Y0-BVX-045

REFERENCE
-
DO NOT TAKE FROM
LIBRARY

Chemical Technology: An Encyclopedic Treatment

VOLUME III

Volume 1 Air, water, inorganic chemicals and nucleonics
2 Non-metallic ores, silicate industries and solid mineral fuels
3 Metals and ores
4 Petroleum, organic chemicals and plastics
5 Natural organic materials and related synthetic products
6 Wood, paper, textiles and photographic materials
7 Vegetable food products and luxuries
8 Edible oils and fats and animal food products
Index

Editors
L. W. Codd, M. A.
K. Dijkhoff, Chem. Drs.
J. H. Fearon, B. Sc., C. Eng., M.I.E.E.
C. J. van Oss, Ph. D.
H. G. Roebersen, Chem. Drs. †
E. G. Stanford, M. Sc., Ph. D., C. Eng., M.I.E.E.,
   F. Inst. P.

General Editor
T. J. W. van Thoor

Foreword

When I read the first announcement of this new eight-volume encyclopedia I immediately recognized it as a marvellously integrated guide to the World's raw materials unlike any other reference work I know.

The encyclopedia is prepared in such a way that its information is accessible to a large number of readers whose knowledge of science and technology is limited. For the first time, the nonspecialist in need of quick, accurate data about materials and processes can refer to a single reference work that correlates technical facts with economic and financial data in a simple yet highly organised and readily understandable fashion. This encyclopedia is a comprehensive guide, broad enough in scope to include information about all the world's raw materials, the primary manufacturing and agricultural processes involved, world output, prices, and other related information, all presented together in such a way that the unknowledgeable reader can rapidly gain an overall view of a subject or a variety of subjects.

The wealth of economic information presented in these volumes should be of immense value to those engaged in business, manufacturing industry, finance, economics, journalism, public relations, research, analysis, government, and indeed to everyone involved in commercial applications and marketing of raw materials.

The encyclopedia will also be a tremendous asset to librarians and students in science, engineering and economics, who can refer to this information as an authoritative introduction to the whole subject of materials and technology.

Although this work has been planned with the nonspecialist in mind specialists will also find it a useful work of reference. Each chapter includes, in addition to fundamental data about a subject, other information usually found in more technical sources. With this combined approach the technical man now has a source he can use to gain knowledge quickly of the functional importance of materials and processes outside his own special field.

This work may be considered a successor to Dr. J. F. Van Oss's *Warenkennis en Technologie (Systematic Encyclopedia of Technology)*. The present encyclopedia, however, has been thoroughly updated and comprises an entirely new work that is in every way a far more ambitious and elaborate project than its predecessor.

The editors have superbly integrated descriptions of operational processes with descriptions of technical and chemical processes. For example, in the first volume, even though most of the industrial processes describe chemical reactions of one kind or another, the first chapter covers the important physical separation processes involved in the manufacture of some of these materials. The editors intend to carry out this same kind of presentation throughout. Subjects are amply cross-referenced, there are numerous line drawings and photographs, and selected bibliographies at the end of each chapter. Each of the eight volumes is to be indexed separately, with a separate final index to the entire encyclopedia.

We in the United States are used to seeing everything from our point of view and tend to forget the tremendous technological and agricultural world that lies outside our country, a world that includes numerous raw materials that play an important part in our own technology and economy. *Chemical Technology: An Encyclopedic Treatment* deals with them all and will be of inestimable value to the user. I recommend it highly.

John J. McKetta, Jr.
Dean of Engineering,
The University of Texas at Austin
Austin, Texas.

Contents

Foreword by John J. Mc Ketta, Jr.	v
Preface	ix
Authors	xi
Acknowledgements	xii
Symbols and Abbreviations	xix
Table of Chemical Elements	xxiii
Units Conversion Table	xxv
General References	xxix

1 **Metals in General** — 1
   1.1 Structure and properties of metals *1*   1.2 Heat treatment *19*   1.3 Economic aspects of metal selection *37*   1.4 Mechanical testing *40*   1.5 Metallographic examination *63*   1.6 Non-destructive testing *72*   1.7 Literature *80*

2 **Ores in General** — 81
   2.1 Ore mining *81*   2.2 Processing of metalliferous ores *97*

3 **Iron and Steel** — 123
   3.1 History *123*   3.2 Iron ores and the manufacture of pig iron *131*   3.3 Steelmaking *151*   3.4 Steels and cast iron *191*   3.5 Heat treatment and hardness of steels *209*

4 **Non-ferrous Metals** — 235
   4.1 Introduction *235*   4.2 Beryllium *236*   4.3 Magnesium *240*   4.4 Aluminium *243*   4.5 Gallium, indium and thallium *252*   4.6 Germanium *255*   4.7 Tin *256*   4.8 Lead *261*   4.9 Antimony *265*   4.10 Bismuth *267*   4.11 Cobalt *269*   4.12 Nickel *274*   4.13 The platinum group metals *280*   4.14 Copper *284*   4.15 Silver *293*   4.16 Gold *297*   4.17 Zinc *301*   4.18 Cadmium *307*   4.19 Mercury *308*   4.20 Titanium *310*   4.21 Zirconium *316*   4.22 Vanadium *319*   4.23 Niobium (Columbium) *321*   4.24 Tantalum *326*   4.25 Chromium *327*   4.26 Molybdenum *329*   4.27 Tungsten *335*   4.28 Manganese *337*

5 Metal Casting — 341

5.1 Introduction *341* 5.2 Patterns *346* 5.3 Moulding and core-making in general *356* 5.4 Sand moulding *359* 5.5 Investment moulding and casting process *381* 5.6 Other precision moulding processes *387* 5.7 Some particular casting equipment and casting methods *392* 5.8 Cores and core materials *396* 5.9 Pouring and feeding of castings *404* 5.10 Metal melting *420* 5.11 Metals cast in the foundry *429* 5.12 Evaluation of castings *446*

6 Metal Deformation — 455

6.1 Metal deformation in general *455* 6.2 Rolling *462* 6.3 Extrusion *473* 6.4 Forging *480* 6.5 Drawing *491* 6.6 Deep drawing and pressing *493* 6.7 Spinning *495* 6.8 Bending *496* 6.9 Recent developments in metal deformation *501* 6.10 Products of mechanical deformation processes *505* 6.11 Literature *521*

7 Working of Metals by Cutting — 523

7.1 History of metal cutting *523* 7.2 Cutting in general *527* 7.3 Turning and boring *538* 7.4 Drilling, reaming and tapping *546* 7.5 Milling *552* 7.6 Shaping, slotting and planing *556* 7.7 Broaching *558* 7.8 Gear cutting *561* 7.9 Sawing *568* 7.10 Grinding, lapping and honing *570* 7.11 Numerically controlled machine tools *576* 7.12 Literature *579*

8 Joining of Metals — 581

8.1 Classification of joining methods *581* 8.2 Mechanical methods of joining *586* 8.3 Welding methods involving pressure *589* 8.4 Fusion welding *601* 8.5 Metallurgical surface bonding *632* 8.6 Design of metallurgical joints *641* 8.7 Adhesive joining of metals *644* 8.8 Literature *665*

9 The Surface Treatment of Metals — 669

9.1 Introduction *669* 9.2 Choice of surface treatment *670* 9.3 Metallic coatings *671* 9.4 Miscellaneous non-metallic coating processes *710* 9.5 Vitreous enamel *717* 9.6 Lacquer, varnish, enamel and paint coatings *719* 9.7 Plastic coatings *727* 9.8 Literature *727*

10 Metal Powders and Hard Metals — 729

10.1 Introduction *729* 10.2 Metal powders *730* 10.3 Hard metals *774* 10.4 Literature *780*

11 The Corrosion of Metals — 783

11.1 Corrosion in general *783* 11.2 Reaction mechanism of corrosion *786* 11.3 Forms of corrosion in the absence of stress *801* 11.4 Corrosion in the presence of stress *813* 11.5 Corrosion of metals *820* 11.6 Cathodic protection *841* 11.7 Corrosion control *845* 11.8 Literature *850*

12 Composite Materials — 853

12.1 Principles of composite materials *853* 12.2 Particulate composites *862* 12.3 Fibre composites *869* 12.4 Planar composites *882* 12.5 The future of composites *886* 12.6 Literature *887*

Index — 889

Preface

Like all volumes of the Encyclopaedia, this one on Metals has been written by specialists for people who require reliable information. Care has been taken to present the information in a way such that it will serve the layman as well as the engineer or technologist. The layman, with only an elementary knowledge of physics and chemistry, will have no difficulty in understanding the text and the engineer and technologist will find it a useful work of reference. Students, in schools, colleges and universities, who are preparing for a career in science or technology will also find the volume helpful in providing background reading in support of the main lines of study.

All the commercially useful metals and alloys are described under the headings of ferrous and non-ferrous metals. Because of the predominance of ferrous metals in commerce and engineering, particular emphasis is given to them by devoting a complete chapter to the subject of iron and steel. A detailed account is given of twenty-five non-ferrous metal systems in another chapter and, in a third chapter, the important development of metallic composite materials is discussed. The information in these three chapters is supported by the remaining chapters which describe: the scientific background to metallurgical technology; the mining of ores and the extraction of metals from them; the processes involved in the manufacture of metals into finished and semi-finished products; the uses to which metals are put, their performance in service and the testing methods which have been developed for the purpose of assessing this performance.

The reader must be warned that some of the information presented is protected by patents and before using it for commercial purposes careful enquiries should be made. Reference in this book to any particular product does not imply approval or recommendation of that product and the use of tradenames in reference to any product does not imply that they are generic terms as they may or may not be protected as registered trademarks.

The publishers have made arrangements whereby they hope to be able to answer any enquiries arising out of the use of this encyclopaedia; these should be sent to Longman Group Limited, Longman House, Burnt Mill, Harlow, Essex. They should be marked specifically 'For the attention of the Editor, Materials and Technology Encyclopaedia'.

THE EDITORS

Authors

Professor C. Bodsworth, and B. Fookes,
Department of Metallurgy,
Brunel University (U.K.)

Ch. 1 Metals in General

R. E. Okell and B. J. Meadows,
The Department of Metallurgy,
The University of Aston in Birmingham
(U.K.)

Heat treatment (1.2)

A. Grierson,
Department of Mining and Mineral Technology,
Royal School of Mines,
Imperial College of Science and Technology, (U.K.)

Ch. 2 Ores in General
Ore mining (2.1)

M. P. Jones, Lecturer,
Department of Mining and Mineral Technology,
Royal School of Mines,
Imperial College of Science and Technology, (U.K.)

Processing of metalliferous ores (2.2)

C. L. Boltz,

Ch. 3 Iron and Steel

Under the editorship of J. H. Fearon,
Alcan Research and Development Ltd.:
N. P. Pinto, J. H. Fearon,
C. J. Thwaites, E. Williams,
P. A. Lovett, J. S. Forbes,
B. W. Smith, R. T. J. Hubbard,
J. H. Schemel, J. C. E. Amos,
W. Fairhurst.

Ch. 4 Non-ferrous Metals

Dr. A. A. DAS,
Industrial Engineering Department,
Loughborough University of Technology,
(U.K.)

Ch. 5 Metal Casting

R. E. OKELL and B. J. MEADOWS.

Ch. 6 Metal Deformation

G. H. RYDER, Principal Lecturer,
Applied Mechanics Branch,
Royal Military College of Science, (U.K.)

Ch. 7 Working of Metals by Cutting

D. SLATER, Chief Metallurgist, and
Dr. C. T. COWEN,
The A.P.V., Company Ltd., (U.K.)

Ch. 8 Joining of Metals

Dr. G. E. GARDAM.

Ch. 9 The Surface Treatment of Metals

Professor C. R. TOTTLE,
School of Materials Science,
Bath University of Technology, (U.K.)

Ch. 10 Metal Powders and Hard Metals

T. H. ROGERS.

Ch. 11 The Corrosion of Metals

Dr. D. CRATCHLEY.

Ch. 12 Composite Materials

Acknowledgements

We are grateful to the following for their kind permission to reproduce the figures listed below;

Fig. 1.12: from *Cold working of Metals* by L. E. Gibbs, by courtesy of The American Society of Metals, Metals Park, Cleveland, Ohio, U.S.A.

Fig. 1.13: from *The Making and Shaping and Treating of Steel,* by courtesy of United States Steel Corporation.

Fig. 1.17: from *Transformation Characteristics of Direct Hardening Nickel Alloy Steels,* by courtesy of International Nickel Company Limited, London.

Fig. 1.18 (a): from *Practical Microscopical Metallurgy* by Greaves and Wrighton, by courtesy of Chapman and Hall, London.

Fig. 1.18 (b): from *The Making and Shaping and Treating of Steel,* by courtesy of United States Steel Corporation.

Fig. 1.19: from *Metallurgy for Engineers* by Rollason, by courtesy of Edward Arnold, London.

Fig. 2.1, 2.2: by courtesy of Boyles Brothers.

Fig. 2.6: by courtesy of A.E.I.

Fig. 2.21: by courtesy of Gunson Sortex Limited.

Fig. 2.23: *De Re Metallica* by Georgius Agricola 1556 (reproduced from edition by Dover Publications Inc. 1950).

Fig. 2.24: by courtesy of the Pacific Tin Corporation.

Fig. 2.28: After M. P. Jones, *Records of the Geological Survey of Nigeria,* 1957, Government of the Federation of Nigeria.

Fig. 2.33: by courtesy of Professor R. B. Fleming.

Fig. 3.2, 3.3: Science Museum, London, Crown Copyright.

Fig. 3.4: by courtesy of Thomas Walmsley Ltd., Atals Forge, Bolton, Lancs.

Fig. 3.7: by courtesy of *The Beneficiation of Iron Ores* by M. M. Fire. Copyright 1968 Scientific American, All rights reserved.

Fig. 3.8: Adapted from a Wallchart, by courtesy of the British Iron and Steel Federation.

Fig. 3.9, 3.10: by courtesy of the British Oxygen Company.

Fig. 3.11: from *The Making and Shaping and Treating of Steel,* by courtesy of United States Steel Corporation.

ACKNOWLEDGEMENTS

Fig. 3.12: Adapted from a Wallchart, by courtesy of the British Iron and Steel Federation.

Fig. 3.13: from *The Making and Shaping and Treating of Steel*, by courtesy of United States Steel Corporation.

Fig. 3.14: from *The Making and Shaping and Treating of Steel*, by courtesy of United States Steel Corporation.

Fig. 3.15: from *The Making and Shaping and Treating of Steel*, by courtesy of United States Steel Corporation.

Fig. 3.16: from Oxygen in Steelmaking by J. K. Stone. Copyright 1968 Scientific American. All rights reserved.

Fig. 3.17: from *The Elements of Steel Making Practise*, by J. D. Sharp, by courtesy of Pergamon Press and the British Iron and Steel Federation.

Fig. 3.19: from *The Elements of Steel Making Practise*, by J. D. Sharp, by courtesy of Pergamon Press and the British Iron and Steel Federation.

Fig. 3.20: from *The Elements of Steel Making Practise*, by J. D. Sharp, by courtesy of Pergamon Press and the British Iron and Steel Federation.

Fig. 3.21: from *Foundations of Iron and Steel Metallurgy* by W. J. Dennis, Elsevier Press.

Fig. 3.22: by courtesy of The British Iron and Steel Research Association.

Fig. 3.26: by courtesy of The British Iron and Steel Research Association.

Fig. 3.27: from *The Journal of the Iron and Steel Institute*, by courtesy of the Iron and Steel Institute.

Fig. 3.31: from *A Simple Guide to the Structure and Properties of Steel*, by courtesy of The British Steel Corporation.

Fig. 3.36: by courtesy of Syndication International.

Fig. 3.37: by courtesy of The London Transport Board.

Fig. 3.38, 3.39, 3.40, 3.41, 3.42, 3.43, 3.44, 3.45: from *Alloying Elements in Steel* by Bain and Paxter, by courtesy of The American Society for Metals, Metals Park, Cleveland, Ohio, U.S.A.

Fig. 3.48: from *A Simple Guide to the Structure and Properties of Steel*, by courtesy of The British Steel Corporation.

Fig. 3.49, 3.50, 3.51: from *Alloying Elements in Steel* by Bain and Paxter, by courtesy of The American Society for Metals, Metals Park, Cleveland, Ohio, U.S.A.

Fig. 3.52, 3.53: from *A Simple Guide to the Structure and Properties of Steel*, by courtesy of The British Steel Corporation.

Fig. 4.1, 4.2: by courtesy of Alcan Aluminium Ltd.

Fig. 4.3: by courtesy of the Tin Research Institute.

Fig. 4.4: by courtesy of the International Nickel Company of Canada Ltd., from the Thompson Nickel Refinery.

Fig. 4.5: by courtesy of the Utah Copper Division, Kennecott Copper Division Corporation.

Fig. 4.6: by courtesy of G. B. Instructional Ltd., Imperial Studios.

Fig. 4.7: by courtesy of the Copper Developement Association.

Fig. 4.8: by courtesy of the Tin Research Institute.

Fig. 5.1: by courtesy of the Conway Library, Courtauld Institute of Art.

Fig. 5.2: by courtesy of Voice and Vision Limited, and Corocraft Jewellry.

Fig. 5.3: by courtesy of Stone Manganese Marine Limited.

Fig. 5.12: by courtesy of Stone Manganese Marine Limited.

Fig. 5.25: from *Modern Foundry Practice,* by courtesy of Odhams Press.

Fig. 5.26: from the *Proceedings of the Institute of British Foundrymen,* Volume XLVII 1954, by courtesy of The Institute of British Foundrymen.

Fig. 5.27: by courtesy of Stirling Metals Limited, Nuneaton, England.

Fig. 5.33: from *Design and Die-casting* by Gustav Leiby, by courtesy of American Foundryman's Society.

Fig. 5.36: by courtesy of *The Foundry Trade Journal,* March 1968, and R. Milliner.

Fig. 5.50: Adapted from American Foundryman's Society Gating and Risering Committee Diagram.

Fig. 5.53: by courtesy of The Inductothern Corporation, U.S.A.

Fig. 5.54, 5.55, 5.56, 5.57, 5.58, 5.59: by courtesy of Loughborough College of Technology.

Fig. 6.2: from *Manufacture of Iron and Steel,* Vol. 2, Page 278 Fig. 145, by Bashforth, by courtesy of Chapman and Hall Ltd. 1951.

Fig. 6.3: from *Science of Engineering Materials* by Goldman, by courtesy of John Wiley and Son, New York, 1957.

Fig. 6.4: from *Proceedings of the Royal Society,* from a paper by Whelan, Hirsch, Horne and Bolmann 1957 A 240 524, by courtesy of the Royal Society and the authors.

Fig. 6.6: from *The Making and Shaping and Treating of Steel,* by courtesy of the United States Steel Corporation Page 421, Fig. 22-1, 1957.

Fig. 6.7: from *Mechanical Treatment of Metals,* by Parkins, Fig. 4/18, by courtesy of Allen & Unwin Limited 1968.

Fig. 6.8: from *Mechanical Treatment of Metals,* by Parkins, Fig. 4/19, by courtesy of Allen & Unwin Limited, 1968.

Fig. 6.9: from *Rolling Mill Design Developed Afresh,* by courtesy of *Engineering* 1963, April 5th.

Fig. 6.10: from *Roll Pass Design,* Fig. 1/3 United Steel Company Limited, Sheffield 1960.

Fig. 6.11: from *The Making and Shaping and Treating of Steel,* by courtesy of the United States Steel Corporation, Page 431 Fig. 23/1, 1957.

Fig. 6.12: from *The Making and Shaping and Treating of Steel,* by courtesy of the United States Steel Corporation, Page 431, Fig. 23/2, 1957.

Fig. 6.13: from *Roll Pass Design,* by courtesy of United Steel Company Limited, Fig. 2/11, Sheffield, 1960.

Fig. 6.14, 6.15, 6.16, 6.17: from *The Extrusion of Metals* by Pearson and Parkins, by courtesy of Chapman and Hall.

Fig. 6.18: from *Cold Forging of Steel* by Feldman, Page 17, Fig. 1, by courtesy of Hutchison and Company Limited 1961.

Fig. 6.19: from *The Extrusion of Metals* by Pearson and Parkins, by courtesy of Chapman and Hall.

Fig. 6.20: from *Cold Forging of Steel* by Feldman, Page 31, Fig. 9, by courtesy of Hutchison and Company Limited, 1961.

Fig. 6.21: by courtesy of De Rotterdamsche Droogdok Mij, Rotterdam, Holland.

Fig. 6.23: by courtesy of the National Association of Drop Forgers and Stampers.

Fig. 6.24: by courtesy of Firth-Derihon Stamping Limited.

Fig. 6.26: from *Hydraulische Pressen* by Müller, Page 29, Fig. 21, by courtesy of Springer, Berlin, 1962.

Fig. 6.27: from *Hydraulische Pressen* by Müller, Page 219, Fig. 117, by courtesy of Springer: Berlin, 1961.

Fig. 6.28: by courtesy of Lamberton and Company Limited.

Fig. 6.30: by courtesy of Warne, Wright and Rowland Limited.

Fig. 6.31: by courtesy of Smethwick Drop Forging Limited.

Fig. 6.32: from *Sheet Metal Industry Conference on Cold Extrusion of Steel* by R. E. Okell, November 1960.

Fig. 6.33: by courtesy of the National Machinery Company Limited.

Fig. 6.35: from *Internat: Colloquium on Forming of Sheet Metals* by Whiteley, Wise and Blickwele, by courtesy of Sheet Metal Industries, London, 1960 Paris.

Fig. 6.38: from *Brass Pressings,* Page 30 Publication 26 by courtesy of Copper Development Association, London.

Fig. 6.40: from A.S.M.E. Handbook, *Metals Engineering-Processes,* Page 147, Fig. 2, by courtesy of McGraw Hill 1958.

Fig. 6.41: from *Press Tool Practice* by Houghton, Page 50, Figs. 230 and 231 (together), by courtesy of Chapman and Hall 1951.

Fig. 6.42: from A.S.M.E. Handbook, *Metals Engineering-Processes,* Page 167, Fig. 1, by courtesy of McGraw Hill 1958.

Fig. 6.43: from A.S.M.E. Handbook, *Metals Engineering-Processes,* Page 175, Fig. 18, by courtesy of McGraw Hill, 1958.

Fig. 6.44: from A.S.M.E. Handbook, *Metals Engineering-Processes,* Page 156, Fig. 8, by courtesy of McGraw Hill, 1958.

Fig. 6.45: from A.S.M.E. Handbook, *Metals Engineering-Processes,* Page 157, Fig. 9, by courtesy of McGraw Hill, 1958.

Fig. 6.46: from *Metal Industry,* 6th June, 1963, Page 802, Fig. 1.

Fig. 6.47: from *Recent Progress in Metal Working,* Page 96, Fig. 3.18, by courtesy of the Institution of Metallurgists, London.

Fig. 6.48: from *Recent Progress in Metal Working,* Page 98, Figs. 3.20/21, by courtesy of the Institution of Metallurgists, London.

Fig. 6.49: from *Bulleid Memorial Lecture 1965,* Vol. IIIB Page (3)-2, Fig. 1, by courtesy of the University of Nottingham, from N.E.L., East Kilbride.

Fig. 6.50: from *Recent Progress in Metal Working,* Page 35, Fig. 2.1, by courtesy of the Institution of Metallurgists, London.

Fig. 6.51: from *Recent Progress in Metal Working,* Page 44, Fig. 2.6, by courtesy of the Institution of Metallurgists, London.

Fig. 6.52, 6.53, 6.54: from *Roll Pass Design,* by courtesy of the United Steel Company Limited, Sheffield, 1960.

Fig. 6.55: from *The Making and Shaping and Treating of Steel,* by courtesy of the United States Steel Corporation, Page 742, Fig. 41.18.

Fig. 6.56: from *Mechanical Treatment of Metals* by Parkins, Page 45, Figs. 1 and 2, by courtesy of Chapman and Hall 1967.

Fig. 6.58: from *Mechanical Treatment of Metals* by Parkins, Fig. 4/52, by courtesy of Chapman and Hall 1967.

Fig. 6.59: from *Mechanical Treatment of Metals* by Parkins, Page 45, Figs. 3 and 4, by courtesy of Chapman and Hall 1967.

Fig. 6.60: from *The Making and Shaping and Treating of Steel,* by courtesy of the United States Steel Corporation, Page 746, Fig. 41-24.

Fig. 6.61: from *The Making and Shaping and Treating of Steel,* by courtesy of the

ACKNOWLEDGEMENTS | XVII

United States Steel Corporation, Page 755, Fig. 41-37.

Fig. 6.62: from film strip *Rotary Piercing* by courtesy of Stewart and Lloyds.

Fig. 6.63: from *The Making and Shaping and Treating of Steel*, by courtesy of the United States Steel Corporation, Page 760, Fig. 41-42.

Fig. 6.64: from *Mechanical Treatment of Metals* by Parkins, Fig. 4/54, by courtesy of Chapman and Hall 1967.

Fig. 7.1: Science Museum, Crown Copyright.

Fig. 7.2, 7.3, 7.4: from *A History of Machine Tools, 1700-1910* by Professor W. Steeds, by courtesy of Oxford University Press and the author.

Fig. 7.19: by courtesy of Société Genevoise.

Fig. 7.20, 7.21: by courtesy of Alfred Herbert Limited.

Fig. 7.25, 7.26: by courtesy of Staveley Machine Tools Limited.

Fig. 7.30: by courtesy of Adcock and Shipley Limited.

Fig. 7.31, 7.32: by courtesy of Butler Machine Tool Company Limited.

Fig. 7.34: by courtesy of Staveley Machine Tools Limited.

Fig. 7.41, 7.42, 7.43: by courtesy of Coventry Gauge and Tool Company Limited.

Fig. 7.44: by courtesy of Pehaka Saws Company Limited.

Fig. 7.46: by courtesy of Kearney and Trecker Limited.

Fig. 8.1: by courtesy of A. I. Welders Ltd.

Fig. 8.2, 8.3, 8.4, 8.5, 8.6: by courtesy of Bristol Siddeley Engines Ltd.

Fig. 8.7: by courtesy of Martons Excelsior Ltd.

Fig. 8.12, 8.17, 8.22, 8.23: by courtesy of CIBA (A.R.L.) Ltd.

Fig. 9.2: by courtesy of W. Canning and Company Limited, Birmingham.

Fig. 9.3: by courtesy of Imperial Chemical Industries Limited.

Fig. 9.7: from *Product Finishing*, by courtesy of Sawell Publications Limited.

Fig. 9.8: by courtesy of G.E.C.-A.E.I. Telecommunications Limited and Harshaw Chemicals Limited, Daventry.

Fig. 9.9: by courtesy of Mono Pumps Limited, London.

Fig. 9.10: by courtesy of Cruickshanks Limited.

Fig. 9.11, 9.12: by courtesy of W. Canning and Company Limited, Birmingham.

Fig. 9.13: by courtesy of Sel-Rex (U.K.) Limited, and Thomas Leigh Advertising Limited.

Fig. 9.14, 9.15: by courtesy of the Tin Research Institute.

Fig. 9.16: by courtesy of Metallisation Limited.

Fig. 9.17, 9.18: by courtesy of Aluminium Laboratories Limited.

Fig. 9.19: by courtesy of Imperial Chemical Industries Limited.

Fig. 10.9 (a) (b): by courtesy of S.M.C. Sterling Limited.

Fig. 12.1: from *Metallurgical Reviews* 1965, Vol. 10, No. 37, page 81, by courtesy of the Institute of Metals, 17 Belgrave Square, London S.W.I.

Fig. 12.2 (a): from the *Proceedings of the 2nd International Materials Symposium* (California 1964); by courtesy of John Wiley and Son Inc., New York.

Fig. 12.2 (b): from the *Metallurgical Reviews* 1965, Vol. 10, No. 37, page 31, by courtesy of the Institute of Metals, 17 Belgrave Square, London S.W.I.

Fig. 12.3: from the *Proceedings of the Royal Society,* 1964 (a) 1 282, 508 by courtesy of the Royal Society.

Fig. 12.4: from *Powder Metallurgy* by G. C. Smith No. 11, page 105, Joint Group of the Institute of Metals and the Iron and Steel Institute.

Fig. 12.5: from *Metal Progress* 1961 80 (2), page 108, by courtesy of the American

Society for Metals, Metals Park, Cleveland Ohio, U.S.A.

Fig. 12.6: from Engineers Digest 1963, Vol. 24, No. 12, page 85, by courtesy of Engineers Digest Limited, 120 Wigmore Street, London, W.I.

Fig. 12.7: by courtesy of the General Electric Corporation of America.

Symbols and Abbreviations

Symbol	Name of unit
*	radioactive
a	are (= 100 m²) (= 119.599 yd²)
A	ampere
Å	angström (= 10^{-8} m) (= 0.003 937 01 μ in)
a.c.	alternating current
at	technical atmosphere
atm	standard atmosphere (= 101.325 kN/m²) (= 14.2233 lbf/in²) (= 14.6959 lbf/in²)
b	bar (10^5 N/m²) (= 14.5038 lbf/in²)
Bé	Beaumé's scale
BP	boiling point
Btu	British thermal unit (= 1.05506 kJ)
bu	bushel (= 36.368 7 dm³)
c	centi (= 10^{-2})
C	coulomb
°C	degree Celsius (= 5/9 (°F − 32)) (temperature value)
cal	calorie (International table)
cd	candela
cg	centigram
Ci	Curie
cm	centimetre (= 0.393 701 in)
c/s	cycles per second
cwt	hundred weight (= 50.8023 kg) (= 112 lb)

Symbol	Name of unit
d	deci (= 10^{-1})
da	deca (= 10^1)
dag	decagram
d.c.	direct current
degC	degree Celsius (temperature interval)
degF	degree Fahrenheit (temperature interval)
dg	decigram
dm	decimetre
dyn	dyne (= 10^{-5}N) (0.224829 x 10^{-5} lbf)
erg	erg (= 10^{-7}W) (= 0.737 562 x 10^{-7} ft lbf)
F	Farad
°F	degree Fahrenheit (= $\frac{9}{5}$ °C + 32) (Temperature value)
fl oz	Fluid ounce (= 28.4131 cm³)
ft	foot (= 0.3048 m) (= 12 in)
ft H₂O	foot water (= 2989.07 N/m²)
g	gram
G	giga (= 10^9)
gal	UK gallon (= 4.596 litres) (= 4.546 09 dm³) (cf. US gallon = 3.78541 dm³)
gr	grain (= 64.798 9 mg)
h	hecto (= 10^2)
H	henry
ha	hectare (= 10 000 m²) (= 2.471 05 acres)

SYMBOLS AND ABBREVIATIONS

Symbol	Name of unit
hp	horsepower (= 745.700 W)
Hz	hertz
in	inch (= 2.54 cm)
in Hg	conventional inch of mercury (= 3386.39 N/m^2) (= 33.8639 mb)
in H$_2$O	conventional inch of water (= 249.089 N/m^2)
J	joule (= 0.737562 ft lbf)
K	kilo (= 10^3)
°K	degree kelvin (= °C + 273)
Kcal	kilocalorie
kg	kilogram
kgf	kilogram force (= 9.806 65 N) (= 2.204 62 lbf)
kJ	kilojoule
km	kilometre
kp	kilopond (= Kgf)
kW	kilowatt
l	litre (= approx 1 dm^3) (= 0.220 0 gal) (= 0.24642 US gal)
lb	pound (= 0.45359237 kg)
lbf	pound force (= 4.44822 N)
lm	lumen
lx	lux (= 1 lm per m^2)
m	metre (= 1.09361 yd)
m	milli (= 10^{-3})
M	mega (= 10^6)
mb	millibar (= 100 N/m^2)
m.g.d.	million gallons per day
mile	mile (= 1.60934 km)
ml	millilitre
mm	millimetre
mmHg	conventional millimetre of mercury (= 133.322 N/m^2) (= 0.0393701 in Hg)
mm H$_2$O	conventional millimetre of water (= 9.80665 N/m^2)
money	£ (= UK pound unless stated to the contrary) $ (= US dollar unless stated to the contrary)

Symbol	Name of unit
M.P.	melting point
μ	micro (= 10^{-6})
μb	microbar (= 0.1 N/m^2)
μ Hg	conventional micron of mercury (= 0.133322 N/m^2)
μ in	microinch (= 0.0254 μm) (= 0.000001 in)
μ m	micrometre (micron) (39.3701 m in)
μ mHg	micron of mercury (= 0.133322 N/m^2)
n	nano (= 10^{-9})
N	newton (= 0.224809 lbf)
n mile	international nautical mile (= 1852 m) (cf. UK nautical mile = 1853.18 m)
oz	ounce (= 28.3495 g)
Oz apoth	apothecaries' ounce (= 31.1035 g) (= oz tr)
ozf	ounceforce (= 0.278014 N)
oz tr	troy ounce (= 31.1035 g) (= oz apoth)
Ω	ohm
P	pico (= 10^{-12})
P	poise (= 0.1 N s/m^2) (= 2.08854 x 10^{-3} lbf s/ft^2)
Pl	poiseville (= N m^2/s)
PS	Pferdestärke (ch)
pH value	measure of acidity/alkalinity
p.p.m.	parts per million
p.s.i.	poundweight per square inch (= 6894.76 N/m^2) (= 68.9476 mb)
pt	pint (= 0.568261 dm^3)
pz	pieze (= 10^3 N/m^2)
PVC	polyvinylchloride
q	quintal (= 100 kg)
qt	Imperial quart (1.13652 dm^3)
°R	degree Rankine (°F + 459.67)
rad	radian
s	second
sp.gr.	specific gravity

Symbol	Name of unit	Symbol	Name of unit
St	stokes ($= 10^{-4}$ m^2/s) ($= 558.001$ in^2/h)	tonf	tonforce ($= 9964.02$ N)
		V	volt
t	metric ton ($=$ tonne) ($= 1000$ kg) ($= 0.984207$ tons) ($= 2204.6$ lb)	W	Watt ($=$ J/s)
T	tera ($= 10^{12}$)	Wb	Weber
ton	Imperial ton ($= 1016.05$ kg) ($= 2240$ lb) ($=$ long ton) (*cf.* US ton $= 2000$ lb $=$ short ton)	wt	weight
		w/w	weight for weight
		yd	yard ($= 0.9144$ m)

Table of Chemical Elements

*These elements, like the transuranic elements (see below), have been produced by artificial means and do not occur naturally (at least, not in any appreciable amount).

Element	Symbol	Atomic Number	Atomic Weight	Element	Symbol	Atomic Number	Atomic Weight
Actinium	Ac	89	227	Gallium	Ga	31	69.72
Aluminium	Al	13	26.98	Germanium	Ge	32	72.60
Antimony	Sb	51	121.76	Gold	Au	79	197
Argon	A	18	39.944				
Arsenic	As	33	74.91	Hafnium	Hf	72	178.5
*Astatine	At	85	(210)	Helium	He	2	4.003
				Holmium	Ho	67	164.94
Barium	Ba	56	137.36	Hydrogen	H	1	1.008
Beryllium	Be	4	9.013				
(Glucinium)	(Gl)			Indium	In	49	114.76
Bismuth	Bi	83	209	Iodine	I	53	126.91
Boron	B	5	10.82	Iridium	Ir	77	192.2
Bromine	Br	35	79.916	Iron	Fe	26	55.85
Cadmium	Cd	48	112.41	Krypton	Kr	36	83.80
Caesium	Cs	55	132.91				
Calcium	Ca	20	40.08	Lanthanum	La	57	138.92
Carbon	C	6	12.01	Lead	Pb	82	207.21
Cerium	Ce	58	140.13	Lithium	Li	3	6.940
Chlorine	Cl	17	35.457	Lutetium	Lu	71	174.99
Chromium	Cr	24	52.01	(Cassiopeium)	(Cp)		
Cobalt	Co	27	58.94				
Copper	Cu	29	63.54	Magnesium	Mg	12	24.32
				Manganese	Mn	25	54.94
Dysprosium	Dy	66	162.46	Mercury	Hg	80	200.61
				Molybdenum	Mo	42	95.95
Erbium	Er	68	167.3				
Europium	Eu	63	152	Neodymium	Nd	60	144.27
				Neon	Ne	10	20.183
Fluorine	F	9	19	Nickel	Ni	28	58.7
Francium	Fr	87	223	Niobium	Nb	41	92.91
				(Columbium)	(Cb)		
Gadolinium	Gd	64	156.9	Nitrogen	N	7	14.008

TABLE OF CHEMICAL ELEMENTS

Element	Symbol	Atomic Number	Atomic Weight	Element	Symbol	Atomic Number	Atomic Weight
Osmium	Os	76	190.2	Thallium	Tl	81	204.39
Oxygen	O	8	16	Thorium	Th	90	232.1
				Thulium	Tm	69	168.94
Palladium	Pd	46	106.7	Tin	Sn	50	118.70
Phosphorus	P	15	30.974	Titanium	Ti	22	47.90
Platinum	Pt	78	195.23	Tungsten	W	74	183.92
Polonium (Radium F)	Po	84	210	(Wolfram)			
Potassium	K	19	39.100	Uranium	U	92	238.07
Praseodymium	Pr	59	140.92				
*Prometheum	Pm	61	(145)	Vanadium	V	23	50.95
Protactinium	Pa	91	231				
				Xenon	Xe	54	131.3
Radium	Ra	88	226.05				
Radon (Niton)	Rn (Nt)	86	222	Ytterbium	Yb	70	173
				Yttrium	Y	39	89
Rhenium	Re	75	186.31				
Rhodium	Rh	45	102.91	Zinc	Zn	30	65.38
Rubidium	Rb	37	85.48	Zirconium	Zr	40	91.22
Ruthenium	Ru	44	101.1				
Samarium	Sm	62	150.43				
Scandium	Sc	21	44.96				
Selenium	Se	34	78.96				
Silicon	Si	14	28.09				
Silver	Ag	47	107.873				
Sodium	Na	11	22.997				
Strontium	Sr	38	87.63				
Sulphur	S	16	32.066				
Tantalum	Ta	73	180.88				
*Technetium	Tc	43	(98.91)				
Tellurium	Te	52	127.61				
Terbium	Tb	65	158.9				

TRANSURANIC ELEMENTS

Atomic Number	Element	Symbol
93	Neptunium	Np
94	Plutonium	Pu
95	Americium	Am
96	Curium	Cm
97	Berkelium	Bk
98	Californium	Cf
99	Einsteinium	E
100	Fermium	Fm
101	Mendelevium	Mv
102	Nobellium	No

Units Conversion Table

This table has been compiled with reference to units and magnitudes appropriate to the technical and commercial operations with which these volumes are concerned, and in the light of the present transitional situation. It therefore includes many units which, though widely employed in imperial or metric usage, fall outside the strict Système International des Unités.

IMPERIAL/US AND METRIC/SI

Length

1 mile	=	1.6093 km	1 km	=	0.6214 mile
1 yd	=	0.9144 m	1 m	=	1.0936 yd
1 ft	=	0.3048 m	1 cm	=	0.3938 in
1 in	=	2.54 cm	1 mm	=	39.37 'thou'
1 mil or 'thou' (1/1000 in)	=	0.0254 mm			

Area

1 mile²	=	2.590 km²	1 ha	=	2.471 acres
	or	259 ha		or	0.386 mile²
1 acre	=	4047 m²	1 km²	=	247.1 acres
	or	0.4047 ha	1 m²	=	1.196 yd²
1 yd²	=	0.8361 m²	1 cm²	=	0.1550 in²
1 ft²	=	0.0930 m²			
1 in²	=	645.2 mm²			

Volume, Capacity

1 yd³	=	0.7646 m³	1 m³	=	1.3079 yd³
1 ft³	=	0.02832 m³		or	35.315 ft³
1 in³	=	16.387 cm³	1 dm³	=	0.0353 ft³
1 gal	=	4.546 l	1 cm³	=	0.0610 in³
1 US gal	=	3.785 l		or	0.0351 fluid oz
1 pint	=	0.5682 l	1 l	=	0.220 gal
1 fluid oz	=	28.413 cm³		or	1.760 pints
				or	0.2642 US gal

Velocity

1 mile/h	=	1.6093 km/h	1 km/h	=	0.6214 mile/h
1 ft/min	=	0.00508 m/s	1 m/s	=	3.2808 ft/s
1 ft/s	=	0.3048 m/s	1 mm/s	=	0.0394 in/s
1 in/s	=	25.40 mm/s			

UNITS CONVERSION TABLE

IMPERIAL/US AND METRIC/SI

Mass

1 ton (2240 lb)	=	1016	kg	1 tonne (1000 kg)	=	2204.7	lb
					or	0.9842	ton
1 short ton (2000 lb)	=	907.19	kg		or	1.1023	short ton
				1 quintal (100 kg)	=	220.47	lb
1 cwt (112 lb)	=	50.802	kg				
				1 kg	=	2.2047	lb
1 stone (14 lb)	=	6.350	kg	1 g	=	0.0353	oz
1 lb	=	0.4536	kg				
1 oz	=	28.349	g				

Mass per Unit Length

1 ton/mile	=	631.3	kg/km	1 tonne/km	=	1.584	ton/mile
1 lb/yd	=	0.4961	kg/m		or 1.774	short ton/mile
1 lb/ft	=	1.488	kg/m	1 kg/m	=	2.016	lb/yd
				1 tonne/m	=	0.900	ton/yd

Length per Unit Mass

1 yd/lb	=	2.016	m/kg	1 m/kg	=	0.490	yd/lb
1 in/oz	=	0.7005	cm/g	1 cm/g	=	0.0721	in/oz

Mass per Unit Area

1 ton/mile²	=	392.3	kg/km²	1 tonne/ha	=	892.2	lb/acre
1 ton/acre	=	0.2511	kg/m²				
1 lb/ft²	=	4.882	kg/m²	1 kg/m²	=	0.2048	lb/ft²
1 lb/in²	=	70.31	g/cm²	1 kg/cm²	=	14.22	lb/in²

Area per Unit Mass (Specific Surface)

1 mile²/ton	=	2549	m²/kg	1 ha/tonne	=	2.511	acre/ton
1 yd²/ton	=	0.823	m²/tonne	1 m²/kg	=	0.542	yd²/lb
1 ft²/lb	=	0.205	m²/kg				

Volume per Unit Mass (Specific Volume)

1 ft³/ton	=	0.0279	l/kg	1 m³/tonne	=	1.332	yd³/ton
1 ft³/lb	=	62.428	l/kg				
1 gal/lb	=	10.022	l/kg	1 l/kg	=	0.995	gal/lb
1 in³/lb	=	36.127	cm³/kg				

Mass Rate of Flow

1 ton/h	=	1016	kg/h	1 tonne/h	=	0.984	ton/h
1 lb/h	=	0.454	kg/h				
1 lb/s	=	0.454	kg/s	1 kg/s	=	2.2047	lb/s

Volume Rate of Flow

1 ft³/s (1 cusec)	= or	28.32 1019	l/s m³/h	1 l/s	=	0.353	ft³/s
1 gal/min	=	272.76	l/h	1 l/h	=	0.220	gal/h
1 US gal/min	=	227.10	l/h		or 0.264	US gal/h

UNITS CONVERSION TABLE | XXVII

Density

1 ton/yd³	=	1.329 tonnes/m³		1 tonne/m³	=	0.7525 ton/yd³
1 lb/ft³	=	16.018 kg/m³		1 kg/m³	=	0.0624 lb/ft³
1 lb/in³	=	27.680 g/cm³		1 g/cm³	=	0.0361 lb/in³
1 lb/gal	=	0.10 g/cm³				
1 oz/gal	=	6.236 g/l		1 g/l	=	0.1600 oz/gal

Force

1 pound force (lbf)	=	0.454 kgf *or* 4.449 N		1 kgf (kp)	=	2.205 lbf
1 poundal (pdl)	=	0.1383 N		1 N	=	0.225 lbf *or* 7.233 pdl
1 ton force (tonf)	=	1.016 tf		1 tonne force (tf)	=	0.984 tonf
1 tonf	=	9.964 kN		1 kN	=	0.1004 tonf

Pressure

1 lbf/in²	=	0.0703 kgf/cm² *or* 6.8947 kN/m²		1 kgf/cm²	=	14.22 lbf/in²
				1 kN/m²	=	0.145 lbf/in²
1 lbf/ft²	=	47.880 N/m²				
1 tonf/ft²	=	10.936 tf/m² *or* 0.1072 N/mm²		1 tf/m²	=	0.914 tonf/ft²
1 atm (760 mm Hg)	=	101.325 kN/m²		1 mb	=	2.088 lbf/ft²
1 in Hg	=	3386.4 N/m²		1 torr (mm Hg)	=	0.019 lbf/in²
1 in H₂O	=	249.1 N/m²				

Energy, Work, Heat

1 Btu	=	0.252 kcal		1 kcal	=	3.9683 Btu
1 therm	=	105.51 x 10⁶ J		1 thermie (10⁶ cal₁₅)	=	0.309 therm
1 ft lbf	=	1.3558 J *or* 3.766 x 10⁻⁷ kWh		1 J	=	0.7375 ft lbf
				1 kWh	=	2.6552 ft lbf
1 ft pdl	=	0.0421 J				
1 hph	=	2.6845 x 10⁶ J		1 J	=	37.25 x 10⁻⁶ hph

Power

1 hp	=	1.0139 met hp		1 met hp	=	0.9863 hp
1 hp	=	0.7457 kW (1 cv)		1 kW	=	1.3410 hp
1 ft lbf/s	=	1.3558 W		1 kW	=	737.56 ft lbf/s

Various Heat Factors

1 Btu/h	=	0.293 W		1 W	=	3.4121 Btu/h
1 Btu/lb	=	2326 J/kg		1 J/kg	=	0.4299 x 10⁻³ Btu/lb
1 therm/gal	=	2.321 x 10⁴ J/cm³		1 J/cm³	=	4.3089 therm/gal
1 Btu/ft³	=	0.0372 J/cm³		1 J/cm³	=	26.839 Btu/ft³
1 Btu/lb deg F	=	4.187 x 10³ J/kg deg C		1 J/kg deg C	=	0.239 x 10³ Btu/lb deg F

General references

Although a list of references is included in most chapters, additional information may be found in the following major reference works.

L. AITCHISON. *History of Metals,*
Macdonald and Evans, 1960

W. ALEXANDER and STREET. *Metals in the Service of Man,*
Harmondsworth (Middlesex), Penguin, 1968

AMERICAN SOCIETY OF MECHANICAL ENGINEERS. *Metals Engineering Design Handbook,*
New York, McGraw-Hill Book Company Inc., 1965

Metals Handbook
American Society for Metals, Cleveland, Ohio, 1964
London, Chapman and Hall

R. ESCHELBACH. *Taschenbuch der metallischen Werkstoffe,*
Stuttgart, Franckh'sche Verlagshandlung

C. H. FRITZSCHE. *Bergbaukunde* (2 volumes),
Berlin, Springer Verlag, 1961-1962

G. W. C. KAYE and T. H. LABY. *Tables of Physical and Chemical Constants,*
London, Longman, 1966

R. E. KIRK and D. F. OTHMER. *Encyclopedia of Chemical Technology,*
New York, Interscience, 1963-1969

C. L. MANTELL. *Engineering Materials Handbook,*
New York, McGraw-Hill Book Company Inc., 1958

GENERAL REFERENCES

R. PEELE and J. A. CHURCH. *Mining Engineers Handbook* (2 volumes),
New York, John Wiley and Sons Inc., 1944

R. B. ROSS. *Metallic Materials*,
London, Chapman and Hall, 1968

C. J. SMITHALLS. *Metals Reference Book*,
London, Butterworth, 1967

H. F. TAGGERT. *Handbook of Mineral Dressing*,
New York, John Wiley and Sons Inc., 1960

Ullmanns Encyklopädie der technischen Chemie,
München, Urban und Schwarzenberg, 1951-1969

U. S. ATOMIC ENERGY COMMISSION. *Reactor Handbook: Materials*,
New York, Interscience

F. W. WILSON and F. D. HARVEY (ed.). *Tool Engineers Handbook*,
New York, McGraw-Hill Book Company Inc., 1959

K. WINNACKER, L. KÜCHLER. *Chemische Technologie Band 5 Metallurgie*,
München, Carl Hanser Verlag, 1961

A. VAN ZEERLEDER. *Technology of Light Metals*,
Amsterdam, Elsevier, 1940

Most of statistical data in this work has been taken from the following references. The most recent editions of these references should be consulted for more up to date statistics. In addition in most countries there is a government bureau of statistics which publishes annual volumes containing data on production, consumption and trade.

Periodicals (annual) of international production and trade statistics:

Statistical Summary of the Mineral Industry. World production, Exports and Imports.
Mineral resources division of the Institute of geological sciences.

Minerals Yearbook. Vol. I and IV.
Washington, U.S. Department of the interior. Bureau of mines.

Industrie und Handwerk Fachserie D.
Reihe 8 Industrie des Auslandes
    I Bergbau und Energiewirtschaft
Stuttgart, W. Kohlhammer.

Statistical Yearbook of the United Nations.
New York.

Statistiques.
Paris, Minerais et métaux S.A.

Statistische Zusammenstellungen.
Frankfurt am Main, Metallgesellschaft A.G.

Engineering and Mining Journal (each Feb. number).
New York, McGraw-Hill Inc.

CHAPTER 1

Metals in General

1.1 STRUCTURE AND PROPERTIES OF METALS

1.1.1 Metals and non-metals

In general, metals may be distinguished from non-metals by their physical and mechanical properties. Metals exhibit the characteristic properties of moderately high melting points, good electrical and thermal conductivities and ease of formability, but there are numerous exceptions, such as cadmium and tin, which possess low melting points and have poor formability. In contrast, the nonmetallic elements carbon and silicon have high melting points, so a general classification in terms of physical properties does not readily differentiate between the two groups.

All the solid elements are crystalline, so that their atoms are bonded to form a regular, symmetrical, three-dimensional pattern. The nature of the bonding forces controls the physical properties of the element, so that elements can be classified according to the type of chemical bond displayed by the crystal; differences in properties can be explained on the basis of differences in the bond character. In the pure elements the interatomic bonding forces may be described as being either covalent or metallic, while in chemical compounds a third type, the ionic bond, may be established. To appreciate the differences in the bonding forces, it is necessary to consider the structure of the atom.

An atom consists of neutrons, which carry no electrical charge, protons, which carry a positive charge, and electrons, which have an equal but negative charge. The electrons are in orbit round the atomic nucleus in a number of distinct energy 'bands' or 'shells'. The lowest order shell, nearest the nucleus, can contain two electrons, while the successively higher order shells may contain 8, 18, 32 etc. These higher order shells are, in turn, subdivided into 'energy levels' with larger energy gaps separating groups of eight electrons into 'stable octets'. These octets of electrons have a marked effect on the chemical bonding which the element portrays.

The periodic table of the elements (table 1.1) expresses the way in which the electrons occupy the energy shells, the horizontal periods revealing the regular increase in electron numbers and the vertical groups depicting those elements which have a similar number of 'valence' or outer shell electrons. This feature accounts for the similarities in the chemical behaviour of elements from the same group. The regularity in the sequence in which electrons occupy the lower order shells, before filling the

Table 1.1 The periodic table of the elements

PERIOD	I	II											III	IV	V	VI	VII	0
	1	2	1										3	4	3	2	1	0
1	**H** 1.008																	2 **He** 4.003
2	3 **Li** 6.940	4 **Be** 9.013											5 B 10.82	6 C 12.01	7 N 14.008	8 O 16	9 F 19.00	10 Ne 20.183
3	11 **Na** 22.997	12 **Mg** 24.32											13 **Al** 26.97	14 Si 28.09	15 P 30.974	16 S 32.066	17 Cl 35.457	18 A 39.944
4	19 **K** 39.100	20 **Ca** 40.08	21 **Sc** 44.96	22 **Ti** 47.90	23 **V** 50.95	24 **Cr** 52.01	25 **Mn** 54.93	26 **Fe** 55.85	27 **Co** 58.94	28 **Ni** 58.69	29 **Cu** 63.54	30 **Zn** 65.37	31 **Ga** 69.72	32 **Ge** 72.60	33 As 74.91	34 Se 78.96	35 Br 79.916	36 Kr 83.80
5	37 **Rb** 85.48	38 **Sr** 87.63	39 **Y** 88.92	40 **Zr** 91.22	41 **Nb** 92.91	42 **Mo** 95.95	43 **Tc** (98.91)	44 **Ru** 101.7	45 **Rh** 102.91	46 **Pd** 106.7	47 **Ag** 107.873	48 **Cd** 112.41	49 **In** 114.76	50 **Sn** 118.70	51 **Sb** 121.76	52 Te 127.61	53 I 126.91	54 Xe 131.3
6	55 **Cs** 132.91	56 **Ba** 137.36	57 **La** 138.92 †	72 **Hf** 178.6	73 **Ta** 180.88	74 **W** 183.92	75 **Re** 186.31	76 **Os** 190.2	77 **Ir** 193.1	78 **Pt** 195.23	79 **Au** 197.2	80 **Hg** 200.61	81 **Tl** 204.39	82 **Pb** 207.21	83 **Bi** 209.00	84 **Po** (210)	85 At (210)	86 Rn 222
7	87 **Fr** (223)	88 **Ra** 226.05	89 **Ac** 227 ‡	(104)	(105)	(106)	(107)	(108)										

← METALS —— NON-METALS →
GROUPS

Electropositive valency ← → Electronegative valency

← Metals with variable valency, i.e. transition groups →

†Lanthanide or rare earth series
‡Actinide or transuranic series

Elements in bold figures are metallic, the others non-metallic
Elements in crosshatched area display properties characteristic of metals and non-metals
Atomic number above symbol of elements
Atomic weight below symbol of elements
Reproduced from *Institution of Metallurgists Year Book* 1959/1960

higher order shells, is interrupted by a series of transition elements. This is due to the overlap in the energies of the higher order shells and causes the variable valence of these elements.

If we exclude the rare earth transition series, the periodic table reveals that elements may possess one of three types of electronic structure, depending upon how the electron shells are occupied, namely:

1. Those elements which have filled electron shells, representing the most stable electronic configuration, which are typified by the chemically inert gases helium and neon. In addition, the inert gases argon and krypton have filled lower order shells but possess a stable octet of valence electrons in the outer shell.
2. Those elements in which the outer shell only is partially filled.
3. Those elements in which the two outer shells are partially filled, typified by the transition elements.

When the pure elements with partially filled valence shells crystallize, the atoms attempt to establish the stable electronic structure shown by the type 1 elements. In establishing the bonding forces with neighbouring atoms in the crystal lattice, the atoms take the line of least resistance.

In general, the elements which exhibit non-metallic properties form a crystal structure in which the bonding is by covalent forces. In forming this type of bond, each atom achieves the stable configuration of eight electrons in the outer shell by sharing valence electrons with neighbouring atoms. Fig. 1.1 depicts the nature of the bond. In this example, the central atom donates an electron to form an electron pair with each of four neighbour atoms.

This behaviour accounts for two significant features in the crystals of these elements. All the valence electrons are firmly bonded to the atoms. Also, the number of near neighbours surrounding any one atom, termed the co-ordination number, is given by the (8-N) rule, where N is the number of valence electrons possessed by the element. Thus carbon, with four valence electrons, has four covalently bonded neighbour atoms in the diamond crystal (fig. 1.2(a)). Bismuth has five valence electrons and, in order to achieve the stable configuration, it adopts a crystal structure with three neighbours (fig. 1.2(b)).

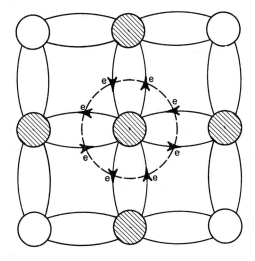

Fig. 1.1 Covalent electron pairs inducing a stable octet in outer valence shell.

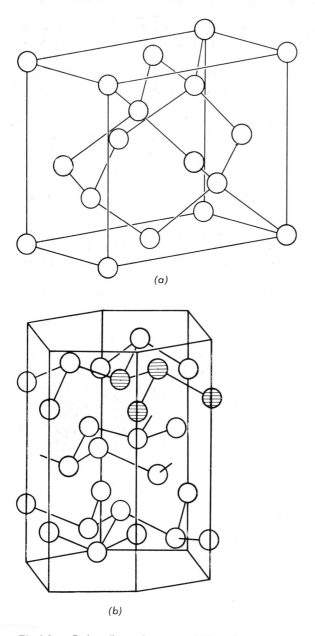

Fig. 1.2 a Carbon diamond structure. *b* Bismuth structure.

Elements having pure covalent bonding possess four or more valence electrons, since it is not possible to form the stable electron octet by electron sharing when the atom has less than this number of bonding electrons.

The periodic table shows that the elements which are wholly metallic in nature possess one or two valence electrons in the outer electron shell. This feature may arise from the regular filling of the outer shell with increasing atomic number, or as a result of electrons occupying a higher order shell before the lower shells are completed, i.e. the transition series. The crystals of these elements have metallic bonding forces, each atom contributing its valence electrons to the formation of an electron cloud that permeates the entire crystal. This action disturbs the electrical neutrality of each atom, which becomes a positively charged ion.

In consequence the metal ions establish an equilibrium spacing on the crystal lattice to balance the repulsive ion–ion forces and the attractive electron–ion bond. In this manner each metal ion achieves the stable electron structure either by the donation of valence electrons to the electron cloud or by the free association of the cloud with any ion. In contrast to the covalent bond, the significant features of metallic bonding are the free movement of the valence electrons within the crystal and the non-dependence of the number of near neighbour atoms on the valence of the element. In fact, metallic elements usually possess closely packed and highly symmetrical crystal structures with co-ordination numbers of eight or twelve.

Whilst differences in the crystal bond clearly distinguish between the non-metallic and metallic groups of elements, it is not easy on this basis to classify a number of elements which lie intermediate in the periodic table between the two groups. The elements with three valence electrons may adopt metallic bonding, as with aluminium, or may possess a structure bonded partly by metallic and partly by covalent bonding forces. The gradual transition from one type of bonding to the other is reflected in the physical and mechanical properties of these elements.

An ionic bond is often formed when two pure elements combine to produce crystalline compounds. This bond type never occurs in the pure elements, since it is created by atoms of one element donating valence electrons which are accepted by the atoms of the other element, so that electron stability is achieved by both species. The exchange of valence electrons results in one element creating positive ions and the other negative ions – hence the term 'ionic bonding'. Fig. 1.2(c) is a schematic illustration of the ionic crystal of the compound sodium chloride, in which the single valence electron of the sodium atom is donated to the chlorine atom, which has seven valence electrons. This action results in both ions achieving the stable octet of electrons in their outer electron shells. In each ion the electronic structure is highly stable, and the crystal is bonded by the electrical attraction between the differently charged ions.

The distinction between metallic and non-metallic properties is derived from the way in which the valence electrons contribute to the different types of bonds. Materials possessing a strong covalent bond are, like ionically bonded solids, hard and brittle and have high melting points. The resistance to chemical change is often a feature of covalent bonding and emphasizes the stability of the bond. Most of these properties are characterized by the elements silicon and carbon (diamond).

In contrast, the free nature of metallic bonding accounts for the marked chemical reactivity, and only moderately high melting points, observed in metals. However, these properties display anomalies which arise as a result of the complex electron structure of the transition element series. The unique character of the metallic bond is reflected more clearly in the physical and mechanical properties of metals (table 1.2).

The high electrical and thermal conductivities displayed by metals depict the free association of the valence electrons in contributing to the metallic bonding. Since the electron cloud is not firmly bound to a given ion in the metallic lattice, the presence of an electric field causes these electrons to flow through the crystal, restricted only by collisions with the ions. These collisions give rise to the resistance in metallic

6 | METALS IN GENERAL

Table 1.2 Properties of the Metallic Elements[1]

Element	Symbol	Atomic Weight[20]	Atomic No.	Melting point[14] °C.	Boiling point[14] °C.	Density[2,3] g/ml.	Specific heat[2,4] cal/g/°C.	Coefficient of Linear Expansion[2,5] × 10^6 cm/cm°C.
Actinium	Ac	(227)	89	(1200)	(3300)	—	—	—
Aluminium	Al	26.9815	13	659	(2440)	2.7	0.215	23.03
Americium[10]	Am	(243)	95	(995)	—	13.7	(12)	6.2, 7.5
Antimony	Sb	121.75	51	630.5	(1635)	6.67	0.049	11.4
Barium	Ba	137.34	56	710	1620	3.6	0.068	(18)/0–100°
Berkelium	Bk	(247)	97	—	—	—	—	—
Beryllium	Be	9.0122	4	1283	2480	1.85	0.42	11.54/25–100°
Bismuth	Bi	208.980	83	271.3	1660	9.8	0.034	17.5, 11.7[15]
Cadmium	Cd	112.40	48	320.9	765	8.65	0.055	29.8
Caesium	Cs	132.905	55	28.6	690	1.95	0.05	—
Calcium	Ca	40.08	20	850	1480	1.53	0.149/0–100°	25/0–21°
Californium	Cf	(249)	98	—	—	—	—	—
Cerium	Ce	140.12	58	804	3500	6.8	0.045	6[(24)]
Chromium	Cr	51.996	24	1900	2665	7.1	0.11	6.2
Cobalt	Co	58.9332	27	1492	2900	8.7	0.099	12.3
Copper	Cu	63.54	29	1083	2570	8.9	0.092	16.6
Curium[10]	Cm	(247)	96	(1340)	—	—	—	—
Dysprosium	Dy	162.50	66	(1500)	(2600)	8.5	0.04	9
Einsteinium	Es	(254)	99	—	—	—	—	—
Erbium	Er	167.26	68	(1500)	(2600)	9.0	0.04	9.5
Europium	Eu	151.96	63	826	(1500)	5.27	0.04	32/50°
Fermium	Fm	(253)	100	—	—	—	—	—
Francium	Fr	(223)	87	—	—	—	—	—
Gadolinium	Gd	157.25	64	(1350)	(3000)	7.9	0.07	(4)
Gallium	Ga	69.72	31	29.8	2250	5.9	0.079	11.5, 31.5, 16.5[(7,9c)]
Germanium	Ge	72.59	32	937	2870	5.4	0.073	(6)/0–100°
Gold	Au	196.967	79	1059	2810	19.3	0.031	14.2
Hafnium	Hf	178.49	72	2220	(3100)	13.1	0.035	5.9
Hoimium	Ho	164.930	67	(1500)	(2700)	8.76	0.04	9.5/400°
Indium	In	114.82	49	156	2070	7.32	0.057	33
Iridium	Ir	192.2	77	2454	4390	22.4	0.031	6.5
Iron	Fe	55.847	26	1536	2875	7.9	0.11	11.7
Lanthanum	La	138.91	57	920	3360	6.15	0.048	5
Lead	Pb	207.19	82	327.4	1740	11.3	0.031	29.1
Lithium	Li	6.939	3	180	1324	0.545	0.79	56
Lutetium	Lu	174.97	71	(1700)	(1900)	9.8	0.04	12.5/400°
Magnesium	Mg	24.312	12	650	1105	1.74	0.25	26.0/–100°
Manganese	Mn	54.9380	25	1244	2050	7.4	0.115	23

STRUCTURE AND PROPERTIES OF METALS

Element	Thermal Conductivity[2,4] cal cm/ cm² sec°C.	Resistivity,[2,5] microhm. cm.	Crystal Structure[6]	Atomic Radius[3] (Shortest Distances) Å	Standard Potentials[3] Volts	Ion
Actinium	—	—	—	—	—	—
Aluminium	0.53	2.69	f.c.c.	1.43	−1.66	Al^{+++}
Americium[10]	—	—	c.p.h., f.c.c.	1.73	—	—
Antimony	0.045	42	r.	1.45	+0.18	Sb^{+++}
Barium	—	50	b.c.c.	2.17	−2.93	Ba^{++}
Berkelium	—	—	—	—	—	—
Beryllium	0.53[(9a)]	3.8[(9a)]	c.p.h.	1.125	−1.7	Be^{++}
Bismuth	0.020	116	r.	1.56	+0.28	Bi^{+++}
Cadmium	0.22	7.4	c.p.h.	1.486	−0.402	Cd^{++}
Caesium	—	21	b.c.c.	2.62	−2.92	Cs^{+}
Calcium	0.3	4.1 (soft) 4.37 (hard)	f.c.c. c.p.h.	1.975	−2.83	Ca^{++}
Californium	—	—	—	—	—	—
Cerium	0.026	78	c.p.h. (α) f.c.c. (β) f.c.c. (γ)	1.82	−2.335	Ce^{+++}
Chromium	0.21[(11b)]	13.5[(11b)]	b.c.c.	1.246	−0.71	Cr^{+++}
Cobalt	0.165	6.3[(1)]	c.p.h. (α) f.c.c. (β)	1.25	−0.27	Co^{++}
Copper	0.94	1.673	f.c.c.	1.275	+0.52 +0.34	Cu^{+} Cu^{++}
Curium[10]	—	—	c.p.h. f.c.c.	—	—	—
Dysprosium	0.024	90[(18)]	c.p.h.	1.75	−2.2	Dy^{+++}
Einsteinium	—	—	—	—	—	—
Erbium	0.023	85[(18)]	c.p.h.	1.73	−2.1	Er^{+++}
Europium	—	90	b.c.c.	1.98	−2.2	Eu^{+++}
Fermium	—	—	—	—	—	—
Francium	—	—	—	—	—	—
Gadolinium	0.021	130[(18)]	c.p.h.	1.80	−2.2	Gd^{+++}
Gallium	0.098, 0.21, 0.038[(7,9c)]	17.4, 8.1, 54.3[(7,9c)]	orthorh.	1.33	−0.52	Ga^{+++}
Germanium	0.14	(100 × 10³)	d.	1.22	−0.15	Ge^{++++}
Gold	0.71	2.3	f.c.c.	1.435	+1.7 +1.42	Au^{+} Au^{+++}
Hafnium	0.056	(35.1)	c.p.h., b.c.c.	1.57	−1.7	Hf^{++++}
Hoimium	—	80[(18)]	c.p.h.	1.74	−2.1	Ho^{++++}
Indium	0.205[(9j)]	9.0	f.c.t.	1.45	−0.34	In^{+++}
Iridium	0.355[(9d)]	4.71/0°	f.c.c.	1.35	(+1.0)	Ir^{+++}
Iron	0.18	9.71	b.c.c. (α) f.c.c. (γ) b.c.c. (δ)	1.24 1.26	−0.44 −0.04	Fe^{++} Fe^{+++}
Lanthanum	0.033	80[(17)]	c.p.h. f.c.c.	1.86	−2.4	La^{+++}
Lead	0.083	20.6	f.c.c.	1.745	−0.126	Pb^{++}
Lithium	0.17	9.35	b.c.c. c.p.h., faulted/ −195° f.c.c., strain-induced	1.515	−3.01	Li^{+}
Lutetium	—	55[(18)]	c.p.h.	1.73	−2.1	Lu^{+++}
Magnesium	0.367	3.9	c.p.h.	1.60	−2.38	Mg^{++}
Manganese	—	260	c. (α) c. (β) f.c.t. (γ) b.c.c. (δ)	1.13 1.18 1.29	−1.05	Mn^{++}

8 | METALS IN GENERAL

Table 1.2 (continued)

Element	Symbol	Atomic Weight[20]	Atomic No.	Melting point[14] °C.	Boiling point[14] °C.	Density[2,3] g/ml.	Specific heat[2,4] cal/g/°C.	Coefficient of Linear Expansion[2,5] × 10^6 cm/cm°C.
Mendelevium	Md	(256)	101	—	—	—	—	—
Merucry	Hg	200.59	80	−38.87	356.58	13.5459[21]	0.033	61
Molybdenum	Mo	95.94	42	2620	4650	10.2	0.066	5.35/0−20°
Neodymium	Nd	144.24	60	1024	(3100)	7.00	0.05	6.5
Neptunium	Np	(237)	93	640	—	—	—	—
Nickel	Ni	58.71	28	1455	2890	8.9	0.105	12.8
Niobium	Nb	92.906	41	2468	4730	8.6	0.064	7.1
Nobelium	No	(256)	102	—	—	—	—	—
Osmium	Os	190.2	76	3045	4300	22.5	0.031	6.1
Palladium	Pd	106.4	46	1552	(3000)	12.0	0.058/0°	11.8
Platinum	Pt	195.09	78	1769	3825	21.45	0.032	8.9
Plutonium[13]	Pu	(242)	94	640	3235	19.81/25°	0.05	46.8/ −186−100°
Polonium	Po	(210)	84	246	965	—	—	—
Potassium	K	39.102	19	63.2	758	0.88	0.177	83
Praseodymium	Pr	140.907	59	935	(3000)	6.8	0.045	4.5
Promethium	Pm	(147)	61	—	—	—	—	—
Protactinium	Pa	(231)	91	—	—	—	—	—
Radium	Ra	(226)	88	700	1140	5	—	—
Rhenium	Re	186.2	75	3180	(5650)	21.05	0.033	6.6
Rhodium	Rh	102.905	45	1966	3780	12.4	0.059	8.4
Rubidium	Rb	85.47	37	38.8	684	1.53	0.080	90
Ruthenium	Ru	101.07	44	3400	4100	12.4[26]	0.055[26]	9.1
Samarium	Sm	150.35	62	1072	1800	7.5	0.05[11]	—
Scandium	Sc	44.956	21	1540	2950	3.0	0.13	12/25−100°
Silicon	Si	28.086	14	1410[23]	3280	2.34	0.162/0°	2.8−7.3
Silver	Ag	107.870	47	960.8	2164	10.5	0.056/0°	18.9
Sodium	Na	22.9898	11	97.7	883	0.97	0.295	71
Strontium	Sr	87.62	38	770	1375	2.6	0.176	23
Tantalum	Ta	180.948	73	3000	5000	16.6	0.034[12]	6.6
Technetium	Tc	(99)	43	(2200)[20]	—	—	—	—
Terbium	Tb	158.924	65	1365	(2500)	8.25	0.04	7
Thallium	Tl	204.37	81	304	1490	11.85	0.031	28
Thorium	Th	232.038	90	1700	4850	11.5	0.028	11.55/20−100°
Thulium	Tm	168.934	69	(1600)	(2400)	9.3	0.04	11.6/400°
Tin	Sn	118.69	50	—	—	—	—	—
White		—	—	231.9	2620	7.3	0.054	—
Grey		—	—	—	—	—	—	—
Titanium	Ti	47.90	22	1660	3260	4.51	0.125	8.5/25°

STRUCTURE AND PROPERTIES OF METALS | 9

Element	Thermal Conductivity[2,4] cal cm/ cm² sec°C.	Resistivity,[2,5] microhm. cm.	Crystal Structure[6]	Atomic Radius[3] (Shortest Distances) Å	Standard Potentials[3] Volts	Ion
Mendelevium	—	—	—	—	—	—
Mercury	0.0201	95.8	r.	1.50	+0.798	Hg^{++}
Molybdenum	0.34	5.65[11f]	b.c.c.	1.36	(−0.2)	Mo^{+++}
Noedymium	0.031	64	c.p.h.	1.81	−2.24	Nd^{+++}
Neptunium	—	—	—	—	—	—
Nickel	0.22	6.84	f.c.c.	1.25	−0.23	Ni^{++}
Niobium	0.125/0°	16.9[19f]	b.c.c.	1.43	−1.1	Nb^{+++}
Nobelium	—	—	—	—	—	—
Osmium	0.21[9e]	8.3/0°[9e]	c.p.h.	1.34	+0.7	Os^{++}
Palladium	0.18	9.93/0°[9e]	f.c.c.	1.37	(+0.83)	Pd^{++}
Platinum	0.17	9.85/0°[9e]	f.c.c.	1.38	(+1.2)	Pt^{++}
Plutonium[13]	0.013	145	monocl (α) b.c. monocl. (β) f.c. orthor. (γ) f.c.c. (δ) f.c. tet. (δ′) b.c.c. (ε)	1.58	−2.03	Pu^{+++}
Polonium	—	50/30° (α) 80/30° (β)	c. (α) r. (β)	(1.40)	—	—
Potassium	0.24	6.86	b.c.c.	2.31	−2.92	K^+
Praseodymium	0.028	68	c.p.h. f.c.c.	1.82	−2.2	Pr^{+++}
Promethium	—	—	f.c.c.	—	—	—
Protactinium	—	—	—	—	—	—
Radium	—	—	—	—	—	—
Rhenium	0.115[9h]	18.8	c.p.h.	1.37	—	—
Rhodium	0.36[9d]	4.33/0°[9e]	f.c.c.	1.342	—	—
Rubidium	—	12.5	b.c.c.	2.485	−2.95	Rb^+
Ruthenium	0.25[9e]	7.13/0°[9e,22]	c.p.h.	1.34	+0.45	Ru^{++}
Smarium	—	105[17]	r.	1.80	−2.2	Sm^{+++}
Scandium	—	66	c.p.h.	1.51	—	—
Silicon	0.25/100°	85 × 10³	d.	1.175	—	—
Silver	1.0/0°	1.6	f.c.c.	1.44	+0.80	Ag^+
Sodium	0.32	4.6	b.c.c.	1.855	−2.71	Na^+
		22.76	f.c.c.	2.145	−2.9	Sr^{++}
Strontium	—		orthorh. (α)	1.06	−0.51	S''
		2 × 10²³				
Tantalum	0.138	14.5[9f]	b.c.c.	1.425	−1.12	Ta^{+++++}
Technetium	—	—	—	—	—	—
					+0.56	Te^{++++}
Terbium	—	115[18]	c.p.h.	1.76	−2.2	Tb^{+++}
Thallium	0.093	16.6	c.p.h. (α) b.c.c. (β)	1.70	−0.335	Tl^+
Thorium	0.090/100°	(13–14)	f.c.c. (α) b.c.c. 1450° (β)	1.80	(−2.1)	Th^{++++}
Thulium	—	65[18]	c.p.h	1.73	−2.1	Tm^{+++}
Tin	—	—	—	—	—	—
White	0.16	12.8	b.c.t.	1.508	−0.14	Sn^{++}
Grey	—	—	d.	—	—	—
Titanium	0.052[9g]	42.1	c.p.h. (α) b.c.c. (β)	1.46	−1.75	Ti^{++}

Table 1.2 (continued)

Element	Symbol	Atomic Weight[20]	Atomic No.	Melting point[14] °C.	Boiling point[14] °C.	Density[2,3] g/ml.	Specific heat[2,4] cal/g/°C.	Coefficient of Linear Expansion[2,5] × 10⁶ cm/cm°C.
Tungsten	W	183.85	74	3380	5500	19.3	0.032	4.6[25]
Uranium	U	238.03	92	1130	4200	19.05	0.028	23, −3.5, 17/ 25–300°[7] 4,6, 23.0/ 20–720°[8]
Vanadium	V	50.942	23	1910	3380	6.15	0.120	9.12/18–100°
Ytterbium	Yb	173.04	70	824	1600	7.0	0.035	25
Yttrium	Y	88.905	39	1530	3300	4.47	0.07	(10)
Zinc	Zn	65.37	30	419.5	911	7.14	0.0915	33
Zirconium	Zr	91.22	40	1850	4400	6.4	2.5, 0.066	2,5, 14.3[8]

[1] All temperatures in °C.
Brackets, indicate approximate values.
[2] At 20°C. unless stated otherwise.
[3] Numerous data for densities, atomic radii and standard electrode potentials with respect of hydrogen were supplied by Dr. O. Kubaschewski, National Physical Laboratory.
[4] Based on *Metals Handbook*, pp. 46–49 (ASM, 1961).
[5] Based on pp. 695–696 of *Metals Reference Book*, by C. J. Smithells (Butterworth Scientific Publications, London, 1962) and pp. 101–107 of *Lange's Handbook of Chemistry* (Handbook Publishers Inc. Sandusky; Ohio, 1956).
[6] Based on pp. 136–137 of *Metals Reference Book* (b.c.)c. f.c.c. = (body-centred) cubic, face-centred cubic; d. = diamond structure; (c.p.)h. = (close-packed) hexagonal; (f.c.)t. = (face-centred) tetragonal; (f.c.) orthorh. = (face-centred) orthorhombic; monocl. = monoclinic; r. = rhombohedral.
[7] Parallel to a, b and c axes respectively.
[8] Parallel and vertical to c-axis respectively.
[9a] R. W. Powell: *Phil. Mag.*, 1953, **44**, 645.
[b] R. W. Powell and R. P. Tye: *J. Inst. Met.*, 1956/57, **85**, 185.
[c] R. W. Powell: *Proc. Roy. Soc.*, A, 1951, **209**, 525. – R. W. Powell, M. J. Woodman and R. P. Tye: *Br. J. Appl. Phys.*, 1963, **14**, 432.
[d] R. W. Powell and R. P. Tye: *Proc. IXth International Congress of Refrigeration*, 1955, **1**, 2083.
[e] R. W. Powell, R. P. Tye and M. J. Woodman: *Platinum Metals*, 1962, **6**, 138.
[f] R. P. Tye: *J. Less Common Metals*, 1961, **3**, 13.
[g] R. P. Tye: *Ibid.*, p. 226.
[h] R. W. Powell, R. P. Tye and M. J. Woodman: *Ibid.*, 1963, **5**, 49.
[i] R. W. Powell, M. J. Woodman and R. P. Tye: *Phil. Mag.*, 1962, viii, **7**, 1183.
[j] R. W. Powel: *Cobalt*, 1964, (24), 145.

STRUCTURE AND PROPERTIES OF METALS | 11

Element	Thermal Conductivity[2,4] cal cm/ cm² sec°C.	Resistivity,[2,5] microhm. cm.	Crystal Structure[6]	Atomic Radius[3] (Shortest Distances) Å	Standard Potentials[3]	
					Volts	Ion
Tungsten	0.45	5.45[(9f)]	b.c.c.	1.41	−1.1	W^{++++++}
Uranium	0.064	29 (α)	orthorh. (α) t. (β) b.c.c. (γ)	1.38	−0.82	U^{++++++}
Vanadium	0.074/100°	19.5	b.c.c.	1.313	−1.5	V^{++}
Ytterbium	—	30	f.c.c.	1.93	−2.1	Yb^{+++}
Yttrium	0.035	80	c.p.h.	1.80	—	—
Zinc	0.27	5.92	c.p.h.	1.33	−0.763	Zn^{++}
Zirconium	0.05	44.6	c.p.h. (α) b.c.c. (β)	1.59 1.57	−1.5	Zr^{++++}

[10] B. B. Cunningham: "Thermodynamics of the Actinides", pp. 61–70 of "Thermodynamics of Nuclear Materials" (IAEA, Vienna, 1962)– D. B. McWhan, B. B. Cunningham and J. C. Wallman: *J. Inorg. Nucl. Chem.*, 1962, **24**, 1025.
[11] L. D. Jennings, E. D. Hill and F. H. Spedding: *J. Chem. Phys.*, 1959, **31**, 1240.
[12] K. F. Sterrett and W. E. Wallace: *J. Am. Chem. Soc.*, 1958, **80**, 3176.
[13] The majority of values for plutonium were provided by Dr. R. J. Wakelin, United Kingdom Atomic Energy Authority.
[14] The data for melting and boiling points of metals are based on "Selected Values of Thermodynamic Properties of Metals and Alloys" by R. Hultgren, R. L. Orr, P. D. Anderson and K. K. Kelley. (Wiley, 1963).
[15] E. F. Cave and L. V. Holroyd: *J. Appl. Phys.*, 1960, **31**, 1357. Values for c and a axes of hexagonal triple cell.
[16] E. Anderson, R. A. Buckley, A. Hellawell and W. Hume-Rothery: *Nature*, 1960, **188**, 48.
[17] J. K. Alstad, R. V. Colvin, S. Legvold and F. H. Spedding: *Phys. Rev.*, 1961, **121**, 1637.
[18] R. V. Colvin, S. Legvold and F. H. Spedding: *Phys. Rev.*, 1960, **120**, 741.
[19] A. G. Knapton, J. Savill and R. Siddall: *J. Less Common Metals*, 1960, **2**, 357.
[20] International Union of Pure and Applied Chemistry Table of Atomic Weights, 1961; based on nuclidic mass of $C^{12} = 12$: *Pure and Appl. Chem.*, 1962, **5**, 258; by permission of IUPAC and Butterworths Scientific Publications.
[21] P. H. Bigg: *Br. J. Appl. Phys.*, 1964, **15**, 1111.
[22] R. J. Tainsch and G. K. White: *Can. J. Phys.*, 1964, **42**, 208.
[23] L. D. Lucas: *Rev. Mét., Mém. Sci.*, 1964, **61**, 22.
[24] R. D. Smith and E. Morrice: U.S. Bureau of Mines, 1964, R.I. 6480.
[25] B. N. Dutta and B. Dayal: *Phys. Stat. Solidi*, 1963, **3**, 2253.
[26] International Nickel Co. (Mond) Ltd.: "Ruthenium", 1963.

Reproduced from The Institution of Metallurgists Year Book 1966–1967.

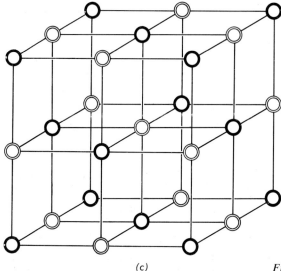

(c)   *Fig. 1.2 c* Sodium chloride structure.

conductors. The motion of the electrons through the crystal does not disrupt the bonding force, since the number of electrons surrounding any ion remains unaltered. In solids subject to covalent or ionic bonding, the valence electrons are held in stable configurations so that electronic conduction is not possible. For this reason, the non-metallic elements tend to be insulators.

Even though the conduction of thermal energy in all solids is achieved partially by electron flow, it is apparent that the mobility of the electron cloud contributes substantially to the high values of thermal conductivity observed in metals.

A feature of metallic bonding is that each positive ion is highly symmetrical and, as a result of this symmetry, metals form close-packed crystal structures. The high degree of formability observed in metals is attributed to the ease with which the close-packed ions can rupture and re-establish the 'loose' bonding forces with the free electron cloud within the crystal. In addition, the close-packed structure of metals is a cause of their high density and promotes their high elastic moduli.

1.1.2 Metallic alloys

The interplay of such properties as strength, formability and high electrical conductivity accounts for the widespread application of metals. However, this combination of properties is not always desirable. While all three would be necessary for the manufacture and use of transmission cables, for heating elements the mechanical properties must be coupled with high electrical resistance. In addition, since heating elements operate at high temperatures, the chemical reactivity of the pure metals would result in catastrophic failure due to oxidation by the atmosphere. It is apparent that the limited range of properties offered by the pure metals does not fulfil the requirements of many industrial uses. In practice this problem is overcome by using specially prepared alloys.

Alloys consist of two or more elements and are formed either when other elements remain in a metal during the extractive process, or by the deliberate addition of other elements to a pure metal. The object of alloying metals is to achieve specific chemical, physical and mechanical properties, in addition to promoting economies in the manufacture and use of components. The properties displayed by alloys vary widely, since they depend upon the nature of the atomic structure of the constituent elements.

The atoms of a metal are arranged in a regular pattern or space lattice. When a second element dissolves in a pure solid metal, it is possible for some of the atoms of the parent metal, the solvent, to be replaced by the atoms of the dissolved element, which is the solute. Alternatively, if the atomic size of the solute is small compared with that of the metal, then the atoms of the solute may be positioned in the spaces between the parent atoms. An analogy with a familiar liquid solution may make this clear. When sugar is added to water, the molecules of sugar disperse among those of the water, and the result is a solution. This is comparable with the dispersion of foreign atoms among the parent atoms of a metal, and such a dispersion is itself a solution, but since the metal is solid it is called a 'solid solution', and the metal is said to possess solid solubility. The extent of solid solubility in alloys is governed by the ability of the crystal to maintain the metallic bonding forces. Thus, the addition of elements with lower valence to a solvent of high valence would reduce the average valence electron-atom ratio and the bonding forces would not be preserved. Similarly, solid solubility is restricted when two elements possessing markedly different atomic sizes are alloyed, since the crystal lattice becomes distorted and the uniform bonding forces are disturbed.

Solid solutions of metals are very common. Most alloys in everyday use consist of one or more solid solutions, of which some of the most important are composed of non-metals in the parent metal. Carbon, nitrogen, oxygen and phosphorus are often involved.

When an alloying element is added to a given metal in an amount which exceeds the limit of solid solubility, a second 'phase' appears together with the primary solid solution. The second phase may be another solid solution formed by the alloying addition. More often it is an intermediate compound, and is so called since its composition lies intermediate between two solid solutions. The type of compound formed at high solute concentrations is governed by the same factors controlling the solubility limits of solid solutions. Interstitial compounds, which are formed when the interstitial solid solubility limits are exceeded, usually form by covalent bonding and exhibit the characteristic high hardness and high melting points. The valence compounds form when the alloying elements obey the normal rules of chemical valence and are bonded by ionic and covalent forces. Metallic bonding is observed in compounds arising from the alloying of two metals of different valence so that, at definite valence electron-atom ratios, phases with complex crystal structures are found in association with the solid solution.

The structure and hence the properties of alloys are essentially determined by the atomic nature of the constituent elements. For this reason, the solid solution alloys formed between two metals display properties which are not dissimilar to those of a pure metal. In alloys containing more than one phase, the properties may be varied, depending upon the amounts and types of phases present in the microstructure.

The addition of small amounts of carbon to pure iron produces an interstitial solid solution whose properties are wholly metallic. When the amount of carbon added to the alloy exceeds the limit of solid solubility, the excess carbon separates from solid solution to form the covalently-bonded compound, iron carbide. The physical properties of the aggregate can be attributed to the mixture of the two bond types. However, the mechanical properties of alloys do not conform with those predicted by the simple atomic theory and, by virtue of their importance in metal-forming processes, merit a more detailed examination.

1.1.3 Metallic crystal structures

The feature common to all true solids is the regular arrangement of atoms which constitute their crystal structures. This regularity affects almost all the properties of solids, so that it is important to describe it quantitatively. The science of crystallography has adopted a system of classifying all possible crystal structures as belonging to one of 14 space lattices. This convention states that a space lattice arises when a distribution of points in three dimensions is arranged so that each point has identical surroundings. The portion of the general space lattice shown in fig. 1.3 reveals that it is built up by the repetition of a simple geometric unit, called the unit cell, and that the geometry of the unit cell is wholly defined by three lattice constants a, b and c and the interaxial angles α, β and γ.

The crystal structure adopted by an element when it crystallizes is commonly named after the geometric shape of the unit cell. Most metals possess non-simple cubic structures, termed face-centred cubic (f.c.c.) and body-centred cubic (b.c.c.) types, and the unit cells of these are shown in fig. 1.4 together with the lattice constants. The elements iron, chromium and vanadium typify the metals which possess body-centred cubic lattices, in which the atoms are arranged at the corners and centre of a cube and each atom has eight near neighbours. Copper, nickel and aluminium are typical of elements forming the face-centred cubic lattice, in which atoms are situated at the corners and the centre of the cube faces. This structure represents one of two ways in which atoms of equal size may achieve the closest possible packing, with each atom having 12 near neighbours. The other method of close packing atoms is seen in the close-packed hexagonal (c.p.h.) lattice, which is adopted by elements such as magnesium, zinc and cadmium (fig. 1.5). This structure is characterized by having atoms in

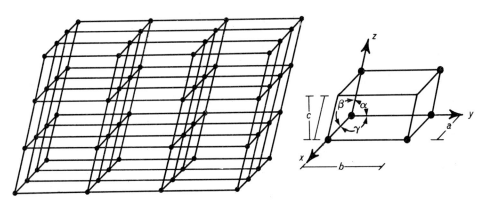

Fig. 1.3 General space lattice and unit cell defined by lattice 'vectors' a b c and interaxial angles α β γ.

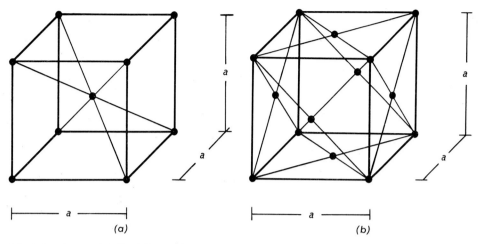

Fig. 1.4 *a* b.c.c unit cell, *b* f.c.c. unit cell.

alternate layers situated at equivalent lattice points, with atoms in the intermediate layer occupying the interstice created by the alternate layers. The layer structure is responsible for the unusual deformation characteristics of the metals having hexagonal structures.

Specific atomic planes and directions within the lattice result from differences in atomic packing. A quantitative description of these is provided by the Miller index. This notation uses the smallest integer reciprocals of the intercepts made by a plane with the axes of the unit cell, in terms of the cell constants as unit dimensions. Thus, the index of the atomic plane normal to the x axis of a cube is given as (100) (fig. 1.6(a)), since it makes intercepts at a distance a along the x axis and at infinity with the y and z axes. Fig. 1.6(b) denotes other important atomic planes in the cubic lattice, and reveals that the Miller indices express the orientation of planes with respect to an arbitrary origin.

Directions in the lattice are represented by the integer values of the coordinates, in terms of the lattice constants, at which a line from the origin intersects the unit cell.

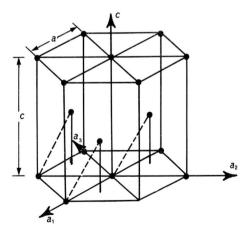

Fig. 1.5   c.p.h. unit cell.

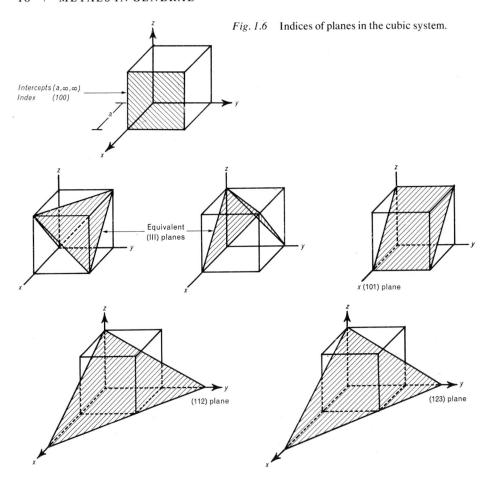

Fig. 1.6 Indices of planes in the cubic system.

Thus the [111] direction in a cubic lattice would denote the cube diagonal, the line from the origin intersecting the unit cell at coordinates (a, a, a). In order to distinguish between planes and directions, the latter are enclosed in square brackets. Examples of directions in the cubic lattice are given in fig. 1.7.

1.1.4 Deformation of metals

The ease with which metallic solids plastically deform is explained not only by the nature of their bonding forces but can be attributed also to the symmetry of their crystal lattices. For plastic deformation to take place, the large-scale movement of adjacent atomic planes must occur by 'slip'; the atoms change their relative positions but the crystal lattice structure is preserved. Each of the different crystal lattices has atomic planes along which slip can take place in specific directions under a smaller applied stress than on other planes and/or directions. The slip plane and the slip direction constitute a slip system. In the face-centred cubic and close-packed hexagonal lattices, where atoms are arranged in the most closely packed arrays, slip occurs along the planes having the closest atomic packing, due to their large interplanar spacing, and the slip direction is the direction of closest atomic packing. A single crystal pos-

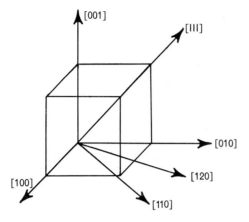

Fig. 1.7 Lattice directions in cubic metals.

sessing the close-packed hexagonal symmetry has only one type of slip plane, although slip may occur within this plane in one of three directions (fig. 1.8). The small number of easy slip systems is the cause of the low ductility in metals and alloys which form this structure.

In contrast, the face-centred cubic crystals have four slip planes, each of which has three slip directions, giving a total of 12 possible slip systems. This explains the high ductility of metals such as copper, silver and gold which possess this crystal symmetry. In the body-centred cubic structure there are no true close-packed atomic planes, although the [1̄11] direction is the most closely packed one. Iron, containing carbon in solid solution, is unique among body-centred cubic materials in exhibiting 42 slip systems. Fig. 1.9 depicts the usual slip systems available in normal crystals of this symmetry. An alternative mechanism of deformation may occur when the slip process is inhibited, for example, in zinc at normal temperatures, and is termed 'twinning'. Unlike the slip process, twinning involves the cooperative movement of successive atomic planes about a particular (twin) plane, so that the lattice on each side of the twin plane forms a mirrored image.

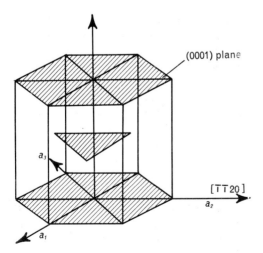

Fig. 1.8 Slip planes (shaded) and equivalent slip directions (arrowed) in the close-packed hexagonal system.

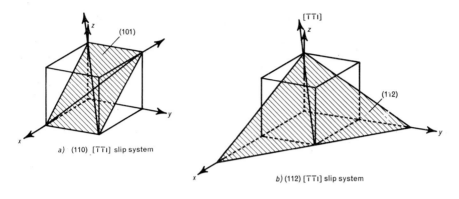

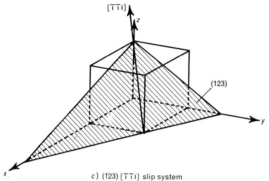

a) (110) [1̄1̄1] slip system

b) (112) [1̄1̄1] slip system

c) (123) [1̄1̄1] slip system

Fig. 1.9 Slip systems in b.c.c. lattices.

The simplest mechanism by which slip could occur in a single crystal would involve the simultaneous movement of an entire plane of atoms relative to the adjacent plane. It is possible to calculate the shear stresses necessary to induce slip in this manner from knowledge of the atomic spacing in the slip plane, the distance apart of adjacent slip planes and the elastic shear modulus of the crystal. Since the atomic positions in a crystal represent a balance between the attractive binding forces and the repulsive ion–ion forces, the atoms occupying these sites must possess minimum energy. The slip plane can be represented schematically as in fig. 1.10, where each atom has minimum energy at the interatomic spacing d. When a crystal deforms elastically each atom is pulled out of the minimum energy trough by the applied stresses. The limit of elastic deformation results when each atom attains the peak of the energy hill. If the applied stresses are removed before the peak is passed, each atom slides down the energy hill and is restored to its former site. The shear stress to induce this movement is given by the relationship:

$$\tau = G/\theta$$

where G is the elastic shear modulus, τ the shear stress and θ the shear strain. For face-centred cubic metals, the common slip planes are spaced at a distance approximately $2d$ apart, so that the limit of the elastic shear strain is $\frac{1}{2}d/2d = \frac{1}{4}$. Hence, the

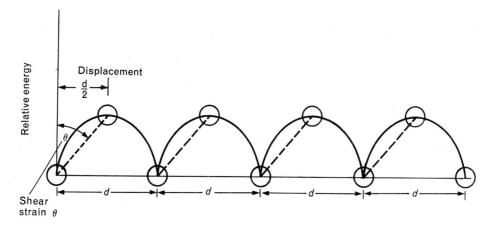

Fig. 1.10 Schematic representation of energy variation along a slip path.

shear stress corresponding to the onset of plastic flow is $\frac{1}{4}G$ or 7×10^6 p.s.i. In practice, metals deform by slip under shear stresses several thousand times lower than the theoretical value, so the atomic theory does not adequately explain the mechanical properties of solids.

If the atoms in a slip plane move consecutively and not simultaneously during slipping, then the lower shear stresses could be explained. By analogy, consider the force exerted by an engine in moving a row of railway wagons if the engine shunts only the first wagon but this transmits its movement down the entire row. The force necessary to accomplish this action is considerably lower than that required to move the entire row simultaneously. In a crystal, at any stage of slip the individual movement of atoms in the slip plane would give rise to a boundary separating the slipped and unslipped portions of the crystal. This boundary would constitute an imperfection in the crystal lattice and would extend over the entire slip plane. In real crystals such imperfections exist and are termed dislocations. The magnitude of the dislocation is measured by the degree of misfit it introduces into the lattice and is termed the Burgers vector. The direction of the dislocation line during slip and its Burgers vector classify it as an edge or screw dislocation type, and both are shown in fig. 1.11.

Since the concept of the dislocation was introduced by E. Orowan, M. Polanyi and G. I. Taylor in 1934, considerable progress has been effected in the explanation of many phenomena which occur in metals as a result of imposed stresses. These effects will be discussed in some detail in the section on Tensile testing, 1.4.2, p.

1.2 HEAT TREATMENT

1.2.1 Introduction

One of the reasons why metals play such a prominent part in everyday life is because they possess a wide range of properties which may be varied and modified according to the intended application and use. Thus, in any one metal or alloy, it may become necessary to alter the strength, increase the workability in order that further deformation may be imposed (see ch. VI), increase the toughness to withstand

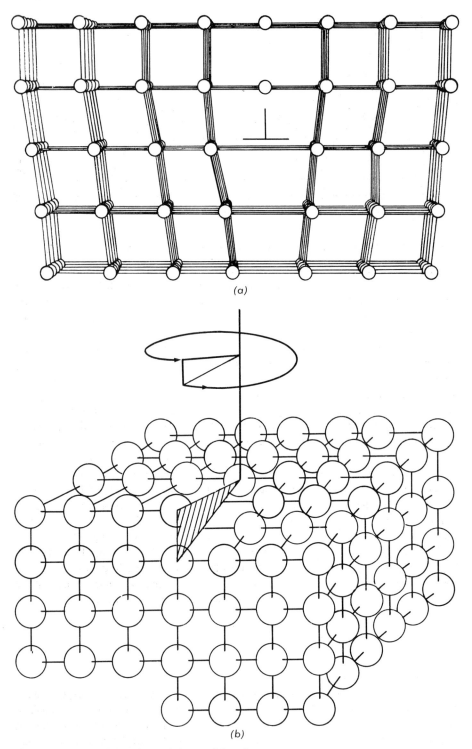

Fig. 1.11 a Edge dislocations *b* Screw dislocation.

service conditions or improve the corrosion resistance. The fatigue resistance of a metal may be significantly improved by suitable treatment and the surface condition of the metal may be modified to further improve the fatigue life, enhance corrosion resistance or increase wear resistance. One of the ways in which these changes may be brought about is by heat treatment in which the metal is heated, possibly at a controlled rate, to a predetermined temperature. It is held for a given time at that temperature before being cooled at a carefully selected rate to room (or even a lower) temperature. The heat treatment may have to be performed in a specific environment and for certain purposes the metal may be immersed in a molten salt or placed in a furnace with a mixture of such gases as carbon monoxide, hydrogen and methane. The heat treatment may require more than one heating and cooling cycle.

In order to appreciate the effect of heat treatment on metal properties reference has to be made to the structure of metals as observed by a microscope. Basically the properties developed in a material are a reflection of the structure of that material and if the structure can be changed then the properties will alter. Pure metals and simple alloys may be regarded as being of a uniform nature throughout and consist of a mass of grains whose size and shape may be altered by deformation and heat treatment. Fig. 1.12 shows the microstructure of brass after cold rolling, and then after heat treatment at 450°C and 700°C. The structures are revealed by microscopic examination of a suitably polished and etched sample. The grain size and shape determine the properties, which are listed beneath each photograph.

However, the more complex alloys cannot be regarded as simple uniform substances. Thus, just as concrete is a mixture of several ingredients, so are the vast majority of alloys used industrially. To the naked eye they may appear homogeneous but on microscopic examination it can be seen that two or more constituents exist. The way in which the second and subsequent constituents are dispersed in the basic metal will determine the properties developed, and this dispersion can be very considerably influenced by heat treatment. If we consider steel to be a mixture produced by adding carbon to iron, then we can anticipate that the form of this carbon will affect such properties as the strength and toughness of the steel. Microscopical examination will reveal that the carbon (as iron carbide) can exist in a plate-like form (fig. 1.13(a)) or as very fine or coarse globules (figs. 1.13(b) and 1.13(c)). These vastly differing structures have been induced by heat treatment and have enabled the range of properties shown beneath fig. 1.13 to be developed.

In order that a proper understanding may be achieved of the control of structure by heat treatment it is necessary first to have an appreciation of alloy theory and of the way in which the dispersion of a second metal in the first may be altered by temperature, time at a temperature and by heating and cooling rates.

1.2.2 Alloy theory and the principles of heat treatment

We have seen from Section 1.1.2 that an alloy consists of two or more elements, at least one of which is a metal; the other element or elements may be metals or non-metals, but the alloy has the characteristic properties of a metal.

Under the microscope a simple uniform solid solution is indistinguishable from a pure metal unless a colour change is apparent, as with copper alloys. Since there are now at least two separate elements involved, however, uniformity of distribution may not necessarily prevail. Even in a liquid solution segregation may occur, e.g. sugar, when added to water, may be concentrated at the bottom unless it is stirred. Similarly,

22 | METALS IN GENERAL

(a)

(b)

(c)

Fig. 1.12 Microstructure of 70/30 brass
a after cold rolling
b after rolling and annealing at 450°C
c after rolling and annealing at 700°C.

Typical Mechanical Properties	a	b	c
Tensile Strength (t.s.i.)	47	28	21
% Elongation	4	50	70
Hardness (Vickers)	220	85/90	65/70

with metals composition gradients may exist and can be observed under the microscope.

Stirring of a solid metal is impossible but a mechanism exists within the metal which can lead to the same effect and if encouraged will result in uniform dispersion. Atoms are in a constant state of vibration as a result of their thermal energy, which increases as the temperature rises. In addition, a metallic lattice contains many faults including vacant sites where an atom is missing. Where such a vacancy exists it is possible for a neighbouring atom to vacate its own site and move in. The newly vacated site is then available for another atom to move. Such movement is known as diffusion and takes place continuously in solids even at room temperature. At higher temperatures the number of imperfections in the lattice increases, as does the thermal energy, and diffusion is progressively encouraged. The mechanism acts for all the atoms present

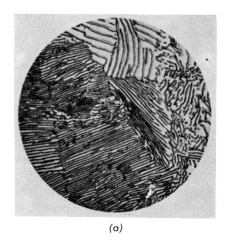

(a)

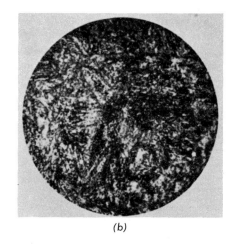

(b)

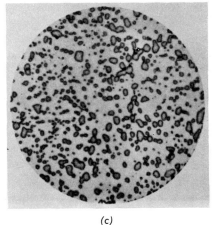

(c)

Fig. 1.13 Microstructure of 0.80% carbon steel showing effect on carbide form and distribution of different heat treatments.
a slowly cooled from 760°C, ×500
b quenched from 760°C and tempered at 550°C, ×1000
c quenched from 760°C and tempered at 700°C, ×500

Typical Mechanical Properties	a	b	c
Tensile strength (t.s.i.)	60	75	37
% Elongation	15	25	40
Hardness (Vickers)	250	350	170

including the solute atoms and random dispersion can be closely approached with sufficient time. For most metals it is necessary to use quite high temperatures for there to be any noticeable effect.

It is common experience that with a solution of a solid in a liquid there comes a point when no more solid will enter the solution, which is then said to be saturated. Similarly, with solid solutions there is often a limit to solubility. The solubility may be either very small to the point of being undetectable, or very extensive. In some cases complete solid solubility exists, i.e. both metals dissolve readily in one another as, for example, copper and nickel. The behaviour of these two metals in the solid state may be compared with that of alcohol and water which also mix in all proportions, whereas oil and water clearly do not.

Of greatest interest is the case where solubility is limited. Adding excess solute results in one of the mixtures mentioned in Section 1.1.2, p. 13, and the nature and distribution of the constituents profoundly alters the properties of the mixture. A further analogy with liquid solutions lies in the dependence of the saturation point on the temperature. If hot water is used it dissolves more sugar than cold. This is illustrated by the solubility curve of fig. 1.14(a). Fig. 1.14(b) shows, similarly, the solubility of aluminium for copper as a percentage of copper against temperature. However, in an alloy the effect of change of solubility with temperature differs from the case of sugar and water. Whereas sugar rejected from solution on cooling falls to the bottom, the solute rejected from an alloy remains dispersed throughout the metal. This dispersed solute may markedly influence the properties of the alloy.

A solid solution is not the only possible way in which two or more metals or non-metals may be associated. Compounds are common. Tungsten carbide is an example of a compound between the metal tungsten and the non-metal carbon in which the compound shows distinct metallic properties. Many intermetallic compounds exist, for example copper aluminide, and most of them are hard and brittle. These, together with the carbides, nitrides and similar compounds provide the possibility of varying structure and properties (see fig. 1.13). They often result when excess solute is deposited from a cooling solid solution; on reheating they will dissolve in that solution.

During rejection of excess solute, whether as a compound or not, the atoms do not deposit suddenly, completely dispersed, but segregate into substantial aggregates. The phenomenon is termed precipitation. The clustering of atoms towards a given deposition point takes place by diffusion. At the instant of initial deposition a nucleus of atoms deposits at a preferred site, e.g. a grain boundary. Growth of the precipitate follows by deposition of further atoms within the sphere of influence of the nucleus. Time is needed for such 'nucleation-growth' processes. If time is allowed as, for example, by slow cooling, diffusion distances can be great – the precipitate particles

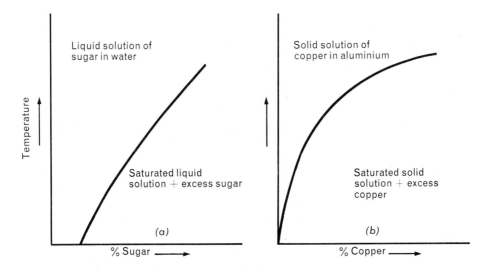

Fig. 1.14 Diagramatic solubility curves: *a* sugar in water, *b* copper in aluminium.

grow large and there are correspondingly fewer of them. Thus, the rate of cooling through a solubility change governs the size of the precipitate.

Since time is needed for precipitation, rapid cooling can depress the temperature at which rejection of the solute would normally occur. The structure obtained is then said to be supersaturated. If the temperature is not too low rapid deposition may occur at many nuclei, giving a fine well-dispersed precipitate. If, however, the temperature rapidly reached is too low, there may be so little opportunity for diffusion that precipitation is completely suppressed. Such a situation is an example of a metastable state, i.e. a state which is not truly stable but where nothing will happen unless a change is triggered off. An analogy is that of an explosive where detonation is required to enable a stable state to be reached albeit with enormous energy change. Many useful conditions in metals owe their existence to a state of metastability.

Not all changes can be suppressed even by the most rapid cooling (quenching). Sometimes the type of change is drastically altered, as in the case of steels where a change involving no diffusion replaces precipitation. On the other hand, suitable alloying can make a change so sluggish that even slow cooling will not result in the truly stable product.

When the constituents of an alloy are those which are truly stable under the given conditions, a state of equilibrium is said to exist. Equilibrium diagrams show the equilibrium products of a given system (see fig. 1.15). The solubility curve of fig. 1.14(*b*) forms the part AB of fig. 1.15, which is the aluminium-rich end of the aluminium–copper system.

A further aspect which has to be taken into account is the possibility of phase change. Each metal has a characteristic atomic structure. Some metals can exist in two or more forms with distinctly different atomic structures. Such metals change from one to the other at a definite temperature. They are said to have undergone a phase change, often indicated by a volume change. Iron, for example, changes at about 900°C from alpha to gamma form and its volume decreases slightly. The reverse change is an expansion and if cooling is rapid can lead to severe stresses and cracking. In the two forms the solubility of iron for carbon is different and the iron–carbon system has vast possibilities for variation in structure depending on whether the carbon is in solution or not. The iron–carbon alloys are the basis of steels. Phase changes also rely on

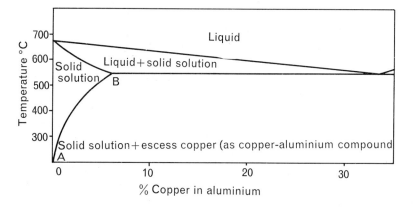

Fig. 1.15 Aluminium-rich end of the aluminium-copper equilibrium diagram.

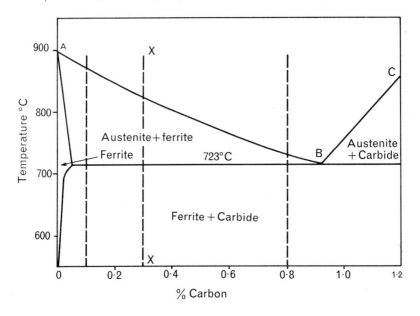

Fig. 1.16 Section of the iron–carbon diagram.

nucleation and growth and can often be suppressed by quenching or rendered sluggish by alloying. New phases may also appear as the alloy content increases.

1.2.3 Hardening of steel

 a. Quenching. Steel may be hardened by heating to a temperature of at least 760°C followed by cooling (usually referred to as 'quenching') at a rate sufficiently fast to produce the required structure. The temperature to which the steel is heated will depend on its carbon content. It is necessary for the steel to reach a temperature at which it will undergo a phase change. Fig. 1.16 is a section of the iron–carbon equilibrium diagram showing the limits of stability of the phases with respect to composition and temperature, i.e. for a 0.1% carbon steel a temperature of 900°C would be required for the steel to be entirely in the gamma state, whereas for a 0.8% carbon steel a temperature of 760°C would suffice (see line AB in fig. 1.16).

The carbon content required in a particular steel, and therefore the temperature selected for the hardening treatment, is dependent on the hardness required in the quenched component. This is shown in fig. 1.17. The hardness range measured on the Vickers scale is from approximately 200 to 850, whereas in the soft state the maximum hardness will be between 100 and 200.

The ability of steels to harden in such a spectacular way is dependent on the fact that quenching may suppress the normal equilibrium phase change from gamma to alpha. Thus, instead of a nucleation process which would lead to the formation and growth of alpha iron and iron carbide, the gamma phase, known as 'austenite', is retained in a metastable state down to a temperature of 300°C or less. Below this critical temperature (M_S) a remarkable structural change occurs. The supersaturated austenite (gamma) begins to transform *instantaneously* to a product having a highly distorted atomic structure with the carbon atoms still forcibly retained within it. This structure

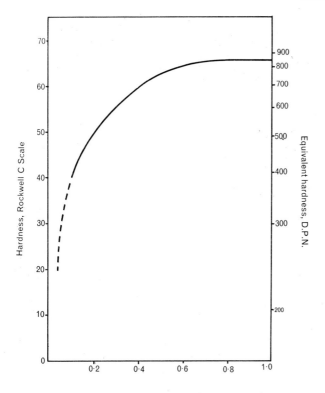

Fig. 1.17 Relationship between the carbon content of a steel and the hardness after quenching.

has a needle-like (acicular) appearance under the microscope and is known as 'martensite'. It is easily distinguishable from the equilibrium structure (fig. 1.18) of alpha iron and iron carbide.

In order that nucleation and growth may be suppressed so that martensite may form, diffusion must be prevented and it is for this reason that quenching is used. However, the rate of diffusion will vary from one steel to another and therefore the actual cooling rate will be selected according to the composition and mass of the steel. Thus a low-carbon low-alloy steel will need a relatively fast quench in water, whereas a highly alloyed steel in which diffusion rates are slow may be hardened by cooling even in air without the risk of nucleation occurring. Such a steel can cool quite slowly down to its M_s temperature before the instantaneous martensite reaction occurs. The ease or otherwise with which a steel hardens and transforms to martensite is a measure of its 'hardenability'. Highly alloyed steels generally have high hardenability and the hardenability that a particular steel possesses will determine the severity of the quench required to harden a given section. Alternatively the hardenability will determine the thickness of section which will harden under given conditions.

At any one temperature below M_s only a certain percentage of austenite will have transformed and there is a second lower temperature (M_f for 'martensite finishing temperature') at which the transformation from austenite is complete. At any one temperature between M_s and M_f the amount of austenite that transforms does so instantaneously. In some steels the M_f temperature is below normal room temperature

 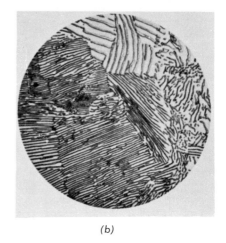

(a) (b)

Fig. 1.18 Showing the micro-structure of carbon steel: *a* after quenching ('martensite'), *b* after slow cooling under equilibrium conditions ('pearlite').

and a structure of martensite and some retained austenite is then observed. However, sub-zero refrigeration would enable the transformation of the residual austenite to take place.

b. Tempering. The as-quenched steel, although very hard, will be very brittle. Some degree of toughness and ductility may be achieved without undue loss of hardness (and strength) by a 'tempering' treatment. In this treatment the hardened steel is reheated to a temperature within the range 100°C to 675°C and soaked for a period of time sufficient to allow the desired structural change to occur. The martensite structure, which is a metastable one at room temperature, will on tempering begin to revert towards the equilibrium structure. If the temperature were high enough and sufficient time were allowed complete equilibrium could be achieved and alpha iron and carbide formed. However, careful control of temperature and time allows a 'tempered martensitic' structure to be obtained (see figs. 1.13(*b*) and 1.13(*c*). The range of properties which may be achieved by tempering is shown in fig. 1.19.

c. Secondary hardening. In certain highly alloyed steels of the 'high-speed tool' type, a process of secondary hardening is employed. On quenching most of the austenite present changes to martensite but some is retained. When the steel is tempered at around 600°C the remaining austenite transforms to martensite and the steel develops additional, or secondary, hardness. A second tempering treatment is usually carried out in order to impart some degree of toughness to the freshly created martensite.

d. Isothermal treatments. The hardening process described above relies on the continuous cooling from the austenitizing (i.e. hardening) temperature down to room temperature. Quenching in this way may lead to an unacceptable level of distortion or even to cracking due to a volume expansion. To avoid these dangers an isothermal treatment is sometimes used. Essentially such a treatment consists, as before, in heating to the temperature at which austenite forms followed by cooling to a temperature slightly in excess of M_s. As will be seen from fig. 1.20, at that temperature the super-

HEAT TREATMENT | 29

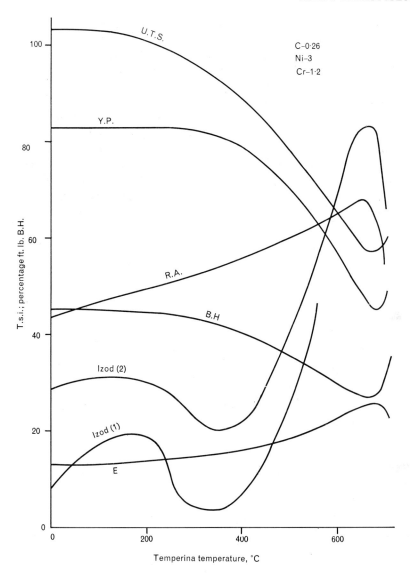

Fig. 1.19 Effect of tempering on the mechanical properties of nickel-chromium steel hardened in oil quenching from 830°C. Izod(2) is for a steel with 0.25% molybdenum added.

saturated austenite can exist for a significant period of time without transforming. Thus, from fig. 1.20, the austenite at 350°C will not begin to transform until a time period of 2 min has elapsed, whereas at 550°C transformation begins after only 10 s. It is therefore possible to obtain much greater uniformity of temperature through the section by this isothermal treatment, i.e. holding at a constant temperature. If the steel is then cooled to room temperature the martensite reaction proceeds uniformly without so much danger of distortion or cracking. This process is known as martempering. It must be stressed that if the maximum hardness is to be obtained the time period during which equalization of temperature is occurring must not exceed that indicated in fig. 1.20. The diagram shown is referred to as a 'time–temperature–transformation' (TTT) curve.

Another isothermal treatment is that of austempering. In this process the object is not to form a martensitic structure. Instead the steel is quenched to a temperature above the M_s line (say AB in fig. 1.20) and held there. After a short induction period transformation begins. However, as the temperature is relatively low, diffusion is limited and the normal alpha iron plus carbide structure cannot be produced. Instead a constituent 'bainite' is formed having a distorted atomic structure resembling alpha iron into which carbide from the supersaturated austenite has precipitated. The hardness of the bainite depends on the temperature at which it is formed but it never reaches the level of hardness achieved in a martensitic structure. Bainites are in many respects similar to tempered martensite and have similar properties. In microscopic appearance there is also a great similarity. Bainite structures enable the required properties to be

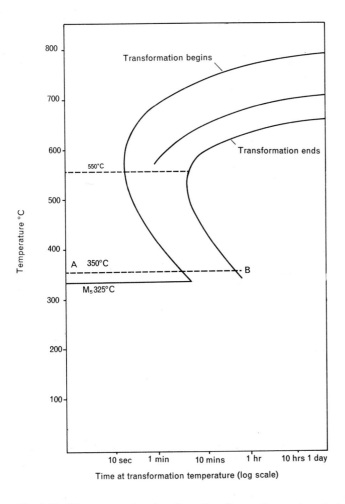

Fig. 1.20 Time-temperature transformation diagram for a carbon steel.

imparted whilst minimizing cracking and distortion. Considerations of quenching speed limit the application of this treatment as it is essential that equilibrium products are not formed during the quench.

The transformation to bainite requires time even after it has once been initiated and the TTT curve shows not only the time required before transformation begins but also the time that is required for the process to go to completion. When the process is carried out in industry it is important, therefore, to know and control these time intervals accurately.

e. Surface hardening, carburizing, nitriding. Many engineering components require a hard wear-resisting surface supported by a softer but tougher core. Such a combination of properties may best be achieved by a special heat treatment which affects only the surface of the steel. The simplest case would be that of a medium to high carbon steel which has its surface rapidly heated by intense heat to the hardening temperature and is then quenched. The heating must be at a sufficient rate so that the core does not reach the hardening temperature. In this way only the surface will be hardened. The heating may be achieved by a bank of burners with flames impinging directly on to the surface, or by high-frequency induction techniques.

Alternative ways of surface hardening involve a change in the chemistry of the steel at its surface. This may best be seen by considering the process of carburizing. It has been shown earlier (fig. 1.17) that the hardness of a steel is dependent on its carbon content. If a low carbon steel can have its surface carbon content increased to approximately 0.8%, the hardness at the surface will be that appropriate to a 0.8% carbon steel, namely, 800 on the Vickers scale, whilst the core will have the hardness and toughness of the lower carbon content. In carburizing processes carbon in the atomic form is presented to the surface of the steel at a temperature of approximately 900°C. The carbon dissolves in the steel to form a solid solution containing 0.8% or more of carbon at the surface; away from the surface the carbon content is controlled by the rate of diffusion of the carbon inwards. When the steel is quenched and tempered a surface layer or case of hard steel is produced, the thickness of which largely depends on the time of treatment. The process is often described as 'case-hardening'.

The carbon that is supplied to the surface may be provided by placing the steel in contact with hot granular charcoal, by heating in a gaseous mixture rich in carbon compounds such as carbon monoxide or hydrocarbons, or by heating in contact with molten cyanide. Whichever medium is used a chemical reaction occurs which allows atomic carbon to be formed. This atomic carbon is then available for absorption by the surface and subsequent diffusion.

A modified form of carburizing is the carbo-nitriding process. This is performed in the same type of gaseous mixture as would be used for carburizing but between 2 and 10% of ammonia is fed into the atmosphere. At the process temperature of between 850°C and 900°C this ammonia cracks and the atomic nitrogen so formed can be absorbed along with carbon by the surface of the steel. Hardening follows when the required depth of case has been produced but, because of the increased hardenability resulting from the nitrogen, the desired martensitic structure may be achieved by oil rather than water quenching. This, together with the slightly lower processing temperature, reduces the amount of distortion that would occur.

A much more specialized process is that of nitriding in which certain special alloy steels are subjected to an ammonia atmosphere at a temperature of approximately

500°C. Ammonia dissociation occurs and the resulting nitrogen is absorbed by the surface and diffuses, somewhat slowly, inwards. The nitrogen forms complex alloy nitrides within the steel matrix. These nitrides, precipitated in a very fine submicroscopic form, strengthen the structure and increase the hardness to values of approximately 800 on the Vickers scale. A useful characteristic of a nitrided steel component is its increased resistance to fatigue and to certain corrosive environments.

1.2.4 Hardening of non-ferrous alloys

Certain non-ferrous alloys may be hardened by heat treatment and, although reactions similar to the martensite reaction in steel do occur in some alloys (e.g. aluminium bronzes), the major heat treatment process which is used is that of 'precipitation hardening'. Many aluminium alloys (e.g. duralumins) as well as certain copper base, nickel base and titanium base alloys are known to respond to precipitation hardening heat treatment. In some ways the actual process appears similar to that used for steel in that the alloy is quenched from an elevated temperature and then reheated to a lower one. However, the mechanism operating is vastly different, as will be seen by the description given below for an aluminium copper alloy.

The equilibrium diagram for the aluminium-rich end of the aluminium–copper system is set out in fig. 1.15. The significant feature of this diagram is that, whereas at room temperature aluminium can dissolve only very small amounts of copper, at approximately 500°C 5.5% copper may pass into solution. Thus, at room temperature the equilibrium structure of an aluminium-5%copper alloy is one of a weak solid solution plus a massively precipitated compound of copper and aluminium. In contrast, at 500°C the structure is a uniform solid solution, all the copper being dissolved in solution. If this alloy is quenched from 500°C, the nucleation and growth process by which the compound is rejected from solution is prevented, and at room temperature a supersaturated solid solution is obtained. This part of the treatment is known as 'solution treatment'. If the alloy is now reheated even to mildly elevated temperatures the supersaturated solid solution will tend towards its equilibrium state and will begin to reject copper atoms. These will be clustered at certain specific sites in conjunction with aluminium atoms, but because the temperature is low, diffusion over large distances is prevented. The clusters so formed are therefore small and still within the lattice. Their effect is to distort the lattice structure of the alloy, which strengthens it, making it more resistant to deformation, i.e. harder. This controlled 'precipitation' is known as 'aging' because in some alloys it takes place at room temperature, and the time required is dependent on the temperature used. As shown in fig. 1.21, if sufficient time is allowed at the higher temperatures, the hardness passes through a peak and begins to decrease. This 'over-aging' effect is due to the increased diffusion giving rise to precipitation and agglomeration into more massive particles, and if allowed to proceed to completion would produce the equilibrium structure.

The solution treatment and aging of other precipitation-hardening alloys may be explained on similar lines although solution treatment temperatures and aging times and temperatures will vary from alloy to alloy.

1.2.5 Annealing

Many heat-treatment processes are described under the term 'annealing'. The common link is that an attempt is made to reach a state nearer true stability (equilibrium) than the initial condition. This condition is commonly associated with maximum

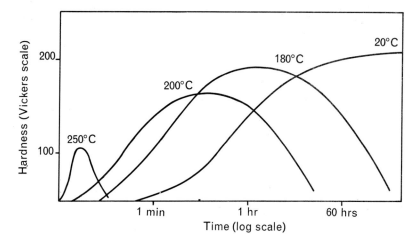

Fig. 1.21 Typical hardness against time curves for an aluminium alloy quenched and aged at various temperatures.

softness and freedom from internal stress. Usually the treatment involves heating to a predetermined temperature, holding for a time, and cooling at a given rate. Usually the active periods in this treatment are the 'soak' at temperature and the cooling period. Since equilibrium is required, cooling is often slow and diffusion has sufficient time to allow processes within the metal to go to completion.

a. Homogenization. Possibly the simplest annealing process is that of homogenization. Cast metals, particularly, suffer from non-uniformity resulting from the solidification process and this may not be desirable. If the cast metal is annealed for a long time, diffusion allows the atoms to redistribute themselves and achieve a random dispersion. The metal is now of uniform composition throughout and is said to be homogeneous. Complete homogenization may take so long that it becomes impracticable but the state is approached in many commercial alloys. Each element under given conditions has its characteristic rate of diffusion and accordingly the time taken for attainment of uniformity varies with the alloy.

b. Recrystallization. Recrystallization implies the replacement of one set of crystals by another. An annealing treatment designed to promote recrystallization may be required to soften a work-hardened metal or to change the grain size. Recrystallization is yet another example of a nucleation-growth process and the size of the new grains can be controlled by the conditions of heat treatment.

In many metals the only possibility of changing the grain structure is by recrystallization after cold work which is described in ch. VI on mechanical working. This type of annealing is often called process annealing. In some other metals there is another possibility. When a solid phase change occurs the new phase forms new grains, i.e. the metal recrystallizes.

The most common example of phase change recrystallization is that of iron and steel. As iron alloys are heated through the ferrite to austenite change, new grains of austenite are created, and if heating is continued they increase in size. On cooling again the change is reversed and new ferrite grains form whose size is dependent in part on the austenite grain size. Iron alloys may thus be treated to alter the grain size. Usually 'grain-refining' is the objective of this treatment and castings and hot forgings may be so treated.

c. *Structural changes, full annealing, normalizing.* Annealing may also be employed to ensure that, in a two-phase alloy, the second phase is in its equilibrium state and that each constituent contains the minimum possible concentration of other elements dissolved in it. Usually this condition is the softest that can be achieved. The important stage in such a case is the cooling period when solutes are rejected. Two possible cases exist. In the first, solute atoms slowly precipitate as cooling proceeds. In the second, a phase change results in a sudden change in solubility and precipitation is completed over a narrow temperature range.

In the first case conditions are simple and the rate of cooling determines the size of precipitate particles as previously outlined. The second case is the more interesting and may be illustrated by reference to carbon steels. On cooling, gamma iron changes to alpha at about 900°C. If carbon is present, the temperature of change is suppressed, as shown by the line AB of fig. 1.16. Consider a steel of 0.3% carbon. In fig. 1.16 the line XX represents this composition The temperature at which the gamma to alpha change takes place is 837°C, where XX cuts AB. When this point is reached some of the iron changes its atomic arrangement and converts to virtually carbon-free alpha iron, called ferrite. The remaining carbon is concentrated in the remaining gamma iron (austenite). Such concentration effectively turns the remaining austenite into (say) 0.31% carbon steel. Steel of this carbon concentration will not transform to ferrite until a temperature of (say) 835°C is reached. Further cooling is therefore essential before any further change takes place. As long as cooling is continued more and more alpha iron is formed and the remaining carbon concentrates into the slowly disappearing austenite. Eventually the carbon concentration reaches a critical figure of approximately 0.86% at a temperature of 723°C when, under equilibrium conditions, all the remaining austenite transforms and all the carbon is rejected as iron carbide. The rejection is not into massive lumps of carbide but into a structure appearing as alternate plates, or lamellae, of carbide and ferrite when viewed under a microscope. This structure is labelled 'pearlite' and is the product of alternate deposition. If a steel with a composition of 0.86% carbon had been chosen to begin with, only pearlite at 723°C would have been formed (fig. 1.13(*a*)). If a steel of more than 0.86% carbon had been chosen, the austenite would have already been of a composition above the critical 0.86% and the initial change on crossing line BC of fig. 1.16 would have been rejection of excess carbide with the remaining austenite gradually reducing in concentration down to 0.86% carbon. In any event the final stage of carbon rejection would occur at 723°C from an austenite containing 0.86% carbon.

Under conditions of extremely slow cooling the pearlite transformation, as the last stage is called, takes place at the equilibrium temperature of 723°C. When a steel has been annealed under these conditions it is said to be 'fully annealed', i.e. it has been austenitized (heated to above the alpha to gamma change temperature) and slowly cooled through the changes outlined above.

Since the reactions described involve the formation of nuclei and the addition of further atoms by diffusion the processes can be modified by altering the rate of cooling. Faster cooling has the effect of depressing the transformation to a lower temperature and consequently reducing the size of the pearlite plates. Ultimately they are so small and finely spaced that even microscopic examination cannot distinguish the structure, which then appears simply dark in colour. The properties of pearlitic steels will vary from soft, as annealed, to strong and tough. When carbon steels are cooled in air they are said to be 'normalized' and have a pearlitic structure of intermediate fineness.

An alternative treatment to achieve pearlitic structures of controlled spacing is that of 'isothermal annealing' in which the temperature of transformation is kept constant. The austenite is rapidly cooled to the required temperature below 723°C. In this state it is metastable and transforms, after an incubation period, to the pearlite of spacing corresponding to the temperature chosen. The times needed for such transformation are incorporated in the time–temperature–transformation curve of fig. 1.20, where the first curve along the time axis represents the beginning of ferrite formation and the second and third curves show the beginning and end of the pearlite transformation.

d. Spheroidization. If pearlite is heated and held at temperatures around 700°C it is found that the carbide plates lose their shape and form globules. Again diffusion has taken place and surface tension acts to form a spherical shape. In this state the steel is even softer than when fully annealed and is said to be 'spheroidized' (fig. 1.13(c)). Spheroidized steels are used when extreme ductility is needed for mechanical working. Spheroidized carbide can be formed directly by an isothermal treatment – by continued holding at temperatures only slightly below 723°C.

Spheroidizing is an example of agglomeration of a structure which will take place in most cases where a precipitate is present and where conditions facilitate diffusion.

e. Malleablizing. Cast iron is usually regarded as a brittle substance but with suitable irons a heat treatment can be applied to produce tough iron castings. Carbon in *grey* cast iron forms flakes of graphite which interrupt the structure and make it weak in the same way that a notched bar is weak. Basically, grey iron can be regarded as steel with graphite dispersed in it. Flake graphite in grey iron cannot be modified but its formation can be suppressed and the resulting iron can be treated. The treatment known as 'malleablizing' produces iron containing graphite in a less harmful form.

Suitable modifications to the composition and cooling rate enable *white* iron to be cast in which massive iron carbide is present and graphite is absent. Such iron is very hard and brittle but it may be made ductile and soft by removing the carbide or by converting it to graphite of near spherical shape. In this form the effect on ductility is slight compared with that of the flake form. Hence the ease of casting of cast irons is combined with strength and ductility.

'Whiteheart' malleable iron is traditionally produced by annealing the casting in contact with iron oxide (haematite ore), which oxidizes the surface carbon of the casting at a temperature of about 1050°C. Fresh carbon diffuses to the surface from the centre and over a period of two or three days the iron becomes decarburized all through. The process is the reverse of carburizing and is diffusion controlled so that only limited thicknesses can be treated. Gaseous atmospheres can replace the ore as the decarburizing medium.

'Blackheart' malleable irons contain 'rosettes' of graphite within the structure, which is then a mixture of rounded graphite and iron (ferrite). The process is shorter than whiteheart annealing and relies on careful control of composition to render the carbide sufficiently unstable to breakdown *in situ*. A neutral atmosphere or packing obviates scaling simply by excluding air. Diffusion is essential to enable carbon from the dissociating carbide to precipitate on to convenient nuclei to form rosettes of 'temper' carbon. During cooling graphite continues to deposit and a fully malleablized blackheart iron has a very soft carbon-free matrix. Cooling can be speeded to retain sufficient carbide to ensure a pearlitic structure which is capable of being heat treated in the same way as steel. Blackheart iron can be malleablized in quite substantial sections.

The terms blackheart and whiteheart are derived from the colours of the fractures of specimens of the irons.

Of recent years processes have been discovered in which nodular graphite may be made to form direct from the melt and little heat treatment is involved. Additions of cerium and magnesium are active in this respect.

1.2.6 Equipment

The basic equipment required for heat treatment consists of a furnace in which the metal is heated, possibly some separate means of providing a controlled atmosphere and arrangements for cooling at a controlled rate on removal of the metal from the furnace.

The actual furnace used for a specific job will depend on:
1. Shape, size and weight of components.
2. Throughput required.
3. Batch or continuous working.
4. Temperature of treatment.
5. Necessity for controlled atmosphere.
6. Economics – fuel used, efficiency, capital cost, maintenance etc.

Furnaces are normally classified according to the type of fuel used, the physical construction and the method of handling stock into and out of the furnace. Solid fuel firing is rarely used nowadays as it is difficult to control compared with gas, oil and electric heating. The running costs of these furnaces may differ widely and are very dependent on the availability or otherwise of natural gas, cheap electricity etc.

The actual construction of furnaces varies considerably. The simplest are merely refractory-lined boxes with a door for charging and discharging the stock, whilst more complex designs can include moving hearths to progress the stock through the furnace. Examples of such devices include the roller hearths, belt conveyers, moving (or walking) beam furnaces and shaker hearths. In all of these the hearth has become a means of handling the metal being heated so that it may be processed through the furnace at the desired rate and emerge, usually through a separate door, at the required temperature. In some instances the unit may be combined with a quenching tank. A shaker hearth furnace, for example, will automatically shake the components at the end of their journey through the furnace from the hearth into the quenchant.

The actual choice of quenchant, whether it be an integral part of the furnace or not, is dependent on the structure and properties required in the metal. In essence, when it is required to suppress a nucleation process (as in hardening of carbon steel) the necessity arises to bring the steel through the critical temperature range as quickly as possible. Oil, water or even brine solutions will then be utilized. When this necessity does not arise, air or even furnace cooling will suffice. The exact choice of quenchant thus depends on the desire to suppress, encourage or control the structural changes governed by nucleation and growth phenomena. When the nucleation and growth occurs at a slow rate, the cooling has to be correspondingly adjusted. Thus an air quench could suffice for hardening alloy steel but a water quench might be required for carbon steel. In contrast, when annealing these steels, an air cool could produce the desired structure and properties in the carbon steel but in the alloy steel a furnace cool might be demanded.

Most metals used industrially tend to oxidize if the surface is exposed to the air even at room temperature. At higher temperatures the rate of oxidation increases and may lead to massive scale formation, as in the case of steel. In order to minimize such effects, many heat-treatment operations will be performed in atmospheres other than air. Immersion of the metal in molten salt provides a satisfactory way of heating the metal efficiently yet reducing to negligible proportions the amount of oxidation that can occur. In other more conventional furnaces a controlled atmosphere is introduced into the heating chamber. This atmosphere consists of one or more gases which are unreactive to the surface of the metal and will be produced in a separate atmosphere generator. When specially reactive metals are heated, vacuum furnaces or inert gas atmospheres may be necessary. On the other hand, controlled atmospheres may be chosen to be deliberately reactive, as in carburizing or nitriding treatments.

1.3 ECONOMIC ASPECTS OF METAL SELECTION

Our knowledge of material properties enables the correct material selection to be made for a given component in a specific environment. However, any material selection must be based upon sound economic practice for, in specifying materials for a particular use, we are in fact costing the manufacturing process and the final properties of a component. A variety of materials may meet the physical, chemical and mechanical requirements specified by a design engineer, but each material may demand different fabrication and heat-treatment processes in order to meet these specifications. On economic grounds it may be expedient to alter the material to meet the optimum manufacturing requirements. An example of this type is provided by the automobile industry in which camshafts, formerly made by forging rough blanks from alloy steel billets, required rough machining prior to heat treatment and a further final machining process to complete the component. This processing route was wasteful of material and required high capital investment. By employing a high-strength cast iron for this component, it was possible to cast the camshaft within close limits of the final tolerances, to heat treat the casting and to conclude manufacture with a machining operation. This alternative manufacturing route permitted the use of a lower cost material and a much less expensive processing route and was less wasteful in terms of material lost during processing. Although such problems lie within the realm of the production engineer, they must be considered as contributing to the metallurgical costs of the material. In view of the difficulties posed by studying the economics of component manufacture, only the basic costs involved in producing materials of standard engineering forms will be considered.

The basic costs of all metals are largely determined by their natural availability and the stability of the chemical compounds from which they are extracted. These factors control the expenditure involved in the winning and concentration of the ores, in extracting a crude metal from the concentrate and in refining the crude metal to a degree necessary for its commercial exploitation. Since materials require costly refining techniques to attain high purity, this stage may constitute a considerable part of the basic material cost. Fortunately, high purity materials are rarely needed in industrial uses, but the tolerable impurity level, residual from the refining stage, is different for each metal.

The basic prices per ton of several metals, of commercial purity, in the cast ingot form are listed in table 1.3. Included in the list are the prices of cast iron and low carbon steels, and the difference in price denotes the additional refining costs involved in producing steel from the base iron.

The addition of alloying elements may reduce the basic cost of an expensive material, as well as conferring upon it specific properties of industrial importance. It is of interest to compare the various costs of copper based alloys, since they embrace a wide range of alloying elements of different basic prices. In table 1.4 the composition, expressed as the weight percentage, is given for five copper alloys, together with their prices. This list clearly demonstrates that it is the differences in properties which are costed since, on the basis of cost alone, the brass would always be employed.

An additional cost arises due to the need to fabricate material into specific usable forms such as plate, sheet, rod, wire etc. and it is apparent that the greater the degree of fabrication required, the higher the cost per unit weight of material.

During the fabrication process, considerable metallurgical control of the material structure is necessary, in addition to control over the dimensional tolerances of the product. Most metals become hard and lose their formability as a result of processing at ambient temperatures and require softening by reheating at high temperatures. This annealing treatment can be used to control the size of the individual crystals or 'grains' within the product, and may modify the anisotropic pattern of mechanical properties induced by the forming process. To avoid surface reactions with the atmosphere, the annealing treatment may be conducted in an inert atmosphere. In view of the many variables which have to be controlled during the mechanical fabrication process, it is

Table 1.3 Basic prices† and world production‡ of some common metals

Metal	Average concentration as percentage of the earth's crust	World production 1965 in millions of tons (2000 lbs)	Cost £1 ton
Pig iron	5.0	359.7	23
Steel	—	501.4	30
Lead	0.002	3.1	93
Zinc	0.004	4.3	112
Aluminium	8.0	7.4	223
Magnesium	—	0.17	252
Copper	0.01	6.3	538
Nickel	0.002	0.47	902
Tin	0.0001	0.17	1316
Silver	0.00001	*251 × 10^6 ounces	0.8 £1 oz
Gold	0.0000001	*47.7 × 10^6 ounces	14.6 £1 oz

†Foundry Trade Journal Jan. 1968
‡Metal Statistics 1967 (The American Metal Market Co.)
*fine ounces

ECONOMIC ASPECTS OF METAL SELECTION | 39

Table 1.4 Cost comparison for copper based alloys. (cast form) B.S. 1400

Material	Composition w/o	Specification	Cost £1 ton
Copper	99.9 w/o		537
Brass	Cu 70.0 w/o, Zn 30 w/o	B.2	450
High Tensile Brass	Cu 55 w/o, Zn 34 w/o, Mn, Al, Fe, Sn	HT B1	400
Gunmetal	Cu 85 w/o, Zn 5 w/o, Sn 5 w/o, Pb 5 w/o	L.G.2	450
Tin Bronze	Cu 90 w/o, Sn 10 w/o, P 0.5 w/o	P.B.1	630
Aluminium Bronze	Cu 90.5 w/o, Al 9.5 w/o	A.B.1	518

not surprising that wrought materials have an apparent high cost in comparison with the initial cast form (fig. 1.22).

Considerable economies can be effected by producing components by casting techniques, thereby avoiding the costs incurred during mechanical fabrication.

The purpose of employing wrought rod and bar products in component manufacture stems from the special properties which the working operations confer upon materials. A wrought material possesses a more uniform structure than a cast product, and defects such as inclusions and porosity are rendered innocuous by the working operation. Thus, wrought materials may be subjected to greater imposed stresses in service than could be tolerated in an equivalent cast form. The sheet, strip and wire products are the basis of many component manufacturing routes since, in addition to their ease of manipulation, they permit great flexibility in design, for example, in weight saving.

Since ferrous alloys form the bulk of all engineering structures in which strength and ductility are of prime importance, it is of interest to study the costs of achieving

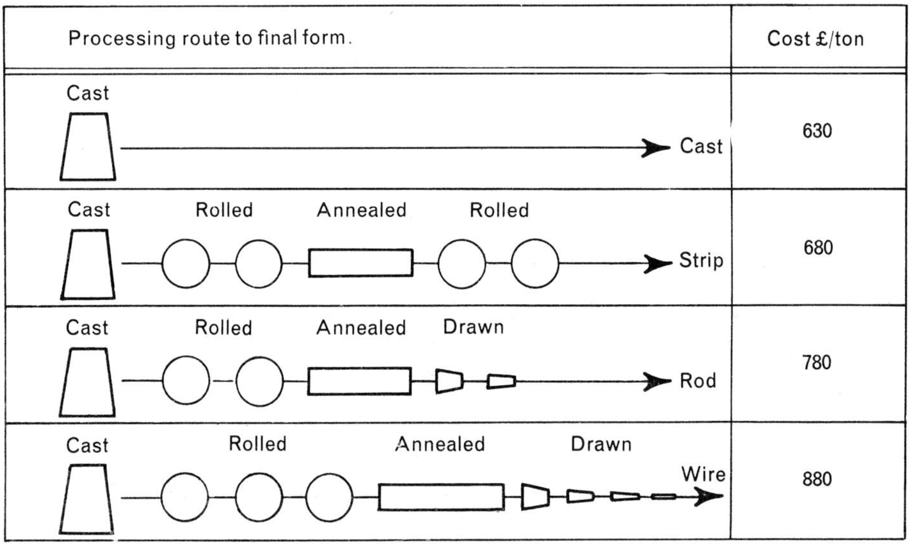

Processing route to final form.	Cost £/ton
Cast → Cast	630
Cast → Rolled → Annealed → Rolled → Strip	680
Cast → Rolled → Annealed → Drawn → Rod	780
Cast → Rolled → Annealed → Drawn → Wire	880

Fig. 1.22 Cost comparison of cast and wrought phosphor bronze (B.S. 1400 PB1).

these requirements. The graphs in fig. 1.23(*a*) relate the costs of attaining strength in various steels in three different metallurgical conditions, while fig. 1.23(*b*) depicts the variation of tensile strength with the property of ductility, here expressed as the elongation experienced by a steel prior to fracture. In selecting a steel to have a specific tensile strength, together with optimum ductility, the metallurgist often has to resort to using hardened and tempered material. Thus, for any particular material, the properties required largely determine its cost.

1.4 MECHANICAL TESTING

1.4.1 General

The purpose of mechanical testing is to evaluate the suitability of materials either for a given manufacturing process or when, as finished components, they enter the service environment for which they were designed. In general, the mechanical properties of materials can be utilized in three basic ways, namely:

1. As a control technique, to ensure that the metallurgical quality of a product conforms, at each stage of manufacture, with the specifications required.
2. To furnish design data for engineering projects.
3. To assess the intrinsic properties of materials, so that the optimum manufacturing method may be employed in their processing route.

The need for mechanical testing as a quality control technique arises from the fact

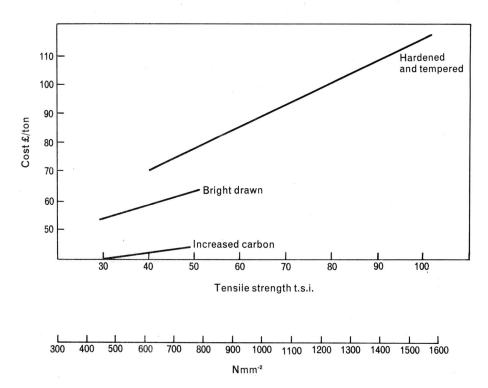

Fig. 1.23 a Relations between cost and strength in steels.

that the chemical composition is only a very crude criterion. The mechanical properties depend markedly upon the distribution of the various phases which may be present in the microstructure and the types, densities and distributions of the crystal defects in the microstructure. Preferential growth of crystals in one or two directions in cast materials and preferential metal flow in mechanically deformed (i.e. wrought) materials may also result in variation of the properties with the direction of testing, but no indications of such variations are apparent from the composition.

For various reasons, the mechanical properties measured on a small test-piece may differ significantly from those of a larger mass of the same material, and this aspect must be considered carefully in interpreting the test results. For example, particularly in cast components, the material may contain gas or shrinkage porosity, cracks and non-metallic inclusions. These are rarely distributed uniformly and a small section used for testing may contain a non-typical concentration. A large defect may occupy almost the entire cross-section of the test-piece and the properties measured would bear no relation to those of the bulk material. The microstructure, and indeed the composition, may vary from surface to centre of a component, in which case test-pieces cut from selected positions over the entire cross-section would be needed to obtain representative data. Changes may also result from the relief of residual stresses when the test-piece is cut from an actual component.

In contrast to the component design criteria, the study of the properties associated with the plasticity of metals is important in all working processes. Thus, the behavior of metals under the complex stress systems experienced in rolling, forging and extrusion techniques may be assessed in hot torsion and compression tests.

1.4.2 Tensile testing

The tensile test is used universally to measure the mechanical properties associated with the strength and ductility of materials. The test is conducted upon a

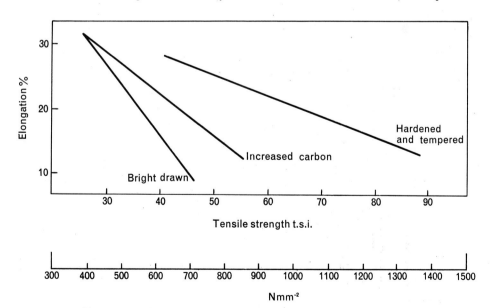

Fig. 1.23 b Relations between strength and ductility in steels.

specimen of suitable geometry by fixing the ends of the specimen in grips, which are attached to the two crosshead members of the machine (fig. 1.24). The specimen is subjected to increasing axial loads, applied via a moving crosshead member, and the extension of the specimen is measured as a function of the applied load until the specimen is fractured. Although apparently a simple test to perform, it is in fact complex and has to be carefully controlled if useful results are to be obtained.

In modern tensile testing machines the electric motor, which controls the hydraulic or mechanical screw loading mechanisms, operates over a wide range of speeds. This feature permits the rate of loading to be controlled, which is of major significance in obtaining accurate test data. The load applied to the specimen is measured by a transducer load cell, attached to one of the crosshead members, and may be displayed potentiometrically or by a mechanical dial indicator. The load capacity of test machines varies widely and this feature determines the maximum cross-section of test specimen which can be fractured. It is essential that the grips holding the specimen are designed to ensure that the specimen is axially aligned with the loading axis of the machine.

It would be useless to extend a prismatic test specimen gripped at its ends by the jaws of the machine because the localized stressing at such positions would inevitably result in failure at the grips. For this reason it is essential that the specimen has enlarged ends suitable for fixing in the grips. Similarly the localized extension associated with the fracture should be confined within fixed positions on the specimen so that the extension experienced during loading can be easily measured. The transition in the specimen cross-section between the enlarged ends and the gauge length should be gradual, and a British Standard (ref. 1) has been formulated to govern the geometry of test specimens. In practice all specimens are circular or rectangular in cross-section and, in order that the test results obtained from specimens of different geo-

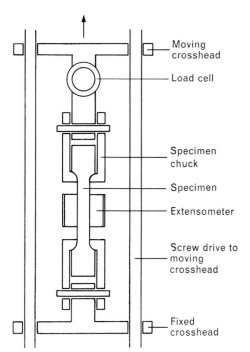

Fig. 1.24 Diagrammatic representation of a tensile testing machine.

metry may be compared, the gauge length and cross-sectional area of specimens are related (table 1.5).

The elongation or strain experienced by the specimen, either during the test or after fracture, is measured by an extensometer. A simple extensometer merely records the over-all extension by denoting the change in length of the specimen reassembled after fracture. Other types of extensometer measure the strain during the test, possibly up to the point of fracture of the specimen. The extensometer is firmly clamped to the ends of the gauge length and may employ a simple micrometer screw, which is reset after each observation of strain and load. An alternative type uses a transducer to continuously record the extension under load, so that the extension-load characteristics of a material may be electrically recorded on an X-Y recorder. An important application of the continuously recording extensometer stems from the fact that it may be used to control the rate at which the specimen deforms, thereby enabling a constant rate of straining of the specimen to be achieved. In this application the electrical signal from the transducer extensometer is measured as a function of time, and is used to control the speed of the motor applying the load to the specimen. The need to control the rate of straining during the test arises from the tendency of materials to display different mechanical properties, particularly at high temperatures, when tested at differing strain rates.

In fig. 1.25 schematic load-extension and true stress-true strain curves are shown for a pure metal tested at a low strain rate at room temperature. The initial portion of the load-extension curve is linear up to point A, which denotes the limit of proportionality for the material. Within this portion of the curve the material deforms elastically and the

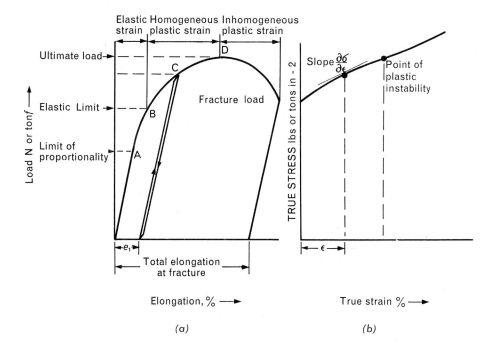

Fig. 1.25   *a* Load extension diagram. *b* True stress/true strain diagram.

44 | METALS IN GENERAL

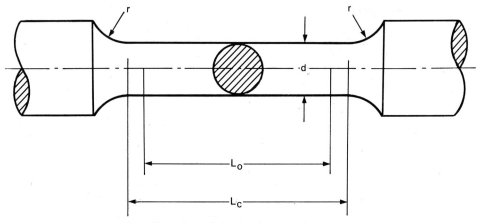

Test piece of circular cross-section

Table 1.5 Tensile specimen dimensions B.S.S. 18. (1966)

Gauge length $= 5.65\sqrt{S_0}$

Cross-sectional area S_0	Diameter d	Gauge length L_0 (See Note 2)	Minimum parallel length $L_c = 5.5d$	Minimum radius at shoulder r		Tolerance on diameter (See Note 1)
				Wrought metals and cast iron	Other cast metals	
A. METRIC DIMENSIONS						
mm²	mm	mm	mm	mm	mm	±mm
500	25.23	125	140	25	50	0.150
400	22.6	113	124	23.5	47	0.125
314	20	100	110	22	44	0.100
200	16	80	88	15	30	0.080
154	14	80	88	1·5	30	0.080
100	11.3	56.5	62	10	20	0.055
78.5	10	50	55	9	18	0.050
50	8	40	44	7.5	15	0.040
38.5	7	35	38.5	6	13	0.035
25	5.64	28.2	31	5	12	0.030
19.6	5	25	27.5	5	11	0.025
7.07	3	15	16.5	4	8	0.015
B. INCH DIMENSIONS						
in²	in	in	in	in	in	±in
1	1.128	5.65	6.25	1.00	2.00	0.006
¾	0.977	4.90	5.40	0.86	1.72	0.005
½	0.798	4.00	4.40	0.70	1.40	0.004
¼	0.564	2.80	3.15	0.50	1.00	0.003
⅕	0.505	2.50	2.75	0.44	0.88	0.0025
⅛	0.399	2.00	2.20	0.35	0.70	0.002
1/10	0.357	1.80	2.00	0.31	0.62	0.0015

Table 1.5 (continued)

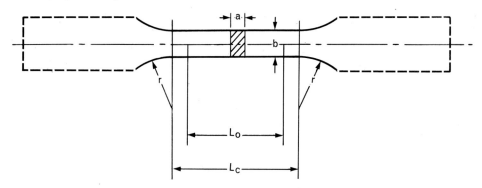

Test piece of rectangular cross-section

Cross-sectional area S_0	Diameter d	Gauge length L_0 (See Note 2)	Minimum parallel length $L_c = 5.5d$	Minimum radius at shoulder r		Tolerance on diameter (See Note 1)
				Wrought metals and cast iron	Other cast metals	
$\frac{1}{16}$	0.282	1.40	1.55	0.25	0.50	0.0015
$\frac{1}{25}$	0.226	1.15	1.30	0.20	0.40	0.001
$\frac{1}{50}$	0.160	0.80	0.90	0.14	0.28	0.0008
$\frac{1}{80}$	0.126	0.65	0.75	0.11	0.22	0.0006

Width b	Gauge length L_0	Minimum parallel length L_c	Minimum radius at shoulder r	Approximate total length L_t
A. METRIC DIMENSIONS				
mm	mm	mm	mm	mm
40	200	225	25	450
20	200	225	25	375
25	100	125	25	300
12.5	50	62.5	25	200
6	24	30	15	100
3	12	15	10	65
B. INCH DIMENSIONS				
in	in	in	in	in
$1\frac{1}{2}$	8	9	1.00	18
1	4	4.5	1.00	12
$\frac{3}{4}$	8	9	1.00	15
$\frac{1}{2}$	2	2.5	1.00	8

from *Methods of Tensile Testing of Materials* B.S. 18 (1962) Part 2 (Amended 1966), British Standards Institution.

extension is proportional to the applied load. Young's modulus of elasticity is defined within this region as:

$$E \text{ (in p.s.i.)} = \frac{\text{Stress}}{\text{Strain}} = \frac{\text{Load/Original cross-sectional area}}{\text{Extension/Original gauge length}}$$

Up to point B on the curve the elastic extension disappears when the applied load is removed. This point is called the yield point of the material. On continued loading of the specimen, beyond the elastic limit, the entire gauge length deforms homogeneously by the plastic deformation mechanisms of slip and twinning. If the load is released at point C, the specimen experiences recovery of the elastic strains but is found to have suffered a plastic strain equivalent to e_1. Upon reloading the specimen the elastic strains are reimposed upon the specimen but, since the material has 'strain' or 'work' hardened, the load required to induce further plastic deformation is that shown by point C and not that originally required to initiate plastic flow at point B.

The load has to be increased over the range of homogeneous deformation due to the increased resistance to flow of the material caused by the 'strain' or 'work' hardening phenomenon. Since insignificant volume changes accompany the plastic deformation, it is clear that marked contractions in the diameter of the entire gauge length are experienced during the specimen extension. Point D depicts the maximum load which the specimen withstood during the test and this load value is termed the ultimate tensile load of the material. At this point the increased load required to overcome work hardening is balanced by the fall in load due to the contraction in specimen diameter, and the material attains the state of plastic instability. Unrestricted extension or 'necking' then occurs at the weakest point along the gauge length and all subsequent extension is localized within the 'neck' of the test-piece. The rapid thinning at this point accounts for the apparent fall in load over this region of inhomogeneous deformation and fracture takes place within the 'necked' portion of the specimen.

The point of plastic instability, which is observed in the tensile test, detracts from the value of such a test to predict the behaviour of metals during metal forming processes. The true stress-true strain curve, derived from the load-extension experiment, depicts the real behaviour of the material during testing, and may be used in the corrected form to measure the work-hardening characteristics of materials. True stress (σ) is defined as the load divided by the cross-sectional area of the specimen at the instant when the load was measured. Similarly, true strain (ϵ) is defined as:

$$\epsilon = \int_{l_0}^{l} \frac{dl}{l} = \ln \frac{l}{l_0}$$

where l is the extended and l_0 the original gauge length. Because the measurements are related to the specimen dimensions at each instant during the test, the σ–ϵ plot does not show a maximum corresponding to the ultimate tensile stress but rises continuously to the point of fracture. The rate at which the material work hardens is given by the slope ($\partial\sigma/\partial\epsilon$) of the curve. In general, the slope varies with the strain and the value of the strain at which the work-hardening rate is measured should be quoted.

The data values determined from the results of the tensile test are given below:

a. Yield stress. The yield stress can be defined as the effective stress at which plastic deformation is initiated in a material.

$$\text{Yield stress (p.s.i.)} = \frac{\text{Yield load}}{\text{Original cross-sectional area of the gauge length}}$$

b. Proof stress. Many materials do not possess a clearly defined yield point and an arbitrary 'proof stress' is quoted which denotes the stress required to induce a plastic strain of 0.1%, 0.2% or 0.5% extension of the gauge length. Fig. 1.26 illustrates how this proof stress is obtained. A line is drawn on the load-extension curve from the required strain increment on the extension axis parallel to the linear elastic portion of the curve. The point of intersection of this line with the curve gives the appropriate proof stress value.

$$0.1\% \text{ proof stress (p.s.i.)} = \frac{\text{Load required to induce 0.1\% plastic strain in gauge length}}{\text{Original cross-sectional area}}$$

c. $\quad$ Ultimate tensile stress (p.s.i.) $= \dfrac{\text{Maximum load}}{\text{Original cross-sectional area}}$

d. $\quad$ Fracture stress (p.s.i.) $=$

$$\frac{\text{Load at fracture}}{\text{Minimum cross-sectional area of the necked region of the gauge length}}$$

e. Percentage elongation $= \dfrac{\text{Total extension}}{\text{Original gauge length}} \times 100$

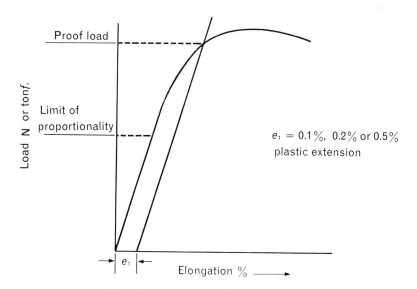

Fig. 1.26 Determination of proof stress.

48 | METALS IN GENERAL

f. Percentage reduction in area $= \dfrac{\text{Original area} - \text{final area at the neck}}{\text{Original area}} \times 100$

The yield and proof stresses differentiate between the elastic and plastic properties of materials, and these parameters form the basis of the design criteria in most engineering components. The ultimate tensile strength of a material describes the stress limit to which it may be loaded prior to ultimate failure. The elongation of the testpiece is a measure of the ductility of a material and is composed of two components: (a) the uniform extension, which is proportional to the specimen gauge length, and (b) the localized extension caused by necking. The latter is not proportional to the gauge length but depends upon the cross-sectional area of the gauge length. Thus it is unwise to compare the elongation values of specimens of different geometry unless the ratio gauge length/area is constant.

The properties recorded during tensile testing depend upon the metallurgical condition of the material, as well as upon the test variables of specimen geometry, temperature and strain rate. Fig. 1.27 depicts characteristic load-extension curves for materials in three different metallurgical conditions. The curve for the single crystal (of the pure metal magnesium) reveals that the initial elastic strain is small and that the yield stress is low in comparison with polycrystalline metals and alloys. Also, a single crystal exhibits a pronounced elongation but displays a low rate of work hardening, since the rise in load beyond the yield load is small. Virtually all the plastic strain is uniform over the gauge length of the specimen and the necking phenomenon is absent. In contrast, the tensile properties of the polycrystalline specimen reveal a high yield load, a high rate of strain hardening and a pronounced reduction in the elongation value.

The differences in these properties of the two pure metals can be explained in terms of the way in which each experiences atomic disturbances during testing. The elastic strains in both specimens result from the displacement of atoms from the equilibrium positions on the crystal lattices under the action of the applied loads. The yield load corresponds to the stress necessary to induce slipping and/or twinning deformation by

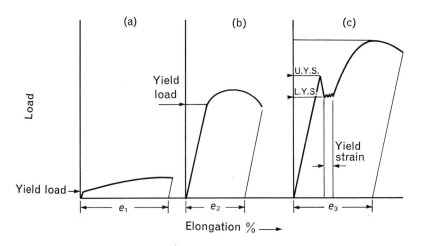

Fig. 1.27 Typical load extension curves for *a* Single crystal, *b* Polycrystalline specimen and *c* Mild steel specimen.

the generation and movement of dislocation lines. For a single crystal, the yield load (F) is related to the critical shear stress (τ) required to move dislocations in the slip plane and in the slip direction by the expression:

$$\tau = \frac{F}{A} \cos \lambda \, \cos \theta$$

where λ and θ are the angles shown in fig. 1.28 and A is the cross-sectional area.

Work hardening is caused by barriers to the movement of dislocations. These barriers may be grain boundaries, precipitates of intermetallic compounds, inclusions and other dislocation lines. The last mentioned are the major source of work hardening in single crystals of pure metals. If, as is frequently the case, the dislocations are moving on only one slip system, they are able to continue moving until they emerge at a free surface. Further slip is then dependent only upon the generation of additional dislocations, and the single crystal shows a low rate of work hardening. If slip occurs simultaneously on more than one slip system, the dislocation paths may intersect. When they do so the dislocation lines become 'stepped' or 'jogged' at the point of intersection. Further movement of these dislocations then requires the imposition of a higher stress and the rate of work hardening is correspondingly increased.

Grain boundaries in polycrystalline materials are very effective barriers to dislocation movement. The dislocations 'pile up' at the grain boundary and exert a back

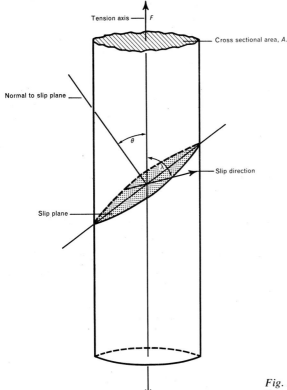

Fig. 1.28 Indices used in the determination of critical shear stress (see text).

stress on the motion of other dislocations in the same plane. It is for this reason that the yield stress is raised by decreasing the grain size.

In addition to the characteristics already discussed, the stress-strain curve for a low carbon (i.e. mild) steel exhibits a discontinuity and irregular strain associated with the yield load (see fig. 1.27 (c)). The upper limit of the first part of the curve is called the upper yield stress (U.Y.S.) and the subsequent, nearly horizontal, portion is termed the lower yield stress (L.Y.S.). This phenomenon arises from the fact that mild steel is a mixture of iron and carbon with some of the carbon in interstitial solid solution. The interstices between the iron atoms are strained to accommodate the carbon atoms and consequently the carbon segregates to the larger interstices in the vicinity of the dislocations. These 'atmospheres' of carbon immobilize the available dislocation lines and, when the metal is strained, either these must be unlocked or fresh dislocations must be generated before the metal can be deformed. Once these dislocations are introduced the metal deforms rapidly and the stress necessary for further deformation falls to that necessary to maintain the dislocations in motion (the L.Y.S.). A similar phenomenon is displayed by other metals when the essential requirements of locking and rapid multiplication of dislocations can be met. This is particularly pronounced when defect-free metal 'whiskers' are strained.

1.4.3 Hardness testing

Hardness is not a fundamental property but is a comparative measure of the resistance to abrasion and indentation offered by one material with respect to another. Hardness tests measure some combination of the elastic and plastic properties of materials, and the results have been employed empirically to assess the tensile strength of materials. Since the behaviour of a material under stress depends upon the type of stress system imposed, the hardness results obtained upon a given material must be interpreted in relation to the type of test conducted. For example, glass will scratch steel but glass would fracture more readily than steel under an indentation test. Similarly, the rebound hardness of rubber, given by the scleroscope method, is approximately the same as that of steel, whereas under indentation loads rubber would distort to such an extent that it would not be possible to obtain valid results.

Hardness testing is used widely in industry as a quality control technique and often the results recorded are the determining factor in the acceptance or rejection of both materials and finished components.

a. Scratch hardness. The first systematic hardness scale was proposed by Friedrich Mohr and it provides a rapid and effective method of assessing the scratch hardness of material. Although not used directly in the testing of metals and alloys, it finds application in the field of mineralogy. The scale consists of ten standard minerals listed in order of increasing hardness, each mineral being numbered according to its position in the series. The standard Mohr scale is given below.

1. Talc
2. Gypsum
3. Calcite
4. Fluorite
5. Apatite
6. Orthoclase feldspar
7. Quartz
8. Topaz
9. Corundum
10. Diamond

The hardness value of a material scratched by topaz (8) but not by quartz (7) would lie between 7 and 8.

A derivation of this method, the file hardness test, has been evolved for determining the abrasion hardness of metals, usually in massive forms, which are difficult to list by other methods. The material is subject to the cutting action of a standard file and the hardness assessed by noting whether or not a visible cut is produced. These tests do not give a sufficient degree of discrimination, as most metals and alloys would fall within the range 5 to 7 on the Mohr scale and the file test results are dependent upon the manual skill and visual interpretation of the operator. The routine tests which are now employed are designed to ensure reasonably reproducible results in the hands of any operator and distinguish differences in hardness levels very much smaller than can be assessed by purely visual methods.

b. Indentation hardness. The three principal methods employed to measure the indentation hardness of materials are the Brinell, Vickers and Rockwell tests. In the first two tests the hardness number is defined by the load and the actual surface area of the impression produced by the indenter, whilst in the Rockwell test the hardness number is defined by the depth of penetration of the indenter under standard loads.

In the Brinell test, the indenter is a hardened steel ball which is impressed into the surface of the specimen under a known load for 15 s. The diameter of the indentation produced is measured with the aid of a low-powered portable microscope fitted with a calibrated eyepiece. The Brinell hardness number (HB) is given by the expression:

$$\text{HB} = \frac{\text{Load (kgf)}}{\text{Surface area of impression (mm}^2\text{)}} = \frac{P}{\frac{1}{2}\pi D\{D - (D^2 - d^2)^{1/2}\}} \text{ kgf/mm}^2$$

where P is the applied load (kgf), D the diameter of the ball indenter (mm) and d the mean diameter of the impression (mm). The major disadvantage of the Brinell test is that the hardness number is markedly dependent upon the value of the load applied to the indenter, since different impression geometry is caused by different loads in the same material. The variation in hardness number with increasing applied loads, for a fixed indenter diameter, is shown in fig. 1.29. Consistent hardness values are obtained only over the horizontal portion of this curve, which restricts the value of applied load. Similar sized impressions are obtained in a given material if the relationship $P/D^2 =$ constant is obeyed, the constant being 30 for steel, 10 for copper and 5 for aluminium. Thus, when the standard 10 mm diameter ball is employed in testing, loads of 3 000, 1 000 and 500 kgf respectively are used upon steel, copper and aluminium.

False readings are obtained if the hardness of the material under test approaches the hardness of the indenter, since part of the applied load is then expended in deforming the indenter.

In the Brinell test, as in all hardness testing procedures, the surface of the specimen to be tested must be smooth and planar to avoid producing non-symmetrical impressions of the indenter. Also, the thickness of the specimen should be at least ten times the depth of the impression, to ensure that the indenter load is supported solely by the material under test. Two features specific to the Brinell test may obscure the true diameter of the impression, and are termed 'piling' and 'sinking'. Both these effects

52 | METALS IN GENERAL

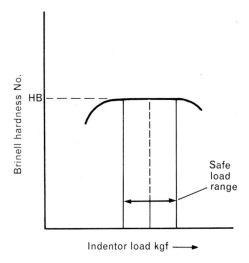

Fig. 1.29 Variation of Brinell hardness number with indenter load.

originate from the plastic deformation of the material around the indenter, resulting from the applied load. The plastic deformation causes the material to work harden and the degree of work hardening determines, in turn, the impression depth. 'Piling' occurs with materials which exhibit a low capacity for work hardening. In this case the material is extruded around the indenter to form a raised lip at the edge of the impression. Conversely, materials which work harden appreciably exhibit 'sinking', in which the area of plastic deformation extends beyond the area of contact with the indenter. In this case, the rim of the impression is not clearly distinguished from the surrounding depressed area.

The Vickers test is performed upon a machine in which the rate of loading and the duration of loading are controlled by a hydraulic dashpot. The indenter is a square-based pyramid diamond, of 136° included angle, which maintains a constant impression geometry under a variety of loads. A microscope is used to measure the lengths of the diagonals of the square impression. The Vickers hardness number is defined by

$$HV = \frac{2P \sin 68°}{d^2} = \frac{1.859P}{d^2} \text{ kgf/mm}^2$$

where P is the applied load (kgf) and d the mean length of the impression diagonals (mm).

The Vickers test has several advantages over the Brinell test, since the geometry of the impression is not affected by the depth of penetration of the indenter. Although the 'sinking' and 'piling' phenomena cause some distortion in the impression, these effects do not affect the measurement of the diagonals. A wide range of loads may be employed, enabling materials of high hardness to be tested without inducing distortion in the indenter. The method has been modified to permit microhardness measurement, which can be employed effectively to discriminate between the separate constituents within a microstructure.

The Rockwell test employs a conical, diamond indenter of 120° angle. The method of testing differs from the usual indentation tests in that a minor load of 10 kgf is applied initially to provide a precise zero. The indenter load is increased to 150 kgf

(RC scale) and the depth of penetration is expressed as a Rockwell hardness number (RC). Alternatively a ball indenter is used, with a total load of 100 kgf, to record the Rockwell B scale hardness. This test, developed in the USA, is particularly useful for rapid and routine control techniques, since the hardness number is conveniently indicated by a dial gauge.

All of the above tests require test specimens of restricted dimensions, since the material has to be manipulated in the testing machine. An alternative testing method is required, which may be used *in situ* upon massive components without conducting a destructive examination. The Shore scleroscope is one type which is used for this purpose, and measures the rebound hardness of the material. In the instrument a small diamond-pointed 'tup' or 'hammer' weighing 1/12 oz is allowed to fall from a height of 10 inches inside a glass tube graduated into 140 equal divisions. The height of the first rebound of the 'tup' is recorded as the index of hardness (HS). This test is also used for measuring the surface hardness of plated components.

Table 1.6 shows the comparative hardness values obtained in all four tests performed upon different materials. The discrepancy between the Brinell and Vickers hardness numbers, at high hardness values, can be attributed to the indenter distortion in the Brinell test.

Meyer analysis. The Meyer analysis utilizes the hardness numbers obtained in the Brinell test to give a measure of the work-hardening capacity of metals. The Meyer constant n, representing the work-hardening index, is related to the load P and impression diameter d by

$$P = ad^n$$

The impression diameters, produced by a given indenter, are measured at two or more loads and the constants a and n are obtained graphically. The index n varies from 2.50 for soft materials to about 2.0 for heavily work-hardened metals. The constant a is some measure of the mean hardness level of the material and does not represent any measure of work-hardening capacity.

1.4.4 Impact testing

In service, many components are subjected to high bending stresses, which result from the sudden application of loads, and under these conditions may fracture

Table 1.6 Comparative hardness values

MATERIAL	Brinell	Vickers	Rockwell		Shore
	HHB	HV	RC	RB	HS
Annealed 70/30 Brass	58	61	–	–	–
Annealed Mild Steel	133	134	–	73	20
Annealed 0.8% C Steel	240	240	23	98	35
White Cast Iron	412	435	42	112	56
Nitrided Steel	750	1050	67	–	100

in a brittle manner. This brittle behaviour of materials could be assessed in tensile tests if the attendant difficulties of reproducing the necessary high strain rates and measuring the property values could be overcome. Various impact tests have been evolved, using tension, torsion and bending techniques, to provide a comparative measure of a material's ability to resist fracture, and this property is termed 'toughness' or impact resistance. Specific tests are used to discriminate between those materials of either exceptionally high or exceptionally low ductility. The impact tensile test is used for materials generally classified as ductile, while the impact torsion test is chosen for brittle materials.

The term 'notch bend' test is applied to the Izod and Charpy methods of impact testing, since the presence of the notch in the test specimen increases the tendency of materials to suffer brittle fracture. In the Izod test, a cylindrical specimen, 0.45 inches in diameter, having a notch of specified dimensions (see table 1.7), is held in a vice so that the root of the notch is level with the top surface of the vice (fig. 1.30). A weighted anvil, attached to the end of a pendulum, is arranged to swing from a fixed height to strike the specimen on the same side as the notch. Fracture of the specimen occurs in a manner depending upon the composition, microstructure and temperature of the material, and the geometry of the notch. The Izod value is a measure of the energy required to fracture the specimen, recorded as the difference between the initial and final energies of the swinging anvil, and is displayed by a pointer at the top of the pendulum. The higher the Izod value (ft lb) the greater is the notch toughness of the material.

A Charpy V-notch test is similar in principle to the Izod test, but the specimen is more easily located in the machine. This facilitates testing specimens at temperatures differing from ambient. The Charpy V-notch specimen, of 10 mm square section and 60 mm length, is supported to form a bridge of 40 mm span so that the anvil strikes it centrally on the face opposite to the notch.

An examination of the impact fractures may reveal a ductile, brittle or mixed mechanism of failure of the specimen. Thus the notch-brittleness of a material can be assessed both in terms of the impact value and by that percentage of the fracture area which displays a characteristic crystalline, brittle appearance.

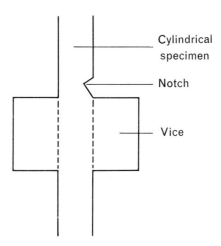

Fig. 1.30 The Izod 'notch bend' impact test.

Table 1.7 Methods for Notched Bar Tests; the Izod Impact Test on Metals. From BS 131 Part I (1961), British Standards Institution.

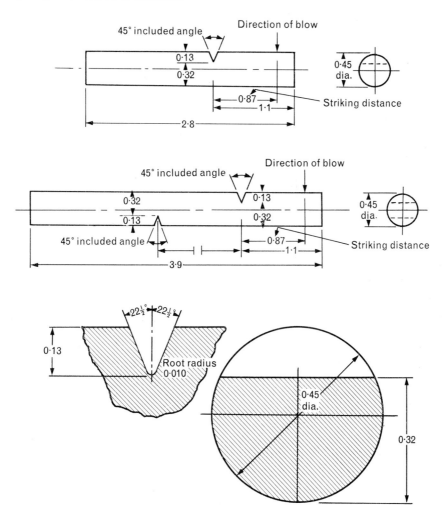

Materials having body-centred cubic crystal structures, which include most structural steels, are notch-brittle below specific 'transition temperatures', where the term transition describes the fairly abrupt change from a ductile to a brittle fracture mechanism under impact loads. Fig. 1.31 shows the temperature dependence of the impact properties of an unalloyed structural steel and emphasizes the value of such tests to predict material behaviour in low-temperature environments, but it must be recognized that the transition temperature is dependent on the specific test conditions. In service the material may undergo the transition at either higher or lower temperatures than those indicated in the test, depending on the imposed condition.

1.4.5 Creep testing

In certain industrial environments such as those encountered in steam generation and chemical refining plant, and in aircraft turbines, the components operate

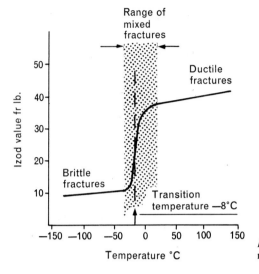

Fig. 1.31 Typical Izod temperature transmition curve for mild steel.

under stress for long periods at high temperatures. Since the safe and efficient operation of such plant depends upon the individual components maintaining dimensional stability, the high-temperature strength properties of materials are the principal design criteria. Although the environmental factors may require the materials to be corrosion-resistant and to possess a stable microstructure, only the strength properties will be considered in this section.

At low temperatures the mechanical properties of metallic materials are virtually independent of the rate of straining or the duration of the test. At high temperatures this condition is changed and the yield and ultimate tensile stresses are markedly dependent on these factors. The material may then plastically deform or 'creep' under the sustained application of loads which would induce only elastic distortions in a short-time tensile test. The temperature range in which this transition occurs is related to the melting point of the material. Thus lead, which melts at 327°C, exhibits creep even at ambient temperatures (c.f. the sagging of domestic hot-water pipes between supports) whereas some nickel base alloys (melting at about 1 400°C) do not undergo significant creep deformation at temperatures below about 650°C.

The design engineer uses two criteria in specifying the life of high-temperature components. Creep test data are used for conditions where small dimensional changes may seriously impair the efficiency or safe operation of the plant. Stress-rupture data are employed in those cases where dimensional stability is less important but where ultimate failure would constitute a hazard. A simpler form of apparatus may be employed to obtain the information in the last case.

In both test methods the specimens are usually tested in tension, but compression and flexure tests are sometimes employed. For the tension creep tests the specimens are similar in shape to those used in the conventional tensile test and are located by screw threads in heat-resisting adaptors linked to the loading mechanism. An extensometer is attached to the specimen and the assembly is enclosed in a furnace which maintains a constant temperature over the entire specimen length. The furnace temperature is controlled within specified limits during the test by a thermostatic control system. The specimen is stressed, usually under constant load, and the resulting plastic strain over the gauge length is measured as a function of time, either to a specified strain or until the specimen fractures. Since the creep strain is measured regularly during the long test period, a highly sensitive method of measuring the extension is required. This extension has to be measured by mechanically transferring the relative movement of the ends of the extensometer to a position outside the furnace, where the displacement is determined by optical, mechanical or electrical techniques (fig. 1.32).

In the stress-rupture test, the time at which the load induces rupture of the specimen and the strain at rupture are noted for fixed temperature and load conditions.

A set of creep curves, in which the creep strain is plotted as a function of time for specimens tested under different tensile loads, at a fixed temperature, is given in fig. 1.33(a), and in each curve the strain is seen to occur in three distinct stages, namely

1. *The primary stage*

The specimen experiences an initial strain on loading, which may be wholly elastic or partly plastic, and then creeps at a decreasing rate.

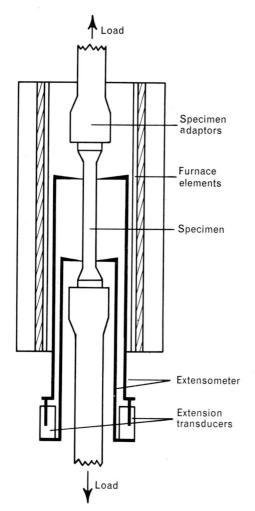

Fig. 1.32 Arrangement of specimen and extensometer for a creep test.

2. The secondary stage
The creep strain continues at a constant rate and is uniform over the gauge length.
3. The tertiary stage
The creep strain increases with time, due to localized plastic flow, until the specimen fractures. This stage corresponds to the fall in load after the ultimate tensile load in the conventional tensile test and is due to the diminished cross-sectional area of solid, load-bearing material.

At any stress level, the minimum permissible creep strain in a component can be computed as the sum of the primary extension and the plastic strain occurring during the secondary stage of creep. For design purposes the creep data is plotted as in fig. 1.33(b), which gives the relationship between the creep strength and the stressing time required to induce fixed values of strain.

Typical data from stress-rupture tests are given in fig. 1.34 which shows the time for rupture under various applied loads.

The mechanisms of plastic deformation in the creep process are clearly different from those experienced by the same material subject to tensile testing and this can be attributed to the temperature of testing and the slow rate of straining which the material undergoes. At elevated temperatures, strain-hardened materials experience recovery and softening as a result of the rearrangement in the 'piled-up' dislocation networks. This is facilitated by the diffusion of unoccupied lattice sites, termed 'vacancies', which allow dislocation 'pile-ups' to climb into free slip planes. The dislocations then migrate to form 'subgrain networks' surrounding volumes of material with very low dislocation densities.

In the primary creep stage, materials strain harden as a result of slip dislocations

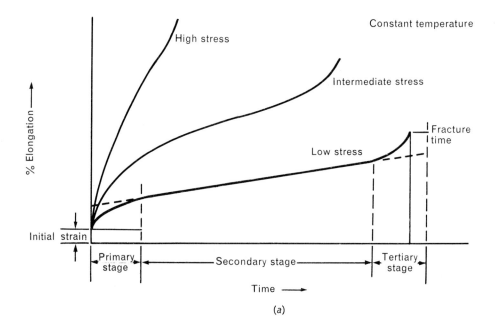

Fig. 1.33 a Creep elongation time curves.

MECHANICAL TESTING | 59

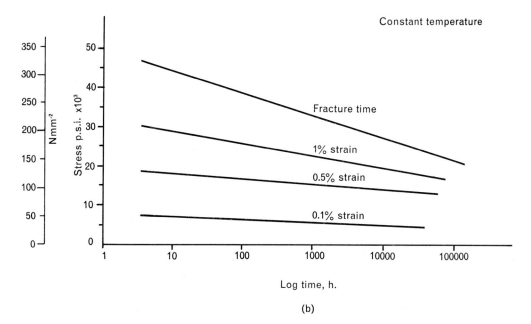

Fig. 1.33 b Creep-stress time curves to achieve specified strains at constant temperature.

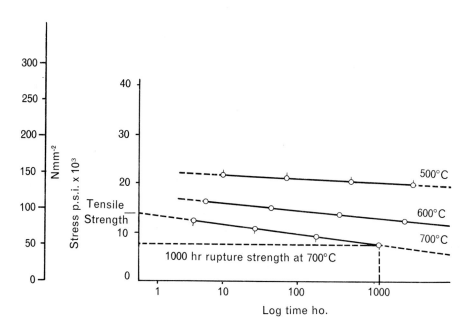

Fig. 1.34 Typical relation between creep-rupture stress and time of exposure at fixed temperatures.

encountering barriers to their movement. An equilibrium is established between the hardening and recovery processes, resulting in a substructure within the original 'grains'. This mechanism accounts for the decrease in the primary creep rate, since presumably the new substructure offers a resistance to plastic flow. The constant secondary creep rate is controlled by the vacancy-induced climb of dislocations into free slip planes and then to the boundaries of the subgrains, causing localized distortion of the lattice. Ultimately, voids are formed at the subgrain boundaries and the tertiary creep stage is initiated, culminating in fracture of the specimen.

Ideally, a material selected for high creep strength purposes should possess a microstructure containing stable precipitates which pin dislocation lines, to restrict the thermally induced climb processes.

1.4.6 Fatigue tests

The cumulative damage to materials resulting from repeated or cyclic stress which ultimately causes fracture is termed 'fatigue'. Since many engineering components such as shafts, bearings and springs experience multiple stress cycles in operation, a tolerable level of stress must be chosen in their design which will avoid failure in service. Numerous manufacturing and environmental factors present difficulties in achieving the component fatigue life of laboratory tests. The size and geometry of the components, the presence of stress raisers in the form of notches, keyways in shafts, machining marks and the presence of certain types of residual stresses, together with corrosion and wear processes, act to reduce the effective fatigue life of components. For these reasons many components are themselves subjected to fatigue tests in order to obtain more realistic estimates of their probable service life.

In laboratory tests, specimens can be subjected to a variety of stress systems, including pure bending, torsion and alternate tension/compression, in addition to combinations of these. Some types of test apparatus used are shown diagrammatically in fig. 1.35. The Wöhler test (fig. 1.35(a)) uses a rod specimen, with a central gauge length, which can be loaded as a cantilever with one end of the specimen attached to the shaft of an electric motor and the other end supporting the load via a bearing. The outer layers of the specimen are subjected alternately to longitudinal compressive and tensile stresses as the shaft rotates and a counter records the number of reversals of stress which the specimen withstands before it fractures. Identical tests can be conducted upon specimens which are 'notched' around the central gauge length, thereby determining the effect of stress concentrations upon the fatigue life of the material.

The complete fatigue data for a material is given by a S/N curve, where S, the applied stress, is plotted against the number of reversals of stress, N, required to fracture the specimen. The typical form of the S/N curve for a ferritic steel is shown in fig. 1.36(a). The horizontal portion of the curve gives the 'endurance limit', which is defined as the maximum stress which will not induce failure in the material at an infinite number of reversals of stress. However, in corrosive environments, ferritic steels display S/N characteristics more typical of non-ferrous materials, and this arises due to the phenomenon of 'corrosion fatigue'.

The majority of non-ferrous materials and the austenitic steels do not display an endurance limit, although discontinuities arise in the S/N curves (fig. 1.36(b)) which have been attributed to changes occurring in the fatigue mechanism.

In many applications the components are subjected to a number of stress cycles of differing amplitude, and the concept of cumulative damage was proposed by Milner as a basis for estimating the useful life of the material. If a material experiences n cycles at a stress S_1, for which the fatigue life is N_1, then the expended fraction of the fatigue life is n_1/N_1. Failure of the material would be anticipated when $\Sigma n/N = 1$, where

$$\Sigma \frac{n}{N} = \frac{n_1}{N_1} + \frac{n_2}{N_2} \ldots \text{etc.}$$

The results of such fatigue tests approximate to this criterion.

The frequency of the stress reversal has little effect on the fatigue characteristics, except where the temperature of the material is raised by the mechanical work

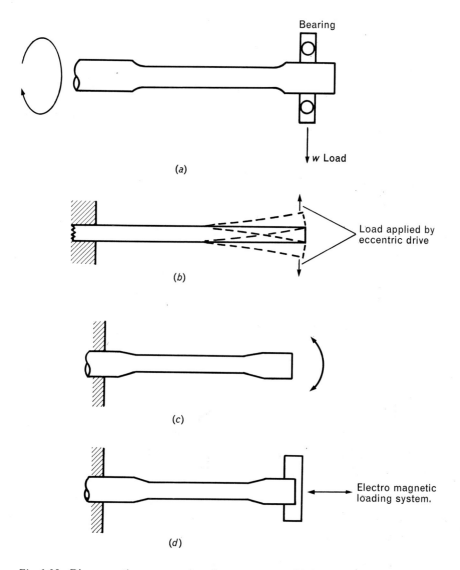

Fig. 1.35 Diagrammatic representation of common types of fatigue tests. *a* Wöhler test (rotating bending) *b* Planar bending fatique *c* Torsion fatigue *d* Push-pull fatigue.

62 | METALS IN GENERAL

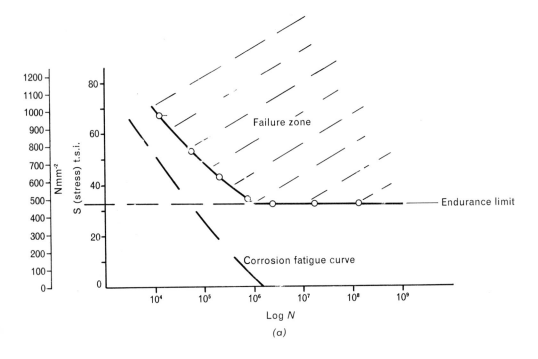

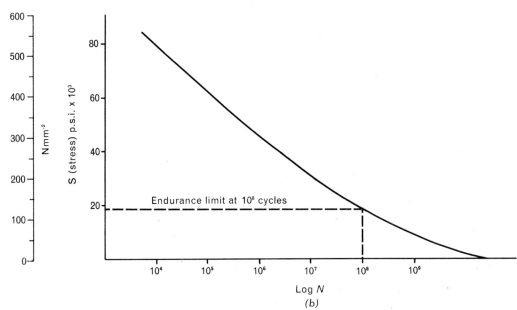

Fig. 1.36 a Fatigue curve for ferritic steels. b Fatigue curve for austenitic steels and non-ferrous alloys.

expended in the test. This causes a decrease in the number of cycles to failure for a given stress. In general, the effect of temperature upon fatigue properties follows that of the yield strength, so that at increasingly high temperatures the fatigue life falls appreciably.

The fatigue of material occurs by three consecutive stages, comprising the nucleation of surface fissures which coalesce to form a single fatigue crack, the growth of the fracture due to the tensile component of the stress cycle and fracture of the specimen when the imposed stress level exceeds the tensile strength of the material.

Fatigue cracks are nucleated in those regions of the specimen which rapidly strain harden as a result of dislocation locking. The material experiences a loss of ductility and the fissures develop within the surface slip bands, which act as points of local stress concentration.

The intermittent progagation of the fracture is revealed by a series of fine striations on the fracture surface, which indicate also the point of initiation of the fracture. Ultimately the unfractured area becomes too small to support the load and the striations cease at the boundary from which simple tensile failure progagates. A typical fatigue fracture is shown in fig. 1.37, where failure nucleated from a keyway in a shaft.

1.5 METALLOGRAPHIC EXAMINATION

The field of metallography encompasses all methods which are used to study the constitution and internal structure of metals and alloys and their relation to the mechanical and physical properties.

The properties of a metallic solid are only partly determined by the chemical composition. In alloys containing more than one crystallographic phase, particularly, the

Fig. 1.37 Fatigue fracture.

64 | METALS IN GENERAL

properties may be varied over a wide range by varying the mechanical and thermal treatments to which the material is subjected during manufacture. With the aid of metallographic techniques of examination, the metallurgist can usually say whether or not the particular properties obtained from an alloy are anywhere near to the optimum. If not, he can suggest alternative processing routes to produce the microscopic structure which will give the desired properties.

The metallographic techniques most commonly used centre around the use of a microscope. The conventional microscope, using light optics, was first adopted for the study of metals by Henry Clifton Sorby during the second half of the nineteenth century. More recently the electron microscope has been applied to study the internal structure in very fine detail, almost at the atomic level. In some metals the location of individual atoms can be revealed with the field ion microscope, but in most cases the detail of the atomic arrangement cannot yet be examined with the microscope and recourse must be made to X-ray diffraction techniques.

1.5.1 Light microscopy

The metallurgical microscope closely resembles those used in biological studies, but with one important difference. Since metals are opaque, light cannot be transmitted through the specimen to the lens system. Instead, the specimen must be viewed with light reflected from the specimen surface, and the optical system is modified to incorporate a suitable means of vertical illumination (fig. 1.38).

If the true structure of the material is to be examined, it is first necessary to prepare a flat surface, free from all extraneous marks, scratches etc. and free from any dis-

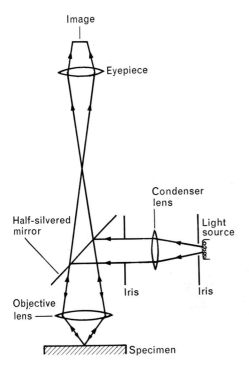

Fig. 1.38 Light paths in the metallurgical microscope.

tortion which may be produced in the early stages of specimen preparation. This is usually obtained by filing or linishing, grinding on abrasive papers of progressively finer grade, and finally polishing with corundum or diamond paste or by electrolytic means.

Examination under the microscope of a specimen prepared in this way will reveal the presence in the surface of cracks, some types of non-metallic inclusions and other defects but very little else, for all the defect-free areas of the specimen generally reflect the light with equal intensity. The internal structure is revealed by subjecting the prepared surface to a controlled corrosive attack with a suitable chemical solvent. This process is called etching.

The atoms at a grain boundary are less regularly bonded to their neighbours than those in the interior of the grains. Hence, when the solvent is applied to the surface, the grain boundaries are dissolved preferentially and when viewed under the microscope they appear as dark lines due to the non-vertical reflection of light from the resultant grooves (fig. 1.39(a)). Some solvents show preferential attack for specific crystallographic planes. In such cases the etched surface of each grain is tilted with respect to the original surface, according to the orientation of the grain, and the microscope image shows contrast which varies from grain to grain (fig. 1.39(b)). When two or more phases are present the anodic potentials of the phases will be different because of their different compositions. By suitable selection of the solvents it is then possible to cause one or more phases to be attacked preferentially and the phase distribution is revealed by the resultant contrast when viewed under the microscope (fig. 1.39(c)). Potentiostatic control of electrolytic cells is now being developed to give very precise control of the etch attack and better differentiation between phases with similar anodic potentials.

The composition of the etchant solution is usually selected to develop the desired microstructure after immersion of the specimen for 10 to 60 s, although in some cases this may extend to about 10 min. If the solution is too strong the attack is vigorous and it is difficult to obtain reproducible results. With very weak solutions the structure is imperfectly developed and lacks contrast. When the contrast has developed sufficiently, the solvent is washed off the surface with water, followed by a rinse in alcohol and the specimen is dried in a blast of warm air. These operations must be performed quickly to prevent the structure being masked by surface oxide stains.

1.5.2 Specialized microscopic techniques

Whilst much useful information can be obtained with the simple uses of the microscope described so far, a wide variety of ancillary techniques are also employed either to increase the contrast or to reveal other features of the structure which are not readily observed in direct examination.

Surface relief is only detected in normal examination when the differences in height are at least 100 Å, or when the surface tilts are several degrees. In some cases this can be improved by illuminating the surface with an oblique beam of light, but better resolution is obtained with an opaque stop to form a dark field image. The stop is a disc, with an annular opening, which is located in the back focal plane of the objective. Each inclined area of the specimen surface produces a separate image in this plane and, by lateral movement of the stop, the light from the various images is allowed to pass the stop and reach the eyepiece. The image of the inclined area is then seen as a bright area on a dark background.

66 | METALS IN GENERAL

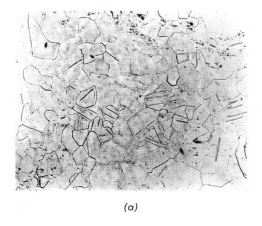

(a)

(b)

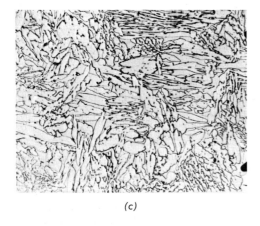

(c)

Fig. 1.39 *a* Pure zinc, incident illumination (Magnification × 250) *b* 70/30 brass, grain contrast in incident illumination (Magnification × 250) *c* $\alpha\beta$ brass, incident illumination (Magnification × 250).

Phase contrast techniques reveal more clearly the differences in height between different areas of the etched surface. The light diffracted from the specimen surface is one-quarter of a wavelength out of phase with the directly reflected light. A glass phase-retarding ring is placed in the back focal plane of the objective and advances or retards the directly reflected light passing through the ring, but the refracted light is not brought to a focus in this plane and is not affected. Projections above the surface are thus made to appear either light or dark, depending on the type of phase plate used.

Interferometry is used for quantitative measurements of the step heights. In principle, the image is produced by interference between two beams of monochromatic light, one of which is reflected from the surface under examination and the other from an optically flat surface. The two beams reinforce where the optical path difference between them is a multiple of one half-wavelength of the light and interfere at other path lengths. The image thus shows optical fringes or contour lines, delineating the outline of areas of the specimen which are at the same level.

Polarized light is very useful for examination of anisotropic (e.g. non-cubic) phases. The wave motion of a light beam contains vibrations in all directions normal to the beam. If the beam is passed through a sheet of polaroid the emergent beam vibrates in only one direction. Insertion of a second sheet of polaroid with the polarizing direction at right-angles to the first (i.e. crossed polars) then extinguishes the beam. In the microscope the first sheet (the polarizer) is placed immediately in front of the light source and the second sheet (the analyzer) is placed below the eyepiece. In the crossed polar position an isotropic phase will appear dark for all positions of rotation of the specimen stage. However, the optical properties of an isotropic phase depend on the crystallographic orientation and, in effect, the reflected light is polarized by the specimen. Thus the light reaching the analyser is uncrossed for certain orientations of the specimen and a bright image is obtained. The grain structure and other features of anisotropic specimens can be examined in this way without the necessity of etching the specimen first (fig. 1.40).

High-temperature microscopy is used to observe microstructural changes occurring in specimens during heating and cooling or on holding at elevated temperatures. Contrast can be obtained by thermal etching, i.e. preferential evaporation of atoms.

Fig. 1.40   Zinc in polarized incident illumination (Magnification × 250).

The heated specimen is usually housed in a vacuum chamber and viewed through a silica window with a microscope using long working-distance objectives.

Microhardness measurements can be made with a miniature form of the diamond hardness indenter. The indenter is attached to the microscope and the optical system is used to locate it on any microstructural feature of interest.

The Quantimet Image Analysing Computer gives a quantitative evaluation of the amounts of phases, inclusions etc. The optical microscope image is scanned using the flying spot principle of the television camera. A photoelectric sensor detects changes in the reflectivity of the specimen surface and is set to trigger the channels of an electronic counter.

1.5.3 Electron microscopy

The limit of resolution (R) of a microscope is given by the relation:

$$R = \lambda/2\text{N.A.}$$

where λ is the wavelength of the light used and N.A. is the numerical aperture of the objective lens. The largest numerical aperture normally obtained, using an oil immersion lens, is about 1.3. With light of the shortest wavelength (ultraviolet, $\lambda \sim 2\,000$ Å) the theoretical limit of resolution is therefore about 800 Å, but in practice it is nearer to 1 μm. This is very large compared with the size of the individual atom. The wavelength of electrons is very much shorter, varying from about 0.05 Å to 0.037 Å as the accelerating voltage is increased from 70 to 100 kV, and, with an electron beam, the limit of resolution is only a few atomic diameters.

The electron microscope resembles an optical transmission microscope, the light source being replaced by an electron gun and the optical lenses by electromagnetic lenses (fig. 1.41). The electron beam is focused through two condenser lenses on to the specimen, which lies in the focal plane of the objective lens. A magnified image is produced at the first projector lens. Part of this is magnified to produce a further image at the second projector lens, where the final stage of magnification produces a visual image on a fluorescent screen. The magnifications of the lenses are varied by altering the current flowing in the electromagnets.

An electron beam can penetrate only a short distance into a metal and, to produce an image of sufficient intensity in transmission, the specimen is limited to a thickness of about 1 000 Å. Foils of this thickness are produced by cutting a thin slice from the bulk metal, taking care to avoid mechanical damage and heating effects which could change the internal structure. The slice is progressively thinned mechanically and by chemical or electrolytic means until small areas of the requisite thickness are obtained. These areas are cut – or punched – out and mounted on 100 mesh copper grids for examination in the microscope.

Very fine precipitates, the density and distribution of dislocations and many other features can be studied *in situ* with the foil technique, but the preparation of specimens is laborious. Much useful information about the nature and distribution of the phases can be obtained more simply by preparing a transparent replica of the etched surface of the specimen. This is prepared in the same way as for light microscopy examination and a thin layer of an electron-transparent substance (usually plastic or carbon) is formed on the surface. Plastic films are stripped off with Sellotape, whereas carbon films are removed by dissolving the underlying metal away, leaving fine insoluble precipitates

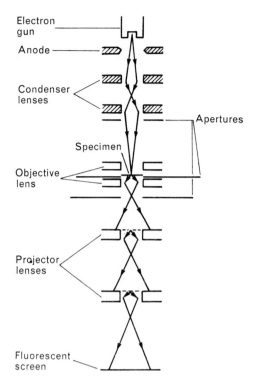

Fig. 1.41 Schematic diagram of lens arrangement in an electron microscope.

attached to the film. The replica is smooth on one side and a 'negative' of the metal surface on the other. When viewed in the microscope contrast appears due to variation from place to place in the path length of the beam, the thinner areas appearing light and the thicker areas dark. Precipitate particles attached to the film increase the adsorption and add to the contrast.

A number of ancillary stages are available to extend the versatility of electron microscopes. Heating and cooling stages can be fitted for examination of foil specimens at temperatures other than ambient, and to observe the change of structure with temperature. Straining stages are used, for example, to study the formation, movement and annihilation of dislocations. With a tilting stage the specimen can be inclined while under observation to obtain the best image contrast, and electron beam tilt devices are used for high resolution, dark field examination.

The electron beam emerging from a thin foil comprises a direct and a diffracted beam. For normal, structural examination the diffracted beam is eliminated by one of the field stops from the final image but, by manipulation of the stop, the direct beam can be suppressed and the diffracted beam used to form an electron diffraction pattern. This pattern can be indexed to determine the crystallographic structure of the diffracting phase. When two phases contribute to the pattern any relationship between the lattice orientations of the two phases can be ascertained. In these respects, the electron microscope fulfils a similar function to X-ray diffraction crystallography but, whereas the latter is more efficient for studying bulk phases, the electron diffraction technique is superior for *in situ* studies of very fine dispersions of phases.

1.5.4 Electron probe microanalysis

The first successful instrument for electron probe microanalysis was described by Castaing as recently as 1951, but it is now regarded as an indispensable tool of the metallographer. Both electron and X-ray pictures of a specimen surface are obtained by scanning a small area of the surface with an electron beam. Crystals or diffraction gratings are used to resolve the spectra of the diffracted beam and a qualitative image of the distribution of specific elements in the area under examination is obtained by selecting the position of the sensor in the spectrometer chamber. Alternatively, the beam can be focused on one particular feature of the microstructure and a quantitative spectrographic analysis of the composition obtained.

The EMMA instrument (combined electron microscope and electron probe microanalyser) is a transmission electron microscope which is modified to allow the electron beam to be focused on the feature of interest. The emergent X-radiations are collected and analysed in a spectrometer chamber in a similar manner to the microanalyser.

1.5.5 Field ion microscopy

The resolving power of the present generation of electron microscopes is not sufficient to identify the sites of individual atoms in metallic solids. Larger microscopes, now under construction, with very high accelerating voltages will bring the resolution nearer to the atomic level and it is possible that this goal will eventually be reached. However, the field ion microscope is the only satisfactory means at present available of revealing the atomic lattice of metallic materials.

The specimen is usually in the form of a fine wire, one end of which is electropolished to form a hemispherical tip with a radius of a few hundred ångströms. The other end is attached to tungsten cathodes which are cooled with liquid nitrogen. The assembly is mounted above a fluorescent screen in a vacuum chamber (fig. 1.42). The chamber is evacuated and then filled to a pressure of about 10^{-3} mm Hg with purified helium gas. A positive potential of about 5 kV is applied to the specimen, relative to the screen, which draws the free electrons in the metal away from the outer surface. A helium atom in the vicinity of the surface gives up an electron to an exposed surface atom and, in so doing, becomes a positively charged ion. The ion is accelerated down the line of force radiating from the metal to the screen and causes the screen to fluoresce. The very high-speed repetition of this process from all the metal atoms exposed above the surface level produces an image of the ionization centres at the metal tip.

The application of this instrument was limited initially to metals such as molybdenum, platinum and tungsten, in which the atomic bonding forces were strong enough to withstand the high voltages imposed. More recently the range has been extended down to the metals of the first transition series by using ions of other gases to form the image and by the development of image intensifiers which allow lower accelerating voltages to be used. With careful manipulation the surface of the specimen can be evaporated while under observation, facilitating a three-dimensional study of the structure.

1.5.6 Relations between microstructures, composition and mechanical properties

At the beginning of section 1.5 it was stated that the mechanical properties of any alloy are markedly dependent on the microstructure. This statement is elaborated in the succeeding chapters but the following examples illustrate the importance of this aspect.

METALLOGRAPHIC EXAMINATION | 71

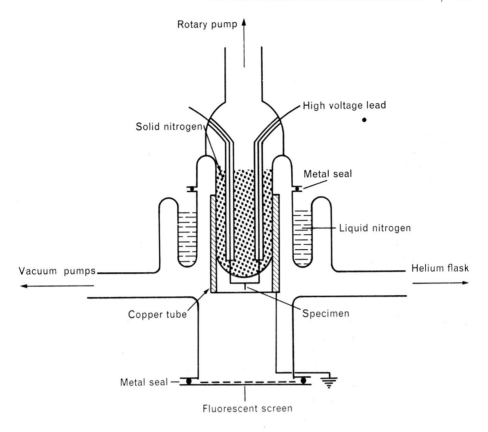

Fig. 1.42 Field ion microscope.

a. Single phase alloys. When a metal solidifies the size of the grains is determined by the rate of heat removal, being much larger with sand-mould castings than when the metal is poured into a metal (chill) or water-cooled mould. Since all the common metals contract during solidification, the solid structure frequently contains voids, which are points of weakness, where the last metal solidified.

A marked decrease in grain size and closure of non-oxidized voids can be obtained by mechanically deforming the metal at temperatures above its recrystallization temperature. At lower temperatures the grains are deformed but do not recrystallize. However, heavy low-temperature deformation followed by reheating to above the recrystallization temperatures can produce very marked grain refinement. If the metal is allowed to cool in the furnace at a slow rate after the heat treatment, the thermal process is called annealing. Table 1.8 illustrates the variation in properties caused by these treatments.

b. Multiphase alloys. Pure copper, or copper with up to 33% zinc added, forms a soft, ductile single phase (α) at all temperatures. If 40% zinc is added (60:40 brass) the microstructure is vastly modified. This alloy solidifies as a hard and brittle Cu–Zn(β) phase. As the temperature falls, this phase becomes supersaturated with copper and, with slow cooling, at 770°C the α phase starts to separate out. At ambient temperature

Table 1.8 Typical mechanical properties of 70/30 brass

	0.1 Proof Stress % Tsi	UTS Tsi	El %	Hardness VPN
As cast (chill mould)	7	13	65	70
Hot worked (wrought) 80%	12	25	45	75
Cold worked 20%	13	23	18	145
Cold worked 80%	36	47	4	220
Cold worked 80% and annealed	7	21	7	70

the structure is a mixture of the soft α and hard β phases, the relative amounts of which can be controlled by the rate of cooling. With very rapid cooling (quenching) of thin sections, separation of the α phase can be suppressed and the β phase completely retained. The mechanical properties are closely related to the relative amounts of the α and β phases.

Very small amounts of precipitates often have a disproportionally large effect on the strength of an alloy. This can be illustrated by reference to the duralumin-type alloys, which are basically aluminium with about 4% copper added. When the alloy is rapidly cooled from about 500°C (solution-treated) the copper is retained in solid solution. The alloy is very soft and readily deformed. In this condition it can be shaped, for example, to form a household utensil or decorative article. On heating between 100° and 200° for a short time (aging) the copper is precipitated as a very fine dispersion. The hardness and strength of the alloy is then very much greater and the ductility is low. With continued holding at this temperature the precipitate particles coarsen and decrease in number. The alloy is then said to be over-aged and the strength decreases towards the solution-treated condition (table 1.9). It is interesting to note that at the peak strength the precipitates are very small and difficult to detect with an optical microscope but they can be examined with the greater resolving power of the electron microscope.

1.6 NON-DESTRUCTIVE TESTING

A manufactured metal component may contain various types of physical defects, such as non-metallic inclusions, voids, seams and cracks, which can act as points of weakness under strain and give rise to premature failure when the component is put into service. Some means is required, therefore, of detecting and rejecting those components containing flaws which may adversely affect the service life. This is particularly important if the component is to be highly stressed, as in high-pressure steam plant, or where premature failure would have serious consequences on human life or expensive machinery. Ideally, the defects should be identified as early as possible in the manufacturing cycle to avoid fabrication and machining costs on components which will ultimately be rejected.

Where a large number of articles are being made by the same processing route, a sample number can be tested to destruction or cut up for detailed examination. Whilst much useful information is obtained in this way, if all the components tested are found

Table 1.9 Typical mechanical properties of an Al–4% Cu (duralumin) alloy

	0.15 P.S. Tsi	UTS Tsi	El%	Hardness VPN
Solution treated	20	30	25	124
Heated 1 hour at 175°C	27	32	14	146

to be satisfactory there is no guarantee that one or more of the remaining components does not contain potentially dangerous defects. For adequate safeguard, testing techniques are required with which all the products can be examined at one or more stages in the manufacturing route and a wide range of non-destructive tests have been devised for this purpose. Some of the techniques most commonly encountered are described in the following sections. It must be emphasized, however, that such tests do not, in themselves, discriminate between good and bad material. Practically all metallic articles contain defects of one type or another, and expert knowledge is required to distinguish those which may adversely affect the service life of a component under a particular set of imposed conditions from those which are probably innocuous.

1.6.1 Crack detection

Cracks which extend from the outer surface of an article may be detected by visual examination after all the scale, grease and other contaminants have been removed from the surface. The chances of detection are very much enhanced, however, by the application of very simple procedures. In the liquid penetrant test, the cleaned component is immersed for a short time in paraffin or warm oil. The surface is then rinsed, dried and coated with a thin layer of an absorbent powder, such as chalk. The liquid retained in any surface cracks slowly exudes on to the surface and stains the powder in the vicinity of the crack (fig. 1.43(*a*)). A better contrast is obtained by adding a coloured dye to the liquid bath. The contrast is further improved by immersing the component in a bath of a fluorescent substance. After removing superfluous fluid from the surface, the component is viewed under ultraviolet light and the cracks are vividly shown up by the liquid retained in the cracks.

The magnetic dust test is suitable only for readily magnetized materials. The article is magnetized, or placed in a strong magnetic field, and a detecting fluid comprising finely divided iron or magnetite (Fe_3O_4) powder suspended in paraffin is applied to the surface either by spraying or by immersion. The two sides of a crack lying across the magnetic flux become magnetic poles and collect the iron dust. This technique will also detect flaws just beneath the surface but, even with surface cracks, the ease of detection decreases as the angle between the crack and the lines of force decreases to zero (fig. 1.43(*b*)). It is usually necessary to demagnetize the article after the test and prior to putting it into service.

Internal flaws are more readily detected by the following techniques.

1.6.2 Ultrasonic tests

The railway wheel tapper used his ear to detect the pitch of the sound wave generated when the wheel was struck a sharp blow. In this way he was able to detect the

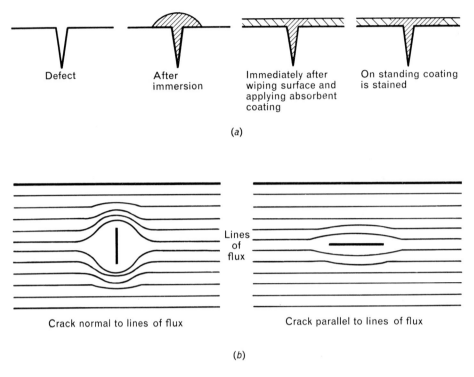

Fig. 1.43 Diagrammatic representation of non-destructive tests. *a* Liquid penetrant test *b* Magnetic test.

presence of a crack, but had to rely on visual examination to locate it. This principle is now utilized in a more scientific way to reveal not only the presence but also the position and dimensions of the flaw. It is not limited to cracks but can be made to reveal any metallic discontinuity at the surface of, or within, the article.

A quartz or barium titanate crystal is used normally to apply pulses of high-frequency sound waves to the surface of the article. The wave is transmitted through the material, reflected from the opposite surface and the signal received at the transmitting surface is shown on a cathode-ray oscilloscope. As the probe is traversed over the surface, any structural discontinuity in the path of the wave will reflect a portion of the signal and this reflected wave arrives at the transmitting surface in advance of the wave reflected from the opposite surface. The time interval between the arrival of the two reflected waves at the receiver is proportional to the difference in the path lengths traversed by the two waves. Hence, by connecting a time base to the oscilloscope, the location of the defect relative to the surface of the article can be determined (fig. 1.44).

Some indication of the size or type of defect can be obtained from the relative intensities of the reflected waves, but it should be realized that these are very much dependent also on the shape and orientation of the flaw. A flat-faced crack normal to the path of the wave reflects the signal more or less completely, depending on the dimensions of the crack, whereas a similar crack parallel to the beam would not be detected. In practice, cracks are generally irregular and give some signal reflection but, wherever possible, it is advisable to test the article in more than one direction.

NON-DESTRUCTIVE TESTING | 75

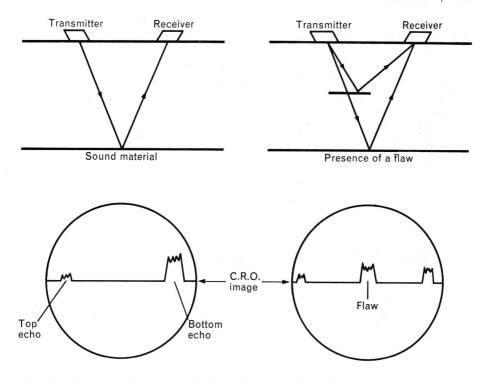

Fig. 1.44 Diagrammatic representation of non-destructive ultrasonic test.

The sound is transmitted as a short train of waves, typically one pulse every 0.02 s, the interval being selected to ensure that the return signal is received before the next pulse is transmitted. The frequencies normally employed are in the range 0.2 to 5.0 Mc/s, giving a wavelength in steel of about 0.05 to 0.20 inches. The angle of spread of the beam is related directly to the wavelength and, consequently, the dimensions of the smallest size of flaw which can be observed are decreased as the wavelength is decreased. With very short wavelengths, however, spurious signals are obtained due to reflection from the grain boundaries, and this places a lower limit on the wavelengths which can be usefully employed.

Acoustic waves are almost completely reflected at an air–metal interface and the contact between the probe and the metal surface is critical. Thin films of various liquid media, such as oil, soapy water or liquid soap, can be interposed to reduce the attenuation of the signal. The surface roughness of the article is also very important. To maintain good contact, the height of the surface irregularities should be less than half the wavelength of the signal and, particularly with castings, it is sometimes necessary first to surface-grind the article. Alternatively, the article and the probes can be completely immersed in a suitable fluid, thus permitting greater separation of the probe from the surface under test.

1.6.3 Radiography

Various types of radiation are employed for this purpose, but X-ray and γ-ray sources are most commonly encountered.

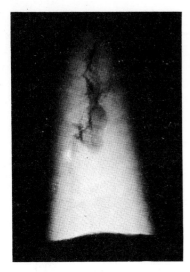
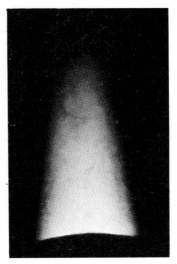

Fig. 1.45 Radiographs of aluminium ingots showing central porosity and surface cracks (left) and surface cracks (right).

The X-ray technique is very similar to that used in medical examinations but a very much harder beam is used, with wavelengths of about 0.4 ångström units (0.4×10^{-8} cm) or less, to give the necessary penetrating power. At these short wavelengths refraction and reflection are negligible and the X-ray beam can be considered as parallel radiation which is not affected by local electric or magnetic fields.

The beam is decreased in intensity by absorption as it passes through a solid, the rate of decay being dependent upon the characteristic absorption coefficient (μ) of the solid. Mathematically the rate of decay is expressed as:

$$I_x = I_0 \exp(-\mu x)$$

where I_x is the intensity after penetrating a distance x and I_0 is the intensity at $x = 0$. For a homogeneous solid, the strength of the signal emerging from the opposite side of the article is proportional to the distance traversed through the solid. In general, the value of μ for the metal is different from that for voids and non-metallic inclusions, and the emergent beam varies also in intensity according to the type, volume and distribution of defects in the path of the beam. An image of the intensity distribution can be obtained by placing an X-ray photographic film in the path of the emergent beam (fig. 1.45).

The penetrating power of the X-ray beam is increased and the value of the absorption coefficient is lowered by increasing the voltage used to generate the X-rays. For radiographic purposes this is typically in the range 50 to 2 000 kV. The operating voltage is selected, on the basis of the thickness and composition of the article, to give a suitable contrast on the photographic film. The contrast obtained is also affected markedly by the focal size of the X-ray beam (resolution increases with decreasing focal size) and the amount of scattered radiation which reaches the film.

The image can be intensified by placing fluorescent screens (e.g. of cadmium tungstate) or thin lead foils in contact with the film, but even when all precautions are taken to ensure maximum intensity, several minutes are needed to obtain each picture. Consequently, the technique of 'fluoroscopy' has been developed, in which the X-ray beam falls upon a phosphor fluorescent screen. A visual image is obtained which can be observed directly or recorded photographically. In principle this resembles the mass miniature chest X-ray technique. The detail obtained is usually less than that from a photographic exposure but with fluoroscopy the article can be moved about during radiation and a more detailed examination can be made.

'Xeroradiography' produces an image of the beam on a flat brass or aluminium plate which is coated with a thick layer of vitreous selenium. The selenium is first charged electrostatically and then partially discharged by exposure to the emergent X-ray beam. The leakage of the charge is proportional to the intensity of the incident radiation. An image of the defects in the article is then obtained by allowing an electrostatically charged powder to settle on the plate, the amount of powder adhering at any point being proportional to the residual charge on the selenium coating.

In gamma radiography a radioactive substance is used which emits γ-rays. Typical sources are radium, cobalt-60, caesium-137 and iridium-192. The technique is very similar to X-ray radiography, the main advantages being that the source is more easily transported for 'on site' examinations and less capital cost is involved. Energy is radiated in all directions from the source and a number of articles can be radiographed simultaneously with one source. The major disadvantages are that the strength of the beam cannot be adjusted, the exposure time is very much longer and the energy of the beam is too high to operate a fluorescent screen.

The sensitivity of a radiographic exposure is determined by placing a 'penetratmeter' (typically a step wedge of accurately known dimensions) at a suitable point on the surface of the article under examination. The image of the penetratmeter is recorded on the radiograph and the dimension of the thinnest visible step, expressed as a percentage of the thickness of the article, is called the radiographic sensitivity. With care, sensitivities better than 1% can be achieved.

Radiographic techniques are used extensively for the examination of welds and for all types of components where rough surfaces or irregular contours present difficulties in the application of other, less expensive, techniques. Rigorous safety measures must be adopted, however, to avoid exposure of any persons to the high-energy X-rays and γ-rays.

1.6.4 The use of radioactive isotopes

Apart from radiographic examination, radioactive isotopes are widely used for such purposes as the continuous measurement of metal thickness during rolling and the determination of the thickness of surface coatings and the wall thickness of hollow-sectioned components. These applications depend on correlation of the rate of attenuation of the beam with the dimension of the material in the direction the beam is transmitted. The emergent beam is usually detected by a Geiger or scintillation counter and converted into a visual signal which can be calibrated to describe variations in the thickness of the material as it is fed past the source.

Amongst other common uses of isotopes are wear testing, where a radioactive substance is incorporated in one of the surfaces and the debris of the other surface is monitored, and locating the position of metallic components embedded in opaque materials. For example, the continuity of metallic reinforcing in concrete columns is readily ascertained in this way.

1.6.5 Electromagnetic (eddy current) tests

These tests depend on observation of the interactions between a metal article and an electromagnetic field. When an alternating current is fed through a coil of wire it creates a magnetic flux. In the same way that a transformer operates, an alternating current of the same frequency is induced by the flux in a second (detector or search) coil which is placed in the magnetic field. If an electrically conducting article is placed in the field, the primary flux induces an eddy current flow in the article which, in turn, gives rise to a secondary flux in opposition to the primary. The current induced in the search coil is thereby diminished. Any change in the dimensions, composition, soundness or microstructure which affects the electrical conductivity of the article will cause a change in the eddy current loss and hence a change in the current flowing in the detector coil. Relatively small changes in composition (particularly with pure metals), grain size, grain shape and dislocation density can cause significant changes in the signal detected. Ferromagnetic materials produce an additional hysteresis loss which is dependent on the domain pattern in the material.

The arrangements of the coils and specimen are illustrated in fig. 1.46. The absolute method (fig. 1.46(a)) gives low sensitivity since signal changes due to flaws etc. in the article are usually small relative to the change when the metal article is placed in the field. Much better discrimination is obtained with the differential method (fig. 1.46(b)) in which the article under test is compared with a standard specimen of similar shape and composition but free from defects. Alternatively, with long cylindrical components, two search coils can be located at different positions on the same article (fig. 1.46(c)). As the component is fed through the coils, the difference in the signals induced in the detector coils is due only to structural differences along the length of the article.

When an alternating current flows through a conductor, the current flow is greatest at the surface and diminishes progressively towards the centre, the gradient increasing with the current frequency (i.e. the skin effect). This applies also to eddy currents and has an important bearing on the location of flaws which are detected in the tests. With radio-frequency signals, the induced current flows in the surface layers, and surface and near-surface defects only are examined. Conversely, low frequencies down to 1 c/s can be used to locate deep-seated flaws. In practice it is often possible to select individual or groups of variables for examination at different depths below the surface of the article by varying the amplitude, frequency and voltage of the input signal and the design of the coils.

1.6.6 Strain gauges

The use of strain gauges to determine the direction and magnitude of the strains which are developed when large engineering structures are loaded is a different, but very important, aspect of non-destructive testing. A strain gauge is basically a transducer whose electrical properties are very sensitive to small changes in dimensions. The two ends of the transducer are cemented on to the structure in the direction in which the strains are to be measured and the change in its electrical resistivity is measured when the load is applied. From a previous calibration of the gauge, this change can be equated to the strain in the structure at the position of the gauge and in the direction tested.

Strain gauges are also employed to ascertain the size and directions of residual strains (i.e. strains which are not relaxed spontaneously) in forgings and other fabricated components. When used for this purpose in the non-destructive sense, only

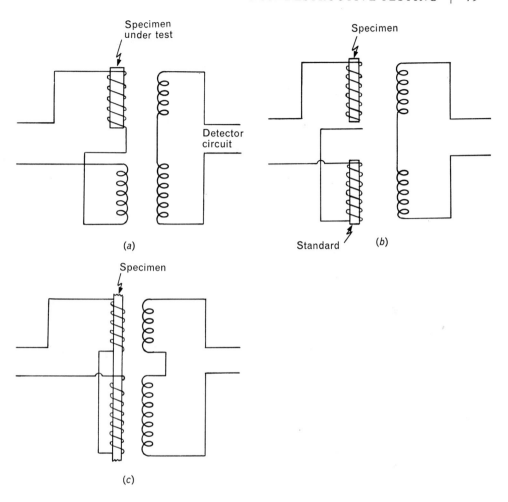

Fig. 1.46 Arrangement of coils and specimens for electro magnetic tests. *a* Absolute method *b* Differential method with standard. *c* Differential method with two coils on specimen under tests.

surface strains are measured. The gauge is cemented to the surface in the direction in which the major strain is expected to lie and a small sliver of metal, to which the gauge is attached, is removed from the article. The change in the reading of the gauge is taken to be proportional to the residual strain, assuming that the strains are completely relaxed when the sliver is removed. There is a danger, however, that the sliver may be plastically deformed during removal, in which case the gauge reading does not give a true measure of the residual strains. A better procedure is to arrange a number of gauges radially around a point on the surface. The change in the gauge readings is noted when shallow holes of increasing diameter are drilled successively at that point and, from previous calibration, both the direction and magnitude of the surface stresses can be determined. Both these techniques involve some damage to the surface. Where this cannot be tolerated, an X-ray diffraction technique is employed to determine the strains.

1.7 LITERATURE

1. *Methods for Tensile Testing of Materials* BS 18, Part 2. London, British Standards Institution. 1962 (amended 1966).
2. A. R. BAILEY, *Textbook of Metallurgy*. Macmillan, London, 1960.
3. C. S. BARRETT and T. MASSALSHI, *Structure of Metals*. McGraw Hill, New York, 1967.
4. BRITISH STANDARDS INSTITUTE *Methods for Tensile Testing of Materials*. B.S. 18 Part 2. London, 1962.
5. A. H. COTTRELL, *Theoretical Structural Metallurgy*. Edward Arnold, London, 1955.
6. J. GORDON PARR, *Man, Metals and Modern Magic*. American Society for Metals, New York, 1958.
7. W. HUME ROTHERY, *Atomic Theory*. Institute of Metals, London, 1962.
8. W. HUME ROTHERY and G. V. RAYNOR, *The Structure of Metals and Alloys*. Institute of Metals, London, 1962.
9. R. F. MEHL, *Physical Chemistry of Metals,* McGraw Hill, New York.
10. B. A. ROGERS, *The Nature of Metals*. American Society for Metals, New York, 1951.
11. A. STREET and W. O. ALEXANDER, *Metals in the Service of Man*. Pelican, London.

CHAPTER 2

Ores in General

2.1 ORE MINING

2.1.1 History of Ore Mining

T. A. Rickard, in his book *Man and Metals* states, 'Civilization did not begin until metals became the material of tools, implements and machines; by aid of metals man emerged from savagery.' Certainly no one can quarrel with this statement for it is indisputable that mining was (and indeed still is in many parts of the world) the precursor of civilization. The Age of Metals is thought to have begun in Europe some 5000 years ago although mining activity in the broader sense of the word had, in fact, antedated this event by several millennia. During the long period of the Stone Age primitive man exploited flint deposits, first on the surface and then underground. The flints were roughly fashioned into tools and weapons and throughout the world many archaeological discoveries point to the widespread mining of flint. Some of these early flint mines consisted of vertical shafts up to 2 m in diameter sunk through the ground to a depth of 10 m. From the foot of these shafts tunnels were driven somewhat haphazardly out into the deposit. Old mining implements included deerhorn picks and the shoulder blades of various large animals. These bones were probably employed as shovels.

Neolithic man not only mined minerals such as flints and stones but also ornamental minerals and rocks. He used mineral pigments and ochres to decorate his dwellings and his body. Clays of many sorts were mined fairly extensively for the manufacture of primitive utensils.

There can, of course, be no sharp demarcation in time between the Stone Age and the Age of Metals, for in some parts of the world metals were in use many thousands of years before their introduction into less advanced communities. Historians divide the Age of Metals roughly into the Bronze Age and Iron Age but no one culture prevailed over the whole earth at the same time. The Syrians, for example, had learned how to obtain and fashion iron whilst Europe still lingered in the bronze phase of development and indeed, in Scandinavia, stone implements were still in universal use. Until fairly recent times certain tribes in the remoter areas of the world could with justification have been classified as of Stone Age culture.

Probably the first metal to attract the attention of man was gold. This substance is one of the few metals found in the elemental or free state in nature, and being so widely distributed and not prone to tarnishing it is not surprising that gold antiquities are common throughout the world. No doubt attracted by its colour and the subsequent discovery that this shining yellow substance could be easily shaped, primitive man soon learned to prize gold as an ornament.

Primitive gold mining merely consists of gathering nuggets from placers, i.e. unconsolidated alluvial deposits. The weathering and erosion of gold bearing rock often yielded gold concentrations of surprising extent and richness and doubtless these miners of prehistory soon discovered the way to wash gold out of gravel by the simple process of panning. Another metal much prized by the miners of the late Stone Age was silver, and archaeologists have excavated burial places in which gold and silver ornaments were equally prominent. In ancient Egypt, silver was known as 'white gold' and, as with gold, the metal was prized chiefly because of the ease with which it could be fashioned into ornaments.

The Age of Metals did not truly begin until men discovered how to smelt ores, although considerable evidence exists that primitive man had learned to fashion tools from native copper found as nuggets on the surface. How the art of smelting was first discovered has produced a multitude of apocryphal stories but it seems likely that the science of smelting was born at the edge of camp fires widely separated both geographically and in time. Early man in building stone enclosures around his fires must have eventually noticed that certain stones appeared to melt and form beads resembling the native copper with which he was familiar. From here the next advancement in knowledge was the casting of molten copper and eventually, by accident, came the significant discovery that particular ores when smelted produced a harder wearing metal. As tin and copper ores are frequently found in close proximity in many parts of the world the primitive metallurgist, in searching for a better product, learned through time to smelt a copper rock together with an adjacent tin-bearing rock, and thus the Bronze Age evolved. With this development the way was open for man to start metalliferous mining in earnest which led to the eventual discovery of the metal that in the form of its steel alloys dwarfs all others in importance.

It is perhaps surprising that the releasing of iron from its ores so long post-dated the smelting and working of copper and tin. The probable reason for this long delay is the fact that, apart from meteoric iron, native iron is very uncommon and thus early man did not have the advantage, as he did with copper, of having first worked with the native metal and then progressing logically to the ores. So highly prized was iron in many parts of the world that this metal had greater value than gold. In many predynastic Egyptian tombs small iron ornaments have been found, but the sparseness of this metal compared with the lavish use of gold and silver suggests that the possession of iron was the prerogative of the rich. Even stepping forward in time some several thousand years, Cortes, when he invaded Mexico, found that only the Aztec chiefs possessed iron knives whereas gold was abundant but of little functional value because of its softness.

Although, as previously discussed, even approximate dating is impracticable, it is generally considered that the Iron Age commenced about 1 000 B.C. Since that time the tonnage of iron metal taken from the ground vastly exceeds the combined tonnage of all other metals put together. Fortunately, iron is abundant and there is no sign yet that the Iron Age is drawing to a close despite the fact that by 1967 annual world production of iron ore was 624 million tons.

2.1.2 Early mines and mining

The earliest records of underground mining are those of the Egyptians some 5 000 years ago. Mining for gold, silver, copper and turquoise was carried out on an appreciable scale although most of these mining operations were confined to near surface deposits. Probably the most graphic and harrowing account of ancient mining is that given by the Greek historian Diodorus Siculus, who visited Egypt in 50 B.C., when he describes the hapless plight of the thousands of prisoners labouring in the old underground gold mines on the Ethiopian border. In this description Siculus mentions fire-setting–the universal method at that time of driving tunnels through hard rock. Basically, fire-setting consisted of lighting brushwood fires against the solid face of the rock and then, when hot, the rock was rapidly cooled by throwing cold water against the exposed face. This abrupt cooling had the desired effect of inducing cracks which enabled the rock to be broken down by primitive picks and wedges.

The next great race of people to advance the spread of mining were the Romans. Although not themselves miners, they had learned mining from the many different nations they had conquered and, in consequence, accumulated considerable experience which had been acquired in the course of centuries by many different races. They mined on a large scale and evidence of their industry–or more correctly that of their slaves–is to be found throughout Europe and Africa. Mining continued following the collapse of the Roman empire and by the Middle Ages the metalliferous mining industry was strongly established in Europe. The famous book *De Re Metalica* by Agricola, published in 1556, contains a complete account of mining at that time and may be considered to be the foundation of mining literature.

The successive application of gunpowder, steam, compressed air and electricity to mining operations brought the industry to a state of development undreamed of by Agricola. Automation of mining activities is now commonplace and many underground operations are remotely controlled by operators sited on the surface. In the past 50 years greater technical development of the mining industry has taken place than in the preceding 5 millennia; it is probable also that the total tonnage of ore obtained in the last half century is greater than in the rest of mining's history.

2.1.3 Prospecting and exploration

Although these two terms are often used interchangeably, more correctly prospecting is the preliminary search for ore deposits whereas exploration involves the delineation of the deposit so that its tonnage and potential value can be ascertained. Nowadays it is usually the mining geologist who is responsible for the discovery of a mineral deposit although there are still several recent examples of the old type of prospector-cum-miner finding profitable ore bodies. Such instances are, however, becoming less frequent, particularly as the discovery of mineral deposits which outcrop at the surface is largely an occurrence of the past. In general, significant new deposits are only found after the expenditure of considerable time and effort by scientifically trained geologists supplemented by the resources of modern technological laboratories.

It is now widely accepted that most minerals originated as molten rocks which, when working upwards to the surface, sequentially deposited the minerals in the form of crystals. Mineral deposits formed directly from magmas are called primary deposits and are classified either as syngenetic or epigenetic depending on whether the deposit is contemporary with the surrounding rocks or younger than those rocks. In the latter case the molten magma has intruded into older rocks and deposited minerals.

84 | ORES IN GENERAL

Primary deposits, through the action of erosion and weathering, give birth to secondary deposits which may be considerable distances away from the position of the original deposition. The gold placer mentioned previously is a good example of a mineral that has been released by weathering and subsequently transported and concentrated by natural agencies into a valuable deposit.

An elementary study of geology will help to indicate the improbability of mineral deposits in certain areas and thus narrow the field for a more comprehensive search elsewhere. A much more rigorous study of geology is necessary to interpret the results of such a search.

The simplest of all prospecting techniques is an examination on foot of the chosen area. Search is made for outcrops, placer deposits, indications of particular types of vegetation, rocks etc. Further detail may be provided by the prospector digging trenches or putting down shallow drill holes (see fig. 2.1). The prospector in this case is likely to be a combination of geologist, surveyor, map maker, miner and financier. Changes of success with this type of prospecting become less each year.

Nowadays large-scale prospecting requires considerable capital and expertise. Mapping of the area is frequently done by aerial photography and stereoscopic photographs are used to build up a fairly complete picture of the terrain. Having obtained effective topographical data the geologist can now proceed to use one or more of several methods for the location of mineral deposits.

a. Geochemical prospecting for minerals. This includes any method of mineral exploration based upon the systematic and precise measurement of the chemical properties of a naturally occurring material. Generally, as applied to mining, this

Fig. 2.1 Small prospecting drill.

involves the determination by analyses of the trace content of some element or group of elements in rock, soil, vegetation, stream sediments or water. Where such measurements show considerable difference from the normal value in the general area the presence of a geochemical anomaly is indicated and thus the prospect of an adjacent mineral deposit may be signified. The normal percentage of the trace element(s) in an area free from mineral interference is referred to as 'background'. These average or background values vary widely over the surface of the earth but can be obtained for the specific area which is being prospected and, in fact, tables of background values are often available as rough guides.

The accuracy of the chemical analyses necessary to determine the minute changes in trace element content can be appreciated when it is considered that contrasts in chemical composition of trace elements as low as 2 p.p.m. in vegetation have been considered sufficient cause for further investigation.

b. Geophysical prospecting for minerals. The science of the physics of the earth may be broadly defined as geophysics. Geophysical prospecting is the application of geophysics to the task of locating mineral deposits or geological structures concealed beneath the surface of the earth. As with geochemical prospecting, the science of geophysical prospecting is not so much concerned with absolute values but with deviations from average values. There are many geophysical methods which can be applied to prospecting.

1. Magnetic method. This relies upon changes in the earth's magnetic field as measured by a sensitive magnetometer. Because of the wide distribution of magnetite in the earth's crust it is possible to locate magnetic ore bodies by measuring magnetic anomalies.

2. Gravitational method. With this system gravimeters are used to measure small variations in the gravitational pull of the earth. In a simple case, if a lead deposit is buried beneath the ground a gravimeter held immediately above this spot would indicate an anomaly by recording an increase in the gravitational pull. In many recent surveys gravimeters have been carried by aircraft, although the use of this method is limited to relatively level country. A further disadvantage is that unless the deposit is large there is insufficient effect to be measurable on the instrument.

3. Seismic method. This involves placing small explosive charges in holes drilled in the ground and measuring the velocity of the resultant shock waves travelling through the ground to strategically placed seismographs. The physical properties of the rocks between the explosion source and the seismograph affect the velocity of the seismic waves. If the normal velocity is known, any deviation will indicate an anomaly.

4. Electrical methods. These are essentially dependent on measuring the differences in electrical conductivity between the ore body and the surrounding rocks or between adjacent geological formations. As metals are good conductors of electricity such methods find wide application in prospecting for ores.

5. Radioactivity method. With this method a simple Geiger counter (or a scintillation counter) is carried by the operator who slowly traverses the area on foot. Usually the radioactivity method is limited to thorium and uranium ores, although some radioactive elements occur in minerals which are often in association with valuable ores and thus the detection of radioactivity acts as a pointer.

All of the foregoing geophysical methods, as with geochemical methods, can merely act as guides to possible deposits. They are by no means infallible and merely serve to indicate areas which justify a further and more positive examination. Such an examination demands exploratory shafts, tunnels or boreholes. Exploratory shafts or tunnels are confined to shallow deposits whereas exploratory boreholes have been put down to depths exceeding 7 000 metres to prove the existence or otherwise of mineral deposits.

2.1.4 Borehole exploration

The function of an exploratory borehole is to provide indication of the depth to the ore body together with the thickness, quality and extent of the deposit. The borehole will also provide data from which the nature of the intervening strata can be determined and this may reveal the presence of faults or other geological irregularities. The amount and direction of the dip of the strata can also be calculated. Normally, several boreholes in a predetermined pattern are put down covering the area where preliminary prospecting has indicated the likelihood of an ore body. However, the cost of deep borehole drilling is high and may reach £20 to £50 per metre of hole drilled and so considerable care is taken when siting the boreholes to ensure that maximum information is provided by the minimum number of exploratory holes.

a. Percussive drilling. Broadly, there are two principal techniques used in drilling deep exploratory boreholes. The oldest system is termed percussive drilling and evidence exists to show that this method was employed in 2000 B.C. by Chinese prospectors searching for oil. The principle of operation is to strike a series of continuous blows on a chisel which is turned slightly after each blow, thus forming a circular hole generally about 6 to 10 cm in diameter. The chisel is attached to a steel rod and when the steel rod has been forced into the earth a new rod is fitted to the top of the first rod and thus the chisel can be forced to a greater depth into the earth. During the boring operation the chisel is attached to the bottom of a long line of steel rods which stretch to the surface. At the surface, a mechanism is arranged to alternately raise and drop the drill string (as the line of steel rods and the attached chisel is known) and give the necessary twist to the string after each blow. After a few hundred metres of hole have been drilled the weight of the steel rods is sufficient to impart a strong blow, for in a deep hole the weight of the drill string is several tons. In the simplest form the percussive drill is operated purely by manpower but for deep borehole drilling diesel driven engines are used to lift the rods.

After every few metres of hole have been drilled it will be apparent that the detritus at the bottom of the hole would prevent the chisel from cutting further rock unless the chippings were removed. To remove these chippings the hole is filled with water and the drill string removed by successively unscrewing 3 to 10 m lengths as the rods are withdrawn to the surface. The greater the height of the derrick on the surface, the longer is the section of rod that can be withdrawn before unscrewing. The chisel is unscrewed from the bottom rod length and is replaced by a baler, which is a hollow section of rod of the same diameter as the borehole and closed at the lower end by a simple valve. When the drill rods are lowered back down to the bottom of the hole the valve opens and allows the wet chippings and sludge to enter the baler. When this is full, the entire string of rods is again withdrawn whilst the valve retains the sludge inside the baler. At the surface the baler is unscrewed and replaced by a sharpened chisel and the sequence of operations resumed. The detritus inside the baler is examined and provides the necessary data on which the geologist and mining engineer can base their assessment of conditions. It will be apparent that, when drilling deep holes, the time spent raising and lowering the drill string far exceeds the time spent on actual cutting operations. For this reason cable tool percussive drilling finds favour for deep hole drilling. With this system the rigid steel rods are replaced by a steel cable, from the bottom of which the cutting chisel is suspended via heavy steel rods attached immediately above the chisel to provide the requisite weight. When withdrawing the chisel from the hole or running back down again the cable can be reeled on or off the driving drum at speeds of up to 200 m per minute.

Percussive drilling has the advantages of cheapness and simplicity compared with other systems and cable tool boreholes have been put down to a depth of 5 000 m. For some exploratory operations, however, percussive drilling is not sufficiently accurate neither in the actual taking of the rock samples nor in the directional control of the borehole. Then again, percussive drilling is limited to vertical or near vertical downward drilled holes. On many occasions it may be necessary to drill holes inclined from the vertical or even bore horizontal exploratory holes into the side of a hill. In actual underground operations it is often necessary to drill holes from underground up into the overlying rocks. In such circumstances rotary drilling is used. When long exploratory boreholes are required, by far the most common rotary system is diamond drilling.

b. Diamond drilling. This system was first employed in the mid-nineteenth century and is now the most widely used method of drilling exploratory boreholes. There are two main techniques, namely, non-coring and coring. With both techniques the cutting action is obtained by means of small inferior diamonds called borts firmly embedded in a specially designed drill bit. Borts are a form of diamond not suitable for gem stones because of various impurities and range in colour from clear to opaque. The size of the stones varies widely depending on duty but in general the harder the rock the smaller the borts used. Average duty diamond drill bits would have borts of total weight 5 to 10 carats with anything between 10 to 100 stones per carat.

For normal mineral exploration purposes the diameter of borehole ranges between 3 to 8 cm although some diamond drill bits have been manufactured up to 35 cm in diameter. Special duty large diamond drill bits cost several hundred pounds and in extremely hard rocks may have a working life equivalent to only a few metres of hole drilled.

c. Non-coring diamond drilling. In operation the solid-ended drill bit is attached to the bottom end of a string of hollow steel rods stretching down the hole. The rods are caused to rotate at speeds ranging between 1 000 to 2 000 rev/min and pressure is applied to the rod string which forces the drill bit into the face of the rock at the bottom of the hole. The borts cut away the face of the rock and the abraded detritus is immediately washed back up the hole by a circulating fluid (water, mud or air) which is passed down the hollow drill rods and returns to the surface via the annular space between the inner face of the borehole and the outer surface of the drill rods. The detritus flushed out of the hole is examined and gives an indication of the strata passed through.

d. Coring diamond drilling. Very broadly, the coring diamond drill bit may be considered as a short section of hollow steel rod with borts set into the cutting edge (see fig. 2.2). As the drill rods are rotated the borts in the cutting end grind away an annulus of rock. As the drill bit descends further into the hole under the action of

Fig. 2.2 Diamond drill coring bit.

downward pressure the narrow cylinder of rock thus cut out intrudes into the hollow drill bit. As the drilling proceeds this cylinder of solid rock passes through the drill bit and into a special length of drill rod called the core barrel. This may be up to 6 m in length. When the core barrel is filled, a special device enables the intruded cylinder of rock to be broken off the parent rock mass. This device, known as a core lifter, then holds the length of snapped off core firmly in the core barrel whilst the drill string is withdrawn from the borehole. At the surface the core is taken from the barrel for close examination. The lengths of core are numbered in sequence and with this system a very precise cross-section of the strata can be obtained. The detritus at the foot of the borehole is continuously flushed up and out of the borehole as with the non-coring system and can be used to provide a check on the core data if this is considered necessary.

An improved coring technique enables drilling to be carried on continuously and eliminates the need for the drill string to be withdrawn when the core barrel is full. This is termed the reverse flush system. The arrangement differs from normal coring in that the circulating flush passes down the outside of the drill string and returns through the drill bit. No core barrel is needed, for, as the drilling proceeds, the core is broken off every few centimetres and is carried to the surface by the circulating fluid as it rushes up the inside of the hollow drill rods.

2.1.5 Sampling of mineral deposits

Estimates of ore grades are based on the assays of samples obtained from exploratory boreholes from the surface, perhaps supplemented by analyses of samples taken from excavations within the ore body. The value of these estimates is governed by the care taken in obtaining the samples, the accuracy of the assays, the number of samples, the weight given to the individual assay results and the statistical treatment of the results, particularly where erratic high or low ore values are encountered. Natural conditions have a great bearing on the degree of accuracy possible. The regularity of the deposit and the size and distribution of the particles of the valuable constituents will affect the results of any given pattern of sampling. In a massive deposit of low-grade ore such as is found in open-pit mining in Australia and North America an error of a fraction of 1% in the assessment of ore grade may make the difference between a profitable mining venture and an unprofitable undertaking. Slight errors in a sampling programme are not quite so significant where high-grade ores are concerned.

A standard sampling principle is that the finer the size of the mineral particles in the ore, the more even the mineral distribution and hence the smaller need be the size of individual samples and the less the frequency of sampling. The converse is also true and indeed in some deposits only mill tests of very large bulk samples can provide any meaningful indication of the average grade of the ore.

The taking of samples by means of percussive drilling has already been considered. Misleading samples may result from caving of the upper wall of the borehole, sludge leaking away through cracks and crevices, and from inaccurate assay techniques. More accurate samples can be obtained with diamond core drilling, although even with this system real precision can only be achieved if core recovery is 100%. Given this ideal state of affairs it must still be realized that the assay value of a perfect core may differ widely from another section of the ore body only a few feet away.

Taking samples from within the excavation is known as face sampling. This involves the cutting or breaking off of sections of ore from the exposed face of the mineral body. In certain circumstances tunnels are driven through the mineral body purely to enable face samples to be taken. This may prove a very expensive undertaking if results are disappointing. For shallow deposits lying near the surface pitting or trenching is carried out to expose sampling faces. In face sampling there are two basic techniques. Small projecting pieces may be broken off more or less at random over the whole exposed face or, alternatively, grooves or channels of uniform width and depth are cut across the face.

Once a mine has started producing it is still essential to maintain continuous sampling. In some mines, particularly gold mines, it is often necessary to ensure that the average grade of ore being mined throughout the entire mine remains fairly constant over long periods. This helps to achieve a balanced economy over the life of the mine and also means that the surface mill or treatment plant deals with a reasonably uniform and consistent input. To this end it is necessary to know if the grade varies, and where it varies, so that if one section of the mine enters a zone of particularly low-grade ore the output from this section can be deliberately reduced and the output from a higher grade section commensurately increased. The outputs from the two sections are then blended prior to despatch to the mineral treatment plant.

With certain mining methods it is necessary to maintain an effective sampling procedure covering the actual mining operations to ensure that a proper balance is achieved between the amount of ore taken from each working place and the amount of waste rock that may have to be mined in conjunction with the ore. If the proportion of waste is inadvertently high the mined ore is considered to have been 'diluted'. Representative samples of broken ore and rock in the working places are obtained by simple 'grab sampling'. If assays reveal that excessive dilution has occurred remedial measures are instigated.

2.1.6 Mining in general

The major differences between coal mining and ore mining are that, in general, ores are harder, often occur in irregular shaped deposits and may vary significantly in grade. These three factors largely account for the different mining techniques and associated equipment that are employed. With most ores, mechanized long-wall mining as practised in coal mines is not possible, although a variety of room and pillar methods (compare Bord and Pillar method in the mining of coal, vol. II, chapter 9) are available for the mining of tabular ore deposits. Indeed, it is impossible to designate any one mining method as specifically a coal mining technique or solely a metalliferous mining method. It is, however, valid to state that certain mining methods are predominantly used in the mining of ores.

Because of the much greater spread in the types of ore deposits it follows that the number of different mining techniques is much greater than in coal mining. Indeed, textbooks are in existence listing several hundred ore mining methods, some of which are, in fact, tailor-made to fit a set of conditions peculiar to one mine. Before considering mining techniques it is expedient to briefly consider the scale of mining operations as carried out at the present time. Table 2.1(a) and 2.1(b) shows the world production of selected metals and minerals for the year 1965.

Table 2.1(a) Tonnages mined in 1965

Gypsum	47 785 000 metric tons
Iron ore	615 351 000 metric tons
Manganese ore	17 612 000 metric tons
Potash	13 500 000 metric tons
Salt	107 590 000 metric tons

When it is considered, for example, that cupriferous ores as mined may contain less than 1% copper or that gold ore is profitable to mine with less than half an ounce of gold per ton of ore, then it is apparent that the figures giving actual metal content represent only a fraction of the tonnage of ore that must be mined to obtain the pure metals.

Table 2.1(b) Actual metal content of ore mined in 1965

Copper	5 075 000 metric tons
Gold	47 700 000 oz
Lead	2 700 000 metric tons
Tin	200 000 metric tons
Zinc	4 310 000 metric tons

Further, the production figures do not give any indication of the volume of waste rock that must be moved to expose the ore deposit. In some open-cast operations, i.e. where the mines are worked on the surface, it is common to find that the volume of overburden or waste rock overlying the mineral deposit is ten times greater than the volume of the ore revealed.

ORE MINING | 91

2.1.7 Alluvial and open cast mines

Mines may be considered under three categories: alluvial, open-cast (or open-pit) and underground. In alluvial mining, operations are fairly straightforward. The ore is taken from the floor of the river or ocean by specially designed dredges or pumps. These dredges often incorporate treatment plants and the dredged material is processed on board and the waste material immediately deposited back into the water. Special gravel pumping equipment is also used in alluvial mining, and hydraulicing or the breaking away of ore on land by means of high-pressure water hoses is common. Alluvial mining is applied in many countries of the world, for example, in Malaysia there are well over 1 000 small alluvial mines operating at the present time. Dredging for diamonds is taking place off the coast of South West Africa and there are trends suggesting that mining the ocean floor will increase in importance in many parts of the world.

Open-cast mining has increased significantly since the end of world war II. Loading equipment is very much bigger and the overburden to mineral ratio is several times that formerly considered feasible. Large-scale operations have now made it possible to mine low-grade deposits at depths well in excess of 500 m below the surface (see fig. 2.3). At the Bingham Canyon copper mine in Utah, USA, which is believed to have the largest annual output of any mine in the world, 225 000 tons of waste are stripped each day and 90 000 tons of 0.7% copper ore are mined. The floor of the open pit is almost 700 metres below the ground level. Huge bucket wheel excavators have been developed for use in open pits capable of stripping nearly 10 000 tons per hour. A walking dragline is currently being built with a loading bucket of 160 m³ capacity mounted at the end of

Fig. 2.3   Large open pit mine in Spain.

a loading boom 90 m long. This giant machine will lift 500 tons of overburden in one bite. The weight of this excavator is almost 15 000 tons. It is estimated that more than half of the world's mineral production now comes from open pits.

2.1.8 Underground mining of ore deposits

Having decided that a deposit should be mined by underground methods the question of how best to gain access to the ore body remains to be resolved. There are three basic methods of providing access to the mine workings from the surface and these comprise: a near horizontal tunnel or adit, an inclined tunnel and a vertical shaft. The choice of opening depends on many factors amongst which are: depth to the deposit, output required, topography of the surface, haulage facilities to be installed, amount of water expected, relative cost of openings, geological considerations etc.

If the mine is wet (some mines pump anything up to 300 000 m^3 of water a day) and the workings lie above the water table it may be advisable to use an adit so that the mine may be de-watered with minimum cost and trouble. If drainage adits are impracticable, or not required, the choice between a vertical shaft or an inclined tunnel is often dictated by the depth to the deposit. In general, the deeper the ore body the more likely it is that a vertical shaft will be chosen. As mining is an extractive industry, so then are the readily available ore reserves reduced each year and man must continuously go deeper. During the past decade mines have been operated at depths formerly believed unrealistic. In South Africa the world's deepest mine, Western Deep Levels gold mine, has already reached a depth of 3 300 m. Even deeper workings are projected despite the difficult problems of extreme rock pressure and high temperature associated with mining at depth.

As previously stated it is difficult to mechanize the actual process of breaking down the ore from the parent mass underground because of its hardness and the irregular shape of many deposits. Thus most mining methods involve dividing the ore body into working sections and subsequently employing high explosives to blast the ore away from each working face exposed by the underground development tunnels. In the blasting process, holes some 3 to 6 cm are drilled 2 to 3 m into the face of the ore. Explosive charges are fed into these holes and then detonated, so bursting away some of the ore ready for loading into transport media for despatch to the surface. Although the amount of explosive varies widely with conditions, an average 3 m hole charged with 2 kg of explosive would yield about 3 tons of ore. Sometimes many explosive charges are detonated simultaneously or in rapid sequence and the use of a ton or more of explosive in one blast is not uncommon. Blasting is usually done between shifts so that men are not endangered and sufficient time is given for noxious fumes to be dispelled.

2.1.9 Stoping methods

In its broader sense the term 'stoping' is employed to mean the operation(s) of excavating ore by means of a series of horizontal, vertical or inclined workings in veins, beds or large irregular bodies of ore. It embraces the breaking and removal of ore from the 'stope' and the timbering or filling of the resultant waste area but excludes operations in tunnels driven for the purpose of exploration and development. Essentially, the stoping method(s) that can be applied to any particular ore body are dependent on the type and disposition of the supports that may be necessary to prevent collapse of the walls of the excavation either temporarily, i.e. during such time as men are working in the stope, or permanently. In ore mining parlance the terms hanging wall, foot wall and side walls relate respectively to the roof, floor and sides of the excavation.

As the choice of mining or stoping method is largely dictated by support considerations, most classifications of methods are based upon the various systems of support that are available. In the simplest case, open stoping, the ore is removed from wall to wall without leaving any support pillars, although occasional props may be left in localized areas of weak ground. This method is only applicable to relatively small ore bodies as there is a limit to the length of span that will stand without support even in the firmest and strongest rocks. Where the ore body is large a number of open stopes one above the other are mined. This is termed sublevel stoping and the method is widely used because it permits almost complete extraction of the ore body. Another popular mining method is the so-called shrinkage stoping system (see fig. 2.4). Here the ore is mined in successive flat or inclined slices working upwards from the haulage level at the bottom of the deposit. After each slice is taken from the 'stope back' or working face, just enough broken ore is drawn off through the ore chutes to provide a working space for the miners on the top of the loose ore filling the lower part of the stope. Usually about one-third of the broken ore is drawn off during the working life of the stope due to the fact that solid ore occupies approximately two-thirds of the space of broken ore. When the ore block has been completely broken down the miners are transferred to another working place and the broken ore (which now completely fills the stoped out area) is drawn off as required. This system has the merit of providing a form of support within the stope during the time when men are working therein.

Cut and fill stoping is typical of another widely used metalliferous mining method (see fig. 2.5). In this system, instead of filling the waste area with broken ore for subsequent removal, as in shrinkage stoping, sand or broken rock material is introduced into the stope from an upper haulage level. This fill consolidates and forms the floor on which the miners stand as they take successive slices of ore. The broken ore is immediately removed from the stope via ore passes. These ore passes are in effect small enclosed shafts or chutes which are maintained within the filled waste area and are progressively built up so that their mouth is always a short distance above the level of the fill in the stope. The essential advantage of cut and fill stoping over many other

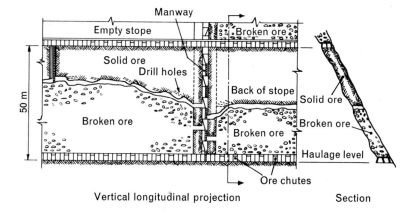

Fig. 2.4   Shrinkage stoping.

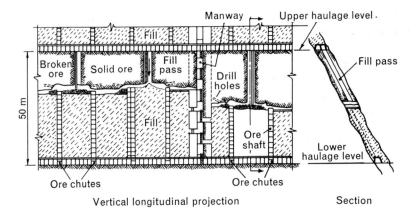

Fig. 2.5 Cut and fill stoping.

mining methods is that the fill acts as a permanent support and thus minimizes surface subsidence. With open stoping or with systems where supports are withdrawn when work is completed within the stope, there is always the danger that the rocks overlying the excavation will cave or break right up to the surface so causing damage to property or mine plant. At some mines where ore bodies large in all three dimensions have been totally extracted, the surface has lowered by several metres over a considerable area.

With all mining methods it is sometimes necessary to leave pillars of ore for all time as support. These pillars may be regular in size and pattern, as when mining stratified and near horizontal deposits, or they may be left purely according to the localized needs of the situation as mining proceeds. The use of wooden, steel and concrete supports is often decided *in situ* but, in general, the need for such support is far less in metalliferous mining than in coal mining. Nevertheless, the cost of maintaining effective support often accounts for a significant proportion of total mining costs. Thus it is important to choose a mining method which is safe and efficacious and yet incurs minimal support costs.

2.1.10 Mine haulage

It is often asserted that mining is essentially a large-scale materials handling exercise. Like most generalizations this is only true up to a point but there can be no doubt that the movement of the ore from the stopes to the surface constitutes one of the most important problems facing the mining engineer. Bulk handling underground is much more difficult than on the surface, for mining, more than any other industry, suffers from a fundamental lack of conformity both as regards processes and products. Such widely varying conditions militate against straightforward application of standard automated techniques. There can be few industrial undertakings on the surface where the working area is spread over several square kilometres, where there are upwards of twenty widely scattered points of production, and where the entire operation cannot readily be observed and supervised. The task of moving the ore to the surface grows more difficult in each mine as it grows older and the distance from stope to shaft increases. There are many mines in operation where the ore must be moved 10 km underground before arriving at the foot of the shaft to be lifted a vertical distance of

2 000 to 3 000 m. Such mines often maintain upwards of 100 km of underground haulage tunnels, although it is a far cry from the days when a large mine might have up to 1 000 men underground engaged solely in moving the ore from the working place to the surface. The same tonnage is often handled now by one-tenth or even one-hundredth of this labour force.

In underground mines, haulage can generally be considered as three separate operations. First, the ore must be collected at the working places. Secondly, the output from a number of such producing units must be gathered together and delivered to the bottom of the shaft. The final operation involves the vertical lifting of the ore up the shaft whence it is despatched to the mineral treatment plant. Although this is the usual sequence of operations, there are exceptions. As discussed previously it is common in some mines to drive an inclined tunnel from the surface down to the working level. If conveyor haulage is installed underground it is then often expedient to continue the conveyor system up the incline and thus eliminate one transfer stage in the sequence of haulage operations.

In ore mining, the choice of haulage medium usually lies between locomotives and conveyors. The fact that in metalliferous mining haulage tunnels are driven horizontally has meant that locomotives have traditionally enjoyed greater popularity than conveyors. With the recent trend towards concentration of operations and high mechanization this situation is changing and many new metalliferous mines have been planned to utilize conveyor haulage on a wider scale. There is little doubt that where high outputs are contemplated the high capacity potential of conveyors makes this system very attractive. There are many underground conveying systems in use dealing with outputs in excess of 1 000 tons per hour, although it is true to say that conveyors find less application in underground ore mines than in collieries.

The increasing size of open-cast mining operations has brought about a significant increase in open-pit conveyors. Some of the these surface mines employ conveyor belt systems handling more than 5 000 tons an hour. Many of these large conveyors have driving units of more than 1 000 hp and single conveyors more than 6 km in length have been installed in mines, both above and below ground.

In modern underground mines the entire conveying system (and this may consist of a dozen or more large conveyors) is continuously scanned and monitored by television cameras. Automatic devices immediately register untoward occurrences. From a central control room on the surface the haulage supervisor can contact maintenance men by radio telephone and direct them immediately to the trouble point and also provide an indication of the cause of the trouble.

Locomotives used underground are usually electrically powered or are of the diesel engine type. They range in size from a few horsepower up to the 200 hp 28 ton giants used in South African gold mines (see fig. 2.6). The locomotive haulage system has the important advantage of being a much more flexible transport facility than conveyor belts. This mobility is of prime importance in metalliferous mining where there may be a score of working levels and it may be necessary to change the disposition of the locomotives considerably from time to time, as ore grades vary or demand fluctuates. The most widely used underground locomotive is the electric trolley wire type. With this system the locomotive picks up current from an electric power line suspended from the roof of the haulage tunnel. The electric locomotive lacks the mobility of a completely self-contained unit such as the diesel locomotive but has not the latter's disadvantage regarding noxious fumes emitted from the engine exhaust.

Fig. 2.6 A trolley electric locomotive in a South African gold mine.

An interesting development of recent years is the application of automatic control to locomotive haulage systems. At Kiruna iron ore mine in Swedish Lapland, the world's first underground electronic data processing equipment was installed in the mid 1960s. This equipment was installed to effect the control and operation of a full-scale underground electric locomotive system. The data equipment is combined with a conventional central traffic control (CTC) system for supervising the movements of 18 trains each consisting of a locomotive and 12 wagons holding 300 tons of ore. The entire operation is automatic, including the addressing of trains to the loading and discharging stations. Information on the position of each train, train weights, ore analysis, ore level in loading chutes etc., is automatically fed into the data centre and the locomotives are thus utilized to maximum efficiency.

At many mines the problem of effectively controlling an extensive locomotive system has been solved by the use of radio receiver-transmitters. The locomotive drivers are in constant touch with each other and also the loading and discharge stations. The overhead trolley wire is used for radio transmission.

The final phase in the haulage operations is to hoist the mineral in the shaft. These high-speed hoists frequently travel at speeds up to 15 m s^{-1} and can lift up to 50 tons of ore at a time. Because of the strength and design limitation of the steel cables used to hoist the ore conveyance in the shaft, the longest single lift in one stage is about 2 000 m. As many mines are operated below this depth the ore must be lifted in two vertical stages using different shafts. The driving units of mine hoists sometimes exceed 8 000 hp. In many modern mines all hoisting operations are completely automatic. The ore conveyance is loaded, hoisted, discharged and returned to the shaft bottom completely under the control of the automatic hoisting gear. Usually there are two conveyances in a shaft and as the full conveyance rises to the surface the empty conveyance descends at the same speed. At very large mines, several hoisting shafts are required to maintain the flow of ore from the mine. At Kiruna, the world's largest underground mine, eventual production will be 15 million tons of ore a year. To handle this massive tonnage ten adjacent shafts have been sunk to the workings.

Improvements in mining techniques and equipment are currently proceeding at such a rate that one may confidently look forward to the time when machines largely take over from men the difficult but essential task of maintaining the world's mineral supplies.

2.1.11 Literature
Mining Annual Review, London, Published yearly in May. U.S. Bureau of Mines, *Information Circulars*, Washington, Published at regular intervals.

2.2 PROCESSING OF METALLIFEROUS ORES

2.2.1 Definitions and history

An ore may be defined as a natural accumulation of minerals that is capable of being exploited for the benefit of mankind: a metalliferous ore is one from which the valuable product is eventually prepared in metallic form. It is difficult to provide a precise definition of the term mineral but it is normally applied to naturally occurring, homogeneous materials: in most instances these possess definite crystalline structures and have compositions that are controlled by well-defined chemical principles.

Most ores consist of complex aggregates of a number of minerals: in many ores the valuable mineral (or minerals) amounts to only a small proportion of the total bulk and this small amount is frequently distributed throughout the mass of the ore in an irregular manner. Unfortunately, minerals that are potentially valuable are of little interest to the purchaser when they occur in this form and, in order that they may be used, these minerals must be separated from the unwanted rock and converted into suitably concentrated products.

The sequence of operations undertaken to prepare a mineral raw material so that it becomes an acceptable primary consumer product is often called ore dressing. Nowadays, however, the terms mineral technology, mineral processing or mineral engineering describe more adequately the complex scientific, engineering and economic aspects that are involved in the treatment of ores.

There are no fundamental differences between the treatment of metalliferous ores and other types of ore but a distinction has traditionally been made and this exists in some measure even today. The mineral technologist obtains his raw material from mining operations and, in the case of metalliferous ores, he sells the products that he makes to the extraction metallurgist. These final products are prepared so that they contain as high a proportion of the desired metal as possible whilst, at the same time, the amount of deleterious impurities is reduced to a minimum. All the mineral processing stages that are required to achieve these aims must, of course, be carried out at a reasonable cost and with an acceptable recovery of the valuable component.

It is probably true to say that the metalliferous ores treated in the historic and prehistoric past were always simple in character and contained comparatively large amounts of the desired minerals. Despite these advantages, the proportion of a valuable mineral that was actually recovered during the ore treatment processes was often very low.

The earliest metalliferous ores to be utilized were those in which elements occurred in metallic form e.g. gold or copper. The metal was doubtless picked out by hand in many instances although there are early references to the use of oiled feathers to pick out gold from granular deposits and the 'golden fleece' of ancient times may also have been a device for collecting fine-grained gold.

The traditional metalliferous ore deposits of the Middle Ages, and even the majority of those exploited in the nineteenth century, consisted of narrow, high-grade veins containing the minerals of copper, lead, zinc, tin, silver and gold. These veins were worked by selective mining methods that removed only the richer portions and it was frequently possible to smelt this material directly or after only a minimum amount of pre-treatment. Simple washing techniques were used for this pre-treatment and the methods that were used relied on the high density of the metalliferous minerals compared with the matrix of worthless material which is called gangue.

In modern terminology the term 'metalliferous ore' includes, in addition to the ores of copper, lead, zinc, gold, silver and tin, the ores of other elements such as nickel, cobalt, aluminium, titanium, vanadium, niobium, iron, molybdenum, tungsten, etc. The ore bodies that are now available often consist of low-grade disseminations of fine-grained metalliferous minerals in a worthless matrix and these ores can only be mined by bulk mining techniques. Such ores never contain metals in the form and degree of purity required by the metallurgist and, consequently, must be up-graded by mineral technology methods before they become acceptable.

The mineral treatment operations reduce the costs of transporting and smelting metalliferous ores, reduce losses of metal in smelter slags, and allow the mineral technologist to meet stringent purchaser specifications. The cost of the mineral treatment operations can be considerable and serious losses of valuable material can occur during the treatment processes. However, the advantages to be gained from mineral processing almost always greatly exceed the disadvantages and a mineral treatment stage is almost invariably used in the winning of metalliferous ores.

During the last few decades there has been a remarkable increase in the volume of mineral products required by man and mineral production during the twentieth century has been greater than the total production during all other ages. With this rapid depletion of the easily found, easily mined and easily smelted mineral deposits it has become increasingly necessary to use low-grade and complex ores. Such ores become economically viable only after scientific evaluation and appraisal, careful process design, and treatment in modern, large-scale, low-cost mineral processing plants, e.g. it is not unusual for a large plant to treat up to 20 000 tonnes per day.

In a few instances the complete operation of metal extraction from mine to smelter has been integrated under single management so as to achieve maximum profitability.

2.2.2 Mineralogy

About 2 500 different mineral species are known and the mineral technologist is generally concerned with, perhaps, 15 to 20 of these in any single ore. This means that the number of combinations of minerals that can exist in an ore is very large and each ore differs materially from all others. Consequently, the best treatment process for any particular ore is closely related to its mineralogy and can only be designed after a careful mineralogical examination of that ore.

The proportion of the desirable element that an ore contains is an important factor in deciding whether a mineral deposit can become a viable economic project but it is equally important to determine the nature, size, and spatial distribution of the mineral species in which that element is found. In addition, it is desirable (if not always essential) to establish the nature, size, etc., of the unwanted minerals in an ore and also to determine certain critical physical and chemical properties for each mineral. The collection of such information is an essential preliminary step in the design of separation procedures which are all based on the exploitation of differences in the properties of the minerals involved (see table 2.2).

Table 2.2 Mineral properties and available separation techniques for a typical ore.

Property	Mineral					Separation techniques available
	A.	B.	C.	D.	E.	
1) *Size*	Large	Small	Small	Small	Small	Screening or classification
2) *Colour*	White	Brown	Black	White	Brown	Optical sorting
10) *Radioactivity*	no	no	no	no	yes	Radiometric sorting
6) *Magnetic susceptibility* (relative units)	-0.2	2.0	162	-0.3	15	Magnetic separation
7) *Surface electrical conductivity* (relative units)	10	10^6	10^6	10	10	Electrostatic or electrodynamic separation
3) *Hardness* (Mohs scale)	7	6	6	7	5	Differential size reduction
4) *Specific gravity*	2.7	4.2	4.8	4.7	5.1	Gravity methods, e.g. dense medium separation, sluice, jig, shaking table, etc.
8) *Surface chemistry*	Each mineral can be selectively waterproofed in the correct chemical environment					Flotation
9) *Mass Chemistry* Acid solubility	V. Low	low	high	low	low	Leaching
Base solubility	low	low	low	low	high	Leaching

100 | ORES IN GENERAL

2.2.3 Liberation in general

Liberation is the process whereby the various mineral species in an ore are freed from one another and is an essential step in the production of high grade products (see fig. 2.7). It is followed by a separation procedure that aims to collect the various freed species into individual fractions. In order to simplify the subsequent separation operations it is usually desirable that the mineral species be liberated from each other whilst, at the same time, retaining their maximum possible size. Consequently, every effort is made to minimize the production of 'under-size' products and this is done by removing material from the liberation process as soon as possible after it reaches the required degree of liberation (see figs. 2.8 and 2.9).

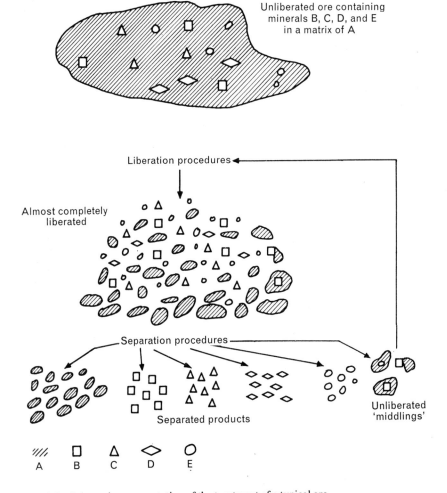

Fig. 2.7 Schematic representation of the treatment of a typical ore.

As a broad generalization it can be stated that ores currently being exploited require to be broken into fragments smaller than 100 μm* in order to achieve effective liberation. This liberation is usually effected by a combination of comminution devices that break the ores into small, irregular fragments. The first step in the liberation procedure occurs, of course, during the mining of the ore when it is broken away from the main mass of rock. The pieces thus obtained vary in size from boulders 2 to 3 m in size to minute dust particles. It is possible that the larger fragments may have to be reduced in size before leaving the mine in order to reduce the transportation difficulties.

2.2.4 Crushing

The coarse breakage of rock is termed 'crushing'. Rocks may be broken by applying a relatively slow compressive force to the individual pieces, or by impact or attrition.

a. The jaw crusher, in which the rock is crushed by compressive pressure, resembles a large robust nutcracker. The rock is pinched between two steel plates and it is broken when these two plates are forced together. The form of the steel plates and the way in which the force is applied varies in the different types of crushers. For example, the Blake type jaw crusher consists basically of two heavy wear plates (jaws) positioned between fixed side plates. The jaws enclose a certain volume which is made smaller when one of the jaws, which is pivoted at the top, is moved inwards. In the open position the cavity between the jaws is filled with large lumps of minerals. These lumps are crushed by the closing movement. Crushed pieces fall deeper into the narrowing cavity and are repeatedly broken until they can pass through the bottom opening between the jaws.

*1 μm = 1 micrometre or 1 millionth of a metre.

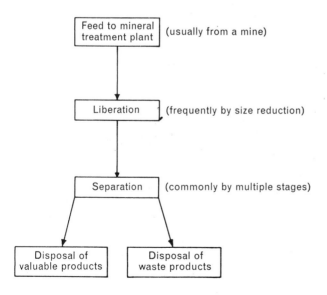

Fig. 2.8 Diagrammatic representation of major mineral treatment stages.

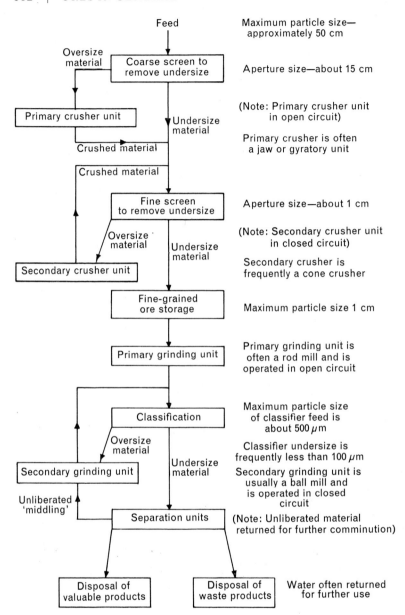

Fig. 2.9 Simplified flowsheet showing liberation by succesive stages of comminution.

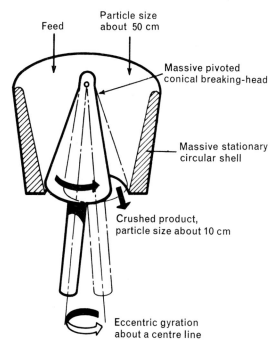

Fig. 2.10 The principle of the operation of a gyratory crusher.

b. The gyratory crusher. In this crusher the pieces are also crushed by applying pressure. The gyratory crusher comprises a conical crushing member placed with its narrow end upward in a vessel (see fig. 2.10). Thus the cone and the vessel enclose an annular cavity which is narrower toward the bottom. The top of the inner cone is fixed but at its base it is free to move so as to produce a gyrating motion, (i.e. the free end of the shaft is moved in a circular path by means of an eccentric without rotating around its own axis.) In this way the cone periodically approaches each point of the vessel wall and the lumps fed between the cone and the vessel wall are crushed.

c. The cone crusher (see fig. 2.11) is basically similar to the gyratory crusher, but the design of the cone has been modified so that the cone is also free to rotate about its own axis. This crusher is a secondary crusher (i.e. it is used for further crushing of pieces which have already been crushed to a certain size).

d. The hammer crusher is an example of a crusher in which rocks are crushed by impact. This crusher consists of a horizontal rotating shaft on which free-swinging hammers are mounted. The hammers crush the rocks by impact and also by projecting the fragments onto a breaker plate.

2.2.5 Grinding

The size range of the material fed to conventional crushers varies from about 2 m to a few centimetres and the size of the final product is commonly less than 1 cm. The amount of liberation achieved by breaking metalliferous ores to this size is often negligible and a further stage of size reduction is necessary. This stage is called 'grinding'.

It is usually carried out in *ball* and *rod mills* consisting of cylindrical metal drums

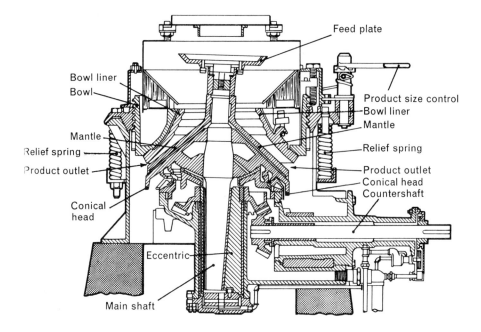

Fig. 2.11 Symons standard cone crusher.

or tumbling mills, wherein the ore (now less than 1 cm in size) is broken by being struck by steel balls or rods (see fig. 2.12). The size of the finished product from these mills depends on the nature of the ore, and in very many modern operations the material is

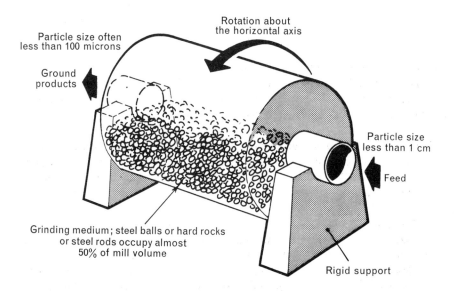

Fig. 2.12 A typical grinding mill.

reduced to less than $100\,\mu$m in size. Once again, great care is taken to minimize the number of particles that are reduced to sizes below that at which they are liberated.

In many plants it is possible to use the larger and harder fragments of an ore to grind the main bulk of material. This practice is called *autogenous grinding*. It is also increasingly common in modern practice to find that both crushing and grinding are combined into a single operation: this has only become possible with the development of very large diameter tumbling mills in which the kinetic energy developed by the tumbling ore is sufficient to break even very large or very hard fragments. The mills for this form of comminution are sometimes more than 10 m in diameter and lifter bars are fitted inside these units so that the ore is lifted to the full height of the mill before being dropped and broken.

2.2.6 Separation in general

During the separation procedures the individual, discrete grains of the required mineral(s) are separated into distinct products by the exploitation of differences in the chemical or physical properties of the various species. These properties may include size, shape, colour, specific gravity, thermal conductivity, surface chemical effects, chemical reactivity, etc. (see table 2.2 and fig. 2.13).

Many of the separating processes that are available require almost complete liberation of the components of an ore, for instance, those processes based on colour or specific gravity, but others such as chemical attack, can proceed satisfactorily even if only a part of a grain is presented to the attacking reagent. See Fig. 2.13d and h.

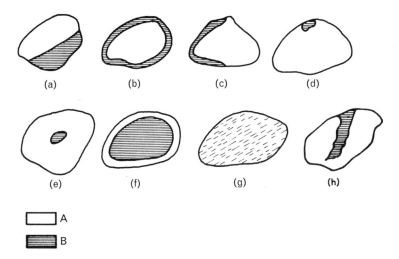

Fig. 2.13 The various types of 'middling' product made up of two mineral phases
A = high density, magnetic and chemically unreactive
B = low density, non-magnetic and chemically reactive
(a) Improved liberation possible by breakage to slightly finer size – (b) In many processes, e.g. flotation, cyanidation, this particle would react like a particle of mineral B – (c) This might behave as a particle of B in a flotation process – (d) This would slightly increase the losses of B in tailings in many processes. – (e) In most processes this would react like a particle of A – (f) In flotation this would react like a particle of A – (g) This would increase the losses of B in tailings – (h) this would be 'liberated' for chemical treatment.

All the known methods of mineral separation can be applied to the treatment of metalliferous ores. However, only the methods that rely on the specific gravity, surface chemistry or chemical reactivity of the minerals are in common use. In specialized applications it is also possible to find a number of the other separation methods being employed e.g. those based on colour, magnetic properties or surface conductivities.

2.2.7 Feed preparation

It is imperative to prepare the feed material for every separation method in such a way that the maximum possible use can be made of the separation characteristics involved. For example, it is obvious that if two minerals are to be separated by a manual sorting action then the mineral fragments must be large enough to be seen and handled by the person carrying out the separation. This simple principle must also be applied to more complex separation procedures such as gravity or flotation methods.

2.2.8 Sizing and classification

One of the most important of the separation characteristics in mineral treatment is the size range of the mineral particles. This range frequently determines the degree of liberation of the various minerals and, at least partly, controls the response of the particles to the separation forces employed. Because mineral grains are, most often, highly irregular in shape it is difficult to define the 'size' of such grains. It is, however, possible to determine an acceptable size parameter either by screening or by classification.

In screening the particles are sized according to their ability to pass through the apertures of a screen (or sieve). The apertures may be round, square or rectangular and the size parameter actually measured varies with the aperture shape. The material used in the construction of a screen is commonly steel but nylon-mesh screens are also used in some applications. Screens are almost always shaken or vibrated as this increases the throughput and reduces the tendency of the screen to 'blind', i.e. for the apertures to become blocked.

When a separation process requires a narrow size range for optimum results then the feed material is prepared by passing it through a succession of screens. Screening can be used for treating mineral grains within the size range of a few centimetres to about 50 μm but the usual range of application is, from about 4 cm to approximately 0.5 mm.

When it is necessary to control the size of material below 0.5 mm it is usual to use a classification device. This produces 'size' fractions consisting of particles which have similar settling velocities in a fluid. Such a device does not, in general, provide a true sizing action because it relies on differences in mass rather than on the linear dimensions of the particles. In this way small particles of high specific gravity are collected along with larger fragments of a lower density. Classification methods are applicable to particles from about 1 mm in size to about 10 or 15 μm and can, consequently, span the range of sizes commonly encountered during the separation processes carried out on metalliferous ores.

The types of plant-scale units employed for size classification are numerous and varied. In most of these the fluid is water and this flows upwards through the classifier at a uniform speed. Particles having settling velocities greater than the upward velocity of the water can settle through the liquid whilst the smaller and lighter grains are carried away by the water flow (see fig 2.14). The classification devices range from simple settling tanks where the settled material is intermittently removed to complex units from which the settled material is continuously removed by a spiral screw, reciprocating rakes, conveyor belts etc.

A hydrocyclone is a modern type of classifier in which the gravitational settling forces employed in earlier classifiers are amplified by centrifugal motion of the mineral pulp and the rate of classification of slow settling particles is considerably increased. The mineral pulp is fed tangentially and under pressure into the cylindrical portion of a cylindro-conical vessel. This produces a fast rotation of the pulp and induces the heavier solid particles to more radially outward whilst the lighter particles and the liquid move radially inward. The heavier particles are constrained by the convergence of the walls to discharge at the apex of the conical portion of the cyclone whilst the lighter material and the liquid is discharged from the opposite end through a pipe that protrudes axially into the cylindrical part (see fig. 2.15).

2.2.9 Hand and mechanical sorting

This method was widely used in the past to separate fragments larger than about 4 cm when there was a marked optical difference between the ore and extraneous material collected with the ore during the mining operation. Hand sorting is now expensive and is being replaced by mechanical sorters that rely on differences in the physical or chemical properties of the various components within an ore. For instance, optical sorters examine an ore fragment from a number of viewpoints and determine whether an optical property, such as colour, complies with a certain standard (see fig. 2.16). Unwanted fragments are then automatically rejected by mechanical plungers, air jets or by electrostatic means. Other mechanical sorters use the natural or the induced radioactivity of various components in an ore as the separation criterion.

2.2.10 Separation by gravity methods

At the turn of this century the separation of metalliferous ores was almost entirely carried out by methods that relied on the comparatively high specific gravity

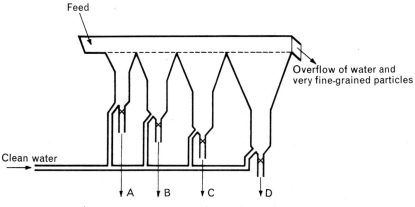

A = coarsest product
B, C & D = progressively finer products } Individual products for separate treatment processes

Fig. 2.14 Hydraulic classifier

108 | ORES IN GENERAL

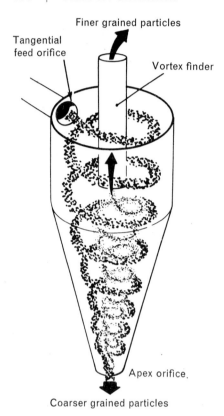

Fig. 2.15   Cyclone classifier

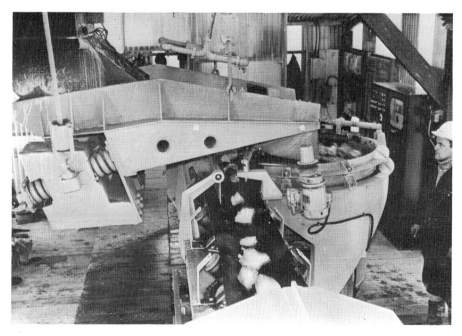

Fig. 2.16   Aligned feed entering the optical chamber of the sortex 811M optical separator.

of the valuable minerals as compared with that of the gangue. In some parts of the industry, e.g. copper treatment, these techniques have been replaced by the flotation process but a number of gravity devices are still widely used in the treatment of tin, tungsten, niobium and other ores.

a. The sink–float method. One of the simplest methods of separating two materials of different specific gravity is by immersing them in a fluid of intermediate density; for example, a mixture of sand and sawdust can be readily separated into its component parts by placing the mixture in a can of water. Unfortunately, almost all minerals are denser than water and the technique, in its simplest form, cannot be used in the treatment of metalliferous ores. However, by using a mixture of water and a finely-divided, slow-settling solid it is possible to produce a 'dense medium' that has an apparent density much higher than water and many useful 'sink–float' separations can be carried out. Mineral grains that differ in specific gravity by only 0.1 of a density unit can be separated by this method. The solids commonly used in the preparation of heavy media are magnetite or ferro-silicon. These solids are employed because they can be recovered readily from the various products of the separation and can then be re-used. (see fig. 2.17). Dense medium plants are used for the separation of mineral particles ranging from a maximum size of a few centimetres to a minimum of about 0.5 mm.

The hydrocyclone is also used as a dense-medium separator for separating fine-grained material since the gravitational settling forces are amplified by the centrifugal motion in the cyclone.

In most other gravity separation techniques it is essential that a considerable difference of specific gravity exists between the various grains that have to be separated (often as much as 2.0 specific gravity units). Consequently, these methods are chiefly applicable to the treatment of fully liberated mineral species where the specific gravity difference between various grains is at a maximum.

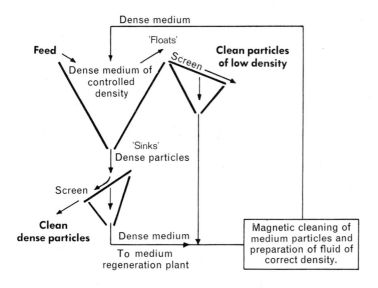

Fig. 2.17 Simplified diagram of a dense medium unit.

110 | ORES IN GENERAL

b. The sluice. A very old method of gravity separation is the sluice in which minerals of low mass are separated from those of higher mass by a stream of water flowing in a small channel (see fig. 2.18). Woodcuts showing various forms of sluice are given in a sixteenth century book on mining* and similar units are still used in many parts of the world.

c. Some other washing devices. A number of other washing devices are used for the treatment of metalliferous ores. Some of these units rely on the separating action produced by vertically flowing currents and others use very shallow, almost horizontal, flows of water.

The behaviour of solid particles in a very shallow stream of water is controlled by the considerable differences of fluid velocity that exist at various depths within the fluid. The velocity at the base of such a liquid film is zero and the speed of the liquid

* G. Agricola, *De Re Metallica*.

Fig. 2.18 Woodcut c. 1556 showing ore being treated in sluices
A   Workman carrying broken rock in a barrow
B   First chute
C   First box
D   Its handles
E   Its bales
F   Rope
G   Beam
I   Second chute
K   Second box
L   Third chute
M   Third box
N   First table
O   First sieve
P   First tub
Q   Second table
R   Second sieve
S   Second tub
T   Third table
V   Third sieve
X   Third tub
Y   Plugs

reaches its maximum near the top of the film. Such a flowing film of liquid has very little effect on very small or flat particles but larger particles tend to be rolled downstream. This differential effect on various sizes and shapes of grains can be used to separate mineral particles and is the major separating action in devices such as shaking tables, buddles, vanners, etc. Some of these devices were widely used from medieval times to the beginning of this century but at present their use is restricted almost entirely to the treatment of some iron and tin ores and alluvial-type deposits, that contain tungsten-, zirconium- or titanium-bearing minerals.

d. The jig. A jig is essentially a water-filled box within which a bed of mineral grains is supported on a perforated surface. A dense particle will fall faster in a fluid medium than a less dense particle of the same size: consequently, stratification of the mineral grains, with the heavier below and the lighter above, will take place either when the screen (and the mineral bed) is moved up and down or when water is pulsed through the screen. This jigging action stratifies the solid particles so that the bottom layers eventually contain all the dense particles and the less dense grains are concentrated in the upper layers. In order that the separating action shall be continuous, provision is made for the various layers to be removed and kept separate. In the most commonly used type of jig the bottom layers of stratified minerals are allowed to pass downward through the perforated surface within the jig and the upper layers are crowded out of the top of the jig by incoming, untreated material.

For a long time jigs were the principal device for processing metalliferous ores but, except in the treatment of alluvial tin and gold ores, they are now virtually obsolete. However, in the few specialized applications mentioned above, jigs are still widely used and their performance is constantly being improved; for example, the modern

Fig. 2.19 Circular jig 7 m. diameter, treating an alluvial tin ore in Malaysia.

112 | ORES IN GENERAL

jig is now often circular, instead of the traditional rectangular shape, and the maximum size of such a circular jig has been recently increased to over 7 m diameter (see fig. 2.19). The maximum size treated by jigging is about 1 cm and the smallest particle size is claimed to be about 0.1 mm.

e. The Humphreys spiral concentrator. This concentrator comprises a spiral channel having three to five turns and a curved cross section (see fig. 2.20). The feed slurry travels down in the spiral channel. If different sized particles of the same material are passed over a spiral, the smallest particles concentrate at the outer rim and the largest particles concentrate at the inner rim of the channel. If particles of different specific gravity are treated the particles with high specific gravity concentrate at the inner rim. Ports set at intervals in the bottom of the channel draw off the various fractions.

f. Pneumatic separators. In pneumatic separators air is used as the separating fluid. In the pneumatic jig particles are separated by passing a vertical flow of air through an inclined vibrating screen carrying the dry particles: the lighter particles are carried down the screen whilst the heavier grains remain near the feed point.

2.2.11 Magnetic separation

Various mineral species have widely different magnetic permeability and this property has become the basis of a useful method of mineral separation. This form of treatment is commonly applied to the separation of iron-rich, ferromagnetic materials, such as magnetite, from a 'non-magnetic' matrix like quartz. However, it is possible by using very strong magnetic fields, to separate weakly magnetic minerals from each other or from a 'non-magnetic' matrix, e.g. columbite from monazite or quartz. The magnetic characteristics of some minerals are shown in fig. 2.21.

In a magnetic separator a continuous stream of the feed material is passed through a strong, converging, magnetic field and the more highly magnetic materials are deflected toward the zone of maximum field intensity (see fig. 2.22 and 2.23).

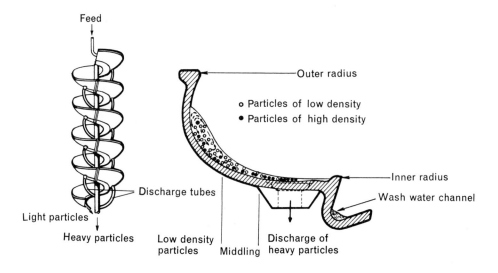

Fig. 2.20 Cross-section through a channel of the Humphreys Spiral Concentrator.

PROCESSING OF METALLIFEROUS ORES | 113

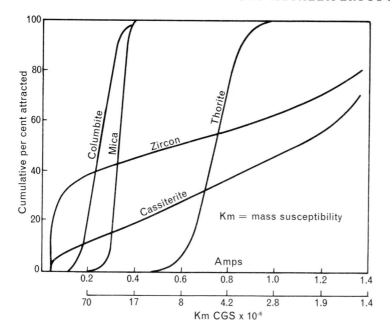

Fig. 2.21 Magnetic behaviour of heavy minerals from a Nigerian Columbite ore.

Magnetic guard units are frequently used to protect mineral treatment machinery, such as crushing units, from damage by fragments of iron tools, nuts and bolts, etc. that accidentally fall into the ore. The major use of magnetic separators, however, is in the separation of the component minerals in iron ores. If the iron-rich mineral is the highly magnetic magnetite and if this exists in a 'non-magnetic' matrix then the separation

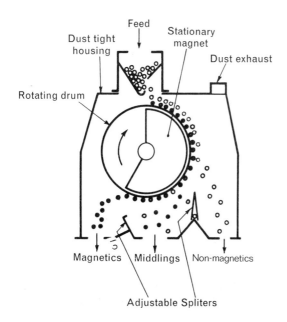

Fig. 2.22 Dry, drum magnetic separator.

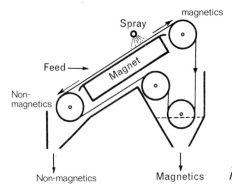

Fig. 2.23 Wet, belt magnetic separator.

presents no difficulties. An iron ore containing a large proportion of the weakly magnetic mineral hematite must, however, be roasted in a reducing atmosphere to convert the mineral to magnetite before magnetic separation can be utilised effectively. Particles ranging from many centimetres to a few micrometres in diameter can be separated by magnetic methods. Separations of coarser-grained ore are usually carried out with dry material but, small dry particles tend to agglomerate by an electrostatic bunching action and it is essential that material smaller than about 0.2 mm is separated as an aqueous suspension.

When a magnetic solid, such as magnetite or ferrosilicon, is used in a dense medium operation (see above) then a magnetic separator is used to clean the medium before re-use.

2.2.12 Electrical separation

This method is not widely used in the treatment of metalliferous ores but it does find application in the processing of certain alluvial deposits and beach sands for the recovery of tungsten, tin, niobium, hafnium, etc. Electrical separation methods are based on the attraction of unlike electrical charges for each other and the repulsion of like charges. The most common form of electrical separation is the high tension or electrodynamic method whereby minerals having different surface electrical conductivity may be separated. The variations in electrical conductivity between mineral species can be very large, but, unfortunately for the mineral technologist, the variation within a single species may also, on occasion, be great. The reasons for this variability within a species are not well understood and, consequently, the method has, for the present, only limited application.

In a high tension (electrodynamic) separator the dry mineral grains are fed as a layer, one particle deep, onto the top of a rotating, grounded, metal roll where all the grains are charged by being subjected to an electrical spray discharge from an adjacent electrode. The particles that are good conductors of electricity quickly lose their charge to the roll and fall off freely whilst the poor conductors retain their charge and, since this is of opposite polarity to that on the roll, they adhere to the roll and have to be mechanically removed by a brush (see fig. 2.24).

It is also possible to use an electrostatic separating process where the mineral particles in a feed material are charged by induction, conduction or friction from a charged surface. The conducting particles and the charged surface repel each other whilst the non-conducting particles are not affected. If the feed is passed over a charged, revolving, roll then the poorer conductors will fall in a trajectory determined only by

PROCESSING OF METALLIFEROUS ORES | 115

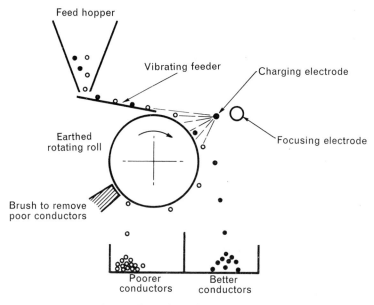

Fig. 2.24 High tension (or electrodynamic) separator.

the size and shape of the particles and the speed of rotation of the roll: the better conductors, however, will be thrown clear of the path of free fall followed by the non-conductors (see fig. 2.25).

2.2.13 Separation by flotation

This method is, probably, the most widely used separation technique in the processing of metalliferous ores. It is very extensively used in the copper, lead, zinc,

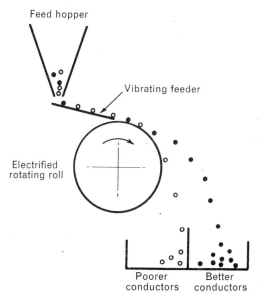

Fig. 2.25 Electrostatic separator.

nickel, cobalt, and molybdenum sections of the mineral treatment industry and is used to a lesser extent in gold, and iron production.

Some flotation operations were attempted between 1860 and 1890 but the process was not successfully applied to the treatment of ores until the first decade of this century. Flotation assumed great importance in the 1920s and since then a very large proportion of all sulphide minerals has been treated by this process.

The flotation process is versatile and selective in discriminating between minerals on the basis of differences in their surface chemistry. Most minerals in the natural state are hydrophilic, that is, they are readily wetted by water or aqueous solutions. In the flotation process selected minerals are rendered hydrophobic (water repellant) by the use of surface-active compounds. The most essential of the compounds are called 'collectors'. These collectors are sparingly soluble in water and have polar (charged) and non-polar (uncharged) groups within a single molecule. If the polar group can be made to react with a mineral surface then the non-polar group, which is hydrophobic, protrudes into the surrounding fluid and the particle becomes water-repellant. The collector layer need not be more than one molecule thick but, because of irregularities in the chemical composition or physical condition of most minerals, the layer that is formed is usually non-uniform and highly irregular.

The great value of the flotation technique is based on the fact that it is possible, by careful control of the chemical environment, e.g. the pH, the concentration and type of ions, etc., to make some mineral species hydrophobic whilst retaining other, associated minerals, in their natural hydrophilic state. Subsequently, when bubbles of air are blown through an aqueous pulp that contains such artificially hydrophobic particles these particles will tend to adhere to the air/water interface by surface tension effects and be carried to the surface of the pulp. If a trace of a froth-making reagent (frother) is added

Fig. 2.26 View of a large modern mineral treatment plant, Bute, Montana.

to such a system the floated grains accumulate in a transient froth and a good separation of minerals is obtained when this froth is collected.

By judicious control of the acidity of a mineral pulp and by adding reagents in the correct sequence and in the correct concentrations it is possible to produce a series of consecutive mineral separations from a complex ore. For example, when treating an ore containing a mixture of copper, lead and zinc sulphides in a quartz gangue it is possible to produce, first a copper-rich concentrate, then a concentrate rich in lead and finally a zinc-rich product. Like most separating procedures the flotation process works best on a limited size range of material. The optimum particle size range for the recovery of sulphide minerals is from about 10 to 300 μm. Grains larger than 300 μm are generally too heavy to be floated in plant-scale flotation units and there is a serious loss of selectivity when floating material smaller than about 10 μm.

The most widely used group of flotation reagents is the xanthates (xanthates are metal salts of dithiocarbonic acid derivatives represented by the formula $[ROC(=S)S]^- M^+$

in which M^+ is the metal ion) which, under controlled conditions and in the presence of oxygen, form coatings of hydrophobic metallic xanthates on a number of important sulphide minerals. Other collectors such as amines, dithiophosphoric acid esters, fatty acid salts, alkyl sulphates and alkyl sulphonates are also occasionally used in the treatment of metalliferous ores, for example, the amines (organic ammonia compounds) are used in the flotation of iron ores and fatty acids are used in the treatment of tungsten ores.

Many flotation plants treat over 20,000 tonnes of ore per day and the scale of such units is continually being increased.

2.2.14 Agglomeration (including pelletising and sintering)

This procedure is used for changing the size distribution of the metalliferous concentrates that are produced by a number of treatment plants. As shown below, these concentrates generally consist of dry, fine-grained particles and such material can present serious difficulties during transportation and during smelting operations. For example, loss by dusting can be severe and, to reduce this, the material is agglomerated by pelletising or sintering.

During pelletising the slightly dampened mineral concentrates are compressed or rolled into pellets. The strength of such 'green' pellets is low but it may be increased by adding a small amount of a 'binding' agent and/or by firing the pellets at elevated temperature.

In some plants, especially iron ore plants, the mineral concentrate (without pelletising) is heated to a temperature at which incipient fusion occurs and the small mineral grains tend to stick together and form a tough, but porous, mass that is very suitable for metallurgical treatment. This process is termed sintering.

2.2.15 Hydrometallurgical methods or chemical treatment

During chemical treatment a mineral is selectively attacked by acids or alkalis in order to recover one or more elements. Traditionally these processes have been the responsibility of the extraction metallurgist and are described in greater detail in the appropriate chapter. However, some of the applications of these methods, notably the cyanidation of gold and the leaching of copper ores, are increasingly considered to fall within the province of the mineral technologist and are briefly reviewed below.

If an ore, whilst in its original position in the earth's crust, is sufficiently porous it may be practicable to leach out some component by passing a suitable solvent through the mass of ore. This procedure is called *in situ* leaching. The solution is later collected and the valuable component (usually a metal) is recovered by controlled precipitation, ion exchange, solvent extraction, or electrolysis. Broken ore that has been stacked into large mounds can sometimes be treated in a similar manner and the procedure is termed dump or heap leaching. Another form of leaching which is widely practiced is the treatment of fine-grained tailings from conventional copper flotation plants to recover small amounts of residual copper.

The rate of solution depends on the reactivity of the mineral and can sometimes be increased by bacterial action e.g. certain copper minerals are readily attacked by the acid ferric sulphate which is generated by bacterial attack on sulphide minerals.

Occasionally the reaction rates between minerals and reactants are increased by the use of high pressure and/or high temperature. Such an operation is expensive and is commonly used only for the leaching of comparatively small quantities of rich concentrates or other high-cost products, e.g. the alkali leaching of uranium ores.

The list of elements obtained by chemical treatment of ores is continually lengthening and at present includes copper, aluminium, manganese, nickel, tungsten, zinc, gold, silver and uranium.

2.2.16 Amalgamation

This is a very old process for the treatment of gold and silver ores that has now been largely replaced by the cyanidation process. The method is only applicable to ores in which the elements occur in metallic form. The crushed ore, which contained comparatively coarse metal particles almost completely freed from the gangue, was passed over, or mixed with, liquid mercury. The gold or silver formed an amalgam and the elements were later recovered in metallic form by distilling the low-boiling point mercury.

2.2.17 Cyanidation

A large proportion of the current production of gold is obtained by this process in which the gold is dissolved by a weak, aqueous solution of cyanide in the presence of oxygen.

The ore is finely ground to expose the gold particles in order that they can react with the cyanide solution. The reaction time between ore and solution is often as long as 48 hours and the mixture is aerated and agitated continually by a current of air. This prevents the solid materials from settling and forming a dense impermeable mass and also ensures that adequate amounts of oxygen are available for the gold-cyanide reaction. When all the gold has been dissolved the solute is filtered off and the gold in solution is precipitated by the addition of zinc dust. This process is very widely used in South Africa where the gold ores do not contain appreciable amounts of cyanicides, i.e. worthless minerals which also react with the cyanide and tend to use up excessive amounts of the solution.

The cyanide solutions used for gold treatment are very weak, but nevertheless, form a potential hazard and extreme care is taken to minimise the danger to plant personnel.

2.2.18 Product disposal

Most of the saleable products from the treatment of metalliferous ores are sold as dry, solid products. However, many of the separation procedures, as outlined above, require that an ore be treated in the form of an aqueous pulp that may contain much more water than solid material. In addition, the waste products, which also contain large proportions of water, must be discarded and it is usually necessary to reduce the water content of this material before disposal. Consequently, it is obvious that the separation of solid particles from large amounts of water constitutes an important operation in the treatment of metalliferous ores and this procedure can be difficult when the solid occurs as particles of very small size. The mechanical procedures employed for this 'de-watering' stage are thickening or filtration or a combination of these processes.

In the thickening process, an aqueous suspension which contains a few per cent solids is allowed to settle so as to give an upper layer of clear water and a bottom layer that consists mainly of solid but which may still contain up to 30–40 per cent water. The clear water is decanted away and the settled solids may be further 'de-watered' by filtration and thermal drying. The process of thickening is carried out in shallow cylindrical tanks where the new feed is introduced at the centre; the clear water overflows at the periphery and the solid collects in the bottom and is continuously discharged by a sludge pump. When very fine-grained solids are being thickened the rate of settlement of the small particles is low and large thickeners are needed to ensure adequate capacity. It is not unusual, for instance, to find thickeners up to 100 m in diameter.

The rate of settlement of fine-grained solids can be materially increased if the small grains can be aggregated together into large clusters. Such a process is called 'flocculation' and can be achieved by correct control of the chemical environment within a thickener; for instance, by control of the pH value, by the addition of electrolytes or by the addition of polymeric flocculating agents.

The minimum water content that is achieved in a thickener underflow product is approximately 30 per cent. Further de-watering can be carried out by a filtration process in which the liquid is forced through a membrane which is impervious to the solid component, e.g. a filter cloth made of cotton or nylon. The process may be intermittent, using filter presses, or continuous as with drum or leaf filters. The filtering action produces a 'cake' of solid that contains only 8 to 12 per cent water.

Any further reduction in the amount of water must be achieved by thermal drying where the remaining water is removed by evaporation. This operation is expensive and is always preceded by thickening and/or filtration. If the material is to be shipped or transported the moisture content of the product of a drying stage is usually about 4 per cent: a reduction below this value may lead to serious dusting problems.

2.2.19 Tailing disposal

The largest product from almost all mineral treatment plants is the waste material which is generally called the 'tailing'. This may amount to 99.9 per cent of the plant feed (e.g. gold or tin treatment) and the disposal of enormous quantities of worthless tailing is sometimes both difficult and expensive.

As stated above, the tailing almost invariably consists of a mixture of finely divided solid particles and large amounts of water that may also contain a variety of chemical reagents. Great care is taken in the disposal of such material so that it does not constitute a danger to the health or the physical well-being of the mining and metallurgical community. Coarse-grained material is readily drained and can be stacked in steep mounds. Dilute pulps of fine-grained tailing, however, must first be thickened and impounded behind strong dams. A tailing pound may be a natural depression, such as a lake, or, more usually nowadays, the tailing is deposited within a retaining wall made from the tailing itself. The coarse material, which is separated from the 'fines' by hydrocyclones, forms the main part of the wall, whilst the finer fractions are impounded within the wall. The solids slowly settle to the bottom of the pound and the clarified water is drawn off through specially prepared sluices – probably for re-use since water is often scarce and always expensive.

It is sometimes possible to replace tailings in the excavation from which the ore was obtained, e.g. almost all alluvial workings and a few underground mines.

2.2.20 Sampling

The materials passing through a metalliferous mineral treatment unit are uncommonly difficult to sample. The marked variations in the shape, size and density of the various components tend to produce marked segregation in the flowing mass of material and extreme care must be taken when choosing a sampling position or method of taking a sample.

2.2.21 Process control

The lowering of the grade of ores, the increasing stringency of product specifications, the need to obtain maximum recoveries and the growing complexities of modern treatment plants necessitate careful control of every stage of mineral treatment. Consequently, process control methods are assuming ever greater importance.

Great improvements have recently been made in the speed at which the necessary data from an operation can be collected. It is now possible to obtain accurate, instantaneous and continuous measurements of the weights of various products (by a totalizing continuous weighing device); of the chemical composition of pulps (by on-stream X-ray analysis), of acidity (by pH meter) and of pulp density (by densitometer). In addition it is possible to obtain a reasonably accurate but continuous measure of the size distribution of the solid particles in a pulp.

Such rapidly obtained data can, in many instances, be used as the basis of an automatic control system. Unfortunately, the majority of mineral treatment processes are not, as yet, sufficiently well understood for full automatic control to be practicable. Consequently, the data obtained by pH meter, X-ray fluorescence analysis, etc. are, more usually, used to give automatic warnings when conditions in a plant stray from predetermined values. However, small automatically controlled loops within a major process are already in use, e.g. the control of a grinding circuit by monitoring the intensity of the sound produced by the mill and then using this information to control the amount of feed entering the unit.

2.2.22 Concluding remarks

The mineral processing of metalliferous ores is a stage in the recovery of metals that is intermediate between the mining of the ore and the extraction of metal by conventional metallurgical methods and it is used because it permits greater efficiency in the exploitation of an ore body. In the past, mineral processing was considered to include only those treatment methods that did not significantly alter the characteristics of the ore minerals. Nowadays, however, it is accepted that mineral processing may also include certain methods that involve major changes in the nature and composition of the ore components.

It can be confidently predicted that greatly increased annual production of most metals will be required in the future. The ores that will be available as sources of these metals are likely to become progressively poorer in grade and of increasing mineralogical complexity. In order that such materials can become economically viable the mineral technologist is developing better methods of assessing the potential of an ore, as well as improved treatment processes, better equipment, more instrumentation and automatic control. These improvements will result in increased efficiency and will reduce the costs of mineral treatment.

In addition, the fundamental principles of the technology are slowly being understood and will allow more rapid and systematic advances than have been possible in the past.

2.2.23 Literature

A. M. Gaudin *Principles of Mineral Dressing,* McGraw-Hill Book Co., New York, 1939.

Werner Gründer *Aufbereitungskunde.* Band I, Allgemeine Aufbereitung, Goslar, Hermann Hubener, 1965.

E. J. Pryor *Mineral Processing*, Elsevier, London, 1965.

A. F. Taggart *Elements of Ore Dressing*, John Wiley & Sons, New York, 1951.

CHAPTER 3

Iron and Steel

3.1 HISTORY

Steel and cast iron are the cheapest structural metals in the world and the most widely used, constituting some 94% of all used metals; they are the metals of choice for structures, machines and tools – in immense quantities too, for example, ten million tons for the British railway-line system alone. This paramount position will not be challenged in the immediate future for, though world production has reached over 500 million tons a year and is still rising, the estimated reserves of ore in known deposits, however published figures differ, suggest that there is enough for at least another two centuries. Increase in scrap from the high production and the discovery of new ore deposits may well extend this estimate further still. The top producers are the United States of America and Russia, with Japan catching up at a phenomenal rate. The output of the top five producers is shown graphically, for two decades, in fig. 3.1.

The production of iron and steel has a long history. Archaeologists refer to an Iron Age when man found that iron was more effective as a material for weapons and tools than copper and bronze. It could just as aptly have been called the Steel Age, in which we still are, because there is little doubt that there was carbon present in some primitive iron implements or they would not have been hard and capable of sustaining a sharp edge. But it was uncontrolled steel or iron production, varying according to the acquired skill of the individual artificer, the type of ore and fuel available and the sort of furnace the early iron-worker could produce. This variation is shown in two examples. History records early battles in which the swords actually bent and were nearly useless; also a find of iron nails of Roman times in Scotland has been analysed and they are shown to range from high-carbon steel to nearly pure iron.

Smelting (which is a word cognate with melting) was known in early prehistoric times and gave us the so-called Bronze Age. But iron presents problems. It melts at the high temperature of 1535°C, which was never achieved in any primitive furnace (compare the melting point of copper, 1082°C, and pure tin, 231.9°C). Indeed the attainment of a sufficiently high temperature has been one of the problems of steel-making and iron-making even in modern times. Iron with a high carbon content, however, melts at a considerably lower temperature (e.g. at 1150°C with 3% carbon) and it has thus been easier to produce cast iron in a furnace than pure iron. But cast iron is very hard and brittle and cannot be worked or wrought. The early history of the

124 | IRON AND STEEL

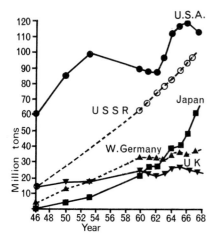

Fig. 3.1   Output of top five producers for 2 decades

industry is therefore milestoned by the achievements of inventive men in finding compromises between the easier melting of cast iron, which could not be worked, and the ease of fabrication of pure iron, which could not be melted.

Iron-making must at first have been sporadic and accidental when stones used in fires produced a metallic material. By 1500 B.C. an iron culture had developed in the Caucasus and it then spread slowly east and west. As far as the western world is concerned this was the beginning of the Iron Age. A number of centres became renowned in India; in Europe the break-up of the Hittite empire in about 1200 B.C. (it gave the name 'chalib', for steel, to the Greek language, from the Chalibes) led to a dispersion of artificers and eventually a noted centre was established in western Austria. Iron-making crossed the English Channel to Britain in the fifth century B.C., and crossed the Atlantic to America in the early seventeenth century A.D. In the course of centuries, iron-making was established in many parts of Europe and Britain, such as the traditional German Siegerland and the British Weald, where ore and forests (for charcoal) were together.

Most of the iron-makers of the Christian era, right through the Middle Ages and much later, produced an impure iron, malleable to such an extent that ornamentation could be done and suits of armour hammered into shape. The iron ore was mixed with charcoal and fired with the help of a forced draught of air. Thus the iron oxide was reduced to iron and the carbon created carbon monoxide and carbon dioxide, both gases. Mixed with the iron were many sorts of impurities, for nothing was known of other elements in the ore, and doubtless there was a demand for the best ones and a mystique built about their origins. The furnace was not hot enough to make the iron melt; it was a spongy mass of red-hot iron with much extraneous matter, the non-metallic part called the gangue (pronounced 'gang'), which has to be beaten out or in. The product was wrought or malleable iron, valuable for many uses.

The prized material, however, was steel, because it could be made into weapons and edged tools. Many primitive craftsmen must have succeeded in melting such a material accidentally in very early times. Not only was it prized because it could make weapons, but because early smiths learned how to harden it by heating it until it was red-hot, quenching it in water and then tempering it to remove brittleness.

Homer, many centuries B.C., referred to the hissing sound of such an activity. This knowledge gives a technological overtone to the Arthurian legend of Excalibur.

From a very early time two main sources of steel became very well known, and remained so for probably two thousand years or more. One was in India, where a type of steel called 'wootz' (a corruption of the word for steel in one language of India) travelled the trade-routes across oceans and over deserts and mountains to provide smiths with material for weapons. The famous saracens' swords of Damascus, (see fig. 3.2) which amazed and dismayed the crusaders, were made from wootz steel, and fashioned so well by the smiths that the blade, with its 'damask' pattern of sheen, could be bent round for the tip to touch the handle. One such sword has been analysed and shown to have 1.3% carbon, well in the high-carbon region for steel. It is doubtful if the Indian craftsmen ever produced steel of much lower carbon content.

Wootz steel was first made as lumps or 'blooms' of impure iron. Each lump then had dried wood and leaves added and the whole was totally enclosed in a clay crucible. The crucibles were then fired in a charcoal furnace in which a high temperature was attained by means of a forced draught achieved with bellows, the operation taking several hours. The iron in each crucible was hot enough to take in carbon from the charred organic matter and then melt. Thus cakes of steel were produced. These were again fired. but not melted, in contact with charcoal. The result was wootz steel which became renowned in lands far away from India. Two points about it are important: (a) it was the first time *melted* steel had been produced, thus ensuring uniformity; and (b) it was the first time crucibles were used.

The process can be seen now to have been one of adding carbon to a contaminated but nearly carbon-free iron. The fact that it was carbon that was being added was unknown as was the truth that it was carbon that effected the change from soft iron to hard steel. Steel-making was a valued and secret craft and a mystery, as was the fashioning of the steel into weapons by skilled smiths. Wootz was famous, but over the centuries many other centres contrived to make steel by adding carbon to soft iron. In Glamorgan, for example, in the Middle Ages, steel was made by passing iron through a fire of charcoal, horn, hooves and bone! One can imagine an incantation accompanying this procedure. Sheffield (Sussex) knives were mentioned by Chaucer.

Another steel-making centre became famous in Styria in Austria, important because of its relation to present-day practice. In the Middle Ages much of its steel was imported into Britain, though Scandinavian sources using the same technique followed and became famous because of the rich iron ore from Sweden. Styrian scythes were much prized. The method depended first on the use of a furnace called the stückofen, which was developed from the early stone furnace so fequently called the Catalan forge. The stückofen (which means 'lump oven') was made 10 to 16 feet high, and the draught was forced by bellows driven by water-power. The demand for steel was increasing and the stückofen method was in response to the demand for bigger blooms and greater economy in fuel. It could and did produce iron but its historical importance is that in it the temperature was high enough to melt iron if enough charcoal was present, and the liquid high-carbon metal ran out of the bottom. Eventually this was allowed to run into parallel channels and solidify, producing lumps called pigs. This was pig iron. It was a cast iron, not quite in the modern use of the term, but nevertheless castable in shaped moulds if desired, a very valuable property. The importance for history, however, is not just that pig iron was produced but that it was then refined, which means that carbon was removed from it, to make steel or malleable iron according to

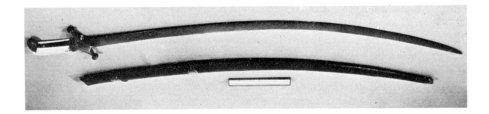

Fig. 3.2 Persian sword; portion of Damascus blade showing watered pattern.

what was required. The refining consists of firing the pig iron with charcoal and air-blast, to create the high temperature, excess carbon being oxidized away until a pasty mass of steel was formed. Improvements in refining followed over the years.

It may seem that the stückofen method, being less direct, was inferior to the wootz method. But there were several advantages. For one thing greater quantities were produced at one smelting and the method was less secretive and guarded. It was more modern. No longer were smiths limited to small lumps made in family-run bloomeries. The weight of iron produced per smelt was up to 6 times that in the best of Catalan forges, and the fuel consumption was about two-thirds. But the outstanding advantage was that the cast iron *melted*. Moreover, the pig iron was more consistent in constitution because of the melting and flowing. Finally, the stückofen was, as can now be seen, the forerunner of the blast furnace.

The Styrian method eventually ousted the Catalan forges and wootz steel ovens and other methods that depended on a lot of chance to get a usable product, but the change was not sudden. No doubt many users stuck loyally to small local bloomeries and in any case communication was poor. In more remote parts of Europe and other continents primitive methods thus stayed until the nineteenth and even twentieth centuries. But the blast-furnace method of making pig iron and the subsequent refining of this to make steel or wrought iron continued to advance. The use of cast-iron as a material for making castings of any shape, such as gun barrels, increased even though pig iron was the raw material for wrought iron. As countries grew more commercial and industrial especially Britain, the founder of the industrial revolution, the demand for steel for weapons and cutting tools was less than that for structural iron, whether wrought or in castings. Invention piled on invention in Britain, especially after the Restoration of Charles II and the revolution of 1688, which saw established, within a few years, cabinet government and party politics, the Bank of England and the National Debt. Many of the inventions are of interest only to historians of technology who wish to follow the somewhat sterile line of investigation into priorities. None of the inventors had any fundamental grasp of the chemistry and metallurgy of iron and steel, but many were ingeniously practical, intuitively inventive, and economically conscious. The history of the iron-and-steel industry from the seventeenth to the mid-nineteenth century was largely British, with a few outstanding developments relevant to the progress to modern times.

There was the blast-furnace problem. Charcoal was too fragile to withstand heavy loads and the blast obtainable was too weak to be effective through such loads. Yet there was increased demand for iron. The first stage of solution of the problem was to use coke instead of charcoal. Coal was useless. Abraham Darby the First (1676–1717) is credited with this development in 1709. He was a remarkable inventor and he patented a way of casting cauldrons and pots in dry-sand moulds, the progenitor of today's iron foundries. He formed the Coalbrookdale Ironworks in Shropshire, and his son, Abraham Darby the Second (1711–1768) carried on the business and the invention, so that at last the industry moved from forests to coalfields. His grandson, Abraham Darby the Third (1750–1791), built with John Wilkinson the first iron bridge in history in 1779, that over the river Severn at Coalbrookdale (see fig. 3.3). It was made of cast iron and still exists. The Wilkinson family was also remarkable. The original was Isaac Wilkinson (1705–1784) from near England's Lake District. His son, John (1728–1808), reorganized his father's ironworks at Bersham in N. Wales and invented a way of making cannon by casting in the solid and then boring them accurately, an achievement which made James Watt's steam engine practicable and provided cannon for the Peninsular War and Russian and Turkish wars. His chief achievement in the history of the industry was the application of steam to power the blast for blast furnaces. He also made the first iron boat, a barge which plied between his ironworks and Bristol. His brother, William (1738–1808), originated Silesia's iron-smelting industry by starting coke-fired blast furnaces, and was the founder of modern ironworks in France. He invented the cupola for making cast iron of a better and more uniform quality than that obtained as pigs from the blast furnace. John Wilkinson's development of steam-driven blasting nearly completed the modernization of the blast furnace, the two remaining steps being the warming of the air first and then the use of waste heat

Fig. 3.3   First iron bridge: Iron Bridge, built 1779.

of the furnace itself to do the warming. This later development did not come until the middle of the nineteenth century.

Another problem was that of producing wrought iron in quantity. The solution to this came from Henry Cort (1740–1800) in his 'puddling' furnace. It was the first mass production of wrought iron, though the term must be related to the times and not to the industry of today. In his new reverberatory furnace Cort overcame two disadvantages of previous processes for refining pig iron. First of all he separated his fire from the iron, and so could use coal as a fuel instead of charcoal. A shallow wall separated the fire from the hearth of sand in which the puddle of molten pig iron was produced. The flues of the fire were on the side of the furnace away from the fire so that flames and gases had to pass over the load that was to be refined. The roof was low and curved and so the heating flames and gases were reflected down on to the load. The construction allowed ordinary air induction and so the second disadvantage of old refining methods was removed—that of needing a powerful draught by means of bellows or blowers. Improvements to the puddling furnace came with the addition of 'hammer scale', an oxide of iron which speeded up the oxidation of carbon into carbon monoxide which escaped and burned with a blue flame. As the mass lost carbon so its melting point rose and eventually the heat was not enough to melt the nearly pure iron; it was pasty and spongy. The puddler had to keep up strenuous stirring with iron rods, working the mass of spongy iron into balls of about 80 lb (35 kg) each. This he kept up

Fig. 3.4 A modern puddling furnace.

for more than an hour and then started another heat, getting in perhaps six heats in a twelve-hour shift. About one ton of iron could be produced per shift. Despite the physical strain of the work there was no shortage of puddlers, and by the mid-nineteenth century there were some 8000 puddling furnaces in Britain, producing 3 million tons of wrought iron a year. Today, as far as is known, all the puddling furnaces except one have disappeared. That one is in England at the works of Thos. Walmsley in Bolton, producing some 8000 tons a year of wrought iron for which there is still a demand for chains, ornamental work, etc., largely because wrought iron is more resistant to corrosion than mild steel.

Henry Cort also made another important contribution to the industry. Nothing has been said here of the need to hammer out lumps of iron or steel to make bars and rods for use. Early hammering by hand was succeeded by water-power tilt hammers, but Cort made grooved rollers· by means of which blooms could be squeezed into bars, plates and so on. This did not come at once and the first rolled joists were made in France in 1847, followed in England in 1855. Meanwhile hammering was further mechanized and the great step forward came in 1842 when James Nasmyth (1808–1890), an inventor and discoverer in many fields, patented his steam hammer, which cut down wastage and removed the need for a steel works to be near a source of water power. Thus iron-making and steel-making, having tended to move from the forests to the coalfields, could now move to them without the need to find a place near a flowing stream. The steam hammer held its place in the forming of iron and steel until well into the twentieth century but was eventually replaced by the hydraulic press.

At about this time, the mid-nineteenth century, wrought iron and cast iron were the materials of choice. In Britain the output of wrought iron was about 3 million tons a year and that of steel 40000 tons a year. These figures may not be precise but all sources agree about the ratio. By 1870 Britain was producing 220000 tons of steel, the U.S.A. 40000 tons, Germany 130000 tons and France 80000 tons. The production of steel was slowly rising everywhere but was not comparable with that of cast and wrought iron. Steel, though needed for weapons and edged tools, was an expensive luxury at over five times the price of wrought iron and ten times that of pig iron. But this situation was about to change. The demand for iron was increasing for the expanding railways and the growth of steamships (some iron ships had existed for half a century) as well as structures and engines and machines. Iron production was nowhere near enough, crude and slow as it was.

The method that supplied the small amounts of steel necessary in the seventeenth and eighteenth centuries in Britain was that of cementation. Iron bars, each completely surrounded by charcoal, were packed into troughs, made as airtight as was possible at that time. The troughs were inside a furnace designed so that the heat passed all round them. The red-heat condition, once attained, was maintained for days. Trial bars were taken out occasionally to test the state of carbonization and when judged correct for the sort of steel required the fire was allowed to die out over a period of days. The surface of each bar had blisters on it from carbon-monoxide bubbles and so was known as blister steel. About 400 tons a year could be made in the cementation kiln. The resulting steel varied between firings and within the bars themselves, because of inadequate and inconsistent penetration of the carbon, and there was still occluded slag present. But by careful cutting-up and hammering together of the bars a bloom suitable for shears and other cutting tools could be produced. However, some manufacturers of tools were already importing steel of better quality, especially from

Sweden where the high-quality ore available and the skill of the craftsmen produced steel that has remained famous ever since. Germany was another country with cementation kilns. Other methods were still in use in different places but blister steel provided all that was needed in Britain for a very long time. The last cementation kiln was closed down in Sheffield as late as 1951.

Cementation was ousted by a method invented by Benjamin Huntsman (1704–1776) in 1740. It was a crucible method, and its outstanding virtue was that, for the first time since wootz, steel was made in lumps no bigger than one's hand, the steel was melted, thus freeing it of slag and impurities, and could be cast into ingots by pouring from each crucible into an ingot mould. The achievement was not so much in the use of crucibles but in their successful manufacture, for they had to be strong and withstand high temperatures without cracking or breaking up. He used local clays and china clay and spent years developing the crucibles. Each barrel-shaped crucible was about 12 inches (0.3 m) high and wide, containing 60 to 80 lb. (25 to 35 kg) of steel. The crucibles, completely sealed, were put into a coke-fire until the steel melted. It was then separated from the floating slag and poured into ingot moulds. It was 'cast' steel. Thus large ingots of steel could be produced without hammering and the steel was consistent and clean. Huntsman at first used blister steel and relied on plain melting, but later on steelmakers used wrought iron, especially Swedish iron, with added carbon or pig iron and ferromanganese (to remove oxygen). It is recorded that Krupps of Essen once exhibited a 25-ton ingot made from crucible steel. Such large ingots were increasingly needed, though in the first place crucible steel was for tools. In the mid-nineteenth century there were some 1500 crucible furnaces in Sheffield alone, producing 60 000 tons of steel every year. By 1919 there were 45 and by 1962 only one was left, and that is now gone.

The reason for their disappearance was that new methods of steelmaking came in the latter half of the nineteenth century. The need of the industrial world for iron and steel could not be met by the wrought iron and cast iron being produced so crudely in small units. Steel was very expensive, but it was valued as a superior product – cleaner, more uniform and workable into sheet or bar or girder or any shape required. Some forms of it were indeed valued and used because of their hardenability for edged tools and weapons, but this was a small part of the market and cost was not so important. What was needed was 'tonnage' steel, which means steel in vast quantities at a higher speed and smaller costs. It is with tonnage steel that the industry is concerned today. It was tonnage steel that Bessemer and Thomas, and Siemens and Martin produced in the latter half of the nineteenth century, and they produced it from pig iron from the blast furnace. Their methods should be treated more as history than as belonging to modern technology because new methods are rapidly taking over the whole industry, but some of these methods are modifications of theirs and their methods did produce the tonnage steel needed. So they will be considered as part of the technology rather than history. It has been said by Sir Charles Goodeve that the whole U.S. railway system of the nineteenth century owed itself entirely to Bessemer, which is an indication of his importance in the technology.

Until late in the nineteenth century, and in some part well into modern times, the iron-and-steel industry was successfully based on ignorance and invention. However, Sidney Gilchrist Thomas (1850–1885), who made the Bessemer process widely usable, and Sir William Siemens (1823–1883), who realized the importance of economy of heat energy and so invented the open-hearth method, both showed greater understanding of the problems, and Thomas was trained in chemistry. Britain was at the time the Mecca of scientists and engineers, and Sir William Siemens, born a German, became a naturalized Briton – he was the younger brother of the man who founded the famous firm of Siemens Halske.

Fundamental understanding of the processes of manufacture and the behaviour of iron and steel was at last beginning. Chemistry had emerged from its dark alchemic womb; metallurgy as a subject came into being as the nineteenth turned into the twentieth century – it merited less than two pages in the 11th edition of the *Encyclopaedia Britannica* in 1911. Physical chemistry did not exist as a separate study until well into the present century. Crystallography earned the Braggs a Nobel prize as recently as 1915. H. C. Sorby (1826–1908) laid the foundations of metallurgical microscopy in the second half of the nineteenth century. Sir William Roberts-Austen (1843–1902) designed his carbon-iron diagrams during the same period. The industry has thus gradually grown research-minded, a fact shown in Britain by the incorporation in 1944 of the British Iron and Steel Research Association, now part of the nationalized British Steel Corporation, and research establishments exist in many countries. The International Iron and Steel Institute was formed in 1967 to collect and circulate information on the steel industry of the whole world. Though the industry is, and must be, chiefly concerned with the economics of production, just as any other industry is, it is becoming one that is science-based. For a long time research followed practice (and still does to the extent that fundamental understanding is necessary); but much of the more recent research has led to practical modifications and innovation. The computer-controlled steelworks is no longer merely a dream.

Scientists are now very concerned with the chemical, metallurgical and thermodynamical problems of iron and steel. Some of the problems have been solved. It must have occurred to many that there might be a way of converting ore directly into steel because, looked at naively, the chemistry seems simple. There is no logical obstacle to the direct and continuous production of steel from the ore, but there are considerable practical problems. There are some plants where iron is produced directly by special methods from the ore, but the main raw material for steel is still pig iron (and some scrap steel from waste and used products), and this iron is produced in the blast furnace.

3.2 IRON ORES AND THE MANUFACTURE OF PIG IRON

3.2.1 Ore supplies

Iron is produced from rock in which iron is combined with other elements to form compounds. This is iron ore. It varies from a richness of over 65% of iron to less than 20%. It is widely distributed in the world in various forms. Intensive geological exploration has been carried out in recent years to find more and more deposits to satisfy the existing and expected demands, the manufacture of steel having risen so fast that fears were once expressed that supplies would soon be exhausted. Estimates of the reserves in the world, a result of the exploration, vary widely and methods of classification differ, but a round figure of 300 000 million tons of iron ore can be taken as fairly representative, with more yet to be found. However much these immense quantities of ore may vary in quality and availability, and however much the iron-and-steel industry may vary in magnitude there is clearly enough iron in the world for another two centuries, and exploration continues.

132 | IRON AND STEEL

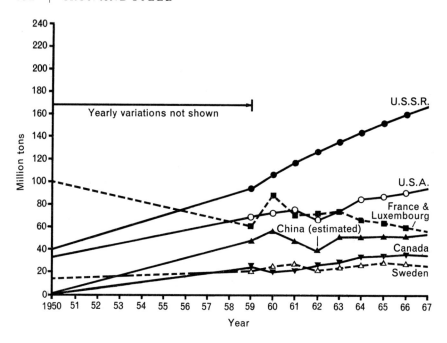

Fig. 3.5 Iron ore production.

Many of the deposits thus discovered have now been developed, especially since the 1950s. Exploitation has changed the world iron-ore situation, bringing prosperity to some small countries and increase in trade to others. Liberia started exporting in 1951 and now ranks among the five major exporters, with a very notable rise since 1962. Canada, which produced no iron ore in the period 1924–1938, is now the world's largest exporter at over 30 million tons a year, and immense new deposits were discovered recently in the north of Baffin Island, inside the Arctic Circle. Before 1951 Venezuela produced no iron ore; now it produces over 17 million tons a year. Fort Gouraud in Mauritania started exporting only a few years ago and has reached more than 7 million tons a year; its out-crop of black haematite is there for the removal and transportation. Swaziland in Africa started to produce iron ore in 1964. Other mines in Africa have stepped up production. Brazil is producing some 20 times as much as it did in 1950. Other South American countries have emerged as exporters. India is now a big producer and, including Goa, is among the major exporters. The outstanding development has been in Australia where continued discoveries in Western Australia have put the country fifth in the list of reserves of iron ore.

Russia and China and the eastern block in Europe, as well as North Korea, remain a special case. Russia is said to have a third of all world reserves but her trade is limited, so far, to exports to her neighbours in Europe. Her own production of iron ore doubled in the decade from 1956–66 as did her production of steel. Little is known about China. Estimates put her in the fourth world position as an iron-ore-producing country, but she covers an immense area and has a vast population, so that she uses all she produces.

World trade in iron ore involves transport over long distances, most of it by sea.

IRON ORES AND THE MANUFACTURE OF PIG IRON | 133

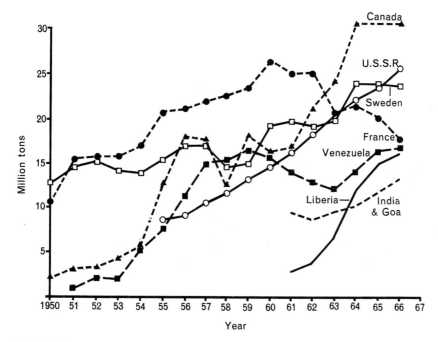

Fig. 3.6 Iron ore exports.

Table 3.1 World iron ore production 1965–66

(All production greater than 1 million tons† per annum)

Country	Production (million tons) 1965	1966	Country	Production (million tons) 1965	1966
Russia	151.0	157.6	Malaysia	6.9	5.8
U.S.A.	87.5	90.2	Spain	5.7	5.1
France	58.6	54.2	Austria	3.5	3.4
China (estimated)	50.0	50.0	Poland	2.8	3.0
Canada	35.7	36.3	Roumania	2.4	2.7
Sweden	28.9	27.6	Bulgaria	1.8	2.6
India and Goa	23.4	26.1	Mexico	2.6	2.5
Brazil	17.9	22.9	Yugoslavia	2.5	2.5
Venezuela	17.4	17.6	Japan	2.5	2.4
Liberia	15.1	16.3	Norway	2.4	2.4
U.K.	15.4	13.7	Czechoslovakia	2.4	2.2
Chile	12.0	12.0	Sierra Leone	2.3	2.2
Australia	6.7	11.4	East Germany	1.6	1.7
Peru	10.1	7.7	Turkey	1.5	1.6
South Africa	6.6	7.6	Algeria	3.1	1.5
West Germany	7.8	7.1	Phillipines	1.4	1.4
Luxembourg	6.2	6.4	Tunisia	1.1	1.3
North Korea (estimated)	5.8	6.0	Morocco	0.9	1.0

†1 ton = 1016 kg = 1016 tonne; clearly, the difference between the British ton, which is equivalent to the U.S. 'long ton', and the metric tonne is so small that throughout the chapter the figures quoted in tons can be taken as representative of those for metric tonnes.

Land transport involved is limited to exports from Russia to its eastern bloc neighbours; France to Belgium, Luxembourg and the Saar; and the U.S.A. to neighbours on the American continent. Sea-borne traffic has increased accordingly, the average distance the ore is carried ranging from 2000 to 5000 miles (3000 to 8000 km), the longest being for iron ore imported by Japan, where native resources are negligible. The importance of this is that sea freight is far cheaper than land freight.

It is more expensive to transport ore from Immingham dock on England's north-east coast to Scunthorpe's iron-making works, a distance of 20 miles, than it is to bring it across hundreds of miles of the North Sea to Immingham.

In addition competition has made ore cheaper. The combination of cheaper freight and cheaper ore has meant that numbers of local sources have been closed down in steel-producing countries, especially where the domestic ore is of low grade. The Salzgitter mines in West Germany no longer supply any ore to the Ruhr, and Salzgitter steelworks use some imported ore. In the USA imported ore accounted for 8% of production in 1950, whereas it is now 40%, a five-fold increase. Japan depends on imported ore for nearly all her rapidly expanding production. In the United Kingdom, the use of imported ore has outstripped that of home ore since 1964; the price of home ore has continued to increase since 1957, and that of imported ore has continued to decrease.

With the growth of this trade there has come a change in shipping. Prices go down as the size of ore-carrying ship goes up, but the rate of decrease gets less as the capacity goes up. As the size of ship increases so does the cost of unloading it in port, but very gradually. There is thus a tendency to build bigger ore-carriers and this brings with it a need for deep-water port facilities. Ports are anticipated in Australia to take ships of 65000 tons deadweight, and in Brazil the need for 100000 ton ships has been realized. Canada is amply supplied with ports, and Liberia can already accommodate ships of 65000 tons. Mauritania can do the same but is planning to take 100000 ton ships. Sweden, Peru and Venezuela can take ships of 60000 to 70000 tons. Sierra Leone hopes to take 85000 ton ships at Pepel. Other countries have similar plans for expansion. There are some ports in the United Kingdom taking ships of 65000 to 100000 tons, and others are being planned.

It is thus seen that since the mid-1950s the pattern of the iron and-steel industry has changed radically. Steelworks and ironworks once had to be in the forests. Then they moved to the coalfields. Now they move to the coast where facilities exist to unload iron ore in great quantities. In this the countries not committed to the old traditional pattern have the advantage. Japan, for example, has most of its steelworks near those ports used for unloading iron ore. The industry, subject like any other industry to the winds of change brought by factors not always technological, and open to the rises and falls of commerce, has changed permanently. The old traditional industry has gone with the wootz.

3.2.2 Composition of iron ores and their mining

Any rock or earth or sand that contains an iron compound is potentially an iron ore, but its use depends on its accessibility and whether iron can be made from it economically. Such materials are so widespread that very few countries are without some, however little and however unprofitable to use. They exist in a range of colours – black, grey, red, yellow, green and brown. They have a wide range of physical forms from soft easily-breakable earth, to hard rock that must be blasted to make moveable boulders. The most common iron minerals and their main iron compound in each case are shown in table 3.2, the proportion of iron being calculated from the chemical formulae.

Of these magnetite and haematite are the two of highest grade. The more widespread is haematite. Magnetite, theoretically the richest, is also magnetic – the material of the traditional lodestone. Goethite, hydrogoethite and limonite contain combined water. Pyrites (and pyrrhotite) and ilmenite are not usually mined for their iron, but for their sulphur in the first case and titanium in the second, iron being a by-product in the residues, but they can in some circumstances be used economically as iron ore.

Table 3.2 suggests that the chemistry of iron production is simple, for it appears that all that is necessary is reduction of the oxide to pure iron. Iron ore technology, however, is in reality complex, mainly because of three factors:
1. the metallurgy of iron and the physical-chemistry of iron-making;
2. the evolution of the earth's crust;
3. the growth of a traditional industry in separate countries, based on empirically-derived methods of production in mass, and leading to an involved terminology.

The first of these will be considered in more detail later (see section 3.2.4) and all that it is necessary to say here is that iron melts at a very high temperature, which means that all the chemistry is high-temperature chemistry, and that, unlike any other metal, it combines easily with carbon.

The second factor is very important. The earth's crust consists of three types of rock – igneous, sedimentary and metamorphic. The last named rocks have been changed by physical and chemical processes from the original igneous rocks. Iron exists in all three types. Probably iron oxide in a pure state existed in igneous rocks originally, but over millions of years carbon dioxide and water have created carbonates and hydrated oxides. Further changes have in some areas oxidized a top cap of richer oxide over these carbonates and hydrated oxides. Other reactive constituents of the earth have created different iron compounds, such as iron chlorite. Silica, a very insoluble and durable material, is widespread in the earth and is mixed with the iron compounds in many different ways, ranging from separate veins in iron-bearing rock to small crystals intimately bound up with iron-compound grains. Silica has also combined with iron to form iron silicate, and has combined with aluminium to form clay or rock material such as feldspar. The result of all this change is that iron rarely

Table 3.2 The most common iron minerals and their main iron compound

Name	Chemical formula	Iron content%
Haematite and martite	Fe_2O_3	70
Magnetite	Fe_3O_4	72.4
Magnesio-ferrite	MgO, Fe_2O_3	56 to 65
Goethite	Fe_2O_3, H_2O	62.9
Hydrogoethite	$3Fe_2O_3, 4H_2O$	60.9
Limonite	$2Fe_2O_3, 3H_2O$	60.0
Siderite	$FeCO_3$	48.3
Pyrites	FeS_2	46.6
Pyrrhotite	$Fe_{11}S_{12}$	61.5
Ilmenite	$FeTiO_3$	36.8
Iron Silicate ('if low in iron')		10 to 35
Iron Silicate ('if high in iron')		35 to 40

exists as pure iron oxide, and where it does the rich ore is being rapidly used up by the growing industry. Almost any combination of the compounds listed in table 3.2 can exist.

Iron ore can thus be a very impure material, and there are two sorts of impurity. The first consists of elements other than iron. These are often in the crystal structure and cannot be eradicated be mechanical methods. Those most frequently found are manganese, phosphorus, sulphur, titanium, copper, chromium, vanadium and nickel. Not all are harmful when they get into the pig iron, but they complicate the chemistry of steel-making. Manganese exists in iron ore in different degrees, and has given rise to a small but separate smelting industry to produce ferromanganese (see chapter 4). Its presence assists in reducing the amount of the undesirable element, sulphur, in steel. Phosphorus is also undesirable because it makes steel cold-brittle and its elimination is never completely achieved. It gets into the raw material, the pig iron, in the blast furnace. Too much copper can make steel hot-brittle. Chromium too, gets into the pig iron and causes subsequent trouble in the production of steel. Even when some of the impurities improve steel they are not easily controllable when they are introduced by the pig iron.

The second sort of impurity is the collection of compounds that constitute the waste or gangue. These are mainly silica, lime, magnesia and alumina. When successfully melted they form a sort of glass which floats on the molten iron. This slag is essential to the chemistry of the making of iron and steel by the traditional methods, and its constitution and amount affects this chemistry. There are two kinds of slag–acid and basic. The first consists chiefly of silica and alumina, while the second is mainly lime and magnesia.

Another important property of iron ore, also the result of geological evolution, is its physical form. Some exists as hard material which has to be broken up. Some is soft. Some is in small lumps mixed with powdery material. Some is just sand–the golden colour of seashore sands is due to iron. Some ore exists in separate bands so narrow that it cannot be excavated separately from the rest of the rock. All these properties of physical form are important commercially in determining how the ore can be mined and handled.

Although iron ore is the raw material for iron and steel, there are other materials needed by the industry that constitute a considerable industry. These are mainly dolomite (magnesium calcium carbonate), limestone, fireclay, alumina and sand. they are needed for refractory linings for blast furnaces and steel furnaces. Some are important in slag production.

Two-thirds of the iron-ore mines are on the surface, and the older deep mines are going out of use. The needed ore, after removal of the overlying earth, is blasted or dug out by great machines. The proportion of this open-cast type of mine is increasing and deep mining of the traditional type will disappear. Such surface mining is cheaper than deep mining, but it has also increased because the old rich 'pure' ores are being exhausted and cheaper low-grade ores can be used.

The third point, contributing to the complexity of iron ore technology, is that the terminology of the iron-ore industry has developed over the years, and though it is less important in its effect on the technology, it does make understanding more difficult for the inventive technologist to understand. Names have come from geographical and geological descriptions. For example, there is the ore of high quality at Kiruna in northern Sweden, the existence of which ore established the high quality of certain

Swedish iron and steels. These ores are mainly magnetic, the magnetic ore existing, in some places, as solid blocks. Other parts of the deposits contain haematite as well as a rock called apatite, which is mainly calcium phosphate. There is the Lake Superior type, originating in the enormous deposits of Minnesota. These were originally mined very carefully in order to ship only high-grade ores through the Great Lakes to the blast furnaces of the United States. Such selective mining is long since finished in this area, and in most other parts of the world. Thus the adjectives Kiruna, Lake Superior, Minette (from the Lorraine deposits of France), Bilbao (from the deposits of North East Spain) and some others have had wide acceptance and can have meaning only for the iron-mining engineer. Classifications of iron ores according to their minerals, their geology, their chemical characteristics, their use for steel making, have all been made, differing from country to country.

A more scientific way of classifying ores is an analysis into constituents. This is regularly done in statistical summaries as shown, in a small extract, in table 3.3 (compiled from the United Nations book: *Economic Aspects of Iron-Ore Preparation*, 1966.) It can be seen that the iron percentage is far less than that shown by the simple formulae of table 3.2. What such analyses do not do is to indicate the physical form of the ore, which is of great importance in relation to their mining, transport and usefulness in the blast furnace. A different analysis has therefore been attempted by dividing iron ores into seven groups:

> Magnetites,
> Haematites (or martites)
> Brown haematites (limonites, oolitic ores)
> Siderites,
> Titano magnetites
> Laterites (goethitic)
> Pyritic

Two of these have not been mentioned earlier. The brown haematites are worked mainly in France, Western Germany and the United Kingdom. They include iron hydroxides, siderite and, in some cases, a haematite. Laterites come from a hard top-cap of ore, arising from the migration upwards of iron after the leaching out of silica. They have in general a high alumina content. Classification into these groups should also include data on the iron content, the nature of the iron-bearing minerals, and the nature and quantity of other minerals.

This sort of classification is an attempt to make a system that is more realistic for the iron technologist than the purely mineralogical one. However, ores vary so much in structure, location, and treatability, that no simple classification is possible. The chemical reducibility has also been tried as a basis for the classification of ores, but methods and standards vary from country to country and no dependable comparison can be made.

The phenomenal growth of the steel industry has necessarily brought into use the medium-grade and low-grade ores, once considered uneconomic but now shown to exist in enormous reserve deposits. They are cheaply obtained from open-cast mines, and are cheaply transported by sea. These are often unsuitable for direct use in the blast furnace, and must be prepared in various ways so that the smelting operations shall remain efficient and economically competitive.

Table 3.3 Specimen analyses of iron ores

Country	District	Ore	Iron (Fe) %	Silica (SiO$_2$) %	Lime (CaO) %	Alumina (Al$_2$O$_3$) %	Sulphur (S) %	Magnesia (MgO) %	Phosphorus (P) %	Manganese (Mn) %
France	Sancy	Calcareous	32.0	6.4	17.6	4.0	0.05	1.6	0.73	0.3
France	Mont-St-Martin	Siliceous	37.3	22.3	3.8	5.4	0.08	1.2	0.66	0.3
India	Cuddapah	Haematite	50.7 to 61.2	3.45 to 6.51	—	1.75 to 5.91	traces	—	0.07 to 0.093	—
Russia	Krivoi-Rog	Haematite	61.0	10.43	0.06	1.04	(SO$_3$) 0.03	0.10	(P$_2$O) 0.07	(MnO) 0.06
Russia	Kerch	Brown Haematite	40.2	19.16	1.85	5.09	(SO$_3$) 0.30	1.13	(P$_2$O$_5$) 2.01	(MnO) 4.6
Sweden	Kiruna	Magnetite with haematite	65 to 68	2.0	1 to 3	0.5	—	—	0.01 to 0.02	—
United Kingdom	Frodingham	Brown ores	18 to 24	4 to 12	15 to 30	3 to 8	0.05 to 0.90	—	0.1 to 0.8	—
United States	Wacootah (Lake Superior) Mesabi	Limonite, Martite, Magnetite	59.37	4.36	—	—	—	—	0.087	—

3.2.3 Ore preparation

With the exception of high-grade pure ores, which merely need to be broken up and sorted into uniformity, all ores must be prepared for use in the blast furnace. The aim is to produce from the ore, as mined, a product richer in iron, more suitable physically for the blast furnace, more easily reducible to iron, and cheap enough to be economic in the production of steel. All preparation increases the cost of ore, but the cost per unit of iron obtained and steel produced can be reduced because of the increased over-all efficiency of the whole process from ore to steel. Ores differ in texture, accessibility, purity and richness from deposit to deposit and these themselves differ in character and geographical situation. There is therefore no single best way of preparation, and many devices have been invented, some of which suit certain local circumstances better than others.

For many years, and even from early times, there was preparation which involved the washing away of clay and other undesired material, and the crushing of large boulders into lumps which could be sorted for uniformity. Such crushing is still an important part of today's preparation of ores. Two further stages, combined under the one term *beneficiation*, have followed in more recent times. These two stages are especially necessary for ores of lower grade. First there is the grinding of the lumps into fine particles, and the separating of these into those that contain iron and those that do not. A more concentrated ore in fine particles is thus obtained, but fine particles and dusts are not suitable for the blast-furnace. So they must be agglomerated into suitable lumps.

Beneficiation of one sort or another has increased in the past twenty years to such an extent that by 1970, it is estimated, all ore used in advanced countries will be beneficiated. No directly-shipped raw ore will be going to the blast furnaces. Even now high-quality concentrates produced from the poorest ores in the largest and most modern beneficiation plants are economically competitive with the richest natural ores. Ores containing in their natural state less than 30% of iron can go to the blast furnaces enriched to 60% and more. There are processes not yet widely used which could produce an iron content of over 90%.

a. Crushing, grinding, and separating according to particle size. Crushing and grinding are the most expensive part of the preparation process. This is especially so for fine-grained ores in which the iron-compound crystals are intimately bound up with crystals of unwanted materials, especially silica. Obviously, a mere breakdown into smaller lumps in this case does nothing to increase the iron content; a small lump has the same proportion as a big lump. Only when such a rock is down to grain dimensions is it possible to sort out iron-rich fragments.

Crushing is done with jaw-crushers or impact-type machines and may have to be done in several stages if the ore is hard, as for example with kidney-stone ore, the lumps of which are like flint. Grinding is done by rolling the ore inside a rotating cylinder mill containing steel balls or rods. The impact of these on the ores breaks them down. A more recent development is to use larger pieces of ore in the grinding media instead of steel balls or rods, which are rapidly worn away by their own grinding action. The special advantage of ball-less grinding is that instead of a very hard material smashing against ore there is a material of the same physical character. The result is that ores tend to break up along the cleavage-lines of crystals so that iron-rich parts are soon released. One such device invented in the USA by D. Weston, and called the Aerofall, claims to grind lumps as big as a metre in diameter down to less than 0.075 mm (0.003 inch) in one pass. It can grind 1700 tons in a day and produce 80% of the output at this small size.

Most mills, however, are ball mills or rod mills and must be used in stages. The grinding can be wet or dry, the former giving no dust and the latter needing no subsequent drying. Each has its special use. Grinding in itself, as a separate operation or on 'open circuit', is a costly operation, and most grinding today is done in association with separation, particles too big being fed back for further grinding. This is closed-circuit grinding. In this way particles of the correct size are taken off, and there is less wastage of useful materials as dust, due to over-grinding. This type of separation is called classification, its function being to separate particles of different sizes. One form of it, the cyclone, allows the slurry from the grinder to enter the top of a hollow cone tangentially. It then whirls round and down and heavy particles go to the periphery and sink to the bottom, while the finer particles are caught up in the vortex and ascend the centre. Another type allows the slurry to move down a spiral trough under gravity, the heavier particles collecting on the inner edge and the lighter ones moving to the periphery.

b. Ore concentration. Coarse-grained ores, in which the iron-bearing mineral is in much larger grains than the quartz and other minerals, can be treated by crushing and washing and perhaps separation to form an ore very much upgraded and suitable for the blast furnace. In general, however, and to make sure of the immense reserves claimed, further preparation is necessary. In these cases the separation techniques are used not merely for classification but for concentration, whereby the iron-containing particles are separated from the unwanted sand etc., and so the proportion of iron is higher than it is in the raw ore. The difference between the iron-rich particles and the others may be that of density (iron oxide being twice as dense as quartz), magnetic susceptibility (magnetite is magnetic, quartz is not and some other ores are weakly magnetic), electrical conductivity (iron compounds are better conductors than quartz etc.) or surface quality (preference for air rather than water in a foam).

The several methods that depend on differences in density vary in the range of particle-size they handle and the rate of separation. The oldest used method is jigging, which takes coarse particles in the range between 3 mm and 50 mm. (Particles are often sized according to a mesh number but there are different systems not strictly comparable.) Upward pulses of water go through the bed of mixed particles and thus force the lighter ores upwards and these are mainly the quartz. The bed is thus stratified and the lighter unwanted particles washed over the sides at the top. Another method, called heavy media separation, depends on the sinking or floating of particles in a fluid which has a density less than that of the quartz and other particles. The most widely used fluid is a suspension of fine ferrosilicon in water. The size of the feed ranges from 6 mm to 76 mm and the method is thus used for fairly coarse particles. If a cyclone similar to that already mentioned for classification is added much finer particles can be separated, in the range from 0.25 mm to 6 mm. A more recent system gaining favour was invented by I. B. Humphreys in the USA. It is the spiral concentrator, with no moving parts and therefore needing no maintenance, which separates particles in the range 0.074 mm to 2 mm (see fig. 3.7). A spiral trough winds round a vertical shaft, and as particles slide down in the slurry the heavier particles collect on the inner part of the trough, and are directed to fall down the central shaft. This is the needed concentrate.

IRON ORES AND THE MANUFACTURE OF PIG IRON | 141

The lighter particles continue down the outer rim of the trough and are collected as waste. Many such spirals, some thousands, are needed for an operation of commercial size.

Magnetic separation is widely used, especially for magnetite, and can be wet or dry. Many devices have been invented. In general magnets are used to create an attractive surface on a drum or belt. The mixed particles flow on the outside of the drum, and the magnetic ones stick while the others do not. As the drum or belt moves beyond the magnetic region the adhering particles are taken off. As the grinding must be fine to get maximum concentration, a previous stage of coarse separation is often used, called magnetic cobbing. In the final magnetic separator the particles are very fine, perhaps as small as 0.04 mm. The method applies especially to magnetite, but with very strong magnetic fields it is possible to concentrate on more weakly magnetic iron ores. It is also possible to roast haematite and thus reduce it to magnetite, and make it suitable for magnetic concentration.

The high-tension electrical system depends on the fact that iron oxide is more electrically conductive than quartz. So if a mixture is made to flow on to a cylinder charged electrically by a high-tension field, the quartz particles retain their charge and stick to the cylinder whereas the iron-oxide particles quickly lose their charge and fall free.

With very fine particles another method, already used with non-ferrous ores, has come into use. The mixed particles are in a liquid which contains a surface active material that makes the iron particles hydrophobic and attracted to air.

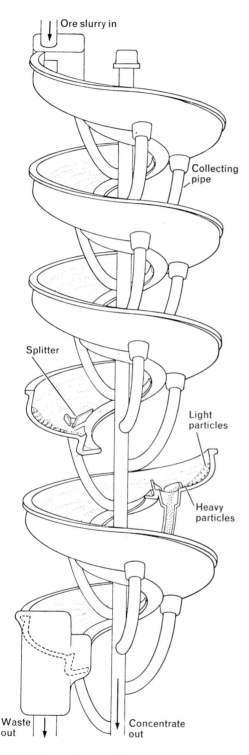

Fig. 3.7 Humphreys spiral concentrator.

When air is blown through the liquid bubbles are formed to which the iron-oxide particles adhere and, thus, these particles rise in the foam. Finally, it should be remembered that screening or sieving is still a practised method of separating particles of different sizes, but for concentration this separation must involve differences in iron content. With certain ores a fine grinding does lead to a product in which the larger particles have a higher silica content than smaller ones, and screening thus serves as a concentration device.

None of the concentration methods is perfect and consideration of which to use depends on the type of ore, the capital cost of equipment, the throughput, the market and the degree of enrichment achieved. A flow-sheet has to be established to show how an integrated system can be made to function efficiently and economically. Furthermore, research continues into the hydrodynamics of separation, the efficiency of roasting and other related problems previously solved in a practical way by empirical methods, so that newer, improved methods of crushing, grinding, classification and concentration can be devised for commercial use. Examples will indicate the differences of usage. Some Canadian examples are given in table 3.4. Electrostatic separation of quartzites from the Olyenegorsk deposits in Russia has enriched ore of 38.5% iron to concentrate of 68.6% iron. Flotation by foam is now widely used in the USA, Russia and China. The Grängesberg plant in Sweden has a flow-sheet combining preliminary screening, magnetic separation, jigging, spiral concentration and screening. The main product consists of lumpy and fine concentrates (from mixed haematite-magnetite ores) containing 61–62% of iron, and some 10% of the output is a super-concentrate containing 71% iron.

c. Sintering and pelletizing. Some ore needs only a few of the processes and can, because of its high quality, be shipped directly as material consisting of small lumps ready for the blast furnace. In general, however, the result of all the processes provides

Table 3.4 Methods of preparation in Canada

Location	Source	Ore	Iron, % crude	Iron, % prepared	Methods
Wawa, Ontario	Siderite	Blocky, fine-grained	32.5	50.4	Crushing, washing, heavy-media separation, washing, cyclone concentration
Vancouver Isle, British Columbia	Magnetite	Medium to fine grain	32.7	56.5	Crushing, magnetic cobbing, magnetic concentration
Wabush Lake area, Labrador	Specular Haematite	Granular, specular haematite and some magnetite	36.10	63.7	Crushing, grinding, Humphreys spiral concentration
Atitokan, Ontario	Haematite Goethite	Earthy	50.93	54.2	Crushing, scrubbing, heavy-media separation, jig concentration, Humphreys spiral concentration

a concentrate of high iron-richness but very small size, some being as fine as dust. Such a product cannot in this condition be used in a blast furnace, because it would make a fused solid impervious to gases and at the same time it would be too weak to sustain the weight of material above it. In early days, when it was necessary to do something about waste dust and fine residues from the mines, methods were evolved to build the particles into lumps suitable for use. Those methods of agglomeration have been taken up and improved in recent years for a different reason – to get suitable ores rich in iron from lean mining ores.

The fine particles can be agglomerated into nodules, briquettes, sinter or pellets. The smaller the formed lumps are, the greater is the ratio of surface area to volume, and this is important for access of the blast furnace gases to as much surface as possible. But the agglomerate units must also be big enough to fit together in such a way that there are plenty of passages between them for gases to penetrate. They must be hard enough to withstand transport, and strong enough to sustain many tons of burden of ore and coke above them. They should also preferably be porous enough to allow penetration of gas. In practice nodules range irregularly from 12 mm to 30 mm in diameter, and briquettes are regular and cushion-shaped, 50 mm by 35 mm by 20 mm deep in the middle. These two forms of agglomerate are obsolescent, though newer methods of briquetting hold promise. The methods of today are sintering and pelletizing. Sinter lumps go up to about 25 mm in diameter, and pellets range between 10 mm and 20 mm. In the decade 1954–64 sinter production was more than doubled in the USA, trebled in the UK, and quadrupled in the USSR. World pelletizing capacity has risen from 10 million tons a year in 1960 to 90 million today. Of the 21 blast furnaces listed in *Economic Aspects of Iron-ore Preparation* (U.N. 1966) as giving optimum performance, only five use ore at all and then in only a small proportion. All the rest take 100% pellets or sinter or both.

Sintering is the sticking-together of particles by means of heat, and is used extensively in powder metallurgy (see ch. 10, vol. III). For iron-ore particles, without any natural binder in the form of fusible material, and without distributed fuel which will burn (as in sulphides), some finely-divided fuel must be provided in the form of coke particles and flue dust. The method used to-day is mainly that due to A. S. Dwight and R. L. Lloyd, who in 1906 produced a continuous sintering machine. It has been enlarged and its ignition modernized, so that the Dwight-Lloyd machine of to-day can sinter thousands of tons every day. The raw materials travel on moving pallets, are ignited by a gas flame and then have air sucked down through them. This fires the fuel and creates adhesive material to stick the particles together to make a porous mass, which moves through rollers for crushing, and is then cooled.

Pelletizing was devised for making use of very finely ground ores (down to less than 0.04 mm) too small for sintering, but the advantages of good pellets as burden for the blast furnace have made it a process that is deliberately pursued, even to the extent of grinding down ores already well beneficiated. The powdered ore is mixed with water and bentonite, or other clay, and rotated in a drum. The particles cohere and the rolling action creates near-spherical pellets, though their correct formation is a matter of exact arithmetic in speed and time of rotation. They emerge as soft pellets usually between 10 mm and 18 mm in diameter. They must then be heated to make them hard. There are three types of heater: the shaft furnace, the continuous grate and the grate-kiln pelletizer. They are first coated with powdered coal, then dried and then hardened by heating enough to sinter but not to fuse. The tendency is to separate the various processes, making pelletizing more expensive. Even so the savings effected in the iron-making are two or three times the expenditure involved in beneficiation and pelletizing. About 60 pelletizing plants are now in operation in the world, most of them in the USA and Canada.

An investigation of the performance of pellets was carried out in 1963 at the Mannesmann works at Huckingen. It lasted some two months, and four thorough tests were made with different mixtures of pellets from sources in Sweden and the USA. The blast furnace output stood at 1 400 tons a day when sinter and lump ore was used. During the tests it rose to 2 040 tons a day in one test and 2 091 in another, both tests being with a burden of 83% pellets. When a considerable proportion of sinter was used with pellets, production went down to 1 856 tons a day in one test and 1 974 in the other. From then on pellets have been increasingly used in Europe. The tests did not decisively prove that pellets are better in every way than sinter, but it was concluded that for a similar performance the sinter would have to be carefully graded for toughness and size, an operation involving the return of unused sinter.

3.2.4 Pig iron production in the blast furnace

The blast furnace is the most widely used device for reducing iron ore to iron; 99% of all the world's pig iron, the raw material for steel and cast iron, is at present so produced. This domination is being challenged, admittedly rather weakly, by other processes but, though they are important in some contexts and may increase in importance in the years to come in countries where electricity or oil or natural gas and ore are cheap and the market limited, today they are not comparable with blast-furnace processes in output or cost. Nor will they be in the foreseeable future in the main steel-producing areas of the world.

Essentially the design of the blast furnace has not changed for a century but there has been much development. There is fuller understanding of the physico-chemical processes involved so that modifications have enabled steel men to drive the furnaces harder. The burden or charge has been modified and the heating gas often enriched. Thus the modern blast furnace is a very different device, in practice, from that of even a couple of decades ago. In the United Kingdom, for example, the consumption of solid fuel (coke) per ton of iron produced, fell from some 3 tons a hundred years ago to 1.2 tons in 1932 and 0.65 tons in 1967, a five-fold increase in fuel efficiency. The output per furnace has increased in the past 30 years from 68 500 to 275 000 tons a year in the UK, and this is small compared with some blast furnace production in the USA. The average number of furnaces in blast, because of the great improvements, has fallen in a decade in the UK from 88 to 55. In the USA, even in a year, the number has been reduced by eight. Bigger and better blast furnaces are being designed and built.

A representative furnace is shown in fig. 3.8. It is a very strong structure of welded steel on deep foundations. At the bottom is the hearth of strong refractory, and on this the molten iron eventually rests and is withdrawn through the notch opening. Several feet up is the notch for the withdrawal of slag. The main volume of the furnace holds coke, ore and limestone. Air, enriched in some cases with oxygen and steam, natural gas, liquid fuel or powdered coal, is blown in through water-cooled pipes called tuyeres to create the combustion. This air is heated beforehand in stoves in which the hot gases from the top of the furnace are utilized. The shape of the furnace is characteristic, being wider in the part called the bosh some feet above the tuyeres, in order to with-

IRON ORES AND THE MANUFACTURE OF PIG IRON | 145

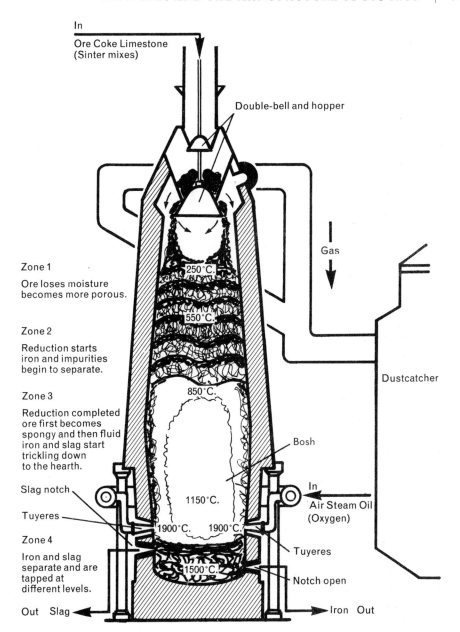

Fig. 3.8 Cross-section through a blast furnace.

stand the expanded volume of components, and then tapering gently towards the top. The loading of the burden is effected by an ingenious double-bell-and-hopper arrangement. Metal plates cooled by water are inserted at intervals through the shell and into the lining.

The smelting process, like many other processes in the iron-and-steel industry, involves very high temperatures, ranging from about 2000°C in the neighbourhood of the tuyeres down to a few hundred degrees towards the top of the shaft. So the linings of steel walls must resist heat, i.e. they must be highly refractory. They must therefore resist melting and spalling–the breaking-away of surface pieces to present new surface for corrosion. They must be strong and resistant to abrasion. They should be as impervious as possible. They must resist chemical action by gases involved in the smelting process. Such gases at high temperatures are very reactive. Even iron oxide could combine with the refractory material, to make it melt and flow at a high temperature.

There is therefore an involved technology of refractories in the industry, part of the technology being concerned with the mining of suitable raw materials needed in bulk. What might seem ideal in a small laboratory experiment cannot be applied in a blast furnace because of the cost of processing, a not inconsiderable part of the cost of smelting. There are some rocks, such as sandstone, which can be hewn into blocks for the refractory lining of furnaces but in general the unit used is manufactured brick or block. Improved science and technology has led to the design of linings in modern blast furnaces which will last long enough to produce millions of tons of iron, which means that the linings last for years. Though this seems a long time, it is finite, and there does come a time when a furnace must be relined.

The refractory linings are several feet thick. The commonest material is brick made of fireclay but there are variants for different zones. At the top of the shaft super-duty brick is used, the densest and least porous of all fired fireclay bricks. Below this there may be high-duty brick down to the broadest part of the bosh. The bosh lining is frequently high-duty brick, but carbon blocks based on anthracite have been used. The hearth wall is usually of this carbon block. The hearth itself may be super-duty fireclay or large carbon blocks, 12 ft deep or more. Cooling efficiency of the whole furnace will vary with the type of lining and some technology has been devoted to this problem. Refractory bricks of high alumina have been considered for parts of the lining. The most troublesome part of this lining is the wall of the bosh, especially near the tuyeres, and of the upper hearth region.

The blast furnace is only the main building of a complex whose purpose is the production of pig iron. The fuel of the operation is coke, which is made from coal in coke ovens, which produce a number of by-products to be used elsewhere. Limestone is used to make the unwanted gangue melt and flow. The effluent gas from the top of the furnace is cleaned and then used as fuel to heat the blast air produced by powerful blowers. The ore is sintered or pelletized. The ore, coke and limestone are mixed. The complex can thus consist of the blast furnaces, coke ovens, gas-cleaners, blower house, blast-heating stoves, sinter or pelletizing plant, mixers and conveyor systems for bringing the raw materials to the mixers, conveying the charge continuously to the top of the blast furnace and taking away the pig-iron and slag. When new complexes are built on the coast there is the added machinery for unloading and conveying. Not all complexes have all the parts listed; sintering and pelletizing are frequently done at the ore field.

The physico-chemical processes in the blast furnace are not completely understood. The overall operation is to reduce iron oxide to iron. A number of the reactions that occur in various zones of the furnace can be represented by chemical equations, but these do not give much information. They are *post facto* symbolizations of what is observed to happen and they thus tell only the quantitative relationships. Consider the equation:

$$2C + O_2 = 2CO$$

This seems simple and definite. What it does not show is that the reaction is exothermic, i.e. it gives out heat. So does the reaction shown by

$$C + O_2 = CO_2.$$

The reaction between steam and carbon, however, needs heat to be supplied; it can be shown thus:

$$C + H_2O = CO + H_2$$

The reaction is endothermic.

The essential points are that the reaction of carbon (in the coke) and water, though endothermic, produces reducing hydrogen as well as reducing carbon monoxide. The two combustion reactions produce reducing carbon monoxide and non-reducing carbon dioxide, which at high temperature reacts with carbon to produce more carbon monoxide.

The reduction changes the magnetite and ferric oxide to ferrous oxide FeO. The reduction equations then become:

$$FeO + CO = Fe + CO_2 \quad \text{exothermic}$$
$$FeO + H_2 = Fe + H_2O \quad \text{endothermic}$$

These reactions occur in the upper parts of the furnace where the temperatures are comparatively low. They are sometimes called indirect reduction because the reducing gases have to be produced by earlier reactions. Lower down in the furnace the carbon reacts directly with ferrous oxide according to the equation:

$$FeO + C = Fe + CO \quad \text{endothermic}$$

There is debate about which is the preferred type of reduction, and general opinion is that there should be considerably more indirect reduction than direct reduction for the more efficient operation. As the metallic iron is produced it drips downwards towards the hearth.

The charge obviously contains a number of other compounds beside iron oxide. There are alumina, phosphate, manganese oxide, magnesia, silica, and compounds of other elements that may be present in some ores, e.g. titanium, copper, chromium, nickel, or even vanadium, tin, selenium, tellurium, zinc, barium or arsenic. All exist eventually as oxides. Sulphur is introduced in the coke. Lime is formed by dissociation of the limestone. Carbon is present as coke. The fate of all these in the fiery furnace where there is so much fierce heat and movement, affects the quality of pig-iron produced.

The lime reacts with the silica and other oxides to make the mixture melt at a high temperature to form a sort of molten glass. This is the slag. Like the metallic iron it drips downwards towards the hearth to form a floating slag on the surface of the iron. However, carbon does not dissolve in the slag but it does dissolve in the molten iron, which therefore carries carbon down with it. At the same time in the hottest parts of the furnace some part of some of the oxides is reduced. A large amount of manganese oxide is reduced and gets into the iron as metallic manganese, the rest remaining in the slag. All the phosphorus oxide is reduced and elemental phosphorus gets into the iron. So does some of the silica, in the form of elemental silicon. All these reductions occur low down in the furnace where the temperature is high. Sulphur combines with iron higher up and descends with the slag and is reduced at hearth level, so that calcium sulphide gets into the slag and sulphur into the iron. Some other elements present also get into the iron.

Efficient action of the blast furnace depends on many factors. There is the temperature of the blast air. The higher this is the less coke is necessary to make the exothermic combustion reactions to create heat. There is the physical state of the burden, which if it fuses makes irregular passages for the reducing carbon monoxide. Pellets provide surface and gaps for the gas to do its work and are porous enough for the gas to penetrate. Sinter is very porous and the pieces do not fuse together. There is also the speed of air. By the introduction of a throttle valve in the gas outlet it is possible to build up top pressure inside the furnace, making a denser and therefore more effective gas and slowing it down so that more is used and less escapes at the top. Fuel oil, powdered coal or natural gas added to the blast intake have been tried in some furnaces about the world with a consequent reduction in the amount of coke, which is expensive, and an increase in the iron-ore capacity.

The reactions of the slag involve complicated physical chemistry. Some compounds, such as manganese oxide, increase the viscosity so that it does not readily flow. Furthermore the diffusion rates in the molten slag are reduced and the chemical actions are consequently slowed down. If there is an excess of silica and alumina over magnesia and lime the slag is said to be acidic. An excess of magnesia and lime makes the slag basic (in the chemical sense), in which case more sulphur is removed from the iron into the slag.

Every few hours the blast furnace is tapped, the slag being removed first and the iron next. The slag is a molten mass of glass, and when it cools is a very hard material which can be used for a number of purposes. Broken up, it makes ballast for railway tracks and metal for road making. Granulated, it makes material for cement. It can be used as a filter-bed for cement. Melted and blown into wool or allowed to form filaments it makes an acoustic insulator. It makes soil conditioners.

The white-hot molten iron is either poured into cars to be carried to the steelmaking plant, or into receptacles to cool and make pigs. It is obviously not pure iron. It can contain, depending on the iron ore used and the smelting conditions, from 3.0 to 4.0% of carbon, 0.15 to 2.5% of manganese, up to 0.2% of sulphur, 0.025 to 2.5% of phosphorus and 0.5 to 4.0% of silicon, and sometimes a fraction of 1% of other elements. It is a very hard material that is brittle, and in former days was used as a structural material and called cast iron. Today, however, cast iron is the result of further treatment on the pig iron. When the pig iron is used for steel-making, as most of it is, certain types, according to composition, are better for some processes than others. If its silicon content is low and its phosphorus high, the iron is said to be basic, i.e. usable in basic steel manufacture. If the silicon content is high it is called foundry iron. The blast furnace is also widely used to produce ferro-alloys, which are addition agents for steel-making. They consist of iron, rich in certain elements such as manganese. The two most produced are ferro manganese and spiegeleisen, the latter having more silicon and less manganese. These and other ferro-alloys for steel-making are also made in special furnaces.

3.2.5 Other iron-producing processes

Many pioneers have contested the dominance of the blast furnace for a number of reasons. First, the blast furnace process is complex and crude, and it cannot be said that all the physical chemistry is fully understood. Secondly, coke-making is expensive; so where there is cheap natural gas or oil or electricity there are economic attractions in alternative processes. Thirdly, there are local conditions where the type of ore available is not best treated by the blast furnace.

The aim is to produce iron from ferrous oxide (made by reduction of haematite and magnetite) by one or more of the routes shown by the equations:

$$FeO + C = Fe + CO$$
$$FeO + CO = Fe + CO_2$$
$$FeO + H_2 = Fe + H_2O$$

The routes are often called direct-reduction processes, but in several instances, though the final reduction is as shown (as it is in the blast furnace), the reducing agents are produced secondarily from other primary fuels or materials. The distinction between direct and indirect reduction thus ceases to have meaning. However, the term direct-reduction is still very widely used to mean the processes that are used without a blast furnace to produce iron from iron ore.

Probably about 40 systems have been invented over the years, some of them quite a long time ago. About half of them were tried and abandoned and belong merely to history, Of the remainder many produce only tens of tons a day, and need not be considered as serious challengers to the blast furnace. A certain number have survived and are increasingly used commercially. There are others still in the development stage.

One technique uses electrical energy for smelting and the reduction is effected by solid carbon in the form of coal, coke or charcoal. In the Tysland-Hole process three electrodes constitute the upper conductors, leading the electricity into the slag. The melting with carbon effects the reduction, and the product is molten pig iron. There are several variants though the main process is the same. The Dwight-Lloyd-McWane (DLM) process in the USA combines the making of sintered pellets with electric smelting. The Strategic-Udy process developed also in the USA combines a rotary kiln with the electric smelter. In the smelter the main gas produced is carbon monoxide, which is used for prereduction of the ore in the kiln. A number of countries in Europe and Asia make use of the Tysland-Hole electric-smelting furnace, which was invented in Norway.

Many processes produce solid iron whether in the form of sponge or small particles. The oldest-established is the Krupp-Renn process which originated in Germany and is used in several countries. It was invented to make use of low-quality siliceous ores. The ore, in small particles, is fed with the reducing carbon into the upper end of a slightly-inclined rotating kiln heated by the burning of fuel at the lower end. A purity of over 90% iron is claimed but there can be considerable sulphur in it according to the quality of the ore. The product is iron and a highly-viscous slag in lumps. These are cooled and crushed, and then the iron is magnetically separated from the slag. Sometimes the product is used as feed for a blast furnace. The RN process (Republic Steel Corporation and National Lead Company) is a modified version in which a basic slag is used, and there is fuel economy that allows use of reducing gas as well as carbon. The final product is again particulate, and it has a low sulphur content. The SL/RN system is a combination of the Stelco-Lurgi and RN systems of rotary kiln, and the first commercial version was built in Korea using anthracite as the reducing solid. Several companies in Canada and Japan have expressed intention of using the system.

The well-known Swedish Höganäs and successful Mexican HyL (Hojalata y Lamina Steel Company) processes make use of retorts as their reducing vessels. The former uses continuously-moving retorts containing coke breeze as the reducing agent. It is a slow low-temperature process but the product is a very pure iron with very low phosphorus and sulphur. The HyL uses gas as the reducing agent, a mixture of carbon monoxide and hydrogen produced by the catalytic reforming of natural gas. The product is sponge which has carbon added to it to make it suitable for steel-making. The process, which has one of the highest production rates, at 500 tons a day, of any process other than the blast furnace, is said to have been developed to produce a substitution for steel scrap, which is very expensive in Mexico.

The Swedish Wiberg-Soderfors process takes place in a shaft furnace not unlike a small blast furnace. The ore is pelletized and the reducing gas, made in ancillary carburettors, consists of carbon monoxide and hydrogen. Sulphur is removed from it by passing it through limestone or magnesia-bearing stone. The gases are heated before use and are regenerated when they have become oxidized as a result of their reducing action. The product is sponge iron low in phosphorus and sulphur and suitable for steel-making in arc furnaces. The several furnaces in use produce some 500 tons a day altogether. The Purofer system, supported by the European Coal and Steel Community, also uses reformed natural gas or coke-oven gas in a shaft furnace.

The 'fluidizing' of fine particles by blowing gas up through them is a known engineering technique and has been utilized in several iron-making processes, especially in the USA. One has been developed by the Bethlehem Steel Company and Hydrocarbon Research and is known as the H-iron process, the reducing agent being hot hydrogen produced from natural gas or coke-oven gas. This gas moves upwards under pressure through successive beds of small-size ore. The United States Steel Corporation produce Nu-iron by hydrogen reducing in a fluidized bed. The ADL process, devised by Arthur D. Little and Esso Research in the USA uses reducing gas made from air and natural gas. All these processes produce a finely divided iron which is agglomerated into briquettes. They have been considered together because of the one common feature of fluidization, but they differ in the pressure of gas used and the details of fuel economy.

Simple chemistry shows that reduction of oxide is not the only route to metallic iron. One new system being tried in prototype turns iron-bearing raw material, whether in scrap or ore, into iron chloride in solution. This is crystallized and reduced by hot hydrogen. The product is a sponge iron. Yet another novel system utilizes electrolysis in an iron solution. A new method has been developed by the United Kingdom Atomic Weapons Research Establishment in England, as a result of its experience in the making of nuclear fuels as oxide pellets. The iron compound is thrown out of solution as a gel, which is dried and then reduced by gas. A wide range of particle sizes and shapes can be produced.

Many of the processes briefly outlined above do not produce molten pig iron. They produce sponge or powder. The sponge is sometimes crushed into powder for magnetic separation, and sometimes is used directly as a raw material for steel, especially in electric furnaces. The powder is in some cases briquetted for remelting for steel-making. There is an increasing use of powder metallurgy for the direct use of iron powder to make final products by sintering. Some of the processes included under the heading of direct reduction are beneficiation techniques, and give a product which is used in improved blast furnaces. The proportion of iron and its purity varies from process to process, and these processes vary according to the reducing agent, the type of product, the economy of fuel and the quantity and speed of output. Several already produce 500 tons a day, and novel processes claiming even practically 100% iron purity are invented from time to time. The final test, apart from suitability for local needs and conditions, is economy. As far as is known at present the blast furnace is the only producer of pig iron on a scale usable by the world steel-making industry.

3.3 STEEL-MAKING

3.3.1 Steel-making in general

There is no one material called steel but there are hundreds of steels. They are all basically iron and contain carbon that ranges in different steels from less than 0.1% to 2.0%. They differ in their microscopic grain structure, the proportion of carbon and how it is combined, the crystalline form of the iron and other metals, the tiny amounts of impurity present (silicon, sulphur, phosphorus etc. and 'trace' elements) and the quantities of alloying elements (chromium, manganese, nickel etc.). These differences in structure and content account for differences in properties such as tensile strength, hardness, toughness, cold-working, response to heat treatment, resistance to creep, fatigue limits, ability to withstand corrosion, behaviour at high temperatures, magnetic behaviour, hot-rolling characteristics and so on. Many countries have standards and codes to express these properties, such as the British Standard Specifications and the American Iron and Steel Institute's tabulations, but they are not easily comparable except by detailed inspection, so that even the most knowledgeable of steel experts has to refer to tables for the properties of steels other than those he regularly works with. Many steels also have proprietary trade-names that further confuse the general picture.

The main raw materials for steel-making are pig iron and scrap steel. The latter is an important part of the industry. In the United Kingdom nearly 14 million tons of scrap were used in 1968. In the USA the annual consumption is well over 50 million tons a year. There are three sources. First there is the waste steel involved in the industry itself—rejected ingots, cuttings from plate, ends from bars, swarf from turning operations, ends cut from blooms and billets. There is also the wastage from industries that make steel products. Finally there are obsolete and discarded products such as washing machines, motor cars, tin cans, building steel and so on. All the scrap must be carefully graded in order to avoid the adding of unwanted elements to the steel melt. It must also be cut and squeezed into suitable sizes for the steel furnaces. As waste is normally a loss, the use of scrap is obviously a necessary economy because it is cheaper than pig iron in industrialized countries. The steel industry thus consumes its own waste, and in countries without a cheap source of scrap, relatively pure iron is sometimes produced separately to make, so to speak, synthetic scrap for steel-making. In certain steel-making processes scrap also serves to reduce the temperature of the melt.

Additive agents are also used in steel-making, the best known being ferromanganese and the one that appears so much in history is spiegeleisen, which has less manganese and more silicon. There are many such agents, and because they were originally produced in the blast furnace and contained considerable iron they are often referred to as ferro-alloys (see chapter 4). They include ferrosilicon, ferroselenium, ferrochromium, ferrotitanium and ferromolybdenum. A considerable number of additive agents, however, contain no iron. Their purposes vary according to what is required. Some are deoxidizers added to remove oxygen created by too much oxidation. Others add certain qualities of hardenability, corrosion resistance, machinability, or grain size.

Other important chemicals involved in steel-making are those that are used as fluxes, slags and refractories. The purpose of a flux is to affect the physical-chemistry of compounds coming out of the molten metal so that they flow as a molten glass. The slag is the rejected material that floats on the liquid metal. Refractories are materials that withstand very high temperatures and are used for the linings of furnaces. There is a continued series of chemical reactions, during steel-making, between fluxes, iron salts and refractories. It is involved and much of it is still so little understood fundamentally that it is the subject of controversy. The action of a flux can be most easily understood by the realization that solutions have lower melting points than the individual constituents, a known fact of elementary physics. In the case of oxides used in steel-making, for example, lime (CaO) melts at 2 580°C and silica (SiO_2) at 1 710°C, but a mixture of them is liquid at temperatures well below 2 580°C, and at one mixture (about 64% silicon) the whole is liquid at 1 436°C, some 274°C below the melting point of silica. The fundamental action of a flux is in breaking up the formation of very large molecules so that the ions are left free to move in a random way. Not only must a flux aid flow so that slag can float with a clean surface between it and the metal, but the viscosity of the floating slag must be such that diffusion of ions and radicals is enough to ensure the desired chemical action. The molten slag has powers of solution to hold the compounds formed. It also serves as a heat insulator, and if thick enough can prevent heat getting to the metal below. The physical chemistry of fluxes and slags is very involved and has developed mostly on empirical lines, but much research in the past several decades has clarified many of the mechanisms, not the least of which are chemical and phase equilibria. The first is concerned with reversible chemical actions which cease to progress in one direction or the other when equilibrium is reached, and equilibrium depends on concentrations and temperature. The second type of equilibrium is that between the phases of mixtures, i.e. in the solid, liquid or gas phases. This type of equilibrium has been much studied in connection with the oxides that make up slags and refractories, each study leading to phase diagrams which are an essential part of the physical chemistry of steel-making. Refractories come into the same field of study, for their action, which seems simply that of remaining solid at high temperatures, is also complex and a small amount of another oxide can act as a fluxing agent to make the refractory bricks actually flow. Iron oxide, for example, has a fluxing effect on silica. Furthermore, the slags react chemically with the refractories. Steel-making indeed is a multi-component, multi-phase, many-temperatured complex, and it can safely be said that had steel-making depended on fundamental knowledge of the reactions involved it would never have developed.

One of the principles that have emerged from the complexity is that of the significance of 'acid' and 'basic' fluxes, slags and refractories. Practical knowledge of these is essential to steel-making. The acid components are those which act in melts just as they would in water, to make acids, and salts of those acids. Thus silica (SiO_2), phosphorus pentoxide (P_2O_5), alumina (Al_2O_3) and others form silicates, phosphates, aluminates and so on. They are the acid components. Other oxides, however, such as lime (CaO), magnesia (MgO), manganese oxide (MnO) and ferrous oxide (FeO), would make hydroxides in water and be alkaline, and so are basic components. In melts they break down the molecular complexes of the acid components. Lime is made from natural limestone; magnesia can come from the sea, dolomite, magnesite and forsterite. Alumina occurs in a number of natural rocks. The basicity of a slag is given commonly by the ratio of the percentage of lime to the percentage of silica, though this simple ratio must be modified by bringing in magnesia and phosphorus pentoxide where the quantities of these are important.

A material important in steel-making though rarely concerned in the chemistry is water. Something like 40 000 gallons (180 000 litres) of water are required in the making of 1 ton of steel, chiefly for cooling purposes. Fuels such as coal, coke, natural gas, oil and electricity enter importantly into the economy of steel-making, and, in a general consideration of the industry, the local availability of certain fuels and their cost affect the decision about which methods of iron-making and steel-making are to be used. The cost of refractories, especially when made into bricks, is another factor in the economics of steel-making.

The material which has had the most dramatic effect on steel-making is oxygen. Before the 1950s the quantity used was small; it came from bottles of liquefied gas and was expensive. As a result of the production of oxygen by the ton by plants erected at the steelworks, the price has come down by more than a factor of ten. The consumption in the USA in 1967 was 185 000 million cubic feet (5 240 million cubic metres), and in the UK 37 000 million cubic feet (1 050 million cubic metres), representing 7.5 million tons and 1.5 million tons respectively. In the UK of the 1.5 million tons used in 1967 about 11% went to blast furnaces, 17% to open-hearth steel furnaces and 40% to converters for steel-making. It is in this last usage that the great revolution is taking place, changing the whole industry.

The fall in the cost of oxygen by the ton followed the design by oxygen-producing companies (such as the British Oxygen Company in the UK and companies of similar status in Europe and the USA) of large plants producing hundreds of tons of oxygen a day. Currently plants of 600 tons/day are operating in the UK and plants of over 1 000 tons/day are under construction in the USA. A schematic diagram of a British Oxygen Tonnox low-pressure plant is shown in fig. 3.9. The principle of the system is to liquefy air by cooling it by sudden expansion using an expansion engine (see section 2.1.2, vol I). The components of the liquid air boil off at different temperatures, a fractional distillation, called in the industry, rectification. The nitrogen goes to waste and the oxygen is pumped round a pipe system to where it is needed. Sometimes systems that involve liquefaction and storage of the oxygen are used to act as a stand-by supply in case of failure, but such internal-compression plants are in general more expensive. A plant installed at a steelworks is shown in fig. 3.10.

The overall chemistry of ordinary steel-making is the removal of carbon to the extent desired by oxidizing it to carbon monoxide. At the same time other unwanted elements such as silicon and phosphorus are oxidized, and the resulting silica and phosphorus pentoxide go into the slag. As already stated the process is a complex one in which the temperature, gas-pressures, slag nature and volume, refractory nature, and the composition of the raw material all enter. The first need is to have the raw materials molten, after which the refining of the iron begins in order to lead to steel. The heat can be supplied by the combustion of fuel or by electricity, and it can be supplied by oxidation reactions which are exothermic and give out energy.

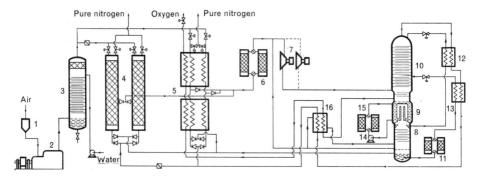

Fig. 3.9 Tonnox low-pressure plant. 1. Air filter, 2. Air turbo-compressor, 3. Direct cooler, 4. Air/nitrogen regenerators, 5. Reversing exchanger, 6. Air purifier, 7. Expansion turbine, 8. Lower column, 9. Condenser, 10. Upper column, 11. Adsorbers (liquid air steam), 12. Liquid nitrogen sub-cooler, 13. Rich liquid sub-cooler, 14. Liquid oxygen circulating pump, 15. Adsorbers (liquid oxygen), 16. Liquefier.

Before considering in detail the different methods in use today in steel-making, it is important to realize their place in the steel-making industry, for though it seems obvious from first principles that the most direct method of oxidation of carbon is to supply oxygen to the hot molten raw material, this usage is quite modern and is only now beginning to oust the traditional methods from the industry. The most widely used method is that of the open-hearth (O.H.) furnace. Half the world's steel is still produced in such furnaces, but there are signs of their decline. As recently as 1961 world O.H. production was 62% of the total. Now it is down to about 50%. Practice varies from country to country. In the UK between 1961 and 1967 the proportion produced in O.H. furnaces fell from 82% to 60%. In the USA it fell over the same period from 85% to 64%. It is now about 55%. In Russia it has changed very little and the O.H. proportion of the whole is still over 80%. In West Germany the O.H. proportion has fallen slightly from 47% to 40%.

Historically the first of the so-called pneumatic methods in which air or oxygen is used for oxidation was that invented by Henry Bessemer and improved by Sidney Gilchrist Thomas; in some countries it is still called the Thomas process. The container in which the Bessemer process goes on is a converter, and so are all the other vessels used today for the pneumatic refining of iron to produce steel. For various reasons, mainly the nature of the available ore, the Bessemer process has been more used in some countries than in others. In 1967 there were only 2 Bessemer converters left in the United Kingdom. In Russia, Bessemer production remains only a very small proportion of the whole. In France in the same period the proportion has fallen, but it still accounts for half the country's steel production.

Production by electric furnace or direct-oxygen process has on the other hand increased noticeably in the past decade. Production by basic-oxygen processes is expected to equal that by the open-hearth furnace in the USA very soon, and then go on to surpass it. Oxygen-process production of steel increased 10-fold there from 1961 to 1966, while its electric production was doubled. In West Germany in the same period oxygen-process production increased 7-fold. In the UK, oxygen-process production increased 3-fold between 1964 and 1967.

STEEL-MAKING | 155

Fig. 3.10 An Oxygen plant installation at a steelworks.

In most of the steel-making processes there is a distinction between those that are basic and those that are acid. The terms have the same meaning as previously explained above and refer to the refractory linings and slag used. Thus there are the basic open-hearth and acid open-hearth processes, the basic Bessemer and the acid Bessemer, the basic oxygen furnace and so on.

It is worth while to consider generally the two that began the tonnage-steel revolution – the Bessemer and open-hearth, or Siemens-Martin, processes. Both arrived at nearly the same time, but there is a long and complicated story of litigation and wrangling about the priorities and patents involved in each. The importance of Sir Henry Bessemer and Sir William Siemens can be judged by the figures. In 1855 steel production in the UK was not more than 60 000 tons and it cost about £75 a ton. In 1890 production was 3 million tons, about half from the Bessemer process and half from the open-hearth process, and the price was about £6½ a ton. In the same year US production just exceeded UK production for the first time, and the combined production was four times that of the rest of the world. These facts are the measure of the importance of the two men.

Both were extraordinary inventors. Bessemer filed 117 patents between 1838 and 1883, covering copper-plating, a perforated die, glass production, sugar-cane machinery, bronze powder machinery, the rifling of gun barrels and diamond cutting. Siemens filed 113 patents covering a differential governor for engines, glass-making devices, telegraphic communication, electric lighting and power, railway apparatus and explosives. Bessemer patented the use of air to refine iron, and a tilting converter in 1855, and his device for converting pig iron to malleable iron in 1856. In August of the same year he delivered his sensational paper to the British Association on 'The Manufacture of Iron without Fuel'. Siemens patented his application of the regenerative principle to furnaces in 1856, and built his own first experimental furnace in 1857. In 1867 he patented 'Improvements in Furnaces and in Processes and Apparatus in connection therein.'

There were essential differences between the two processes. Bessemer used air, with its natural supply of oxygen, for the oxidation of molten pig iron, to reduce the carbon and silicon. He aimed at producing malleable iron. He blew air in through the bottom of a steel vessel containing molten pig iron. The temperature was high enough to allow oxidation reactions, which were exothermic, and thus the temperature rose rapidly. It was all over in twenty dramatic minutes some of which he himself describes thus: 'A roaring flame rushes from the mouth of the vessel, and as the process advances it changes from violet colour to orange and finally to a voluminous white flame. The sparks, which at first were large like those of ordinary foundry iron, change to small hissing points, and these gradually give way to soft floating specks of bluish light as the state of malleable iron is approached.' He was describing the acid Bessemer process and anyone who today sees a Bessemer 'blow' would agree with his description.

Bessemer and his process immediately ran into trouble. An American metallurgist, William Kelly, claimed priority with some justification but was eventually disowned by his compatriot steelmasters. And technically there was trouble. For one thing the licencees had nothing like the success demonstrated by Bessemer himself. The steel was said to be 'red short', i.e. producing cracks when being forged red-hot. This difficulty was overcome by Robert Forester Mushet (1811–1891) in Wales, who added spiegeleisen. The effect was due to oxygen present in the Bessemer product. It was over-oxidized. The spiegeleisen reacted with the oxygen and iron oxide. Through various mishaps and something near sharp practice on certain people's part, Mushet failed to get a patent but Bessemer tacitly admitted his indebtedness by giving him a pension of £300 a year for life. Mushet died unhonoured by any society or establishment. But there was another more serious trouble with Bessemer steel: the presence of phosphorus. In ignorance of the physical chemistry involved nobody understood this; Bessemer had been lucky with his first low-phosphorus iron and his process was successful when, as in Sweden, there was low-phosphorus ore. The man who solved the problem, an achievement much appreciated by Bessemer, was Sidney Gilchrist

Thomas. He worked out a way of making the convertor 'basic' by using lime, and eventually magnesia, for the lining as well as using lime in the slag. The Bessemer process used from 1857 to 1880, or thereabouts, was in fact an acid process which would not work with high-phosphorus pig iron, but afterwards the basic Bessemer process was used widely in many countries.

3.3.2 The open-hearth furnace

This type of furnace has for nearly a century been the main source of steel in the world, but its use is slowly decreasing and as furnaces arrive at the stage where the capital investment has been repaid and they are ageing they are phased out of the steel-making cycle, to be replaced by more modern steel-producing devices. In normal use open-hearth furnaces are built in a row, the working platform for the steel-maker being at hearth level, but the floor of the main building behind the furnaces is considerably lower, and carries rails and overhead cranes for the ladles that are filled when a furnace is tapped. Sizes of furnaces have increased to a capacity of even up to 600 tons but these are rare.

There are two essential features of the open-hearth furnace. First it is reverberatory: there is a shallow curved roof which reflects flames and gases downwards as well as emitting heat radiation. The second is more important: it is that the furnace is regenerative. It is this conservation of heat energy that allows a burning fuel to create enough heat to melt the metal and slag. It is a simple idea, but in practice it makes the whole furnace complex a massive affair. The burning gases do not escape up the chimney and so waste energy by warming up the atmosphere; instead they are made to go, after traversing the furnace over the melting metal on the hearth; through an open-work structure of bricks, called checkers, which absorb heat and store it. Then the next intake of air and fuel gas is made to go into the furnace by way of these checkers, which give up heat and thus preheat the air and fuel gas. A similar set of checkers exists at the other side of the furnace to absorb heat from the outgoing flames and gases. Then the direction of intake of fuel gas and air is reversed again. Thus by reversing the direction in which the flames pass over the 'heat' or metal at regular, calculated intervals, enormous quantities of heat are conserved and combustion, which itself consists of exothermic reactions and supplies heat, is made easier. The checkers are under the charging floor and, as the furnace operator faces the front charging door of the open-hearth furnace, the direction of flow is from his left to his right and then from his right to his left and so on.

The furnace is a robust structure of steel, concrete and brick. And it emits heat everywhere; about a third of the heat supplied by combustion and chemical action is lost through doors and other parts of the structure by escape of hot gas and by radiation. (In the old days when the steel-maker was the key craftsman working long shifts, beer was constantly supplied to make up for sweat lost by the men). About one sixth of the supplied energy is allowed to escape through the chimney if there is a by-product usage of waste heat to raise steam. If there is no such use the wastage is about 40%. This leaves 25% as the efficiency in the utilization of heat energy for the actual steelmaking.

The structure is illustrated in fig. 3.12, which is a sketch of an open-hearth furnace, showing the reversal of flow and the regeneration. As with all steel-making processes, one serious problem is the maintenance of the lining structure, especially the roof, at temperatures that range between 1300°C and 1600°C. The inside is in fact an inferno

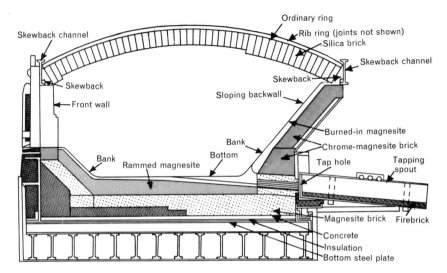

Fig. 3.11 Vertical cross-section through the tap-hole of a modern basic open-hearth furnace (with silica brick roof) indicating materials of construction of the various parts. Buckstays and other binding elements are not shown.

of heat and chemical action and very few linings last long. None lasts indefinitely. The lining serves several purposes. First it is an insulator. No unlined steel furnace could be used; it would be red-hot and melting in a very short time. Second, it is refractory, that is to say a material that will be solid at temperatures well above red heat. Thirdly, it is involved in the chemistry of exchange between molten metal slag and lining. A silica lining, for example, being an acid one, would be immediately attacked by a highly basic slag.

a The basic open-hearth furnace. The linings of the hearth furnace are shown in fig. 3.11, in the biggest furnaces the hearth can be several feet thick. The structure of the hearth being comparatively shallow, means that a very large area is in contact with the slag, and this high ratio of surface to volume is an essential part of the action. The steel charging doors are water-cooled. There are openings for the removal of slag. Many open-hearth furnaces of today can be tilted for the pouring of metal and slag. The fuel used to provide the heating flame, originally coal, has passed through various changes from the use of producer gas to coke-oven gas and natural gas and hydrocarbon fuels, which can produce many more units of heat energy per unit volume or weight. Practices vary throughout the world according to the economics of fuel available. Oxygen can be used with the fuel.

Open-hearth practice varies so much according to the quality of steel required and the quality of raw material economically available, that it is impossible to give detailed comprehensive descriptions of the process. It is a process in which individual variation can happen even within a single steelworks. (What is called 'floor practice' covers, in every steel process, the individual variations in charging, testing, pouring etc.)

The raw materials consist of pig iron, scrap steel, limestone and iron ore or other oxide of iron such as mill scale. The constitution of the pig iron must be accurately known, so that the right amount of other constituents can be calculated in order to

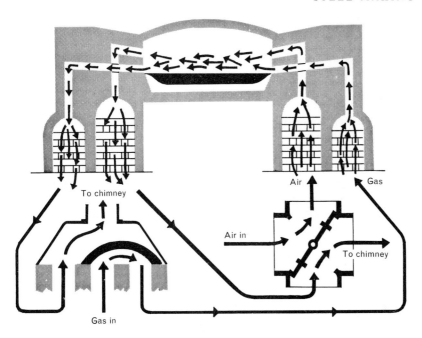

Fig. 3.12 Reversal of flow through an open-hearth furnaee.

refine the charge, to produce a steel that is economical and as nearly as possible a pure iron with a known percentage of carbon. (This knowledge of the constituents applies to all steel-making processes, in order to produce the hundreds of different sorts of steel.) Five possible sorts of charge are possible, though in general two are used. One charge is an entirely molten metal, but this is rare, for where there is excess blast-furnace capacity it is preferable to use some in a converter and then add this 'blown' molten metal to molten pig iron for the open-hearth charge, which thus becomes the second type of charge: liquid pig iron and liquid steel. The charge may also be wholly scrap metal in solid form, in which carbonaceous material is added to provide enough carbon for the final product. This practice is followed only where there is a shortage of pig iron.

The two common charges for the basic open-hearth furnace are: (1) solid scrap and molten pig iron; (2) solid scrap and solid pig iron, this method being compulsory at a steelworks without a blast furnace. Ore is not always used but when it is, it is a way of supplying oxygen, ore being iron oxide. If the scrap is light, such as turnings and sheet edgings etc., it will oxidize quickly in the furnace and present a large surface area, and its low mass demands more pig iron or less ore, whereas heavy scrap having a high mass needs less pig iron or more ore. Light scrap takes longer to charge in sufficient quantity while heavy scrap takes longer to melt. These considerations in the preliminary charge are some indication of the variables that make the process impossible to describe comprehensively and accurately in detail. The quantity of limestone depends on the amount of silicon, sulphur and phosphorus in the charge. The limestone is charged first, then the ore, if any, then the scrap. If the rest of the charge is solid pig iron it is added either immediately afterwards or when the scrap is partly melted.

If the rest of the charge is molten pig iron, however, it must be added at an accurately-judged time, when the scrap is sufficiently oxidized to be a valuable source of oxygen, and hot enough not to chill the molten additive, and not over-oxidized to get rid of too much carbon in which case there is not enough finally to constitute the steel required.

The first stage is the melting down. If the charge includes cold pig iron the whole charge has to be melted. If, as is more usual, the pig iron is to be added when molten, the scrap, perhaps 100 tons of it in a 200 ton furnace, begins to melt and oxidize to iron oxide, the source of oxygen being in the flame oxygen. When the molten pig iron is added, reactions increase. Silicon is oxidized to silica, which rises to the surface and is the first stage of slag. Manganese, if any, forms manganese oxide which also goes into the slag. By this time the carbon is being oxidized as well and this process speeds up, so that carbon monoxide bubbles to the surface, creating the condition known as the 'ore boil' or 'iron-oxide boil', because the source of oxygen is largely in the iron oxide, whether this is in ore or scale, though other sources of oxygen exist in the oxides of the slag. If the proportion of scrap in the charge is high the ore boil is somewhat suppressed. According to the contents and proportions of the original charge the slag may foam, or in any case get too thick for heat to penetrate it quickly enough. In this case, there is a preliminary run-off of slag, removing much silica and manganese oxide and iron oxide. When the heat penetrating to the bottom of the molten charge is enough, the limestone dissociates to form lime and carbon dioxide, which bubbles violently and creates the 'lime boil'. If the whole charge is cold to begin with, the melting takes longer and the lime boil is delayed. The ebullition mixes the charge, and lime rises to the slag, making it basic, while carbon dioxide oxidizes some carbon ($C + CO_2 = 2CO$). The charge becomes wholly molten and the refining period begins. Phosphorus is oxidized to phosphorus pentoxide, which is captured and held in the basic slag. Sulphur is removed by combination with lime. The carbon is still further reduced, making the melting point higher, which necessitates an increase in fuel rate and a more frequent reversal of air and fuel, in order to increase the heat supplied and maintain the charge molten. It is necessary at this stage for the carbon content to be higher than that needed in the final steel, in order to allow time for control of the process. During the various stages additives are introduced and changes in heating effected to get the correct slag and molten metal. One of the old sayings in the mythology of steel is: 'Look after the slag and the steel will look after itself.' Deoxidants are needed, either in the furnace or the ladle, to remove excess oxygen from the melt. At the correct time the tapping spout is cleared (no small operation in itself) and the steel is poured into ladles which transfer the metal to ingot moulds.

The process thus briefly described above is that in which iron ore is used, and is faster than one in which no ore is used at all. Practices vary, a common one being the use of mill scale in the charge to provide iron oxide for the oxidation. Representative timings for the whole heat, using ore practice as above, for the hot-metal process are:

Melting down	2.5 h
Molten pig-iron addition	0.5 h
Ore or iron-oxide boil	3.0 h
Lime boil	1.5 h
Refining or working period	2.5 h
Total:	10.0 h

This is not the time from tapping one melt to tapping the next, because some time is used in cleaning-up and preparing the hearth where necessary – the fettling time. So a tap-to-tap time for the above process could well be 12 hours or more. The time can be reduced somewhat by skill in the melting-down period and by the use of oxygen with the fuel.

The chemical actions of the basic open-hearth are indicated in the following equations.

$$C + O = CO$$
$$C + CO_2 = 2CO$$
$$Fe + O = FeO$$
$$3FeO + 2C = CO + CO_2 + 3Fe$$
$$FeO + C = CO + Fe$$
$$2FeO + Si = SiO_2 + 2Fe$$
$$FeO + Mn = MnO + Fe$$
$$5FeO + 2P = P_2O_5 + 5Fe$$
$$P_2O_5 + 4CaO = 4CaO \cdot P_2O_5$$
$$S + CaO = CaS + O$$
$$CaCO_3 = CaO + CO_2$$

These are merely representative of some of the reactions and it would be naive to think they are definitive, for many of the reactions inside the slag and between it and the melt and the refractory are excluded. Some are not even known. The whole process is seen to be a complex one, and yet the extraordinary truth is that the process as it has evolved has produced most of the world's steels of all qualities, from soft low-carbon steels to hard high-carbon steels – steels to reinforce concrete, make wire, build bridges and battleships. It is a flexible process subject to some control and variation, but with a tap-to-tap time of about 12 hours it is a slow process. The capacity of the big open-hearth furnaces, however, brings the rate of production up to some 17 tons and more per hour.

b. The acid open-hearth process. This is still used in a small way. In the UK in 1932 the steel produced in acid open-hearths was about one-fifth of total steel production; in the USA the proportion of acid open-hearth steel was down to 0.2% of the whole in 1966. The reason that it is still used at all is partly historical and economic, in that in the early days acid steel acquired a reputation for quality and some users still insist on it, and in that acid furnaces still represent capital investment and will not be demolished while they earn returns. There is some validity in the quality claim, in that the end of the process allows the correct carbon content to be maintained for some while without further oxidation, and this carbon content can be accurately controlled.

The structure of the furnace is different from that of the basic open-hearth furnace only in the hearth and linings, which are siliceous, the hearth being built up of fused sand on the refractory base. This means that care must be used in the process to see that basic oxides, such as lime and especially iron oxide, get no chance of reacting with the walls and hearth. No phosphorus is removed in the process, for if any gets as oxide into the slag, equilibrium is quickly established so that any further action sends phosphorus back into the metal. No sulphur is removed in the process. The pig iron used must therefore be very low in phosphorus and sulphur to start with. The care needed in the selection of the charge was one factor that led to steel of high quality in the early days. The product was said to be clean steel.

The charge consists of cold pig iron and steel scrap, which must also be carefully selected. A molten or hot-metal charge is not used. The fuel too is restricted because some fuels contain sulphur. During the melt-down, taking several hours, oxidation of silicon and manganese and iron takes place to make an acid slag, and some carbon is removed. The final state of this is the 'dead' melt. The source of oxygen is mainly in the fuel. A steady boil follows while carbon is consumed, iron ore and lime being added sparingly to assist oxidation and make iron oxide to make the slag less viscous. The steady bubbling effects some mixing and the removal of unwanted inclusions. Eventually deoxidants such as spiegeleisen or ferro-silicon or silico-manganese are added to remove oxygen. This 'blocks' the boil. As with basic open-hearth procedure, the skill of the operator in testing and controlling the various stages is very important. When the skill is high the resulting steel is of high quality, with fewer inclusions and less hydrogen than with the basic open-hearth process. Furthermore, some alloying of other elements such as molybdenum and nickel and silicon can be done.

The open-hearth furnace is sometimes used in combination with another furnace, such as a Bessemer converter, in a *duplex* process. Metal from a Bessemer has lost a lot of its silicon and can be used as hot metal in the open-hearth furnace, as already stated above. Such a combination can sometimes be economic and reduce the total time of steel production. It is favoured in some countries but is unlikely to increase as the open-hearth process is phased out. Any combination of two steel-making processes as one process is of course a duplexing procedure, such as the combination of a Bessemer and an arc furnace.

All the open-hearth procedures so far discussed are traditional, however much they have been improved and automated. The one more fully described above is that of 'ore practice'. A more recent development is the use of tonnage oxygen, which is a concentrated oxidant and can speed up an open-hearth process. The furnace has to be specially designed or considerably reconstructed, though the details need not be discussed here. One aspect, however, must be mentioned because it applies to a greater or lesser extent to all oxygen-lancing steel-making processes. It is the creation of dust and iron-oxide by intense oxidation. A fume collector must therefore be incorporated in the furnace complex. The oxygen is brought in through water-cooled tubes called lances. These may project through the roof or end-walls or back-walls, and much experimentation has been done to improve their design. One special method used by the Appleby-Frodingham Steel Co. in the UK is called AJAX. It is the only one in the world and named after the inventor Albert Jackson. In this the charge is entirely of molten metal and the oxygen lances are through the end-walls. The tap-to-tap time for a 200 ton furnace has, by this method, been reduced to an average of about 7 hours and the output increased. This increase in the speed of production has been accompanied by a reduction in fuel consumption, maintenance, and services. In Canada a 450 ton open-hearth furnace has been fitted with oxygen lances through the roof, and the tap-to-tap time is from 4 to 5 hours, giving a production rate of 90 to 100 tons per hour. In Japan there has also been development in the use of oxygen-lancing leading to claims of tap-to-tap times of less than 4 hours.

3.3.3 Converters

Modern methods using oxygen in large quantities involve vessels called converters. The original converter was that of Bessemer, who made one of the most dramatic impacts ever felt by the world's steel industry, when he showed that many tons of steel could be produced in a matter of minutes and without any fuel to provide heat. The bulk of steel produced in the world in the latter third of the 19th Century was Bessemer steel. In most countries it was outpaced by open-hearth furnaces, and in recent years its use except in some countries – e.g. France, Japan and Belgium – has rapidly declined. There are no basic Bessemer converters left in the U.K. and one firm only is keeping its two acid Bessemers. In the USA there have never been basic Bessemers, and the acid Bessemer production is down to 0.2% of the total. No new Bessemers have been built there since 1949.

   a. The Bessemer converter. A Bessemer converter is shown in section in fig. 3.13. It is a steel vessel, cylindrical for most of its length and lopsidedly coned towards the top. It is thickly lined with refractory. It rotates on heavy trunnions, one of which is hollow in order to convey air from a compressor to the wind box at the base of the vessel. The air-inlet is through vertical channels in the refractory base, channels called, as with the blast furnace, tuyeres. The action is fast and fierce and the refractories near the tuyeres and at the open mouth at the top are quickly eroded, an average life for the bottom being 25 to 30 heats or blows. The floor and linings in the acid process are siliceous, with underlinings of firebrick where practicable, but in the basic process, dolomite brick is used. The common size of acid Bessemer, perhaps 15 feet (4.5 m) in diameter and 20 feet (6 m) in height, will hold 25 tons of material. Bigger basic converters exist.

The procedures in the basic and acid Bessemer production of steel differ somewhat, though they have much in common. The shared fact is that air is blown in under pressure

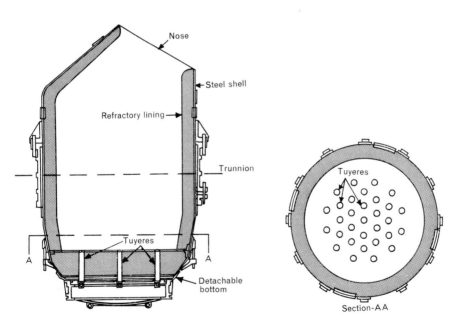

Fig. 3.13 Plan of bottom and section through a 25-ton accentric Bessemer converter, so-called because of the location of the opening in the nose. Concentric converters have the nose opening parallel to and concentric with the bottom. The total area of the openings in the 31 tuyeres of the bottom shown (each tuyere containing seven $\frac{7}{8}$-inch diameter holes) is 66.6 square inches

Fig. 3.14
A Bessemer converter 'blowing'.

through the bottom, and the heat and oxygen effect oxidation reactions which are exothermic, and so give out heat to accelerate other actions and the melting of the batch at higher and higher temperatures as the carbon is removed. The acid procedure starts with some scrap followed by molten pig iron, poured in when the converter is horizontal. The silicon content must be fairly high and the phosphorus and sulphur contents very low. The blast is turned on and the converter turned upright, whereupon oxidation of the silicon begins straightaway, creating a short transparent flame at the mouth. The rate of removal of silicon is high at first while the carbon is hardly removed at all, but gradually the carbon removal accelerates and the flame at the mouth lengthens and becomes whiter, and reaches eventually perhaps thirty feet into the air. This flame then gets shorter when the carbon content is getting very low and fans out and red appears in it. Soon the whole flame is reddish and short. Keen observation of the flame tells the skilled operator what is going on inside the converter, and he must decide on action within seconds. If the temperature is too high coolants are added such as pig iron, scrap or ore. Alternatively steam is added to the air being blown in. Over-blowing leads to dense brown fumes of iron oxide. The process is fast and keen and does not admit of the more leisurely control possible with the open-hearth furnace. The point when the white flame changes to a short pinkish flame is called the end point, and sometimes an afterblow is continued for some seconds. The time of a blow for a 25 ton converter ranges from 11 to 15 minutes. At the right moment the converter is turned nearly horizontal, and the molten steel is carefully poured into ladles, the slag being poured off afterwards.

The *basic* Bessemer process, also known as the Thomas process, is the one which can remove phosphorus, and so it can be used with iron produced from a less restricted range of ores. It might be inferred that the lower the phosphorus-content of the pig iron, the easier the process should be, but this is not so. There *must* be enough phosphorus to allow the exothermic reaction of its oxidation to provide heat at the right time. The charge must therefore be relatively high in phosphorus and low in silicon. If there is too much sulphur in it this is often removed before charging, by treating the molten pig iron with anhydrous sodium carbonate. In the basic converter process the silicon blow is shorter and the carbon blow starts earlier. Eventually sufficient iron oxide is produced to flux the lime and make it into basic slag. The phosphorus blow then starts *after* the carbon blow, and the timing of it is sometimes done by stopwatch.

Both processes involve the same chemical reactions as in the open-hearth processes and the equations need not be repeated. Both processes involve details not described above and concerned with the nature and timing of additions, whether to converter or ladle, the sequence of operations, deoxidation, and the precise quality of steel to be produced. However, because air is 79% nitrogen, some of this gas gets into the metal, the more so if this is very hot, and this affects the quality of steel produced. By a fortunate accident this does not matter with steels for certain purposes, such as those for railway lines, for example, which were almost all made from the acid Bessemer process for the vast American railroads of the nineteenth century.

b. Converters operated with gas mixtures with a low nitrogen content. Various ways have been devised to restrict the nitrogen content of Bessemer steel. These include side blowing (below the surface of the metal) to give a shorter path through the melt; double-blowing by the introduction of part of the molten charge later in the blow; and the addition of ore or oxide scale just before the end point to speed-up oxidation other than by air. A lower temperature also helps to reduce the entrainment of nitrogen. Some makers have tried mixtures of carbon dioxide or steam with oxygen. One firm in the UK still uses a very-low-nitrogen (VLN) converter process in which an oxygen-enriched air is succeeded by an oxygen-steam mixture, but these converters will be phased out. In the Tröpenas converter the air intake is blown on to the surface of the melt and gives a hotter process, and is used most by foundries to provide molten steel for castings. It is not a major steel producer. All the bottom-blown basic converters that produce low-nitrogen steel are comparable in quality of output with basic open-hearths, but the process is more limited in the range of steels that can be produced. It is impossible, for example, to make high-carbon steels because, to get the phosphorus out, the carbon content must be reduced to about 0.04%, and so additional processing must be done to put back carbon to produce even medium-carbon steels.

c. The LD process. It is natural that there has been for a long time a great interest in pure oxygen as the prime oxidizer, though this was obviously too expensive for large-scale steel-making until the oxygen manufacturers, keeping up with technological developments, showed how to get cheaper oxygen in the quantities needed. Use in the open-hearth has already been mentioned. Early attempts were made with bottom-blown converters, so widespread in use in some countries. But it was soon found that the use of pure oxygen under pressure eroded the bottoms very rapidly. So, though the use of pure oxygen alone is successful in its limited use in specially-designed or converted open-hearth furnaces it has never been successful with Bessemers. A new technique had to come before tonnage oxygen was brought into general use.

This development came with the LD process. It was thus called because it came from Linz in Austria and it used a nozzle procedure or Dusenverfahren, though the D is usually taken to refer to Donawitz where development of the process also took place. The term BOF, for basic oxygen furnace, is used loosely for it though this expression more correctly includes the Kaldo process as well. It is the most rapidly developing steel-making process and will replace the open-hearth. It wil take the place of Bessemers, VLN or any other, as these are phased out. In the USA it produced 4% of the amount of steel made in open-hearths in 1961. The percentage was up to 60% by 1967 and it will catch up and surpass the open-hearth production in the early '70s. What will happen in Russia, heavily committed to open-hearth production, nobody knows, but some oxygen converters are planned. Japan, West Germany, Canada, the China People's Republic and many others — the record is the same so far as the transfer to oxygen converters is concerned.

The original LD process was mothered by necessity because of the quality of the Austrian ore, very low in phosphorus and high in manganese, and so unsuitable for both acid and basic Bessemer processing, and because there was a shortage of scrap, thus making the open-hearth process uneconomic. After success with a small converter

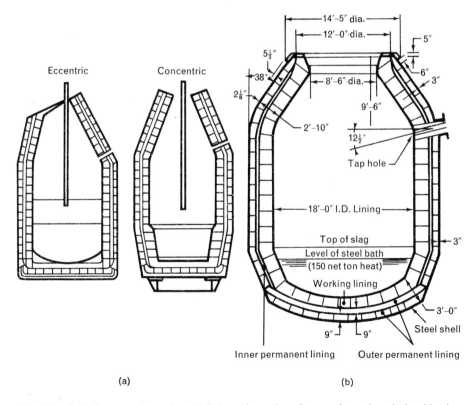

Fig. 3.15 L.D. Converter in section. (a) Schematic section of types of vessel used when blowing oxygen vertically downwards on to the metal through a water-cooled lance (b) Diagrammatic section of an actual concentric type of vessel.

the first 35 ton LD converter was installed at Linz in 1952. Early attempts had eroded the bottom of a Bessemer converter. By using top-blowing, reducing the oxygen pressure and moving the jet further away from the surface of the melt success was achieved. Though it excited a great deal of interest, it was not immediately foreseen as a major new development of the twentieth century until modifications increased its versatility.

The basic LD converter of today is shown in fig. 3.15. It is similar to a Bessemer converter but with a solid bottom; its lining is basic. In the original there was no tap-hole and the charge was low in phosphorus. A later modification enables iron with more phosphorus to be used and is known frequently as the LD-AC process, where the A is the first letter of a Belgian firm, ARBED, and the C is the first letter of a Belgian research organisation CNRM. The main modification is the use of powdered lime in the oxygen stream to achieve greater dephosphorization. The water-cooled lance to supply the oxygen jet has also been modified, since the first one was made, to have several orifices to spread the gas and the pressure. The Japanese trident lance has become almost standard with its three orifices jutting at an angle with the lance. Modification of jet pressure has improved the blow. However, in diagrams such as that of fig. 3.15 a single channel is normally shown. The size of LD converter has increased beyond all comparison with the first Linz installation. Development continues.

The key point of the LD process is the use of top-blowing with 99.5% pure oxygen at a pressure of 130 to 200 psi. (900 to 1400 kilonewtons/m^2). In a converter with an open top some distance away, the furious combustion produces less roof erosion than there would be in an open-hearth with the lance point high above the metal. And the same position of the lance ensures that the action at the bottom of the melt is not disastrously erosive. The pressure of oxygen produces a depression in the molten surface and engages, owing to lance construction and liquid turbulence, a large area. The lance-end is several feet (a metre or so) above the surface.

The charge is scrap steel – the use of which has been improved by the development of the process, thus removing one of the early objections to it in comparison with the open-hearth – and molten pig iron, both put in while the converter is tilted nearly horizontal, and then lime when it has been turned upright. The oxygen is then turned on. The immediate oxidation of much iron produces iron oxide, which fluxes the lime to make a basic slag at an early stage. The AC modification involves no fundamental change but does necessitate a fine lime supply connected to the lance. The pure oxygen is jetted and then powdered lime and oxygen are jetted together, forming a great quantity of basic slag, which is poured off. The lime and oxygen jetting is then restarted and continued until just before tapping the molten steel into ladles. It is thus a double-slagging procedure. Owing to the early creation of enough basic slag (and its removal before phosphorus equilibrium is over-reached) the phosphorus comes off early and so the steel is not decarbonized as quickly. Thus steel of higher carbon content can be made from phosphoric iron. The LD process and the LD-AC process thus combine to make the one converter provide a versatile process for many types of steel (see figs. 3.16 and 3.17). There is no entrained-nitrogen problem and there is adequate removal of sulphur from most pig irons. There is no need to restrict the charge to pig iron with enough phosphorus to produce an exothermic reaction, as there is with the basic Bessemer. The chemical actions are the same as those in the basic open-hearth or the Bessemer. One source of economy not previously mentioned is that there is a store of iron oxide in the slag retained after the tapping of the melt.

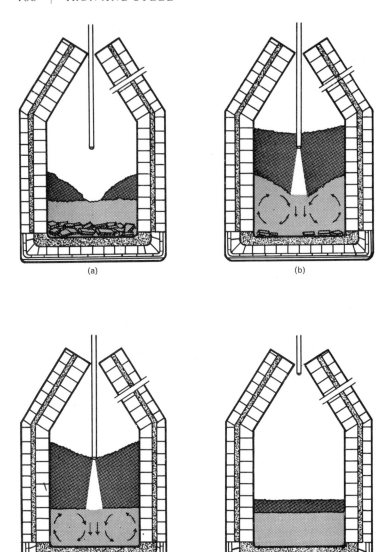

Fig. 3.16 Stages of LD process; the refinement of a mixture of scrap and iron into a batch of steel are represented as they appear inside a furnace during a blow. The stages are (a) about 15 seconds after the beginning of the blow, (b) approximately midway in the 22-minute operation (c) A minute or two before the oxygen is shut off and (d) 30 seconds after the end of the blowing operation. The lance is being withdrawn so that the furnace can be tipped to pour off first the steel and then the slag.

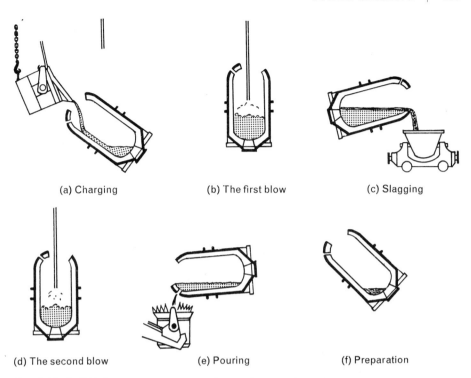

(a) Charging (b) The first blow (c) Slagging
(d) The second blow (e) Pouring (f) Preparation

Fig. 3.17 L-D/AC converter sequence of operation. (a) The ladle pours molten iron into the mouth of the converter which tilts forward to receive the charge. (b) The lance descends into the mouth of the converter directing a jet of oxygen and powdered lime on to the surface of the molten iron. (c) Tilting backwards the converter pours the primary slag, which contains most of the impurities, from the surface of the metal into a ladle. (d) The converter returns to its position shown in (b) and a second oxygen/lime injection further purifies the metal. (e) Tilting forward completely the converter pours the steel into the waiting ladle. (f) The residual slag remains in the converter which returns to its position shown in (a) for the next charge.

The product is a clean steel, low in gaseous inclusions, which can be made in a range of qualities comparable with those from open-hearth furnaces. The converter is relatively inexpensive to make and is fast in its action, though not as fast as the Bessemer, (See fig. 3.18 for a representative interpretation of the difference.)

d. The Kaldo converter process. The second most widely used oxygen-lanced converter process is the Kaldo process, from the names of Swedish professor Bo Kalling and Domnarvets Jernwerk the establishment where it was first installed. A 30 ton version was built in 1956. The converter, which is basic, is shown in fig. 3.19.

The first notable innovation is that the converter is rotated on its longitudinal axis when in action, in which case it is at an angle to the horizontal. It can also be tilted on a horizontal axis in the same way as the LD and Bessemer converters. The oxygen is lanced also at an angle with the molten-metal surface. The procedure is to charge with scrap and lime or sintered ore and lime, this charge acting both as coolant and slag builder, and then with molten pig iron, after which the oxygen is turned on for the first blow while the whole massive converter, weighing hundreds of tons, is rotated at

170 | IRON AND STEEL

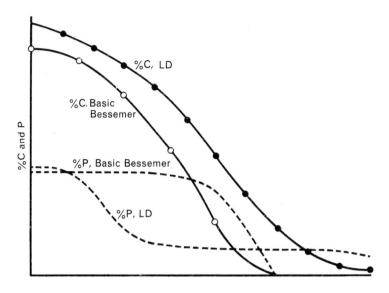

Fig. 3.18 Graph contrasting removal of C and P in basic Bessemer and LD.

speeds that can go up to 30 rotations a minute. The first basic slag (and its phosphorus compounds of course) is poured off and more lime is added, after which the second blow (and rotation) takes place until the molten steel is ready for pouring.

A treble slagging can be performed if the phosphorus content of the pig iron is very high – it is this utilization of high-phosphorus iron that is one of the outstanding characteristics of the Kaldo process. Control is possible by varying the speed of rotation. When it is slow the slag builds up for the removal of non-metallic elements; when it is fast the carbon removal is accelerated. The purpose of the rotation is the mixing of slag and metal on a large-surface basis. Owing to the combustion of carbon monoxide to carbon dioxide, an exothermic reaction, great heat is evolved, allowing the melting of higher proportions of scrap than in the LD process. The oxygen pressure is much lower than that for the LD process, only some 45 psi (310 kilonewtons/m^2), from a

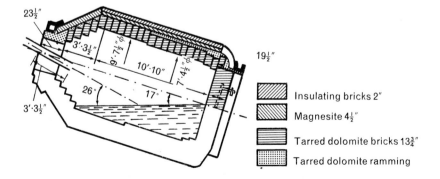

Fig. 3.19 Kaldo converter in section.

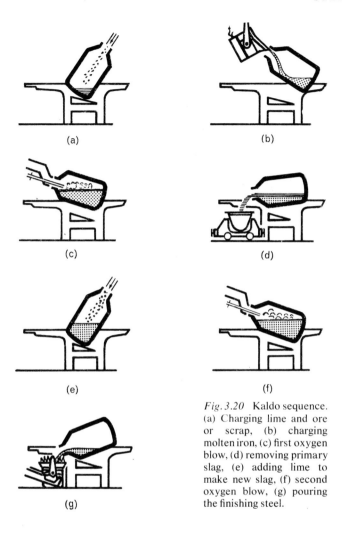

Fig. 3.20 Kaldo sequence. (a) Charging lime and ore or scrap, (b) charging molten iron, (c) first oxygen blow, (d) removing primary slag, (e) adding lime to make new slag, (f) second oxygen blow, (g) pouring the finishing steel.

third to a quarter of that in the LD converter, so that there is less turbulence. The process is a little slower than the LD process, a tap-to-tap time ranging from $1\frac{1}{2}$ to 2 hours, according to size. This is still fast compared with the open-hearth time, even when oxygen-lanced.

Kaldo steel, like LD steel, is low in dissolved gases. It can be made in a higher-carbon range from high-phosphorus pig iron, and it is versatile in its use of raw material and the range of qualities of steel produced. Low carbon steels are by far the greatest part of steel production, so that the LD process is in wider use, but the two processes together, LD and Kaldo, constitute the basic oxygen furnaces of today's steel-making industry. Although a number of open-hearths are successfully lanced with oxygen (apart from enrichment of fuel) the smaller capital cost and high speed and versatility of converters as they are now developed, indicate that further open-hearth conversions are less likely, though in this respect as in many others, local conditions of capital investment, ore quality availability, and market demands affect decisions of this sort.

e. The Rotor. There is one other basic oxygen converter in use, the Rotor, invented by R. Graef in Germany and built as a 60 ton unit soon after the first Kaldo. The converter is cylindrical and horizontal and rotated slowly at up to 2 rotations a minute. Two oxygen lances are used, entering through one end, which is also the charging end. One lance is above the melt and at an angle to it as in the Kaldo converter. The other lance dips below the surface of the melt and uses about a quarter of the oxygen of the first lance. At the opposite end is a tap-hole low in the periphery.

Ore and lime are charged, followed by molten pig iron. The oxygen lances are then inserted and blowing and rotation begin. Double-slagging is used, for the phosphorus is removed long before the carbon, there still being 2.5% carbon when the phosphorus is down to 0.25%. After the second blow the metal is tapped. The effect of the double lancing is that the submerged oxygen effects quick decarbonization and production of carbon monoxide as well as oxidation of the non-metals, while the upper oxygen changes the carbon monoxide into carbon dioxide, an exothermic reaction that produces extra heat, and also produces quick iron oxide to flux the lime as in the LD and Kaldo processes. Though the Rotor was first devised as a pre-refining unit to get rid of phosphorus it is used for steel-making, being able to deal with high-phosphorus iron and produce high-carbon steel. The process seems theoretically as effective as the others but its use has not spread to the same extent, largely because of economics and also because of the difficulties in rotating heavy structures weighing hundreds of tons. These difficulties apply also to the Kaldo process.

Though it is not possible to get precise figures for all the oxygen furnaces used, some comparison can be made between them in relation to oxygen consumed, number of heats before repair, and capacity. The facts are listed in table 3.5.

3.3.4 Electric furnaces

Electricity as a source of heat energy for steel-making has attracted inventors

Table 3.5 Comparison between various types of oxygen furnaces.

Type	Capacity (tons)	Oxygen per ton of steel		Tap-to-tap Time	No. of heats between repairs to linings
		ft^3	m^3		
Basic open hearth	200 to 450	AJAX 1250 to 1600	35 to 45	7.0 to 7.5 h*	300*
Oxygen-lanced		Canada 1150	33	4.0 to 5.0 h	
		Japan 925	26	4.0 h	
		Others		9.0 h*	270*
LD converter	35 to 250	1600 to 2000	45 to 56	35 min	450*
LD-AC					280*
Kaldo	30 to 100	2000	56	1.5 to 2.0 h	90*
Rotor	30 to 100	2500	71	2.0 h	130 to 220*

For comparison the basic Bessemer tap-to-tap time is 20 min. *British latest average figures. All figures are approximate and there are many variables to make the comparisons imprecise, though they are of the right order. Much depends on shop practice, the need to keep production going, the use of costlier linings etc. Comparison of production rates in tons per hour is not made because differences in size of furnace and the number of heats possible make such a comparison meaningless.

for many years, doubtless in the very early years because of its novelty. The economics of the method held it back, while open-hearths and Bessemer converters were functioning satisfactorily. The increased need for very special sorts of steel (e.g. stainless steels) brought its use to the fore, and it has advanced steadily so that today, with oxygen steel-making, electric steel-making of all types of steel constitutes the progressive part of the industry, the two processes being likely to replace all other methods. When the British firm of Steel, Peech and Tozer a decade ago replaced 21 open-hearths at its Templeborough works by 6 130 ton arc furnaces for plain carbon steels, and so went all-electric, it was the first major European concern to herald the new age.

In 1952 electrically-produced steel was about 1% of total production in the UK. In 1956 the proportion was 5%. Today it is nearer 15%. The same progress is seen in the USA where the proportion increased from about 9% in 1961 to 12% today. In the world as a whole, electrical production is about 10% of the total, amounting to some 50 million tons of steel. The rise has not been as dramatic as that of oxygen steel-making but it is steady. Wherever there are plans for new plant it consists of basic oxygen converters or electric furnaces.

The advance has come with the general rise in the use of electricity everywhere and its transmission on national networks at high voltage and power, as well as with the progress in heavy electrical engineering that has led to components which are readily available or which can be made by known techniques. Transmission and distribution has meant availability of electricity. Electricity has in general risen in price, but in countries such as Britain this rise has been less than that for most other fuels. It is a clean source, easily available at the throw of a switch, and involves no gas producer or fuel storage and avoids the contamination of steel by elements such as sulphur. It can produce temperatures higher than that of any combustion process. Its capability in melting large quantities of scrap, which is cheap in many countries, makes the economics attractive.

There are several ways of using electricity as a source of heat. It can merely pass through metal, as it does in electric heaters, producing the resistive form of heating. It can induce currents in metals not electrically connected. It can radiate heat on to the surface of a metal.

a. The electric-arc furnace. The only method that has emerged as a major steel-making process is that of the electric arc passing between electrodes and the metal. This electric-arc furnace (see fig. 3.21 and 3.22) is known under many proprietary names but is still based fundamentally on the furnace of Dr Paul Heroult, a French mining engineer who first made an arc furnace for steel-making in 1900 at the age of 27. Three large electrodes, made of graphite in the larger furnaces but cheaper carbon in furnaces of smaller power, are connected to the three phases of an electrical 3-phase a.c. supply. The metal to be melted and refined, lying on the hearth, constitutes the second electrode (looking at the three graphite electrodes as one) and the arc is struck between the electrodes and the metal, similar in nature to that of any arc lamp, producing a plasma temperature up to 4000°C or so. The electrodes are of graphite or carbon because this material is an electrical conductor and has a very high melting point. It is chemically inert and does not dissolve in slag. Graphite is more expensive than carbon but will carry a higher current. Lengths range from 24 inches to over 100 inches (0.6 m to over 2.5 m) and diameters from ¼ inch to 40 inches (5 mm to 1 m). They are massive and expensive and special threading devices must be built on to them to allow new lengths

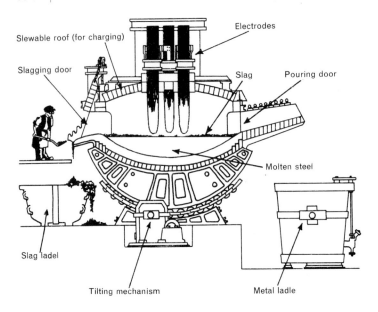

Fig. 3.21 Cross-section through an electric arc furnace.

to be added as they wear away. The furnace is cylindrical with shallowly curved hearths and tops. It is made of steel and lined with refractory bricks, which have an upper layer that is dolomite or magnesite in basic furnaces, and is siliceous in acid furnaces. The diameter of the inside ranges from 7 ft to 25 ft (2 to 7.5 m) for steel capacities of 4 to 200 tons. The furnace may be charged through doors, as with the open-hearth furnace, or, more normally, from the top, the massive roof and retracted electrodes being swung away for the purpose. The furnace can be tilted and has doors on opposite sides, one for slagging and one for pouring. Some furnaces can be rotated through a small arc to speed up the melting.

The heavy electrical engineering involved is essential to the action of the electric arc furnace. Electrical contacts must be firm on the electrodes and large transformers designed, cooled by water and oil and with tappings to allow for varying the voltage. An electric arc does not behave like a plain resistive circuit. As the current increases the voltage-drop decreases, and this change is fast at currents below a certain volume. This causes instability so that steps must be taken to include enough reactance, but not too much, to cause the arc to act more like a normal resistance. Then, as the molten metal falls away from the arc its resistance increases and the electrode must be lowered to maintain the current. This special behaviour necessitates regulating circuits, that will switch on motors to raise or lower the electrodes automatically, to maintain constant power consumption, and will move the connections to tappings on the transformer. An inductive device to cause stirring of the melt is frequently added. The usual electrical switchboard with its instruments and heavy-duty circuit-breakers is part of the complex.

STEEL-MAKING | 175

Fig. 3.22 An electric-arc furnace.

The metallic charge is wholly cold scrap steel, and in this respect the electric-arc furnace is different from all other major-production furnaces. The scrap must be very carefully selected in order to have the right elements in it and no others, and in some cases well over 60 different segregations have been found necessary. Shop practices, as in other steel-making procedures, vary according to the nature of the scrap available and the steel needed. Thus if a low-carbon steel is needed and the carbon content of the scrap is too high there must be oxidation of the carbon. Oxidation of silicon and other elements may also be necessary. On the other hand, if the scrap has too little carbon for the final product, carbon must be added. Some alloying elements needed can be added if not oxidizable, or they may already exist in the scrap. Scrap must thus be separated out, which is simple when it is produced in the steelworks because its exact specification is known, but it is far from simple when scrap is bought from outside. In this case special chemical spectrographic and magnetic tests are added to, or used instead of, rough tests by the skilled steel-maker.

By control of the slag composition the *basic* electric-arc provides the most flexible of the fast methods of making steel. Lime may be added first. Oxidation may or may not be needed. If it is, it comes from the oxygen of the air or from added oxides such as ore or mill-scale. It can also come from injected pure oxygen, *via* a lance, an increasing practice. The first stage, when the furnace is closed after charging, is the melt-down, achieved by lowering the electrodes and switching on. At first the current must be kept low to avoid danger to the refractories but it is soon raised and many thousands of kilowatts of power produced, involving currents of thousands of amperes at voltages not very different from normal mains voltage. The melting is fast and at the same time oxidation, if this is needed, takes place and there is a carbon 'boil'. Phosphorus is removed and, if necessary, silicon, and the slag increases. A process of double-slagging is usual. In this case, when the melting and oxidation have gone far enough, in accordance with chemical reactions similar to those in the open-hearth, the first slag is removed. It is, as all slags in other furnaces previously considered, an oxidizing slag containing oxides that could yield oxygen. The second slag, however, and this makes the electric-arc procedure different from any other, is a reducing one (in the chemical sense), and by care in the choice of its constituents control can be effected. A common one consists of lime, fluorspar and silicon to which carbon must be added, usually as coke dust, whether with the slag charge or afterwards. The extremely high temperature of the arc furnace then produces calcium carbide, which reacts with iron sulphide and lime to produce iron, calcium sulphide (which goes into the slag) and carbon monoxide. Thus sulphur is removed more efficiently than in other steel-making methods. The reduction also effects the removal of oxygen from the steel, an action which may also be expedited by the addition of an agent such as ferro-silicon in the melt before the second slag materials are added. Oxides of iron, manganese, chromium and so on, which could try to form in the slag are reduced to metals by the calcium carbide and return to the melt. The slag may also have lime-silicon or lime-alumina added to act as a deoxidant and desulphurizing agent, in which case the reducing nature of the slag can be maintained by further additions of metallic aluminium. Practices vary according to the product needed and the nature of the raw material. Alloying elements may be added. About 15 minutes after the slag has formed, whatever the additional procedures, the molten steel is poured off. The slag is lighter in colour than oxidizing slag and disintegrates into powder when cool. Double-slagging (sometimes indeed treble-slagging) is not essential to every procedure, for it is sometimes possible to maintain one slag and make it a reducing one if necessary by suitable additions.

Clean steels with low oxygen and sulphur and phosphorous of any desired carbon content and with alloying elements for special types, can thus be produced in the basic electric-arc furnace. The process is fairly fast, especially with oxygen lancing, but times vary with shop practice, the size of furnace and the type of steel made, so that tap-to-tap times can be as low as 3.0 hours (average about $3\frac{1}{2}$ hours at the Steel, Peech and Tozer steelworks) or less, and as high as 6 to 10 hours. With adequate care in charging etc. the wear on refractories is low and the consumption of electricity some 400 or so kilowatt-hours per ton of product. Where electricity is cheap the fuel cost is thus very low.

Little need be added about the *acid* electric-arc furnace used by the foundry part of the industry, though it came into use, as with the old Bessemer and open-hearth, before the basic variety. There are several procedures, such as that of complete oxidation and then reduction of silica to silicon, which re-enters the metal as needed, for though normally silicon is regarded as an element to be brought down to small proportions it is needed in certain steels for sheets for electrical devices such as transformers. The commonest practice, however, is complete oxidation. The charge must be carefully chosen in order to produce the required steel. The slag is silica and iron oxide. As previously stated, the acid electric-arc furnace is used mainly by foundries where the high fluidity of the melt, because of the high temperature of the arc, is an advantage for casting the metal into moulds.

b. The induction electric furnace. The induction electric furnace as used today is confined to small units capable of dealing with up to 3 tons of steel, or 5 tons at the most. Though classed as a steel-making device it is hardly more than a melter in the most common usage, though occasionally there is slag and some chemical action. The principle is simple. When a coil carries alternating electric current it creates varying magnetic fields and if there is an electrical conductor in the field, currents called eddy currents are included in it. The flow of current through a resistive conductor creates heat. With the induction furnace the scrap to be melted is the conducting core inside a coil of copper tube, through which water flows to cool it, carrying a heavy current. The eddy currents create heat and movement, which effects stirring. The frequency is about 2 000 hertz and upwards, and the voltage may be of the order of 1 000 V.

The melting rate is high and the power consumed approaches 1000 kilowatt-hours per ton in small furnaces, falling to some 700 kilowatt-hours per ton in the largest. The metal container can be a separate crucible or be made by working refractory material into the current-carrying coil.

The commonest type of induction furnace utilizes the high frequencies mentioned, but more recently some furnaces have come into use at the mains frequency of 50 hertz in the UK and 60 hertz in the USA. As the furnace is primarily a melter, many sorts of special alloy steels, not needed in large quantities, can be economically made using this method.

3.3.5 Secondary refining methods

With the possible exception of the induction furnace all the methods so far described have been for steel-making, and chemical actions take place to oxidize unwanted elements to slag oxide, get rid of carbon to a predetermined level, reduce phosphorus and sometimes sulphur to acceptable levels, and reduce oxygen entrained in the melt or present as iron oxide. For a century and more crude steel made in the open-hearth furnace or Bessemer converter supplied what was needed, and large structures such as bridges (the Forth bridge was built in 1890), built with a large factor of safety, have survived satisfactorily. There are still many applications where the steel specification is adequate for its task. In more recent years, however, demands have come for much more stringent specifications, partly dictated by economics, in that large factors of safety can no longer be afforded and consequently steels are used much nearer to their limits, and partly dictated by special applications where small quantities of impurities become more important in the behaviour of the steel. Larger and larger electric generators, jet engines for aircraft, nuclear-power-station pressure vessels and fuel cladding, heavier moving parts everywhere, and many smaller developments demand steels that will withstand very high temperatures or sometimes very low temperatures, extremely corrosive conditions, and so on. And it must still be possible to machine, forge, cast or extrude these steels. Steels made under old conditions, though still in bulk constituting the greater part of the market, cannot satisfy these increased demands.

It is a notable fact that unwanted damaging elements are often present in very small quantities; a slight change in these is important. An example will show the small changes that are considered worth making by a refining technique. The problem was to find, for one refining method (to be considered later), the correct slag composition to refine a 5% chromium steel. The composition before refining and afterwards is shown in the table:

	C%	Si%	Mn%	S%	Cr%	O_2%
Before remelt	0.41	0.92	0.27	0.01	5.10	0.004
After remelt	0.40	0.69	0.28	0.005	5.00	0.002

The important reduction is in sulphur and oxygen without affecting the proportion of manganese and chromium to any important extent.

The reduction of sulphur from 0.01% to 0.005% seems negligible but is important in practice. The importance of oxygen has already been mentioned. It was a serious cause of trouble in early Bessemer steel. The importance of phosphorus in all steels has already been emphasized. Nitrogen and hydrogen get into steel from air and moisture respectively. Gases are less soluble in solid metal than in molten metal, and their harmful effects begin to show when the steel has solidified and is being heat-treated or rolled or forged or used. Too much oxygen produces 'red shortness' which means the steel can hardly be forged without troubles. Hydrogen produces flaking and cracking. Nitrogen produces work-hardening. Phosphorus produces brittleness. Sulphur causes 'red shortness'. These factors have been known for a long time and the basic process was evolved to reduce phosphorus, while many deoxidizing devices have been used. To get rid of hydrogen has necessitated long heating procedures, for there is no metal which will combine with it and so remove it as deoxidants remove oxygen. Furthermore, the contaminants are never reduced to nothing; it is a matter of compromise and economics to reduce the proportions sufficiently to make the steel usable and reliable. Another important property in steel is lack of homogeneity and the presence of unmeasurable quantities of other materials which make the steel unclean.

Refining to some degree has been part of the steel-making process for a century, but for modern special steels further, secondary refinement is necessary. There are several ways of doing it and the best known are vacuum degassing, vacuum-arc remelting (VAR) and electroslag refining.

a. Vacuum degassing. Vacuum degassing is the removal of gases, especially hydrogen, from the molten metal. It started seriously in the mid-1950s, and hundreds of plants now exist throughout the world and there are many variants on the same theme. This is merely the placing of the molten metal in a gastight chamber and then pumping the air out. Under this reduced gas pressure the occluded gases can escape into the space above the metal and are pumped away. It is basically a simple notion, but satisfactory commercial methods have come only as a result of years of research. The simplest application is where the ladle of steel is put inside the vacuum chamber and removed afterwards for pouring into moulds. Stream degassing is the method wherein two ladles are used, a full one outside the vacuum chamber and the empty one inside. As the metal pours into this ladle it exposes a big surface to the vacuum because it breaks into drops. For large ingots the ingot mould itself can be inside the vacuum chamber. Another modification is to cast the metal directly into shaped moulds inside the vacuum, thus producing the finished casting directly. All these are batch methods degassing a certain number of tons at a time, but the British Iron and Steel Research Association has pioneered one continuous degassing technique and development of this goes on. Whichever the variant, vacuum degassing is an achieved secondary process used, especially in combination with electric-arc steel, for such steels as those required for aircraft, ball-bearings and large alloy-steel forgings, in all of which failure because of hydrogen would be disastrous.

Such vacuum degassing is done on material that is already molten. Other methods of refining melt the metal inside the vacuum chamber. A common way to do this is to have a small induction furnace inside and melt the material electrically. The capacity of such melting devices is usually small but there are some with a capacity up to several tons.

b. Vacuum arc remelting. This method is more widely used. The principle of it is shown in fig. 3.23. The electrode is a bar of the steel to be refined and it extends downwards towards a small amount of the same metal, as turnings or as a plate, which is put at the bottom of the crucible fitted to the bottom of the vacuum chamber. The air is pumped out and an arc is struck from a d.c. source between the electrode and the bottom metal. The metal of the electrode melts, as does the bottom metal, and builds up from the bottom of the crucible into an ingot. An electronic circuit maintains a constant voltage across the arc, so that the power consumed is directly proportional to the current. The circuiting and gear needed make the whole installation much less simple. The current is measured in thousands of amperes, up to 25 000 A for an ingot of diameter 25 inches (0.65 m), and the choice of current is important in relation to the quality of ingot. The degree of vacuum is higher than in vacuum degassing, as high as 0.001 mm of mercury (0.001 tons) in some cases.

The result of this treatment is the elimination of hydrogen, the reduction in quantity of other gases, the removal of inclusions and the reduction of what are left to microscopic size, and a general cleaning-up. A tiny proportion of some alloying metals, such as manganese, is lost. The general effect is thus to produce a cleaner steel with high fatigue characteristics, high tensile strength, greater toughness and longer life for some uses such as bearings. In general, there is a greatly improved reliability, forgeability and consistency from batch to batch.

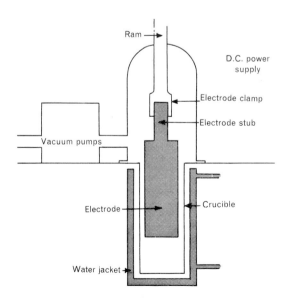

Fig. 3.23 A basic diagram of a consumable electrode vacuum arc remelting furnace.

c. *Electroslag refining.* A non-vacuum process gaining ground widely is electroslag refining (ESR). The principle has been known for a long time in Russia and the U.S.A. (where there is a similar process called Electroflux) but it has gained ground in commercial use only since the mid-1950s. The British Iron and Steel Research Association gave the method a boost when, after years of research, it set up the Electroslag Refining Technology unit in 1966 to develop the process. A dozen or so plants are used in Britain. Other countries are interested in Europe, and in the United States there is now an Electroslag Institute. The method has many variants and development continues at increasing speed, with advantages in eases of operation, quality of product and lowering of cost claimed by numbers of inventors. Electroslag refining may oust VAR processes from the refining field. Direct current or alternating current can be used, the latter in single-phase operation. Ingots of refined steel up to 4 feet (1.2 m) in diameter, weighing 20 tons, can be made, but common practice is to make ingots smaller than this, of the order of tons rather than tens of tons.

The general principle is that an electric arc is struck between an electrode of the steel to be refined – though this can be interpreted broadly – and the base (or in one variant between two such electrodes) on which a special slag is placed. When general steel-making is considered the development to electroslag refining is obvious, for the main factors in steel-making are the temperature achieved, the complex physico-chemical reactions between slag and metal, and the degree of contact between the two to facilitate these reactions and spread them through the melt. In electroslag refining the very high temperature (2000°C or more) is achieved by the electric arc, the slag is carefully calculated, and intimate contact between metal and slag is achieved by the dispersion

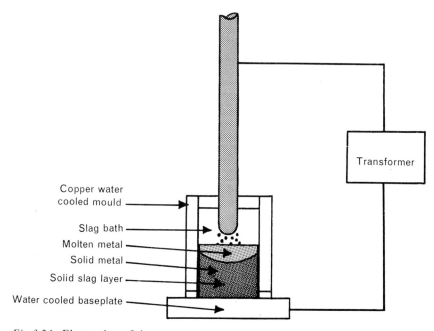

Fig. 3.24 Electro-slag-refining.

of the steel into drops, so that there is a very high ratio of surface area to volume. A simplified diagram, ignoring the complexity of all the engineering and circuiting is given in fig. 3.24 though since 1966 modifications such as steel moulds sprayed with cooling water have been used, and in some cases graphite moulds, as well as devices for continuous withdrawal of the ingot.

The formulation of the slag is central to the process. In general calcium fluoride is included to increase fluidity and the remaining constituents are lime and alumina. Excess lime will make the slag basic and the composition can be chosen to achieve just what refinement is needed. One has already been mentioned designed to remove sulphur and oxygen. Magnesia and silica are other components sometimes used. Depending on the composition of the raw steel and the need in the refined ingot, the slags are varied, and much research has been done on the correct proportions for any one purpose. Examples will demonstrate this point. Sulphur in a high-speed tool steel has been reduced from 0.013% to 0.005% with a slag of 80% calcium fluoride, 15% lime, 5% silica. Oxygen in a silicon alloy has been reduced from 0.006% to 0.002% by a slag of 50% calcium fluoride, 20% lime, 30% alumina. A slag of 70% calcium fluoride, 10% lime and 20% alumina, used on a carbon-chromium bearing steel, reduced the silicon from 0.31% to 0.10%, sulphur from 0.017% to 0.005%, oxygen from 0.006% to 0.004%, and phosphorus from 0.038% to 0.032%, without changing the proportions of carbon, manganese, chromium and nitrogen to any important extent. The slag not only does the chemistry; it is also a conductor, and so gets hot, and it protects the top of the ingot from the atmosphere.

The start of the process has undergone some modifications, such as the use of a tablet containing chemicals to start an exothermic reaction when the electricity is switched on, a tablet of ceramic that rapidly melts, and the pouring-in of molten slag. Electrodes have been modified by a ceramic coating to prevent oxidation of iron, and in the development a plasma torch has been used to avoid the need for a solid electrode.

Results already obtained show that very good refining can be obtained, not only in the reduction of unwanted elements but also in the reduction of non-metallic inclusions and in the production of a good ingot surface. The metal is clean and has desirable mechanical qualities such as higher tensile strength and toughness and greater homogeneity.

All the refining methods described are applied to special steels, whether of low-alloy or high-alloy specification. These steels constitute only a fairly small proportion of the total steel production, but this proportion is rising – approaching 10% in the USA and over 6% in the UK – and they are important in their applications to advanced engineering, where reliability, cleanliness (absence of non-steel inclusions) and improved mechanics, anti-corrosion and heat properties are essential.

3.3.6 Preparation for use

Steel-making processes produce steel in liquid form, and this has to be handled sometimes in hundreds of tons at temperatures that range above 1500°C. Mishandled, such white-hot metal could be a lethal weapon and would lead to mechanical difficulties as well as serious financial loss. The mechanical handling involved is considerable and well-established in its techniques. It is a variable operation because crude steel is not stored; it is produced to satisfy a demand which can fluctuate from day to day and melt to melt.

All crude steel is poured into ladles. These are refractory-lined steel buckets that range in capacity from about a hundredweight (about 50 kg) to a hundred or more tons. They are brought along on overhead cranes. The smallest can be tipped over to pour their contents into whatever receptacle is waiting, but the common practice is to have a stopper in the bottom, and this can be removed by a hinged and levered device to allow metal to flow out of the bottom.

If the steel is further to be refined the ladle may go directly to a refining unit for stream or ladle degassing. Otherwise the molten metal is transferred (poured or teemed) to moulds in which it solidifies to make ingots, which can also range from a few hundredweights to many tons in weight. The steel ingot is the crude material for further treatment.

The ingot moulds may be special ones to make electrodes for vacuum-arc remelting or electroslag refining. In general, for the majority of steels, the moulds are cast iron containers at least twice as long as they are wide, with rectangular or square cross-sections, and tapering slightly along the length to facilitate the withdrawal of the white-hot solid ingot. When the liquid metal is poured into the mould it can splash, like any liquid, and form irregular portions on the mould wall that make 'scabs' on the surface of the ingot. So several methods have been evolved to avoid harm to the ingot when this is economically justified. Thus the simple method of teeming is to pour directly into the top of the mould, but this process can be modified, by an extra container, called a tundish, which takes up turbulence and allows the steel to pour more quietly. Furthermore, moulds can be filled from the bottom upwards by means of an intermediary channel, the so-called 'uphill' teeming.

Moulds can be used in several different ways. The first obvious variation is to use them either with the wide end at the top or the other way up, i.e. narrow end up. There is one sort with a narrowed opening at the narrow end, called the bottle-top mould. These shapes have evolved as a result of experience, and the chief reason for the variation is the need to get the greatest possible yield from the steel. Because of the changes during cooling and solidification some steels may eventually lose as much as 15% or more, and with expensive steels this would represent serious financial loss. Decisions as to the type of moulds to be used is made at the time when the sort of steel to be produced in a furnace has been decided.

A device called a feeder-head is sometimes used with an ingot. This is a structure, placed on top of the ingot mould, which can be kept hot by various means in order to influence the cooling of the ingot.

A small proportion of steel is cast into its final form for use. In general casting is done in a foundry, which gets its crude steel or scrap and does its own melting, and foundries can be quite small establishments. But some casting is done in steelworks direct from the ladle. The mould is shaped to a desired pattern and the final product is taken out fully fashioned. On the whole, only very large castings are made at a steelworks. In general, steelworks produce ingots, the crude steel, which is later made into semi-finished steel for further fabrication.

The ingots for the next operation must be heated to the right temperature and be as homogeneous as is possible. Traditionally the heating places are called soaking pits but the old types of these hardly exist today. Instead, reheating furnaces are used.

From the soaking pits the ingots go to be made into blooms or billets or slabs. These are traditional terms without precision but the bloom and billet have square or circular cross sections, the bloom being considerably bigger than the billet. A slab has a rectangular cross-section that is much wider than it is thick.

STEEL-MAKING | 183

These semi-finished products are usually produced in enormous rolling mills, about which there is a highly developed technology which is still expanding. The system is to squeeze the ingots, while still incandescent, between rollers driven by motors in what are called 'coggling' mills. This hot-rolling procedure is not always followed, for blooms and billets can be forged into shape by hammering, with power-operated hammers no different in principle from that wielded by the blacksmith. During the roling stages of steel production there is some wastage in the rejection of ends of blooms and billets and slabs as a result of defects. These rejects constitute scrap for use in steel furnaces. The amount of them is related to the furnace practice, ladle practice, and ingot practice that produce the crude steel. All the first rolling constitutes the primary mills.

The semi-finished steel can be taken away as the raw material for the fabricator elsewhere, in which case re-heating is needed, although cold working is used in making certain products. Energy, and therefore, money, can be saved if the semi-finished steel can be used at once, as hot and plastic bloom or billet or slab. Some large steelworks therefore produce finished products such as joists, beams, rails, girders and the like by hot-rolling with specially-shaped rollers, and some produce rod, plate, strip and sheet. (The difference between the last three is in the thickness.) Some large steelworks also make heavy forgings as finished products.

The semi-finished steel blooms, billets, slabs and bars are the raw material for the makers of final products. The finished steel consists of rails, plates, wire rods, light rails, light rolled bars, bright steel bars, strip (both hot-rolled and cold-rolled), sheet, tinplate, blackplate, tubes, pipes, tyres, wheels, axles, forgings and castings, beams, angle-iron, H-bars, etc. These are the products that go into ships, bridges, power stations, turbines, motor cars, wire cables, electrical and mechanical machinery of all sorts, bolts, screws and nuts, domestic ware, railways, tools; just some of the things in which steel is used, from heavy structures to the most delicate of instruments and devices.

3.3.7 Deoxidation

As the overall action of making steel is the oxidation of carbon and unwanted elements, it is not surprising that steel as made in furnaces has too much oxygen in it. The amount of this depends on the sort of steel being made, for it increases as the carbon is removed, so that low-carbon steel is richer in oxygen than high-carbon steels and alloy steels. Procedures have been evolved over the years for dealing with excess oxygen by a final process of deoxidation, which is done in the furnace or in the ladle or even in the ingot, according to empirical rules learned by experienced steel makers. Such is the physical chemistry of steel-making that it is never possible to get all the oxygen out in the furnace – it is a matter of chemical equilibrium and activity. The constitution of a steel to be made is decided in advance, i.e. the percentage of carbon, silicon, phosphorus, sulphur, etc., and the responsible official choses his raw materials – the proportion of scrap, the amount of additives, the composition of slag, and so on. Analyses are made from time to time in the laboratory, and if all goes well the correct proportion of constituents and the correct pouring (or tapping) temperature will be reached. The most important figures in the last analyses are the carbon and iron-oxide contents. The operator then known how much deoxidizing to do.

The deoxidants used have already been mentioned, namely ferromanganese, ferrosilicon and aluminium. Ferromanganese and ferrosilicon exist in several grades according to the proportions of constituents (see chapter 4). Ferromanganese, for example, can contain carbon, as well as iron and manganese, and some silicon and phosphorus. Other deoxidants include anthracite and pig iron. The exact nature and quantity of deoxidant is decided by the operator, according to the analysis and the desired constitution and type of steel. However, whichever deoxidant is used and when and where, the purpose is to remove oxygen from the steel. But the quantity to be removed depends on yet another specification of the steel, a specification that determines the behaviour of the molten steel in the ingot mould. This behaviour leads to several different qualities of the final steel.

When there is a fair amount of oxygen present (and a large amount would be merely 0.1% by weight) it fizzes out of the molten metal as it is poured. All the time it is being teemed from furnace to ladle and ladle to ingot, the metal is losing heat and falling in temperature, and as soon as it touches the cold cast iron of the ingot-mould wall it cools very rapidly and begins to solidify on the outside and bottom. But the whole process of solidification, or freezing as the steel-maker calls it, is a long one occurring over a range of temperature, and many changes take place in the steel during the process. For one thing the physico-chemical equilibrium, which depends on concentration and temperature, is changed, and elements in solution in the molten metal react according to their activation energies. Thus with enough oxygen present this reacts with carbon to form carbon monoxide gas, which acts as it would in any liquid: it bubbles towards the top. However, as this is happening the steel is cooling from the outside inwards and an action takes place similar to that of zone refining, already known for the extreme purification of metals in today's technology. Substances in solution migrate inwards, leaving a rim of nearly pure iron on the outside with a chilled skin of steel in the extreme inside. This action is called rimming and the steel thus produced in the ingot is called rimming steel, or rimmed steel.

This rimming action brings a bonus, which must have been an accident in the early days of tonnage steel but soon became part of accepted practice. This bonus is that the rim of nearly pure iron and good surface, when the steel is rolled, provides a softer outer skin, more corrosion resistant than the steel inside. Sheet made from it can be pressed into shapes without breaking this outer skin. It is very suitable therefore for car bodies. The action of rimming can be stopped when desired by capping the ingot with a metal plate, which cools the top, turning it solid and preventing the evolution of gas. Thus it is possible to have rimming steel with chosen thicknesses of rim. If the plate to be made from it is to be used for car-body pressings or tinplate, for example, the rim must be thick to withstand the extreme reduction to plate thickness while still retaining the strength of steel.

The rimming action eventually stops naturally if the ingot is not capped, because as the steel gets more viscous (less fluid) and solidifies so the movement of gas bubbles is stopped, and the bubbles are frozen into place as blowholes. It is good practice to ensure that these blowholes are not near enough to the surface at the top end of the ingot to break under rolling and allow oxidation from the air. The blowholes also tend to collapse at the bottom under the pressure of steel above. Where they exist and are not too near the surface the metal edges of the blowholes are clean and will weld together when under the great pressure of rolling. The amount of oxygen is related to the amount of carbon, as previously stated, in such a way that the less carbon there is the more oxygen there can be, and the more carbon there is the less the oxygen that can co-exist with it. As a result of this steel with more than 0.2% carbon cannot be rimmed because there is not enough oxygen to cause evolution of gas, rapid mixing and quiet migration of elements like sulphur, phosphorus, carbon, silicon and manganese to produce the purer rim. The degree of deoxidation, if any, needed to produce the right amount of rimming is thus another factor in the preparation of the steel, and the operators will know in advance not only the composition intended in the final steel but also whether the steel is to be rimmed. The ingot operative judges from the first ingot cast whether it is satisfactory or whether more or less deoxidation is needed.

When there is not enough oxygen, gas is not evolved rapidly and the zoning action does not take place. So there is no rim of softer, clean-surfaced iron. As near complete deoxidation as possible 'kills' the steel so that it lies quietly in the mould. An intermediate state of deoxidation produces semi-killed or balanced steels. It is possible to kill all steels, but with low-carbon steel this is an expensive procedure beacuse of the amount of deoxidation needed. It is not therefore usually done. Steels containing up to 0.3% carbon can be produced as balanced steel economically. There is no rimming because of the less violent movement of the molten metal. Such balanced steel is the inexpensive maid-of-all-work of the carbon–steel industry, and balanced steel of 0.2% carbon is a general purpose mild steel for sections for building and structures, and for plate for shipbuilding, and for rod to make nuts and bolts and so on.

It is obvious that with the zone-refining action of rimming steel and with the evolution of enough gas in balanced steel, some of the elements (and compounds such as oxides) move towards the top and centre. This segregation changes the properties of the steel in different parts of the ingot, and subsequent rolling is a necessary part of the procedure, contributing to the quality of the final product. At the same time gas that is occluded swells the metal and compensates the shrinkage due to solidification and cooling. Thus the yield from an ingot mould is affected. In a balanced steel there should be less segregation than in rimming steel and enough gas just to make a slight dome on the top surface, giving a high yield.

Steel is fully killed when a more homogeneous steel of forging or casting quality is required. This is too costly to be done with low-carbon steels and so killed steels are of higher carbon content and often have alloying elements in them as well. In general, steels with more than 0.2% carbon cannot be rimmed and steels with more than 0.3% carbon cannot be balanced. Killed steels must therefore invariably have more than 0.3% carbon. With no gas evolution there is less movement and no rim refining and, therefore, less segregation. There is also, inevitably, shrinkage and this takes the form of a funnel-shaped hollow from the top downwards, called a pipe. This leads to lower yield and quite a sizeable part of the ingot has to be removed (and used as scrap).

Considerations of yields and the type of steel to be made affect the choice of ingot mould and the way it is used. Thus killed steel is made in ingots with the wide end up, and with a feeder-head that either ensures heat or has heat provided, so that a reservoir of molten steel is maintained to fill up the pipe as it forms. Thus a smaller piece of the ingot can be discarded than if the mould were used with the narrow end up and without a feeder-head. Capped rimming steel on the other hand is used with the narrow end up and has a bottle-top head so that a small area has to be chilled by the cap. Balanced steel is made in moulds with the narrow end up. These different practices, so much a part of the operative's expertise are suggested crudely in fig. 3.25 where the black marks indicate hollows.

3.3.8 Continuous casting

If the ingot-making, with all its complexities, and the reheating and primary

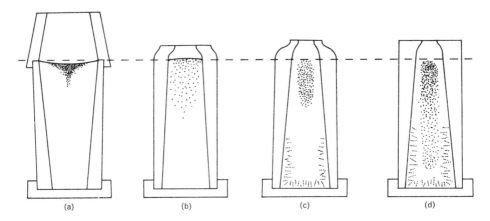

Fig. 3.25 Ingot usage (a) Killed steel, (b) Balanced steel, (c) Capped rimmed steel, and (d) Rimmed steel.

milling could be omitted from the steel making sequence, there would obviously be a saving of energy and time and, therefore, money. If also billets and slabs could be made continuously, the first stage would be reached of making steel continuously instead of in batches. Development on these lines since World War II and especially in the past decade has led to continuous-casting installations that are increasing yearly. More than 200 machines have been installed throughout the world (including Russia and China), with a capacity which can be roughly estimated at between 20 and 30 million tons a year, representing a five-fold increase or thereabouts in well under ten years. The size of casting has increased so that today blooms and large slabs as well as billets are being made by this method. This continuous casting is part of the revolution going on in the steel industry.

The first machine of all was installed in Germany in 1943 and ceased working in 1954. The first-ever production plant of over 2 tons ladle capacity was installed in the UK in 1946, mainly for stainless steels. Siegfried Junghaus in Germany and Irving Rossi in the USA were the chief pioneers, though development has continued in many establishments such as that of the British Iron and Steel Research Association in the UK.

The molten metal is poured from a special ladle into a tundish from which it flows gently into a vertical mould, which can be up to a yard or so (1 metre) in length and is of the cross-section needed in the final casting. The mould can be of thick copper and steel and the molten steel must solidify sufficiently on its way down through it. At the beginning, the bottom of the mould is closed by a dummy bar of the appropriate size, and the bar is withdrawn when the mould is full, and the machine starts the rolling action. The billet in the watercooled mould, which is lubricated, moves downwards and more molten metal flows in at the top. At emergence the billet has a solid, though incandescent, skin which holds the shape. The billet moves downwards between rollers and is then bent round by passing between moveable rollers pushed forward to the correct extent. Thus it bends round until horizontal and moving over a roller track. It is sprayed with water for further cooling after leaving the mould. At a predetermined stage along the horizontal track it is cut off to form separate billets.

This over-simplified account divulges nothing of the technical problems solved in

STEEL-MAKING | 187

Fig. 3.26 A beam being continuously cast.

the development of the machine to successful commercial application. The temperature of the poured metal must be accurately determined, and all parts of the machine before the mould must be adequately heated to preserve fluidity. One major problem was to prevent the billet from sticking to the mould, thus leading to a ruptured skin and consequently transverse cracks in the billet. One well-known technique is to reciprocate the mould down and up to prevent sticking, though the precise way of doing this varies from one design to another. The speed also must be low enough to allow adequate cooling and high enough to make the process economically competitive. One way of achieving this economical competitiveness is to make the machine with several billets running simultaneously, or having several 'strands' as the steel maker describes it. Thus the tons-per-hour production is increased. Careful control of all the factors involved is necessary, and because of this a continuous-casting machine produces steel of more consistent quality. So far balanced steels cannot be produced, but killed steels have been successful from the beginning, and rimming steels can be produced, but are not of the same quality at those produced in ingots because the rim is not mainly iron. Nevertheless these effervescing high-oxygen steels can be continuously cast. The yield in all cases is high.

The normal machines have to be very high and it is the custom to build them in some countries with the final product well below ground level. One firm has produced a curved mould which does away with the need for great height. A number of these more compact machines has been installed. The continuous casting of shaped cross-sections has also been achieved, by the Algoma Steel Production Corporation in Canada using a technique evolved at the Sheffield laboratory of the British Iron and Steel Research Association. This achievement suggests that joists, beams, girders and the like will eventually be continuously cast, thus avoiding the need for costly mills. A stage further will be *continuous* continuous casting with steel delivered to the casting ladles all the time. Such continuously produced steel, acting as bloom, billet or slab could then be fed continuously to the re-rolling mills.

3.3.9 Development and the future

The manufacture of iron and steel is an industry, and as such is open to competitive challenge. To answer this challenge research and innovation are necessary so that productivity in quantity and quality can be increased, and so that the improved steels can be made more cheaply than the materials challenging them – other metals such as aluminium, to say nothing of less common metals for special uses, and other materials such as plastics. (The enamelled steel bowl and the galvanized bucket have disappeared from the home to be replaced by articles of polythene. In this sphere stainless steel is making a challenging return.)

Research continues to understand more of the fundamentals of steel making, for many details of empirically-based practices are unknown and many of the equations so frequently quoted are presented with diffidence. Even the action of deoxidants is not thoroughly understood. The aim of the research is that fundamental knowledge shall lead to better processes more precisely predictable and controllable. Other research is the development of techniques that will one day perhaps replace the blast furnace and the steel furnace, with all the high-temperature problems involve, and lead to continuous processes that can be controlled automatically, and eliminate the

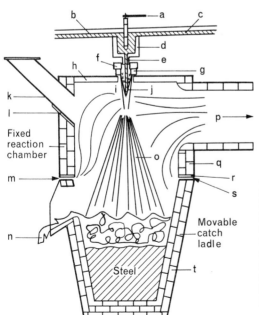

Fig. 3.27 Cross-section of Millom prototype spray steel plant. a. Stopper rod. b. excess overflow iron. c. molten iron from blast furnace. d. tundish. e. metal flow regulating nozzle. f. flux ring. g. water-cooled oxygen injection ring. h. water-cooled toplate. i. flux injection. j. oxygen injection. k. chute for additions. l. water-cooling. m. air entrainment. n. slag overflow. o. spray. p. fume extraction. q. brick lining. r. water-cooling brick retaining ring. s. air entrainment. t. brick lining.

pseudo-mystical procedures of the traditional steel maker. There is indeed a spate of innovation, much of it widely publicized only to disappear from discussion within a few months or years. Some of it will survive and contribute to the revolution that is making the industry one that is science-based and economically sound. A few of these developments can be discussed.

a. Spray steel making is one. Pioneered by BISRA, it has achieved an unprecedented rate of removal of carbon from pig iron, to such an extent that pig iron becomes low-carbon steel while falling through a distance of about six feet (two metres). The principle is simple. It is a way of achieving higher surface contacts between reactants than in any known steel making process. This is due to the splitting up of molten pig iron into droplets of a millimetre or so in diameter. This atomizing is done by the pressure of oxygen injected against the stream. The droplets fall in the oxygen stream with a basic slag material in powder form. The great heat created by the exothermic oxidation reactions melts the slag and any scrap steel used, while the oxygen removes phosphorus, silicon, carbon etc. The molten mass falls into a collecting vessel and the slag floats as foam and is allowed to overflow. The steel runs into a container which can be on wheels for easy transportation. The principle is illustrated in fig. 3.27, while fig. 3.28 compares the rates of carbon removal in different processes. It can be seen that the rate moves right off the diagram for spray steel-making.

In the pilot installation in the United Kingdom the molten pig iron came directly from the blast furnace nearby. The first commercial installation, with a capacity of 50 tons an hour, was made in the United Kingdom in 1967. The method could become part of changes to make steel-making continuous from the blast furnace to spray steel-making to continuous casting to re-rolling mills.

b. Direct steel-making has been dreamed about for years and methods of achieving it are being investigated. One United States concern has a device for using crushed ore and limestone with oxy-fuel heating, and then reducing the metal to 99% iron, to which additions are made to create steel. This direct reduction has already been referred to in the section on iron-making and it appeals to every physical chemist as the most

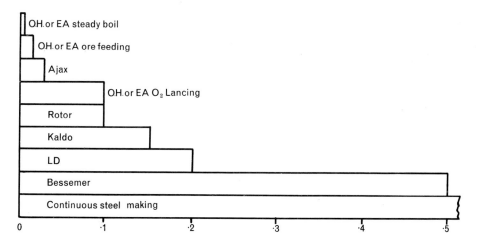

Fig. 3.28 Histogram of carbon removal.

obvious procedure i.e. make pure iron and add the needed carbon and other elements. Pelletizing has been described also in the section on iron-making. By reducing these pellets to nearly pure iron they can be used in an electric furnace for melting and refining. Japanese companies have practical schemes for achieving steel by this method. Sponge iron has also been referred to elsewhere. A Canadian Company is interested in using such sponge in an electric furnace. Whether any of these methods will get beyond the laboratory or pilot plant remains to be seen. They would achieve what forward-looking steel-makers have wanted, i.e. they would get rid of the blast furnace. However, in the world as a whole there is little doubt that this blast furnace will be part of steel-making for the immediately foreseeable future.

c. Powder metallurgy. Considerable research and development is going on into the use of techniques of powder metallurgy in the steel industry. If molten steel is atomized into fine spray, dried as a powder, and then formed into final customer product by pressure and sintering, the ingot and primary-milling stages are avoided. Steel powder is indeed used to make some components which can more readily be made this way than by casting or machining. Some confusion exists between 'iron powder' and 'steel powder,' the former term being used sometimes when the latter is meant. Sweden exports steel powder made by the Höganäs direct-reduction process. A new process (commissioning year 1969) in Canada, in which a British firm has a share, can produce up to 50000 tons a year of steel powder made from scrap. The process is a chemical one of turning scrap steel into iron chloride and reducing this with hydrogen, creating a sponge which is crushed. Carbon in the scrap is still in the powder, so it is in fact a steel powder. This process is versatile and not all the products will be steel powder. The British Iron and Steel Research Association has succeeded in making stainless steel strip directly from steel powder, and mild steel strip suitable for tinplate at a rate that is promising. The use of steel powder is increasing and production in the USA is in the tens of thousands of tons a year, so there is interest in the production of it as a route for certain steel products. One establishment in Britain is applying a gel precipitation technique, already successful in the production of nuclear-fuel pellets. Another process uses electrolysis to produce iron powder. How any of this development will affect the steel industry depends on its success in comparison with conventional processes, both qualitatively and economically.

d. Integrated steelworks. The changes already in existence are leading to the concept of the integrated steelworks. This is at the coast where ore is taken in, either in raw form for sintering or already sintered or pelletized, and goes to the blast furnace, from which the molten pig iron goes to the steel furnace and thence to the mills. By-product gases are integrated into the heating system and fumes of iron oxide into the sintering plant. Thus ore comes off the ships and steel products come out of the mills, foundries and fabricating shops. The oxygen-producing plant is integrated into the whole as, indeed, a nuclear power station could be to make the works self-supporting. In such an integrated works continuous casting and continuous steel production would fit naturally.

e. Instrumentation. Another important change is in instrumentation. The skill of the steel-maker has been very much a matter of personal experience and expertise closely guarded and valued. The open-hearth operator could sample his melt, judge slag fluidity and steel composition, and decide on change in fuel and reversals and additives.

The converter operator could judge with the flick of his eye the precise moment to stop a blow. The ingot operator knew just when to thrust in a handful of aluminium pellets for further deoxidation. And a man of one skill would never reveal his mystique to a man of another. All this expertise is a dying art, for instrumentation can effect analysis in a few moments. Spectrometers, pyrometers and flowmeters can do it all, and it is a short stage from such off-line analysis and measurement, to on-line automatic measurement, and thence to automatic feedback to control mechanical movement. Finally there is the computer to control a whole section of the steelworks. It has been done with tinplate production and the blast furnace. It can come with continuous steel-making. It can eventually control the whole of an integrated steelworks. The manpower structure of the industry is thus also in the process of change, the traditional work force diminishing to make way for the white-collar technician, technologist and computer operator.

3.4 STEELS AND CAST IRON

3.4.1 Definitions

Iron is the basic element of all steel and cast iron. Pure iron is silvery-grey and fairly soft. The specific gravity (relative density) is 7.87, melting point 1535°C, atomic weight 55.85 (based on oxygen as 16), atomic number 26. Four stable isotopes are known, the abundance of Fe^{56} being 92%. It has the outstanding property of being easily magnetized; so much so that the term ferromagnetic has been invented to describe this property, shared, to a smaller extent, only by two other elements, cobalt and nickel.

All steels and cast irons consist essentially of iron and carbon, one great difference between them being that the percentage of carbon present is less than 2% in steel and more than 2% in cast iron; there are other important structural differences, especially in the way in which the carbon is present. The cast iron produced in the world today, though considerable, is about one-seventh of that of all types of steel, which therefore stands out as the leading world metal. What is called malleable iron has a higher proportion of carbon than steel, but a lower proportion than that of industrial cast iron. Wrought iron, a term once more widely used than it is today, is more nearly a pure iron, nearly all the carbon having been burnt away. Its making involves beating and shaping, so that the archaic word 'wrought' meaning shaped or worked-on has retained its use. A rough qualitative line of progress can thus be made from pure iron to cast iron according to the carbon content: pure iron → wrought iron → steel → malleable iron → cast iron. However, the percentage of carbon is by no means the only factor in the composition of steels and irons; nor is there a figure to give the word 'steel' a precise meaning. Other important factors are the proportion of alloying elements, the treatment given in the production of the crude metal and the heat treatment carried out to produce desired properties. All of these combine to make the technology of iron and steel very complex and the vocabulary extensive and variable within the industry and between countries, so that no expert could know the whole of it without reference. Some terms are purely descriptive, some refer to the metallic content, others refer to the grain structure, and many are trade-names.

3.4.2 The phase diagram of iron and steel

Most of the properties of irons and steels are related to their crystal and grain structure, and these are related to the changes through which iron and carbon go during manufacture. Some knowledge of these changes is therefore necessary.

An established tool of the metallurgist and physical chemist is the phase diagram. This shows the different phases in which components can exist in equilibrium, the simplest example frequently given being that of H_2O, which can exist in three phases: solid, liquid, gas. If there are two components the diagram can be simply drawn showing how the phases change with temperature and the proportions of one component. If there are three or more components the complete phase diagram would be 3-dimensional and impossible to illustrate. What is done then is to take sections in the forms of triangles and many such diagrams have been made for the slags, fluxes and refractories of the steel-making process. The most generally-used diagram for iron and carbon plots temperature along the ordinate axis and carbon percentage along the abscissa axis. Lines are then drawn to show the boundaries between phases. Such a diagram, very simplified, is shown in fig. 3.29. Above the thick line everything is liquid and it can be seen that pure iron melts at 1537°C, whereas an iron–carbon combination of 4.27% carbon melts at 1152°C, and with higher-carbon content the melting point of the whole rises again. It is this near-4.27% carbon lowest melting temperature that determines the constitution of pig iron from the blast furnace. It is the eutectic composition that is retained until cold. The presence of other elements changes this temperature of eutectic composition. There is another similar point at about 0.8% carbon where a stable composition is formed, and as the material is solid above this, the word eutectoid is used to describe it. The point B corresponds to 2.0% carbon and all steels lie to the left of this. The symbols α (alpha) and γ (gamma) refer to the basic crystalline form of the iron, which can have three forms α, γ and δ (delta), but the last is formed only in nearly pure iron at temperatures above 1400°C and is of no interest in the context of this discussion. The alpha form of iron is based on what is called a body-centred cube, which can be imagined as a cube consisting of eight atoms at the corners, and one right in the centre. The gamma form of iron is based on a face-centred cube, which has the same eight atoms at the corners, but also has an atom at the centre of each face of the cube. This phenomenon of change of crystal lattice is called allotropy. [The word crystal is often used to mean different things. The basic lattice structure is truly the crystal lattice, and will determine how a crystal grows to form its final shape if conditions are perfect. The crystalline result is often called the crystal, but for clear discussion of microscopic forms of iron and steel it is better to use the term grain.] The important point about this change is that the face-centred lattice is bigger than the body-centred lattice, and it can take in more atoms of carbon. In other words gamma iron can hold in solid solution more iron than alpha iron. This gamma iron with carbon in solution is called *austenite*. The alpha iron with (much less) carbon in solution is called *ferrite*. On the extreme left of the phase diagram there is no carbon and the ferrite is pure alpha iron and the austenite pure gamma iron, which is, incidentally, non-magnetic.

3.4.3 The structure of steel
  a. *Austenite, ferrite, cementite and pearlite.* The phase diagram (again very simplified) of most interest in the discussion of steel is a small portion of fig. 3.29 to the left of B. As there is no change below the line HG, the diagram is not usually carried on to 0°C but is terminated at about 600°C. A modified phase diagram of this sort is shown in fig. 3.30. In this all the phases are solid, however incandescent the material

STEELS AND CAST IRON | 193

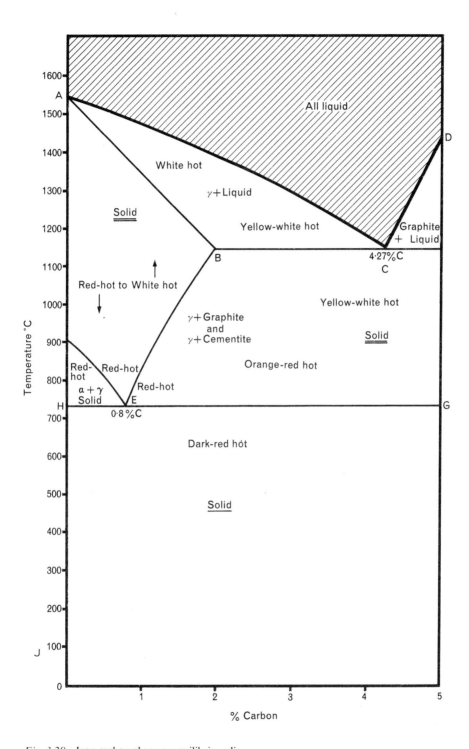

Fig. 3.29 Iron-carbon phase or equilibrium diagram.

194 | IRON AND STEEL

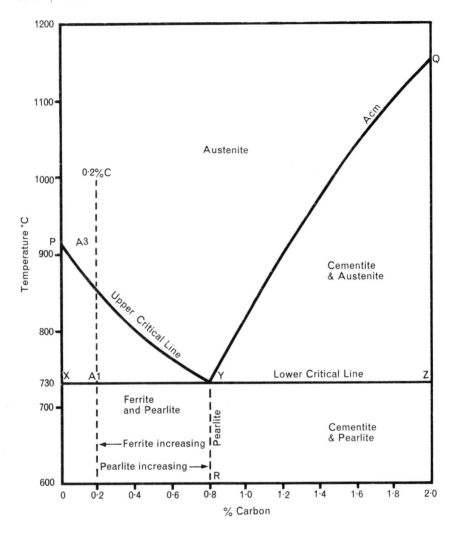

Fig. 3.30 Iron-iron-carbide phase diagram.

may be in reality. The point Y in this diagram does not represent a melting point. It must further be remembered that the diagram is an equilibrium diagram, and is idealized in that it represents what happens when molten iron has cooled very slowly indeed to about 1152°C and is cooled in the solid state also very slowly indeed. It does not represent what happens in practical steel making because alloying changes it and other things are done deliberately to change it, such as heat treatment (see section 3.5). Nevertheless it gives a basic understanding of the changes that would take place if not interfered with. One example will amplify this point. It can be seen that below 730°C, whatever the carbon proportion, no austenite exists. Yet there are austenitic steels in everyday use. How this is achieved will be discussed later.

Above the lines PYQ (or the A3 and austenite-cementite (Acm) lines) all the steel is in the form of austenite, the solid solution of carbon in gamma iron. It exists as polyhedral grains the size of which, unless interfered with, increases with the temperature. Now consider a carbon content of 0.2% in the austenitic phase. As it cools the grains get smaller, until at about 880°C the austenite starts rejecting iron and gets richer in carbon in the remaining grains until it reaches 730°C. The ferrite or iron grains remain but the carbon-enriched austenite changes suddenly. The carbon combines with iron to form a compound called iron carbide or cementite (Fe_3C) and forms with the remaining iron a structure called pearlite. This consists of very thin plates of iron and cementite. They are so thin that they allow some light to go through and be reflected back, causing a slight interference effect. This effect, which is the same as that occurring in pearls and mother-of-pearl, gives the somewhat iridescent effect that has provided the name pearlite. On further cooling nothing else happens; the steel is a mixture of ferrite grains and ferrite plates with cementite plates. If the same procedure is followed for 0.8% carbon the austenite changes at 730°C wholly into pearlite. At carbon contents above the 0.8% eutectoid composition, austenite changes first into cementite grains and austenite grains, but at the temperature of 730° ceases to be austenite and changes into pearlite as before, but this time there are cementite grains and pearlite, and the cementite 'grains' form the boundaries of the pearlite.

b. The grain size. As shown above the microstructure depends on temperature, so heating and cooling can affect this microstructure and therefore, the properties of the steel. Thus the hotter the steel the larger are the austenite grains. The temperature to achieve austenitic structure is called the austenitizing temperature, which is an essential feature of heat treatment. The grain size of the austenite influences the development of grains in the cooled steel, so that the correct choice of austenitizing temperature is important. In one example, from experiments with medium-carbon steel, it was found that the grain size at about 1070°C was 0.2 mm. diameter. At 900°C it was 0.127 mm. and at 760°C 0.045 mm. The conditions were such that the grain size was due to nothing but the heating. The thickness of pearlite plates also depends on the temperature. As steel cools below 730°C the first pearlites have a coarser structure, i.e. thicker plates, but as the temperature falls so the layers get thinner. The words 'coarse' and 'fine' applied to these layers must not obscure the fact that even in the coarse structure the layers can be of thicknesses comparable to that of the wavelength of visible light, giving the pearly lustre from which the term pearlite comes.

Grain size has an important bearing on the properties of steel. A metal is crystalline, which means that rows and layers of crystal lattices are connected together. Sir Lawrence Bragg used to demonstrate in a most elegant way the mechanical effects of distorting this structure. He had an array of bubbles on a liquid, the edges of the raft of tiny bubbles being two rods. He could distort the raft by moving one rod parallel to the other and he showed that as long as the rows stayed together the raft would resume its original shape. This corresponds to the elastic behaviour of a metal. At a certain amount of strain however, he showed that one row of bubbles would slide quickly over the other. The stress to achieve this slipping was small. Thus a single crystalline array of crystal lattices has weaknesses along the slip planes. He went further, and showed that if the orientation of the rows of bubbles was broken up into groups in different directions, the amount of sliding was less because the moving row hit a boundary of differently-aligned bubbles. This corresponds to the formation of grains in metals, randomly orientated. The boundaries between the grains allow only smaller slippings.

Thus the metal is strengthened by the formation of grains. Furthermore, the smaller the grains the greater the number of boundaries. The similarity between a raft of bubbles and a real metal must not be pushed too far, but the simple illustration does show the advantage of fine graining for greater strength and toughness, though the strength also depends on the individual strength of the grains themselves. Certain elements added to austenite can change the grain size. So can heat treatment.

c. Martensitic steel. Another important type of steel not yet mentioned is the martensitic, but as this is produced only as a result of heat treatment, it belongs to section 3.5. All that need be said here is that it is a distorted form of alpha iron (body-centred cube) in which cementite is in solid solution. It is very hard and brittle. Thus there are three forms of steel related to the grain material (but not grain size) viz. ferritic (which has ferrite grains of alpha iron with a little carbon), austenitic (which has grains of gamma iron containing carbon in solid solution, i.e. with carbon atoms in the interstices of the lattice and in the place of iron atoms) and martensite (which consists of needle-like grains of a solid solution of iron carbide in a distorted body-centred type of iron). These descriptive terms occur frequently in the steel literature.

d. The influence of working on the structure. The grain structure can be changed by cold-working or hot-working. The latter case is familiar: it is the rolling of ingots to make blooms, billets or bars. In this process the steel is red-hot and plastic, and the stresses of rolling under great pressure contribute a great deal to undoing some of the unwanted changes that have taken place in the ingot. The rolling squeezes the steel and breaks down needle-like structures and forces the grains into greater homogeneity. This obvious way of making semi-finished steel from the ingot is thus again a lucky development for the steel-maker, who would otherwise have to go to other time-consuming techniques. Forging has a similar effect: the blacksmith of old was not only fitting a horse shoe to the horse; he was producing a tougher steel. Cold-working is quite different. This can be rolling, drawing or extruding. The steel can be squeezed cold between rollers, drawn forcibly through hard orifices smaller than the original billet or bar, or pushed through an orifice under a force great enough to produce plastic deformation. The effect of such cold-working is to elongate the grains in the direction of movement, and without subsequent heat-treatment this grain structure will remain. The effect is to increase the hardness, tensile strength, Young's Modulus, and elastic limit, and decrease the ductility, so that when wire is being drawn it is necessary, depending on the amount of reduction in diameter effected, sometimes to give heat-treatment to restore the original ductility before proceeding with more cold-drawing. A familiar everyday example of the effects of cold-working is the breaking of a flexible steel wire by bending it quickly backwards and forwards in the fingers until it gets brittle enough to snap. The complex yield point of steel sheet can be removed by cold-rolling in one pass. This is done when a fairly weak steel is needed for pressings.

Enough has been said already, and more will be said in section 3.5, to show that steels in use are often abnormal in their microstructure. The behaviour indicated in fig. 3.30 is interfered with in order to produce special properties. In many steels the result is a solid solution forced into permanency, so it is natural to expect that it may change with time. And some do, especially carbon steel. This ageing appears as a hardening at normal temperatures, accelerated by heating the steel and by straining it. It is due to the precipitation from solid solution of elements such as carbon, oxygen and nitrogen.

Obviously well deoxidized steels will not age-harden much, with the inference that rimmed steels *will* age-harden. If the ageing is due to strain the effect is still some hardening but also some embrittlement. Designers must therefore give some thought to this phenomenon, when the usage can produce strain of tension or compression.

3.4.4 Metallurgical examination

Methods have been evolved to make the grain structure visible and so subject to observation. The piece of metal is first polished very finely to a mirror finish. It is then treated with a chemical which effects a selective dissolving of different constituents. There are several reagents for this purpose, the commonest being a solution of nitric acid in alcohol. The grains have preferential directions of chemical action, the edges of pearlite, for example, being more eaten away than the flat surfaces, and are unequally susceptible to erosion. Thus cementite is attacked more than ferrite. The result of all this etching is that the previously polished surface is very uneven, with surfaces of unequal reflectivity and with edges that cast shadows. Such a surface can be examined in a microscope up to about 1 000 magnification or, with a suitable replica made, in an electron microscope up to hundreds of thousands of magnification. Photographs of etched specimens figure prominently in discussions of iron and steel but they need experience for their understanding. Moreover photographs of different magnifications can be misleading. Figs. 3.31-3.35 show such photographs, fig. 3.35 being made in an electron microscope. Fig. 3.31 is a micrograph of plain ferrite showing the grain boundaries. Fig. 3.32 shows clearly the structure of pearlite where the overlapping edges of plates are alternating light and dark. This is the characteristic photograph of pearlite. Fig. 3.33 shows the structure of a high-carbon steel when cooled slowly. The white cementite boundaries to grains show clearly in the predominantly pearlitic structure. Fig. 3.34 shows a grey cast iron with some of its carbon as flakes of pure graphite. Fig. 3.35 is a much more enlarged picture, made with an electron microscope, of the sort of pearlitic structure shown in fig. 3.32.

3.4.5 Steel classification

Steel is usually described as an alloy of iron and carbon, the term cast iron being kept for the materials with more than 2% carbon and the term steel for those with less than 2% carbon. This description, however, tells very little, for steels vary widely in their composition, grain structure and behaviour. Description by words fails to specify every type, and the most detailed specification gives the type under terms to be discussed later as well as the composition in percentage, usually omitting the fact that iron is still the major element present. These specifications are coded, and as the demands change and different sorts of steel are made and deeper understanding of them is obtained, so these specifications have to be re-examined, re-coded and made official by bodies such as the British Standards Institution in Britain, the American Iron and Steel Institute in America and comparable authoritative organizations elsewhere.

The several standard specifications are not easily comparable. Thus the British En_2 steel (1955 coding) is described as having less than 0.2% carbon, less than 0.8% manganese, less than 0.06% phosphorus, less than 0.06% sulphur; whereas the AISI C1017 specifies 0.15–0.20% carbon, 0.30–0.60% manganese, 0.04% phosphorus maximum and 0.05% sulphur maximum. These two are seen to be very similar in composition and would have comparable properties, but the specifications are not exactly the same. This lack of strict comparability is not normally important to the steel-maker,

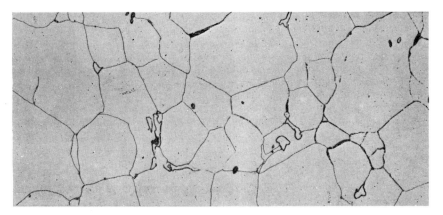

Fig. 3.31 Ferrite micrograph.

Fig. 3.32 The microstructure of pearlite 500x. The microstructure of slowly cooled, high-carbon steel showing pearlite with cementite in the grain 500x.

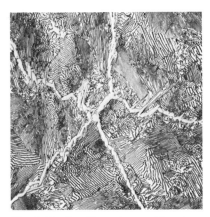

Fig. 3.33 Pearlite and cementite micrograph.

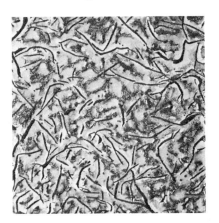

Fig. 3.34 The microstructure of gray cast iron showing graphite flakes.

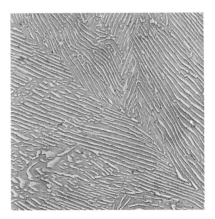

Fig. 3.35 Teletron micrograph showing the microstructure of fine pearlite formed at 1 100°F 750x.

because he is working in the context of his own national specification and market, but it is important in international trade and makes it necessary for anyone not in the industry to use great care in describing steels.

Many ways exist for qualitative description of the many steels that are made. There is the end-use method in which terms exist such as locomotive structural steels, bridge-building structural steels, armature steels, etc. A bridge is a static structure but is subjected to recurrent irregular stresses and vibrations as well as being open to the air and moisture, whereas an armature endures high temperatures and high-speed vibrations and centrifugal forces. The same steel would not be suitable for both uses. A method of description based on the melting procedure, such as 'basic open hearth', 'acid Bessemer' and so on, must eventually vanish as oxygen steel-making and electric steel-making take over in mass production, and methods of analysis that are more precise come into use.

The terms austenitic, martensitic and ferritic are also used to typify steels and refer them to the nature of the metallurgical composition. Another descriptive term commonly used is one indicating the amount of alloying element present, such as, for example, 18-8 steel, meaning ore with 18% chromium and 8% nickel, a common stainless steel. Mechanical properties such as hardness, toughness, strength and ductility are also used in the description of steels and are important in decisions about their use. A steel that will be used under conditions where it can suffer rubbing stresses (abrasion) must be hard on the surface, whereas one that has to withstand impact shocks must be tough. The meaning of these terms can be qualified according to certain methods of testing. Even a steel of one composition will change its properties according to the treatment it is given such as hot-rolling, cold-working and heat treatment (to be discussed in Section 3.5).

There is thus no one way of describing every aspect of any particular steel, but discussion cannot go on without a method of classification, however crude. The most general way is first to distinguish between carbon steels and alloy steels, then subdivide the carbon steels according to the carbon content and divide the alloy steels into groups that indicate the limits of alloying elements as well as the purposes of certain of the alloy steels. Such a simplified classification is widely used. Carbon steels are specified as if there were no other elements present, whereas as soon as some of these elements reach a certain proportion the steels become alloy steels. Carbon steels can be divided thus:

Low carbon steels	(a)	less than 0.07%C
	(b)	0.07% to 0.15%C
Mild steels		0.15% to 0.25%C
Medium carbon steels		0.20% to 0.60%C
High carbon steels		0.61% to 1.40%

The first two are frequently considered together as mild steels and constitute the greater part of structural steels. Not stated in this classification is the amount of other elements present. Thus a mild steel could have the composition:

0.15%C   0.03%Si   0.05%S   0.05%P   0.5%Mn.

Iron, unstated, thus constitutes 99.22% of this steel. The very small percentage of sulphur and phosphorus are taken for granted, many methods (as already discussed in Section 3.3) being used to reduce them as much as is economically and technically possible. Ideally there would be none at all. The British Steel Corporation, for statistical purposes, specifies that high carbon steel contains by weight not less than 0.60% carbon and not more than 0.04% phosphorus or sulphur and not more than 0.07% of the two added together. For the purposes of the Trade and Navigation Accounts in Britain, steels containing less than 2.0% of manganese or 0.3% of tungsten or cobalt or 0.5% of chromium or nickel or 0.1% of molybdenum or vanadium are classed as carbon steels. Steels with more than these proportions of any one of the elements would be classed as alloy steels. In the USA, however, there are some steels classed as high-strength low-alloy steels which would be called carbon steels in Britain. When such borderline differences are ignored, steels with elements added as alloying material in order to produce special properties are again subdivided into alloy steels, stainless steels, heat-resisting steels, tool steels, etc.

3.4.6 Carbon steels

Steels with under a certain small proportion of elements other than carbon are carbon steels. They invariably have manganese in them from the deoxidizing they have undergone in the steel-making, but as long as this does not exceed 2% and other elements are present in negligble amounts, they are still classified as carbon steels in the UK for the purposes of the Trade and Navigation Accounts for trade figures. For general purposes of definition, if other elements are negligible, the percentage of manganese can go up to 10% for classification as carbon steel. The manganese not only acts as a deoxidant; it also makes sulphur into manganese sulphide instead of iron sulphide and thus nullifies the embrittling effect of sulphur. It is thus a beneficial element in facilitating forging and hot-rolling. It increases tensile strength for a given carbon content. When present in higher proportions it improves the hardening effected by cold-working.

There are many of these carbon steels, depending on the carbon content, manganese content and treatment given to modify their properties. A few of them can be considered here, though they make by far the greatest output of steel in the world.

a. Low carbon steel. As ferrite is soft, tough and ductile, steel with very little carbon, falling on the left of the iron-iron carbide diagram (fig. 3.30) has these properties. Normally its tensile strength is low but the material is suitable for tubes and pressings. Such a low-carbon steel does not behave well when machined because it tears under the impact of the hard cutting tool, and thus is not suitable to have a screw thread cut on it. There is a variety with higher manganese and much more sulphur. It is called free-cutting or rapid-machining steel. One composition is 0.12%C, 0.85%Mn, 0.23%S, 0.005%Si, 0.03%P. The easy machining makes it a general-purpose soft and inexpensive steel for nuts, bolts, studs, etc. It has a yield point that ranges from 16.8 to 26 tons per square inch (259 to 401 meganewtons per square metre) according to whether it is used as hot-rolled or cold-drawn. The corresponding maximum stress or tensile strength is from 24.8 to 26.8 tons per square inch (383 to 413 meganewtons per square metre). With slightly more carbon a mild steel (with up to 0.25% carbon) is suitable for forged, machined or cold-worked parts.

Mild steels such as these account for thousands of small engineering and domestic devices familiar in everyday use. Soft mild steels make the bulk of sheets and tinplate and are used for a considerable amount of steel wire. As the proportion of carbon goes up and as therefore, steels are considered that move to the right along the bottom of fig. 3.30 so the hardness increases in proportion. The tensile strength and yield strength also increase up to the eutectoid 0.8%, and then increase only slightly. The ductility goes down as the carbon increases to 0.8%, and then decreases more slowly. The toughness also decreases very rapidly to the same eutectoid composition and then much more slowly. The importance of the eutectoid point in separating the fast change from the slow change, is such that the steels are frequently described in some contexts as hypo-eutectoid (0–0.8% carbon) and hypereutectoid (0.8–2.0% carbon). What is happening is that as the steel composition moves to the right along the bottom line of fig. 3.30, so the proportion of pearlite is increasing, and being of relatively small grains, is stronger and harder. Past the eutectoid point, however, the hard cementite is increasingly formed, and the effect is to slow down the rate of increase of toughness and strength and the rate of decrease of ductility, whereas the hardness continues to increase at the same rate. Hardness, therefore, can be said to be directly proportional to the carbon content. So is the strength up to eutectoid composition. Ductility and toughness are, up to the same point, inversely proportional to carbon content. This is why carbon steels are commonly classified according to carbon content. With this the manganese content is often stated while the content of silicon, phosphorus and sulphur is assumed to be below a certain small percentage.

Composition, especially the carbon content, is not the only factor in determining the properties of steels. The other is the microstructure – the size and nature of the grains. These grains range from about 4 per square millimetre to 160000 per square millimetre, that is from an average diameter of 0.5 mm to 0.0025 mm. However, in working steels, a material with an average grain diameter of 0.032 mm would be called fine-grained. The very coarsest grains are just about visible to the naked eye, but in general they are seen only by means of a microscope, and the very fine structures only by means of the electron microscope. In carbon steels made from melt to ingot to billet, nothing much can be done about this microstructure except in the killing of an ingot and in the rolling. To change the microstructure effectively alloying elements must be introduced or there must be heat treatment, or both. These will be discussed later.

There are many mild steels besides these detailed above; the USA lists about a dozen. They sometimes have lead added in a very small amount, less than 0.25%, distributed in sub-microscopic particles, to make them more machinable. Copper is sometimes added to improve the resistance to atmospheric corrosion and improve strength and hardness. It is also added to medium-carbon steels for the same purpose.

b. Medium carbon steel. The USA lists twice as many medium-carbon steels in the range 0.25% to 0.60% carbon. They make most of the structural steel commonly used. They can be forged and rolled and machined. In the lower-carbon range they are suitable for small forgings, in a higher-carbon range for large forgings, and in the form of bars are used for a wide range of parts. With somewhat higher manganese content they are suitable for helical springs. Their tensile strength ranges from 20 tons per square inch to 40 tons per square inch (310 to 620 meganewtons per square metre). They make structural steels for bridges, buildings, ships, locomotives and cars, and rivet steels for all purposes. They are used for strong steel wire and rods for concrete-reinforcement.

c. High carbon steel. High-carbon steels (from 0.61% carbon upwards) are of course the hardest of plain carbon steels and they have high tensile strength. They used to be the only materials for tools. One very important use for these steels is in the making of wire for wire ropes and cables, for valve springs for engines, for spokes for wheels and for pianos. In the last-mentioned the tensile strength needed may be up to 390000 lb or 174 tons per square inch (2690 meganewtons per square metre), though this must be related to the small diameter of the wire to see that the total force involved in much smaller. This tensile strength is higher than for any other steel use. Another important use is in rails for railroads and other laid tracks and for heavy-duty wheels.

3.4.7 Special steels

Steels with elements other than carbon deliberately introduced into them have different properties from those of carbon steels. They have increased in scope and volume and are still increasing and more varieties can be expected as innovation brings new needs. They are often described as alloy steels, but this expression is now frequently used for a specific range of special steels which can be considered separately. Special steels include corrosion-resistant, or stainless steels, heat-resistant steels, tool steels, spring steels, maraging steels (see section 3.4.7(e)), valve steels, electrical steels, etc. All these constitute only a small fraction of the world's steel tonnage but in economic value and technological importance they count for more.

All carbon steels contain elements other than iron and carbon. These include gases, manganese, phosphorus, sulphur, silicon–all of which have been referred to elsewhere–as well as perhaps nickel, copper, molybdenum, chromium and tin picked up from scrap, and deoxidation elements such as aluminium, titanium, vanadium or zirconium. A valid enquiry, therefore, is how to distinguish such steels from special steels. Is there a certain proportion above which the steel is alloyed and below which it is not? Attempts have been made to resolve this difficulty. Thus, as already stated, for the purposes of trade statistics in the UK the Trade and Navigation Accounts stipulate 2% of manganese or more as taking steel out of the carbon-steel category, and there are specified proportions of other elements, but in the USA the comparable limit for manganese is 1.65% so that there is a fringe area where a universal definition of special steels is impossible. What can be said is that when the amount of another element present is enough to modify the iron-iron carbide diagram of fig. 3.30 to any significant extent, the steel ceases to be carbon steel and becomes special steel. The person least worried by difficulties about definition is the steel-maker himself, who decides on a composition, the method of making, and the treatment to be given so that the final product shall satisfy the customer.

R. F. Mushet first used tungsten in steel in the nineteenth century, and Sir Robert Hadfield, in about 1913, made steel with more than 7% manganese and went on to introduce other steels with alloying elements in them. People were by this time working on similar techniques in other countries. In 1902 what is now the United States Steel Corporation supplied nickel steel for the Queensborough Bridge connecting Manhattan to Long Island. Harry Brearley in Sheffield introduced chromium, in 1913, and so made the first corrosion-resistant steel, but the present common range of these stainless steels was developed by Benno Strauss in Germany very soon afterwards.

The most commonly used alloying elements are manganese, nickel, chromium, and molybdenum. Others are tungsten, cobalt, copper, aluminium, silicon, niobium (or columbium), titanium, vanadium, beryllium, boron, tantalum, selenium and zirconium. No enquiry into common fundamental properties of all these reveals a guiding principle that decides their behaviour in steel. What they all help to do, in sufficient quantities (and slight differences in their individual properties makes one special steel differ in properties from another), is modify the equilibrium diagram of fig. 3.30. Furthermore this can no longer be drawn as one diagram because of the innumerable combinations possible, but series of such diagrams can be drawn, one for each type of alloy. Their final effect, for the cold steel, is a change in the microstructure—the size and shape and nature of the grains–and this is bound up with the hardenability, which depends on heat-treatment behaviour. It is this microstructure that determines the yield point, tensile strength, toughness, hardness and fatigue strength of the used product, and it is these mechanical properties that determine the sphere of usefulness of the steel in general, though there are other properties for special purposes such as high resistance to corrosion, or resistance to high or very low temperatures.

A few of the effects of alloying can be stated more specifically. Thus silicon and manganese harden ferrite; nickel both hardens and toughens; manganese and chromium refine pearlite to smaller grains; and tungsten, vanadium, chromium and molybdenum are good carbide makers (with carbon), so that when these are present in large proportions all the carbon of steel is bound up in complex carbides. More generally, some alloying elements go into solid solution in ferrite, some form carbides, some change the distribution of cementite in ferrite, some favour the formation of austenite, some favour the formation of ferrite.

a. Alloy steels. For the purpose of this discussion alloy steels are restricted to steels of wide general use and having comparatively small amounts of alloying elements. Every country has a range of such steels. In general the percentage of any one alloying element is less than 2%, with the exception of some with nickel up to more than 4%. The actual alloying elements can be varied, because in this usage of them (to increase hardenability and improve microstructure), one element is as good as another, and choice rests on cost and availability. There is one group of *low-carbon alloy steels* in which the carbon does not exceed 0.2% and yet, because of the alloying and heat treatment, they have the toughness and weldability of low-carbon steel combined with increased strength and hardness. Such steels have been used for pressure vessels and mining equipment as well as in structures. Such steels have been made with tensile strengths of up to 60 tons per square inch (925 meganewtons per square metre). Their resistance to atmospheric corrosion is several times that of carbon steel of comparable carbon content. They can frequently be used in smaller sections and so be economically preferable to carbon steel. A typical composition would be: 0.1%C, 0.60%Mn, 0.15%Si, 0.70%Ni, 0.40%Cr, 0.40%Mo, 0.03%V, 0.002%B, 0.15%Cu. This represents US Steel's 'T-1' steel.

Another group has slightly higher carbon, and slightly more of one or more constitutuents, and makes what are called *high-strength low-alloy steels*. The most famous of these is COR-TEN from the US Steel Corporation. It was the first of the range and was introduced in 1933. Since then these high-strength low-alloy steels have been greatly improved. They are more springy, because of the high yield point, than carbon steel and need more force for cold-working. Their abrasion resistance is somewhat higher than, and their fatigue resistance superior to, comparable carbon steels. They are tougher and they can be welded. These steels have been used to save weight in transport equipment, and have found applications in bridges, television towers and containers for liquified petroleum gases. Because of its superior resistance to atmospheric corrosion COR-TEN has been used in bare conditions for exposed parts of high-rise buildings, a fact that Dr. H. M. Finniston of the British Steel Corporation underlined as a pointer for the future, when he addressed the International Iron and Steel Institute in Los Angeles in 1968. The composition of COR-TEN for plates under 0.5 in(1 cm.) in diameter is: 0.12%C, 0.20–0.50%Mn, 0.07–0.15%P, 0.25–0.7 %Si, 0.25–0.55%Cu, 0.65%Ni, 0.30–1.25%Cr.

The final group of these alloy steels has somewhat higher carbon contents extending into the medium-carbon range: these are called *medium-carbon alloy steels*. It is the largest group of the alloy steels and the one that most people would think of under this heading. In them small amounts of copper, nickel, chromium and molybdenum under the respective proportions of 0.35%, 0.25%, 0.20%, and 0.06% are considered incidental, and the proportions that count are well above these limits. The low-carbon steels of the group are intended to be case-hardened and used for cam shafts, clutch fingers and other parts needed in the motor industry. Those with higher alloying are for gears, piston pins, universal joints, aircraft-engine parts, even rotary rock-bit cutters etc., and all are intended to be case-hardened. There are many varieties for different purposes but in general it can be said that it is the growth of the motor industry that has brought them into being, though they are used as constructional steels for such things as machinery axles and shafts.

All the alloy steels briefly described above make the bulk of production of special steels. The economics of them has changed so much with demand and with the deeper understanding of the metallurgy of alloying, that they may yet challenge the carbon steels as oxygen steel-making, electric-arc steel-making, and other modern aspects of the industry take over from the old furnaces and indeed the blast furnace.

b. Corrosion-resistant, or stainless, steels. These depend mainly for their special property on chromium, which forms a very tough oxide film on the metal surface and this acts as a protective coating against the environment. Alloying, however, affects the microstructure of steel, and other elements have been added to provide certain desirable properties, the commonest composition being that of 18% chromium and 8% nickel in a low-carbon steel, the one known throughout the industry as 18-8. The most familiar environment is the atmosphere, which rusts and destroys steel (because surface rust peels off and presents fresh metal), so that large and intricate structures have to be perpetually painted. Industry also supplies its own corrosive elements, whether acid or alkaline, oxidizing or reducing, so that ordinary steel with all its advantages of strength and low cost cannot be used. Thus corrosion-resistant steels are needed for a variety of environments, which include the temperature at which the steel is used. In addition, for different sorts of components and structures, the mechanical properties(tensile strength, hardness etc.) must be suitable. The properties needed for processing during manufacture–whether the steel is to be hot-rolled or cold-worked into bar or plate or sheet or wire, or forged or cast into shape make further variants demanding a more extensive range of steels with the needed corrosion-resistant properties. Availability and cost of alloyant metals also affect the composition as when, during a nickel shortage, substitutes were tried which led after considerable research, especially in the USA, to the use of manganese and nitrogen together instead of nickel. A consequence of all this is that there is no one stainless steel; there are dozens of stainless steels. They make up some 20% of the special steels produced.

The composition of some stainless steels is given in table 3.6, taken from US data.

Table 3.6 Composition of some stainless steels

Type	C %	Ma % max.	Si % max.	P % max.	S % max.	Cr % max.	Ni %	N %
Martensitic	<0.15	1.00	1.00	0.04	0.03	12.5	—	—
Ferritic	<0.12	1.00	1.00	0.04	0.03	16.0	—	—
Austenitic	<0.15	2.00	1.00	0.045	0.03	18.0	8.0	—
Austenitic	<0.15	7.5	1.00	0.06	0.03	17.0	4.5	0.25

These have been chosen to illustrate three main types of stainless steel, and a fourth which shows the use of increased manganese and some nitrogen and a decrease in nickel. The first two are, apart from the usual Mn, Si, P and S, iron-chromium-nickel steels. The second has more than 12.5% chromium. They are representative of groups called martensitic, ferritic and austenitic because of their predominant structure at working temperature, the third being a solid solution of carbon in gamma iron. (How this can be made to exist at normal temperatures will be discussed in section 3.5). These are important groupings because their properties and uses are different. The martensitic stainless steels can be made very hard by heat treatment, and their limit of chromium is about 12.5% as shown in the table. With more chromium the steels become ferritic and when nickel is added they are austenitic. Neither ferritic nor austenitic stainless steels can be hardened by heat treatment. Austenitic stainless steel, however, can be hardened by cold-working.

Austenitic grades of stainless steel are the most widely used. They have excellent corrosion resistance because of their high chromium content. They are tough and duc-

Fig. 3.36 Goonhilly satellite station. The 'dish' is made of stainless steel.

tile and so can be deep-drawn (i.e. formed into hollow containers by pressing). They are used for kitchen equipment and utensils, dairy and food-processing gear, and in the oil-processing and chemical-processing industries. Varieties with lower carbon contents are welded with less subsequent trouble. If these also have higher chromium they are suitable where there is severe corrosion risk. Ordinary austenitic stainless steels cannot be used for articles that cannot be cold-worked by rolling, a motor-engine valve for example. Special stainless steels that can be age-hardened have therefore been developed such as the US 'Stainless W.' Austenitic types, unlike the other two, can be used successfully at cryogenic temperatures.

Ferritic stainless steels are much less tough than the austenitic, but they have certain advantages such as increased resistance to some types of corrosion, for example, that due to nitric acid. So they are widely used in chemical plants. They can be drawn easily and so have uses in motor-cars and buildings.

Martensitic stainless steels can be made with a wide range of mechanical properties because they can be hardened by heat treatment. They are used in the oil industry for ballast trays and liners, for blades and buckets in steam turbines, for valves, valve-seatings and shafting. They are the grade that provides surgical instruments, and cutlery of all sorts, and it is interesting that one formulation for this purpose is just about the same as that used by Harry Brearley in 1913.

Stainless steels are not in general resistant to every possible corrosion and new corrosive environments are created by innovation. They present welding problems because of the depletion of the protecting chromium near the weld, and these problems have to be anticipated by special formulations. But they are made now in massive electric-arc furnaces and they have increased in range and quality. So they are replacing older steels and will continue to do so if the cost can be reduced. It is unlikely that any housewife or hotel manager would choose to go back to the old high-carbon steel for cutlery. There has been a minor social consequence of the change brought about by corrosion-resistant steels, in that specialized tableware such as silver fishknives and forks are now an elegant anachronism.

c. Steels for use at high temperatures. High or elevated temperatures present difficult problems for the steel-maker, because of the variety of environments in which the steel must retain its designed properties. These environments can vary from steam boilers to the red heat of an engine exhaust-valve or the inferno of a gas turbine. Added to the heat hazard there can be a corrosion hazard from liquids or gases. As so many steels owe their properties to the controlled heat treatment given them before they leave the steelworks, it is difficult to forecast just what will happen to these same steels subjected to an uncontrolled heating. As a result there are many attempted solutions to the problems but there is no one simple solution.

Above about 540°C there is the problem of corrosion by oxidation. Below this temperature carbon steels can be used in air without much risk, but above it the corrosion increases at an increasing rate, so that at about 700°C it is twice what it is at 650°C. The element most effective in preventing this is chromium, and the more there is the better is the resistance to oxidation. If a smaller amount of silicon is added the result is still better. The effect is about the same in several other environments experienced in industry such as sulphurous gas, flue dust and hydrogen. Nickel, however, is sensitive to SO_2 and H_2S, and aluminium with chromium and silicon have been found useful against H_2S attack at 1 000°C.

The chief general effect, however, is due to creep, the slight plastic flow at high temperatures that weakens steel and distorts it. In carbon steels the carbon can be kept down to less than 0.2% with good results at temperatures that are not too high. The additive metals that improve creep strength are mainly molybdenum, tungsten and vanadium. The higher the temperature the greater is the tendency to upset the microstructure of the steel, by depositing some of the carbon as graphite and turning the iron carbide into spheroidal granules. In some steels chromium carbide is deposited at the grain boundaries, but this does not weaken the metal. Oil-cracking distillation tubes have been made of steel in which this happens without damage. On the other hand, there are steels in which chromium–iron compounds are deposited at the grain boundaries and make the metal brittle. There are carbide stabilizers such as titanium, niobium or tungsten.

By careful manipulation of the right alloys (whether in alloy steels, carbon steels or stainless steels) in the right amounts, and with suitable heat treatment steels can be produced which are suited to the particular use. Some of them are given separate names such as valve steels and boiler steels. One famous class of such alloys has wide use, especially for turbine blades. They are the nimonic alloys (and their casting equivalent nimocast) in which the iron is as low as 5% while the nickel is up to 50% and more. The composition of one of them shows iron as low as 1%, hardly meriting the name steel. It has 15% chromium, 4% titanium, 5% aluminium, 15% cobalt, molybdenum from 3.6% upwards, the rest being iron, copper, manganese and carbon in very small amounts and the balance nickel.

d. Tool steels. Steels and other metals are used for thousands of products of all sizes. Some are made by casting into a mould, some by forging, some by rolling. Extruding, die-stamping and other manipulations are used. One very widespread method of shaping is done by a cutting tool fixed on a machine such as a lathe. This tool must be harder than the metal it cuts, and it must retain its hardness and cutting edge when made very hot by the friction of cutting at high speeds. Tool steels are designed for such work and there are many sorts. They are all high-carbon steels to start with, and can have small amounts of alloying elements or they can have enough to classify them as alloy tool steels. Some have less carbon if the tool must withstand impact. The heat-treatment of these tool steels is of the greatest importance. There are tool steels that must resist wear and keep a smooth cutting edge. These have intermediate amounts of alloy. High-speed tool steels have to retain their hardness when very hot. They are the ones containing up to 18% tungsten and 4% chromium with some vanadium.

e. Maraging steels. A remarkable series of steels called maraging steels has been developed by the International Nickel Company. In composition their outstanding feature is the reduction of carbon content nearly to nothing. This means that when they are changed to martensite, by heat treatment, they are ductile and relatively soft, but when they are aged deliberately they become hard, strong and tough – hence the name from *mar*tensitic *age*ing. Thus they can be fabricated into shape while ductile. The ageing is believed to be due to the precipitation of nickel-molybdenum and nickel-titanium alloys on grain boundaries. They contain up to 0.01 carbon and nickel in fairly high proportions, with molybdenum, cobalt or chromium, titanium and aluminium. The main class of maraging steels contains 18% nickel and there is a class of tougher steels having 12% nickel. They all have very high tensile strengths – 200 000 to 300 000 pounds per square inch (1400 to 2100 meganewtons per square metre)–and good notch-toughness.

f. Other special steels. There are special steels for permanent magnets, developed over the past half a century. They are alloys which vary according to the magnetic and steel qualities required and many have proprietary names such as the Alnico steel containing iron, nickel, aluminium and cobalt, developed at Sheffield in 1934 and Ticonal (nickel-aluminium-cobalt). Many new ones are being produced. And the special steels for transformer cores, containing silicon to make the steel electrically resistant so that eddy-current losses are extensively reduced, must be mentioned. Some steel-makers distinguish steels for special uses, e.g. *spring* steels. Among the thousands of steels made only a few have been discussed here. In many of them the heat treatment is as important as the alloying.

3.4.8 Cast iron

In the iron-and-steel industry steel is by far the chief product but cast iron, though a fairly small proportion of the output, is widely used and remains the second structural metal in world use. It is the material for all engine blocks, crank-shafts, brake drums and transmission housings – the automotive industry consumes nearly a million tons a year in the UK alone. The machine-tool industry consumes nearly 200 000 tons a year; some two-thirds of a million tons a year go into piping; 150 000 tons are used for manhole covers and other gratings and covers. Its excellent properties for accurate casting, with little need for finishing afterwards, make it a preferred material in thousands of engineering applications.

Pig iron is obviously a cast iron, but the lack of fine control over its constituents makes it crude, serving as the raw material for steel and usable cast irons. Pig iron has a high carbon content of 3.5 to 4.5%. This carbon can exist in several physical forms, and it is the business of the maker of cast iron to control the form of carbon to produce what sort of cast iron he requires. In order to do this he remelts the pig iron with additional materials, and cools it in the right way. The commonest remelting kiln is a cupola, which is a simplified blast furnace. Other methods include the use of an air furnace, not unlike a puddling furnace, and the electric furnace.

A cast iron is obviously a metal that can be cast. That is to say, the molten metal can be poured into moulds and allowed to solidify. To achieve such casting is the work of the iron foundry. It is a direct, inexpensive way of achieving complex shapes. However, the composition can vary considerably and the technique of cooling is very important – several sorts of cast iron can be produced from the same molten metal by different rates of cooling. The main property desired is machinability (drilling etc.) and this depends on the structure. A very hard cast iron, for example, cannot be machined because it wears out the cutting tools. Another property is the tensile strength. There is also the brittleness. This is measured by a sudden blow in a testing machine and tabulated as the impact strength. These and other properties of various grades of cast iron are covered in leading countries by standard specifications, those in the UK being the British Standard (BS) specifications.

Cast irons are divided into groups that depend on the composition and microscopic structure. The most commonly used are the grey cast irons. In these the carbon exists as flakes of graphite and the iron as pure iron (ferrite) or pearlite (a layered structure of ferrite separated by iron carbide boundaries). Too much silicon present has a harden-

HEAT TREATMENT AND HARDNESS OF STEELS | 209

Fig. 3.37 London Transport Victoria line underground tunnel, lined with knuckle-jointed cast-iron segments.

ing effect and reduces machinability. The grain-size and proportions of pearlite to ferrite have an important bearing on the properties. White cast irons have the carbon present only as the compound iron carbide. They are extremely hard and unmachinable but are excellent for crushing and pulverising components. Appropriate heat treatment to break down the carbide and precipitate some carbon produces malleable cast irons, which have enough ductility to replace steel for some uses. Nodular, or spherical-graphite, cast irons have the carbon in the form of near-spherical nodules. These are produced by the addition of other metallic elements, and they have qualities comparable to those of the malleable cast irons with greater potential for variation. By adding certain elements cast irons can be made to have special properties such as resistance to corrosion.

3.5 HEAT TREATMENT AND HARDNESS OF STEELS

3.5.1 *Heat treatment*

The versatility of steels, with varieties to suit all purses and purposes, is due mainly to two factors. One is alloying, which has added a new dimension to the steel industry. The second is heat treatment. Historically this was known, for one special application, in ancient times when smiths made Damascus blades from wootz, and more recently when the cutlers with their jealously-guarded skills produced swords and scythes and sickles. In the past half a century much, but not all, of the mystery has been made clear, so that heat treatment today is a complex technology of its own which can be discussed only in general outline here (the basic concepts are discussed in Chapter 1).

Heating and cooling are part of the processes of steel-making from blast furnace to billet, but heat treatment as such refers to a subsequent process. It is a way of arriving at a desired microstructure – in order to get the required properties – by the controlled manipulation of heating and cooling at known temperatures and prescribed rates.

Understanding of the process begins with interpretation of the phase or equilibrium diagram (see fig. 3.30 and section 3.4.3) which, it must be remembered, relates only to iron and carbon – alloys will be considered later. All the discussion of heat treatment will in the first place be concerned with plain carbon steels.

The temperature at which a microstructure forms is already important, but it is not the only factor in steel transformations. Steel deoxidized by a very small amount of aluminium, for example, contains aluminium nitride formed in the solid by the combination of aluminium with nitrogen. This nitride, formed in fine dispersion, prevents grains from enlarging when the steel is heated.

The one unavoidable factor not yet considered is *time*. The diagram of fig. 3.30 is relevant only to very slow change. It takes time for carbon to diffuse through its surroundings, though this time may be small. E. C. Bain, formerly vice-president of the United Steel Corporation, and an acknowledged authority on alloying, has given a figure, for one case examined, for the migration of carbon atoms from iron carbide in a ferrite surround. The carbon moved 0.0003 inch (0.0076 mm.) in 5 seconds in solid steel at 746°C; this diffusion time also depends on the temperature. Chemical reactions take time, as is seen in the chemical industry when catalysts are essential to reactions that otherwise would take far too long. The establishment of lattices takes time, however short. Time enters into the transformations of heat treatment in yet another way. It is obvious that when a piece of steel is heated the heating time depends on the *heat conductivity*, the heat transfer between the steel and its surroundings, and the *size* of the specimen. A small piece can be heated through much more quickly than a large piece. The same applies to cooling. In data on heat treatment, therefore, the limiting size of the article treated is frequently stated and the temperature and conditions are set to suit the size. It is frequently stated by authoritative experts that because of the time factor the equilibrium diagram of fig. 3.30 is irrelevant, yet it is an essential starting point for an understanding of steel, although by manipulation of the time factor (as well as the temperature) the steel-maker can interfere with the steel and control the microstructure. Thus austenite according to fig. 3.30 does not exist below about 730°C. But by making things happen fast it can be made to exist even at room temperature as seen in the austenite stainless steels.

The starting condition for all heat treatment is austenite, the solid solution single phase, though this condition must of course first be created by heating. Much of the forging and rolling of steel is done on the material in the austenite condition, and it is then allowed to cool on the bed after the working, and this cooling though not accelerated is fairly fast. So 'as-rolled' steel has had heat treatment, though this was not recognized by older steel-makers – one more example of the lucky accidents that established good steel practice before there was scientific investigation of the processes. Usually, however, the as-rolled steel is not considered to have had heat treatment and the expression is mean specifically to apply to the deliberate heating to austenite temperatures and the controlled cooling, which for the most important type of heat treatment is followed by reheating to a temperature below 730°C.

HEAT TREATMENT AND HARDNESS OF STEELS | 211

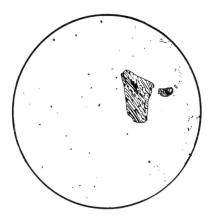

Fig. 3.38 Transformation to pearlite in a eutectoid steel arrested at its beginning. Note nodular growth from nuclei and simultaneous formation of both ferrite and carbide. 500x.

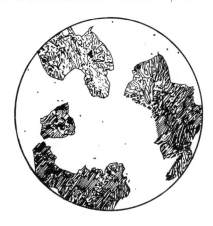

Fig. 3.39 Transformation to pearlite arrested when 25% of the austenite had transformed. 500x.

Fig. 3.40 Transformation to pearlite arrested when 50% of the austenite had transformed.

Fig. 3.41 Transformation to pearlite arrested when 75% of the austenite had transformed.

The first time-effect to be noted is how long it takes to transform austenite into carbide and ferrite. According to fig. 3.30 these transformations start on the A3 or Acm line as the steel cools, but they take time. If not enough time is allowed the transformations are slowed down and incomplete. So the faster the cooling rate the lower is the temperature at which the transformation starts. In other words, whereas according to fig. 3.30 all the steel below 730°C is ferrite and carbide in one form or the other (according to carbon content) it is possible to supercool austenite much below this before the atoms have time to diffuse and reassemble. The lower this starting temperature is, the thinner are the plates of carbide in the pearlite formation. This means

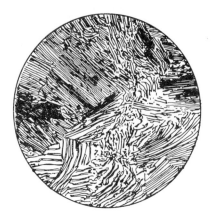

Fig. 3.42 Coarse pearlite formed at about 1325°F (720°C.) Hardness Rockwell C5, 174 BHN 1250x.

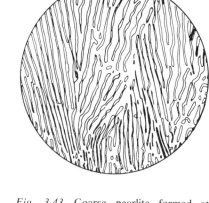

Fig. 3.43 Coarse pearlite formed at about 1325°F (720°C.) Hardness Rockwell C5, 170 BHN 1250x.

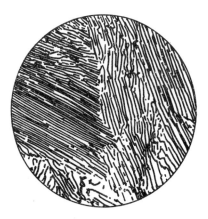

Fig. 3.44 Coarse pearlite formed at about 1300°F (705°C.) Hardness Rockwell C15, 208 BHN 1250 x.

Fig. 3.45 Medium pearlite formed at about 1225°F (665°C.) Hardness Rockwell C.30, 296 BHN 1250x.

that the hard carbide is more finely dispersed and this leads to a harder material. Thus by speeding up the cooling rate one can increase the hardness of the final product. Diagrammatic sketches of etched specimens as pearlite grows are shown in figs. 3.38 to 3.41, showing that the process is time-dependent. Further micrographs in figs. 3.42 to 3.45 show how the thickness of the plates depends on the temperature of formation. The grain boundaries can also be seen separating one group of pearlite plates from another. Examples of the hardness achieved by transforming at different temperatures are given in table 3.7, the austenitizing temperature in all cases being 760°C. See section 3.5.2 for a further discussion on hardness.

It can be seen from the table that, in this limited experiment, the fastest cooling brings the lowest transformation temperature and makes the hardest steel. It is also seen that the slowest cooling rate corresponds to a temperature of transformation of

HEAT TREATMENT AND HARDNESS OF STEELS | 213

Table 3.7 Hardness achieved by transforming at different temperatures

Carbon percentage	Cooling rate °C per min.	Transformation temperature °C	Brinell hardness number
0.75	2.8	707	210
	128.0	671	315
	1000.0	543	415
1.00	1.8	704	217
	55.6	682	262
	222.2	674	302

just over 700°C, but not 730°C, which is the approximate figure for the full equilibrium condition. The table does not show the effect of more carbon on the hardness because the cooling rates are not strictly comparable.

The rate of transformation of austenite is studied by cooling it suddenly to a certain temperature and holding it there, providing an isothermal transformation which of course takes time. The time of beginning and ending the transformation is noted. When this is done for many specimens [made small to bring the experiment within the time-scale of the laboratory] of the same steel the results can be shown on an isothermal drawing frequently called a Time Temperature Transformation (or TTT) diagram as given in fig. 3.46 for a eutectoid steel (0.83%C). It can be seen from this that even at 730°C the austenite takes between 100 and 1000 seconds even to start transformation. At below 300°C the delay of starting is more than 1000 seconds. The austenite is therefore less unstable near these temperatures. At about 550°C, however, the transformation starts very quickly, in 2 or 3 seconds. In the neighbourhood of this temperature therefore, the austenite is most unstable. When the steel is not eutectoid the TTT curve has a double line at the top, the upper line starting above 730°C, because ferrite is being deposited from hypo-eutectoid steel and carbide from hypereutectoid steel (see Section 3.4.6(a)). This extra curve joins the TTT curve well above the minimum-stability knee. For simplified discussion fig. 3.46 is sufficient.

Experiments such as these on isothermal transformations on austenite have revealed some interesting facts on the resulting structure. It is found that over the range from 730°C down to about 540°C the structure is pearlite for eutectoid steel, that is, the familiar lamellar structure of ferrite and cementite plates. This was to be expected. Over the range 540°C to about 230°C the structure has another name. It is called bainite. It has a feathery look (in micrographs) at the upper temperature but has needle-like constituents at the lower part of the range. It does not appear when cooling is continuous and slow because pearlite starts forming much earlier. Bainite is not therefore shown in the equilibrium diagram, based on ultraslow cooling, but it comes into the discussion of heat treatment.

The TTT diagram relates to isothermal transformation but cooling curves can be drawn on it to give a guide to what happens during continuous cooling, which is the practical reality in heat treatment. Some cooling curves are therefore drawn on fig. 3.46 to show a few different procedures. [The straight lines ought to be curves, but as long as one does not attempt to use fig. 3.46 quantitatively no error will enter into the discussions.] The first curve, r_1, is for the slowest cooling shown. The rates in fact are in increasing order: r_1, r_2, r_3. [The one shown at r_4 is a special case to be mentioned later.]

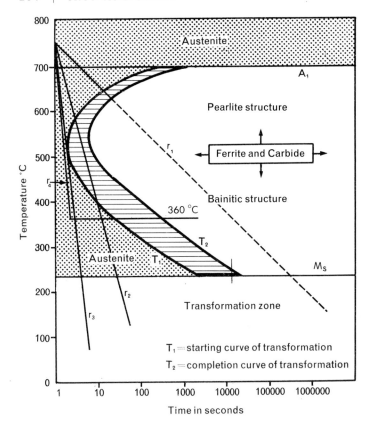

Fig. 3.46 Time-temperature transformation diagram for Eutectoid Steel.

Consider first r_1. The austenite starts transforming at 650°C and continues until the temperature 620°C is reached, when all the austenite has changed into pearlite [for a eutectoid steel]. Any cooling curve that crosses both T_1 and T_2 does represent a set of changes of complete transformation, though as the temperature is falling all the time the result is a mixture of structures. If the austenite cools faster as shown by curve r_2 it starts transforming at 580°C, but because of the fast cooling the transformation is not complete. This is shown by the fact that the curve r_2 does not cross T_2. So some transformation into pearlite and bainite takes place, but the rest is still austenite as it cools towards the temperature M_s which also has a special meaning to be discussed later. The rate of cooling represented by r_3 is important, for it never crosses the transformation curves. The austenite is not transformed at all in such rapid cooling and goes on cooling towards M_s. The cooling curve r_4 is different. It represents rapid cooling, without transformation, until the austenite reaches 360°C, when the cooling is halted and the steel is allowed to change isothermally, ending up as bainite. All these transformations, it should be noted, are in the solid steel.

It can be seen that, starting with austenite, the grain-size of which depends on the austenitizing temperature (being finer at the lower temperatures), the rate of cooling determines the final structure at room temperature, if no further heat treatment is given. Control of the rate of cooling – and the decision when to stop the action if necessary – thus provides the steel-maker with a remarkable means of arriving at a wide variety of microstructures in steel. And it is microstructure that determines properties. Traditionally many ways of effecting this control have been evolved so that it can be done in the works. First and most obvious is air-cooling, whereby the red-hot metal cools naturally in still air (there must be no draughts to interfere uncontrollably.) For small specimens this gives a cooling rate of from 1° to 3°C per second, a moderate rate. Another method is to let the steel cool normally in the furnace in which it has been heated. This is a very slow cooling. For isothermal change the steel is put into a suitable hot fluid such as molten lead, which can be kept at a variety of temperatures above its melting point Another technique is to clamp the steel between copper plates, which conduct heat away to the air. Fused salts such as potassium nitrate, sodium chloride and sodium nitrate are also used as cooling media.

One especially important case must be considered. If the cooling is so fast that the cooling curve does not cross the T_1 line of fig. 3.46, i.e. if the curve is such as r_3, and if the cooling is not halted, an entirely different transformation occurs below the 230°C line marked M_s. It is a very fast transformation involving no diffusion of atoms. The gamma face-centred cubic lattice is distorted by the stresses imposed, and to such an extent that planes of atoms slip and the lattice adjusts itself to be a body-centred type, not a cube, but having one dimension greater than the other two, with carbon atoms as part of the structure. The more carbon there is the greater is the ratio of the long dimension to the other two. The slipped planes create long needle-like structures when seen in micrographs. The carbon is thus in solid solution in the iron and the structure is called *martensite*. Steel that has this martensitic microstructure is very hard, the hardest of all, and brittle. The hardness increases with carbon content, reaching as much as 700 BHN, which can be contrasted with the 415 BHN already shown in table 3.7 for rapidly-cooled austenite transformed into pearlite with a carbon content of 0.75%.

For eutectoid steel the needles start forming at the temperature of 230°C, marked M_s in fig. 3.46, and they go on increasing in number until the finish of the transformation. Both the martensitic-forming temperature M_s, and the finishing temperature, M_f, decrease with increasing carbon content and for steels with more than 0.6% carbon M_f is in fact below 0°C. This means that if the transformation is stopped at room temperature, which is the normal thing to do, there is still untransformed austenite, compressed inside the martensite, and extra cooling must be provided to effect full transformation. The martensite may not always be pure because one can have the condition shown by r_2 in fig. 3.46, when austenite with partly-formed pearlite cools towards M_s. To avoid this sort of complexity the cooling must be very fast indeed to bring the cooling curve to the left of the nose of the cruve T_1, and such cooling is not always easy to achieve with carbon steels, because the nose of the curve is so near the zero-line axis.

a. Quenching. To achieve this extra-rapid cooling the traditional method is quenching, which means the plunging of the red-hot steel into a much colder liquid. It has been known for 3000 years at least, for Homer refers to it. The cold medium can be anything non-corrosive but the usual range of media include water, oil, brine, and aqueous solutions of other salts. These do not all give the same severity of quench, and variations include the stirring of the liquids to make faster cooling. Brine provides the severest quench having a higher latent heat and better conductivity than water. Water comes next. Oil because of lower latent heat, lower conductivity, and higher viscosity, provides a less severe quench. The quench gives the metal a tremendous thermal shock which can cause damage in splitting or cracking. It can distort the steel because martensite has a higher specific volume than austenite. So there are modifications of quenching such as the 'through water or oil' procedure, a two-stage process. There is interrupted quenching. The cooling can be at two rates, the first very high and the second lower, as long as the cooling curves do not cross the T_1 line of a TTT diagram. All the methods have one aim–to produce martensite. Quenching is thus a *hardening* process.

b. Annealing. When austenitized steel is cooled slowly in the furnace it follows a course more comparable to that of fig. 3.30. A fairly coarse pearlite and ferrite, or pearlite and cementite (according to whether it is hypoeutectoid or hypereuctectoid) is then produced, though if the austenitizing temperature is low enough the grain is finer than it would be if the steel were overheated. This very slow cooling is called annealing, and aims at producing a softer steel. It is very time-consuming for an industrial process and so isothermal annealing is often done instead. This involves rapid cooling to 600°C or thereabouts and then isothermal cooling in the upper part of the transformation area to produce the coarse pearlite needed, i.e. pearlite with thick ferrite layers. If the austenitizing temperature used is low enough there is also enough free carbide to form more pearly globular grains–the process of spheroidization. This type of structure is even softer than the coarse pearlite structure. The softness aimed at is to make the steel more easily cut by tools on lathes etc., or more easily formed in other ways. Annealing is also used for steel castings. There is also *sub-critical* or *process annealing*. When this is done the heating is not carried through to austenitization. It is really a re-crystallization procedure for steel which has hardened or produced stresses by cold-working. Sometimes, however, this type of annealing is done to effect spheroidization, but the process is very slow.

c. Normalizing. Cooling in still air is more rapid than in the furnace but not as rapid as quenching. It is too fast for the transformations of fig. 3.30 with their slow growth of ferrite. Instead the ferrite does not get time to grow and settles in the grain boundaries in the form of very small particles. Thus a fine-grained structure is obtained. A somewhat higher austenitizing temperature is used. This is the process called normalizing, which is really a grain-refining process. It is used very much for steels that have been rolled at austenitizing temperatures, which are found to vary in a steelworks. The process, which also ensures greater uniformity, can be used on a continuous basis for such materials as strip, sheet, plates and bars. It is a hardening and toughening procedure.

d. Patenting. This is a process used in making wire and rod from medium-carbon and high-carbon steel. It is a continuous process wherein the rod or wire passes through a heating tube to bring it to austenitizing temperature, and then through a cooling bath maintained at a fixed temperature. The aim is to produce toughness and strength in wire already over hardened by drawing. There are several different ways of achieving the patenting; which leads to higher tensile strengths than annealing. It is an isothermal process and the structures produced depend on the isothermal temperature-range. Thus any isothermal procedure in the 450°C to 550°C range produces sorbite, a finer form of pearlite, whereas somewhat below this range the structure is bainite, and is harder and tougher than pearlite. According to the temperatures chosen in patenting, wire can be made with tensile strengths up to 375 000 psi (260 kg/mm^2) and yet be very tough as well.

To achieve austenitizing temperatures, the heating is done in special furnaces, and there is of course a difference in design according to whether the process is part of continuous production or is for batch production, and according to the size of the steel product being heated. There is considerable engineering involved but not much fundamental technology. Once the heating is done the cooling by various methods provides a great variety of microstructures and properties. Which method is to be used depends not only on the final microstructure required, but also on what has happened to the steel before the heat treatment is applied. Thus it is not possible here to list which methods must be used for any particular steel product.

e. Tempering. Hardening by quenching produces a very hard product indeed, based on martensite though it may not be 100% pure. As such the steel is unusable, except for magnets that have to endure no mechanical stresses and for glass-cutting bits; for other uses it is far too brittle and has high stresses in it. The hardening by quenching is done not for the final hardened product but because it is the first of a two-stage procedure which is widely used. Items such as rolled bars, joists, girders, rails, sheet and strip, are often used in the as-rolled condition, though some may be normalized, but a multitude of steel products are first quenched and then immediately tempered. Tempering is rather similar to sub-critical annealing in technique, but more controlled and sophisticated. It is a process wherein the martensite steel is reheated very carefully to a chosen temperature below the A_1 line of the equilibrium diagram, i.e. below about 730°C. Recirculating-air furnaces are used for large-scale tempering, and baths of molten salts or metals are frequently used. The cooling to room temperature afterwards is not critical for carbon steels.

One purpose of tempering is to relieve stresses, and when this is done, not after quenching but after cold-working or welding or some other stress-inducing procedure, it is called *stress-relieving*. This is not such a careful process and can be done on site with large structures such as pressure vessels: merely by heating them electronically and allowing them to cool slowly. True tempering is much more than this. The purpose of it is to re-establish some ductility and toughness in the martensite, and this involves softening, which is not the purpose but is inevitably associated with increases in ductility and toughness. These changes vary with the tempering temperatures and the time for which the steel is maintained at this temperature. Thus a 1.2% carbon steel, with a hardness approaching 700 BHN, when kept at the tempering temperature for 1 hour shows a BHN of about 600 if the temperature is 250°C, about 520 if the temperature is 370°C, 420 if the temperature is 480°C and 380 at 530°C. There is thus a progressive softening with increase in tempering temperature. There is in general a similar progressive improvement in toughness and ductility, though there are anomalies. It is notable that most of the hardness numbers quoted in this example are still well above what is achieved in steels that do not go through the quenching-tempering process. This procedure can thus produce steels of greatly enhanced engineering properties. It should be

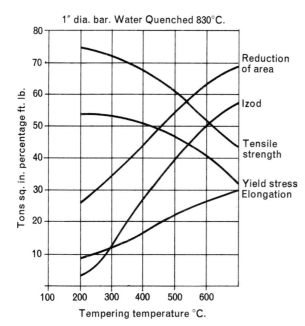

Fig. 3.47 Tempering at different temperatures.

noted also that the hardness figures quoted are for 1.2% carbon steel. Those for 0.35% carbon steel are about 240 below the ones quoted.

Martensite is rarely pure, even if quenched to well below freezing point. There is some austenite present and this has its effect on the structure, for the austenite can turn to bainite or one of the other structures with special names, such as troostite or sorbite. The tempering procedure for the martensite itself is the precipitation of carbon out of the solid solution in the form of iron-carbide plates, and the creation of ferrite. The more ferrite created, the softer is the material, but the carbide maintains its hardness, and as the tempering temperature is raised so the carbide tends to form spheroidal grains; this change improves the ductility but reduces the strength. Careful control of the tempering can thus create almost any desired combination of tensile strength, yield point and toughness, and in general steels put through this double heat treatment are stronger, harder and tougher than steels made in any other way or heat-treated merely by controlled cooling. The number of variations in cooling from austenite temperature allows an enormous range of microstructures to be created to produce steels with a wide variety of properties. In fig. 3.47 the results of tempering at different temperatures are shown for 0.5% carbon steel quenched from 830°C. Note the high yield strength ranging from over 112 000 pounds per square inch to 67 000 pounds per square inch (about 800 to 450 meganewtons per square metre).

Anyone can do crude quenching-tempering in his workshop. He can take, for example, a screwdriver and heat the end to bright red heat and plunge it into cold water. It will then become glass-hard and brittle. He can then reheat it gently and judge the temperature by removing the screwdriver from the flame and looking at the surface in the light. A faint colouring can be detected, an effect due to interference of light in the oxide film. The colouring varies with the thickness of this film, i.e. with the temperature. A light straw colour corresponds to a temperature of about 220°C, yellowish-brown to about 250°C, violet to 280°C and blue to about 300°C. If the screwdriver is held at any of the chosen temperatures for a minute or so and then cooled – slowly or quickly, it does not matter much – the glass-hardness is corrected and the steel is hard and tough. This sort of procedure has been practised by craftsmen for hundreds of years.

There are variants on these quenching-tempering heat-treatment procedures. With some steel products, such as tools, bearings, dies and so on the distortion and cracking of severe quenching cannot be tolerated. An alternative form of cooling, called *martempering* or *marquenching* has been used. The sudden quenching does not go to room temperature. Instead it is halted at just above the martensitic-transformation temperature, M_s (230°C in fig. 3.46). This is done by plunging the austeritized steel into a bath of suitable fused salts, and keeping it there until it is all at the chosen temperature. It is then allowed to cool in air and tempering then follows.

Austempering is another variant. In this there is the same quenching in a molten salt at a temperature above the M_s line, but not far above it, but the steel is kept in the bath for a much longer time so that complete isothermal transformation takes place. At such a temperature the structure produced from the austenite is bainite, which is tougher than, and as hard as, some tempered martensite. Austempered steel can be further tempered if desired. The use of the isothermal transformation, a slow process for the fairly stable austenite, ensures that there is very little distortion and no cracking. Austempering is used mainly for high-carbon steels in small sizes, though it is sometimes used to minimize distortion in cast iron and some alloy steels. The disadvantage of it and martempering is that the slower cooling in the salt bath allows some austenite transformation at higher temperatures, and this effect is all the more evident in larger sections. So steels should be used with a greater hardenability, which will be discussed later (Section 3.5.3).

f. Age-hardening. Ageing has already been mentioned as the inevitable process with time in some steels. It is speeded up at higher temperatures. So it can be deliberately used as a procedure to produce hardness by soaking the steel for a long time at an elevated temperature. For example, a 0.06% carbon steel if soaked at 60°C attains a hardness after 10 hours some 15% greater than its initial hardness though after 30 hours this decreases somewhat and remains fairly constant. If soaked at 41°C it achieves a hardness 30% or more greater than its initial hardness after 100 hours. This process is age-hardening and is used by some steel-makers when suitable. The effect is due to the precipitation of elements from solid solution, so it is a case of *precipitation-hardening* a phenomenon of deposition of particles in the grain boundaries, whether these are intermetallic compounds or carbides or nitrides or anything else.

3.5.2 Hardness of steel

It is not always easy to grasp the significance of terms, used so freely in steel-making, such as hardness, toughness, ductility, strength etc. Very great hardness has been associated with brittleness, but adequate treatment – usually heat treatment of one sort or another – can make some steels of a fairly high hardness without brittleness. And the important thing here is that hardness is closely correlated with strength. Thus a steel with a Brinell hardness number (BHN) of 495, has a tensile strength of 113 tons per square inch (1745 meganewtons per square metre). Table 3.8 gives a few representative comparisons.

Tensile strength is also closely related to yield strength, which is the maximum stress

220 | IRON AND STEEL

Table 3.8 Values of hardness and tensile strength for various carbon steels

Hardness (Brinells)	Tensile strength tons per square inch	Meganewtons per square metre
495	113	1745
401	91.5	1413
302	69.0	1066
202	45.7	706
153	34.9	539
109	24.8	383
101	23.2	358
80	18.4	284

well below which all steel structures and components must work. Thus strength is the most-desired property. The correlation with hardness means that the Brinell Hardness Number gives some idea of the mechanical properties. This hardness can be achieved in several ways. There is heat treatment, but there are other aspects as well. For example, a steel can be cooled from austenitizing temperature in such a way as to turn into any of the structures of pearlite or bainite, and the lower the temperature of transformation the harder the resulting fine pearlite or bainite.

Hardness by itself, however, is not enough for most working steels. The material needs to be tough as well. That is to say, it must resist sudden stresses. Fine grain structure contributes to this. Thus normalizing is seen to be a process for getting the right combination of strength and toughness for much of the structural steel used, for it cools the steel from the austenitizing temperature at such a rate that it forms fine pearlite or bainite. Rolling or forging the steel when red hot is also nearly a normalizing procedure though more erratic, for when it cools in the air it transforms to pearlite or bainite. This is the procedure for a vast range of steels sold in the as-rolled condition. The grain structure has been broken down in the process of squeezing under immense pressure. Work-hardening by rolling or drawing or extruding is another way of reaching a certain degree of hardness. Indeed it is the only way that austenitic stainless steel can be hardened. Age-hardening or precipitation-hardening is another route to a certain desired hardness, accelerated by treatment at higher temperatures. All these routes to hardness have their special advantages in cost, time, convenience and quality of product. Annealing, it should be noted, is the exception, for it is a softening procedure, though the word 'soft' must be related to the material itself. A 'soft' steel is by ordinary standards still a hard material. In alloying, a certain amount of strength is added in the alloying procedure itself. Thus nickel, manganese, chromium, vanadium and some others go into solution in ferrite, and make a contribution to its strength while others are combined in the iron carbide. This is the basis for high-strength low-alloy steels replacing carbon steels for so many structural uses. This material is not heat-treated, but some of the alloying such as copper and chromium produces corrosion-resistance so that the fairly high strength [yield point 50 000 pounds per square inch or 350 meganewtons per square metre] and corrosion-resistance combine to enable thinner sections to be made.

Steels are of course made for use, and the uses involve forming procedures. Many must be shaped by mechanical tools, which means they must be machinable. This may seem a simple matter but it brings some difficulties. If a steel is too hard it cannot be machined because it wears the tool out too easily. On the other hand if it is too ductile it cannot be machined either because the pieces taken off by the tool do not break – a continuous spiral of swarf is made which rubs on the cutting tool, raises its temperature, and interferes with its movement. So machinability is a special property and one of the things that improves it is the presence of non-metallic inclusions such as oxides and sulphides – manganese sulphide is an example well known. These allow the cut steel to break off. It is one more of the ironies of the evolution of steel-making that in fact the inclusions that the steel-maker of the past half a century has done so much to eradicate are necessary in some steels and have to be put back into too-clean steels if there is not enough oxygen or other element to form the inclusions. Another method of forming is welding, and this brings problems as well because it is a form of heat treatment to make joints. The steel is heated to austenitizing temperature and melted together and then left to cool. In this process the cooling rate may change the structure and in the case of alloy steels leave corrodible or brittle joints. In the low-alloy high-strength steels the elements and their proportions have been chosen to ensure that welding shall not bring ill effects. Some welding is done under circumstances where stress-relieving is possible in which case the problems are not so great.

3.5.3 Hardenability of steel

The size of a steel product which is subjected to quenching–tempering heat treatment is obviously important. When a piece of steel is red-hot and it is plunged into water the first effect is that water is vaporized and escapes as hissing steam. This cools the part of the steel immediately in contact with the water and soon it is not red-hot at all. Within a short time the temperature of the skin is very near that of the medium. But further inside the steel the conduction of heat towards the outside depends on the immediate temperature gradient, and at the very middle this is less than it is towards the outside. So the inside cools much more slowly. Its cooling rate may be so much less than that of the outside that the cooling curve would cross T_1 in fig. 3.46. In this case the transformation would include structures other than martensite – pearlite, ferrite, bainite etc., according to the speed of transformation and where the cooling curve crosses T_1 and T_2. Thus the whole of the steel would never turn to martensite and the piece of steel would not harden all the way through. This is not always a disadvantage for certain uses because a tougher, softer interior may be useful. But if high strength and toughness are needed throughout the section this heat-treatment would fail.

This brings in the concept of *hardenability*, which is not a good term because of its normal wide connotations. It does not mean merely the ability to make a steel hard, nor does it indicate how hard a steel can be made. It means rather the ability of steel to avoid transformations to softer steel when put through hardening treatments. And it means the ability to do this all through the steel section. Thus steel with good hardenability is not so limited in the size that can be hardened at usable cooling rates. Hardenability for different steels can be measured in terms of the ideal diameter of bar which could be hardened to the centre. It is also measurable by the depth inside a bar at which a certain specified hardness exists. A profile can be drawn of the limit of this specified hardness. Diagrams of this sort figure in all detailed discussions of hardenability. It is not an easy concept to keep clear in one's mind, but a crude picture of it can be obtained by saying that the more hardenable a steel is, the bigger is the section that can be hardened by quenching procedures.

a. Hardenability of carbon steels. When the concept of hardenability is applied to carbon steels their limitations are realized. So far all the discussion on heat treatment has indeed been concerned only with carbon steels. For years these were the materials for engines, bridges, battleships and railways, and they worked satisfactorily, and indeed still do. But when enhanced properties are needed the treatment is quenching and tempering. [It is in fact the only treatment recognized as such by some steel workers – when they say 'heat-treated' they mean quenched and tempered.] When this is necessary for carbon steels the size of product that can be completely hardened is limited to small sections, and larger products would have to be made of thin pieces welded or rivetted together. The limitation of carbon steels for this heat treatment can be seen in another way by means of fig. 3.46, which shows that the cooling rate for hardening is very high. If the steel had very low carbon content it would be impossible to harden it at all by quenching, because the nose of the curve gets nearer the zero-time axis as the carbon content is reduced. Fortunately, this is a minor limitation for the low carbon would restrict the hardness obtainable in any case, and low-carbon steels are the mild steels used as-rolled for so much of the steel-using industry. High-carbon steels, however, can be, and are, heat-treated but again only in small sections because of the low hardenability. It was lucky that swords and sickles and knives had thin sections that allowed quenching and tempering when the only steels available were carbon steels. It was lucky too that the slight amount of phosphorus was there to increase the hardenability, despite all the steel-makers' efforts to reduce it because it made steel brittle. Today small cutting tools, gears, shifts and axles, twist drills, dies, screw taps, wood saws and so on are made of high-carbon steels heat-treated by quenching and tempering.

There is a way of getting better hardenability in a carbon steel, which means that somewhat bigger sections could be heat-treated. It is to heat the austenite to a higher temperature, such as 980°C, for example, giving a grain diameter of 0.18 mm. which is a coarse grain. By severe quenching this can be hardened and the hardness penetrates quite deeply into a 1 inch (25 mm) diameter bar. The reason for this dependence on grain size is that a fine grain provides many nucleation centres for transformation, much in the way that the inclusions in an oyster start off the formation of a pearl, and that rough protrusions inside a vessel allow the formation of bubbles when a liquid is heated to boiling. The consequence of this in the austenite with a fine grain is that it is very sensitive to a fall in temperature. The effect is just as if the T_1 curve of fig. 3.46 were brought much nearer the zero-time axis. Conversely, the coarser the grain the less sensitive the austenite is. So a fine-grained austenite, when quenched, has an interior which transforms to pearlite easily at a temperature not far below A_1; whereas a coarse-grained one, less sensitive, can withstand a greater range of cooling inside it and so resist change of structure. Thus the coarse-grained austenite is more hardenable. The result, however, is a coarse-grained martensite and this is very brittle and creates big stresses and tends to crack. These defects, already mentioned in connection with quenching, cannot be tolerated for the sort of machine part for which hardenability is essential. A gearwheel in a car, distorted in shape and with cracks in it, is unthinkable. Moreover the tempering does not produce much refinement in the coarse grain, and the advantages of a fine grain for mechanical properties has already been stressed. Thus the limit to the use of a carbon steel is reached by using a very high austenitizing temperature and a very severe quench.

b. Hardenability of alloy steels. The way out of the difficulty is the introduction

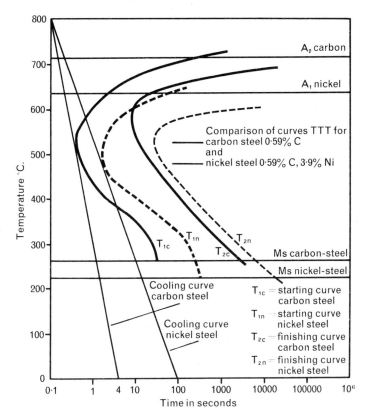

Fig. 3.48 Time temperature transformation diagram, Tempering curves 0.50% carbon.

of other elements, that is, *alloying*. Again the lucky accident of deoxidizing steel with ferro-manganese has helped the steel industry by making so much steel, as traditionally provided, contain some manganese in small properties. Its effects are hardly measurable, but they must be there, so that the industry can be said to have had a mildly alloyed steel for a century. However, this is not what is meant by alloying in the modern sense.

The end result of alloying as it is now understood is a modification of the microstructure of steel in such a way as to extend the range of usable steels. It has done this to such an extent that the low-alloy structural steels are edging their way into the market previously held by ordinary carbon steels. A great many of the modifications are brought about by the changes due to heat treatment of one sort or another. Indeed, apart from certain special properties, such as the resistance to corrosion, many of the benefits of alloying can be obtained only in connection with heat treatment. Heat-treatment technology has, as a result, been elaborated and investigated to a greater extent than ever before.

The first effect of some alloying elements is shown by an examination of the TTT diagram. In fig. 3.48 two such diagrams are superimposed for comparison. One is for a 0.59% carbon steel (i.e. a medium-carbon steel) and the other for a steel of the same carbon content but with 3.9% nickel added. (The curves are simplified for easier understanding, in that the separation of the upper parts of T_1 into two parts, because the steel is hypoeutectoid, is not done). It is seen that the effect of the nickel addition is that the curve T_1 has been pushed to the right, and the nose of the curve for minimum stability shows a delay of about 4 seconds before transformation begins, whereas the comparable start for the carbon steel is within a second. The critical cooling rate is thus much less for the nickel steel than for the carbon steel. Quantitative inferences from fig. 3.48 should not be drawn, but the difference is too great to be mistaken: the critical cooling rate for the carbon steel is many times that for the nickel steel. This means that less severe quenching can be done, well within the scope of steelworks practice. Thus there is less stress induced in the nickel steel when quenched, which means that cracking and distortion can be avoided.

This is of course an advantage, but the result is more specific than this. It is that the 3.9% nickel has *improved the hardenability*. This merits some discussion in more detail than the previous short account of hardenability. Consider what happens inside a steel bar 1 inch (25 mm) in diameter in both cases, for the carbon steel and the nickel steel. At any moment there is a temperature gradient from the centre outwards. Let it be supposed that the temperatures can be measured at 6 spots of equal distances from centre to the outside, and at a point one fifth of the distance from the centre the temperature is 700°C. There are two temperature gradients, one in space from the inside outwards, and one in time because the bar is cooling by losing heat to the cooling medium. Both gradients are less for the nickel steel than for the carbon steel because the former has a milder quench. From the TTT diagram of fig. 3.48 it can be seen that if there were no further heat flow the carbon steel would in 20 seconds or so start to transform to pearlite, but more than 1000 seconds would be needed for the same isothermal transformation in the nickel steel. However, there *is* heat flow and, taking the period of 20 seconds the specimens may have cooled, at the same place inside the bar, to 500°C in the carbon steel and 600°C in the nickel steel. Examination of such a cooling on fig. 3.48 shows that the carbon steel will have started to transform into pearlite in this 20 seconds, because the cooling curve from 700°C to 500°C in 20 seconds has crossed T_{1c}. The nickel steel, however, has not transformed at all because its cooling curve, of 100°C in 20 seconds (from 700°C), has not crossed T_{1n}. If one assumes that the cooling rate stays constant (it slows down in fact) then in the next 20 seconds the cooling curve in the carbon steel will still be inside the transformation region between T_{1c} and T_{2c}, so more transformation, into pearlite and bainite, has taken place, whereas in the nickel steel no transformation has taken place. Thus the inside of the carbon-steel bar cannot become entirely martensite, whereas the same spot in the nickel-steel bar does become martensite. Now, if the bar were 2 inches (50 mm) in diameter, the spatial temperature gradient would be halved. In the 1 inch bar the spot considered would be one-tenth of an inch (2.5 mm) from the central axis, but in the 2 inch bar it would be one-fifth of an inch (5 mm) from the central axis. Thus in this hypothetical example, the depth of penetration of martensite would be greater in the nickel-steel bar than in the carbon-steel bar, in which a still larger volume would convert to pearlite and bainite and never be martensite. In other words, a larger size of bar can be hardened completely if made of nickel steel but not if made of carbon steel, even when the quenching is milder. This is the significance of hardenability. The discussion is simpler if one keeps to the critical rate of cooling, but this more involved picture has been built here because the concept of hardenability is difficult, and yet is fundamental to the discussion of heat-treatment by the quenching-tempering route. It is also of great practical importance in the applications of steel to engineering. Had alloying not been done, there are many steel parts, for example, in the motor industry, which could not have been given the properties needed. This would have meant greater ingenuity of design to overcome the difficulties and, of course, greater cost.

Nickel, therefore, increases hardenability. This does not mean that the nickel steel is necessarily harder. Indeed nothing harder than effectively hardened high-carbon steel has been made. But hardness, though a necessary property in some applications of steel, is in heat treatment only a means to an end – enhanced mechanical properties. It has been shown that the quenching–tempering route gives much better properties than isothermal cooling to lead to, for example, a bainitic microstructure. Comparisons of three treatments of a medium-carbon steel, with 0.95% chromium and 0.65% manganese, are shown in table 3.9, taken from a Russian source.

It can be seen how much greater the tensile strength and yield strength are for the quenching-tempering process. Yet with these enhanced strengths the ductility shown by the elongation and reduction in area, is also improved. So is the notch toughness, at least twofold, though the figures, based on Russian measurements, which do not give results in Izod or Charpy numbers, are not included in table 3.9. The yield strength of 49.19 tons per square inch (750 meganewtons per square metre) is very high indeed for a medium-carbon steel of the modest carbon content of 0.4%. [An as-rolled carbon steel of this percentage would have a yield strength of about 21 tons per square inch].

With one important exception all alloying elements improve hardenability. The exception is cobalt. All the others push the T_1 curve to the right, and as has been shown, this means improved hardenability. They do not all contribute to the same extent. Thus vanadium, titanium, molybdenum and tungsten need be present in very small amounts, and when they are they increase hardenability considerably. The elements most used are nickel, chromium, manganese and silicon. They are inexpensive and their hardenability influence is cumulative. Thus from the point of view of hardenability there is a choice of compositions and, up to a point, it does not matter which of them is used.

Hardenability can be illustrated quantitatively by drawing a diagram of the profile of hardening achieved in a bar of stated diameter. Thus fig. 3.49 shows four such profiles for a carbon steel (0.35% carbon) and three values of added chromium. The way to interpret this sort of diagram can be understood by looking at the 2% chromium profile. It can be seen that at $\frac{3}{16}$ inch from the outside the hardness is 50 Rockwell C, at $\frac{5}{16}$ inch it is 47.5 Rockwell C, at $\frac{7}{16}$ inch it is 44.5 Rockwell C and at the centre 43 Rockwell C. With this understanding of the hardness chart the effects of alloying with chromium can be seen. The pure carbon steel has at the centre a hardness of only about 17 Rockwell C and on the outside 23 Rockwell C, a difference of about 24%. The addition of 0.5% chromium improves this somewhat and the difference between out-

Table 3.9 Comparison of three heat-treatments of a medium–carbon steel.

	Tensile strength tons per square inch	Yield strength tons per square inch	Elongation %	Area reduction —
Annealing	41.7	23.12	21	53.5
Normalizing	47.87	28.57	20.9	56.0
Quench-tempering	55.62	49.19	22.5	67.5

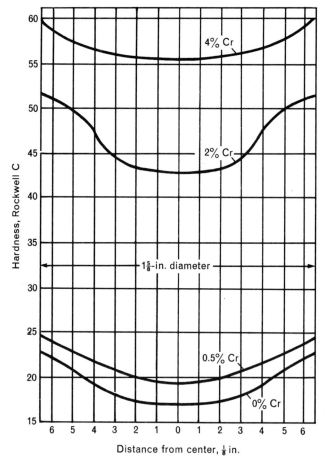

Fig. 3.49 Hardenability. Distribution of hardness across 1⅜ in rounds of four 0.35% C. steels with chromium as indicated. The bars as quenched in oil were free from undissolved carbide.

side and centre is slightly less. At 2% chromium the hardness is more than doubled, and the difference between outside and inside is about 18%. At 4% chromium the central hardness is up to 56 Rockwell C and the difference between outside and centre is down to 6%–the whole bar is very hard. Fig. 3.50 shows the effect of adding 3% chromium to 1% nickel. Fig. 3.51 shows that 2% molybdenum is not as effective in improving hardenability as the same amount of chromium, but the addition of only 0.5% molybdenum has improved the hardness by 50% or so, and increased the hardenability in that the difference between centre and outside is only 9%.

The most effective way of improving hardenability is to combine the alloying elements rather than use more of any chosen one. This is already seen in fig. 48. Another example is a steel of carbon content 0.45% with manganese content 0.75%. The critical diameter, that is the one that will have 50% martensite at the centre, for this steel is 1 inch (25 mm). The same quenching procedure for a constructional alloy steel of about the same contents of carbon and manganese, but with 1.25% nickel and 0.6% chromium allows a bar 2 inch (50 mm) in diameter to reach the same degree of hardness at the centre. The hardenability is thus doubled by the addition of these small amounts of nickel and chromium. These two steels are those classified in the USA as SAE 1045 and SAE 3140.

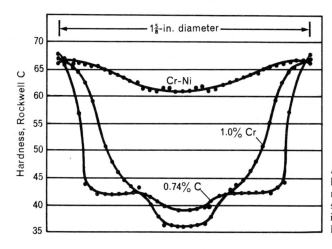

Fig. 3.50 Distribution of hardness across the 1⅝ in diameter of three 0.75% C. steels. Effect of alloy may be inferred by comparison with lower curve for carbon steel.

The alloying elements act in different ways when added to the steel and only the most general statements can be made here. Most of the elements dissolve in ferrite and all of them can be forced to dissolve in austenite – even the least soluble – by raising the temperature high enough. There is one important behaviour that is relevant to the heat treatment already discussed. It is the readiness with which some of the elements form carbides by combining with carbon, some of the compounds being complex carbides containing more metals, such as, to quote one example, the molybdenum-iron carbides. These carbides are hard and brittle and they have several important effects in alloy steels. The most important is their influence on grain size, a matter already

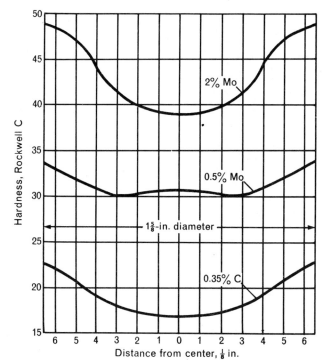

Fig. 3.51 Distribution of hardness across 1⅝ in rounds of three 0.35% C steels with molybdenum as indicated. The bars as quenched in oil were free from undissolved carbide.

mentioned in connection with carbon steels. Austenite increases its grain size as it gets hotter but the effect of carbides is to restrict this grain size. The elements that most readily combine with carbon are (in decreasing order of carbide-forming tendency): titanium, zirconium, niobium, vanadium, tantalum, tungsten, molybdenum, chromium, manganese. Iron comes next, and does form carbide as already described. Less effective than iron are nickel and cobalt. The higher up the list a carbide-forming element is, the more easily it will rob steel of carbon and segregate it in carbide particles. Nickel and cobalt and silicon are not found in the carbide form.

The argument still holds that the coarser the grain the more stress and creation of defects there are when a steel is quenched. Unfortunately the argument also holds that the finer the grain the more difficult it is to quench the steel completely into martensite. However, differences are shown between the behaviour of alloy steels and carbon steels in this respect. The behaviour of vanadium is much-quoted example, because it is high on the list but not as strong a carbide-former as niobium, zirconium and titanium (or uranium, though the use of this in steel is not common), while being more effective than tungsten, molybdenum, chromium or manganese. A comparison of hardenability [as done in figs. 3.49 to 3.51 above] in two steels, heated to the same temperature and quenched, one a carbon steel and the other with a small proportion of vanadium added, gives for equal hardenability a grain size in the range 0.06 to 0.09 mm. diameter in the carbon steel, and 0.02 to 0.03 mm. diameter in the vanadium steel. The carbon steel is three times as coarse as the vanadium steel for equal hardenability, when the austenitizing temperature is about 900°C. When the austenitizing temperature is higher (982°C) the average grain size of the carbon steel is 0.2 mm. whereas that of the vanadium steel is 0.076 mm. Thus the carbon steel remains three times as coarse as the vanadium steel even at the higher temperature, which gives complete martensite across a 1 inch (25 mm) bar.

It must not be thought that the effect of carbides is always, and at all temperatures, grain refinement. It depends how finely the carbide remains in the grain boundaries and how readily it can be made to coalesce or break up, and the carbide-forming elements differ in this respect. As carbides they do not contribute to hardenability, which is due to solution of them in austenite. It might be thought that the aim should therefore be to get rid of carbides by heating to higher and higher temperatures, and for the most part this is true; but free carbides can be an advantage in steels that need to be abrasive, such as those used in machine tools, especially when used at high speeds. A further inference from these examples is that the degree of separation between the solution of the carbides in austenite and the retention of them as free carbides, can be controlled by the choice of correct austenitizing temperature. Thus alloying has widened the choice of the steel maker, in suiting the composition to the heat treatment in order to get a wide variety of properties in the final steel.

The two main effects of alloying in heat treatment are thus seen to be an increase in hardenability, allowing bigger articles to be deep-hardened and milder quenched to achieve this, and a restriction on grain size by the formation of carbides (as well as intermetallic compounds in some cases) which allow finer grain size at austenitizing temperatures comparable to those used with carbon steels. There are other effects, many of which cannot be considered here, but one merits discussion. It is that all alloying elements, again with the exception of cobalt, depress the temperature at which martensite starts to form, the one marked M_s in TTT diagrams. It is shown for a small amount of nickel in fig. 3.48, at about 240°C. Larger amounts of alloy achieve so much lowering of this temperature that M_s may be below room temperature. When this happens, if quenching is done in the normal way, martensite does not form. The steel is still austenitic at room temperature. The effect of the alloying is thus to stabilize austenite at room temperature. This is what is achieved in the austenitic stainless steels, which obviously cannot be hardened by heat treatment.

For complex alloys containing carbide-forming elements the TTT curves have very different shapes from those already considered. They often have two minima of delay-time and a maximum in between, shaped almost like a figure 3 backwards. One such is shown in fig. 3.52 for an alloy containing nickel, chromium and molybdenum. From this it can be seen that with normal rates of cooling the austenite would never transform to the coarse pearlite but only to the harder pearlite or bainite. The critical cooling rate to go directly to the martensite would be comparatively small. It is not proposed to follow this complex isothermal transformation curve in any more detail, but anyone can judge from it how useful this type of curve is in arriving at a knowledge of the final structure when heating or cooling is done.

Tempering, which is the process that immediately follows quenching, is done to order to get any desired structure. Martensite is too hard and brittle, and some modification of microstructure is needed to give a measure of ductility and toughness to a steel that is to be used. In the tempering behaviour, alloys also play their special parts. The first effect is a general slowing-down in the softening due to a rise in temperature. The softening, in carbon steel, is caused by a diffusing of carbon and some transformation of the depleted iron into ferrite, while iron carbide is formed which turns into more spheroidal particles instead of the platelets of pearlite. These dispersed particles give added ductility. Thus the heating procedure to temperatures which range from not far above the M_s value to perhaps 600°C or more, and of holding the steel at the pre-determined temperature, relieves stresses induced by quenching and increases ductility and toughness at the expense of hardness. The carbide particles get bigger with

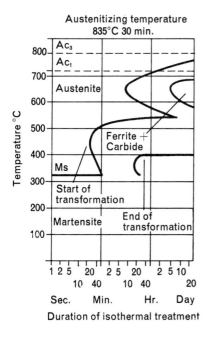

Fig. 3.52 Time-temperature transformation diagram for alloy steel. This illustrates the effect of alloying elements which could delay transformation at some temperatures to such a degree that it is not complete in many hours. In heat treatment the metallurgist uses his knowledge of these diagrams to obtain particular structures and properties in the steel.

increasing tempering temperature and eventually add nothing to the hardness, so that what is left is ferrite as the only structure. The first effect of alloys is to defer this, so that for example, a steel with 0.5% silicon may have a hardness of 630 BHN on quenching, and after tempering at 425°C may have a hardness of 380 BHN, but if the silicon content is 3.8% the hardness is about 520 BHN. In effect the steel is harder after tempering than if there had been negligible silicon present. Nickel and manganese act similarly. It must be remembered that softening is not the prime purpose of tempering and the hardness afterwards is still high, though not as high as the 700 BHN or more of martensite.

Carbide-forming alloying elements contribute another phenomenon. Instead of the gradual softening there is a levelling-out so that the hardness stays constant over a range of tempering temperatures, and in addition there is an *increase* in hardness, the so-called secondary hardening. This effect is enhanced by raising the temperature of austenitization, showing that the degree of solution of the carbide-forming element in the austenite is important in relation to the correct tempering temperature. Not only is the halt in hardening related to the tempering temperature, it is related to time as well and experiments have shown the effect of maintaining steel at the tempering temperature for up to 1500 hours. When this temperature was 450°C the hardness started at 600 BHN, went down to 485 BHN after 6 seconds, was a little softer after 1 minute, was increasing after 10 minutes and was up to 530 BHN after 1000 hours. The actual hardness achieved was 550 BHN after 1 minute when the tempering temperature was 650°C. By adding molybdenum to chromium and tempering at 450°C the hardness, originally 650 BHN or higher, went down to 530 BHN and stayed there for nearly 1000 hours. It has now been shown, by means of electron-diffraction techniques, that in fact the process is one of depositing or precipitating carbide particles, but at temperatures above 400°C the rods of carbide that have been growing are replaced by much smaller ones.

It must not be forgotten that alloys affect the temperature at which martensite starts to form, and the temperature at which it ceases, M_f. If the latter is below room temperature all the austenite in a quenching procedure cannot be turned into martensite. It is still austenite. When heated at tempering temperatures this precipitates carbides and the austenite left, being purer, will transform to martensite more readily when it cools. Then a second tempering is necessary to transform this martensite. This has to be done with some high-alloy steels used for making high-speed tools.

All this discussion is relevant to the very practical need of steels that can be used at elevated temperatures. First there is the phenomenon of creep resistance. Obviously any heat treatment that allows considerable hardness at higher tempering temperatures will produce a steel which withstands stress at such temperatures with much less yield than a steel produced by a treatment that allows only moderate tempering temperatures. It is alloying that allows this. Secondly there are the tool steels already mentioned, which must not lose hardness even when at the dull red-heat caused by the speed of cutting. The same type of secondary hardening at a high temperature is necessary and the chief carbide-forming elements that are used for this purpose are tungsten, vanadium, molybdenum and chromium. One of the best known of these is the so-called 18-4-1, meaning 18% tungsten, 4% chromium and 1% vanadium. They are high-alloy steels.

The above discussion shows how complex the effects of alloying can be, and some of them are not yet thoroughly understood. The most general statement that can be made is that alloying increases hardenability, thus extending the range of sizes hardenable and simplifying quenching procedures, and provides a whole new range of properties from correct tempering. Alloying brings with it the need for much greater understanding of the importance of microstructure as the master of all properties, and the need for greater control in various procedures. The reward for these is an extension of the range of steels already in use, and other new ones still to come, and to such an extent that even ordinarily structures that have relied on well-produced carbon steels are now being made of low-alloy steels.

3.5.4 Surface hardening

Some steels that have been hardened and toughened by full heat treatment still need a harder surface to resist wear. There are also steels that are innately softer and tough and do not need the full heat treatment right through, and yet must have a hard surface to resist wear. There are uses that would not suit a steel which is hard right through and yet demand a hard-wearing surface. If, at the same time, the surface can have some qualities of resistance to chemical attack from the environment then this is desirable as well. There are several ways of converting the surface to the sort that is required. One obvious way would be to effect a measure of work-hardening by firing shot rapidly at the cold surface. This procedure, known as shot-peening, is indeed done. In general, however, heating and chemical treatments are used, for what is called *case-hardening*, meaning the casing of the steel in a hardened skin.

One group of techniques uses heating only, subjecting the surface to a localized heat treatment. One way is to rotate the steel article and subject it to a hot flame directed at the surface. This is widely done because it is applicable to different shapes. It carries the risk of overheating the surface and coarsening the grains, but it is simple and effective. Quenching is done by directing a spray immediately behind the heating ring of flames. Another method of heating is to make the steel part the cathode of an electrolytic cell, in such a way that the evolution of hydrogen bubbles at the cathode creates local high electrical resistance. If the current density is high enough the high-resistance surface is heated very rapidly. The electrolyte solution does not heat up and so acts as the quenching medium when the current is switched of. A third method of surface heating is that provided by induction heating with alternating current. The steel part is inside a coil through which alternating current flows. Eddy currents set up in the

Fig. 3.53 Gear—Flame hardened.

steel part create heat – an effect unwanted in transformers – and the skin effect is such that the surface is heated more than the interior, and the higher the frequency of the current the more pronounced this skin effect is. So the depth of hardening can be controlled by the correct choice of frequency, the range being between 150 hertz and 250 000 hertz. The shape of the coil can be varied to suit the shape of article to be treated. Cooling by, for example, spraying can often be done on a continuously-moving part passing at the correct speed through the magnetic field. In all such hardening procedures the surface must be made austenitic and the cooling procedure decides the degree of hardness to be achieved. Heating procedures of this sort can be used for such parts as crankshafts, gears, ploughshares, cylinder liners for engines etc.

The other methods of case-hardening depend on the diffusion, through the surface of the steel, of materials which have a hardening effect. At high temperatures carbon and other elements diffuse through the steel and their diffusion rates increase with the temperature. The two commonly used elements are carbon and nitrogen. When the first is used the process is called *carburizing*, the oldest and most widely used method of case-hardening in the industry. When nitrogen is used the process is *nitriding*. When both are used together the process is *cyaniding* (see chapter 1).

3.6 LITERATURE

EDGAR C. BAIN and HAROLD W. PAXTON. *Alloying elements in steel.* American Society for Metals, 1962.

G. R. BASHFORTH. *The manufacture of iron and steel;* Vol. I Iron production 3rd edition 1964; Vol. II Steel production 2nd edition 1959. London, Chapman and Hall, 1962.

A. K. BISWAS and G. R. BASHFORTH. *The physical chemistry of metallurgical processes.* London, Chapman and Hall, 1962.

THE BRITISH IRON AND STEEL RESEARCH ASSOCIATION. *Research and development reports and annual reports.* London. (Now The Inter-Group Laboratories of the British Steel Corporation.)

BRITISH STANDARDS INSTITUTION. *Specifications relating to the steel industry latest revisions.* London.

BRITISH STEEL CORPORATION. *Iron and steel: annual statistics for the United Kingdom, 1967.* 1968.

BRITISH STEEL CORPORATION. *A simple guide to basic processes in the iron and steel industry.* London.

BRITISH STEEL CORPORATION. *A simple guide to structure and properties of steel.* London.

BRITISH STEEL CORPORATION. *A simple guide to technical developments in the iron and steel industry.* London.

INSTITUTE OF GEOLOGICAL SCIENCES (Mineral Resources Division). *Statistical summary of the mineral industry 1961-66*. London, Her Majesty's Stationary Office, 1967.

SECRETARIAT OF THE ECONOMIC COMMISSION FOR EUROPE. *Economic aspects of iron-ore preparation*. United Nations, Geneva, 1966.

UNITED STATES STEEL CORPORATION. *The making and shaping and treating of steels,* 8th edition. 1964.

B. ZAKHAROV. Translated by N. Ivler. *Heat treatment of metals*. Foreign Languages Publishing House, Moscow, 1963.

CHAPTER 4

Non-ferrous Metals

4.1 INTRODUCTION

There are more than 100 known elements of which about 80 are metals, but only 8 of these (iron, copper, lead, tin, aluminium magnesium, zinc and nickel) are sufficiently abundant and cheap to provide common engineering materials. About 15 other metals (the most important are chromium, manganese, titanium, cobalt, tungsten and vanadium) fulfil specific, often indispensable applications. Other groups of metals are the alkalis, the precious metals and the rare earth metals. Materials which are based on iron are known as 'ferrous'; all the other metals are styled 'non-ferrous'. Ferrous materials are made in quantities which exceed all the non-ferrous metals together.

Although the usage of non-ferrous metals is so much less than that of ferrous metals, their practical importance is very great. For example, aeroplanes require metals of high strength and low weight, and electricity supply depends on the high electrical conductivity of copper and aluminium and the special properties of lamp filaments. Modern developments in aero-space engineering and nuclear power depend upon these materials, many of which have been specially developed in recent years.

4.1.1 Properties

Any one non-ferrous metal will have a wide range of properties which depend on its purity and method of production. In addition, most non-ferrous metals are deliberately alloyed with small quantities of other metals, which again profoundly alter their properties.

In the space available in this chapter, therefore, only an indication can be given of the properties of the various non-ferrous metals. If the properties of the pure element are quoted, they will be of value to the chemist but of little interest to the engineer wishing to use the metal for structural purposes. On the other hand, space does not permit the publication of engineering properties for the hundreds of different materials commercially available. Readers wishing to obtain detailed information on the properties of a specific type of non-ferrous metal are referred to table *1.2* and to the books listed, on page xxix or the National Standards published for many common metals in most countries, or to the manufacturers themselves who produce detailed tables of properties for their materials. Statistical data in this chapter marked with * have been compiled from *American Bureau of Metal Statistics Yearbook, 1969*, those marked with a † from *World Metal Statistics*, 1969.

4.2 BERYLLIUM

4.2.1 History

Beryl in the form of emerald was mined for centuries before Vauquelin, in 1798, first identified it as a compound of a new element, then called glucinum but later changed to beryllium. Although the metal beryllium was extracted in 1828, commercial potential was not recognized until the 1920s when beryllium–copper alloys were produced from beryllium oxide reduced by carbon in the presence of copper, and production of these alloys began in 1932 in the USA and shortly thereafter in Germany. Beryllium X-ray windows were made in the 1930s, and because of the nuclear characteristics of beryllium, production in the USA was increased to 100 lb quantities for the atomic bomb project during world war II. To satisfy increasing nuclear requirements, tonnage production began in 1947 at a USAEC-built plant and became commercial in 1957 at two privately financed refineries in the USA. During the same postwar period, a French refinery started production of electrolytic flake and semi-finished mill products. Plants for refining ore and making beryllium oxide and metal have been operated in the UK. At present, beryllium and beryllium-copper are produced principally in the USA, with smaller industries in France and Japan; beryllium–copper master alloys and mill products are also made in the UK.

4.2.2 Occurrence

Beryl occurs as a mineral deposit, extracted as a by-product of other minerals such as feldspar, spodumene and mica. It can be concentrated only by hand picking. Amongst the principal producers are Brazil, India, USSR, Malagasy Republic, Rwanda, Mozambique, Argentina and Uganda. Beryl contains 10 to 11% of beryllium oxide (BeO).

Deposits located in western USA contain bertrandite ($4BeO.2SiO_2.H_2O$). These contain less than 1% beryllium oxide but can be concentrated and refined using already developed methods when future demands justify the costs of constructing the refineries. Beryllium occurs in about 60 other minerals, notably phenacite and helvite, but workable deposits or recovery methods have not been developed.

4.2.3 Extraction

Adequate supplies of beryl are available and industrial recovery plants are based on beryl. In the USA, beryl is rendered soluble by melting and quenching and then reacting with sulphuric acid. The sulphate solution is purified by precipitation or fractional crystallization, and beryllium sulphate ($BeSO_4$) is converted to hydroxide or oxide. Alternatively, crushed beryl may be reacted with sodium fluorosilicate and soda ash to form a soluble fluoroberyllate, which is water leached and the solution purified by precipitation. Hydroxide from either process is then reacted or dissolved in ammonium bifluoride, to be decomposed thermally to beryllium fluoride. This is converted to beryllium by reducing with molten magnesium.

Metallic beryllium is manufactured in France by the electrolysis of a mixed chloride bath, which includes beryllium chloride produced by reacting beryllium oxide with chlorine in the presence of carbon.

4.2.4 Fabrication

Either magnesium-reduced or electrolytic beryllium is melted into ingots, lathe-turned into chips, attrition milled and screened into powder. Billets are hot-pressed at 1 900°F (1 040°C) from powder and are generally machined into parts. Some billets (a small fraction) are rolled into sheet or foil, and a still smaller number are extruded or forged.

Hot-pressed blocks weigh up to several thousand pounds. Block is extruded into rod up to 5 inches in diameter, round tubes up to 7 inches in diameter and square tubes. Experimentally, structural shapes, including angles and channels, have been extruded. Block is also rolled into plate up to $\frac{1}{2}$ inch thick, and sheets 36×96 inches are available down to 0.020 inch; smaller sheets of foil are rolled to 0.001 inch. Wire down to 0.002 inch in diameter is drawn from cast and extruded rod.

In producing parts, powder metallurgy is preferred. Sound ingots have not yet been cast in very large sizes and beryllium ingots have coarse grains which are difficult to refine by rolling. With a sufficient reduction, beryllium wire and thin foil are best made from ingots, and sheet made from ingot is more formable than that made from powder.

4.2.5 Alloys

Alloying is limited since few metals have solid solubility in beryllium. Copper additions increase the strength, but these alloys are difficult to roll. An alloy of 0.4% calcium has improved the resistance to corrosion, particularly by carbon dioxide. None of these is produced commercially.

An alloy containing 62% beryllium and 38% aluminum has interesting engineering properties, including elasticity, lightness and machinability, with relatively good formability. It is available commercially and is one of a family of alloys called Lockalloy.

4.2.6 Properties

Beryllium is attacked slowly by moist air and more rapidly at sea. It is often anodized before being put into service. Parts may also be passivated (see chapter 11, p. 799) to resist high temperature oxidation. Several metals, including copper and nickel, can be plated on parts.

In comparison with other engineering metals and alloys, beryllium is a brittle material. All shaping operations, except machining, are performed hot; most forms and compositions have low elongation and poor formability at room temperature. Between room temperature and 1 400°F (760°C) (the temperature for complete recrystallization) there is a range in which sheet may be formed, wire may be drawn and bars may be forged without loss of strength. The chosen temperature depends on the form, composition and prior processing. Wire is drawn at 750°F (400°C) ingot sheet is formed at 500 to 900°F (260 to 480°C), powder sheet is formed at 1 350°F (730°C) and bars are forged at 1 200 to 1 400°F (650 to 760°C). The block produced by hot-pressing powder has a yield strength of about 17 tons/sq in. and is the starting material for most other mill products. Block is rolled or extruded into sheet with 27 tons/sq in. yield strength or bars with about 40 to 45 k.s.i. Forgings are difficult to produce and must have simple contours but have yield strengths between 45 and 90 k.s.i. Very heavily worked beryllium, in the form of wire of 0.050 inch diameter and finer, shows 60 to 70 tons/sq in. yield strength. These strengths may be decreased substantially, almost to the strength of block, by annealing at over 1 350°F (730°C).

Beryllium is toxic and inhalation of dusts or powder must be avoided. Ordinary handling and simple manipulation may be performed without ventilation. Operations which generate chips or dust or which employ high temperatures require special precautions which have been developed with experience. Beryllium is generally machined in specially equipped and ventilated shops and is processed in plants safeguarded against excess beryllium in the atmosphere. The properties of beryllium are summarized in table *4.1*.

4.2.7 Joining

Beryllium may be joined to itself and to other metals by soldering, brazing, resin bonding or mechanical fastening. Zinc solders are satisfactory, but most beryllium joints are made by brazing using aluminum–silicon brazing metal. Joints with excellent strength at room and elevated temperatures are made by brazing with silver or silver alloys. Fusion welding is not satisfactory since cracks develop in the weld during cooling. Mechanical joints using Monel metal, stainless steel, titanium or aluminum rivets, pins or screw fasteners are used widely. Resin bonding is well developed and is sometimes used in conjunction with mechanical fasteners to provide strength at room temperature and up to the point at which the resin weakens and the fasteners take over. In this temperature range beryllium is less notch sensitive and has higher fatigue properties.

4.2.8 Uses

Applications of beryllium rely on its stiffness, lightness and strength, together with its nuclear characteristics. Its use for certain parts of aircraft, ship and missile

Table 4.1

Physical and mechanical properties of beryllium

Density, g/cm^3	1.85
Electrical conductivity, % IACS	40
Melting point, °C	1285
Boiling point, °C	(2970)
Modulus of elasticity, p.s.i.	42×10^6
Poisson's ratio	0.025
Latent heat of fusion, cal/g	260
Latent heat of vaporization, cal/g	5390

Typical mechanical properties of beryllium mill products

	Ultimate tensile strength p.s.i.	Yield strength p.s.i.	Elongation %
	Powder metallurgy products		
Hot-pressed block	48 000	35 000	2
Sheet	77 000	55 000	9
Extruded rod	77 000	40 000	7
Forgings	75 000	45 000	5
	Cast and wrought products		
Sheet	49 000	40 000	2
Wire	150 000	125 000	3

guidance systems is virtually unchallenged, because of its lightness and because the parts retain extremely precise dimensions. Aerospace programmes require large amounts for re-entry vehicles of intercontinental ballistic missiles and for structural parts of satellites. A third major use is in the nuclear field, especially for reactors and weapons which depend on the ability of beryllium to moderate or reflect neutrons. Beryllium is transparent to radiation and subatomic particles and is therefore used for windows in X-ray and radiation-measuring equipment.

From a wide range of development projects, future applications in several fields may be forecast. *Instrumentation* will rely more heavily on beryllium for both guidance systems and computers. Parts for *space optics* require precise dimensional control for mirrors, cameras and telescopes. Some missile and satellite programmes are considering beryllium for *heat shields*. Generally, *aerospace structures* utilize its unique combination of mechanical properties and light weight. Scope for use in *space and flight hardware* includes aircraft brakes, antennae, nuclear power sources, rocket nozzles, radiation-measuring instruments and sheet metal fasteners. Beryllium has potential applications as an additive to *rocket fuels* and as a wire *reinforcement* in composite materials.

Beryllium is most widely used as an alloying addition to other metals discussed in this chapter The greatest tonnage is used in copper base alloys containing 98 or 99% copper and $\frac{1}{2}$ to 2% beryllium. It is a potent strengthener of copper, nickel, iron, molybdenum, titanium and a number of other metals. Beryllium is used in magnesium-bearing aluminum alloys and in other magnesium alloys to minimize the loss of magnesium through oxidation or burning. Beryllium oxide is used in the electronics industries because, unique among ceramics, it combines electrical insulation with good thermal conductivity.

4.2.9 Statistics

Beryllium sells for about $54 to $62 per lb in raw form and as semi-finished products up to $300 per lb.

Metal Statistics 1967 indicates that world mine production in recent years has been about 6000 tons per year, which is somewhat lower than the 1960 and 1962 levels.

Most beryllium production, well over 90%, is concentrated in the USA where most of the output is consumed; minor quantities are exported to Europe. French and British production is used principally in Europe, with some consumption in the USA. Japanese products are used locally.

Production of beryl ore concentrates

	World production of BE concentrates (tons)	USA imports of BE concentrates (tons)
1960	12300	10430
1961	12900	8510
1962	11000	8550
1963	7700	6240
1964	5200	5425
1965	5700	7790

4.3 MAGNESIUM

Although magnesium does not occur free in nature, it is the eighth most abundant element in the earth's crust and in quantity is second only to aluminium among metals. Its compounds are widely distributed as minerals and it forms about 0.13% of sea water – an almost limitless source.

4.3.1 History

The first metallic magnesium was extracted by Sir Humphrey Davy in 1808. He electrolysed a mixture of magnesia and mercuric oxide to give an amalgam of mercury and magnesium and then boiled off the mercury leaving metallic magnesium. Bussy in 1828 developed a method in which anhydrous magnesium chloride was heated with potassium. After dissolving out the chlorides, he obtained magnesium as a powder which could be melted to give globules of relatively pure metal. Faraday in 1833 electrolysed a fused magnesium salt and his method is, in essence, the modern electrolytic process. Further work on the electrolytic production of magnesium was carried out by Bunsen and subsequently by Matthiessen, who developed the use of anhydrous salts for feeding the electrolytic cell.

Commercially magnesium can be said to date from 1866 when it was produced in quantity in Germany using a modified Bunsen cell. Most of the metal used was produced in Germany until 1915, when wartime needs led to its production in the UK, USA, France and Canada. Eight companies were producing magnesium at this time, but by 1927 only one, Dow Chemical Company, was left.

Little development occurred until the outbreak of world war II but from 1939 onwards progress has been rapid. New alloys with improved properties have been developed, methods of fabrication have been developed and improved and the price has been reduced to a level at which, on a volume basis, magnesium is the cheapest of all common metals. Expansion of production during world war II was phenomenal; fifteen new plants with a total capacity of nearly 300 000 tons per annum were built in the USA alone.

4.3.2 Occurrence

The most important magnesium-bearing minerals are: magnesite, dolomite, hydromagnesite, brucite, carnallite, kieserite, kainite, serpentine and olivine. Of these, magnesite ($MgCO_3$), brucite ($Mg(OH)_2$) and dolomite ($CaCO_3MgCO_3$) are the ones normally used commercially in the production of metallic magnesium. Magnesite is found in many countries including Austria, Greece, Russia and the USA. Brucite, although rich in magnesium, has the disadvantage of being very localized, the only deposits of significance being in North America. Dolomite is very widely distributed but it is only used as a source of metal when magnesite is not available.

Other sources of magnesium are brines from salt beds and sea water or from the waste materials from salt and potash production. Magnesium can constitute up to 15% of the total dry salt from brines and up to $3\frac{1}{2}$% of salt from sea water.

4.3.3 Production

There are two main production methods for metallic magnesium: one electrolytic and the other thermal. The electrolytic process is usually cheaper and is widely used in the USA. Its main disadvantage is that the raw material – fused magnesium chloride –

must be very pure. In the thermal process, magnesium oxide is reduced by heating with ferrosilicon or carbon and, because it is a batch process, is usually more expensive than electrolysis.

Electrolytic process. Sea water is the most important source of material for electrolysis although other natural or artificial brines and magnesium-bearing minerals are also used. Whatever the source, the first stage is the production of magnesium chloride which is then electrolysed in the second stage. Sea water and other brines are first converted to magnesium hydrate by treatment with lime and then converted to chloride by the addition of hydrochloric acid. Alternatively, the magnesium hydrate precipitate may be calcined and treated with chlorine gas to give magnesium chloride.

Mineral ores are first crushed and then calcined to convert them to magnesium oxide. This again is treated with chlorine gas to convert it to the chloride.

The electrolytic reduction cell is an iron or steel box with carbon electrodes. The charge consists of the anhydrous magnesium chlorides together with alkaline earth chlorides. In the reduction process, chlorine is given off at the anodes and is collected for reuse in the production of more magnesium chloride. The liberated metal floats on the surface of the cell and is removed for refining in a melting furnace and casting into ingots. The electrolytic process gives metal of about 99.9% purity.

Thermal process. The thermal reduction process employing ferrosilicon normally operates with magnesite or dolomite as raw materials. The latter has the advantage of containing lime which acts as a slag to combine with any silica that is present. The ores are pulverized, mixed with pulverized ferrosilicon and charged into steel retorts. The retorts are evacuated and heated to about 1200°C. Magnesium vapour is formed which condenses in the cool end of the retort. On conclusion of the cycle the metal is removed, refined in melting furnaces and cast into ingots of 99.5% purity or greater.

The thermal reduction process can also be operated with carbon as a reducing agent and in this case an electric arc furnace is employed. Magnesium vapour is given off and condensed on cooled surfaces in the form of dust, also containing magnesia. The dust is reheated to distil off the magnesium metal with a resulting purity of 99.5% or greater.

4.3.4 Properties

The most noteworthy property of magnesium is its low specific gravity (1.74 compared with 2.71 for aluminium and 7.85 for steel). In addition, it can readily be machined, requiring less power than any other common metal, and can also be cast by all the normal foundry processes. In the form of alloys, it has a very high strength-to-weight ratio, leading to its widespread use in structural applications, particularly for transport and other moving equipment.

Because of a protective oxide layer on the surface, magnesium has good resistance to atmospheric exposure and it is also resistant to alkalis and to pure water. Acids, however, will attack it.

The principal properties of pure magnesium (99.9%) are shown in table *4.2*.

4.3.5 Alloys

Alloys of magnesium are widely used because they have, in general, superior properties to the pure metal. Alloys with additions of aluminium and zinc, manganese, zinc and zirconium are extensively used in industry for anodes, castings, sheet and extrusions. Where higher strength and better performance at elevated temperatures are

Table 4.2

Symbol	Mg
Atomic number	12
Atomic weight	24.32
Crystal structure	close-packed hexagonal
Mass numbers of the isotopes	24, 25, 26
Density at 20°C (68°F), g/cm^3	1.74
Electrical resistivity at 20°C (68°F), microhm-cm	4.46
Mean coefficient of expansion (20° to 500°C)	0.0000299
Mean specific heat (0° to 100°C), cal/g	0.249
Latent heat of fusion, cal/g	88.8
Btu/lb	159.8
Thermal conductivity 0° to 100°C (32 to 212°F), cal/cm degC s	0.376
Btu ft^2/in. degF h	1090
Modulus of elasticity, kg/mm^2	4570
p.s.i.	6.5×10^6

required, alloys with thorium and some of the rare earths are used.

Table *4.3* gives some typical alloy compositions but is far from exhaustive; one major manufacturer's catalogue lists more than 25 compositions.

4.3.6 Uses

The light weight and good strength of magnesium alloys have led to their widespread use for structural purposes in the aircraft industry, for such components as landing gear, engine parts and equipment. Magnesium is also used to a great extent in the automobile industry, a noteworthy example being the German Volkswagen car, in which the air-cooled engine is cast mainly from magnesium. Wheels for sports and racing cars are often made from magnesium. Other instances where its light weight is an advantage are found throughout industry, for example, in portable tools like electric drills, in bobbins and spools in the textile industry and in lawn-mowers. In addition to its widespread structural uses, magnesium finds application for sacrificial protection of other metals in an application known as cathodic protection (for further details see ch. Corrosion of metals, p. 841). Briefly, a magnesium casting weighing from 1 to 20 lb is attached to the structure (usually steel) to be protected either underground or under water. The slow disintegration of the magnesium anode generates an electric current which protects the steel structure and prevents corrosion. The magnesium is eventually

Table 4.3

Alloying elements	Uses
3% Al, 1% Zn	Sheet
6% Al, 3% Zn	Castings. Anodes.
9% Al, 2% Zn	Castings
6½% Al, 1% Zn	Extrusions
6% Zn, 0.5% Zr	Extrusions
1.2% Mn	Extrusions
3% rare earths	Castings

all used up and has to be replaced. Anodes are used in oil tankers and other ships and attached to structures such as piers and jetties.

Much magnesium is also used for alloying with other metals, such as aluminium, and it is also used in metallurgical production processes as a reducing agent and as a purifying material. In the production of special ductile forms of cast iron, magnesium is the agent employed to modify the iron structure. In powder form it is used in fireworks and signalling equipment and also in the production of some types of organic chemical by the Grignard reaction.

The use of magnesium is growing rapidly in all industrialized countries (see below) because of its useful combination of properties for many modern applications.

4.3.7 Statistics

Magnesium production (tons of 2 000 lbs)*

	1965	1966	1967	1968
Canada	10 108	6 723	8 887	9 878
Italy	6 959	7 180	6 996	7 267
Japan	8 763	5 832	7 438	6 236
Norway	29 100	31 195	31 400	34 500
USSR	36 000	40 000	45 000	
UK	6 083	4 143	4 059	3 901
USA	81 361	79 794	97 406	98 375
World	183 100	179 800	206 800	

Magnesium price in USA for selected years

	US cents/lb
1940	27.0
1945	20.5
1950	22.0
1955	30.9
1960	36.0
1965	28.8

Magnesium end uses in USA in 1964

	Short tons
Castings	8 000
Sheet and plate	5 000
Extrusions and forgings	4 500
Powder	1 500
Alloying and metal treatment	26 500
Cathodic protection	5 000
Others	5 000

4.4 ALUMINIUM

4.4.1 History

Aluminium is a modern metal, unknown before the nineteenth century and used in large quantities only in the last 50 years. Its existence was first postulated by Sir Humphrey Davy in 1807 but his attempts to separate it from alumina (aluminium oxide) by electrolysis were not successful. In 1825 the Danish physicist H. C. Oersted isolated small quantities by heating potassium amalgam with aluminium chloride and removing the mercury from the resulting aluminium amalgam by distillation. Frederick Wöhler produced aluminium in a similar way in Berlin in 1827 using metallic potassium instead of amalgam. Aluminium remained a curiosity, however, until 1854 when Henri St. Claire Deville improved Wöhler's methods by using the much cheaper sodium in place of potassium. Eventually the price came down to 200 francs per kilogramme, but this still kept it in the precious metal class and its uses were confined to such things as jewellery, cutlery and works of art.

A major breakthrough occurred in 1886 when, independently of each other, the Frenchman Paul L. T. Heroult and the American, Charles Martin Hall devised the modern method of extraction, by electrolysing alumina dissolved in a bath of molten cryolite ($3NaFAlF_3$). Hall founded the Pittsburgh Reduction Company (now the Aluminum Company of America) to exploit his invention and Heroult became associated with Aluminium Industrie A.G. at Neuhausen, which changed its name recently to Swiss Aluminium Company. An original associate of Hall's, A. V. Davis, was still active in the management of the Aluminum Company of America in the 1950s.

Another important name in the development of aluminium is that of Alfred Wilm, who discovered in 1909 the strong heat-treatable aluminium–copper alloy known as duralumin, and thus laid the foundations of the aircraft industry.

In a little over 100 years aluminium has progressed from a curiosity to one of the essential materials of modern civilization, produced at the rate of many millions of tons a year.

4.4.2 Occurrence

Aluminium is the most abundant metal in the earth's crust, making up about 8% of the total volume. It is a constituent of many rocks, particularly clay, shale, slate, schists and granite, and also occurs fairly commonly in the form of hydrated oxides, known as bauxites. This name is derived from the village of Les Baux in the south of France where the mineral was originally worked.

Bauxites are the principal ores of alumina and although they vary greatly in appearance and composition they are all mixtures of hydrated aluminium oxides, with impurities in the form of compounds of iron, silicon, titanium and other elements. Most of the high-grade bauxite deposits, suitable for aluminium smelting, occur in tropical or semi-tropical regions, notably Jamaica, Guyana, Surinam, West Africa, India, Malaysia and Northern Australia. Deposits are, however, worked in many other regions for economic or strategic reasons. Bauxite deposits are of three main forms: blanket, interlayered and pocket deposits. Most of the tropical deposits are of the blanket type, about 20 ft thick, and are mined by open-cast methods. Interlayered deposits occur, for example, in France and Greece and are mined by both open-cast and underground methods. Pocket deposits, which occur in Jamaica, Istria and Dalmatia, are mined by open-cast methods.

Mined bauxite is commonly subjected to a crushing, washing and drying process known as 'benification' to improve its quality.

4.4.3 Extraction

The first step in the extraction of aluminium is to reduce the bauxite to pure alumina (Al_2O_3). There are several methods, but the Bayer process is the most used and usually the most efficient. The bauxite is treated under pressure with hot caustic soda

ALUMINIUM | 245

Fig. 4.1 Aluminium: Mining bauxite for the manufacture of aluminium. The large crane or 'dragline' is removing the top surface, known as over burden, to expose the bauxite. This mine is in Guyana.

which dissolves the alumina to form sodium aluminate. The residue, known as 'red mud' and containing iron oxide, titania and silica, is filtered off and discarded. Aluminium oxide trihydrate ($Al_2O_3 3H_2O$) is precipitated from the liquor, washed and calcined in rotary kilns at about 980°C to give alumina in the form of a white powder. The yield in the Bayer process is approximately 1 lb of alumina for each 2 lb of high-grade bauxite.

Metallic aluminium has a strong affinity for oxygen and cannot readily be reduced from alumina by the traditional smelting methods. Although small quantities were produced in the nineteenth century by reducing an aluminium compound with sodium or potassium, mass production of aluminium did not become feasible until the development of the Hall–Heroult reduction process.

Roughly 10 kWh and 2 lb of alumina are consumed for each pound of aluminium produced. Because of the need for large quantities of electric current, aluminium smelting is almost always located in centres where cheap power can be provided. By far the greatest capacity is in areas with access to water power, for example in Canada, Scandinavia, Scotland, Switzerland, Austria, USSR, India and Japan. Some smelters, notably in the USA and Western Germany, are powered by generating stations fired with brown coal, and others are associated with natural gas deposits. In the late 1960s nuclear power stations were being investigated. All, however, depend on electricity costing less than $\frac{1}{2}d$ per unit. This leads to another characteristic of aluminium smelters:

they are often located in what might be termed 'frontier' regions. Once the power source has been developed and used for smelting for a few years, other industries tend to move in, which results in an increase in power charges. This in turn forces the smelters to move on to other undeveloped territories and the process is repeated.

Because high-grade ore and cheap power are vital to the existence of the aluminium industry much effort is expended in the exploration and development of new sources of raw materials and further attention is devoted to possible new sources of energy such as magnetohydrodynamic or atomic power. Recent development work on direct methods of reducing aluminium ores has not been commercially successful and the industry is looking to cheap atomic power as its future source of energy.

A reduction plant for aluminium consists of one or more 'pot lines' comprising a number of large steel boxes, or pots, lined with carbon about 6 inches thick forming the cathode. One or more carbon anodes which dip into each pot are either prebaked from petroleum coke and pitch or are of the 'self-baking' type in which coke and pitch are continuously added as the electrode burns away. Each cell takes a direct current of from 10000 to 100000 A according to size and operates at about 6 V. It is usual to connect pots up to a total of about 150 in series. The d.c. power is fed from banks of rectifiers. The electrolyte employed is cryolite, a double fluoride of sodium and aluminium ($3NaFAlF_3$). Although natural cryolite, obtained from Greenland, was formerly used synthetic cryolite, made in an extension of the Bayer process for alumina,

Fig. 4.2 Potline of aluminium reduction cells at Kitinat B.C. Canada. The carbon lined steel cells or "pots" are seen on either side. In the foreground (right) molten aluminium is being extracted by vacuum into a crucible.

is the usual material today. Powdered cryolite is loaded into the cell and melted by electric current. Alumina is then dissolved in the molten cryolite and decomposed by electrolysis into oxygen and metallic aluminium. The latter falls to the bottom of the cell. When the electrical resistance of the cell increases suddenly to about 10 times its normal value, more alumina is added.

Metallic aluminium is siphoned from the pots and either cast into blocks for remelting and alloying, or continuously cast in large slabs for subsequent rolling or extrusion.

4.4.4 Properties

Pure aluminium is a silvery coloured metal whose main characteristic is lightness. Its specific gravity of 2.7 (0.1 lb/in^3) is about $\frac{1}{3}$ that of other common metals such as steel, copper and zinc. Although chemically it is highly reactive, it is protected against deterioration by a thin tough oxide skin that forms spontaneously on the surface and accounts for the high durability of aluminium exposed to the atmosphere. This oxide skin can be thickened artificially by an electrolytic process known as 'anodizing' or 'anodic oxidation' and the resulting film can be coloured artificially. Anodized aluminium is much used in architecture and decorative applications, as well as in chemical plant and other situations where protection is needed.

Although pure aluminium is not very strong (about 6 t.s.i compared with 28 t.s.i for mild steel) its strength can be increased by up to 50% by cold-working in a rolling mill.

However, much greater increases in strength can be obtained by the addition of small quantities of other elements such as magnesium or silicon, a process known as 'alloying', which produces aluminium alloys. The term aluminium as used in industry is normally taken to mean the whole range of aluminium alloys which are used commercially. Different alloying elements give different properties and a wide range of materials can be produced with properties suitable for industrial applications. For example, one class of alloy has very high strength for use in aircraft, another has high resistance to sea water and is used in ships and another takes a high polish and is much used in architecture.

Although all the alloys made in this manner can be increased in strength by cold-work, a number of them can be made stronger still by heat treatment, in which the metal is heated to a temperature below its melting point and then suddenly quenched. It is usual to have also a second heat treatment or 'aging' process. Heating the metal causes changes in the internal structure and the quenching locks these changes in the individual grains. Aging then gives a stable structure of high strength, which can only be destroyed by reheating to a high temperature.

The melting point of aluminium is about 650°C and it is has a very high capacity for heat absorption. Thus, although the melting point of steel is about double that of aluminium, the heat required to melt a ton of each is about the same. At the same time, aluminium has very high heat conductivity, hence its widespread use in heat exchangers in industry and also in domestic cooking utensils (in which use, as in many others, chemical inertness is another advantage).

Aluminium is a good electrical conductor, surpassed only by silver and copper among the common metals. On a weight basis, aluminium is a better conductor, pound for pound, than any other metal; this accounts for its widespread use in overhead conductors and, in recent years, for underground cables and transformer windings. The price of copper has also been a factor encouraging the use of aluminium.

Aluminium will take a high polish and then is a good reflector of light, hence its free use in decoration and thermal insulation.

4.4.5 Alloys

Commercially pure aluminium, as delivered from the smelter, is relatively soft and of value only in situations where high strength is not required, for example, decoration, packaging, electrical conductors, roofing and domestic utensils. Several million tons a year are consumed in this form.

Addition of small quantities of other elements produces drastic changes in properties and a number of such alloys are of great commercial importance. Some of the principal alloy systems are listed below, but many others have been developed for special purposes.

1. *Alloys for making into sheet, plate and extruded sections*

   (a) *Aluminium–magnesium.* Alloys containing up to 5% magnesium have good strength and notable resistance to corrosion. They are used in shipbuilding and chemical engineering. The strength of these alloys can be increased by cold-working but not by heat treatment.

   (b) *Aluminium–magnesium–silicon.* This system, sometimes known as 'the magnesium silicide group' gives a range of medium-strength heat-treatable alloys used largely in engineering, structures and architecture.

   (c) *Aluminium–copper.* Addition of up to 4% copper, with other elements in smaller quantities, gives alloys of very high strength used principally in the aircraft industry. Resistance to atmospheric corrosion is not good and the material is often supplied with a surface coating of pure aluminium.

   (d) *Aluminium–zinc–magnesium.* There are two classes of heat-treatable alloy in this range. One gives medium-strength readily weldable alloys much favoured in Europe, while the other produces ultra-high-strength material whose tensile strength can exceed that of steel. This latter class is used almost exclusively for aircraft.

2. *Alloys for casting to shape in the foundry.*

   (a) *Aluminium–silicon.* The vast majority of general purpose castings are made from aluminium with up to 12% silicon added. These alloys are very fluid when liquid and give sound castings.

   (b) *Aluminium–silicon–copper.* Small quantities of copper increase the strength of the material and improve its pressure-tightness. They also enable the strength to be further increased by heat treatment.

   (c) *Aluminium–magnesium and aluminium–copper.* These two systems have high strength but are not particularly easy to cast. For this reason their uses are normally restricted to uncomplicated shapes.

4.4.6 Available forms

Remelt ingot. Aluminium from the smelter is supplied in the form of ingots of different weights and shapes. The first step in the production of commercial forms is the remelting of the ingot, followed by degasification and other purification treatments and the addition of alloying elements as required. The metal is then cast, usually by a semi-continuous chilled mould process, into ingots for rolling or extrusion.

Sheet. Aluminium sheet is of thicknesses from 0.01 to 0.25 inch and is supplied in flat or coiled form. It may be in pure aluminium or any of a large number of alloys and can also be supplied with different surface finishes, such as lacquered or stove-enamelled, or coated with high purity aluminium to increase its durability or decorative properties. Corrugated and profiled aluminium sheet is much used for roofing and siding.

Foil. Sheet material below about 0.008 inch is known as foil and has widespread application in the packaging industry for wrapping products such as chocolate and cigarettes. It is used for paper-insulated and electrolytic capacitor electrodes in the electrical industry and it is also being used increasingly in place of round wires for coil winding in transformers and magnets.

Plate. 'Plate' is flat material over $\frac{1}{4}$ inch thick up to a normal maximum of 6 in. With modern machinery plate can be produced in large pieces up to 10 ft wide and 50 ft long and up to $2\frac{1}{2}$ tons weight. Plate is supplied in several alloys.

Extruded sections. Aluminium is amenable to extrusion, that is, long lengths of shaped sections can be produced by forcing the metal through a shaped hole or die. The process, one of the most versatile manufacturing methods available to the engineer, enables shapes to be designed with built-in features for special applications. Any section that will fit within a 16 inch diameter circle can be extruded. The normal maximum length is 38 ft but longer lengths can be supplied. The sections may be solid or hollow and are available in a wide range of alloys, as with other aluminium products.

Typical examples of extrusions are window sections, skirting boards and aircraft wing spars.

Tubes and other hollow sections may be either extruded or made on a draw bench.

4.4.7 Wire and rod

Aluminium wire is the commodity of principal interest to the electrical engineer and it is produced by conventional wire-drawing methods. The starting point is circular rod about $\frac{3}{8}$ inch in diameter, which is produced by extrusion, by rolling and drawing or by continuous casting.

The rod is reduced to wire by pulling it through hardened steel dies of successively smaller diameters.

Paste and powder. Aluminium can be reduced to very small particles by a process known as ball-milling and these particles may be round or flat. The powder made from round particles is used in making fireworks. The flat particles are mainly used as pigments for aluminium paint, generally supplied mixed with a solvent as 'paste'.

4.4.8 Working of aluminium

Aluminium can be handled by all the traditional metal-working techniques. Some of the more important methods are described below.

Casting. Although much aluminium is cast in sand moulds, more is cast in metal dies, filled either under gravity or by pressure. Die casting is used to produce repetition parts for industry, for example, pistons for cars and door handles for buildings. The cost of a metal gravity die may be justified by quantities of 500 to 2000, but pressure dies are more expensive and the economical quantity rises to about 10000. A recent development is low-pressure die casting which has some of the advantages of both gravity and pressure die casting.

Aluminium foundries are moving towards complete automation and much use is being made of casting machinery that can be programmed to carry out a given sequence of operations.

Machining. Aluminium is easily cut by saws of all types, particularly band saws, and can also be shaped by turning, drilling, planing and all the usual machine-shop operations. For repetition parts a free-machining alloy would be recommended.

Pressing and forming. Being a ductile metal, aluminium is readily pressed, spun and formed to shape in conventional machinery.

Forging. High-strength parts such as aircraft undercarriage parts are often hot-forged to shape; repetition parts are made under the drop-stamp. The strong alloys need heavy machinery for such deformation.

Joining. By far the most important joining process for aluminium is welding, which is carried out by electric arc under a shield of inert argon or helium gas. The arc may be struck with a tungsten electrode (t.i.g. welding) or with an aluminium consumable electrode (m.i.g. welding) and fast reliable welds can be made in metal from 0.008 to 6 inches thick. Aluminium can also be spot welded, and thin material is joined by electron-beam and ultrasonic welding techniques.

Soldering is readily carried out but is used almost exclusively in electrical joints. The soldering process may employ a flux and tin-lead solders containing a little zinc, or a fluxless tin-zinc solder which is merely rubbed on the hot metal.

With the development of welding, riveting has lost some of its importance as a joining method but it is still used to a large extent in aircraft construction and in heavy engineering. The aircraft industry uses special rivets that can be driven from one side only.

4.4.9 Uses

Aluminium is used in virtually every facet of modern life and in every country in the world. The major uses are very similar in the industrialized countries but the emphasis varies slightly in different parts of the world as shown in table *4.4*.

The principal uses in each category can be summarized as follows.

Building. Roofing and siding sheet is by far the largest individual use, but more and more components of buildings are being factory-made in aluminium as the demand for housing increases. Aluminium windows in many countries now account for as much as 25% of the total window usage. Other uses include doors, door furniture and metallic trim.

Transportation. Aircraft still account for a large part of the transport usage, although with the decline of military aircraft it is decreasing. Large passenger ships are being made with aluminium superstructures to save weight and space, and railway rolling stock, particularly for urban services, is often all-aluminium. In road transport, commercial vehicles have all-aluminium bodies to save weight, while the amount of

Table 4.4 Percentage of total usage consumed by different industries

End use	North America	Europe	Rest of world
Building	26	12	16
Transportation	24	29	15
Consumer products	10	11	21
Electrical	14	14	19
Packaging	9	10	8
Engineering and machinery	8	10	10
Others	9	14	11

aluminium in private vehicles increases every year. This is mainly in the form of trim, window frames etc. and in cast engine components. Several cars have aluminium cylinder blocks with steel liners and a large increase in this use is expected if an aluminium cylinder can be developed that does not need a steel liner.

Consumer products. These include domestic equipment of all kinds, including hollow ware, and other decorative and utilitarian goods found in hardware stores. Table 4.4 shows that such uses are relatively more important in the less industrialized countries.

Electrical. The principal electrical use is for overhead line conductors, in which stranded aluminium cable is used, often with a steel core to impart extra strength. In the UK and Europe underground cables with stranded or solid aluminium conductors are becoming of increasing importance and form a market comparable with that for overhead conductors. Such cables may have tubular aluminium outer sheaths, corrugated for flexibility. Transformer windings in aluminium are becoming more common and these may be of wire or of aluminium flat wide strips. Aluminium seems likely to be used increasingly for small wiring cables for houses, shops and offices.

Packaging. Aluminium packaging may be divided into rigid containers such as food cans, and flexible containers such as collapsible tubes and foil wrappings. Both of these uses employ large quantities of aluminium, but the flexible market is the larger. In most countries the rigid container in aluminium is restricted to non-sterilised foods, such as jams, preserves, coffee etc. and other materials such as cigarettes and pharmaceuticals. However, in some European countries, notably Norway and West Germany, there is a large usage of aluminium rigid containers for fish, vegetables, meat and other processed foods.

A recent innovation is the 'easy open' can end which is nearly always made of aluminium. Originally for beer, it is now used for other packs, including processed foods.

The market for foil and foil packs in aluminium is very extensive, as also is that for toiletries, cosmetics, food and other pastes.

Engineering. Aluminium is a normal engineering material both for load-bearing parts and for casings for machinery. It is chosen for all uses where light weight and strength are required, including machine tools and portable equipment. Chemical engineering uses vast quantities of aluminium for tanks and equipment when high resistance to attack is needed. It is also widely used in low temperature (cryogenic) applications because its properties do not fall as the temperature drops and it is not subject to brittle failure.

Domestic ware. The familiar aluminium cooking utensil needs no description. A recent development has been plastic coatings for a non-stick surface. Composite aluminium–copper and aluminium–stainless steel utensils have been made, but they do not have any apparent advantages over the pure metal.

4.4.10 Statistics

Alumina production from bauxite: capacity in 10^6 tons of 2000 lbs

North America	7.4
Caribbean and South America	2.7
Europe	3.5
Asia (including USSR)	4.0
Australia	1.3
Total	19.0

World aluminium production: capacity in 10^6 tons of 2 000 lbs (1968)

North America	4.4
South America	0.15
Europe	2.4
Asia (excluding China and USSR)	0.7
Australia	0.2
Africa	0.15
Total	8.0
USSR and associates	1.5
Total world capacity	9.5

Aluminium production (tons of 2 000 lbs)*

	1965	1966	1967	1968
USA	2 754 476	2 968 366	3 269 259	3 255 041
Canada	830 505	889 915	963 343	979 171
W. Germany	262 562	268 837	278 763	283 764
France	375 364	400 698	398 166	403 159
UK	39 911	40 934	43 042	42 066
USSR	1 150 000	1 185 000	1 270 000	1 350 000
Norway	303 802	364 075	397 912	518 169
Japan	321 947	369 414	421 189	531 200
Netherlands		22 422	35 881	54 170
World Total	7 125 470	7 741 284	8 501 536	9 042 120

Aluminium consumption (tons of 2 000 lbs)*

	1965	1966	1967	1968
USA	3 093 400	3 308 700	3 310 900	3 684 000
Canada	210 000	209 000	217 500	220 000
W. Germany	427 000	462 400	459 400	594 500
France	274 000	328 800	324 100	324 100
UK	400 800	407 200	397 300	430 566
USSR	985 000	995 000	1 050 000	1 150 000
Norway	42 400	43 000	45 700	70 500
Japan	370 400	466 500	552 800	696 200
Netherlands	21 100	23 000	24 000	34 600

Yearly average price of aluminium in USA cents/lb

1930	1935	1940	1945	1950	1955	1960	1965	1966*	1967*	1968*
23.3	20.5	18.69	15.00	17.69	23.67	26.00	24.5	24.5	24.98	25.58

World production of aluminium by years (in 10^6 tons of 2 000 lbs)

1930	1940	1950	1956	1958	1960	1962	1964	1966*	1967*	1968*
0.25	0.80	1.66	3.72	3.87	4.95	5.58	6.73	7.74	8.50	9.04

4.5 GALLIUM, INDIUM AND THALLIUM

These three metals belong to group III of the periodic system, which also contains boron and aluminium.

4.5.1 History

All three are of comparatively recent discovery and all owe their identification to the use of the spectroscope. Lecoq de Boisbaudran, a French chemist, discovered and isolated gallium in 1875 while searching for new elements predicted by Mendeljeff in his periodic table of the elements. It was the first new element to be discovered after the publication of the table. Indium was discovered in 1863 by Reich and Richter who were examining sphalerite ore with a spectroscope. The discovery was accidental; they were searching for new sources of thallium, which had been discovered in 1861 by Crookes, the well-known English scientist. (Here again the discovery was accidental; Crookes was in fact searching for tellurium). A Frenchman, Lamy, discovered thallium at about the same time, but Crookes was generally accepted as the first.

4.5.2 Occurrence

Gallium occurs abundantly in nature, in about the same degree as lead, but it is very scattered and even the richest sources only contain 0.5 to 1% of the metal. These rich ores, of which germanite is typical, are very scarce but small quantities are found in South Africa. Commercially usable quantities are found in some zinc ores and in coal but the commonest source of exploitation is the residues from the Bayer process in aluminium extraction in which bauxite is converted to alumina.

Indium is also found widely distributed in minerals but only occasionally is it in appreciable quantities. Some pegmatite ores in the USA are said to contain over 2% indium. Ores of other metals such as tungsten, magnesium, tin and iron also contain indium, usually in minute quantities but sometimes in larger amounts. Sphalerite is the commonest ore used commercially. Indium is also present as a by-product in the lead, zinc and aluminium industries.

Thallium sources are comparatively scarce, the most important ores being crookesite, occurring principally in Sweden, and lorandite. It is also a constituent of several ores of iron, copper, zinc, cadmium and aluminium but not usually in any workable quantities. Chimney soots are a rich source of both indium and thallium. None of the three metals occurs in a native state.

4.5.3 Extraction

Gallium can be extracted in several ways. The residues from zinc and germanium ores are treated with aqua regia and unwanted metals precipitated by the addition of zinc. They are then filtered off and gallium precipitated as a basic salt. This is dissolved in hydrochloric acid and converted to oxide which is then reduced by passing carbon dioxide into the solution. This process is used in the USA and Germany.

Bayer residues in the aluminium industry (sodium aluminate lyes) are concentrated to $\frac{1}{2}$% gallium which is then precipitated by carbon dioxide or chalk. This process is operated in France and the USA. An electrolytic process for extraction of gallium from Bayer residues is reported to be under development in Germany.

Several methods are in use for purification of the metal, including acid treatment, filtration, electrolysis and distillation.

Indium is extracted from sphalerite ore by roasting it, treating it with sulphuric acid, and then placing zinc plates in the neutralized solution. Indium precipitates on the zinc, which is removed by treatment with acid. The residue is dissolved in nitric acid and indium oxide precipitated by adding barium carbonate. The oxide is converted to metal by heating in a stream of hydrogen residues. Waste liquors from many industrial processes are treated to recover indium and it is also recovered from chimney soots in industrial plants.

Thallium can be extracted electrolytically from solutions of various salts including perchlorate, sulphate or carbonate. Metallic thallium is also prepared by fusing the chloride with sodium carbonate and potassium cyanide. The fusion product is treated with water and the metal remains undissolved.

A third method is to heat thallous sulphate with pure zinc plates, leaving a deposit of thallium sponge in the zinc from which it can easily be separated.

It will be seen from table 4.5 that, despite the close affinity of these metals, they have widely different physical properties.

Gallium is a greyish-silvery metal which preserves its lustre. The most notable features of gallium are its very low melting point (it will sometimes melt in the hand) and high boiling point. Its melting point is next above that of mercury and when molten it resembles this metal. Gallium shares with antimony the property of expanding on solidification.

Indium is a very soft silvery metal resembling tin, and like tin it gives out a creaking sound when bent. It is resistant to most influences except alkalis. Because it cold-welds, pieces are usually supplied coated with grease.

Thallium is a very soft metal which can easily be cut or deformed. It is very toxic and must not be allowed to come into contact with the skin. It oxidizes rather quickly in air and so is often supplied in sealed containers.

4.5.5 Uses

The low melting point of gallium makes it useful in fusible alloys and also as a liquid in high-temperature instruments. It is widely used in semiconducting devices as gallium arsenide, and is also a constituent of some superconducting alloys. Another important use is as an additive in the analysis of uranium, to increase the sensitivity of the measurement. Because its thermal conductivity varies with the crystal structure, it is also used in special types of heat exchangers.

Indium is used as an electrodeposited coating on other metals and is often heat-treated to fuse it to the base metal. When fused in this manner, it is resistant to flaking and peeling and gives high corrosion resistance to the underlying metal. It is widely used as a sealing material for high vacuum apparatus.

Indium is sensitive to neutron radiation and for this reason is used in safety monitoring badges for workers in the atomic industry. It is used in batteries, bearings and in fusible alloys and is also added to other metals such as lead, copper and aluminium to strengthen them.

Table 4.5

	Gallium	Indium	Thallium
Density, g/cm^3	6	7.33	11.85
Melting point, °C	30	156	303
Boiling point, °C	2337	2480	1475
Thermal conductivity, cal/cm degC s	0.08	0.20	0.993
Resistivity, microhm-cm	26	9	18

An important use of thallium is as an alloying addition to other metals to lower their freezing point. An example of this is mercury for use in low-temperature thermometers. It is also used in fusible alloys, electrical rectifiers and as an addition to glass where special optical properties are needed. Its salts find application as pesticides and fungicides. Care is needed, however, because they are very poisonous.

4.5.6 Statistics

None of these three metals is used in quantity and no reliable information is available on output and consumption. It is probable that the total usage of all three does not exceed a few tons a year.

4.6 GERMANIUM

4.6.1 History

This metal was discovered by Winkler in Germany in 1886 in a mineral known as argyrodite (silver germanium sulphide).

4.6.2 Occurrence

Germanium also occurs in silver-tin sulphides to the extent of about 1% and in other tin and zinc ores, these last being the most important commercial sources of germanium. It also occurs in flue dusts from which it can be recovered.

4.6.3 Extraction

The residues from zinc extraction processes are the most common source of germanium. These residues are first roasted to remove arsenic and sulphur and then treated with hydrochloric acid. This gives a volatile tetrachloride which is distilled off, converted to dioxide and reduced with hydrogen, charcoal or potassium cyanide at 900°C. The germanium is recovered as a powder which is then fused. An alternative method is to heat the residues with sulphur in a reducing atmosphere to give germanous sulphide. This is roasted to convert it into dioxide, which is reduced in the manner described above.

4.6.4 Properties

Germanium is similar in properties to silicon. It is a stable metal only slowly attacked by acids but it is brittle, with a hardness of 6.25 on Mohs' scale. Its thermal conductivity is 0.014 cal/cm degCs and its resistivity 2 to 4 ohm-cm as normally produced with impurities of about 1 p.p.m. After careful refining the resistivity may increase to about 60 ohm-cm.

4.6.5 Uses

Almost the only widespread use of germanium is in semi-conducting devices such as transistors. For this application it must be of exceptionally high purity and a process known as 'zone refining' is used to remove impurities. A bar or rod of metal is passed slowly through a coil carrying a high-frequency current which raises a small cross-section of the metal to its melting point. This cross-section or zone thus travels along the bar, taking impurities with it to the end of the bar which is then cut off. The process may be repeated several times. The germanium is made semi-conducting by the addition of minute traces of other materials such as arsenic or indium.

4.6.6 Statistics

No reliable information is available on the production and use of germanium as much of it is employed in military equipment. The estimated free world production is about 100 short tons per year, most of which is produced in the Congo, South West Africa, USA and Japan. The price of unrefined germanium is of the order of 25 to 30 US cents per gramme.

4.7 TIN

4.7.1 History

There is clear evidence that tin was one of the earliest metals known to man. Its ability to harden copper and lower its melting point made it easier to cast useful implements and this was probably discovered accidentally when copper and tin ores were smelted together. Deliberate alloying of copper with tin produces bronze, hence the Bronze Age, which began about 4000 B.C. in South West Asia and 2000 B.C. in Europe. The alloy normally contained about 10% tin and this is still the most commonly used composition. The Romans and Chinese made speculum metal, a white bronze containing at least 30% tin, which was cast in thin plates and polished as mirrors.

For many centuries from pre-Roman times Cornwall in South West England was an important world source of tin. In the thirteenth century other sources of tin appeared in Gascony and Bohemia, but it was not until the nineteenth century that the present sources in Malaysia were developed. The other major producers have all appeared within the last century. As recently as 1870, one-third of the world's production of tin was from Cornwall, but today this fraction is down to about 0.5%.

Fig. 4.3 Tin: Floating dredge for tin ore.

4.7.2 Occurrence and mining

The major deposits of tin are worked in Malaysia, Bolivia, Thailand, China, USSR, Indonesia, Nigeria, Congo and Australia. Of the world production of about 160 000 tons per year (excluding Communist countries) 43% comes from Malaysia and 15% from Bolivia. The USSR and China are believed to be substantial tin producers but official production figures have not been released. Other countries produce from a few tons up to 2 000 tons per year; the UK's output is around 1 300 tons.

The major tin mineral is cassiterite (tin oxide, SnO_2) but the sulphide and mixtures with other metals such as lead and antimony also occur. In Malaysia and Indonesia, tin ore is found in alluvial deposits and can be won by dredging. Highly sophisticated floating dredges have been developed for this purpose (fig. 4.3). The tailings remaining after separation of the cassiterite from the ore are dumped back from the stern of the dredge. Alluvial deposits on ancient, raised sea beds as in Nigeria are washed down into troughs known as launders by high-pressure water jets. In Bolivia and Cornwall, the cassiterite forms irregular lodes and veins in rocks such as granite and must be deep-mined.

The richness of tin ores varies widely but they are generally lean, not over 1% tin.

4.7.3 Extraction and refining

The tin ore is concentrated by conventional mineral dressing techniques such as crushing mills, classifiers, shaking tables, hydrocyclones, flotation and so on. Initial roasting or leaching may be necessary. Much research is carried out by the tin producers to discover better methods of 'benefication' and the normal highgrade concentrate from the preparation plant contains about 70% tin.

The concentrated ore is smelted to tin metal by reduction with coke in a reverberatory, rotary or electric arc furnace. Low-grade concentrates may be smelted in a blast furnace. Fire refining is carried out in stages at various temperatures by liquation (melting off unwanted ingredients) and by settlement of impurity particles from the molten tin. Blowing air or steam through the tin and stirring in sulphur or other additives are common treatments which are carried out in large heated cast-iron kettles. Electrolytic refining is also used and tends to be more efficient for the removal of certain impurities. The highest purities of tin are obtained by zone-refining (see germanium, 4.6.3).

The most commonly used quality of ingot tin is 'refined', containing at least 99.75% tin. Standards exist for tin of various degrees of purity. The impurity elements present and their amounts depend on the nature of the original ore deposit and on the extraction and refining process employed.

4.7.4 Properties

Tin is normally a white or silver-like metal (white or 'β' tin) which is the stable form above 13.2°C and has a density of 7.29 g/cm³. It becomes a superconductor at very low temperatures. Grey (α) tin has a density of 5.77 g/cm³ and is a semi-conducting material like selenium and silicon. The β to α change is normally prevented by the presence of small amounts of antimony, bismuth or lead. Unusual conditions of low temperature, stress etc., may enable the transformation to occur, when purple-coloured powdery patches appear on the surface having the appearance of a disease – hence the name 'tin pest'.

White tin melts at 231.9°C and has a boiling point of about 2 300°C. The vapour pressure of tin is remarkably low: 1 mmHg at steel melting temperatures. The crystal structure of white tin is body-centred tetragonal and of grey tin is cubic. White tin has an electrical resistivity of 11.0 microhm-cm at 0°C (i.e. about 13% of the conductivity of copper).

The atomic weight of tin is 118.7, the most abundant natural isotopes having mass numbers of 116, 118 and 120. The most useful radioisotope that can be produced in sufficient quantity by irradiation of bulk white tin is ^{113}Sn which emits γ radiation and has a half life of 118 days.

Tin is a ductile, soft metal, having a hardness of about 4 Brinell units or 1.8 on the Mohs' scale. The tensile strength is strongly dependent on the rate of test and is given as 1.5 kg/mm² at 15°C with an elongation of 75%. The shear strength is 1.26 kg/mm² and the fatigue endurance ± 0.3 kg/mm² for 10^7 stress reversals.

When rods of cast pure tin are bent, there is a crackling noise, popularly known as the 'cry' of tin.

Tin oxidizes only slowly at room temperature, but at temperatures above 150°C a yellow oxide film forms fairly quickly. Hydrogen does not react with tin. Dilute acids attack tin very slowly while concentrated nitric acid forms a surface layer of hydrated stannic oxide which stops further dissolution.

Metallic tin has insufficient strength to be used extensively in the bulk and unalloyed form. Pure tin coatings are, however, widely used, particularly because of their non-toxicity. Tin is alloyed with other metals such as lead to form a range of low melting alloys and solders, with antimony to harden it, or tin may be the minor constituent as in copper-tin alloys (bronzes) or in cast iron.

Tin salts of commercial interest are the sulphate, fluoroborate, chloride and sodium or potassium stannate, all of which are largely used for electroplating solutions.

Tin and inorganic tin compounds are generally considered to be non-toxic, hence the wide use of tin coatings in the food-handling industry. The recent rapid development of organic tin compounds has led to a variety of new uses in the plastics and biocidal fields.

4.7.5 Uses

a. As the pure metal. For economic reasons pure tin is not now widely used in massive form but foils for cheese, brewery piping, water stills and organ pipes are still made from the pure metal. For organ pipes pure tin has largely been replaced by tin–lead alloys. The mirror-like surface of molten tin when kept free from oxidation by an inert atmosphere is used for making plate glass by the 'float' process. Costume jewellery is centrifugally cast in rubber moulds from nearly pure tin. Collapsible tubes for medical products are impact extruded from virtually pure tin slugs, but tin-coated lead and aluminium have claimed the rest of this field.

b. As an alloy

Solders. The soft solders widely used in electrical applications normally contain about 60% tin and 40% lead and are chosen because they have a low melting temperature close to the eutectic point (183°C). General machine and bit soldering is often done with a 50:50 alloy. Alloys with 30% tin and lower are used for plumbing and cable jointing and motor-car body filling. Tin with 5% antimony is a solder having better high-temperature strength than tin–lead solders, as also are lead–tin–silver alloys.

Bearing metals. Tin alloyed with antimony (7 to 9%) and copper (3 to 4%) are the highest quality metals used extensively as bearings for large marine diesel engines and generating plant. These alloys are frequently called by the name of the originator, Isaac Babbitt. There are a number of other alloys containing varying amounts of lead. Lead-rich bearing metals for less exacting applications contain 12 to 14% antimony, a small percentage of tin and a little copper.

Small, high fatigue-strength bearings for the modern internal combustion engine consist of a thin layer of bearing metal bonded by a rolling process to a steel backing or shell. They are made as a continuous strip and are subsequently cut to size, shaped and machined to the required bore. Aluminium-20% tin-1% copper alloy is widely used for this purpose. A 6% tin in aluminium alloy is also used as a shell or as a solid bearing.

For less heavily loaded engines, shell bearings may consist of thin layers of tin base or lead base white metal cast on to the steel.

Bronzes. Copper–tin alloys containing 5 to 15% tin with additions of lead and/or zinc (e.g. gunmetal – 5% each of tin, lead, zinc) are widely used for gear wheels, bearing bushes etc., where corrosion resistance and good frictional properties are required. The majority of bronze parts are now machined from rods or tubes made by a continuous casting process rather than from sticks cast individually in metal chill moulds.

Sintering of bronze powder is the technique used for oil filters where porosity of the product is required. Small bearing bushes made by sintering may be made self-lubricating by impregnation with oil, graphite or polytetrafluoroethylene.

Bronze is the traditional metal for architectural devices and statuary. High-tin bronze (20 to 25% tin) is the alloy used for making bells and cymbals because of the peculiarly resonant properties of the metal.

Pewter. There is a resurgence of interest in pewterware, which is an alloy of tin with about 6% antimony and between 0 and 2% copper. Tankards and vases are usually spun from rolled sheet and thicker sections such as plates and candlesticks may be cast. The alloy develops a dark patina with age.

Others. Type metals are lead–antimony–tin alloys. Fusible alloys containing tin, lead, bismuth, cadmium, indium etc., have melting points ranging down to room temperature and are used for sealing purposes, safety fuse links and as casting patterns. Dental amalgams contain tin, silver, copper and zinc and are plasticized with mercury for filling cavities in decayed teeth; they rely on the formation of intermetallic compounds by reaction of the mercury. Tin with 20 to 40% antimony is used widely as a die casting alloy for numerical counter wheels and ornaments. 0.1% of tin added to cast iron promotes a uniform high hardness and hence good wear resistance and gives a heat-stable structure. The niobium–tin alloy Nb_3Sn is a superconducting material and has the highest superconducting temperature (20°K) known at the present time.

c. As a coating

Tin-plate sheets. The major outlet for tin is in the manufacture of tin-plate. This consists of high quality low-carbon mild-steel sheets, having a thickness of 0.15 to 0.5 mm and coated with a layer of tin between 0.4 and 2.5 μm (millionths of a metre) thick. The coating thickness is internationally specified in grammes per square metre (g/m^2) of tin-plate (i.e. per 2 m^2 of surface). A commonly used tin-plate for can making has a tin coating 0.77 μm in thickness (11.2 g/m^2).

About 90% of the world's tin-plate is made by the electrolytic process. Steel slabs are hot-rolled in continuous tandem mills to produce coils of material about 2.5 mm thick. These coils are welded end to end, passed through a continuous pickling line and are then passed through a line of 4 or 6 cold rolling mills to produce the required final thickness. After annealing in an inert atmosphere to prevent oxidation and temper-rolling to the necessary hardness and surface finish, the steel is fed to the tinning lines. It passes as a continuous strand at speeds up to 2500 ft/min through electrolytic degreasing and pickling vats and is then tinplated using current densities of several hundred amps per square foot. The matt grey/white plated coating is then momentarily melted by resistance or induction heating to produce the traditional bright finish and to impart good solderability and resistance to handling. This 'flow-brightening' process produces a thin layer of the intermetallic compound $FeSn_2$ at the tin-steel interface, which has an important influence on the corrosion behaviour of the tin-plate. The continuous band of tin-plate is sometimes given a brief chemical treatment to provide additional corrosion resistance and is electrostatically coated with an oil film before being cut into the required size of sheet.

A small proportion of tin-plate is still made by the older hot-dip process in which the coil of steel from the rolling mills is cut into sheets which are individually cleaned, acid pickled, fluxed and passed through a bath of molten tin covered with hot palm oil. Only the heavier coating grades can be made by the hot-dip process.

Tin-plate is used largely for the manufacture of cans, containers and boxes for packaging foodstuffs and other merchandise. Jar closures, toys and a variety of small brackets are among other things often made in tin-plate. Tin-plate sheets for can manufacture are frequently printed on one face for brand presentation and may be lacquered on the inside face for especially corrosive packs.

Thicker steel sheet and heavier coatings either of pure tin or tin–lead alloys may be produced for manufacturing fuel tanks and roofing sheet.

On fabricated articles. Ferrous and non-ferrous fabricated articles are easily coated by dipping in a bath of molten tin after a suitable preparation. Tin provides a corrosion-resistant, hygienic and non-toxic coating for domestic hardware and food-handling equipment. Hot-dipped tin coatings facilitate soft soldering and the bonding of bearing metals cast on steel or bronze bearing shells. Tin–lead alloy coatings containing normally 30 to 70% tin may be used as a cheaper coating where non-toxicity is not important, as, for example, on soldering tags and heat exchangers.

Pure tin or alloys with lead (40 to 70%), zinc (25%), cadmium (30 to 50%), copper (58%), nickel (35%) may be electrodeposited onto fabricated work from suitable aqueous solutions to provide coatings having a variety of properties and applications. The first three listed generally provide corrosion protection and solderability, while the last two are hard, decorative coatings which are lustrous when polished.

d. As a chemical. Tin is widely used in the chemical field. Stannic oxide has long been used as an opacifier in enamels or as an abrasive. Films of this oxide on ceramic are used as high-stability electrical resistors. Tin oxide was used for 'weighting' silk but this usage has diminished with the spread of man-made fibres.

Organic tin compounds are finding increasing usage for high quality polyvinyl chloride (PVC) where transparency or pale colours require to be maintained. There is a large consumption of dibutyltin compounds for this purpose and now about 1% of dioctyltin is also in use in fruit squash bottles and other food containers. The antifouling properties of certain paints for ship's hulls and wooden boats are dependent on the incorporation of organo-tin compounds. Tributyl tin oxide is a powerful disinfectant used for cleansing hospital operating theatres from dangerous bacteria. There is an increasing general use of such compounds in fungicides and pesticides.

4.7.6 Statistics

Production of tin ore in terms of metal – tons of 2 240 lbs*

	1966	1967	1968
Malaysia	68 886	72 121	75 069
Bolivia	25 522	27 283	28 883
Thailand	22 565	22 430	23 601
Indonesia	12 526	13 601	16 632
Nigeria	9 534	9 340	9 649
Congo (Dem.-Rep.)	6 925	7 013	7 377
Australia	4 838	5 600	6 921
S. Africa	1 745	1 764	1 837
UK	1 272	1 475	1 798
Brazil	1 320	1 600	1 600
Argentina	1 300	2 040	1 840
World[1]	164 000	171 900	182 600

[1]Excluding Communist countries. An estimated production for China and the USSR is 50000 tons per annum

Consumption (long tons) of primary tin by country excluding Communist countries*

	1968
USA	58 120
Japan	22 242
UK	17 172
Fed. Rep. Germany	11 582
France	9 900
Italy	6 200
Canada	4 950
Australia	2 600
Netherlands	4 783
Belgium/Luxembourg	2 701
India	3 000
Brazil	2 100
All others less than	
World consumption	169 700

Consumption of primary tin by uses (1966)

	USA	UK	France
Tin-plate	28 584	9 089	4 800
Solder	15 245	1 577	2 425
Bearing metals	2 145	2 981	560
Bronze	5 339	2 218	1 030
Tinning (fabricated work)	2 589	1 281	545
Chemicals and oxide	1 213	1 464	340
Tubes and foil	1 215	380	470
and others to make *totals of*	60 209	18 425	10 300

4.8 LEAD

4.8.1 History

Lead was known to the earliest civilizations. There is evidence that it was used for vessels and ornaments in Mesopotamia about 3500 B.C., and in predynasty Egypt of about the same period. By Roman times it was in widespread use for water pipes. The methods of production then used survived until the eighteenth century.

Lead is not found free in nature but its ore, galena (lead sulphide), was used in ancient times for ornaments and probably cosmetics. It seems likely that metallic lead was discovered by the accidental smelting of galena in a charcoal fire.

4.8.2 Occurrence

Lead is present in the earth's crust to the extent of only about 0.002%, but it is concentrated in various parts of the world in deposits rich enough to warrant mining. By modern methods minerals with a lead content as low as 3% have been worked profitably.

The principal lead-mining countries are Australia, USA, Mexico, Canada and Peru, which together with the USSR account for about 50% of world production. Several other countries have important mine productions.

Some of the older industrialized countries, which at one time met their lead requirements entirely from home mine production, have now replaced or augmented these supplies by imports of refined metal supplied direct to the consuming industries, or of ores and bullion to feed the smelters and refineries. In this category are notably the UK, USA, Germany, France and Belgium. The principal large-scale producers with an exportable surplus are Australia and Canada, which are the main sources of supply for the UK.

Total world consumption of lead is three million tons annually, a large proportion of which is recovered from scrap.

4.8.3 Ores and their preparation

The most important lead ore is galena (lead sulphide). Others, such as cerrusite (carbonate) or anglesite (sulphate), may be regarded as weathered products of galena, and usually occur nearer the surface.

Mined rock containing galena is crushed and concentrated. The traditional method was some form of water-gravitation treatment such as jigging or the action of a riffle-table, and these devices are still sometimes used. The modern technique is froth flotation in which, by violent air agitation, the ore is suspended in water to which a small amount of pine oil (and/or chemicals) has been added. The rock remains in the body of the liquid, while the sulphide ore is caught in the froth which can be skimmed off and dried.

4.8.4 Smelting

The first stage in the smelting of galena is a roasting process, called sintering, in which most of the sulphur is removed. Some flux is introduced and the resulting product (sinter) is delivered in the form of coarse lumps. Fine powder would choke the blast furnace into which the charge is next transferred. The charge for this furnace consists of roasted concentrated ore, coke, limestone and other suitable fluxes, and scrap iron is often added to help to reduce any residual lead sulphide to lead.

A process is widely used in which a mixed concentrate of lead and zinc sulphides is smelted in a special type of blast furnace. The two metals are produced simultaneously but separately, the lead being in a form suitable for conventional refining.

4.8.5 Refining

At this stage the lead contains many impurities, which may include antimony, arsenic, bismuth, copper, gold, silver, tin, zinc and other metals in traces. In the refining process the copper is the first to be removed. The lead is held just above its melting point and the copper, which is solid at this temperature, rises to the top and is skimmed off. Sulphur is added to help the copper to rise.

The lead is further refined by one of several methods. In the Betts electrolytic process it is made the anode in an electrolyte of acid lead fluorosilicate. The lead which deposits on the cathodes still contains tin and a small amount of antimony and these are removed by subsequent melting and oxidizing with air. For many years the Betts process was the only one which effectively removed bismuth.

Another electrolytic process, developed in Italy, uses a sulphamate electrolyte which, it is claimed, has several advantages over fluorosilicate solutions.

A more widely employed method of refining is the Parkes process, discussed below, but the lead must first be freed from antimony, arsenic and tin. These metals are removed in a softening furnace or by the Harris process. Softening is carried out in a reverberatory furnace in which the metal is melted and agitated with air, the impurities being oxidized and skimmed off. The resulting lead is now soft – hence the name given to this method. In the Harris process the lead is melted and pumped through a molten mixture of sodium hydroxide and sodium nitrate, or some other suitable oxidizing agent. After several hours processing the impurities leave the lead and are suspended in the alkali melt as sodium antimonate, arsenate and stannate and as zinc oxide. The alkali and the metals are then separated and the alkali reused.

The Parkes process is used to extract any silver and gold. The molten lead is agitated with molten zinc in which the impurities are preferentially soluble. The zinc, carrying these metals, rises to the top in the form of a crust and is skimmed off. The small amount of zinc which is left in the lead can be removed by oxidation in a reverberatory furnace, by preferential oxidation with gaseous chlorine, or by vacuum distillation. The Betterton process (an alternative method for the removal of bismuth) is similar to the Parkes process for the removal of silver, except that magnesium and calcium are used instead of zinc.

Combinations of these methods can yield a very high degree of purity. Lead is offered to the market and regularly supplied in bulk quantities of guaranteed purities exceeding 99.99% lead (i.e. impurities about $3\frac{1}{2}$ oz per ton). Purity as high as 99.9999% lead can be produced for special applications (e.g. for electronic engineering).

4.8.6 Properties of lead

Atomic number	82
Atomic arrangement	face-centred cubic
Interatomic distance, Å	3.49
Atomic weight	207.2 (lead derived from radioactive sources may have other atomic weights)
Density at 20°C	11.34 g/cm^3 (708 lb/ft^3)
Melting point, °C	327.4
Latent heat of fusion,	6.26 cal/g (11.27 BTU/lb)
Thermal conductivity, cal/cm degC s 0°C	0.083
100°C	0.081
Specific heat (0°C to 100°C)$_{av}$, cal/g degC	0.031
Coefficient of linear expansion, per degC	0.000029
Increase in volume from 20°C to liquid at melting point, %	6.1
Shrinkage on casting taken in practice at 7/64 to 5/16 inch per ft.	

4.8.7 Uses as unalloyed metal

As a pure metal or a dilute alloy lead is largely used for the sheaths of underground cables. The main function of a cable sheath is to prevent moisture from reaching insulating materials and conductors.

As sheet and pipe, lead is an important material in building and in water supply systems. Lead sheet provides a tough covering, with exceptional resistance to atmospheric corrosion, which can be readily shaped, lead being the most easily worked of the common metals.

Lead is used widely in chemical plant because of its corrosion resistance, particularly in the manufacture and handling of sulphuric acid.

In the atomic energy field, because of its density and high atomic number, lead provides an effective barrier to the dangerous γ-rays emitted by many of the radioactive isotopes. Similarly, it is used as a shield against X-rays to ensure that operators and bystanders are protected from the harmful effects of irradiation when X-ray machines are in operation. Lead sheet is an effective sound-barrier and is used in buildings for this purpose.

4.8.8 Uses as alloys

The electric storage battery or accumulator, used in vehicles and generally in industry, consists basically of grids of antimonial lead alloy into which a paste of lead compounds has been pressed. This is the major use of lead throughout the world.

When alloyed with tin, lead forms a wide range of soft solders. These have fairly low melting points and therefore do not adversely affect the metals being joined. A solder containing 62% tin and 38% lead melts suddenly at 183°C, becoming a freely flowing liquid. For this reason it is used in tinsmith's work where penetration into small openings is often required. A solder of 30% tin and 70% lead is pasty over a fair range of temperature and is better for plumbers, who have to shape the metal while it is setting. Soldering machines use large amounts of lead–tin solders, since by adjusting the compositions and introducing other metals, if necessary, a wide range of properties can be obtained and a great variety of work undertaken.

Lead has been used for many years for anti-friction bearings either as the basic metal or as a major or minor constituent. Lead based white metals, alloys containing antimony and sometimes other elements, are widely used for bearings in many types of machinery.

In unlubricated bearings, as used in washing machines, lead is combined with plastic to work for years without oil and without attention.

Some alloys of lead shrink very little or not at all on casting. This property is valuable in type metals where small individual castings must retain very fine details of the letters to be printed. Alloys of lead, tin and antimony are commonly used for this purpose (see antimony, 4.9.7).

Lead and its alloys are often used as coatings for other materials, mainly to give corrosion resistance. These coatings are applied by spraying, hot-dipping or electroplating. Sprayed coatings may be of almost any composition. Dipped coatings are usually lead alloys containing between 2% and 15% of tin or 6% antimony. Electroplated coatings will usually be pure lead although it is possible to electroplate lead–tin alloys.

To resist severe corrosion, and for some radiation shielding requirements, a much greater thickness of lead may be applied to another metal, usually steel or copper.

This material is known as homogeneous lead, since by correct pretreatment and fluxing of the backing metal a metallurgical bond may be obtained with the lead cladding. The lead used may be either pure or alloyed according to requirements.

Large quantities of lead compounds are used as anti-knock additives in petrol, and as constituents of paints, ceramics and glass. More detailed information will be found under these headings.

4.8.9 Statistics

The principal lead producing countries in order of importance are:
- Lead ores.   Australia, USA, Canada, Mexico, Peru, Yugoslavia.
- Refined lead.   USA, Australia, West Germany, UK, Canada, Mexico, Japan, France, Belgium.

Output and consumption figures for 1966, 1967 and 1968 are given in table 4.6.

Table 4.6 World refined production – 1 000s of metric tons †

	1966	1967	1968
Europe	967.4	1059.3	1096.1
Africa	128.5	125.8	118.8
America	1018.7	938.4	1019.6
Asia	137.1	170.2	180.8
Oceania	226.7	216.9	203.2

World refined consumption – 1 000s of metric tons†

			(Jan–Sept)
Europe	1184.9	1174.6	897.0
Africa	40.0	41.0	32.5
Asia	211.5	233.8	215.8
Oceania	69.4	72.2	52.5

4.9 ANTIMONY

4.9.1 History

Stibnite, the principal ore of antimony, has been known from antiquity. The preparation of the metal was described by Valentine in the seventeenth century. Naturally occurring antimony was recognized about A.D. 1750.

4.9.2 Occurrence

Small quantities of antimony are found occurring naturally and also as a compound with arsenic, AsSb, known as arsenical antimony or allemontite. However, the principal source of commercial significance is stibnite, Sb_2S_3. Other minerals are the sesquioxide, Sb_2O_3, cervantite, Sb_2O_4, and stilbite, $Sb_2O_4H_2O$.

Stibnite occurs in many countries of which the most important in order of output are China, South Africa, USSR, Bolivia and Mexico. Workable deposits, however, are found in many other countries including the UK.

4.9.3 Ore treatment

Some sulphide ores are rich enough to be smelted without further treatment but the majority are concentrated by flotation (see lead, 4.8.3) and then further purified by a process known as 'liquation' in which the ore is heated to a temperature at which the sulphide melts and is run off, leaving the residues. The concentrated ore may be reduced to metallic antimony or converted to the volatile trioxide which is drawn off and condensed. This process is often used when the residues contain commercial quantities of gold and silver. Reserves of antimony ores greatly exceed the demand and only the most economic sources are worked.

4.9.4 Extraction

Several methods are practised commercially for the extraction of antimony:
1. High-grade stibnite ore is crushed into small (about ½ inch) pieces and heated for 2 to 3 hours with iron, salt and slag. The charge is then poured into moulds and the antimony solidifies underneath the slag. The crude metal is broken into pieces and reheated with salt and liquated sulphides giving a metal of about 99% purity. It is sometimes known as 'bowl metal' or 'star bowls'.
2. Sulphide ores are also reduced with coal and soda in reverberatory furnaces.
3. Oxides, particularly the trioxide, can be reduced directly in roasting furnaces, using reducing agents such as producer gas or acetylene.
4. Metal from the above processes, which may be of 85 to 95% purity, is refined by slagging with fluxes such as Glauber salts and charcoal.

4.9.5 Properties

Antimony is a grey lustrous metal, very brittle, with a low thermal conductivity – only 5% of that of silver. Its principal properties are given in table 4.7.

Antimony is closely related to nitrogen, phosphorus, aresenic and bismuth and has many similarities to arsenic. It is stable at normal temperatures but oxidizes when heated and starts to burn at red heat, evolving dense white fumes. It is attacked by nitric acid and by strong sulphuric and hydrochloric acids. An allotropic form produced by electrolysis is known as 'explosive antimony' because its instability causes it to explode violently on scratching.

Perhaps the most important practical attribute of antimony is that it expands on solidification; so do its alloys.

4.9.6 Alloys

Antimony is an alloying constituent in type metals, where its expansion on solidification gives a sharp outline to cast type. Typical type metals have compositions in the following ranges: lead 55 to 85%, antimony 12 to 30%, tin 2 to 15%. Added to lead in small quantities it gives 'antimonial lead' which is stronger and less liable to cracking than pure lead. Large quantities of this material are used in lead–acid batteries, ammunition, cable sheathing and pipes.

Table 4.7

Specific gravity	6.6
Melting point, °C	630
Boiling point, °C	1 440
Atomic weight	121.77
Mass numbers of the isotopes	121, 123
Thermal conductivity, cal/cm degC s	0.043
Resistivity, microhm-cm	39.1

Antimony is alloyed with tin and copper to give 'Britannia metal' used for tableware, utensils etc. Pewter is a similar type of alloy.

Antimony is also an important constituent of most anti-friction or bearing metals, often known as 'Babbitt metals' (see tin, 4.7.56) after their discoverer.

4.9.7 Uses

Antimony is little used commercially in a pure state compared with the extensive uses in alloys already noted. It is used as a metallic coating on brass or copper and also as an electrolytic coating on metals. Its compounds are used widely in medicine, particularly in a form known as 'tartar emetic', but in larger doses it is poisonous.

4.9.8 Statistics

Antimony – production (tons of 2 000 lbs)*

	1960	1965	1967	1968
China	16 500	16 500	13 200	
S. Africa	13 478	13 796	13 664	18 532
USSR	6 300 (estimate)	6 800 (estimate)		
Bolivia	5 872	9 663	12 650	12 254
Mexico	4 664	4 924	4 190	4 116

US consumption (tons of 2 000 lbs) (primary)*

1966 – 19 681
1967 – 17 350
1968 – 18 224

4.10 BISMUTH

4.10.1 History

Bismuth has only been known from the Middle Ages. It gets its English name from the German 'wismuth'. The nature of bismuth appears to have been imperfectly understood until the work of Pott and Geoffrey in the eighteenth century.

4.10.2 Occurrence

Bismuth is found in its native state and also as the sulphide Bi_2S_3, 'bismuth glance' or bismuthinite. World demand is low and is met from the more productive ores, principally in Peru, Mexico, South Korea and Japan. Many other countries have deposits, some of which are mined for economic or strategic reasons. It is often found associated with tin and precious metals; this makes the working of the ores more attractive.

4.10.3 Extraction

The ores are crushed and concentrated by hand sorting or gravity methods. The bismuth is then extracted by a variety of reduction processes.
1. With reverberatory furnaces, the concentrates of native bismuth and sulphide are roasted to remove sulphur and oxide and then reduced by carbon. The metal from this process is often refined by melting it on an inclined plane so that the metal runs off leaving the residues behind.

2. Cylindrical column furnaces are also used, the charge being ores, fuel, iron and fluxes.
3. Another method is to put a charge of ore with iron turnings, sodium carbonate, lime and fluorspar into a clay crucible which is then fired in a furnace.

In all the above methods, flue dust may be treated to recover bismuth.

Metal from these processes is purified by remelting and fluxing. Valuable by-products such as gold and silver are recovered.

4.10.4 Properties

Bismuth is a grey-white metal with a red tinge. It is very brittle and easily powdered, with a hardness of 7.3 Brinell. It melts at 271°C, boils at 1477°C and has the important property of expanding on cooling. Its electric resistivity (106 microhm-cm) and thermal conductivity (0.019 cal/cm degC s) are both very poor, worse than almost any other pure metal.

Bismuth is diamagnetic; a bismuth rod suspended between two opposite magnetic poles would take up a position at right-angles instead of in line with the two poles.

Chemically, bismuth is stable in dry air but oxidizes in moist air or water. It burns in air giving off yellow fumes. Nitric acid will attack it readily but other acids do so only slowly.

4.10.5 Uses

The pure metal has little commercial use. Almost all the applications are for alloys or compounds. Bismuth wire is used in thermocouples, galvanometers and similar apparatus.

4.10.6 Alloys

Bismuth acts as a hardener for lead, tin and cadmium. Lead battery plates normally contain bismuth, while tin is given greater strength, hardness and lustre by the addition of this metal.

In copper even small quantities of bismuth are disastrous, reducing the strength drastically and making it brittle.

Probably the most important use of bismuth is in the class of materials known as 'fusible alloys', i.e. those which melt at very low temperatures. They are widely employed as safety devices in fire prevention systems, solders, fuses and as temporary fillers for tubes that have to be bent.

All these alloys contain about 50% bismuth, together with lead, tin and cadmium. The best known is 'Woods metal' (with tin and cadmium) which has a melting point of 71°C. There are many others named after their discoverers, including Newton's, Rose's and Lipowitz's. Compounds of bismuth are widely used in pharmacy as sedatives, astringents and antiseptics.

4.10.7 Statistics

Production (short tons)	1963	1965
Peru	622	856
Mexico	470	550
Japan	412	556
South Korea	124	550
Bolivia	280	291
World	3400	4700

US consumption in 1966 (short tons)

Fusible alloys	437
Other alloys	225
Miscellaneous	20
Pharmaceuticals	820

4.11 COBALT

4.11.1 History

The name cobalt appears to be derived from the German 'kobalt' or 'kobold' meaning a mischievous spirit that bedevilled the miners and rendered their ore worthless. Cobalt or variations of it became associated with impure material, difficult to refine, which did not necessarily contain the element itself. It was isolated as a metal in the eighteenth century but only during the last threequarters of a century or so has it assumed industrial importance.

The ore was used in China, Persia and Egypt from about 2000 B.C. to impart a beautiful blue colour to glass and ceramics. The oxides are so used today.

4.11.2 Occurrence

Cobalt minerals are fairly widespread, invariably associated in small amounts with other metal ores such as those of copper, nickel, iron or silver. For technical reasons the sulphide, arsenide and oxidized minerals form almost the entire economic source of cobalt. Its production is restricted to a few countries, the most important being the Katanga area of the Republic of the Congo, Zambia, USA, Canada, Germany, Morocco and Finland. The principal minerals and their occurrence are listed in table *4.8*.

4.11.3 Mining and recovery

Cobalt-containing ores are mined in a similar way to those of copper and nickel, either by open-cast methods or, more usually, by normal underground working. The ore is ground prior to concentration, the optimum size being about $\frac{3}{8}$ or $\frac{1}{2}$ inch. The prepared ore is then either heat treated to form a speiss or matte, or subjected to

Table 4.8

Mineral name	Type	Composition	Occurrence
Cobaltite	Sulphide	$CoAsS$	Canada
Linnaeite	Sulphide	Co_3S_4	Katanga, USA, Finland
Carrollite	Sulphide	$CuCo_2S_4$	Zambia, Katanga, USA
Siegenite	Sulphide	$(Co, Ni)_3S_4$	Katanga
Safflorite	Arsenide	$CoAs_2$	Canada, Morocco
Skutterudite	Arsenide	$CoAs_3$	Canada, Morocco, USA
Smaltite	Arsenide	$CoAs_3$	Canada, Morocco, Germany
Asbolane	Oxidized	–	Canada, Zambia, Katanga
Heterogenite	Oxidized	$CoO \cdot OH$	Katanga, Zambia
Sphaerocobaltite	Oxidized	$CoCO_3$	USA, Katanga, Zambia
Erythrite	Oxidized	$Co_3(AsO_4)_2 8H_2O$	Canada, Morocco, Germany

flotation prior to hydrometallurgical or electrolytic treatment. The extraction processes are varied and complex because the metallurgical properties of cobalt differ insufficiently from those of associated metals and also because the cobaltiferous raw materials comprise the arsenide, sulphide, oxide or a mixture of these. Firing techniques are not in themselves sufficient and although in almost all cases some preliminary heating is necessary various liquid processes followed by electrolysis have been adopted. Three of these processes are briefly described below.

Rhokana process – sulphate roasting, hydrometallurgy and electrolysis:

The cobalt flotation concentrate – containing about 2.8% cobalt, 17.0% copper and 16.0% iron – which is produced from the sulphide ores is roasted so that a maximum proportion of the cobalt is converted to soluble form, leaving most of the copper and iron insoluble. The roasted residue is leached with hot water and, under suitable conditions, the greater part of the soluble copper is precipitated as hydroxide, from which copper is later recovered. After purification of the filtrate, the cobalt is precipitated as hydroxide and separated as a filter cake. The cake is then dissolved in spent electrolyte, with sufficient sulphuric acid to bring the solution, which at this stage contains about 20 to 30 g/l. of cobalt, to the correct acidity for electrolysis. High purity cobalt is deposited on mild-steel cathodes, and then stripped off by hand and either melted in an electric furnace and granulated in water, or broken into suitable sized pieces, burnished and packed into steel drums for shipment. Either form contains well in excess of 99% cobalt; the granules contain less zinc as this volatilezes during the melting, but have the disadvantage of containing de-oxidation residuals, since aluminium and calcium silicide are used as de-oxidants.

The Katanga–Shituru process – hydrometallurgy and electrolysis:

The material treated for copper recovery consists of oxide concentrates which are dissolved in acid and electrolysed. The cobalt remains in solution and this cobalt-rich electrolyte, discarded from the copper circuit, forms the starting point for cobalt recovery; it still contains impurities such as copper, phosphate, alumina, magnesium, manganese, zinc, iron and nickel. There are 15 to 25 g/l. of cobalt and 20 to 25 g/l. copper some of which is removed electrolytically; the solution is used to leach further ores and concentrates, more copper is electrolytically extracted, and iron and alumina are precipitated by lime. The remaining copper is now removed by cementation on cobalt granules. Lime is again used to precipitate cobalt hydroxide from the solution as a pulp which is dissolved in acid. Electrolysis can be carried out only in neutral solution; excess of cobalt hydroxide must therefore be ensured. The pulp is agitated by air blown through the anodes to avoid the harmful effects of the solids during electrolysis.

The cobalt cathodes produced contain 93.5 to 95.0% cobalt, and traces of nickel, iron, copper, manganese, zinc and sulphur. Melting of the cathodes with lime under oxidizing conditions removes manganese and sulphur, and zinc is eliminated by poling as with copper. High purity cobalt remains, with average composition 99.10% cobalt and 0.50% nickel, and is granulated to shot in water. The marketed cobalt is 3 to 12 mm in diameter. About 4500 tons of electrolytic cobalt are produced annually by this means.

In the newly constructed Luilu works with a planned output of 2500 tons of cobalt a year, electrolysis is carried out in clear solution instead of in a suspension of the hydroxide.

Port Colborne process – hydrometallurgy and electrolysis:

The electrolytic method used in the extraction of nickel and copper from ore mined at Sudbury, Ontario, by the International Nickel Company of Canada Limited, yields both high purity cobalt and cobalt oxide. Some of the oxide is reduced to the metal and marketed in the USA, and some is exported to Clydach in South Wales for further refining.

At Port Colborne, cobalt is obtained from the electrolyte remaining after nickel recovery, iron being removed by precipitation. The almost iron-free solution is treated with chlorine and nickel carbonate, and cobalt is precipitated in a cobalt-nickel slime. This is treated first with sulphur dioxide to make the slime acid-soluble and then with sulphuric acid to dissolve it. Any remaining copper or iron is removed from the solution and cobalt hydroxide is precipitated with sodium hypochlorite. This is purified and roasted to a black oxide containing about 70% cobalt, most of which is reduced to metal, cast into anodes and electrolysed.

4.11.4 Properties

Cobalt, a silver white metal with a slight bluish tinge, exists in two allotropic forms. The close-packed hexagonal form ϵ (or α) is stable below 417°C and the face-centred cubic form γ (or β) is stable at higher temperatures up to the melting point. Table 4.9 lists the currently accepted values of the various properties.

The purity of the cobalt used for measuring these properties is important and accounts for the scatter normally found in published results. The mechanical properties are based on results obtained with vacuum melted, de-oxidized metal and hot-rolled powder metallurgical products.

Finely divided cobalt is pyrophoric, but the metal in massive form is not attacked by air or water at temperatures below approximately 300°C; above this temperature it is oxidized in air. Cobalt combines readily with the halogens to form the respective halides. It combines with most of the other metalloids when heated or in the molten state. However, it does not combine directly with nitrogen, but decomposes ammonia at elevated temperature to form a nitride. It reacts with carbon monoxide above 225°C to form the carbide, Co_2C.

Metallic cobalt dissolves readily in dilute sulphuric, hydrochloric or nitric acids to form cobaltous salts. Like iron, cobalt is passivated by strong oxidizing agents, such as the dichromates. It is attacked by ammonium hydroxide and sodium hydroxide.

Cobalt is not very resistant to oxidation. Scaling and oxidation rates of unalloyed cobalt in air are 25 times those of nickel. The scale formed on unalloyed cobalt during

Table 4.9 Properties of cobalt

Melting point, °C	1 495
Boiling point, 760 mmHg, °C	2 802
Specific heat 15°C to 100°C, cal/g degC	0.1056
Coefficient of thermal expansion, per degC at room temperature	12.5×10^{-6}
Thermal conductivity, 25°C, cal/cm degC s	0.215
Electrical resistivity, 20°C, microhm-cm	5.8
Young's modulus, p.s.i.	30.6×10^6
Poisson's ratio	0.32
Hardness, Vickers scale	215
0.2% proof stress, p.s.i.	47 000
Ultimate tensile strength, p.s.i.	122 000
Elongation, %	20

exposure to air or oxygen at high temperatures is double layered. In the temperature range 300 to 900°C the scale consists of a thin layer of cobalto-cobaltic oxide (Co_3O_4) on the outside and a cobaltous oxide (CoO) layer next to the metal. Above 90°C cobalto-cobaltic oxide decomposes to cobaltous oxide.

4.11.5 Alloying

Cobalt metal was until recently generally available only in the form of granules, broken electrolytic cathode and powder. More recent developments have now brought on the market a wide range of wrought forms such as plate, strip, sheet, rod and wire.

Few of the technological uses of cobalt alloys involve binary alloy systems. The important industrial alloys are multicomponent systems that may be partly interpreted in terms of ternary and quaternary systems. An extensive literature exists on this subject.

4.11.6 Uses of the metal and alloys

Cobalt is mainly used in metallurgy and about 20 to 25% for chemicals. Relatively little use has been made of the pure metal, the most important being as radioisotope cobalt 60 for teletherapy and radiation processing units. As cobalt has recently become available in wrought forms its intrinsic properties may suggest it for magnetic devices, wear resistance at elevated temperatures, high-temperature bearings and heterogeneous welding rods for hardfacing.

Major uses of the metal may be classified as follows.

High-temperature materials. The need to produce alloys capable of reliable performance under the arduous conditions created by the gas turbine engine has led to the successful development of a wide range of alloys based on cobalt and cobalt containing iron and nickel.

The improvement in properties at high temperature has been brought about by a combination of solid solution hardening and precipitation hardening. To the oxidation-resistant cobalt-chromium matrix varying amounts of refractory metals, mainly tungsten, molybdenum, niobium and tantalum, are added to strengthen the matrix further and promote the formation of stable carbides which make a substantial contribution to the high-temperature strengths. Many of the earlier cobalt base alloys could be air melted and cast but the later alloys demand vacuum melting and casting facilities to ensure optimum properties and component reliability.

Magnetic materials. Cobalt and iron are the two most magnetic elements and consequently the manufacture of permanent magnet alloys represents a major application for cobalt. The discovery by Mishima in 1931 of the high coercivity iron–cobalt–nickel–aluminium alloy led to the development of a range of successful alloys.

Hardfacing and wear resistant alloys. The original cobalt–chromium–tungsten–carbon alloys (stellites) were developed at the turn of the century and brought into commercial use for cutting tools during the 1914–18 war. The ability of these alloys to retain high hardness at elevated temperatures and their outstanding abrasion resistance led to their use on the cutting edges of rotary drills used in oil-drilling. This technique of imposing a highly alloyed heat, wear and corrosion resistant surface on a cheaper substrate has now been widely adopted in general engineering.

Tool steels and cutting alloys. The addition of cobalt to a high-speed tool steel was one of the earliest uses of cobalt but it is only relatively recently that so-called super high-speed tool steels containing from 6 to 12% cobalt have begun to be more widely used. The cobalt base alloys retain a higher hardness through the red-heat range (500 to 900°C) than any other available tool steels and as such are recommended for cutting operations which require large roughing cuts, particularly under conditions where lubrication may be difficult.

In structural engineering steels cobalt has little advantage over cheaper alloying elements, but it is used in the more specialized alloy steels.

Miscellaneous alloys. A wide range of cobalt base and cobalt-containing alloys finds use in such applications as constant modulus alloys, magnetostrictive alloys, dental and prosthetic alloys, spring materials for use up to 400°C, glass to metal seals, electrical resistance heating elements and soft magnetic alloys.

4.11.7 Statistics

The production by countries for the years 1966 to 1968 is shown in table *4.10*. Table *4.11* is based on the statistics available from the US Bureau of Mines. There were no detailed consumption figures for European countries but it is likely that the over-all pattern is similar although varying by countries. For example, a relatively high proportion of cobalt is used in the production of high-temperature alloys in the UK.

Table 4.10 Cobalt production (in pounds)*

	1966	1967	1968
Republic of Congo	24 818 000	21 424 000	
Zambia	3 566 000	3 608 000	2 960 000
Germany	2 304 000		
Canada	3 511 169	3 603 773	3 488 656
Morocco	4 298 000	4 254 000	3 346 000

Table 4.11 Cobalt consumption in the USA, by uses

	1964		1965	
	s.t.	%	s.t.	%
High-temperature alloys	1 231	23.1	1 631	24.0
Cutting and wear-resisting alloys	169	3.2	207	3.1
Magnet materials	1 105	20.8	1 368	20.1
Hardfacing rods	400	7.5	528	7.8
Cemented carbides	215	4.0	265	3.9
High-speed steels	153	2.9	152	2.2
Other tool steels	77	1.4	56	0.8
Other alloy steels	282	5.3	403	5.9
Non-ferrous alloys	163	3.1	165	2.4
Other metallic uses	213	4.0	446	6.6
Total Metallic	4 008	75.3	5 221	76.8
Salts and driers	639	12.0	838	12.3
Ground-coat frit	299	5.6	268	4.0
Pigments	105	2.0	129	1.9
Other non-metallic uses	274	5.1	342	5.0
Total Non-metallic	1 317	24.7	1 577	23.2
Total	5 325		6 798	

4.12 NICKEL

4.12.1 History

Nickel was identified as a separate element by the Swedish chemist Axel Cronstedt is 1751, but for thousands of years it had been incorporated in alloys used for making swords, ornaments, cooking utensils and coins. Meteoritic iron-nickel was raw material for the swords of many an ancient warrior.

Uncertainties about the validity of Cronstedt's discovery were removed by Richter, a German chemist, who published a paper in 1804 'On Absolutely Pure Nickel...'. At this time the sources of nickel were the cobalt and copper ores in Sweden and Germany, but around 1850 rich deposits were discovered in Norway and these became the main source until the discovery in 1867 of vast quantities of nickel silicate ore in the Pacific island of New Caledonia, and the erection of a smelter plant there in 1875. This French possession became, and remains to this day, a major supplier of the world's nickel.

In 1883 work on the eastern end of the Canadian Pacific railroad had reached a place now known as Sudbury, Ontario, when a highly mineralized outcrop was exposed by one of the workmen. News that the outcrop was rich in copper soon brought prospectors to the area, but many claims were abandoned when the sulphide ore proved uneconomical to smelt and refine into copper because it was heavily contaminated with nickel. The ore is in fact extremely valuable and fifteen different elements are now recovered from it.

The solution to the extraction problem came in three ways. In 1891 J. L. Thompson, Superintendent of the Orford Copper Company, reported that, by adding sodium sulphate to the molten mixture of nickel and copper sulphide, the copper sulphide would float on top of the nickel sulphide and when solid the two could be separated easily.

This 'tops and bottoms' process solved the smelting problem and made it possible to produce large quantities of nickel oxide for steel making at an economic price. The production of refined metallic nickel was still an expensive process.

In 1890, at the time when Thompson was developing his 'tops and bottoms' process, Dr. Ludwig Mond and his chemist, Dr. Carl Langer, announced the discovery of a completely new chemical reaction between carbon monoxide gas and solid nickel. The reaction produced a metallic gas, nickel carbonyl, which could then be decomposed by heat into nickel and carbon monoxide. This discovery paved the way for a nickel refining process which is operated to this day by the International Nickel Company in South Wales. The process was offered to the Canadian mine owners but rejected after protracted negotiations. Instead the Canadians adopted an electrolytic refining process invented by a Swedish chemist, Hybinette. This process is used today in Canada and also in Norway, where Hybinette set up his own company to refine nickel concentrates from Canada. Thus, by three inventions, the riddle of the Sudbury ores was solved.

4.12.2 Nickel ore bodies

Nickel is present in certain rocks found in many parts of the world, for example, in peridotite, 0.20%, gabbro, 0.016%, diorite, 0.004%, and granite, 0.0002%, but the content is too low to make mining and extraction worth-while. The most important operational ore bodies are those located in Canada, Russia, Cuba and New Caledonia. New ore bodies in Western Australia, Guatemala and the Philippines are likely to become major sources. Ores worth mining have undergone natural concentration, of which two different kinds result in the two types of nickel ore most widely mined, lateritic in New Caledonia and Cuba and sulphide in Canada and Russia.

Lateritic ores. The concentration process begins with peridotite, a rock composed chiefly of olivine, a silicate of magnesium and iron. Water, rich in carbon dioxide from the atmosphere and decaying vegetation, decomposes the olivine, and the iron, magnesium and nickel go into solution. The iron oxidizes readily and precipitates iron as ferric hydroxide to form the minerals goethite, $FeO(OH)$, and haematite, Fe_2O_3. The magnesium, nickel and silicon remain in solution and are carried downwards. The solution is oxidized by reaction with the surrounding rocks and soil and tends to precipitate a hydrous silicate. Because nickel is less soluble than magnesium, the nickel content of the precipitate is higher than in the solution. This, very briefly, is the mechanism of concentration, which can produce ores with nickel contents of 1.5 to 9%.

Sulphide ores. As with the lateritic ores, nickel in the sulphide form is found in association with rocks high in iron and magnesium and relatively low in silicon. These are igneous rocks which have ascended molten from great depth and forced their way among rocks nearer the surface. There are many ideas about how the nickel was concentrated; one theory is that when the rock was still molten and cooling, nickel sulphide was precipitated and sank to the bottom of the molten area. This would account for the pockets of rich ore found at Sudbury. Another theory suggests that sulphur selectively removed the nickel from the solidified rocks and concentrated it. This would account for the presence of nickel sulphide in Thompson, Manitoba, only where it is alongside peridotite, a nickel-bearing rock.

4.12.3 Mining, smelting and refining

Lateritic ores. In New Caledonia the ore is predominently of the silicate type, as distinguished from the limonite of Cuba. The nickel content runs at about 3%–twice that of Cuba–and the ore bodies are up to 40 ft deep, below an iron-rich overburden. In one large mine the ore is removed by a combination of earth moving machines and (for difficult areas of rich ore) hand mining. It is graded by machines and by hand picking and then sent by aerial ropeway to the local smelter or to the docks for shipment to overseas smelters.

The raw ore is a hydrated nickel–iron–magnesium silicate containing about 2.8% nickel, 0.06% cobalt, 13% iron, 24% magnesia, and 24% silica. The ore is dried, crushed and screened and then sintered. The sinter is smelted in a blast furnace with coke and gypsum ($CaSO_4 \cdot 2H_2O$) to produce an iron–nickel sulphide. The smelter matte comprises on average nickel plus 27% cobalt, 63% iron and 10% sulphur. The matte is mixed with a silica flux in horizontal converters and air is blown through to oxidize the iron and cause it to form an iron silicate slag which can then be tapped off. The cooled and crushed converter matte contains 78% nickel and 22% sulphur and in this form is shipped to refineries at Le Havre, France, and in Japan and Canada.

Another product, by a different process which omits the gypsum addition, is ferronickel, which is solid directly to steelmakers.

At the Le Havre refinery of Société Le Nickel, the nickel sulphide is calcined to remove sulphur
$$2Ni_2S_2 + 7O_2 = 6NiO + 4SO_2$$
first in a fluid bed roaster and then in a kiln. The nickel oxide is crushed, mixed with water plus a binder and briquetted. These are dried, mixed with charcoal and charged

into a retort where the oxide is reduced and a final product called a rondelle, containing 99.25% nickel, is produced.

Sulphide ores. The famous Sudbury ore body occurs in an intrusive (due to an inflow of molten rock) which is roughly spoon shaped, some 37 miles long by 17 miles wide, with veins of mineral-rich ore rising to the surface at intervals around the edges.

The main method of mining is to undercut vertical slices of ore called stopes 200 ft high by $30\frac{1}{2}$ ft wide, leaving 19 ft wide pillars of ore between stopes to support the mine. The stope is mined by drilling and blasting the underside, and then drawing away some of the broken ore so that the drillers can return to work on the newly exposed face. This process is repeated – drill, blast, remove some ore, drill and blast again, until the top of the stope is reached. Then the broken ore is drawn out from the bottom and the empty space filled with tightly packed sand. The pillars are then removed. This basic method, with variations to suit the strength of the ore and rock, is widely used in Canada.

The broken ore from the mining areas falls down ore shutes into waggons and is drawn away to an underground crusher which reduces it to a size suitable for hoisting to the surface.

The ore is reduced to a fine powder by successive crushing and grinding operations until all the mineral particles have been split away from the rock. The pulp is then treated in banks of flotation machines, the particles being exposed to chemicals which selectively condition their surfaces in such a way that they attach themselves to air bubbles, induced in the pulp by agitation and frothing agents, and so float to the surface; the mineral-rich froth is then skimmed off. The remainder of the pulp flows into the next flotation cell and the process is repeated until all the mineral particles have been removed leaving only a tailing of rock particles which are pumped back to the mine as fill. By the use of different chemical flotation reagents it is possible to float or depress the different minerals in succession and a high degree of concentration of chalcopyrite, $CuFeS_2$, pentlandite, $(Ni, Fe)_9S_8$, and pyrrhotite, $Fe_{n-1}S_n$, is achieved.

After dewatering and filtering the nickel concentrate is mixed with a silica flux and roasted in multi-hearth furnaces. Of the three elements, nickel, copper and iron, iron is the most readily oxidized,

$$3Fe_7S_8 + 38O_2 = 7Fe_3O_4 + 24SO_2$$

so, by restricting the amount of oxygen entering the furnace to roughly half that required to oxidize the iron content, very little of the nickel and copper sulphide oxidizes.

Smelting. The roaster product is fed directly into reverberatory furnaces where the iron oxide forms an iron silicate slag which is tapped off and discarded. Any nickel and copper oxides would go out with the slag but for the presence of the remaining iron sulphide, which reacts with the two oxides to convert them back to sulphides, changing the iron to iron oxide

$$9NiO + 7FeS = 7FeO + 3Ni_3S_2 + SO_2 \quad Cu_2O + FeS = Cu_2S + FeO.$$

The reverberatory furnace matte is taken to horizontal side-blown converters where compressed air is blown through the matte to complete the oxidation of the iron sulphide and its combination into a slag which is periodically removed and returned to the reverberatory furnace. By the successive additions of fresh matte and flux for 24 hours a converter product containing about 47% nickel, 30% copper, 0.5% iron and 22% sulphur is achieved.

The controlled slow cooling and matte separation method of treating the converter matte is an invention of the International Nickel Company which completely replaces the old Orford 'tops and bottoms' process which served the industry so well in its early days. The matte is poured at about 980°C into large flat moulds capable of holding 25 tons. An insulated lid is placed over the mould and under these conditions the matte slowly solidifies, and cools for three days to about 480°C. During cooling the nickel and copper atoms concentrate as distinct sulphide crystals of high purity but, since the sulphur content of the converter matte is below that needed to produce 100% nickel plus copper sulphide crystals, some of the nickel and copper solidifies as a metallic alloy into which minute quantities of gold, silver and the platinum-group metals dissolve preferentially. After a final cooling for one day the ingot is broken up, crushed and ground, and the pulp passed through a magnetic-belt separator to remove most of the metallic alloy. The rest of the pulp is treated by flotation to separate the nickel and copper sulphide. The nickel sulphide is once more given a magnetic separation treatment to clean out more of the valuable metallic alloy and the final product contains some 73% nickel, 0.6% copper and virtually no precious metals—a remarkable technical achievement.

The nickel sulphide is pelletized and fluid-bed calcined to nickel oxide for reduction either by electrolytic refining or by refining using the carbonyl gas method. The two processes will next be described.

Electrolytic refining. Most of the feed material to International Nickel's Port Colborne refinery is nickel oxide containing about 74% nickel, 0.7% cobalt, 3% copper, 0.7% iron and 0.2% sulphur plus other impurities. It is reduced to metal with petroleum coke in reverberatory furnaces, reduction and melting taking place simultaneously at about 1550°C.

The metal is cast into anodes which are then transported to the electrolytic tanks. The tanks take between 30 and 40 anodes interspersed with thin sheets of pure nickel which form the cathodes. The cathodes are placed inside canvas-sided boxes into which purified electrolyte is continuously fed at a head of pressure sufficient to ensure that the liquid flows *outward* through the canvas, thus preventing impure electrolyte outside the canvas from reaching, and contaminating, the purity of the cathode. The electrolyte is a solution of nickel, sulphate, sodium and chloride ions and boric acid. The sequence of operation is as follows: under the influence of the applied voltage the anode dissolves into the electrolyte; the electrolyte is continuously withdrawn from the tank and sent to a large purification unit where the copper, iron, cobalt and other impurities are removed. The purified electrolyte pours into the catholyte around the cathode and deposits its nickel on the cathode; after 10 days a cathode reaches about 135 lb and is removed, washed and cut into 4 inch squares for marketing.

Refining by the carbonyl gas process. The starting material is the fluid-bed calcined nickel oxide which emerges from the smelting process. The oxide is reduced to crude metallics in multi-hearth reducer columns. The hearths are heated to 425°C by flue ducts, while water gas, which contains about 51% hydrogen, is passed through the column to effect the reduction

$$H_2 + NiO = Ni + H_2O.$$

The metallic powder is cooled conveyed, without exposure to air, to the top of a similar tower where treatment with carbon monoxide occurs. A stream of carbon monoxide flows countercurrent to the powder and after reacting with the nickel emerges at the top of the tower rich in nickel carbonyl. The reaction

$$Ni + 4CO = Ni(CO)_4,$$

Fig. 4.4 Nickel: charging first starting sheets to electrolytic tanks (The International Nickel Co. of Canada Limited) Thompson Nickel Refinery.

generates heat and the tower must be water-cooled to maintain 40 to 60°C. The gas emerging from the tower contains as much as 15% nickel, but not all the nickel is removed in a single tower. A number of volatilizers are used and the exit gas from the last one may carry as little as 0.5% nickel.

After dust filtering, the combined exit gases, containing about 8% nickel, are passed to the decomposers. A pellet production unit (or decomposer) is a gas-tight tower containing a shell-and-tube preheater, below this a reaction chamber and at the bottom a collection area where the pellets are removed by a bucket elevator and transported to the top of the preheater for a further passage through the unit. At the top of the preheater is a sizing device which removes any pellets above 5/16 inch in diameter.

Some 30 tons of nickel pellets are kept in continuous circulation, to prevent them adhering to each other, and in the reaction chamber carbonyl gas encounters the heated surfaces of the pellets, depositing its nickel on the pellets. Nickel powder is introduced to nucleate the pellets. It can take three months to produce a pellet of marketable size. The released carbon monoxide, containing about 0.1% nickel, is recycled through the volatilizer units.

4.12.4 Uses

Nickel unalloyed is used for coinage–the French 50 cent and 1 franc coins are good examples – in chemical plant, in electronic valves for the electrode assemblies and pins, and particularly as an undercoat for chromium plating where layers of nickel provide most of the protection to the underlying metal.

Nickel base alloys belong to four alloy systems:
1. Nickel-copper alloys, used for general purpose corrosion resistance, particularly against sea water, alkalis and certain acids.
2. Nickel-iron alloys, which, depending on composition, have high magnetic permeability, low curie point and low thermal coefficient of expansion.
3. Nickel-chromium alloys, which are used for heat resistance and range from the alloys used for electric heating elements to the complex alloys chosen for their strength and creep resistance for parts of gas turbine engines.
4. Nickel-chromium-iron alloys, used for heat resistance and corrosion resistance, particularly in furnace components.

The greatest use of nickel as an alloying element is in making stainless steel. 18/8 stainless steel containing 8% nickel is widely used in chemical plant, domestic ware, architecture and in many large road and rail vehicles. Other stainless steels containing more nickel are used for chemical plant and for furnace components.

Nickel in medium-carbon steel is important because it gives the steel a consistent response to heat treatment so that components of large size can be hardened in depth to a high strength. Typical applications are highly stressed gears and shafts and aircraft undercarriages.

Nickel in copper base alloys produces cupro-nickel for coinage, ships' condenser tubes and sea water piping; nickel silver (an alloy of copper, nickel and zinc) for telephone relay springs, buttons, cutlery and hollow ware (electroplated nickel silver (EPNS)); and nickel aluminium bronze (an alloy of copper, nickel, iron and aluminium) for ships' propellers and other high strength corrosion-resisting applications.

Nickel gives aluminium base alloys high strength at moderately high temperatures, for example, pistons and cylinder heads in internal combustion engines.

4.12.5 Properties

The properties of Nickel of 99.94% purity.

Atomic number		28
Atomic weight		58.69
Crystal form		face-centred cubic
Density, g/cm^3		8.908
Melting point, °C		1455
Specific heat, 0°C, cal/g degC		0.1025
Thermal expansion, 20 to 200°C, per deg C		13.3×10^6
Thermal conductivity, 100°C, cal/cm deg Cs		0.198
Electrical resistivity, 20°C. microhm-cm		6.844
Tensile properties:		
Annealed bar	Maximum stress, t.s.i	24 to 26
	0.2% proof stress, t.s.i.	6 to 8
	Elongation, %	40 to 50
	Reduction of area, %	60 to 75
Modulus of elasticity, p.s.i.		25 to 30×10^6
Hardness, annealed bar, HV		80 to 110

4.12.6 Statistics

Consumption of nickel in 1966,* excluding Communist countries, in 10^6 lb

USA	410
Canada, Europe and Japan	400
Other areas	20
Total	830
Stainless steel	285
Nickel-plating	130
High-nickel alloys	120
Constructional alloy steels	95
Iron and steel castings	85
Copper base alloys	35
All others	80
Total	830

* From *1966 Annual Report of The International Nickel Company of Canada, Limited.*

World production of nickel in 1965,* including estimates of Communist countries' production, in 10^6 lb

Nickel production (tons of 2 000 lbs)*

	1965	1967	1968
Canada	267 308	246 954	265 000
USSR	90 000	105 000	114 000
New Caledonia	53 054	67 829	88 100
Cuba	20 200	26 000	40 000
USA	13 510	14 615	15 154
S. Africa	3 300	6 300	6 500
Others	19 898	14 571	24 246
Total (estimated)	468 270	481 269	553 000

4.13 THE PLATINUM GROUP METALS

The platinum metals fall into two sub-groups: platinum, iridium and osmium, sometimes referred to as the heavy group because of their higher densities; and palladium, rhodium and ruthenium – the light group. Together with gold and silver they are often called the precious or noble metals.

4.13.1 History

Platinum, the most important of the six metals, was known by the inhabitants of South America prior to the Spanish conquest. Primitive articles of adornment dating from this time and made from platinum–gold alloy have been discovered in Ecuador. The early Spanish colonists encountered platinum in their search for gold but considered it of little value and in fact a nuisance, since its separation from gold was difficult. They did, however, fabricate such articles as sword hilts and buckles in the platinum–gold alloy which they called 'platina', meaning little silver. In the mid-eighteenth century, samples of the alloy were sent to Europe where they attracted the curiosity of scientists. A sample was brought to England by Charles Wood of Jamaica and given to Dr. Brownrigg, a scientist, who reported his own and Wood's observations to Sir William Watson who in turn reported them to the Royal Society in 1750. Further research was carried out in England by William Lewis and by workers in Sweden, Germany and later in France, Spain and Russia. This work was aimed at evolving methods, not only of purifying platinum, but also of producing it in a malleable form and melting it.

The first platinum industry arose in Spain in the latter part of the eighteenth century, using the results of research in Spain and other countries, particularly by Chabaneau who succeeded in producing platinum in a malleable form. A chalice made from his metal is now in St. Peter's, Rome. The industry in Spain declined at the end of the eighteenth century.

Woolaston was the first to refine platinum on a truly scientific basis and to prepare it in a suitable state for commercial use. In the course of his researches he discovered two new metals, palladium and rhodium, which he separated from the aqua regia solution of the native impure platinum. At about the same time (1804) Tennant, who was working on the aqua regia insoluble material from native platinum, discovered the new metals iridium and osmium. The last member of the platinum group, ruthenium, was discovered by the Russian chemist Claus in 1844.

4.13.2 Occurrence

Native platinum alloyed with gold and iron was first found in any quantity in South America in what is now Colombia. The main sources of platinum at the present day are:
1. The Merensky reef, part of the Bushveld igneous complex at Rustenburg in Central Transvaal, where it occurs partly as native platinum alloyed with iron and partly as sulphide (e.g. cooperite) and arsenide (sperrylite) with sulphides of iron, nickel and copper.
2. The Sudbury area of Ontario, Canada, where it is present mainly as arsenide in sulphide ores of nickel and copper.
3. The Ural mountains and Eastern Siberia in USSR, occurring in alluvial deposits and in nickel ores.
4. Alaska, where it is present as native platinum.
In all areas the platinum is associated with other platinum metals. It is understood that palladium constitutes about 60% of the Russian production of platinum metals. In the other deposits, platinum normally predominates.

4.13.3 Extraction and refining

The South African deposits are mined primarily for the platinum metals, whereas the Sudbury ores are worked chiefly for the nickel and copper. The output of platinum metals from both these sources is large and supplies the greater part of the market in Britain.

Crushing and preliminary concentration of the ores from the Rustenburg mines is carried out in South Africa. The product, partly in the form of gravity concentrates and partly as a matte, is shipped to Britain for further extraction and refining. The nickel fraction of the matte is separated from the copper fraction by layering with molten sodium sulphide. It is roasted to oxide, then reduced to nickel and formed into anodes. These are dissolved electrolytically yielding pure nickel and anode slime. Both the anode slime and the gravity concentrates from South Africa contain a high proportion of platinum metals and are subjected to chemical separation and refining processes. The platinum, palladium and gold are taken into solution in aqua regia and, after removal of the gold by precipitation with ferrous chloride, the platinum and palladium are precipitated as complex salts, the platinum as ammonium chloroplatinate and the palladium as palladium diammino dichloride. After recrystallization the precipitates are calcined to yield the metals in the form of sponge. The other platinum metals are recovered from the aqua regia insoluble material by fusing with alkaline oxidizing fluxes and dissolving in water. The ruthenium and osmium are distilled as oxides and the metals recovered by precipitation of complex salts and calcining. The iridium and rhodium are recovered from the solution remaining after distillation by precipitating as complex salts and calcining. All the metals are melted in high-frequency induction furnaces and are produced in a high degree of purity.

During the treatment of the Sudbury ores the platinum metals with some gold and silver are mainly concentrated in the anode slime which is formed in the electrolytic refining of the nickel. The concentrates are sent to Britain for separation and refining of the precious metals. The methods differ only in some details from those used in the treatment of the concentrates from South Africa.

4.13.4 Properties

The principal properties of the platinum metals are shown in table *4.12*.

All are white metals, exceptionally resistant to corrosion and most chemical reagents. Platinum is not attacked by any of the common acids but is slowly soluble in aqua regia yielding chloroplatinic acid (H_2PtCl_6). Palladium is resistant to tarnish in ordinary atmospheres but is less resistant to acids and chemicals reagents than platinum; it is readily soluble, for example, in nitric acid. Iridium, rhodium and osmium are exceptionally corrosion resistant, being unattacked by most common acids and chemical reagents, including aqua regia. Osmium and ruthenium form volatile oxides when heated in air.

Pure platinum and palladium can be worked in the cold, rhodium at temperatures above 800°C but ruthenium can only be forged with difficulty at about 1 500°C. Platinum can be readily welded and this method of joining is normally used for industrial equipment. It can also be soldered, high melting point alloys such as gold/platinum or palladium alloys being generally employed.

The platinum metals are transition elements with variable valencies and are capable of forming a large number of coordination compounds; in consequence their chemistry is complicated.

4.13.5 Alloys and Available Forms

Platinum and palladium form solid solution alloys with each other and with the remaining precious metals, also with a number of base metals. Alloying additions

Table 4.12

	Platinum	Iridium	Osmium	Palladium	Rhodium	Ruthenium
Chemical symbol	Pt	Ir	Os	Pd	Rh	Ru
Atomic number	78	77	76	46	45	44
Atomic weight	195.09	192.2	190.2	106.4	102.905	101.07
Density, 20°C, g/cm³	21.45	22.65	22.61	12.02	12.41	12.45
Melting point, °C	1769	2443	3050	1552	1960	2310
Boiling point, °C	3800	4500	5000	2900	3700	4100
Specific heat, 0°C, cal/g degC	0.0314	0.0307	0.0309	0.0584	0.0589	0.0551
Resistivity, 0°C, microhm-cm	9.85	4.71	8.12	9.93	4.33	6.80

are frequently made to platinum and palladium in order to increase hardness. Nickel has the greatest hardening effect on platinum, followed by ruthenium, iridium and rhodium in that order. Ruthenium and nickel are the most effective hardeners for palladium.

The pure metals are available as sponge or powder, sheet, wire etc. For catalytic uses they can be supplied in finely divided state supported on alumina. The alloys are also available in a variety of forms. Platinum, palladium and rhodium salts are marketed for plating baths.

4.13.6 Uses

Owing to their unique properties the platinum metals and their alloys have many important industrial applications. The first major one was the use of platinum for the fabrication of boilers for concentrating sulphuric acid manufactured by the lead chamber process. It was this application which firmly established the platinum industry in Britain. The original prototype standard metre and standard kilogramme were made in platinum by Janety, a Frenchman, in 1795. A new official standard metre was required in 1870 and this was made in platinum/iridium alloy by George Matthey in London on behalf of the French government; subsidiary standards were also made. Between 1828 and 1846 over $1\frac{1}{4}$ million platinum coins were minted and circulated in Russia. Platinum salts were much used in the early days of photography, when the so-called 'platinotype' process produced prints of great beauty and permanence. The process is, however, no longer in common use.

The principal current applications of the platinum metals and the metal or alloys used are as follows.

1. Catalysts: pure platinum, pure palladium and platinum alloys such as platinum/rhodium. They are employed both industrially and in the laboratory for a variety of reactions. They may be either unsupported, for example in the form of gauze or sponge, or supported on a material such as alumina. Among the many important industrial processes relying on platinum catalysts are:
(*i*) hydrogenation processes in the chemical industry;
(*ii*) reforming processes in the petroleum industry for converting low octane hydro carbon fuel into high octane fuel and aromatics;
(*iii*) the oxidation of ammonia for the production of nitric acid; and
(*iv*) the manufacture of sulphuric acid by the oxidation of sulphur dioxide.
Palladium is much used for (*i*) and platinum for (*ii*) to (*iv*).
2. Nozzles in glass-working equipment: 10% platinum/rhodium alloy. This alloy is particularly resistant to molten glass.
3. Spinnarets for rayon production: 10% platinum/rhodium or gold/platinum.
4. Thermocouples for pyrometers: typical combinations are platinum with 10% platinum/rhodium or 15% platinum/rhodium with 5% platinum/rhodium for higher temperatures.
5. Platinum resistance thermometers: high purity platinum (99.99%).
6. Furnace windings: platinum/rhodium (between 10 and 20% rhodium). Windings of these alloys have a long life in high-temperature resistance furnaces.
7. Sparking-plug points for aero engines: platinum/iridium, platinum/tungsten or pure iridium.
8. Electrical contacts: platinum/iridium, platinum/rhodium, platinum/palladium, pure palladium or palladium alloys. This is one of the principal uses for palladium.

9. Laboratory apparatus (crucibles, electrodes etc.): pure platinum, platinum/rhodium or platinum/iridium.
10. Electrodes for industrial processes: pure platinum or platinum alloys.
11. Safety bursting discs: pure platinum or platinum alloys. The discs are for protection of pressure vessels where corrosive chemicals are involved.
12. Dental appliances: platinum and palladium alloys. They are also used as minor alloying elements in gold alloys for the same purpose, e.g. wires and frames for dentures.
13. Brazing alloys: palladium/silver/copper alloys. These are used in the manufacture of special purpose thermionic valves.
14. Permanent magnets: cobalt/platinum alloys.
15. Tips for pen-nibs; osmiridium (naturally occurring alloy consisting principally of osmium and iridium).
16. Jewellery manufacture; platinum/copper, platinum/iridium, platinum/ruthenium and to a small extent palladium/nickel and palladium/ruthenium. They are principally used for wedding rings, gem-set rings, watch cases and settings for brooches. Most European countries have adopted a minimum standard of 95.0% for platinum jewellery. Palladium is used as alloying ingredient in 18 and 14 carat white golds.
17. Coatings. Platinum, palladium and rhodium can be electrodeposited and such coatings have some applications, particularly in the electronics industry and in the jewellery trade. Silver jewellery and white gold articles are frequently rhodium-plated, to give the former a tarnish-free surface and to improve the colour of the latter; it is normal practice to employ only a very thin deposit. Silver electrical contacts are also frequently rhodium-plated, partly to prevent tarnish and partly to give greater resistance to abrasion.

Silver or base metal alloys clad with platinum are used for the fabrication of equipment in the chemical industry.

4.13.7 Statistics

The estimated figures in table *4.13* for the world production of platinum metals in 1965 are quoted in the *US Bureau of Minerals Year Book*.

Table 4.13

Country	Total production in troy ounces
Canada	452 000
South Africa	750 000
USSR	1 700 000
Other countries	58 000
Total	2 960 000

In 1967 the estimate for Canada was 401 263, for South Africa 825 000 and for USSR 1 700 000.

4.14 COPPER

4.14.1 History

Copper, the first base metal used by man, was probably discovered in Armenia and worked in Sumeria and Chaldea more than 6 000 years ago. It appeared in predynastic Egypt *c*. 4 000 B.C., mainly as malachite and native copper, drawn first from the Eastern Desert and Sinai and later (about 3 000 B.C.) from Cyprus. After another

2 000 years, during which copper was widely used in the Middle East, the Phoenicians discovered it in Southern Spain, a region which came under Roman control during the Punic wars. In Huelva the Romans opened large mines with vertical shafts and waterwheels, as well as a smelter of high efficiency. At a very early date tin from Asia Minor was mixed with copper, perhaps accidentally at first; this alloy, bronze, being harder, quickly supplanted the pure metal for tools and weapons. Around 800 to 600 B.C. iron, which had long been known as a curiosity, was extracted from ore, notably at Hallstadt, Austria. Abundant and cheap, iron quickly supplanted both the older metals, despite its corrodibility. Thus there was first a Copper or Chalcolithic Age, then a Bronze Age, and then the Iron Age. The older metals continued to be used for special purposes throughout.

Copper was known to the Greeks as *Chalkos*. The Romans called it *Aes Cyprium*, whence the modern name is derived. The origin of 'bronze' is more doubtful but is possibly from the old German *bronce* or *brown*.

In Britain the Cornish tin mines long preceded Caesar, the tin being taken east by the Phoenicians through Gaul and Cadiz; some copper must have been involved in the mining. Irish copper from Avoca may also have been bartered abroad. There were Roman and pre-Roman mines in Cheshire, Anglesey and the Welsh borders. On the Continent equally ancient mines existed at Madjenpek in Jugoslavia. By the thirteenth century A.D. a very noticeable band of copper ore was being worked near Mansfeld, southern Germany, and an even older mine in Sweden has been operating continuously for eight centuries. An Elizabethan nobleman discovered copper in Cumberland and worked the mine by imported German labour. This find and the Cornish mines, together with the rise of English brass wire manufacture, led to the building of smelters and refineries at Swansea, a centre that dominated the world's copper industry until mid-Victorian days. The great modern copper mines, with their gigantic tonnages of low-grade ores, date mainly from the latter half of the nineteenth century in the USA, Canada, Peru and Chile, but those of Zambia are more recent, dating only from the 1930s.

4.14.2 Principal copper ores

The metal occurs in nature in a bewildering variety of conditions, even within a single mine. Like most metals copper was formed as hot vapours or liquids, arising in the earth's interior, which flowed into cracks and veins in the crust, especially in places where great intrusive bodies like granite and greenstone had forced their way up into the layered or bedded rocks, shattering their walls and consolidating far below the surface. The solutions permeated the rocks on both sides, replacing the mineral grains and occasionally giving rise to so-called 'copper mountains'. Of the wall rocks, limestones proved particularly liable to this kind of replacement. From near the earth's surface downwards the pore spaces in rocks are filled with slightly acid ground water, the slow creep of which has often transported mineralized grains and then redeposited them on others, thereby creating the enriched zones which miners most prize. The still larger pore space above the ground-water zone permits free oxygen to circulate, causing more active acid change, and in this oxidized zone we find the green malachite (57% Cu), blue azurite (55% Cu) and the rarer but very valuable reduction products cuprite (90% Cu) and native copper, which is almost pure. The enriched zone below yields copper glance (chalcocite, nearly 80% Cu), a grey or sooty sulphide. The primary deposits still farther down are (besides some chalcocite) peacock ore or bornite ($55\frac{1}{2}$% Cu), the brilliant blue hues of which occur in tarnished specimens, and the yellow copper or chalcopyrite, which contains one-third each of copper, sulphur and iron and is by far the most widespread of all copper ores. More than 20 other copper ore minerals are known, but only enargite (sulpharsenide, 48.3% Cu) and the less abundant covellite, tetrahedrite, atacamite and famatinite, are commercially valuable in a few mines. Copper-nickel association occurs on a large scale in South West Ontario, where it has given rise to the world's largest nickel mines. Many copper mines contain some zinc, lead, arsenic, cobalt and small but valuable amounts of gold and silver.

4.14.3 Chief mines

The USA, the world's largest producer, obtains about $1\frac{1}{4}$ million long tons annually from Butte (Montana), Bingham (Utah) and Chino, Inspiration, Ajo, Jerome, Globe and other large mines in Arizona, Nevada and New Mexico. The Canadian mining district of Sudbury, north of Lake Huron, includes Copper Cliff, Creighton (the deepest copper mine in the world, over 8000 ft) and Falconbridge; other important mines are Noranda in North West Quebec, Gaspe, Flin Flon, Lynn Lake in Manitoba and active workings in British Columbia. On the high semi-desert interior plateau of Peru, home of the Incas, are the famous mines of Cerro de Pasco and Morococha. For to the south in Chile, between the western Andes and the sea, is Chuquicamata, the largest open-cast mine in the world, and beyond that a great new mine at El Salvador, 9 500 ft up in the arid Atacama Desert. Much farther south still, near Santiago de Chile, the Braden (El Teniente) copper mine rests inside the crater wall of an old volcano; the town of Sewell is built on the outer face of the crater in a system of terraces, with steps up and down instead of streets. The African copperbelt straddles the Katanga-Zambia border, the surface being a flat open forest with almost no signs of mineralization. The Zambian mines, including Nkana, Nchanga, Mufulira, Roan Antelope (Luanshya) and the newer Chibuluma, Chambishi and Bancroft mines, produce over 600000 tons annually, the Katanga mines about 280000 tons. Uganda has a mine overlooked by Ruwenzori, the famous Mountains of the Moon. South and South West Africa include mines at Messina and Palabora, O'okiep and Tsumeb. There is also copper in Botswana and Algeria. In Asia the USSR has large mines in Kazakstan and Daghestan and older ones in the Urals, with a total production stated to be 690000 tons in 1966. Japan obtains over 100000 tons annually from numerous small mines. Erzerum in Turkey, Outokumpu in Finland, Bor in Yugoslavia, Rio Tinto in Spain and Mount Isa in Queensland are also important. Britain no longer produces new copper but the ancient workings of Avoca in the Irish Republic are still being operated.

The world has abundant copper reserves for very many years to come.

4.14.4 Mining and extraction

As most copper today runs only 1.0% or 2.0% of ore at most, 998 tons in 1 000 are wasted and the fundamental problem of producers is to extract the metal economically. It is often quarried on the bench system after stripping off any overburden, the face being then blasted off and the shattered rock removed in trucks or hugh bucket excavators. After preliminary crushing it reaches the smelter, where it is further crushed in machines working on the principle of a pestle and mortar, and is then ground to a fineness of about 200 mesh. Exactly the same method is employed for the material mined, which may be by the ordinary gravity system of successively deeper horizontal tunnels or cross-cuts, or by block-caving, which involves loosening the walls of the

Fig. 4.5 Copper mining: Loading ore cars. Utah Copper Division, Kennecott Copper Corporation.

ore body and excavating beneath it, so that in falling it shatters through its own weight. The underground workings in a large mine may total hundreds of miles; those at Luanshya (Zambia) are over 1 000 miles.

At the smelter the ore may or may not be given a preliminary roasting, to get rid of some of the sulphur; or heaps may be laid out in the open leached with acids and the liquor collected into tanks, where the metal is precipitated over scrap iron to yield 'cement copper'. This method is also employed for much oxidized ore. The sulphide ores, in finely divided particles, are fed at the smelter into froth flotation tanks, an invention of the 1920s which alone has made the extraction of such low-grade metals practicable. The flotation plant comprises batteries of cylindrical tanks holding oily reagents. Air is injected upwards through the tank, carrying the mineralized grains (which resist wetting) to the surface as a froth that overflows, the great bulk of the waste stone falling to the bottom. The ore, now copper concentrate, is partially dired and then passed on to the reverberatory furnace, a chamber up to 130 ft long, 25 ft wide and 10 or 12 ft high, lined with refractory bricks. The piles of ore, together with a suitable flux, are subjected from above to intense heat provided by a flame of compressed air and pulverized coal and oil. Two liquid layers result, slag above and a mixture of copper and iron sulphides called *matte* below. The exact details vary not only from plant to plant but also from day to day according to the nature of the ore at that moment. In one or two very modern plants preheated air under pressure soon provides its own heat, thereby dispensing with the coal; this is known as 'flash smelting'. The furnace works continuously for perhaps a year, after which the lining has become so shattered that the furnace, which may have treated 1000 tons daily, has to be relined. In the furnace wall are two sets of openings; from the upper one the slag is withdrawn and from the lower row the matte. Hot matte passes in great ladles to a converter (fig. 4.6), much like the old Bessemer steel converter. Along the converter wall runs a series of tubes called tuyeres, through which pressurized air passes. A large converter handles 60 to 70 metric tons of liquid and a 'blow'-a spectacular operation-lasts about 3 hours. First all the iron is oxidized and removed as slag; then the sulphur is blown out. The remaining green copper is next poured into water-cooled mounds (themselves made of copper) either on a casting wheel or an endless chain. Although over 98% pure, the metal still cannot be worked as it is full of blow-holes, hence its name of blister copper. It must finally be refined.

These operations take place as near to the mine as possible, but refining can be done anywhere. Much Far Western blister is refined in New Jersey; formerly almost all the world's supply had to go to Swansea. The nature of the refining depends upon the grade involved, but in general there are two types of process.

1. Blister is remelted and fire-refined in smaller reverberatories, an interesting but essential detail being 'poling' or the immersion of green wood into the hot liquid so as to remove the excess oxygen by hydrogenation. The product–fire-refined tough pitch copper–may be cast direct into wire bars, cakes or slabs and billets, from which copper wire, sheets and plates, coiled strip, tubes and sections are fabricated.

2. The electrical conductivity of all metals is seriously affected by very slight impurities

Fig. 4.6 Copper converter being charged.

and although fire-refining produces high-grade metal, the highest purity necessitates electrolytic refining. In this process lugs cast onto slabs make then into anodes. Usually about 36 anodes are hung in a tank, being interleaved with thin starting sheets (the cathodes) of pure copper. The tanks contain dilute sulphuric acid and dissolved copper sulphate. Through this solution a current having an average density of about 15 A/ft² at the cathode dissolves the anodes and deposits pure copper on the cathodes, waste falling to the bottom as anode slime. After about 14 days each cathode has grown from 11 lb to some $2\frac{1}{4}$ cwt and half the anode has been consumed. A second set of cathodes completes the operation. The sludge contains any gold and silver which may have remained in the metal since it left the mine; this 'waste' is therefore dried and retreated for this recovery.

4.14.5 Properties

Symbol Cu	
Atomic weight	63.54
Atomic number	29
Melting point, °C	1083
Electrical conductivity	100
Tensile strength	About 14 t.s.i., which by work-hardening rises to 25 t.s.i.
Purity	Up to 99.90% regulated by seven British Standards
Specific heat (N.T.P.), cal/g degC	0.092
Density, 16 ft³	555/558

The metal is ductile, malleable and very corrosion-resistant at moderate temperatures and pressures. Once it has acquired the familiar green patina corrosion is inhibited. The high heat conductivity of copper makes it an important means of heat exchange.

4.14.6 Fabrication and uses

Economic considerations such as availability, price, appearance and publicity strongly influence a manufacturer's choice of materials. Copper, like other metals, is in and out of favour from time to time, but its extreme adaptability, especially in the creation of new alloys to meet a particular need, gives it an advantage in new projects. The world production of newly mined copper has been climbing steadily since world war I, reaching 5 million tons in 1966; about a quarter of this comes from British Commonwealth countires. A very large additional tonnage is derived from remelted scrap, the bulk of both plain copper and its alloys being non-consumable.

Roughly two-fifths of all manufactured copper goes into electrical engineering, mainly as wire; Britain alone consumes over 300 000 tons yearly. The wire is derived from copper wire bars which generally weigh about $2\frac{1}{2}$ cwt and are the basis of most exchange prices. After heating, wire bars are reduced by repeated passes through a rolling mill to coils of wire rods 1/4 or 5/16 inch in diameter, which are then drawn through tungsten carbide dies to the diameters demanded. Diamond dies enable the smaller sizes to be drawn down to 1/1 000 inch diameter.

Copper wire is used in cables of many patterns and for telegraph and telephone wires, in dynamos and alternators for generating stations and for electric motor windings. In proton synchrotrons and other giant atom-smashing plant the windings

Fig. 4.7 Preparing to load starting sheets of pure copper into electrolytic refining tanks.

of the electromagnets are copper. Multicore copper cables may require up to 169 copper wires each 0.107 inch in diameter. Modern telephone cables are even more complex and may include up to 18 'units' each with 101 pairs of copper wires, 1818 pairs of wires in all. Small wires carry current in buildings and vehicles, and are in a vast range of instruments like oscilloscopes, voltmeters, ammeters, navigational aids and many other dial systems. The bonds for conductor rails of electrified railway lines are made of stout copper wire, and in overhead systems British Railways employ both copper and cadmium copper, an alloy with about 0.9% cadmium which is widely used for catenaries. Every 100 miles of electrified track uses some 2 500 tons of copper and cadmium copper. Copper rods and bars are important in electrical engineering. Hollow copper extruded sections are used for waveguides. For busbars, the vital connectors for heavy currents in power stations, copper has no rival except aluminium. Telephone contacts in millions are made from nickel silver, an alloy of copper, nickel and zinc. Brass wire is the basis of one of England's oldest industries, the manufacture of pins; today great quantities of pins and other small fastenings are made of plated brass wire.

Copper and copper alloy plate, sheets, strip and foil are produced in ordinary rolling mills. Copper plates are widely used for brewing vats and whisky stills; Guinness's brewery has 19 huge coppers, each holding 23 400 gals of beer. The slotted false bottoms of brewery mash tuns are made of bronze or copper. Round or oval coiled tubes called attemperators, through which cold water or brine circulates in the fermenting vessels, are of copper because of its high heat conductivity, as are the steam coils in the brewing copper. Jam boilers, varnish kettles and similar chemical plant are also of these materials and for the same reason.

Copper distilling columns for the production of industrial alcohol, fatty acids, essential oils etc. are of special interest. The great penicillin plant at Speke, near Liverpool, is an example. Its fractionating columns are 23 ft high, 5 ft in diameter, and built of copper sheets. Each column is in seven sections, six of which have riveted copper bubble plates, and each plate incorporates 60 copper bubblers and up-tubes. Such is the modern version of the Arab practice of a thousand years ago, when they distilled essences in gourd-shaped copper vessels.

Copper sheet and strip is important in architecture, whether on large structures like the domes of the London Planetarium and the British Museum reading room, the roofs of Coventry, Guildford and Liverpool Cathedrals, or on domestic dwellings. Copper for wall sheathing, long used overseas, is now being more widely adopted in the U.K. Copper flashings, weatherings, dormers, cills and so forth are to be seen everywhere. Many handsome modern doors are of bronze or of copper bronzed; the surfaces of other metals are often coated to look like them. Copper water tanks and boilers are commonplace.

With the addition of a minute amount of silver or arsenic to raise its softening temperature, copper is widely used for printers' process plates, especially half-tones. The artist etcher uses the same medium. Copper and copper alloy strip is stamped or pressed into blanks that are then cupped and drawn into numerous objects in daily use. Britain's present annual production of copper and copper alloy sheet etc. is about 150 000 tons.

A recently developed use is for printed circuits in electrical and electronic engineering. Elaborate wiring layouts for radio and television sets and computers are produced by photographing the wiring diagram on a copper sheet, which is then etched so as to leave the diagram in relief. By punching holes through the points of connection, through which terminal wires or tags pass, and soldering the tags to the plate, the circuits are established; or with dip-soldering all the joints can be made in one operation.

Copper tubes are made from cylindrical cast billets or tube shells. They may be drawn through a mandrel and then through dies on a draw bench or forced through extrusion presses to the size required. Extrusion is the opposite to drawing; the rod, called a slug, is rammed by hydraulic pressure through a press holding a die of suitable size and shape. By the modern process of continuous casting, rods and bars can be cast, cooled, and cut into lengths while being fed continuously from a reservoir of molten metal.

Copper tubes are now standard for gas, water and steam to British Standards. Common sizes range up to 4 or 6 inches in diameter, but solid drawn copper tubes are available for special needs up to 16 inches, or exceptionally even 24 inches in diameter. Small-bore copper tubes in long lengths are admirably adapted to central heating of existing houses, since they can be fitted along skirting boards and connected to tasteful radiators, whether the heat source be coal, oil or gas. In new dwellings and in public buildings nowadays panel-heating with copper tubes is employed; for many years prefabricated copper tube panels have been the practice in the USA and Canada. The recent Chase Manhattan Plaza in New York, which is 60 stories high and the sixth tallest building in the world, has about 60 000 ft of copper tubes for plumbing services alone. Tubes are easily joined either by capillary soldered joints or by a range of compression joints; where copper joins a cast-iron pipe it is bronze-welded.

Small sizes of copper tube are widely used in refrigeration. Copper, because of its high thermal capacity, is used for heat exchangers of many kinds, a noteworthy instance being the large conical assemblies of aluminium brass tubes for the large compartments on giant oil tankers. The world's first nuclear-powered merchantman *Savannah* has over 30 tons of cupro–nickel condenser tubes. The main condensers of the liner *Empress of Britain* each comprise 5 949 cupro–nickel tubes, with tube-plates and baffles of naval brass, a copper–zinc–tin alloy (62/37/1). Copper and copper alloy tubes are also used for the feed-water, salt-water and fresh-water cooling systems on ships, as well as for evaporators, low-temperature steam lines and other items. For coils in a large variety of hot-water cylinders, as well as in essential oil and spirit stills and other chemical plant, copper is chosen for its high heat conductivity, resistance to corrosion and ease of joining and manipulation. Tubes subject to high duty may be in Monel metal, which is a natural nickel–copper alloy. Special sizes of brass tubes, drawn to fine limits or machined, are used for microscopes and other optical instruments. Copper base flexible metallic tubes, commonly phosphor bronze, are called for where corrosive fluids have to be handled. Flexible brass tubes are used for connections in aero-engines.

Rods of machining brass (copper about $58\frac{1}{2}\%$) are to be found in work shops all over the world, the alloy being easily worked, non-magnetic and fairly corrosion resistant. A good deal of small brass rod is turned into screws by thread-rolling. The pinion wheels of clocks and watches are commonly made of small pieces cut off extruded bars, in a special alloy known as clock brass. A considerable quantity of high-conductivity copper rod finds its way into spot-welding electrodes.

Copper (and also some of its alloys, particularly aluminium bronze) is employed for a large range of castings, from the 35 ton manganese bronze propellers of an ocean liner or from a ship's stern-port down to the smallest die castings. Many types of bushes and bearings are also cast in these metals, and sintered copper powder bushes are self-lubricating. Bells of all sizes have been cast in tin-bronze for perhaps 1 500 years: the early English industry may be remembered by London's Billiter (Bilyeter) Street, where they were made. Intricate castings can be obtained in the same versatile metal by electrodeposition. Powdered copper and some of its salts provide the rich ruby-red and other fine glass hues, and copper glazes go back to the superb work of the early Egyptians. Gold having disappeared from circulation, coins are now in most countries either bronze or cupro-nickel; the present (1967) British 'silver' contains 75% copper and 25% nickel, the penny $95\frac{1}{2}\%$ copper, 3% tin and $1\frac{1}{2}\%$ zinc, and the threepenny piece 79% copper, 20% zinc and 1% tin. Production of money all over the world absorbs much copper; the Royal Mint uses 7 000 tons a year. Among many other uses the following are of special interest. Copper sulphate is mixed with lime to make Bordeaux Mixture, which is used for spraying potatoes and other crops. The fungoid diseases of many soft fruits are also sprayed with copper-bearing compounds. Copper is one of the two known bases (the other is mercury) of anti-fouling compositions, which effectively keep ships' bottoms free from barnacles and other growth by poisoning the organisms.

In many cases copper is a valuable additive to alloys of other metals. Small quantities, for instance, improve certain cast irons and aluminium alloys for high-duty work, and are also used as a flash in electroplating.

4.14.7 Statistics
Copper production by countries of origin of ore (tons of 2 000 lbs)*

Region	1961	'62	'63	'64	'65	'66	'67	'68
Europe	141	163	176	177	171	173	184	214
Africa (total)	1 070	1 052	1 074	1 155	1 222	1 249	1 293	1 373
Zambia	634	620	648	710	767	687	731	804
Republic of the Congo	325	323	298	305	318	349	353	358
S. and S.W. Africa	81	77	92	101	99	175	173	174
Asia (total)	232	248	267	259	265	291	297	333
Japan	106	114	118	117	118	123	130	132
USSR	524	550	600	675	710	770	850	905
Australia	97	119	119	110	96	116	94	111
America (total)	2 495	2 597	2 616	2 705	2 819	2 932	2 598	2 875
Canada	439	465	462	487	510	508	603	608
USA	1 160	1 224	1 208	1 251	1 356	1 408	950	1 203
Chile	603	646	662	685	645	701	728	726
Peru	218	184	196	192	199	194	212	235
World (total)	4 662	4 834	4 974	5 228	5 438	5 691	5 479	5 983

4.15 SILVER

4.15.1 History

Silver, one of the precious or noble metals, was known and used by man in ancient times. Its recovery from ores such as galena is simple and could have occurred accidentally in the first place during forest fires; the metallic silver would have been easily recognizable. In Asia Minor silverware and silver jewellery were common by 3000 B.C., at which date controlled mining and refining operations were employed. Silver-lead mines in Greece were worked from about 500 B.C. until the first century A.D. The Spanish mines then continued to supply most demands until the eighth-century. Then the main source was Central Europe until the Spanish discoveries of silver in Central and South America in the mid-sixteenth century. Silver was discovered in the USA in the early part of the nineteenth century and that country was the world's largest producer from 1871 to 1900. Other countries have contributed to total production.

The craft of the silversmith stretches from early times to the present day and examples of the art of each period may be seen in museums. Methods of raising, embossing, casting, engraving etc. have changed little for special pieces, but modern techniques such as spinning and stamping have partly replaced them for quantity production. Silver was used as a coating on base metal at an early date, being applied as a thin foil and soldered (close-plating). The discovery of the method of cladding known as Sheffield plate first popularized plated wares. From about 1830 this process was itself superseded by electroplating, which has since been used extensively for domestic articles.

Silver was used as a medium of exchange by early civilizations, first in bulk form and later as coinage. Units such as the Babylonian talent and the Hebrew shekel referred to a definite weight of silver. It was adopted as a coinage material by the Greeks and later by the Romans. To check inflation, governments usually kept a certain ratio of gold to silver coins issued and this ratio remained remarkably constant between about 1:10 and 1:15. Gradually, however, countries dropped bimetallism in favour of gold as a single standard. This occurred in England in 1816. In 1870 the ancient right of the public to bring silver bullion to the Mint and exchange it for coin after payment of a fee called 'seigniorage' was abolished. Silver continued for many years to be used as a coinage material, but is no longer used in Britain and in most other countries. Industrial uses of silver are now the more important.

4.15.2 Occurrence

Silver is found in almost every country in the world; it is present in trace amounts in sea water. It occurs in the native state and as sulphide, arsenide, antimonide, telluride, bismuthide and halide. Typical ores are argentite (Ag_2S), proustite (Ag_3AsS_3), pyrargyrite (Ag_3SbS_3) and cerargyrite or horn silver (AgCl). These ores are usually associated with ores of base metals such as copper, lead or zinc.

The principal sources at the present day are in Mexico, USA, Peru, Canada, USSR, Australia and Japan.

4.15.3 Extraction and refining

The greater part of the world's output of silver is now a by-product of the working of base metal ores such as those of copper and lead. A smaller quantity is produced from ores mined primarily for their precious metal.

There are many methods of extracting silver and innumerable variations have been used over the centuries. One of the earliest was smelting argentiferous lead ores to give a lead/silver alloy, followed by cupellation, i.e. heating with free access to air in a special furnace having a bed of bone ash. The resulting litharge, together with other base metal oxides, was removed partly by volatilization and partly by absorption in the bone ash, leaving the silver behind. This was the principal method prior to the sixteenth century. Cupellation is still used for refining crude bullion.

Recovery from silver ores

Amalgamation methods. Between the sixteenth and nineteenth centuries amalgamation processes were principally used. Of many variations the original patio process, first used in Mexico, is typical. The crushed ore was mixed with salt and water and thoroughly ground. Burnt pyrites and mercury were added and the mixture trampled by mules on a stone floor. The silver salts were reduced to metallic silver which formed an amalgam with the mercury. This was collected, washed, squeezed through canvas bags and distilled to remove the mercury.

Leach processes. Various leach processes were used. Augustin's process consisted of roasting the ore with salt to convert the silver to chloride. This was dissolved in hot brine and the silver precipitated on copper. In the Patera process the silver chloride was leached with sodium thiosulphate and the silver precipitated as sulphide.

Cyanide method. The amalgamation and leach processes were almost entirely replaced at the end of the nineteenth century by the cyanide process. The ore is crushed and ground with water into a slime, mixed with sodium cyanide solution and aerated.

The silver together with any gold present dissolves in the cyanide solution and is recovered by reduction with zinc dust.

Recovery from base metal ores

The current production of silver is mainly from the treatment of argentiferous base metal ores, particularly lead, copper and zinc, the silver being recovered as a by-product in the extraction of the base metal. The ore is usually concentrated by flotation (see lead, 4.8.3, and copper, 4.14.4). Copper and lead concentrates are smelted to produce argentiferous copper or lead bullion from which the silver is subsequently recovered. One of the following processes is usually used.

Parkes process. This process makes use of the fact that silver is more soluble in molten zinc than it is in lead. Zinc is added to the molten argentiferous lead bullion; the silver/zinc alloy which is formed floats to the surface and is skimmed off. The zinc is distilled off and the residue containing the silver, together with any gold originally present and a small quantity of lead, is then cupelled to recover the gold and silver.

Pattinson's process. When molten argentiferous lead is cooled, crystals of pure lead separate first leaving the residual melt richer in silver. Repetition of the process results in a lead low in silver and a lead alloy containing 2 to 3% of silver. The latter is cupelled to recover the silver and any gold present.

Electrolytic processes. In the Betts process impure lead is cast into anodes which are electrolytically dissolved in an acid lead fluorosilicate bath; pure lead is deposited on the cathode. The silver, gold and other impurities form a layer on the anode which is removed and the precious metals recovered.

In the electrolytic method for refining copper bullion, the silver and other precious metals form an insoluble sludge or 'slime' at the anode. The slimes are smelted, cast into anodes and refined electrolytically, or they can be cupelled with lead to remove the base metals.

Refining impure silver or doré metal (silver containing gold)

This was formerly carried out by boiling with sulphuric acid (parting). The silver is dissolved leaving the gold as a residue.

Electrolytic refining is now widely employed. In the Moëbius process the impure silver is cast into anodes which are enclosed in cotton or linen bags and dissolved electrolytically in acid silver nitrate solution. Pure silver is deposited in a loosely adherent form on stainless-steel cathodes and is continuously removed by wooden scrapers, after which it is washed and melted.

4.15.4 Properties

Silver is a white ductile metal with a melting point of 960.8°C a density of 10.49 g/cm^3 at 20°C and an atomic weight of 107.87. Its electrical and thermal conductivities and its optical reflectivity are higher than those of any other metal. Molten silver readily absorbs oxygen which is given up on cooling, giving rise to a phenomenon known as 'spitting'. Silver is unattacked by pure air but tarnishes in town air owing to the presence of traces of hydrogen sulphide and sulphur dioxide. It is also tarnished by contact with sulphur-containing substances such as rubber. It is resistant to a large number of chemical reagents but is soluble in nitric acid and hot concentrated sulphuric acid yielding silver nitrate and silver sulphate respectively. It is univalent in most of its compounds. The best known of these are the nitrate, sulphate, chloride, bromide, iodide, cyanide and the double salts with sodium and potassium cyanide. The silver halides are all insoluble in water and are photosensitive.

4.15.5 Alloys and available forms

Silver forms alloys with many other metals, the most important industrially being the silver-copper, the ternary silver-copper-zinc and the silver-copper-cadmium alloys.

Pure silver and silver alloys are supplied in ingot form or as sheet, tube, wire or grain. Silver anodes are manufactured for the electroplating trade. Among silver salts available for special purposes are silver nitrate for the photographic industry and silver cyanide and potassium argentocyanide for the electroplating industry.

4.15.6 Uses

The two principal industrial uses of silver are for photographic materials and electrical contacts. Other applications are in the production of brazing alloys and manufacture of jewellery and silverware, including electroplated wares. It is also used extensively for silvering mirrors, for chemical plant and as a catalyst in oxidation reactions.

A large quantity of silver is used annually by the photographic industry for the manufacture of plates, films and printing papers. The photographic process is based on the sensitivity to light of silver halide emulsions in gelatine. The greater part of this silver is never recovered.

Pure silver is excellent for silver contacts owing to its high electrical conductivity and freedom from oxidation. Alloying additions are made for certain applications where its resistance to wear, strength and other properties are desired. Silver-copper, silver-cadmium, silver-gold, silver-platinum and silver-palladium are most frequently used.

Silver-copper or silver-copper-cadmium alloys are used for the manufacture of silverware and jewellery, pure silver being too soft for this purpose. In Britain all such wares are required by law to be at least of sterling standard (not less than 92.5% silver), and most articles weighing a quarter of a troy ounce and above have also to be assayed and hallmarked at one of the official Assay Offices before they are offered for sale. Some other countries have similar requirements but the minimum standards vary. Solders used in the manufacture of silverware usually consist of silver-copper-zinc alloys.

Sterling silver (92.5% silver, 7.5% copper) was used in British coinage prior to 1920. Between 1920 and 1946 it was replaced by the ternary alloy containing a 50% silver. An alloy consisting of 90% silver and 10% copper was formerly used in the USA for coinage and is known as 'coin' silver.

Industrial silver brazing alloys usually contain silver and copper with zinc and/or cadmium.

In the manufacture of mirrors the glass surface is cleaned and treated with stannous chloride. It is then covered or sprayed with an ammoniacal silver solution containing a reducing agent such as formaldehyde or Rochelle salt. Reduction to metallic silver occurs on the surface of the glass.

Silver is used as a protective lining in industrial plant for handling certain chemicals, dyes, vinegar etc.

Electroplated coatings of silver on base metal are used widely for decorative and other industrial purposes as a cheaper alternative to silver. Nickel silver (see nickel,

4.12.4) is the usual basis metal for domestic articles and the final product is referred to as EPNS (electroplated nickel silver).

Silver salts are used in medicine for their antiseptic properties, for example, silver nitrate, known as lunar caustic.

4.15.7 Statistics

The world silver production for 1965, 1966, 1967 and 1968 is shown in table 4.14.

Table 4.14

Silver production (fine ounces)*

	1965	1966	1967	1968
Mexico	40 332 075	41 983 529	40 172 285	41 206 768
USA	39 000 000	42 500 000	31 000 000	32 437 000
Canada	31 917 243	32 824 514	37 206 023	45 389 141
Peru	36 470 353	32 841 241	32 703 668	36 019 599
Other Central & S. Americas	14 282 182	15 672 941		
Australia	17 312 716	18 278 000	19 765 000	21 618 000
Japan	16 672 795	18 327 079	22 173 015	25 874 819
USSR	27 000 000	33 000 000	35 000 000	
World Total	249 247 733	262 778 042		

4.16 GOLD

4.16.1 History

Gold, which occurs naturally in its metallic state, may well have been the first metal to be discovered by man. It would have been easily recognized and its separation would have been a simple matter. There are many fine examples of ancient gold jewellery which owe their survival to the fact that the metal is completely resistant to corrosion. Some of Egyptian origin date from as early as 3000 B.C. and ever since this time the craft of the goldsmith has been practised with great skill in many countries. The attractive appearance of gold, its permanence and its scarcity have ensured its pre-eminent position. For many centuries it has held a unique place amongst metals not only in the goldsmiths' craft but also because it has been accepted throughout the world as a convenient medium of international exchange. As a coinage material gold has been used from early times. The Greeks in particular developed coining to a high level of skill. Another craft, gold beating to make gold leaf, is at least 5 000 years old.

4.16.2 Occurrence

Gold is widely distributed in the earth's crust but usually it is present in such small quantities that its extraction is not an economic proposition. Sea water has a minute gold content. In the principal deposits it normally occurs in the native or free state, usually alloyed with silver and occasionally with mercury or tellurium. The earlier finds, in the form of nuggets or flakes, were located in streams or alluvial deposits and are known as placers. They are still one of the sources of the metal. Gold in quartz veins or lodes in rocks, often at great depths, is the main source of the world's production. It is also to a lesser extent found in the metallic state associated with the sulphides of other metals such as copper, silver, nickel and lead.

The principal mining areas at present are South Africa, California, Eastern Canada, Russia, Western Australia and Ghana. Gold is also produced on a lesser scale in many other countries throughout the world. It was formerly mined in small amounts in the British Isles, principally in Wales.

4.16.3 Extraction and refining

The methods used for the recovery of gold depend on the mode of its occurrence. Placer deposits require little treatment other than sluicing with water to remove the gravels with which the gold is associated. The primitive method was hand panning, which was also used in prospecting. The ore which has been mined from vein or lode deposits requires crushing before the gold can be extracted. Stamp batteries with mechanical stamps weighing up to 1 ton were formerly employed for this purpose but crushers and ball-mills have now generally superseded them. The gold was originally recovered from the crushed ore by sluicing but this method was later assisted by the addition of mercury with which it forms an amalgam. This was collected on amalgamated copper plates or by other methods. Part of the mercury was squeezed out and the remainder removed by distillation in retorts. Mercury was also used in a similar fashion in recovery of gold from placer deposits. As an alternative to the amalgamation method, the cyanide process first introduced in 1890 is now widely used. The crushed ore is treated with dilute sodium, potassium or calcium cyanide solution which dissolves the gold and silver. The metals are recovered by the addition of zinc dust. In the treatment of base metal ores containing small quantities of gold the metallic components are first separated from the gangue (rock) by flotation, the gold being then recovered by smelting and refining.

The impure gold from these operations contains silver and usually some base metals. The gold is refined either 1. electrolytically (Wohlwill Process), 2. by chlorine (Miller Process) or 3. by treatment with hot nitric or sulphuric acid. In method 1 the impure gold is made the anode in a gold chloride solution containing hydrochloric acid. Gold of 99.95% purity is collected on the cathode. An alternating current is superimposed on the direct current to prevent the formation of silver chloride on the anode. In the Miller Process (2) chlorine gas is bubbled into the molten impure gold. Base metals and silver are converted into chlorides which float to the surface and are removed. Method (3) is termed 'parting'; the silver and base metals are dissolved in the hot acid leaving the gold unaffected. It is only suitable, however, for separating the gold from a gold-silver alloy (doré metal) containing less than about 33% gold. If the impure metal contains gold in excess of this figure it must be alloyed with silver before parting. This method is also used to separate small quantities of gold which are frequently present in silver recovered from silver ores and in silver obtained as a by-product in the treatment of base metal ores.

Formerly, a gold/silver alloy was separated from base metals by heating with lead under oxidizing conditions on a hearth of bone ash. In this process, known as cupellation, the litharge and base metal oxides are partly volatilized and partly absorbed by the bone ash, leaving the gold and silver unaffected. Separation of the gold from the silver was a more difficult operation before the advent of the methods described above.

4.16.4 Properties

Pure gold is a soft yellow metal which has exceptional malleability and ductility. It can be beaten into leaves less than 1/300000 inch thick or drawn into very fine wire.

Densities ranging between 19.2 and 19.4 g/cm^3 at 20°C have been reported, the difference probably being due to variations in the physical state. It has a melting point of 1063°C, a boiling point of 2808°C and an atomic weight of 196.967. It is a good conductor of heat and electricity and forms alloys with many other metals, their colours ranging from white through yellow to red. It is unattacked by air and most chemical reagents including the strong acids but is readily soluble in a mixture of hydrochloric and nitric acids (aqua regia) yielding a solution of tetrachloroauric acid, H(AuCl$_4$). Gold has valencies of one and three, yielding aurous and auric salts respectively. Compounds of gold are readily reduced to the metal. Colloidal gold can be prepared by reducing solutions of gold salts.

4.16.5 Available forms of gold and its alloys

Gold and gold alloys are supplied in different forms by bullion dealers. Pure gold is available in bars and as sheet, wire or grain. Gold bullion as produced for monetary reserves is usually in the form of bars of 99.99% purity. Gold bars of 99.9% and 99.7% purity are also manufactured. Gold alloys for the jewellery trade and other industrial purposes are available as sheet, wire, tube etc. A variety of alloys are produced of differing gold content and physical properties. Alloys suitable for casting and a range of gold solders are also supplied. Some jewellers and goldsmiths make up their own alloys but most prefer to buy them in the required form from bullion dealers.

Rolled gold (see under 'Uses') is also sold in many forms for the jewellery trade. Gold salts are manufactured for electroplating baths, the commonest being potassium gold cyanide. Other forms of gold commercially available are 'liquid' gold and gold salts for medicinal purposes (see under 'Uses').

4.16.6 Uses

Gold alloys. The gold used by ancient civilizations for making articles of adornment was the native gold which in its natural state was invariably alloyed with silver. The name electrum is given to this pale yellow alloy of gold and silver which usually contains between 15 and 35% silver. Later in history alloys of definite composition were prepared. Pure gold is too soft to stand up to wear and accordingly copper and silver and, in modern times, zinc, nickel and palladium, are added to harden the metal and at the same time to achieve the desired colour. Addition of copper produces a red colour, the higher the content the deeper the colour. Additions of silver or zinc lighten the yellow colour of the pure metal. The yellow or red alloys used in the manufacture of jewellery usually contain copper or silver or both of these metals with sometimes a small percentage of zinc. Nickel or palladium are added to produce a white gold. An alloy of 75% gold, with the remainder principally palladium and nickel, is pure white and is extensively used in jewellery as a cheaper alternative to platinum. The percentage of gold in a gold alloy is usually designated by the carat. A carat is 1/24 parts by weight. Thus 18 carat gold contains 18/24 parts or 75% gold. In Great Britain there are four legal standards of gold, 22, 18, 14 and 9 carat. Most gold articles are required by law to be assayed and hallmarked at one of four official Assay Offices before they are offered for sale. Some, but not all, other countries have similar requirements.

Many items of jewellery formerly made by hand are now manufactured by the 'lost wax' casting method or by machine stamping.

Gold has been used as a coinage material for many centuries. The gold soverign was withdrawn from circulation in Britain during World war I but in recent years some tens of millions have been minted and sold abroad. The soverign contains 91.66% gold (22 carat). The US gold dollar contained 90% gold. The minting of sovereigns is a highly skilled operation, the weight and composition being controlled within very close limits. Until the gold standard was abolished, the pound sterling was defined in terms of gold, and paper currency and was freely convertible into gold coins. There are restrictions in some countries on the personal holding of gold coins or bullion.

Apart from jewellery and coinage, gold alloys today have many important industrial and other applications. For example, they are used in the electronics industry, in space research, for chemical equipment, for medical appliances and in dentistry. High-content gold alloys are particularly suitable for electrical contacts, being resistant to oxidation and good electrical conductors. Gold alloys are used in dentistry for inlays produced by 'lost wax' casting and for wrought wires for various appliances. Fourteen carat gold is much used for fountain pen-nibs as it is resistant to corrosion by ink.

Gold coatings. Electroplating of gold on base metal dates from 1840 and is now used extensively both for jewellery and in the electronics industry. Formerly, pure gold coatings were used for jewellery, but more recently alloy coatings have been developed which can be varied in colour and are harder. Thicknesses vary from flash or 'gilt' coatings, which are decorative only, to coatings up to 20 μm which will withstand many years of wear. Gold-plated base metals are used as an alternative to gold alloys to reduce cost, for example, in the manufacture of electrical contacts. Printed circuits are often produced by electrodeposition of gold.

An alternative to gold-plating in the jewellery trade is 'rolled gold' which consists of base metal or silver with a covering of gold alloy of 9, 10, 14 or 18 carat. A thin sheet of gold alloy is soldered to one or both sides of a bar of the base metal or silver, the whole being rolled down together to form a composite sheet of the required thickness. Tubing and wire are also produced in rolled gold. The resulting material is used in the manufacture of jewellery, cigarette cases, watch cases etc. In USA such articles are referred to as 'gold filled'.

Before the advent of electroplating, gilding was carried out by a process known as fire or mercurial gilding. An amalgam of gold and mercury was brushed on the article and the mercury subsequently volatilized by heating. This finish was usually applied to silver articles. The coating has exceptional permanence but the process is no longer used in Great Britain owing to the injurious effects of mercury fumes.

'Liquid' gold, consisting of organic gold compounds in suitable solvents, is used to decorate glass and porcelain. It can be applied by brushing or spraying. On firing, the material decomposes leaving a film of finely divided gold.

Gold leaf has a variety of uses mainly in the decorative arts.

Other forms of gold. If gold chloride solution is treated with stannous chloride a colloidal product known as 'Purple of Cassius' is formed, which has the property of imparting a deep ruby colour to glass. Ruby glass has been made by this method since 1685.

Gold salts have been used in medicine for the treatment of rheumatoid arthritis.

4.16.7 Statistics

The quantities of gold produced in the Western world are shown in the table 4.15.

Table 4.15

Gold Production (in fine ozs.)*	1965	1966	1967	1968
S. Africa	30 553 874	30 879 700	30 532 880	31 094 466
Canada	3 587 168	3 273 905	2 961 999	2 688 018
USA	1 675 500	1 801 600	1 525 500	1 038 600
Australia	877 643	916 985	801 009	815 560
Ghana	755 191	684 394	762 609	727 122
World	46 642 302	47 066 365		

4.17 ZINC

4.17.1 History

Although metallic zinc was probably known to the ancients, its early history is uncertain. Zinc sheet dated to the fourth century B.C. was found in Greece and there are also one or two other ancient references but these are much too slight to provide any evidence of widespread use. The Romans knew brass, but this was made from zinc ores, not metallic zinc.

From the sixteenth century onwards, zinc was imported into Europe from the East under the names of 'Indian Tin' or 'Pewter'. It was also known as 'Spelter' in the UK.

In 1721 Henckel isolated metallic zinc and Champion patented a method of production by distillation in 1738.

4.17.2 Occurrence

Zinc is contained in ore deposits widely distributed throughout the world. The most important mines are found in Australia, North and South America, Japan, North and Central Africa, Scandinavia, Germany, Poland, Spain, Italy and the USSR. Zinc is often smelted or refined near the mines but large quantities of concentrates are also sent abroad for treatment, notably to Belgium, the UK, Germany, Japan and the USA. Over 250 000 tons of primary slab zinc is used each year in the UK; the total annual world consumption is over $3\frac{1}{2} \times 10^6$ tons, excluding the Communist countries.

4.17.3 Ores and their preparation

Zinc has been extracted from a number of ores, and was once mainly produced from calamine, the carbonate which the Romans mixed with copper ores to make brass before zinc itself was recognized as a metal. The word 'calamine' is still used in the UK to describe zinc carbonate, and in the USA (where zinc carbonate is known as smithsonite) it covers any oxidized zinc ore and very often refers to a silicate. Zinc is now nearly all obtained from sulphide materials which usually also contain lead (and often silver). The most plentiful of these is blende or sphalerite, though marmatite – which contains iron – is also important.

Although it is still possible in a few mines to pick out sufficiently pure zinc material by hand, zinc ores nearly always need to be concentrated before the metal can be extracted. Two methods are in use: the first, a wet gravity method which takes advantage of the differences in density between mineral particles in the ore and the worthless sandy gangue material, and the second, the more widely used and efficient flotation process which relies mainly on the reluctance of water to wet the mineral sulphide particles.

In the flotation process, the finely crushed ore is agitated with water containing certain chemicals and a suspension of air bubbles, and the mineral particles – which tend to attach themselves to the air bubbles – are carried to the surface to form a froth which is skimmed off. The gangue material is readily wetted and, though lighter, sinks to the bottom. By adding suitable reagents it is possible to make some constituents float and others sink, thus permitting the treatment of complex lead and zinc sulphide ores and the economic separation of each mineral in a concentrated form.

The first stage in the smelting or refining of zinc sulphide concentrates is a roasting process, called sintering, which is carried out on a hearth or in a flash roaster. Most of the sulphur is thereby converted to sulphur dioxide, which is used for the production of sulphuric acid. The crude zinc oxide product is then treated either by thermal smelting or electrolytic refining. The sulphuric acid is often used at the zinc works to make fertilizers.

Thermal smelting. The original Champion process for zinc production was a thermal one, and in a modified form still accounts for about one-third of the metal made. The roasted concentrates are heated to a temperature of about 1 100°C with anthracite or a similar carbonaceous material in banks of small horizontal fireclay retorts. The thermal reduction of zinc, which is formed as a vapour and caught as liquid metal in condensers outside the furnace, may be represented chemically as follows:

$$ZnO + C \rightleftarrows Zn(gas) + CO.$$

There are, however, two stages:

$$ZnO + CO \rightleftarrows Zn + CO_2 \qquad (1)$$

and

$$CO_2 + C \rightleftarrows 2CO. \qquad (2)$$

The reduction only proceeds at temperatures above the boiling point of zinc and, since reaction (1) tends to reverse as the temperature drops, the proportion of carbon dioxide must be kept as low as possible (i.e. a strongly reducing atmosphere must be maintained). As a consequence excess carbon has to be added to the charge so that reaction (2) is not reversed. A typical retort will produce a daily output of between 50 and 70 lb of metal containing about 1% of lead. This is known in the UK as Good Ordinary Brand (GOB) zinc and in North America as 'Prime Western' zinc. Zinc dust in the finely divided form known as 'blue powder' is a by-product of this method of smelting.

New Jersey process. An important modification of the thermal method was made by the New Jersey Zinc Company in the late 1920s. The company developed a process in which a briquetted mixture of roasted concentrates and anthracite is heated in a large vertical retort made of silicon carbide bricks, which allows the reduction to proceed during the steady descent of the charge. Although the initial cost of this continuous plant is much higher than that of a hand-operated retort, the labour costs are greatly reduced and daily output of around 7 tons of metal is possible. Fractional distillation of vertical retort zinc gives a metal of more than 99.99% purity, which is suitable for modern die casting alloys and a number of other applications.

The ISF process. In 1957, the Imperial Smelting Corporation Ltd. announced the successful development at Avonmouth of a blast furnace for making zinc. The furnace is fed with a preheated mixture of roasted zinc concentrates and coke, and is supplied with hot air blasts above and below the central take-off point. Zinc vapour is removed in a stream of gas containing carbon dioxide and carbon monoxide and condensed rapidly by a spray of molten lead. The lead circulates continuously through heat exchanger– for cooling–and a separating chamber where molten zinc of GOB quality can be run off. Since the charge is heated internally, the ISF process can produce much more zinc daily than a vertical retort. Lead and zinc can be smelted simultaneously if roasted mixed lead-zinc concentrates are fed in. Low-grade concentrates can also be treated by this process and a single unit can made 100 000 tons of zinc in a year.

The reduction in a blast furnace follows the same course as in a retort, with the difference that oxygen (from the air blast) provides the heat by burning the excess carbon and some of the carbon monoxide. Zinc vapour is thus formed in a much greater volume of gas than in a retort, and the special condenser is necessary because carbon dioxide is also present. Waste gas from the condenser is utilized to preheat the charge and air blast. There are two ISF blast furnaces working in the UK at present, and some twelve others are in operation or under construction overseas.

Electrothermal process. An electrothermal refining process of Swedish origin is also used on a limited scale in America. Basically this is a vertical retort process in which a mixture of crude oxide and coke is heated internally by passing a heavy electric current through the charge. A similar plant has also been adapted for the production of zinc oxide. One of the drawbacks of this method is the very close control of charge composition necessary to maintain the electrical conductivity at the required value.

Electrolytic process. More than half the world's zinc is refined by an electrolytic process first used commercially during world war I. Roasted concentrates are first dissolved in sulphuric acid, then–after an intensive purification of the solution which includes treatment with zinc dust to precipitate nobler metals–zinc is deposited electrolytically on aluminium cathode sheets from which it is stripped off, melted down and cast into slabs. Metal produced in this way has a purity greater than 99.95%, and exceeding 99.99% when required, and the process provides most of the special high-grade zinc needed throughout the world for die casting and other alloys, and for anodes for zinc-plating and cathodic protection. It was the successful development of the electrolytic process that first led to the large-scale production of zinc outside Europe and the USA, although the process is used there in places where cheap electricity is available (usually from hydroelectric schemes).

4.17.4 Properties

The most important properties of pure metallic zinc the widest-used die casting alloy, and commercially-available rolled zinc, are as listed in table *4.16.*

4.17.5 Uses

About 35% of the world consumption of zinc is used in protective coatings for iron and steelwork. Die casting accounts for 25%, brass-making for 20% and sheet zinc for 10%.

Zinc coatings. Because of their excellent resistance to corrosion in most atmospheres, in fresh and salt water, and in contact with many natural and synthetic substances, zinc coatings are widely used for the protection of finished products ranging from structural steelwork for buildings and bridges to nuts, bolts, strip, sheet, wire and tube. The following are the five principal methods of application.

Table 4.16

A. Pure metallic zinc

Melting point, °C	419.5
Boiling point, °C	906
Specific gravity at 25°C	7.13
at 419.5°C (solid)	6.83
at 419.5°C (liquid)	6.62
Thermal conductivity (18°C), cal/cm degC s	0.27
Linear coefficient of thermal expansion (polycrystalline)	
from 20°C to 250°C, per degC	39.7×10^{-6}
Specific heat (50°C), cal/g degC	0.094
Heat of fusion (419.5°C), cal/g	24.09
Heat of vaporization (906°C), cal/g	425.6
Electrical resistivity (20°C) microhm-cm	5.92
Electrical conductivity, % IACS	28.27
Atomic weight	65.38
Electrochemical equivalent, mg/C	0.3388
lb/1000 Ah	2.6886
Crystal structure	close-packed hexagonal

B. Pressure die casting alloy

BS 1004: alloy A – made up of 4% aluminium, 0.04% magnesium and the remainder special high purity (minimum 99.99%) zinc.

Specific gravity	6.7
Density, lb/in³	0.24
Melting point, °C	387
Solidification point, °C	382
Specific heat	0.10
Solidification shrinkage, %	1.2 (0.14 in/ft)
Linear coefficient of thermal expansion, per degC	27×10^{-6}
Electrical conductivity, % IACS	27
Thermal conductivity, cal/cm degC s	0.27
Tensile strength (21°C), t.s.i.	18.5
Elongation (21°C), %	15
Impact strength (unnotched on Charpy machine with 40 mm gap)	
on ¼ in. × ¼ in. bar, ft lb	42
Brinell hardness, kg/mm²	83

C. Rolled zinc

Commercial zinc of	*Pack-rolled*		*Strip-rolled*	
98.5% to 99% purity	*Parallel**	*Perpendicular*	*Parallel**	*Perpendicular*
Tensile strength, t.s.i.	12	18	12	16
Elongation, %	23	14	40	23
Brinell hardness, kg/mm²	45 to 50		48 to 51	

Table 4.16 (Continued)

Zinc–copper–titanium alloy**	Parallel	Perpendicular
Tensile strength, t.s.i.	14	19
Elongation, %	45	28
Brinell hardness, kg/mm^2	50 to 55	

*Parallel to the direction of rolling.
**Made up of 0.6 to 1.2% copper, 0.1 to 0.2% titanium and the remainder zinc.

Hot dip galvanizing. Goods to be treated are dipped, after pickling and fluxing, in a bath of molten zinc at a temperature of about 430°C to 460°C. The zinc reacts with the iron or steel to form a series of alloy layers on the surface, each successive layer containing a higher proportion of zinc until, in the outer layer, the coating consists of ductile unalloyed zinc. The zinc is thus effectively bonded to the basis metal to form a protective coating possessing excellent resistance to corrosion and rough handling.

Zinc spraying. Atomized particles of molten zinc are projected on to a grit-blasted steel surface from a special pistol fed with wire or powder. The process is often applied to structural components too large to be dipped in a galvanizing bath, or to large structures which cannot withstand heating or which require a thicker coating than can economically be applied by other methods.

Zinc plating. (Electrogalvanizing) Zinc is electrodeposited on the prepared steel from a solution of zinc salts. The process is used to protect small articles such as nuts and bolts and light pressings – which require a finer finish than galvanizing can normally provide – and can be specially adapted to provide thin coatings on continuous strip and wire.

Sherardizing. Prepared iron or steel articles are heated with zinc dust and sand in a slowly rotating drum until the zinc has formed an alloy coating over their surfaces. The process, which gives a matt-grey coat, is limited to fairly small articles because of the difficulty of heating large containers evenly.

Zinc-rich paints. These consist of fine zinc dust suspended in a vehicle allowing very high pigmentation and the formation of an electrically conductive dry film. These paints can be applied to any rust-free and scale-free steel surface by brushing, spraying or dipping, and are mainly used to protect factory steelwork, ships' hulls and parts of car bodies, and to repair damage to other types of zinc coatings. Automatic grit-blasting and zinc-dust paint spraying equipment is now widely used in shipyards to protect plates during storage.

Rolled zinc. Rolled zinc sheet and strip is mainly used in dry batteries and in the printing and building industries. The ordinary dry battery operating on the Leclanché principle uses zinc as the current-producing element. Individual cells are made in various shapes and sizes depending on the purposes of the battery; the cylindrical type (used in torch batteries) and the more compact rectangular layer type (for higher voltage applications) are the best known.

Another long-established use of zinc sheet is in printing, as plates for photo-engraving and lithographic work. The sheet is rolled to close thickness tolerances and produced with a controlled uniform fine-grain structure.

Zinc sheet and strip are widely used both in the UK and on the continent for roofing, gutters, rainwater pipes, flashings and weatherings, and – provided the proper techniques are followed – give long maintenance-free service.

4.17.6 Alloys

Certain alloys to *BS 1004* based on special high-grade zinc of minimum purity 99.99% are used for pressure die casting, a fast mass production process for making strong accurate components with excellent corrosion resistance and able to take a variety of decorative finishes. Molten alloy is injected under pressure into a steel die made in two or more parts to permit removal of the casting, and runs of many thousands of identical castings can be made at a high production rate, depending on the size and complexity of the part and the tolerances to be held. The alloy most used (alloy A) contains 4% aluminium and 0.04% magnesium; a slightly harder alloy (alloy B) is obtained by adding 1% copper. To avoid any risk of intercrystalline corrosion under warm, humid conditions, very close limits are imposed on the impurities – tin (less than 0.002%), lead and cadmium (each less than 0.005%) – in the die castings.

It is estimated that over half the total weight of die castings made are used in car manufacture, for example, as plated exterior components and trim. Die castings are also widely used in the manufacture of locks, door handles, bathroom fittings and other items of builder's hardware, for a variety of engineering components where dimensional accuracy and high strength are required, and for scale model toys and many other products.

Brasses are copper–zinc alloys with a zinc content ranging from 20% to 45%, and sometimes containing additions of other metals. Alpha brasses with 28% to 37% zinc content are single-phase alloys suitable for cold-working (i.e. rolling, pressing and drawing) and for small castings. The alpha-beta brasses with 40% to 45% zinc content consist of two phases and are suitable for casting, hot-pressing and extruding. Brasses are easily recognized by their yellow colour and, because of their ease of working, high corrosion resistance and good electrical conductivity, are widely used for plumbing and electrical components.

Zinc alloys containing small additions of copper and titanium have recently been developed and are now also available in sheet and strip form. These alloys show better creep strength than unalloyed zinc and are being used for prefabricated roofing, ventilation ducting and various pressings.

4.17.7 Statistics

Production of metallic zinc in 1960 and 1966 in thousands of metric tons is shown in table *4.17*.

Table 4.17

	1960	1966
Europe	923	1 157
Africa	83	88
North and South America	1 131	1 380
Asia	181	405
Oceania	122	178
World	2 439	3 007

4.18 CADMIUM

4.18.1 History
Cadmium was discovered in 1817 when Stromeyer extracted it from the oxide, which he had found in zinc carbonate. About the same time Hermann obtained cadmium sulphide from impure zinc oxide.

4.18.2 Occurrence
Small quantities of the sulphide, known as greenockite, occur in association with zinc ores. The principal deposits are in Scotland, Czechoslovakia and the USA. It also occurs in zinc sulphide ores and is obtained as a by-product in the production of zinc. Some metal is recovered from lead and copper ores during refining. Flue dusts and some mine waters also contain cadmium.

4.18.3 Extraction
Zinc residues are the principal source of cadmium. The concentrates are roasted and then leached with sulphuric acid. Cadmium is precipitated from the solution by the use of metallic zinc powder and is obtained in the form of a sponge. This is either melted in an arc furnace or distilled to give pure metal.

4.18.4 Properties
Cadmium is a bluish-white metal. It is soft, ductile and malleable, and can be drawn into fine wires and rolled to thin sheet. It is readily attacked by most acids.

Melting point, °C	321
Boiling point, °C	767
Density, g/cm^3	8.65
Hardness, Brinell scale	21
Thermal conductivity, cal/cm degCs	22
Resistivity, microhm-cm	6.85

4.18.5 Uses
Pure cadmium has applications in secondary cells, in solders and as a plating metal. The secondary cells or 'accumulators' are of the alkaline type and are generally more robust and lighter than lead-acid cells, but also more expensive. Cadmium-plating is extensively employed as a protection for steel, because cadmium is widely separated from steel in the galvanic series and confers sacrificial protection (see ch. 9, p. 699).

Cadmium is also widely employed as an alloying element in other metals. Added to copper it gives 'cadmium copper' which is much stronger than pure copper and is therefore used for overhead electric lines of long span. One of the earliest of these was across the Thames at Dagenham in the 1920s. Alloys with silver are used, for example, in the jewellery trade, and cadmium is added to lead cable sheaths to increase their strength.

As cadmium vapour and cadmium salts are poisonous, special precautions are needed in industries using them. In recent years the use of solders containing cadmium has been discouraged, particularly in the electrical industry.

4.18.6 Statistics

Cadmium production – in pounds*

	1962	1965	1967	1968
Canada	2 605 000	1 756 000	2 418 000	2 719 000
USA	11 136 729	9 671 000	8 699 000	10 651 000
USSR	3 500 000 (estimate)	4 200 000 (estimate)		
Japan	1 948 000	3 254 000	4 185 000	4 874 000
Belgium	1 854 000	849 000	410 000	

U.S. Price – 1966 – 255 cents/lb
1967 – 265 cents/lb
1968 – 265 cents/lb

4.19 MERCURY

4.19.1 History

Mercury or quicksilver was known to the ancients and was in use in the first millennium B.C. Aristotle refers to 'liquid silver' and, at a later date, Theophrastus describes its extraction by rubbing cinnabar (mercury sulphide) with vinegar in a metal mortar. In the first century A.D. Dioscorides describes a distillation method using a clay retort.

There is no record of what use the Greeks made of mercury but the Romans used it for the extraction of gold from its ores and the recovery of gold from woven fabrics. In the Middle Ages it was a favourite with alchemists.

4.19.2 Occurrence

Mercury is widely distributed but in low-grade ores. Many of the important sources occur in areas of volcanic activity, for example, near hot springs. Some of the principal producing countries are Spain, Italy, USA, USSR and Mexico. The Almaden mine in Spain, the richest in the world, has been in production for more than 2 000 years.

The principal ore, cinnabar (HgS), when in a pure state is of a brilliant red colour and is used for the pigment vermilion, but impurities change the colour to red-brown and grey. A black variation, metacinnabarite, is also worked. Other ores include livingstonite, in Mexico, a grey sulphide of mercury and arsenic. Mercury sometimes occurs as native metal mixed with cinnabar, but the quantities are not sufficient to work by themselves and its presence is a health hazard to the miners. Natural amalgams with gold and silver are also found occasionally.

Both underground and open-cast methods of mining are employed. The ore is rarely found below 1 000 to 1 500 ft.

4.19.3 Concentration

Although most ores rarely exceed 5% mercury content, and may be as low as $\frac{1}{2}$% or $\frac{1}{4}$%, concentration is not widely practised. This is because cinnabar is friable and crushing gives rise to powders which are difficult to handle. Some use is, however, being made of froth flotation methods.

4.19.4 Production of metal

Decomposition of a cinnabar is relatively simple, being effected by direct roasting at fairly low temperatures. Developments in extraction technique are mainly related to plant design and ore treatment, but most of the metal produced is still extracted from untreated ores. The reduction process can be expressed by the equation:

$$HgS + O_2 = Hg + SO_2.$$

Other processes, carried out without air, use lime or iron:

$$4HgS + 4CaO = 4Hg + 3CaS + CaSO_4$$
$$HgS + Fe = Hg + FeS.$$

Kilns have been used since ancient times and are still employed, but the trend is towards continuous furnaces of two types.
1. Rotary kilns, similar to those used in the cement industry, usually about 3 to 5 ft in diameter and 40 to 80 ft long.
2. Multiple hearth or Herreschoff furnace which is also used in copper smelting. For mercury, heat must be added, usually by oil burners.
In both these processes, escape of dust from the flues must be prevented as it is poisonous; dust extractors are fitted.

In all extraction processes mercury is vaporized and must be condensed. This is done in inverted U tubes, dipping into water, in which the mercury collects and is continuously removed. Careful sealing of the tubes is necessary to prevent leakage. The pipes are periodically opened and scraped to remove the flue dust from which mercury is extracted by agitation. Metallic mercury produced in this way is about 99.7% pure and may be still further purified by redistillation. When high purity is required further treatment with nitric acid and caustic potash is used. Mercury is commonly sold in cast iron 'flasks' containing 76 lb of mercury.

4.19.5 Properties

Mercury is a silvery white heavy metal which freezes at $-39°C$. When solid it is white, soft and malleable. The boiling point is 357°C. It has a thermal conductivity of 0.0148 cal/cm degC s at 0°C which is 1/20 that of silver. Its electrical conductivity is also poor being 1/56 that of copper. Mercury is used for the international standard of resistance, an ohm being defined as the resistance to an unvarying current at 0°C of a column of mercury 106.3 cm long and 14.4521 g in weight.

Mercury is resistant to cold hydrochloric and sulphuric acids but is attacked by nitric acid and most hot acids. An important property of mercury is its ability to amalgamate with other metals such as gold and silver.

Both the metal and its salts are very poisonous and strict precautions are necessary in mines and workshops where they are handled.

4.19.6 Uses

Only about 30% of the mercury produced is used in metallic form, the remainder being in salts and compounds.

Metallic mercury is used in thermometers, barometers and many other instruments. It is a component of certain types of electric switches and is widely used in mercury vapour lamps and in arc rectifiers. It is also used as a heat transfer medium in heat exchangers.

Mercury compounds such as calomel (mercurous chloride) are used in pharmacy. Others are in pesticides, anti-fouling paints, paints and inks and in the textile industry.

Mercury fulminate ($Hg(ONC)_2$) is a highly explosive substance used in munitions and fireworks.

These are only a few of the more important industrial applications of this widely used metal.

4.19.7 Statistics

US production in various years (in flasks of 76 lb)

1861	35 000
1876	72 716 (greatest year)
1900	28 317
1940	37 777
1950	4 535
1960	33 223
1966	20 210

Major uses in USA in 1966 (in flasks of 76 lb)

Preparation of chlorine and caustic soda	11 578
Electrical apparatus	11 461
Paints	7 947
Instruments and control	3 106
Pharmacy	3 762
Agriculture	2 367
Dentistry	1 153

US price in various years ($ per flask of 76 lb)

1955	291
1960	210
1966	441
Min and max	186 to 570

4.20 TITANIUM

4.20.1 History

The element now known as titanium was first discovered by Reverend William Gregor in 1790, who found its oxide in a sample of Cornish sand. Five years later the oxide, rutile, was found by the Austrian chemist Klaproth, who called it titanium after the Titans – gods of ancient Greece who were credited with great strength.

During the nineteenth century there were several attempts to extract the metal from its oxide. In 1825 Berzelius produced very impure metal; Dumas in the following year converted the oxide to its tetrachloride and in 1887 Nilson and Petersson reduced the tetrachloride with sodium to make metal of 95% purity. At this time also titanium in the form of an impure alloy with iron was being used as a deoxidizing agent for steels.

In 1910 Hunter of the American General Electric Company made titanium of 99.5% purity by sodium reduction of the tetrachloride in an enclosed bomb. His hope that the new metal would be suitable for electric lamp filaments was short-lived when the melting point was found to be only 1 800°C instead of the expected 6 000°C. Hunter noted that this titanium was ductile when hot but brittle when cold, and interest in the metal waned. Fifteen years later van Arkel and de Boer dissociated titanium tetraiodide on a hot wire and made enough very pure titanium to show that the metal was, in fact, ductile at room temperature.

Between 1928 and 1938 Dr. Wilhelm Kroll of Luxembourg produced small quantities of titanium by Hunter's bomb process and later patented a method for manufacturing pure ductile titanium by magnesium reduction of titanium tetrachloride. Kroll followed this discovery by successfully consolidating his titanium sponge product by electric arc melting under vacuum using a tungsten electrode.

The invention of the Whittle jet engine led to a search for new materials to operate in the hot turbine and titanium was intensively studied during the 1940s in England, the USA and in Australia. Titanium did not prove a suitable replacement for nickel alloys but sufficient was learned of its metallurgy and technology for its rapid development as an engineering metal, and titanium was first produced in commercial quantities in 1948.

4.20.2 Occurrence

Titanium is the fourth most abundant engineering metal in the earth's crust after aluminium, iron and magnesium. It almost always occurs in nature as its oxide. Rutile (TiO_2) is found in beach sand in Australia, Mexico and in Travancore, South India, and rock deposits of the mixed oxides ilmenite (($FeTi)O_3$) and titanomagnetite (($Fe_2Ti)O_4$) are extensive in Norway, the Transvaal, the Ilmen mountains of the USSR, and in the USA, Canada and Brazil.

4.20.3 Purification of the ores

The mined ores are separated from siliceous gangue by the conventional techniques of froth flotation, specific gravity methods and, for ores containing magnetite, magnetic separation. Typical ore concentrates contain 95% titanium dioxide (TiO_2) from rutile, 40 to 80% TiO_2 from ilmenite and about 10% TiO_2 from titanomagnetite. Smelting of ilmenite ores to produce pig iron leaves a high titanium slag containing about 70% TiO_2.

4.20.4 Metal production

Titanium dioxide is chemically very stable and cannot be reduced to the pure metal by common reducing agents such as carbon or hydrogen. Furthermore, oxygen, nitrogen, carbon and hydrogen are soluble in titanium metal, making it brittle, and must be excluded if ductile titanium is to be produced. Metal is produced by converting titanium oxide to titanium tetrachloride which is reduced with magnesium or sodium to form a titanium sponge or finely divided granules.

Titanium tetrachloride is manufactured by heating an intimate mixture of titanium dioxide and coke, bound with coal tar to form briquettes, at 800°C in a chlorine atmosphere. Reaction temperature is maintained by resistance heating with carbon blocks and by the heat generated by the reaction:

$$TiO_2 + 2C + 2Cl_2 \rightarrow TiCl_4 + 2CO.$$

The chlorination temperature can be lowered by prefiring the titanium dioxide with carbon or with carbon and ammonia at high temperature to form titanium carbide or titanium cyanonitride. The impure product of the chlorination process is refined by filtration, fractional distillation, and by refluxing with copper or hydrogen sulphide to remove vanadium oxychloride, an impurity with a boiling point similar to that of titanium tetrachloride itself. Titanium tetrachloride is a yellow liquid boiling at 136°C which hydrolyses readily on contact with moist air.

The next stage in metal production is reduction of titanium tetrachloride. In the Kroll process titanium tetrachloride is fed into a clean steel reaction vessel containing molten magnesium at 800 to 900°C in an inert atmosphere of argon. The liquid magnesium reduced the tetrachloride to form spongy masses of titanium metal. Reaction temperature is maintained by controlling the rate of addition of the tetrachloride.

$$TiCl_4 + 2Mg \rightarrow Ti + 2MgCl_2.$$

Sufficient titanium tetrachloride is added to react with 85 to 90% of the magnesium present; then as much magnesium chloride as possible is tapped off and the products are allowed to cool in the inert atmosphere.

The spongy mass of titanium containing residual magnesium and magnesium chloride is removed from the reaction vessel into a dry atmosphere and the titanium sponge is separated by leaching with cold hydrochloric acid and by vacuum distillation at 900°C.

Further refining of the raw titanium metal is not usually necessary but can be achieved by the van Arkel iodide process or by electric refining in fused salt baths.

The sodium reduction process for producing titanium metal granules is essentially similar to the Kroll magnesium process, although final vacuum distillation is not needed. The major disadvantage of magnesium or sodium reduction is that it can only be operated as a batch process.

Electrolytic reduction of molten salts containing titanium chlorides and fluorotitanates can produce very pure metal and is being developed as a commercial process.

4.20.5 Melting and conversion of titanium ingot

Because of its high reactivity and melting temperature titanium cannot be melted in air using conventional refractory materials. Production of titanium and titanium alloy ingot is therefore achieved by arc melting (in vacuum or an inert atmosphere) an electrode of titanium sponge or granules compacted with a uniform distribution of the required alloy additions. The melting crucible is made of copper and is vigorously cooled with water or with liquid sodium–potassium alloy. Ingot size is from 1 to 10 tons.

Conversion of titanium ingot into useful shapes follows conventional metal working practices, equipment used for steel being generally more suitable than that used in other non-ferrous metal production. Titanium and its alloys are supplied for manufacture of components as billet and bar for forging or machining and as plate, sheet, wire, extruded sections and tube.

4.20.6 Properties

Titanium is a silver-grey transition metal with an atomic number of 22 and an atomic weight of 47.9. The density of pure titanium is 4.51 g/cm^3 (0.162 lb/in^3) and its melting point is 1668°C. The melting point is raised by impurities such as oxygen, nitrogen and carbon. The specific heat of titanium is about 0.12 cal/g degC and the coefficient of linear thermal expansion is about 8.3×10^{-6} per degC at room temperature. Thermal conductivity of pure titanium is 0.04 cal/cm degC s; alloying lowers the thermal conductivity to 0.015 to 0.030 cal/cm degC depending upon the alloy addition.

The strength and ductility of titanium depend upon purity. Iodide-refined metal (99.9% Ti) has a tensile strength of about 15 t.s.i.; commercially pure titanium melted from sponge or granules has a strength of 20 to 30 t.s.i. which can be increased to 50 t.s.i., at the expense of ductility, by the addition of oxygen. The modulus of elasticity (E) is about 16×10^6 p.s.i.; the fatigue limit for titanium is 45 to 60% of the ultimate tensile strength.

4.20.7 Corrosion resistance

The excellent corrosion resistance of titanium to many chemicals and industrial liquids results from the formation of a thin passive surface oxide film. Titanium is resistant to oxidizing acids, most organic acids, chloride solutions and moist chlorine gas, sulphur and dilute mineral acids and alkaline solutions. Hydrofluoric acid and strong or hot solutions of hydrochloric, sulphuric, phosphoric, formic and oxalic acids attack titanium, although in many cases resistance can be regained by anodic passivation or by addition of traces of oxidizing agents in the corrosive liquid.

At temperatures above 700°C titanium oxidizes rapidly in air, and oxygen dissolves into the metal causing hardening and ultimately embrittlement. Oxidation of titanium at temperatures of 300 to 500°C forms temper colours only and protection in service is not required. Hydrogen also will be dissolved rapidly at high temperatures and embrittle titanium and care must be taken when heating titanium to ensure a neutral or oxidizing atmosphere.

4.20.8 Titanium alloys

Although strengths of up to 50 t.s.i. can be obtained with commercially pure titanium containing high impurity levels of oxygen, carbon and nitrogen, these impurities tend to reduce ductility. It is necessary to alloy titanium with other metals to achieve higher strength with good ductility and in particular good creep resistance at temperatures of 150 to 500°C.

Aluminium is the most important alloy addition because it strengthens alpha titanium (see table *4.18*) and also reduces density, thus improving an already favourable strength-to-weight ratio; aluminium additions are limited to about 6% because higher levels can cause embrittlement. Additions of beta stabilizing elements improve the forgeability of titanium alloys and lead to a duplex structure of alpha and beta phases which, by suitable heat treatment, can be modified in composition and distribution to provide titanium alloys varying widely in strength, ductility, creep resistance and toughness. With larger additions of beta stabilizing elements it is possible to retain the high temperature beta allotrope on cooling to room temperature; limited use has been made of heat-treatable metastable beta alloys, but they have higher density than alpha and alpha-beta alloys. Complex alloys containing two or more alpha stabilizers with one or more beta stabilizers or compound-forming additions generally have superior mechanical properties to the simpler alloys.

Titanium alloys are classified by the phases present at room temperature and table *4.18* indicates the principal characteristics of commercially available alloy compositions.

4.20.9 Manufacturing techniques

Titanium components are manufactured by conventional metal working processes.

314 | NON-FERROUS METALS

Table 4.18 Titanium alloys

Alloy type	Alpha	Alpha	Alpha age hardening	Near-alpha	Alpha-beta	Metastable beta
Chief alloy additions	O_2	Al, Sn, Zr	Cu	Ai, Sn, Ar, Mo, V, Si	Al, Sn, Mo, V, Mn, Si	Mo, V, Cr, Al
Typical ultimate tensile strength, t.s.i.						
Annealed	20 to 45	50 to 65	35 to 42	60 to 70	55 to 70	45 to 60
Heat-treated			45 to 53	65 to 75	70 to 85	70 to 90
Maximum service temperature, °C	200	400 to 500	350	450 to 500	400	
Density, g/cm³	4.51	4.45 to 4.8	4.56	4.4 to 4.9	4.4 to 4.9	4.8 to 5.0
Forgeability	Good	Fair	Good	Fair/Good	Good	Good
Formability	Good	Poor	Good	Fair	Fair/Poor	Good
Weldability	Good	Good	Good	Fair	Fair/Poor	Fair
Oxidation resistance	Good	Good/Fair	Good	Good/Fair	Fair	Poor
Examples of commercial alloy compositions (balance titanium)	Commercially pure titanium 99.8 to 99.2% Ti	5% Al, 2½% Sn	2½% Cu	8% Al, 1% Mo, 1% V 11% Sn, 2¼% Al, 5% Zr, 1% Mo, 0.2% Si	6% Al, 4% V 4% Al, 2% Sn, 4% Mo, ½% Si 11% Sn, 2¼% Al, 4% Mo, 0.2% Si, 6% Al, 6% V, 2% Sn	15% Mo 13% V, 11% Cr, 3% Al

Forging is carried out at preheat temperatures from 700 to 1 100°C depending upon the strength and metallurgical requirements of the particular alloy. Highly strengthened alpha and near-alpha alloys are more difficult to forge than other alloy types unless high temperatures can be used. After hot-working operations surface contamination is removed by machining, grinding or vapour blasting followed by pickling in hydrofluoric acid and nitric acid solutions. Titanium sheet alloys of up to 50 t.s.i. tensile strength can be formed cold; hot-forming or sizing at temperatures up to 700°C is preferred for stronger alloys. Alpha alloys and some near-alpha and alpha-beta alloys can be joined satisfactorily by fusion welding, using inert-gas shielded metal arc processes; the stronger alpha-beta alloys are generally not weldable although the use of electron-beam welding can improve weld quality significantly. Resistance and spot welding and diffusion bonding can be used to join titanium. Brazing is rarely satisfactory.

Titanium is readily machined with slow cutting speeds, heavy feed rates and correct tool design. There is a tendency for fretting and titanium alloys have poor wear resistance. Surface hardening treatments and coatings to improve wear resistance have been developed but they tend to impair other mechanical properties, particularly fatigue resistance.

Production of titanium shapes by casting is possible by arc-melting the metal in a water-cooled copper crucible and pouring under vacuum into moulds of non-reactive ceramic materials, usually lined with graphite.

4.20.10 Uses of titanium and titanium alloys

Titanium is used in applications where its particular properties of high strength-to-weight ratio, good creep strength at temperatures up to 500°C and excellent corrosion resistance offer design efficiency and economic advantages over more conventional and cheaper metals. Major outlets for titanium are in the aerospace and chemical engineering industries.

Titanium alloys are used for compressor wheels, blades and casings of gas turbine engines and in airframe and space vehicle structures for high strength and heat resistance. Low-temperature pressure vessels for rocket motors frequently use titanium, for its good strength and toughness at sub-zero temperatures.

In rotating and reciprocating machinery such as steam turbines, high-performance internal combustion engines and ultracentrifuges, use of titanium alloys allows higher operating speeds and efficiency.

In chemical plant, titanium valves, pumps, pipework and vessels can outlive by several times similar components in stainless steel and other corrosion-resistant metals, thus eliminating frequent replacement of parts and loss of production. Titanium anodizing jigs and nickel-plating baskets have excellent service lives. Compatibility with body fluids permits the use of low-density titanium for surgical implants and artificial limb joints.

4.20.11 Statistics

In the years 1952 to 1966 world titanium consumption (excluding USSR) increased from 1 000 tons per year to about 12 000 tons per year, while median prices fell from £10 per lb to £2 per lb. The cost of titanium semi-finished products varies according to alloy grade and form, from £1 to £2 per lb for forging billet up to £8 per lb for high-strength alloy sheet.

Titanium has become the ninth major engineering metal, after iron and steel, aluminium, copper, lead, tin, zinc, nickel and magnesium.

4.21 ZIRCONIUM

4.21.1 History

The element zirconium was discovered by Klaproth in 1789 but elemental metallic zirconium was not produced even in the laboratory until 1824, when Berzelius first prepared the metal in brittle form. Ductile zirconium was first made by Van Arkel, De Boer and Fast who developed the iodide decomposition process at the University of Leyden and the Philips Lamp Works at Eindhoven, Holland, in 1925. The 'iodide crystal bar' process survives today as a method of refining sponge metal to very high purity. A simpler process using magnesium metal to reduce zirconium tetrachloride was developed by W. J. Kroll at the US Bureau of Mines station at Albany, Oregon, in 1947. About the same time Kaufman and Utermeyer discovered that early measurements of the thermal neutron capture of zirconium were in error because the metal used contained about 2% of hafnium. When the hafnium was removed zirconium proved to have a very low capacity for absorbing low velocity neutrons produced in atomic reactions. This relative transparency to thermal neutrons coupled with excellent corrosion resistance and strength combined to make zirconium the ideal metal for cladding the fuel in atomic reactors like those of nuclear-powered submarines. The impetus of this use made zirconium a useful commercial metal. In about 1958 the metal became available for other than military purposes and the chemical process industry chose it for a wide variety of uses. Early commercial power reactors used stainless steel to clad the uranium fuel but by 1965 zirconium had replaced it as the prime material for water-cooled systems.

Zirconium is produced commercially in the USA, France and the USSR. Plants were operated for a time in the UK and Japan.

4.21.2 Occurrence

The primary ore used for the production of zirconium metal is the mineral, zircon ($ZrSiO_4$). Normally it contains 1 part hafnium for each 50 parts zirconium, although some ores contain as much as 1 part in 15. It occurs widely in beach sands in nearly every country of the world. These beach sands are mostly silica but contain ilmenite, rutile and other valuable minerals as well as zircon. Deposits in Australia and the USA supply most of the commercial ore; smaller deposits are being worked in India, Nigeria and Brazil.

Baddeleyite (ZrO_2), found in Brazil, is also a possible ore but it has not been exploited for the production of metal. This ore contains less than 1 part hafnium for each 100 parts zirconium.

4.21.3 Mining and beneficiation

The beach sand deposits are mined by the floating dredge technique; sometimes the early stages of preliminary treatment known as beneficiation take place on the dredge. The valuable minerals are much heavier than the silica sands and are separated in two stages using spiral concentrators. The heavy sand concentrate from the spirals containing the ilmenite, rutile and zircon along with other minor minerals is dried for further separation. The ilmenite is removed with a magnetic separator leaving the rutile and zircon to be dealt with by an electrostatic separator. The purpose of these separations is to reduce the amount of other metals left in the zircon concentrate, since they have to be removed in later processing. The purest concentrates are usually set aside for metal production and the rest used for refractory, ferro-alloy and foundry work.

4.21.4 Metal production

The zircon concentrate is first converted to a carbonitride in a submerged arc furnace similar to those used for ferro-alloy production. The zircon reacts with the coke in the presence of air at very high temperatures to form the gold-coloured carbonitride, the silica being driven off as a fine fume. The crushed carbonitride is treated with chlorine to produce zirconium tetrachloride ($ZrCl_4$) with evolution of heat. This crude tetrachloride contains the hafnium that was also present in the zircon ore. To produce metal of reactor grade with low hafnium content, the crude tetrachloride is put into aqueous solution and complexed with ammonium thiocyanate. The hafnium is extracted from this solution with hexane in a complex liquid–liquid extraction system. The separated metals are precipitated from the respective solutions and calcined to the pure oxides. The oxide is mixed with carbon and again chlorinated to produce the tetrachloride, which is then reduced with magnesium in the absence of air to produce the sponge metal and magnesium chloride. The reduction products are vacuum distilled to remove the left-over magnesium and magnesium chloride, leaving the sponge metal which is crushed and blended for melting. The reduction process can also use sodium metal in place of the magnesium.

The sponge metal is pure enough for most uses but can be further purified by the iodide decomposition process to produce crystal bar exceeding 99.9% zirconium.

The sponge metal, together with alloying agents and scrap, is melted into ingots in consumable-electrode vacuum arc furnaces. These furnaces are like giant welding machines. A d.c. arc is struck between the electrode of compacted charge resembling the welding rod and the ingot being built up in a water-cooled copper pipe mould. The metal solidifies quickly against the copper and thus the metal pool is actually contained in a crucible of the zirconium itself. This type of melting was developed to prevent contamination of reactive metals with the refractories used in conventional melting furnaces. The metal is melted twice to achieve the desired homogeneity. Ingots are about 40 cm in diameter and normally weigh over a ton. The ingots are converted to all forms of wrought product by forging, rolling, extruding and other conventional metal-working techniques.

4.21.5 Properties

Zirconium is a ductile, silvery metal that has an appearance much like stainless steel. The pure metal and many alloys are allotropic. The alpha phase, stable at room temperature, is hexagonal close-packed. Above 862°C the structure transforms to a body-centred cubic lattice designated as the beta phase. This phase transformation and the solid-state reactions initiated by it play a major role in the metallurgy of most alloys.

The physical properties of the unalloyed metal are given in table *4.19*. These properties are largely a function of the oxygen content; the values given are for arc melted sponge metal with an oxygen content of about 0.1%.

Apart from the nuclear properties mentioned earlier, zirconium is notable for its resistance to a wide variety of chemical solutions. It is completely resistant to caustic solutions in all concentrations up to the boiling points; even fused caustic does not attack the metal. Zirconium resists most common acids to the limits shown in table *4.20*.

Table 4.19 Typical properties of zirconium

Density, g/cm^3	6.5
Melting point, °C	1845
Coefficient of linear expansion, per degC	5.8×10^{-6}
Thermal conductivity, cal/cm degC s	0.05
Specific heat, cal/g degC	0.067
Electrical resistivity, microhm-cm	40
Modulus of elasticity, kgf/mm^3	21 000
Ultimate strength (20°C), kgf/mm^2	45
Yield strength, 0.2% offset (20°C), kgf/mm^2	31
Elongation in 5 cm (20°C), %	21
Reduction in area (20°C), %	34
Hardness, Vickers scale (50 kg)	175

Table 4.20

Acid	Concentration	Temperature
Sulphuric	Up to 55%	Boiling
Nitric acid	All concentrations	Boiling
Hydrochloric acid	Up to 30%	Boiling
Phosphoric acid	Up to 60%	Boiling

It is attacked by hydrofluoric acid, most fluorides, aqua regia and wet chlorine gas. The metal will react with atmospheric gases starting at about 500°C.

4.21.6 Alloying and heat treatment

Most of the zirconium used in water-cooled reactors is in an alloy, called zircaloy, containing 1.5% tin and lesser amounts of iron, chromium and sometimes nickel. Other alloys in commercial use are with $2\frac{1}{2}$% niobium and with $\frac{1}{2}$% molybdenum. A family of alloys based on a combination of chromium and iron is being studied for possible use in steam. These alloys all depend for optimum properties on proper distribution of an intermetallic compound in the structure. They are very sensitive to heat treatment, particularly in the range near the allotropic transformation.

The elements oxygen, nitrogen, aluminium and tin tend to stabilize the alpha or hexagonal phase; most other elements tend to promote the early formation of the body-centred cubic or beta phase.

The recrystallization or annealing temperatures for unalloyed zirconium or the zircaloys are 650 to 750°C.

4.21.7 Fabrication

Wrought forms of zirconium can be made and machined with conventional machines and techniques. Bending should be done with care allowing bend radii of 3 to 5 times the metal thickness and forming warm where possible.

Machining is best done using sharp tools, slow speeds and relatively heavy feeds with a flood of water-soluble oil lubricant. Tool relief should be greater than for other metals as zirconium has a tendency to stick and gall on the tool. Drilling and tapping require the most care. Machines should be kept free of chips as they ignite spontaneously. Chips should be burned in small batches rather than accumulated.

Grinding is done on standard wet grinders using silicon carbide (carborundum) abrasive. Open grit wheels or belts perform best and a generous supply of coolant is essential.

Power sawing is best done with a very coarse (3 to 5 teeth to the inch) high-speed steel blade. Medium stroke rates are recommended and again a generous supply of lubricant is needed.

Welding is the most common method of joining zirconium. Since the liquid metal in the weld pool will react with atmospheric gases during welding, an argon gas shield is used. Nonconsumable electrode welding with tungsten, and consumable electrode welding using zirconium wire or rod, are both widely used. Zirconium can be welded to itself, hafnium, titanium and a few other metals but not to most common metals. Welds properly made are ductile and strong but if the argon shield is not complete the weld will be brittle. A black or white deposit on the weld indicates that the gas shield was not adequate.

4.21.8 Uses

The largest use of zirconium is as cladding for atomic fuel in water-cooled reactors. The usual arrangement is a lattice of 1 cm diameter tubes filled with pellets of uranium oxide. The zirconium protects the uranium oxide from the hot water or steam and contains the fission gases generated in the fuel.

The next largest use is as construction material for chemical plants. Here the resistance of zirconium to corrosion by process solutions makes it attractive where other more common materials are unsuitable.

A surprisingly large quantity of zirconium is used in the manufacture of photo flashbulbs. Very small bulbs were made possible through the use of shredded zirconium foil, which burns with a great burst of white light. Foils are also used to exhaust all traces of gas from electronic vacuum tubes – a process known as 'gettering'.

Zirconium metal powder is used in some fireworks and has been considered as a fuel for solid propellant rockets.

4.21.9 Statistics

Accurate records of metal production are not kept but it is estimated that usage in 1967 was about 1 500 tons; it is expected to double by 1970. The cost of the sponge metal is about £5 per kilogramme. Sheet and plate cost about £10 to £15 per kilogramme according to size.

4.22 VANADIUM

Vanadium, niobium and tantalum are closely related and have similar properties. They are characterized by high melting points and the need for high energy to reduce their oxides to the metallic form.

4.22.1 History

The discovery of vanadium is generally credited to Sefström in 1830, although it had previously been found, but not properly identified, in lead ores from Mexico. The

German chemist Wöhler showed that the two new substances were the same. Further investigation was carried out by Berzelius in the mid-nineteenth century and later by Roscoe, who was the first to prepare the pure metal.

4.22.2 Occurrence

Vanadium does not occur in a native state and its ores are not widely distributed. There are, however, rich deposits in the USA, Peru, Rhodesia and South Africa. The principal ores are the sulphides, such as patronite, found in Peru, silicates, such as roscoelite in Colorado, uranium-vanadium ores in the USA, and metallic vanadate ores including vanadinite, mottramite and descloizite.

4.22.3 Extraction

Extraction calls for many different processes because of the varied types and qualities of ores. In several processes the ores are decomposed with acids and the metal precipitated as insoluble vanadates which are then reduced by zinc. There are also fusion processes in which the ore is fused with fluxes and vanadium salts leached out with water. In the more modern 'matte' process the ore is smelted in a furnace until all the impurities are converted into a matte or mixture of sulphides, leaving the vanadium in the slag, from which it is extracted as a ferro-alloy. (This process is described under copper.)

Most of the vanadium extracted today is obtained by a thermite reaction (self-generated heat in mixture with powdered aluminium) or by a similar process in which the heat is supplied electrically.

4.22.4 Properties

Vanadium is a bright ductile metal with a high melting point. It is difficult to obtain in its pure form, but the following are accepted data.

Density, g/cm^3	6.0
Melting point, °C	1 800
Atomic weight	50.95
Resistivity, microhm-cm	26

Vanadium is generally resistant to cold acids and alkalis but is readily attacked when they are hot. It forms a large number of complex compounds and in this respect is similar to carbon.

4.22.5 Uses

It is mainly used as an alloying element in steel. Vanadium is added to cast iron, cast steel and alloy steels to increase tensile strength and hardness. It is often used in combination with other alloying metals such as chromium or manganese. The amounts of vanadium added usually vary between 0.1 and 2% according to the properties required.

It is also employed as a minor alloying addition to some non-ferrous metals such as bronze and aluminium.

Vanadium salts are generally poisonous but they have found some medical applications.

4.22.6 Statistics

World production (short tons)

	1960	1965
South and South West Africa	1 444	2 794
Finland	625	1 100
USA	4 971	5 226

4.23 NIOBIUM (COLUMBIUM)

4.23.1 History

Niobium is a high melting point metal of growing industrial importance. In the mid-1950s interest in niobium as an industrial metal was stimulated by the desire to take advantage of its low-neutron absorption coefficient, high-temperature strength and good corrosion resistance to liquid metals, properties which favoured its use as a cladding material for fuel elements in nuclear reactors. About 1959 this interest was extended to niobium base alloys for high-temperature applications in gas turbines and space vehicles. Since 1965 there has been increasing use of niobium alloys as superconductors in high field magnets.

In mineral form niobium is almost always found with tantalum. This and their chemical similarities caused some confusion at the time of their discovery. Niobium was first observed in 1801 in columbite from Connecticut by the English chemist Charles Hatchett, who called the new element columbium. In 1844 another chemist, Heinrich Rose, discovered what he considered to be a new element in Bavarian tantalite which he named niobium. Identification of the elements known as columbium tantalum and niobium aroused some controversy at the time but it was eventually established that there were only two elements, tantalum and one other. The other element still has the two names. In Europe it is known as 'niobium' (symbol Nb) and in America as 'columbium' (symbol Cb).

None of these early investigators succeeded in isolating the element and it was some years before pure niobium was produced. It is likely, although not quite certain, that the first pure ductile niobium was produced around 1905 by von Bolton at the same time as he produced pure tantalum. Until the advent of nuclear power virtually the whole of the world's niobium production was used as an alloying addition for steels. In recent years there have been an increasing number of uses for niobium itself.

4.23.2 Occurrence

Niobium and tantalum are very rarely found apart in nature and never in the free state. Most minerals containing niobium as a major constituent are complex oxides which may contain other elements including the rare earths, uranium and thorium. In general they are chemically complicated and variable in composition. The bulk of the world production of niobium is derived from minerals of the columbite–tantalite series, but in recent years large low-grade sources of niobium have been found in pyrochlore deposits. Geologically columbite is derived from ordinary granite sources such as pegmatic dykes and placer deposits. Pyrochlore is found in carbonites which are associated with alkalic igneous rocks.

Columbite and tantalite form a continuous isomorphous iron-manganese series with the general formula $(Fe, Mn)O.(Nb, Ta)_2O_5$. High-grade columbite contains up to 77% niobium pentoxide (Nb_2O_5) and high-grade tantalite up to 84% tantalum pentoxide (Ta_2O_5). Between these values there is a whole series of minerals of varying grades. The iron and manganese contents also vary. Columbite and tantalite are usually found as opaque minerals, iron black to brown in colour, with a Mohs' hardness value of about 6. The specific gravity ranges from 5.3 to 8.0 according to the proportions of niobium and tantalum present.

Pyrochlore is a complex niobate of calcium and sodium which forms an isomorphous series with microlite, a complex tantalate. The general formula is $(Na, Ca)_2O.(Nb, Ta)_2O_5$-$(ON, F, O)$. Although pyrochlore may contain as much as 73% niobium pentoxide the mineral varies widely in composition. Pyrochlore is brown, yellow, green or black in colour and has a Mohs' hardness value of about 5.5 and specific gravity about 4.3.

The principal producers of niobium minerals are Nigeria and Canada. In 1965 they supplied between them about 90% of the world's output. Niobium minerals are also found in Norway, Brazil, Malaysia, the Congo Republic and Burundi-Ruanda.

4.23.3 Mining and extraction

Columbite and pyrochlore are generally mined from open pits by various dry earth-moving methods. This is followed by crushing, grinding and gravity flotation to produce a concentrate of over 50% niobium pentoxide. Concentrates from Nigerian placer deposits, which are worked primarily for cassiterite (tin), are dried and treated by magnetic separators to remove the magnetic constituents of columbite, magnetite, ilmenite, monazite and magnetic cassiterite. A further magnetic separation enables the magnetite and ilmenite to be removed. Columbite is then separated from the remaining minerals by air flotation and electrostatic means.

The up-grading of pyrochlore concentrates has proved to be a somewhat difficult problem, but by selective flotation and chemical beneficiation a concentrate containing around 52% niobium pentoxide is obtained. Pyrochlore is generally used for the production of ferroniobium; virtually all niobium metal produced originates from columbite.

Many methods of extracting niobium from the concentrates have been proposed and several have been developed on a commercial scale. In all these processes there are essentially three stages. First the concentrate is decomposed to form either pure niobium and tantalum fluorides or chlorides. Next these compounds of niobium and tantalum are separated. Finally the pure niobium compound is reduced to metal powder. If the halide is used the final reduction can be carried out by hydrogen, magnesium or sodium or by fusion electrolysis. When the oxide is used, niobium metal is produced either by thermic reduction with aluminium or by reduction with carbon.

4.23.4 Consolidation

Because of its high melting point and extreme reactivity, conventional methods of consolidation and refining cannot be used for niobium. Consolidation is carried out in an inert atmosphere or in vacuum. High purity metal is required in order to obtain the necessary ductility for subsequent fabrication. There are three processes which can be used to consolidate niobrium.

(i) *Vacuum sintering.* In this process bars of compacted powder are gradually heated in a vacuum bell, either directly by resistance heating or indirectly by radiation. As the temperature increases volatile impurities are removed and densification of the bar occurs. A final temperature of about 2 300°C is maintained for about 2 hours to produce a bar of about 90% theoretical density. Full density can be obtained by cold-rolling and then resintering. Although high purity niobium can be obtained by vacuum sintering, the main disadvantage is the restriction on weights which can be economically produced. A typical analysis is given in table *4.21*.

(ii) *Vacuum arc melting*. Large piece-weights of niobium can be consolidated by arc melting an electrode of compacted and partially sintered niobium bars welded together. An arc is struck between the tip of the electrode and a small amount of niobium powder in a water-cooled copper crucible. The electrode gradually melts off and an ingot is built up in the crucible. Since the process takes place in vacuum some purification of the niobium occurs, but as the metal is only molten for a short time the degree of purification is not as great as with vacuum sintering. Double melting techniques are therefore generally used.

(iii) *Electron-beam melting*. The electron-beam melting process overcomes the disadvantages of sintering and arc melting in that very large piece-weights of high purity niobium can be produced. Melting is carried out by the bombardment of a highly accelerated beam of electrons on feed-stock of compacted and sintered niobium bars welded together. The molten niobium drips into an open-ended water-cooled copper crucible in which is a block or 'soleplate' of niobium. As the crucible fills up, the soleplate is gradually retracted and an ingot produced. A typical electron-beam furnace is shown in figure 4.8. Two melts are generally required; the first is carried out slowly in order to achieve the required degree of purification and the second fairly fast to produce a homogeneous ingot of good surface quality.

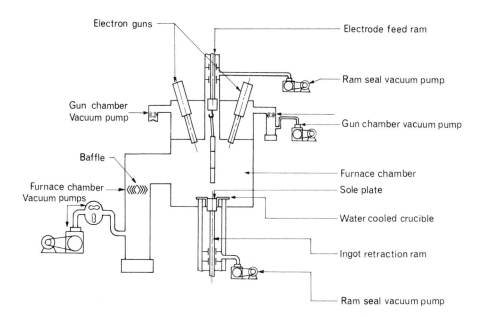

Fig. 4.8 Diagram of a typical electron beam furnace for melting niobium.

324 | NON-FERROUS METALS

Table 4.21 Analyses of niobium consolidated by different processes

Element p.p.m.	Sintered		Arc melted		Electron-beam melted	
	Before	After	Before	After	Before	After
O_2	5 000	50	800	650	1 570	38
C	2 500	100	300	230	280	70
N_2	700	30	300	240	200	60
H	80	4	14	11	21	1

4.23.5 Properties

Niobium is a transition element with the atomic number of 41 and an atomic weight of 92.91. The general physical and mechanical properties of unalloyed niobium are given in table *4.22*. The mechanical properties vary according to the interstitial content and the figures quoted are for niobium containing about 150 p.p.m. oxygen, which is a typical commercial grade.

Generally niobium has a higher resistance to corrosion than most other metals. It has good corrosion resistance to acids except hydrofluoric acid and hot concentrated sulphuric and hydrochloric acids. Resistance to alkalis is poor.

Unalloyed niobium has good resistance to molten sodium, potassium, lithium, calcium, cerium, bismuth, lead and silver up to 1 000°C.

Niobium is embrittled by hydrogen and to some extent by nitrogen, and has poor oxidation resistance above 600°C.

4.23.6 Alloys

The available niobium alloys fall into three main groups.
1. Moderately strengthened alloys for nuclear applications at 1 000°C; typical compositions are Nb1%Zr and Nb11%Ti. These alloys are available as sheet, bar and tube but are not widely used at the present time.
2. High-strength alloys for use at 1 200/1 300°C. Typical alloys are SU16-Nb11%W, 3%Mo, 2%Hf, 0.08%C and Cb752-Nb10%W, 2.5%Zr. These alloys are not yet in wide commercial use but their primary potential application is for jet engine turbine blades and other aerospace components. These alloys can be made into sheet and bar and have strengths of around 12 t.s.i. at 1 300°C.

Table 4.22 Physical and mechanical properties of unalloyed niobium

Crystal structure	body-centred cubic
Melting point, °C	2 415
Density, g/cm³	8.6
Young's modulus, 20°C, p.s.i.	12.4×10^6
Thermal conductivity, 0 to 100°C, cal/cm deg C s	0.13
Linear thermal expansion coefficient, 0 to 1 000°C, p.s.i.	6.9×10^6
Thermal neutron absorption cross-section, barn/atom	1.1
0.2% proof stress, t.s.i.	7
Ultimate tensile strength, t.s.i.	15
Elongation, %	55

3. Superconducting alloys such as Nb44%Ti, Nb25%Zr, Nb_3Sn. These alloys have excellent superconducting properties in that they remain superconducting in the presence of high magnetic fields. Many metals become superconducting at or near absolute zero but most lose their superconductivity in the presence of high magnetic fields in which they must operate for useful applications. The best superconductors are at present clad in copper, which has a stabilizing effect. They are available as wire or strip.

4.23.7 Fabrication and coatings

Niobium and niobium alloys can generally be fabricated on conventional metallurgical equipment. Primary breakdown of unalloyed niobium and some alloys is achieved by forging at moderate temperatures to prevent excessive oxidation. Breakdown of high-strength niobium alloys is carried out by extrusion at around 1400°C; forging, rod rolling or sheet rolling can then be performed on these alloys. Heat treatment of niobium and its alloys is always carried out in vacuum in order to avoid surface contamination.

Because of the poor oxidation resistance of niobium base alloys, they have to be coated before advantage can be taken of their good high-temperature mechanical properties. Much research work on the production of oxidation-resistant coatings for niobium base alloys has been carried out, and although some coatings give a satisfactory performance under isothermal conditions, a great deal remains to be done. The aims of current research are to produce a coating with a satisfactory combination of thermal shock resistance, resistance to erosion and impingement, and oxidation resistance at both elevated and intermediate temperatures. Some of the best coating systems are based on niobium disilicide modified with additions of chromium and titanium or aluminium. Even more promising are the newer composite coatings consisting of a disilicide plus a glaze overlay.

4.23.8 Joining

Because oxygen contamination embrittles welds, welding is carried out either in vacuum or in an inert atmosphere of argon or helium. Niobium is welded commercially by three different methods: resistance welding, inert-gas shielded arc welding and electron-beam welding. The third method produces the best welds but the size of workpiece is often limited by the need to use a vacuum chamber.

4.23.9 Uses

Unalloyed niobium is used primarily as a constructional material in nuclear engineering, where use is made of its low neutron absorption coefficient and good resistance to liquid sodium at 600°C. Niobium is also used for heat shields in high-temperature vacuum furnaces.

The application of high-strength creep-resistant niobium alloys in gas turbines and space vehicles will not become a practical proposition until the problem of oxidation-resistant coatings has been solved.

Superconducting niobium alloys are already being used in the construction of high field magnets for research purposes, and their application in power transmission and electric motors is under investigation.

A large quantity of niobium is used for ferroniobium in the steel industry, as an addition to certain stainless steels and nickel base alloys.

4.23.10 Statistics

Estimated production (in lb) of niobium concentrates* (other than Communist countries) Source: US Bureau of Mines Report

Country	1962	1963	1964	1965
Canada	1 909 433	2 692 935	4 222 424	4 510 182
Brazil	38 164	42 767	24 643	675 168
Norway	769 405	782 633	410 056	187 391
Portugal†	42 565	4 465	21 527	
Spain†			14 610	
Malaysia	246 400	197 120	125 440	103 040
Congo and Burundi Ruanda	55 846	163 437		42 125
Mozambique	34 000	33 000	40 000	32 187
Nigeria	5 066 880	4 506 880	5 239 360	5 707 520
South West Africa	1 116	418	447	1 080
Uganda		22 488	8 717	18 199
Estimated free world total	8 133 809	8 446 143	10 106 214	10 676 892

* Figures do not include niobium in the form of complex niobium–tantalum–tin concentrates.
† US imports.

4.24 TANTALUM

4.24.1 History and occurrence

As tantalum is almost always coexistent with niobium, the reader is referred to the previous section on niobium.

4.24.2 Extraction

Extraction of tantalum is identical to that of niobium to the stage where the ore concentrates are decomposed to form fluorides or chlorides. The tantalum salts are then separated from the niobium salts and reduced to pure metal using sodium or calcium. This gives powdered metal which must be heated in a vacuum or in an inert gas such as helium or argon, because of its high reactivity.

4.24.3 Properties

Tantalum is a silvery metal of very high melting point (2 996°C) and high density (16.6 g/cm^3). It is ductile and can be rolled into sheets or drawn into wire. Chemically it is reactive but is protected by a resistant oxide layer. As with aluminium, this oxide layer can be thickened by electrolysis or 'anodizing'.

4.24.4 Uses

Tantalum is used to make an extremely small capacitor, in which a roll of anodized tantalum sheet or sintered tantalum powder is employed, in midget electronic apparatus for civilian and military use. The major use of tantalum is in the chemical industry for heat transfer equipment, coils, condensers etc., for example, in hydrochloric acid manufacturing plant. Because of its high melting point it is used in furnace equipment and for crucibles in metallurgical operations.

4.24.5 Statistics

The production of tantalum is small compared with that of niobium, and reliable information is lacking. It is believed to be of the order of 500 tons per year.

4.25 CHROMIUM

4.25.1 History

Metallic chromium was first isolated in 1797 in France by Vauquelin, who obtained the metal by reducing chromic acid with carbon. He produced chromic acid from natural lead chromate ($PbCrO_4$), known as crocoite. Shortly afterwards the same method was developed independently by M. H. Klaproth. Vauquelin also demonstrated that emerald owed its green colour to the presence of chromium.

4.25.2 Occurrence

Although chromium occurs widely in the earth's crust, it is not found in its metallic state. Only one ore, chromite ($FeO.Cr_2O_3$) is of commercial importance as a source of the metal, but a large number of other minerals contain chromium. Some of the more important are crocoite, mentioned above, phoenicite ($3PbO.2CrO_3$), vauquelinite ($2(PbCu)CrO_4(PbCu)_3(PO_4)_2$) and chromitite ($((FeAl)_2O_3.2Cr_2O_3)$). Chromium has been found as a constituent of meteriorites and it is also present in a number of silicate minerals to which it gives a characteristic green colour. Chromite is found in many countries and is extensively mined in USSR, Turkey, South Africa, Rhodesia, Philippines, Yugoslavia and Cuba; lesser quantities are mined in a number of other countries. Of the major powers, only the USSR is self-sufficient in chromite; all others are net importers. The USA has some low-grade ore which could possibly be worked in an emergency

4.25.3 Mining

Chromite is worked by quarrying, open-cast mining and underground mining according to the type of deposit. The ores usually worked are of a high enough grade not to require further treatment before melting.

4.25.4 Metal production

There are two principal methods of production, according to the end use. For alloying with steel to make chrome steels, an iron-chromium alloy is used and this can be produced direct from chromite without first separating metallic chromium. If metallic chromium is required, the ore is reduced by a thermal process, using either aluminium or silicon as the reducing agent. A third method is employed when very high purity metal is required.

Ferro-alloys. These may be of two types: high-carbon or low-carbon. High-carbon ferrochromium contains about 70% chromium, 4 to 6% carbon and the remainder iron. It is produced in an electric arc furnace consisting of a steel shell, loaded with chromite ore and coke, having carbon electrodes submerged in the charge. An arc is struck between the electrodes and the ferrochromium collects at the bottom of the furnace from which it is tapped into moulds. Low-carbon ferrochromium is also produced in arc furnaces, but using silicon as the reducing agent. This gives a substantially carbon-free ferrochromium with a high silicon content. While still molten it is mixed with a synthetic slag obtained from another arc furnace in which chromite is reacted with lime. This gives a low-silicon low-carbon ferrochromium, the quantities of each being less than 1%.

4.25.5 Commercially pure chromium

Chromite may be reduced to chromium by an exothermic or heat-generating process using either aluminium or silicon. This use of aluminium powder was described by Wöhler in 1859 but was not adopted commercially until aluminium became cheap following the development of the Hall-Hercult process at the end of the century. No external source of heat is required. First chromic oxide is produced from chromite by a two-stage process. It is then mixed with aluminium powder. The charge is placed in a cylindrical container, about 5 ft in diameter, and ignited, the reaction proceeding in accordance with the equation:

$$Cr_2O_3 + 2Al = 2Cr + Al_2O_3.$$

Great care is required to prevent premature ignition of the charge by sparking.

The resulting slag and metal may be either poured from the container or allowed to solidify and removed by dismantling the cylinder. The ingots, weighing up to 1 ton, are broken into lumps and the slag picked out by hand.

Silicon reduction is also exothermic but the amount of heat generated is not sufficient by itself, and the process is carried out in an electric arc furnace:

$$2CrO_3 + 3Si = 4Cr + 3SiO_2.$$

Although external heat is needed, this is offset by the use of silicon; it is cheaper than aluminium and less is needed. High purity chromium can be obtained by the electrolysis of very pure chromic acid and small quantities are made commercially in this way.

4.25.6 Properties of chromium

Chromium is a transition element, allied to vanadium and manganese with which it has many points of similarity. Its principal properties are listed in table 4.23.

4.25.7 Uses

Almost the only commercial uses of chromium are as a coating on other metals and as an alloying element, but these two uses between them have had a major impact on our civilization.

Table 4.23

Atomic number	24
Atomic weight	52
Mass numbers of the isotopes	50, 52, 53, 54
Crystal structure	body-centred cubic
Melting point, °C	1850
Density, 20°C, g/cm³	7.19
Coefficient of linear expansion, per degC	6.2×10^6
Young's modulus, p.s.i.	36×10^6
Resistivity, 20°C, microhm-cm	12.9
Specific heat, 20°C, cal/g degC	0.11
Thermal conductivity, 20°C, cal/cm degCs	0.16

4.25.8 Chromium plating

Chromium can be electrically deposited to impart decorative or wear-resistant properties to a variety of other metals. Decorative chromium plating is applied to steel, brass, aluminium and other metals to give a hard bluish-tinged coating which will take a high polish and is very resistant to deterioration. Typical uses of chromium-plate are on automobile trim, tableware, ornaments and household metalware. It is usually applied over a nickel or copper undercoat or sometimes a combination of the two. Chromium-plate is usually deposited from a solution of chromic acid. Thick plated coatings, known as 'hard chrome-plating' are deposited direct on the base metal either to give protection against wear, as, for example, on press tools, or to repair worn surfaces. Another method of protecting metal surfaces with chromium is known as 'chromizing' and consists of exposing the metal to vaporized chromium in a carrier gas such as chlorine. A chromium salt is often used instead of metallic chromium. The chromium diffuses into the surface layers of the metal and gives a hard durable coating. The process is particularly applicable to steel, giving a product known as 'chromized steel'. For a more detailed discussion of plating see Chapter 9.

4.25.9 Alloying

The production of ferrochromium alloys for use in making alloy and stainless steels has already been described. Metallic chromium is also used as an alloying addition to nickel, giving a range of nickel-chromium alloys noted for their resistance to oxidation, making them very suitable for electric heating elements. Chromium also forms similar heat-resistant alloys with cobalt, and these are generally used in the form of castings. Chromium is used as a minor alloying addition in a number of aluminium alloys.

4.25.10 Statistics

Because chromium is not used to any extent in metallic form, statistics for its production are difficult to obtain. As a guide, however, the figures for production of chromite ore may be cited.

Production of chromite (short tons)

Country	1960	1965
Albania	318 650	347 000
Iran	74 957	165 000
Philippines	809 579	611 288
Rhodesia	668 401	624 500
Turkey	530 676	625 078
South Africa	850 921	1 038 498
USSR	1 010 000	1 565 000
World	4 885 000	5 370 000

4.26 MOLYBDENUM

4.26.1 History

The Greeks and Romans used the word molybdos to describe a mineral that was soft and lead-like in appearance. The word 'molybdenum' was introduced around 1816.

Molybdenite (MoS_2) was first identified by the Swedish chemist Karl Wilhelm Scheele in 1778; and in 1782 P. J. Hjelm isolated the metal itself.

For the next hundred years little was done with the metal except to use it in certain chemicals and dyes. Then, in 1893, two German chemists, Sternberg and Deutsch, obtained a 96% pure metal by reducing molybdate of lime with carbon and then removing the lime with hydrochloric acid. The impure molybdenum may have been used in experiments to find a substitute for tungsten in tool steels. In 1894 an attempt was made to produce molybdenum in an electric furnace but the carbon content of the product – 9% – made it unsatisfactory for use.

The first recorded use of molybdenum as an alloying element in steel occurred in 1894 when the French Schneider Company produced molybdenum-bearing armour plate at its Creusot works. A short time later the French chemist Henri Moissan obtained 99.98% pure metal by reduction in an electric furnace. The metal then achieved commercial importance for the first time.

Production of molybdenum ore was sporadic until 1900; since then there has been commercial production every year. Among the early producing areas were the Knaben mine in southern Norway and several mines in Australia and the USA. Production in the USA was suspended in 1905 but resumed during world war I; it was this stimulus that brought the fabulous Climax deposit in Colorado USA into production. Despite lean years between 1919 and 1933, the position of molybdenum as an additive and as a pure metal was assured.

At the outbreak of world war I, world production totalled some 200 000 lb. The shortage of tungsten and a widespread rumour that the Germans were using molybdenum in their armament steels (which proved later to be untrue) combined to favour an intensive search for molybdenum deposits. By 1918 1.8×10^6 lb was being produced.

The close of world war I left large amounts of ferromolybdenum in the hands of USA consumers who had accumulated the stocks in anticipation of military orders for 'Liberty' aircraft engines and 'baby' tanks. These large stocks, along with stocks at the mines, were more than sufficient to meet the small peacetime demand. As a result, the USA molybdenum industry collapsed: production ceased completely in 1920.

Since 1925, when the first industry-wide specification for chromium-molybdenum steels was approved by the American Society of Automotive Engineers, there has been a chain reaction of metallurgical research and development. In 1933 world molybdenum production was approximately 5×10^6 lb in 1938 over 20×10^6 lb and in 1966 production in non-Communist countries was near 130×10^6 lb, principally in the USA and Canada.

The principal molybdenum minerals and their properties are shown in table *4.24*.

4.26.2 Production

Virtually all molybdenite concentrate is roasted in Nichols-Herreshoff type furnaces to technical-grade molybdic oxide (MoO_3) – also known as roasted concentrate – which is the starting material for manufacturing almost all other molybdenum products. The molybdenum oxide plant of the Climax Molybdenum Company, which came on stream in 1966, has also made it possible to recover economically the residual molybdenum oxide formerly wasted.

Products produced from the molybdic concentrate are as follows.

Technical molybdic oxide. This grade is the raw material for producing all types of molybdenum salts and compounds and is also added to iron and steel, either with the charge or to the molten bath. The technical molybdic oxide contains most of the residual gangue present in the mine concentrate. It is also referred to as roasted concentrate.

Table 4.24 Physical and chemical properties of molybdenum minerals

Name	Formula	%Mo	Usual colour	Specific gravity
Molybdenite	MoS_2	60.1	Lead-grey	4.62 to 4.73
Wulfenite	$PbMoO_4$	26.1	Ranges: orange-yellow, yellowish grey, grey-white, olive green, reddish brown, to bright red	6.5 to 7.0
Powellite	$Ca(Mo,W)O_4$	39 to 48	Straw yellow, brown, pale green, grey	4.22 to 4.53
Ferrimolybdite	$Fe_2(MoO_4)_3.8H_2O$?	40	Canary yellow, straw yellow, greenish yellow	2.99 to 4.5
Chillagite	$3PbWO_4.PbMoO_4$	10.6	Yellow to brownish	7.5
Ilsemannite	$MoO_2.4MoO_3$? $Mo_3O_8.xH_2O$? $MoO_3.SO_3.5H_2O$?	Variable	Black to blue-black	
Koechlinite	$(BiO)_2MoO_4$	14.9 to 15.6	Greenish yellow	8.29 (calculated)
Lindgrenite	$Cu_3(MoO_4)_2(OH)_2$	36	Yellowish green	4.26

Purified molybdic oxide. Made by volatilization of roasted concentrate, the molybdic oxide content generally exceeds 99.5%. The roasted concentrate is fed as a thin layer on to a 'doughnut' type sand-hearth electric furnace with hearth temperature of 2 000 to 2 200°F (1 095 to 1 200°C). Air drawn over the heated surface removes the oxide through ports leading to bag filters. Since only 60% of the charge is converted to molybdic oxide, the remainder is collected and reclaimed for other processing. Other methods of production include chemical processing, sublimation and distillation of the roasted concentrate.

Briquettes. A mixture of technical molybdic oxide and carbon, in the form of pitch. Used in the manufacture of iron and steel, each package of four briquettes holds either 10 lb or 5 kg of contained molybdenum.

Ferromolybdenum. Both the thermite process and electric furnace are used to produce ferromolybdenum.

Ammonium molybdate, prepared from purified molybdic oxide, is the major starting material for the production of molybdenum powder. Hydrogen reduction of the ammonium molybdate produces a high purity molybdenum powder which can be consolidated either by powder metallurgy techniques or by arc casting.

Molybdenum powder is compacted mechanically or hydrostatically into ingots or bars, which are sintered by resistance heating. The sintering is done in vacuum or a hydrogen atmosphere, with the maximum temperature about 4000°F (2 200°C). Large ingots do not lead themselves to resistance heating and are sintered by radiation or induction heating; ingots weighing several hundred pounds are handled in this way. The primary processing of sintered ingots is done by hot-rolling, swaging or forging.

Commercially, arc casting is the main means of obtaining cast molybdenum ingots. The arc-casting process comprises vacuum or inert-gas arc melting in a water-cooled copper mould. The melting takes place in an arc, which operates between a vertical consumable molybdenum electrode and a pool of liquid metal in the mould. The electrode, composed of molybdenum powder, a de-oxidizing medium (usually carbon) and sometimes alloying additions, is compacted and sintered before melting. Ingots weighing more than a ton can be made in this way. The arc-cast ingots are extruded to billets, which may then be rolled or forged to the desired size.

Special production methods, such as electron-beam melting, skull melting, zone melting, slip casting, direct rolling and vapor deposition from carbonyl or pentachloride, have been tried experimentally.

4.26.3 Physical properties of molybdenum

Atomic number	42
Mass numbers of the isotopes	
Natural	92,94,95,96,97,98,100
Artificial	90,91,93,99,101,102,105
Atomic weight	95.95
Lattice type	body-centred cubic
Melting point, °C	2610
Heat of fusion (estimated), kcal/mol	6.7
Boiling point, °C	5560
Specific heat (0 to 500°C), cal/g degC	0.065
Electrical conductivity, 0°C, % IACS	34
Density at 20°C	10.22 g/cm^3, 0.369 lb/in^3

4.26.4 Chemical properties

At temperatures over about 500°C unprotected molybdenum oxidizes so rapidly in air or oxidizing atmospheres that its use in these conditions is impracticable. The rate of oxidation of solid molybdenum is not so extreme, however, as to cause combustion of the metal. There is no grain boundary weakening or internal oxidation; the molybdenum oxidizes evenly although there may be some preferential attack of corners or protruding sections. Uncoated molybdenum is being used satisfactorily where very short lives are involved (as in some missile parts) or where the surrounding atmosphere is non-oxidizing.

The oxidation tendencies of the molybdenum depend on the amount of oxygen and water vapour in the atmosphere. In vacuum, uncoated molybdenum has unlimited life at high temperatures. This is also true in the vacuum-like conditions of outer space. For example, the oxidation rate of molybdenum at altitudes approaching 200 000 ft may be as low as 1% of that at sea-level.

Pure hydrogen, argon and helium are completely inert to molybdenum at all temperatures. Molybdenum is also relatively inert in carbon dioxide and nitrogen up to about 2000°F (1090°C). A thin carbide case may be formed in carbon monoxide at these high temperatures and a thin sulphide case in hydrogen sulphide. The type of stress in service will determine whether these cases will impair the usefulness of molybdenum parts. In some oxygen-deficient combustion gases, uncoated molybdenum appears feasible for temperatures up to 2 500°F (1 370°C) where a short life is acceptable.

Although much work has been done on coatings for molybdenum, no one coating – whether metallic or ceramic – has been found that provides long-term protection at temperatures in excess of 2 500°F (1 370°C). Manufacturers of missile components have found that silicide coatings are effective for up to 10 hours protection at 2 500°F (1 370°C).

Molybdenum has particularly good resistance to corrosion by mineral acids provided oxidizing agents are not present.

4.26.5 Applications of molybdenum

In the USA in 1966, the steel industry took 72% of the molybdenum used, alloy cast iron almost 9%, high-temperature alloys 6%, chemicals and miscellaneous approximately 9% and molybdenum metal somewhat over 4%. This pattern of usage appears to be generally true of other nations.

The extensive and growing use of molybdenum in steel and cast iron arises from a combination of factors including economy, availability and technical advantage. Except for relatively short periods of great demand, users of molybdenum have been assured of a supply more abundant than either tungsten or vanadium and less subject to price fluctuations (especially relative to tungsten). Another important economic consideration is the high recovery rate of molybdenum additions to the steel or iron; unlike boron, chromium, manganese, titanium and vanadium, the molybdenum addition is completely recovered in the ingot, and substantially complete recovery is obtained from the molybdenum contained in scrap.

The form in which molybdenum is added to cast and wrought steels depends largely on the steel-making process, on local conditions and on the proportion of molybdenum to be added. Ferromolybdenum is adaptable to any steel-making process but molybdic oxide additions are usually cheaper.

Molybdenum is used in nearly all kinds of steel, with many and varied effects on its properties. For example, it is added to constructional steels for better strength and ductility, stainless steels for better corrosion resistance, elevated temperature steels for its remarkable contribution to creep strength, and tool steels for better strength and hardness at elevated temperatures. In general, molybdenum is used extensively in irons and steels because it enhances their hardenability.

Chemicals. The earliest commercial use of molybdenum was probably in the production of colours and dyes. Other applications today include reagents, ceramics (white pigments and as an agent when applying vitreous enamels to iron and steel) and catalysts (especially in oxidation-reduction reactions). In the production of high-octane gasoline, molybdenum is one of the preferred catalytic agents because of its resistance to sulphur and other poisons.

Lubricants. Molybdenum disulphide was first used commercially as a solid lubricant in the early 1940s. With an operating temperature range of -450 to $+750°F$ (-268 to $+400°C$), molybdenum disulphide is replacing both graphite and conventional hydro carbon lubricants in conditions of extreme pressure and/or high vacuum. It has a low coefficient of friction, resists extreme pressures, shears readily and bonds to metals and plastics. It is proving valuable in prolonging the life of parts where conventional lubricants are inadequate.

Selected applications by industry. Many applications (such as in the glass and missile industries) have depended on the availability of large ingots. Ingots of molybdenum and molybdenum base alloys up to 3 000 lb weight are produced in consumable-electrode vacuum arc furnaces using a device for continuous electrode production from molybdenum powder.

Missiles and aircraft. Arc-cast molybdenum alloys have been used for leading edges of control surfaces (as in the ASSET and Dyna-Soar space gliders) for rocket nozzles, and for support vanes where high resistance to erosion is required at elevated temperatures. For jet engines the alloys are used for guide vanes, blades, combustion liners and after-burner parts.

Glass. Resistance to the action of molten glass, high-temperature strength and good electrical properties are the reasons for the use of molybdenum in resistance heating electrodes, stirring devices, pumps and wear parts.

Nuclear. Most of the molybdenum going into nuclear energy is used as an addition to conventional steels and alloys. Some of the most important power installations in the USA for example, have reactor vessels made of ASTM 302 Grade B alloy steel which contains, in percentages: C, 0.20 to 0.35; Mn, 1.15 to 1.50; Si, 0.15 to 0.30; and Mo, 0.45 to 0.60.

Die-casting. Of considerable interest is the programme sponsored by INCRA (International Copper Research Association) to develop a better die material for die casting copper base alloys than the commonly used chromium–tungsten–molybdenum hot-work steels. In a production mould made of TZM alloy, 26 000 parts have been run with no evidence of heat checking. A hot-work steel die is generally considered unusable for a part of this type after 6 000 shots.

Molybdenum compounds. Molybdenum forms compounds in which it has the valence numbers 0, +2, +3, +4, +5 or +6. The chemistry of molybdenum compounds is complex because of the following factors: 1. Molybdenum compounds readily yield mixtures of compounds in which molybdenum occurs in different valence states. 2. Shifts between several coordination numbers (4, 6 and 8) of molybdenum atoms result from only minor differences in controlling conditions. 3. Molybdenum has a strong tendency to form complex compounds. With the exception of the halides and sulphides, almost no simple salts of molybdenum are known.

Any system containing molybdenum, therefore, must be extremely complex, since two or more variables must each reach equilibrium independently to define a specific condition. These equilibria are independent of each other and are easily distributed individually. The product of an apparently simple reaction must thus consist of a mixture of compounds whose identities and proportions are determined by temperature, pressure, concentration, pH, redox potential and proportions of the several original constituents and are changed by even minor variations of any of these.

4.26.6 Biological aspects

Acute toxicity and acute percutaneous toxicity tests on molybdenum disulphide have demonstrated that it falls in the 'relatively harmless' class, the highest classification for describing the safety of a material. As a result the US Food and Drug Administration raised no objection to the use of molybdenum disulphide lubricants for bakery oven chains. *US Public Health Bulletin No. 293* reports tests on several molybdenum compounds and concludes: 'A careful study of the histopathological changes produced by long continued exposure of animals to the fumes of dust of molybdenum compounds and of animals fed these substances indicates that molybdenum compounds in general are of a low order of toxicity both from the point of view of observed chemical effects as well as from the histopathological point of view.'

Molybdenum is an essential trace element for plant growth and is considered by some scientists to be essential for animal nutrition as well. Although minimum daily requirements for molybdenum have not been established, it is contained in some of the mineral and vitamin supplements now on the market. A molybdenized iron sulphate has been prescribed in the USA for many years to combat anaemia in pregnant women. Current investigations indicate also that molybdenum in the diet contributes to healthy tooth enamel.

Apart from laboratory tests designed to learn the possible effects of extremely high

TUNGSTEN | 335

levels of molybdenum, the only deleterious effects of molybdenum that have been observed occur in cattle, sheep and other ruminant animals grazing on certain abnormal soils. A number of research organizations have demonstrated the interrelation of molybdenum and copper in the nutrition of these animals. Where molybdenum in the forage is low, sheep suffer chronic copper poisoning, which can be prevented by treating the soil with molybdenum. On the other hand, where molybdenum in the forage is high, cattle and sheep suffer from copper starvation. This can be corrected by feeding the animals supplemental copper, by treating the soil with copper salts or by injections of copper glycinate suspensions. Both conditions have been found only in local areas with very limited soil types.

4.27 TUNGSTEN

4.27.1 History

Tungsten (sometimes also known as Wolfram) was first identified by the Swedish chemist Scheele in 1781, but he was not able to produce it in metallic form. The first to do so was Bergmann who, in 1783, reduced tungsten oxide with charcoal. Shortly afterwards tungsten was also obtained by reducing the oxide with hydrogen – the first example of what was to become an important method of reducing metallic oxides. 'Tungsten' is Swedish for 'heavy stone'; it is one of the heaviest substances known.

Occurrence. Minerals containing tungsten are very widely distributed, but commercially workable deposits are much rarer. The USSR and China produce about 60% of world requirements. Other important producing countries are Bolivia (28%), South Korea (26%) and Portugal (14%); smaller quantities are turned out in Australia, Argentina and Burma.

The most important of the minerals are wolframite ($(Fe\ Mn)WO_4$) and scheelite ($CaWO_4$). Others which are mined include ferberite ($FeWO_4$), powellite ($Ca(MoW)O_4$) and huebnerite ($MnWO_4$).

4.27.2 Mining

Some tungsten ores occur on the surface and, because of their high density, can be separated by 'placer' operations or water separation, also known as panning. However, most ore is won by underground mining.

An important attribute of many tungsten ores is that they fluoresce under ultraviolet light and this helps in prospecting and underground mining. The principal ore, wolframite, is magnetic because of its iron content and can be separated and concentrated with electromagnets.

4.27.3 Production

Tungsten is produced either as ferrotungsten, an alloy with iron, or as pure metal. Ferrotungsten is reduced direct from wolframite in an electric arc furnace in which the ore is mixed with iron oxide or scrap iron together with carbon, glass and quartz. If a purer metal is required, tungsten trioxide is used instead of wolframite.

Metallic tungsten is produced in a number of ways including reduction of the trioxide with hydrogen, electrolysing a fused mixture of tungstates and reducing the trioxide with zinc.

4.27.4 Properties

Tungsten is a grey metal with a metallic lustre. It is one of the heaviest metals known; its specific gravity of 19.3 is about the same as that of gold. It also has the highest melting point (3387°C) of any metal. It is resistant to atmospheric exposure and to most acids except a mixture of nitric and hydrofluoric acids. It tends to be attacked by hot alkalis, and oxidizes very rapidly when heated above 400°C.

The principal properties of tungsten are listed in table 4.25.

4.27.5 Alloys

Tungsten forms an important series of alloys with steel which are hard and heat resistant, hence their use for cutting and other tools. The tungsten content varies from 1 to 18%.

Tungsten carbide, as an alloy of tungsten with carbon, is one of the hardest substances known, second only to diamond, but it is extremely brittle. Other commercially significant alloys of tungsten are those with cobalt and chromium, known as 'stellites' and used for cutting tools.

4.27.6 Uses

Nearly half the total tungsten consumption is in the form of tungsten carbide for cutting and other tools. Wire, used for electric lamps, takes 20%, metallurgical and powder uses 18%, high-speed and special steels 18%.

Because of its brittleness tungsten carbide is usually embedded in cobalt and is then known as 'cemented tungsten carbide'. This material is used for the cutting tips of tools such as drills or lathe tools. Small portions of the carbide are brazed to the steel body and provide a very hard cutting edge that can be operated at high temperature and high speeds without blunting. When sharpening is required it must be done with a diamond or silicon carbide wheel.

At one time tungsten wire for lamp filaments was made by extruding tungsten powder. Later, a method was developed by the General Electric Company in the USA by which very pure powdered tungsten is compressed and heat treated to produce a ductile form that can be drawn through dies. Other methods have been developed of making ductile tungsten by addition of alloying elements such as rhenium, but they are expensive.

Table 4.25

Atomic weight	183.92
Atomic number	74
Specific gravity	19.3
Melting point, °C	3387
Coefficient of thermal expansion, 20°C, per degC	4×10^{-4}
Resistivity microhm-cm	5.48
Specific heat, cal/g degC	0.034

4.27.7 Statistics

The estimated world production of tungsten in 1967 was 66×10^6 lb of which 41×10^6 came from China and the USSR and 25×10^6 from the rest of the world. The USA stockpile in that year was estimated at 192×10^6 lb, compared with the USA 1967 consumption of 16.8×10^6 lb (4.2×10^6 lb of this was imported).

4.28 MANGANESE

4.28.1 History

Manganese was first reported as a new metal by Scheele in 1774, although various ores had been known for centuries before. The name is of uncertain origin, but is probably a corruption of magnesia. Scheele was unable to isolate the metal but a contemporary, J. G. Gahn, succeeded in producing small globules of manganese from ores supplied by Scheele. In the early days manganese was often known as 'Braunstein metal'.

4.28.2 Occurrence

Manganese is widely distributed in the form of oxides, hydroxides, silicates and carbonates, the oxides being the most important commercial source. Some of these are: pyrolusite (MnO_2) and a colloidal form, psilomelane, manganite ($Mn_2O_3H_2O$), braunite ($3Mn_2O_3MnSiO_3$), hausmannite (Mn_3O_4).

Manganese ores are mined in many countries of the world but the most important sources in tonnage are the USSR, India, South Africa and the USA.

4.28.3 Mining

Most deposits are near the surface as their usual mode of formation is by secondary deposition after having been dissolved out of crystalline rocks. Mining is therefore by open-cast methods, although there is some underground mining, for example, in India. The ore as mined is often pure enough for sale without treatment. Some low-grade ore may need to be crushed and washed to remove earth, or it may be concentrated by a flotation process.

4.28.4 Production

Manganese is produced either as ferro-alloys or as pure metal. Ferro-alloys have been made since the early nineteenth century but production methods have been improved in recent years.

4.28.5 Ferro-alloys

High-carbon ferro-manganese, containing about 80% Mn 13% Fe and 6% C, is made in a blast furnace or electric arc furnace. The blast furnace method is very similar to that used for pig iron (q.v.) but the fuel requirements are much greater. High-grade manganese ore and coke form the charge for the blast furnace. If cheap electric power is available, the electric arc furnace is a suitable method for producing ferro-alloys. The two methods yield alloys of very similar compositions.

Another form of ferro-alloy known as 'spiegeleisen', iron with about 20% Mn, 5% C and 1% Si, is also made in the blast furnace but its production is tending to drop in favour of high-carbon ferromanganese. Low-carbon ferromanganese, (containing less than 0.1%C), which is required in certain steel-making operations, is made in an electric arc furnace in which the charge consists of manganese ores and silico-manganese, without any carbon fuels.

Silico-manganese, which contains about 20% Mn, 22% Si, 6% Fe and less than 1% C, is made by the reduction of manganese ores containing silica or with added silica.

4.28.6 Pure metal

There are three principal methods for the production of commercial grades of manganese: reduction of ores with silicon, electrolysis and thermite reduction. The reduction process with silicon in an electric arc furnace is very similar to the process for low-carbon ferro-alloys described above. The manganese ore is reduced with a silico-manganese specially produced with low carbon and iron contents. Electrolytic production was developed in the USA. Ores are roasted to convert all the manganese to oxide which is then leached out with sulphuric acid leaving insoluble impurities behind. The liquid is then purified and electrolysed in a cell operating at about 5 V with a current of about 50 A. Electrolytic metal is usually purer than that produced by the other methods. The thermite reaction employs aluminium powder mixed with powdered metal oxide in carefully measured proportions. On ignition, the aluminium combines with the oxygen with an exothermic, or heat-generating, reaction that is sufficient to reduce the ore to metal. The charge is loaded into cylindrical refractory vessels which produce about 1 ton of metal. The contents of the furnace are broken up after cooling and the slag is separated from the metal by hand.

4.28.7 Properties

Manganese is a transition element with many similarities to chromium. Its principal properties are listed in table 4.26.

It exists in three allotropic forms known as α, β and γ manganese.

4.28.8 Alloys and uses

The principal uses of manganese are in steel making, where it is used to refine the steel and remove impurities. About 14 lb of manganese can be used in the production of 1 ton of steel. For this purpose, it is normally used in the form of one or other of the ferro-alloys already described. Although steel made from ferro-alloys is only about half the cost for the same manganese content, pure metal is said to have technical advantages for some special steels.

Although manganese appears to be plentiful, there is some anxiety about the rate at which it is being consumed in the steel industry and steps are being taken to recover it from slags and low-grade ores.

Manganese is employed as an alloying element in a large number of commercial

Table 4.26

Atomic number	25
Atomic weight	54.93
Melting point, °C	1 244
Density, 20°C, g/cm^3	7.2
Coefficient of linear expansion, per degC	22×10^{-6}
Resistivity, ohm-cm	200×10^{-4}

alloys, both ferrous and non-ferrous. Among the more important are an aluminium-1¼% manganese alloy and an electrical resistance alloy of manganese and copper known as Manganin, which contains about 12% Mn. Others are the manganese bronzes, which are copper–zinc alloys in which some of the copper is replaced by manganese.

There has recently been some interest in the use of manganese coatings electroplated on to other metals, principally for protective purposes. The coating is hard and also more electronegative than zinc, so that it confers a greater degree of sacrificial protection to more noble metals such as steel.

Manganese ores are used in electric dry batteries, in glass manufacture, as pigments and in fertilizers.

4.28.9 Statistics

Figures are not available for the production of manganese metal but tables *4.27* and *4.28* are a useful guide to the principal sources. Almost every major industrial country, with the exception of the USSR, is a net importer of manganese.

Table 4.27 World production of manganese ore (short tons per annum)

	1960	1965
Mexico	171 400	202 800
Brazil	1 101 387	1 296 987
Hungary	135 888	194 000
Rumania	192 872	110 000
USSR	6 080 300	8 598 000
India	1 321 411	1 657 874
Japan	357 131	338 409
Ghana	600 261	665 821
South Africa	1 316 132	1 727 822
Congo	420 671	416 205
Morocco	532 508	414 337
World total	14 989 000	19 406 000

Table 4.28 US consumption of manganese (short tons per annum)

	1960	1965
Ore	1 946 389	2 866 079
Ferromanganese	755 804	992 788
Silicomanganese	98 634	105 893
Spiegeleisen	32 128	20 042
Metal	15 733	19 408

CHAPTER 5

Metal Casting

5.1 INTRODUCTION

A casting may be defined as a metal object which is obtained by directly pouring liquid metal into a mould or cavity and allowing it to solidify and take the shape of the mould. Founding or casting is the manufacturing process for the production of castings and a foundry is a commercial establishment where the manufacturing process is carried out.

Practically all metal is initially cast, either directly into the final shape (castings), or into intermediate shapes or ingots which may then be mechanically worked by such processes as forging, rolling and extrusion to obtain the desired shape (wrought products). There are, of course, other methods of producing metal components, e.g., machining, fabrication and welding, stamping and numerous other techniques, but the starting material for most of these processes is a cast shape such as an ingot.

The basic strength of the foundry industries lies in the diversity of casting techniques available and it is rare for a completed engineering product not to include several cast components in its assembly. Examples are shown in figs. 5.1 to 5.3.

The basis for choosing casting in preference to other shaping processes in a particular case is dependent on many – sometimes conflicting – economical and technological factors. Some inherent advantages of the metal casting process are, however, easily recognized:

1. Any metal or alloy that can be melted or poured is capable of being cast, often into the most intricate of shapes, both external and internal. No other processing technique can match this capability.

2. Many refractory and reactive metals which cannot be economically hot- or cold-worked by rolling, forging, extrusion or forming and welding may be cast using suitable techniques. Recent developments in this field are the castings used for elevated-temperature applications for the aerospace industries.

3. Almost limitless varieties of shapes and sizes can be cast as single pieces, which if processed any other way may need complicated assembly of many parts, thus increasing the cost of manufacture.

4. Casting methods are easily adapted to mass production. The bulk of the components for the motor-car industry are cast.

5. Extremely large and heavy metal parts, weighing up to 200 tons, may be made by

Fig. 5.1 Investment casting: Perseus by Cellini A.D. 1582.

casting. This is rarely possible by any other processing technique. On the other hand, thin sections and extremely light metal parts can also be cast.

6. Metal components can generally be cast into shape with reasonably close dimensional tolerances, thus eliminating or minimizing machining operations, e.g., spheroidal graphite iron cast crankshafts for the automobile industry.

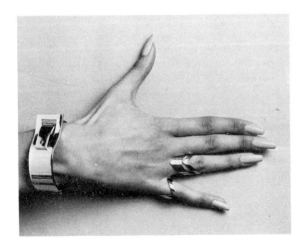

Fig. 5.2 Fashion jewellery cast in rubber moulds.

INTRODUCTION | 343

Fig. 5.3 Sand cast 31 ton marine propeller in copper-manganese aluminium alloy.

7. Some specific engineering properties are found only in cast metals:
(*i*) Damping capacity of cast irons.
(*ii*) Minimum directionality in mechanical or physical properties.
(*iii*) Special properties of SG (nodular) and malleable cast irons.
(*iv*) Composite structures produced by differential rates of solidification, e.g., chilled iron rolls.

5.1.1 The foundry industry

Metal-casting techniques are known to have existed as early as 5000 B.C., and many of the basic techniques developed between 5000 B.C. and A.D. 1700 are still applicable. Modern investment castings technology, for instance, stems directly from the ancient 'lost wax' process and differs from it only in details. Permanent mould castings, the forerunner of contemporary pressure and gravity die castings, can be traced back to Chinese 'money moulds' of 2000 B.C.

The foundry industry has always been a major part of the national economy of a country. However, after world war II, considerable progress was made in cast metals technology, concurrent with developments in transportation, communication and other areas of major growth. Castings are now used in transportation, communication, construction, agriculture, power generation, petro-chemical industries, in aerospace and atomic energy applications, and many other fields including domestic appliances and ornamental fittings. For defence and the military market, investment castings are in great demand in the manufacture of small-arms and rocketry.

The industry in the UK has a turnover of about £600 million, employs 150 000 workers and produces 5 to 6 million tons of saleable castings annually, and sustains the subsidiary business of foundry equipment and material supplies. Comparable figures for the USA are $8.5 billion, 475 000 workers and 18 to 20 million tons of saleable castings annually.*

Foundries may be classified as ferrous and non-ferrous, or more specifically as iron, steel and malleable, or brass and bronze or light metals (aluminium and magnesium) foundries. Alternatively classification based on the process is not uncommon, e.g., investment casting, die casting and centrifugal casting foundries. Further classification based on the nature of products and processes may also be adopted. A jobbing foundry is one where a small number (of each type) of a variety of castings may be produced, while a production foundry is usually a highly mechanized plant where very large numbers of a limited variety of castings may be made, in order to achieve a low unit cost of production. A production foundry is often specifically made for a definite range of products and is therefore less flexible with regard to the sizes and shapes of castings that may be produced, while a jobbing foundry is particularly adapted for variety at the expense of unit cost and production rates. Very often, foundries are mixtures of the two, i.e., a section of the works is organized on a jobbing basis while the rest is on a mass production basis. A captive foundry is a subsidiary of a larger manufacturing organization. The castings from such foundries are consumed solely by the parent organization in their finished products. Examples are the captive foundries of machine-tool manufacturers and motor-car manufacturers. The classification system for foundries is summarized in fig. 5.4.

5.1.2 Basic steps in making castings

For readers unfamiliar with the subject, the basic steps required to produce a casting are given below. In subsequent sections, the different methods used are explained in more detail.

*For latest statistics consult *The Foundry Trade Journal* – UK, *The Foundry* – USA.

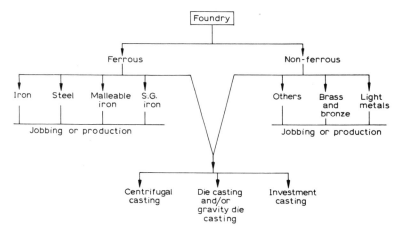

Fig. 5.4   Usual classification system for foundries.

The basic steps are:
1. Pattern making.
2. Core-making.
3. Moulding.
4. Melting and pouring.
5. Cleaning.

a. Pattern making. A pattern is basically a replica of the exterior of a casting. If the casting is to be hollow, additional patterns referred to as core boxes are used to form shapes, from sand or other material, to create these cavities or hollow regions.

To make the moulds plastic refractory material such as moulding sand is packed around the pattern, as illustrated in fig. 5.5. The mould cavity produced when the pattern is withdrawn is eventually filled with metal to become the casting. Patterns may be made from wood, metal, waxes or plastics, depending on the method of manufacturing the casting. Patterns are not required for die casting processes – the mould cavity is in fact a metal die – although in some cases the dies themselves may be initially cast rather than machined.

b. Core-making. Cores are forms, usually made of compacted refractory material such as sand, which are placed in a mould cavity to form the interior surfaces of castings. The space between the core and mould-cavity surface is then filled with molten metal to produce the casting. Core prints are the extensions or supports for the core in the mould cavity. Cylindrical or other simple cores for die casting are often permanent metal shapes.

c. Moulding. Moulding involves placing a moulding aggregate around a pattern held within a supporting frame, called a box or flask, withdrawing the pattern to leave the imprint or mould cavity, setting the cores in the cavity (if necessary) and closing the mould (usually two parts). The mould is then ready for pouring.

d. Melting and pouring. The preparation of molten metal for making castings is referred to as melting. It is carried out in suitable furnaces in a specified area in the foundry and the molten metal is transferred to the moulding area, where the moulds are poured. The channel or channels through which the molten metal enters the mould cavity is known as the running system. Often, risers or feeder heads are provided in the mould which act as reservoirs of molten metal and compensate for the shrinkage that takes place when liquid metal solidifies.

e. Cleaning. Cleaning includes all operations necessary for the removal of sand, scale and runners and risers from the casting. Defective castings may be salvaged by welding or other repair if practicable. Inspection of the castings for defects and general quality follows. The castings are then ready for shipment or further processing, e.g., heat treatment, surface treatment and machining.

The cast metals industry has advanced further in the last 20 years than in its entire history. This progress has been due to the application of the sciences of physics, chemistry, metallurgy and ceramics to foundry problems, and the adoption of basic engineering criteria for product design and processing techniques. The contributions from science, technology and engineering that enter into the manufacture of a typical casting are illustrated in fig. 5.6.

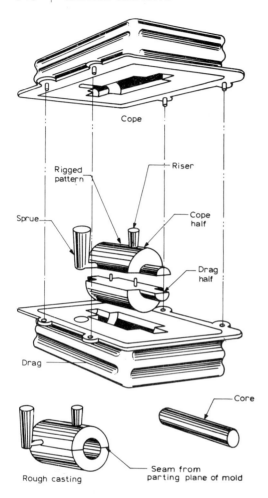

Fig. 5.5 Basic components in sand mouldings.

In subsequent sections a summary is given of the most important features of cast metals technology, with the object of presenting a broad general picture to the non-specialist reader.

5.2 PATTERNS

Except in the case of die castings or other permanent mould castings, the construction of suitable pattern equipment is the first step in making castings. Not all foundries, however, have pattern-making departments. Small and medium sized organizations which do include pattern departments are concerned mainly with modifying and repairing existing patterns (rigging) rather than producing new patterns. The vast majority of patterns are made by pattern shops which are independent of the foundry and operate as separate businesses. In general, larger foundries are more likely to have pattern shops organized on a production basis and may include both wood and metal pattern-making facilities as well as arrangements for making plastic patterns.

PATTERNS | 347

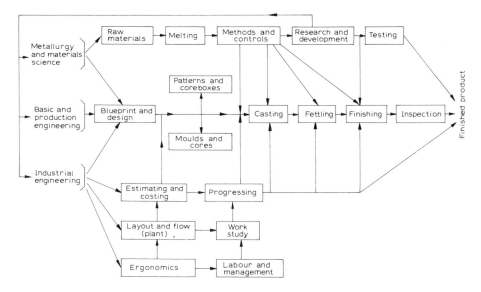

Fig. 5.6 The interdependence of technologies in a production foundry.

Investment casting foundries almost invariably have their own arrangements for making wax and other consumable patterns.

The detailed technology of making patterns which will produce the finished casting is beyond the scope of this book. However, certain principles which are applicable to patterns should be clear to all who may be concerned with castings.

5.2.1 Types of patterns

Although a great many castings can be made from a single pattern, it must be realized that a pattern is necessary even if only one casting has to be made. Several types of patterns are therefore used in foundries, depending on the number of castings to be made and the casting technique involved.
1. Single or loose patterns.
2. Loose patterns incorporating running and feeding systems – gated patterns.
3. Match-plate patterns.
4. Cope and drag patterns.
5. Skeleton, sweep and other special patterns.
6. Consumable patterns.

a. Loose patterns. Loose patterns are single copies of the external shape of the casting but incorporating the allowances in dimensions and the core prints necessary for producing the casting. These patterns are generally wooden but may also be of metal, plaster, plastics or any other suitable material. Relatively small numbers of castings are made from loose patterns since moulds can only be made by slow and costly hand-moulding processes. The running and feeding systems are also hand-cut and the patterns are withdrawn from the mould by hand. Casting dimensions are therefore difficult to reproduce and depend largely on the skill of the operator (moulder). A loose pattern is shown in fig. 5.7. Such patterns are cheap to produce and

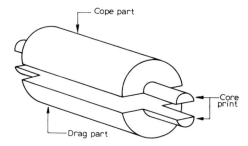

Fig. 5.7 Loose pattern for a hollow cylindrical casting.

are used when a small number of castings are to be produced (which does not justify more costly patterns). They may also be used for producing prototype castings.

b. Gated patterns. A gated pattern (shown in fig. 5.8) is better than a loose pattern because the running and feeding arrangements are integrated into the pattern, thus eliminating hand cutting. More rapid moulding of small quantities of castings is possible but gated patterns are more costly to produce than loose ones.

c. Match-plate patterns. Moulds are usually made in two parts, the top part known as the cope and the bottom known as the drag (see fig. 5.5). In match-plate patterns, the cope and drag portions of the pattern are mounted on opposite sides of a wood or metal plate forming the parting line, i.e., the plane which separates the two parts of the mould. Fig. 5.9 illustrates match-plate patterns incorporating the running and feeding systems. Match-plates are also integrally cast, i.e., the patterns and plate are cast as one piece in sand or plaster moulds. They are generally used in conjunction with moulding machines using snap flasks; the cope and drag halves of the mould are simultaneously moulded by placing the match plate between two moulding boxes (filled with moulding material) and squeezing from the top (see fig. 5.10).

The main advantages of mounting patterns on plates are that machine moulding can be carried out and excellent location and alignment of the patterns (with respect to the moulding boxes) are possible – this improves the dimensional accuracy of the finished castings.

The improved rate of production possible usually compensates for the increased cost of these patterns compared with loose or gated ones. Match-plate patterns are ideally suited for the quantity production of small castings. Size is limited by the weight

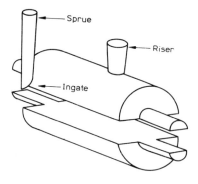

Fig. 5.8 Same pattern as in fig. 5.7 now shown rigged, i.e. incorporating the gate and riser.

PATTERNS | 349

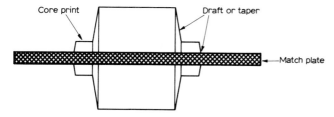

Fig. 5.9 Pattern for cylindrical casting mounted on a matchplate.

of mould and flask which can be handled by the moulder, since the moulding machines used are not easily adapted for mechanical handling, although fully automated systems incorporating match-plate patterns are now available for mass production.

d. Cope and drag pattern plates. In these patterns the cope and drag parts of the pattern are mounted on separate plates, usually incorporating the running and feeding arrangements as well. The cope and drag halves of the mould may then be made separately by operators on different moulding machines, making it possible to achieve high rates of production. The moulding of medium and large castings on moulding machines is almost universally carried out using this type of pattern equipment. Since high-speed automated or mechanized moulding techniques are used, the pattern and plates are generally made of metal to avoid rapid wear. The technique of moulding using separate pattern plates requires accurate alignment of the cope and drag by means of locating pins and bushings in the flasks, so that perfect matching occurs on closing the mould. All this adds to the cost of tackle but is fully justifiable when mass production is desired. Cope and drag pattern plates are shown in fig. 5.11.

e. Special patterns and equipment. Specialized pattern equipment is necessary when none of the previously mentioned types are suitable.

For very large castings *skeleton* patterns, illustrated in fig. 5.12, may be used. Such patterns are suitable for unusually large castings where the mould is manually constructed, using the skeleton as a format. These castings are generally produced singly and a solid pattern may be uneconomical or impractical to handle. Moulds for large

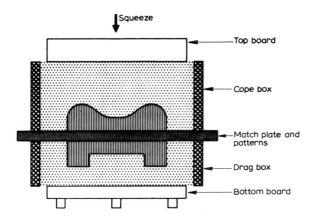

Fig. 5.10 Match plate pattern.

Fig. 5.11 Cope and drag parts mounted on plates.

castings of symmetrical shape are sometimes made with *sweep* patterns. These moulds are also manually constructed.

Loose patterns which do not have well-defined parting lines or which have irregular parting lines may require a *follow board* or *match* to support the loose pattern during moulding of the drag part of the mould, and also to establish the parting surface when the match is removed. The frame and bottom of the follow board may be wooden, but very often the actual matching contour is formed by rigidly compacted sand or plaster.

A *master pattern* is generally constructed of wood or plastic and used as the original for casting metal patterns. Several metal patterns may be cast from the master pattern and finally mounted on pattern plates. The master pattern must incorporate certain dimensional allowances to produce the metal patterns, which in their turn incorporate allowances to produce the required dimensions of the casting.

f. Consumable patterns. Consumable patterns are usually made from wax or plastic for the investment casting process and, as the name signifies, each pattern serves to produce one casting. The patterns are usually mass produced by injecting molten

Fig. 5.12 Skeleton pattern of a marine propeller blade in position for moulding.

wax or plastic into metallic dies, allowing solidification to occur and then withdrawing the pattern from the die.

A more recent development is the use of consumable patterns made from polystyrene bead boards in the full-mould process (see section 5.4.8).

5.2.2 Pattern allowances

The phenomenon of expansion and contraction that takes place when a solid, liquid or gaseous substance is subjected to increasing or decreasing temperatures is well known and dealt with elsewhere in this Encyclopedia. In casting technology, in addition to the normal expansion and contraction phenomenon, additional changes in dimensions, as a result of phase changes, have to be taken into account. A phase change occurs when liquid metal changes to solid or vice versa; phase changes may also occur due to changes in the metallurgical structure of metals on heating or cooling in the solid state.

Moulds are usually prepared and stored at room temperature so that when liquid metal is poured into the mould there is a rise in the temperature of the mould with consequent expansion. This reaches a maximum and then the whole assembly, mould and casting, gradually cools down to room temperature with the mould undergoing contraction. The metal in the mould cavity undergoes contraction in three stages with decreasing temperature: liquid contraction as the temperature decreases from the pouring temperature to the solidification temperature; solidification shrinkage (phase change) as it freezes*; and finally, contraction in the solid state as the temperature decreases from the solidification temperature to room temperature. A further change is not uncommon due to phase change in the solid state resulting in expansion or contraction. (Phase change in the solid state will usually depend on the alloy system as well as the rate of cooling and other metallurgical factors.) The expansions and contractions in the mould do not coincide with the contraction of the metal in the mould either in sequence or in degree.

It will be clear, therefore, that for metallurgical and mechanical reasons a number of allowances must be made on the pattern used to produce the mould if the resulting casting is to be dimensionally correct. A pattern is used to produce a casting of the required dimensions but *it is not* dimensionally identical with the casting.

5.2.3 Shrinkage allowance

Shrinkage allowance on patterns is a correction for the resultant contraction of the casting upon cooling down to room temperature from the pouring temperature of the metal, as already explained. The total contraction is volumetric but the correction for it is usually expressed in linear terms. Pattern shrinkage allowance is the amount by which the pattern must be made larger than the casting to provide for total contraction. It may vary from a negligible amount to about $\frac{3}{8}$ inch/ft, depending on the metal, the nature of the casting and the technique used in making the casting. Typical shrinkage allowances for castings made in sand moulds are shown in table 5.1.

It must be realized that such tables are only guides and many special conditions apply with different metals and different techniques. White cast iron, for example, contracts about $\frac{1}{4}$ inch/ft as cast, but if annealed after casting it expands about $\frac{1}{8}$ inch/ft, so that the net shrinkage allowance for castings to be annealed may only be $\frac{1}{8}$ inch/ft.

---

*There are exceptions to this rule, e.g., bismuth expands on freezing.

Spheroidal graphite (nodular) iron may contract $\frac{1}{4}$ to $\frac{1}{8}$ inch/ft depending on the degree of graphitization which it undergoes during freezing (i.e., the more the graphitization the less the total contraction). From these examples it will be evident that accurate determination of pattern allowances for different applications is a highly specialized subject, and a competent cast metals technologist should be consulted before specifying pattern shrinkage allowances.

The *shrink rule* is a special scale which eliminates the calculation of the amount of shrinkage allowance necessary on a given dimension during construction of the pattern. For example, an $\frac{1}{8}$ inch shrink rule is one on which each foot is $\frac{1}{8}$ inch longer and each graduation is proportionately longer than on a regular scale. Thus, 2 ft on this scale is an actual length of 2 ft $\frac{1}{4}$ inch, while 6 inches on this scale is equivalent to an actual length of $6\frac{1}{16}$ inch. Shrink rules are available with different standard allowances such as those shown in table *5.1*.

Double allowances may be necessary when making a master pattern. Many other special conditions apply when making patterns for investment castings and other castings requiring precise dimensions.

5.2.4 Machining allowance

Castings which require machine finish either on external faces or in the bore, or other internal surfaces, must be made oversize in these dimensions to provide stock for machining. In such cases allowances are made in the pattern, depending on the metal, the design of the casting and the method of casting and cleaning. Table *5.2* is a guide to pattern (machine-finish) allowances used for castings made by conventional and moulding techniques.

For other casting processes, such as investment castings and shell moulded castings, different sets of machining allowances are specified. Therefore, each casting must be considered individually. Properly specified allowances will not only produce more accurate dimensions but will also achieve considerable savings in cost of production.

5.2.5 Pattern draft

Draft is the tapering necessary on vertical faces of a pattern to allow it to be withdrawn from the sand or other moulding medium without damaging or tearing the mould-cavity surfaces. A taper of about $\frac{1}{16}$ inch/ft is common for vertical walls on patterns withdrawn by hand. However, small machine-moulded castings may not require any taper if suitably smooth and well-aligned patterns are available. In general, tapers of from 1° to 3° are common, depending on the depth of mould cavity, the pattern tackle, the moulding medium and the technique of moulding and withdrawal of pattern. In die castings, similar considerations apply in the case of the vertical walls of the die, which must be tapered to permit ejection of the casting and withdrawal of cores.*

5.2.6 Size tolerances

The permissible variation on any given dimensions of a casting is called the tolerance. In general, the maximum tolerance is usually half the shrinkage allowance. For close tolerance castings, i.e., tolerances of only a few thousandths of an inch, a

---

*For quantitative details consult *Design of Die Castings*, American Foundrymen's Society, Des Plaines, Illinois, USA.

Table 5.1*

Casting alloys	Pattern dimension	Type of construction	Section thickness (inch)	Contraction (inch/ft)
Grey cast iron	Up to 24 inch	Open		1/8
	From 25 to 48 inch	Open		1/10
	Over 48 inch	Open		1/12
	Up to 24 inch	Cored		1/8
	From 25 to 36 inch	Cored		1/10
	Over 36 inch	Cored		1/12
Cast steel	Up to 24 inch	Open		1/4
	From 25 to 72 inch	Open		3/16
	Over 72 inch	Open		5/32
	Up to 18 inch	Cored		1/4
	From 19 to 48 inch	Cored		3/16
	From 49 to 66 inch	Cored		5/32
	Over 66 inch	Cored		1/8
Malleable			1/16	11/64
Cast iron			1/8	5/32
			3/16	19/128
			1/4	9/64
			3/8	1/8
			1/2	7/64
			5/8	3/32
			3/4	5/64
			7/8	3/64
			1	1/32
Aluminium	Up to 48 inch	Open		5/32
	49 to 72 inch	Open		9/64
	Over 72 inch	Open		1/8
	Up to 24 inch	Cored		5/32
	Over 48 inch	Cored		9/64 to 1/8
	From 25 to 48 inch	Cored		1/8 to 1/16
Magnesium	Up to 48 inch	Open		11/16
	Over 48 inch	Open		5/32
	Up to 24 inch	Cored		5/32
	Over 24 inch	Cored		5/32 to 1/8
Brass				3/16
				1/8 to 1/4

*Pattern Makers' Manual. American Foundrymen's Society, Des Plaines, Illinois, USA.

complete analysis of all the processing factors (e.g., the behaviour of the mould, the design of the running and gating systems, mould dilation due to pressure of the liquid metal) and of the behaviour of the metal must be carried out. Such analysis is beyond the scope of the present discussion.

5.2.7 Distortion allowance

Due to the complex mechanics of non-uniform forces working on a casting during the process of solidification and the contraction taking place in the solid state while it is cooling down to room temperature, some castings are distorted. This is

Table 5.2 (pattern) machine allowances*

Casting alloy	Pattern size	Allowance – Bore (inch)	Allowance – external surfaces (inch)
Cast iron	Up to 12 inch	1/8	3/32
	13 to 24 inch	3/16	1/8
	25 to 42 inch	1/4	3/16
	43 to 60 inch	5/16	1/4
	61 to 80 inch	3/8	5/16
	81 to 120 inch	7/16	3/8
	Over 120 inch	Special instructions	Special instructions
Cast steel	Up to 12 inch	3/16	1/8
	13 to 24 inch	1/4	3/16
	25 to 42 inch	5/16	5/16
	43 to 60 inch	3/8	3/8
	61 to 80 inch	1/2	7/16
	81 to 120 inch	5/8	1/2
	Over 120 inch	Special instructions	Special instructions
Malleable iron	Up to 6 inch	1/16	1/16
	6 to 9 inch	3/32	1/16
	9 to 12 inch	3/32	3/32
	12 to 24 inch	5/32	1/8
	24 to 35 inch	3/16	3/16
	Over 36 inch	Special instructions	Special instructions
Brass, bronze and aluminium alloys	Up to 12 inch	3/32	1/16
	13 to 24 inch	3/16	1/18
	25 to 36 inch	3/16	5/32
	Over 36 inch	Special instructions	Special instructions

Pattern Makers' Manual, American Foundrymen's Society, Des Plaines, Illinois, USA.

inherent in the design; preferably, designs which are prone to distortion should be avoided. However, in cases where alterations in design are not possible, the pattern may be intentionally distorted in the opposite direction to the casting distortion, so that when finally solidified and cooled down to room temperature the casting will be free from distortion. Considerable experience is necessary to make calculations of distortion allowances with any degree of accuracy because of their complexity. Special stays may also be cast-in in some cases to avoid distortion of the casting.

5.2.8 Particular provisions in pattern making

a. Running and feeding systems. In addition to being a replica of the external shape of the casting, the pattern will generally incorporate the running (channels through which metal enters the mould cavity) and feeding (reservoirs to compensate for solidification shrinkage) systems. It will be seen in subsequent sections that proper design of running and feeding systems is essential for producing sound castings, and the incorporation of the running and feeding system is therefore a most important part of pattern making.

PATTERNS | 355

b. Parting line. In some cases, the parting line cannot follow a single plane because of the shape of the casting necessary for providing core support or for the proper design of the running and feeding system, i.e., in cope or drag plate patterns or match-plate patterns the surfaces of the plates may not be on a single plane. In loose patterns, multiplanar or curved parting lines may be cut by hand during moulding or a follow board may be used for successive moulds.

c. Core prints. Core prints are parts of the pattern (and therefore also of the mould cavity) which are necessary to support and locate the cores in proper positions in the mould. The core prints do not appear in the casting since they are blocked off by the core (see fig. 5.13).

d. Locating points. Locating points may be used in the pattern tackle for establishing points on the casting to check its dimensions, or for ensuring proper positioning of the cores and matching of sections. Machining operations may also use locating points on the pattern to establish the position of machined surfaces relative to the rest of the casting.

5.2.9 Core boxes

For all castings with internal cavities or surfaces, core boxes are an essential part of the pattern equipment. In conventional casting techniques, core boxes are constructed of wood or metal, depending mainly on the number of cores required and the core-making method employed. The simplest type of core box is a dump-box (see fig. 5.14), the core being removed by placing a plate over the box and then inverting it. A split box is usually a two-piece box; the individual box parts may be separately removed (fig. 5.15).

For simple small cores, a multiple or gang box may be used for making several cores simultaneously. Cores which do not have any flat surfaces need core driers,

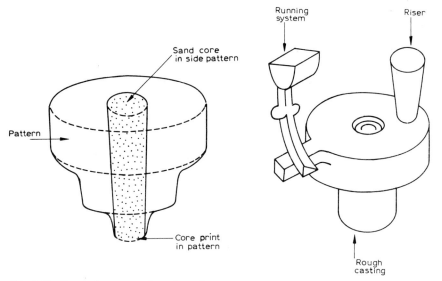

Fig. 5.13 Core print, core and rough casting.

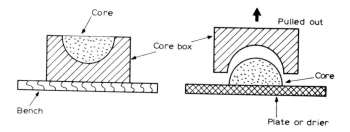

Fig. 5.14 Simple dump core box.

contoured to support the cores until baked hard. Newer processes, i.e., carbon dioxide, shell, hot-box and air-set core-making, have obviated this problem since the cores in those processes are set hard before being removed from the box and therefore need no special support for baking.

5.3 MOULDING AND CORE-MAKING IN GENERAL

5.3.1 Classification

It was realized even in the very early days of foundry practice that a casting of the required quality could not be produced without a good mould (although this is only one of the prerequisites). Owing to this historical importance, casting processes and castings are often described by the materials and processes used in making the moulds. Casting processes are therefore not easy to separate from moulding processes, and the terminology may appear somewhat confused to the non-specialist.

The basic casting processes are as follows:

1. Sand castings generally include all moulding processes employing 'granulated compact' moulds and require individual moulds for individual castings, i.e., the mould cannot be reused but the mould material is usually reused.
2. Permanent mould castings, as the name suggests, use metallic or other moulds which can be reused. One mould will generally produce a large number of castings, e.g., gravity die castings.
3. Die castings also involve metallic permanent moulds or dies but the difference from other permanent mould castings lies in the method of introducing the metal into the mould. In die casting, or pressure die casting as it is often called, the metal is intro-

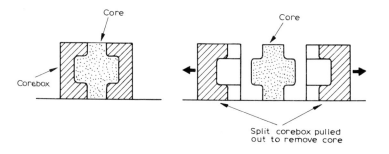

Fig. 5.15 A split core box.

duced into the mould under pressure to improve the mould filling characteristics.
4. Centrifugal castings utilize centrifugal force in channelling and distributing the metal in the mould. The mould is usually rotated when the metal is being introduced, and the metal is forced against the outer periphery of the mould. Both metallic (permanent) or aggregate (sand) moulds may be used.

The various moulding processes differ primarily in the method of making the mould, in the granular refractory aggregate used and in the method of bonding the aggregate. The most important moulding processes are as follows:
1. Sand moulding or sand casting:
   (*i*) Green sand moulding.
   (*ii*) Skin-dried sand moulding.
  (*iii*) Dry sand moulding.
  (*iv*) Core sand moulding.
   (*v*) Loam moulds.
  (*vi*) Cement moulds.
 (*vii*) Chamotte and compo moulds.
(*viii*) Air-set or self-curing moulds.
  (*ix*) C process or shell moulding.
   (*x*) Pit and floor moulding.
  (*xi*) Carbon dioxide (CO_2) process.
 (*xii*) Full-mould, fluid sand and other special processes.
2. Investment or precision moulding.
3. Ceramic moulding.
4. Plaster moulding.
5. Graphite moulding.

A brief description of the casting and moulding processes is given later, together with core-making processes.

Before going into the specific cases, it is necessary to consider the various types of moulding materials used and the principles underlying the bonding of aggregate material.

5.3.2 Choice of moulding and casting methods

There are many factors controlling the choice of a moulding process for any particular product. The designer, engineer or others who use castings are not expected to be fully aware of the most favourable moulding method for producing particular castings. Consultation with the foundry technologist leads to a more efficient use of castings through the selection of the most suitable moulding and casting process. In general the most important factors to be considered are:
1. Casting size and shape.
2. Dimensional accuracy and tolerances required.
3. Surface finish required.
4. Metallurgical properties.
5. Nature and choice of alloy to be cast.
6. Quantities to be produced.

How these several factors impose limitations on the choice of moulding and/or casting processes is summarized in table 5.3.

Even in the manufacture of components for the aerospace industries, quality at any cost can only be a temporary expedient. In the ultimate analysis, once the minimum requirements with regard to quality have been determined, the cheapest method of production which will ensure quality will obviously be adopted for quantity production. The economics of the casting process, therefore, do not end at an analysis of the design features of the component and the corresponding moulding or casting method, but must also take into account the relation of design to the process, i.e., economics is also dependent on producibility. Fig. 5.16 shows what is implied by 'producibility'. In fig. 5.16(*a*), the cooling fins on the electric motor casing are radial and therefore it is not possible to produce this casting by a simple two-part moulding (divided by the parting line). In fig. 5.16(*b*) the casting fins have been redesigned (to serve the same function) to make simple two-part moulding possible.

Fig. 5.17 is another example of a slight modification in the design (without affecting function) which will improve producibility by eliminating cores. Not all designs, of course, can be modified to improve producibility without at the same time affecting the functional aspects of the component, but considerable economies are possible in most cases as a result of joint consultations between the design engineer and the foundry technologist before the component design is finalized. Producibility is not restricted to a simplification of moulding and core-making operations alone but will include such features as the location of the feeders, the running system and mould filling, core venting and support, distortion and residual stresses (which may be vitally affected, depending on the design of the component). Again, considerable economies are possible in a majority of cases, with suitable alterations to the design.

Assuming that the casting design has been examined from the point of view of producibility and suitable modifications made where possible, and that the correct choice has been made in specifying such details as casting alloy, surface finish, dimensional tolerances, inspection standards and packaging, the cost of production is largely dependent on the nature of the pattern tackle and the moulding (or casting) technique used. If loose (hand-moulded) patterns are used the purchaser cannot expect the low prices attainable by machine moulding for large quantities. On the other hand, insistence on machine moulding when only a few castings are required will greatly increase the cost.

In order that the foundry technologist can select the most satisfactory casting technique and offer the most economical price, inquiries from buyers of castings should include the following essential information.

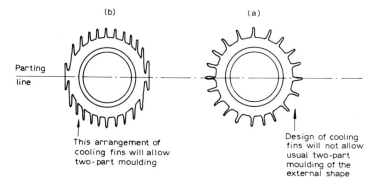

Fig. 5.16 Schematic representation of design problem encountered in castings such as motor castings.

SAND MOULDING | 359

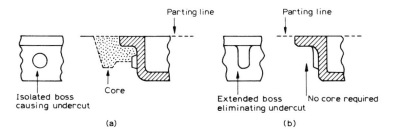

Fig. 5.17 Design of boss in a cast housing: (a) Isolated boss requires coring of the external shape. (b) Redesigned boss to eliminate core.

1. Metal specification. Apart from the actual price of the casting alloy, metal cost can depend on the range of chemical and physical properties. Unwarranted rigidity can add to casting cost. Wherever possible, the relevant standard specification should be quoted.
2. Drawing. Detailed drawings should include dimensional tolerances, machine-finished surfaces, cored sections, locating points on chucking surfaces and the actual surface finish required. Special requirements, such as heat treatment and surface treatment (plating, anodizing, galvanizing etc.) should also be indicated.
3. Quantity. The number of pieces required, with delivery dates and schedules, must be stated on the inquiry.
4. Pattern equipment. In some cases the purchaser may have existing patterns which he wants to use for making the castings. Since this imposes a limitation on the methods of production, accurate description of such pattern equipment must be submitted, together with details of material and construction. It is advisable to let the foundry technologist specify the type of pattern tackle even if the purchaser is paying for it separately.
5. Machining. If the casting is to be machine-finished by the producer, full details of tolerances and finish must be supplied with the inquiry.
6. Inspection. Specifications requiring pressure tests, non-destructive tests (e.g., radiographic, ultrasonic, magnetic particle inspection), metallographic tests and other special tests must be clearly indicated. All these add to the cost and should be specified under expert advice.
7. Packaging. Requirements of packaging, crating and marking for shipment or storage should also be clearly stated. Incomplete details can add to the cost.

It is general practice to quote for castings either in terms of price per piece or price per unit weight (per lb or kg). From the purchaser's point of view, price per piece is usually more satisfactory since price per pound or kilogramme may involve him in paying for extra weight of metal arising from poor foundry practice.

5.4 SAND MOULDING

5.4.1 Introduction

Sand castings generally include all moulding processes employing a granular aggregate refractory mixture to form the mould cavity. This granular aggregate may be

360 | METAL CASTING

Table 5.3 Limitations on the choice of casting processes

	Shell process	Permanent mould	Pressure die casting	Plaster moulding	Investment casting	Centrifugal casting
Weight	1 oz to several tons	1 oz to several hundred pounds. Up to 50 lb for shell more common	Few ounces to 50 lb, occasionally up to 500 lb	Few ounces to 50 lb occasionally more	Less than 1 oz to 100 lb	Less than 1 oz to several hundred pounds. Most commonly under 10 lb	Few pounds to several hundred pounds
Shape						
External	No limit	Limited by pattern drawing	Limited by casting ejection	Limited by casting ejection	No limit	Limited by wax pattern ejection from die	Favoured by circular periphery but any shape can be centrifuged
Internal	No limit	No limit	Limited. No limit if sand cores are used	Limited by metal cores	No limit	Limited by assembly of wax patterns that can be ejected from dies	Hollow circular, unless centrifuged in a cored mould
Number						
Minimum	One	Not less than 200	1 000 to 5 000	1 000 to 5 000	One	500 to 5 000	One or more depending on whether sand or metal moulds are used
Maximum	Pattern life limited according to material used for pattern making	Pattern life limited	Mould life limited. 1 000 to 100 000	Mould life limited 5 000 to 500 000	Pattern life limited according to material used	Die life for pattern making limited	Pattern (if sand moulds) or mould life (metallic) limited
Alloys that can be cast	1,2,3,4,5,6, 7,8,9,10,11	1,2,3,4,5, 6,9	1,3,4,5,6, 7,8,10,11	4,5,7,8, 10,11	4,5	3,4,5,6,9	1,3,4,5,6,9

SAND MOULDING

	Patterns	Mould or core constraint	Best tolerances	Minimum Section thickness	Maximum	Minimum cored hole diameter	Chilling power relative to green sand
	Wood, metal, plastic etc.	Mould and core can be made collapsible	±0.005 inch/inch	1/8 inch cast iron	No limit	1/4 inch	1.0
	Metal	Mould and cores are collapsible	0.003 inch/inch	1/8 inch	No limit	1/8 inch	0.2 to 1.0
	Machined mould	Rigid mould and core. Constraint can give rise to tearing of casting during freezing	0.015 inch/inch	3/32 inches	2.0 inch	3/16 inch	200 to 500
	Machined mould	Same as permanent mould	0.003 inch/inch much lower in some	0.015 inch Zn	5/16 inch. Normally 0.05 inch	1/32 to 3/16 inch	200 to 500; depends on mould temp.
	Metal patterns (polished)	Not as easily collapsible as sand	0.005 inch/inch or less	0.040 to 0.060 inch	—	0.500 inch	0.2 to 4.0
	Metal die for casting wax or plastic pattern	Not as easily collapsible as sand	0.002 inch/inch or 0.001 inch/inch	0.025 to 0.050 inch	0.500 inch, may be more	0.020 to 0.050 inch 1.0 (heated mould)	
	Pattern for centrifuged casting. Usually metal mould	No constraint due to circular periphery, unless central core is used which is not collapsible	—	—	—		

Alloys

1. Grey iron
2. Malleable iron
3. Steel
4. Aluminum
5. Copper
6. Nickel
7. Zinc
8. Magnesium
9. Refractory alloys
10. Tin
11. Lead

any refractory material (not necessarily sand) provided that it has the necessary stability and strength at molten metal temperatures. Sand is the most common material because of its ready availability and low cost compared with other materials.

The various moulding processes described under sand castings have one feature in common, i.e., they tend to produce porous moulds. The porosity allows for the free escape of gases and moisture when the molten metal is introduced into the mould. It also allows the air in the mould cavity to be replaced by the metal without excessive back-pressures being built up to prevent mould filling (special venting arrangements are necessary in permanent mould castings to displace the air from the mould cavity).

The pores should never be large enough, however, to allow metal penetration. In practice this is unusual because of the relatively high surface tension of molten metals compared with liquids like water. The maximum size of pore into which metal cannot penetrate is given by:

$$r = \frac{2\tau \cos \theta}{p}$$

where r is the pore radius in centimetres, τ the surface tension of the metal in dynes per centimetre, θ the contact angle between the metal and the refractory mould, i.e., the angle the surface makes at the point of contact with the mould and p is the pressure head of the metal.

Considering a steel casting about 2 ft deep, the values will be: $\tau = 1500$ dyn/cm (this is an average value quoted in literature); $\theta = 120°$ (minimum); $p = 4.6 \times 10^5$ dyn/cm² (2 ft head). Thus r is calculated to be about 0.3 mm, which is larger than the pore size usually obtained in moulds used for steel castings. The pore size is increased if a coarse-grained sand or other refractory aggregate is used for moulding. The importance of grading the refractory grains will therefore be readily appreciated. An adverse effect arising out of high surface tension is that it is difficult to reproduce very fine details in the casting.

The many moulding processes classified under sand castings are usually distinguished from one another by one or more of the following factors:
1. The method of forming the mould, i.e., by compaction, by free-flow of dry aggregate or by free-flow of slurry or fluid aggregate.
2. The nature of the refractory material used.
3. The method of bonding the aggregate.

The brief descriptions of moulding processes that follow are based on these.

5.4.2 Mould and core materials

In the final analysis, the type and severity of reaction between the molten metal and the refractory or other material forming the mould will decide the usefulness of the mould. In other words, the usefulness of the mould is largely dependent on its behaviour at elevated temperatures.

Behaviour of the mould depends not only on the chemical composition of the mould material but also on the type of bonding and the arrangement of atoms or molecules, i.e., the physical structure. For example, a common moulding material such as silica (SiO_2) can exist at room temperature as quartz, vitreous silica and several other forms. Quartz will disintegrate easily on rapid heating and cooling, while vitreous silica of the same composition can be heated or cooled rapidly without shattering. For a complete understanding of the ceramic and organic materials used in cast metals technology, a full review of the crystal chemistry of these materials – including bonding – is necessary. Further, an understanding of the principles of polymerization and cross-linking (phenol or urea formaldehyde resins, and the oxidation of core oils) is also important. These and many other relevant principles are discussed elsewhere in the Encyclopedia, in appropriate volumes and chapters.

A brief outline is included here so that the reader can readily appreciate the principles underlying the casting and moulding processes described later. The bibliography at the end of the section will help the reader in selecting literature for further reading.

a. Silica and silicates. Silica and silicates are by far the most important and extensively used mould and core materials in the foundry industry due to their stability at elevated temperatures (an essential requirement for mould materials).

Silica, the main constituent of foundry sands, can exist in four common modifications differing in atomic structure and density. These modifications are:

Type	Equilibrium temperature ranges
Quartz	Up to 870°C
Tridymite	870 to 1470°C
Cristobalite	1470 to 1710°C
Vitreous silica	Above 1710°C

Fortunately, the transformation of one form to the other is extremely sluggish – the long life of vitreous silica at room temperature is well known. During heating and cooling, however, each of the types listed undergoes various changes (without rearrangement of bonds), e.g., volume changes (see fig. 5.18).

In all but vitreous silica, the curves representing volume change show vertical discontinuities. Quartz (the commonest form) shows a large increase in volume at about 573°C, this has an important bearing in foundry sand practice since the mixture of aggregate and bond and the packing density must be such that the mould does not crack or collapse before the metal has solidified (i.e., when the metal is poured and the temperature of the mould rises above this point). For the same reason, it is more difficult to reproduce casting dimensions accurately from moulds made from common sand. Investment moulds made from quartz sand must be poured at temperatures (mould temperatures) above the α to β inversion point (573°C) to avoid mould shattering, which would prevent reproduction of fine details. In special moulds where dimensional accuracy and reproduction of delicate details are important, vitreous silica or other suitable material (see later) may be used.

b. Clays. Clays are essential ingredients for moulds in the majority of sand casting practices. They provide the bond between individual silica particles, thus allowing the silica (sand) aggregate to be compacted. Clay minerals used as bonding additions to sands are usually of the following types:
1. Montmorillonites (bantonites).
2. Kaolinites (fire clays).
3. Illites, and other special clays.

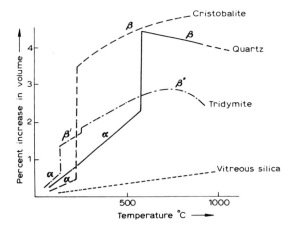

Fig. 5.18 Volume changes with temperature in the four types of silica.

Montmorillonites and kaolinites are most commonly used.

Clays are composed of very small crystals of hydrous aluminium silicates which have the form of sheets. The sheets possess a surface charge which is neutralized by exchangeable ions such as calcium, sodium or hydrogen ions. Due to this polar character clays are able to absorb strong polar molecules, especially water molecules. These form thin layers of water around the clay particles and are responsible for the plasticity of the clay.

c. Clay-water-silica bond. Moulding sands consist of water, clay and sand. The water molecules are attracted to the clay as already described. Since water molecules are polar, the first layer adjusts to the electric field of the clay and a second water layer may be attracted to the first and so on. The attraction is weaker for this second layer and is progressively weaker for all successive layers. Attraction is also present between the water molecules and the fractured surfaces of the quartz (silica). A linkage, such as quartz-water-clay-water-quartz (or clay), is set up throughout the moulding sand. To effect a proper and uniform distribution of sand and to develop this bond of sand, clay and water, mixing or mulling is necessary in suitable mix-mullers. The pH of the sand is also important since under acid conditions (low pH) flocculation or belling together of the clay occurs (explained by hydrogen ions being bonded to the clay particles and attracting them together). For optimum distribution, therefore, slightly basic conditions are desirable.

d. Hydraulic bond. Although water may participate in this type of bonding as water of crystallization, the actual binding together occurs due to some reaction which crystallizes a network of a new material. The decomposition of ethyl silicate to produce silica and ethyl alcohol is an example (fig. 5.19).

In preparing an investment mould, a slurry consisting of activated ethyl silicate and silica is poured round an expendable pattern (usually wax). A network of silica is gelled round the silica (sand) grains by hydrolysis, providing the bond. Surplus water is eliminated from the gel by heating, which strengthens the bond further.

In the carbon dioxide (sand) process, a similar reaction takes place. The mould or core is made with a sand mixed with hydrated sodium silicate. Carbon dioxide (CO_2)

gas is then passed through this assembly which, by forming a weak acid (H_2CO_3), hydrolyses the silicate to form silica gel, producing the bond. The sodium salts unfortunately remain in the sand, lowering its refractory properties; a special mould or core wash is generally used to obviate this difficulty.

Another hydraulic bond of commercial importance in the foundry industry is from the hydration of plaster of paris to form gypsum. If gypsum ($CaSO_4 \cdot 2H_2O$), a naturally

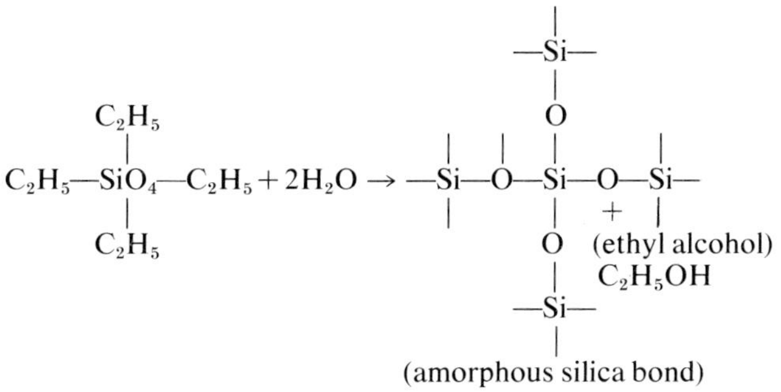

$$\begin{array}{c} C_2H_5 \\ | \\ C_2H_5-SiO_4-C_2H_5 + 2H_2O \\ | \\ C_2H_5 \end{array} \rightarrow \begin{array}{c} -Si- \\ | \\ O \\ | \\ -Si-O-Si-O-Si- \\ | \quad + \quad | \\ O \quad (\text{ethyl alcohol}) \\ | \quad C_2H_5OH \\ -Si- \\ | \end{array}$$

(amorphous silica bond)

Fig. 5.19. Hydrolysis of ethyl silicate.

occurring mineral, is heated to a temperature of 88 to 93°C it is converted to plaster of paris ($CaSO_4 \cdot \frac{1}{2}H_2O$). When the plaster of paris is again mixed with water, growth of a network of new gypsum crystals occurs, forming a compact mass.

Complex phosphide bonds are also used in investment moulds. One example is the bond structure brought about by the interaction of monobasic sodium and ammonium phosphate and magnesia in water (low pH, i.e., acid).

The formation of a gel or a new structure is accompanied in most cases by an expansion in volume which helps to reproduce the pattern details accurately (by pressing against the pattern). This is an important consideration in choosing moulding materials.

e. *Miscellaneous silicates.* Although sands and clays may be said to constitute the most important foundry materials (at least in terms of bulk consumption), a large number of other silicates are used for special purposes, zircon ($ZrSiO_4$) and olivine (Hg_2SiO_4) being the most common. Zircon (due to its radius ratio of 0.67) forms a very tightly bonded, high melting point refractory and is used in moulding for castings made from metals with higher melting temperatures.

f. *Organic bonds.* In a large number of applications, such as shell moulding and core-making, organic bonds are used in addition to or in place of the types already described. This is achieved by polymerization or by cross-linking. Condensation polymerization, where two different types of molecules react to give the polymer and a condensation product such as water or alcohol, is well established in the cast metals industry in the phenol formaldehyde or urea formaldehyde reactions in shell sands.

Cross-linking, i.e., the joining of two molecules at double-bond points by an atom, such as an oxygen atom in air-setting linseed oil, has also become quite common.

5.4.3 Green sand moulds

Green sand moulding is undoubtedly the most popular and widely used process known to the foundry industry, the majority of small miscellaneous castings being made in these moulds. As moulds become larger the problem of attaining the required mould properties becomes progressively more difficult, and other methods of making moulds are generally adopted. Although castings weighing up to 3 to 4 tons have been made in green sand moulds, the process is really more suitable for mass production of smaller castings.

a. General properties and ingredients of moulding sands. For green sand moulding the refractory aggregate consists of sand grains, clay, water and other additives, mixed and milled together to produce a plastic mixture. The moulds are known as green sand moulds because of the moisture content (which is usually up to about 8%).

The testing procedures adopted for the quantitative evaluation of the properties of moulding sands are set out in the *Foundry Sand Handbook*, published by the American Foundrymen's Society. However, some of the more important requirements are summarized below.

1. Green strength. This is a property of the sand which determines its suitability for the construction and subsequent handling of the mould, and refers to green sand, i.e., with the water added.
2. Dry strength. When a mould is filled with molten metal, the sand adjacent to the hot metal loses its water very quickly (as steam). The dried sand must retain sufficient strength to resist the metallostatic pressure and should not become too friable and thus liable to erosion.
3. Hot strength. Water may be expelled from the sand as steam at temperatures above 100°C but the actual metal temperature may be much higher, e.g., 1 600°C for cast steel, so that the dry strength and also the strength at the actual temperatures involved need to be known. Pressure of the liquid metal against the mould walls may cause dilation of the mould (enlargement), or cracks and breakages may occur, unless the sand possesses adequate hot strength.
4. Permeability. Heat from the metal causes evolution of a great deal of steam and other gases. The mould must be permeable (porous) to allow the gases to escape or a defective casting will result due to gas holes in the metal. Permeability also helps to prevent air-locking in the mould cavity when the metal is run-in.
5. Thermal stability. When casting the metal, a rapid expansion of the sand surface at the metal/mould interface takes place. The mould surface may then crack, buckle or flake off unless the moulding sand is relatively stable (dimensionally) under conditions of rapid heating.
6. Mouldability. For compaction of the mould the individual sand grains must be able to slide past each other and thus pack as closely as possible during the moulding operation. Suitable mouldability is therefore essential.
7. Refractoriness. The ability to withstand high temperatures without melting or softening is particularly important for sands used in moulds for casting steels and other alloys at temperatures of 1 200 to 1 700°C.
8. Collapsibility. After solidification, when the casting has cooled down sufficiently, it

is necessary to remove it from the mould. Sand which becomes very hard on heating is difficult to remove and may also cause the contracting metal to tear or crack due to mould constraint. The ability to give or collapse under pressure at a certain stage during the process is therefore important. Satisfactory collapsibility does not necessarily imply unsatisfactory dry or hot strength.

9. Fineness or grading. This may include average grain size, grain size distribution and grain shape. Apart from its effect on the strength, permeability and other properties of the sand, grading is most important from the point of view of metal penetration and surface finish of the castings.

10. Reusability. Most sands are used over and over again, more or less in a continuous cycle from mould making to casting to removal of castings and then to reconditioning and back to moulding again. Not all sands can be so used and the rate of deterioration for different sands may be different.

Most of these requirements apply to other refractory aggregates as well.

Moulding sands are mixtures of three or more ingredients. A green sand contains clay and water as well as the principal constituent, silica (SiO_2). These provide the basic requirements for moulding. Other materials may also be added to the mixture to improve or develop some of the properties mentioned.

Granular particles of sand, mainly silica, make up from 50 to 95% of the total material in a moulding sand. The actual sand particles may differ from sand to sand depending on source and origin. The main differences are in: average grain size, grain size distribution and grain shape; chemical composition; and refractoriness and thermal stability.

In general, sand grains which consist of pure silica, (SiO_2 99.8% +), are considered to be the most refractory and thermally stable. The presence of excessive amounts of iron oxide, alkali oxides and lime can lower the fusion point of the grains to undesirable levels. The shape of sand grains may be rounded, angular or subangular, depending on the geological history. In moulding sands used in foundries, the sand grains tend to be of mixed origin and are often agglomerates of angular or subangular grains. This is a natural consequence of mixing and repeated exposure to moisture and high temperatures.

Moulding sands also contain from about 2 to 50% of clay. With a suitable water content, it is the main source of the plasticity and strength of the sand. In other words, clay is the binder for moulding sands, the bond being developed by the mechanism already described. In some mineral deposits, clay and sand occur mixed in suitable proportions so that the sand can be used directly for moulding with very little adjustment. Such sands are known as 'natural moulding sands'. Sands which require additions of clay to give them bonding properties are known as 'synthetic moulding sands'. The types of clay minerals used as bonding additions to sands have been described earlier (see Section 5.4.26).

Water is the third principal constituent of moulding sands and is present in amounts of about 1.5 to 8%. It is sometimes referred to as 'tempering water'. The water is adsorbed by the clay up to a limiting amount (for the particular clay), and excess water can lower the strength properties. Additional water, however, acts as a lubricant and makes the sand more mouldable at the expense of strength of the bond.

Besides the three principal ingredients, other materials may be added to moulding sands to develop specific properties. The most important of these 'additives' are listed below.

1. Cereals – finely ground corn flour or gelatinized and ground starch. They are used for increased green or dry strength or for improving collapsibility. Up to 2% may be added for these purposes.
2. Ground pitch. A by-product from the carbonization of coal, pitch is generally used in amounts up to 3% to improve hot strength of the mould or to improve surface finish of the casting. Asphalt, a by-product of the distillation of petroleum, is also used for the same purposes.
3. Coal. Soft coal is used mainly in the cast iron foundries to improve surface finish of the castings and to make cleaning easier. The amount added may vary from 2 to 8%.
4. Graphite. Both synthetic and natural varieties may be used in amounts from 0.2 to 2.0% to improve the mouldability of the sand or to improve the surface finish of the castings.
5. Fuel oil. This is sometimes used in small quantities – not exceeding about 0.2% – for improving the mouldability of the sand.
6. Cellulose. Ground wood flour, sawdust or other cellulose substances such as cob flour, cereal husks and carbonized cellulose may be added in amounts up to 2% to improve collapsibility and mouldability of the sand. By burning out at higher temperatures, they may help to control the expansion of the sand and also stop it becoming too hard on exposure to higher temperatures, thus preventing tearing or cracking in the casting as the metal contracts.
7. Silica flour. Powdered silica, usually finer than 200 mesh, is used in varying amounts up to about 30% to improve the hot strength of the sand. It may also be used to increase the density of less dense sands (i.e., by decreasing the porosity) to enable it to resist metal penetration.
8. Perlite. This is an aluminium silicate mineral with special heat-insulating properties and is sometimes used in amounts up to 2%. It is claimed that it gives a better thermal stability to the sand.
9. Molasses and dextrine. Both dextrine and unrefined molasses may be used in moulding sand mixtures in amounts up to 2 to 3%. The main purpose is to improve the dry strength of the mould and prevent tendencies towards friability of the mould surface. It is also used in slurries applied to mould surfaces, i.e., in mould coatings or mould paints.

When moulding sands are in continuous use, fresh additions are necessary during each sand-preparation cycle to replenish the material which suffers deterioration by burning or from the heat of the metal. Sand in a foundry system must inevitably consist of burned or partially burned substances and particles of metal and/or slag, in addition to the fresh additions of clay, water, sand and other additives discussed. Without going into the detailed technology of sand control, reconditioning and reclamation, the flow diagram in fig. 5.20 will give a general idea of the usual cycle for moulding sands in a production foundry.

Green sand moulding is carried out by compacting the prepared plastic sand mixture round the pattern. As the sand is rammed or packed, it develops strength and becomes rigid within the box or flask. When relatively few castings are required, compaction may be carried out by manual ramming using loose patterns. Both the cope and the drag parts are moulded the same way except that the cope must provide for the sprue and the running and feeding systems. With the cope and drag parts of the mould made and the pattern withdrawn, cores are set into the mould cavity to form the internal surfaces of the casting. The cope and drag parts are then closed and the cope

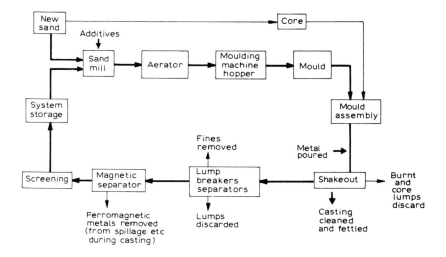

Fig. 5.20 Flow diagram for the sand system in a mechanised foundry.

either clamped or weighted down to prevent it from floating when the molten metal is poured into the cavity.

For mass production or when a large number of castings are to be made, green sand moulding is carried out with match-plate or cope and drag patterns, the compaction of the sand being achieved by moulding machines which may be partially or fully automated. The cope and drag may be closed manually or with the help of machines.

Green sand moulding is normally the least costly method of making moulds and is probably the most flexible production process. Mechanical equipment may be used in all stages of production and mass production units may be easily automated. The moulding sand itself can be reused many times by reconditioning it with water, clay and other additives. The moulding process in mechanized plants is rapid and repetitive.

The limitations in the use of green sand moulding are:
1. In casting design. Thin long projections of green sand in a mould cavity are likely to collapse and wash away with the molten metal.
2. Certain casting alloys must be cast in moisture-free moulds to prevent undesirable reactions resulting in defective castings.
3. In dimensional accuracy. A dimensional variation of $\pm 1/64$ inch on small castings and up to $\pm 3/32$ inch on larger ones may be encountered, which may preclude the use of this process for some applications.
4. Large castings require high mould strength and resistance to erosion (by the metal), which is not easily achieved in green sand moulds.

5.4.4 Compaction of green sand moulds

In green sand moulds the bonding or holding together of the moulding aggregate is achieved by the clay-water-silica bond; special additives are also used to impart desirable properties such as easy breakdown of the compact after casting (i.e., collapsibility) and other specific properties.

In order to develop the bond, however, it is necessary to pack the moulding sand

as close as possible round the pattern in the box or flask so that a hard rigid mould results, which is resistant to dilation and erosion by the molten metal. This is generally referred to as compaction. Except in the case of hand-moulding with loose patterns compaction is carried out by moulding machines. The most important (from a very wide range of designs) of these machines are:
1. Squeeze machines.
2. Jolt machines.
3. Jolt-squeeze machines.
4. Slingers.
5. Blowers.
6. A combination of numbers 1 to 5.

The bulk density of a freshly mixed moulding sand mixture may be as low as 50 lb/ft.^{3} The object of using these machines is to compact the sand round the pattern in the box or flask so that the bulk density of the compacted mass reaches a value of between 90 and 115 lb/ft.^{3} or more.

Even when compacted, the solid particles only occupy about 60 to 65% of the total volume. For a granular material the limiting bulk density is reached when compaction has occurred to the point of grain to grain compact throughout the mass. For typical foundry sands the limiting value of the bulk density is 100 to 115 lb/ft.^{3} Once this limiting value has been reached application of higher mechanical forces will not cause further compaction. Most moulding machines are designed to produce a compaction which will give a bulk density equivalent to at least 90% of the limiting value.

a. Squeeze machines. Squeeze moulding machines, as the name suggests, use pressure as a means of compacting the sand. The maximum squeezing force of a pneumatically operated machine is limited by its piston diameter and the air pressure available (usually 90 to 110 p.s.i.), since

$$\text{Squeezing force} = P \times \frac{\pi d^2}{4} - W$$

where P is the air pressure in the squeeze cylinder in p.s.i., d is the piston diameter in inches and W the weight of the pattern, flask, sand and any other items on the worktable of the machine.

Although squeezing force is constant for a particular machine (and air pressure), the moulding box or flask may be of a different size for different castings, so that the squeeze pressure may be quite different and will depend not only on the total squeezing force but also on the surface area of the moulding box subjected to this force, i.e.,

$$\text{Squeeze pressure} = \frac{\text{Squeezing force}}{\text{Surface area of sand in the box in square inches}}.$$

Since squeezing pressures of 100 to 150 p.s.i. are required to approach the limiting value of the bulk density, the size of the moulding box also becomes a limiting factor for a particular machine.

It must be realized, however, that the pressure applied to the surface of the sand is not uniformly transmitted or distributed throughout the body of the mould (since sand aggregates do not behave as perfect fluids). To overcome this difficulty and to ensure satisfactory compacting in areas further away from the surface, contour squeezing or diaphragm squeezing is used (fig. 5.21). More recently, multiple squeeze head machines have been designed to obtain the necessary uniformity of compaction throughout the body of the mould. In general, however, squeeze-type machines will tend to produce greater compaction near the squeeze surface and progressively less at greater depths, and so less compaction is possible in the mould near the pattern surface. There is, therefore, a limit to the depth of the moulding box or flask that can be successfully used in combination with squeeze-type machines.

 b. Jolting machines. In these machines, the work-table with pattern, flask and sand is raised by a pneumatically operated piston and allowed to fall against the base of the machine under the influence of gravity. Packing of the sand is accomplished by the work done by the kinetic energy of the falling sand. Compaction results from the momentum of the falling sand being converted to work (in ft lbf/s) when the momentum is instantaneously arrested by the jolt table.

$$\text{Power of jolting} = \frac{Mv}{A} \text{ ft lbf/s ft}^2$$

where M is the weight of sand in lb, v is the velocity at the instant of jolt $\{=(2gd)^{1/2}\}$ in ft/s, A the jolt area in ft^2 and d the jolt stroke in ft. Power for moulding is therefore largely determined by the jolt stroke, which is a characteristic of the machine. The number of jolts is, of course, also a factor. It is found that the limit (compaction) is usually reached with about 20 to 30 jolts.

In this type of compaction, the maximum effect is at the pattern surface (see fig. 5.22) and it is progressively less further away from the surface. It may be necessary to finish off the top or back of the mould by squeezing or ramming. Only the first 1 to

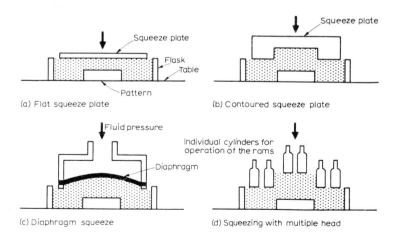

Fig. 5.21 Four methods of squeezing sand for compaction.

$2\frac{1}{2}$ inches of the sand above the pattern is well compacted. The limiting capacity of a jolting machine is the total weight which it is able to lift and let fall, defined as:

$$W = \frac{\pi d^2}{4} \times P$$

where W is the total weight in lb which can be lifted, i.e., flask + pattern + jolt table + sand, d is the diameter of the jolt cylinder in inches and P the air line pressure in p.s.i.

c. Jolt-squeeze machines. These machines use a combination of jolting and squeezing for compaction. Satisfactory compaction by jolting alone is limited to the sand volume up to 2 inches above the pattern surface, whereas squeezing alone tends to produce insufficient compaction near the pattern surface in deeper moulds. It is therefore easier to obtain desirable uniformity of compaction by using both methods either in sequence or simultaneously.

d. Slingers. Sand slingers are used to compact moulds by discharging the sand at high velocity onto the pattern surface inside the box. Sand is conveyed by a belt into the slinger head which contains a rotor. This rotates at 1 200 to 1 800 rev/min and the rotor blades pick up the sand as it falls into the head and throw it against the mould. The sand, which can achieve velocities of about 10 000 ft/min, is stopped instantaneously by the pattern plate, thus compacting the mass. The main disadvantage of this method is the wear caused on the metal pattern, due to the blasting effect of the sand, and the comparatively slow rate of production. However, slingers are almost universally used for making moulds for larger castings. Many types of slingers are available, including mobile slingers which can move up and down the foundry floor and compact the mould *in situ*, thus cutting down on the handling of large mould boxes. A stationary slinger is shown in fig. 5.23.

e. Mould blowers. Mould blowers are machines which blow the sand into the mould box pneumatically. If the green shear strength of the sand exceeds about 1 p.s.i. it becomes increasingly more difficult to blow and for this reason blowers are more commonly used for making cores, core sands being easier to handle in this respect. However, some of the disadvantages have been overcome by introducing special additives to the sand mixture and this method of moulding is becoming increasingly

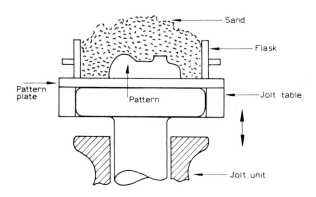

Fig. 5.22 Diagramatic arrangement for jolt moulding.

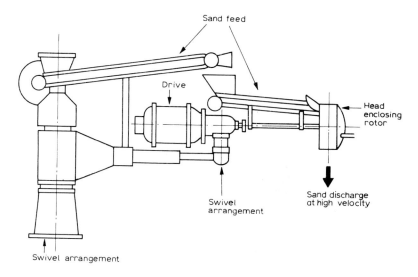

Fig. 5.23 Stationary sand slinger.

popular. The bulk density of the compact as blown is usually of the order of 80 to 85 lb/ft^3. The additional compaction necessary is usually done by squeezing the mould against the blower head. Blow squeeze moulding is considered to be the fastest method of moulding but it is generally restricted to moulds which are comparatively shallow, i.e., up to about 5 to 6 inches in depth. Mould blowing requires large volumes of air at 100 p.s.i. (about 60 p.s.i. in the magazine). A variation of the process known as core shooting requires a smaller volume of air since it depends on extrusion of the sand into the mould or core box by the instantaneous pressure generated in the magazine and not on the continued supply of large volumes of air at 100 p.s.i. Special additives are usually required when using these machines for green sand moulding.

5.4.5 High-pressure moulding

This is a variation of green sand moulding developed with the object of improving the dimensional accuracy and surface finish of the castings. The sand consists of a mixture of sand grains, clay and water, often with additions of organic and inorganic binders. The sand mixture needs close control to obtain the desired results. The procedure is similar to green sand moulding except that improved compaction is obtained by using machines capable of producing high pressures for squeezing; the pressures used may vary from 150 to 400 p.s.i. Distortion of the pattern and tackle may take place at such high pressures and so carefully designed metal pattern equipment is used. As with green sand moulding, the mould does not require any further treatment, such as baking, and the sand is reusable. The main advantage claimed is improved dimensional tolerance of the castings produced, due to a more rigid mould and less mould dilation from metal pressure.

5.4.6 Dry sand moulds

Dry sand moulds are initially made with the refractory aggregate (similar to green sand) in the green or moist condition. The sand mixture is modified by the addition of binders, such as dextrine or other cereal powders, to develop good strength and other properties when the mould has been dried. For small castings, dry sand moulding is carried out in the same way as green sand moulding. After making the mould, the mould-cavity surface is usually coated or sprayed with a mixture to improve the hardness or refractoriness of the surface. The entire mould is then dried either in an oven at temperatures of from 150 to 300°C, or by circulating heated air through the mould. The additional handling and the time-consuming drying operation are the main disadvantages of this process.

Dry sand moulds are used only when green sand moulds are found to be unsuitable for reasons of inadequate strength or insufficient resistance to erosion by the metal (see Section 5.4.3). They are also used for casting certain alloys which have a tendency to react with the moisture in green sand moulds. Copper and magnesium base alloys are susceptible to these reactions and some steels also fall within this category.

The properties of a dry sand mould may be partially obtained by drying the mould surface to a depth of only $\frac{1}{4}$ inch to 1 inch. This may be effected by means of flame-throwing torches or by radiant heating units directed to the mould-cavity surface. These moulds are known as skin-dried moulds. The main disadvantage is that they must be poured shortly after drying so that the moisture remaining in the body of the mould will not have time to penetrate into the dried skin.

Floor and pit moulding. The production of large intricate castings, generally weighing from a few tons to over 100 tons, requires special technological effort and control in construction and handling of the moulds. Ordinary machines and mould boxes are obviously not suitable for such purposes. When the castings are not exceptionally large, they are usually moulded on the floor in large mould boxes using sand slingers. The moulds are cast and sometimes even dried *in situ*. When the casting is too large to be moulded in a box or flask, the moulding is done in pits. Moulding pits are concrete-lined box-shaped holes in the moulding floor. The pattern is lowered into the pit and the moulding material is tucked and compacted under the pattern and up the side walls to the parting surface. This then forms the drag part of the mould. The cope is usually made in a box and closed over the drag after the cores have been set.

Due to the special requirements of strength and rigidity, both floor and pit moulds are invariably dried, i.e., dry sand moulding is used. Many different types of refractory mixtures or aggregates are used for making these large moulds; the most important described below.

a. Cement-bonded moulds. The moulding material is a mixture of sand, 8 to 12% hydraulic cement with high early strength and 4 to 6% water. This sand develops great hardness and strength by the setting action of the cement. The sand must be allowed to set or harden before the pattern can be withdrawn. The mould is then allowed to cure, or continue setting, for about 72 hours before the mould can be closed (i.e., cope and drag assembled for pouring the metal). When the metal is poured, steam is evolved due to the escape of the water of crystallization of the cement, and therefore special care is necessary to ensure that the mould is sufficiently porous to allow the steam to pass through the mould. The main advantage of these moulds is the improved dimensional accuracy possible in the larger castings.

b. Compo and chamotte Moulds. A mixture often known as 'Sheffield Compo' is probably one of the oldest synthetic moulding mixtures to have been used in England.

The original mixture consisted of crucible clay pots, used firebricks and gannister, crushed and bonded with fireclay and carbonaceous material such as blacking or graphite. Later, high-alumina firebricks (scrap) were used after crushing and a crushed and calcined bauxite clay was added to the main firebrick base. The bauxite clay occurred naturally in Ayrshire and was similar to that found in Europe – a calcined refractory clay-grog called 'chamotte'. Chamotte moulding has been popular in Germany and Czechoslovakia for producing large steel castings.

The composition of present day compo is:

Crushed scrap firebricks	65–70%
Crushed chamotte (calcined)	15–22%
Rough-ground Derby fireclay	12–16%
Blacking or graphite	1%
Water	9–11%

The equipment required to produce compo in the foundry consists of:
1. Pulverizer or mill for crushing the bricks and the chamotte.
2. A screening unit to grade the crushed material.
3. A heavy duty muller for mixing the ingredients.
4. A ball-mill for the production of fine powder (for a chamotte mould coating or paint).

Pit moulding techniques are usually employed but only about an 8 to 10 inch layer next to the pattern consists of compo. The rest of the mould body up to the side walls of the pit is filled with tightly compacted backing sand. First the backing sand is rammed up in the pit to the approximate shape of the mould cavity but larger all around by about 10 inches. A layer of compo about 4 inches thick is then rammed into this sand cavity. The pattern is positioned and the remaining 4 inches or so of space is filled with compo and rammed hard (fig. 5.24). After removing the pattern, the mould cavity is painted with a fine slurry of chamotte paint. After allowing the excess moisture to be removed by air drying for 3 to 4 hours, the mould is dried with portable high-temperature gas driers. The drying cycle consists of a slow initial phase at 200°C followed by a final stage of 500°C. The high temperature of drying removes chemically combined water from the mould-cavity surface. Vertical faces of the mould are sometimes propped or tied up with tie-rods before drying to prevent possible mould collapse.

The cope is made in a similar way except that bolted box parts are used to hold

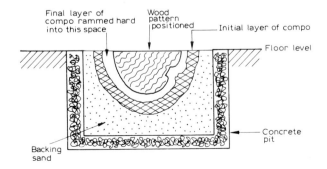

Fig. 5.24 Section through a compo moulding pit.

the mould together. The whole assembly is eventually closed over the drag in the pit.

The main advantage of compo moulds is their relative stability under the varying temperatures encountered during drying and casting. Silica (sands), on the other hand, show a wide variation in thermal expansion during drying and casting with the result that cracking and mould collapse may occur during the pouring of large castings where the effect of heat from the metal is particularly severe.

c. Loam moulding. When a large mould of regular and simple shape can be made in multiple piece boxes or by bricking up a large portion of the mould, loam is used as the moulding material. Loam is a wet plastic mixture of about 50% sand and 50% clay, together with additives such as sawdust, straw or other cellulose materials. The rough contours of the required mould cavity are usually constructed of brickwork ether in a pit or a box (drag part) and the loam is then trowelled or plastered on to the brickwork surface and brought to the pattern dimensions by using skeleton patterns, sweeps or templates (fig. 5.25). The cope part is constructed in a similar way in a box or flask. Both parts of the mould are thoroughly baked to drive out moisture. A refractory wash is usually applied to the warm mould for better surface finish and to prevent erosion.

This process has the advantage of being economical, particularly for one-off jobs, but is generally restricted to large or heavy castings (made in iron) of a circular or similar shape, such as large cylinders and machinery rolls. Dimensional accuracy of the castings produced is not good and usually large machining allowances are required.

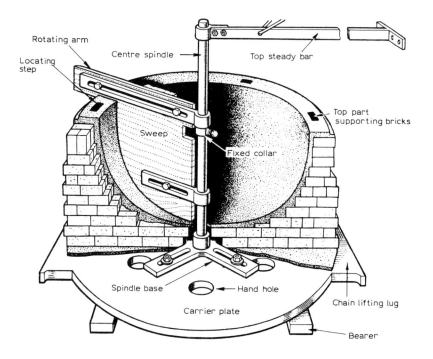

Fig. 5.25 Loam moulding.

The cores to form the internal surfaces of the casting are also made in a similar way by plastering loam on brickwork.

5.4.7 Self-curing moulds

Large castings, particularly those made from high melting point alloys such as steel, require the special moulding methods and materials already described under dry sand moulding. The main drawback of moulds such as compo moulds and cement moulds is that they take considerable time to construct and present special problems in handling and drying. These moulding materials are also difficult to ram or compact round the pattern, due to their poor flowability. In order to increase productivity for large castings, three conditions must be satisfied.
1. Handling must be minimized.
2. Flowability of the refractory mixture for moulding must be improved.
3. Drying or stoving must be avoided.

The self-curing binder process meets these requirements.

The lubricative effect of the oil, which permits easier sliding of one sand grain past another during the compaction of the mould, is a further advantage. The method is also known as the cold-setting, air-setting or air-curing process.

The moulding aggregate consists of dry sand, a drying oil and a catalyst to promote oil gelation (by oxidation of the oil). Self-curing drying oils usually contain metallic driers which take part in the catalytic action between the accelerators (oxygen donors) and the oil itself. This results in oxidation of the oil to a gel, thus providing a strong bond for the sand grains. A typical sand mixture may consist of 1 000 lb dry sand, 20 lb self-curing oil and from $\frac{1}{2}$ to 2 lb accelerator.

Unfortunately, control is difficult since the process is dependent upon a chemical reaction and the kinetics of the process are significantly affected by such factors as the ambient temperature and the temperature of the sand itself, which has a high capacity for heat retention. A further problem is that the oxidation reaction is exothermic during the development of air-set strength. Even the heat generated during the mixing of the ingredients may have significant effects. Premature gelling of the oil round the sand grains will prevent development of a bond. Consistent strength is therefore difficult to attain – especially hot strength, which is particularly important in larger castings. The process is still in its infancy and a great deal of systematic research is necessary to establish it on a firm footing.

5.4.8 The full-mould process

This may involve any of the usual sand practices except that the pattern is constructed from expanded polystyrene bead boards and is not withdrawn from the mould cavity. The pattern is an exact replica (internal and external surfaces) of the casting to be produced, with the necessary contraction allowances. It is positioned in a suitable mould box and sand packed round it. The polystyrene pattern also incorporates the running and feeding systems. When the hot metal is poured through the pouring cup it melts and volatilizes the polystyrene and thus replaces the polystyrene in the mould. The volatile products escape through the permeable sand mould which is well vented for this purpose. The casting is removed from the mould as usual and then cleaned and fettled. A separate pattern is required for each casting. For quantity production, patterns must be mass produced, which can be done by injecting expanded polystyrene into metallic dies. The main advantages are that the mould can be made in

5.4.9 The fluid-sand process

In this process a slurry is made from sand grains, a detergent, a binder such as sodium silicate and setting agent such as dicalcium silicate. The mould is made, by free-flow of the slurry around the pattern in the flask or box, allowed to harden and then the pattern is withdrawn. The cope and drag parts are assembled and the metal poured in the usual manner. After mixing, the foaming agent (i.e., the detergent) produces the free-flowing characteristics in the sand. The main advantage of the process is that moulding is carried out without the aid of any machines, the sand being allowed to discharge into the box straight from the mixer. Mass production of ingot moulds and other large castings by this method is reported from Russia. At this stage of development it is debatable whether the moulds produced are sufficiently dense and rigid to withstand metallostatic pressure to the extent necessary for close dimensional tolerances. However, considerable development work is in progress.

5.4.10 The carbon dioxide process

The principle of the carbon dioxide (CO_2) process was described in a British Patent of 1898 taken out in the names of two Lancashire chemists, Hargreaves and Poulson. Apparently, it was too radical for the foundrymen of the time and no trials were carried out until after world war II.

The process, also known as the sodium silicate process, uses a mixture of sand and between 2 to 6% of liquid sodium silicate (waterglass). The sand mixture is compacted around the pattern in the usual way but before removing the pattern carbon dioxide gas is passed through the body of the mould. This hydrolyses the silicate to form a silica gel and the mould sets hard. The pattern can then be removed and the mould processed in the usual way, i.e., the cores are set and the mould closed before pouring.

The main advantage of the process is that there is no necessity to bake or dry the mould if a suitable sand mixture is employed. Other advantages resulting from the elimination of drying or baking are a quick turnover, decreased capital cost (stoves not required) and a much healthier atmosphere in the foundry due to a considerable reduction in the volume of fumes generated.

From the technological point of view, CO_2 moulds and cores have lower volatile content than oil-sand or resin-sand mixtures and consequently there is less chance of defects due to entrapped gases in the casting. With higher temperatures at the mould/metal interface CO_2 moulds have higher plasticity so that they yield under expensive and compressive stresses, thus minimizing defects arising out of buckling or cracking of the mould surface. For the same reason fusion of the mould and metal is also more common, particularly for higher temperature alloys like steel, and so CO_2 moulds are invariably coated with an alcohol base refractory mould paint.

The collapsibility of CO_2 moulds can be improved by one or more of the following additives: asphalt emulsion, cellulose fibres, cereal binder, graphite and coal powder. One or more of gilsonite compound, silica flour, iron oxide, graphite and coal powder, may be added to improve surface finish.

5.4.11 The D process

The D process employs an accurately machined metal pattern mounted with gates on a metal blow plate, which in turn is bolted to the blow chamber of a mould or core blowing machine. A hot or warm metal plate (drier) which is contoured around the pattern surface leaving a small space (usually $\frac{3}{8}$ inch) is placed under the pattern (fig. 5.26). The two are clamped together and the sand is blown through holes in the pattern plate, thus filling the space between the pattern and the drier. The pattern is cold and the drier hot. The drier with the blown shell is placed in an oven and baked. After baking, two such moulds, i.e., the cope and the drag, are clamped together to form a rigid mould. The mould is then ready for pouring.

Several sand mixtures have been used for the D process, among them:
1. Fine sand with D process oil.
2. Fine sand and D process oil with cereal and water.
3. Fine sand (precoated) plus resin and liquid.

Two processes for precoating the sand are in use, a hot and a cold. The cold process employs a liquid resin and a catalyst which are added to the sand at room temperature in a mixer. As mixing proceeds the resin envelops the individual sand grains.

In the hot process the sand is heated to a temperature of about 350°C and a dry powdered resin is added. The mixture is mulled and, as the resin softens due to the heat, the sand grains are coated and a liquid catalyst is added. The sand is immediately discharged into a cool sand mixer and thoroughly mixed to break up any lumps.

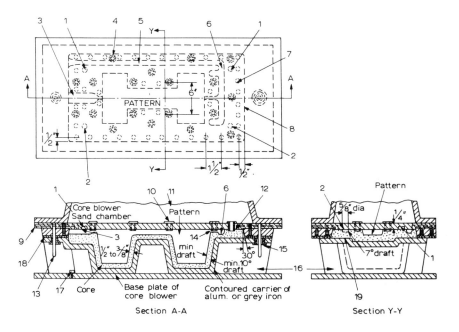

Fig. 5.26 D process (1) male locator, (2) female locator, (3) header, (4) blowing holes 9/16 in. i.d. (5) sprue, (6) runner, (7) vent holes $\frac{1}{2}$ in diameter. (8) outer edge of core, (9) pattern and blowing plate, (10) blowing holes, (11) sand, (12) vent holes s.s. cloth 150 mesh. (13) pine, (14) ingate, (15) elongated bushing, (16) rib, (17) carrier locator, (18) machine or disco-grind carrier, (19) sand-cast file finish.

380 | METAL CASTING

This method has not found much favour in mould making due to the many controls necessary and the high cost of the sand mixture. The main problem is to ensure that the sand possesses blowability and minimum stickiness. This process has been used mainly in the USA for making cores rather than moulds.

5.4.12 Shell moulding

The essential feature of shell moulding is that the mould is in the form of a thin shell round the mould cavity which is sufficiently strong and rigid to withstand and contain the pressure from the liquid metal after pouring.

A fine sand (i.e., small grains) is used together with about 3 to 10% of synthetic resin binder. Resins generally used are the phenolformaldehydes, urea formaldehydes, alkyds and polyesters. The resin must be a thermosetting plastic and is used as a powder in dry mixtures. More often, however, the sand grains are precoated with the resin by mixing either by the cold or hot processes. The mixture must be dry and free-flowing for moulding.

The steps in shell moulding are best illustrated with the help of fig. 5.27. The sand mixture is dropped or dumped on a metal pattern which is heated to a temperature of

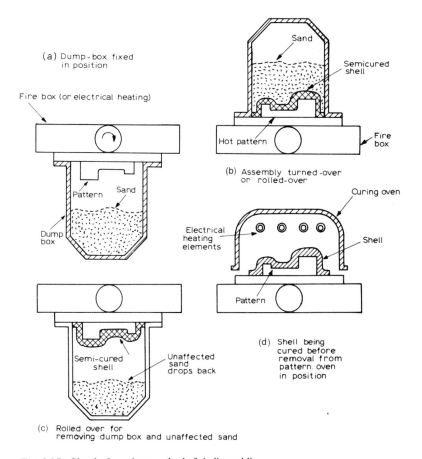

Fig. 5.27 Simple dump box method of shell moulding.

170 to 350°C. The plastic partially thermosets and builds up a coherent sand shell next to the pattern. The thickness of this shell is $\frac{1}{4}$ to $\frac{3}{4}$ inch and depends on the pattern temperature, the time elapsed and the nature of the resin and the sand mixture, i.e., on the rate of heat transfer from the pattern through the sand and the temperature required to make the resin plastic and tacky.

To restrict the thickness of the shell the whole assembly may now be inverted so that the sand not affected by the heat (usually all sand at a distance of over 1 inch from the pattern face) will drop back into the dump-box, leaving the shell next to the pattern. The shell remaining on the pattern is then 'cured' by heating it for 1 to 3 min to a temperature of between 230 to 350°C, again depending on the type of resin mixture employed. The shell is very strong and grips the pattern lightly. A mould-release agent or parting agent is used (usually silicone solutions) to obtain clean stripping when the ejector pins push the shell off the pattern. The cope and drag parts of the shell are then assembled by pressing together while still hot (or by using adhesives) and the shell mould is ready for pouring.

Advantages claimed for the process, such as exceptionally good surface finish of the castings, dimensional accuracy (reduction or elimination of machining) and reduced cleaning cost, appear to be largely substantiated by the rapid growth in shell moulding over the last decade. Although not comparable with small investment castings, dimensional tolerances have been held to 0.002 inch in some cases. The precision appears to be dependent on the shape and size of the mould and the type of alloy being cast.

In general the cost of the pattern equipment and the cost of the actual sand (which cannot be reused without expensive treatment) would tend to discourage the growth of shell moulding. However, when this is offset against the reduction of machining and/or finishing costs the economic advantages become apparent. The process is therefore primarily used for smaller castings where machining can be drastically reduced or eliminated. Completely automatic shell moulding machines are now available with rated outputs of up to 100 shells per hour. The collapsible nature of the organic bond in shell moulds limits their use for heavier castings.

A variation of the process, known as the hot-box process, employs a heated core box. The moulding mixture contains 1.5 to 4.0% resin of the furane or furfuraldehyde type. Heat from the core box causes the catalysts to start an exothermic polymerization process, and the mass hardens with rising temperature.

Shell moulding is not restricted to resin-sands. Shell moulds made by the carbon dioxide process are also used.

The ceramic shell process will be described under investment castings. Recently, blow squeeze machines have been introduced for shell moulding (fig. 5.28). Better dimensional stability of the moulds is claimed.

5.5 INVESTMENT MOULDING AND CASTING PROCESS

5.5.1 *Principles*

Modern investment casting processes have been developed from the ancient 'lost-wax' process and have gained steadily in popularity and importance with the development of the aerospace industries and small-arms weaponry and rocketry. The main advantage of this process in the manufacture of small metal components

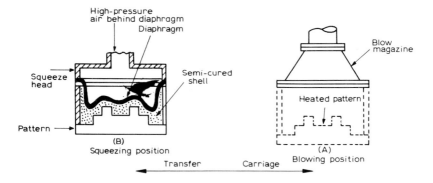

Fig. 5.28 Blow-squeeze shell moulding: Squeezing by a diaphram squeeze head.

is the dimensional accuracy attainable and the precise reproduction of finer details. This applies particularly to the more reactive and refractory metals and alloys which are difficult to machine or fabricate. For this reason where high precision is required investment castings are almost universally employed e.g., for special alloy turbine blades. Investment castings are often referred to as precision castings.

A number of processes have been used for making investment casting to suit different applications but the main features may be summarized as follows.

1. Disposable or expendable patterns are used, i.e., individual patterns must be made for each casting and gating system.
2. Moulding is carried out with a fluid aggregate or slurry which flows round the pattern and is allowed to set with a 'hydraulic' bond. The setting reaction often results in an increase in volume of the moulding aggregate, thus helping to reproduce fine details from the pattern by pressing against it.
3. Organic substances or binders are not used in the refractory aggregate.
4. The mould is heated to drive off all gases and the metal is poured with the mould preheated to a controlled temperature, which varies according to the alloy being cast, the minimum thickness of section of the casting and the type of refractory aggregate and binder employed.

The sequence of operations necessary to produce an investment casting is illustrated in the flow diagram in fig. 5.29.

A metal die is made for casting the wax patterns, with the appropriate allowance for shrinkage of both wax and metal casting (about 0.011 to 0.015 inch/inch). The pattern may include the gating system if the design will allow; otherwise separate wax patterns may have to be produced for it. The technique of producing the patterns is by wax injection into the metal dies. The waxes used are generally blended from beeswax, carnauba, ceresin, acra wax, paraffin and other resins. The wax is injected into the mould at temperatures between 65 and 90°C and at pressures between 100 to 500 p.s.i. Polystyrene is used if patterns require temperatures of between 140 and 280°C and pressures up to 12 000 p.s.i. Mercury may be used in place of wax patterns but requires deep freezing to prevent it from melting and running out during preparation of the mould. Separately prepared patterns and gating systems are assembled together by heating the surfaces to be joined (in the case of wax patterns). For polystyrene patterns the surfaces to be joined are moistened with carbon tetrachloride.

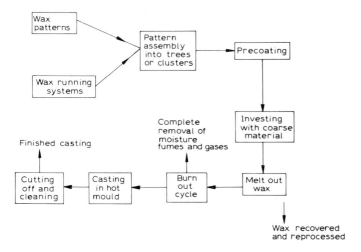

Fig. 5.29 Flow diagram for typical investment casting process.

The assembled pattern is now precoated by dipping into a slurry of a refractory coating material. A suitable slurry may consist of silica flour (300 mesh) suspended in ethyl silicate solution of controlled viscosity to produce a uniform coating after drying. Other suitable coating mixtures are listed in table 5.4. After application of the slurry the assembly is sprinkled with 40 to 50 mesh silica sand and allowed to dry. In some cases precoating is not used and the wax pattern is directly invested in the moulding material, in which case vacuum de-airing is necessary to prevent air bubbles from sticking to the pattern.

The next step is investment of the mould. The wax assembly is inverted on a table within a paper-lined steel flask and the moulding mixture is poured around the pattern. Settling of the material is helped by vibrating the table. The moulds are then allowed to air-set. Some well-established moulding mixtures are given in table 5.5.

After the mould has hardened the wax is melted out by heating it in an inverted position at temperatures between 95 and 150°C. The wax is allowed to run into receptacles and is usually reused. Moulds with polystyrene patterns in them are usually dried at about 70°C. The moulds are then heated at a rate of about 70°C per hour to temperatures of from 850 to 1050°C for ferrous alloys and 560°C for aluminium alloys. The preheat temperature must be controlled to suit the particular alloy being cast. It is also most important to ensure that all wax and other gas-forming materials are completely eliminated from the mould during this heating cycle.

When the mould is at the desired temperature metal is poured into the pouring cup. Air pressure is sometimes applied to the sprue to assist mould filling. The entire investment mould may also be placed in a centrifuge during pouring to assist filling of thin sections.

Cleaning operations are carried out after the casting assembly has cooled down sufficiently.

With the rapid growth in the investment casting industry, a very wide range of refractory materials and binders is being used at present–too many to be discussed individually. However, a summary of the basic principles underlying the choice of materials may serve as a guide.

Table 5.4 Some investment casting precoating mixtures

Type	Approximate composition		
Precoating	Tetraethyl silicate	60%	For alloys with
	Water	7%	high melting points
	Ethyl alcohol	33%	
	Hydrochloric acid	a few drops	
Precoating	325 mesh silica	94 parts	For alloys with high
	325 mesh alumina	56 parts	melting points
	40 mesh silica	37 parts	
	Sodium silicate	64 parts	
	2% polyvinyl alcohol	16 parts	
Plaster moulding	Plaster of paris	60%	For low melting point
	50 mesh silica	25%	non-ferrous alloys
	Talc	15%	
	Water to give creamy consistency		
Precoating	Silica	90%	For high melting
	Magnesia	5 to 6%	point alloys
	Monobasic ammonium phosphate	2 to 3%	
	Monobasic sodium phosphate	1%	
	Water or 10% HCl or HNO_3, for required consistency		

For precision casting of metal components, the investment moulding material must have the following characteristics:
1. Stability to withstand contact with metal at the pouring temperatures involved.
2. Ability to harden or set from a workable pouring consistency (fluid slurry) within a reasonable period of time. Practical considerations of processing preclude mixtures which set over a very long period.
3. Simplicity and stability during mixing and pouring of the slurry.
4. A minimum (and reproducible) dimensional change during the entire cycle, i.e., setting, curing and metal pouring.
5. Cost within practical economical limits.

In general, materials are classified as high-temperature or low-temperature investment material. In the high-temperature category are materials used for the casting of the various types of irons and steels and some other metals such as titanium and zirconium.

Table 5.5 Investment-moulding mixtures

Refractory	Binder		Water	Application
Sand	Alumina cement	5%	As required	Investment moulding
Sand	Ethyl silicate or sodium silicate	3%	—	Investment moulding
Sand	Calcium phosphate	6.5%	As required	For ceramic or
	Magnesia	2.5%		investment moulding
Sand	Ammonium phosphate	6%	As required	For ceramic moulding or investment moulding

5.5.2 High-Temperature materials

Many aggregates will stand the temperatures involved, including silica (SiO_2, 1720°C), alumina (Al_2O_3, 2050°C), chromium oxide (Cr_2O_3, 2430°C), magnesia (MgO, 2800°C), zirconia (ZrO_2, 2690°C), mullite ($3Al_2O_3.2SiO_2$, 1810°C), fosterite ($2MgO.SiO_2$, 1890°C), zircon ($ZrO.SiO_2$, 2500°C) and spinel ($MgO.Al_2O_3$, 2135°C). There are also at least 14 nitrides and 16 carbides with melting or decomposition temperatures above 1650°C. The melting point, however, is not the only criterion in the selection of investment moulding material; many other requirements have to be met.

The non-uniform thermal expansion behaviour of the various forms of silica has been discussed earlier. In recent years, the tendency has been to replace the silica-type aggregate with materials which have a more uniform thermal expansion curve. Materials such as aluminosilicates, zircon, alumina or the non-crystalline low-expansion aggregates like silica glass are finding favour.

Availability in the proper particle-size ranges is another important requirement. This eliminates many of the synthetic materials such as the carbides, borides or nitrides.

The thermal conductivity of the investment mould material has also assumed increased significance in recent years. Because of the high hot strength of these materials (and therefore low collapsibility) compared with conventional types of dry or green sand moulds, rapid cooling of the casting would result in hot tears or cracks. Fortunately, the thermal conductivities of investment moulds are much lower than conventional moulds. For example, the thermal conductivity of dry sand is approximately 0.0037 cal/cm degCs, whereas for an ethyl silicate investment it is 0.0009 cal/cm degCs. A slow rate of solidification of the casting is thus possible, allowing the contraction stresses to be dissipated within the still-plastic body of the casting without adverse effects such as cracking or tearing. The solidification times for ordinary sand moulds and investment moulds are compared in fig. 5.30.

5.5.3 High-temperature binders

The esters of silicic acid were the first organic silicon compounds to be used. Monomeric ethyl silicate and its polymers are now large tonnage industrial chemicals manufactured in a continuous process by reaction between silicon tetrachloride and ethanol. Three types of ethyl silicates are available: tetraethyl orthosilicate $\{(C_2H_5O)_4Si$ (28% $SiO_2)\}$, condensed ethyl silicate, which is a mixture of the tetraethyl orthosilicate and some polysilicates, and ethyl silicate, which is a mixture of polysilicates and contains about 40% Silica (SiO_2). With ethyl silicate (relatively cheap) as a starting

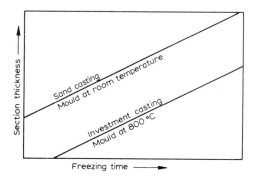

Fig. 5.30 The freezing times of castings in sand moulds and in investment moulds.

material, the solution must be hydrolysed and caused to react with water in order to produce a solution which will deposit the adhesive form of silica desirable for bonding refractory aggregates. The reaction may be written as follows:

$$\begin{array}{c}
OC_2H_5 OC_2H_5 OC_2H_5 OC_2H_5 \\
| | | | \\
C_2H_5O-Si-O-Si-O-Si-O-Si-OC_2H_5 + 12\,H_2O\,(HCl) \\
| | | | \\
OC_2H_5 OC_2H_5 O OC_2H_5 \\
| \\
OC_2H_5-Si-OC_2H_5 \\
| \\
OC_2H_5
\end{array}$$

$$\begin{array}{c}
OH OH OH OH \\
| | | | \\
HO-Si-O-Si-O-Si-O-Si-OH + 12\,C_2H_5OH. \\
| | | | \\
OH OH O OH \\
| \\
OH-Si-OH \\
| \\
OH
\end{array}$$

To effect hydrolysis the following mixture may be used:

                      Ethyl silicate (40% SiO_2)   37.6% by volume
                      190° proof ethyl alcohol   59.8% by volume
                      3% hydrochloric acid   2.6% by volume

Complete hydrolysis takes 2 to 3 hours and the 'anhydrous' solution can be stored for a few weeks without gelling provided that evaporation of the alcohol is prevented.

To promote gelling the pH is controlled and excess water added. The pH of the non-gelling solution is about 1.2 and as this value is increased quicker gelling takes place until, at a pH of 6, the gelling time is 60 to 90 s. In practice the pH is controlled to give a setting time of about 1 h. Small amounts of magnesia (MgO) may be added to the mixture to act as a neutralizing agent for hydrochloric acid, gradually raising the pH and causing the precipitation of $Si(OH)_x$. Even distribution of the magnesia is necessary to prevent hard and soft areas in the investment.

The setting time may also be controlled by the addition of salts of strong bases and weak acids. Both sodium hydroxide (NaOH) and sodium carbonate (Na_2CO_3) may be used but ammonium salts such as ammonium hydroxide (NH_4OH), ammonium carbonate (($NH_4)_2CO_3$) or the amines are preferred since they afford better control.

Phosphate binders are also quite common and a number of phosphates have been used, including aluminium, magnesium, calcium, zinc and lead phosphates. The magnesium phosphate bond is probably the most important.

$$3\,MgO + 2\,H_3PO_4 + H_2O \rightarrow Mg_3(PO_4)_2 \cdot 4\,H_2O.$$

Due to the hazards involved in using phosphoric acids, phosphates such as dihydrogen ammonium phosphate are generally used.

$$MgO + NH_4H_2PO_4 + 5\,H_2O \rightarrow MgNH_4PO_4 \cdot 6\,H_2O.$$

Setting time is controlled by the grain size, surface area and degree of calcination of the magnesia. Proprietary brands are available with different setting times within the range 15 to 40 min. Curing by heating is necessary to produce the pyrophosphate bond before pouring the metal.

$$2MgNH_4PO_4 \cdot 6H_2O \rightarrow Mg_2P_2O_7 + 13H_2O\uparrow + 2NH_3\uparrow.$$

At about 1310°C the magnesium pyrophosphate ($Mg_2P_2O_7$) decomposes giving phosphorus pentoxide (P_2O_5) gas. This is therefore the upper limit for the use of this bond. The aggregate material generally preferred with this type of bond is alumina (Al_2O_3).

The calcium aluminate cements, $CaO.Al_2O_3$ and $CaO.2Al_2O_3$, are stable at 1600°C and 1720°C respectively and may be used for bonding (hydraulic bond). The main problem is the long setting time but also the hydration reaction is exothermic.

In the acetate process, by using a soluble aluminium acetate and raising the pH, alumina gel ($Al(OH)_3$) is formed which acts as a binder for the aggregate. Similar techniques may be used to form oxide bonds such as zirconium oxide (ZrO_2) from zirconium acetate or other metallic oxides from soluble metallo-organic compounds.

5.5.4 Low-temperature materials

The most important material for investment moulding to produce castings from low melting point alloys is plaster of Paris. The setting time of plaster of Paris may be controlled by additives such as cations of aluminium (Al^{+3}), magnesium (Mg^{+2}) or terra alba.

$$2CaSO_4 \cdot \tfrac{1}{2}H_2O + 3H_2O \rightarrow 2CaSO_4 \cdot 2H_2O.$$

For investment castings, plaster of paris is seldom used alone. The more common additives are:
1. Silica (SiO_2) to improve thermal conductivity.
2. Asbestos, glass fibre or talc to improve strength and eliminate cracking tendencies.
3. Inert materials or foaming agents to improve permeability (also by autoclave treatment).
4. Additives to overcome shrinkage of plaster when heated.

Under ideal conditions these moulds may be used at temperatures up to about 1/100°C.

5.6 OTHER PRECISION MOULDING PROCESSES

5.6.1 Ceramic-shell moulds

Ceramic-shell moulding is essentially the same as the investment moulding process except that the mould is in the form of a thin shell enveloping the mould cavity. These moulds are produced by alternately dipping the consumable pattern (usually wax) in a coating slurry and a dry aggregate such as silica. The dry aggregate may be conveniently applied by suspending the slurry-coated pattern in a fluidized bed containing the dry refractory aggregate. A shell of $\tfrac{1}{4}$ inch or more is built up in this way. The pattern is then melted out and the mould cured and processed in the same way as in investment moulding.

5.6.2 Ceramic moulding

Ceramic moulding is also a development derived from the investment moulding process but reusable patterns are used as in the case of conventional sand moulds. The moulding aggregate consists of a slurry made up from refractory aggregate and a ceramic binder. This slurry has a similar composition to the slurry for investment castings, i.e., it may consist of silica grains, ethyl silicate, water, alcohol and hydrochloric acid. Other refractory aggregates and bonding agents are also used (similar to the investment moulding slurries). The slurry is poured around the pattern contained in a suitable box or flask and allowed to gel for about 10 min. The pattern is then removed and the mould fired by igniting the alcohol in the aggregate. After the cope and drag parts have been prepared, the mould is assembled and preheated to a suitable temperature (depending on the alloy being cast) before the metal is poured.

Like the investment casting process the main advantages of ceramic moulds are:
1. High pouring temperature alloys may be cast to accurate dimensions. (An accuracy of ± 0.003 inch/inch is possible.)
2. Machining is reduced or eliminated for alloys which are difficult to machine.
3. Excellent reproduction of intricate details is possible.
4. Because of the heated moulds, extremely small sections can be cast.

5.6.3 Plaster moulds

The general principles involved in the use of plaster of paris moulds have been discussed in the section dealing with investment moulding with plaster. For conventional moulding the patterns used may be similar to those used for sand moulding, i.e., they consist usually of the cope and drag parts. A plaster slurry of creamy consistency is poured around the pattern placed in a suitable box or flask and allowed to set. The

Fig. 5.31 Plaster moulded turbine charger impeller in aluminium alloy.

pattern is then removed and the mould parts processed and assembled in the usual way before pouring. Where autoclaving treatment is carried out, the assembled moulds, after such treatment, are allowed to dry in air for about 12 h and finally in an oven for up to 20 h at about 230°C before pouring.

The main advantage of plaster moulds is in the casting of comparatively low melting temperature non-ferrous alloys. Good dimensional tolerances and surface finishes are possible. The method is popular with manufacturers of aluminium metal patterns. Moulds for foam rubber and other rubber products, such as tyres, are also cast in plaster moulds.

5.6.4 The Shaw process

The Shaw process, developed in England, is a variation of the investment moulding process and uses reusable patterns instead of the expendable wax or plastic patterns.

Split moulds (cope and drag) are made by pouring a refractory slurry over the patterns enclosed in boxes or flasks. The mould can be stripped from the pattern within a few minutes and then 'flamed' by setting fire to the alcohol in the mould mixture. Assembled moulds are fired in a furnace and metal can be poured when the moulds are red-hot or at lower temperatures.

This method has been used for casting aluminium, magnesium, brasses and bronzes, cast iron, low- and high-alloy steels and high-temperature super alloys. The largest casting made by this process is reported to be a steel die block weighing 700 lb.

The refractory slurry is a mixture of coarse and fine sillimanite grains, hydrolysed ethyl silicate and a liquid catalyst such as ammonium hydroxide (NH_4OH). ammonium carbonate (($NH_4)_2CO_3$) or the ammonium-organic salts (amines). These are mixed together quickly (mixing time less than 1 min) into a slurry and poured over the pattern, the gelation being controlled by the amount of catalyst used. During the gel stage the mould is flexible, which facilitates the stripping of the pattern without damage to the mould.

After stripping, a high-temperature gas flame is applied to the mould, which continues to support combustion due to its alcohol content. Flaming causes the mould surface to develop a network of fine cracks (craze-cracks) throughout the mould body. This network of fine cracks results in a higher permeability and a better thermal shock resistance for the mould.

After assembly, the mould is heated at a temperature of about 900°C for periods depending on the thickness of the mould walls. This is claimed to further strengthen the bond and to produce an inert, gas-free and collapsible mould. Metal patterns are preferred in all applications using a slurry aggregate.

Dimensional accuracy and surface finish are reported to be excellent in castings made from the Shaw process moulds. Additional advantages are:
1. Patterns need not have draft or taper, since at the time of stripping the mould is flexible or rubbery to a certain extent and therefore not likely to be damaged during pattern withdrawal.
2. A wide range of casting sizes can be made.
3. Simple pattern construction and moulding procedure allows short lead times and significant economies when small numbers of castings are being produced.

The Shaw process has found application in the manufacture of high-alloy steel jet-engine blades and manifolds. Other components such as landing-gear struts, shock-absorber housings, fuel system components, stainless-steel pump impellers and filter-press frames, blades and vanes for gas and steam turbines are among castings which have been made by this process.

Larger castings such as glass moulds and dies for forging, rolling or extruding are also being developed.

5.6.5 Graphite moulds

Recently, with the need to cast reactive metals such as titanium, inert moulds, both permanent and aggregate, made from graphite have come into prominence. The expendable aggregate moulds are usually prepared in the same way as sand moulds except that the moulding mixture consists of about 70% graphite grains and 30% binders, composed of pitch, carbonaceous cement, starch and water. After preparing the mould by squeezing the aggregate round the pattern in the usual way, the mould is dried and heated in a reducing atmosphere at a temperature of about 1000°C. The assembled mould is transferred while still hot (about 200°C or more) to a vacuum chamber where the metal is melted and poured.

Permanent graphite moulds are made by machining the mould cavity in solid blocks of graphite. Graphite shows signs of oxidation above 400°C and the mould can become unusable after a few castings have been made unless a special coating is applied to it. A mould coating of ethyl silicate, which deposits silica on heating, usually helps to minimize mould deterioration. Permanent graphite moulds are used in the centrifugal casting process for casting brass and bronze bushings, sleeves and other similar shapes. They are also used for permanent mould castings when comparatively small numbers are required (they are cheaper than metal dies and therefore more economical for short runs). Recently, railway-car wheels have been cast in graphite moulds, with dimensions accurate enough to eliminate machining.

5.6.6 Permanent moulds

Castings made in moulds which can be reused many times are known as permanent mould castings or gravity die castings. The mould cavity or die cavity is frequently cast to approximate shape and size and then machine-finished to its final dimensions. Dies are generally made of cast iron or steel, although some non-ferrous metals and alloys such as bronze and aluminium are sometimes used for specific applications. The mould or die also includes the running and feeding systems.

Permanent moulds are most frequently used for **aluminium, magnesium, zinc, lead and copper base alloys, followed by the cast irons. The high melting temperature of steels usually leads to a rapid deterioration of the dies, so that gravity die casting of steel is rare.** The temperatures involved in casting the **more common alloys are reproduced below:**

Metal	Pouring temperature, °C	Mould life	Mould operating temperature, °C
Cast irons	1260 to 1475	Up to 20000	300 to 450
Aluminium alloys	700 to 760	Up to 100000	325 to 425
Copper alloys	1025 to 1150	Up to 20000	125 to 260
Magnesium alloys	650 to 710	Up to 100000	150 to 320
Zinc alloys	390 to 430	Over 100000	200 to 260

Due to the cost of moulds or dies the process is generally limited to production in large quantities, usually carried out as a continuous cycle of mould preparation, metal pouring and ejection of casting. This is carefully timed in sequence so that the temperature of the mould can be strictly controlled within the operating limits. The operating temperature of the mould is the most important single factor in permanent mould castings. It not only determines the proper filling of the mould but also affects mould life. The expansions and contractions of the mould are kept to a minimum to avoid excessive thermal shock and consequent 'heat checking' (cracking of the mould surface). To improve mould life the cavity may be coated with lampblack or soot. If this is not advisable due to the nature of the alloy being cast, the mould cavity may be sprayed with a fine suspension of refractories in a suitable liquid carrier. The thickness of the mould coating also affects the rate of heat transfer from the casting to the mould.

Moulds may be assembled from two or more parts (depending on the shape of the casting) to help in ejecting the casting from the mould. Ejector pins may be incorporated in the mould for this purpose. Both metallic and/or sand cores are used when necessary and these are set before the mould parts are closed or assembled. The metal is usually fed into the die by gravity (hence gravity die casting) but in some cases air pressures up to 10 p.s.i. may be applied to the metal in the pouring cup after the casting has been poured. The cycle of operations for gravity die castings is illustrated in fig. 5.31.

Refractory permanent moulds have also been used. One method using granular silicon carbide, bonded with bentonite clay, and water containing sodium carbonate has been used for this purpose. The mixture can be moulded like green sand moulds and has the advantage that no machining is required. The moulds are fired at temperatures approaching 800°C to develop a stable hard mould. The thermal conductivity of silicon carbide moulds is superior to sand moulds but not nearly as good as metal moulds. The number of castings that can be made from the same mould is also not firmly established.

Graphite as a permanent mould material has been described earlier.

Dimensional tolerances of ±0.010 inch are possible with gravity die castings. The chilling action of the metallic mould also helps to produce sound castings in many alloys. The mass production of a wide range of engineering components by this process has become quite common – carburettor bodies, hydraulic brake cylinders, connecting rods, oil-pump bodies, components for washing machines, refrigerators, typewriters and numerous other mass produced components.

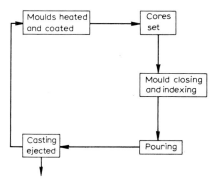

Fig. 5.32 Cycle of operation in gravity die casting.

5.7 SOME PARTICULAR CASTING EQUIPMENT AND CASTING METHODS

5.7.1 *Die casting*

Die casting or pressure die casting is a technique whereby the molten metal is introduced into a metallic die or mould cavity under pressure. This method is particularly suitable for the production of small thin-walled castings of complicated shape, since the metal moves through the die with high velocity and is therefore able to fill the thin sections of the mould cavity, particularly around bends and corners. The air present in the die has to be displaced by the entering metal and, since the mould walls are not permeable, special venting methods have to be employed. Arrangements for displacing the air from the die could be eliminated by pouring the metal into a die which is under vacuum. However, the cost of such installations has limited their use to small parts made of low melting point alloys.

Castings produced by the die casting processes are distinguished by their superior dimensional accuracy and surface finish. Compared with other metal-working processes, the die casting process also shows a higher yield of finished castings (i.e., a higher ratio of casting weight to weight of metal melted), and in general they are accurate enough in size to minimize or eliminate machining altogether.

The bulk of die castings are made from lead, zinc, tin, magnesium and aluminium base alloys, although copper base alloy die castings are also quite common. The types of alloys most suitable for die castings are discussed in subsequent sections.

There are two basic types of die casting machines: the-chamber and the cold-chamber machine. In the hot-chamber die casting process, molten metal is forced under pressure into a permanent mould or die made of steel. The chamber, where pressures of about 400 to 5000 p.s.i. may be produced, also contains the molten metal, i.e., the chamber is heated to a temperature not less than the melting temperature of the metal or alloy to be cast. A plunger is generally applied to produce the pressure but compressed air or some other unreactive (to the metal) gas under high pressure may be used.

In the cold-chamber method the metal or alloy to be processed is not melted in the pressure chamber but melted separately and transferred to the chamber for individual castings. The chamber is therefore separated from the melting pot or furnace and is not heated. The requisite amount of metal is placed in the chamber and a plunger applied to force the metal into the die cavity. At the time of applying the pressure the metal or alloy may be in a semi-solid plastic state. The high pressures involved, 3000 to 30000 p.s.i.,* make the plastic mass fluid enough to fill the die cavity. Pressure could, of course, be applied to the alloy when it is completely in the liquid state.

The surface quality and the mechanical properties of castings made by either the hot-chamber or the cold-chamber processes are very similar. However, certain molten alloys such as aluminium and zinc base alloys, which tend to pick up iron from the chamber with consequent deterioration in mechanical properties, are best cast by the cold-chamber process, where there is only limited contact between the molten metal and the chamber walls. The reported improvement in mechanical properties of the castings processed in extra-high-pressure cold-chamber machines have not yet been fully explored.

* Much higher pressures have been employed in some instances to improve the mechanical properties of the castings produced.

The dies can be constructed to allow their use in either hot-chamber or cold-chamber machines, provided that the die clamping devices in the machines are similar. Slight differences in shrinkage are inevitable from the two processes but the effect is not significant enough to upset dimensional accuracies of small or medium sized castings.

The principle of operation of a hot-chamber gas-pressure die casting machine is illustrated in fig. 5.33.

Cold-chamber machines have been developed which incorporate automatic filling of the chamber instead of manual filling for each casting. Completely automatic die casting machines have also been installed for mass production.

The die parts, usually known as the ejection die and the cover die, must include such features as venting, provision for ejector pins, and the metal running system (gate) and location points for proper closing of the dies. Provision must also be made for cores where needed.

Due to the inherent mechanics of the process it is not possible to provide for the usual risers or feeders to compensate for the shrinkage when the casting solidifies. For this reason, heavy section castings cannot be successfully made by this process, although efforts are being made to develop mechanical feeding systems by using multihead plungers.

The usual wall thicknesses for die cast components rarely exceed 0.2 inch and as far as possible uniform section thickness is maintained. The lower limit of wall thicknesses may vary between 0.025 inch and 0.075 inch, depending on the alloy being cast and the surface area of the die cast part.

Another important point to note is that the casting design must be such that the mould cavity and the cores allow the casting to be ejected. This is a fundamental limitation of metal moulds which does not apply to sand casting processes.

The most important advantage of the die casting process is the dimensional accuracy that can be obtained. The saving possible due to elimination or minimization of machining operations is usually between 30 and 50% and may be as high as 90% in some cases. To summarize the main advantages of the process:

1. High production rate – up to 500 shots per hour may be possible.
2. Dimensional tolerances between $+0.001$ and $+0.003$ inch may be obtained in commercial castings.
3. Very thin sections down to about 0.015 inch can be cast (for small castings).
4. Accurate coring and core location can be maintained.
5. Good surface finish may allow direct buffing in most cases.
6. The rapid cooling rate in metal dies results in superior mechanical properties in many die cast alloys, e.g., zinc alloy.

5.7.2 Centrifugal casting

Centrifugal force is employed in many ways to distribute metal in the mould, and centrifugal casting refers to the method of metal distribution rather than any specific moulding process. Metal is poured into spinning moulds, either of the permanent or the expendable type. The mould spinning axis may be either horizontal or vertical.

The pipe industry and various tubular and sleeve industries have been using this method for mass production of their products. Since the cylindrical core can be eliminated and high production rates achieved in permanent metal moulds, the cast-iron pipe industry (pipes for water, sewage etc.) relies almost entirely on this method of

394 | METAL CASTING

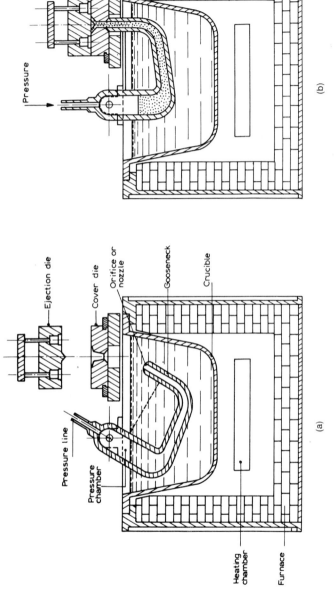

Fig. 5.33 Hot chamber gass-pressure type pressure die-casting arrangements.

SOME PARTICULAR CASTING EQUIPMENT AND CASTING METHODS | 395

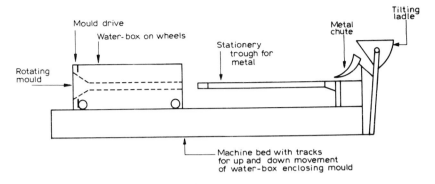

Fig. 5.34 Schematic diagram of De Levaud centrifugal casting machine.

production. Fig. 5.34 is a schematic representation of the De Levaud process of pipe manufacture.

Centrifugal force has the effect of rapidly forcing the molten metal against the mould walls (periphery), thus avoiding miss-runs due to solidification. Properly designed moulds will also promote directional solidification and a section free from shrinkage cavities. Gears, pipes, brake drums, fly wheels and similar castings are produced by this process.

Shorter and smaller castings are often made in conventional sand or permanent moulds clamped to a spinning table during pouring. The spinning axis in such cases is usually vertical; the object of spinning the mould is to effect a better liquid-metal distribution within the mould cavity.

Centrifuging differs from true centrifugal castings in that the entire mould cavity is spun off the axis of rotation as shown in fig. 5.35. The main advantage is in easier filling of the mould cavity and to some extent forced feeding to compensate for the solidification shrinkage.

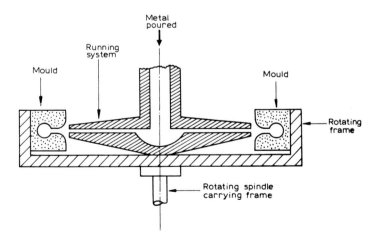

Fig. 5.35 Centrifuging of castings.

5.8 CORES AND CORE MATERIALS

5.8.1 Introduction

The basic materials used for making moulds are generally suitable for making cores as well. The methods for bonding aggregates and other processing techniques are also similar in principle. One may therefore describe cores as green sand cores, dry sand cores, metal cores, ceramic cores, carbon dioxide (CO_2) cores, shell cores, air-set cores and so on, in much the same way as moulds are described.

There is, however, an important difference between moulds and cores, and this lies in the fact that in most cases the core is totally submerged or surrounded by molten metal when the casting has been poured, while only one face of the mould is so affected. The conditions giving rise to erosion, breakage, metal penetration and dimensional instability are therefore much more severe in the case of cores. Moreover, the problems regarding collapsibility and permeability are also more acute. It follows that the physical and mechanical properties specified for cores must be, in the majority of cases, superior to those specified for moulds.

One direct result of this difference is reflected in the choice of method for the production of cores since the moulding method used to produce a casting does not necessarily predetermine the type of core to be used; i.e., each individual case has to be assessed separately. In this connection the more important considerations are:
1. Dimensional accuracy and surface finish required in the cored region.
2. Metallostatic pressure and buoyancy forces acting on the core.
3. Pouring temperature and reactivity of the metal.
4. Type of mould used.
5. Shape, size and location of the core in the mould cavity.

The geometry and location of the core, together with the pouring temperature, are probably the most important factors to be considered in specifying the property requirements such as collapsibility, permeability and hot strength.

The choice of the core-making method therefore reduces to an assessment of the mechanical and physical properties desirable and the most economical way of attaining them. It is therefore not uncommon to incorporate dry sand cores, shell cores, carbon dioxide cores or other suitable cores with green sand moulds. Similarly, a metallic mould or die used in gravity die casting may well include a sand or ceramic core instead of a metallic core.

5.8.2 Core-making methods

Cores may be made of metal, plaster and investment and ceramic materials as well as core sand. Metal cores, used in permanent mould castings and die castings, do not possess the important property of collapsibility and therefore have shape limitations.

The function of the cores is to form the internal cavities in castings and provide the casting process with the flexibility which is unique among all known methods for producing metal components. In fact, shapes are commercially mass produced which would be impossible to machine or fabricate in any other way. The water-cooling chamber in internal combustion engines and the exterior and interior of air-cooled engines require intricate coring and they can only be mass produced because it is possible to make these shapes by coring.

Most cores are made by compacting a mixture of sand, or other refractory grains, and organic binders, which provide the necessary strength and other properties.

Although some small cores are still made manually, most of them, particularly in quantity production, are made in machines which are becoming progressively more automated.

a. Bench work. When small cores are made by hand-filling core boxes with the aggregate mixture by placing the core box on a bench, this is usually referred to as bench work. The only equipment necessary is the core boxes and core plates. The core box is essentially a wooden or metal mould for producing the external shape of the core and may be constructed in two or more parts to allow for the withdrawal of the core after compaction. In this respect the relation of the casting to the mould may be equated to the relation of the core to the core box – at least in the geometrical sense. The core box is filled with the mixture, rammed and the core removed and placed on the core plate for baking if necessary.

The simplest box is a dump-box, which consists of a single part only; the core is withdrawn from the box by placing the core plate over the box, inverting the whole assembly and then drawing the core box away from the core.

Cores which are to be baked must have a flat surface to rest on the core plate. Cores with no flat surfaces must be supported on core driers, contoured to support the cores, until the cores are baked and hardened. For this reason the baked core process is being replaced by newer techniques such as the hot-box, shell or carbon dioxide processes. The great advantage of these processes is that the core is set or hardened before being removed from the core box, so that uneconomical handling, baking and supporting with core driers during the manufacturing process are eliminated.

b. Machine work. The operating principles of core-making machines are very similar to those used for making moulds, i.e., compaction to the required degree is achieved by jolting, squeezing, slinging or blowing or by a combination of these methods. Shell core-making machines differ from shell moulding machines merely in the geometry of the shapes handled and not basically in the principle of operation. The design of the core boxes will obviously be different from the design of the pattern plates and so the machines employed are suitably altered to accommodate the different geometry. There are, however, certain minor differences, the most important of which are outlined below.

Jolt machines. Unlike the jolting machines used for moulding, machines used for core-making are usually simple jolt tables suitable for the jolt compaction of cores in simple dump-boxes. The core material used is usually a mixture of sand, drying oil and other organic material such as dextrin. These cores are subsequently baked to develop the necessary strength properties.

Shell core machines. The basic principles of shell moulding already discussed are also applicable to shell core-making. In this process a metal core box is heated by gas burners or electrical heaters to a controlled temperature and a resin–sand mixture is fed into the box either by gravity or by blowing. After a dwell period which establishes the thickness of the shell, the unheated sand from the centre of the core is drained out and reused in subsequent cycles. The shell thickness built up in the core box cavity is usually between $\frac{1}{4}$ and $\frac{1}{2}$ inch. The core is then stripped from the box and may be placed directly into the mould without baking or further processing. In some cases the stripped core may be further cured by baking in an oven to develop superior strength and dimensional stability. Completely automated units are also available.

Sand slingers. These machines are suitable for ramming up larger sizes of conventional sand cores. Separate facilities are provided for stripping the cores from the core boxes.

Core blowers. The core blower is the most important equipment for the rapid mass production of small and medium sized cores. The basic principle of the machine involves filling the core-box and ramming simultaneously by an air-stream carrying sand.

A core blower usually consists of a movable sand reservoir from which the sand is blown into the core box. The reservoir has an opening at the top which admits sand when moved into the fill position under a sand hopper. A plate, called the blow plate, covers the bottom of the reservoir and prevents the sand from dropping through. Special holes in the plate (blow holes) allow sand to be blown out of the reservoir. The filled reservoir slides into the blow position over the core box and the core box and table are raised tightly against the blow plate of the reservoir, which in its turn is raised against a sealing gasket in a blow valve so that the full line air pressure may be built up in the sand reservoir. This sequence is set in motion by a hand-operated valve (usually pneumatically powered). There is also a horizontal clamping arrangement (air chuck) which is used on vertically split core boxes so that the air pressure will not blow them open. When air pressure is applied to the sand reservoir through a hand-actuated valve, the sand is forced into the core box through the holes in the blow plate. The air is vented through specially designed vent holes in the core box. The blowing action is very rapid and less than 1 or 2 s are required to fill and compact the sand in the box, even for fairly large core boxes. Blowers are available in a wide range of sizes capable of producing cores weighing only a few ounces to ones weighing 300 lb or more. The essential features of core blowing are illustrated in fig. 5.36.

Core blowers are usually designed to operate from air delivered at 90 to 110 p.s.i. For maximum density of compaction of the core the air pressure must be more than 90 p.s.i. The actual volume of air required for the blow varies with the size and design of the machine. However, an average value of 25 ft^3 of air for a core weighing 100 lb is not uncommon. The air delivery system must cope with this volume flow without any significant drop in the pressure during the blow cycle, and many core blowers are equipped with pressure-tank air reservoirs to overcome this problem.

The movement of the sand through the blow hole occurs when the air pressure reaches about 5 p.s.i. in this region. Pressures of up to 50 p.s.i. inside the reservoir keep the sand moving into the core box, forming a channel in the sand. Sand from near the wall of the reservoir collapses into this channel (air stream). The process is aided by the design of the reservoir walls and by the aspiration effect set up by the movement of the air through the channel.

The sand mixture is most important for the successful operation of a core blower. Good flowing characteristics are desirable and the strength (in compression) should not exceed about 2 p.s.i. Stronger mixtures will tend to pile up by compaction within the reservoir unless special agitators are employed.

Core boxes used in conjunction with core boxes as almost invariably made of metal, usually aluminium or cast iron. Adequate venting of the core boxes is most important and the walls should be strengthened sufficiently to prevent distortion during clamping and blowing. Venting may also be used to direct the movement of the air–sand stream within the core box so that awkward corners can be filled (see fig. 5.36). Areas near vents and under the blow holes may wear rapidly, thus affecting the di-

Fig. 5.36 Complex core arrangements for internal combustion engine, ready for closing.

mensional accuracy of the cores made. When cores are produced in large numbers the core boxes need to be carefully checked for this.

The only real limitation in the use of core blowers arises from the nature of the sand mixture. However, with reasonable free-flowing characteristics this problem is reduced to negligible proportions. Almost all types of sand cores, i.e., oil-sand cores, carbon dioxide cores, shell cores, air-setting cores etc., can be produced with the help of a core blower and this is therefore the most popular method of core-making at present.

Stock core machine. Small cores of cylindrical, hexagonal, rectangular or other simple cross-sections may be produced by extrusion through suitable dies. The machines usually operate on the same principle as a meat-grinder. These cores may be made in many standard shapes and sizes, stored, and then cut to suitable lengths before use.

Core shooters. These are a development of core blowers and differ from them principally in the operation and design of the air valve introducing air into the reservoir or magazine. The snap opening of the air valve generates instantaneous explosive pressures within the magazine, which forces the sand out through the blow holes in the blow plate by a mechanism similar to extrusion. The action is similar to the shooting of a bullet from an air rifle. The great advantage over core blowers is that large volumes of air are not required and venting of the cores is much simpler. Core-box wear is also minimized since the abrasive action of large volumes of sand-carrying air is not involved. This method has largely superseded blowing for small and medium sized cores.

c. Core ovens. Either continuous or batch-type ovens are employed for baking cores. The mode of heating may be by gas or oil firing or by electricity. Dielectric baking is a relatively recent development. It should be noted, however, that baking of cores is necessary only for the conventional sand core processes and the general trend is to adopt newer processes such as the carbon dioxide, shell or air-set processes where baking is eliminated.

d. The hot-box process. One of the methods employed in core-making which eliminates baking is the hot-box core-making process. Sand is transferred into a heated core box by gravity, in the case of simple boxes, or by blowing, complex boxes. The boxes are usually cast iron and are heated to a temperature of about 210°C. A thermo-setting resin-sand mixture is used. When furan resins are used, exothermic polymerization continues even when the core is withdrawn from the box (after an initial shell has formed) so that the entire body or cross-section of the core may harden. In contrast, in shell core-making using urea formaldehyde or phenol formaldehyde, only the sand directly affected by the heat from the box is set around the core-box cavity and the unreacted sand may be drained out leaving a hollow shell.

The main advantage of the hot-box process is that the core is hardened in the box and has sufficient strength (immediately) for handling and use in the mould. The rapid rates of production possible are another advantage – one finished core per minute is not uncommon.

5.8.3 Core finishing

In the case of baked cores, the baking operation may involve changes in shape and size or distortions of various types. It may be necessary, therefore, to inspect the cores and check for dimensional accuracy and make the necessary corrections, where possible, or reject the core if such corrections are not possible. Cores are often coated with refractory or protective materials which improve their resistance to molten metals. These coatings are generally applied by spraying, dipping or swabbing.

Cores which are not made in one piece have to be assembled before they can be used. Core assemblies may be held together by core paste or other suitable adhesives or by bolting together (usually in larger cores). For small cores anchoring of the pieces is sometimes accomplished by pouring molten lead into matching cavities in the pieces and allowing the lead to solidify, thus holding the pieces together.

All these operations – cleaning, sizing, coating, assembly and inspection before use – come under the heading of 'core finishing'.

5.8.4 Core setting

Core setting is the operation of placing cores in the moulds. The positioning is possible due to the provision of core prints. Unless securely anchored the cores will be displaced by the buoyancy forces when molten metal is poured into the cavity. When a number of cores are used in a mould, dimensional errors are likely to be additive so that some fixtures and gauges may be necessary for proper location. Positive location in three directions (i.e., one vertically and two horizontally) are necessary in any case, even when a single core is used. Location or reference points may thus be required both in the mould and the cores to ensure proper placement of cores.

The buoyancy force in a core is equal to the weight of liquid displaced by the core minus the weight of the core. Thus, if 1 ft³ of metal (say cast steel) is displaced by the core, the buoyancy force will be:

$$\text{wt. of 1 ft}^3 \text{ liquid iron} - \text{wt. of 1 ft}^3 \text{ core} = 490 - 100 \text{ lb} = 390 \text{ lb}.$$

Obviously, the buoyancy force will depend on the density of the liquid metal. For a light metal like aluminium there are no buoyancy forces since the core is heavier than the liquid aluminium it displaces. The total buoyancy forces have to be supported by the core prints, so careful design of the core prints is essential. Where adequate support cannot be provided by the core prints, chaplets may have to be used. Chaplets are metal forms placed between the mould and core surfaces to prevent movement of the core when metal is poured into the cavity. The chaplets, however, may not fuse completely into the body of the casting, thus giving rise to lack of pressure tightness or other local defects. For this reason the use of chaplets is usually restricted to cases where proper core support is not possible with the core prints or where thin long cores are involved. Efforts should be made during the design stage of the casting to avoid their use.

5.8.5 Other core applications

Apart from their principal function of forming the internal cavities or shapes in castings, cores are also used for many other purposes in castings technology, e.g.,
1. As strainer cores used in the running system.
2. As the running system itself.
3. As the pouring cup.
4. As risers (feeders).
5. As core assembly moulds.
6. To enable moulding of undercuts or other awkward shapes which cannot be moulded easily any other way.

These functions of cores are best understood from the diagrams in fig. 5.37.

5.8.6 Materials

It has been stated earlier that the core must resist erosion, breakage, thermal shock and metal penetration while submerged in the molten metal. It must also retain dimensional location and stability and yet must be collapsible enough to prevent casting defects such as 'hot tears' and to enable it to be removed from the casting (when the casting is solid) without undue effort.

Although cores may be made of metal, ceramics or sand, the most versatile materials are core sands, from which the bulk of cores are made.

The core sand mixture is prepared so that it combines some of the important properties listed below:
1. Adequate green strength.
2. Response to baking (for conventional baked sand cores).
3. Adequate strength for handling in core setting and for the retention of dimensional accuracy.
4. Resistance to the action of molten metal, i.e., erosion, fusion, thermal shock.
5. Venting ability to allow the considerable amount of steam and other gases (generated due to the rise in temperature when the metal is cast) to escape.

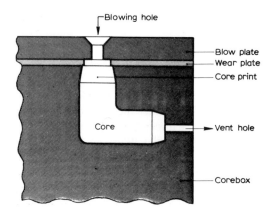

Fig. 5.37 Typical arrangement for blow hole, blow plate and core box, showing location of vent hole to direct filling of core round the bend.

6. Collapsibility and ease of removal from the solidified casting.
7. Retention of desirable properties in storage.

Cores and mixtures such as shell, furan resin, air-set and carbon dioxide (silicate) sands do not necessarily require baking response since they may not be baked.

a. Core sands. Silica is the most common form although zircon, olivine and other granular refractory materials are also used. In selecting these refractory materials, the most important considerations are their refractoriness, dimensional and chemical stability with temperature and heat transfer characteristics and the particle size distribution (sieve analysis).

b. Binders. Binders used for core-making may be classed as organic, inorganic and metallo-organic.

Organic binders are combustible and are destroyed by heat, so that they usually help to satisfy the property of collapsibility. Inorganic binders are not combustible and may therefore retain considerable strength at elevated temperatures and may be virtually non-collapsible at the temperatures involved.

Organic binders. Core oil, cereal, resins, plastics, pitch, dextrin, molasses, rosin, rosin oil, lignin, casein and gelatin are generally used as organic binders.

Core oils are prepared from a blend of several ingredients such as linseed oil, soy oil, fish oil, petroleum oils and coal tar. They also contain drying agents and/or extenders (polymerized by heat or oxidation) to convert the oil to a solid. The thin film of oil on the sand grains is converted to a solid by baking in conventional oil–sand cores.

Air-set oil binders contain catalysts which accelerate polymerization and may induce it at room temperature.

Driers are used in oil–sand cores to hasten the curing process. They function either as catalysts to the polymerization process or as providers of additional oxygen and heat for the reaction. Ammonium nitrate in amounts up to 0.15% may be used for this purpose. Sodium perborate, manganese dioxide, manganese oleate and other metallo-organic compounds are also used with liquid oils and resins to cut down the baking time by as much as 80%.

The amount of core oil used in the sand mixture varies widely depending on the type of oil and other ingredients in the mixture. However, amounts of between 0.5 to 3.0% by weight are most common.

Cereal is mainly used for green strength in amounts up to 2.5%. Starches, dextrin and other binders such as casein also provide some green strength. These are all water soluble and the baked strength depends to a large extent on the amount of water used to develop this potential. Sugar, molasses and dextrin require about three times their weight in water to develop optimum properties on baking.

Lignins, better known as sulphite binders, are water-soluble compounds of wood sugars produced as a by-product in paper-making. Lignin binders are used in green or dry sand core making since they can be used in sands containing clay. They reabsorb water on standing and therefore cores containing lignin cannot be stored.

Pitch is produced as a by-product in the manufacture of coke and is used as a ground powder in amounts up to 3%. The pitch portion fuses and distills (partially) at temperatures of 150 to 300°C during baking, leaving a solid film which binds the sand grains together. After casting, the pitch in the core changes into coke and thus develops hot strength to resist the action of the molten metal.

Ground hardwood cellulose, wood flour, is used in amounts up to 3% to reduce the tendency of cores to set 'rock-hard' after casting and also to improve collapsibility.

Organic hydro carbon fluids are used in amounts of less than 0.10% as releasing agents, to reduce the sticking of sand in core boxes and to ease the removal of the core from the box. Silicones and various waxes dissolved in solvents are also used for this purpose.

Phenol formaldehyde and urea formaldehyde are thermosetting plastics and, as partially polymerized liquids or powders, may be used in core-making (binders). When heated they quickly polymerize, at temperatures between 100 and 250°C, to a strong solid.

Thermosetting plastics used in shell core-making are the same as those for shell moulding. To prevent sticking to core boxes, small amounts of kerosene, light fuel oil or other proprietary 'release agents' may be added to the sand.

Inorganic binders. Fireclay, bentonite, silica flour, sodium silicate and iron oxide are the most common inorganic binders used in core sand mixtures. Fire clay and bentonite are generally added to improve green strength, baked strength or hot strength, while the main function of iron oxide appears to be the prevention of 'finning' (cracking). Silica flour is used to improve hot strength and to resist fusion on contact with the metal. The main use of sodium silicate is in the carbon dioxide process.

c. Surface coatings. Core coatings usually consist of a liquid carrier, a refractory material and a binder and may be applied as liquids by spraying, dipping or brushing and as solids by dusting. The object of a coating is to produce a superior finish on the casting surface by improving the core's resistance to metal penetration and fusion. In some cases, it is claimed that the gas generated from the coating during pouring acts as a barrier (thin layer) between the molten metal and the core, thus protecting the core from the severe action of the molten metal.

Coatings are sometimes used for metallurgical effects. Tellurium-bearing coatings have the effect of producing a chill (white iron) in grey cast iron for a considerable depth below the coated surface so that special wearing surfaces may be cast by applying such coatings to them.

It is important to note, however, that not all coatings are suitable for all types of metals and alloys, since undesirable reactions may occur. A competent technologist should be consulted before specifying a coating for a casting made in any particular alloy.

5.9 POURING AND FEEDING OF CASTINGS

If suitable moulds and cores are available to reproduce the external and internal shapes of castings accurately, the soundness of the castings will depend primarily upon the mode of entry of the liquid metal into the mould cavity and the solidification behaviour of the metal or alloy being poured.

Metals and alloys are not always chemically inert with respect to the moulds and cores, nor are they free from gas absorption, shrinkage on cooling and erosive tendencies. Further, the specific gravities of different metals and alloys vary widely. A sound knowledge of the behaviour of the various alloys in their molten state, together with the mechanics of liquid metal flow and solidification characteristics, is therefore necessary for the production of satisfactory castings.

A comprehensive treatment of the many factors involved is not possible here and the discussions that follow must be supplemented with further reading (see bibliography) to appreciate the importance of these factors in cast metals technology.

5.9.1 Solidification of metals in general

Solidification of metals occurs by a process of nucleation and growth. Nucleation involves the formation of minute particles of solid surrounded by liquid. To determine how and when nucleation takes place, it is necessary to approach the problem from both the thermodynamic and kinetic aspects. For our purpose it will be sufficient to realize that, from the thermodynamic point of view, there is only one temperature at which a pure solid metal can exist in equilibrium with the pure liquid metal. At this temperature the free energies of the two phases are equal, as shown schematically in fig. 5.38, while below this temperature the free energy of the solid phase becomes progressively smaller than the free energy of the liquid phase. This decrease in free energy therefore provides the driving force to transform liquid to solid.

Unfortunately, once a particle forms, a solid/liquid interface is established and this interface has energy associated with it which makes a contribution to the free energy change shown in fig. 5.38, with the result that some cooling below the equilibrium temperature is necessary before the free energy of the solid phase becomes low enough (compared with the liquid phase) to support nucleation.

In practice, however, most commercial metals contain a large variety of insoluble impurities and the mould walls also provide other foreign substances. Under these conditions, and provided that the surfaces of such impurities are at least partially wetted by the liquid metal, the atoms in the liquid metal readily form the solid metal 'embryos' on the surfaces of these foreign substances (substrate). In other words, the substrate helps to overcome the opposition to nucleation resulting from the energy associated with the solid/liquid interface mentioned earlier.

Due to the presence of these impurities, most commercial metals can readily nucleate at temperatures of 1 to 10°C below the equilibrium temperature.

If the number of effective nuclei is insufficient a nucleating agent (a substance

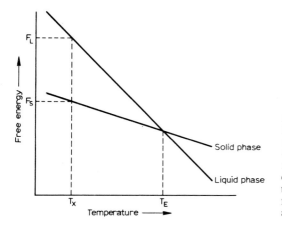

Fig. 5.38 Free energies of the liquid and solid phases in pure metals. T_E = thermodynamic equilibrium temperature, i.e. melting point, F_S = free energy of solid at temperature T_X, F_L = free energy of liquid at temperature T_X.

which can act as a nucleation catalyst) may be added to the molten metal. It may be a compound that is insoluble in the melt (but can be wetted by the liquid metal), or it may be a substance that will react with the liquid to form a nucleation catalyst.

Controlled nucleation is extremely important during the solidification of castings since the amount of nucleation controls the structure and therefore the properties of the solidified castings. Each grain or crystal grows from a single nucleus and the number of nuclei available during solidification determines the final grain size in the as-cast metallurgical structure. The rates of cooling are almost equally important, to allow the potential nuclei to act and grow. Controlled cooling and the addition of nucleating agents are standard methods for producing the correct grain size in castings.

If a casting is vibrated during solidification, a fine grain size can result. The mechanism, however, is not clearly understood at present.

Once a stable nucleus has formed, it grows by acquiring atoms from the liquid. The rate of growth is controlled by the amount of cooling below the thermodynamic equilibrium temperature, i.e., undercooling below the melting point. The growth rate increases with the degree of undercooling, reaches a maximum and then drops off on further undercooling.

When the molten metal is introduced into the mould cavity, the metal next to the mould walls starts to solidify first. This is because this region is rapidly undercooled by contact with the cold mould wall. The nucleation potential of the mould wall will also tend to produce a very large number of nuclei in this region. As solidification proceeds next to the mould walls, the latent heat of solidification is released, with the result that the remaining liquid is not undercooled to the degree necessary for further nucleation. Growth can, however, continue from the grains already formed, controlled by the rate of heat transfer from the casting through the mould walls, the direction of growth being in a direction opposite to the direction of heat flow. Growth is also favoured in certain crystallographic directions, so that not only conditions imposed by the direction of heat flow but also those imposed by crystallographic direction must be satisfied before the grains can continue to grow towards the centre of the casting. Only those grains which are favourably oriented will therefore grow towards the centre, resulting in columnar grain growth (illustrated schematically in fig. 5.39).

406 | METAL CASTING

In pure metals these columnar grains will extend to the centre of the casting but in commercial alloys they may be partially or wholly replaced by equi-axed grains, as shown in fig. 5.39. In commercial alloys, growth occurs most commonly in a dendritic manner, i.e., like trees, and this applies to columnar grains as well as the equi-axed variety. Since most castings are made from commercial alloys, it is more important to consider the freezing of alloys than pure metals.

5.9.2 Freezing of alloys

Freezing of alloys occurs over a temperature range, the extent of which is determined by the alloy system and the composition of the alloy. This point is illustrated by considering a simple hypothetical binary phase diagram for metals A and B (fig. 5.40). Alloy of composition X has a short freezing range (i.e., between temperatures T_x and T_z) while alloy Y has a long range (i.e., between T_Y and T_E). The only exceptions are in the cases of a composition of 100% A or 100% B and the composition E (eutectic), where freezing takes place at a constant temperature.

In the freezing of alloy X or Y the precipitating solid differs in composition from the liquid. This results in a concentration gradient being set up in the liquid which may have far-reaching effects on the freezing process and result in what is known as constitutional undercooling. A schematic illustration of constitutional undercooling is given in fig. 5.41. The instability arising out of this condition promotes the growth of spikes extending into the liquid. Continued growth of these spikes into the liquid in a direction opposite to the direction of heat flow, as well as in a lateral direction, results in the typical dendritic or tree-like structure illustrated in fig. 5.42.

If the cooling rate eventually results in a temperature gradient represented by the line OL (fig. 5.41), fresh nucleation is possible in the liquid, ahead of the solid/liquid interface. As these new nuclei grow, the process is repeated and fresh nuclei grow further into the liquid and so on. These grains grow in a dendritic equi-axed manner.

Referring back to fig. 5.39, alloy X freezes over a range of temperature and freezing may take place almost simultaneously throughout the cross-section in a 'mushy' manner, as opposed to the growth of solid from the mould wall towards the centre of the casting in the case of pure metals.

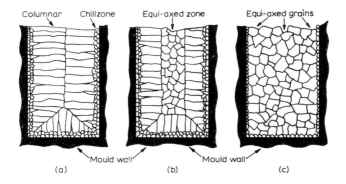

Fig. 5.39 Progress of solidification in pure metals. (a) Wholly Columnar, (b) Part columnar equi-axed, and (c) Wholly equi-axed. In alloys it generally indicates absence of thermal gradient and/or presence of nucleation catalyst.

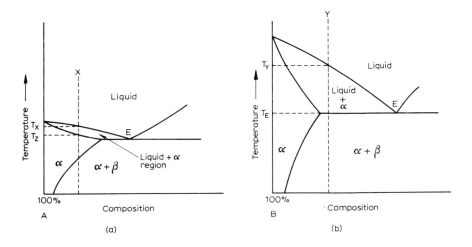

Fig. 5.40 Hypothetical phase diagrams to illustrate (a) short freezing range (b) long freezing range binary alloy systems.

The width of the region where solid and liquid may coexist within a casting poured from a commercial alloy at any one time during solidification is affected by a large number of factors, the most important of which are:
1. The solidification range of the alloy.
2. The thermal properties of the mould.
3. The thermal properties of the solid and liquid alloy.
4. The solidification temperature of the alloy.
5. The type of phase diagram and composition of the alloy.
6. The pouring temperature.

The shrinkage effects in these alloys are also complicated. This is becase localized

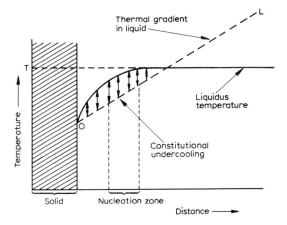

Fig. 5.41 Schematic representation of constitutional undercooling: Effect of composition gradient on the liquids, temperature, with the thermal gradient in the liquid superimposed.

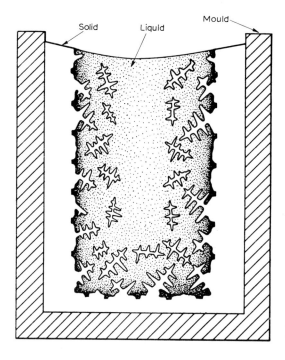

Fig. 5.42 Schematic illustrations of dendretic (tree-like) growth during the solidification of a cast alloy.

shrinkage in the liquid trapped between dendritic grains gives rise to small voids distributed throughout the cross-section, referred to as microporosity.

The solidification of a casting poured from an alloy of eutectic composition (alloy E in fig. 5.40) takes place in a manner very similar to that of pure metals. However, some differences exist:

1. The grain size of the two precipitating phases is much finer than that of a pure metal.
2. The shape of the precipitating particles can be quite varied and may include structures described as lamellar, rodlike, globular or polyhedral (fig. 5.43).
3. Eutectic alloys are known to freeze either exogenously (from the surface to the centre of the casting) or endogenously (random nucleation of cells, or colonies consisting of individual clusters of the two phases growing in the liquid in more or less spherical masses). In fact, only modified eutectics, e.g., Al + 12% Si (sodium modified), freeze in an exogeneous manner in commercial practice.

5.9.3 Related effects

a. Fluidity. In cast metals technology the term fluidity does not mean the reciprocal of viscosity. It refers to the ability of the liquid metal to fill the mould. Since all metals in their liquid state have the ability to fill all but the narrowest of mould cavities, fluidity in this sense reduces to an estimation of the 'fluid life' of the metal on being poured into the mould, i.e., the time lapse before the metal begins to solidify. 'Fluid life' is therefore largely controlled by the solidification characteristics of the metal and these in turn are dependent on such factors as metal composition, amount of superheat in the metal and the rate of heat transfer through the metal and the mould.

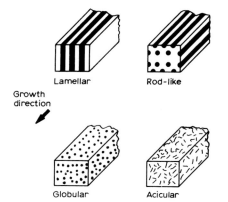

Fig. 5.43 Schematic cut-away view to illustrate eutectic morphology.

Other factors such as the metal viscosity, surface tension, surface oxide films, adsorbed gas films and suspended inclusions may also affect the fluidity of the metal to a certain extent, but these are of more or less academic interest in the practical problem of mould filling. The essential point is to make sure that the metal remains liquid until the entire mould has been filled.

b. Hot tearing. If the casting is restrained from contraction during the solidification process due to mould or core geometry or other constraints, the resulting stresses may cause cracking in the casting. Cracking occurs during the later stages of freezing when solidification is almost complete except for a thin film of liquid surrounding the grains (dendrites). The condition is aggravated by the presence of low melting point constituents in these regions and by a coarse grain size. It also appears to be made worse by the presence of small amounts of eutectic (believed to form interdendritic liquid films). Cracking is therefore associated with the solidification behaviour of the alloy in general and the freezing ranges and cooling rates in particular.

c. Evolution of gases. The solubility of gases dissolved in most of the more common casting alloys shows a sudden drop at the solidification temperature of the alloy, with the result that gases may be evolved during solidification due to supersaturation. The gas bubbles, formed by a process of nucleation, are sometimes unable to escape from the casting, thus giving rise to gas porosity.

d. Inoculation effects. The addition of a small amount of another metal or alloy before pouring may modify the structure and properties of the parent alloy as-cast. For example, the addition of sodium to a eutectic (12% Si) aluminium-silicon alloy and of magnesium to cast iron results in modified structures. None of the explanations advanced so far to explain these phenomena have been universally accepted, although it is recognized that these additions affect the nucleation and/or growth processes and thus alter the solidification patterns.

5.9.4 Gating system

The system through which molten metal flows to reach and fill the mould cavity is called the gating or running system.

The main objectives of a gating system may be summarized as follows.
1. The metal should flow through the gating system with as little turbulence as possible to prevent mould gases and air being trapped in the metal stream and also to limit the erosion of the mould and core surfaces.
2. The metal should enter the mould cavity in a manner that will promote directional solidification (i.e. solidification from remote points of the casting towards the feeder or metal reservoir which compensates for the shrinkage during solidification).
3. Metal, free from slag and dross, should be delivered to the mould cavity at a rate sufficient to fill the cavity completely before any freezing can occur.
4. The gating design should be easy to incorporate during the preparation of the mould and should be economically practical.

Some of these requirements are conflicting, so that any good gating system is the result of compromises.

Metal and mould compositions also affect the choice of design for a gating system. For example, a more elaborate system is necessary to prevent the formation of dross (oxides, etc.) in an easily oxidized metal such as aluminium than for cast iron, which can withstand considerable turbulence without drossing. Similarly, the characteristics of a heated ceramic mould (e.g., investment casting) will permit certain variations in design.

The basic requirement for any discussion on gating is an understanding of the mechanics of flow of molten metals in vertical and horizontal channels or passages.

5.9.5 The flow of molten metal

In the foundry, the term 'fluidity' means the ability of the metal to fill a mould. Fluidity, as mentioned earlier, is affected by several factors, the most important of which is superheat. The degree of superheat is indicated by the temperature of a metal or alloy above its melting point. Superheating, in effect, means that it will take a longer period of time for the metal to cool down to the temperature at which it will begin to solidify, i.e., superheating increases the length of time it will remain molten and flow. Care must be taken when increasing superheat, however, since excessively high temperatures can damage both the mould material and the as-cast metallurgical structure of the solidified casting.

The composition of the alloy also affects fluidity by determining the manner in which the metal in the channel freezes. Long freezing range alloys solidify in a 'mushy' manner with the growth of dendrites, while short freezing range alloys solidify by a gradual inward growth of solid metal from the mould wall (fig. 5.44). It will be apparent that metal flow in channels will be most hindered by the mode of solidification characterized by long freezing range alloys.

Gravity (i.e., acceleration due to gravity) affects the velocity of a stream of liquid metal, so that the height through which the metal stream is allowed to fall before entering the mould cavity will determine the velocity of entry for any particular set of channel dimensions for the gating system. The further a stream of molten metal is allowed to fall the greater its velocity.

The flow of liquid can be either turbulent or laminar (smooth). Turbulent flow is characterized by the irregular movement of the 'particles' of liquid, both across the stream and in the direction of flow. In laminar flow, the 'particles' of liquid metal proceed smoothly parallel to the direction of flow.

Some metals are inherently more sensitive to turbulence than others, e.g.,

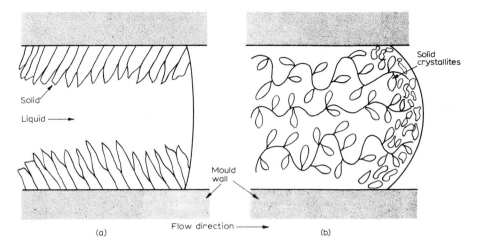

Fig. 5.44 Freezing in mould; channels (running system) (a) short freezing range alloy or pure metal. Columnar growth from the walls will leave central channel open for flow until the entire cestion is solid. (b) Long freezing range alloy. Mushy freezing will cause crystallites to be swept ahead of the matal stram and choke up the channel at an early stage.

High sensitivity	Medium sensitivity	Low sensitivity
Light alloys, such as aluminium and magnesium.	Low carbon steels, ductile iron.	Irons, other than ductile.

However, it must be recognized that laminar flow is not achieved in practice since the velocities necessary to ensure this are too low (impractical for mould-filling purposes) for commercial casting alloys. The importance attached to the velocity of entry arises from the efforts to control the degree of turbulence. The greater the velocity, the more turbulent is the flow of the metal for any given set of conditions.

The basic laws applicable to the flow of liquid metals may be summarized as follows.

a. The law of continuity states that for a system (shown in fig. 5.45) with impermeable walls and filled with an incompressible liquid,

$$Q = A_1 V_1 = A_2 V_3$$

where Q is the rate of flow, A the area and $V =$ the velocity, and $V_2 = \sqrt{2gh}$ at the exit point.

Therefore, the velocity of a stream is increased when it enters a smaller channel.

b. Bernoulli's theorem which is based on the first law of thermodynamics, states that the sum of the potential energy, the velocity energy, the pressure energy and the frictional energy of a flowing liquid is a constant, i.e.,

$$wH + wPV + (wv^2/2g) + wF = K$$

where w is the total weight of liquid flowing (lb), H the height of liquid (ft), P the

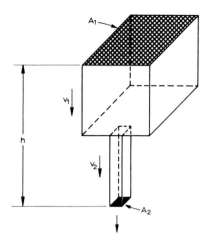

Fig. 5.45 The law of continuity Q = rate of flow $= A_1 V_1 = A_2 V_2$, and $V_2 = \sqrt{2gh}$ at the exit point.

static pressure in the liquid (lb/ft²), V the specific volume of liquid (ft³/lb), g the acceleration due to gravity (32 ft/s²), v the velocity (ft/s) and F the frictional loss (ft).

If this equation is divided by w, all the terms have dimensions of length and the equation reduces to

$$H + Pv + v^2/2g) + F = K.$$

The potential energy of the metal can be considered to be a maximum as the metal enters the pouring cup (or pouring basin). This energy is then rapidly changed to velocity energy (i.e. kinetic energy) and pressure energy as the metal passes into the mould through the gating system. Once flow is established, the height of the liquid and frictional loss are virtually constant, so that velocity is high when pressure is low, and vice versa, according to this equation.

While metal is flowing, there is a loss of energy in the form of fluid friction between the metal and the channel walls. The loss of energy through heat loss, which eventually leads to the solidification of the metal, is not included in these considerations.

   c. *Newton's law of motion* states another characteristic of fluid flow, i.e., that an object (or fluid) in motion remains in motion in a straight line and at the same speed until some other force is exerted on it to change its speed or direction.

   The consequences of this principle applied to the flow of metals may be summarized as follows:
1. Any sudden change in the direction or speed of flow will create turbulence.
2. A stream of liquid flowing round a sharp corner will contract as it makes the turn and will cause the liquid to separate from the inside corner of the channel and create a low-pressure space. The result is a sucking action (known as the aspiration effect) which may suck in air or gas through the permeable walls of the mould, with detrimental effects on the soundness of the casting. Erosion may take place at the high-pressure point at the outside corner.
3. Another effect may be 'vortex' formation in the pouring basin, which is similar to the vortex formation in bath tubs when the water is allowed to run out.

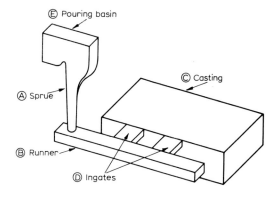

Fig. 5.46 The basic elements in a gating system.

5.9.6 Gating design (horizontal)

The illustration in fig. 5.46 shows the basic parts of a gating system. The sprue or down-gate (A) is a vertical passageway through which the molten metal enters the runner (B). The pouring basin (E) is the enlarged offset portion at the entrance to the sprue into which the molten metal is first poured. The runner (B) is the horizontal passageway through which the molten metal flows to the in-gates (D). The gate (D) connects the runner to the mould cavity. That part of the runner (C) which extends beyond the last gate as a blind end is known as the runner extension.

Having identified the parts we can turn our attention to the functions performed by these parts.

a. Pouring basin. The function of the pouring basin or cup is to initiate the proper flow conditions by acting as a reservoir of molten metal, thus keeping the gating system filled with metal during the pouring operation. Slag and dross are usually lighter than the molten metal and will float on the surface; thus, if the pouring basin is properly filled, slag and dross can be prevented from entering the sprue and hence the mould cavity. Excessive turbulence of the metal in the gating system can also be prevented by a properly designed basin with smooth contours.

b. The sprue. The design of the sprue is extremely important in establishing desirable flow conditions. When the metal first enters the sprue from the basin (by free fall before the system is filled) the stream of falling metal is not parallel but has a narrower cross-section at the bottom, as a result of the increased velocity at the lower level, so that, if the sprue walls are parallel, air and mould gases may be drawn into the stream. For this reason, the sprue is usually tapered to fit the natural contours of the falling metal (fig. 5.47). The flow rate through a tapered sprue is about 15% greater than the flow rate through a parallel sprue of equal exit area. Provided that there are no other constrictions in the runner or in-gates, the cross-sectional area at the bottom of the sprue will control the flow rate and the pouring rate for a particular sprue height. The bottom of the sprue, which usually has the smallest cross-sectional area in the entire gating system, is referred to as the 'choke' and the velocity of the metal stream is a maximum here. To decrease the velocity (and thus the kinetic energy and turbulence of the molten metal and the entrapment of air and dross), the junction between the sprue and runner is usually enlarged as shown in fig. 5.47. This region is known as the sprue base.

414 | METAL CASTING

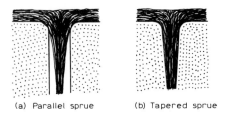

(a) Parallel sprue   (b) Tapered sprue

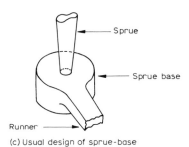

(c) Usual design of sprue-base

Fig. 5.47 Contour of falling metal (a) in parallel sprue, (b) sprue tapered to fit and (c) usual design of sprue base.

c. The runner. A gating system may have one or more runners, depending on the shape and size of the casting. To reduce velocity in the runners appreciably, the cross-sectional area of the runners must be greater than the outlet or choke area of the tapered sprue. Runners should also be designed to avoid abrupt changes in the direction of flow of the molten metal. The runners are usually moulded in the drag part of the mould to make sure that they are filled before the metal enters the in-gates, which are in the cope part (see fig. 5.48). Runner design, therefore, is aimed at reducing turbulence and air entrapment by reducing the velocity of flow and ensuring that the runner is properly filled before any metal enters the mould cavity through the in-gates. The reduction in velocity also allows lighter impurities and inclusions to float and stick to the top surface of the runner channel. The runner extension serves the same purpose by entrapping the washed-down sand and other impurities.

d. Gates. Gates or in-gates are the final sections of the gating system connecting the runners to the casting. In the multiple gate system shown in fig. 5.49, back-pressure

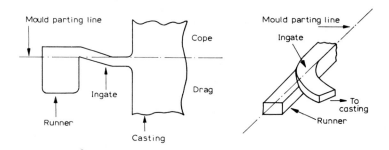

Fig. 5.48 Two examples of a split level arrangements for runners and ingates, designed to ensure that the runners are full before metal enters the mould cavity (casting proper).

is created when the molten metal reaches the end of the runner. This pressure works its way back towards the sprue base, causing molten metal to flow through the gates marked C, A, B, in that order. Therefore, the amount of metal entering through the gates will not be equal even though the cross-sectional areas of the gates C, A, B may be the same. If the total cross-sectional area of the gates is less than the cross-sectional area of the runner, a more or less even flow will result after the initial period, but if the total cross-sectional area is greater, the condition of uneven flow will persist. It is therefore most important to determine the most suitable area of cross-section of the in-gates in relation to the rest of the system, as well as to look at the particular geometry to ensure even flow and avoid pressure differentials.

In designing gating systems, the most important objective is to maintain a flow at constant velocity and pressure by considering the proportional relationship between the cross-sectional areas of the sprue, runners and gates through which the metal flows. This ratio is known as the gating ratio.

5.9.7 Gating design (vertical)

The discussions so far have been related to horizontal gating systems where the in-gates are located at the parting line (between cope and drag) of the casting, since this is the most common method of introducing metal into the mould cavity. However, depending on the shape and size of the casting and the type of alloy being cast, many different types of vertical running systems are also used in order to satisfy technological or economical requirements. It is not possible to go into these cases individually but fig. 5.50 will serve as a guide.

5.9.8 Feeding of castings

It has been mentioned in section 5.2.2 that when liquid metal in a mould cavity is allowed to cool down to room temperature, shrinkage takes place in three stages, i.e., contraction in the liquid metal, contraction during the liquid to solid transformation and finally contraction in the solid state. Of these three stages, only the second, i.e., contraction during the liquid to solid transformation, is important with regard to feeding.

From a knowledge of the amount of solidification shrinkage in different metals and

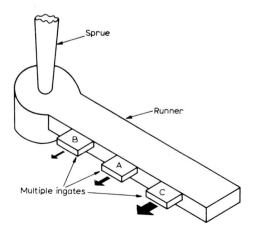

Fig. 5.49 Unequal flow through multiple ingates when the running system is not properly filled in the initial stages.

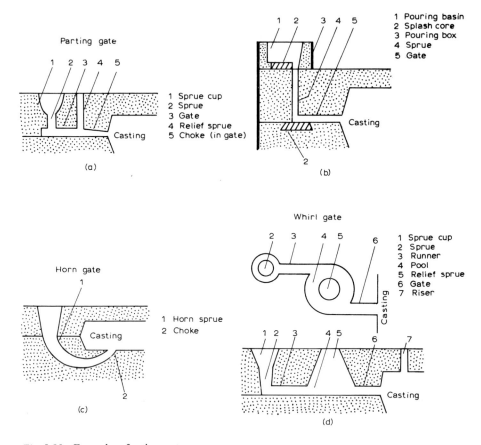

Fig. 5.50 Examples of gating systems.

alloys it is clear that voids will result within solidifying metals restricted a container such as a mould, unless ways are found to compensate for this shrinkage by introducing the requisite volume of liquid metal into the container. For example, if molten steel is allowed to solidify in a container with a volume equal to 1 ft^3, upon solidification the void is expected to have a volume of 0.03 ft^3 (solidification shrinkage of steel is about 3% by volume).

The primary function of a feeder or riser is therefore to compensate for the shrinkage in volume during solidification by feeding liquid metal to the casting.

Table 5.6 shows the approximate volume shrinkage during solidification for some common commercially cast metals.

The important point to note is that the contraction can take place at constant temperature or over a range of temperatures, depending on the composition and the phase diagram for the alloy system. However, the actual amount of contraction *per se* is not the determining factor in riser design, although in the case of grey cast iron, where graphitization occurs during the final stages of solidification, the resulting expansion counteracts volume shrinkage and therefore feeding requirements for this type of iron are not critical.

Contraction takes place during solidification due to a density change accompanying

Table 5.6

Metal	Percentage volumetric solidification shrinkage
Low-carbon steel	3
High-carbon steel	4
White iron	5
Grey iron	Usually expansion up to 2%
α brass	4.5
Aluminium bronze	4
Aluminium-silicon alloy (12% Si)	3.8
Magnesium alloys	4
Zinc alloys	6

the more orderly and densely packed atomic structure that results as the metal solidifies, and feeding requirements are largely controlled by the mechanism of solidification of the metal.

Two classes of alloys, typified by their basic modes of solidification, have been briefly described earlier:
1. The skin freezing type of alloy in which solidification proceeds from the mould wall towards the centre of the casting, generally described as short freezing range alloys.
2. Alloys which solidify in a 'mushy' manner virtually throughout the cross-section of the casting, classed as long freezing range alloys.

Provided that the metal in the riser (feeder) stays liquid until the casting has solidified and that solidification occurs from the furthest point away from the riser, towards the riser, it is comparatively easy to ensure satisfactory feeding in short freezing range alloys. In long freezing range alloys, however, due to the presence of intertwined dendrites and liquid metal throughout the body of the casting, the passage of feed metal to deficient areas through the capillary spaces between the dendrites becomes increasingly more difficult as solidification proceeds. It is almost impossible to ensure completely satisfactory feeding in these alloys (see 5.51).

Table 5.6 indicates that a relatively small amount of feed metal is necessary to compensate for shrinkage in common casting alloys. The problem is therefore not due to the actual amount of metal required but to the difficulty of keeping the riser liquid and keeping the feed channels open during the entire solidification period of the casting. Many factors are involved; the following discussion outlines the most important ones.

a. Riser shape. Heat is transferred from the casting to its surroundings by radiation, conduction and convection. Without going into the relative importance of these three modes of heat transfer, it can be shown that the ratio of the surface area of the casting to its volume controls the rate of heat transfer for a given set of metal/mould conditions. In other words, the solidification time is proportional to V^2/A^2, where V and A are the volume and surface area of the casting respectively. This relationship was first demonstrated by Chvorinov and, although somewhat oversimplified, indicates that for a riser to have a solidification time greater than that of the casting, the ratio for the riser should be greater than the one for the casting. Hence, the most favourable ratio would be obtained if the riser were spherical in shape. Prac-

418 | METAL CASTING

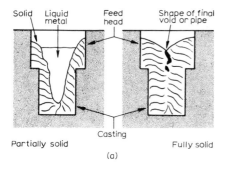
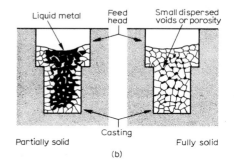

Fig. 5.51 (a) Progressive solidification from the mould walls allows feed metal to reach all areas through the central channel. (b) Mushy solidification resulting in tortous feed channels. Difficulty in feed metal to all regions may result in dispersed porosity.

tical difficulties in moulding spheres dictate that the next best shape, i.e. a cylindrical shape, be used. Except for some specially designed risers such as blind risers (see later), the cylindrical shape is almost universally adopted.

b. Riser size. Various procedures have been developed to calculate riser size and the most important consideration appears to be the 'shape factor' of the casting, which is determined by a simplification based on its volume/surface area ratio. It will be realized that the use of a riser does not eliminate shrinkage voids altogether, but restricts them to regions outside the casting proper, i.e., in the riser itself. The shape of this shrinkage void, or 'pipe' as it is called, will therefore also influence the choice of a particular riser size. If the pipe tends to extend into the body of the casting, it may be necessary to enlarge the riser sufficiently to overcome this. The shape of the pipe is dependent on complex metallurgical factors beyond the scope of the present discussion.

c. Location of riser. For an alloy with a wide freezing range, the casting will require a closer spacing of risers than for other alloys. This is because the effective feeding distance of a riser in a wide freezing range alloy is restricted for the reasons already mentioned. This problem is further aggravated if the riser is located on a section of the casting which is smaller than adjacent sections. In such a case, the smaller section will solidify first, thus cutting off feed metal to the larger section. The actual distance feed metal can travel through a section is limited by the solidification characteristics of the alloy concerned even where only uniform sections are involved. Empirical methods have been developed to calculate the feeding distance in different alloys through different sections, and the reader is referred to the appropriate literature, listed in the bibliography, for further details.

d. Riser neck. The connection between the riser and the casting, known as the neck, is important for two reasons. First, it influences the riser's ability to feed the casting, and secondly, it determines how easily the riser can be detached from the casting. A comparatively recent development to facilitate removal of the riser without sacrificing efficiency of feeding is the 'washburn core'. A hard wafer core with a small hole in the centre, sometimes made of ceramic material, is interposed between the riser and the casting, restricting the area of contact between them so that removal is comparatively easy. The thermal properties of the core ensure that the small channel is kept open until the casting has solidified so that feeding is not restricted in any way.

In general, however, the riser neck should be so designed that it is able to control the depth of the shrinkage cavity by solidifying just before the riser freezes, thereby preventing the shrinkage cavity from extending into the casting.

e. Chills. It has been stated earlier that directional solidification, i.e. solidification towards the riser from the remote areas, is essential for producing sound castings. This process can he helped by chilling the metal in those portions of the casting that are more remote from the liquid metal source in the riser. Both external and internal chills can be used for this purpose. External chills are metal inserts of steel, cast iron or copper that are placed at appropriate locations in the mould to increase the freezing rate of the metal at those points. Chills may be of standard shapes or specially shaped to conform to the mould-cavity contour. An example of an external chill is shown in fig. 5.52(a).

Internal chills are sometimes placed in the mould cavity at locations that cannot be effectively reached with external chills, as shown in fig. 5.52(b). The use of such chills is not recommended since the chill may not fuse properly with the casting, thus endangering the soundness of the casting in that region. However, bosses and lugs that are to be drilled or bored are suitable locations for internal chills.

f. Insulators and exothermic compounds. A riser can be made more efficient by adding some exothermic compound to the top of the riser immediately after the metal has been cast. These compounds are added for two purposes: to generate heat and to act as radiation shields. The combined effect is to keep the riser molten for a longer period. A shield against radiation may be provided by substances such as powdered

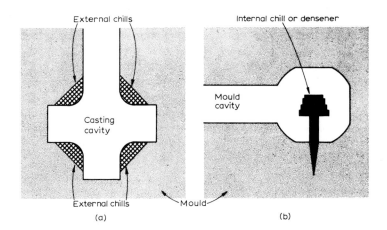

Fig. 5.52 (a) Appropriate external chills may help to eliminate shrinkage porosity in sections such as 'X' and 'T' and 'Y' junctions. (b) Internal chills may be placed in thicker sections of the casting cavity to promote rapid solidification. Normally only in regions not easily covered by external chills.

graphite or charcoal, rice or oat husks and refractory powders. An exothermic reaction may be provided by a mixture of aluminium and iron oxide powders which reacts according to the equation

$$2Al + Fe_2O_3 = 2Fe + Al_2O_3.$$

Oxidation of the metal in the riser by a stream of oxygen and heating with an electric arc are other means of keeping the riser molten in very large castings.

Insulating sleeves and/or exothermic compounds may also be used to form the side walls of the riser. Mouldable exothermic compounds are generally used for this purpose.

Insulating pads are sometimes incorporated in the mould in the thinner sections so that the freezing is delayed in these regions to keep the feeding channels open.

It must be noted, however, that not all insulating materials and/or exothermic compounds can be used with all types of alloys, since undesirable contamination of the metal may occur with consequent deterioration in the quality of the casting. A careful analysis of the metallurgical factors involved must be made before recommending the use of these devices for any particular application.

5.10 METAL MELTING

In the early days of the foundry industry, the supply of molten metal for casting was obtained directly from the primary furnaces used for the extraction of metals from their ores (e.g., the blast furnace for the production of pig iron). With the growth of the industry, the demand for metals and alloys of different chemical compositions, 'tailor-made' for specific applications, increased rapidly. It was soon realized that the primary extraction furnaces were no longer versatile enough to meet this demand. Today, the methods used in the foundry to ensure supplies of molten metal rarely, if ever, involve primary extraction. In this sense, the melting furnaces used in the foundry may be described as secondary units.

The basic metals are purchased from primary extractors or dealers in scrap and then melted with suitable alloying and refining in the foundry to obtain the molten metal to the required specification. This operation is quite often carried out in separate establishments (manufacturers and suppliers of secondary ingots) so that the foundry can purchase ingots ready made to the required specification. This is particularly advantageous for small foundries where the technical knowledge or the capital resources necessary to establish a refining and alloying plant are not available. Some of the furnaces in the foundry can therefore be classed as simple melting furnaces, although it must be recognized that even in these operations, where extensive alloying or refining are not involved, adequate controls are necessary to produce molten metal free from undesirable contamination from gases and other impurities.

The design and operation of the many types of furnaces used in the metallurgical industry and the thermochemical principles involved in melting, refining and alloying, are well documented and many authoritative books and other literature are also readily available (see bibliography and suggestions for further reading).

In this section, therefore, the proper selection of a furnace for a particular process and the principles controlling such choice are briefly discussed. Limitations in space preclude a detailed discussion on the design and operation of the very large number of furnaces currently available.

5.10.1 Choice of furnace

Although furnaces differ greatly in design, in the kinds of fuel used and in capital and operating costs, there are only a few major factors which, in addition to volume requirements and metal specifications, control the choice. The most important factors are as follows.

a. The shape, size and compostion of the raw materials available. The design of the furnace required to carry out substantial refining and alloying will obviously not be the same as the one required to melt secondary ingots. However, the control of metal chemistry is not limited to the control of the standard elements present in the alloy. It also involves the control of gases and other impurities (sometimes present in traces only). For example, the action of trace impurities is thought to be responsible for the difference in the chilling tendencies exhibited by cast iron melted in cupolas and electric furnaces respectively. These so-called 'hereditary' effects, resulting from melting in different furnaces, are observed in other alloys, manifested as differences in machinability, even though the nominal chemical compositions and other details of casting procedure were identical.

b. Metal temperature. In section 5.9.5. it is mentioned that the degree of superheat in the metal at the time of casting is paramount in determining the mould-filling capacity of the metal. Excessive temperatures, on the other hand, can have damaging effects, e.g., undesirable oxidation, altered freezing characteristics, reaction with the furnace refractory and mould material, excessive gas pick-up and so on. To add to the problem, the method of metal transfer from the furnace to the moulds and the time elapsed between the pouring of the first mould and the last will also influence the temperature required in the melting furnace. Thus, the optimum temperature of the metal, as tapped from the furnace, has often to be controlled within narrow limits, so that a consistently obtainable tapping temperature from a given furnace may be of primary importance in the selection of the appropriate furnace type.

c. Volume and rate of metal delivery. Smooth and economical operation in the foundry demands that two vital aspects of production planning are thoroughly understood. These are the rate of metal delivery and the sequence or mode of metal delivery.

When large castings are made the batch process of melting is preferable, since the time needed to accumulate the required amount of metal from a continuous melting unit such as a cupola might seriously affect control over the pouring temperature (i.e., the metal may be too cold by the time it is poured into the mould).

On the other hand, if a batch-type melting unit is employed in a mechanized foundry producing a large number of small castings, it is necessary to provide a heated receiver (holding furnace) to provide a continuous supply of hot metal for casting. If consistent temperatures and compositions can be obtained, a continuous melting unit will obviously be more suitable for this type of foundry.

Variable (i.e., part of the time continuous and part intermittent) demands are not uncommon and many foundries therefore choose their furnaces solely on the basis of economical and technological advantages, regardless of whether or not continuous or batch melting is involved. The necessary flexibility is built into the system by providing a heated holding furnace (receiver), into which the metal from the melting furnace is transferred and held at the appropriate temperature. Given that the capacity of the receiver has been properly calculated, fluctuations in demand do not create disruptions or down time in the production cycle. Many foundries are using computers to facilitate this vital aspect of production planning.

However, not all metals and alloys can be held in a receiver without deterioration and therefore (even disregarding the cost) the receiver cannot be considered a panacea.

d. Melting cost. The balance between capital investment and operating cost is often critical. Capital invested in electric melting units is usually much higher than for units using conventional fuels such as coal, coke, fuel oil and gas. On the other hand, if the demand for molten metal is substantial and continuous, depreciation based on the full capacity of the plant may be a nominal amount on the cost per ton, and need not inflate the operating cost significantly. If, however, a reduction in demand occurs, depreciation assumes prohibitive proportions and the operating cost increases accordingly. Where large volumes and continuous demands are predicted, electric melting units will obviously be more economical in the long run, owing to their low operating cost. A reasonable prediction regarding the metal requirements over a period of 5 to 10 years is therefore necessary before a satisfactory choice can be made.

5.10.2 *General features of melting furnaces*

Some of the principal types of melting furnaces are represented by the schematic diagrams in fig. 5.52. They may be further subdivided into two main groups:
1. Furnaces for batch melting.
2. Furnaces for continuous melting.

a. Batch processes. In all batch processes, the metal, in a bath or similarly shaped refractory container, is heated by electricity or by the combustion of oil, gas or solid fuels like coal or coke. The oldest type of batch furnace is the crucible furnace, which is also the most common method for the melting of non-ferrous alloys. The crucible is heated from the outside and transfers the heat to the metal through the side walls. Conventional fuels, i.e., coke, gas or fuel oil, are used. The melting of ferrous alloys in crucible furnaces is considered uneconomical and the desirable metal temperatures are also difficult to obtain, so that in spite of the long history of melting iron and steel in crucible furnaces, this practice is no longer favoured.

The modern substitutes for crucible-type furnaces for ferrous alloys are the induction melting furnaces, which are fast superseding all other methods of melting in the foundry in popularity. In this furnace, the heat is developed directly in the metal instead of being applied externally to a container. Many different designs incorporating either high, medium or low frequencies of a.c. power are utilized. The basic principle, however, is simply that eddy currents are induced in the metal by means of an induction coil positioned outside and encircling the containing crucible. The resistance of the metal to the passage of this induced current produces the heat necessary to melt and superheat the metal. In the induction melting of non-ferrous metals, a conducting crucible of graphite or silicon carbide is often used because the resistance of the solid metal or the melt is not always sufficient to generate the heat necessary for melting and superheating. The induced current in the crucible can heat it up rapidly, and this heat is then transferred to the metallic charge within it.

Reverberatory, open-hearth and air furnaces are also batch-type furnaces, in which the flame from the combustion of the fuel heats the bath both directly and by radiation from the roof. Some oxidation and contamination of the metal by contact with the products of combustion is inevitable. These furnaces sometimes find favour in non-ferrous foundries for the melting and refining of foundry returns and other scrap in bulk for the production of secondary ingots. These ingots are subsequently remelted in crucible or other furnaces for the production of the actual castings. Open-hearth furnace melting used to be quite common for the production of large steel castings, up to 100 tons in weight or more. However, with the availability of large electric arc furnaces, very few open-hearth furnaces are in use now for this purpose.

There are two basic types of electric arc furnaces: the 'direct arc' and the 'indirect arc' designs. In direct arc furnaces the heat energy is generated in the metal in the immediate vicinity of the electrodes which arc through the metal. In the indirect arc furnaces, the heat of the arc is radiated to the metal surface and the roof and side walls. The efficiency of heat transfer to the metal is improved by rocking the furnace about its horizontal axis so that the metal is alternately brought into contact with the heated side walls. Often, the initial stage in melting in indirect arc furnaces is due to electrical resistance from the cold charge of metal placed between the electrodes.

The side-blown converter also finds limited use in the foundry as a batch unit for the production of general purpose steel castings in small quantities. The converter is not really a melting furnace but only a reaction vessel. In this process, cast iron is melted in a cupola or other suitable furnace and then transferred to the converter. Air or oxygen is blown into the converter and causes the combustion of silicon, carbon and other elements present in the cast iron, reducing the level of these elements and thus converting the metal into low-carbon steel. Suitable alloy additions may be made before casting to bring the metal up to the necessary specification. The combustion processes are exothermic so that the metal is heated up rather than cooled down by the blow process. The blow is stopped as soon as the combustion of the elements is complete.

b. Continuous processes. The most widely used continuous melting unit is the cupola, which has no real bath. The metal charge is melted by the hot gases resulting from the combustion of coke and the metal droplets gain additional heat as they percolate through the incandescent coke bed.

The coke consumed by combustion and the metal melted are continuously replaced by adding fresh charges from near the top of the vertical shaft. Limestone and other fluxing materials are added in requisite quantities together with the coke charge to form a fluid slag which effectively entrains most of the impurities and oxides.

The cupola is used mainly for melting cast iron but occasionally for melting brass and bronze. The average melting rate for iron is about 20 lb/min per square foot of cross-section of the melting zone. Air for combustion of the coke is delivered through tuyeres via a wind-box surrounding the outer shell of the cupola from fan-type or positive-displacement-type blowers. The thermal efficiency of the cupola may be improved by heating the combustion air, which may also result in other improvements in metal quality, notably the tapping temperature. Such installations are known as 'hot blast cupolas'.

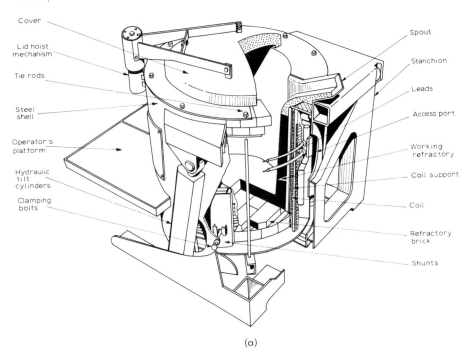

(a)

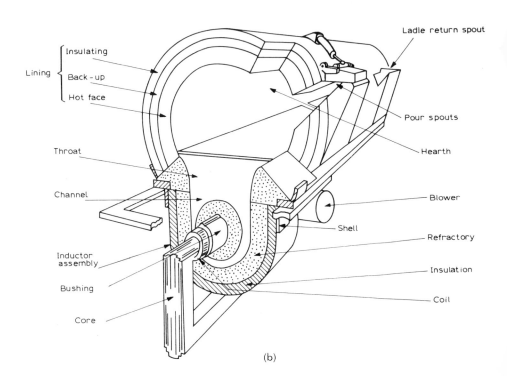

(b)

METAL MELTING | 425

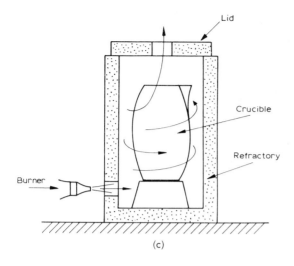

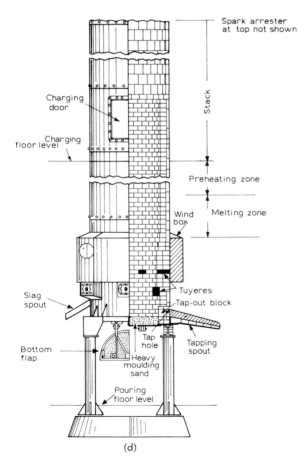

Fig. 5.53 (a) Cut-away view of line frequently coreless induction furnace. (b) Cut-away view of line frequently channel-type induction furnace. (c) Gas-fired lift out type crucible furnace arrangement. (d) Sectional view of the basic elements of a conventional cupola.

5.10.3 Metallurgical features

An understanding of the complex metallurgical factors controlling the design, selection and operation of furnaces requires detailed knowledge of thermochemical principles, reaction kinetics and chemical equilibria, as well as some acquaintance with the behaviour of refractory and other materials used in furnace construction. A knowledge of metallurgical calculations involving mass and energy balances is also essential. Such considerations are beyond the scope of the present discussions. However, an outline of some of the basic features will help the non-specialist reader to appreciate the problem at hand. The bibliography and suggestions for further reading are intended for readers who require comprehensive details regarding the processes.

a. Non-ferrous. The melting of non-ferrous metals and alloys in the foundry is somewhat different from that of ferrous metals, since the chemistry of non-ferrous metals is largely controlled by the selection of melting materials of the proper specification and quality. Extensive refining operations are rarely indicated and once a satisfactory source of secondary ingot or scrap has been found, the main problems arise from the solution of gases during melting and alloying. The most undesirable gas is usually hydrogen, derived from the dissociation of water vapour in contact with hot metal. Protection from excessive oxidation may also be a problem during melting.

For light metals like aluminium and magnesium alloys, the refining process is accomplished by the use of gaseous or solid fluxes which entrain the dross and other impurities and also remove the amount of gases to a safe level. Crucible melting, which is the commonest method for melting non-ferrous alloys, suffers from a disadvantage in that the source of heat is derived from the combustion of coke, fuel oil or gas, so that the combustion products will invariably contain hydrogen gas and water vapour in addition to carbon dioxide, sulphur dioxide and other gases. This may be a source of contamination for the metal.

Crucible melting in medium- or low-frequency induction melting units does not suffer from this disadvantage but certain other problems associated with this type of melting preclude its use in many cases. For example, for complex metallurgical reasons, copper alloys are generally slightly oxidized (intentionally) during melting and then subsequently deoxidized before pouring the metal into the casting. Oxidation during melting is better achieved by controlling the combustion to produce a slightly oxidizing atmosphere in a fuel-fired furnace. Because of the low electrical resistivity of copper, induction melting may be uneconomical or it may limit the choice of crucible materials. Molten magnesium reacts with siliceous refractories and for this reason is best melted in crucibles made of steel. Steel crucibles cannot be used for induction melting of magnesium since the crucible will tend to melt before the magnesium alloy.

Induction melting furnaces, are very suitable for melting aluminium alloys the only special requirement being some provision for removing sludge from the bottom of the crucible, since some of the oxides and other impurities are actually heavier than the molten metal and therefore sink to the bottom. This occurs in the melting of magnesium alloys as well.

Copper base alloys are occasionally melted in reverberatory furnaces when large volume production is desired. Oxidizing conditions are easily produced and the process can be economical.

Brasses and bronzes are sometimes melted in indirect arc electric furnaces but the bulk of production is from crucible furnaces.

Each method must therefore be assessed on the basis of the chemical composition of the metal required and the reactivity, thermal or electrical conductivity of the metal and many other considerations, and in general, except for aluminium where induction melting has many advantages, the crucible melting technique is the most flexible for the melting of non-ferrous metals.

b. Ferrous. Melting of ferrous metals like cast iron and steel has received much more attention because the majority of castings are made in these metals and also because considerable refining is sometimes necessary when melting them.

Until recently the cupola functioned as a cheap melting unit only and no serious attempts were made to control the composition of the resulting metal beyond ensuring that the average composition of the charge material was appropriate. This meant that expensive raw materials had to be used in most cases, since the successive remelting of foundry returns (runners and risers) and other scrap gradually decreased the carbon, silicon and manganese contents of the metal while increasing the sulphur content. With the development of spheroidal graphite iron (iron with the graphite distributed in the form of nodules making the iron ductile – in contrast with conventional cast irons where the graphite is distributed in the form of flakes), the need for closer control of the composition, particular the carbon, phosphorus and sulphur content, became more urgent. This led to the development of the basic cupola.

In the basic cupola the usual fireclay refractories are replaced by either magnesite brick, monolithic magnesite rammed lining or by a water-cooled steel shell which develops a thin layer of basic slag from the limestone and other fluxes. A cold blast can be used with magnesite lining but a hot blast is usually required for water-cooled cupolas. With suitable slag control, i.e., control of the amount of ferrous oxide (FeO) and the basicity of the slag, it is possible to obtain sulphur contents of less than 0.02%. Sulphur removal is also helped by high operating temperatures and high carbon contents in the metal. Unfortunately, due to the lower operating temperatures necessary for the effective removal of phosphorus, simultaneous removal of sulphur and phosphorus is not practicable.

The solution of carbon from coke by molten iron depends on the composition and temperature of the iron, the coke/iron interface and the type of coke used. In general, both hot-blast and basic cupolas operate at higher temperatures than the conventional firebrick-lined cupolas, and carbon solution is therefore increased. In basic cupolas, due to the nature of the slag, metal-coke contact is also superior, so that increased carbon pick-up is possible.

Hot-blast cupolas are often used to economize on coke and to allow a higher proportion of steel scrap to be used in the charge. Steel scrap is usually cheaper than pig iron so that further economies are possible. The hot blast reduces the silicon loss in the cupola, while at the same time increasing the carbon content of the metal. Carbon and silicon are the most important elements in determining the properties of cast irons and effective control of these elements is essential for producing castings of the required quality. Many different systems of preheating the blast are in use, the optimum blast temperature being in the neighbourhood of 500 to 600°C.

Carbon-lined cupolas (neutral) have also been used. The main advantage is that these cupolas can operate under either acid or basic slag conditions.

In the melting of special cast irons, which include spheroidal graphite irons, malleable irons and alloy cast irons, the control possible in cupolas may not be adequate to produce the molten metal to the rigid specifications called for unless selected scrap and other raw materials are available for the furnace charge. In such cases the metal may be initially melted in a cupola and then the molten metal treated in an electric arc furnace to adjust the final composition, make special alloy additions and superheat the metal to the required level before casting. Such processes are known as duplex processes. Occasionally the iron may be melted and refined entirely in an arc furnace, but the process is more expensive and high production rates are difficult to achieve.

The induction furnace is an ideal melting unit for cast irons as long as the charge can have the same chemical composition as the casting. Apart from gas elimination, very little refining is practicable because the slag is colder than the metal and the refractory lining or crucible walls have to be quite thin for efficient melting. Slag erosion, accelerated by the stirring action of the induced current, can cause early failure of the furnace walls if efforts are made to utilize refining slags to any great extent. Low-frequency or mains-frequency induction furnaces are becoming increasingly popular, particularly as holding furnaces for metal initially melted in the cupola.

Air furnaces are often used for the melting of malleable iron and to some extent grey cast iron. Except for a small decrease in the carbon content, no significant changes take place in the charge analysis so that no refining or other alterations in the composition are practicable. It is claimed, however, that the chilling tendency of the iron is markedly increased by melting in an air furnace. This effect may be related to the solution of the furnace gases, notably hydrogen, or to the destruction of potential nuclei for graphitization.

Melting of steel in foundries is almost invariably carried out in either direct arc furnaces or in induction melting units. When the volume of metal required is comparatively small and suitable scrap is available to make up the charge, induction furnaces are generally used. Induction furnaces are preferred for steels highly alloyed with nickel, chromium and other elements for the production of heat- and corrosion-resistant castings. Induction melting is also resorted to in the manufacture of special quality heat- and corrosion-resistant alloys for investment castings, particularly when melting under vacuum is needed. Fig. 5.54 illustrates an actual installation of a small induction melting furnace.

For larger volumes of general purpose steel melting in the foundries, the direct arc furnaces are generally used. These may be lined with acid or basic refractories. Basic-lined furnaces are preferred because they are more resistant to attack by the feed charges and refining constituents. The greater chemical stability allows a greater flexibility in charge composition and a considerable refinement with poor or contaminated charges. A double slag practice, i.e., oxidizing slag to oxidize the impurities followed by a reducing slag to reduce the iron oxide back to iron, is usually resorted to in such cases. Acid-lined arc furnaces are cheaper to operate and generally function on single slag practice, provided that reasonably clean scrap of appropriate composition is available.

Open-hearth furnaces for melting are useful in the foundry only when batch melting of large quantities is required. The chemistry of the basic and acid open-hearth processes is exhaustively described in the literature and needs no further mention here.

Finally, small side-blown converters are sometimes used in foundries where the demand for molten steel is so small or irregular that the installation of conventional steel melting furnaces is not justified. These foundries are primarily manufacturers of iron castings and usually have the necessary installations for melting cast iron. The

Fig. 5.54 Tapping on a small frequency induction furnace.

converter, in effect, is a reaction vessel which converts molten iron into steel by the oxidation of carbon, silicon and manganese by a blast of air or oxygen. Only small acid-lined converters are used since basic lined converters require high phosphorus (about 2%) irons for operation. No refining is therefore accomplished and in general the resulting steel may contain objectionable amounts of nitrogen unless pure oxygen is used for the blow. Metal loss is also quite high and may be up to 15% or more.

5.11 METALS CAST IN THE FOUNDRY

The preceding sections contain a brief description of the general principles and methods used in the production of castings. In this section, some consideration is given to the selection of appropriate alloys for different casting applications.

The combination of properties necessary to meet the conditions for successful processing, together with the mechanical, physical and chemical properties required of the casting to withstand the service conditions, form the basis for alloy selection.

A very large number of alloys are currently available for casting purposes, and each of these alloys, or groups of alloys, possesses certain metallurgical and other characteristics that necessitate giving them special consideration in the foundry. Accordingly, emphasis is placed on the foundry characteristics of the alloys, to enable the reader to have some insight into the many complex factors that control the selection of suitable alloys for specific applications.

Full details of the chemical composition limits and the range of physical and mechanical properties are to be found in the relevant standard specifications, published by standards institutions, e.g., BS, ASTM, DIN and so on, including the International Standards Organization (ISO).

5.11.1 Classification of casting alloys

Iron base alloy castings are usually known as ferrous castings, while non-ferrous castings refer to castings made from other alloys, such as copper base, aluminium base and so on. Subdivisions of these two major groupings can be made as follows.

a. Ferrous
1. Steel
(i) Plain carbon steels
(ii) Low-alloy steels
(iii) High-alloy steels

2. Cast iron
(i) Grey cast irons or flake graphite irons
(ii) Spheroidal graphite or ductile irons
(iii) White cast irons
(iv) Malleable irons

b. Non-ferrous
1. Aluminium base alloys
2. Copper base alloys
3. Lead base alloys
4. Magnesium base alloys
5. Tin base alloys
6. Zinc base alloys
7. Nickel base alloys
8. Miscellaneous alloys

The non-ferrous alloys may be further subdivided into light alloys (i.e., aluminium and magnesium base alloys) and heavy alloys (i.e., copper, nickel, lead, tin and zinc base alloys).

There are also cast alloys which do not fall into any of the classifications given so far, and while it is not practicable to include the entire list, some of the more important groups are mentioned below.

a. Alloys for high-temperature applications
Complex heat-resistant alloys are often cast by precision (e.g. investment) casting methods. Elements such as cobalt, chromium, niobium, tantalum, tungsten, nickel, titanium and zirconium are used in a wide variety of different combinations. The aerospace industries, among others, have helped to create a demand for these castings.

b. Alloys for atomic energy applications
These include many special alloy castings for structural and other applications. Control rods, cans and sheaths, as well as uranium fuel elements may be cast.

c. Alloys used in dentistry
Dental applications usually call for complex alloys cast by precision casting techniques.

d. Precious metals

Parts made of gold, silver and platinum or their alloys are cast either for their exceptional resistance to oxidation and corrosion or for their aesthetic appeal in jewellery. Precision casting techniques are usually used.

Individual groups of alloys will generally include many subdivisions and variations within the groups, so that a comprehensive list of casting alloys will literally run into thousands. The reasons for this proliferation in the number of alloys cast in the foundry are to be found in the many and varied properties demanded of castings.

5.11.2 Selection of alloys

The factors that determine the selection of a casting alloy will generally include specified physical, chemical and mechanical properties of the finished casting, in addition to the properties required of the metal to produce a sound casting, e.g., fluidity.

These several considerations may be summarized as follows.

a. Properties which may be specified for finished castings
1. Corrosion resistance and oxidation resistance.
2. Strength.
3. Toughness.
4. Resistance to high or low temperatures.
5. Weight and strength-to-weight ratio.
6. Electrical properties.
7. Magnetic properties.
8. Wear resistance and abrasion resistance.
9. Damping capacity (vibrations).
10. Weldability.
11. Rigidity.
12. Fatigue resistance.
13. Creep resistance.
14. Thermal properties (e.g., conductivity).
15. Machinability.

b. Properties important in processing (casting)
1. Fluidity.
2. Solidification characteristics.
3. Feeding range.
4. Freedom from hot tearing.
5. Reactivity (i.e., freedom from undesirable reactions with furnace and mould materials and with gases).
6. Melting temperature and the degree of superheat possible without serious damage.
7. Response to heat treatment.
8. Response to surface treatment, coating and other finishes.
9. Section sensitivity.
10. Toxicity of vapour.
11. Other properties of the liquid state such as surface tension and drossing tendency.
12. Density.

c. Miscellaneous considerations
1. Cost
2. Appearance
3. Tradition and personal preferences.
4. Dimensional stability in service, i.e., freedom from aging or other solid-state transformations which may give rise to volume change.

This is by no means a comprehensive list and many other factors may influence the choice from time to time.

The first step in selection is to make a critical analysis of the properties desirable in a casting and then (as far as is practicable) to set quantitative limits for these properties. In general, the physical and mechanical properties are comparatively easy to specify in this way. Difficulties may arise in some cases due to the method of evaluation followed. For example, it is well known that both aluminium and copper have superior electrical conductivity compared with iron. Further, of the two, the electrical conductivity of copper (measured on a volume basis) is higher than that of aluminium. However, if this comparison is made on a weight basis, the conductivity of aluminium will be found to be superior. Similar considerations may be applied to strength and other properties. Weight may be a significant factor in certain applications so that evaluation on a weight basis may be important.

Greater difficulties are encountered in evaluating some of the chemical properties such as corrosion resistance: (*a*) because they are generally long-term effects and therefore comprehensive data are difficult to tabulate; (*b*) because the environmental factors are not easy to determine in advance; and (*c*) because minor differences in metallurgical structure (e.g., segregation) may have important effects on corrosion within the same alloy system.

Other properties, such as wear resistance and abrasion resistance may prove intractable, since the mechanism of wear in metals is not clearly understood. Experience may therefore be the decisive factor in these cases.

Some of the properties listed may also pose conflicting metallurgical problems. For example, improvements in strength may be obtained by sacrificing corrosion resistance and vice versa.

Once the requirements under group *a*. have been analysed, attention must be directed towards the casting properties of the alloy (group *b*.) in order to determine whether or not a sound casting will result. Satisfactory mechanical and other properties of the alloy do not automatically predetermine the casting properties, and therefore the choice will be narrowed down further by applying this process of selection.

The next stage of selection will be applied by an evaluation or analysis of the requirements under group *c*., cost being the predominant factor in this group in most cases.

It must be recognized, however, that the selection process does not necessarily proceed in the order mentioned since there is a great deal of interdependence or interrelationship between the many factors listed. Methods based on 'value engineering' or 'value analysis' may be found useful in this respect.

5.11.3 *General properties of cast metals*

It is not the intention here to provide fully comprehensive data on the many properties of cast metals discussed in the preceding paragraphs. Instead, only such general properties are discussed as are thought pertinent to the actual use of castings.

Structural (metallurgical) control, and therefore control of the properties of the resulting casting, may be achieved in a number of ways:

1. The properties of the casting are largely determined by the properties of the base metal. Thus, aluminium base alloys would be expected to have properties basically different from iron base alloys like steel.

2. Further adjustments in the properties of the base metal are made through the addition of suitable alloying elements. Theories on alloying are well documented in books on basic physical metallurgy and need not be discussed here, but the fact remains that different alloy additions can alter the properties significantly. For example, the addition of aluminium to copper will generally result in an increased strength, whereas addition of lead may improve machinability and some casting properties.

3. Treatment of the alloy before or during casting, and the actual casting technique adopted, may also have a vital effect on the structure and therefore the properties. Inoculation (affecting nucleation and growth) may be carried out or the cooling rate of the casting may be controlled in order to modify the metallurgical structure, with consequent effects on the properties. Alterations in structure as a response to alterations in the cooling rate are generally described as 'section sensitivity'. The casting technique, e.g., die casting, sand casting, centrifugal casting and so on, may also have an important effect on the metallurgical structure.

4. The casting may be subjected to a thermal cycle generally referred to as 'heat treatment', to further alter the properties in desirable directions. This method of control is a very important factor in the utilization of many of the casting alloys, particularly those based on iron, aluminium and copper. The heat treatment of metals cannot be dealt with adequately in a discussion such as this, and the reader is referred to standard texts on the subject.

Clearly, strict metallurgical control is necessary prior to and during and after casting to obtain the desired metallurgical structure (microstructure). The microstructures of some common casting alloys are shown in figs. 5.55 to 5.59.

General properties of the important groups of casting alloys are summarized in tables 5.7 to 5.9.

5.11.4 *Properties of ferrous alloys*

A very wide range of properties may be obtained from iron base alloys.

1. Grey cast irons or flake graphite irons are characterized by low cost, good wear resistance and machinability and excellent vibration damping capacity. The over-all casting properties are also most favourable. Carbon, silicon and manganese are the principal alloying elements, with sulphur and phosphorus in small amounts. The sulphur and phosphorus originate from the basic raw material (pig iron) and are very rarely added intentionally. The properties of cast irons are largely determined by the chemical composition (namely, the carbon and silicon contents) and the cooling rate. The alloy exhibits pronounced section sensitivity, so that considerable variations in the graphite morphology are possible by alterations in the cooling rate. The strength and other properties are therefore generally controlled by controlling the type and distribution of the graphite flakes in the structure. Although grey cast irons have little or no ductility, large tonnages are produced throughout the world. Many types of alloy irons are also produced, the more important alloy additions being nickel, chromium and

434 | METAL CASTING

(a) As cast 100x.

(b) Normalised 100x.

Fig. 5.55 Microstructure of steel casting for general engineering purposes.

(a) Grey iron as cast 50x.

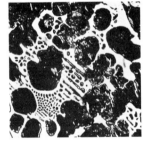

(b) White iron as cast 50x.

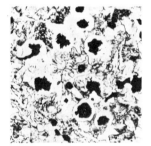

(c) S.G. iron as cast 50x.

Fig. 5.56 Microstructure of different types of cast irons.

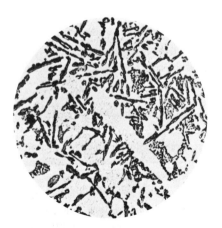

(a) As cast, unmodified 100x.

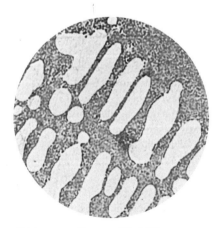

(b) As cast sodium modified 100x.

Fig. 5.57 Microstructure of Aluminium Silicon Eutecyic alloy.

METALS CAST IN THE FOUNDRY | 435

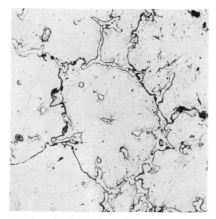

(a) Mg + 10% Al alloy 500x.

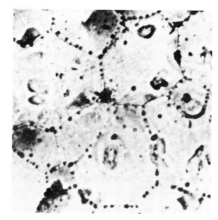

(b) Mg – Zn – Zr – alloy 500x.

Fig. 5.58 Cast magnesium alloys.

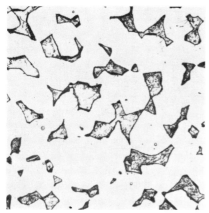

(a) 60/40 brass annealed 200x.

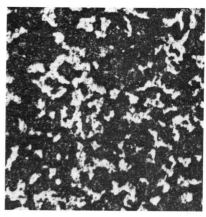

(b) Manganese bronze 500x.

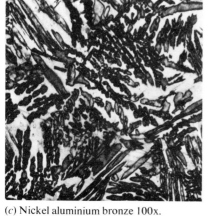

(c) Nickel aluminium bronze 100x.

(d) Tin bronze 50x.

Fig. 5.59 Some cast copper alloys.

Table 5.7 Cast ferrous alloys – range of mechanical and physical properties

Property	Grey cast irons	Malleable irons	S.G. (nodular) irons	Carbon and low alloy steel
Tensile strength, p.s.i.	20000 to 80000	48000 to 120000	60000 to 160000	60000 to 200000
Yield strength, p.s.i.	Nearly same as T.S.	30000 to 95000	40000 to 135000	30000 to 170000
Compressive strength, p.s.i.	3 to 5 × T.S.	48000 to 120000	40000 to 135000	60000 to 200000
Elongation (2 inch) %	0 to 3	1 to 26	1 to 26	5 to 35
Reduction of area %	0	0 to 23	0 to 30	5 to 65
Brinell hardness number	135 to 350 TS/HB = 0.16 to 0.21	125 to 285	140 to 330	130 to 750 for case hardened parts. TS/HB = 0.47 to 0.58
Modulus of elasticity, p.s.i.	12000000 to 22000000	25000000	24000000 to 26000000	30000000
Endurance limit, p.s.i.	0.4 to 0.6 × T.S.	0.4 to 0.6 × T.S.	0.4 to 0.55 × T.S.	0.4 to 0.5 × T.S.
Impact resistance, ft lb Charpy V	Low	1 to 20	1 to 20	3 to 65
Density, g/cm³	6.95 to 7.35	7.15 to 7.60	7.15 to 7.50	7.81 to 7.86
Thermal conductivity, g cal/cm³s°C	0.07 to 0.14	0.138 to 0.151	—	0.12 to 0.14
Specific heat, cal/g°C	0.13	0.122	—	0.11 to 0.12
Melting range, °C	1090 to 1280	1090 to 1380	1090 to 1295	1410 to 1500
Casting temperature range, °C	1186 to 1540	1380 to 1540	1190 to 1470	1540 to 1700
Heat treatment temperature ranges, °C				
Stress relief	425 to 675	425 to 675	425 to 675	425 to 675
Annealing	815 to 980	870 to 955	870 to 955	900 to 930
Hardening	815 to 925	815 to 900	815 to 925	815 to 900
Tempering	180 to 590	180 to 650	180 to 650	425 to 730
Normalizing	815 to 980	—	—	870 to 925
Maximum temperature in service, °C	450	450	450	450

Table 5.8 Cast non-ferrous alloys – range of mechanical and physical properties

Property	Al-base alloys	Cu-base alloys	Mg-base alloys	Ni-base alloys	Zn-base alloys
Tensile strength, p.s.i.	19 000 to 53 000	21 000 to 125 000	22 000 to 45 000	50 000 to 145 000	25 000 to 52 000
Yield strength, p.s.i.	8000 to 43 000	11 000 to 100 000	11 000 to 30 000	25 000 to 118 000	—
Compressive strength, p.s.i.	19 000 to 53 000	8000 to 60 000	22 000 to 45 000	18 000 to 80 000	55 000 to 93 000
Elongation in 2 in., %	0 to 22	0 to 52	1 to 12	1 to 45	0.5 to 10
Reduction of area, %	—	4 to 40	—	1 to 47	—
Brinell hardness number	40 to 140	47 to 425	45 to 84	100 to 390	75 to 100
Modulus of elasticity, p.s.i.	10 300 000	9 100 000 to 20 000 000	6 500 000	22 500 000 to 28 000 000	—
Endurance limit, p.s.i.	6500 to 23 000	4000 to 15 000	9000 to 13 000 based on 500 000 000 cycles	—	6875 to 8500
Impact resistance, ft lb					
Charpy unnotched	0.5 to 35	—	—	—	1 to 48
Charpy keyhole notch	0 to 8	—	0.5 to 10	4 to 70	—
Izod	—	0.5 to 40	—	3 to 85	—
Density, g/cm³	2.57 to 2.95	7.3 to 9.5	1.79 to 1.83	7.80 to 9.24	6.6 to 6.7
Thermal conductivity, g cal/cm³sec°C	0.21 to 0.40	0.166 to 0.942	0.16 to 0.27	0.027 to 0.142	0.25 to 0.27
Specific heat, cal/g°C	0.23	0.12	0.23 to 0.25	0.09 to 0.136.	0.10
Melting temperature range, °C	650 to 770	910 to 1050	460 to 645	1290 to 1520	385 to 420
Casting temperature range, °C	700 to 780	960 to 1270	650 to 845	1460 to 1610	400 to 480
Heat treatment temperature ranges, °C					
Stress relief	345	—	260	—	—
Solution treatment	500 to 540	870 to 900	390 to 565	650 to 1150	—
Ageing treatment	150 to 260	535 to 650	160 to 315	590	75 to 100

438 | METAL CASTING

Table 5.9 Cast ferrous and non-ferrous alloys – some properties important for processing (qualitative)

Property	Grey cast irons	Malleable cast irons	Nodular irons	Carbon and low alloy steels	Al-base alloys	Cu-base alloys	Mg-base alloys	Ni-base alloys	Zn-base alloys
Machinability	Good	Good	Good	Inferior compared to the irons	Good to excellent	Fair to good	Excellent	Comparable to steel	Excellent
Section sensitivity (i.e., variation in properties depending on section size or cooling rate)	Yes	No	As cast – yes Annealed – no	To a limited extent	Yes	To a limited extent	To a limited extent	To a limited extent	No (usually die-cast)
Suitability for Joining									
Brazing	Yes	Yes	Yes	Yes	Yes for a few alloys	Yes	No	Yes	No
Soldering	Yes	Yes	Yes	Yes	Yes	Yes	Yes	Yes	Yes
Welding	With difficulty, special precautions	With special precautions	With special precautions	Yes, with ease	Yes	Yes	Yes	Yes	Yes
Mould Filling Capacity	Excellent	Good	Excellent	Not as good as cast irons	Excellent	Fair to good	Good to excellent	Comparable to steels	Excellent

METALS CAST IN THE FOUNDRY | 439

Property									
Susceptibility to hot tearing	No	Yes	No	Yes	Depends on composition	Depends on composition	Depends on composition	Comparable to steels	No
Pressure tightness	Yes	Yes	Yes	Yes	Depends on composition	Depends on composition	Depends on composition	Yes	Yes
Properties altered by heat treatment	Yes, limited	Yes, production involves 100% heat treatment	Yes, fair response	Yes	Yes, for most alloys	Yes, for a few alloys	Yes, for the majority of alloys	Yes, limited to a few alloys	Not for the usual alloys, except dimensional stability
Response to molten metal treatment i.e., control of properties through inoculation, modification etc.	Yes, good	Not marked	Yes, treatment necessary in all cases to produce graphite in nodular form.	Not marked.	Good for a few alloys	Not marked	Yes	Yes, only in carbon alloys.	No

molybdenum. High-silicon irons (more than 6% Si) find application for their resistance to acids.

2. White cast irons usually contain less silicon and carbon than grey cast irons, with the result that when the metal solidifies the metallurgical structure is free from graphite, the carbon being present in the form of carbides. This gives the characteristic white appearance when the metal is fractured and hence the name.

White cast irons are generally used for their excellent wear or abrasion resistance and low cost.

3. Malleable irons may have tensile strengths approaching 100000 p.s.i. together with appreciable ductility. They are therefore superior to grey cast irons in this respect. Malleable iron castings are produced by a special cycle of heat treatment, carried out on castings initially made in white cast iron, to break down the carbon in the form of carbide and cause it to precipitate as graphite, which takes the form of clusters or nodules. The fractured surface of the metal (after heat treatment) appears dark and hence is generally called 'black-heart' malleable iron.

In another process of heat treatment, the carbon is partially oxidized (by diffusing out of the body) and partially precipitated as graphite. Due to the loss of carbon, the fractured surface may appear comparatively bright and these irons are known as 'white-heart' malleable irons.

The malleable irons offer the advantage of a more precise control of the properties through heat treatment and are therefore less subject to property variations than alloys used in the as-cast condition. Large tonnages of 'black-heart' malleable castings are produced annually for the transport industry (motor cars, tractors, trucks, earth movers etc.).

4. Spheroidal graphite (SG) irons, also known as nodular irons, are examples of cast irons where the graphite morphology has been modified by inoculating magnesium and/or cerium into the liquid metal before pouring the casting. In general, the mechanical properties of SG irons are similar to those of malleable irons, the difference being that the graphite morphology is obtained in the as-cast condition rather than after heat-treatment. SG irons, suitably alloyed, may be heat treated (like steel) to produce pearlitic or martensitic matrix structures, in addition to a ferritic structure. The graphite nodules retain their form after such treatment. The SG irons (and malleable irons) fill the gap between grey iron and cast steel and have replaced ordinary steel castings for a large number of applications.

5. The large variations in the properties of steel castings are possible through control of the chemical composition (carbon and other alloying elements) and heat treatment. Steel castings offer a high degree of ductility, toughness and strength, coupled with excellent weldability. Suitably alloyed stainless steels, and other temperature-resistant and corrosion-resistant steel castings, constitute a major part of alloys used for applications requiring such properties. The range and versatility of steels is too well known to require further comment. Steel castings are particularly responsive to heat treatment and hardening is carried out by rapid cooling from an elevated temperature (quenching), followed by reheating to an intermediate temperature (tempering). In this respect the heat treatment of steel castings does not differ from the heat treatment of steel components fabricated by any other process. There is one difference, however, in that steel castings are rarely, if ever, used in the as-cast condition. Some form of annealing is invariably carried out for homogenization of the as-cast structure and to relieve residual stresses introduced during the casting process.

METALS CAST IN THE FOUNDRY | 441

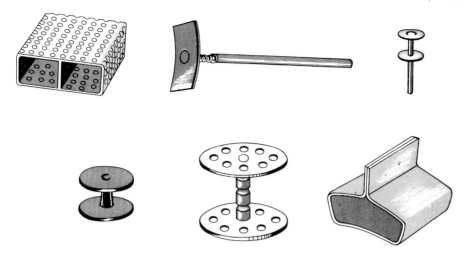

Fig. 5.60 A selection of chaplets used in the foundry industry.

5.11.5 Properties of non-ferrous alloys

a. Aluminium base alloys. The popularity of aluminium alloy castings is derived largely from the characteristic properties of the alloys, such as strength-to-weight ratio, non-toxicity, corrosion resistance, thermal or electrical conductivity and castability. These characteristics also provide a useful basis for the classification of aluminium alloys:

1. Alloys with outstanding casting properties, which include Al-Si, Al-Si-Cu and Al-Fe-Mn alloys.
2. Alloys developing high and intermediate mechanical properties through heat treatment (zone forming alloys), e.g., Al-Mg, Al-Cu-Mg, Al-Si-Mg and Al-Zn-Mg alloys.
3. Alloys with superior corrosion resistance, e.g., the Al-Mg, Al-Mg-Mu and Al-Si systems.
4. Alloys with superior heat resistance, exemplified by the complex Al-Cu-Mg, Al-Cu-Mg-Ni and Al-Cu-Si-Mg-Ni systems.

It will be apparent that the most important alloying elements are silicon, copper, magnesium, zinc, nickel and manganese. However, several other elements, notably iron, chromium, cobalt, titanium, boron, niobium, sodium and phosphorus, are also used in small controlled amounts to develop special features in the basic alloy systems. For example, small additions of sodium may be made to modify the structure of the Al-12% Si eutectic alloy and thus improve the mechanical properties, while titanium may be added to some alloys to modify the grain size.

The heat treatment of aluminium alloys largely centres round the so-called zone forming alloy systems mentioned in group 2 in the classification. Some of the elements alloyed with aluminium exhibit what is known as retrograde solid solubility. For example (see fig. 5.61), the solid solubility of copper in aluminium is 5.65% at 548°C but negligible at room temperature, so that if alloy K is rapidly cooled (quenched) from a temperature of 548°C the copper is retained in solid solution even at room temperature.

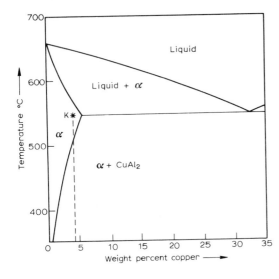

Fig. 5.61 Retrograde solid solubility in the Al-Cu phase diagram (Aluminium-rich alloys).

This, however, is an unstable state and the copper complex tends to precipitate out upon reheating or prolonged holding at room temperature. The strength properties are developed according to the nature, size and distribution of the precipitate; the reheating and/or holding process is known as the aging process, and the heat treatment process (i.e., quenching from solution temperature and then aging) is known as the age-hardening process.

The name 'zone forming alloys' is derived from the identification of the early stages of precipitation by two workers in the field, Guinier and Preston, who carried out X-ray diffraction analysis, and the zones affected by the initial stages of precipitation came to be known as Guinier-Preston zones.

b. Copper base alloys. The majority of the copper base alloys cannot be hardened by heat treatment. Exceptions to this rule include the alloys containing aluminium, silicon and beryllium (e.g., aluminium bronze, silicon bronze etc.). The heat-treatable alloys also represent the group which accounts for the highest strengths obtainable in copper base alloys. Corrosion resistance, thermal and electrical conductivity and architectural value are the most important properties in copper base alloys.

The most common alloying elements in copper alloys are tin, lead, zinc, iron, nickel, manganese, aluminium, silicon and phosphorus. When zinc is the major alloying constituent, copper alloys are usually known as brasses, and when tin, aluminium, silicon, etc. form the major constituents, the alloys are generally known as bronzes. This terminology is not strictly followed in all cases so confusion may arise from time to time.

Zinc and tin are appreciably soluble in copper in the solid state and represent the most common additions for strengthening purposes.

Lead is almost completely insoluble in all cast copper alloys and forms small round globules throughout the matrix of the solidified casting. Small amounts of lead may be used to improve machinability. The main reason for using lead is generally to improve bearing properties, although it is debatable whether it serves any useful purpose.

In aluminium bronzes there are two or three distinct solid phases, known as the alpha, beta and gamma phases. The alpha phase is soft and ductile, whereas the beta phase exhibits the opposite properties. The gamma phase is very hard and decreases the ductility of the alloys considerably. Since this phase forms quite rapidly by the breakdown of the beta phase when the alloy is slowly cooled through the critical range (400 to 600°C), castings should be removed from the sand moulds as soon as possible. Iron is added to aluminium bronzes to decrease grain size and thus improve the mechanical properties to some extent.

The amounts of alpha and beta in the structure (and therefore the properties) may be varied through heat treating by a process which is similar to that used for carbon steel. The casting is first heated to a temperature of 900°C (approximately), for a period of time dependent on the bulk or section thickness of the casting, and then quenched in oil or water. It is next tempered or aged for varying times and at varying temperatures from 400–650°C, depending on the properties desired. This aging precipitates the harder constituents and gives added strength and hardness at the expense of ductility. Nickel is added to aluminium bronzes in amounts up to 5% to improve the mechanical properties and to retard the transformation to gamma during cooling.

Manganese bronzes are copper-zinc alloys with additions of iron, nickel, manganese, silicon, aluminium and sometimes tin and lead. These are alpha-beta alloys (i.e., soft alpha + hard beta) and control of properties depends on controlling the amounts of alpha and beta in the structure. All alloying additions except lead and nickel tend to increase the beta phase. Iron is also used to decrease grain size. Lead is completely insoluble while nickel increases the alpha phase. Small amounts of tin may improve corrosion resistance.

Silicon bronzes are basically copper-silicon alloys with addition of zinc. These alloys are mainly used for their corrosion resistance and high strength. Surface quality and appearance may be affected by the presence of lead, so this element must be held to a very low level in silicon bronzes.

Pure copper is used for applications where the thermal or electrical conductivity is of prime importance. Examples are the castings made for blast furnaces, such as tuyeres, and some castings made for the electrical industry.

c. Magnesium base alloys. Some of the magnesium base alloys, although lighter than aluminium alloys, can be heat treated to almost the same strength. Machinability is usually excellent. The largest portion of magnesium alloy castings are used in aircraft in the permanent mould cast, die cast or sand cast form.

The most important alloying elements are aluminium, manganese, zinc, the rare earths, thorium, zirconium, silicon, copper and nickel.

Four general types of magnesium alloys are used for casting. Those of the Mg-Al-Zn and Mg-Zn-Zr series are used in applications where the service temperatures are below about 120°C. Alloys containing rare-earth elements and zirconium, with or without zinc, are used in applications where the service temperatures are up to about 260°C. Alloys containing thorium, with or without zinc, have been used at temperatures approaching 330°C.

The Mg-Al-Zn series usually contain 6 to 10% aluminium, 0 to 3% zinc and 0.10 to 0.30% manganese. They are generally heat treated in a manner analogous to aluminium base alloys, i.e., solution and quenching followed by aging, although the aging treatment is not carried out in all cases.

Alloys containing the rare earths or thorium are of special interest due to their resistance to creep. Some of these alloys are heat treated by solution treatment followed by aging. Alloys in these series containing zinc are usually heat treated by aging only. The alloys may contain up to 3% thorium or about 3 to 4% rare-earth elements. Most of these alloys must contain zirconium to the saturation point (i.e., beyond the solubility limit of zirconium in magnesium, which is usually 0.6%) in order to ensure the desirable properties.

The Mg-Zn-Zr series combine high yield strength with good ductility. The zinc content is about 4.5 to 6% and zirconium content 0.7% (saturation) to develop the optimum properties. Alloys with the lower zinc content are generally heat treated by aging alone, while those containing higher amounts of zinc may require a solution heat treatment followed by aging.

d. Nickel base alloys. **Nickel base alloy** castings range from commercially pure nickel to complex corrosion- and heat-resistant alloys.

For the nickel-copper alloys, improved strength and hardness may be obtained in some cases by heat treatment involving an aging process. Alloying with molybdenum, silicon, aluminium or boron usually strengthens the alloy. Improvements in properties are also possible, in some cases, by heat treatment. Ni-Cr-Fe alloys provide outstanding heat- and corrosion-resistant castings.

Nickel alloy castings are made in several grades. The general purpose alloys used in the textile, chemical and drug industries and in food-handling operations may have the following composition range:

$$Cu - \text{up to } 0.30\%$$
$$Fe - \text{up to } 0.25\%$$
$$Si - 1.5 \text{ to } 6.0\%$$
$$Mn - 0.75 \text{ to } 1.5\%$$
$$C - 0.25 \text{ to } 2.50\%$$
$$Ni - \text{balance.}$$

In applications such as food-handling where contamination with elements other than nickel is undesirable, an alloy containing 1.5% silicon (Si) and 0.50% carbon (C) is generally used.

The high-carbon and high-silicon grades are bearing or metal-to-metal contact materials, suitable for shaft bearings, pump seals and other similar applications. The carbon may appear in the structure as flake graphite or spheroidal graphite depending on treatment. The high-silicon grades are useful where hard metals or hard-faced surfaces are encountered in metal-to-metal contact operations. In general, nickel alloys are specified wherever complete elimination of 'copper contamination' is needed in the chemical or food-processing industries.

The nickel-copper alloys are characterized by complete solid solubility throughout the composition range. Coefficients of expansion and magnetic properties change with composition, which makes it possible to use these alloys for castings required for electrical, thermostatic or instrumentation equipment. Other nickel-copper alloys such as the Monel group (containing silicon and other elements in addition to about 32% copper) are used, for their high strength and corrosion resistance, in the chemical, textile, paper and oil industries. Food processing, pharmaceutical, laundry, soaps, caustics as well as household and architectural uses are common. Other grades of Monel castings are used for valves and pumps handling steam and chemical process work. The Monel type of alloy resists most mineral acids, most organic acids and almost all alkalis.

Nickel-chromium alloys containing about 14% chromium are noted for their exceptional heat and corrosion resistance (e.g., Inconel). They are frequently cast by the investment process.

Tin imparts bearing properties to nickel alloys and is used in amounts of up to 10% for this purpose.

Molybdenum alloys containing nickel (Hastelloy, Chlorimet) are known for their superlative resistance to hydrochloric acid and oxidizing acids at high temperatures. The only other known material to match this performance is titanium.

Aluminium in amounts of 17% has been used for elevated-temperature properties. However, this is still in the experimental stage.

Alloys containing up to 10% silicon may be used for applications in the handling of hot corrosive liquids, e.g., boiling sulphuric acid.

e. Zinc base alloys. The zinc base alloys are used primarily for die casting although some sand castings and investment castings are also produced.

The principal alloying elements are aluminium, magnesium and copper. Aluminium greatly increases the strength and hardness. It also improves the casting properties, allowing fine and intricate details to be accurately reproduced. With more than 4.5% aluminium, however, the alloy begins to lose its impact strength and therefore most alloys contain between 3 and 4.5% aluminium.

Improved strength and hardness coupled with corrosion resistance result from additions of copper. Alloys containing over about 1% copper are, however, unstable, showing deterioration in impact strength and dimensional change due to room temperature transformations. The use of copper is therefore limited to a maximum of about 1%.

Most zinc base alloys contain magnesium in addition to aluminium and copper. Small amounts of magnesium retard some structural changes that take place in the alloys. The practical effect is to prevent intergranular corrosion. The composition range for magnesium is most important, since below the range rapid intergranular corrosion may take place, while above the range casting properties such as fluidity and resistance to hot tearing may be affected. The usual range for magnesium is 0.03 to 0.08%.

The main applications for zinc alloy castings are in automotive parts, household utensils, office equipment, building hardware, typewriter and other machinery and instrument frames, padlocks, toys, lighting fixtures and novelties of all descriptions.

f. Miscellaneous alloys. Lead base alloys may contain antimony, tin and arsenic in varying amounts to produce the desired physical and mechanical properties. Casting properties are very good so that gravity die, centrifugal or pressure die casting methods may be employed. Although lead base alloys have a wide variety of applications, the most important application is in bearings. Good frictional and running-in properties, ease of bonding to steel or copper alloy shells and satisfactory behaviour under conditions of poor lubrication are the principal properties sought in bearing applications and lead alloys are especially suitable.

Lead by itself is quite soft so that antimony and/or tin is added in amounts up to 20% to improve the strength and hardness.

Arsenic is added to Pb-Sb-Sn alloys to improve strength and hardness at elevated temperatures and to produce a finer grain structure. Fatigue properties are also improved.

Lead base alloys used in printing (type metal) contain 3 to 20% tin and 3 to 20% antimony, depending on whether a soft backing metal or hard printing surface is required.

The important tin base alloys are based on the Sn-Sb-Cu system. Antimony and copper may each be used in amounts up to 8.5% to obtain suitable strength. Broadly speaking there are three general classes of these alloys: babbit bearing alloys, die casting alloys and Britannia metal and pewter, which are used for a wide variety of purposes including pewter mugs, vases, and tea and coffee services. Lead is sometimes added to the babbits and the die casting grades, where toxicity is not a problem, mainly to reduce the cost. However, die castings produced for food-handling applications cannot tolerate lead and it must be controlled to below 0.35% in the alloy.

Titanium base alloys are classified into three groups related to the metallurgical structure of the alloys. These groups are: alpha alloys (hexagonal close-packed); beta alloys (body-centered cubic); and alpha plus beta alloys. Unalloyed titanium, Ti-Al and Ti-Al-Sn alloys belong to the alpha group. The addition of aluminium decreases the density and strengthens and toughens the alloys in this group. Heat treatment has no appreciable effect.

The beta alloys contain alloying elements such as vanadium, chromium or molybdenum. These, however, increase the density and therefore are less attractive on a weight basis. These alloys are potentially capable of providing a combination of good strength and ductility.

The best combination of strength and ductility is developed in the alpha-beta alloys. Aluminium and tin are the elements that tend to stabilize alpha, while all other elements tend to stabilize the beta phase.

The introduction of titanium to the field of casting is still in the developmental stage. Potential applications are as fittings in chemical and petroleum plants, on ships, in aircraft and in airborne equipment. These applications would be based on the excellent corrosion resistance of titanium, which is equal to or better than stainless steels in most media, and on the favourable strength-to-weight ratio of titanium alloys at temperatures up to 380°C. Titanium is a 'reactive' metal so that processing problems are enormous – the entire melting and casting process has to be carried out under vacuum. Further, titanium reacts with all known refractories in its liquid state and melting has therefore to be carried out in vacuum furnaces of the consumable electrode skull melting type.

5.12 EVALUATION OF CASTINGS

5.12.1 General

Testing and inspection may be described as the science of examining raw materials or manufactured products in order to determine their fitness for certain specific purposes.

Any casting or assembly of castings, whether simple or complex, is invariably subjected to mechanical loads of varying types and magnitudes, e.g., steadily applied loads (dead loads), fluctuating loads (live loads), suddenly applied or shock loads, or loads due to impact in some form. These loads may also be applied under different environmental conditions such as high or low temperatures and corrosive atmospheres.

In every case a casting must fulfil three basic requirements. It must perform its function effectively, operate safely and be able to withstand the service conditions for a sufficiently long time to make it worth-while to use or install.

To attain these objectives, the foundry technologist must have a thorough working knowledge of all the properties of a casting relevant to the load intensities and durations under the ambient conditions to be met in service.

The assessment of mechanical properties (mechanical testing) is based on certain assumptions:
1. that, under the same conditions of loading, a casting of a particular external shape will always behave in the same way with respect to the applied load.
2. that the material is homogeneous throughout its entire section.
3. that load distribution within the material follows simple geometric principles depending on the shape of the casting.

Clearly, these conditions cannot hold in practice, even when the casting has been produced under exacting standards of quality control.

A large variety of defects and flaws may result from the casting operation, e.g., cracks, holes and cavities, lappings, pipes and inclusions of slag and other foreign matter, and residual stress. Inhomogeneities may result from the solidification characteristics of the alloy and the thermal and other properties of the mould material, e.g., columnar and equi-axed grains, variations in the grain size, micro- and macro-segregation, and dispersed shrinkage and gas porosity. The metallurgical structure derived during the solidification process is variable to some extent, e.g., the distribution and orientation of the phases, solid-state reaction products present (largely dependent on the chemical composition of the original alloy and the direction and rate of heat transfer after casting the alloy into the mould). All these variables may affect the mechanical properties in an unpredictable manner. It will be clear, therefore, that the foundry technologist must devise other methods of testing, in conjunction with mechanical testing, if he is to succeed in making a reasonable prediction regarding the suitability of a particular casting for a specific application.

Some of these variables can be checked by non-destructive methods of testing, others by determining the chemical analysis (conventional or instrument techniques) and the remainder by using metallographic and allied techniques and measurements of physical properties.

Until recently, tests to determine the 'soundness' of castings were either neglected or carried out by destroying the casting, e.g., cutting it up to examine internal flaws. Destructive testing is necessarily a wasteful process and not a practical proposition for commercial products. Non-destructive methods of examination of castings have therefore been gaining popularity and the progress of these techniques has been phenomenal over the last few years, both in the number of methods of non-destructive testing and in the scope of their applications.

Apart from the purely utilitarian aspect of non-destructive testing (i.e. determination of flaws and discontinuities on the surface or within the casting), these methods are being used increasingly as development tools (for developing satisfactory running and feeding systems, for example) and for comparing or assessing average metallurgical structure, and segregation variations in the chemical composition and, in some instances, for determining the mechanical properties.

Some other aspects of testing, such as determination of physical properties and metallographic techniques, may also require special attention in cast metals technology.

5.12.2 Standard specifications

In order to guide the manufacturer in evaluating the quality expected of his product and give an indication to the customer of the expected performance or utility of his purchase, most industrialized countries have their organizations or institutions for standards. The primary function of these institutions is to formulate or state the standards of quality required in the raw materials, as well as in the semi-finished and finished products. However, specifications cannot be expected to cover the entire range of products, nor can they be expected to anticipate new developments in materials and techniques. The standard acceptable today may no longer be acceptable tomorrow and this is more especially so with the trend towards mass production and the increasing use of automation in industrial processing.

A standard specification, therefore, should not be taken to be a static comprehensive manual for a particular product, but should be examined objectively for what it is meant to be – a useful guide to the producer and consumer alike for assessing the quality of the product in the light of our present knowledge in the various branches of science and technology.

A specification formulated for a certain type or group of castings may take into consideration one or more of the following requirements:

1. Design, 2. tolerance – size, 3. chemical composition, 4. mechanical properties, 5. surface finish, 6. corrosion resistance, 7. physical properties, such as magnetic and electrical properties and coefficient of expansion, 8. resistance to oxidation, 9. low-temperature properties, 10. high-temperature properties, 11. structure, including grain size, phase type and distribution, 12. special requirements such as pressure tightness, wear resistance, 13. heat treatment.

All these requirements are not necessarily explicitly stated in the standard specifications since, quite often, stating one set of requirements will predetermine another set. As an example, let us consider a steel casting which, from the design point of view, must have a yield strength of 18 t.s.i. under normal environmental conditions. If these two factors alone were stated in the specification, a steel founder may be able to supply the casting without annealing and satisfy the yield strength requirement by a good margin. However, if, in addition, the specification stated a ductility property such as 20% elongation, the casting must be annealed since the elongation in the as-cast state is unlikely to exceed 5%. Therefore, without specifically stating that annealing must be carried out in such and such a way, stating the percentage elongation required together with the yield strength will not only predetermine the range of chemical analysis of the steel but will also ensure that the casting is annealed.

Another requirement in the specification may be that the casting must be sound. Soundness cannot be determined by carrying out mechanical tests on a test bar representing the casting, and unsoundness may range from fine dispersed porosity to gross shrinkage cavities within the casting or any of the usual discontinuities and flaws. Unsoundness may even include segregation, inclusions, grain size and structural variations which are not necessarily reproduced in the test bar. The standard specification cannot possibly include all the different types of unsoundness that can occur in different castings produced under different conditions. Much will therefore depend on the integrity of the foundry technologist and his knowledge and ability to recognize and interpret the results of his findings in deciding whether or not a particular casting is suitable for the application. The progressive foundryman must also be in a position to use the results of these tests in an intelligent manner, not only to improve the quality of his products but also to project the data in developing new techniques and products.

5.12.3 Soundness and unsoundness

'Sound' as defined in the dictionary would imply complete freedom from defects. Strictly, the word 'sound'–to the foundry technologist–is in the same category as 'safe'. It cannot be used without qualification or expression of degree. For example, what is sound for a cast-iron block used as a counterweight is certainly not sound for a cast-iron engine block. In general, therefore, soundness or unsoundness must be applied with degrees of qualification in the context of the actual application of the casting.

Different degrees of unsoundness may be indicated by the following, for example:

1. Cracks and breaks. A crack may be a comparatively fine discontinuity in the body of the casting which may or may not cause the casting to fall apart with the application of a small load, whereas a break would imply that falling apart is highly probable.

2. Holes, cavities, voids, blow holes, gas holes are all words in common usage for describing certain types of defects found within the casting. They may be further qualified by the words large, small, spherical, elongated, branching etc. A hole is normally connected to the surface of the casting, whereas a cavity is not so connected. Similarly, porosity indicates an agglomeration of small holes normally connected to the surface, which means that the casting will not be pressure-tight, and may be further qualified by the term 'spongy'. The non-communicating type may be termed (depending on the size and shape) pin-hole porosity or simply porosity.

3. Lack of fusion. A 'cold shut' may appear in several forms. It may look like a crack, a hole, or it may develop as an overlap, as unfused chaplets (studs) or internal chills (densers). In practice, chaplets rarely fuse completely and may be a source of all sorts of defects and weaknesses.

4. Foreign matter. Solid inclusions of a non-metallic nature may consist of slag, sand and other refractory material and mould dressings washed down by the metal. They may be described as sand pockets, inclusions etc.

5. Pipes. A pipe may be a hole or a cavity. Strictly applied this should only indicate the shrinkage defect at the point where a casting or ingot is fed. A primary pipe will then be a hole and a secondary pipe a cavity. Shrinkage holes or cavities may, however, occur in other parts of the casting and be designated as such. Shrinkage defects may also take the form of dispersed porosity.

It will be clear from the foregoing (by no means a comprehensive list) that it is extremely important to describe the type of defect in detail, possibly with sketches and photographs, to leave no room for ambiguity in interpretation. Another point to note is that 'defect' implies that the casting is unsuitable for the application, whereas 'flaw' merely indicates the presence of some form of discontinuity or inclusion which may or may not render the casting unsuitable for use. Typical examples of unsoundness are included in Fig. 5.62.

5.12.4 Mechanical properties generally specified

1. *Tensile stress* is that type of loading in which the two parts of the casting, on either side of a typical stress plane, are tending to pull apart (see Fig. 5.63a).
2. Compressive stress is the reverse of tensile stress (see Fig. 5.63b).
3. Shear stress exists when the two parts tend to slide across each other in any typical plane of shear (see Fig. 5.63c).

Strain: Whenever a stress, no matter how small, is applied to a material, a proportionate dimensional change takes place in the material. Such a change is called strain. For all practical purposes, it may be assumed that the volume remains the same, i.e., if the length increases the transverse section will decrease to keep the volume constant. (This is not always strictly true, e.g., strain-hardening.) Strain may take two forms: elastic strain and plastic strain.

Elastic strain: is taken as a transitory dimensional change that exists only as long as the initiating stress is acting and disappears on removal of the stress. The phenomenon is referred to as elasticity.

Plastic strain: is dimensional change that does not disappear when the stress is removed. It is usually accompanied by some prior elastic strain. The phenomenon is referred to as plasticity.

Most metals have some elasticity at room temperature, and some plasticity as well. At elevated temperatures, metals are likely to have more plasticity and less elasticity. The state of stress at which plastic strain begins to appear is known as the elastic limit and is defined by the magnitude of the stress and the corresponding value of elastic strain.

Poisson's ratio: is the relationship, or ratio, of lateral to axial strain.

True strain (or logarithmic strain): for a given load increment, is the ratio of the change of dimension (resulting from the load increment) to the magnitude of the dimension immediately prior to the load increment. For example, if a bar has been deformed uniformly to a length l from an initial length l_0.

$$\text{True strain} = \ln(l/l_0)$$

(ln signifies natural logarithm.)
Similarly, for area,

$$\text{true strain} = 2\ln(D_0/D)$$

where D_0 is the original diameter and D the Final diameter of the specimen.

Nominal strain (or simply strain): is the ratio of the change in dimension after loading to the original dimension before loading.

$$\%(\text{Nominal}) \text{ strain} = \frac{l - l_0}{l}$$

where l is the length after the load is applied and l_0 the length before the load is applied. Similarly,

$$\% \text{ Reduction of area} = \frac{A_0 - A}{A_0} \times 100$$

This method of finding the strain is quite accurate for all practical purposes and only ceases to be sufficiently accurate when plastic strain becomes the dominant type.

EVALUATION OF CASTINGS | 451

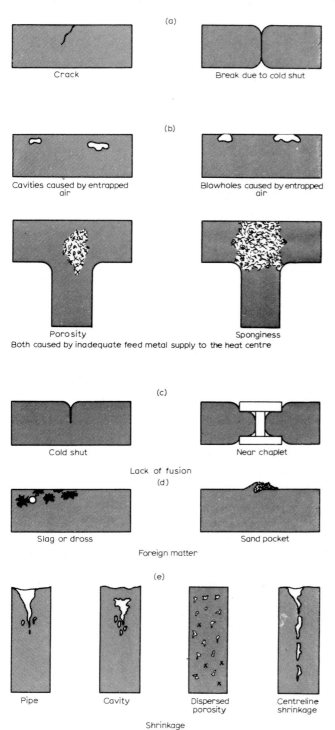

Fig. 5.62 Typical examples of unsoundness.

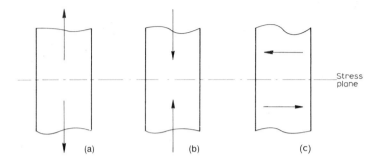

Fig. 5.63 (a) Tensile stress. (b) Compressive stress. (c) Shear stress.

True stress: is found by dividing the load applied to a member at a given instant by the true cross-sectional area supporting that load at the same instant. In finding out the true stress, therefore, it is necessary to take into account the changes in cross-section taking place during continuous loading.

(Nominal) stress: is the load at any instant divided by the original cross-sectional area.

Stress-strain diagrams: The relationship between stress and strain cannot be expressed by a simple formula. It is customary to indicate the relationship in the form of a diagram in which stress is plotted on the y axis and strain on the x axis (see Fig. 5.64). The area under the curve, bounded by the x axis and an ordinate dropped from the curve to the relevant value of strain, gives a measure of the strain energy absorbed in causing that amount of strain. These diagrams are generally drawn for nominal stress and nominal strain.

True-stress true-strain diagrams: For more accurate work and for fundamental investigations, the true stress and true strain are plotted in such diagrams.

Malleability: is the plastic response to compressive force.

Ductility: is the plastic response to tensile force.

Loading Methods: can be classified under five headings:

1. Steady loads are loads that do not change in intensity (or change so slowly that they may be taken as steady).
2. Fluctuating loads are loads which change their intensity and/or direction but at relatively low speeds.
3. Cyclic loads are loads which change value by following a regularly repeated sequence of change at significantly high speeds (or very high speeds).
4. Shock loads are suddenly applied loads in which there is a very rapid build-up of stress (e.g., motor-car piston).
5. Impact loads are specially severe shock loads such as those caused by the instantaneous arrest of a falling mass (e.g., mechanical hammer).

The determination of mechanical properties may be desirable for various reasons.

1. *For design:* Unless the loading conditions and the response of the material to these conditions are clearly evaluated, it is not possible to design a component on a safe and competitive basis. For example, if, under a particular system of loading, a casting has to withstand steady loads of 5 t.s.i. a large number of alloys could be found which would satisfy the requirements. To be competitive, however, one must produce the

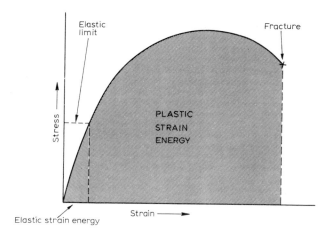

Fig. 5.64 Stress-strain diagram.

casting in the alloy which will be the cheapest to produce and yet will ensure a safety margin.

Very few castings are of a type in which solidification takes place in a simple and uniform manner. Generally, there is some restraining force present, due to the geometric constraint from the mould or core, or to unequal shrinkage or uneven shrinkage caused by thermal gradients. These forces may cause cracking or deformation or may simply leave residual stresses. Any of these conditions may result in failure in service.

It is therefore not sufficient to have a knowledge of the appropriate mechanical properties; the foundry technologist must also understand how these properties can be affected by solidification and cooling down to the room temperature.

2. *For minimum standards:* It is desirable to be able to state to the customer that the casting will withstand a specific load under specific conditions – this may help to ascertain the suitability of the casting for a particular application.

3. *Quality Control:* Even when tests are not specified, mechanical properties may be determined for controlling and improving the quality of the product.

4. *Research:* Study of individual properties of particular fundamental or practical interest is of importance for developing new materials and techniques.

A systematic study of mechanical properties will therefore not only involve the study of individual properties but also the interrelationship of properties and the devising of effective test methods for routine testing. This may be quite complicated in the foundry industry and quite often 'tailor-made' tests have to be devised for particular applications.

5.12.5 Chemical composition

Although the chemical composition of a foundry raw material such as sand or a casting alloy will predict some of the properties to be expected in service, it is only a guide and factors such as the cooling conditions in the mould can produce completely unpredictable results. The other complicating factor is the presence or the absence of trace elements which are not normally determined in routine chemical analysis. Some of these trace elements are known to affect the properties drastically. Similarly, the chemical analysis of a foundry sand will rarely predict the behaviour of the sand unless it is carried out in conjunction with a large number of other tests. One must therefore consider chemical analysis as one of the tests and never, by itself, a sufficient test to be able to predict the properties of the material.

In the present discussions it has not been possible to include descriptive details regarding the machinery and methods used in carrying out the various types of tests, and the aim has been to introduce the most important factors that influence the methods of evaluation of castings.

CHAPTER 6

Metal Deformation

6.1 METAL DEFORMATION IN GENERAL

6.1.1 History

The earliest records of metal deformation take us back to the Stone Age. Small pebbles of native gold are known to have been hammered with stone implements to make amulets and primitive jewellery. Even in that age copper, discovered also in its native form, was being used for elementary tools, being crudely hammered by stone to knives, arrow-points, chisels and even needles. The ability of the metal to become harder as it was hammered to shape produced, for those times, a reasonable cutting edge. The advent of the Bronze Age was associated with more sophisticated shaping and many examples of fine decorative bronzeware produced by hammering have been found, particularly in Mesopotamia and the surrounding area. The royal tombs of Ur have provided some notable examples. Iron was developed in a similar manner and, being harder than copper, was found to be a better material for tools and weapons. Brass is known to have been in use by the Romans for coinage but was not commonly used for shaping until the eleventh and twelfth centuries when it was used for containers. In the fifteenth century bronze and iron were being deformed to produce ewers, fire-irons, bedwarmers, vases, mortars and decorative ware.

6.1.2 Elastic and plastic deformation

Most metals in the solid state are capable of undergoing a change of shape through the application of an external force. The processes which alter the shape of a metal are known as metal deformation processes, and take advantage of the fact that metals, although appearing to be rigid and solid, possess both elastic and plastic properties. This may best be illustrated by the simple example of a metal bar subjected to an external tensile force in order to increase its length and decrease its cross-sectional area. The change in shape may be illustrated graphically (see fig. 6.1). In the section AB the metal is undergoing elastic deformation, i.e. on the removal of the load the metal reverts to its original dimensions; its behaviour is similar to the springing back of a rubber band when stretched and released. However, once the load exceeds the value at B (the yield point, elastic limit) some of the extension becomes permanent and the metal is said to have been plastically deformed. Thus, on loading to a point represented by B', both elastic and plastic deformation are obtained. If the load is removed

456 | METAL DEFORMATION

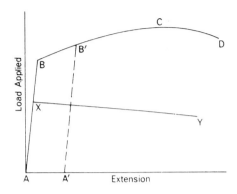

Fig. 6.1 Lond–Extension curve showing the influence of temperature.

springing back still occurs, as can be seen from the slope of the line A'B', but a permanent extension AA' remains.

It is common for the behaviour of a metal subjected to a tensile load to follow the curve BCD; beyond the point C very localized extension occurs with falling load until at the point D the metal finally fractures owing to the corresponding localized reduction in cross-sectional area (necking).

A more conventional representation of the effect of increasing load on the extent of deformation is given by a stress–strain curve, in which

$$\text{stress} = \frac{\text{load}}{\text{original area}} \quad \text{and} \quad \text{strain} = \frac{\text{change in length}}{\text{original length}}.$$

Of even more fundamental significance are the true stress and true strain which relate the load and length changes at any one stage to the actual area and length at that stage.

It is significant that when a metal is subjected to a stress within its elastic range (AB), the elastic strain is proportional to the stress applied. That is,

$$\frac{\text{stress}}{\text{strain}} = \text{constant } E.$$

This constant E is known as Young's modulus of elasticity and may be considered constant for any one metal. It is of particular significance in the metals used in tools for plastically deforming other materials, since these tools distort elastically under the working stresses. The amount of elastic strain that a deforming tool undergoes affects the dimensional accuracy of the plastically deformed material and will obviously be dependent on the slope of the elastic part of the stress–strain curve, i.e. on Young's modulus E. This elastic property is also of importance in engineering structures which deflect or bend under their own weight or under other forces. Some elastic behaviour may be tolerated but permanent or plastic change of shape would be unacceptable. Once the value at B (the elastic limit) has been exceeded the plastic strain that is incurred is no longer proportional to the applied stress.

The form of a stress–strain curve may be profoundly modified by such variables as the temperature and speed of deformation. Thus at high temperatures continuous plastic deformation may occur at a constant load as shown by the curve AXY (see fig. 6.1). This temperature-dependent behaviour leads to the concept of 'hot working' and 'cold working' which will be amplified in later sections.

6.1.3 Reasons for metal deformation

Almost any metal can be deformed either in a cold or hot deforming process. Metals are deformed in order not only to change the shape to a more desirable form but also to alter and generaly improve the mechanical properties. It is difficult to envisage the production of sheet metal for car bodies, silver paper or curtain runners without resort to a deforming process. Many other shapes could be made by machining but this would involve considerable wastage of metal. Other instances may be cited where resort might be made to casting techniques for the manufacture of a component (e.g. a crankshaft) but where a deformation process is preferred. In such cases deformation is used because of the superior properties that are normally attained in wrought shapes. A comparison of strength and elongation values for a number of cast and wrought metals is given in table 6.1, clearly demonstrating the improvement resulting from the deformation of a cast structure.

In order that the reasons for these improvements may be fully understood, reference must be made to the structure of a typical cast ingot (see fig. 6.2). The grain structure of a cast ingot is seldom uniform, generally consisting of a layer at the surface of very fine rapidly chilled crystals, below which lie large elongated (columnar) grains pointing towards the centre. Near the centre there may be a zone where grains of moderate size exist, which exhibit no obvious directionality (equiaxed crystals). The presence of columnar grains may give rise to planes of weakness in the metal, along which fracture might easily occur. Large holes (porosity), resulting either from gas trapped in the metal as it solidified or from the shrinkage and contraction that all metals undergo during the solidification process, may also be present in a cast structure. The presence of such porosity obviously weakens the metal and renders it more prone to failure. When an alloy is being produced, segregation of certain of the alloying constituents may be detected in cast structures. Such segregation may be on a large scale (macroscopic), as typified by 'tin-sweat' on the outside of tin-bronzes, or on a microscopic scale, detectable only by the microscope or similar instrument. Such segregation, whether macroscopic or microscopic, is generally regarded as detrimental and may give rise to inferior mechanical properties.

Plastic deformation of a cast metal significantly changes the structure outlined above and leads to an enhancement of the metal's properties. Thus, the highly directional crystal structure may be changed to a randomly oriented one in which no planes of weakness exist, whilst the porosity, if the deformation is sufficient, may disappear altogether. The extent of segregation may be reduced (although seldom removed completely) if deformation is carried out at elevated temperatures. The higher temperatures assist in the diffusion of one metal into another until (ideally) a uniform composition

Table 6.1 Comparison of mechanical properties of Aluminium, Copper and Brass in both cast and wrought states.

	Aluminium– 3–6% Mg		Copper		Brass	
	Cast	Wrought	Cast	Wrought	Cast	Wrought
Ultimate Tensile Strength – t.s.i.	9–11	14–17	10	14	16	20–23
% Elongation	3–5	18	25	50	60–70	65–75

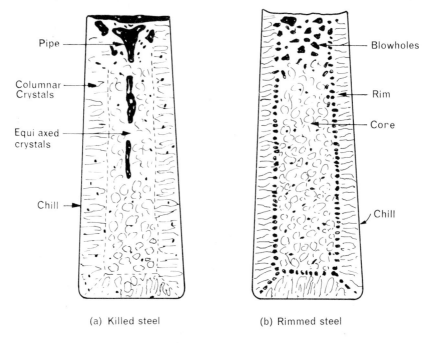

Fig. 6.2 Structure of typical cast steel ingots. (a) killed steel (b) rimmed steel.

exists in all places. It is normal practice for the initial stages of the metal deformation process to be performed at an elevated temperature. As may be seen from fig. 6.1 the stresses (i.e. loads) required to deform a metal are lower at higher temperatures and thus the energy required to perform the process is less. This obviously has an economic significance: less powerful (and therefore less costly) equipment is required. The 'hot-working' processes leave the metal in a soft state with a randomly oriented grain structure.

In the later stages of deformation, the process is more likely to be performed at ambient temperature – 'cold working'. This may be necessary to obtain accuracy of dimensions (difficult in hot-working processes when allowance would have to be made for thermal contraction and scale) as well as improved surface finish. It is common practice after the hot-working stage for the material to be pickled in some acid liquor in order to remove surface oxide and scale prior to final cold working. The cold working then serves to smooth out surface irregularities and may enable a highly polished mirror-like finish to be achieved, as with aluminium sheet. One other important effect of cold working is its influence on the grain structure and mechanical properties. Because of the phenomenon of 'work hardening' most metals may be strengthened by the simple device of deformation at room temperature. This change in strength is accompanied by a progressive change in grain structure and, in conjunction with annealing, the grain size and mechanical properties of the metal can be controlled.

6.1.4 Mechanism of deformation

In order to understand the ability of a metal to deform yet retain its coherency, or to work harder, some consideration must be given to its fundamental structure.

METAL DEFORMATION IN GENERAL | 459

Metals are essentially simple crystalline substances built up of atoms arranged in a regular manner in space. Most common metallic objects consist of a large number of irregularly shaped crystals (called grains by metallurgists) which may be large enough to be seen without aid. The crystals of zinc on galvanized iron are one example. There are many metals, however, which need to be microscopically examined for evidence of their grain structure. The essence of a crystal is that it consists of its unit parts arranged in regular order in a 'space lattice' (see fig. 6.3). A metal may then be visualized as an arrangement of layer upon layer of atoms built up, usually as closely packed as possible, to form a solid. The atoms are kept together by attraction forces resulting from a shared electron bond. It is the peculiarly metallic electron bond which results in many special metallic properties.

When a metal deforms, the layers of atoms may be thought of as sliding one over the other, breaking and remaking bonds in the process. Since the atomic lattice is three dimensional it is easy to see that movement may also be in many directions in most cases. Alternatively the effect may be visualized rather as the sliding of a pack of cards. Atomic slip is the term used to describe this type of movement. Some metals do not have the facility to deform equally well in several directions and there are other processes involved besides the simple sliding described here. For a more detailed treatment the reader is advised to study an introductory metallurgical textbook.

6.1.5 Cold working a metal

When a metal is deformed cold the analogy of a card pack is too simple. As slip in a metal proceeds the force necessary to bring it about increases progressively, i.e. the metal work hardens and becomes stronger, whereas in a card pack the force required to distort it sideways remains the same no matter how many times the cards of the pack slide over one another. Clearly a mechanism must be operating to explain the effect.

Calculations of atomic bonding forces have established that the stress necessary to slide a layer of metal atoms simultaneously over another layer is up to a thousand times greater than the load at which the metal is observed to deform. Here again is an anomaly. Both the unexpectedly low load for deformation and the work hardening may be explained on the basis of the theory of dislocations. Atomic lattices of metals are far from perfect. Examples of defects are vacant sites (or vacancies), where an atom is missing, and the presence of a foreign atom. In this latter case the foreign atom may be

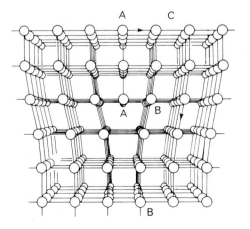

Fig. 6.3 Space lattice with simple edge dislocation.

an alloying element (e.g. zinc atoms interspersed in a copper lattice) or an impurity. When a defect is present the surrounding atoms rearrange themselves slightly to accommodate the strain and reach a compromise on the atomic forces which are now out of balance. A common defect takes the form of a part row or part plane of atoms missing. Such a defect is a dislocation. Fig. 6.3 illustrates the simplest form of dislocation–the edge dislocation. The lower part of row AA is missing and the resulting distortion is accommodated within the neighbouring few rows. A careful look at the row of atoms BB in the figure will show that it is in such a position that very little force would be necessary to push it to the left, in which case the dislocation row would now be row CC. The process could be repeated so that the dislocation would be moved through the lattice in a rippling movement, rather as a ruckle in a carpet can be trodden across a room. When the dislocation reaches a surface there is a step which represents a deformation. With the aid of millions of dislocations visible deformation is brought about. It is estimated that some 10^8 dislocations are present on a square centimetre of metal surface in annealed metal. As working proceeds, more and more are formed as a consequence of atomic strain. Hence there is no shortage of dislocations. The stress required to move one plane of atoms over another is thus merely the stress needed to move a dislocation step by step and one anomaly is explained.

The second anomaly, work hardening, is similarly explained by the dislocation theory. Slip takes place by the piecemeal movement of atomic planes through the agency of dislocations. If dislocation movement is hampered for any reason, then slip is hindered. A dislocation represents a distortion of its parent atomic lattice and if two similar defects exist together the distortion becomes even more pronounced, i.e. the lattice strain is increased. Accommodation by neighbouring atoms becomes more and more difficult. For this reason the approach of dislocations towards one another so that their surrounding strain fields interact is strongly resisted by the atomic forces of the parent lattice and similar dislocations effectively repel one another. On the other hand, if fig. 6.3 is viewed upside down there appears no reason why a dislocation should not consist of a row of atoms missing from the top half of the lattice. If such a dislocation encounters a dislocation of the first type then clearly they complement one another and lattice strain disappears. Unlike dislocations, then, effectively attract one another. The attraction and repulsion exist even when direct confrontation is impossible and when the two are on entirely separate but adjacent planes. Other defects such as the vacancies or foreign atoms previously mentioned also possess strain fields which interact with those of dislocations, attracting or repelling respectively depending upon whether the net effect would decrease or increase lattice strain. Such interactions serve to anchor dislocation movements if one of the sources of strain is itself stationary. For example, a foreign atom cannot readily move, and a dislocation on a plane at right angles to the moving one is immobile. As deformation proceeds more and more dislocations are rendered immobile and movement in the planes is progressively blocked. Large-scale observation records work hardening.

A very effective block is the grain boundary, which is an area of intense mismatch between neighbouring grains and hence is associated with very high lattice strains. Fig. 6.4 shows a 'pile up' of dislocations against a grain boundary, revealed under an electron microscope. As a consequence of the over-all change in shape, each individual grain is also distorted. A grain is bonded to its neighbours and cannot distort in isolation but must accommodate its shape by plastic and elastic movement to match surrounding grains, which are themselves similarly constrained. Heavily worked grains are distorted

METAL DEFORMATION IN GENERAL | 461

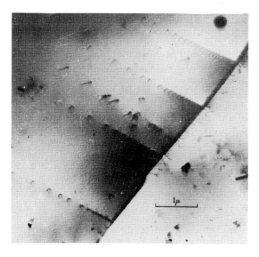

Fig. 6.4 A 'pile-up' of dislocations at a grain boundary, revealed by the electron microscope.

in external shape and, in lattice form, are elastically stressed and contain many locked dislocations. They are storing strain energy which may be released by heat treatment. One reason for wishing to restore the grains to their former situation might be to enable working to proceed, for it is clear that work hardening can only continue to the point where either the metal is as strong as the deforming tool or the ability of the metal to deform is exhausted.

In time most cold-worked metals will recover some of their original properties and the rate at which they do so is increased by the application of heat. Electrical conductivity, for example, may be partially restored even while the metal remains hard. Similarly, the level of residual stress in the metal will be reduced by lattice accommodation.

6.1.6 Annealing

Annealing is the term used to cover many processes which involve heating a metal (sometimes other substances) to a high temperature, retaining it at that temperature for a period and then slowly cooling it.

Complete restoration of properties is possible only by converting the old heavily strained grains into new ones by a process called recrystallization, which is a pseudo-spontaneous change in which new grains grow from nuclei and absorb the old grains. However, even very heavily cold-worked metal will not, in most cases, recrystallize without some encouragement. Energy is needed to start the change and is supplied as heat when the metal is raised to its annealing temperature. This temperature varies widely from metal to metal and also depends on the conditions. For example, a heavily deformed metal recrystallizes at a lower temperature than a lightly deformed one.

The new grains grow until all the old ones are gone and the new ones touch to fill all the available space. For given conditions, the larger the number of grains which start growing the smaller are the final ones. Nuclei form at sites of high energy and the number of such sites available will depend on the energy imparted to the metal by deformation. The size of the final grains then will depend on the degree of cold work, which leads to the common observation that heavy working leads to fine annealed grain size. The size will also depend on annealing temperature since the rate of growth depends on temperature. At high temperature potential nuclei may be absorbed by rapidly growing grains before they even begin to grow.

The end effect of annealing is a return to the original properties of the metal, which now consists of strain-free grains and possesses its original hardness and ductility.

6.1.7 Hot working

When metal is heated, it becomes softer, weaker and capable of sustaining a higher degree of deformation than when cold. Hot working is conducted to take advantage of these trends. For each metal there is a temperature above which no work hardening can be detected. This means that not only is the strength of the metal low but also that it stays low in spite of even heavy working. Distortion is still taking place by dislocation movements in the lattice but the high temperature enables dislocations to disperse and disappear. New grains form by recrystallization simultaneously with working. Clearly the lowest temperature at which this effect occurs is strongly related to the recrystallization temperature. In fact, the lower limit of the range of hot-working temperatures is the same as the recrystallization temperature in annealing. Below this temperature changes take place slowly and recrystallization does not occur. Hot working can be regarded as simultaneous cold working and annealing.

Theoretically a metal could be hot worked right up to its melting point, but in practice other factors limit the upper working temperature. Segregation of alloys or impurities may cause sections of the structure to melt before the apparent bulk melting point is reached. Heat may be generated by the working process itself and raise the metal temperature to above the safe limit. Chemical reactions with the atmosphere are rapid at high temperatures and may limit the permissible temperature. All of these things, together with the need for a safety margin in furnace control, combine to determine the upper limit of the hot-working range.

Table 6.2 shows recrystallization temperatures, melting points and upper limits to hot working for a number of metals and alloys.

In subsequent sections individual working processes and their products are described. In each of them metal is being deformed hot or cold and its behaviour and properties depend on the basic mechanisms discussed briefly in the preceding pages.

6.2 ROLLING

6.2.1 Principles of rolling

The plastic deformation of a metal by rolling consists essentially of passing the material between two cylinders (the rolls) revolving at the same speed but in opposite directions. The distance between them is somewhat less than the thickness of the metal to be deformed. The rolls thus grip the metal, process it through the space between them (the roll gap) and consequently reduce its cross-sectional area, thereby increasing the length. The process is shown schematically in fig. 6.5. The gap may be altered manually or automatically. By passing the metal through progressively reduced roll gaps it may be rolled to any desired thickness. This may be achieved by passing backwards and forwards through one mill or by processing through several when high-output demand justifies it. Some sideways or lateral spread may occur, which is dependent on the geometry of the material being rolled, the roll size, the temperature of rolling, the composition of the metal and the amount of reduction being effected. In general,

Table 6.2 Temperatures of recrystallization, melting and hot-working upper limit.

Temperature °C

Metal/Alloy	Recrystallization	Melting	Hot-working upper limit
Mild Steel	600	1520	1350
Copper	150	1083	1000
Brass 60/40	300	900	850
Aluminium	100	660	600

however, most spread will occur in the early stages of rolling when the thickness/width ratio of the ingoing material is at its greatest.

In order that rolling may take place it is necessary for the rolls to 'bite' the material and process it through the roll gap. This ability to bite requires the presence of a force acting tangentially to the roll surface at the plane of entry. This force, F, will be the frictional force between the roll surface and the surface of the metal that is being deformed. It is related by the coefficient of friction μ to the radial force N applied by the rolls to the material being deformed, and $\mu = F/N$. For rolling to be possible, i.e. for bite to occur, it may be shown that μ must be greater than $\tan \alpha$ (see fig. 6.5). Rolling is thus the one deformation process which requires the presence of some frictional force. In all metal-working processes the presence of friction increases the amount of work required above that needed to overcome the resistance of the material itself to deformation. Rolling relies on friction for its action but, in order to avoid unnecessarily high pressures, the absolute minimum value of μ is desired. The minimum value of the coefficient of friction to ensure bite is of more significance in the rolling of ingots than in the rolling of sheet and strip. The angle of bite is necessarily greater in the former case because of the thickness of the stock. It is not uncommon for difficulty to be experienced in getting the rolls to bite a large ingot and to facilitate this

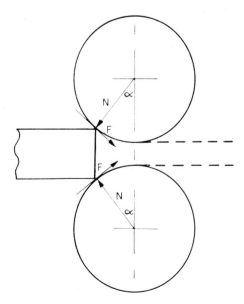

Fig. 6.5 Diagram showing rolling forces and angle of bite.

the roll surface may be deliberately roughened (ragged). As the product thickness decreases with increased rolling, the angle of bite α also decreases and it becomes easier for the rolls to grip the surface of the metal to be deformed.

6.2.2 Types of mill

The principal types of rolling mill in use are shown in fig. 6.6. The nomenclature, two-high, three-high and four-high mill, refers to the number of rolls in the mill housing. In a two-high mill with non-reversing rolls, it will be necessary to return the metal over the top of the rolls if re-rolling is required. Such a mill is known as a two-high pull-over mill and although much used at one time is mainly utilized nowadays for

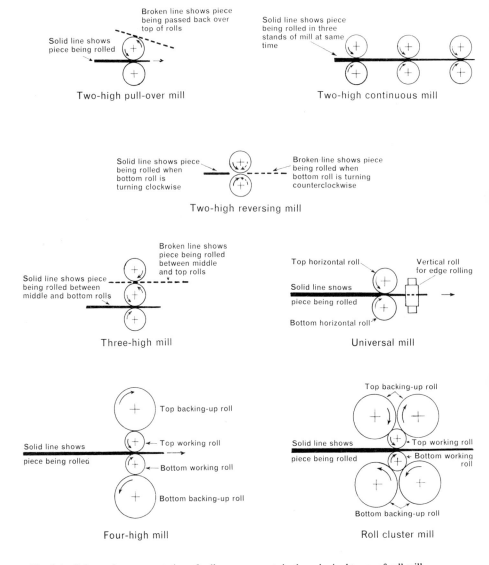

Fig. 6.6 Schematic representation of roll arrangements in the principal types of roll mill.

finishing operations on sheet. A more modern version of the two-high mill is the one in which the direction of rotation of the rolls may be reversed, thus allowing rolling to take place in either direction. In this way, much longer and heavier lengths of stock may be handled than would be possible in a non-reversing pullover mill. In a three-high mill, the top and bottom rolls rotate in the same direction with the middle roll moving in the opposite direction. This allows the material to be processed continually back and forth, passing from one side between bottom and middle rolls and from the other side between middle and top rolls. For the rolling of heavy stock such a mill would have to be used in conjunction with lifting tables, mechanically operated.

The use of large diameter rolls in two-high and three-high mills inevitably means a large area of contact between the rolls and the material being deformed. With large contact areas, the load required to effect deformation must be correspondingly great, which may create problems in mill design. In an effort to overcome this problem, small diameter work rolls are now used in many mills, supported by one or more backing-up rolls. Such smaller diameter rolls, giving a reduced contact area, result in lower rolling loads. More significantly, the pressure is decreased since the reduction in contact area in turn reduces the restraint of friction on metal movement. The stresses imposed on the mill housing are therefore less than for corresponding two-high and three-high mills, and the rolls themselves are subjected to less (elastic) deformation. Thus, the use of four-high mills has enabled thin stock such as sheet and strip to be produced to greater dimensional accuracy as well as facilitating the production of foil of 0.001 inch thickness or less. The necessity for backing-up support of the work rolls in such a mill has led to the development of cluster mills, culminating finally in the Sendzimir-type mill (see fig. 6.7) in which two work rolls, which may be less than 1 inch in diameter are each supported by as many as ten backing-up rolls. Such a mill may be used for rolling material with high resistance to deformation.

In recent years two specialized mills have attracted considerable attention. In one, the planetary mill (see fig. 6.8), the deformation is caused by a large number of small (approximately 2 inches in diameter) work rolls revolving planet-like around a large inner backing roll. Apart from its construction, which differs radically from any previous conception, it is capable of reducing material from approximately 1 inch thickness to less than 0.1 inch thickness in one pass, a reduction of more than 90%. In conventional mills it is unusual for the reduction to be greater than 30% in any one pass. Output rates from the planetary mill are reported to be of the order of 50% higher than

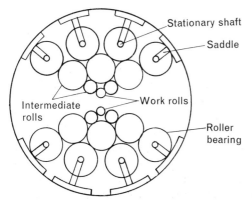

Fig. 6.7 Schematic concept of the Sendzimir cluster mill.

466 | METAL DEFORMATION

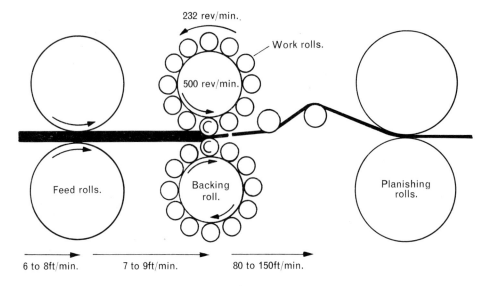

Fig. 6.8 Schematic representation of the Planetary mill showing rolling action.

for a jobbing mill and it may be that the mill will be of particular value to small and medium size undertakings rolling a highly varied range of material.

The second specialized mill of interest is the pendulum mill (see fig. 6.9). Two work rolls oscillate backwards and forwards as a pendulum while the metal moves slowly through them under the action of a set of driving rolls. The pendulum rolls oscillate rapidly (500 to 2600 c/min) in relation to the forward movement of the material but are not power-rotated, the rotation which does occur being due to contact with the work material. The mill is capable of effecting much larger reductions in one pass than conventional two-high, three-high and four-high mills and has a particular field of application in the rolling of hard and expensive metals.

6.2.3 Mill layout

A set of rolls is termed a 'stand'. Probably the most common arrangement is for one stand, be it reversing or not, to be operated independently as a single unit. However, with the advent of mass production techniques, more and more mills are being operated in groups. The continuous tandem mill, for example, contains from two to possibly 20 stands laid out in one long line (train). In such an arrangement a continuous length of metal can be processed simultaneously through all stands with its front end issuing from the last stand whilst its back end is still to enter the first. Clearly, in unit time each stand must pass an equal volume of material. The cross-sectional area decreases as the stock proceeds through the train and therefore each stand must roll faster than its predecessor to keep pace with the ever increasing length. Rolling speeds are matched to control tension or looping between stands, i.e. a 'speed balance' is maintained. Where flexibility is required or space limitations prevent individual stands from being laid out in one long line to give the tandem arrangement, the mills may be placed side by side, or even staggered (see fig. 6.10). The metal is transported or bent round (by means of guides) to feed into the next stand. Such an arrangement

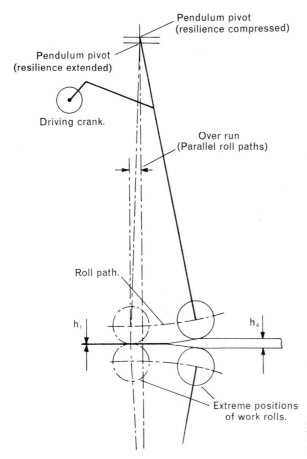

Fig. 6.9 Diagrammatic illustration of the basic principles of the Pendulum Mill.

would not be feasible for sheet and strip rolling but is often encountered in rod and bar mills. Where bending is used, such mills are generally known as looping mills, but individual arrangements of stands have been given such titles as the Belgian mill, cross-country mill and guide-mill. The initial stages of rolling are designated 'roughing' and the final stages 'finishing'. Ingot working is frequently termed 'cogging'. These terms are coupled with specific layouts, e.g. a roughing stand, roughing train, finishing train, cogging mill, etc.

6.2.4 Roll design

The principal parts of the rolls themselves are shown in fig. 6.11. The deformation of the metal being worked is caused by the roll body or barrel, on either side of which are the roll necks which act as the bearing surfaces for the rotation of the roll. The wobblers at the end of the roll serve to connect the roll to the motive power of the mill, the connection being via coupling boxes, spindles and pinion box. The roll barrel may be either flat (nominally) or grooved (see fig. 6.12) according to whether flat, round or angular products are to be rolled.

Proper roll design plays a most critical part in producing a correctly shaped product

468 | METAL DEFORMATION

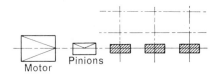

(a) Three stands in open train

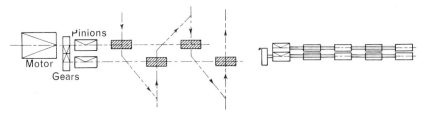

(b) Four Stands cross-country — split-drive

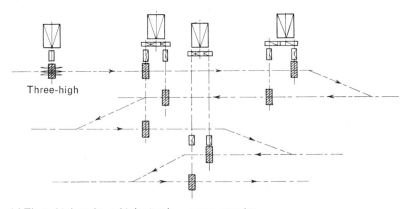

(c) Three-high and two-high stands — cross-country

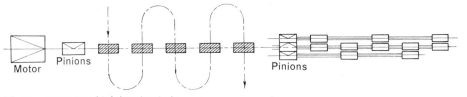

(d) Alternating two-high looping train

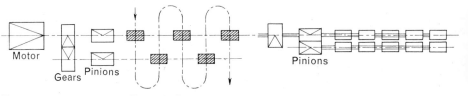

(e) Two-high looping train-offset stands

Fig. 6.10 Diagrammatic representation of the layout of stands.

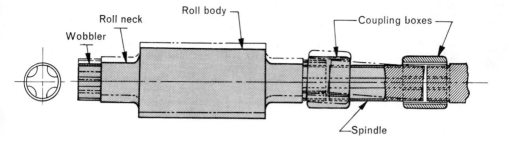

Fig. 6.11 Nomenclature of the parts of a rolling-mill roll and the units that connect it to the motive power driving the mill. The broken lines indicate the position of the parts when the mill is raised.

as well as influencing the life of the roll itself, whilst rolling load must be limited to avoid overstressing the mill housing. Despite the application of much scientific and technological knowledge, experience and art play a part in roll design.

In sheet and strip rolling the most important feature is to include a correct camber on the roll barrel face to ensure the production of a flat product. If this were not done and a flat parallel roll face were used, the forces established during the process would give rise to some (elastic) bending of the rolls. Such bending would cause the edges of the strip to be reduced more than the middle, the effect showing up as buckled (over-rolled) edges. Roll bending can never be altogether prevented but its effects can be

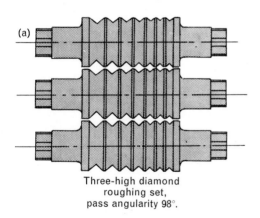

Three-high diamond roughing set, pass angularity 98°.

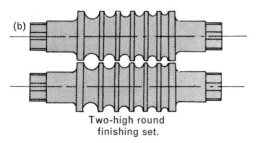

Two-high round finishing set.

Fig. 6.12 (a) Set of three-high rolls with diamond passes, showing the seven passes used in roughing down a billet to a round cornered square section.
(b) Two-high set of rolls with grooves for finish rolling of rounds.

mitigated by giving to the roll barrel surface a camber which compensates for the bending and allows the same reduction to be carried out across the whole width of the strip. The extent of this camber may be calculated from a knowledge of the magnitude of the rolling load, the elastic characteristics of the roll material, the width of the material to be rolled and the dimensions of the roll itself.

The design of rolls for rod, bar and section rolling is much more complex, the designer having to create the correctly shaped grooves in the rolls so that the right balance is obtained between elongation of the metal and lateral spread. In the production of this type of shape, some spread is usually desired and the amount of reduction together with the groove profile has to be such that the metal spreads just the right amount, to fill the pass completely. Too much spread would result in excess metal forming fins at the junction of the two rolls.

Reaction between the material and inclined parts of the roll groove will lead to a force which tends to push one roll sideways over the other. Such a force if unopposed would give mismatch and incorrect shape. By correct balance of groove geometry this 'end thrust' can be minimized (see fig. 6.13) and the forces on the bearings and roll

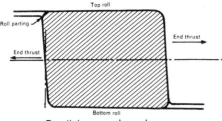

Parallelogram shaped pass

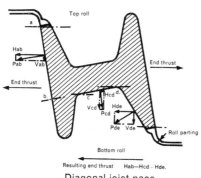

Diagonal joist pass

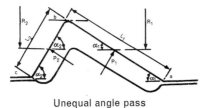

Unequal angle pass

Fig. 6.13 The causes of end thrust.

collars reduced to a minimum. If the reductions effected on different parts of a rolled section are not similar the material issuing from the groove will be bent and its path unpredictable. This is again a roll design problem and, as perfect balance is very difficult to achieve, a slight deflection is often aimed at so that the product leaving the roll bears on guide devices which then control the subsequent path.

6.2.5 Roll materials and manufacture

Rolls may be made from cast iron or steel, typical compositions being shown in Table 6.3. Steel rolls may be produced directly by casting or indirectly by the forging of an ingot. The selection of the correct material and method of manufacture is dependent on the use to which the roll will be put. For rolls which are required to effect heavy reductions on ingots, billets and slabs, strength and the ability to withstand mechanical shock and high temperatures are of the greatest importance. Rolls for cold-finishing sheet and strip require an excellent surface finish which has to be maintained over long periods. Such rolls should have a very hard wear-resistant surface, shock and thermal resistance being relatively unimportant. For these reasons the trend is to use cast carbon or alloy-steel rolls for the hot rolling of steel, iron or forged alloy-steel rolls for the intermediate stages and cast iron for the finishing rolls. Non-ferrous metals such as copper and aluminium-base alloys are hot worked at lower temperatures than steel and the necessity for the rolls to be resistant to thermal shock is obviously less. Aluminium is therefore often hot rolled with forged alloy-steel rolls but would be cold finished on a mill with similar rolls to those used for cold rolling steel.

A method of obtaining a roll with a very hard surface and an underlying tough core is by double pouring. Such rolls are known as 'composite' or 'duplex' rolls and are manufactured by firstly filling the (chill) mould with a highly alloyed white iron which, as solidification begins, forms a very hard shell. When solidification has produced a shell of the desired thickness, the remaining still-molten white iron is flushed out by a softer (often grey) iron. Solidification of this second iron produces the tougher core. A further advantage of this technique is in the ease of machining the necks of the rolls as they need not be chill moulded and therefore little or no solidification of the white iron occurs. The neck is therefore cast in the second iron which is easily machinable.

The development in recent years of ductile or nodular iron has led to its introduction as a roll material. Such rolls possess (for iron) a remarkable combination of strength and toughness due to the nodular or spheroidal shape of the graphite in the structure, which contrasts with the flake form that graphite has in grey irons. Spheroidal graphite cast-iron rolls are easy to cast and can be readily machined to give a good surface finish. Although more expensive than both iron and steel rolls, nodular iron rolls may be particularly useful where a normal iron roll has insufficient strength or where steel rolls would have a short life due to excessive wear.

For specialized rolling of such products as foil it is necessary to minimize roll flattening which is caused by the roll separating force. Such flattening opens up the roll gap and, if significant, makes the further reduction of thin stock such as foil difficult. Rolls are therefore sometimes made from material having a high elastic modulus such as tungsten carbide. These rolls are, however, very expensive and must be installed in highly specialized mills which are constructed to ensure that the rolls do not suffer bending and consequent breakage. In a given case the choice of roll material is governed by the operating conditions, the quantity and quality of material to be rolled and economic considerations.

Table 6.3 Typical compositions, hardnesses and applications of iron and steel rolls

Type	Carbon	Silicon	Manganese	Phosphorus	Sulphur	Nickel	Chromium	Molybdenum	Hardness (shore)	Typical Applications
Iron Rolls										
Grain Rolls	2.75	0.75	0.6	0.5 max	0.1 max				30–38	Roughing and intermediate in plate and rod Mills
Alloy-Grain Rolls	3.00	1.00	0.5	0.25 max	0.08 max	1.0	1.0	0.25	50–55	Intermediate and finishing in strip and rod Mills
High Alloy-Grain Rolls	3.40	1.00	0.5	0.15 max	0.06 max	4.5	1.75	0.25	75–85	Finishing stands in strip mills
Spheroidal Graphite Rolls	3.30	2.00	0.4	0.15 max	0.02 max	2.0	0.75	0.50	45–80	Roughing or intermediate stands in rod, bar and strip mills
Chill Iron Rolls	3.0	0.70	0.25	0.45 max	0.08 max				58–70	Sheet, merchant, plate and rod mills
Nickel Chill Rolls	3.30	0.50	0.25	0.10 max	0.10 max	4·5 max				
Steel Rolls										
Cast-Steel Rolls	0.5				0.06		1.00	0.25	80–90	Finishing stands in strip mill
Cast-Steel Rolls	0.8				0.06				28–32	Cogging, slabbing, heavy roughing
Cast Alloy-Steel Rolls	0.9		0.75		0.06	2.5			32–36	Billet roughing stands
Cast Alloy-Steel Rolls	1.0			0.06	0.06		1.0	0.3	30–42	Roughing and intermediate stands
Forged-Steel Rolls	0.45				0.06		1.75		35–55	Backing up rolls in four-high mill
Forged-Steel Rolls	0.70	0.30	0.3	0.06	0.06				24–27	Cogging, slabbing, heavy roughing
Forged Alloy-Steel	1.00	0.3	0.3	0.06	0.06	0.5	1.75		27–30	Intermediate mills
Forged-Steel Rolls									50–55	Large backing-up rolls
									80	Small backing-up rolls
									90–100	Work rolls in cold rolling

6.3 EXTRUSION

6.3.1 Principles of extrusion

The process of extrusion is one in which a cylindrical block of metal is forced to flow out through an orifice by the application of pressure. This orifice is contained in a flat plate which is called the die. A long length of metal is thus produced, the shape of which corresponds to the shape of the die orifice. The operation is shown schematically in fig. 6.14, both the direct and indirect (inverted) processes being illustrated. In the direct process (by far the most common) the metal block (billet) is placed in a container at one end of which is the die assembly. A mechanically or hydraulically operated ram then moves into the container at the other end and applies pressure to the billet. After initially upsetting the billet to fill the container the pressure causes the metal to flow through the orifice. Extrusion proceeds until merely a thin plate of metal (discard) is left in the container. The die holder assembly is then separated from the container unit and the extruded length cut free from the discard in the container. The ram moves further forward to push out this discard. After reassembly of the die and container, the operation may be repeated. The process is thus essentially a batch one, unlike rolling which may be made fully continuous. The die orifice may be as simple or as complex as required and this makes the process equally suitable for bar, tube or other shapes. The production of tube by extrusion requires the presence of a centrally aligned mandrel to form the tube bore and determine the wall thickness. This mandrel may be part of the ram assembly, moving through the billet into the die orifice, or alternatively be incorporated as part of the die assembly (bridge die).

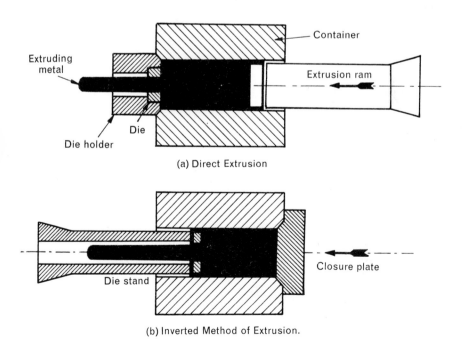

Fig. 6.14 Schematic representation of direct and indirect processes of extrusion.

In the indirect extrusion process the container is sealed at one end and the die is part of the ram. The extruded product thus passes back through the ram before emerging. The design of the press and the general inaccessibility have restricted the use of this form of extrusion.

A specialized branch of the extrusion process is that used for the sheathing with lead (or aluminium) of electrical cable for power and telegraphic transmission. A schematic picture of a vertical hydraulic cable-sheathing press is shown in fig. 6.15. Molten lead is poured into the container and when solidification is complete the ram applies pressure to the hot metal causing it to flow into a chamber in which there is a hollow mandrel. The cable to be sheathed is contained inside the hollow mandrel and passes out through the die. The lead divides into two streams which flow round the cable and weld together on its underside. It is then forced out through the aperture between the cable and the die to form a sheath. As the sheath emerges it grips the cable sufficiently tightly to draw it off a drum behind the press.

When most of the lead in the container has been extruded on to the cable the process is stopped and a fresh charge of molten metal is poured in. The cycle then repeats itself. During the break for recharging the cable will remain in the die-block and in this way a very long length of cable may be sheathed with many charges of lead successively extruded on to it.

Aluminium has been similarly extruded as a covering for cable. More technological problems are encountered with aluminium than with lead because of the greater resistance to deformation and the higher temperatures which are required. To prevent possible thermal damage to the cable insulation, die boxes of complicated design have become necessary. The introduction of molten aluminium into the container may also lead to difficulties due to excessive container wear and it is therefore more usual in the aluminium sheathing industry for the metal to be introduced into the container as a solid billet.

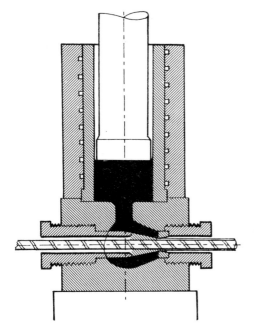

Fig. 6.15 Diagrammatic section through the container and die-block of a vertical cable press.

6.3.2 Types of press

Extrusion presses are usually classified as hydraulic or mechanical. In the former type the press is driven by water under pressure, which may be supplied directly by pumps or indirectly via an accumulator system installed between the pumps and the press. An accumulator is a device for storing energy, in this case liquid under pressure. The liquid may then be released at a much higher rate than would be possible by direct drive from the original pump. Direct pump drive is usually preferred when a prolonged period of time is required for extrusion whereas an accumulator-driven hydraulic press system is recommended when higher extrusion speeds are used and particularly with large presses. A further advantage of the accumulator system is that a number of presses may be operated simultaneously and the unit evens out any irregularities arising from the pump. Mechanical presses are cheaper and occupy less over-all space than hydraulic presses but the stroke is limited and the energy available varies with the position of the ram. If overloaded, mechanical presses are more easily damaged than the hydraulic type. Most mechanical presses operate in the vertical position whereas the hydraulic type may be vertical or horizontal. Larger size presses in excess of 3 000 tons are generally designed for horizontal operation which reduces erection and siting problems. A further difficulty with vertical presses is the limitation in length of extrusion which can be produced, unless the press is raised above ground level or a pit constructed below it. However, when extrusion is being used for tube production, vertical presses are often preferred in order to ensure concentricity in the product.

6.3.3 Tool materials

In an extrusion operation certain tools are subject to the effects of heat and pressure. The ability of the die, the container liner, the mandrel (if present) and the pressure pad in front of the ram to withstand these effects will determine tool life and influence the economics of the process as well as the ease with which the required dimensions can be obtained in the product. Typical temperatures and pressures for the extrusion of a number of metals are given in Table 6.4.

It is therefore important that materials able to withstand these conditions should be selected. Where very high temperatures are encountered, as in copper and its alloys, nickel-base alloys and steels, it becomes imperative for extrusion to proceed rapidly to prevent gross overheating of the tools and loss of strength. In general, alloy steels are used as they can retain their strength up to several hundred degrees centigrade. A further attribute required in the tool material selected for the die, and to a lesser extent, the container, is resistance to abrasion. Any abrasive material (such as scale or oxide)

Table 6.4 Typical temperatures and pressures for extrusion

Metal	Temperature range °C	Specific pressure t.s.i.
Copper	850–900	30–40
60/40 Brass	690–730	15–20
18% Nickel silver	850–900	50–60
Aluminium	400–550	20–40
Duralumin	430–450	50–65
Magnesium	350–430	50

on the surface of the billet is likely to cause wear as it moves over the tool surfaces during the extrusion stroke.

The composition of typical steels is shown in Table 6.5 together with the main field of application. In all cases the steels would be in the hardened and tempered state. There is a significant trend towards the use of nickel-based heat-resisting alloys for the tools subjected to arduous conditions.

6.3.4 Tool design

Possibly the most important aspect of design of extrusion tools is that associated with the die. The profile of the die orifice will determine the dimensions and shape of the extruded product. However, several factors, including frictional restraint and the over-all geometry of the die, combine to ensure that metal flow does not occur uniformly (see fig. 6.16). For example, the surface of the billet is partially restrained by frictional forces between it and the container. The central elements of the billet thus tend to flow in advance of the peripheral layers and in the later stages a central cavity extending into the product may be formed. Under the worst conditions, flow may be such that towards the end of the stroke the original surface is located in the centre of the remaining billet. The last material to issue will thus contain an annular ring of oxide (see fig. 6.17).

It is the task of the designer to modify bearing lengths (die land) and angles to achieve uniform flow. With irregularly shaped sections, in which marked differences in cross-section occur, the problem is particularly difficult as the metal will tend to flow easily through the thick sections where the die orifice is greatest and with difficulty where the section is thinnest. By the use of frictional restraint in some parts of the die and control of the natural tendency of the billet material to flow easily from the centre and less so from the outside, the die designer aims at uniformity and a straight product.

In certain instances, multihole die extrusion is practised in which several extrusions are obtained from one billet by the expedient of including more than one orifice in the die. In this form of extrusion, the positioning of the openings relative to the cross-section of the billet is important in order to obtain a balanced flow in thick and thin sections.

Another aspect of tool design concerns the dimensional clearance between the ram–pressure pad unit and the container. This may be at a minimum compatible with the avoidance of seizure, or may be a significant amount up to $\frac{1}{4}$ inch smaller than the diameter, which will cause a thin shell or skull to be left in the container. The pressure

Table 6.5 Composition and uses of extrusion tools

Carbon	Silicon	Manganese	Chromium	Nickel	Tungsten	Vanadium	Molybdenum	Uses
0.2	0.2	0.25	2.5	2.0	10	0.5		Dies, containers at high temps.
0.35	0.25	0.25	3.0	–	9	–		Dies, pressure discs
0.40	1.3		12	12	2			Dies for high temperature use
0.35	1.0	0.5	1.3		4.25	0.35		Mandrels, pressure pads
0.35	0.25	0.25	2.5		9.0			Mandrels
0.40	–		1.5	–	2.5	–	0.5	Containers for low temp. use
0.35			1.25	3.5			0.25	Containers for low temp. use
0.35			5.0					Dies, mandrels

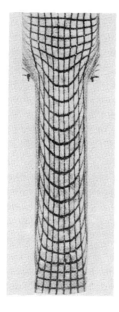

Fig. 6.16 Non-uniform flow in a billet during extrusion, the surface layers being restrained by the frictional forces present.

pad effectively cuts through the billet leaving the skull between the ram and the container; the process is called 'skull cutting'.

A perfect scale-free billet, if well lubricated, should flow easily and produce a satisfactory surface on the extrusion. This is attained in the hot extrusion of steel with a vitreous lubricant, and no skull need be left behind.

When a scaled billet surface is present and lubrication is not completely effective, the extruded product may have an unsatisfactorily rough surface derived from the scale or contain an internal defect near the rear end due to the flow of the scale inwards, and skull cutting may be necessary. Skull cutting is practised when the conditions of billet surface, lubrication and temperature are such that free sliding of the billet through the container is hindered sufficiently to result in a defective product. The cutting of a skull effectively leaves the defective surface in the shell.

Periodic removal of the skull is necessary and this normally takes place after every cycle of the press.

Fig. 6.17 Extrusion defect of the rear end of the product caused by the flow of oxide from the surface towards the centre.

6.3.5 Impact extrusion

The extrusion of softer metals at room temperature by means of a sudden blow dates back to a patent taken out in 1841 for the production of collapsible metal tubes. Such tubes are still produced by this technique, which is illustrated in fig. 6.18, the process being known as 'impact extrusion'. A slug of metal, usually cold, is placed in a shallow die and subjected to an impact from a punch which causes the metal to flow up over the punch through the annulus between it and the die. The backward flow of metal over the punch necessitates the most careful polishing of this tool in order to prevent damage to the tube and facilitate its easy removal.

In contrast to this form of impact extrusion the Hooker process, dating from the early part of the twentieth century, produces similar shapes by forward extrusion. The technique is demonstrated in fig. 6.19. A cold blank is placed in a container at one end of which is the die. The punch fits the container very accurately and prevents escape of metal around its sides. On descent the punch shoulder exerts pressure which causes the metal of the blank to flow out through the annulus between the punch stem and die land. An elongated thin-walled tube is thus formed, as shown in fig. 6.19. The tube may be open-ended or closed since the base takes no part in the actual process (c.f. deep drawing, p. 493). Substitution of a completely pre-pierced shell for the blank shown in fig. 6.19 would enable an open-ended thin-walled tube to be produced.

Although drop hammers have been used, the process of impact extrusion is usually carried out on mechanical presses of the crank, eccentric or toggle type, and was

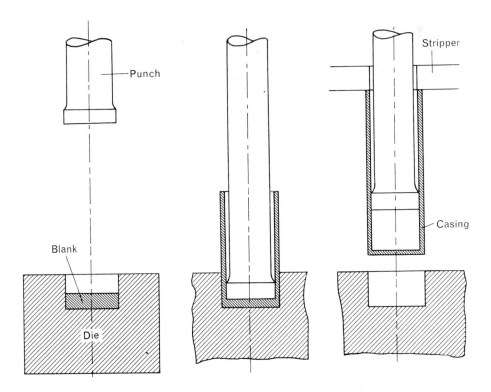

Fig. 6.18 Extrusion of a casing from a blank of soft metal.

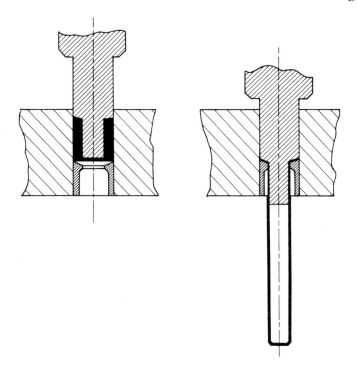

Fig. 6.19 Hooker process of impact extrusion.

originally applied to the metals lead, tin, zinc and aluminium. Later, brass cartridge cases were produced by this method and more recently it has been utilized for the manufacture of copper tubes, reductions of 95% to 97% being reported.

The cold extrusion of steel was initiated in Germany just prior to the second world war for the production of cup-like shapes and has led to the development of the general technique of cold forging (see p. 490) for a vast variety of shapes. With the harder metals 'impact' is no longer a feature of the process but essentially similar operations are carried out at normal press speeds. Backward and forward impact extrusion may be combined to produce shapes such as those shown in fig. 6.20.

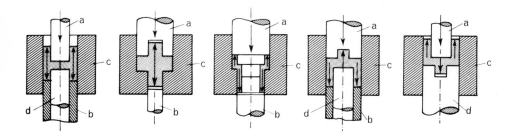

Fig. 6.20 Combined forward and backward extrusion. a = punch; b = stripper; c = counter-punch or ejector.

480 | METAL DEFORMATION

6.4 FORGING

6.4.1 Principles of forging

Primitive man extracted iron from oxidized ores and, because his furnaces were incapable of melting the metal, a sponge of iron particles resulted. By beating on stone anvils the mass was consolidated and forged into shape. Forging can claim to be the oldest mechanical working process. Development has led to great refinements in technique but forging is still characterized by bringing about bulk change of shape through the agency of pressure exerted between tool faces which are brought together either in impact or in slower squeeze.

Most forging is conducted as a hot-working process but significant amounts of metal are forged cold, particularly in the production of such items as nuts and bolts. Of all the metal-deforming processes, forging spans the widest range in size from tiny pins and lock parts to giant forgings weighing hundreds of tons.

The blacksmith forges iron to the shape he wants by applying the energy of the hammer through appropriately shaped tools. These tools impart their shapes to the metal and are chosen to have a wide range of application. Thus there are tools for wedging and cutting, piercing, flattening, rounding and grooving and many others. By increasing the available energy, similar techniques can be employed to forgings as large as that shown in fig. 6.21. Such a technique is described as 'open-die' or 'free-form' forging. A development from this technique uses preshaped tools in the form of a die containing a cavity of a shape such that material forced into it will take up the desired form. This technique is described as 'closed-die' forging. The early die forgings were always made on gravity 'drop' hammers and so received the name of drop forgings

Fig. 6.21 Open-die forging of a large ingot.

or drop stampings. Much of the preliminary shaping can be done on parts of the dies, or even in separate tools, which deform only part of the metal at a time. The final shaping die must, however, totally enclose the forging to achieve the finish and accuracy required, which means that a die-block substantially larger than the forging itself must be used. Hence die costs increase rapidly as the size of forging increases. The hammer or press, which has to be capable of accomodating the dies, becomes itself prohibitively large and expensive since the forging load increases in proportion to the increase in the area of contact between die and metal. Practical and economic considerations, therefore, limit the size of closed-die forgings.

Basically similar equipment is used for both types of forging with variations in tool fixing arrangements, accuracy of guiding or accessibility. As the size of the forged parts decreases machinery of more and more sophistication can be introduced since the forces involved become smaller.

6.4.2 *Grain flow*

Mechanical working results in alignment of zones of segregation, and any foreign material such as slag particles or oxides, in the direction of working. When pronounced the effect is termed mechanical fibre or grain and it is observed in most wrought steels. Consequently, if properties are measured in different directions, different values are obtained; this is illustrated in fig. 6.22. One of the aims of forging practice is to align the fibres in such a way that the optimum properties are developed in the direction in which they are required. Examples of good grain flow are illustrated in fig. 6.23.

6.4.3 *The Hammer*

In principle a hammer is an instrument which gains energy by accelerating a moving mass and imparts it by impact. Hammers used in industrial forging employ gravity or steam or compressed air acceleration. Height of lift or pressure of steam or

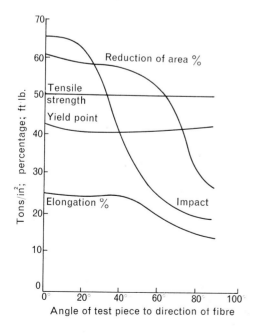

Fig. 6.22 Effect of mechanical fibre on properties.

482 | METAL DEFORMATION

Fig. 6.23 Diagrammatic representation of metal flow lines in (top) single-throw crankshaft, (bottom) bor stock, intermediate stage and = final forging of gear blank.

air can be controlled by the operator to enable him to vary the force of the blow. Early types of mechanical hammer copied the action of hand-wielded hammers in that a mass of metal fixed to a wooden beam was made to rise and fall. Such hammers are often preserved and may still occasionally be found working.

The moving mass of a modern hammer (the tup) is always lifted vertically so that the two striking faces are parallel at all positions. In British and American practice the size of a hammer is described by the weight of its tup, or the equivalent if the tup is artificially accelerated. On the Continent the energy of the tup is quoted, i.e. weight multiplied by height of fall for gravity hammers or $\frac{1}{2}$ mass × (velocity)^2 for steam or air hammers. Within the two main categories of gravity and artificially accelerated hammers lie many variations which result in important practical differences in speed of working and control of blow.

a. Gravity hammers. The simplest form of gravity hammer or stamp still survives in jewellers' shops and consists of a tup which is connected to a rope passing over a pulley on an overhead shaft. The tup is raised by hauling manually on the rope, and its downward fall is guided by a frame. A looped rope enables the operator to actuate the hammer with his feet and many hundreds of such 'kick stamps' were formerly employed in the English Midlands lock trade. This type of hammer was developed by applying mechanical power to rotate the pulley and lift the tup. The rope was replaced by a wide belt so that tups of several tons in weight could be lifted. A clutch was added to actuate the lift and to control the height. Modern 'belt hammers' are efficient and flexible pieces of equipment used to produce intricate closed-die forgings and are widespread in the UK (see fig. 6.24). An American development designed to increase rate of stroking uses rotating rollers which grip maplewood boards attached to the tup. In operation, a trip mechanism causes the rollers to approach one another, grip the boards for a predetermined time and release them. Hence the boards and the attached tup are lifted and then allowed to fall. The mechanism operates automatically whilst the hammer pedal is depressed and rate of stroking is very rapid. In this design, height of fall is fixed and and to some extent flexibility is sacrificed. The latest types of single acting compressed air or steam hammer go some way to combining rapid rate of stroking and flexibility, e.g. air lift, gravity drop.

Fig. 6.24 Eighteen ton bridge-type friction drop hammer.

b. Accelerated hammers. More energy can be imparted to a hammer tup by increased acceleration. The most frequent means employed is to attach the tup or ram, as it is then frequently called, to a piston forced to move up and down in a cylinder by steam or compressed air. This is a double acting steam/air hammer. Single acting steam hammers are not accelerated since the steam pressure is used merely to lift, and fall is under gravity. The valves of a steam/air hammer can be operated to give blows of varying speed and energy and the machine is therefore very flexible. It is also smaller than a gravity hammer for the same equivalent energy, but much more ancillary equipment is needed to supply the motive power. An interesting variation is the counter-blow hammer in which both tup and anvil move. In this design the anvil or countertup is connected to the tup by flexible steel bands in such a way that when the tup is actuated, the countertup moves simultaneously towards it. By this means dissipation of energy into the anvil block and foundations is much reduced and efficiency thus improved.

c. The hammer as a tool. Hammers are generally regarded as being highly flexible tools in their range of application. Blows are rapid and often controllable, and metal may be shaped and made to flow into intricate forms by a skilled forger. They are simple machines and are cheap compared with other sources of energy. Unfortunately the hammer is an antisocial device not popular in centres of population and destructive both to itself and to surrounding property. A metallurgical disadvantage is that the action of an impact blow is to work the surface of the metal preferentially. Thus in large pieces the centre may not be sufficiently deformed to develop optimum properties.

Energy is wasted by conversion into shock waves and noise. The size of hammer which can be tolerated both by the materials of construction and on social considerations is thus obviously limited, and most hammers do not exceed 5 tons in tup weight.

Beyond this capacity, presses become increasingly more common.

6.4.4 The press

A press is essentially a device for exerting and maintaining a pressure between platens for a length of time sufficient to accomplish a given purpose.

Presses are variously designated by the means used to apply the pressure. For example, a screw press uses a screw thread, a crank press uses the throw of a crank and the hydraulic press uses liquid pressure. According to the means, the dwell or maintenance of a pressure may be almost infinite, or merely momentary. In metal working a short dwell is usually sufficient. As a metal-working tool the press does not suffer from some of the disadvantages of the hammer. The action is slower, the energy is contained within the frame and hence little noise and shock are produced. Furthermore, the deformation of the metal is more uniform.

a. Hydraulic presses. The hydraulic press was developed in England during the nineteenth century. Liquid under pressure is fed to the working cylinder where it exerts its pressure on the piston. The piston is forced to move and actuates the ram of the press. The cylinder and anvil block are coupled externally by tie rods forming part of the press frame so that a pressure or load is developed between anvil and ram (see fig. 6.25). Fig. 6.26 shows a large free-form forging press capable of exerting a forging load of 6000 tons. In such large presses the energy required is supplied by a pump, which pressurizes the liquid. The pump would not be capable of supplying energy at a rate sufficient to operate the press at normal speeds and a system is adopted whereby energy is continuously put into store to be used intermittently at each forging stroke. Such an energy store is called an accumulator and commonly takes the form of a bank of reservoir cylinders containing the liquid and air or nitrogen under pressure. Most of the energy is stored in the compressed gas since the liquid is almost incompressible. On tapping the liquid, the gas expands. Fig. 6.27 shows a hydraulic accumulator with its electrical controls. With small presses direct drive from the pump becomes feasible.

b. Mechanical presses. The use of mechanical mechanisms to exert pressure anticipated the use of hydraulic pressure by thousands of years. The lever and screw were possibly the first mechanisms employed. Both are still used today in forging presses. The coupling of the energy source to the press ram is less direct with mechanical presses than with hydraulic presses and consequently the construction of the moving parts is more complex and less robust. Mechanical presses are rarely built in large sizes and tend to be used for light forging work. Such presses range in size from a few tons capacity to perhaps 3000 tons with a vast majority under 1000 tons. Their stroke is generally fixed, which makes them unsuitable for open-die forging. Mechanical presses are found in closed-die forging shops producing small components for the mass production industries. It is in this type of work that a high rate of stroking is an advantage. The die designer cuts his dies so that the mating parts reach the correct distance apart even though the stroke is fixed.

c. Forging machines. Normal forging presses exert vertical pressure only. A forging machine, or horizontal upsetter, is effectively a press laid on its side in which the

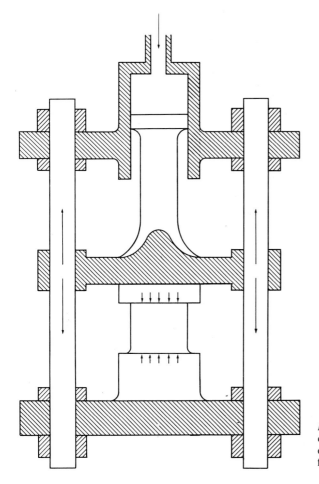

Fig. 6.25 Diagram of a Hydraulic Press showing the forces exerted on the forging and in the frame.

dies are closed by a toggle action to grip the stock and the ram is actuated at right-angles. Hollow forgings, flanged rods and large bolts are typical products of such machines. Related to these machines are those used to produce small parts from coils of wire. A typical horizontal forging machine is shown in fig. 6.28.

6.4.5 Forging tools and operations

Although simple geometrical shapes may be obtained by forging between flat faces, more complex forms require the application of auxiliary tools. A selection of tools used in hand forging is shown in fig. 6.29. The sett is used for cutting, the fuller for shaping, the flatter for finishing and the swage for rounding. These tools are a selection of hand tools but the shapes are essentially similar even when the scale of operations is so large that the tools themselves have to be handled mechanically. When a number of identical forgings of moderate size are required a complex tool called a die may be made. The die commonly consists of a number of impressions, the last of which forms the final shape. It is rare for the finished shape to be obtained in one operation. The initial piece of metal is simple, e.g. a square bar, and progressive

486 | METAL DEFORMATION

Fig. 6.26 Forging press of 6000 tons at the beginning of a stroke.

displacement is needed to thin down or gather material into the required positions. Frequently a rough or dummy impression precedes the final one. Stages in the production of a lever forging using such a multi-impression die are illustrated in fig. 6.30, and a selection of typical drop forgings is shown in fig. 6.31.

Forging dies are cut ('sunk') in steel blocks which may be provided with inserts of special steels for more than usually exacting applications. Typical hot-die steels are given in Table 6.6. Die sinking is a highly skilled trade although much hand work has been superseded by automatic copy-milling machines and even electro-erosion techniques. The blocks are commonly alloy steel, hardened and tempered to withstand the arduous duty. When a set of impressions are worn they may be renovated by re-sinking, until the block is too shallow for further use. Dies are fixed to the hammer or press by wedges (keys) and are 'set' so that the impressions are in true relationship to one another. In action the dies are lubricated with light oil, sawdust or salt solution. The prime purpose of the lubricant is to ease removal of the forging from the die and to this end cavities are tapered with several degrees of draft.

Hand and free-form forgings represent the product of individual skill and are subject to variation in sizes and shape according to that skill. Drop forgings on the other hand reflect faithfully the shape of the impression. Once cut the dies will produce thousands of very nearly identical shapes. Wear occurs as work proceeds, thickness

Table 6.6 Typical forging die steels

Element	Composition mean weight %			
Carbon	0.55	0.4	0.7	0.35
Silicon	0.3	0.2	0.2	1.0
Manganese	0.7	0.6	0.4	0.3
Nickel	1.5	0.3	—	—
Chromium	0.7	3.2	3.5	5.0
Molybdenum	0.3	1.0	—	1.5
Vanadium	—	0.2	—	0.5
Tungsten	—	—	—	1.5
Use	General hammer dies	Heat resistant press dies, hammer dies	Heat resistant press dies, forging machine dies	Very heat resistant heavy duty press or machine dies and punches

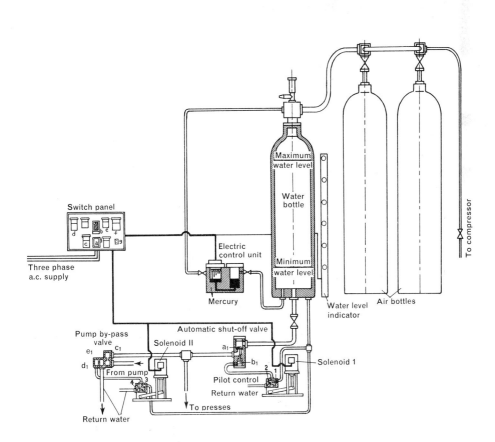

Fig. 6.27 Hydraulic Accumulator with electrical controls.

488 | METAL DEFORMATION

Fig. 6.28 Horizontal forging machine.

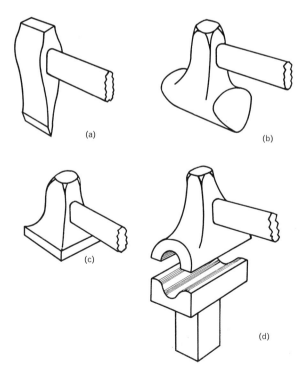

Fig. 6.29 Hand forging tools: (a) Sett (b) fuller (c) flatter (d) swages

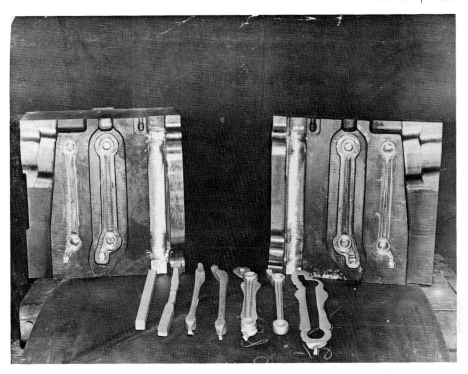

Fig. 6.30 Dies and stages in the production of a lever forging.

Fig. 6.31 A typical group of general drop forgings.

may vary according to blow intensity, and mismatch is caused by die movement. Of these, wear can be observed and followed and the die rejected when an unacceptable stage is approached.

6.4.6 Cold forging

Although forging is frequently thought of as a hot process, metals are capable of plastic deformation in the cold-working range (see introduction, p. 455). Naturally the stresses involved are higher than in hot working and the ability to flow is restricted. Forgings made cold tend to be small and simple in shape. Many thousands in the form of bolts, nuts, studs, rivets and nails are made from coiled wire or rod up to about 1 inch in diameter and from straight rod in sizes above this. On the other hand cut pieces are individually forged in sizes up to some 6 inches in diameter, at the present time, on a production basis. The reader is referred to specialist texts on the subject, e.g. H. D. Feldmann's *Cold Forging of Steel*. Fig. 6.32 shows a selection of cold-forged components.

By far the greatest tonnage of metal which is cold forged is converted into fasteners on automatically fed heading machines. The wire is fed, cut and transferred from station to station in the tools in a multistage machine. At each station the metal is progressively formed towards its final shape. The whole operation takes place under a flood of oil. Quite complex parts can be made on the multistation machines (see figs. 6.32 and 6.33). Simpler machines have one or two stations, but are still suitable for many fasteners.

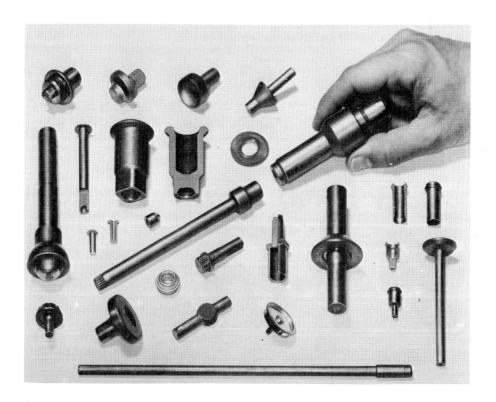

Fig. 6.32 Selection of Components made on a multistation cold-heading machine.

Fig. 6.33 Multistation Cold-Heading Machine.

Nails also are made on similar machines in a series of operations which form the head, feed wire to the required length, cut off and point and eject. The head is formed by upsetting and flattening a short length of wire projecting from a die. Wire is then fed to give the correct length by pushing through the die and the nail is finished by pointing and severing.

6.5 DRAWING

An important method of reducing the cross-section of a metallic object is that of drawing. As the title implies, drawing operates by the application of a tensile load to draw the work-piece through a die, which consists of a hole in some suitable tool material. Drawing thus differs from extrusion in that the product is pulled through the die rather than pushed. Fig. 6.34 illustrates diagrammatically the essentials of a drawing operation.

Drawing is the simplest method of reducing the cross-section of a piece and the process is an adaptation of simple stretching where the die serves to suppress local necking. For example, when the load is applied in a tensile test, the test piece elongates uniformly at first. Some materials elongate steadily but metals rapidly reach a point where instability sets in and a neck develops, and rupture soon follows. In drawing, care is taken to ensure that necking is avoided by limiting the reduction at each stage and hence

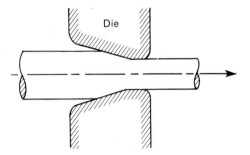

Fig. 6.34 Schematic representation of drawing.

the drawing load. The drawn material must be capable of withstanding this load, which is deforming the undrawn material. Since the section of the drawn material is less than the undrawn it follows that the issuing metal has to be stronger than the metal entering the die. Hence work hardening is essential for successful drawing.

In addition to the resistance to deformation exerted by the metal, friction caused by the metal sliding over the die surfaces has to be overcome, which adds to the drawing load. The limiting drawing ratio (i.e. the ratio of initial to final area which may not be exceeded in a given step) is governed by the strength of the drawn material and in most cases amounts to about 1.4.

It is an advantage to reduce friction to as low a value as possible and lubrication during drawing is essential to minimize both the drawing load and die wear. The most important function of a lubricant in drawing is to form a film between metal and die. This ensures that contact is avoided and, in the case of metal dies, welding (pick up) is suppressed. Drawing die pressures are such that liquid lubricants are squeezed out if applied on their own unless the drawing speed is sufficient to cause the work-piece itself to pump liquid into the die by a hydrodynamic effect. Most low speed drawing, therefore, uses solid lubricants such as lime, soaps, soft metal coatings and phosphates, which if porous may themselves act as vehicles for carrying a liquid lubricant into the die. In wire drawings speeds of several thousand feet per minute enable wet drawing to be successful in producing a product of close tolerances and good surface. Bars, rods, tubes and wire are the normal products of drawing, apart from those sheet metal articles which are deep drawn from sheet. Deep drawing can be regarded as a specialized example of tube drawing. Drawing is a cold-working operation and its products have accuracy and good finish coupled with enhanced properties resulting from work hardening. Bright drawn bars, for example, are used extensively by machinists since chucking is precise and the drawn material is more easily machined than soft metal. Drawing operations differ only in the type of apparatus employed. In general, products which can be coiled are drawn by reeling and products which cannot or may not be coiled are drawn straight. The importance of this differentiation is reflected in the length of the product. Wire may be miles long without a break but a tube in excess of 100 ft would be exceptional.

The apparatus on which tubes, rods and bars are drawn is the draw bench, which consists of a heavy frame on which the die head is mounted. An endless chain runs on a track along the centre of the frame and this chain provides the motive power. The front end of the work-piece is sufficiently reduced by rotary hammering (swaging or tagging) to pass through the die. The tag so formed is gripped in a set of jaws mounted on a trolley (the dog) which is provided with a hook to engage in the links of the moving chain. Drawing takes place as the dog is pulled along the bench by the chain. As the end of the work-piece leaves the die the hook disengages from the chain, leaving the dog free to begin a new cycle.

Coilable products are drawn on 'blocks' consisting of a drum, around which the product is wrapped, and a fixed die. When the drum is rotated it reels the product and in so doing exerts tension which pulls the work-piece through the die, i.e. the drum acts as a capstan. Wire is always drawn in this manner. Heavy gauges are drawn on individual blocks but, as the wire becomes finer, drawing machines consisting of many blocks are used. The wire is then passed through each stage in a continuous strand.

6.6 DEEP DRAWING AND PRESSING

Deep drawing and pressing are secondary cold-working processes applied to sheet and strip. The basic elements of each are shown in fig. 6.35. In deep drawing a blank sheared from sheet or strip is made to flow in a controlled manner by the descent of a punch. The metal is forced to bend over the die radius and is then drawn into the space between the punch and die. As the blank is drawn in, its diameter steadily decreases. When, for example, the segment A (see fig. 6.36) moves to position B the area in the plane of the paper must decrease. As the volume of segment A remains constant, the displaced metal must be accommodated. There are two possibilities. The metal may wrinkle as shown in fig. 6.37(a) or it may thicken (see fig. 6.37(b)). Wrinkling is usually undesirable and may be suppressed by applying pressure through the blankholder, but it must not be so great as to prevent the blank sliding across the die. The increase in thickness which must then occur may be subsequently modified by ironing between the punch and the die.

In contrast, when pressing or stretch-forming, the flow of metal in the periphery of the blank is prevented by firm clamping and the metal is merely stretched over the nose of the punch as it descends. The deforming material is subjected to tensile stresses, in

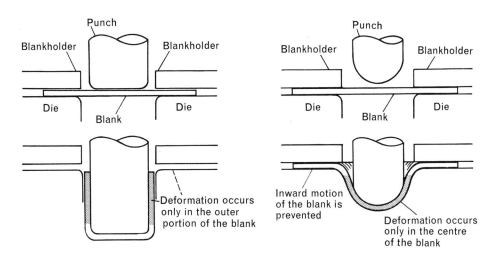

Fig. 6.35 Deep drawing and pressing.

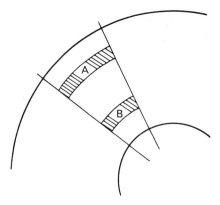

Fig. 6.36 Diagram showing the decrease in area of a segment flange as drawing proceeds.

contrast with the complex stress system operating in deep drawing involving both compression and tension. In practice most shapes are produced by a combination of deep drawing and pressing in which the distinction between the two processes is difficult to define.

As with most cold-working processes there is a limit to the amount of reduction that may be achieved in one step. The load required to draw in the flange must be carried by the wall of the cup, which will neck and fracture if too heavy a load is required. In deep drawing the critical reduction may be regarded as the maximum ratio of the blank diameter to punch diameter. A corresponding measure in pressing is the depth to which the punch can descend without fracture. When smaller diameter or deeper shapes are required two or more stages of drawing or pressing may be used. Typical redrawing operations are shown in fig. 6.38.

Not all metals may be deep drawn or pressed. For example, lead, although soft and malleable, is particularly difficult. Considerable research effort is being made in order to establish the mechanical properties necessary for successful shaping of sheet and strip and the tensile test has revealed much useful data on this matter. Thus the rate at which a metal work hardens appears significant, as does the ratio of thickness strain to width strain, which is governed by the degree of alignment of the individual crystals of the sheet (anisotropy). Hence anisotropy in the correct direction may be of great advantage. Hardness, strength and ability to elongate are in themselves of little value in predicting the deep drawing behaviour of a metal and are at best only useful as quality control guides for a material whose ability to draw or press is known.

The material properties may also cause a defect known as 'earing' to be produced in the component. In the production of sheet and strip, unless adequate care is taken, the metal may develop undesirable anisotropic properties from the alignment of the

Cross section of segments A & B showing accommodation of length A by wrinkling and thickening.

Fig. 6.37 (a) Drawing in producing wrinkling.
(b) Drawing in producing thickening.

SPINNING | 495

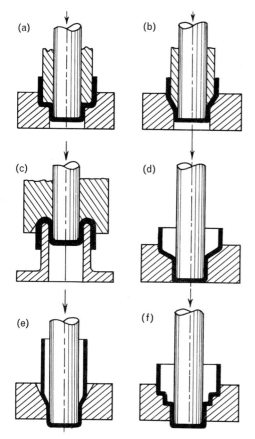

Fig. 6.38 Redrawing operations showing the different methods which may be adopted.
A and B double action redrawing;
C – reverse drawing;
D, E and F – single action redrawing.

individual grains along common axes. Different mechanical properties may then be observed according to whether the metal is tested in the rolling direction or at 90° to the rolling direction. When a material is sufficiently anisotropic, metal flow will occur more easily in one direction than another and this manifests itself in the pressed shape by the presence of the 'ears' (see fig. 6.39). Correct rolling and annealing should prevent this defect being formed.

6.7 SPINNING

Spinning is a method of forming hollow shapes from sheet metal by folding a blank around a former whilst both are being spun in a lathe. The blank may be a flat disc or a preformed pressing which needs a feature more readily spun than pressed, as, for example, a beaded rim.

Spinning is traditionally applied to the more ductile metals and has two advantages over pressing: it is capable of forming re-entrant shapes (using collapsible formers) and the equipment is considerably cheaper than presses and press tools. However, the skill needed is high and spinning is slow compared with pressing. The process, therefore, is favoured for the production of small quantities of articles where press tool

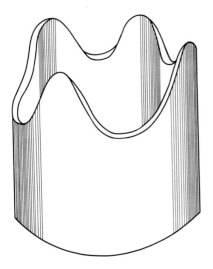

Fig. 6.39 Ears on a deep drawn cup produced from sheet with anisotropic properties.

costs would be prohibitive. Prototype shapes are often spun to achieve minimum cost and delay since the spinning tools are simple.

The equipment consists of a lathe in which the blank is held between the former, held in the chuck, and the tailstock. The forming tools are usually simple rods of various shapes with rounded ends. They are brought to bear on the spinning blank which is gradually folded round the wooden or metal former. The manipulation is a hand operation with pressure being applied by leverage against pins in the tool rest. During the process the metal thickens as its diameter collapses, exactly as in deep drawing. Consequently the spinner encourages longitudinal flow to reduce the metal to the required thickness. Considerable plastic working is involved and the metal must be ductile. Lubrication is essential and soaps and waxes are commonly employed.

In hand operation, which is traditional, the thickness of the blank is limited by the pressure which a human hand can apply and spun articles do not often exceed about $\frac{1}{8}$ inch in thickness but may be several feet in diameter. The usual products include ornamental metalware, dished ends for cylinders, cooking utensils and the like.

Power spinning has been employed for some time and the hard metals, e.g. steel, can be spun in thicknesses of up to $\frac{1}{2}$ inch using equipment which is considerably more expensive than the traditional apparatus.

6.8 BENDING

A great variety of articles in everyday use are fabricated by bending and folding sheet metal. This is sometimes followed by welding or soldering when seams are required. Tin-plate work is an obvious example, welded pipe is another. The first illustrates the technique applied to individual articles; the second to the fabrication of a continuous product.

Bending tools exhibit as great a variety as their products. Presses, press brakes, draw dies and rolls are all used in high-quantity production applications and special automated machines make tin cans in vast quantities. Hand tools, presses and press

6.8.1 Press-brake bending

The press brake is essentially an elongated press, i.e. one where the ram is wide from side to side compared with its depth from front to back (10 ft × 2 in is typical). Press-brake operation traditionally uses tools of simple and often general purpose form to fabricate sheet metal by bending to shape. Skill on the part of the operator is required and in this sense press-brake forming is comparable in some degree with hand forming. Simple bends are formed in tools such as those sketched in fig. 6.40. The angles of the tools are about 85°, i.e. less than a right-angle, so that spring of the bent metal can be anticipated by overbending. Heavy loads developed by the contact of tools over a large area are also avoided.

With such tools, sheet metal boxes and casings, U-section channels, angles etc. can be readily made with very low tooling cost. This makes press-brake forming very attractive for small batch production and prototypes where sophisticated tooling would far outweigh the cost of the extra labour used on the press brake. Once a design has been established the press brake may still be used but special purpose tools may be made to remove some of the need for skill in use. The press brake is not limited to bending and can perform other press operations such as notching, piercing and punching, all of which increase the versatility of this tool.

6.8.2 Press bending

Bending tools fitted to ordinary presses can work in exactly the same manner as in press-brake production, but tooling is usually smaller and may perform several operations simultaneously or in follow-on tools. Production rates may be high when, as is usual, a tool set is made specifically for a given item. In such a case the tool designer can make sophisticated locations and transfer stops or even special feed devices and the operator exercises little skill.

Auxiliary operations may be incorporated as for the bracket shown in fig. 6.41.

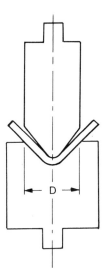

Fig. 6.40 Vee bending dies.

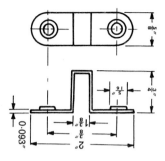

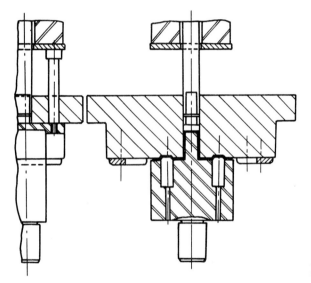

Fig. 6.41 Bending and indenting a bracket.

6.8.3 Roll forming

Roll forming is the term used to describe the shaping of a continuous strip of metal into a continuous bent section by a series of rolls which progressively deform the strip. Roll forming can easily produce 5000 ft of a section per hour and is thus a mass production technique useful for pipes, constructional sections, car trim sections and so on where the quantity required justifies the rather expensive rolls and machine. Two examples commonly encountered are those of electrical conduit tubing which is hot formed and corrugated iron. Corrugated iron is progressively bent, one flute at a time, to allow the metal to flow laterally inwards.

These examples are simple and involve large-scale equipment by virtue of the size of the product. A selection of typical but more complex sections which can be roll formed is shown in fig. 6.42.

In most cases the edges of the strip are shaped first and this is followed by bending up to form the section as the strip passes progressively through the rolls. Fig. 6.49 shows the stages in forming a lock-seamed tube. Such a tube may be further shaped by follow-on rolls, as with some examples of fig. 6.43.

BENDING | 499

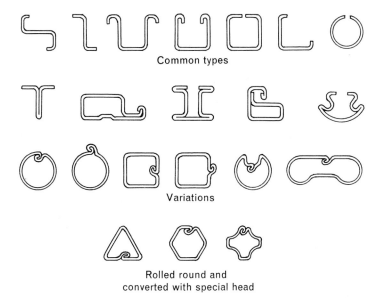

Fig. 6.42 Typical Roll-formed sections.

Roll-formed sections are often in competition with extruded sections.

6.8.4 Draw-die forming

Drawing dies can also be used as a bending tool if the bending is not too severe. The use of bell dies for the folding of strip into tubing is an example in which the strip, in passing through the die, is forced to bend up and close to form a tube. Other sections are possible but frictional restraint and resistance to deformation limit the process. In effect the rolls of a roll-forming process act as a die but, instead of the strip being pulled through, each roll provides the power for its own increment of work and by moving with the strip eliminates much of the friction. In this way large tensile loads are avoided and there is no danger of breakage.

The prime advantage of draw-die bending of strip into sections is the simplicity and cheapness of the tools and equipment.

6.8.5 Section bending

There are many structural members which are composed of solid or tubular sections which are not straight. Conveyor lines or ships' frames, for example, contain many such items. The bending of sections is not as straightforward as the bending of sheet materials since distortion of the section has to be kept to acceptable limits. Hollow sections can collapse and channels flatten during bending and usually both must be avoided.

Bending machines enable heavy sections to be bent and are basically of two types: the ram-type and the rotary-type. The ram-type machine is in essence a horizontal press, usually hydraulic, which advances its ram between two vertical pillars which hold movable dies. The section is positioned across the two dies and the ram applies a load to its centre, causing it to bend. Much skill is needed to avoid wrinkling and output

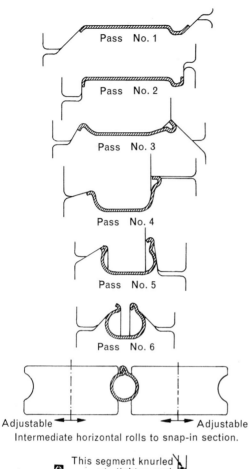

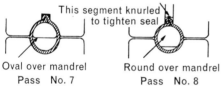

Fig. 6.43 Roll-forming a lock-seam tube.

is low. However, the machine is inherently robust and capable of being built in large sizes to handle heavy sections.

The rotary machines wrap the section round a former by restraining one end at the start of the former surface and rotating the other. In larger machines the former may be rotated instead, with one end of the section attached to it, and will thus 'coil' the section through part of a turn. A mandrel may be used to prevent collapse of hollow sections. This mandrel is located at the point where bending is taking place and is withdrawn into the straight section as bending proceeds.

Both machines use tools which are designed to fit the section and to provide the finally required angle of bend. Figs. 6.44 and 6.45 show examples of tooling.

If bendings is confined to one plane, various arrangements off rolls may be employed; for example, the rolls can be arranged to wrap a complete circle to form a rim.

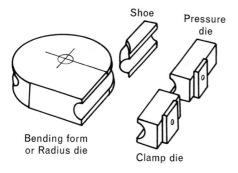

Fig. 6.44 Typical forms of die.

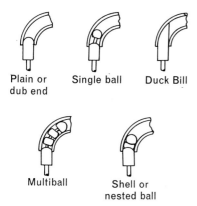

Fig. 6.45 Bending mandrels.

6.9 RECENT DEVELOPMENTS IN METAL DEFORMATION

6.9.1 High-energy-rate forming

The methods for metal deformation described in preceding sections have been known and used for many years. Within the last ten years new methods have been developed which have in common the feature of shaping the metal by the application of energy delivered in a relatively short interval of time. These 'high-(energy)-rate forming' processes have the advantage that they may provide energy in excess of that normally available in conventional processes such as rolling, forging or extrusion. The high-rate forming processes fall into the following categories:

1. Chemical explosives.
2. High-voltage discharges.
3. Electromagnetic fields.
4. High-rate machines.

The first three processes are usually applicable to sheet material only and a particular advantage is the reduced tool cost. The sheet material is forced into a shaped cavity by the shock wave and so a punch is not required. The die may often be manufactured from cheap easily moulded materials, particularly if only a few components need to be made.

a. Chemical explosives. An explosive is an unstable compound which, by the action of heat, either directly applied or induced by a shock wave, breaks down to a more stable form. Under the correct conditions the process occurs in a very short space of time, producing a large mass of gas at a very high pressure. It is this high pressure which is utilized to deform sheet metal. The pressure developed may be used directly to shape the material, or alternatively may be allowed to pressurize some other medium such as water which in turn produces the required deformation. Another way of harnessing the pressure generated by the explosion is by allowing it to operate a piston which in turn applies pressure either directly or through another medium.

The use of explosives to deform sheet material is shown schematically in fig. 6.46. The thin sheet metal blank is firmly clamped at its edges and the space beneath it evacuated. The stand-off distance (between charge and blank) as well as the size of charge are critical if the correct shape is to be achieved.

The essential tool in explosive forming is the die, which may be made from a variety of materials, depending on how many components (i.e. explosions) it has to make. Clay, wood, concrete lined with an epoxy resin, plaster and mild steel have all been used. Plaster dies will probably only withstand one explosion, a concrete die may manufacture several components and a reinforced plastic die may last indefinitely.

b. High-voltage discharge forming. If a large voltage is applied across two electrodes, an electrical spark will be established between them. If the system is immersed in water, the current passing through the liquid medium causes it to ionize and form a gas bubble. This bubble by its fast rate of expansion creates a pressure sufficient to deform sheet material. An apparatus for doing this is depicted in fig. 6.47. The mechanism of deformation by shock waves caused by a high-pressure gas is thus similar to that in explosive forming.

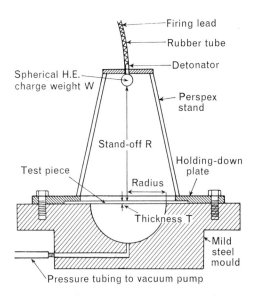

Fig. 6.46 Arrangement for explosive forming.

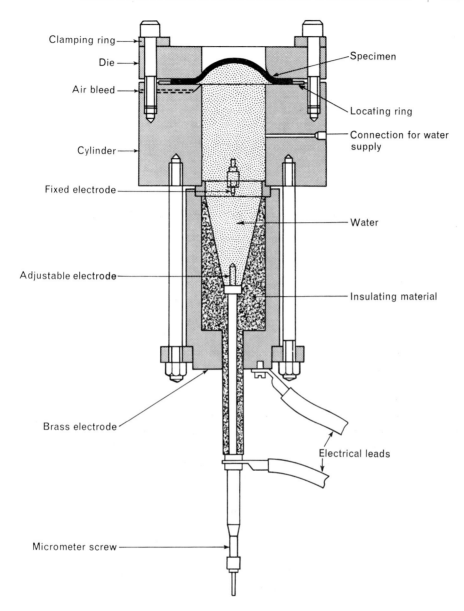

Fig. 6.47 Electrical-discharge forming.

c. Electromagnetic forming. In contrast, electromagnetic forming allows the discharge of very high electrical energy, not across a spark gap but through a coil. In this way a voltage and current may be induced in the metal to be formed. The arrangement for shaping of tube by this method is shown in fig. 6.48. The electromagnetic fields in the coil and metal are in opposition to each other and thus a force acts on the outside of the tube causing it to be deformed according to the shape of the inner insulating mandrel. The intensity of the field is such that the metal is flung against the

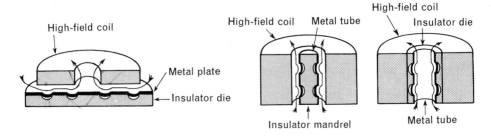

Fig. 6.48 Electro magnetic forming.

shaping tool in much the same way as an electric induction motor is caused to rotate. The process is particularly suited to crimping or expansion of tubes and has also been used for indenting sheet material. The advantage of forming by spark discharge or electromagnetic forming compared with explosive techniques is that the processes are more suitable for mass production and are very readily controlled by specification of the electrical characteristics.

d. High-rate machines. A number of other devices for forming depend on the activation of a forging hammer by the explosion of a petroleum mixture ('petroforge') or by the sudden release of the energy stored in a gas under high pressure (e.g. the 'dynapak' machine).

6.9.2 Hydrostatic extrusion

Failure which occurs in a brittle manner is essentially tensile failure. In working processes, even if the stress system is primarily compressive, secondary tensile stresses develop as a consequence of inhomogeneity of deformation and friction. If the process could be conducted under sufficient hydrostatic pressure so that tensile stresses were suppressed, brittle failure would not occur. The effect may be so marked as to enable brittle materials to be plastically deformed. Examples are found in nature in extrusions of rocks into crevices under extremely high pressures.

Extrusion of metals under such conditions has been studied and in some cases is being applied in production. The apparatus used is shown in diagrammatic form in fig. 6.49. The conventional ram of an extrusion press is replaced by pressure exerted by the fluid. Friction between billet and container is eliminated as there is no contact, and the billet can be of any proportions. The billet may be replaced by a reel of wire which will uncoil and pass through the die just as readily as a straight billet. Differential pressure extrusion is a development which involves extruding a material into a second chamber which is itself under pressure (lower than that of the extrusion chamber). In this way the product is maintained under pressure as well as the billet.

The industrial application of hydrostatic extrusion is accompanied by many difficulties since the pressures needed may exceed 100 t.s.i. and consequently the process will be an economic one only for applications where conventional processes are inadequate. There are hopes that some of the materials which are virtually unworkable, but which possess desirable properties in other respects, might be manipulated by this technique.

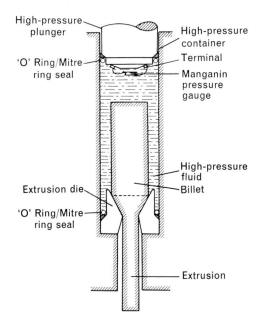

Fig. 6.49 Schematic representation of hydrostatic extrusion.

6.10 PRODUCTS OF MECHANICAL DEFORMATION PROCESSES

6.10.1 Plate, sheet, strip

Whether a product is termed plate, sheet or strip is largely convention based on dimensions. The definitions in such an arbitrary scale naturally vary but in general plate is thicker than sheet and strip, and strip is generally narrow and of greater length in comparison with either. All of them are commonly produced by rolling, beginning with slabs of metal which are reduced in stages until the desired dimensions are reached.

Much sheet and strip is finished by cold rolling to improve its surface finish and attain high dimensional accuracy and in some cases to strengthen by work hardening. Temper rolling is the term used to describe the deliberate achievement of work hardening by rolling. A temper-rolled sheet is designated fully hard, half hard or quarter hard, which reflects the degree of cold reduction it has received. Mild-steel sheet and strip is frequently lightly rolled by a 'skin pass' resulting in 1 or 2% reduction. The object is to suppress the characteristic yield point which would otherwise cause surface blemishes on subsequent pressing as a result of discontinuous yielding (Lüders lines or stretch-strain markings). The suppression is temporary and recovery usually occurs within days or weeks depending principally upon the temperature of storage.

a. Plate. In the production of plate, ingot is rolled in several passes in a cogging mill to long slab which is then sheared to lengths suitable for the plate which is finally required. If the slab in stock has the same width as that required in the plate, rolling takes place directly. If this is not the case slab width can be adjusted by one of two techniques. The first is shearing the length of the original slab to the desired width of the plate, followed by rolling the original width into the new length. The second is by cross rolling the original width to a new width before further rolling in the original direction.

The edges may be controlled by a set of vertical rolls operating on them as the plate issues from the horizontal rolls; this also helps to suppress edge cracking (universal mill). Alternatively the edges may be left uncontrolled and sheared to size.

Plate is rolled mainly on two-high or four-high mills. Three-high mills are used but with modern developments in motors such units are becoming rarer. Single stand, open train, semi-continuous and fully continuous layouts are used depending upon the flexibility required and the output. In open train or semi-continuous sequences it is common to employ a two-high roughing stand through which the slab is passed several times. Subsequent units are two-high or four-high stands through which the material passes only once. Similarly, continuous mills consist usually of a two-high roughing train followed by a four-high finishing train. Such a layout enables a high output to be attained. Accurate control of dimensions and good surface finish are achieved by limiting the work of the final stands so that roll wear and mill distortion are minimized.

Plates are usually finished as a hot-rolled product.

b. Sheets. Sheets are rolled in a similar manner to that used for plates but on a smaller scale. However, as a consequence of the high rolling loads developed as the gauge decreases the mills must be more rigid. Pack rolling, whereby several sheets were rolled together, was an expedient used to alleviate the rise in roll pressure experienced as sheet thickness diminished. Formerly, all sheet iron (and later steel) used for tin plate was rolled in packs which were separated into single sheets after rolling. Modern mills are capable of rolling wide thin continuous bands of metal which can then be sheared into sheets, and much sheet material is now cut from such 'wide strip'. As a typical example, a strip of perhaps 2 000 ft in length may be produced on a continuous mill consisting of four-high stands in roughing and finishing trains.

Sheets are frequently cold finished.

c. Strip. Strip is rolled from slabs or, in the smaller sizes, from sheet bars and billets which have themselves been hot-reduced from ingots. Normal roughing and finishing sequences are used and both hot-rolled and cold-rolled strip are coiled to facilitate handling. If cold finishing is to be employed, the hot-rolled coils are cooled and cleaned by acid pickling to remove oxide prior to cold rolling. The coiler following the last cold stand recoils the strip. Coiler and decoiler speed may be varied to apply tension to the strip for such purposes as controlling flatness, preventing wander across the roll face, reducing rolling load and in one method of gauge control. Where output is low, as with some non-ferrous metals, the strip may be simply passed back and forth in a single reversing stand, coiling and decoiling. However, strip is often produced in vast quantities and many strip mills have become highly mechanized. Fig. 6.50 shows a continuous strip mill in straight line layout which may operate at several thousand feet per minute. At such speeds manual control is quite impossible. Dimensional control is particularly important as strip users demand accuracy of gauge. Formerly, gauge was controlled by hand measuring and manual adjustment of the roll gap. At modern high rolling speeds strip thickness is measured by absorption or scattering of X-, β- or γ-rays, or by measuring the variation in stretch of the mill housing caused by changes in gauge,

PRODUCTS OF MECHANICAL DEFORMATION PROCESSES | 507

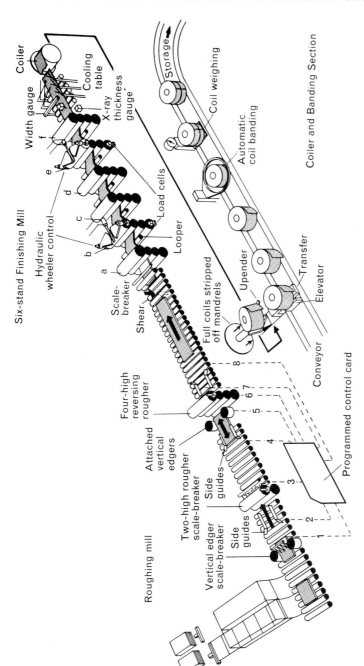

Fig. 6.50 Continuous strip mill.

i.e. the mill itself is used as a 'micrometer'. Time delay between measurement and response must be a minimum as otherwise false correction may be applied to conditions which may already be different. Rapid response may be achieved by computer control of the whole operation.

Sensing devices feed information relating to dimensions, temperature and roll forces to the computer, which selects responses and in turn feeds them to the mill controls which make adjustments to roll speed and gap. Receipt, digestion and response take place in a tiny fraction of the time a human brain needs to carry out the same sequence. Fig. 6.51 illustrates the logic of computer reaction.

6.10.2 Rolled sections

The rolling of sections is accomplished by progressive reduction and shaping in grooved rolls of an ingot or a previously rolled piece. Shapes produced by rolling are inherently simpler than may be formed by extrusion as a consequence of the necessity for the shape to be continuously withdrawn in two directions from the roll grooves. Nevertheless, rails, angles, building sections and so on are rolled in vast tonnages as well as such simpler shapes as squares, rounds and hexagons. The rolling of the more complex sections requires great experience in roll pass design but in principle the metal is gradually displaced into the position it occupies in the final shape by reduction in height and control of spread. Fig. 6.52 shows a typical sequence employed in the rolling of a rail section. A more detailed consideration of the rolling of the simple sections will elucidate the techniques involved.

Starting with the cast ingot the products resulting from rolling, in progressively diminishing sizes, are designated blooms, billets, bars and rods.

a. Blooms. Traditionally, ingots are treated in stages of which the primary stage results in bloom. Blooms are rolled in single stand mills of the reversing three-high type. The cast ingot has to be treated gently until it has been consolidated and refined by the rolling, after which reductions in height (draught) of up to 25% in one pass may be imposed. During the initial sequence care is taken to avoid forming defects on the surface. The bloom is turned frequently so that each face is rolled and cracks do not develop. Corner cracking is suppressed by the use of grooved rolls which support corners, although economy in roll space may not allow this in the earlier larger sizes. A parallel length of roll known as a bullhead may be used to accommodate all the initial sizes of stock which are too big for the grooves. Fig. 6.53 illustrates a typical blooming sequence using this system.

A recent development is the continuous casting process which produces ingots of great length and small cross-section. The initial cogging of large ingots may thus be eliminated since the ingot size as cast approximates that of a bloom or billet.

b. Billets. Billets may be rolled in a similar manner to that used for blooms but unless the requirement is very low it is not economic to do so. As rolling progresses the work required to reduce the stock increases because of the greater length. One blooming mill can produce enough output to justify a separate billet mill consisting of several stands. Billet mills are usually open train, cross-country or continuous mills (see fig. 6.10) and the design of pass becomes critical in order to achieve the highest reduction and control spread, so that material does not enter the roll parting (overfilling). Fig. 6.54 illustrates part of a typical sequence in a continuous billet mill. The

PRODUCTS OF MECHANICAL DEFORMATION PROCESSES | 509

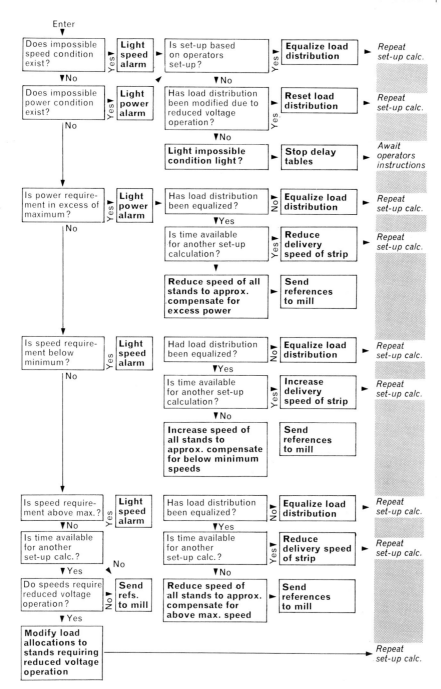

Fig. 6.51 Hot strip mill computer logic diagram.

510 | METAL DEFORMATION

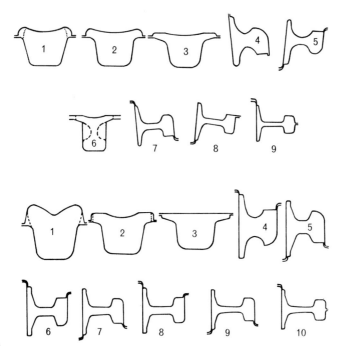

Fig. 6.52 Sequence employed in rolling a rail section.

sequence shown is known as 'diamond-square' and twisting through 90° is necessary between the diamond and square grooves. In this way the diamond section can enter the square pass and spread can occur without overfilling.

c. *Bars and rods*. Bars and rods are produced in the third stage of rolling from sheared lengths of billet. Reheating is usually necessary. The methods employed range from hand-operated open train mills to the most sophisticated automated continuous mills, consisting of up to 25 stands. Rods may be coiled as they leave the finishing stand. Since heat loss is rapid from small sections it is essential to roll as quickly as possible and speeds of up to 7000 ft/min are reached in modern mills rolling several strands at a time in continuous sequence. Pass sequence is similar to that used for billet rolling and many combinations are favoured for reduction to the final shape. Round products are mainly finished by rolling a vertically guided oval to round.

The problems of maintaining speed matching are great and the effects of a 'cobble', as the breakdown in the progress of stock through the rolls is termed, may be imagined. Tensions and guiding are automatically controlled and in the event of cobbling the following material is automatically chopped up to avoid further congestion whilst the mill is being stopped.

6.10.3 Extruded section

Bar, rod and other shapes may also be produced by extrusion as described previously (see extrusion, p. 473). The die orifice is cut to the profile required and the desired shape is produced in one operation when the ram forces the metal through the

PRODUCTS OF MECHANICAL DEFORMATION PROCESSES | 511

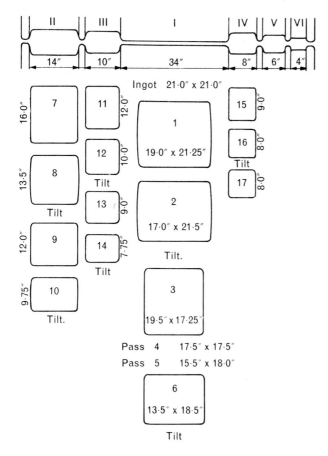

Fig. 6.53 Typical sequence for rolling ingot to bloom.

orifice. The output from an extrusion press is usually less than that from a rolling mill but the process has the merit that shapes with re-entrant angles may be produced, which would be impossible by rolling. Extrusion is sometimes preferred when many small orders require a large number of tool changes, which are easily accomplished in the extrusion process.

6.10.4 *Drawn sections*

Drawing is frequently used to produce sections which possess greater accuracy of dimensions than rolling can achieve and in metals where extrusion is difficult or uneconomical. Most simple shapes are obtainable in drawn stock and machinists employ drawn bars to eliminate chucking problems. The accuracy in form is often acceptable for a finished part. Hexagonal bar, for example, is used when a nut section is needed on an otherwise cylindrical piece. More complex sections are produced for special products but drawing cannot compete with extrusion for freedom of shape.

Economic factors dictate that drawing is the favoured process for producing long products of small cross-section; wire illustrates this point as very little is made by any other method. The cold work of drawing serves to strengthen the product and drawn

512 | METAL DEFORMATION

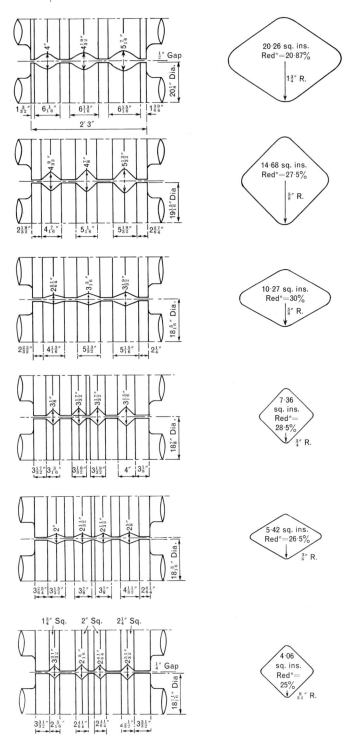

Fig. 6.54 Diamond-square sequence for rolling billet.

wire is probably the strongest common form of metal available. Piano wire drawn from 0.8/0.9% carbon steel, previously heat-treated to give optimum drawing properties, may have a tensile strength of 175 t.s.i. and still be ductile enough to allow an eye to be formed.

6.10.5 Tube production

a. Tubes in general. Tubes are generally 'welded' or 'seamless'. In the former, rolled strip is bent (or wrapped) to give the required tube dimensions, and then welded. The wall thickness is the thickness of the strip. Further information on this is given in Chapter 8. Seamless tubes are made from solid ingots or blooms and because of the absence of a welded joint are considered safer for pressure applications than welded tubes. The production of seamless tube may be considered a two-part operation. The first requirement is to create a hole in the solid ingot to produce a relatively short, thick-walled tube shell. The shell is then further deformed to increase the length and reduce the outside diameter and wall thickness. The history of tube making is such that nowadays there are a wide range of possible processes which may be used for shell formation and for the subsequent working of the shell. The principal ones in use are described below.

b. Shell formation. The process most commonly used to create a hole in a solid ingot is that of rotary piercing. This was originally developed towards the end of the nineteenth century by Mannesmann in Germany and the process is often known by his name. The technique is shown in fig. 6.55. Two rolls are placed side by side but positioned so that their axes are inclined at opposite angles (6° to 12°) to the horizontal centre line. Both rolls rotate in the same direction and present to the billet a converging and then a diverging gap. Critically positioned near the maximum roll gap is a plug mounted on a mandrel which assists in the piercing operation. The inclination of the rolls, their rotational movement and the profile of the gap are such that the billet, when gripped by the rolls, is caused to spin round in a helical manner – thus moving forward. The compressive forces acting at any one time on the billet cause secondary tensile stresses to be established. These act on the centre of the billet and encourage cavity formation. This stress system is illustrated in fig. 6.56. Even without the presence of a plug a hole could be made to develop, but as shown in fig. 6.57 the bore surface would be very rough and unsuitable for a tube. The plug therefore anticipates natural cavity formation and enables a smooth bore surface to be produced.

Other processes relying on the same principle but using a different roll profile are the Stiefel disc and the cone piercers shown in fig. 6.58. The most recent development in rotary piercing is the three-roll piercer which by the presence of the third roll creates an entirely compressive stress system. As the secondary tensile forces are entirely suppressed, the piercing has to be effected by the plug (see fig. 6.59).

Another process used for tube-shell production is that of extrusion. Here the initial piercing may be done by a mandrel which is part of the ram assembly. The mandrel is forced through the billet and finally enters the die orifice. Difficulties will be encountered if the bore is small and if the material to be extruded is strong. Thus, in the hot extrusion of steel tube, mandrel piercing of a solid billet is never attempted. A hollow billet is presented to the extrusion press. Softer alloys such as brass may be successfully pierced in the press and then extruded. Irrespective of the method used

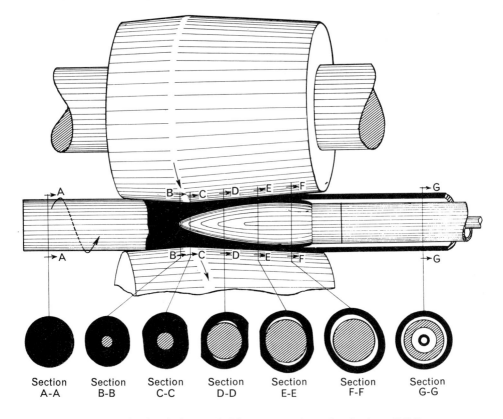

Fig. 6.55 Diagram showing the layout of a Mannesmann piercer for piercing solid billet.

for establishing the hole in the billet, a mandrel is then positioned concentrically within the die aperture and extrusion of the billet takes place. The metal is forced into the space between the mandrel and the die and a tube shell emerges.

c. Tube Formation. Once the shell has been formed, further work is usually performed on it to obtain the final dimensions. The principal processes used are:

1. Plug rolling.
2. Continuous mills and stretch reducing mills.
3. Assel mill.
4. Rotary forging (Pilgering).
5. Reducing.
6. Drawing.
7. Push bench.

In the plug-rolling process, a pair of rolls with semi-circular grooves are used, similar to those required for bar rolling. A plug, located on a support bar on the exit side of the mill, is positioned in the roll gap. The tube shell when pushed into the mill has its outside diameter reduced by the rolls. The internal diameter, and hence the wall thickness, is controlled by the presence of the plug. A schematic picture of the process is shown in fig. 6.60.

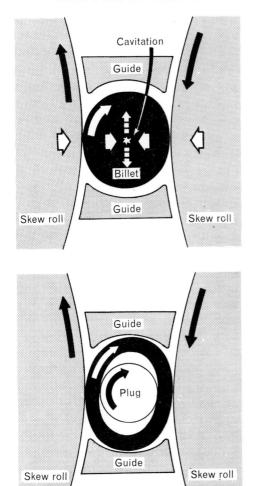

Fig. 6.56 Cross-section through a Mannesmann piercer showing the forces operating.

Greater continuity and output may be achieved in continuous mills. The pierced shell is placed on a mandrel, which fixes the final internal diameter, and is then passed through a number of mills in tandem which gradually reduce the outside diameter and wall thickness. The stretch reducing process also relies upon a rolling action to change the tube-shell dimensions. A large number of mills – possibly 20 to 30 – are arranged in tandem and the shell is progressively rolled through each, similar to strip passing through a continuous strip mill. No internal support is used. The tube is in more than one stand at any one time and, by the use of a speed differential, tension may be developed between stands. Reduction in the outside diameter is accomplished by the rolls in successive stands whilst the tension on the tube between stands is sufficient to effect a small reduction in wall thickness.

When thick-walled shell needs to be reduced an Assel mill may be used (see fig. 6.61). This consists of three rolls set at an angle of approximately 8° with a profile not dissimilar to that used in Mannesmann piercing rolls. The tube shell is mounted on a mandrel and on entering the rolls is spun forward into the decreasing gap. The outside

516 | METAL DEFORMATION

Fig. 6.57 Surface of bore produced without the presence of a plug.

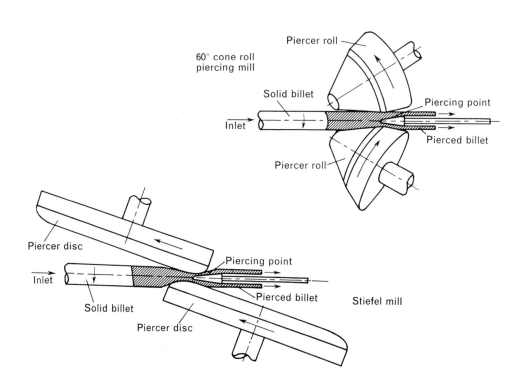

Fig. 6.58 Stiefel and cone piercers for tube shell production.

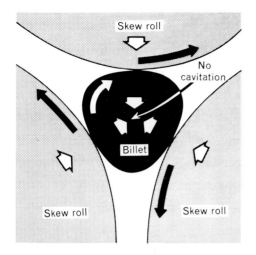

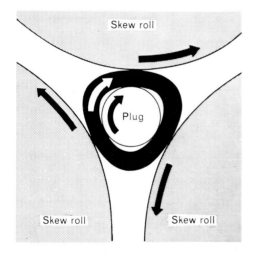

Fig. 6.59 Three-roll piercer showing the completely compressive stress system.

diameter is fixed by the minimum gap whilst the mandrel controls the internal diameter (and hence wall thickness).

The Pilger process is also used for reducing thick-walled shells but utilizes a forging action. However, because a pair of rolls is used the method is often described as rotary forging (see fig. 6.62). The use of forging by rolls enables much greater reduction to be achieved than would be possible by conventional rolling in one stand. The profile cut into the roll surface varies in size and thus as the rolls rotate the pass presented to the tube shell will change considerably. At the maximum (called the 'gap') the pass is larger than the ingoing tube shell. At the minimum the pass is substantially smaller. The direction of rotation of the rolls is such that they will tend to reject the ingoing metal stock (in contrast to conventional rolls which grip and carry the materials through the pass). Thus the tube shell mounted on a mandrel can only enter the space between the rolls when the 'gap' is presented, i.e. when the pass is at the maximum. If this is done, as the rolls rotate further to the 'bell mouth' position, a wave of metal is trapped on the exit side of the rolls and forward movement of the tube shell is pre-

518 | METAL DEFORMATION

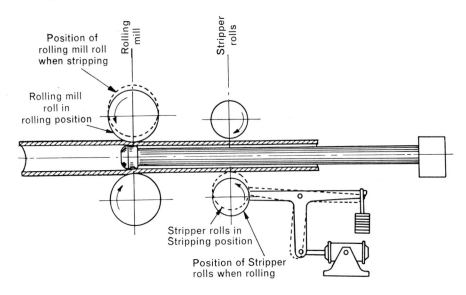

Fig. 6.60 Plug rolling process.

vented. The rolls then rotate to the 'parallel section'. At this stage the rolls reject the shell, expelling it backwards, whilst ironing out on to the exit side the wave of metal trapped by the bell mouth section. For further working to occur the mandrel on which the shell is mounted has to be advanced by a carriage. The shell is then pushed into the gap prior to the trapping of another wave of metal when the bell mouth position comes round again. The working of the tube shell thus proceeds incrementally until the complete length has been processed by the Pilger rolls. After each bite the shell is twisted 90° to obtain uniformity.

The tube-reducing method is a cold-working process but is similar to hot rotary forging and is sometimes known as 'cold pilgering'. The set-up is shown schematically in fig. 6.63. The specially shaped rolls, as well as rotating, move along the axis of the tube tracing out a gradually converging pass, and thus progressively decrease the diameter of the tube. The tube is mounted on a stationary tapered mandrel so that the final dimensions of the finished tube are controlled by the profile of the rolls, the minimum diameter of the mandrel and the space between it and the rolls at the end of their stroke. As in the rotary-forging process, the tube has to be fed into the cold-reducing machine in stages and the over-all reduction is achieved incrementally.

Tube drawing, again a cold-working operation, uses a draw bench or for the smaller sizes of tube, a bull block (see drawing, p. 521). The outside of the drawn tube will be the size of the die orifice and the process may be operated with or without support to the bore of the tube. If no support is used the process is known as 'sinking' and the internal diameter will collapse inwards as the outside diameter decreases under the influence of the compressive stress from the die. Some wall thickening is likely to occur and it may be difficult to obtain the accuracy of dimensions normally required in cold-worked products. In addition the level of residual stress in the tube may be unacceptably high. For this reason tube drawing is often practised with support given to the bore by means of a plug or mandrel. Mandrel bar drawing is now largely of historical interest.

PRODUCTS OF MECHANICAL DEFORMATION PROCESSES | 519

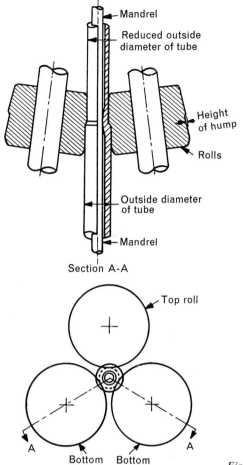

Fig. 6.61 Assel mill.

The mandrel and the tube are drawn through the die together and then separated by, for example, reeling. The need to separate tends to reduce output and has led to the increasing use of plug drawing techniques. The plug may be positioned in the die by attachment to a long thin mandrel or be allowed to 'float' in the die orifice. In the latter case, the plug position is maintained by control of the geometry of plug and die in conjunction with the forward acting forces pulling the tube through the die. The various types of drawing operation are shown in fig. 6.64.

A process similar to that of mandrel bar drawing is the push-bench process and is used particularly for producing closed tubes such as are required for gas cylinders. An incompletely pierced tube shell (bottle) is mounted on a mandrel which determine the final bore diameter. Mandrel and shell are then pushed (as opposed to the pull applied in the drawing processes) through a series of ring dies of progressively decreasing diameter until the desired dimensions are achieved. Care has to be taken to ensure that the mandrel is not pushed through the end of the bottle. The reductions achieved by each die and the forces established have therefore to be related to the thickness and strength of the metal at the bottle end.

520 | METAL DEFORMATION

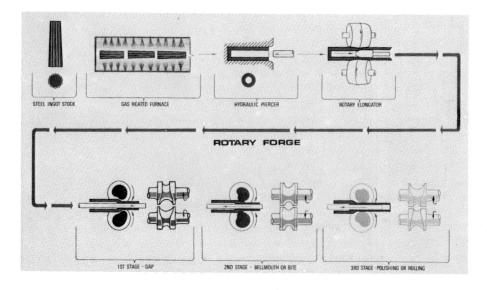

Fig. 6.62 The rotary forge process.

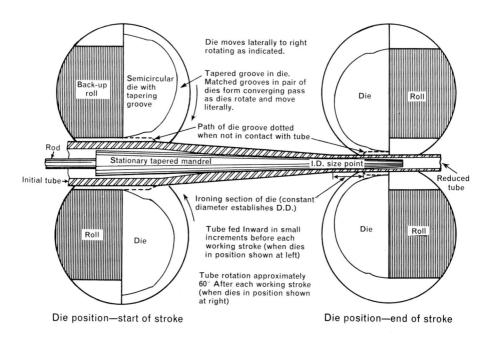

Fig. 6.63 Vertical section through a tube reducer pass showing dies at the start and end of the stroke.

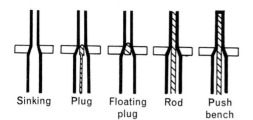

Sinking Plug Floating Rod Push
              plug   bench

Fig. 6.64 Types of operation for cold-drawing tube.

6.11 LITERATURE

H. D. FELDMAN, *Cold Forging of Steel*. Hutchinson 1961
C. E. PEARSON and R. N. PARKINS, *Extrusion of Metals*. Chapman and Hall 1950
UNITED STATES STEEL CORPORATION. *Making, Shaping and Treating of Steel*.
UNITED STEEL COMPANY OF SHEFFIELD. *Roll Pass Design*.

CHAPTER 7

Working of Metals by Cutting

7.1 HISTORY OF METAL CUTTING

The earliest requirements of man in his struggle for existence and for domination of the animal kingdom were basically twofold. Firstly, he required weapons for hunting, and these also proved useful for family and tribal defence against hostile neighbours. Indeed, the military aspect has provided a continuous spur in the development of workshop techniques lasting to modern times. Secondly, early man required domestic equipment, initially for cooking purposes and subsequently for providing other essentials such as shelter and clothing. Following these initial stages came the development of simple hand-powered tools by the early civilizations of Greece, Rome and Egypt for the creation of ornamental works in wood and stone. At the same time, the development of wheeled transport was greatly assisted by simple forms of the drill and lathe.

The forming of implements from metals can be traced back to roughly 5000 B.C. (Bronze Age) when the method of working was generally by casting and forging. However, after 1000 B.C., it is certain that, with the advent of the Iron Age, the process of grinding was developed for the sharpening of weapons and tools. The process was a hand operation, using a flat grindstone, and it probably remained at this stage of development for many centuries, even after the introduction of a grinding wheel for the grinding of grain.

The first documentary evidence showing the use of a grinding wheel for sharpening purposes is contained in the *Utrecht Psalter* of A.D. 850. This shows a man holding the work against a stone wheel which is being hand rotated by a second man. Apart from the replacement of the second man, firstly by a treadle drive and subsequently by a power drive, this method of tool sharpening has remained basically unchanged to the present day.

The early lathes were operated by two men in a similar manner to that of the grindstone. The turner controlled the cutting tool against the work-piece while his assistant rotated the work-piece by means of a simple cord drive. The introduction of the pole lathe in the thirteenth century A.D. (fig. 7.1) dispensed with the help of an assistant but featured a reciprocating motion of the work-piece. This meant that cutting could only be carried out for one stroke, the work-piece being withdrawn for the return stroke. The end of the fifteenth century saw the introduction of continuous rope drives and simple forms of tool holder. This enabled the cutting operations to be performed much

Fig. 7.1 Pole lathe.

more quickly. During the same period the use of horsepower and water-power for driving the lathe enabled cutting to be achieved on harder materials and metals. Other developments around these Middle Ages, namely those of screw cutting and gear cutting machines, resulted from the requirements of the clock makers.

The industrial revolution of the eighteenth century brought about great advances in the design of machine tools. Machine beds were more rigidly constructed, the wooden spar framework being replaced by iron bars; the accuracy of slideways and lead screws was improved; tool holders with a cross-slide motion were developed, and materials having a greater bearing strength were used for supporting the work-piece.

The manufacturers of navigational instruments and clocks were continually trying to improve their products, while on the military side it was appreciated that a machined bore would improve the accuracy and firepower of cannons. Thus, there was a general requirement for improved machining accuracy and early development work on the steam engine, by pioneers such as Newcomen and Watt, pressed this requirement still further because effective operation of the engine depended upon the close dimensional accuracy of the moving parts.

This led to the adoption of power drive for machine tools by the end of the eighteenth century. The first machine shop of the modern era, the Soho Foundry, was built by Boulton and Watt in a suburb of Birmingham, England. They installed an impressive array of lathes, boring machines and drilling machines and concentrated on the manufacture of steam engines. Considerable development took place during this same period in screw cutting machines and measuring instruments. Maudsley produced a micrometer having 100 threads per inch with 100 graduations, each division therefore representing 0.0001 in. This accuracy marked the beginning of standardization and interchangeability which was so essential for quantity production.

HISTORY OF METAL CUTTING | 525

The milling machine is believed to have been developed around 1820 by Eli Whitney, of America, for the production of small arms weapons (fig. 7.2). The design was a natural development of the lathe, the headstock being used to hold a rotary cutter while the tool holder and cross slide were used to hold the work-piece. This machine enabled non-cylindrical shapes and contours to be cut. There was now a requirement to enable the accurate machining of flat surfaces to be achieved and this led to the development of the planing and shaping machines.

During the first half of the nineteenth century, Whitworth, a follower of Maudsley, appreciated the necessity for a standard screw-thread form over a range of sizes. Following a study of the products of various manufacturers he proposed a thread form which was adopted as the British national standard in 1860. This thread form is named after him. Whitworth was also the first large-scale producer of machine tools to rigid specifications and performance.

Although British engineers had successfully mechanized production, rapid progress was also being made in America. The general requirements of the pioneers, who were expanding westward, and later the armoury requirements of the civil war all aided this progress. The methods employed in the production of the 'Colt' revolver were subsequently employed in the production of sewing machines and bicycles for the consumer market. The successful exploitation of such a mass market depended upon a standard machine accuracy which would ease assembly and ensure interchangeability of spares and replacements.

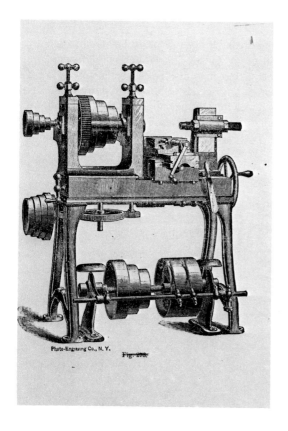

Fig. 7.2 Milling machine. Developed by Eli Whitney for the production of small arms.

In the second half of the nineteenth century multi-tool machines were developed and progress was particularly marked in the field of lathes and drilling machines. The lathe development of this period can be recognized as the beginning of turret lathe design (fig. 7.3). The successful manufacture of a bonded grinding wheel improved precision grinding which led to higher standards of dimensional accuracy. Thus, inspection gauges improved and hence closer tolerances for machine parts and bearings were achieved. In America, Brown and Sharpe designed a universal milling machine and also standardized gear tooth dimensions. Methods of generating gears had been developed by the end of the nineteenth century, the invention of Fellows being particularly outstanding (fig. 7.4).

Whereas the improvements in machining precision and in the rigidity and versatility of a machine have been a continuous process, the whole development process is marked by a number of major advances, each of which produced, at the time, a great technological stride forward. The harnessing of steam-power to drive machinery ended the era of handheld cutting tools while the use of cast iron and steel for machine construction made possible the large-scale production of interchangeable parts at low cost. Following these came three developments, at the end of the nineteenth and beginning of the twentieth century, which paved the way for modern mass production techniques.

Firstly, the introduction of individual electric motor drives for each machine produced greater flexibility in cutting speeds and simpler and quicker gear changes than with the earlier pulley drives from common overhead shafting. The pulley drives were replaced by gear-boxes, giving greater reliability and independence of operation, and the overhead shafting was dispensed with, so making the space available for overhead handling gear and travelling cranes.

The second development was a major breakthrough in tool steel materials. Tungsten and chromium alloy steels were produced by Taylor and White and perfected at the Bethlehem Steel Works by a long series of experiments. These steels, now known as high-speed steels, enabled much higher cutting speeds to be achieved together with a longer tool life than previously. Coupled with this, the development of coolants allowed heavier cuts to be made without tool failure occurring because of overheating. Not until the 1930s was the third advance made. This was the introduction of tungsten carbide as

Fig. 7.3 Turret lathe. This early model (c. 1850) has the turret axis parallel to the spindle.

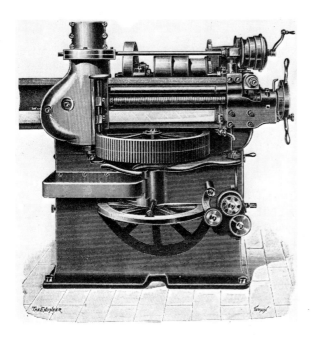

Fig. 7.4 The first Fellows gear-shaping machine, 1897.

a tool material by Krupps of Essen. This material allowed cutting speeds to be doubled, the higher initial cost being rapidly recovered by a greater output of work.

Further technological advances accompanied the age of the motor car, the aeroplane and, more recently, the computer. Gear hobbing, shaving and grinding techniques were perfected, so allowing gear-boxes to be manufactured which transmitted higher loads, ran at higher speeds and were quieter and more reliable than before. Quantity production of interchangeable components was achieved by the development of 'automatics', of transfer machines which could carry out a series of machining operations without a human operator ('automation') and lastly of numerically controlled machine tools to perform a series of operations from instructions computed directly from a design drawing. The modern tendency, which is likely to continue, is for a greater capital expenditure on machinery and control equipment accompanied by reduction in labour requirements. Furthermore, higher production, greater accuracy and repeatability are obtained by the elimination of human errors and, when a machine has an inspection and tool adjustment process incorporated in its design, near perfection is achieved.

References 1 to 4 apply particularly to this section.

7.2 CUTTING IN GENERAL

7.2.1 Principles of metal cutting

In many machining operations the cutting tool is basically wedge-shaped. The cutting edge is straight and moves relative to the original plane surface of the workpiece. In practice, the cutting edge is usually set at an angle to the cutting direction and the operation is referred to as oblique cutting (fig. 7.5(a)). However, in order to analyse the process of cutting, it is easier to consider the simplest case, when the cutting edge

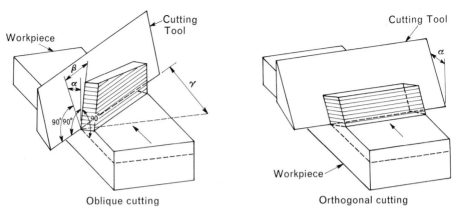

Fig. 7.5 Idealized cutting processes. (a) Oblique cutting, (b) Orthogonal cutting, α Primary rake angle, β Chipflow, γ Angle of obliquity

is at right-angles to the cutting direction as shown in fig. 7.5(b). This is known as orthogonal cutting.

As the tool advances into the work-piece, the metal ahead is severely stressed so that a chip is formed by a process of continuous shear which occurs on a plane extending from the cutting edge to the uncut surface. This plane is termed the shear plane and the angle which it makes with the machined surface the shear angle (designated ϕ in fig. 7.6). The front surface of the tool which is in contact with the chip metal is called the rake face, and the angle it makes with the normal to the machined surface the rake angle (α in fig. 7.6).

The mechanism of chip formation is analogous to the successive displacements of cards in a stack being pushed along a table, where each card is moved forward along the shear plane with respect to its neighbour. If the material is brittle, it will fracture at a low shearing strain and small discontinuous chips will be formed (fig. 7.7(a)). If the material is ductile, it can be strained without rupturing so that the metal is removed as a long continuous chip. Ductile material usually gives good tool life (as measured by the volume of metal removed between successive regrinds) as well as a good surface finish and machinability (fig. 7.7(b)). If there is high friction between the tool face and the chip, the tool edge becomes hot and work-hardened fragments of metal tend to cling

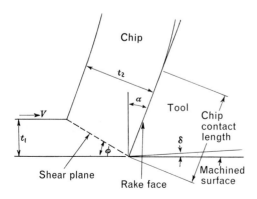

Fig. 7.6 Diagram of work piece/chip/tool interface.

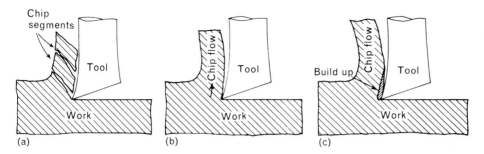

Fig. 7.7 Chip formation, (a) Discontinuous chip, (b) Continuous chip, (c) Continuous chip with built up edge.

to the tool face and build up (fig. 7.7(c)). This build-up breaks away intermittently together with the chip, causing a poor surface finish and poor tool life.

Schematic diagrams of the forces acting at the tool edge during the machining operation are shown in fig. 7.8. The resultant force vector R, which forms the diameter of the force circle, is common to the systems of both the tool and the work. Moving contact between the tool and the chip causes friction along the tool rake face; this is termed the chip friction force, F. Its normal component N, the contact pressure between the rake face and the chip, completes a triangle (fig. 7.8). The angle τ between R and N is called the friction angle, and the chip friction coefficient $\mu = \tan \tau$ or F/N. The chip friction force, which may cause metal build-up on the tool, can be greatly reduced by using suitable lubricants; these are discussed later.

The vector R can also be resolved into the shear force F_S (fig. 7.8), the magnitude of which is determined by the area of the shear plane and the shear strength of the material being cut, and F_N, the compressive force acting normally to the shear plane.

The vector R may be resolved a third time (fig. 7.8) into the cutting force F_C parallel to the plane of tool motion, which represents the total work done in cutting, and its normal thrust force F_T. In practice it is usually important to know the value of F_C as the product $F_C V$ indicates the power absorbed, where V is the cutting speed.

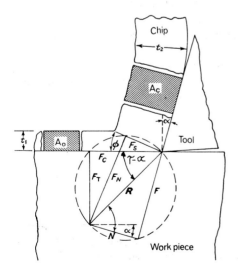

Fig. 7.8 Force diagram showing relationships between components.

The value of the shear angle ϕ depends upon the friction angle τ, the rake angle, α, of the tool, and the material being cut. Merchant (see ref. 5) proposed the following relationship between these quantities when a continuous type of chip is being produced:

$$2\phi + \tau - \alpha = C$$

where C is a constant for a given material. From this expression it can be seen that the shear angle ϕ will increase if the rake angle α is increased, but will decrease if the friction angle τ is increased. In each case the change in ϕ is equal to one-half of the change in the other two variables.

This is a very important relationship as it has been shown (see ref. 6) that the value of the shear angle is indicative of the force required to remove a given chip and of the quality of finish which will be obtained on the work-piece. Typical values for C which have been obtained by experiment are 80° for steel of 200 units of hardness on the Brinell scale, 70° for mild steel, 50° for aluminium and 45° for copper.

From fig. 7.9 it will be observed that although the tool ploughs through the uncut metal to a depth of t_1, the resultant chip has a thickness of t_2. The ratio t_1/t_2 is called the cutting ratio or chip thickness ratio and is directly related to the shear angle and to the rake angle. If the chip cross-sectional area is constant, then the product of the work-piece material length l_1 and thickness t_1 will be equal to the product of the chip length l_2 and thickness t_2, i.e. $l_1 t_1 = l_2 t_2$.

From the geometry of fig. 7.9 it will be seen that:

$$\frac{t_1}{t_2} = \frac{\sin \phi}{\cos (\phi - \alpha)}$$

As $\cos (\phi - \alpha)$ is usually close to unity this expression may be further simplified to:

$$\text{Cutting ratio} = \frac{t_1}{t_2} = \frac{l_2}{l_1} = \sin \phi$$

Thus, by measuring the length of a chip and the length of the path on the material from which it was cut, the shear angle may be calculated.

The factors which influence the magnitude of the cutting force are as follows.

a. *Work-piece material.* The physical properties of a material, such as the tensile and compressive strengths, hardness and ductility, determine the ease with which a material can be machined. These properties are examined later (see Machinability, p. 535).

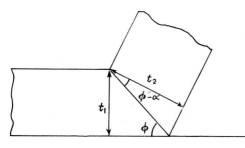

Fig. 7.9 Diagram illustrating chip thickness and length relationship.

CUTTING IN GENERAL | 531

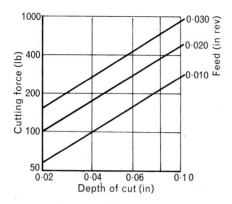

Fig. 7.10 Variation of cutting force with depth of cut and feed rate for turning mild steel with H.S.S. tool.

  b. *Depth of cut.* This corresponds to the undeformed chip thickness t_1 defined previously. The way in which the cutting force varies with the depth of cut is indicated in fig. 7.10.

  c. *Width of cut.* In any machining operation the area of cut is determined by the product of its depth and width; the cutting force is approximately proportional to this value. The width of cut is dependent upon the feed rate between the tool and the workpiece. The direction of feed may be tangential (as in turning) or normal (as in drilling) to the surface being cut and is governed by the motions of the tool holder and the worktable of the machine.

  d. *Tool shape.* The rake angle of the tool has the major influence on the cutting force (fig. 7.11). Other features, such as the shape of the cutting face and the clearance angles, are of secondary importance.

  e. *Cutting speed.* The cutting force is generally independent of the cutting speed. There is, however, a tendency for the force to rise at lower cutting speeds because of an increase in the coefficient of friction (fig. 7.11).

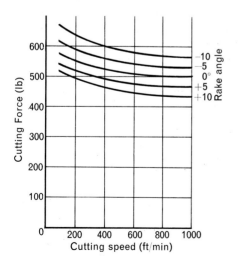

Fig. 7.11 Variation of cutting force with cutting speed and rake angle when turning mild steel with tungsten carbide tool (depth of cut 0.10 in feed 0.020 in/rev).

f. Tool material and condition. The effect of the tool material on the cutting force is dependent on the friction at the tool/work-piece interface and also on the sharpness of the cutting edge. The latter is a function of tool life between regrinds (see machinability, 7.2.5). Cutting tool materials are more fully discussed in a later section.

g. Cutting fluids. These may act either as a lubricant, so reducing the cutting force, or as a coolant to improve the tool life and dimensional accuracy.

7.2.2 Surface finish

As a work-piece is machined by a series of overlapping cuts, its surface cannot be expected to be perfectly smooth. In fact, it will generally consist of a series of 'hills and dales' whose deviation from the average height is a measure of the surface finish.

The root mean square (r.m.s.) deviation* is usually quoted and typical values for various machining operations are:

Turning and milling	100 to 300 μin
Drilling	80 to 200 μin
Grinding	10 to 100 μin

In addition to the contours formed by the tool geometry, irregularities may arise due to wear of the cutting tool, break away of any built-up edge, material inclusions and lack of rigidity of the machine tool structure and drive. Lack of rigidity may also allow vibrations to be set up which produce a wavy surface on the work-piece (fig. 7.12). Such vibrations are known as 'chatter' and may be either a self-excited phenomenon arising from fluctuations in the cutting conditions or forced vibrations set up by resonance from the driving motors or gears.

7.2.3 Tool life

Tool life is a difficult factor to define adequately, but is usually taken as the cutting time before complete failure of the tool occurs or the time necessary to produce a given amount of wear on the tool face. In practice, the tool is seldom allowed to fail completely; chatter vibration and difficulty in obtaining the required finish and dimensional accuracy usually make it necessary to change the tool some time before complete breakdown occurs. In addition, the amount of regrinding necessary to restore a tool which has failed completely is so large that the practice is uneconomical. Thus, the tool is often replaced at regular intervals during the natural breaks which occur in the working day, such as shift changes, meal times etc.

Failure of a tool is normally measured by the amount of abrasive wear on the rake

---

* The r.m.s. deviation is obtained by measuring the deviation from the average height at a number of different points on a line, usually about $\frac{1}{4}$ in. long or less, on the work-piece. It is then calculated by squaring the individual values thus obtained, computing the average and then taking the square root, i.e.:

$$\text{R.m.s. deviation} = \left(\frac{A^2 + B^2 + C^2 + D^2 + \ldots}{N}\right)^{1/2}$$

where A, B, C, D etc. are the deviation measurements obtained and N is the number of points where measurements were taken.

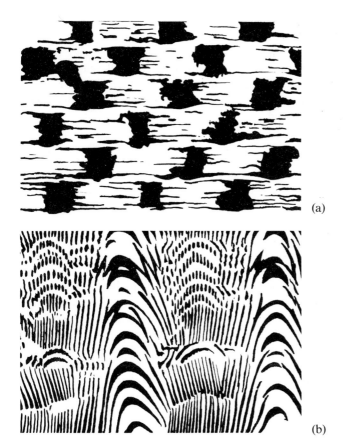

Fig. 7.12 Diagrammatic representation of surfaces cut under vibration, (a) Armour plate (planing machine). Frequency approximately 6 cycles per sec. Cutting speed, 17.5 ft. per min. Depth of cut 7/16 inch. Feed 0.20 inch (approximately $1\frac{1}{2} \times$ full size), (b) Medium carbon steel (lathe). Frequency, 506 cycles per sec. Cutting speed 409 ft. per min. Depth of cut 0.10 inch, Feed 0.013 inch (approximately $1\frac{1}{2} \times$ full size).

face (i.e. 'cratering' at the chip/tool interface) or on the clearance faces (between the tool and the work-piece). If unsatisfactory tool materials are used or the cutting conditions are too severe breakdown may occur due to the fracture or chipping of the cutting edge. It has been determined that where the life of a tool is governed by conditions of wear, a relationship exists between the cutting speed and the useful life of the tool. This relationship is expressed by Taylor's law:

$$VM^n = C$$

where V is the cutting speed, M is the tool life, n an experimental index and C a factor which is constant for a given set of cutting conditions (i.e. tool, work-piece and size of cut).

As a result of experimental work which has been carried out on a range of tool and work materials it has been found that n has a value of approximately 0.16 for high-speed tools, 0.25 for carbide tools and 0.5 for ceramic tools. If M is measured in min and V in

534 | WORKING OF METALS BY CUTTING

ft/min, then lines of constant C, assuming $n = 1/6$, can be obtained as shown in fig. 7.13. Such a chart is used to determine the cutting speed which will give a desired tool life between regrinds. To obtain this optimum speed, a test is run under the required cutting conditions at a chosen cutting speed. The resulting life of the tool will establish a position on the chart. Then, for the same conditions of cut and materials, a line drawn through this point parallel to the plotted lines gives the required relationship between the life of the tool and the cutting speed.

Alternatively, the relationship

$$\frac{V_1}{V_2} = \left(\frac{M_2}{M_1}\right)^n$$

may be used.

7.2.4 Economics of metal cutting

When a component is machined from its rough cast or forged form it is necessary to remove a known volume of metal. The depth of cut and rate of feed will be determined by the capacity of the machine and also by the surface finish and dimensional accuracy required. This usually involves a 'roughing' cut followed by a 'finishing' cut and the rate at which the machining process is completed is then governed mainly by the cutting speed. Any increase of cutting speed will be accompanied by a reduction in the life of the tool and the consequent increase in tool costs may more than offset the decrease in machining costs. The total cost can be minimized by operating at the optimum speed, which is determined as follows.

Let the machining time be T_0, the cutting speed be V_0 and the corresponding tool life be M_0. Then, at a speed of V,

$$\text{Machining time } T = \frac{V_0 T_0}{V}$$

$$\text{Tool life } M = \left(\frac{V_0}{V}\right)^{1/n} M_0$$

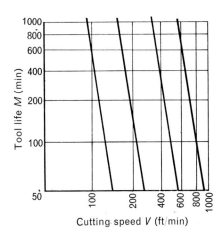

Fig. 7.13 Variation of tool life with cutting speed (assuming n = 1/6).

If the handling time is H (independent of cutting speed), the labour rate L and the overhead rate O, then:

$$\text{Machining cost per part} = (L+O)(T+H)$$

$$= (L+O)\left(\frac{V_0 T_0}{V} + H\right)$$

If the initial tool cost is P, the regrind cost R (including tool changing cost) and the number of regrinds before the tool is discarded N, then:

$$\text{Tool cost per part} = \left(\frac{P}{N} + R\right)\frac{T}{M}$$

$$= \left(\frac{P}{N} + R\right)\frac{T_0}{M_0}\left(\frac{V}{V_0}\right)^{(1/n)-1}$$

These costs are plotted in fig. 7.14 for the following values: $T_0 = 10$ min, $V_0 = 200$ ft/min, $M_0 = 250$ min, $H = 1$ min, $L = 10s$ per hour, $O = 20s$ per hour, $P = 25s$, $N = 10$, $R = 5s$ and $n = 1/6$.

It can be seen that the minimum cost occurs when V_0 is increased to approximately 245 ft/min, and that above this speed the rapid increase in tool costs outweighs any saving in machining costs.

7.2.5 Machinability

This is the allowable rate at which metal can be removed, subject to a satisfactory tool life and surface finish, and in practice it is usually determined by relating the cut-

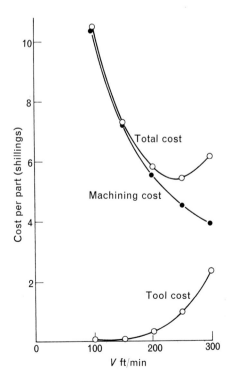

Fig. 7.14 Manufacturing costs.

ting force, for a given area of cut, to the cutting speed to produce a specified tool life.

Dr. Schlesinger carried out a series of tests on different materials to determine the cutting speed corresponding to a tool life of 60 min between regrinds. The machinability of the material was then evaluated as the cutting force required to remove 0.001 in^2 of metal.

High-strength alloy steels are usually machined in the annealed condition, subsequently hardened and then finished by grinding. Improvements in machinability can be achieved by the addition of small percentages of elements such as sulphur and lead.

Corrosion and heat-resistant steels and non-ferrous alloys present particular difficulties in machining because of their high strength at elevated temperatures and their tendency to work harden. This often leads to a built-up edge being formed and hence to high friction forces at the tool/chip interface. The rate of tool wear is then high and may also be aggravated by fragmentation of the cutting point. In order to obtain an acceptable tool life it is usual to machine such materials at low cutting speeds. Also, light finishing cuts should be avoided owing to the skin hardness produced by previous cuts.

Table 7.1 gives average values of cutting force required per unit area of material being cut. Variations from these values occur with changes in speed, feed, depth of cut, tool geometry etc. The power required to remove a given volume of material is directly proportional to the unit cutting force.

References 7 and 8 apply particularly to this section.

7.2.6 Cutting tool materials

The material used for a metal cutting tool must satisfy the following requirements if a suitable tool is to be obtained.
1. It must have a greater hardness than the work material if the cutting edge is to penetrate without distorting.
2. It must be tough and resistant to shock so that any break up of the cutting edge is prevented.
3. It must be resistant to wear so that a good tool life is obtained.
4. It must be possible to grind the material to a sharp edge.

a. High-carbon tool steels. The steel used in the manufacture of hand tools usually has a carbon content of between 0.75% and 1.0%, while for cutting tools such as drills, reamers and taps the content is 1.0% to 1.25%. After the tool has been preformed it is hardened and tempered to produce the required mechanical properties.

Carbon steel loses its hardness at high temperatures and is therefore only suitable for tools which will be used for light work or slow-speed cutting. The hardness at high

Table 7.1

Material	Unit cutting force p.s.i.	Unit power hp/in^3 min
Steel	250 to 400	0.6 to 1.0
Cast iron	200	0.5
Brass	130	0.32
Aluminium alloys	110	0.27
Titanium alloys	250	0.62

temperatures, and also the tool life, is, however, improved by adding small percentages of alloying elements such as chromium, tungsten, cobalt and vanadium.

b. High-speed steels. These are a range of air-hardening tool steels which enable much higher cutting speeds to be achieved. The most common alloy now in use contains from 0.5% to 0.8% carbon, 18% tungsten, 4% chromium and 1% vanadium, and is known as 18-4-1 H.S.S. Molybdenum can also be used as the main alloying element.

High-speed steel is suitable for all types of cutting tool, its additional cost over plain carbon steels being fully justified when higher speeds and heavier cuts can be utilized. For rough machining or for intermittent cutting action, such as in milling, its toughness is advantageous compared with harder but more brittle materials. It can also be formed fairly easily into intricate shapes such as drills, hobbing cutters and broaches.

c. Cemented carbides. These are manufactured in small block form from sintered powders which are mainly composed of tungsten carbide. As their initial cost is very high they are used either as tips for single-point tools, which are brazed onto a carbon-steel shank, or as inserts in multipoint tools such as milling cutters.

Cemented carbides are extremely hard and resistant to abrasion, so enabling high cutting speeds to be used while also giving a long tool life. They are therefore widely used for turning tools and milling cutters in large-scale production runs. They are also rather brittle and therefore tend to break if the cutting action is interrupted, although this tendency can be lessened by using a negatively raked tool cutting face.

d. Ceramics. The basic ceramic material used for cutting tools is sintered aluminium oxide. Although it retains its external hardness at very high temperatures and has a low coefficient of friction, it is very brittle and has a low shock resistance. Consequently, its main application is for high-speed machining under light cutting conditions, e.g. for cutting non-ferrous metals or finish cutting. Ceramic tools are not suitable for intermittent cutting and are easily damaged by 'chatter' vibrations or other fluctuating conditions.

e. Diamonds. The diamond is the hardest material available but, owing to its high cost, its use is usually restricted to light finishing work with single-point tools and to grinding operations. For the latter purposes diamond grit is bonded with metal or resin to form the cutting face of the grinding wheel.

A diamond tool is also used for the 'dressing' of grinding wheels, i.e. truing of the wheel after use, and either a single-point, multipoint or impregnated dresser is used.

The use of a diamond tool allows very high cutting speeds to be used and under ideal conditions a long life is also obtained. Resetting of the tool is usually carried out by the supplier rather than the user. To obtain the maximum benefit from the use of diamond tools the machinery must be sensitive, rigid and free from vibrations.

7.2.7 Cutting fluids

The cutting action of a tool and finish of a work-piece can be improved by applying a cutting fluid to the tool/work-piece area. The reasons for this are given below and are of varying importance depending upon the particular application. The fluid is referred to as a 'lubricant' or 'coolant' depending upon which of these is the dominant function.

538 | WORKING OF METALS BY CUTTING

The functions of a cutting fluid are:

1. To lower the interface friction between the cutting tool and the chip by lubricating action. This reduces the cutting forces and the power required, and hence increases the tool life.
2. To lower the tool and work-piece temperatures by acting as a coolant. The tool life is thus improved, and the reduction of distortion and thermal expansion results in a greater dimensional accuracy. The work-piece may also be easier to handle as a result of its lower temperature.
3. To improve the surface finish of the work, mainly by preventing or reducing the formation of a built-up edge on the tool.
4. To protect the work and the machine against corrosion.
5. To wash away any chips and swarf from the cutting area.

A cutting fluid is usually a soluble oil (emulsion), a mineral oil or a fatty oil.

a. Soluble oils. These are emulsions of oils, additives and water in a ratio varying from about 1:5 to 1:100 oil:water mixtures. Owing to the high specific heat of water and the low cost of the emulsion, soluble oils are widely used for high-speed machining operations in which the cooling effect is more important than lubricating properties.

Typical applications are for finish turning, drilling and grinding.

b. Mineral oils. These are used either in their 'straight' form or with the addition of small quantities of sulphur or organic compounds, when they are known as extreme pressure oils.

Owing to their low specific heat and load-carrying capacity, straight mineral oils are usually restricted to the machining of non-ferrous metals. The additive oils are, however, suitable for heavier duty work on tougher materials in operations such as screw cutting, gear cutting and automatic lathe work.

c. Fatty oils. Despite their expense, lard oil and other fatty oils are widely used in combination with mineral oils for heavy, slow-speed cutting operations, such as tapping and broaching, where their oiliness and penetrative properties are advantageous. Edible fats are not used widely nowadays and it is therefore usual to refer to the fatty ingredient as saponifiable oil.

Table 7.2 indicates the type of fluid suitable for different operations and materials.

7.3 TURNING AND BORING

In these actions the work-piece is rotated by the machine about a fixed axis, which is usually horizontal, while the tool cuts into the work to the required depth. The required shape is produced by feeding or traversing the tool relative to the work and roughly 40% of all machining operations fall within this category.

7.3.1 The tool

The geometry of the cutting point and the performance of the tool are determined by various angles and dimensions which are shown in fig. 7.15.

Table 7.2

	Operation			
Material	Turning	Drilling and milling	Screw cutting and tapping	Grinding
Steel below 35 t.s.i. tensile strength	Emulsion 1:30	Emulsion 1:20	Emulsion 1:10	Emulsion 1:40
Steel up to 65 t.s.i. tensile strength	Emulsion 1:20	Emulsion 1:20	Sulphurized mineral oil	Emulsion 1:40
Steel over 65 t.s.i. tensile strength	Sulphurized mineral oil	Sulphurized mineral oil	Sulphurized mineral/ fatty oil	Emulsion 1:40
Cast iron	Emulsion 1:20 or dry	Emulsion 1:20	Emulsion 1:10	Emulsion 1:40
Aluminium alloys	Mineral oil	Mineral oil	Mineral/fatty oil	Emulsion 1:40
Brass	Emulsion 1:20	Emulsion 1:20	Mineral/fatty oil	Emulsion 1:50
Titanium alloys	Sulphur-chlorinated emulsion 1:3	Sulphur-chlorinated mineral oil	Sulphurized mineral oil	Emulsion 1:25

a. Rake angles. The true rake angle is measured in the direction of chip flow across the face of the tool. It is therefore a combination of the back rake angle (measured in a plane perpendicular to the work axis) and the side rake angle (measured in a plane parallel to the work axis). An increase in the rake angle reduces the cutting force, so improving the surface finish, but weakens the tool by reducing the lip angle. Consequently, low rake angles are used for machining hard materials. The use of negative rake angles with carbide or ceramic tools enables intermittent cuts to be made without damaging the tool.

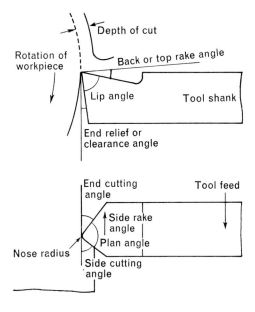

Fig. 7.15 Tool geometry.

540 | WORKING OF METALS BY CUTTING

The rake angle is usually between 5° and 20° for high-speed steel tools and between 0° and 10° for carbide tools. However, a negative rake of up to 10° can be used with a carbide or ceramic tool.

b. Relief angles. To provide clearance between the tool and the cut surface, end and side relief angles are applied to the front and side cutting edges respectively. The end relief angle is shown in fig. 7.15. Both end and side relief angles usually vary between 5° and 10°.

c. Nose radius. To preserve the cutting tip of the tool, the nose is ground to a radius of approximately 1/8 in.

d. Tool shank. This should be of sufficient proportions to withstand the cutting forces, and stiff enough to resist deflections and the build up of chatter. This can usually be achieved by keeping the tool overhang (from tool holder to cutting edge) to a minimum and by setting the tool point at the same level as the work-piece axis.

7.3.2 Turning operations

Turning is a general term for a variety of machining operations, shown in fig. 7.16, which are achieved by different traverses of the tool relative to the rotation of the work-piece.

a. Plain turning and boring. These are the simplest of all turning operations. The tool traverses parallel to the work-piece axis and the cutting speed is proportional to the product of the rotational speed and the diameter of the work. When an internal cylindrical surface is produced the operation is then termed 'boring'.

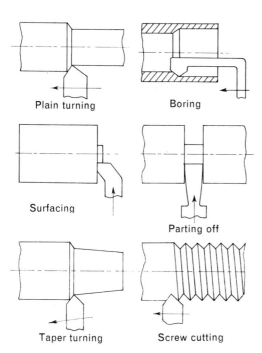

Fig. 7.16 Turning operations.

b. Surfacing and parting off. The tool traverses in a radial direction and produces a cut perpendicular to the axis of rotation. This motion is also used to cut through or 'part off' a section of a bar material when a chisel-shaped tool is used. Unless the spindle speed is increased, the cutting speed becomes lower as the tool approaches the axis.

c. Taper turning. When the motion of the tool relative to the work-piece axis is a combination, in a certain ratio, of motions (*a*) and (*b*) a conical surface is produced. For small axial lengths, this operation can be carried out manually by setting the top tool slide to the required taper angle. For longer lengths, and when a power traverse is used, a taper turning attachment is fitted. This is discussed more fully later (see attachments and accessories, 6.3.3.b).

d. Screw cutting. A helical thread can be cut on a cylindrical section if the plan angle of the tool is made equal to the V angle of the required thread. The tool must be fed axially through a distance equal to the pitch of the thread for each rotation of the work. The required speed ratio between feed and rotation is obtained by selecting the appropriate gear ratio between the spindle drive and the lead screw which drives the tool holder.

e. Copy turning. A non-standard profile, of a desired shape, can be obtained by causing the tool to follow the required path. To achieve this, a tracer follows the template profile and operates the tool slide via a servo-mechanism.

7.3.3 Lathes

The principal working parts of a standard motor-driven lathe with manual controls are shown in fig. 7.17.

The lathe consists basically of a bed, a headstock gear-box, a tailstock and a compound slide. The headstock is fixed on one end of the bed. The tailstock and compound slide can be moved along the upper surface of the bed and can be fixed in any desired position. A constant-speed electric motor drives the headstock gear-box, from which a selection of spindle speeds is obtained. Coupled to this is a feed change gear-box where-

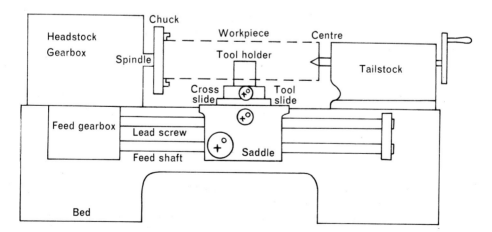

Fig. 7.17 Layout of general purpose lathe.

by the movement of the compound slide along the bed per spindle revolution is selected. For screw cutting, the drive is taken through the lead screw instead of the feed shaft. The tool holder is mounted on the compound slide. The tool slide forms the top section of the compound slide and can be swivelled, when required, to allow the tool holder to be fed in at angles other than normal to the work axis. The cross slide can only move at right-angles to the work axis and it is mounted on the saddle, which can be traversed by hand or driven by either the feed shaft or lead screw. The work-piece is held in a chuck, or on a centre which is driven by the headstock spindle, and is usually also supported by a centre in the tailstock. The depth of cut is set by positioning the cross slide, and the feed is controlled by the saddle motion. The saddle and tailstock can both move axially along guides in the machine bed, which must be sufficiently rigid to withstand the maximum cutting forces and thereby control the dimensional accuracy of the work produced.

a. Rating and accuracy of lathes. The nominal size of a lathe is determined by the maximum diameter ('swing') and length of work-piece which can be accommodated on it. The length of the work-piece is dependent upon the length of the bed, which can vary between 3 ft and 16 ft. Additionally, the horsepower of the drive motor limits the rate at which metal can be removed, and the rigidity of the bed determines the size of cut which can be taken. Floor-mounted workshop lathes normally have a swing within the range of 10 to 50 inches but bench lathes can be obtained in smaller sizes.

Special tool room lathes, constructed with great precision, can be expected to turn to an accuracy of 0.001 inches per inch in the hands of a skilled operator, but workshop lathes normally require a tolerance of 0.010 inches for rough turning and 0.005 inches for finish turning.

b. Attachments and accessories. The principal attachments available for increasing the versatility of lathe operations are those for taper turning and for copy turning.

The many accessories include a variety of chucks for holding either the work or the tool (fig. 7.18). The chucks may be of either the three-jaw or four-jaw type and the jaws, which are operated either manually or by air pressure, can be adjusted independently or in unison on a single thread. Long or overhanging work-pieces are usually given extra support by means of a 'steady'.

The tailstock can be removed and replaced by drills, reamers, taps or dies (see drilling, reaming and tapping, p. 546). These are clamped in position and fed into the work by advancing the tailstock spindle.

c. Boring machines. These are of either the horizontal or vertical type with a single-point or multipoint tool. It is preferable, from handling and rigidity considerations, to support a heavy work-piece on a horizontal table which rotates about a vertical axis. The tool is then fed vertically, at the required radial distance from the table axis, to produce the machined bore.

d. Jig borers. To produce a hole or internal surface to a location and accuracy better than 0.005 inches it is necessary to use a special machine called a jig borer. In fact the jig borer is not a lathe but a drilling machine which can be used for boring operations. The work-piece is clamped to a stationary table and the hole machined by a rotating spindle (see fig. 7.19) to which is attached a drill, reamer (see later section) or boring tool. The actual tool used depends upon the size of hole and the precision re-

TURNING AND BORING | 543

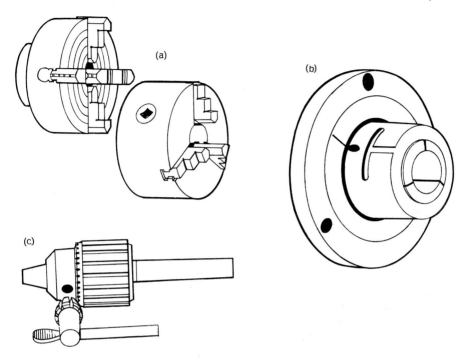

Fig. 7.18 Lathe chucks, (a) Three and four jaw chucks, (b) Collect chuck for bar materials, (c) Drill chuck.

Fig. 7.19 Jig Borer.

quired. To achieve the greatest accuracy a single-point boring tool must be used for the finishing cut.

The work-table is positioned by precision lead screws, which operate in two perpendicular directions, and by a fine measuring system incorporated between the table and the machine bed. Particular attention is paid in the design and manufacture of the machine to ensure that the motions of the table and the spindle axis are truly orthogonal and that there is no free play or distortion. A suitable choice of materials minimizes any errors from differential temperature expansion and from the effects of wear at sliding surfaces.

The capital cost of a jig borer is much higher than that of a production machine and its use is therefore generally restricted to the type of tool room work which demands a high degree of precision, such as the manufacture of jigs, dies and press tools. However, as the machine enables intricate parts to be manufactured without the prior production of costly jigs, it may also be used economically for the manufacture of prototype and small non-recurring batches.

e. Capstan and turret lathes. Where repetitive work and multiple tool operations are to be performed, production rates are usually improved by using a multi-tool holder. This is designed so that the required tools can be selected in sequence and is the principle employed in capstan and turret lathes. To be beneficial, the additional setting-up time must be justified by the quantity produced.

The headstock is provided with either a jaw chuck (for non-cylindrical work) or a collet chuck (for bar work) and an intermediate saddle is fitted with cross slides and traverse as in a normal lathe. However, the normal tailstock is replaced by a capstan or turret in which usually six tools can be fitted in different positions. Each of the six tools can successively be placed in the working position by rotating the turret about a vertical axis and fixing in one of the six indexed positions.

In the case of the capstan lathe this tool holder is carried on a slide which can be hand or power fed into and away from the work. The slide is supported by a saddle which is set and locked on the lathe bed. The tool movement is thereby limited, for any particular set-up, to the length of the slide.

With the turret lathe (fig. 7.20) greater tool travel and rigidity is obtained by mounting the turret directly onto its saddle, which is usually power fed along the machine bed. Consequently, whereas smaller machines may be of capstan design, larger machines are generally of the turret type.

f. Automatic bar machines. The automatic bar machine is basically a turret lathe provided with a means of feeding and turning long lengths of bar automatically, and is designed primarily for the quantity production of components from bar stock.

The sequence of operations, spindle speed and tool feed are automatically controlled by a cam-operated mechanism. The long bar is fed to its working position by passing it through a tubular rotating spindle of the headstock to a stop. The bar is then gripped by a collet chuck mounted on the hollow spindle.

If the machine has one spindle the tool turret can be rotated about a vertical axis to select the desired tool and the sequence of operations is then carried out by indexing from one tool position to the next. The parting-off tool for the final operation is mounted on the cross slide.

With a multispindle machine each bar is passed through a hollow spindle, gripped

TURNING AND BORING | 545

Fig. 7.20 Turret lathe.

Fig. 7.21 Close up of single spindle automatic bar machine.

by a collet chuck and independently rotated by the spindles arranged symmetrically around the main axis of the machine. The tool turret of a multispindle machine, unlike those of the common lathe and single-spindle bar machine, is designed to be rotated about a horizontal axis. The special turret position makes it possible to turn several bars at the same time. A single machining process, in the required sequence, is simultaneously carried out on each of the bars by one of the tools in the turret or on the cross slide. Thus, one component is completed at each index position. The processes of turning, surfacing, drilling, reaming and thread cutting by taps or dies can all be carried out on the machine and, if the quantity of work involved at each index position is approximately equal, thereby minimizing the idling time to that of the tool indexing motion, the maximum efficiency will be achieved.

7.3.4 Speeds and feeds for turning operations

Typical values of cutting speed in feet per minute, applicable to a range of metals and tool materials, are given in table 7.3. The feed per revolution is assumed to be between 0.02 and 0.04 inches for a roughing cut and 0.005 to 0.010 inches for a finishing cut. The actual cutting speed employed depends upon the depth of cut, the rigidity and power of the machine and the economic life of the tool.

7.4 DRILLING, REAMING AND TAPPING

7.4.1 Principles and definitions

Drilling is the process whereby a circular hole is cut into a solid section of material. When an existing hole is to be enlarged with a second drill, this is termed counter-drilling. Reaming is a finishing process which follows drilling and is used when an accurate size of hole and good surface finish are required. Although these operations, using drills and reamers respectively, are often carried out on a lathe (frequently as one of a sequence of operations on a turret lathe) it is generally more economical to use a drilling machine. The essential difference between a lathe and a drilling machine is that in the former the work-piece is rotated and the drill remains stationary, while in the latter the converse applies. The component to be drilled is clamped in a fixture on the work-table and the drill is then adjusted to the correct position. If the component

Table 7.3 Tool material–cutting speeds in ft/min

Work material	High-speed steel		Cemented carbide	
	Roughing	Finishing	Roughing	Finishing
Steel below 35 t.s.i. tensile strength	100	150	250	500
Steel up to 65 t.s.i. tensile strength	40	70	200	350
Steel over 65 t.s.i. tensile strength	30	50	150	250
Cast iron	60	100	175	300
Aluminum alloys	300	500	600	1000
Brass	200	350	500	800
Titanium alloys	20	30	100	175

is small, this is carried out by moving the work over the table, but for larger components a radial drill is used whereby the drill head can be moved over the work. The feed of the drill into the work is normally in a vertical direction and the gap between the work-table and the drill head is adjustable so that different components and hole depths can be accommodated.

Tapping is the cutting of an internal thread in a previously drilled hole by means of a rotating tool, named a tap, which has a similar thread cut on its external surface. The feed of the tap into the hole is equal to the pitch or lead of the thread for each rotation of the spindle drive. After the cutting is completed the tap is extracted from the hole and if the machine is not equipped with a reversing device a special attachment must be fitted.

7.4.2 The tools

 a. Drills. The standard twist drill has two cutting edges and two spiral flutes which form a helix and provide a passage for the cutting fluid and for the removal of the cut metal (fig. 7.22). Drills vary in diameter from 0.010 inches to a maximum of about 3 inches, although larger sizes may be used for special applications. They are normally made from high-speed steel, although carbon steel can be used for light jobbing work and the high cost of carbide-tipped drills may be justified by quantity production.

The helix angle (see fig. 7.22) is usually about 30° but varies from 15° for hard metals to 45° for soft metals. The point angle varies between 100° and 140° for metal drilling but is usually 118° as this is satisfactory for most steels and non-ferrous metals.

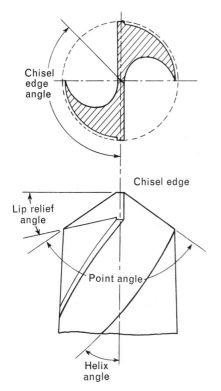

Fig. 7.22 Twist drill angles.

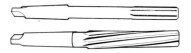

Fig. 7.23 Parallel reamers with straight and helical flutes.

The chisel edge angle is about 130° while the lip relief angle varies from roughly 12° for softer metals, where a high feed rate is used, to 6° for harder metals which necessitate a lower feed rate.

b. Reamers. Typical reamers, which can have straight or spiral flutes, are shown in fig. 7.23. The amount of metal removed by a reamer is generally quite small and most of the cutting occurs on the chamfered edge at its extremity.

c. Taps. A tap has four cutting edges along its length, interspaced by flutes which are normally axial (fig. 7.24). The thread is progressively cut as the tap is rotated in the hole and, to aid this, varying amounts of lead-in at the nose are provided. Taps for cutting threads by hand are supplied in sets of three consisting of a taper, plug and bottoming tap. The taper tap is used to start the thread and the bottoming tap to complete it. The plug tap is most frequently used for machine tapping, but is also used before the bottoming tap when hand or machine threading a blind hole. The tap must always be aligned with the hole and to avoid breakage of the tap the torque should be limited. It is common practice to drill the hole oversize and cut the threads to only 75%, or even less, of the full depth of the thread. This reduces the torque required and eases the swarf removal with only a small loss (around 5%) of thread strength. The use of a suitable lubricant assists in reducing the cutting forces and improving the surface finish.

7.4.3 Drilling machines

Drilling machines, excluding hand portable varieties, can be classified according to the range and type of work that can be accommodated on them. They may be of column or radial form, with either a single-spindle or multispindle drive.

a. Pillar drills. These are general purpose production machines which cover a wide size range. The drill is held vertically in a chuck attached to the end of a driving spindle with the cutting edge downward above the machine work-table. The spindle and work-table are mounted on a vertical column or pillar and the distance between them can be adjusted by moving either the spindle or the work-table or both. Movement can be manual or power operated. During the drilling operation the drill is rotated and gradually lowered onto the work-piece mounted on the table by independent gearboxes driven by an electric motor.

b. Radial drills. The main advantage of a radial drilling machine is that the drill spindle can be easily moved to any required position within the range of the machine. A typical machine is shown in fig. 7.25, and it can be seen that the drilling head can be

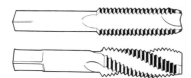

Fig. 7.24 Machine taps with straight and spiral flutes.

traversed along the radial arm, which itself can be rotated about the main column of the machine. The drilling of holes in heavy and bulky components can therefore be easily accomplished.

c. Sensitive drills. These are used for drilling very small holes and a high spindle speed is necessary to achieve the correct cutting speed. The drive for smaller machines is by a V-belt and stepped pulleys, which are driven by a fixed-speed motor, while for larger machines a gear-box or variable-speed motor is used.

A sensitive feed control is usually incorporated, but to prevent the breakage of very small drills the machine is usually hand-operated, so enabling the drill thrust to be gauged by 'feel'.

d. Multispindle drills. A machine having more than one spindle has the following advantages:
1. A separate operation can be carried out on each spindle, as part of a series, so eliminating tool changes. The component is moved by hand or by automatic indexing.
2. A set of holes can be drilled simultaneously with a multispindle drill head (fig. 7.26) and a conversion attachment for a single-spindle drill is frequently available. The drills are usually positioned symmetrically but special purpose machines can be obtained in which the drills can be set to any required position.

e. Deep hole drilling. To obtain long holes that are accurately finished and correctly aligned a gun drill must be used. This was developed for boring gun barrels

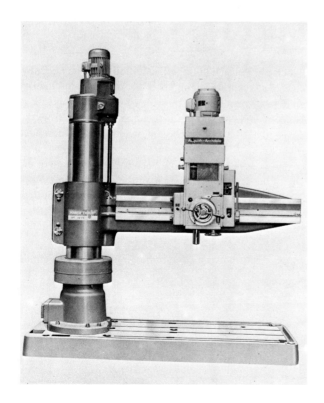

Fig. 7.25 Radial drill.

Fig. 7.26 Multi-spindle drilling machine.

and it has a single cutting edge, which can be carbide-tipped, mounted on a body occupying only half the hole section (fig. 7.27). Accurate alignment of the tool is maintained by using a bearing pad, usually wooden, to bear against the side of the bore opposite to the cutting tip. Thus, the maximum space is provided for the chip flow away from the cutting zone and this is further assisted by the flow of coolant, which is supplied under pressure through a hole in the drill body.

Gun drilling machines are usually mounted horizontally because of the length of the work-piece. It is preferable to rotate the work-piece instead of the drill, but this is not possible where rotational unbalance is present or non-central holes are to be drilled. Higher speeds and lower feed rates than for normal drills are used and guide bushes must be provided to ensure accurate drill alignment.

f. Performance and accuracy. The following conditions are necessary to obtain an accurately machined hole.

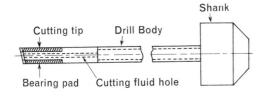

Fig. 7.27 Gun drill.

DRILLING, REAMING AND TAPPING | 551

1. The machine tool must be sufficiently rigid to prevent distortion by the loads produced during the machining operation. The drive spindle must run true and without free play. The feed must be accurately aligned with the axis of the tool and the worktable must be square to the tool.
2. The cutting edges of the tool must be of equal length and at equal angles to the tool axis, otherwise non-circular and oversized holes will result. The part of the tool shank which fits into the drive spindle is also critical upon the concentricity of the tool.

The tolerance on drilled holes varies from about 0.002 inches on a $\frac{1}{4}$ inch diameter hole to 0.010 inches for holes of 1 inch diameter and above. For reamed holes the tolerance can be 0.001 inches irrespective of the hole size, and greater accuracy can be obtained by using a jig borer (see turning and boring, p. 542).

g. Location of the hole. When several holes are required in the same component the accuracy of spacing may be more important than their size. For a small number of components the hole positions can be located by hand marking-out, but this is a tedious process which is not likely to produce a positional accuracy of better than 0.005 inches. The use of a coordinate positioning table would produce a greater accuracy.

For quantity production, however, the hole positions are usually controlled by a jig. Basically, the jig consists of a plate provided with holes, which are jig bored and lined with hardened steel bushes. The component to be drilled is clamped between the jig and the table. The holes in the jig form a guide for the cutting tool and so determine the accuracy of location of the finished holes. The cost of producing a jig can only be justified if a large number of components are required.

7.4.4 Speeds and feeds

Cutting speeds and feed rates are given below for drilling, reaming and tapping.

The quoted cutting speed refers to the peripheral speed at the cutting edge, and to obtain the spindle speed the cutting speed is divided by the circumference of the hole. It should be noted that when drilling solid metal the cutting speed varies from zero at the centre to a maximum at the circumference.

Feed rates are given per revolution of the tool, and in the case of tapping the feed rate is equal to the thread pitch. The time required to cut a hole of given depth is obtained by dividing the length of the hole by the product of the spindle speed and the feed rate.

If a drilled hole is subsequently reamed out, the depth of the reamer cut is usually between 0.005 and 0.010 in.

a. Drilling. For steels and cast irons the cutting speed is 40 to 100 ft/min, depending upon the hardness of the material.

For aluminium and copper alloys the cutting speed is 100 to 300 ft/min.

Feed rates vary according to the hole size and are usually:

0.001 to 0.005 inches for diameters of less than $\frac{1}{4}$ inch,
0.004 to 0.018 inches for diameters between $\frac{1}{4}$ and 1 inch,
0.015 to 0.035 inches for diameters greater than 1 inch.

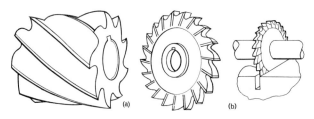

Fig. 7.28 (a) Plain cylindrical cutter, (b) Side and face cutter.

b. *Reaming.* The common practice is to use a cutting speed of roughly two-thirds the drilling speed and a feed rate of twice the drilling feed rate.

c. *Tapping.* The cutting speeds are usually:

20 to 40 ft/min for steels,
40 to 60 ft/min for cast irons,
60 to 100 ft/min for non-ferrous alloys.

7.5 MILLING

Milling is a machining method which is used to produce flat or curved surfaces, slots, keyways, cams and gears by means of a rotating cutter. The milling cutter is of a cylindrical or disc form and has 'teeth', or a number of cutting edges, equally spaced round its periphery. As the cutter rotates, metal is removed from the work-piece by traversing the machine table, to which it is attached, across the cutter. Thus, the cutting action of each tooth is usually intermittent and the chip produced is of non-uniform section.

7.5.1 Types of cutter

There are several types of cutter and these are classified according to their relative length, diameter and number of teeth. Fig. 7.28 shows examples of: (*a*) a plain or slab milling cutter – a general purpose cutter for machining flat surfaces; (*b*) a side and face cutter – this cuts on both the cylindrical and end faces of the cutter and can be used to machine slots, keyways and recesses; Fig. 7.29 shows an end mill cutter – this can be fed parallel to the cutter axis. Form cutters (not shown) are for producing specific contours such as gear teeth.

Milling cutters are made either from high-speed steel or from carbon steel with cemented carbide teeth as tips or separate inserts.

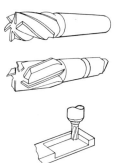

Fig. 7.29 End mill cutter.

7.5.2 Cutting action

For plain milling, the work feed may be either opposing the cutter teeth ('up milling') or in the same direction ('down milling' or 'climb milling') (fig. 7.30).

Although the former method is more frequently used, there is an undesirable rubbing action on each tooth at the beginning of its cut and this can shorten the tool life. Furthermore, the resultant force between the tool and the work is away from the machine table, but in down milling this effect is reversed and the tendency to lift the work off its fixture is removed. The entry conditions of each tooth are also advantageous when down milling, particularly when using carbide-tipped tools. Even so, this method is used less frequently than up milling because the tendency of the cutter to climb onto the work means that it can only be used if the machine is absolutely rigid and there is no backlash in the feed mechanism, in the guiding devices in the table and its supporting members or in any other moving part.

7.5.3 Milling machines

A milling machine consists basically of a milling cutter, which rotates about a vertical or horizontal spindle, and a work-table which can be moved under or along the cutter. The distance between the cutter and the table is adjustable.

   a. Knee-type machines. These are general purpose machines in which the work-table is supported on a knee and the cutter spindle, which can be either horizontal (fig. 7.31) or vertical, is driven from an attachment on the column of the machine. A three-dimensional movement of the work-table is achieved by movement of the knee up or down, the saddle towards or away from the column and the work-table longitudinally on the saddle. On 'Universal' machines the table can be swivelled on the saddle in a horizontal plane through 45° in either direction.

   b. Production machines. These have a fixed bed and, generally, the table can only be traversed in one direction. Consequently, they are less versatile but more rigid and simpler to operate than the knee-type machine. They can have either one or two cutter spindles and, by moving the spindle carrier in the column, a limited adjustment in the vertical and transverse directions can be obtained.

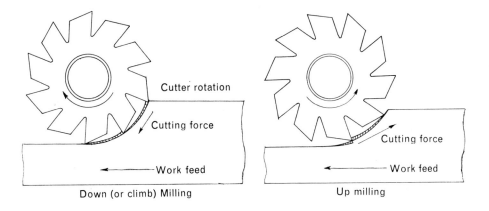

Fig. 7.30 Cutting action in milling.

Fig. 7.31 Horizontal knee-type milling machine.

c. Planar-type machines. These machines are designed to accommodate large work-pieces. A table feeds the work-pieces longitudinally under a cross member. Several milling heads can be set and traversed on the cross member (fig. 7.32) and in addition the whole cross member can be moved up and down to select the desired distance between the table and the milling cutters (compare with planing machines (6.6.4) which are of similar construction).

d. Dividing head attachment. If a series of equally spaced slots is to be milled around the circumference of a component, e.g. a gear blank, then the component must be rotated through an exact angle between each cut. This is known as indexing and is carried out by attaching the component to the spindle of a dividing head. The dividing head is fixed to the machine table and the component is moved through the required angle by rotating a crank on an indexing plate. The crank rotation is transmitted to the component via a worm gear, which usually has a reduction ratio of 40:1.

For example, if a 48-toothed gear is to be cut, the blank must be rotated through 1/48 of a revolution between each cut. The indexing crank must therefore be rotated through $40/48 = 5/6$ of a revolution, and if there are 24 holes on the indexing circle this corresponds to a movement of 20 holes.

7.5.4 Speeds and feeds

Typical cutting edge speeds for high-speed steel (H.S.S.) cutters when machining different materials are given in table 7.4 in feet per minute. The lower speeds should

Fig. 7.32 Planar-type milling machine.

Table 7.4

Steel below 35 t.s.i. tensile strength	80 to 100 ft/min
Steel below 65 t.s.i. tensile strength	60 to 80 ft/min
Steel above 65 t.s.i. tensile strength	40 to 60 ft/min
Cast iron	70 to 75 ft/min
Aluminium alloys	200 to 300 ft/min
Brass	150 to 250 ft/min
Titanium alloys	30 to 40 ft/min

be used for heavier cuts. For carbide-tipped cutters, speeds of approximately three times those for H.S.S. can be used.

Table 7.5 gives the feed per tooth in inches. To calculate the feed per minute multiply the number of cutter teeth by the spindle speed in revolutions per minute.

Table 7.5

	Face mill	End mill
Steel, soft	0.016 in	0.008 in
Steel, medium	0.012 in	0.006 in
Steel, hard	0.008 in	0.004 in
Cast iron	0.014 in	0.007 in
Aluminium alloys	0.022 in	0.011 in
Brass	0.020 in	0.010 in
Titanium alloys	0.005 in	0.003 in

7.6 SHAPING, SLOTTING AND PLANING

7.6.1 Principles and tools

Shaping, slotting and planing are cutting methods for generating flat, and sometimes curved, surfaces by means of one or more single-point tools each having a straight line reciprocating motion relative to the work. Thus a single cycle comprises a cutting stroke and a return stroke. The three cutting methods in question are similar processes using different machine tools.

When shaping or slotting, the work is stationary and the tool moves horizontally on a shaping machine, or vertically on a slotting machine. For planing, however, the table reciprocates horizontally and the tool is stationary.

Because of the reciprocating motion, the cutting speed varies during the stroke, the variation depending upon the type of drive. Consequently, only average cutting speeds can be quoted and optimum tool life conditions are not realized. As the return stroke is non-working it is speeded up by a quick return motion, and to minimize the effect of the tool rubbing on the work during this return stroke the tool holder is designed to pivot away from the work. This is usually achieved by mounting the tool box on a hinge so that it can pivot laterally through about 20°. A solenoid may be used to lift the tool during the return stroke and the device is known as a 'clapper box'.

The tool is usually made of high-speed steel or cemented carbide and its shape and cutting angles are similar to those used for turning. However, it is usually more robust to allow it to withstand the impact loads and intermittent cutting action.

7.6.2 Shaping machines

A standard shaping machine is shown in fig. 7.33 and the following features should be noticed.
1. The work-table can be adjusted vertically to bring the work within reach of the tool.
2. The cross feed of the table occurs during the return stroke of the tool and is either manually or automatically controlled.
3. The ram, which carries a tool holder and clapper box at its forward end, is operated either by a link mechanism or by a hydraulic cylinder. The mechanism is driven by a crank which, in turn, oscillates a slotted lever whose end is connected to the ram.
4. The length of stroke is governed by the radial setting of the crank pin.
5. The cutting speed depends upon the product of the crank speed and the length of stroke. The time ratio of cutting stroke to return stroke is usually between 1.5 and 2.

Although a hydraulic drive is more expensive, it provides a wider range of cutting speeds and better uniformity during the stroke. Also, a quicker return stroke is usually obtained, so reducing the cycle time.

A draw-cut shaper is one which cuts on the stroke towards the column, as opposed to the normal push stroke. It provides a greater stability and rigidity for work requiring heavy cuts and long strokes.

The travelling head shaper is used with large work-pieces or for multiple operations. This machine usually has two shaper rams, each attached to a saddle which traverses along a bed. Work-tables can be attached to the side of the bed or can be mounted directly on the floor.

7.6.3 Slotting machines

A slotting machine is similar to a shaping machine except that the tool ram is ver-

Fig. 7.33 Shaping machine.

tical. The work is carried on a circular table which can move in a rotary or straight line motion in a horizontal plane.

The machine is used to cut keyways, slots, curved surfaces etc.

7.6.4 Planing machines

The planing machine is designed for large and heavy components and it has a much longer stroke than the shaping machine. There are usually two tool holders which are mounted on, and can traverse along, a horizontal cross member. This cross member, which is supported on columns at each end of the bed, provides a horizontal feed across the work and during the cutting stroke the work-table is also driven along the bed, under the tools. Any additional tools required are mounted on the columns, on which they can slide vertically.

Alternative designs of planing machines include the open-side planer, which has only one vertical column, so allowing very wide work to overhang the other end of the table, and the divided or tandem planer, which has a two-part table on the same bed. The two tables can be used separately, one table being loaded while the other is working, or coupled together for large components.

The variable speed and reversing action required to drive the planer table is generally obtained from a motor generator set feeding a d.c. motor, with the final drive through a mechanical reduction gear.

The size of a planing machine is specified by the table width, the maximum height of the tool above the table and the length of stroke. The sizes available range from approximately 2 ft × 2 ft × 4 ft to 16 ft × 16 ft × 60 ft.

7.6.5 Relative merits of shapers, planers and millers

Many of the operations performed on shaping and planing machines, using single-point tools and reciprocating action, could be carried out more rapidly on a milling machine using multipoint tools with continuous cutting action. However, for small or medium sized components numbers, the setting time and tool costs, taking into account initial and regrinding costs, are much less than for a milling machine and result in more economical production.

The shaping machine, with its relative simplicity, has a lower initial cost and, for tool room work or production work on small components, the use of the more elaborate milling machine is not usually justified. At the other end of the scale, the planing machine accommodates larger components and allows longer cutting strokes than a standard production milling machine. Only in the field of plan-milling machines is there any comparative alternative.

7.6.6 Speeds, feeds and depths of cut

Similar cutting speeds are used for both shaping and planing. Typical values using high-speed steel (H.S.S.) cutters are:

30 ft/min for hard steel and cast iron,
60 to 70 ft/min for mild steels,
100 ft/min for brass,
150 ft/min for aluminium alloys.

If the cutters are cemented carbide tipped then speeds of three times those given for H.S.S. cutters can be used, so that for non-ferrous metals the limit is usually imposed by the maximum table speed.

The feed rate can be as high as $\frac{1}{4}$ inch using a broad-nosed tool, and as low as 0.010 inches when using a sharp-pointed tool. With a very large planing machine even higher feed rates can be achieved.

The depth of cut depends upon the quantity of metal to be removed and on the capacity of the machine. Two or three cuts are often required, the final one being only 0.001 to 0.005 in.

7.7 BROACHING

7.7.1 Principles and tools

Broaching is a machining process for obtaining internal or external surfaces of a uniform transverse section by a single pass of a long multipoint tool, called a broach, which is drawn through a hole or longitudinally over a surface.

A selection of broaches is shown in fig. 7.34. Each consists essentially of a straight core with a series of protruding teeth which form the cutting edges. They are usually made of high-speed steel, either in one piece or in sections, but for long production runs and where close tolerances are required inserted teeth of cemented carbide are often used.

BROACHING | 559

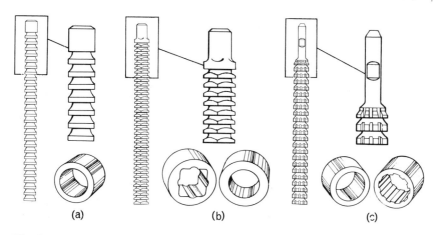

Fig. 7.34 Selection of broaches, (a) Circular hole broach, (b) Square hole broach, (c) Multi-spline broach.

Each section of the work is cut by all the teeth on the broach during each stroke. This is achieved by gradually increasing the dimensions of the teeth or the core, with the smallest teeth or core diameter at the starting end of the broach. The teeth fall into three categories: roughing teeth which remove most of the metal, semi-finishing teeth which complete the shaping, and finishing teeth. Although it is desirable, in order to equalize the load carried by individual teeth, that the relative depth of cut between adjacent roughing teeth should be equal, only the grinding of the finishing teeth is really critical. If the amount of metal to be removed by the finishing teeth is reduced to a minimum, long tool life and low regrinding costs result.

Typical applications of broaching are spline and keyway cutting, rifling, and the machining of non-circular holes. The work-piece is clamped to the machine table and the broach either pulled or pushed past it. The pull method is most frequently employed, as heavier cuts and longer broaches can be used, but for finishing operations and light cuts the push method is preferable, because it can be carried out on a simpler machine with a corresponding reduction in the cycle time.

7.7.2 Broaching machines
These can be classified according to the following features:
1. Horizontal or vertical stroke.
2. Pull or push action.
3. Internal or surface broaching.
4. Hydraulically or mechanically driven.

A horizontal broaching machine (fig. 7.35) works on a pull action, is suitable for either internal or surface broaching and can accommodate long strokes. Furthermore, access to the tool and work is more convenient than with a vertical machine, so making the general handling easier, but a larger floor space is required. The machine is generally hydraulically operated as this is the most suitable method for applying a continuous steady pull of 15 tons or more.

Continuous broaching machines, usually operating horizontally, are used for rapid large-scale production work. The broach is held stationary while the work-pieces, which are held in fixtures, are pulled past the cutters by an endless chain. The loading

Fig. 7.35 Horizontal broaching of keyway.

and unloading is frequently carried out automatically, the capacity being limited only by the loading space on the chain and its speed.

A vertical broaching machine is hydraulically operated and can be designed for internal or external broaching. Those operating on a pull action can cut on either the up or down strokes, but the push action type cuts only on the down stroke.

Electromechanical drive machines have been developed for high-speed surface broaching.

7.7.3 Advantages and disadvantages of broaching

Broaching produces an accurate internal form, tolerances of 0.001 inches being obtainable. This is because the finishing teeth, which are external, can be ground to an exact size. The effective application of a cutting fluid assists in producing a good surface finish.

The process is well suited to long production runs and can be readily automated. The setting-up requires only a simple fixture, while the operation is relatively fast and gives a long tool life.

The disadvantages of broaching are the high initial cost and special nature of the tool, with which only one job can usually be carried out. Furthermore, only through holes can be broached as the tool must pass completely through the work.

7.7.4 Cutting speeds

The maximum speed of most broaching machines is 25 to 30 ft/min. The speed used varies from 3 to 4 ft/min, for high tensile steels, up to 16 to 20 ft/min for mild steels. Cast steel and brass are machined at speeds of up to 25 ft/min, while the maximum speed available is used with aluminium.

High-speed surface broaching machines operate at 50 to 150 ft/min with high-speed steel cutters, and up to 300 ft/min with carbide-tipped cutters.

7.8 GEAR CUTTING

7.8.1 Gears in general

Toothed gears are machine elements which transmit motion and power between rotating shafts. The relative motion is equivalent to that between friction discs which roll together without slipping. The surfaces of these equivalent rolling discs are known as pitch surfaces, and a plane which is transverse to the shaft axis will cut this surface in a 'pitch circle'. The pitch circle is a geometrical property of a gear, and the speed ratio of mating gears is inversely proportional to the ratio of the pitch circle diameters.

The teeth enable an appreciable load to be transmitted without any slip occurring, and the tooth shape must be such that uniform motion is transmitted for all positions of contact between the teeth of a gear pair. It is preferable to use a standard tooth form which can be easily machined and which will allow interchangeability between a set of gears of varying sizes.

The tooth profile which satisfies these requirements and is most widely used in engineering applications is of an involute form. Before machining methods can be discussed some features of gears, shown in fig. 7.36, should be appreciated. An 'involute' can be defined as the path traced out by a point on a cord which is being unwrapped from a circle. For a gear, this circle is known as the 'base circle'. Consequently, in a transverse plane, each side of a tooth is an involute which is formed from opposite sides of the base circle. In operation only one side of a tooth is driving at any time and it can be shown that the point of contact between mating teeth must lie on the common tangent to the base circles. For a pair of external toothed gears, this drive is equivalent to a crossed belt between the base circles.

The size and spacing of the teeth are determined by the radial height, the face width and the pitch. The 'tooth height' is the sum of the dedendum, measured from the root of a tooth to the pitch circle, and the addendum, measured from the pitch circle to the tip of a tooth (fig. 7.36). The 'face width' usually refers to the axial dimension of the gear, though with helical and spiral gears the actual length along the line of the teeth will be greater than this. The 'circular pitch' is the distance measured round the pitch circle between corresponding points on adjacent teeth, but it is usually more convenient to use 'diametral pitch'. This is the ratio of the number of teeth on a gear to its pitch circle diameter. All gears having the same pitch will mesh together.

One feature influencing the tooth shape, and hence the choice of gear cutter, is the pressure angle ψ. This is the angle between the line of action of mating teeth, i.e. the common tangent to the base circles, and the tangent to the pitch circles at their point of contact. The usual value of ψ is either $14\frac{1}{2}°$ or $20°$.

An advantage of choosing involute gears is that the rack form, corresponding to a gear of infinite diameter, has teeth whose sides are straight and inclined at the pressure angle to the centre line. This geometrical property greatly simplifies the manufacture and grinding of the cutters, from which the gears are produced by a generating process.

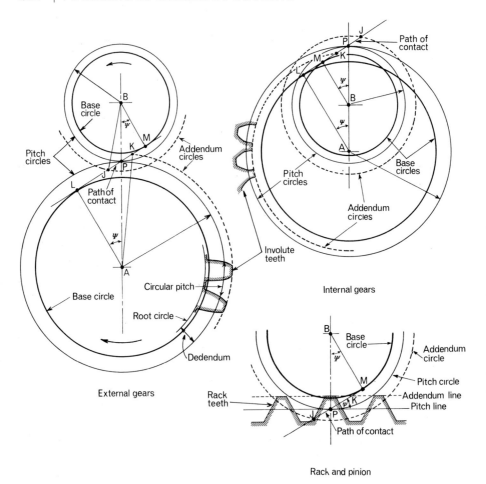

Fig. 7.36 Geometry of involute gear teeth.

7.8.2 Types of gears

Gears can be classified according to the angle between the axes of their shafts when they are put together to transmit motion.

a. Parallel shafts. Straight tooth gears, known as spur gears, are the simplest and most common form, but for heavier loads and smoother running at high speeds helical gears can be used (fig. 7.37). In the latter case, the tooth line is inclined at the helix angle to the shaft axis and, in order to balance out the axial force which is set

Fig. 7.37 Helical gears.

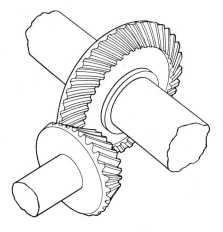

Fig. 7.38 Spiral bevel gears.

up between the teeth, gears for large installations, such as turbine reduction gears, are manufactured in double helical form, i.e. two gears of opposite hand are positioned side by side.

Spur and helical gears can have either external or internal teeth.

b. Intersecting shafts. Bevel gears, with either straight or spiral teeth, are used to transmit the drive between intersecting shafts (fig. 7.38). Their teeth lie on the surface of a cone which forms the pitch surface. A bevel gear having a large cone angle is called a crown wheel.

c. Non-intersecting shafts. Crossed helical gears having unequal helix angles can be used, but a particular form having advantages in gear ratio and load capacity is the worm and wheel pair (fig. 7.39). The sliding velocity between the teeth is high but friction loss and wear can be minimized by using a hardened steel worm and a bronze wheel.

7.8.3 Gear cutting methods

To produce a finished gear the tooth spaces must be cut out of a machined blank, which is in either a disc or cylindrical form. The dimensions of the original blank and the radial depth of cut determine the height and width of the teeth. The number of teeth is related to the pitch of the cutter and the circumference of the gear.

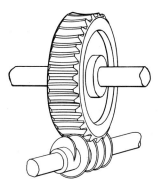

Fig. 7.39 Worm gears.

The following methods can be used.

a. Forming. This term applies to methods where the tool shape corresponds exactly to the tooth space. The tools which can be used are as follows:

Single point. The gear blank is set up on a shaping or planing machine and the tooth spaces are cut one at a time by the reciprocating action of the tool. The gear blank is indexed through one pitch between each completed cut. Spur and helical gears can be formed by this method.

Milling cutters. A quicker cutting process is obtained by using a rotary cutter, though in other respects this method is similar to the single-point method. It is suitable for the cutting of spur, helical and worm gears.

Broaches. These can be suitable for the large-scale production of internal gear teeth.

b. Generating. In these processes the cutter has teeth which resemble those of a mating gear or rack. During the cutting process the blank and the cutter are given motions corresponding to the pitch line speeds of the finished gear meshing with the cutter teeth. All gear types can be cut by a generating process, and the following methods are generally used for quantity production.

Planing. An involute rack has teeth with straight sides and therefore a cutter can be easily made in this form. When such a cutter is given a motion relative to the blank which corresponds to that of a gear engaging with a rack, involute teeth are formed in the blank.

In practice, the gear blank is mounted on the machine about a vertical or horizontal axis, and the cutter reciprocates across the face of the blank (fig. 7.41). Between each cut the blank is given a small rotation and the cutter receives an equivalent tangential movement. When a number of cuts have been taken the total movement will slightly exceed one tooth pitch. The cutter is then traversed back to its starting position and the process continued for the next tooth. Several complete rotations of the gear are necessary to cut the full depth of the teeth.

Spur gears and helical gears having external teeth can be produced by this method.

Shaping. This process is similar to planing except that the cutter is in the form of a pinion, i.e. a small gear. It has either straight teeth, for shaping spur gears, or helical teeth, for shaping helical gears (fig. 7.40).

Although the cutter still has a reciprocating action, comprising a cutting stroke and a return stroke, it does not require resetting after traversing one pitch. The cutting pro-

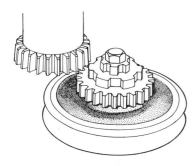

Fig. 7.41 Gear shaping.

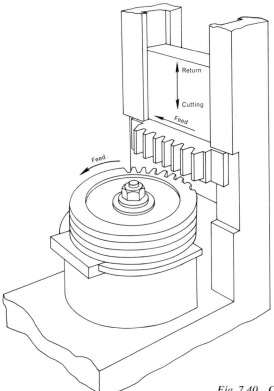

Fig. 7.40 Gear planing using rake cutter.

cess is continuous around the circumference of the pinion cutter, which is fed radially in two or three stages to produce the full depth of tooth.

The shaping process is the only satisfactory generating method for cutting the teeth of internal gears.

Hobbing. A rotating cutter, having teeth of an involute rack form at any diametral section, can be used to generate a gear by a continual cutting action. This type of cutter is called a hob, and it has the appearance of a large screw or worm where the cutting edges have been formed by grinding along axial grooves (flutes) at a number of points around the circumference.

On a hobbing machine, the axis of the hob is inclined at its helix angle to the direction of the tooth spaces to be cut in the blank. It is rotated at the required cutting speed while the gear blank is rotated in the required ratio depending on the pitch, number of teeth and diameter (fig. 7.42). A complete cut is taken by setting the hob to the full depth of tooth and traversing it slowly across the face of the gear.

Spur, helical and worm gears can be produced by hobbing.

Bevel gear generating. There are two processes, one using reciprocating tools and the other rotating tools. The former process is used for straight teeth bevel gears and the latter for spiral tooth gears. In both instances there is a rolling action between the cutter and the work about a line through the apex of the gear, and in this way the whole face of the tooth is formed.

Fig. 7.42 Hobbing cutter producing helical gears.

7.8.4 Comparison of gear cutting methods

Forming processes are comparatively slow and therefore are not used for quantity production. They are also likely to produce an inaccurate tooth shape, as this is derived directly from the cutter shape. A different cutter should be used for each gear size, even though the pitch is the same, but in practice one cutter is usually used for a range of sizes, so producing only an approximate involute form.

With gear generating methods one cutter will reproduce true involute teeth on all gears having the same pitch. The rack-type cutter has a low initial cost and is economical for the production of small to medium quantities. The pinion cutter is similar in operation but can also be used for cutting internal teeth. Double helical gears can be cut by both methods without leaving a gap around the centre of the wheel. For large-scale production the hobbing process is most frequently used, accounting for about two-thirds of all machined gears. As the cutting action is continuous, the cutting force is uniform and the interruptions experienced with reciprocating tools are avoided. This generally eliminates troubles from elasticity and backlash in the machine, and the heat distribution is uniform due to the fairly rapid rotation of the gear.

7.8.5 Gear finishing processes

When a gear is cut by a generating process, such as shaping or hobbing, its profile consists of a series of small flats, due to effects of the feed between successive cuts. A reduction of the feed, while resulting in a closer approximation to a continuous involute, is accompanied by a lengthening of the production time, and it is usually more economical to correct the tooth shape by a finishing process such as shaving.

Gears which are to operate under exacting conditions of heavy loads and at high speeds are generally hardened by heat treatment after cutting. It is then necessary to use a grinding or lapping process to produce the fine tolerances and accurate profile required.

a. Gear shaving. This is a method of finishing gears already cut by forming or generating to eliminate any small residual errors and to improve surface finish.

The cutter used for this process has accurately formed involute teeth with serrated edges. Rotary cutters are more common, although rack-type cutters, which have the advantage of a simple tooth form, can be used for small to medium gears.

The action of the cutter on the gear teeth is similar to hand scraping in that the high spots on the teeth are removed, so bringing them nearer to a true involute form. The cutter and gear are rolled together at fairly high speed and several passes are taken over each tooth, the amount of metal removed rarely exceeding a total of 0.001 in.

To produce the necessary scraping action the axes of the cutter and work are crossed, i.e. set away from their normal meshing positions, at an angle usually within the range 5 to 20°, and one element is driven while the other follows. For larger gears the work is driven, but for smaller sizes either the work or cutter may be driven. A radial or tangential loading must be applied to produce the cutting action. Radial loading is used when it is desired to shave both faces of the teeth at the same time. It is usually a quicker operation than tangential loading, which is applied by means of a band brake and used when only one side of the teeth is to be shaved.

Gear shaving may be carried out on a special purpose machine or on a combined hobbing and shaving machine. The time to shave small to medium size gears can be less than 1 min.

b. Gear grinding. This is a precision finishing process for gears made from hard materials or which have been heat treated. In the latter case it corrects any distortions produced by the hardening of a previously machined gear.

Tooth grinding may be carried out by either a forming or a generating method. As for form milling, the first method employs a grinding wheel whose edge is identically shaped to the tooth space. With the generating method, the cutting edge of the grinding wheel is of a rack tooth shape and one or more cutting edges can be used, depending on the design of machine (fig. 7.43). The gear is rolled in contact with the grinding wheel, as if in contact with a rack.

c. Gear lapping. As an alternative to grinding, and also to produce smooth running between a pair of gears, it is common practice to lap them together after hardening.

568 | WORKING OF METALS BY CUTTING

Fig. 7.43 Grinding a helical gear.

They are run in mesh under load and submerged in an abrasive fluid. A reciprocating axial motion and reversal of rotation are used to ensure that all tooth surfaces are lapped. The time required to lap most gears is approximately 5 min.

7.9 SAWING

Sawing is an economical method, in both power consumption and material wastage, for cutting off from bars, tubes, sheets or other shapes of metal. It can also be used to cut narrow slots, grooves or notches and, by using a band saw, curved profiles can be obtained.

A saw is usually made from high-speed steel. It has a large number of teeth but at any instant only a few of them are actually cutting, each removing a small chip. The removal of heat is assisted by applying a soluble oil coolant, and this enables a high cutting speed and pressure to be maintained.

Five principal methods of power cutting have been developed. Their relative merits and applications are discussed below, together with a description of the machine, where appropriate.

a. Hack saws. A hack saw is a straight steel strip, usually about 2 ft long, which has teeth along one edge which are set to cut in one direction only. Heavy cuts can be made with coarse teeth, but fine teeth are required on thin sections to ensure a continuous contact with the work. The blade thickness varies from 0.05 to 0.01 inches and its width is normally between 1 and 2 inches.

The work is held in a vice and the saw is given a reciprocating motion by the machine, which is similar in appearance to the horizontal band saw of fig. 7.44. A downward pressure is applied on the cutting stroke but during the return stroke the blade is raised clear. Adjustable cutting pressure and blade tensioning are provided by gravity, springs or feed screws, and a quick return motion to raise the blade to its starting position is incorporated to reduce the non-cutting time.

Hack saw machines are cheap, easy to set up and simple to operate, in some instances being automatic. However, due to their intermittent cutting action, they are considerably slower than a continuous band saw machine.

b. Band saws. In this case, the blade is a continuous strip which is driven as a flat belt between two pulleys. Cutting is two to three times faster than with the hack saw and the width of cut is less 1/16 in. The number of teeth varies from 2 to 32 per inch according to the section and the material being cut, while the blade speed varies from 50 ft/min for high tensile steel to 250 ft/min for non-ferrous alloys.

Cutting-off machines are made in a horizontal form (fig. 7.44) for the ease of handling bar and tubular materials. Vertical machines are more versatile, and an automatic feed can be applied hydraulically to the work-table. Additionally, these machines can be used for the contour cutting of internal or external surfaces, such as die blocks or cams. An accuracy of approximately 0.002 inches per inch is obtainable, with a good surface finish, but the best result is obtained by following the main cut with a light finishing cut.

c. Circular saws. A circular saw is a thin disc which resembles a large-diameter milling cutter. It has a large number of teeth, equally spaced around the periphery,

Fig. 7.44 Horizontal bandsaw.

which are alternately set to opposite sides of the disc in order to provide a clearance during cutting. The saw axis mounted horizontally and driven by a variable-speed gear. The work-piece is clamped in position and may be fed into the blade or the blade may be advanced into the work-piece. Feeding may be either by hand or by power. The cutting is fast and accurate results, within a close tolerance and with a good finish, can be obtained.

d. Friction sawing. The principle of friction sawing is that a heavy pressure and high speed are applied to the saw element, thereby causing the material to heat up and soften. In this state the material is easily cut away.

Either a band saw or a circular saw machine can be used, and there is no necessity to maintain sharp teeth on the saw. The cutting speed varies from 10 000 to 20 000 ft/min.

Ferrous alloys are most amenable to friction sawing techniques, and the rate of cutting is mainly dependent on the melting point range and not on the room temperature hardness. It is an excellent method for the cutting of plates and thin-walled tubes, and also for difficult materials such as stainless steel.

Cutting is limited to sections having a small thickness, generally less than 1 inch, so that sufficient heat can be concentrated in the cutting area. Materials, such as aluminium and copper, having a high thermal conductivity are not suitable for cutting by this method.

e. Abrasive wheel sawing. Thin grinding wheels, running at a high speed, can also be used for cutting-off purposes. The most suitable types of wheel are either resinoid bonded or rubber bonded, and the fastest cutting is obtained with a coarse grit. The grade or hardness of a wheel also influences cutting speed as it indicates the strength of the adhesive bond between the particles. For instance, a soft bond allows the particles to be pulled out fairly easily, and ideally the particles should be pulled out as soon as their cutting edges have become dulled to reveal the sharp ones beneath.

The wheels are usually mounted on a swinging frame and manually fed through the work. Their application includes cutting off rolled and extruded sections. They have a reasonable accuracy but some burr is formed, particularly when cutting dry.

7.10 GRINDING, LAPPING AND HONING

7.10.1 *Principles*

When closer tolerances or a finer surface finish than can be achieved by standard cutting processes are called for, it is necessary to use a finishing process such as grinding, lapping or honing. Each of these uses an abrasive material, the first in the form of a wheel, the second as loose particles suspended in a slurry and the third in a stick form. In addition to its application for close tolerance and high quality production, grinding is also universally used for sharpening cutting tools. (Reference 9 applies particularly to this section.)

A satisfactory abrasive material must have the following properties.
1. It must have sufficient hardness to penetrate the material to be cut and to be able to withstand deformation.
2. It must be resistant to damage by fracture under intermittent loading conditions.

3. It must be resistant to wear so that the cutting edge is not rapidly glazed.

As these properties depend jointly upon the abrasive material and the work material, a careful selection of the abrasive material should be made if effective and economical results are to be obtained.

7.10.2 Abrasives used in metal working

a. Aluminium oxide. Although this occurs naturally as emery or corundum, it is not, in this form, sufficiently consistent in grain size and properties for precision work, and is now manufactured in a controlled form from bauxite ore. The resulting grains usually consist of approximately 95% aluminium oxide (Al_2O_3), together with small percentages of titanium oxide, silicon carbide and iron oxide which are also beneficial to the cutting action.

b. Silicon carbide. This material is made from silica sand and coke in an electric furnace. A very high temperature is required to cause the silicon from the sand to combine with the coke. Salt is added to the mix to remove metallic impurities. The result is a mass of silicon carbide crystals of varying size.

c. Diamond. This is a crystalline form of carbon and is the hardest material known. It is used in a grit form, usually for grinding or lapping extremely hard metals, such as tungsten carbide. Small diamonds, which are suitable for grinding purposes, can be synthesized from pure carbon under a very high pressure and temperature.

Abrasive materials are graded according to their hardness, toughness and suitability for different materials. In general, aluminium oxide is most suitable for use with ferrous alloys and hardened steels, while silicon carbide is better used on non-ferrous alloys and brittle materials, such as cast iron.

Abrasives are also classified according to their grain size or grit. This can range from very coarse, screened through a mesh of 4 to the inch, to a fine grit screened at 240 to the inch. Even finer abrasive powders, known as flours, are separated by a flotation process.

7.10.3 Grinding wheels

A grinding wheel consists of abrasive grains held together in a matrix, called the bond, so that the grains form multiple cutting edges on the surface of the wheel. The properties of the wheel depend on the choice of bond and the grit size.

Vitreous bond is a hard, porous, glass-like material which is suitable for most types of grinding wheel. Its strength and stability allow heavy cuts to be made, while a good finish is obtained with a light finishing cut. There is, however, a limitation on the size of wheel that can be made with a vitreous bond, and for larger wheels a silicate bond is used. This is a softer material which gives a cooler cut, so making it suitable for tool and cutter grinding.

For purposes such as cutting-off, when a thin flexible wheel is required, resinoid or rubber bonding is used. A close dimensional accuracy and a high quality surface finish can be obtained with such wheels. Shellac-bonded wheels can also be used for cylindrical grinding where a fine surface finish is required.

The cutting speed is measured at the surface of the wheel and ranges from 4000 to 6000 ft/min, for the vitreous and silicate wheel bonds, to 9000 to 15000 ft/min for the others. It is generally essential to work with a copious supply of coolant in order to maintain the dimensional accuracy of the work-piece and to prevent surface cracking when grinding hard surfaces and brittle materials. The coolant helps to keep down dust and to reduce the fire hazard when grinding magnesium and titanium.

The 'grade' of a wheel indicates the adhesion between the bond and the abrasive particles. A hard grade has strong adhesive properties, while for a soft grade the adhesion is much weaker. This adhesion depends not only on the bond material and the grit size but also on the proportion of bonding to abrasive material and on the grain spacing. The spacing, which is referred to as the structure, can be dense or open, depending upon the porosity of the wheel. This porosity is achieved by leaving voids between the grains and the bond, so providing passages for chip flow and coolant penetration.

The grade of wheel to be used should be chosen to suit the work material and the cutting conditions. It should be sufficiently hard to retain the abrasive grains until their useful life is exhausted, but not to retain them after they have become glazed. A glazed wheel will cause rubbing and so lead to a rapid increase in both cutting force and temperature, with the possibility of burning the work-piece.

To obtain a good dimensional accuracy and a chatter-free surface it is essential that the grinding wheel is balanced. This also avoids the setting up of fluctuating loads on the machine bearings when the wheel is running at a high speed. The wheel and spindle can be statically balanced by the adjustment of weights, which are set in the holding collet.

7.10.4 Grinding machines

Grinding machines can be classified into two main groups, according to whether the surface to be ground is cylindrical or flat. Special machines are also available for specific operations such as thread grinding and gear grinding (see gear cutting methods, 6.8.3).

a. Centre-type cylindrical grinders. The centre-type grinder is one in which a cylindrical work-piece is rotated about its axis by means of a rotating face plate, a chuck or a pair of centres. The grinding of straight and tapered cylinders, shoulders and faces, and cylindrical form work can be carried out with these machines.

External grinder. The work is usually held between centres and rotated at a surface speed of 60 to 100 ft/min. The grinding wheel is mounted on a separate spindle and driven independently at high speeds about a parallel axis.

The depth of cut is determined by the radial movement of the grinding wheel towards the work axis. This is called infeed and varies from 0.002 inches for roughing cuts to 0.0002 inches for finishing cuts. An axial traverse is obtained by moving either the wheel or the work-table. This can be effected by manual, mechanical or hydraulic means, but the last-mentioned is preferable because it produces stepless speed variation and a smooth vibration-free action.

Long work-pieces are ground by axial traverse. The traverse for each rotation of the work is approximately half the width of the wheel but is less for a finishing cut. Each cut is set to the chosen infeed and the length of the work traversed. Thus, the total grinding allowance, which may be 0.010 to 0.020 inches is removed by a series of cuts, a high precision finish being obtained with a light final cut and a slow traverse.

When the section to be ground has a width less than that of the wheel it can be cut by plunge grinding, i.e. entirely by infeed motion.

Internal grinder. With the most common type of internal grinding machine the work is held on a face plate or chuck and rotated at a peripheral speed similar to that used for external grinding. As it is important to avoid any distortion of the work-piece, special holding jigs are often used. The spindle of the grinding wheel is provided with a radial infeed, to set the depth of cut, and with an axial traverse.

The arc of contact between the wheel and the work is much longer with internal grinding than with external grinding and a softer wheel should therefore be used. The width of the wheel is not usually greater than 1 inch and the traverse must be such that there is an overlap between cuts. Because the wheel must have a diameter smaller than that of the hole to be ground, there is a danger of bending occurring in the wheel spindle. This presents a more serious problem as the ratio of length to diameter increases and necessitates very light finishing cuts.

To grind a large bore or an awkwardly shaped work-piece it is often more convenient to hold the work stationary. Thus, the wheel spindle must have a circular motion and the machine is then known as a planetary grinder. The axial traverse required is usually provided by movement of the work-table.

b. Centreless cylindrical grinders. As implied by the name, this machine does not have centres to support the work, although an end stop may be present for this purpose. The work revolves between the grinding wheel and a control or regulating wheel, both of which have axes parallel to that of the work. The height of the work is governed by the position of the work-rest, as shown in fig. 7.45. Pressure is applied between the grinding wheel and the control wheel, both of which can have an infeed control, depending upon the machine design. The control wheel is generally a rubber-bonded wheel of fine grit and it is rotated at the required work speed.

Long components of a uniform diameter are ground by the through-feed method, the axial traverse being provided by slightly inclining the spindle of the control wheel. Components having various diameters, tapers, shoulders and contour lines must, however, be ground by an infeed method. If stepped, contoured or ganged wheels are used, more than one diameter can be machined at the same time.

To grind conical forms an end feed is employed. The grinding and control wheels are set to the required angles and the work is fed axially into the gap between them.

A centreless machine can also be used for internal grinding. The outer surface is ground first, and a very accurate concentricity can be obtained by this method.

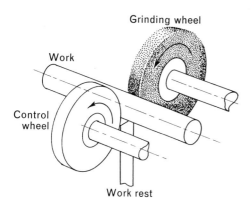

Fig. 7.45 External centreless grinding.

Centreless grinding is most suitable for relatively simple shapes and on long production runs. The longer setting-up time required makes this machine less economical for smaller batches than a centre-type machine. There are, however, several advantages obtained from centreless grinding. These are:
1. The loading and unloading times are short while the cutting time is practically continuous, particularly when using through-feed grinding.
2. The work support is good, so enabling a heavy cut to be made without deflection occurring.
3. The machine can be operated by unskilled operators so reducing labour costs. If the work is very simple it may be possible for one operator to be in control of more than one machine.

c. Horizontal spindle-type surface grinder

Reciprocating table. This has a disc-type wheel which is mounted on a fixed horizontal spindle. The work, which is clamped to a table, traverses under the wheel rim by means of a reciprocating action. The depth of cut is set by radial infeed, and by cross feeding the table a large surface can be ground. The work is often held by magnetic chucks and a wide range of shape and size of component can be accommodated.

Rotating table. This produces a grinding action in concentric circles, which may be desirable for circular components such as pipe flanges.

d. Vertical spindle-type surface grinder. The wheel is cup-shaped, cylindrical or segmental, and the grinding takes place on the lower side of the wheel and in a horizontal plane, as shown in fig. 7.46. A large surface area is cut and the rate of metal removal is much greater than with a disc wheel.

The table again has either a reciprocating or rotating motion, but the latter is most suitable for large-scale continuous production of small or medium sized components. Loading and unloading of the work can be carried out without stopping the machine.

e. Disc-type surface grinder. This is used for rough grinding work, often on

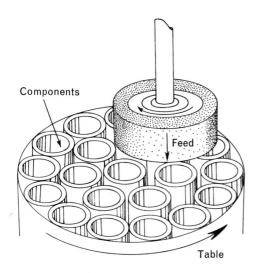

Fig. 7.46 Surface grinding with vertical spindle wheel.

newly cast or forged components. The work is usually held by hand so that a machined surface, which may be required for location purposes, is obtained quickly.

The wheel is generally fairly large and can be mounted either horizontally or vertically. The machine has no work-table and the cutting pressure is applied manually, or by the weight of the component in the case of a vertical machine.

f. Thread grinders. Manufacturing screw threads with thread grinders is advantageous if close tolerances and a fine finish are required or if the thread is to be ground after hardening. For production work grinding compares favourably with other screw cutting methods.

A formed wheel having either a single or multipoint cutting edge is used. A centreless machine is suitable for external threads but a centre-type machine can be used for both internal and external threads.

7.10.5 Lapping

Lapping is a finishing process to correct surface imperfections and to provide the required finish between mating parts. Only a small amount of metal is removed and this is carried out either by hand or machine, using a low rubbing speed and a light cutting pressure. Examples of the application of lapping are valve seatings, metalforming dies and inspection gauge blocks.

The process consists of a random rubbing action between a lap and the work, cutting being due to abrasive particles held in suspension by a fluid known as the vehicle. The lap is a stick, or block, which is made of wood, cast iron, steel or a non-ferrous metal. It is essential that the lap material is soft enough to allow the abrasive to embed itself in the surface, where it is held, to provide a cutting action.

The vehicle which carries the abrasive powder can be water, oil or grease. It is also desirable to have a uniform suspension which can be easily removed from the lapped surface.

Aluminium oxide, silicon carbide and diamond powders can all be used for lapping compounds, but it is usual to confine the use of the latter to work on cemented carbides. The grit size is usually in the range 200 to 500 mesh, but a finer size can be used, according to the finish required. The amount of metal removed does not usually exceed 0.001 inches and for hand lapping may be as little as 0.0001 inches. Lapping machines usually have a simple rotary action and have either a vertical spindle and a horizontal disc lap or are of a centreless type, as in grinding.

7.10.6 Honing

This is a finishing process which is generally applied to internal surfaces such as cylinder bores, bearings and ring gauges. A honing tool is of a cylindrical form, with sticks of abrasive set axially into its outer surface. This tool is inserted into the work and rotated, usually by a machine. The cutting pressure is applied to the abrasive sticks by mechanical or hydraulic means, embodied within the tool. The cutting is continued until the required dimensions and finish are obtained.

7.10.7 Surface finish and applications

The surface finish obtained by grinding varies from about 50 μin when rough grinding to 10 μin when fine grinding. These standards are suitable for components such as plain bearings, gear teeth, keyways, splines etc. but if a better finish is required the surface must be lapped or honed. Cylinder bores are finished by honing, while lapping is used on gauge faces, ball bearings and valve seats. The best commercial finish attainable is a highly polished mirror finish of about 2 μin.

7.11 NUMERICALLY CONTROLLED MACHINE TOOLS

A system where the data relating to the size and shape of a component is transferred as a series of instructions and numerical values to a machine, which then performs the cutting operations automatically, is known as numerical control. It can also be referred to as programme control or automatic control but should not be confused with automation, which implies a production line for processing large quantities of the same or similar components.

Numerical control is most appropriate for the production of small to medium sized batches or when the demand for a particular component is intermittent and recurring. To develop fully the potential of a control system it is necessary that the machine tool has a rigid frame, that backlash is eliminated from moving parts and that friction in the slideways is reduced to a minimum. Although this normally entails a specially designed or modified machine, the additional cost is compensated for by the greater accuracy and repeatability. Thus, such a machine may prove to be suitable for the manufacture of components that would otherwise have to be produced on a special machine, such as a jig borer. The use of a numerically controlled system also offers several quite considerable economic advantages, and provided the system is suitably applied its additional cost could well be recovered within an economical period.

7.11.1 Advantages of numerical control

There are many advantages, of varying significance, of a numerically controlled system and these are divided here into three main groups, depending upon the production stage where they occur.

a. Pre-production stage. The production and cost of jigs and fixtures is often eliminated or greatly reduced as only simple holding devices, which can often be standardized, are required. Complex jigs are replaced by a data storage unit, using either punched cards or tape. Thus, an indirect saving is made in storage space, again reducing costs, while any design modification, at prototype stage or to meet customer requirements, can be easily and quickly incorporated.

The accumulated effect of these features is that the lead time, from specification to commencement of manufacture, is reduced and this, in turn, can be the vital difference between gaining or losing an order.

b. Production stage. The marking-out stage is eliminated and setting time is considerably reduced as it is only necessary to clamp the work-piece at a known position on the table. Tool setting can be carried out away from the machine and tool changes are either manual or automatic. The depth of cut, feed rate, speed, tool selection etc. are controlled by the programme and not by an operator.

These features produce a greater cutting efficiency, because optimum conditions are always used, and better machine utilization. A large quantity output of a high quality is achieved during a shift, while operator fatigue is considerably reduced. Furthermore, as

NUMERICALLY CONTROLLED MACHINE TOOLS | 577

any dimensional variation arises almost totally from tool wear, the inspection required with conventional machining is no longer necessary.

c. Assembly stage. The consistency of the dimensions of components means that selective assembly is avoided. This reduces assembly time and virtually eliminates scrap.

7.11.2 Essential features of numerical control systems

a. Design data. These are usually obtained from a drawing where the dimensions are shown in coordinate form from a convenient origin outside the component. Tolerances are not shown as they are beyond the control of the programmer, but a typical accuracy obtained is 0.001 inches or better.

b. Data processing. The information on a drawing must be converted into machine instructions to set in motion the table traverse, tool feed and speed etc. At this stage the planner, or programmer, has to decide the sequence of operations and the choice of tools.

For simple positioning or sequencing systems, such as drilling or turret lathe instructions, the information can often be fed directly into the machine console, via a plug-board or a set of decade switches. However, for more complicated programmes, or when it is desired to store the instructions for future production runs, punched cards, paper tape or magnetic tape are used. For point-to-point systems (see later), this data conversion can be carried out manually by cutting the tape on a desk machine which is similar to a typewriter. However, when complex contouring operations are involved a digital computer is usually used because of the large number of positions which must be calculated for a curved path.

c. Control unit. The function of this console (fig. 7.47), which is usually positioned close to the machine, is to interpret the information supplied to it manually, or by paper tape etc., into machine instructions. Pneumatic, electric and electronic means are used by different manufacturers, but in all cases a signal is transmitted to the machine via a servo-amplifier to produce the power to operate the machine table, tools etc.

d. Machine movements. Although these can be operated by mechanical or hydraulic means, the latter is most suitable for controlling the table traverse because of its stiffness, low inertia and absence of backlash.

e. Feedback. In more sophisticated systems, the table position is automatically measured at each stage of the cutting cycle and compared with the instructing signal. Any difference is fed back as an error signal and this actuates the table movement until the error is reduced to zero. Thus, the maximum positional accuracy is obtained and a repeatability error of as little as 0.0001 inches can be obtained.

7.11.3 Types of systems

a. Sequential control. This system, usually using plug-board or push-button methods, is the simplest available. The operational sequence, such as table or saddle

578 | WORKING OF METALS BY CUTTING

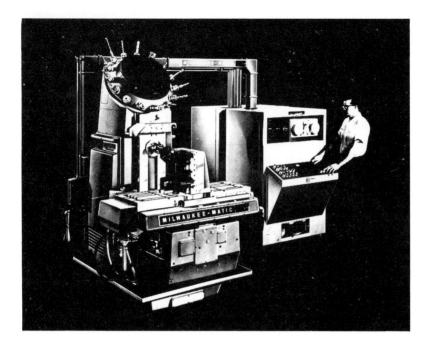

Fig. 7.47 Numerically controlled machining centre. Note control console with tape input, tool input, tool storage magazine and work location.

traverse, tool feed and speed, turret rotation etc., is selected on a control panel. The tool travel is controlled by trip-dogs which operate microswitches.

This method of control is suitable for automatic lathes, turret machines and cycle milling when medium to large quantities are required.

The system can generally be applied to existing machines with the minimum of modification.

The saving of pre-production work and setting-up time is not as great as with a fully programmed system, and the information cannot be stored for future use.

b. Positioning systems. The required machine operation together with the appropriate tool are selected by instructions from the control tape. The tool must be accurately positioned relative to the work, and this normally necessitates controlled movements along two axes, while the third axis is used to provide a power feed for the length of cut.

Positioning systems are particularly suitable for controlling drilling, milling and allied operations, where rotating tools, which cut in straight paths parallel to an axis of the machine, are used. For example, when drilling several holes in a component a considerable saving is achieved in programming, as compared with either jig drilling or hand-setting.

One of the most versatile machines using a positional control system is shown in fig. 7.47. The range of tools that can be held in the rotary magazine includes drills, reamers, taps and milling cutters. The tape provides the instructions for tool changes, which are automatically effected by the swinging arm below the turret. Thus, only one spindle is in use so enabling replacement tools to be prepared while work is in progress.

c. Contouring systems. For the fully automatic machining of curved surfaces and contours, such as cams, turbine blades and dies, a continuous path control must be employed. Except for simple geometric curves, this necessitates the use of a computer-prepared programme and therefore, with the additional complexity of the control system, is very expensive. Thus, contouring systems can only be justified for high quality work in a variety of small batches. To assist users, and to save them the expense of purchasing a machine tool controlled by a digital computer, a programming service is usually provided by the manufacturers.

7.12 LITERATURE

1. ROLT, L. T. C. *Tools for the Job*. London, Batsford, 1965.
2. SCIENCE MUSEUM. *Illustrated Catalogue of Machine Tools*. London, H.M.S.O.
3. WOODBURY, R. S. *History of the Lathe*. Cleveland, Ohio, Society for the History of Technology, 1961.
4. STEEDS, W. *Metal-cutting Machine Tools*. London, I. Mech. E., 1964.
5. MERCHANT, M. E. 'Mechanics of the metal cutting process' in *J. Appl. Phys.*, **16** (1945) 267-75 and 318-24.
6. ERNST, H., and MERCHANT, M. E. *Chip Formation, Friction and Finish*. Cincinnati, Ohio, The Cincinnati Milling Machine Co., 1940.
7. TOURRET, R. *The Performance of Metal Cutting Tools*. London, Butterworths, 1958.
8. SCHLESINGER, G. 'How to measure Machinability' in *American Machinist*, (1947).
9. HOUGHTON, P. S. *Grinding Wheels and Machines*. London, Spon, 1963.
10. *Machinery's Handbook*. New York, Industrial Press.
11. *Machine Tool Specification Manual*. London, Maclean-Hunter, 1963/64.
12. *Technology of Engineering Manufacture*. London, I. Mech. E., 1958.
13. *Tool Engineering Handbook*, A.S.T.M.E. London, McGraw Hill, 1959.

CHAPTER 8

Joining of Metals

8.1 CLASSIFICATION OF JOINING METHODS

Since the early discovery of metals, problems of joining them have occupied the mind of man to an ever increasing extent. Until the present century, available methods were few, but now such a multiplicity of techniques and processes exists as to justify the formation of several technical institutions concerned solely with the development and exploitation of methods for joining metals.

Such methods can conveniently be grouped under three main headings:

1. Mechanical.
2. Metallurgical.
3. Adhesive or non-metallurgical.

Of these groups, the second is the largest and technically the most complex.

8.1.1 *Mechanical methods*

Possibly the oldest form of mechanical joining involved the use of rivets. Many examples of ancient riveted work in iron and bronze are to be seen in our museums and, indeed, it was not until the development of screw cutting machinery in the early part of the industrial revolution that nuts and bolts began to play a significant part in ousting the generally used rivet. Many examples of recent large-scale riveted work spring readily to mind. Riveting of ships has been replaced by welding only in the last two or three decades, whilst welded bridges have an even shorter history. A rivet consists of a shank passing through the parts to be joined and heads at either end formed by plastic deformation. Riveting may be done either hot or cold, depending upon the size of the rivet and the amount of force available for closure of the head. In the case of large work, it is usual to heat the rivets in order to make the formation of the heads more easy to carry out.

The production of agricultural implements, cutting tools, weapons and household utensils in early times by forging is well known. In the case of wrought iron, these forging operations generally resulted in a form of metallurgical bond between the components being joined. In many cases such bonding was neither particularly efficient nor strictly necessary; the metal worker's art was such as to ensure a mechanically reliable joint, even in the absence of any metallurgical bonding. This would be done by forging in such a way that effective mechanical keying of the parts was achieved.

Bronzes were often used in cast form – a notable use in medieval times being for church bells. Indeed, church bells and cannon constituted the acme of the founders' art in terms of weight, soundness and precision of individual castings for many centuries. For correct functioning of church bells, it was necessary to fix a hinged clapper inside the crown of the bell, and in many early instances this was achieved by incorporating an iron clapper staple in the mould during the casting operation, and thus the bell was cast round the staple. This practice of joining by 'casting-in' components is not nowadays favoured in view of the difficulty of ensuring good contact and freedom from blow holes in the casting which might result from the generation of gas from contaminants on the surface of the cast-in component.

Nuts, bolts, screws, split pins and self-tapping screws are well-known means for joining metallic and non-metallic pieces. Screws, however, are fairly costly items to make and their fitting is also time consuming. The search for improved and cheaper methods of manufacture has quite naturally led to the exploitation of simpler mechanical joining devices and also of welding. Spring clip designs of all forms abound and find widespread application in the car manufacturing industry. One common device consists of a relatively soft pin used in conjunction with a pierced spring steel washer having a hole slightly less in diameter than that of the pin. This hole is so arranged that it is easier to push the pin through one way than the other, and thus when once pushed into place the washer bites into the pin and prevents the assembly from loosening. (For a further description of mechanical joining methods, see section 8.2.)

8.1.2 Metallurgical methods

Table 8.1 shows the definition of metallurgical joining processes as given in *B.S.S. 499 Part 1*, and indicates their relationship.

A distinction must here be drawn between welding and other forms of metallurgical joining. The definition of a weld given in *B.S.S. 499* is as follows:

'A union between pieces of metal at faces rendered plastic or liquid by heat or by pressure, or by both. A filler metal whose melting temperature is of the same order as that of the parent material may or may not be used.'

The definition of brazing is:

'A process of joining metals in which, during or after heating, molten filler metal is drawn by capillary attraction into the space between closely adjacent surfaces of the parts to be joined. In general the melting point of the filler is above 500°C but always below the melting temperature of the parent metal.'

Soldering or soft-soldering may be described as brazing in which the filler metal melts at a temperature lower than 500°C and is generally a lead or tin based alloy.

It will thus be seen that welding involves heating some part of the material to be joined to a temperature in excess of its melting point. Surface bonding techniques such as soldering and brazing, pressure welding, forge welding and so on do not involve any melting of the material being joined. In all cases, however, whether or not fusion is involved, the processes falling into this section of the classification involve intimate metallurgical contact between the components being joined, which means that non-metallic surface films such as oxides are not present at the interface. (For a detailed treatment of metallurgical joining methods, see section 8.5.)

Table 8.1

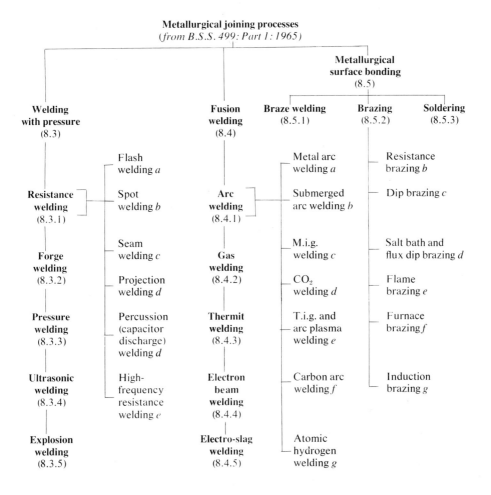

a. Welding methods involving pressure. Although all the techniques in this section involve the use of applied pressure for the production of satisfactory joints, the processes can be considered in three main groups. The largest group is that associated with electrical resistance heating and goes under the general description of resistance welding. In this process and its variants, the passage of an electric current through the components to be joined, and across their interface, results in sufficient local heat for the material to be raised above its melting point. External pressure is applied, usually by means of the copper electrodes carrying the heating current, in order to contain the 'nugget' of molten metal produced by the resistance heating. When the heating current is interrupted the nugget freezes and a metallurgical joint results. The many variants listed in table 8.1 do not differ in principle from this basic outline.

The second group of processes falls under the heading of forge welding. The principle of this method lies in bringing the two components – which are usually heated – into close contact, and then the contact surfaces are substantially increased in area by hammering or rolling whilst, by virtue of their close contact, the atmosphere is effectively excluded. Freshly produced, uncontaminated, contacting surfaces result and a sound metallurgical bond is made. Explosive welding can be considered as an extension of forge welding, resulting as it does in an increased contact area, rapidly produced and within a sufficiently short time for no atmospheric oxidation to take place.

The group of other pressure processes includes hot and cold pressure welding as such, together with ultrasonic and friction welding.

b. Fusion welding. As the name implies, this group of processes depends upon the melting of both parent metal and, in most cases, added filler material in the area of the joint. A source of heat is necessary in order to raise the temperature at the joint sufficiently high for fusion to take place and this source of heat may either be electrical or chemical.

A great deal of welding used to be carried out by means of the oxyacetylene process and this process is still used, particularly for thin sheet work and in the smaller workshop where more expensive electrical welding equipment is not readily available. Oxyacetylene or gas welding has now largely been superseded by one or other of the electric arc processes. In gas welding the source of heat arises from chemical combustion of a gas such as acetylene or hydrogen in the presence of oxygen. Adjustment of the relative amounts of oxygen and combustible gas flowing through the torch allows an oxidizing, neutral or a reducing flame to be produced. These flame conditions are important to the welding of specific materials and their use depends upon the type of metallurgical reaction which is likely to arise between the molten metal and the products of combustion in the flame.

Heat from a chemical source is also employed in thermit welding. This is a relatively old process and the heat arises from reaction of powdered aluminium with iron oxide to form iron and aluminium oxide. The reaction is started by the ignition of a highly oxidizing mixture and the whole process goes off rather like a firework.

All metals with the exception of the noble ones, such as gold, platinum and silver, oxidize when heated in air. The presence of surface oxides seriously interferes with any attempt at metallurgical joining, whether by fusion welding, pressure welding or surface bonding. Such oxides must therefore be removed and prevented from reforming if fusion between the components is to be achieved. In gas welding this is done by chemically dissolving the oxides by means of a suitable flux applied to the joint immediately before heating with the welding blowpipe. The application and maintenance of sufficient flux to keep the metal clean and oxide-free is thus an essential part of gas welding technique.

The main group of fusion welding processes derives heat from an electric arc, which is normally struck between an electrode and the work-piece itself but this is not invariably so. In some cases, such as in atomic hydrogen welding and in one form of the very new arc plasma welding process, the arc is struck between two non-consumable electrodes. Here, the work-piece does not form any part of the welding current circuit. The arc temperature is very much higher than that of the chemical flames used in gas welding, and thus it is even more important to ensure that the arc and surrounding molten metal is well protected from contamination by the air.

The most widely encountered arc process is that in which coated electrodes are used. In this case, the electrode is fixed into a holder to which is attached a heavy cable carrying the welding current from a suitable generator. The other lead from the generator is clipped to the work-piece and an arc is struck between the work-piece and the electrode, which is subsequently consumed. The electrode has a baked-on flux coating which serves a number of purposes, including that of providing a liquid slag film over the molten weld pool which is capable of dissolving any oxides formed, thus keeping the metal surface clean. There are a number of variants of this process, such as submerged arc and automatic open arc welding, which are fundamentally similar and which can all be grouped as flux arc processes.

The other group is comprised of the non-flux arc processes and here protection from atmospheric contamination is assured by the provision of an inert or chemically suitable atmosphere around the arc zone and over the molten weld pool. These are generally known as the inert gas shielded processes, although in some instances, such as in carbon dioxide welding, the gas shroud is not strictly inert.

Other forms of welding involving heating by means of electrical energy include electron beam, induction and laser welding. In the last-mentioned case, electrical energy is converted into short pulses of intensely bright light and the system is suitable for welding very thin sections on a micro scale. Electron beam welding works in the same way as a cathode-ray tube in a television set, except that the power involved is very much greater. In a cathode-ray tube a beam of electrons is caused to move rapidly over the whole screen, whereas in welding applications a higher current beam is focused onto a single spot or line. Impact of the fast moving electrons on the work-piece causes a sharp rise in temperature and adjustment of the beam current may allow sufficient heating for fusion welding to take place. This process is characterized by very deep penetration welds and has a number of specific applications which will be described later. Induction welding utilizes the heating effect of high-frequency alternating electric currents induced in the work-piece by the proximity of current-carrying coils. There is no actual contact between the generator and the work-piece. It is a very rapid process and is usually confined to small components of thin section or, in one specific application, to the production of longitudinally welded tube.

c. Metallurgical surface bonding. If soldering is considered as a special form of brazing, there is one other method of surface bonding which may be considered distinct from brazing and that is bronze welding. In this method any low melting point filler metal, not necessarily bronze, may be used, the surface of the metal to be joined being heated by one means or another to a temperature intermediate between the melting point of the filler material and that of the metal to be joined. The filler metal is then applied to the joint as in welding and the method is distinctly different from brazing which depends upon capillary attraction of the filler metal between the parts to be joined (see bronze welding, 8.5.1, and brazing, 8.5.2).

Diffusion bonding can be considered as a very special form of pressure welding. In this method, cleaned components are heated under slight pressure in a vacuum and a metallurgical bond results from the fact that atoms from one component tend to migrate across the interface into the other component at points where intimate mechanical contact exists. This migration occurs where different quantities of any particular atom exist on either side of the interface and the method therefore works best for the joining of dissimilar materials.

8.1.3 Non-metallurgical adhesive methods

Under the stimulus of necessity in world war II, close attention was paid to methods of producing light-weight yet stiff composite sections in both metals and non-metals. The gluing of wood was first improved by the development of stronger adhesives of the resin type, which set by chemical action following the addition of a hardening gel or solution. This development marked a significant advance over the use of air-hardening glues or evaporative glue solvents.

Wood cannot, in general, be formed or shaped so easily as metal and therefore, as aluminium is both light-weight and easily formed, the development of suitable adhesives for metals became an obvious necessity for effective exploitation of stiff light-weight structures of the honeycomb type. Here again, the impetus came from the aircraft industry.

Glued joints in metals are subject to some limitation in that the mechanical properties of the adhesive seldom approach those of the parent metal. It is necessary, therefore, to design joints in such a manner that direct tensile or cleavage stresses on the adhesive are avoided. Best results are obtained when the adhesive is in shear or compression. Surface preparation is of critical importance if reliable and effective bonding is to be achieved. Service behaviour is often affected by environmental humidity, and temperature variations are particularly important as the thermal expansion characteristics of parent metal and adhesive often differ widely, thus leading to significant thermal stress.

A detailed treatment of the theoretical and practical aspects of adhesive bonding – which may in some senses be considered as a lower temperature extension of brazing and soldering – is given later (see section 8.7).

8.2 MECHANICAL METHODS OF JOINING

Although riveting, bolting and mechanical fastening may seem, in the light of present welding technology, to be old-fashioned and inconsequential means for joining metals, it should be remembered that those who travel risk life upon the efficiency of some mechanical joining device. Despite the fact that welding plays a great part in the construction of jet engines, all civil air frames are, in fact, riveted. Again, major points in car steering and suspension assemblies are joined by means of nuts and bolts. Thus more than passing thought must be given to these old, yet still firmly entrenched, mechanical joining methods.

8.2.1 *Bolting*

The behaviour of mechanically joined assemblies, subject only to static loads, is quite calculable and may be predicted with certainty. Few structures, however, are subject solely to static loading and the designer must in many cases give serious thought to the effects of vibrating loads which might result in structural fatigue. In general, fatigue cracks are likely to start at those points in the structure where some discontinuity exists; notches left by bad machining, bolt holes, or sharp corners forming part of a design detail are typical sites where fatigue cracks may start under vibrating load conditions. Having said this, it follows that most screw-thread forms are likely sources of fatigue trouble, unless careful attention is paid to detailed loading conditions in the design of the structure as a whole. In view of the fact that most aircraft materials have to be strong in order to reduce the weight of the components to a minimum, the ultra-high-strength steels and high-strength aluminium alloys have received a great deal of attention. These materials are not readily welded as the heat of welding disturbs their metallurgical structure and seriously weakens them. Thus much research into the behaviour of mechanical joints has been carried out by the aircraft industry and as aircraft are invariably subject to vibration and fluctuating loads, design against fatigue in aircraft has been very highly developed.

It is usual that the development of an advanced technology benefits industry as a whole and the results of research in the aircraft field are now being applied in other sectors of the transport industry with obviously beneficial results. High-tensile steel bolts are now produced with generous fillets merging the head to the shank, and waisted studs having reduced stress levels in the threads, which are themselves formed with radiused roots to improve fatigue life, are now commonplace.

The increasing use of automation in the production of motor cars has led to some simplification of conventional bolted assemblies. One such simplification involves the use of integral flanges formed under the bolt heads and also on the nut faces, thereby spreading the contact load and eliminating the need for separate washers. Self-locking nuts are widely used and avoid the need for separate spring washers, the object of which is to maintain shank tension should the nut become loosened by a fraction of a turn. Self-locking nuts have a nylon or other plastic insert trapped within one part of the threaded zone and, in use, this insert is compressed into the bolt threads and by its binding action inhibits loosening.

The problems of assembling nuts and bolts by means of hand spanner tightening are not great. However, bolts above about three-quarters of an inch in diameter require greater force than can consistently be applied by hand. It is usual under these conditions to specify tightening to a particular torque level. The designer specifies the torque level which will guarantee the bolt shank tension that he requires for adequate tightening purposes. The difficulty of this method of shank tension control lies in the fact that friction characteristics between nut and bolt tend to vary according to the actual thread shape and degree of misfit. To take an extreme example, it will be appreciated that a very high torque level would be necessary to screw up a rusty nut and bolt assembly. Under these circumstances the specification of a particular torque level would not guarantee adequate tension in the shank, and as a consequence the assembly would be loose. Under vibrating load conditions this could well lead to premature fatigue failure. In these cases, it is usual to carry out detailed inspection of the machine threads and to use some friction-reducing lubricant such as graphite or molybdenum disulphide during assembly.

An unfortunate property of stainless steel is its tendency toward seizure between surfaces, or galling as it is commonly termed. This characteristic is also seen in titanium and aluminium assemblies. In these cases the use of anti-friction compounds in tightening threaded assemblies is essential. A further point to note when tightening threaded assemblies in low-strength materials such as aluminium and brass, for example, is the highly likely risk of over-tightening and thread stripping, due to the low shear strength of these materials. It is even possible to shear $\frac{1}{4}$ inch diameter mild-steel bolts by hand tightening alone.

The problem of nut loosening may well be solved in future by the use of adhesive locking. Again as a result of research in the aircraft industry, a number of entirely reliable non-air-curing resins have been developed. It is possible to apply the resin and hardener to threaded components during assembly and, after an appropriate reaction period during which curing takes place, the assembly is tightly locked and becomes highly resistant to shock and vibration.

8.2.2 Riveting

Riveting is still widely used both for thin sheet work and for heavy sections, although in the latter field–particularly where high stress levels are encountered– welding is usually preferred. Perhaps the principal factor inhibiting the use of welding for thin sections is the inevitably large degree of distortion which accompanies the welding processes. This distortion is relatively more severe in the case of thin compared with thick sections. Although a very large number of rivet forms exist for sheet metal work, possibly those most commonly used are the semi-tubular and tubular types. The tubular or pop rivet type can be fixed where access is limited to one side of the joint. The tubular rivet is flanged at one end, and the other end is pushed through the rivet hole together with a heading tool which passes through the hollow rivet. A suitable device pulls the heading tool through the rivet and, in so doing, a flange is formed on the reverse side, thus locking the assembly. Riveting is a commonly used process for joining aluminium, especially in the aircraft industry. The diameter of a rivet should not be less than the thickness of the thickest part through which the rivet is driven, or greater than three times the thinnest part. When a small-diameter rivet is used in thick material, driving is difficult, and when a large rivet is used in thin material, the driving and heading pressures necessary often result in distortion of the sheet. Rivets are normally pitched at about three times their diameter.

In the case of heavy sections and large-diameter rivets, driving is often done hot. Subsequent thermal contraction of the rivet results in good joint tightness and this is especially advantageous in the case of stainless steel and other materials exhibiting high coefficients of thermal expansion.

8.2.3 Crevice conditions

All bolted and riveted joints, and indeed any joint executed by mechanical fastening methods, result in the production of crevices in the assembly. Such crevices may require special consideration in the case of service conditions involving corrosive environments. Stainless steel is a particular case in point. Where acid conditions are encountered coupled with traces of chloride, rapid selective corrosion in any crevices may almost certainly be expected. Welded vessels can be designed with a complete freedom from crevice conditions at the welded joints; however, most vessels require some form of access, manway or pipe connections and hence flanged joints are involved. These bolted flanged joints form crevices and may be corroded under adverse circumstances.

8.2.4 Special applications

There are many applications in the pressure vessels field where special problems of leak-tightness occur, particularly under service conditions involving temperature variation. It is not easy to avoid leakage at bolted joints, particularly those of the flange type, unless some form of compressible gasket is used. Such gaskets fill the inevitably uneven gap between the flange faces and allow normal manufacturing errors to be tolerated. One of the most common gasket materials is compressed asbestos fibre but rubber, soft metal or sealed thin metal flexible tubes are often met. In general, the more flexible and elastic gasket materials or constructions behave best and are most tolerant of misfit between the components to be joined; however, these materials are often more costly than alternatives such as compressed asbestos fibre.

One solution to the problem is to design the bolting system in such a way that it possesses a high degree of elasticity, small changes of distance between head and nut having no appreciable effect upon tension in the bolt shank. The simple spring washer is a step in this direction and it will be seen that adjustments to the rate and strength of the spring will allow uniform bolt pressure to be maintained on the flange even under some conditions of relative movement between the faces. The flange can then be made to conform to a less elastic gasket, rather than relying upon the gasket to compensate for unequal bolt loading.

In cryogenic applications, the spring washer technique is very valuable as wide differences in temperature are likely to occur between bolt shanks and flanges, especially during cooling and heating periods. In cooling, for example, the flanges contract while the relatively warmer bolts remain near their original length. Shank tension is therefore lost and the flange may leak. Specially designed spring washers or even coiled springs will overcome this problem.

In some cases it is necessary to use flanges and bolts of different materials whose thermal expansion characteristics may not be the same. Where cryogenic or elevated temperature service is required, some form of compensation must be introduced taking into account these differences in thermal behaviour. Although suitable spring systems may be designed, it is usual to introduce thick washers having controlled expansion characteristics. The washers may be made from the nickel-iron series of alloys having zero or even negative thermal expansion coefficients, and compensation becomes a matter of simple arithmetic. Pipelines for the handling and storage of liquid methane at temperatures between ambient and $-200°C$ on test have been made with aluminium alloy flanges, long 9% nickel-steel bolts and thick nickel-iron alloy compensating washers. They have also been made with chromium-nickel stainless-steel bolts and conical spring steel washers. In both cases, satisfactory service behaviour was obtained although the former method was somewhat cheaper.

8.3 WELDING METHODS INVOLVING PRESSURE

In order to facilitate reference, a systematic classification as indicated in table 8.1 has been followed in dealing with the joining processes. Table 8.2 has been included in order to link the available metallurgical joining processes with particular materials of construction.

The welding methods involving pressure are metallurgical joining methods which, at some time or other, use pressure as an integral part of the welding cycle. Resistance welding is a special case involving the production of molten material at the joint, pressure being applied to ensure containment of this molten nugget during the time in which it freezes to form a bond between the two pieces being joined. In the other methods, irrespective of whether heat is applied, the pressure is used to cause intimate contact by plastic deformation of the two components to be joined, usually after some treatment of the surfaces to ensure that the contact points are clean and free from oxides or other contaminants.

8.3.1 Resistance welding

Reference to table 8.1 will show that this group includes six methods; five further subdivisions of spot and seam welding are given in B.S.S. 499.

Table 8.2
Guide to Application of Welding Processes

Material	Thickness (note (i))	Resistance spot and seam	Flash	Projection	Percussion (C.D.)	High-frequency	Forge	Pressure	Ultrasonic	Explosion	Metal-arc	Submerged-arc	M.i.g.	CO_2	T.i.g. and plasma	Carbon arc	Atomic hydrogen	Gas	Thermit	Electron beam	Electro-slag	Comments
Mild steel	<0.1in	A	T	A	A	B	T	T	C	T	C	NA	NA	B	A	C	C	A	NA	C	NA	Weld deposits always stronger but less ductile than parent material, unless heat treated after welding. Electro-slag rating A where thickness exceeds about 2 in.
	0.1–0.5in	B	B	A	T	T	T	T	T	T	A	NA	NA	A	B	T	T	B	T	C	NA	
	>0.5in	T	A	B	NA	NA	B	NA	NA	T	A	T	C	C	C	T	NA	C	B	C	C	
Low-alloy steel	<0.1in	B	T	B	A	B	T	T	C	T	C	NA	NA	B	A	C	C	A	NA	C	NA	Preheat and post-weld heat treatment often necessary to avoid brittle welds and heat affected zone cracks. Heat treatment essential where alloying elements total more than about 2%.
	0.1–0.5in	C	T	B	T	T	T	NA	NA	T	A	NA	NA	A	B	T	T	B	T	C	NA	
	>0.5in	T	B	T	NA	NA	T	NA	NA	NA	A	T	C	C	C	T	NA	C	B	C	C	
Chromium-nickel stainless steels	<0.1in	A	T	A	A	T	T	NA	NA	C	C	NA	NA	T	A	T	T	B	NA	C	NA	Avoid prolonged periods in temperature range 600–800°C if steel is not stabilized against 'weld decay'. Carburizing conditions in gas welding must also be avoided.
	0.1–0.5in	B	T	B	T	T	T	NA	NA	T	A	T	B	T	A	NA	T	T	NA	C	NA	
	>0.5in	T	T	T	NA	NA	T	NA	NA	NA	A	A	B	T	C	NA	NA	T	NA	C	T	
Chromium-nickel heat resisting steels	<0.1in	A	T	A	A	T	T	NA	NA	C	C	NA	NA	T	A	T	T	B	NA	C	NA	Some of the higher carbon steels are subject to hot short cracking. Welding restraint should be kept as low as possible.
	0.1–0.5in	B	T	B	T	T	T	NA	NA	T	A	T	B	T	A	T	T	C	NA	C	NA	
	>0.5in	T	T	T	NA	NA	T	NA	NA	NA	A	T	B	T	B	T	T	T	NA	C	T	
Aluminium	<0.1in	T	NA	C	A	A	A	B	A	A	NA	NA	B	NA	A	NA	NA	B	NA	C	NA	Sources of hydrogen must be avoided in order to minimize weld metal porosity. Gas and submerged arc fluxes are highly corrosive. Metal arc welds usually of poor quality.
	0.1–0.5in	T	NA	C	T	T	NA	B	B	B	T	NA	A	NA	A	NA	NA	B	NA	C	NA	
	>0.5in	NA	NA	T	NA	NA	NA	B	B	T	T	T	A	NA	B	NA	NA	B	NA	C	NA	

WELDING METHODS INVOLVING PRESSURE | 591

This page is a large, complex table with many columns of welding process suitability ratings (A, B, C, T, NA) for different materials at different thicknesses, plus a column of comments. Due to the density and column ambiguity of the table, the content is reproduced below in a simplified row-wise format rather than as a precisely-aligned wide markdown table.

Materials, thicknesses, and comments:

- **magnesium alloys**
  - 0.1–0.5 in
  - >0.5 in
  - Comment: Cracking of the 2% magnesium alloy can be overcome by adding 5% magnesium alloy filler. Gas welding is difficult where the magnesium level exceeds about 2%.

- **Aluminium-magnesium-silicon. Heat treatable alloys**
  - <0.1 in
  - 0.1–0.5 in
  - >0.5 in
  - Comment: The weld heat affected zone is always soft and weak. Cracking can be overcome by adding 5% magnesium or 5% silicon filler alloy. The heat treatable alloys containing copper are virtually unweldable.

- **Nickel and nickel based alloys**
  - <0.1 in
  - 0.1–0.5 in
  - >0.5 in
  - Comment: These materials are seriously embrittled by sulphur contamination. Oils, greases and sulphurous furnace atmospheres must be avoided. Oxygen and nitrogen both give rise to porosity.

- **Copper**
  - <0.1 in
  - 0.1–0.5 in
  - >0.5 in
  - Comment: Gas welds are generally weaker than mig or tig welds but properties may be improved by hammering and annealing the joints after welding. Copper should be phosphorus deoxidized.

- **Bronzes**
  - <0.1 in
  - 0.1–0.5 in
  - >0.5 in
  - Comment: Aluminium bronze is welded with a nickel iron aluminium bronze filler in order to avoid cracking. This deposit is stronger than the parent plate.

- **Brasses**
  - <0.1 in
  - 0.1–0.5 in
  - >0.5 in
  - Comment: Because of the ease with which zinc volatilizes the brasses are virtually unweldable and must be hard or soft soldered.

- **Titanium Zirconium**
  - <0.1 in
  - 0.1–0.5 in
  - >0.5 in
  - Comment: Tig welding is the only process used for large-scale production at present. Air contamination must be carefully avoided to retain ductility in the welds.

Notes:
(i) See text for any restriction of individual processes to particular material forms, e.g. bars, sheet, plate, castings etc.
(ii) 'A' represents most likely general choice, 'B' and 'C' next choices but the rating will be affected by individual economics and technical circumstances.
(iii) 'T' represents technically possible but unlikely production choice.
(iv) 'NA' represents not applicable under normal production conditions.

592 | JOINING OF METALS

a. Flash welding.

Process. In this method, one lead from a source of electricity, usually a very large-capacity transformer, is connected to each of the two components to be joined. They are then brought together until they touch. As soon as this happens, intense heat is generated at the local points of contact by the passage of an electric current between the components. The metal melts where they touch, thus allowing the component faces to approach each other more closely. Further heating and molten metal formation takes place at new points of contact until the whole face is virtually at the melting point. The welding current supply is interrupted and extra pressure is then applied to bring the two components into intimate contact. This 'upsetting force' is sufficiently great to cause some distortion and spreading of the hot metal at the joint face, and as a result a satisfactory hot pressure bond results. There are a number of variants of this method all directed toward the same end, which is to raise the temperature of the butting faces to a suitable level for effective hot pressure welding to take place when the upset force is applied. In one method, a low preheating current is applied while the two components are brought into and out of contact with each other. This method is useful where very heavy sections are involved. After the preheating period, the current level is raised and flashing takes place.

Equipment. As heating in this process depends on the passage of electric current through the components and as the resistance of the components is not likely to be very great, it follows that very high welding currents are necessary in order to achieve an adequate heating rate. The transformers used for flash welding are therefore so designed that they will produce currents of many thousands of amperes but at an electric pressure of only a few volts. Apart from the physical risks of burning from contact with the

Fig. 8.1 Automatic flash butt welder for commercial vehicle rims.

flash, these machines are, because of the low working voltage, inherently safe to handle. Two further features of the equipment are associated with the current conditions and the means of gripping the work-pieces. Because very high currents are involved, it is necessary to have good low resistance contacts between the current conductors and the work-pieces; thus, in many cases, these must be specially shaped to suit the components involved and must be constructed of copper or some other suitable highly conducting material. The work-piece grips must be capable of transmitting quite high upset loads and so it will be appreciated that even small machines for, say, the flash butt welding of concrete reinforcing rods are physically quite large.

Applications. One of the most well-known applications of flash butt welding is in the production of welded railway track. The relatively simple cross-section of the rail makes it very suitable for welding by this process as the upset metal or flash can easily be removed by grinding after welding. Flash butt welding is also used for tubular components in large heat exchangers and steam generating plants and for many structural components in those cases where the upset metal can be readily removed by grinding or where it is of no consequence in the application.

Properties of joints. The application of upset pressure expels virtually all of the molten material from the joint and thus the final metallurgical structure consists of hot-worked material and not cast material as in the case of fusion welding methods. It is generally true that the properties of hot-worked metals are better than those of castings of the same composition, largely because hot working results in a refinement of grain structure with the elimination of brittle grain boundary films and also in an over-all improvement in ductility of the material. Although this is true in the general case, there are some notable exceptions, particularly in the case of very low carbon cast chromium-nickel stainless steels which are very ductile in their as-cast state.

As a result of the hot working, it is usual to find that a properly made flash butt weld has properties equal to those of the parent material in its annealed state. Whilst the method is principally used for joining carbon and low-alloy steels it has been used for high-alloy heat-resisting steels although results are not so good with materials having limited ductility. In these cases plastic deformation is difficult and there is a tendency toward incipient breakdown of the structure and fissuring under the action of the upsetting loads. Some non-ferrous metals could be flash butt welded but, generally speaking, they are so liable to atmospheric contamination during the heating cycle that it is better to adopt other methods of joining. The inert gas shielded fusion welding processes are particularly important alternatives.

b. Spot welding.

Process. As the name implies, a weld is produced between two components at a particular spot. The components are usually sheets of approximately similar thickness and are held together between two electrodes, one on either side of the joint. Pressure is applied between the electrodes continuously throughout the process. Whilst the pressure is applied, a predetermined pulse of electric current is passed between the electrodes. Resistance heating occurs at the interface between the two sheets being joined and a nugget of molten metal forms which subsequently solidifies on interruption of the welding current. This nugget joins the two components and is the spot weld. Its diameter is usually about equal to that of the electrode tips at the point where they contact the work-piece. Variants of the process exist but all have the same principle. In stitch welding repeated spot welding is carried out by means of an intermittent current, and in multiple spot welding a number of pairs of electrodes are connected to the welding transformer either simultaneously or in an automatically controlled sequence, thus allowing two or more welds to be made at the same time. In series spot welding the current is made to flow between one pair of electrodes and then through a further pair in electrical series, again allowing two or more welds to be made simultaneously. In a further variation, the electrodes are replaced by two conducting wheels which apply continuous force to the work-piece whilst rotating. Current is passed intermittently and thus a series of separate spot welds are made in line.

Equipment. As in flash welding, large electric currents at low voltage are involved and thus high current output transformers are required. The control devices are, however, more sophisticated as precise control over the quantity of electric current is necessary in order to form a weld nugget of the desired size in the very short time available. Thyratron or silicon-controlled rectifier systems are used which permit a precisely determined number of cycles of alternating current to be passed through any given spot weld. At the same time, these systems permit a variable and controlled portion of the alternating current cycle to be used and thus afford close control over the total heat input.

Applications. Spot welding is a well-known and indispensable method of joining sheet steel components, especially in the motor-car manufacturing industry. Although generally applied to sheet thicknesses up to about 0.1 inches, occasional applications up to 0.5 or even 1.0 inches in carbon steel are encountered. The two components should be of equal thickness but, in practice, successful welds may be made where the thickness ratio between the components is as much as 2:1. Beyond this ratio, the quality of the joints steadily deteriorates and at a ratio of about 4:1 the bonding efficiency is definitely suspect. Although spot welding is applied in the main to carbon and low-alloy steels, it is also successfully used for the joining of nickel alloys, especially in aircraft engines, the stainless and heat-resisting steels and, albeit with rather more difficulty, some aluminium alloys. It will be obvious that, as heating depends upon the resistance of the interface, greatest difficulty is experienced when dealing with materials having high electrical conductivity – copper, for example, is virtually impossible to resistance spot weld.

Properties of joints. Most spot welding applications result in shear rather than plain tensile loading of the joint. The diameter of a spot weld is about equal to the thickness of the sheet being joined and therefore shear properties of a spot or series of spots may be calculated from a knowledge of the parent-metal properties in shear. It will be appreciated that a very severe notch exists around the weld nugget and therefore, under vibrating load conditions, the problem of fatigue may be serious. This is particularly true in the case of large thin vessels and components which rely for their stiffness and static mechanical strength upon spot welded reinforcing members. Membrane flexure under vibrating load conditions almost always results in failure by fatigue at the spot welds. The type of joint inevitably gives rise to crevice conditions and, here again, care must be exercised in using spot welding where crevice corrosion might be important. This is particularly so in the case of stainless steel.

c. *Seam welding.*

Process. In this resistance welding process, force is applied continuously to the work-piece by two electrode wheels which, by their rotation, cause the work-piece assembly to move between them. The welding current is normally applied intermittently but at such a pulse frequency as to produce a series of overlapping spot welds. Although repetition frequencies of the order of five cycles on and five cycles off are usual, one cycle on, one cycle off, conditions are met and it is also possible to apply a continuous alternating current to the seam.

Equipment. The same sophistication in control apparatus is necessary as for spot welding and the electrical power source arrangements are similar. Because of the rapid pulse repetition frequency, however, the electrical power capacity of the equipment must be suitably increased. For resistance seam welding two 0.2 inch thick sheets together, for example, a power of about 100 kW would be required. The electrode wheels are usually carried on conducting arms of sufficient length to allow substantial sizes of work-piece to pass through the throat of the machine. The longer the arms, the greater their mass and the greater the over-all size and strength of the machine for a given electrical duty.

Applications. Resistance seam welding is now widely used for the production of sheet mild-steel containers such as motor-car petrol tanks, in aircraft engine manufacture and in the production of many domestic appliances. It is also used for stainless-steel constructions, notably in refrigeration plant and in large storage vessels, where heating or cooling of the walls is required. Heating or cooling panels can conveniently be made by resistance seam welding corrugated sheet to the vessel wall, and indeed most of the beer produced in the UK has been fermented or processed in vessels embodying this type of construction.

Properties of joints. These can be calculated by considering a series of overlapped spot welds. The stress concentration and other effects on mechanical and corrosion-resisting properties of the inevitable crevice must be considered, but it is interesting to note that very high pressures may be sustained in constructions of ductile material such as stainless steel where the stress concentration at the crevice may be reduced by deformation of the structure under pressure. In one particular example, 0.07 inch thick corrugated sheet, resistance seam welded at 2 inches pitch to an 0.14 inch vessel wall, withstood 2 500 p.s.i. in the cavity before bursting occurred. Very close control over the electrical parameters in welding is, however, necessary in order to achieve reliable results. This is particularly important where mains voltage fluctuations are likely to occur, as visual examination of the outer surfaces alone cannot give a reliable indication of the degree of fusion. In any case of doubt, an examination of cross-sections taken from production welds is essential.

d. Projection welding.

Process. In this process, the location of the weld is determined by the provision of a projection or number of projections on one or both of the contacting surfaces of an assembly to be resistance welded. Prior to welding, contact is at the projections only and the passage of welding current from one sheet to the other is, of course, by way of the projections. These in turn become preferentially heated and, because of the force applied, resistance welding occurs in the locality of the projection, which collapses during the welding cycle.

Equipment. This is the same in principle as that used in resistance spot welding except that the weld nugget size may not be determined by the electrode geometry but rather by the size and disposition of the projections on the contacting surfaces of the assembly to be joined. It is possible to use a large flat type of electrode and thus to make a number of projection welds simultaneously.

Applications. This method is most suited for mass-produced components where the projections can be raised in the primary forming operation, which might, for example, be die forging.

Properties of joints. These are essentially similar to those obtained by resistance spot welding except that the nugget size tends to be smaller in depth for a given diameter. As a result of this, the over-all distortion tends to be less and thus the process is very suitable where close dimensional control of the welded assembly is important.

e. Percussion (capacitor discharge) welding.

Process. This is a resistance welding process in which the two components are brought together with a rapidly increasing force as, for example, by the release of a spring. A short current pulse is provided by the discharge of a capacitor at the instant the components begin to touch and while the welding force is very low. The discharge of the capacitor creates a transient arc condition between the components and raises their surface in the vicinity of the arc to a suitable temperature for welding to take place under the action of the rapidly applied welding force. The essence of the process lies in the momentary application of force after the heating part of the welding cycle which itself occupies only a few milliseconds.

In order to aid the formation of the transient arc, a small projection on the surface of one of the components is often provided. This projection may be covered with a thin coating of sprayed metal or other substance in order to provide a flux to counteract the effects of atmospheric contamination during the transient arc condition. Due to the strictly limited quantity of electrical energy used in the process, very low distortion is encountered.

Equipment. In contrast with the equipment required for resistance welding methods so far discussed, this process is characterized by the use of light-weight and readily transportable apparatus. It is only necessary to have unhindered access to one side of the joint as relatively small forces are involved. Power is normally drawn from electrical energy stored in a large capacitor. The discharge takes place over a very short period of time but the charging time of the capacitor may be many hundred times greater and still allow high weld production rates. This relatively long charging period allows the use of small electrical components and thus the cost and weight of the power sources are kept to a minimum.

Applications. The most common application for this process is in the attachment of studs for ancillary fixing purposes, particularly to thin sheet components. Distortion introduced by the process is negligible and therefore the fixing of studs to decorative panels and trim where there must be no evidence of welding on the reverse side is particularly well accomplished. The studs are held in a spring-loaded gun in contact with the sheet or panel. Operation of a trigger allows discharge of the capacitor while the stored energy in the spring provides the necessary force on the weld.

Properties of joints. The joint is essentially a hot pressure weld and there is little or no evidence of the formation of a nugget in the sense of spot welding. The small quantity of molten metal formed during the transient arc condition is expelled by the subsequent application of pressure to the joint. In the case of studs, it is usual to provide a substantially larger diameter head at the point of welding in order to increase the area in contact, thus reducing any stresses likely to be applied in service. The size of the head is usually arranged so that, because of its increased contact area, the strength of the joint exceeds that of the stud shank. Freedom from crevices at the stud joints, however, cannot be guaranteed and therefore the same considerations of corrosion in service apply as in the case of spot and seam welds.

f. High-frequency resistance welding.

Process. In this process, a high-frequency electric current passes through contacts to the work, where it concentrates along adjacent edges to produce local heating just before welding force is applied. The frequency of the current is normally greater than 10 kHz and sometimes the current in the work-piece may be induced by appropriate coils placed in close proximity, thus avoiding direct contact. It is a characteristic of alternating electric currents that the higher their frequency, the greater is the degree to which they concentrate on the surface of the metal in which they are present. Thus, appropriate choice of frequency in this method of welding affords control over the location and extent of heating.

Equipment. Most high-frequency current generators consist of a high-power thermionic valve oscillator having circuit components chosen to produce resonance at the desired frequency, which may vary between 10 kHz up to 1 or 2 MHz. This electrical equipment is generally quite separate from the equipment needed to produce the welding force. The form of the mechanical equipment will be governed by the type of product being welded, which may be sheet, strip, bar or tube, and, in each case, particular gripping arrangements will be required.

Applications. Butt welding of bars, using the same type of mechanical equipment as for flash welding, may be successfully carried out. In this case, direct flashing is replaced by high-frequency induction heating. The production of longitudinally welded tube from strip is perhaps the most common use of this process, especially for the manufacture of carbon-steel tubes. It is less satisfactory for stainless-steel tubes unless special attention is given to the removal of the crevices associated with the rough inside bead. On account of the critical duties to which these tubes are generally put, fusion welding methods are preferred. A great deal of aluminium tube is produced by this method, again mainly for light structural purposes.

Properties of joints. The joint characteristics are essentially those of a hot pressure weld and, for the highest quality, depend upon a considerable degree of upset in the pressure stage of welding. In the absence of this significant upset, surface contaminants are likely to remain in the joint line with a consequent weakening effect.

8.3.2 Forge welding

Processes. This most ancient of welding methods is defined as any welding process in which the weld is made by hammering or by some other impulsive force while the surfaces to be united are plastic. The method is usually considered under three headings:

1. Blacksmith welding.
2. Hammer welding.
3. Roll welding.

The first is the original manual method whilst the latter two are machine methods.

Blacksmith welding is principally applied to wrought iron and low carbon steels, although the use of higher carbon steels is not unknown. Successful blacksmith welding depends on heating the steel to a sufficiently high temperature for the surface oxide to become a liquid slag whilst avoiding any melting of the base metal. Having achieved this temperature, the two components are hammered together, thus expelling the liquid slag from the joint and allowing intimate contact between the clean heated steel surfaces. Considerable judgment is necessary in order to work at an adequately high temperature and yet avoid overheating and actual melting of the steel which would then result in spoiling the forged work. The heat for welding may be obtained from a traditional blacksmith's coke fire or from gas or oil fired burners. Silicious fluxes are sometimes used to lower the melting point of the surface oxide and thus make expulsion of slag from the joint easier.

In hammer welding the force is applied by mechanical means after heating the components as in the production of forged parts. Roll welding is brought about by progressively applying forging pressure to the components in mechanically operated rolls after heating.

Equipment. That required for blacksmith welding is well-known and consists of a heating hearth, anvil and hammer. Hammer welding, on the other hand, requires mechanical forging hammers of considerable strength. The tup and anvil may support suitably shaped dies in order to control the shape of the work. In roll welding, the usual equipment found in steelworks for the production of plate is employed, and the method depends upon careful cleaning and close contact of the two components to be roll welded prior to heating and rolling. In general, the greater the degree of reduction in thickness, the more satisfactory is the resulting bond. This follows as a consequence of the greatly increased contact area produced in the virtual absence of atmospheric contamination, thus allowing intimate metallurgical contact to be effected.

Applications. Blacksmith welding is still of some significance as, for example, in the production of heavy chain for ships' anchoring purposes. It is also used in the manufacture of decorative articles such as gates, candlesticks and so on. Two notable uses of roll bonding are met in the production of clad plate and tube-in-strip. Clad plate is widely used where corrosion-resisting vessels of high strength are required but where the expense of construction in solid corrosion-resisting materials cannot be justified. In process circumstances where stainless steel, for example, might be required, it would be appropriate to use mild-steel plate clad with a thin layer of stainless steel, the cladding being effected by roll bonding.

The production of aluminium tube-in-strip is an excellent example of intelligent use of roll bonding. In this method, which is used for the production of refrigerator components, two relatively thick plates are prepared with a screening material sandwiched between them. The screening material is of a form corresponding to the pattern of refrigerant passages required in the ultimate strip. The whole sandwich is rolled and bonding takes place except in those areas where the screening material exists. On completion of the bonding, hydraulic pressure is applied to the non-bonded area which is blown up like a balloon to form the required passage.

Properties of joints. Well-made forge welds, containing no significant inclusions or discontinuities at the interface between the two components, have excellent properties because of the improvement in strength conferred by the forging operation. These properties can exceed those of fusion welds, which are essentially cast structures surrounded by a zone of fully annealed parent material. It must be emphasized, however, that the production of forge welds having no significant defects at the interface is not a particularly easy matter.

8.3.3 Pressure welding

Processes. The general definition of this process is one in which a weld is made by sufficient pressure to cause plastic flow of the surfaces, which may or may not be heated. In none of the variants of this general method is liquid metal produced at the interface, although there is a certain element of doubt with respect to friction welding. Heat may be applied to the joint area by various means, including oxyacetylene flames and high-frequency current induction. Having raised the temperature of the joint to the necessary level, pressure is applied to cause plastic flow and welding.

The difference between this process and forge welding lies in the fact that constant pressure is used rather than a series of hammer blows. Heat is used to reduce the parent-metal strength and to reduce to a reasonable level the amount of force necessary to cause plastic flow at the butting surfaces. In the case of soft materials such as aluminium and copper, heat is unnecessary as plastic deformation may be achieved by the use of relatively small forces. Cold welding or cold pressure welding is therefore the method in which pressure alone is used to effect plastic flow of the surfaces.

Friction welding is a method in which one component is rotated with respect to a mating part with which it is in forced contact. This relative movement of the contacting surfaces induces heat by friction and when the interface has been raised to the appropriate temperature an upset force is applied and relative rotation is stopped.

Equipment. Two types of hot pressure welding may be carried out using the same basic equipment, consisting of a hydraulic ram system for applying pressure to the two components. In one method the components are heated to the requisite temperature and then pressure is applied to cause welding. In the other method, constant pressure is applied during the time the components are heated. In the second case, welding takes place as soon as the required temperature has been reached, and there are marginal advantages over the first method in that slightly less contamination of the joints during heating may result. The application of pressure results in the generation of excess material at the joint which, in many instances, must be ground away or otherwise removed. The size of the equipment is very similar to that employed in flash butt welding but its electrical complexity is not so great.

Cold pressure welding is usually carried out on relatively small components with the simplest of gripping tools, many of which are hand-operated. The two parts to be joined must be cleaned with great care and all visible traces of scale, oxide or other forms of contamination must be thoroughly removed. In order to break up the residual surface oxide films, considerable cold deformation is necessary and, in many cases, more than 60% reduction in total thickness will be needed to ensure effective joining.

Friction welding requires a lathe bed plate type of set-up with a rotating chuck and stationary chuck through which axial force may be applied by means of a hydraulic ram. During the welding cycle, axial pressure is continuously applied to the rotating component and this causes characteristic expulsion of plastic metal from the heated joint area. When the required amount of expulsion has occurred, and clean faces are in contact, the relative rotation is stopped and increased upset pressure is applied in order to produce a sound hot pressure weld.

Applications. Hot pressure welding is used for joining carbon-steel bars and sections, whilst cold pressure welding of aluminium is used in the electrical industry for making joints in flat strip for transformers. Although considerable technical interest has been aroused in the development of friction welding, it has so far not received extensive application. This may result from the poor appearance and economic disadvantage of considerable quantities of waste metal expelled during the welding process which must usually be machined away. Notable examples exist of friction welded joints between unlikely combinations of dissimilar materials. These are, however, to some extent technical curiosities, as the need for some of the more unlikely combinations does not at present frequently arise in practice. Satisfactory dissimilar joints are possible because of the absence of fusion, which might result in the production of brittle compounds should one material become diluted with the other. The metallurgically difficult joint between aluminium and copper is now produced by this method for the electrical industry. Friction welding is restricted to assemblies where one component can be rotated, that is, in general, to bar and tube stock.

Properties of joints. The strength of hot and cold pressure welds depends upon effective dispersal of interface oxides and other contaminants. Provided that sufficient upsetting is carried out, good bonding results and the properties are similar to those produced by other hot-working methods which do not involve the incorporation of fused material in the final joint. Friction welds are normally of excellent quality as the process lends itself to effective removal of contaminants from the mating surfaces. It is necessary, however, to machine away the expelled metal in order to remove the objectionable crevice which would otherwise remain. Dissimilar joints between, for example, aluminium and copper are invariably stronger than the aluminium member and failure in a tensile test occurs well away from the joint, the over-all strength being determined by that of the weaker parent material. The possibility of galvanic corrosion in service should not be overlooked.

8.3.4 *Ultrasonic welding*

In this process, a relatively light pressure is applied between the two components to be joined. By means of a suitable piezoelectric transducer, high-frequency electrical energy is converted into high-frequency mechanical vibrations which are applied to the work-piece. These vibrations cause local disruption of the oxide film between the overlapped components and welding takes place. This is a solid-state joining process, no fusion being involved. Any heat generated as a result of friction between the components is rapidly dissipated and is not essential to the welding process. The method is limited to lap joints in thin materials up to about 0.05 inches thick. It works well for aluminium but has been used for many other materials and particularly for dissimilar metal joints in the electronics industries. Some auxiliary heating may be applied in order to facilitate welding but this is unusual. The generator and transducer usually work at an ultrasonic frequency of about 50 kHz but this frequency is dependent upon the physical size of the transducer necessary for the job in hand. As would be expected from their resonance characteristics, larger transducers work at lower frequencies.

8.3.5 *Explosion welding*

The main application for this interesting process is in the production of clad plate, and particularly in the manufacture of titanium clad mild-steel plate. It is found that an explosive charge detonated over a cladding sheet held at a slight angle above the plate to be clad causes intimate contact of the cladding sheet along a front corresponding with the shock wave produced by the explosion. Impact of the cladding sheet with the plate causes a ripple type of disruption of the surface oxide film and of the surface layers of the plate, with the result that a metallurgical bond is produced. This bonding is further aided by adiabatic heating which occurs at the interface as a result of impact. Once again, no fusion is involved and therefore bonding of dissimilar metals is quite feasible as brittle compound formation, which might result from fusion, is avoided.

Several obvious hazards in explosion welding exist. The process must be carried out under strict supervision and control in a suitably reinforced chamber away from buildings or other vulnerable structures. Various attempts have been made at explosion welding of tubes into tube plates in the construction of heat exchangers and in the welding of socket joints in aluminium pipework. These attempts, however, have not met with whole-hearted acceptance because of the difficulty in predicting weld quality, which may be affected by variations in the detonation characteristics of the explosives used. Perhaps the principal inhibition, however, lies in the obvious physical hazards involved in carrying out the process.

8.4 FUSION WELDING

Fusion welding methods are metallurgical joining methods in which joints are in a molten state without the application of pressure. The most widely used group of processes is that in which the source of heat is derived from an electric arc. Chemical heat sources are also considered.

8.4.1 Arc welding

The general description of arc welding covers the most significant group of welding processes in current use. The essential features of the processes consist of an electric arc or arcs normally struck between the work-piece and an electrode, which may or may not be consumed in the welding process. This arc provides the source of heat necessary to raise the temperature of the parent metal above its melting point in the vicinity of the joint. Protection from the effects of the atmosphere is achieved by the use of gasifying and/or slag-forming fluxes or by the provision of an inert gas shield around the arc zone. The current in the welding arc may vary from 20 A or so in the case of thin sheet material through 500 A, which is about the maximum practical limit for hand processes, up to 2 000 A, which might be used for automatically welding very heavy plate. A number of the many arc welding processes will now be described.

a. Metal arc welding. Of all the arc welding processes this is probably the most important in that it is still the most widely used method for general welding applications.

Process. A source of electrical power, which may be either direct or alternating current, is normally obtained from a rotary generator or from a transformer whose output may or may not be rectified. One terminal of the power supply is connected to the work-piece; the other is connected to a metal electrode carried in a handheld grip. The electrode is normally of a similar composition to that of the parent metal and is covered with a baked-on flux layer, except for one end which is inserted into the holder at the current pick-up point. In use, the other end of the electrode is touched on the work-piece and an arc is drawn. By careful manipulation of the electrode the arc is held at a constant length of about one-eighth of an inch. The heated end of the electrode melts and globules of molten electrode material pass across the arc into the weld pool which forms on the work-piece.

602 | JOINING OF METALS

Fig. 8.2 Metal-arc welding

As the electrode melts, the holder is advanced toward the work in order to maintain a constant arc length as short as possible in order that the effects of atmosphere contamination in the arc zone may be minimized. The temperature in this zone may be several thousand degrees and thus material in transit across the arc is greatly superheated. Reaction with only small traces of atmospheric gases is rapid and, in the case of carbon steel, for example, oxidation of the carbon can readily occur producing carbon monoxide and carbon dioxide gas in the form of bubbles in the weld metal. The presence of this gas might well result in an unacceptably high level of porosity in the weld, which could in turn give rise to mechanical weakness or actual leakage through the joint.

Maintenance of a short arc also allows the production of an uninterrupted slag cover over the weld pool, thus further protecting against atmospheric contamination. A long arc tends to disrupt this slag cover and again quality suffers. Besides protecting against atmospheric contamination, the slag cover, which results from melting of the electrode flux, helps to form the weld bead profile. In the case of electrodes intended for making vertical welds, for example, the fluxes are compounded to produce a fairly viscous slag which assists in holding the weld pool in the desired position. A further important function of the flux is to provide ionizable materials in the vicinity of the arc in order to stabilize and assist in striking the arc in the first instance.

Electrodes are manufactured from cut lengths of wire having diameters between 0.06 and 0.5 inches although the usual range lies between 0.1 and 0.3 inches. The most useful and widely used sizes are 0.1 and 0.13 inches diameter. The electrode core wire usually matches the composition of the parent material, whether it be carbon or low-alloy steel, stainless steel, nickel or nickel alloy. Dissimilar joints between stainless and carbon steel are usually made with high nickel electrodes of the 80% nickel, 20% chromium type. Satisfactory joints may also be made using cheaper stainless-steel electrodes of the 18% chromium, 10% nickel type containing about 4% molybdenum. Although some copper based electrodes are made, it is more usual to weld these materials and the aluminium alloys by the inert gas shielded arc processes. Synthetic or metal-powder electrodes are available having significant quantities of ferro-alloys or other alloying elements present in the flux coating. In the heat of the arc these elements melt and alloy with the core wire, which is often mild steel but may be stainless steel or nickel where very highly alloyed weld deposits are required. This is a useful method of obtaining special deposit compositions where suitable matching core wires may not be readily available.

The cut lengths of wire are passed through an electrode extrusion press containing a wet mix of flux compounds. In passing through the press a concentric envelope of flux is formed round the core wire. Electrodes formed in this way have the flux removed for a short distance from one end, giving a current pick-up point. They are then carefully dried and baked to ensure that the flux forms a hard damage-resisting coating and is free from moisture levels which might adversely affect the welding operation. Moisture tends to decompose in the arc zone causing porosity either by oxidation, as in the case of carbon steel, or by hydrogen solution and subsequent expulsion, as in the case of some aluminium alloys which are occasionally metal arc welded.

The rate at which metal is melted from the end of an electrode is a direct function of the current passing through the electrode. It is therefore desirable that this current be maintained constant in order that adequate control over the shape and penetration of the weld deposit may be achieved. Current flows through the welding arc because of the potential difference between the electrode and the work. The arc column can therefore be regarded as a variable resistance, although its actual characteristics are far from simple. It is, however, reasonably satisfactory to consider the arc resistance as a direct function of its length, increasing with increasing length and decreasing as the arc length is shortened. If the welding current is to be kept constant, it follows that the characteristic of the power source must be such that variations in the resistance across the arc produced by the normal variations in hand welding position do not significantly affect the current flowing in the circuit. Thus, to take an extreme example, if the electrode is touched to the work and held in direct contact, then the current flowing should not greatly exceed the current which flows under normal arc conditions. The characteristics of a suitable power source for this purpose are termed drooping, in that if a plot is made of voltage against current drawn, any increase in current is accompanied by a sharp drop in output voltage. Other power sources have different characteristics which may either be flat, in which case increase in current drawn results in little effect upon output voltage, or even rising, in which case increase in current drawn results in an increase in output voltage. These last two characteristics are used for specific semi-automatic welding processes which will be described later.

Welding transformers and motor generators for metal arc welding are of the drooping characteristic or constant current type, having an open-circuit voltage under non-working conditions of between 70 and 110 V. Short-circuit current of these units will vary between 20 and 600 A according to the capacity of the power source and its actual setting. Generally speaking, the higher the open-circuit voltage, the more readily is an arc struck, and this applies particularly to alternating current conditions.

Direct current is preferred for smooth arc running but electromagnetic effects can give rise to the phenomenon known as 'arc blow' when ferromagnetic materials are welded. The arc column can be regarded as a movable conductor and it will be appreciated that a considerable electromagnetic field surrounds an electrode and arc column carrying a large direct current. The electrode is therefore subject to electromotive force according to Lenz's law and that portion of the circuit free to move, namely the arc column, does so. Deviation is to one side of the joint and may seriously interfere with control of the weld bead. If alternating current is used, the electromotive forces reverse at mains frequency and therefore the total effect upon the arc is negligible.

The higher voltage required with alternating current sources is necessary to ensure immediate reignition of the arc should extinction occur in the zero voltage part of the cycle. There is a disadvantage in using these relatively high potentials of about 100 V in that there is some hazard of electric shock, especially when working outdoors under wet conditions.

Under direct current conditions heat distribution in the arc is about two-thirds at the column end nearest the positive terminal of the power source and one-third at the column end nearest the negative terminal of the power source. In order to ensure effective melting of the electrode, therefore, it is usual to connect it to the positive terminal. This connection is known as reverse polarity in American literature, while the electrode negative connection is known as straight polarity. This latter connection is used in non-consumable electrode arc systems and is seldom used for metal arc welding.

Metal arc welding is normally carried out manually and the greatest electrode length which can conveniently be held steady is about 18 inches. As a consequence, welding by this process is made up of a series of short runs. One interesting improvement in the technique is termed 'fire cracker' welding and, in this process, an electrode of greater length than is practicable for hand welding is either laid up a joint or at a slight angle to the joint in an appropriate fixing device. The arc is started between one end of the electrode and the work and travels along the work as the electrode melts. Constant arc length is maintained by contact of the insulating flux coating with the work and this maintains the constant distance between the work and the core wire. As the arc advances and the electrode burns away, it is kept in contact with the work either by its own weight or by slight pressure.

Hand welding electrodes may also have fluxes compounded in such a way as to allow direct contact with the parent metal during welding and this technique is known as 'touch welding'. This is a particularly useful technique since it demands less skill than is required in maintaining an arc of definite length without recourse to touching.

Equipment. The equipment required for manual metal arc welding is essentially simple, consisting of a power source which is usually a motor generator connected to a three-phase supply or driven by an internal combustion engine, or alternatively a single-phase connected constant-current transformer. The electrodes may be used direct from the manufacturers' packets but are preferably kept in a heated container powered by a suitable voltage derived from the welding generator.

Open arc processes generate considerable ultraviolet radiation and therefore it is essential to shield exposed parts of the body from the direct effect of the arc. This is done by wearing suitable leather gloves and a head shield having a dark filter glass window. The use of this dark filter allows observation of the arc zone and manipulative control of the electrode and weld pool.

After passage of the arc, the deposited weld metal is covered with a layer of solidified slag. In many cases this slag lifts away from the joint or is easily removed by brushing and it is the aim of electrode manufacturers to make slags either self-lifting or easily removable. In some cases, however, metallurgical considerations require the use of viscous slags which stick to the resultant deposit and are not easy to remove. In these cases, the use of specially pointed chipping hammers is necessary and the chipping may be done either by hand or with the aid of pneumatic tools. In stubborn cases or where some correction of the groove profile in multiple pass welding is required, narrow grinding wheels must be available.

Applications. Flux coated electrodes are available for a very wide range of materials: steels, stainless steels, heat-resisting steels, nickel, nickel alloys, aluminium alloys, copper alloys and even copper itself, although not all materials and thicknesses can be so readily welded as, for example, carbon steel. In sheet work, it is usual to select an electrode diameter approximately equal to the thickness of the material being joined; hence it follows that material much less than 0.05 inches thick becomes difficult to weld by this process. Above about 0.15 inches it is usual to make the weld in a series of runs one upon another. Thus a butt joint in 0.25 inch plate might be made in a 'V' groove, laying first a 0.12 inch electrode in the bottom of the groove and following this by a run with a 0.16 or 0.20 inch electrode, grinding the underside of the joint down to sound metal and filling the resulting groove with similarly sized electrodes. In very heavy plate sections of, say, 2 inches thickness, 30 or 40 runs might not be uncommon. There is no real limit to the thickness of plate which can be welded by this process except that arising out of economic considerations when comparing the process with quicker machine methods.

The application of metal arc welding to non-ferrous metals is less widespread than application to ferrous metals and the stainless and heat-resisting steels. Metallurgical problems exist in that the non-ferrous metals are often adversely affected by atmospheric contamination when molten. This is particularly so in the case of aluminium, which not only forms a refractory and tenacious oxide film difficult to remove by chemical fluxing, but also rapidly dissolves hydrogen, which may be formed by the decomposition of moisture in the arc zone, and subsequently rejects it again on freezing to form grossly porous welds. The refractory nature of aluminium oxide makes the use of highly aggressive and corrosive fluxes essential in metal arc welding. These fluxes are very susceptible to moisture pick-up and as a result it is almost impossible to produce defect-free aluminium welds by this process. Much attention has therefore been given to perfecting the techniques of inert gas shielded arc welding.

Nickel and its alloys are readily welded by this process, provided that the use of large-diameter electrodes, high welding currents and large weld-pool sizes is carefully avoided. At these high heat input levels, cracking of the weld deposit is frequently encountered. Suitable electrodes are available for all compositions of stainless and heat-resisting steels in common use and, of course, for carbon and low-alloy steels.

Properties of joints. Welds in carbon steel, if properly made and free from defects, invariably have a tensile strength greater than that of the normalized parent plate. This is because rapid cooling from welding temperatures must result in somewhat harder structures than those which would be obtained by full tempering, normalizing or annealing. In making multiple pass welds, the early runs are, in fact, normalized by the later runs and it is only the capping or final pass which has an unrefined metallurgical structure. Refinement is usually obtained by post-weld heat treatment but if such treatment of the complete joint is impracticable, it might be feasible to remove the unrefined capping pass by grinding. This can often be done without reducing the section thickness of the joint and thus no strength is lost.

Unrefined carbon-steel weld metal has poor impact properties and where shock loading conditions are likely in service, then due caution should be exercised. These low impact and poor ductility properties of 'as-welded' deposits become more pronounced as the alloy content of the weld metal is increased and in the case of, for example, 2½% chromium, 1% molybdenum deposits, not only is preheating necessary in order to prevent cracking of hard brittle deposits during the welding operation but post-weld heat treatment to produce a fully tempered structure or even annealing is necessary in order to preserve some measure of impact resistance. Post-weld heat treatment may result in a slight loss of strength but these chromium-bearing low-alloy steels are principally used for their improved heat resistance in the 500°C region and in these circumstances the slight reduction in strength brought about by post-weld heat treatment is not of great consequence.

Stainless-steel deposits also exhibit greater tensile strength than the annealed parent material, except in the case of very low carbon deposits, where the joint strength may sometimes be a few per cent less than that of the parent material. Many stainless-steel deposits are stabilized by the addition of niobium and niobium carbide which, as a consequence, acts as a marked strengthening agent in the deposit. Excessive niobium additions do, however, result in a lowering of ductility and in extreme cases where the niobium exceeds about 1.2% the weld deposits may crack, particularly in heavy sections.

Ductility is usually assessed by means of a bend test in which a strip sample is cut transverse to a butt welded test joint. The test strip, after suitable machining in order to remove irregularities, is bent over a former which has a radius about twice the thickness of the test strip. A 'U' bend is formed and if the weld metal does not fissure or crack it is reckoned to be satisfactory. This testing procedure has its limitations in those cases where the deposit is significantly stiffer than the parent material in that bending deformation may take place remote from the weld deposit. The test then becomes one of parent-metal ductility in the heat-affected zone rather than weld metal ductility, the practical significance of which is open to question. The test will, however, reveal the presence of excessive hardness and brittleness in weld heat-affected zones such as might result from inadequate or incorrect post-weld heat treatment of low-alloy steel weldments. Indiscriminate application of the test to non-ferrous metals of low inherent ductility and also to some high carbon heat-resisting steels has led to much unnecessary dispute over the acceptability of otherwise sound welds. In these cases, a thorough understanding of the metallurgical characteristics of the materials in question is essential.

It is commonly observed that austenitic stainless steels are non-magnetic and it comes as a surprise to discover that welds in these materials show a marked magnetic response. The reason for this lies in the fact that these stainless steels are hot worked at temperatures at which the non-magnetic phase, austenite, is stable. Weld metals of similar composition, however, are rapidly cooled from a higher temperature at which the magnetic phase, delta ferrite, is stable, and normally between 5% and 10% of this magnetic phase persists in the microstructure at ambient temperature. Its presence is necessary in order to prevent weld cracking, which would occur if a fully austenitic ferrite free weld structure were produced.

This delta ferrite is adequately ductile and has no adverse effect upon mechanical properties of the joint. However, in certain corrosive environments, such as are met in the production of some organic acids, selective attack of the ferrite may occur with consequent failure of the structure as a whole. It should be emphasized, however, that these specific conditions do not often arise and, for the vast majority of applications, the presence of delta ferrite is of no consequence and is of considerable advantage in ensuring freedom from weld cracking.

The properties of joints in nickel and its alloys are similar to those of the parent material, although they may be adversely affected by porosity which might result from excessive atmospheric contamination in welding. Care must be taken to avoid excessive pick-up of mild steel as the resulting dilution might give rise to an intensely air-hardening composition which would prove extremely brittle and unsatisfactory in service. As already indicated, defect-free welds in aluminium and its alloys cannot be produced by this process.

b. *Submerged arc welding.*

Process. For many years, this method occupied the position of greatest importance among the automatic processes. It is essentially a method suitable for joining heavy sections. Of recent years, the inert gas shielded semi-automatic metal arc process (8.4.1) has become, because of its flexibility, the preferred method for some applications, particularly in the welding of thinner sheet and plate. The essentials of the submerged arc process consist of a continuously fed bare wire electrode carrying an arc between its advancing end and the work-piece. The arc zone is surrounded by a finely granulated flux of a composition similar to that which would be used for the production of covered electrodes. The flux is usually fed from a hopper and is deposited along the joint just ahead of the arc. The heat of the arc causes some of this flux powder to fuse and form a protective slag over the deposited metal. The fluxes are compounded in such a way that the resulting slag is either self-lifting or very readily removed from the weld deposit. On completion of the weld, the unused flux is normally sucked away by a vacuum device and passes through a screen which removes any lumps of fused slag. The cleaned flux powder is then returned to the hopper for further use.

Equipment. As the name implies, the arc zone in this process is submerged in the pile of flux and is therefore invisible. This is a most important aspect in considering the type of equipment necessary and the general application of the process. As both the arc and the deposited metal are invisible, great care is necessary in controlling the position of the deposit with respect to the joint line. Thus it is evident that a machine arrangement in which the welding head can be accurately positioned over the joint and

Fig. 8.3 Submerged arc welding a stainless steel pressure vessel 0.75 in. thick.

checked over its whole length prior to making the weld is virtually essential. Recent attempts have been made to produce a manual submerged arc welding apparatus, but its application is limited to relatively non-critical applications such as the reinforcement of heavy section fillets where the operator can more easily sense the position of the arc from the geometry of the work-piece.

In machine applications, the welding head, consisting of the electrode wire driving motor working through spring-tensioned drive rolls, an electrode wire coil carrier and a flux hopper, is mounted either on a motorized tractor running on a suitable railway above the joint or, alternatively, on a welding column and boom in which the boom is driven out at right-angles from the vertical column at a controlled speed over the joint. Some sort of pointer or optical indicator is used in front of the flux nozzle in order to check that the head does not deviate from the intended line.

The welding current is conveyed to the electrode wire by means of a spring-loaded split copper contact about an inch above the arc point. The arc is started by slowly advancing an electrode wire until it touches the work, at which instant intense resistance heating takes place at the point of contact. A globule of molten metal falls away from the end of the electrode wire under the influence both of gravity and of the electromotive forces present when the welding current flows. Due to the heat and partial ionization of some of the flux compounds, an arc may be drawn across the gap as this globule falls away. The electrical circuitry is so arranged that the electrode drive motor is automatically speeded to the right level as soon as the arc is established.

Either drooping or flat characteristic power sources may be used. If a drooping characteristic source is used, there may be some difficulty in starting the arc as the available short-circuit current is limited and the electrode may simply fuse to the work-piece without actually melting. The advantage of the drooping characteristic source, however, is the ease with which motor speed control may be effected. Clearly, if the arc length, that is, arc voltage, increases for any reason, this increase may be used to speed the motor a little, thus shortening the arc once more, and the whole process adjusts itself automatically. The other advantage of a drooping characteristic source is that it provides a constant welding current irrespective of arc conditions and this is often important in the precise control of penetration.

Arc starting with a flat characteristic source is very much easier as high short-circuit currents are available which result in rapid initial melting of the electrode. The wire drive motor system is simpler than that used with a drooping characteristic source. The motor works at a constant speed and the arc is self-adjusting in that any increase in arc length is accompanied, because of the characteristic of the power source, by a marked decrease in welding current. This decrease in welding current results in a lower rate of electrode burn-off and as the electrode is advancing at constant speed the arc length must shorten. Conversely, if the arc length becomes too short, its reduced voltage results in a significantly greater welding current which in turn causes a greater electrode burn-off rate and the system adjusts itself. The advantages of this system in terms of electrical simplicity are obvious but its possible disadvantage lies in the wide variation of welding current, and therefore penetration characteristics, which might occur from time to time.

Because of the need to keep control over the heap of flux powder, the process may be considered as limited to the downhand or flat position only. Any attempt at welding in the vertical position brings the problem of maintaining a flux cover against the effects of gravity. When working in a flat position, however, it is possible to use very high welding currents to produce large weld pools and deep penetration. By this means, plates of the order of 1 inch thickness may be satisfactorily welded by butting the square edges together without any groove preparation, simply making one welding pass from either side. Adjustment of the welding current on the second pass ensures penetration into the first pass and a completely welded joint results. This contrasts with the large number of runs which would be required in manual metal arc welding.

The possible use of these very high welding currents, which may be of the order of 1 000 to 2 000 A, requires the availability of suitably large power sources. Although rotary generators may be used, it is common to find transformer rectifier systems for this type of plant. The welding heads, columns and booms are themselves bulky pieces of equipment and the power sources are of correspondingly large dimensions. As a consequence, therefore, the plant is not normally regarded as portable and it is usual to make provision for its permanent installation in a special part of the workshop where suitable manipulating equipment and cranage exists for the handling of the vessels and structures which may be welded by this technique.

Applications. Probably the most widespread use of this process is in the shipbuilding and heavy pressure vessel fabricating industries. Ship structures frequently require the attachment of stiffeners to plates and bulkheads and submerged arc welding is very suitable for this application, where the components or sub-assemblies can be set up for welding in the flat position. The twin fillet welding machine consists of two submerged arc heads set up on a gantry and simultaneously working on either side of a deck plate stiffener fillet joint. The advantages of simultaneous welding lie both in the speed and economy of the process and also in the reduced level of distortion which results from balanced welding.

For high production efficiency in very thick plate welding, it is possible to arrange for two or more electrodes to feed the same welding run or bead and the electrodes may either be connected to a common power source or to separate power sources. These applications relate to the welding of carbon and similar steels, and, in thick plate welding, electrodes as large as a quarter of an inch in diameter might be encountered.

Corrosion-resisting stainless steels are also welded by this process. For metallurgical reasons, currents as high as those found in carbon-steel welding cannot be used. The upper limit is of the order of 800 A. Higher welding currents require correspondingly greater arc lengths and the difficulty of alloying element losses due to atmospheric contamination in the arc zone is encountered. The composition of the deposit may therefore be unacceptably modified and, in addition to this, there is always an increasing risk of hot cracking trouble with some of the common compositions of stainless steel as the weld-pool size is increased. Those stainless steels which do not contain carbide stabilizing elements are also liable to suffer impairment of corrosion resistance by chromium carbide precipitation in the weld heat-affected zone and, as chromium carbide forms rapidly at red-heat, there is again a metallurgical case for limiting welding current and heat input rate.

Non-ferrous metals such as nickel, and even copper, have been submerged arc welded experimentally but the use of the method for these materials in production is not widespread. It is reported that submerged arc welding of aluminium is practised in Russia but the process involves the use of corrosive fluxes and in the UK inert gas shielded arc welding is preferred, even for very thick plate. At the other end of the scale, thin carbon-steel sheet containers of 0.05 or 0.10 inch thickness are regularly welded by the submerged arc process and attempts have been made to exploit the method for thin stainless-steel work. The reasons behind this attempt lie in the fact that submerged arc welds normally have a smooth ripple-free surface which is easy to polish. This contrasts with the relatively irregular surfaces produced by metal arc welding. There is, however, a minimum electrical energy level in the arc zone which is required to bring about effective ionization and fusion of the flux cover and this level is about 250 A in the case of stainless-steel fluxes. Such current levels are excessively high for thin sheet work and therefore, for most practical purposes, 0.2 or 0.25 inches may be regarded as the minimum sheet thickness for straightforward production application of this process.

Properties of joints. In multiple pass metal arc welding, the heat of successive runs effectively normalizes or tempers preceding runs. The absence of such thermal structure modification in carbon and low-alloy steels adversely affects impact resistance and ductility. Submerged arc welds are generally made at a high rate of metal deposition, particularly in the final stages cf a multi-run weld. Each run refines the previous run or runs but the final run is not refined by the heat of welding. Because of the high deposition rate, the final run is large compared with the final run in manual metal arc welding, for example, and this unrefined final run may constitute a large proportion of a submerged arc weld. As a consequence, submerged arc welds should be given some post-weld heat treatment in critical applications where best impact properties are required. On the other hand, post-weld heat treatment of stainless steels is the exception and should only be undertaken when there is a possibility of stress corrosion arising from the presence of residual stress. It is particularly easy to impair the corrosion resistance of stainless steel by incorrect heat treatment and the requirements for stress relief and best corrosion resistance are to some extent incompatible. Effective stress relief demands slow and even cooling at all stages but, for the preservation of corrosion resistance, it is important to cool the assembly through the critical carbide precipitation range of 800 to 600°C as rapidly as possible. Thus, the detailed metallurgy of the particular circumstance should be carefully considered before undertaking post-weld heat treatment.

As earlier indicated, the characteristics of submerged arc welding are its deep penetration and the difficulty of ensuring accurate conformity with a predetermined line. When making two-pass interpenetrating welds, one pass from either side, it will be obvious that accurate steering down the joint is essential to success. If the steering control is inadequate, the second pass will not interpenetrate the first. The resulting joint will have an unfused zone at its centre line and will be weak, particularly under conditions of dynamic loading. Under these conditions, failure by fatigue crack propagation from the unfused area may be a distinct possibility. This is particularly so in the case of stainless steels which have low inherent fatigue resistance.

c. Metal inert gas welding.

Process. In the inert gas shielded metal arc process, an arc is struck between the end of an advancing electrode wire and the work-piece as in submerged arc welding. Instead of obtaining atmospheric protection and weld bead control by means of flux additions, the metal inert gas (m.i.g.) process, as the name implies, utilizes a shroud of inert gas around the arc zone. In its early development this process was directed toward the welding of aluminium and its alloys and argon was, and is, used as a shielding gas. The process has been developed for use on other non-ferrous metals and stainless steels and various modifications to the shielding gas composition have been investigated. Helium is often used in the USA as a shielding gas and has the advantage

Fig. 8.4 Metal inert gas welding an aluminium alloy pressure vessel 0.5 in. thick.

of producing a hotter arc because of the higher voltage developed across the arc column compared with that produced when argon is used. The cost of helium in the USA is not very much greater than the cost of argon, whereas in Britain the cost has, until recently, been as much as 100 times as great. There are, however, signs that the cost differential will in the future be reduced to a factor of about five, but the use of helium is still likely to remain inhibited on the grounds of cost.

The system is normally operated in the so-called spray-transfer mode in which molten metal is transferred across the arc from the tip of the advancing electrode wire in the form of very small-diameter globules. The frequency with which these globules detach themselves is in the region of 300 times per second and their size may be gauged from the fact that electrode wire one-sixteenth of an inch in diameter might be fed into the arc at a rate of about 250 inches per minute. D.C. electrode positive connections are made to the welding power source and, under these conditions, an intense quasi-stationary electromagnetic field exists around the arc, which in turn is constricted by these forces. The constriction forces increase with increasing welding current and exert a progressively more marked 'pinch' effect upon the material melting from the end of the electrode. If the current, or more particularly current density, is high enough, the pinch forces overcome effects of surface tension and constrain the growing globules to a small diameter causing rapid detachment and transfer across the arc to the weld pool.

If the current density is insufficiently high, globule growth is controlled by surface tension and transfer takes place under the effects of gravity. The point at which electromotive forces become controlling is known as the threshold of spray transfer and current conditions below this point are known as subthreshold conditions. Under these conditions, very large globules are transferred at low frequency and tend to chill rapidly upon striking the work-piece, fusing ineffectively to the work-piece. Above the threshold level and in the spray-transfer range, the increased current allows better preheating and premelting of the work-piece and the fine spray of transferred material allows effective fusion.

At excessively high current densities, the pinch effects and metal transfer velocities become so great that a significant low-pressure zone develops around the arc, leading to entrainment of a proportion of air with the inert gas shroud. In the case of aluminium, severe oxidation of the deposit occurs under conditions of extreme turbulence in the pool and the resulting welds are useless. This condition is known as puckering. For a given wire diameter, therefore, there is a specifically usable range of working current below which large globular transfer takes place with resultant lack of fusion, and above which puckering and general instability can occur. With wire one-sixteenth of an inch in diameter, the usable current range is about 200 to 400 A for aluminium wires in argon. For larger diameter electrode wires, the current range is correspondingly higher, although 600 A appears to be the maximum upper limit for a single aluminium wire system.

Twin or multiple wire systems are possible and are capable of increasing the deposition rates from a maximum of about 7 lb per hour of aluminium for a single wire system to about 25 lb per hour for a three wire system. In the multiple wire system, the electrodes are fed by means of a common drive roller through a common current pick-up point to the arc zone, which is shrouded by inert gas directed by means of a special vaned gas barrel designed to ensure laminar flow. As in single wire m.i.g. systems, power is supplied from a flat characteristic source and, because of the high current surge available, little difficulty is experienced in simultaneous arc striking. Each arc works at a current density within the spray-transfer range and well below the level at which puckering would commence. The arcs are spaced at a distance of about three-eighths of an inch and work in a common weld pool. The low-pressure zone around the individual arcs is not so marked as that which would occur around a single arc working at the same total current and thus the multiple wire system overcomes the disadvantage of entrainment at high current levels and makes the process comparable with high-current submerged arc welding in its applications.

The existence of a minimum useful current density tended to inhibit the application of this process to thin sections as, in order to work with a low current, it was necessary to use fine wires which were difficult to feed consistently into the arc zone. Pulsed arc welding is an extension of the m.i.g. system in which high-current pulses of short duration are superimposed upon a low current arc. The pulse repetition frequency is normally either 50 or 100 Hz and the magnitude of each pulse is sufficient to bring the arc condition into the spray-transfer range. As a consequence, at each high-current pulse, a small globule is detached from the end of the electrode and transferred across the arc. During the low-current part of the cycle no transfer takes place. Thus a pseudo-spray condition is developed at a low mean current which, normally, for a given wire diameter is well below the spray-transfer threshold level. It is therefore possible either to use large-diameter easily fed electrodes for a given application or to extend the range of application by the use of very low currents with the smallest diameter wires which can be fed in practice.

Equipment. The process may be operated either manually or by machine and, apart from the mechanical arrangements necessary for carrying the welding head, the equipment for manual and machine welding is essentially similar. As indicated earlier, power is normally derived from a flat characteristic source although, as in the case of submerged arc welding, it is possible to use a drooping characteristic source and in this case the system is known as 'controlled arc', whereas with a flat characteristic source the description 'self-adjusting arc' is used. The essential difference between the two is that in controlled arc welding the current is held constant and the wire feed speed varied to compensate arc voltage fluctuations, while in self-adjusting arc welding, the wire feed speed is held constant and the burn-off rate is adjusted by current variations induced by changes in arc voltage.

The electrode wires for this process are generally smaller in diameter than those used in submerged arc welding. As a consequence, the mechanical problems of providing constant wire feed conditions are considerably greater. Wire is drawn from a spool by means of a spring-tensioned roller drive system and pushed through a guide, which may be rigid in the case of a machine set-up or flexible in manual welding apparatus. As in submerged arc welding, current is fed to the electrode wire at a point close to the arc.

In manual welding the wire emerges from a holder termed a 'gun' and this is usually provided with a pistol grip and trigger switch to actuate the welding controls. The inert gas is fed by means of a hose to the gun assembly or, in machine welding, to the barrel of the welding head and is directed as a laminar stream around the arc zone. Operation of the trigger switch causes flow of shielding gas, connection of electrical power and flow of cooling water to the gun or head.

Pulsed arc welding equipment is the same as that for m.i.g. welding except that a Duplex power source is provided. The base or pilot arc current of a few tens of amperes is drawn from one section of the power source whilst the high-current pulses are injected from another. Both parts of the power source may be independently set, thus allowing a wide degree of control over transfer conditions and mean current level.

Applications. There is no doubt that the fabricating of aluminium and its alloys on the present scale would have been impossible but for the invention of m.i.g. welding. This process has done for aluminium what submerged arc welding did for steel fabrication. One of the most widely used aluminium alloys is that containing 5% magnesium, and material of this composition is almost impossible to weld on a production basis by any means other than the inert gas shielded processes.

The range of application of m.i.g. welding is from sheet thickness of about 0.15 inches upwards. There is no upper thickness limit and examples exist of 6 inch thick aluminium sections having been welded without difficulty. Pulsed arc welding extends the thickness range from 0.15 inches down to about 0.05 inches. As both the spray and pulsed arc transfer modes are unaffected by gravity, the processes may be operated in any position. Hence, for aluminium welding at least, there is a universal application limited only by the physical size of the welding heads and guns.

This physical limitation is of considerable significance in those cases where the highest quality joints are desired. Aluminium depends for its good corrosion resistance upon the presence of a tenacious refractory oxide surface film; unlike surface oxides of steel, aluminium oxide remains solid over the molten weld pool. It is removed only by electrical dissociation in the arc and, as a consequence, it is necessary for the arc to play upon the whole of those surfaces on which it is intended to deposit weld metal.

If this cannot occur, fusion will be prevented wherever residual oxide particles are encountered and very weak joints will result. It will be obvious that head or gun approach angle is of great importance if effective scavenging of the surface oxide, by the action of the arc, is to be achieved. The arc should be allowed to play equally on both sides of the groove and, as a result of the relatively cumbersome but necessary design features of most welding guns, limitations of access are bound to occur in most practical structures. Designers should therefore have this aspect carefully in mind where joint quality is critical.

Stainless-steel applications are probably next in importance and, in this case, it is usual to add between 1 and 5% oxygen to the argon shielding gas in order to stabilize the arc. This stabilization probably results from the endothermic dissociation and exothermic recombination of molecular oxygen. The greater quantities are added where maximum penetration is required. A problem of oxidation arises both from the use of oxygen arc-stabilizing additions and also from the retention of heat in the weld zone due to the low thermal conductivity of stainless steel. The weld zone remains hot after the arc zone and its protective shroud have passed and post-weld oxidation takes place before the deposited weld has had time to cool. The presence of this thin oxide film on the surface of the deposited metal gives rise to the risk of lack of fusion in multiple pass welding unless very careful attention is paid to its removal between passes.

In many instances, it is easier to obtain good quality smooth welds by means of alternative processes such as submerged arc or metal arc welding, but there are occasions where the increased productivity available by m.i.g. welding, compared with metal arc welding for example, is the determining factor. In examples such as fillet welding of 0.15 inch sheet or fillet welding of 1 inch plate in short runs, where submerged arc welding would be impracticable, m.i.g. welding would be the logical choice. In principle, m.i.g. welding should be considered for any material containing significant quantities of aluminium or other refractory oxide forming element for the reason that the electrode positive connection used in this method is suitable for removing the oxide without recourse to chemically aggressive slags or fluxes.

The case of copper welding is a good example of the application of m.i.g. welding resulting in improved productivity and better weld quality. Until the development of this process copper was oxyacetylene welded at very slow rates and with an ever present risk of weld cracking due to alternate oxidation and hydrogen reduction to form steam at the weld and parent-metal grain boundaries. M.i.g. welding under inert conditions involves neither oxidation nor reduction where the process is carefully executed avoiding draughts which might disturb the inert gas shield. The disadvantage of the process lies in the considerable capital cost of the equipment, which may be three or four times that required for manual metal arc welding, and in the expense of providing an argon gas shield in the arc zone. As a consequence there is little economic incentive to apply the process in those cases where alternative cheaper processes might be equally satisfactory, albeit possibly more time consuming. Because of these economic considerations, m.i.g. welding as such, using an expensive inert gas shield such as argon, has not found great application in mild-steel welding.

In those cases where there is little alternative, as in the welding of aluminium, for example, the process has found ready acceptance as developments, aimed at improving the mechanical reliability of the wire feeding equipment and the electrical characteristics of power sources, have proceeded satisfactorily.

Properties of joints. It is unfortunate that hydrogen readily dissolves in molten aluminium and that there is a marked change in solubility between the solid and liquid states. Traces of water vapour and hydrated oxide decompose readily in the arc zone and the resulting hydrogen enters the weld metal. As the weld pool solidifies, hydrogen is expelled in the form of bubbles at the freezing front, and therefore there is always a tendency for the hydrogen bubbles to be trapped between the growing metal crystals. This is particularly so in overhead welds where the gas cannot readily escape from the weld pool. One of the characteristics, therefore, of a m.i.g. weld in aluminium is its porosity, which varies in degree according to the amount of hydrogen present in the arc zone. Steps can be taken to reduce this level of hydrogen but its complete elimination is a matter of great difficulty. It is fortunate, however, that aluminium and its alloys are, in the main, ductile materials and therefore the effects of fine porosity on the mechanical strength are not very great and only become really significant when the weld metal takes on the appearance of foam rubber.

A more serious problem in aluminium welding is that of incomplete oxide removal leading to lack of fusion. This defect is not readily detected by non-destructive means such as radiography but its effect on tensile strength and on fatigue resistance is very marked, as the planes of weakness can well amount to large percentages of the weld cross-section.

Aluminium weld metal deposited by this process is essentially a fine grained cast structure and its properties seldom exceed or even equal those of wrought sheet and plate of similar composition. They do, however, generally meet the minimum specification requirements for parent materials, whose actual strengths normally exceed these minima by a comfortable margin. The same considerations apply to nickel and copper and, to a lesser degree, to the extra low carbon stainless-steel alloys. With most stainless-steel alloys, it is usual to have small quantities of niobium present in the weld metal, for the purpose of preventing the formation of chromium carbide. The niobium carbide preferentially formed acts as a stiffener and improves the strength of the deposit to a level equal to that of the annealed parent metal.

The corrosion resistance of welds made by this process is not significantly different from that of welds made by other processes using similar materials. In some specific cases, for example, when welding aluminium bronze, it is normal to use non-matching filler material in order to prevent weld metal cracking or other difficulties. In these special cases, choice of filler metal composition must follow detailed consideration of the service conditions to which the weldment may be exposed.

d. Carbon dioxide welding.

Process. This process is very similar to the m.i.g. welding process except that the inert gas shroud is replaced by carbon dioxide gas which, in the context of welding carbon and low-alloy steel, is not inert. An extension of the process to the welding of thin sections has required the development of special power source characteristics not normally required for m.i.g. spray-transfer conditions.

The carbon dioxide process was developed in order to provide a cheap alternative to metal arc and oxyacetylene welding specifically for mild steel. As a continuous electrode wire is used, there is a saving in welding time because the need to change electrodes or filler rods at frequent intervals no longer exists and thus continuous welding for considerable periods of time is possible.

In the presence of the arc and molten weld metal, the gas shroud does not play an inert role, there being a reaction with iron to produce iron oxide and carbon monoxide.

Deoxidation is necessary in order to prevent this reaction and avoid the formation of weld porosity. This is usually achieved by the addition of powerful deoxidants, such as titanium, aluminium and manganese, to the electrode wire. Thus the cost of the electrode wire is increased significantly above the comparable cost for covered electrodes and this must be offset against improved productivity gained by using a continuously fed electrode.

The current density necessary to ensure spray transfer in carbon dioxide is much greater than that required in argon and, because of the high levels of heat input, the process has a limited use under these conditions. In developing the process, it was found that useful conditions resulted from a marked reduction in working arc voltage to a level at which the globule formed on the end of the electrode actually touched the work before detachment, thus establishing a momentary short-circuit condition. The current surge resulting from the short circuit increased the burn-off rate momentarily, thus detaching the droplet from the end of the electrode and reigniting the arc. This condition is known as 'dip-transfer' welding and, because of the low working voltage and reduced heat input, is particularly useful for thin sheet welding applications.

Equipment. The most important feature of carbon dioxide welding equipment is the power source, which is specially designed to have an over-all flat characteristic under open arc welding conditions and at the same time a limited short-circuit current in order to control the extent of the current surge during the momentary short circuit. The force with which the droplet becomes detached from the electrode is therefore limited and hence the amount of spatter is also limited and controlled.

This special characteristic is achieved by the inclusion of a series inductance in the welding power circuit. Under quasi-stationary current conditions the inductance has no effect, but where the rate of current change is high, as in the momentary short-circuit condition, careful choice of inductance value will afford the appropriate measure of control over the rate of current increase. These so-called short-circuit or dip-transfer conditions are workable at current levels considerably lower than those required for spray transfer and thus the process becomes applicable to the welding of thin sheet.

The extent of heat radiation from the short circuiting arc is considerably less than that radiated from a spray-transfer arc and thus it is possible to dispense with water cooling for the large majority of applications. This considerably simplifies the equipment. Due to the occurrence of wire stubbing in the intermittent short circuiting, it is critically important that the wire drive unit has an adequate reserve of power and that the wire is well gripped in the rollers in such a way that slipping is unlikely. This aspect of positive feed is much more important than in spray-transfer welding and has received a great deal of attention from the equipment manufacturers. In all other respects, the equipment is exactly the same as that supplied for spray-transfer m.i.g. welding and indeed much equipment specifically supplied for carbon dioxide welding has been used for spray-transfer m.i.g. welding with complete success.

Applications. The most notable field of application for dip-transfer carbon dioxide welding is in the motor-car manufacturing industry where the process, despite its higher capital and running costs, has superseded metal arc and oxyacetylene welding. Using 0.045 inch diameter wire very satisfactory joints can be produced on steel sections around 0.05 inches thick. Some care is necessary, however, in indiscriminate application of the process to thicker sections. Dip-transfer carbon dioxide welding is characterized by its low heat input and, where sections of, say, 0.15 inches and greater are encountered, therefore, is ideally suited for sheet metal work and, in the thicker sections, other processes such as submerged arc welding offered improved productivity.

Careful attention to technique can result in a useful extension to the range of application. This is particularly so in the case of pipeline welding. Many miles of steel pipeline have been laid using dip-transfer carbon dioxide welding for the butt joints. Such lines may have pipe wall thicknesses in the 0.2 to 0.5 inch range and are readily welded *in situ* using a multiple pass technique. Attempts have been made to apply the carbon dioxide process to the welding of stainless steel but a technical objection exists in that carbon pick-up to a level of about 0.1% is likely and may seriously impair the corrosion resistance of the weld deposit. As a consequence, this application has not found general acceptance.

Properties of joints. As the process is characterized by small weld bead size, it follows that multiple pass welds in carbon steel exhibit well-refined structures with good ductility and impact properties. Due to the active nature of the gas shroud, there is, however, a risk of porosity formation. When using covered electrodes in manual metal arc welding, the formation of a fluxing and refining slag allows a certain tolerance towards the presence of millscale, rust and other contaminants on the steel surfaces. The absence of slag and fluxes in carbon dioxide welding makes preweld cleaning of greater importance if porosity is to be avoided. It is also obvious that disturbances in gas shroud, due to the presence of draughts or adverse weather conditions, are also likely to impair weld quality.

Sources of hydrogen such as oil or grease on the electrode wire reels or plates to be welded should be avoided, especially when welding the higher carbon or low-alloy steels. Any dissolved hydrogen tends to accumulate in weld heat-affected zones and, should these be composed of brittle martensite, cracking may occur as a result of the pressure developed within the metal by the nucleation of diffusing hydrogen. As a general rule, however, it is found that the hydrogen level of carbon dioxide welds is exceedingly low where adequate care is taken in the use of specially dry carbon dioxide gas and in appropriate preweld cleaning procedures.

Some of the metal refining advantages of the presence of slag on the weld-pool surface are obtained in the recently developed carbon dioxide flux process. In one form of this method the electrode consists of a small-diameter tube formed from strip and containing a quantity of flux just sufficient for weld-pool refining purposes. Carbon dioxide is fed to the arc zone in the normal way and continues to protect the pool from atmospheric contamination. In another form, the flux is compounded with magnetic iron oxide and is fed to the arc zone with the carbon dioxide gas stream. In this case the powder clings to the current-carrying solid electrode wire by magnetic attraction and is carried appropriately to the arc zone and weld pool.

In both cases, improved deposition rates and weld metal fusion may be obtained at the expense of dealing with some slag between the passes of multiple run welds.

e. Tungsten inert gas and arc plasma welding.

Process. Historically, the tungsten inert gas (t.i.g.) process precedes the m.i.g. process, which was a development of the former. In the early part of world war II, the possible exploitation of magnesium and its alloys, particularly for aircraft parts, drew close attention to the problems of welding these materials. As with aluminium, magnesium has a tenacious refractory oxide film on any surface exposed to the atmosphere and this surface oxide film is particularly difficult to remove by chemical means.

Fig. 8.5 Double operator tungsten inert gas welding an aluminium vessel 0.5 in. thick. The second operator is working inside the vessel controlling the penetration of the weld.

Complete removal is necessary in welding if effective fusion is to take place.

It was found that the oxide of magnesium could be dissociated by electric arc action if the work surface was made the cathode in a d.c. system and was thus required to provide electrons to maintain the arc current. The work function of an oxide interface is lower than that of a pure metal interface; that is to say, the potential necessary to cause an electron to escape from an oxide surface into the arc column is lower than that required to cause electrons to escape from a clean metal surface into the arc column. This effect is well-known and is exploited in the formulation of oxide cathodes for thermionic valves.

As the oxide work function is lower, the analogy of lower surface resistance may be drawn, and thus it will be seen that current will tend to flow into the arc preferentially where particles of oxide exist on the metal surface. As a consequence, the oxide dissociates into ions and electrons; the electrons escaping from the surface pass through the arc column into the positive electrode. By surrounding the arc column with an atmosphere of inert gas, reoxidation is prevented, and thus wetting and fusion of the now cleaned oxide-free metal surface can occur.

In order that the process should be useful, it was necessary to use a refractory metal for the positive electrode and thus the choice fell naturally upon tungsten which, by the emerging techniques of powder metallurgy, could be produced in rod form. In a d.c. arc system, however, there is a serious disadvantage in the electrode positive connection in that the greater part of the heat is developed at the electrode and thus there is a very great tendency for even the highly refractory electrode materials to melt. Such melting does not so readily occur if the electrode negative connection is used. However, under these circumstances, thermionic dissociation of the surface oxide film of the work-piece cannot take place and thus welding is difficult, if not impossible.

The solution to this dilemma lies in the use of alternating current. During the electrode positive half cycle oxide scavenging takes place, and during the electrode negative half cycle electrode cooling and weld-piece heating occurs. The difficulty of using alternating current, however, lies in the tendency for arc extinction during the zero voltage parts of the current cycle. To overcome this difficulty a high-frequency high-voltage spark discharge is superimposed upon the arc and, should extinction occur, reignition follows immediately due to the ionization created by the spark discharge.

It was further discovered that the emissivity of tungsten could be improved by the addition of rare earth oxides to the metal as is done in the case of thermionic valve cathodes and thus the working temperature of the electrode could further be reduced with a lessening of the risk of electrode melting and inclusion of tungsten particles in the weld.

The electrode positive oxide scavenging mechanism is only necessary in those cases where refractory oxides exist on the molten metal surface. In stainless-steel and carbon-steel welding, for example, where the oxides become molten at or near the melting point of the parent material, no difficulty is experienced in achieving good surface wetting and fusion. The presence of the inert gas shroud prevents undesirable or excessive oxidation. With this type of material, it is possible to use a d.c. system with the electrode negative connection and, as a consequence, little trouble is experienced with the cooler running tungsten electrodes.

In the t.i.g. system, filler material is added separately and does not pass through the arc as in the case of m.i.g. welding. Thus, as filler conditions and welding current can be varied independently, there is a great measure of control possible in the operation of the system.

Arc plasma welding is a relatively new development. In the indirect arc system, the arc is contained wholly within the welding torch. The electrode exists in a hollow water-cooled copper cavity into which is fed the plasma gas, which may be argon with or without additions of nitrogen or hydrogen. An orifice is positioned beneath the electrode and the arc is struck between the electrode and the copper body of the welding torch. The plasma gas passes through the arc column and issues from the orifice at a very high velocity and very high temperature. With the addition of a suitable outer annular shroud of inert gas, the plasma jet may be used as a heat source for welding rather as a blowpipe may be used in oxyacetylene welding. In the direct or transferred arc plasma welding system, the arc, working at a very high voltage, is struck between the electrode and the work-piece and passes through the orifice in the torch, being constricted in so doing. This arc has very great penetrating power and welding speeds considerably in excess of those found in conventional t.i.g. welding are possible. The process is, however, mainly suited for machine welding and cutting applications.

Equipment. The torch for t.i.g. welding consists of a holder and collet arrangement to carry the tungsten electrode rod, which may vary in diameter between 0.03 inches for low-current d.c. applications up to 0.15 inches or even 0.25 inches for high-current a.c. applications. The collet is water-cooled where the torches are designed to carry more than about 200 A and thus welding current water and shielding gas must be fed to the torch by means of suitable hoses. A ceramic or water-cooled metal shield is fitted around the electrode end in order to direct inert gas to the arc zone.

The separately fed filler wires may vary in diameter between 0.05 and 0.25 inches although sizes in excess of 0.17 inches are not common. For hand feeding, the wires are normally in the form of 3 ft straight lengths and for mechanical feeding the wires are spooled as in m.i.g. welding. A great deal of care has to be taken to ensure cleanliness of the filler wire and surfaces immediately prior to welding.

Alternating current power sources consist of a tapped transformer having an open-circuit voltage in the region of 100 V. Lower voltages are unsuitable as intermittent arc extinction becomes a problem. A remotely controlled solenoid-operated current contactor is provided together with a high-frequency spark discharge circuit. This circuit consists of a broad-band spark discharge oscillator inductively coupled to a voltage amplifying coil, the energy from this being fed to the torch. Although voltages of the order of 15 kV may be generated, the welding operator is at no risk as the power involved is very small. The high-voltage discharge occurs between the electrode and the work-piece and provides an ionized path for establishment of the arc. Although continuous injection of this high-frequency voltage is desirable for the maintenance of arc stability in a.c. welding, it is unnecessary in the case of d.c. welding. In these circumstances, the oscillator circuit is automatically switched off by a voltage-sensitive relay as soon as the arc is initiated and the welding power source voltage drops from open circuit to the working level.

In a.c. plants, there is a tendency for partial rectification to occur due to the presence of dissociating oxide on the work-piece side only of the arc and the greater ease with which electrons can move in the one direction from the work-piece. This rectification leads to an imbalance in heat distribution and is corrected by the use of a d.c. component suppressor. This suppressor consists of a very large bank of electrolytic capacitors, usually of the order of one farad capacity. This capacitor bank has negligible impedance to alternating current when connected in series with the welding arc.

D.C. systems use motor generators or transformer rectifier units and both a.c. and d.c. power sources have a drooping electrical output characteristic. This characteristic produces constant current under varying arc length conditions such as are encountered in normal manual welding operations.

Arc length control systems are available for machine welding applications, although complex electronic circuitry is involved in order to provide a system which has a rapid positive yet dead-beat response without any tendency for overshooting or hunting. The circuits sense changes in arc voltage and compare them against a set reference voltage. The differences are electrically amplified and applied to a servo motor system which, by suitable gearing, causes the welding head to move nearer or further from the work to bring the arc voltage back to preset value. In welding, control over arc length is required and it is assumed that there is a simple relationship between arc voltage and arc length. Whilst this assumption is valid for most practical purposes, it must be recognized that slight changes in shielding gas composition, resulting from disturbances due to draughts, can bring about a marked change in the relationship between arc length and arc voltage. Normally the addition of diatomic gases to the inert gas shroud reduces the arc length for a given voltage. In many applications, it is desirable to work at a very short arc length and thus disturbances due to draughts can cause a short arc to become even shorter, to the point where short circuiting and arc extinction occur with ill effects upon weld quality.

Applications. With very few exceptions, it can be said that any metal which is at all weldable is weldable by the t.i.g. process. The process is, however, relatively slow and therefore, although there is theoretically no upper limit of thickness which may be welded by this method, it is seldom economic to consider using the process for thicknesses much in excess of 0.5 inches. Above this thickness, and in many cases below, the m.i.g. submerged arc or even metal arc processes would be considered. There are some exceptions however. In the case of high-alloy heat-resisting steels where the steep thermal gradients induced by m.i.g. welding are likely to give rise to micro cracking and where the alternative of metal arc welding involves difficult fluxes and irregular weld beads, t.i.g. welding is often preferred as its slowness and the cost of argon are of less consequence than the labour and technical risk involved in the alternative processes.

T.i.g. welding is especially suitable for pipework and, where butt joints can be locally purged with inert gas in the bore, excellent control of root penetration characteristics may be obtained. In aluminium pipe welding, however, root oxide films are a problem as the underside cannot be exposed to arc scavenging action. Thus it is usual to find heavier penetration beads as excess aluminium filler metal is often added to ensure that the root oxide films are thoroughly disturbed and do not form a continuous crevice. With aluminium and other materials having refractory surface oxides, it is always preferable to weld from both sides of the joint, removing unsound metal on the reverse side before seal welding.

The t.i.g. method is probably the most widely used process for welding stainless-steel sheet, especially in thicknesses less than 0.12 inches. The advantages of the process lie in the ease with which distortion can be controlled and the neatness of deposit which can be achieved. This is particularly important in the case of polished work such as is widely found in the food and potable liquor industries.

Plasma welding is likely to have an impact on the joining of sheets, especially in stainless steel, around 0.25 inches thickness. At this thickness, conventional t.i.g. welding is slow and submerged arc welding is difficult to apply. The so-called 'keyhole' technique is used in which the plasma arc blows a vertical hole between the close butted plate edges. As the arc moves forward, freshly melted material flows round the sides of the hole to coalesce immediately behind the arc, forming a completely fused weld in a single pass. This has the advantages of speed and minimal distortion. Furthermore, as a square edge may be used, little or no filler material need be added, and the economic advantages are considerable. It is necessary, however, to have a substantial sheet clamp and an accurately positioned welding head. The capital investment in the whole equipment is quite large and is comparable with that required for a high-capacity submerged arc welding plant. As a consequence a considerable volume of work is necessary in order to justify the expense of acquiring plasma welding equipment. The electrode negative connection is normally used in plasma welding torches and therefore there is some difficulty in applying the process to the welding of aluminium and magnesium.

Properties of joints. Some nominally single-phase alloy systems are prone to hot cracking. Alpha aluminium bronze and fully austenitic stainless steel are good examples. This hot cracking may be overcome by the addition of a lower melting point second phase which seals off hot cracks as they form and thus prevents their further development. Thus, for the welding of single-phase alpha 7% aluminium bronze, a complex nickel iron aluminium bronze filler is recommended and, for the welding of aluminium $2\frac{1}{2}$% magnesium alloy, the 5% magnesium alloy rather than parent composition filler is recommended. Being a slow process, t.i.g. welding results in deposits closer to equilibrium than those produced by quicker processes such as m.i.g. welding. Hence t.i.g. deposits are likely to contain less second phase for a given composition and are therefore, under these circumstances, more liable to hot cracking. It is therefore necessary to pay special attention to the precise technique by which multiple pass welds are made, particularly when dealing with materials which are liable to crack. In such cases, it is advisable to avoid the use of very large weld pools which are easily produced in t.i.g. welding by the adoption of weaving techniques.

The absence of flux is implicit in any inert gas shielded process and therefore it is particularly important to see that not only are the butting surfaces clean but the filler is also scrupulously clean. Failure to do this can result in porous or even cracked welds, particularly if sulphur-bearing contaminants are allowed to enter nickel based deposits. The heated end of the filler rod must be kept within the inert gas shroud if the introduction of oxides to the weld pool by this means is to be avoided. The presence of oxide films in the weld can have a very damaging effect upon tensile strength. The filler metal must be added to the edge of the weld pool and normally to the leading edge. It must not be fed into the arc column or disturbance and atmospheric entrainment will occur, resulting in further oxidation of the weld pool.

The grain size of t.i.g. welds tends to be larger than that produced by m.i.g. and metal arc welding and, as a consequence, some alloy systems show a slight reduction in tensile strength of t.i.g. welds compared with those produced by the quicker processes. The double operator vertical process is a very useful technique often applied to aluminium and its alloys, up to 0.5 inches thickness. In this technique, two operators work, one on either side of the joint, simultaneously at a common weld pool and a single pass joint results with very little distortion. The weld is, however, coarse grained and in the case of aluminium 5% magnesium alloy, for example, a tensile strength of about 18 t.s.i. would be expected, compared with 19 t.s.i. for multiple pass m.i.g. welding. The joints, however, do meet the minimum specification requirements for annealed parent plate.

Because of the high degree of control possible in t.i.g. welding, excellent radiographic quality may be obtained. This is particularly so in the case of aluminium welding and the level of porosity encountered is usually significantly lower than that experienced in m.i.g. welding. Little is known at present about the properties of plasma welds but it is not expected that they will differ significantly from those of conventional t.i.g. welds.

f. Carbon arc welding.

Process. In this process a direct current arc is struck between a carbon electrode and the work-piece, the carbon electrode being connected to the negative pole of the power source. The process has been largely superseded by the advent of t.i.g. welding, although it is still in use for thin steel sheet work and in some specialized techniques such as the joining of thermocouple wires. In the latter application an indirect arc would be struck between two carbon electrodes, the wire to be joined being introduced to the arc zone between them. In the direct arc system, the carbon electrode slowly oxidizes in the atmosphere providing a somewhat inefficient carbon monoxide–carbon dioxide shroud, thus allowing steel sheets to be joined without the need for flux additions.

Equipment. A d.c. power source is required and this is usually a motor generator, but any drooping characteristic source would be suitable. Carbon electrodes of about 0.15 inches in diameter are used at relatively low currents, usually of the order of 50 A depending upon sheet thickness, which is not normally greater than about 0.05 inches. The equipment is thus essentially simple and inexpensive and well suited to a strictly limited field of application where very high levels of quality control and metallurgical refinement are unnecessary.

g. Atomic hydrogen welding.

Process. In this process an arc is struck between two non-consumable electrodes normally made of tungsten and the arc zone is surrounded by an atmosphere of hydrogen. The intense heat of the arc causes dissociation of the molecular hydrogen to the monatomic state during which energy is absorbed. The monatomic hydrogen is blown away from the arc zone by further incoming molecular hydrogen and the monatomic gas recombines at the fringe of the arc zone to form molecular hydrogen once again. This recombination is accompanied by a considerable evolution of heat and it is this exothermic reaction which forms the principal heat source for the welding operation.

Equipment. The two electrodes are fixed in a holder so that the gap between them can be readily adjusted whilst the arc is running. Hydrogen is directed to the arc zone through tubes in the holders. In order to produce an arc of useful symmetrical shape it is normal to use an a.c. supply from a drooping characteristic transformer. The process is used mainly for small work with low welding currents and is similar to carbon arc welding in its range of applications and its infrequent use compared with t.i.g. welding. The equipment is somewhat more complex than that required for carbon arc welding in that gas regulators and current contactors are required. An advantage of the process is the relative efficiency of the reducing gas shield compared with carbon arc welding. By careful manipulation, it is possible to produce welds which are scale free and perfectly sound.

The increasing use of mechanically fixed expendable tool tips is likely to result in a further decline in the application of this process, which at one time found considerable use both for hard surfacing and for carbide tool tipping by brazing.

8.4.2 Gas welding

Process. The most important variant of the gas welding process is that in which the source of heat for welding is derived from the combustion of acetylene in the presence of oxygen. The advantage of using acetylene lies in the fact that its flame temperature of around 3000°C, when burnt in the presence of oxygen, exceeds that produced by the combustion of other gases such as town gas, propane, butane and hydrogen. This high flame temperature improves the flexibility of the process, particularly in dealing with thick sections, although the flexibility in this respect is considerably less than that available with the arc welding processes.

There are a number of stages possible in the combustion of acetylene depending upon the relative proportions of acetylene and oxygen available for the reaction. In the presence of excess oxygen, complete combustion occurs with the formation of carbon dioxide and water and such a flame condition is known as oxidizing. In neutral flame conditions, the amount of oxygen is exactly that needed for complete combustion. Further reduction of oxygen to a level below that required for complete combustion results in a reducing stage in which the carbon from the acetylene is only partially oxidized, although the hydrogen, being a very strong reducing agent, remains completely oxidized. The products of combustion under these conditions are therefore carbon monoxide and water. The carbon monoxide burns to carbon dioxide on the fringes of such a flame where it comes into contact with atmospheric oxygen.

The reducing flame is quite useful as the presence of a reducing atmosphere is important when welding some types of heat-resistant steel and when depositing hard surfacing materials.

In the absence of added oxygen, an acetylene flame freely burning in air has access to oxygen only at the edge of the flame zone where it comes into contact with atmospheric oxygen. Under these circumstances, although the acetylene decomposes into carbon and hydrogen, it is only the hydrogen which is oxidized, and thus the products of combustion are water vapour and carbon. Large quantities of soot are formed and the flame is quite valueless as a heat source in welding.

The reducing flame is composed of three zones. The inner zone, or cone, has its base at the blowpipe orifice and defines the space where the issuing gas mixture is being raised to combustion temperature. The length of the cone is primarily a direct function of the orifice size. The middle zone of the reducing flame is known as the 'feather' and forms a luminous envelope around the inner cone. In this middle zone there is a deficiency of oxygen and complete combustion does not occur. The luminosity arises from the presence of finely divided carbon particles at very high temperature. The length of the feather depends upon the proportion of oxygen and acetylene admitted to the blowpipe and increases with greater quantities of acetylene in the mixture. The outer limits of the feather define the point at which combustion of the carbon particles to form carbon monoxide becomes complete. The outer zone is much greater in size and is at a lower temperature. In this zone, final combustion of carbon monoxide to form carbon dioxide takes place, the source of oxygen being the surrounding atmosphere.

Flame conditions have an important bearing upon the success of gas welding since strength or corrosion resistance of the joints can be markedly affected by the conditions adopted. Although the welding of some materials such as carbon steel may be satisfactorily carried out by careful adjustment of flame conditions alone, other materials which have difficult surface oxides, e.g. aluminium, require the use of aggressive fluxes in order to expose clean metal and to allow effective fusion to take place. The formulation of these fluxes is often the result of considerable development work and, as a result, a large number of proprietary liquids and powders are on the market.

Equipment. Gas welding is essentially a simple process. Two cylinders of gas are attached to a common outlet and the issuing mixture is ignited. In order, however, to avoid the risk of explosion or fire certain precautions are necessary. Oxygen is usually supplied in gaseous form under pressure in metal cylinders of suitable strength to withstand 2 to 3000 p.s.i. internal pressure. A gas regulating valve is necessary to reduce this pressure to the few pounds per square inch required for combustion purposes. Acetylene cannot be supplied at high pressure because of the risk of explosion and it is therefore normally supplied at about one-tenth of the pressure found in compressed oxygen cylinders and dissolved in acetone. Regulating valves are again required to reduce the pressure from about 200 p.s.i. to the few pounds per square inch required for flame combustion.

Opening the acetylene cylinder valve allows acetylene gas to escape because of the considerable decrease in solubility of acetylene in acetone as the pressure above the liquid is reduced. It will be obvious that cylinders containing dissolved acetylene must be kept in an upright position if ejection of acetone is to be avoided.

Connection to the blowpipe is made through high-pressure rubber hoses. The oxygen hose fittings have normal right-hand threads whereas the acetylene hose fittings, as indeed any fittings used with combustible gases, are left-hand threaded and conspicuously marked with a groove around the flats. The blowpipe itself is designed for hand use and has two small screw valves which enable the operator to control the proportion of gases admitted to the mixing chamber within the blowpipe handle. The mixed gases pass to the flame through a suitably sized orifice in the blowpipe tip. The tips are replaceable and their size may be chosen in relation to the type and thickness of the work to be undertaken, as the orifice diameter largely determines flame and therefore heat source size.

Filler material may be required in wire or rod form and is often of slightly heavier gauge than that met in similar t.i.g. welding applications. Where fluxes are required, these may be applied in paste form to the joint immediately prior to welding and the filler rod itself may be similarly coated.

Applications. Flame conditions must be carefully considered in the metallurgical context of each application and it is fortunate that visual adjustment of the flame may quite easily be achieved. In starting, it is normal to admit excess acetylene to the flame, which under these conditions takes on a long bright appearance. The acetylene level is then progressively reduced by means of the hand control on the blowpipe until the luminous zone is reduced to a feather about twice as long as the inner cone. This is known as the carburizing condition and is used for hard surfacing or for the welding of high carbon heat-resisting steels where carbon pick-up is preferable to any decarburization of the deposit.

Further reduction in acetylene level until the feather is only slightly greater in length than the inner cone results in a reducing flame condition. In practice, the cone tip is the hottest part of the flame and it is this point which is normally applied to the workpiece. Under reducing conditions, therefore, the weld pool will be enveloped in an atmosphere rich in carbon monoxide. This condition is used for the welding of high-strength steels of the carbon and low-alloy types where carbon losses in the weld metal must be minimized whilst at the same time avoiding excessive carbon pick-up.

Further reduction in acetylene level until the luminous feather just disappears produces a neutral flame condition, which is that most widely used. Here the cone tip zone contains neither an excess nor deficiency of oxygen and this condition is suitable for welding mild steel and the corrosion-resisting stainless steels. Further reduction in acetylene level causes a shortening of the inner cone and a marked reduction in the length of the outer flame, the feather being completely absent. Under these conditions, the cone tip zone contains excess oxygen and is mainly used for the welding of copper alloys, particularly those containing zinc.

The importance of understanding the metallurgical factors involved and of being able to exercise adequate control over flame conditions is well illustrated in the early history of stainless-steel welding. The ductile chromium-nickel austenitic stainless steel came into use around 1930 and contained about 0.1% carbon. Electric arc welding was not common and it was found that gas welding could be facilitated by the use of reducing flame conditions, the neatest smoothest welds of all being produced with a carburizing flame. Although the welds were of excellent appearance, their corrosion resistance was so impaired by the precipitation of chromium carbide, both in the weld and in the heat-affected zone, that disintegration of the joints occurred after a matter of weeks in damp industrial atmospheres.

Because of the difficulties of ensuring supervisory control at all times, inert gas arc welding is now preferred for critical applications of this type. Gas welding is, however, still of considerable significance in the smaller workshop and particularly in sheet metal shops working with mild steel. The advantage of the process lies in its low capital requirements and low operating costs. Its chief disadvantage lies in the higher level of skill and conscientiousness required on the part of the operator if consistently satisfactory results are to be achieved.

Properties of joints. As already indicated, considerable variations in joint properties are possible by simple alterations in gas mixture proportions and the ease with which this can happen and remain undetected makes the application of the process of special concern for the metallurgist. The welding of copper is a good example. Under the action of welding heat, copper oxide will dissolve in a copper weld pool and on solidification a grain boundary network of copper–copper oxide eutectic will form. The formation of copper oxide will obviously be aided by the use of a highly oxidizing flame. Suppose now that a second pass of welding is made over the first but, by accident, a reducing flame is used. With a reducing flame there is always a chance of bringing hydrogen to the weld zone and any hydrogen present will diffuse rapidly into the metal, reacting with the copper oxide to produce water vapour under considerable pressure. This pressure is often great enough to cause splitting of the grains and cracked welds result. The conditions for copper welding must therefore be just slightly oxidizing and must be maintained so.

When welding mild steel, a strictly neutral flame must be maintained if carbon pick-up and embrittlement of the joint is to be avoided on the one hand, or decarburization and weakening is to be avoided on the other. The problem of dealing with stainless steel and the importance of avoiding any carbon pick-up has already been high-lighted. Carburized stainless-steel welds, although having adequate mechanical properties, will have poor corrosion resistance.

In general, design engineers recognize the control limitations of the gas welding process and either specify a careful system of inspection and control or, alternatively, accept the fact that some variation in weld joint properties will be inevitable and derate the stress bearing requirements of the joint accordingly. This is notable in the case of copper where a gas welded joint strength of about 80% that of the annealed parent material is admitted.

8.4.3 Thermit welding

Process. In this interesting process, the heat for welding is obtained from the exothermic reduction of iron oxide and the molten iron so produced is used as the filler metal for the weld. In one method, the heat of reaction may be used to preheat the parent metal in the region of the joint, but the more usual method is to carry out the thermit reaction in a separate crucible and to pour the molten iron into a specially prepared mould as in a casting operation.

The process is really a development of the 'burning-on' technique used in the foundry for the reclamation of defective castings, in which the casting is mounted in a refractory mould and the prepared defective area is exposed to a stream of molten metal of suitable composition. In passing over the defective area, the molten stream heats the casting locally until fusion of the surface takes place and the pouring operation is terminated. In thermit welding, a limited amount of metal is produced and therefore the section to be welded must normally be raised to a temperature close to that required for fusion prior to the introduction of thermit melted filler.

Equipment. This is a simple process and does not require connection to an electricity supply; therefore it is particularly useful for repetitive field work such as in the welding of railway track. The parts to be joined are assembled with a suitable gap between their ends and a wax model is made of the desired weld profile. A moulding box is placed around the joint which is set up with the wax pattern and the box is rammed with foundry moulding sand. Openings are cut in the sand to allow access for the preheating torches. A conical crucible having a refractory lining is positioned above a pouring channel which connects with the mould cavity. As in normal foundry practice, a riser is arranged directly above the mould cavity and any slag formed is allowed to spill through a chute into a catchpot. Thermit powder consisting of a mixture of finely divided aluminium and iron oxide is placed in the crucible, the bottom hole of which is closed by means of a plug and tapping pin. The amount of thermit mixture is calculated in accordance with the size and shape of the joint to be welded.

Exothermic reduction of iron oxide by aluminium occurs only at temperatures in excess of about 1 000°C and therefore it is necessary to start the reaction by means of a quantity of ignition powder placed on top of the thermit mixture in the crucible. The ignition powder is usually compounded of powdered aluminium and a highly oxidizing substance similar to firework mixture. Control of filler metal composition is achieved by making appropriate ferro-alloy additions to the thermit mixture prior to ignition.

In operation the wax model is first carefully melted from the mould cavity by means of a gas or paraffin fired preheater. Preheating is continued until the joint temperature is about 1 000°C, at which point all the mould openings except the pouring gate and riser are closed off with sand stoppers. The thermit reaction is then started and when, after a minute or so, this is complete, the tapping pin is dislodged allowing the molten iron to descend into the mould cavity. The level of preheat and the amount of thermit mixture is judged to cause fusion of the parent metal to take place. If either are insufficient, defective fusion is likely.

In order to ensure soundness in the weld joint, it is essential to keep the riser molten as long as possible and thus promote feeding of any shrinkage which might otherwise develop. This may be done either by covering the exposed riser surface with sand or, preferably, with an exothermic compound similar to that used for igniting the thermit mixture.

Applications. Because of the need for moulding boxes, wax patterns and for careful calculation of preheat levels and quantity of filler metal required, it is obvious that the chances of success are greatest where repetitive work on a standardized type of joint is undertaken. The rail welding case is obviously suitable as equipment may be set up for checking preheat levels and the precise amount of thermit mixture may be pre-packed, thus avoiding weighing errors.

As already indicated, the method is itself a development of the foundary 'burning-on' technique and it follows that thermit welding may be used for the repair of heavy casting sections, particularly in the field where full foundry facilities may not be readily to hand. The process also finds some use in the construction of large items of plant, particularly from castings. In many cases, the piece required is either too large or too complex to be made satisfactorily as a single casting and, when broken down into smaller units, problems of unsoundness may be more readily overcome. In these circumstances, a simple and reliable joining technique is obviously desirable. Electro-slag welding* was developed in Russia for this purpose and it is currently favoured in many quarters. It is, however, limited to joints of uniform section thickness and in this respect thermit welding offers somewhat greater flexibility.

Properties of joints. The weld metal is more akin to a casting in this process than in many others and this is particularly the case when very heavy sections are joined. Carbon-steel welds may therefore require a normalizing treatment after welding if

*See 8.4.5

good ductility and impact properties are desired. Such normalizing treatments are not difficult to carry out as the preheating equipment may be used and, in those cases where thermocouples have been inserted to give an indication of preheating temperature, the task is much simplified. Control of post-welding heat treatment temperature becomes increasingly important with the higher carbon and low-alloy steels, particularly those containing chromium. In some cases a double normalizing or tempering treatment may be desirable in order to ensure complete transformation and tempering of any retained austenite.

The most commonly encountered defects in thermit welds are lack of fusion and cold lapping where insufficient preheat has been applied. Attention is necessary to the precise design of wax pattern in order to ensure smooth blending of the thermit weld with the parent material. Provided that the mould is perfectly dry and adequately heated, porosity is not a problem. Centre line shrinkage may be troublesome if the riser size is inadequate.

8.4.4 Electron beam welding

Process. This process is of recent origin and results from the great improvements in vacuum technology and thermionic cathode design which have taken place since world war II. The physical principles underlying the production of an electron beam are well-known. Electrons from a suitable source, which is usually a heated metal cylinder coated with rare earth oxide, are accelerated in a vacuum toward a cylindrical anode by an electric potential difference of the order of several kilovolts maintained

Fig. 8.6 Electric beam welding chamber

between the anode and the cathode. Because of their momentum, the electrons pass through the hollow anode and may then be subject to further stages of acceleration. A diffuse beam of electrons is thus produced and this is deflected or focused to a spot by electrostatic or electromagnetic means. Any dense object placed in the beam is heated by the impact and deceleration of the electrons. By adjustment of the focus of the beam, intense heating may be concentrated within a very small local area.

The beam of electrons is sometimes known as a 'cathode ray' and, in a television tube, energy released by deceleration of the electrons causes the phosphor coating of the tube end to emit light. The power involved in this case is very small and because of the scanning device the beam is made to cover a very great area with little or no heating effects. In electron beam welding the power involved is of a much greater order and the beam is also stationary. Thus the local heating effect is intense. In an X-ray tube the electron beam strikes a water-cooled target to generate X-radiation and, although the power involved is similar to that used in welding, the beam current is less and the potential difference between anode and cathode is appropriately higher. In welding, high beam currents are used with relatively low potential differences between anode and cathode and soft easily absorbed X-radiation is produced. Screening is therefore not a problem. Most of the X-radiation is absorbed by the welding cabinet walls.

Equipment. A severe limitation of this process will be evident in that, for effective operation, the whole process must be carried out in a vacuum. The presence of any gas molecules in the path of the electron beam will, by collision, reduce the effective energy of the beam as it arrives at the work face and thus the efficiency of the expensively provided accelerating circuits will be seriously reduced. The whole apparatus and the work-piece must therefore be mounted in a chamber capable of being pumped in a reasonable time to a very high vacuum. The chamber must have observation windows and suitable mechanical means to enable the operator to position and move the work-piece under the beam.

The penetration of the electron beam into the work-piece is governed by the speed and energy of impact which, in turn, depend primarily upon the accelerating voltage applied between anode and cathode. Penetration is progressively reduced as the number of molecular collisions in the evacuated zone increases. The weld width is governed by total heat considerations and the important factors are beam current, traverse speed and degree of focus. The electrical equipment is obviously complex and the whole apparatus is costly. In general terms, the capital cost lies between 10 and 20 times that which might be required for other possibly suitable alternative processes.

Some development work has gone ahead in attempts to reduce the difficulty of having a very high vacuum in the whole of the system and it has been found practicable to use a rough vacuum in the main chamber, a very high vacuum in the immediate vicinity of the electron gun and an intermediate stage between the two. As already indicated, welding efficiency in terms of depth to width ratio falls off as rougher vacuum conditions are used in the main chamber. It has proved possible to extend this system to allow the electron beam to emerge into the atmosphere prior to striking the work-piece. Under these circumstances, however, the depth to width ratio of the penetration bead becomes similar to that found in m.i.g. welding, and this, together with the undesirable effects of atmospheric contamination, make this development of limited attraction in normal circumstances.

Applications. In many cases, electron beam welding may be successfully applied to fully machined components with very little risk of unacceptable distortion. This seldom applies when more conventional fusion welding processes are adopted as very much greater amounts of weld metal are required, a depth to width ratio of about two being common. The thermal disturbance in these circumstances is very great and the consequent distortion is much more significant.

Although electron beam welding is not very suitable for jobbing work, it is of considerable value in the mass production of relatively small precision articles. In terms of material compatibility it is an extremely flexible process and it finds greatest use in the joining of exotic materials for critical applications, particularly in the aircraft field. As the process is operated in a vacuum, there is no danger of weld contamination and thus the refractory materials which absorb oxygen and nitrogen can be readily welded. Molybdenum, tantalum and titanium are welded with less metallurgical difficulty by this process than by others. The large depth to width ratio means that, for a given penetration, there is very little thermal contraction stress and thus quite brittle and conventionally unweldable materials such as tungsten may in some configurations be satisfactorily welded by this method.

Properties of joints. The characteristics of the process make comparisons with other methods difficult. The shape of the weld bead is unique. It is characterized by a very high depth to width ratio, which may reach 50, and thus the use of alternative processes in a particular application will involve extensive redesign of joint details and possibly also of manufacturing sequences.

Although startlingly good results have been claimed for this process, hot cracking tendencies still remain an inherent problem with some materials. Single-phase alloy systems which contain small quantities of low melting point second phase can still show cracking even in electron beam welds. As it is impractical to add filler metal of a different and beneficial composition in such a way that homogeneous mixing may occur in the joint, little can be done to improve matters. However, because the contraction stresses are of a significantly lower order than those found in welds made by other processes, less trouble would be expected, and this is confirmed in practice.

The narrow width of the weld makes joint alignment under the electron beam of critical importance. As it is not necessary to add filler metal, joints are machined square and are close butted together. It will be obvious that any slight departure of the narrow electron beam from the joint line will result in non-welding. High-strength materials which gain their strength from solution treatment and precipitation hardening can be welded by this process with least effect upon their mechanical properties, as the over-aged weakened zone is exceedingly narrow. However, materials liable to volatilize when fused in a vacuum are not suitable for welding by this process. This consideration applies particularly to alloys containing zinc and, to some extent, tin.

8.4.5 *Electro-slag welding*

Process. This is a process essentially developed for welding very heavy sections together and the early development work was carried out in Russia in the 1950s. It is somewhat similar to submerged arc welding as there is no visible arc and consumable electrode wires are fed into a weld zone covered by a layer of flux, but beyond this the similarity is not great. The process is operated in the vertical direction in a rectangular gap between the edges to be joined. The weld is started in a cavity defined at its bottom by a starting plate, at its two longer sides by the edges of the material to be joined and at its two shorter sides by movable water-cooled copper shoes. The arc is struck much as in submerged arc welding by allowing the electrode to touch the starter plate beneath a layer of powdered flux. The resulting arc causes the flux to melt and begins heating the surrounding material. Further powdered flux is added until the arc becomes invisible and at the same time the covering of molten slag becomes deeper. After a while, the molten slag pool becomes sufficiently deep for the end of the electrode to be submerged. At this point, the arc is extinguished and further heating arises from the passage of current through the molten slag bath.

The formulation of the flux is adjusted so that the conductivity of the resulting molten slag is sufficiently high to allow the passage of large currents and yet is low enough to ensure the development of adequate resistance heating. Heat generated in the slag bath passes to the parent-metal side walls of the cavity, raising them to fusion temperature. It is normal, in fact, for a considerable quantity of the parent material to be melted into the weld pool. The heat of the slag bath also melts the incoming electrode wire, which in turn has undergone resistance heating by the passage of electrical energy from the current pick-up point along the wire to the slag bath.

Having established the process, small powdered flux additions are made from time to time in order to ensure that the slag depth is maintained at the required level. With the infeed of electrode wire, the weld pool level rises and the water-cooled copper shoes are simultaneously adjusted in order to prevent spillage of the weld metal or the slag bath. A fairly long period of time is necessary to establish equilibrium conditions in which the heat input exactly balances that conducted away by the parent material and the water-cooled copper shoes. As a consequence, the first few inches of a weld are usually defective through lack of fusion; therefore joints are generally started on appropriate scrap plates attached as extensions to the main seam.

Equipment. As in submerged arc welding, it is possible to operate this process with more than one electrode wire feeding into the weld pool and this is often done where very thick joints have to be welded. By adopting multiple wires, heat distribution in large weld pools can more readily be controlled and lack of fusion avoided. It is also possible to feed thick strip electrodes to achieve the same result. The equipment consists of an electrode wire driving unit of somewhat greater power than that used in submerged arc or m.i.g. welding. Greater power is necessary as the electrode wires are driven horizontally from one side of the joint and are turned through a right-angle to enter the slag bath. The water-cooled copper shoes are both fixed to the welding head, the one on the far side of the joint being supported by a bar passing through the gap between the parent plates ahead of the welding point. The whole mechanism is mounted on a rack and pinion allowing it to move vertically up the seam as welding progresses. With the smaller heads, a magnetic crawler device may be used to raise the machine up a work-piece, thus dispensing with the rack and pinion column. Power is drawn from a transformer of suitable capacity, welding currents of up to 2 000 A being required.

In a relatively recent development of the process, the head is fixed stationary at the upper end of the seam and the electrode passes through a guiding tube or fabricated constraining system to the weld zone. The guide tube is fixed and is therefore consumed in the welding operation. This procedure is known as the consumable guide technique and its advantage lies in the simplicity of the apparatus as movement of the wire driving mechanism is no longer required. Control of the copper shoes is, however, still necessary.

Applications. The process was developed and is best suited for the joining of very heavy sections. Problems of non-uniform heat abstraction become pronounced with section thicknesses less than about 1 inch. The process has been used for joining both cast and forged components, notably in the metal fabricating equipment industry itself and one of its first applications was in the building of very large press frames. Because of negligible angular distortion and the high speed of metal deposition compared with multiple pass submerged arc welding, considerable attention has been given to the use of the process for pressure vessel manufacture and therefore techniques have been developed which allow the welding of circumferential seams. In this case, the vessel rotates while the welding head remains stationary and, after the welding has proceeded for some distance round the seam, the weld start is dressed at an angle such that a vertical containment face is presented to the weld pool during the closing stages. Toward the end of the seam, the inner sliding shoe is replaced by a series of stationary shoes, thus allowing the support bar to be withdrawn from the weld gap and the seam to be closed.

Properties of joints. Dilution of the weld pool by melted parent material may be as high as 50% and, as little slag is consumed, control of weld metal composition must be by means of the electrode wires. For mild-steel welding, strongly deoxidizing electrode compositions are used and alloying additions may also be made in this way. Although the deposition rate is very high the travel speed along the joint is not very great and, as a consequence, coarse grained welds result. This is of importance where consideration of impact properties has to be made and, in these circumstances, post-weld normalizing heat treatments are adopted. This can often be done while welding is in process by the use of following gas burners. Various attempts have been made at grain refinement during welding in order to avoid the need for post-welding heat treatment but it is doubtful whether for critical applications the need for such treatments will be entirely eliminated.

The quality of electro-slag welds is normally very high as the weld pool solidifies slowly allowing gas and slag inclusions to rise to the surface and escape from the solidifying weld metal. Where attempts are made to speed the process excessively, by using very high deposition rates in conjunction with a large heat input, there is a risk of inducing hot cracking, particularly where the carbon level of the steel exceeds about 0.4%. Porosity is only a trouble where inadequate deoxidation arises through the use of low manganese or low silicon materials.

8.5 METALLURGICAL SURFACE BONDING

This section deals with joining methods which are metallurgical in nature and yet involve neither melting of the parent material nor the application of pressure to effect the joint.

8.5.1 Braze welding

Process. In this process, the techniques are similar to those used in fusion welding from the point of view of heating the joint and adding filler metal. The distinction lies in the use of a filler metal having a much lower melting point than that of the parent material. Capillary attraction of the filler metal between the parts to be joined does not form part of this process and the parent metal is not intentionally melted. The oxy-acetylene flame is the most usual source of heat but entirely satisfactory results have been obtained using arc techniques and particularly the t.i.g. technique.

Although it is impossible to eliminate parent metal fusion entirely when using arc techniques, the aim is to reduce the arc current level to the point where virtually no fusion takes place, the parent material still being heated sufficiently for melting of the filler to occur when it is touched upon the heated surface. Control is somewhat easier when an oxyacetylene flame is used; the object of the operator is to ensure wetting of the parent metal by rapid and adequate additions of filler, at the same time providing sufficient filler to ensure proper reinforcement of the joint.

The process is often known as bronze welding when specifically copper-rich fillers are used. In this case, however, the fillers usually contain substantial quantities of zinc and are, in effect, brasses. In many cases, a brazing type flux must be applied to the joint in order to ensure effective wetting.

Equipment. This has the simplicity of that required for manual gas, arc or t.i.g. welding, and is in most respects identical.

Applications. One of the most common uses for braze welding is in the bell butt joining of pipes. In this technique, the end of one pipe is swaged out to receive the end of the other pipe and the groove so formed between the butting pipes is filled with a suitable low melting point copper alloy normally applied with the aid of an oxyacetylene blowpipe. The repair of cast-iron parts and joints between them are often effected by this technique, again using a brass filler. Where oxyacetylene flames are used, these should be slightly oxidizing in order to prevent excessive loss of zinc from the filler metal. The provision of an oxidizing flame causes the formation of a tenacious oxide film on the surface of the pool, thus inhibiting further oxidation as the flame is moved about. An interesting application of braze welding is in the joining of galvanized iron sheet. Excellent joints of good rust resistance can be produced by the t.i.g. method using copper 3% silicon alloy filler coated with tin. The copper-silicon alloy is very fluid and wets the iron base material well. The tin acts as a flux and tends to inhibit oxidation of the zinc galvanizing coat, whilst the use of a carefully controlled t.i.g. arc can ensure exactly the right level of heat input so that minimum disruption of the galvanizing occurs.

Properties of joints. The joint between brass and steel or cast iron is more satisfactory in shear than in direct tension and joints should therefore be filled with this in mind. Wider joints are used than those commonly found in fusion welding and, provided that adequate control over wetting has been achieved and that they are well reinforced, the increased section at the joint should compensate for the relatively low strength of the filler metal compared with that of steel.

8.5.2 Brazing

The difference between brazing and braze welding lies in the fact that capillary action of the filler material between the faces of the parts to be joined is an essential part of the brazing process. The general definition of a filler metal for brazing is that its melting point shall be above 500°C and below the melting temperature of the parent metal.

a. Materials. A very large number of brazing alloys exists, the most important group being copper and copper based. Copper is used for the furnace brazing of steel components in an atmosphere of hydrogen and it is necessary to have the furnace running at a temperature of about 1100°C. Lower brazing temperatures are required with the copper alloys, the lowest of all being for silver alloys containing copper, cadmium and zinc, all in roughly equal proportions. The copper-zinc brasses are relatively cheap and are widely used for joints between hard steel components. They require a brazing temperature between 900 and 950°C. The copper-phosphorus alloys are used at a similar temperature and exhibit a fairly wide pasty range in which they are neither wholly liquid nor wholly solid. The 95% copper, 5% phosphorus alloy, for example, has a pasty range of some 200°C. Although they are not as fluid as some of the copper silver zinc cadmium alloys, they are useful where there is difficulty in controlling joint clearance and where wide gaps have to be bridged.

The silver bearing alloys are worked at temperatures around 700 to 800°C depending upon the particular composition. They are very fluid and will readily penetrate narrow joints. They wet and spread well and are almost as easy to use as the tin solders.

Brazing of aluminium is effected by means of the aluminium-silicon alloys, containing up to 13% silicon, which melt in the 540 to 630°C range depending upon composition. This temperature range is normally only about 50°C below the melting range for the parent material and very great care is needed in temperature control during brazing.

For high service temperature applications and in special circumstances where corrosion problems occur, nickel base brazing alloys may be used. These may contain phosphorus, silicon, boron or chromium singly or in combination and according to their composition may melt in the range 875 to 1135°C. One of the highest melting point braze metals is the 60% palladium, 40% nickel alloy which melts at 1235°C. The composition and melting ranges of commercially available alloys are given in *B.S.S. 1845* and *1723*.

Brazing alloys may be obtained in a variety of forms such as rod, for addition directly to the heated joint, or as powder, paste, sheet or special sections for preplacement. Flux is generally necessary in order to clean the parent-metal surfaces of all oxide traces and to allow satisfactory wetting and bonding to take place. Surface bonding results from the long range action of interatomic forces and alloying does not necessarily take place. A number of variants of the basic brazing technique are now described.

b. Resistance brazing.

Process. In this process the filler metal is preplaced in the position required for the joint and the heat for brazing is obtained either by passing an electric current between the parts to be joined using copper electrodes as in resistance welding, or by passing an electric current through two carbon electrodes and the part to be joined. In the latter, more usual, case, much of the heat for brazing is generated in the carbon electrodes and reaches the joint area by conduction. This method has the advantage of more easy control of temperature at the joint face.

Equipment. Lower currents of greater duration are required for resistance brazing as compared with resistance spot welding. Control over and limitation of closing pressure between the electrodes is also of importance. Because of the relative slowness of the heating cycle, visual control is often practicable and therefore the need for complex automatic timers can be avoided. In many cases simple handheld tongs through which the current passes will suffice.

Applications. The method is very suitable for small components which might be distorted or damaged by excessive heating, and thus the process finds use in the electrical and electronics industries for the assembly of electromechanical devices, motor coils and conductor joints of many forms.

c. *Dip brazing.*

Process. In this method, a bath of molten brazing metal is covered by a layer of molten flux and the work-pieces to be joined are first cleaned of excessive surface contamination and then dipped into the bath. In so doing they pass through the molten flux layer and become further cleaned and coated to allow wetting by the braze metal to occur as soon as the work-piece reaches the appropriate temperature. Heat for brazing is derived from that contained in the molten bath and, in the case of large components, some degree of preheating may be necessary in order to conserve heat in the bath.

Equipment. As heat is supplied to the components from the molten brazing alloy, it is important that the bath be of adequate size to prevent large temperature fluctuations resulting from entry of the work-pieces. Graphite crucibles are commonly used for containing the bath and these may be externally fired by means of oil or gas. Automatic temperature control of the bath may be arranged as is done for heat treatment furnaces.

Applications. The cheaper brazing alloys, such as the copper zinc types, are often used in this method and one of the principal uses is in the manufacture of bicycle frames. The process is generally useful for joining steel components where precision in filler material placement is not of great importance.

d. *Salt bath and flux dip brazing.*

Process. This technique differs from dip brazing in that the components are assembled with preplaced filler metal. They are then immersed in a bath of molten salt at a temperature suitably above the melting point of the filler metal, which consequently flows into the joint as soon as the work-piece reaches the bath temperature. The salt bath may or may not act as a flux and the technique derives either of its names accordingly.

ʃ Equipment. The salt bath, which is usually of considerable size and heat capacity, may either be externally fired or heated by the passage through it of an electric current. Components to be brazed must be perfectly dry as most molten salts react violently to the admission of even small quantities of water. In order to ensure complete freedom from moisture in the assembled components, they are often preheated, and this is also useful in limiting heat loss in the salt bath.

Applications. One of the more important applications of flux dip brazing is in the

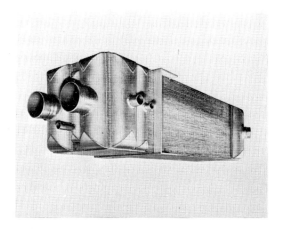

Fig. 8.7 Salt bath brazed aluminium heat exchange assembly for use in low temperature air separation.

manufacture of aluminium honeycomb-type heat exchanger assemblies. These assemblies are made from crimped thick foil placed between thicker flat sheets. Aluminium-silicon brazing alloy is preplaced and, after preheating, the assemblies are brazed at a precisely controlled temperature of about 600°C, which is just above the melting point of the brazing alloy and just below the melting point of the parent materials. A great deal of technical skill is involved in the control of this highly specialized application. The process can be used at higher temperatures for the brazing of mild steel with copper base alloys although the aluminium application is the most common. The major disadvantage of the process lies in the need for complete removal of the highly corrosive fluxes, particularly in the aluminium application.

e. Flame brazing.

Process. This process, sometimes also known as torch brazing, has notable flexibility. Whilst it may be regarded primarily as a manual method, it can be mechanized, in which case brazing material would normally be preplaced. In manual operation, the heat for brazing is derived from a chemical flame which may be either oxyacetylene or some other combustible gas mixed with air or oxygen. The choice of gas and flame size will depend upon the size of the work to be undertaken. Although oxyacetylene flames are the most widely used, there is some merit in the use of less fierce oxypropane flames, particularly when dealing with the lower melting point brazing materials.

As with all brazing methods, a high degree of cleanliness is necessary in the parts to be joined and a suitable flux is essential in order to achieve complete wetting and penetration of the brazing material. The operator should aim at raising the temperature of the components uniformly to a level just above the melting point of the filler metal. If the appropriate flux has been chosen, then an indication that the correct temperature level has been attained is given by the fluidity and free running of the flux at this point. Filler rod is then applied to the joint and, if the temperature has been gauged correctly, should immediately melt and run into the joint. Uniform heating is continued until the joint is seen to be completely wetted by the brazing alloy.

Equipment. The equipment for flame brazing is very simple and flexible in application. It is exactly similar to that required for gas welding except that a greater degree of choice is possible in selecting flame conditions and heat input level, due to the fact that generally lower heat inputs are required than is the case in gas welding.

Applications. These are as varied as those encountered in gas welding. The method is principally used, however, with the copper and silver based brazing materials for joining steel, stainless steel, copper, nickel and their alloys. Some assemblies may be designed in such a way that simultaneous brazing of all the joints is impossible due to the difficulty of holding the components together while the brazing alloy is still molten. In these circumstances, it is often possible to make the joints in sequence using brazing alloys of successively lower melting point and flame brazing is very suitable for this technique.

f. Furnace brazing.

Process. The mechanism of brazing in this process is similar to that in other brazing processes, in that the assembled components must be perfectly clean, fluxed, and have the requisite amount of brazing alloy available to fill the joint. In this variation brazing is effected by placing the fluxed assembly into a controlled atmosphere furnace at the appropriate temperature above the melting point of the particular brazing alloy.

The controlled atmosphere prevents scaling during the relatively long brazing cycle and excellent reproducible results may be obtained because of the high degree of automatic control over brazing time and temperature which is possible with this technique.

Equipment. Although batch furnaces may be used, best production results are obtained where continuous conveyer type furnaces are available. In this type of furnace, completely controlled and reproducible conditions exist. Where a good finished appearance of the work is to be maintained, there is no difficulty in ensuring the essential coverage of controlled atmosphere during the cooling part of the brazing cycle.

Applications. The method is very suitable for awkwardly shaped assemblies and particularly where there is a large imbalance in heat capacity between the components to be joined. Heating in the furnace is more likely to be uniform than by other means and therefore distortion of the work is also minimized. The technique is suitable for brazing with the high-temperature high-strength nickel base brazing alloys, particularly where vacuum furnaces are available.

g. Induction brazing.

Process. In this technique the heat for brazing is obtained by high-frequency induction methods.

Equipment. Power is usually obtained from a high-frequency oscillator circuit coupled to a coil manufactured from water-cooled copper tube and of such a shape that it will fit closely around the part of the component to be joined. The design of fixtures and jigs for holding the components during the brazing operation must be carefully considered in view of the tendency for high-frequency currents to be induced in any other metal components near the induction coil. Close control over the power level and consequent heating effects can readily be achieved in this essentially machine technique.

Applications. The process is most suited for repetitive mass production of similar components as the induction coils must be specifically designed to suit each particular application. The process is a very rapid one and it is possible to control the extent of heating within quite fine limits. Hence, considerable forethought is necessary to ensure that exactly the right temperatures are reached in the very short time cycle. Typical applications might be the attachment of small gears to shafts, brazing of container seams and the attachment of pipe fittings.

Properties of joints. As the brazing process depends upon capillary attraction of the filler metal between the surfaces to be joined, it follows that most joints will be of the lap or socket types. With the exception of aluminium based brazing alloys used for brazing aluminium and aluminium-manganese parent metals, brazing alloys in general tend to be weaker than the parent metals to which they are applied. If, however, assemblies are designed about shear loaded lap joints, it is normally possible to produce sufficient area in shear to compensate for this reduced strength. If direct tensile loads are avoided it is then only necessary to consider the corrosion possibilities where heterogeneous joints are put into service.

Nickel based chromium bearing brazing alloys are often suggested for use on stainless-steel assemblies but where such assemblies are likely to be exposed to nitric acid, in food plant cleaning and sterilizing, for example, attack of the brazing alloy in the joint may be expected. Apart from the possibility of direct attack, as in this example, galvanic effects must also be taken into account. This is especially important in those cases where the braze metal may become anodic to the parent metal, as the very much larger cathodic area will promote a high current density and rapid rate of attack in the small anodic braze metal area. De-zincification or selective attack on the zinc-rich constituents in brass is a good case in point.

8.5.3 Soldering

Process. From a metallurgical standpoint, soldering, being operated at a lower temperature, is simply an extension of brazing. Because of this, a certain amount of confusion exists with regard to the terminology and soldering with tin-lead alloys at temperatures up to about 400°C is commonly called 'soft soldering' to distinguish between 'hard soldering' in which silver bearing or silver based filler alloys are used at temperatures in excess of 500°C. Hard soldering above 500°C is, by definition, a brazing operation.

The soft solders are essentially based on the tin-lead alloy system, which is a simple eutectic system. Lead melts at 327°C, tin at 232°C and the 62.5% tin eutectic at 183°C. Intermediate compositions on either side of the eutectic point show the usual pasty melting range which, in the case of a 40% tin alloy for example, begins at 183°C and extends to about 230°C. The existence of this pasty condition is of great practical importance in the production of certain types of joint by the wiping technique.

Small quantities of other elements may be added to the basic tin-lead alloy in order to improve the mechanical properties of the solder or to modify its fluidity. Among these are cadmium, which improves fluidity, antimony, which increases hardness, and silver, which improves strength at elevated temperature. A small amount of copper may also be added in order to prolong the life of very fine copper soldering bits by inhibiting further solution of copper in the solder.

The basic process of soldering is very similar to that of brazing in that the parent material must be very carefully cleaned of surface oxide films and contamination prior to the application of heat, flux and filler metal. Capillary attraction of the filler metal between the parts to be joined may or may not be an essential feature of the joint design. Where capillary attraction into a lap joint is desired, then a solder having a high fluidity and low melting range is chosen and its composition would be near that of the eutectic. Where reinforced fillet joints are involved, a solder with a long pasty range is preferred and its composition would be of the 40% tin type. The use of a flux is essential in order to remove residual surface oxide films and to reduce the surface tension of the deposited solder, thus promoting good wetting of the parent-metal surface.

Equipment. The use of relatively low temperatures in this process not only simplifies the equipment but broadens the scope of possible heat sources. The most common means of applying solder is by way of a relatively large heat sink at a suitable temperature. Soldering irons are, in fact, made of copper, having good heat capacity, and, because of their high thermal conductivity, are easily heated from a point well away from the soldering tip. Hence they may be heated by gas flame or by means of electrical resistance heating wires suitably embedded in a holder. Copper itself may be soft soldered very readily and thus it is easy to wet the solder bit well, thus allowing rapid heat conduction from the heat sink through the liquid solder to the work-piece.

As in brazing, it is possible to use blowpipe methods of heating work-pieces which are assembled with preplaced quantities of solder. Because of the lower temperatures involved, however, oxy-coal gas or similar relatively cool flames are used in preference to oxyacetylene flames. For mass production work, components may be soldered in batch or continuous ovens, provided that adequate control over the temperature is available. Furnaces and ovens are progressively more difficult to control as their working temperature is reduced and it is often necessary to use forced air circulation at the lower temperatures if reasonable uniformity of heating is to be achieved.

Fluxes may be applied in paste, liquid or powder form, having various degrees of corrosiveness. Where large quantities of surface oxide or contaminates have to be removed, then it is often necessary to increase the corrosivity of the flux and, in these cases, very careful attention must be given to the removal of all flux residues after soldering.

'Killed spirits' is a corrosive acid flux for steel soldering made by adding zinc granules to hydrochloric acid until hydrogen evolution ceases. All traces of this type of flux must be removed from the assembly after soldering if subsequent corrosion is to be prevented.

In certain fields, it is possible to preclean the components sufficiently well for a relatively inert or only mildly corrosive flux such as the resin type to be used. Such resin fluxes are often incorporated as a core in extruded and drawn solder wires. Such wires are useful in the machine soldering operations where a predetermined quantity of solder and hence flux is automatically added to the joint.

One of the easiest materials to solder is tin-coated copper and many small components, particularly in the electronics industry, are pre-tinned by dipping or electrodeposition prior to soldering. Where large contact areas are involved, the components may be precoated with tin or solder and then fluxed prior to assembly. The whole assembly may then be heated in an oven or by other suitable means, raising its temperature uniformly until the tin or solder coating melts. Slight pressure on the joint causes intimate contact between the components and excess solder or tin is expelled. This method is much more reliable than ancillary feeding or even preplacement of solder foil, as the effectiveness of the precoating bond can reasonably easily be checked prior to assembly.

Molten baths to which fluxes have been added are used for precoating and the containers may either be of metal or graphite. A certain amount of alloying of the solder with a metal container may occur with the passage of time and the formation of relatively high melting point brittle compounds, which are washed away from the container walls, may ultimately impair the quality and fluidity of the solder in the bath. This situation is only important where the baths are kept molten for long periods where little work may be passing through. Under normal circumstances, drag-out losses are made up by fresh solder additions and thus the intermetallic compounds seldom reach an undesirable level.

Applications. The electrical and electronics industries could scarcely have arrived at their present advanced states had soft soldering techniques not been available. One advantage of using soldered joints in this field lies in the low temperatures used with the minimum resulting risk to heat sensitive components. Another advantage is the excellent electrical conductivity of properly made joints. The need for close control over soldering efficiency in electronics component assembly is especially important where high reliability in electrical conductivity is required.

Copper domestic water supply pipes and central heating systems are often joined by soldering using a sleeve technique. Fittings such as bends and tees are made up to allow socket joints, and the fittings contain preplaced solder in sufficient quantity to fill the joint by capillary action when the cleaned and fluxed assembly is heated. This is a useful technique and is generally more economical than the use of mechanical couplings. Similar techniques are used for joining bronze or gunmetal pipe fittings to larger diameter copper pipes in the brewing, distilling and chemical industries.

Solders may be obtained in the form of powders or pastes ready mixed with fluxes of varying characteristics to suit particular applications. Active fluxes are often based on zinc chloride and may contain small amounts of excess acid to assist the cleaning action. Organic acids are also used and generally have a milder, less corrosive action. Mild steel, nickel and stainless steel can be satisfactorily soldered, although not quite as easily as copper. Special problems surround the soldering of aluminium on account of its highly tenacious oxide film but fluxes and solders have been developed which permit the joining of aluminium by this method with almost equal ease as in the soldering of copper. The behaviour of soldered aluminium joints in normal humid atmospheres is, however, poor and they invariably corrode unless steps are taken to protect them completely from atmospheric contact. It has been found that the magnesium-bearing aluminium alloys spontaneously crack when exposed to the low melting point aluminium solders, particularly where the sheets or components to be joined are in a work-hardened condition.

In many food industry applications, the use of lead-bearing materials in the construction of process plant is prohibited. Items of plant such as filters and screens made from stainless steel or nickel wire mesh and other thin sections must often of necessity be soldered. Under these circumstances, the use of pure tin, or tin with a small addition of silver where better high-temperature properties are required, is a practical solution.

Properties of joints. Where joint strength is of major importance, it is not customary to consider soft soldering as a possible joining technique unless careful design steps are taken to ensure low stresses in the solder. The use of long socket rather than butt joints for the soldering of pipe fittings is a good case in point. With a socket joint, the area in shear may be increased to a point where a direct tensile pull on a fitting will result in failure of the pipe itself.

Soldering is commonly used either as a method of sealing a container against leakage or to provide joints with good electrical conductivity. In the design of the familiar tin can, mechanical strength of the joints is ensured by a fold over or interlocking mechanical design. The solder seals against leakage, although with some forms of interlocking design where a good area in shear can be arranged, it does contribute somewhat to the strength. Because of the difficulty of controlling the many parameters in soldering, such as actual metal temperature, time of contact, precleaning and efficiency of fluxing, the strength of joints made in production may be expected to show a wide scatter. A further difficulty lies in the fact that tin, lead and their alloys have very poor creep properties, even at ambient temperature. Thus, while a short time mechanical test might show a shear strength in the joint of about 2 t.s.i., stresses of the order of only 2% of this may be sustained for prolonged periods.

In the electronics field most soldered joints are made using non-corrosive resin fluxes, in view of the impossibility of flux removal after assembly. Resin fluxes suffer the disadvantage that they clean the parent-metal surface only slowly whilst at the same time promoting the spread of solder over the parent-metal surface. It is thus possible to obtain a visually acceptable joint which, in fact, has a thin layer of solidified flux between the solder and the parent material. In many cases, this joint is mechanically adequate but, as it has very poor electrical conductivity, its effect upon circuit performance can be quite marked. Such joints are difficult to trace and therefore the condi-

8.6 DESIGN OF METALLURGICAL JOINTS

8.6.1 General

tion should be avoided by careful attention to precleaning techniques and heating time in soldering.

Although the designer will have the mechanical and corrosion-resisting properties of joints firmly in mind when drawing up a construction, it is important that he bears in mind a number of factors concerning the means by which these desirable properties are to be obtained. Not only must the metallurgical characteristics of the material in question be examined, but the joining process, with all its limitations of applicable thickness range and need for unhindered accessibility, must also be carefully considered.

The degree and method of inspection to be applied to the finished joints will also play a part in determining the precise type and disposition of the joints in the structure. Where, for example, leak testing by sensing traces of halogenated hydro carbon gas is to be carried out, then the design must allow enclosure of the test gas on one side of the joint and ready access for the sensing equipment on the other side. If examination is to be by radiography, then it must be possible to place sensitized film material on one side of the joint and a source of radiation on the other in such a way that a clear and undistorted image of the weld is subsequently obtained on the film.

With the growth in the number of alternative joining processes and techniques, it is essential that design engineers thoroughly understand their scope of application and mark drawings with adequate detail, defining the precise technique in mind and the way in which the component edges are to be prepared for joining. Increasing reliance is placed on metal joining techniques and it is no longer acceptable for drawings to be marked with the vague but optimistic instruction 'weld here'.

The scope of this section does not allow highly detailed treatment of this involved and sometimes obscure subject of joint design and further information should be sought from standard codes of practice, individual metal manufacturers, the suppliers of the welding equipment or from the literature indicated for further reading.

8.6.2 Lap joints

Joints made by brazing, soldering or by any of the non-fusion methods such as resistance or pressure welding will almost certainly be lap or overlap joints of one form or another (see fig. 8.8). Such joints will normally be shear loaded in service and, beyond specifying the area to be in shear, the designer has little freedom of action, apart from ensuring that the specified process is suitable for the thicknesses under consideration and that in brazing or soldering, for example, the components can be held together satisfactorily while the filler material is molten.

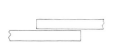

Fig. 8.8 Lap joint

8.6.3 Fillet joints

Brazed or soldered lap joints of the socket type often require a reinforcing filler between the components. In these circumstances, the designer should specify the appropriate filler alloy composition which will allow the development of the desired

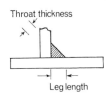

Fig. 8.9 Fillet joint

fillet. Pure metals and eutectic alloy compositions are in general unsuitable for this purpose and alloys having a substantial pasty range should be selected.

The size of fillet welds (see fig. 8.9) should be specified either in terms of leg length or throat thickness as defined in *B.S.S. 499*. These dimensions have some meaning in terms of engineering calculations and, as a general rule, the leg length of a fillet weld should not be less than the thickness of the thinner component at the joint.

In some applications where the risk of leakage through a fillet weld cannot be tolerated, it is often desirable to specify that the weld be made with not less than two runs.

8.6.4 Butt joints

a. Joints between plate and sheet edges. This type of joint is usually made by fusion welding and it is necessary to prepare the edges by machining, grinding or some form of thermal cutting to such a form that they can be properly fused, and where necessary additional filler metal may be deposited in such a way that lack of fusion and slag inclusions are avoided. In material more than about 0.15 inches thick, it is necessary to provide a groove preparation in which the weld metal may be deposited. This thickness of 0.15 inches is quite arbitrary and may be considerably greater where deeply penetrating machine welding techniques are applied. In machine welding, the edges to be joined are often machined square and close butted together prior to welding.

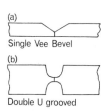

Fig. 8.10 (a) Single vee bevel (b) grooved butt joints.

Where a groove is required this may be either a simple bevel, 'V' shaped, or may be 'J' or 'U' shaped. In order to balance the welding and minimize distortion, a double V, U or J groove may be formed (see fig. 8.10), subsequent welding being from both sides of the sheet or plate. The radius at the bottom of U or J grooves is of sufficient size to allow good fusion and deposition of weld metal without forming slag pockets or voids by weld metal bridging, which sometimes occurs in the bottom of a V groove. The J or U form is normally applied to thicker materials where a V groove would be excessively wide at the top.

In general, a designer should aim at weld cross-sections of minimum area as these will not only prove most economical but will also minimize the problem of welding distortion. Distortion problems generally become more troublesome as the section thickness increases as it is usually possible to correct distortion of thinner sections by mechanical means, e.g. rolling or hammering where cold working is not barred for metallurgical reasons.

Groove angles vary considerably and are determined by the conductivity of the parent material–the greater the conductivity, the wider the groove necessary to allow effective fusion–and the difficulty or otherwise of manipulating the welding equipment. Cumbersome equipment such as that used in m.i.g. welding of aluminium requires the use of wide groove angles in order that the arc may be presented squarely to the surface on which the metal is being deposited. Certain types of t.i.g. weld carried out by machine on materials of low thermal conductivity, such as heat-resisting stainless steel for example, may be made quite satisfactorily in J grooves having parallel sides.

In all cases where joint quality is of importance, it is advisable to carry out a practical test using the proposed joining technique and joint design before welding work on the assembly itself is allowed to begin. Test pieces prepared in this way may then be subject to destructive examination in order to ensure that adequate fusion and weld quality have been achieved.

   b. Vessel penetrations. Because of the difficulty of carrying out non-destructive tests on butt joints between, for example, a pipe and a vessel wall through which it penetrates, such joints have tended in the past to be classed as fillet welds and have often been of quite inferior quality. Searching applications, such as those found in the nuclear power industry, have demanded improved standards for this type of joint and much attention has been given to detailed design. The problem is essentially that of achieving good fusion between two sections of different heat capacity. The same principles apply as in joining plate edges, in that heat conduction and access for the welding equipment must determine the precise configuration of the edge preparation. In general, the groove angle on that side of the joint extracting heat at the greatest rate will be shallower than that on the other side, and this would explain some of the assymmetry seen in pressure vessel joint details. Reference to the pressure vessel codes *B.S.S. 1500* and *1515* will show that a great number of typical details have been included and the need for the many variations can be judged in the light of the principles outlined above. Here again, preproduction tests are invaluable where first-class joint quality is desired.

8.6.5 Pipe butt welds

Most pipe butt welds are made by fusion welding and the difficulties of producing a smooth, even, penetration bead on the inside of a pipe are well-known to those who have tried. T.i.g. welding is probably the most useful process for pipe butt welding and even where large wall thicknesses are involved, the t.i.g. method is often used for the first passes in the root. Filling passes may be made by the metal arc, submerged arc or m.i.g. processes. Where the t.i.g. process is used, the availability of argon makes it possible to arrange local purging of the pipe bore at the point of welding. By displacing the air in a zone between two suitably positioned bungs, oxidation and contamination of the weld root penetration bead can be eliminated.

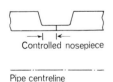

Controlled nosepiece

Pipe centreline

Fig. 8.11 Pipe butt weld preparation.

This technique is particularly valuable in stainless and heat-resisting steel welding as it allows unhindered flow of the molten metal and the production of an exceptionally smooth penetration bead. Here, the butting edges are prepared by machining so that they match exactly. A J groove is provided for thick walled pipes and the bottom of the J may have a closely controlled projecting nose-piece (see fig. 8.11) so that the root pass is in effect made on material of only about 0.05 inches wall thickness. The important features of this joint design are that the butting nose-pieces shall match exactly and shall be of equal thickness. Any variation in diameter or wall thickness between the butting pieces will result in poor irregular penetration.

Carbon-steel pipes are often prepared with a single V bevel, the root pass being made by manual metal arc welding in a gap maintained between the butting edges. Skilled welders can produce quite acceptable results although the smoothness of the penetration bead is seldom as good as that produced in t.i.g. welding.

It will be evident that the material presented in Sections 8.3 to 8.6 is far from adequate where detailed instructions in welding, brazing or soldering are required. An attempt has been made to highlight some of the features and pitfalls, but omissions are bound to be numerous in any condensed treatment of this type. The reader is strongly recommended to pursue the subject by reference to the literature indicated at the end of the chapter. In this respect, the *American Welding Handbook* is to be highly commended.

8.7 ADHESIVE JOINING OF METALS

8.7.1 Introduction

The use of adhesives as a method of joining components has a long history, but one of the first serious applications of adhesives in structural work was their use for bonding metal to wood in the construction of the highly successful de Havilland Hornet aeroplane at the end of world war II. In modern times, advances have been made in adhesive technology and extensive use is still made of this technique in the aircraft industry, exemplified by the use of Redux adhesive in the construction of Aeroweb honeycomb used for the buoyancy tanks and car deck of the SRN4 hovercraft. Needless to say, reliability is of paramount importance when considering the use of adhesives in aircraft construction and the fact that they are so extensively used is a mark of confidence in adhesively bonded joints.

Ideally, any adhesive should be as strong as the material it is bonding but when used in conjunction with metals this ideal has not yet been achieved. In fact, considerable advances must be made before an approach to this situation is realized. A comparison of the tensile strength of a typical aluminium-magnesium alloy, which lies in the range of 20 to 25 t.s.i., with that of an epoxy resin adhesive, which in shear is 2 to 3 t.s.i., adequately illustrates this point.

8.7.2 Theories of adhesion

Before discussing the practical aspects, brief mention may be made of some of the more fundamental principles and theories of adhesion. The science of adhesion

Fig. 8.12 Aeroweb aluminium honeycomb cored panels coated with an Araldite based non-skid formulation (exterior surface of the SRN 4 Hovercraft).

has developed during the past 50 years but even today there is no single unified theory which can adequately explain the behaviour of all adhesives under all conditions. The three widely accepted theories, namely the diffusion (see ref. 1), electrostatic (see ref. 2) and adsorption theories, are all applicable to certain aspects of adhesion within a given field, and persuasive arguments have been advanced by the supporters of each. It is outside the scope of this section to cover all aspects of the different theoretical treatments and only the adsorption theory will be briefly outlined on the basis that it explains many of the more important technological aspects of adhesion.

a. Adsorption theory. The adsorption theory is based on the principle that gaseous and liquid molecules are adsorbed onto solid surfaces, physical forces such as van der Waals and London dispersion forces making contributions to bond formation. Reinforcement of the bond may be effected through permanent dipoles of constituent groups of the adhesive molecule or even by the formation of covalent bonds (chemisorption) which occur with certain adhesive-adherend combinations. The role played by polar groups in the adhesive is therefore of prime importance and a lot of experimental work has been directed at studying molecular orientation at interfaces by workers such as Crisp, Glazer and Sharpe (see refs. 3, 4 and 5). Adsorption of a liquid onto a solid surface involves an energy change and it is on the basis of surface energetics that any discussion of the adsorption theory of adhesion must be conducted. For a detailed account, reference should be made to the survey by Eley and Dunning (see ref. 6) and the A.C.S. Monograph on adhesion (ref. 7).

b. The effect of temperature. It is often considered that failure of an adhesive at elevated temperature is due to the adhesive melting. This is not always the case and consideration of this in terms of the adsorption theory will elucidate the current thinking. Wake (see ref. 8) has pointed out that the adsorbed adhesive molecule loses one and possibly all three of its translational degrees of freedom. In complex polymers the adsorbed section of the polymer has lost rotational freedom and, on increasing the temperature, the increased thermal vibration of the molecule results in a loss of the degree of order which is a characteristic of adsorption. Thus, increasing the temperature, where only physical adsorption is involved, causes increasing disorientation and hence fewer adsorbed molecules. Evidence of this theory is presented by the same author (see ref. 9) who, using data of Kraus and Manson (see ref. 10), shows that failure of polyethylene and polystyrene bonded to steel occurs at temperatures below their melting points and suggests that this may correspond to the onset of some freedom of movement such as rotation of a whole segment of the molecule.

The theory of adhesion based on adsorption is, however, not applicable where chemisorption occurs, i.e., where the adhesive and adherend molecules chemically combine. Strong evidence for this type of bonding is presented by Buchan (see ref. 11), who was able to show that in a rubber–metal bond the strength was increased by brass plating the metal surface. It is postulated that copper in the brass chemically combines with the sulphur atoms in the rubber to form covalent copper sulphide bonds. Failure of the bond in this case is linked with dissociation of the bond rather than with desorption.

c. Viscosity. For an adhesive to be effective a prime consideration is that it must wet the surface, i.e., it must exhibit a low contact angle and spread rapidly. Although surface energetics are important in this respect, the viscosity of the adhesive also influences wetting. For example, with a rough porous surface penetration of the adhesive into the cavities of the surface must occur for the adhesion to be effective and, whilst over-penetration is generally regarded as undesirable in so far as it may cause a break in the glue line, penetration is necessary and the viscosity of the adhesive plays an important role in this respect. A mathematical treatment of the effect of viscosity on joint strength has been made by de Bruyne (see ref. 12) but for a more extensive treatment of the whole aspect of adhesive rheology, reference should be made to the work described by Hoekstra and Fritzius (see ref. 13). A further aspect of adhesive viscosity which is worthy of mention is that, where gap-filling and jointing are simultaneously required, low viscosity Newtonian fluids are of little value, the more viscous and preferably non-Newtonian being more applicable.

8.7.3 Nature of the adherend surfaces

In bonding metals, probably more than other materials, the pretreatment of the bonding surfaces is of paramount importance. This varies from metal to metal and may comprise the steps from simple degreasing to chemical etching. It is also found in practice that the pretreatment necessary will be determined by the nature of the adhesive being used as well as the adherends. In the case of degreasing, wiping over the surface with solvent is only effective in removing gross contamination and does not remove a thin adherent layer which can seriously affect the strength of the subsequently formed bond. If a bond of premium strength is required, it is usually necessary to use a technique such as vapour degreasing. As mentioned, chemical treatment of the surface may be necessary and there are various reasons for the need of this. In the case of iron, for example, the oxide film is very loosely held to the metal surface and therefore the strength of a bond formed on lightly oxidized steel is limited to the strength of the iron–iron oxide bond. On the other hand, the tenaciously held chromium oxide film on stainless steel forms an excellent bonding surface.

Aluminium is a metal to which considerable attention has been paid, due mainly to its extensive use in the aircraft industry and the increased use of adhesives in this field. Whilst it will be recognized that an oxide film rapidly forms on any aluminium surface, a point not often appreciated is that the film so formed is of uncontrolled structure and thickness. In order to form a surface to which adhesives may form strong bonds, the accepted procedure is to remove the naturally formed oxide and reoxidize the surface under controlled conditions. Use of sulphuric–chromic acid mixture at a temperature below 65°C produces a continuous film, controlled in thickness, structure and composition, of bayerite (β-Al_2O_3.$3H_2O$). Above 65°C boehmite, (α-Al_2O_3.$3H_2O$) is formed which is known to possess bonding properties inferior to bayerite. The full etching procedure and composition of the etchants is described in the *Ministry of Aviation specification DTD 915B* which is the accepted procedure for use in the aircraft industry.

8.7.4 Surface pretreatment

The accompanying table outlines the procedure for metal treatment when using the epoxy based resins as adhesives (see ref. 14) but it must be emphasized that, in all cases, simple degreasing and abrading (and degreasing), which frequently may be adequate, always precede the chemical treatment. In general terms, it is only necessary to resort to chemical etching of the surface if bonds of utmost reliability and strength are required.

Metal	Pretreatment
Aluminium	Degrease. Abrade with emery cloth (and degrease). Etch with a solution containing 15% v/v sulphuric acid (s.g. 1.82) and 7½% w/v sodium dichromate at a temperature of 60 to 65°C for 30 min. Rinse and then wash in clean cold running water. Dry in hot air, the temperature of which must not exceed 65°C.
Cadmium	Degrease. Abrade with emery cloth (and degrease). Preferably electroplate with silver or nickel.
Cast iron	Degrease. Grit blast or abrade with emery cloth (and degrease).
Chromium	Degrease. Abrade with emery cloth (and degrease) or preferably etch with 16% w/v hydrochloric acid for 1 to 5 min at 90 to 95°C and follow by washing with cold and then hot water. Dry with hot air.
Copper and copper alloys	Degrease. Abrade with emery cloth (and degrease) or preferably etch with a solution containing 2.6% w/v ferric chloride and 12.5% v/v nitric acid for 1 to 2 min at room temperature. Wash with copious amounts of cold water and dry in a cold air stream in order to avoid staining. For intricate shapes where thorough washing may be difficult a 25% ammonium persulphate solution may be used at room temperature for 30 s. Wash with cold water and dry in a stream of cold air. As with the ferric chloride etch, staining of the metal may occur if the metal is not dried immediately or if hot air is used for drying.
Gold	Degrease.
Lead	Degrease. Abrade with emery cloth (and degrease).
Magnesium	Degrease. Abrade with emery cloth and apply adhesive immediately. Alternatively, immerse in an 11% w/v solution of caustic soda at 70 to 75°C for 5 min. Wash with cold water and etch with a 9% w/v solution of chromic oxide (CrO_3) containing 0.06% sodium sulphate.
Nickel	Degrease. Abrade with emery cloth (and degrease) or preferably etch for 5 s in concentrated nitric acid (s.g. 1.42), wash with cold and then hot water and dry in hot air.
Solder	Degrease. Abrade with emery cloth (and degrease).
Steel – mild	Degrease. Grit blast or abrade with emery cloth (and degrease). A more suitable chemical treatment is: immerse for 10 min in a 33% v/v solution of orthophosphoric acid (88%) in industrial methylated spirits at 60°C, wash with cold water, brush off the black deposit and dry at 120°C for 1 h.
Steel – stainless	Degrease. Remove any surface deposits with alumina-grit cloth or other non-metallic agents. For final degreasing an aqueous metasilicate/phosphate/hydroxide solution is preferable to the usual chlorinated hydro carbon solvent. A typical degreasing agent comprises:

Sodium metasilicate	2 lb
Tetrasodium pyrophosphate	1 lb
Sodium hydroxide	1 lb
*Nana S. Powder	5 oz
Water	10 gal

*Marchon Products Ltd.

For bonds of maximum heat or peel resistance, it is necessary to use chemical etchants. For maximum heat resistance etch the steel surface with a solution containing 10% w/w oxalic acid and 10% w/w sulphuric acid. Treat at 85 to 90°C for 10 min, wash the sample with clean cold water, remove the black surface deposit by brushing and dry with hot air. For maximum peel strength, use an etch of concentrated sulphuric acid (s.g. 1.82) containing 3.5% of a saturated solution of sodium dichromate. Treat at 50°C for 15 min and wash with sodium meta-silicate solution (as described before) rather than a chlorinated hydro carbon.

Tungsten and tungsten carbide — Degrease. Abrade with emery cloth (and degrease). Etch with a 30% w/w solution of sodium hydroxide for 10 min at 80 to 90°C, wash with cold water, then hot water and dry with hot air.

Zinc and zinc alloys — Degrease. Abrade with emery cloth (and degrease) and apply adhesive immediately.

8.7.5 The design of joints

a. Stresses in joints. Before embarking on a description of joint design, an illustration of the various types of stress to which a joint may be subjected will form a useful introduction (see fig. 8.13).

In any system to which an external force is applied, the distribution of stress is rarely uniform. This is particularly the case in a system incorporating a joint, and the difference at either side of the joint, that is, in the adherend and the adhesive, will result in a stress concentration. The significance of this is that different areas of the joint are subject to different stresses and initial failure will occur in the most highly

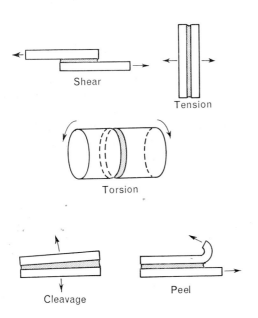

Fig. 8.13

stressed area. If, for example, a simple lap joint is considered (see fig. 8.14), it will be noted that there is an eccentricity of loading in the region of the overlap which produces a bending moment which can produce final deformation of the joint adherends. Therefore, a simple lap joint in shear will, when deformed, give rise to peeling stresses, the strain differential present in the joint being a maximum at the ends of the joints which are therefore the points of initial failure. Since the load transmitted by the adhesive is equal to the applied load, the relationship

$$\tau l w = \sigma_t t w$$

may be deduced where τ is the shear stress in the bend line, σ_t is the direct stress in the metal, l the joint overlap, w the metal width and t the metal thickness. For any system, a curve of failing stress against the t/l ratio may be established which will aid the designer in joint design. McMullen and Garnish (see ref. 15) have presented a series of typical curves for Redux bonded aluminium alloy (see fig. 8.15). Using the technique of photoelasticity with model joints, McLaren and MacInnes (see ref. 16) were able to illustrate that a lap joint modified in the form shown in fig. 8.14 (b) showed a marked reduction in peeling stress whilst Greenwood and colleagues (see ref. 17), using steel adherends, designed a joint with a negative bending moment and were able to show how such joints exhibited the property of a higher load to failure than the more common type of joint.

The main implication of the foregoing is that the strength of a joint cannot be expressed simply by the breaking load divided by the area of the joint. Analyses of the stresses in glued joints have been made by several workers and the reader is referred to publications by Volkersen, Goland and Reissner, Plantema and Sneddon (see refs. 18 to 21) for more detailed information.

Design of joints on a mathematical basis is obviously a complex process but, from a purely practical aspect, the prime objective is to ensure an even distribution of the load over the whole bonded area, thereby minimizing stress concentrations. Furthermore, the conditions approaching an evenly stressed joint are only obtained when the joints are in tension or shear. (The case of torsion will, upon consideration, be recognized as a variant of a system in shear.) In the case of cleavage, for example, the stresses are concentrated along a small section of the adhesive where the bending moment is greatest. The case of peeling stress is perhaps the weakest type of all since the stress is concentrated along the glue line. With these concepts in mind, it will be appreciated that certain designs are acceptable whereas others are most unsatisfactory from a strength point of view.

b. Lap joints. Lap jointing is one of the most frequently used types of bonded joint and there are several modifications to the basically simple joint, all aimed at

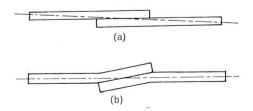

Fig. 8.14

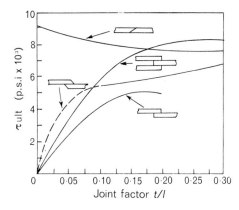

Fig. 8.15 Average ultimate shear stress in Redux bonded aluminium alloy 2.4 S-T tested at room temperature.

reducing peeling and cleavage stresses, thereby increasing bond strength (see fig. 8.16).

(a) The simple lap joint is reasonably good although it has limitations due to the fact that, in shear, the applied forces are not acting in the same plane and hence the resulting bending motion results in cleavage stresses on the joint. The main advantage is that it requires no premachining of the adherends and is therefore extremely simple to apply in practice. It is universally accepted as the configuration for testing adhesives and may be used as a basis for providing much design information.

(b) A modification to the simple lap joint produces the joggled lap in which it will be noted that the stress forces are made to act along the same plane and distribute the load evenly over the bonded area. This type of joint is therefore stronger and requires only the simplest of metal-forming operations to prepare one of the mating surfaces. In very thin sheet metal, the joggle tends to straighten introducing cleavage stresses into the joint.

(c) The scarf joint, which may also be considered to be a variant on the butt joint (see later), produces a similar effect and distributes the stresses evenly across the jointed surfaces. It is generally regarded as being a most satisfactory joint but it suffers

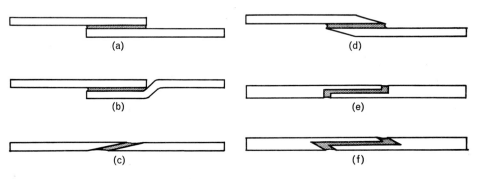

Fig. 8.16 Lap joints: (a) Simple lap joint.
(b) Joggled lap joint.
(c) Scarf joint.
(d) Tapered single lap joint.
(e) Double lap joint.
(f) Double scarf lap joint.

ADHESIVE JOINING OF METALS | 651

two serious disadvantages in so far as it requires accurate machining of the mating surfaces and is not applicable to thin sheet sections.

(d) The tapered single lap is another variant of the simple lap. The principle by which the stressing loads are made to act in the same plane is based on the assumption that the tapered edge of the joint will permit bending in preference to failure of the adhesives. Again, this joint suffers from the disadvantages of the need for machining the edges.

(e) and (f) The double lap and double scarf joints are similar, both being very strong, as would be expected from their construction which maintains the load bearing area in the same plane as the shearing stress. The advantage of the double scarf joint is that it shows excellent resistance to bending forces. However, neither type of joint is suitable for thin metal sections since both necessitate accurate machining of the adherend faces, an essential if the advantages of this type of bond are to be fully exploited.

c. Stiffening joints. The use of adhesives in bonding stiffening members to thin sheet material in order to improve rigidity is assuming increasing importance. The actual form of the stiffener may be one of several types, e.g., a T section, inverted U section or a castellated section (see fig. 8.18). The most important feature is that the thickness of the sheet and stiffener must be matched in order to ensure that any flexing of the bonded structure will not introduce peeling stresses.

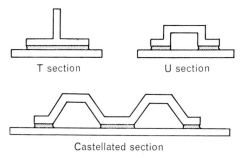

Fig. 8.18 The use of various sections to increase the rigidity of thin sheet material.

d. Cylindrical joints. Although the butt joint can be used successfully for joining cylindrical sections, it has certain limitations. A joint of this type shows excellent strength in pure shear and when subjected to pure torsional stresses, but when subject

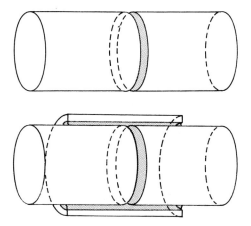

Fig. 8.19 Cylindrical joints.

652 | JOINING OF METALS

Fig. 8.17 Bonnet assembly of Aston Martin D.B. 5 car using Araldite epoxy adhesives. The aluminium alloy double curvature bonnet is strengthened by glueing a positively located inner panel to which hinges and bonnet catches are attached.

to bending forces, the joint has limitations due to cleavage stresses. This defect may be overcome by sleeving, which not only minimizes failure through cleavage but also increases the strength of the joint to shear and torsion (see fig. 8.19).

e. Angle joints. This type of joint (see fig. 8.20) is particularly prone to failure through cleavage stresses and, in the case of thin gauge metal, peeling stresses may contribute largely to failure. The simple angle joints are singularly prone to failure and generally regarded as poor design, although a joint of the type shown in fig. 8.20 (c) is somewhat of a compromise but undoubtedly possesses greater strength than either of the joints in (a) or (b). The rebated angle joint is of better design in so far as it relieves the cleavage stresses. It does, however, necessitate machining operations and is obviously unsuitable for thin gauge sheet metal.

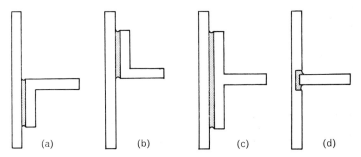

Fig. 8.20 Various types of angle joint:
(a) and (b) Simple angle joints.
(c) T section joint.
(d) Rebated angle joint.

ADHESIVE JOINING OF METALS | 653

f. Butt joints (see fig. 8.21). The failings of a simple butt joint are attributable to the relatively small area which is bonded and to the very poor resistance to bending moments. By tongueing and grooving the butting edges, an improvement in the general performance of this type of joint is achieved, although as with many of the stronger joints they are unsuitable for use in thin gauge metal and necessitate accurate edge machining. By incorporating the straps as in fig. 8.21(e) and (f), joints which may be regarded as two lap joints back to back are obtained. In so far as the stresses are uniformly distributed through the adhesive layer, which lies in the same plane as the

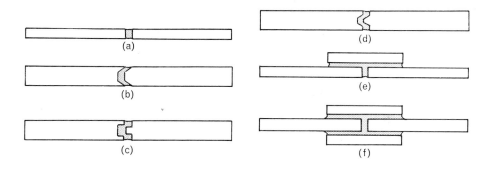

Fig. 8.21 Butt joints of various types:
  (a) Simple butt.
  (b) Scarfed and groove.
  (c) Tongue and groove.
  (d) Landed scarf tongue and groove.
  (e) Single strapped butt.
  (f) Double strapped butt.

applied load, these joints show useful strength in shear. They also have the advantage over the simple butt joint in that they are resistant to cleavage stresses under bending loads, particularly in the case of the double strapped joint.

8.7.6 Adhesive materials

Reference to any treatise on adhesives will show that the variety of materials available to the engineer is formidable. Some of the more important groups of adhesives are given below.

Acrylics	Phenolic neoprene	Rubber–natural
Cyanoacrylate	Phenolic nitrile	neoprene
Epoxy resins	Phenolic polyamide	nitrile
Epoxy phenolic	Phenolic vinyl	Silicate
Epoxy polyamide	Polyamide	Silicone
Epoxy polysulphide	Polyurethane	Special adhesives.
Epoxy silicone	Polyvinyl acetate	

654 | JOINING OF METALS

In general, the demands of industry vary so widely that no strict recommendations can be made for bonding any particular metal or combination of metals. It is frequently advisable to approach the manufacturers of adhesives who have the facilities for evaluating adhesive performance under specific conditions. For example, in structural work, the magnitude and types of stresses to which the bond is to be subjected will dictate the type of adhesive to be used. If the bond is likely to be continuously stressed in peel or shear, a material which exhibits low flexibility and high creep tendencies would obviously be unsuitable. On the other hand, a vibratory motion which would induce fatigue in a rigid bond would be withstood by a flexible adhesive. The effect of moisture must also be taken into account as some adhesives are more prone to deterioration in a damp environment than others. Similarly, as has been shown, all adhesives lose strength at elevated temperature and therefore the maximum working temperature must be considered along with the other criteria. The following descriptions of adhesives are aimed at providing basic information on the characteristics of the main groups and the relative merits and limitations of each. The list is by no means exhaustive and it must be emphasized that, for more fundamental data, the reader should refer to one of the excellent textbooks on the subject (see refs. 22 and 23).

a. Acrylic resins. This group of resins, which have limited use as adhesives, are based on acrylic and methyl acrylic esters,

i.e.

$$CH_2 = CH.COOR \qquad CH_2 = C(CH_3).COOR$$
$$\text{acrylic ester} \qquad\qquad \text{methyl acrylic ester}$$

These materials undergo polymerization under the influence of acids, heat, ultraviolet light and organic peroxides. In general, the acrylic esters undergo polymerization more readily than the methacrylates. The methacrylates are usually polymerized with peroxides (e.g. benzoyl peroxide) and, if in solvents, the nature of the solvent can have a marked effect on the degree of polymerization (see ref. 24) and hence on the properties of the polymer. The range of Loctite adhesives (Douglas Kane (Sealants) Ltd.), which are polyester-acrylate based, have proved singularly effective in many engineering applications. Typical examples quoted include permanent thread locking, bonding non-threaded tubing, retaining bearings and thread locking against vibration. Tensile strengths of up to 6000 p.s.i. are quoted but values in the range 2000 to 3000 p.s.i. tend to be characteristic. Modifications to the polyester–acrylate resins can produce adhesives which show a good degree of flexibility and peel strength but this is at the expense of maximum service temperature which is normally in the range 100 to 150°C.

b. Cyanoacrylates. This range of somewhat unique adhesives is based on cyanoacrylic acid, one such proprietary product (Eastman 910) being the methyl ester

$$CH_2 = \underset{\underset{COOCH_3}{|}}{\overset{\overset{CN}{|}}{C}}$$

ADHESIVE JOINING OF METALS | 655

Fig. 8.22 The use of a cyancrylate adhesive (Eastman 910). Its rapid bond forming properties are fully exploited.

This compound possesses the property that a thin film of the monomer readily polymerizes at room temperature when applied between two surfaces. The bond strength achieved with the majority of materials is high, the exceptions being when applied to low surface energy polymers. To illustrate the speed with which bonds are formed, results published by Eastman Chemical Products Inc. may be considered (see ref. 25 and table 8.3).

The monomer has a very low viscosity and surface tension and therefore high energy surfaces are rapidly wetted even in the complex irregularities of a machined surface and subsequent polymerization ensures a continuous film with a minimum of flaws. The nitrile groups, which possess very high dipole moment, are readily oriented and strongly adsorbed at the interface and possibly account for the high bond strengths achieved. Although very expensive, cyanoacrylates are very economical in use, one or two drops being adequate for many bonding operations. The polymer shows a good resistance to solvents such as alcohol, benzene and acetone but prolonged immersion in water, dilute acids and alkalis reduces the bond strength. The maximum recommended service temperature is 80°C, although short periods of operation at 100°C can sometimes be tolerated – a minimum operating temperature of −65°C is quoted.

Being of low viscosity in the monomeric form, these adhesives have no gap-filling

Table 8.3

Adherends	Time to firm set	Shear strength p.s.i.	Age of bond
Al-Al	2 min	1484	10 min
		2188	1 h
		2700	48 h
		2800 (tensile)	24 h
Steel-steel	2 min	1362	10 min
		2224	1 h
		2800	48 h
		5030 (tensile)	48 h

properties and therefore they necessitate a closely fitting joint or well-machined surface if their advantages are to be fully exploited.

c. Epoxy resins. Adhesives based on epoxy resin are the most important group of general purpose adhesives, the whole subject being recently covered by Lee and Neville (see ref. 26). The discovery of these materials dates back to 1938 (see ref. 27) but the majority of commercial resins are still based on Bisphenol A (bisphenol dihydroxy-diphenyl methane):

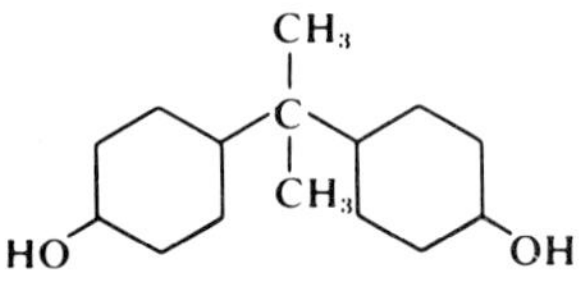

CH3
C
CH3
HO
OH

Bisphenol A

O
CH2.CH.CH2Cl

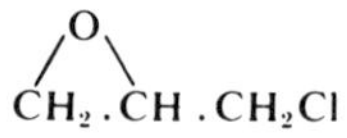

epichlorhydrin

which, on reaction with epichlorhydrin, produces the co-polymer terminated by epoxide groups:

CH3
C
CH3
CH2.CH.CH2
O
CH2
CH2
CH
OH
n
CH2.CH.CH2

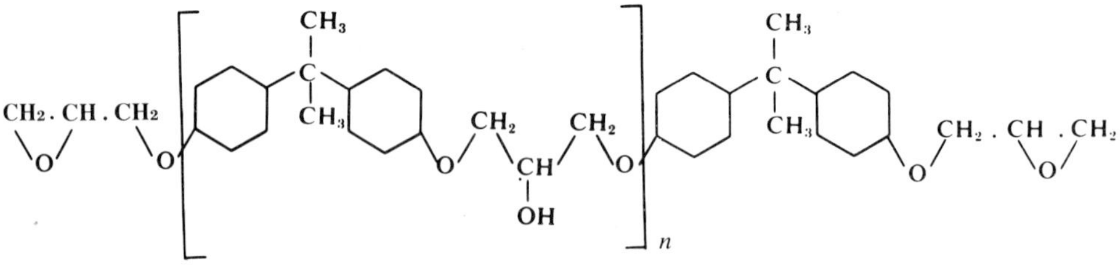

The lower molecular weight polymers (i.e. $n < 10$) are viscous liquids whereas the higher polymers tend to be high melting point solids. The presence and versatility of the hydroxyl and epoxide groups makes the choice of cross-linking and polymerizing agents very wide and hence the properties of the final product can be varied to suit any particular application.

For example, the primary and secondary amines react additively with the functional groups

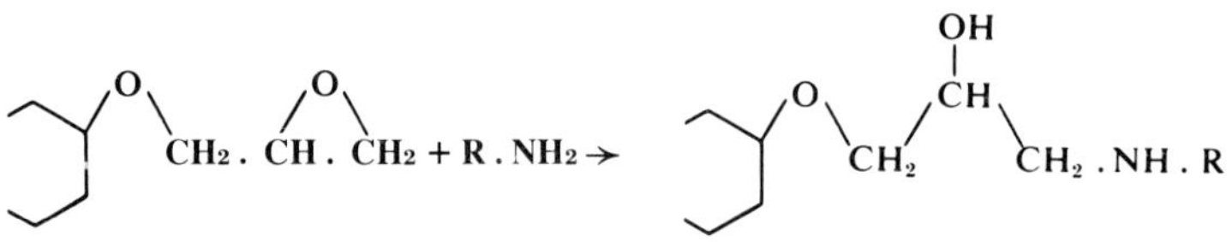

O
CH2. CH. CH2 + R.NH2 →
OH
CH
CH2
CH2.NH.R

while polyfunctional amines can yield cross-linked products represented by

CH
CH
CH2
CH2.NH.R.NH.CH2
OH
CH
CH2
O

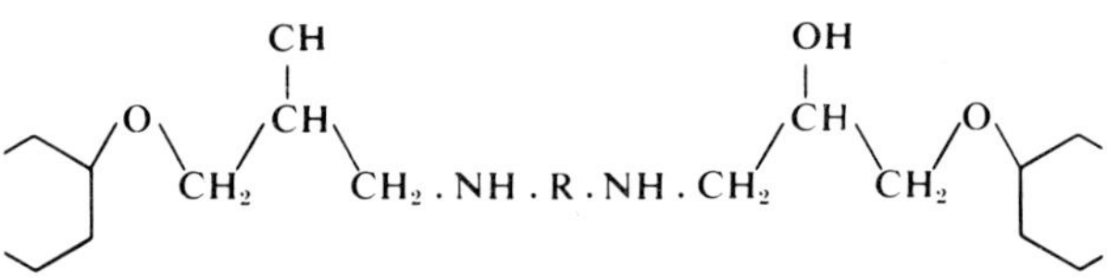

ADHESIVE JOINING OF METALS | 657

Tertiary and quaternary amines catalyse the reaction between hydroxyl and epoxide groups:

$$\text{R-O-CH}_2\text{-CH(OH)-CH}_2\text{-O-R'} + \text{CH}_2\text{-CH-CH}_2\text{-O-R''}$$

$$\downarrow$$

although with a low molecular weight polyamine such as diethylene triamine or tetraethylene pentamine, it is probable that both the above two mechanisms contribute to the polymerization process. There are innumerable amine curing agents in use and amongst these may be listed tridimethyl aminomethyl phenol, 4-4' diaminodiphenyl methane, morpholine derivatives and *m*-phenylene diamine. Complex amine curing agents such as dipentamethylene or tetramethyl-thiuram tetrasulphide (see ref. 28) are claimed to provide cured resins with improved adhesion properties. Other non-amine curing agents are also used including polyamides and phenolic resins, which are discussed later, and fatty acids/anhydrides such as pyromellitic dianhydride (see ref. 29). Isocyanates have been employed to produce resins which, although possessing good flexibility, have rather poor chemical resistance. More recently, materials such as triethanolamine-chelated titanium or phosphorus compounds and a borate ester (see ref. 30) have been suggested in the formulation of epoxy adhesive films.

Examination of the literature and trade brochures shows that a large number of modifications to the epoxy resin systems have been proposed, aimed at improving such properties as flexibility, service temperature and adhesive strength and mention can be

Fig. 8.23 Aluminium window frames being assembled using Araldite epoxy resin.

made of such material as butadiene acrylonitrile co-polymer (refs. 31 and 32), heterocylic nitrogen containing monomers (see ref. 33) and polyvinyl acetal resins (see ref. 34), as well as materials such as polysulphides and silicone resins which are discussed in subsequent sections.

With the unmodified epoxy systems, extremely high bond strengths, 5000 p.s.i. in shear, are obtainable but their drawback lies in their low flexibility. A further disadvantage is due to the decrease in bond strength which occurs at increased temperature and an upper service temperature of 90°C is the usually accepted value. Due to their availability in highly viscous forms, they are useful where gap-filling as well as adhesion is required. They also show good resistance to water and a wide range of solvents and, in spite of their disadvantages, they form an extremely versatile and useful group of adhesives.

Epoxy phenolics. Co-reaction of epoxy and phenolic resins gives rise to a series of extremely useful co-polymer adhesives but, since the reaction is a thermally initiated process, the resins necessitate a heat curing cycle. They do, however, produce bonds with a shear strength of up to 3000 p.s.i. which may be only reduced to 30% of this value at 250°C. It is also reported (see ref. 35) that a resin is available which retains a good bond strength at 300°C whilst possessing good peel strength and is therefore suitable for bonding applications where a high service temperature is operative. Although showing excellent resistance to oils and water, the bonds tend to be very brittle, but this feature may be improved by incorporating further polymers such as nitrile rubber. This type of additive generally has an adverse effect on the maximum service temperature.

Epoxide polyamide systems. The mechanism of cross-linking epoxy resins with polyamides is similar to that in which amines are used. The polyamides, formed by condensing a diamine with a dimeric fatty acid, contain terminal amine groups which may react with the functional groups of an epoxy resin, the product being an epoxide resin cross-linked with polyamide chains:

$$NH_2-R-NH.CO.R'.CO.NH.-R-----R'CO.NH.R.NH_2$$

The size and composition of the polyamide, to a certain extent, determines the properties of the final co-polymer and, in general, the polyamide cured resins possess increased flexibility and toughness compared with the polymeric epoxy resins. The condensation reaction generally occurs at room temperature, the rate of cure being determined by the reactivity of the polyamide. The systems are insensitive to wide variations in the amount of each component and, whilst increasing the polyamide content increases the flexibility of the co-polymer, the temperature resistance is reduced owing to the thermoplastic nature of the polyamide.

These adhesives form excellent bonds with metal and are extensively used in the formation of bonds with typical shear strengths of 3000 to 4000 p.s.i. They do, however, show a rapid fall-off in bond strength at temperatures of the order of 100°C. Being viscous materials in the unpolymerized form, they can be used in jointing procedures which also rely on the gap-filling property of the adhesive. They show excellent resistance to water, oil and some solvents and therefore find widespread application in industrial bonding.

Epoxy polysulphides. Modification of the epoxy resins with a polysulphide results in a material which shows improved elasticity, peel strength and impact resistance. A typical polysulphide modifier is the Thiokol LP 33 (Thiokol Corporation)

$$HS-(C_2H_4-O-CH_2-O-C_2H_4-S-S)_6-C_2H_4-O-CH_2-O-C_2H_4SH$$

which co-reacts with an epoxy resin in the presence of an amine catalyst. The hydrosulphide groups, being labile, can react with the epoxide group and the molecule, being bifunctional, can act as a cross-linking agent. Depending on the type of resin and amine hardener used reaction will occur at room temperature, but for maximum bond strength a heat curing schedule at up to 120°C is often recommended. Bond strengths with a shear strength of up to 3 000 p.s.i. are obtained but the maximum temperature of operation is limited to 90°C.

Since both the polysulphide and epoxy resins are polymerizable materials in their own right, a wide range of polymers is possible possessing properties which vary from the hard epoxy resin to the elastomeric polysulphide.

Table 8.4 The properties of Epikote 828 (Shell Chemical Co.) Thiokol LP-3 co-polymer

	Epoxide resin/polysulphide ratio			
	1:0	2:1	1:1	1:2
Tensile strength p.s.i.	3 500	2 300	1 700	900
Elongation – % at break	0	30	40	60
Impact strength ft lb	1.3	4.7	78	78
Impact strength – ft lb at – 40° C	0.3	2.5	3.5	5.0
Water vapour transmission g/m²/24h	17.5	17.5	2.45	0.92

As shown in table 8.4 (see ref. 36), incorporation of a polysulphide increases the flexibility and reduces the water vapour transmission of a straight epoxy resin. However, these advantages are gained at the expense of reduced tensile strength but, for specific applications, co-polymers can be formulated in which the relative advantages of the two resin systems can be fully exploited. The adhesives show good water and oil resistance and are finding increasing use in architectural bonding.

Epoxy silicone. The high cost of silicone resins has detracted from the more widespread application of the epoxy silicone co-polymers. Incorporation of the silicone resin bestows a high degree of temperature resistance and in environments where bond strengths of up to 1 500 p.s.i. at 300°C are required, these adhesives find an outlet. The cured adhesives are also resistant to oxidation and are therefore extremely valuable materials for operation under arduous conditions.

d. Phenolic resin based adhesives. The interaction of phenol and formaldehyde produces, as initial reaction products, ortho and para substituted methylol phenol:

[Chemical reaction: phenol + CH₂O → ortho-methylol phenol (with -CH₂OH) + para-methylol phenol (with CH₂OH at para position)]

Self-condensation of the ortho or para isomers produces the appropriate dihydroxy-dibenzyl ether. Alternatively further reaction may occur with phenol to give the methylene coupled bisphenol or with formaldehyde to give dimethylol phenol:

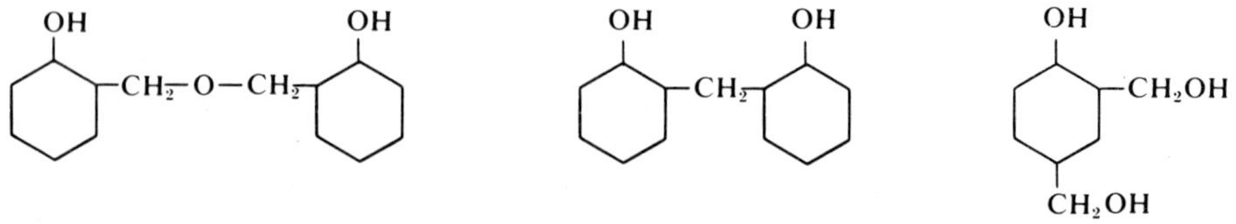

Polymerization can then be effected through the methylol groups giving the polymeric ether or a condensation product in which the linkage of the phenol nuclei is through the methylene groups.

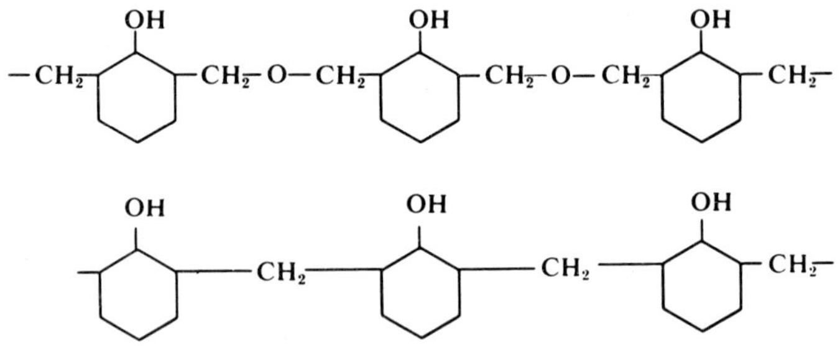

Cross-linking of the chains is the final stage in the development of a resinous product.

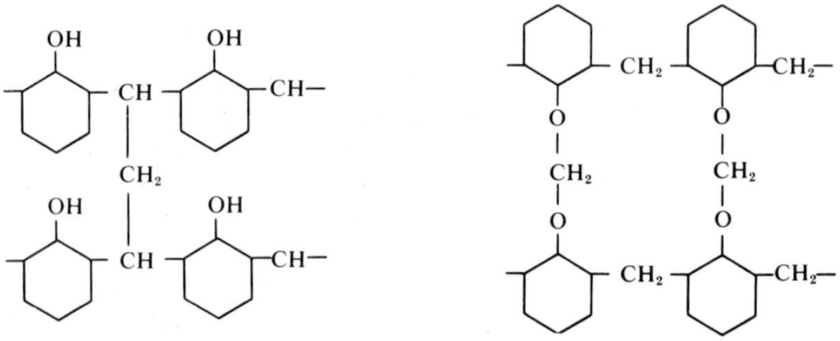

The composition and structure of the final product will be determined not only by the ratio of phenol to formaldehyde but by the conditions under which polymerization is effected and the catalyst used. In the partially polymerized form with terminal methylol groups (the so-called resol stage) polymerization and development of an adhesive bond may be carried out by heat or in the cold by using an acid catalyst, e.g. p-toluene sulphonic acid. The presence of acid in the fully cured bond can cause corrosion and subsequent deterioration of metal adherends and therefore the cold curing phenolic resins are not widely favoured.

The usefulness of the thermosetting phenolic resins is considerably increased by incorporating a nitrile or neoprene rubber. The role of the resin is to cross-link the rubber, and the polymers so produced show excellent metal–metal bonding properties, high impact resistance and a bond which is resistant to fatiguing stresses. Heat curing is necessary if maximum bond strength is to be achieved and, whilst the curing schedule largely determines the ultimate bond strength, values as high as 2 500 p.s.i. shear strength have been reported for the phenolic neoprene adhesive and up to 4 000 p.s.i. for the phenolic nitrile resins. The phenolic nitrile resins are suitable for continuous duty at temperatures up to 170°C although, at this temperature, the bond strength is likely to have deteriorated by 50% or more. Both the adhesives show excellent resistance to water and oil although, like all rubber-containing adhesives, aromatic solvents can cause degradation of the bond.

Replacement of the rubber by a polyamide which is thermoplastic provides a further range of adhesives which show excellent temperature resistance and very high shear strengths.

A further combination which has been extensively used is the thermosetting phenolic resin and thermoplastic vinyl resin. The latter is generally in the form of a powder which is dusted onto the surface of a liquid phenolic resin applied to one surface of the adherends. After mating the adherend surfaces, the resin is cured under pressure at a temperature of up to 150°C and the resulting polymer has a bond strength of up to 5 000 p.s.i. This is the so-called Redux process which was developed by Aero Research Ltd. (Bonded Structures Ltd.) for application to the aircraft industry in the production of bonding structure and honeycomb sandwich structures. The use of the Redux process is covered by the *Ministry of Supply specification DTD 775*.

e. Polyamides. The background to the development of polyamide thermoplastic resins has been outlined by Bolton (see ref. 37). This range of materials, although frequently used in conjunction with epoxy resins, are employed as hot metal adhesives in their own right. A typical polyamide obtained by condensing ethylene diamine with a dimeric fatty acid has a general formula

$$H_2N.CH_2.CH_2.NH(CO.R.CO.NH.CH_2.CH_2.NH)_nCO.R.NH.CH_2CH_2NH_2$$

The inclusion of a small amount of unsaturation helps to promote cross-linking of the polyamide chains during the heat curing cycle. In view of the large number of variants which are feasible, it is difficult to state specific conditions of use. In general, however, these materials are heat cured at 100 to 180°C and, on cooling, they rapidly set to form bonds, the tensile strength of which may be up to 3 000 p.s.i. They show good resistance to water and oils and the maximum service temperature, although determined primarily by the reactants and degree of polymerization, may be up to 300°C.

f. Polyurethane. The polyurethanes are a series of organic polymers produced by the interaction of a polyhydroxy compound and a polyisocyanate. Like many polymeric materials, the final form of the polymer may be varied according to the type and proportion of monomeric components and may take any form from an elastomer to a resinous product. Consequently, bonds of good flexibility, good resistance to shock and vibration and with a good peel strength can be formulated. The polymerization reaction may be represented thus:

$$\text{R—N=C=O} + \text{HO—R}' \rightarrow \text{R—NH—C=O} \rightarrow \text{R—NH—}\underset{\underset{\text{O—R}'}{\|}}{\text{C}}\text{—}\underset{\text{N}}{\overset{\text{R}}{|}}\text{—C=O}$$
$$\phantom{\text{R—N=C=O} + \text{HO—R}' \rightarrow \text{R—NH—C=O} \rightarrow \text{R—NH—C—N—}}\text{O—R}'$$

The initial reaction product is a urethane which undergoes further reaction with the isocyanate to give a cross-linked polymer via the allophonate structure. Use of bifunctional reactants is frequently employed to ensure a high degree of cross-linking:

$$OCN-\bigcirc-CH_2-\bigcirc-NCO + HO-R-OH$$

$$OCN+[\bigcirc-CH_2-\bigcirc-NH \cdot CO \cdot R \cdot NH]_n-\bigcirc-CH_2-$$

In initial work on polyurethane adhesives trouble was experienced due to foaming, attributable to the interaction of the isocyanate group with water and release of carbon dioxide:

$$-N=C=O + H_2O \longrightarrow \left[-NH-C\begin{smallmatrix}O\\OH\end{smallmatrix}\right] \longrightarrow -NH_2 + CO_2$$

but use of specially dried reagents has largely eliminated this trouble. Of the catalysts which are available, metal salts of organic acids and organo-metallic compounds are effective for room temperature curing resins and generally favour the (linear) polymer growth reaction. Tertiary amines tend to favour the water-isocyanate reaction and cross-linking and are therefore more suitable for use when degassing can be effected during polymerization. Typical isocyanates frequently used include 4-4' diisocyanate diphenyl methane and tolylene diisocyanate whilst polyols such as 1-4 butanediol are employed.

Although the polyurethanes are resistant to a wide range of chemical environments, their susceptibility to attack by water seriously limits their usefulness. A bond of reasonable strength at room temperature can be obtained but the upper temperature limit must be regarded as 90 to 95°C. Their use in metal bonding is limited but Mauser (see ref. 38) has described their use in this application.

g. Polyvinyl acetate. This range of adhesives based on polymerized vinyl acetate

$$CH_2=CH.OOC.CH_3$$

suffers from a serious disadvantage in that the adhesives are subject to cold flow when continuously stressed even though the applied load may be relatively low. This is in spite of the fact that the initial bond strength may be as high as 3 000 p.s.i. These are frequently employed as an aqueous emulsion of the polymer which dries to a hard friable material. This material, on heating above the glass transition temperature, produces a brittle but coherent film when cool. Although the bonds show a good oil resistance, they are susceptible to deterioration by water and temperatures above 50°C.

h. Rubber based adhesives. The rubber based adhesives may utilize either natural rubber, based on the isoprene molecule

$$-CH_2-\underset{\underset{CH_3}{|}}{C}=CH-CH_2-$$

or nitrile rubber, an acrylonitrile butadiene co-polymer

$$-CH_2-\underset{\underset{CN}{|}}{CH}- \; + -CH_2-CH=CH-CH_2$$

or neoprene, i.e. chloroprene,

$$-CH_2-\underset{\underset{Cl}{|}}{C}=CH-CH_2-$$

The general advantages of this group of adhesives lies in the fact that they exhibit good specific adhesion for various surfaces and their higher elongation permits bonds to absorb much of the strain although certain specific adhesives are distinguished by high adhesion for metal surfaces rather than their elasticity.

Two basic rubber adhesive systems are in general use: (i) in which a bond is formed by solvent evaporation, and (ii) in which a bond is formed by vulcanizing the rubber. Types which are formulated from natural rubber and which rely on solvent release have maximum service temperatures of 65°C and therefore find only limited application. However, those which utilize vulcanization, which is frequently effected with sulphur and an accelerator, produce adhesives which show improved temperature resistance, a reduction in plastic flow and an increase in elasticity. Vulcanization is usually carried out at elevated temperatures with accelerators such as tetramethyl thiuram disulphide or mercaptobenzthiazole but certain room temperature accelerators such as sulphur chloride are available. The use of this is not general owing to the rapid aging it produces in the vulcanizate and its lachrymatory nature. The systems used for vulcanizing natural rubber are also generally applicable to the acrylonitrile butadiene based cements and produce bonds which are stronger, up to 1 000 p.s.i. in shear, and more temperature resistant, some material being usable up to 150°C.

Neoprene rubber cements are different from the others in that vulcanization can occur through natural aging processes or by the addition of accelerators, but without the need for sulphur, which, in certain applications may be disadvantageous. The advantage of the vulcanized neoprene lies in its resistance to aliphatic solvents and oils but, like the natural and acrylonitrile butadiene rubbers, it is rapidly attacked by aromatic solvents. It has a maximum service temperature of 90°C although certain formulations are capable of withstanding short periods of operation at temperatures in excess of this.

Other types of rubber based adhesives are available and modifiers such as phenolic resins are extensively employed. A detailed description of the use of rubber based adhesives has been given by Wake (see ref. 39).

i. Silicate adhesives. These materials are based on sodium silicate with a $SiO_2:Na_2O$ ratio of 2.5:1 or greater–potassium silicates with a ratio greater than 2.0:1 have also been used. The silicate adhesives find certain applications in which the bonding of metals such as aluminium, copper, steel or lead to an absorbent substrate is required. Since these materials do not possess initial tack, the bonds must be held under slight pressure until the set is initiated. The 'curing' of the adhesive is effected by carbon dioxide which insolubilizes the silicate and forms a water resistant bond. The mechanism of bond formation is considered to be partly mechanical due to penetration of the pores of the absorbent substance and partly due to chemisorption by the formation of hydrogen bonds. A detailed description of sodium silicate as an adhesive has been given by Macfarlane and Sewell (see ref. 40).

j. Silicone adhesives. Silicone adhesives first appeared on the market in 1955, followed in 1958 by the appearance of the single-pack systems. They have been developed for highly specific applications such as high service temperature. Their bond strength is low (250 to 400 p.s.i. in shear) when compared with the more conventional adhesives but they remain serviceable up to 250°C and even higher for short periods. They are generally unaffected by wet and humid conditions and are resistant to oxidation. They are, however, singularly prone to hydrolysis by some acids, e.g. dilute nitric acid, and they deteriorate in the presence of many organic solvents. In such environments the fluorsilicones are more suitable.

k. Special adhesives. With the advent of the space age, a need for highly specialized adhesives has arisen. The list of materials under current investigation is formidable but mention may be made of two types. The polyimide resins are a recent development in this field, and they are remarkable in that, whilst of an organic nature, they are capable of withstanding prolonged exposure to temperatures of 300°C. Their drawbacks as adhesives include the very drastic curing cycles which are necessary to effect polymerization and the brittleness of the cured polymers.

The other group of adhesives to fall under this classification are those of an inorganic nature and based on elements such as boron, silicon, fluorine, nitrogen, phosphorus, titanium and germanium (see ref. 41). Being of a highly sophisticated type and still the subject of extensive research, reference should be made to one of the textbooks on inorganic polymers for further details.

8.7.7 Conclusions

The use of adhesives in structural bonding is an ever increasing field of activity. It spans so many industries that it would be almost impossible to try and enumerate them. As has been pointed out, the application of adhesives as a means of joining materials is by no means the answer to every problem and, although many systems are available in the field of cryogenics, deficiencies at the other end of the temperature spectrum are apparent. It is fairly true to say that research into adhesives will continue until the ultimate has been achieved, i.e., until materials are available which show bonding properties equivalent or superior to the materials being bonded. In their application to the engineering field, much work has yet to be done.

8.8 LITERATURE

8.8.1 Further reading for sections 8.3 to 8.6

a. General

AMERICAN WELDING SOCIETY. *Welding Handbook*, 4th edn in 5 vols. Ed. by ARTHUR L. PHILLIPS. London, Cleaver-Hume, 1958–62. 5th edition in preparation.

E. FUCHS AND H. BRADLEY. (Eds). *Welding Practice*, 3 vols. London. Butterworths, 1952.

J. F. LANCASTER. *The Metallurgy of Welding, Brazing and Soldering*. London, Allen & Unwin, 1965.

J. M. SOMERVILLE. *The Electric Arc*. London, Methuen, 1959.

R. HAMMOND. *Automatic Welding*. London, Redman, 1963.

b. British Standards – welding and brazing

B.S.S. 499. 1965. *Welding Terms and Symbols*.

B.S.S. 638. 1966. *Arc Welding Plant and Equipment*.

B.S.S. 709. 1964. *Methods of Testing Fusion Welded Joints and Weld Metal in Steel*.

B.S.S. 1077. 1963. *Fusion Welded Joints in Copper*.

B.S.S. 1126. 1957. *General Recommendations for the Gas Welding of Wrought Aluminium and Aluminium Alloys*.

B.S.S. 1719. Part 1. 1963. *Classification and Coding of Covered Electrodes for Metal-arc Welding*.

B.S.S. 1723. 1963. *Brazing*.

B.S.S. 1724. 1959. *Bronze Welding by Gas*.

B.S.S. 1845. 1952. *Filler Alloys for Brazing*. (In revision.)

B.S.S. 1856. 1964. *General Requirements for the Metal Arc Welding of Mild Steel*.

B.S.S. 2937. 1957. *General Requirements for (Resistance) Seam Welding in Mild Steel*.

B.S.S. 2996. 1958. *Projection Welding of Low Carbon Wrought Steel Studs, Bosses, Bolts, Units and Annular Rings*.

B.S.S. 3019. Part 1. 1958. Part 2. 1960. *General Recommendation for Manual Inert Gas Tungsten Arc Welding*.

B.S.S. 3571. Part 1. 1962. *General Recommendations for Manual Inert Gas Metal Arc Welding*.

c. Standards for pressure vessels

B.S.S. 1500. Part 1. 1958. Part 3. 1965. *Fusion Welded Pressure Vessels for use in the Chemical, Petroleum and Allied Industries: Part 1 – Carbon and Low Alloy Steels; Part 3 – Aluminium*.

B.S.S. 1515. Part 1. 1965. Part 2. 1968. *Fusion Welded Pressure Vessels (Advanced Design and Construction) for use in the Chemical, Petroleum and Allied Industries: Part 1 — Carbon and Ferritic Alloy Steels; Part 2 — Austenitic Stainless Steel*.

A.S.M.E. *Boiler and Pressure Vessel Code*, Sections VIII and IX. New York, American Society of Mechanical Engineers, 1965. (Revised at about three year intervals.)

d. Materials

H. WIGGIN & CO. LTD. *Welding, Brazing and Soldering Wiggin Nickel Alloys*. Pubn 3238. Hereford, 1966.

H. WIGGIN & CO. LTD. *Welding Dissimilar Metals*. Pubn. 3233. Hereford, 1965.

COPPER DEVELOPMENT ASSOCIATION. *The Welding, Brazing and Soldering of Copper and its Alloys.* Pubn 47. London, 1957.
UDDEHOLMS AKTIEBOLAG. *Stainless Steel Working Instructions – Welding, Brazing, Soldering, Riveting.* Sweden, Uddeholm, 1963.
UNITED STEEL COMPANIES LTD. *Silver Fox Stainless Steels – Notes for Fabricators on Welding.* Sheffield, S. Fox, 1961.
ALUMINIUM FEDERATION, LONDON, INFORMATION BULLETINS:
No. 5. *The Gas Welding of Aluminium.*
No. 6. *Resistance Welding of Wrought Aluminium Alloys.*
No. 19. *The Arc Welding of Aluminium.*
No. 22. *The Brazing of Aluminium and its Alloys.*
No. 23. *Soldering Aluminium.*
TITANIUM METALS CORPORATION OF AMERICA. *Titanium Engineering Bulletin No. 6. Titanium Welding Techniques.* New York, 1962.
IMPERIAL METAL INDUSTRIES (KYNOCH) LTD. *I.M.I. Titanium Fabrication.* Birmingham, 1965.
BRITISH WELDING RESEARCH ASSOCIATION PUBLICATIONS:
Arc Welding Low-alloy Steels.
Arc Welded Machinery Construction in Mild Steel.

e. Processes

INSTITUTE OF WELDING. *Inert Gas Arc Welding Handbook.* London, 1965 (in revision).
BRITISH WELDING RESEARCH ASSOCIATION. CO_2 *Welding*, 2nd edn. Cambridge, 1965.
A. H. MELEKA and J. K. ROBERTS. 'Electron Beam Welding – The Need for Further Development as Revealed from Production Experience.' *Brit. Welding J.* **15**, 1968.

f. Design

INSTITUTE OF WELDING. *Handbook for Welding Design.* London, Pitman, 1956.
BRITISH WELDING RESEARCH ASSOCIATION PUBLICATIONS:
Welded Structural Details.
Design of Welded Joints.
Design of Welded Pipe Fittings.

8.8.2 References particular to section 8.7

1. DERYAGIN, *Nottingham Conference on Adhesion, 1966.* London, Maclaren, 1968.
2. R. M. VASENIN. *RAPRA Trans.* **1005, 1006, 1010, 1075**, R. J. MOSELEY.
3. D. J. CRISP, *J. Coll. Sci.*, **1** (1946), 101.
4. J. GLAZER, *J. Polym. Sci.*, **13** (1954), 355.
5. L. H. SHARPE, *Proc. Chem. Soc.*, **12** (1961), 461.
6. D. D. ELEY. (Ed.) *Adhesion.* Oxford, Clarendon, 1961.
7. *Contact Angle, Wettability and Adhesion.* A.C.S. Series no. 43. Washington DC, American Chemical Society, 1964.
8. W. C. WAKE. *Adhesives.* Lecture Series no. 4. London, Royal Institute of Chemistry, 1966.
9. W. C. WAKE. *Trans. Inst. Rubber Ind.*, **35** (1959), 145.
10. G. KRAUS and J. E. MANSON. *J. Polym. Sci.*, **6** (1951), 625.

11. S. BUCHAN. *Rubber to Metal Bonding*, 2nd Edn. London, Crosby Lockwood, 1959.
12. DE BRUYNE. *Bulletin 168*. Cambridge, Ciba (A.R.L) Ltd.
13. J. HOEKSTRA and C. P. FRITZIUS. *Adhesion and Adhesives*. Ed. by N. A. DE BRUYNE and R. HOUWINK. Amsterdam, Elsevier, 1951.
14. Pubn. A. 15d. Cambridge, Ciba (A.R.L) Ltd.
15. E. B. MCMULLEN and E. W. GARNISH. *Metals and Mats.*, **1** (1967), 398.
16. A. S. MCLAREN and I. S. MACINNES, *Brit. J. appl. Phys.*, **9** (1958), 72.
17. L. GREENWOOD, T. R. BOAG and A. S. MCLAREN, *Nottingham Conference on Adhesion, 1966*. London, Maclaren, 1968.
18. O. VOLKERSEN. *Luftfahrtforschung*, **15** (1938), 41.
19. M. GOLAND and E. REISSNER. *J. Appl. Mech.*, **11** (1944), 417.
20. F. J. PLANTEMA. *Nat. Luchtfahrtlaboratorium*, Report 1181.
21. I. N. SNEDDON. *Adhesion*. Ed. by D. D. ELEY. Oxford, Clarendon, 1961.
22. R. L. PATRICK. (Ed.) *Treatise on Adhesives*. New York, Dekker, 1967.
23. R. HOUWINK and G. SALOMON. (Eds.) *Adhesion and Adhesives*. Amsterdam, Elsevier, 1965.
24. D. E. STRAIN. *Ind. Eng. Chem.*, **30** (1938), 345.
25. Eastman Chemical Products Inc. Bull. No. R-103.
26. H. LEE and K. NEVILLE. *Handbook of Epoxy Resins*. London and New York, McGraw-Hill, 1967.
27. E. PREISWERK. *The Invention of Araldite, 1944–1964*. Basle, Birkhäuser Verlag, 1965.
28. Minneapolis & Honeywell Regulator Co. B. P. 859,573.
29. W. R. Grace & Co. B.P. 864,437.
30. S. H. LANGER, I. N. EBBLING, A. B. FINESTONE and W. R. THOMAS. *J. Appl. Sci.*, **5** (1961), 370.
31. Armstrong Cork Co. B.P. 784,565.
32. Rubber & Asbestos Corp. U.S.P. 2,879,252.
33. Phillips Petroleum Co. U.S.P. 2,898,315.
34. N.V. de Bataafshe Petroleum Maats, B.P. 962,010.
35. Ciba (A.R.L) Ltd. Cambridge, INF. Sheet Hidux 1197A.
36. J. S. JORCZAK and D. DWORKIN. *Res. Soc. of Plastics J.*, **10** (1954).
37. E. BOLTON. *Chem. & Ind.*, **61** (1942), 31.
38. R. A. P. MAUSER and E. MAUSER. B.P. 834,917.
39. W.C. WAKE. *Adhesion and Adhesives*. Ed. by R. HOUWINK and G. SALOMON. Amsterdam, Elsevier, vol.1.1965, vol.2.1967.
40. W. S. MACFARLANE and J. F. SEWELL. *Adhesion*, April, 1952.
41. *Research on Elevated Temperature Resistant Inorganic Polymer Adhesives*. U.S. Dept of Commerce, PB 131,934.

CHAPTER 9

The Surface Treatment of Metals

9.1 INTRODUCTION

Metals are essentially artificial and unstable materials, that is to say, they are not found as such in nature (excepting gold and copper) and they tend, under the influence of the weather, waters and similar corrosive exposure, to revert back to a non-metallic state. It is convenient for reasons of cost, strength and ease of working to manufacture metallic articles without too much regard to their external appearance or to their behaviour against corrosion, and then to deal with these purely surface characteristics by a subsequent treatment, which may therefore be designated in general as surface treatment of metals. Since iron and steel are by far the most commonly used metals, and also unfortunately among the most prone to rust, decay and corrosion, they figure very largely, but not by any means exclusively, in the general subject of the surface treatment of metals.

The main purposes of the surface treatment of metals are predominantly twofold, with emphasis on one or the other to different degrees in different cases, viz:
1. To improve the appearance.
2. To improve resistance to corrosion, tarnish or staining under the conditions of use which are foreseen.

There are other objectives, each important in its own sphere, but these together are only of minor importance compared with the above main purposes. Examples of these minor objectives are to provide resistance to scaling by heat, to facilitate other subsequent surface treatments or to change the surface hardness or other surface properties, or even to change the overall dimensions.

The surface treatment of metals almost invariably consists of applying a coating of some kind to the metal, and thus providing an external skin which can be selected solely for its appearance and corrosion resistance. If such a coating can be applied without any discontinuities, and if it is firmly adherent to the base, then the surface behaviour and appearance become entirely a question of the behaviour of the coating. However, such external coatings must inevitably be thin, unless the whole character of the article is changed. In the normal course of use these coatings therefore entail the risk of penetration through to the basis metal by mechanical damage or flexing; the effect of the small areas of basis metal thus exposed cannot be ignored, not only because of the local corrosion of the exposed metal, but because of the risk that the remainder of the coating may be undermined and thrown off.

In the choice of surface treatment for any particular metallic article, many factors must be considered, such as cost, type of corrosive environment to be withstood, and compatibility with special materials such as foodstuffs or washing materials. In very many cases an additional factor of continuing good appearance must be considered. Metal surfaces, especially when they are highly polished, have a uniquely attractive and aesthetically satisfying appearance. It is obviously appropriate to provide a metallic article with a metallic coating of higher corrosion resistance and more attractive appearance, so that the whole system is a unity. But very few metallic coatings are absolutely resistant to corrosive influences, and these few are the most expensive. In particular, metallic coatings are not completely resistant to some of the many chemical substances now involved in everyday life, such as washing chemicals, fruit acids and many foodstuffs. Nor do they provide much opportunity to introduce colours, including white, which may be desired to conform to decorative schemes or fashions. Thus non-metallic coatings, such as vitreous enamel, paint and lacquer, are applied to metallic articles for decorative as well as protective reasons.

Before any surface coating can be applied it is essential that a clean, sound surface must be provided on the article to be coated. In particular, scale and oxide, grease and dirt remaining from the manufacturing operations, must be removed; the surface usually must be smoothed or even polished. It may also need to be prepared physically or chemically to ensure satisfactory adhesion of the coating. These preparatory treatments are an integral part of the surface-treatment process, and obviously have to be applied at the end of the main manufacturing process. For this reason these cleansing, protective and preliminary decorative treatments are almost invariably applied as the final stage of the manufacturing process of a metallic article, to such an extent that the processes are often collectively referred to under the term 'metal finishing'. In the cases of hot-dip tinning and hot-dip galvanizing of sheet steel the protective films are applied to the semi-fabricated sheet material; here, however, the subsequent processes are limited to cutting, bending and joining; moreover, decorative effects are subordinate to economics.

Although metal finishing is normally considered to be a separate aspect of production technique, the processes entailed are really an integral part of the manufacturing process, and in larger firms they are mostly carried out within the one factory. It was formerly the custom for most electroplating to be carried out by separate specialist electroplating enterprises, and this is still to some extent the case. Similarly, paint and lacquer is usually applied as a logical completion of the manufacturing process; but there are still many processing firms which continue to specialize in this work. The large and specialized plant for galvanizing and hot-dip tinning or continuous electrotinning of sheet and strip steel is often associated with the production of these commodities. For these reasons, it is difficult to extract statistics indicative of the size of the metal-finishing industry as a whole. It must suffice to say that it is an important, very progressive and widely pervasive component of the metal manufacturing industry.

9.2 CHOICE OF SURFACE TREATMENT

Metallic coatings have the advantage of being hard and strong and of resisting moderately high temperatures; they are thus fairly robust. They can usually be chosen to give a high degree of protection against most types of corrosive environment, excluding strong acids and chemicals. Where low cost is important and good protection is required, without much emphasis on appearance, hot-dipped coatings are cheap and very effective. Where a neat, highly polished surface is required costs are much greater and electroplating is pre-eminent. Both hot-dipping and electroplating involve immersion of the articles into a liquid held in a container, so the maximum size of object which can be treated is limited. Metal coatings can be applied by metal spraying to objects of any size, if necessary *in situ*.

Vitreous-enamel coatings can only be applied to a restricted number of metals which withstand the high temperature of application. As they are composed in effect of glass, they are highly resistant to most chemicals, but are extremely brittle and can only be applied to relatively rigid articles.

Paint systems can be applied to articles of any size; if necessary to selected areas only. They can be transparent, opaque or coloured, as desired. On the other hand, even the hardest stoved paints are damaged relatively easily, are not very heat resistant, and tend to be adversely affected by alkaline washing materials and organic solvents; they suffer gradual deterioration from the effect of bright light and the weather, and are seldom absolutely impermeable to water. Although they are fairly cheap as regards raw materials, paints are expensive and wasteful to apply.

The choice of coating is also dictated to some extent by the basis metal or alloy. Iron and steel predominate and, as they are so readily and disfiguringly corroded, they need special protection. Copper alloys, such as brass and nickel-silver, are much less corrodible, but are more expensive and thus justify a more costly finish. The lower-melting point metals and alloys of lead, tin and zinc obviously cannot endure a hot process. Zinc, which is used very extensively in the form of zinc base die-castings, is a very active metal and needs effective protection.

Aluminium is a special case, since both the metal and its alloys have a very high resistance in the uncoated condition to atmospheric corrosion and to water, and do not readily accept coatings of other metals or of paints. This natural resistance is a consequence of an oxide film which forms automatically on the metal. Most schemes for the protection of aluminium therefore utilize this characteristic, and consist of an artificial thickening of the oxide film.

Alone of the cheaper industrial alloys, stainless steel does not need further protection, but its full appearance can only be brought out by suitable polishing, which is made expensive by the hard and tough nature of the material.

9.3 METALLIC COATINGS

9.3.1 *Choice of metallic coating*

In choosing a metallic coating for the protection of a metallic article it is not sufficient to consider the behaviour of the coating metal alone. It is necessary also to have regard to the effect of the mutual exposure of coating metal and basis metal at pores, cuts and other initial or subsequent breaks in the thin film of coating. This is a consequence of the predominantly electrochemical nature of the corrosion of metals by the weather, by water and by aqueous solutions, which must now be briefly considered. All metal atoms in the metallic state are characterized by having one or more rather loosely attached electrons, the mobility of which accounts for the high electrical and thermal conductivity of metals. Metal atoms in the combined state have lost these

672 THE SURFACE TREATMENT OF METALS

electrons and are bound to other atoms only by their resultant positive charge. Corrosion of a metal is therefore essentially the loss of one or more electrons from its atoms. But since both the metals and the aqueous solutions which corrode them are conductors of electricity, the absorption of electrons essential to corrosion may occur elsewhere than at the site of the attack; that is to say, corrosion is electrochemical in nature. The various metals differ in the relative ease with which they part from their outer electrons, and a table, known as the 'electrochemical series' can be arranged to show this (table 9.1). For this purpose hydrogen gas can also be considered to be a metal, because its atoms each have one loosely attached electron. In the combined state hydrogen loses this electron and becomes the 'hydrogen ion', which is present in all aqueous solutions to a greater or a lesser extent, because it is present both in water and in acids.

It will be seen from table 9.1 that hydrogen occupies a somewhat central position in the electrochemical series. The metals below it have a more negative potential, which indicates a greater willingness to part with their outer electrons than that of hydrogen. When, therefore, one of these reactive or base metals, e.g. zinc, is in contact with an aqueous solution, the zinc atoms can reduce their energy by donating their outer electrons to the hydrogen ions, and can thus pass into the combined state, i.e. corrode. The hydrogen ions, by virtue of having each received an electron, become hydrogen gas, but as a result of the electrical mobility of the electrons in the metal or in the aqueous solution, this transfer can quite readily occur at a site on the metal solution interface far removed from the site of the corrosion of the zinc atom. The hydrogen may be evolved as gas, but this seldom occurs except from strongly acid solutions. The aqueous solutions more commonly involved in the corrosion of metallic articles range from films of condensed water (dew), which will usually be slightly contaminated by dissolved atmospheric polluting gases, to quite strong solutions of chemicals, such as sea-water, or dissolved foodstuffs or laundry materials, etc. Under practical conditions these solutions will also be saturated with, and in contact with, the air. The dissolved oxygen of the air can combine with the freshly formed or 'nascent' hydrogen gas to form water, so the corrosion of those metals which are more negative than hydrogen tends to proceed continuously.

Nevertheless, subsidiary factors often conspire in practice to prevent such corrosion proceeding continuously. Thus in relatively neutral solutions, zinc, when it passes into the combined state, immediately forms zinc hydroxide, which is insoluble in water, and coats the metal surface, thus inhibiting further corrosion. Similarly, aluminium

Table 9.1 The electrochemical series for metals

Expressed as the standard electrode potential of the metal related to that of hydrogen gas in an aqueous solution of pH 0 at normal temperature and pressure (0°C and 760 nm Hg)

Gold	+1.42 V	Nickel	−0.23 V
Rhodium	+1.40 V	Cadmium	−0.40 V
Silver	+0.80 V	Iron	−0.44 V
Copper	+0.52 V	Chromium*	−0.56 V
Hydrogen	0 V	Zinc	−0.76 V
Lead	−0.13 V	Aluminium	−1.67 V
Tin	−0.14 V	Sodium	−2.71 V

*Chromium is usually passive, and in this state behaves like gold.

immediately forms a thin, but extremely dense and insulating oxide film on the surface, which entirely stops further attack.

The metals above hydrogen in the electrochemical series, which have a more positive potential than hydrogen, are less prone to progressive corrosion in aqueous solutions. Those metals with the most positive potentials, such as gold and the platinum metals, indeed have practically no tendency to pass into the combined state; accordingly they are designated noble metals.

The position of iron in the electrochemical series should particularly be noted; it is more negative or base than hydrogen, and considerably more negative than copper, tin and nickel, almost identical in position to cadmium, and considerably more positive than zinc.

Where two different metals are in electrical contact with each other and also with an aqueous solution, they form a galvanic cell or battery, because of their difference of potential. The metal with the more negative potential is stimulated to corrode faster, and the metal with the more positive potential is discouraged by the difference of potential. A current flows around the circuit made up from the two metals and the solution; it passes out of the corroding metal (the anode) into the solution and thence back into the protected metal (the cathode). The anode and the cathode may be some distance apart in this action, since both the metals and the solution are electrically conducting to the electron transfer.

This situation of two different metals in contact being exposed together to a corrosive liquid, arises when a protective metallic coating is discontinuous for some reason. It may be inherently porous, or it may be locally deficient, or it may have become cut or damaged in service, or even by former corrosion. The corrosion of the basis or under-metal thus exposed will be stimulated or protected, according to the relative potentials in the electrochemical series of the coating and the basis metal. The more positive metal will always stimulate the corrosion of the more negative metal. If it is the corrosion of the basis metal which is stimulated in this way the effect can be serious, because the area exposed will be small compared with that of the undamaged coating metal, and therefore the current generated by the difference of potential will be concentrated on a small area of reactive basis metal. On the other hand, if the corrosion of the basis metal is inhibited by the adjacent presence of a more negative coating, quite large gaps in the coating may not be very detrimental.

It follows that in the important case of articles of iron and steel, coatings of the metals more positive than iron in the series, such as copper, nickel or tin, should be particularly complete and impervious; otherwise corrosion of the steel at pores and discontinuities will be stimulated, and the resulting rust will be washed out around the pores over the sound coating. On the other hand coatings of zinc, aluminium and even cadmium will actually protect adjacent areas of exposed bare steel, but will themselves become corroded in doing so. These anodic or negative potential coatings are therefore said to give sacrificial protection.

9.3.2 Preparation for metal coating

a. Pickling. Articles which are intended for coating with metal must first be cleaned free from heavy grease and dirt; oxide scale which may have been formed during hot fabrication operations such as annealing, casting and welding must also be eliminated completely. Removal of grease and dirt, where necessary, is usually accomplished by solvent vapour degreasing or by immersion in a hot alkaline metal

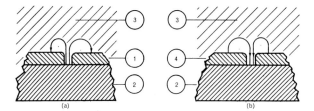

Fig. 9.1 Effect of pores in cathodic and in anodic coatings.
1. Copper coating; electrochemical potential, +0·52 volt.
2. Steel base: electrochemical potential, −0·44 volt.
3. Corrosive liquid, such as salt water or contaminated rain water.
4. Zinc coating,; electrochemical potential, −0·76 volt.

The cathodic copper coating (a) causes a current to flow *out* of the steel, which is thus corroded at the base of the pore. The anodic zinc coating (b) causes a current to flow *into* the steel at the base of the pore and the steel does not corrode.

cleaning solution. These processes are described in the section dealing with preparation for electroplating because they must, in any case, be repeated at that stage, especially if the article has been polished with greasy polishing compounds.

Oxide scale on iron and steel is removed by pickling, that is by immersion in hot dilute (about 10%) sulphuric acid solution or in hydrochloric acid solution. Similar acid solutions are used for scaled copper and its alloys, for brass and for nickel-silver. The dissolution of the oxide in the acid is a purely chemical reaction and is rather slow, taking many minutes to complete if the scale is thick. As the clean metal becomes exposed, it too would become attacked, which is undesirable, if small quantities of complex organic pickling inhibitors were not added. Unlike the dissolution of oxide, the acid attack on the metal is electrochemical and anodic, as already explained, and hydrogen gas is evolved at other cathodic places. Inhibitors are chosen to block either the anodic or the cathodic part of this process. Apart from the loss of sound metal caused by attack of the acid, the hydrogen evolved may be absorbed in part into the metal itself. With hardened high-tensile steel this may lead to dangerous 'hydrogen embrittlement', to such an extent that sections which are already stressed may crack during pickling. Pickling inhibitors, however, reduce the hydrogen absorption. The absorbed hydrogen can be driven out subsequently by a low-temperature baking treatment.

Pickling also removes the casting skin and burnt-in sand residues from castings. Scale and sand can alternatively be removed from rigid objects by sand- or shot-blasting. This process is rather costly and leaves a rough and pitted surface which is unfavourable for electroplated or hot-dipped coatings, but which provides a useful mechanical key for sprayed metal coatings and for paints. Shot-blasting can, however, be applied locally *in situ*, which is especially useful in the case of very large objects, such as ships' hulls and bridges, which could not be treated by the wet method.

Steel, and most of the higher melting point metals, can be efficiently and rapidly descaled, without danger of attack on the solid metal, or of hydrogen embrittlement, by treatment in molten caustic soda (350 to 370°C) containing about 2% of sodium hydride. The process is expensive and is usually only applied where wet treatments are undesirable, e.g. because of risk of hydrogen embrittlement.

b. Pre-polishing. Where a final smooth or polished metallic surface is intended, most of the smoothing must be applied to the uncoated article, since the metallic coating would be unduly and irregularly thinned by the smoothing process. All these processes are loosely described as 'polishing', but as applied at this stage, they consist in fact only of careful grinding with progressively finer abrasives. The articles are pressed against the abrasive-loaded rim of rapidly rotating flexible wheels made of leather or coarse cloth discs. Two or three successive processes are usually used with progressively finer abrasives; grease or oil lubricant is applied during the final cuts, and the surface is brought to a satin smooth but entirely lustreless condition. The grease not only lubricates the operation, but supports and to some extent masks the abrasive particles; it also removes the frictional heat and prevents burning or scorching of the surface. These processes are known in the trade by various names such as bobbing, scurfing and buffing. In bobbing, the abrasive emery is glued to the rim of a leather or felt wheel; in scurfing or buffing a finer and milder abrasive is applied to a sisal or cloth wheel from a grease-bound cake. In sand-buffing, which is especially used for high-class silver and plated ware, loose oiled sand is fed between the article and a rotating leather-rimmed wheel. The abrasive used in the final stages is often Tripoli, or diatomaceous earth (see silica minerals and rocks in Vol. II). Of recent years the coarser grinding has often been performed on long endless belts of emery cloth, supported between a soft-faced pulley, against which the work is pressed, and another pulley some distance behind. This process is called back-stand emerying.

Because of the irregular shape and contour of most articles sent for decorative electroplating, they mostly have to be presented to the revolving wheels by the hands of operatives, who turn and twist them to ensure that the whole area is smoothed, and who also visually check that all blemishes and roughnesses are ground out. This is heavy, dirty and tedious work, which nevertheless requires considerable skill and experience. Its cost is therefore high; often as great as, or greater than all subsequent processes. It is moreover a serious bottleneck in production because the work is unattractive to labour.

Mechanization of the presentation of the articles to the scurfing and buffing wheels is possible where these are of somewhat regular shape, and are required in sufficient numbers to justify the considerable cost and complexity of the machines. For example, motor-car hub caps, saucepans and bumper bars (or fenders) can be held mechanically for scurfing, but of course the machine cannot distinguish between defects to be removed and inequalities which are a part of the design.

Another long practised, but crude method of overcoming the need for mechanization is known as barrel polishing. Originally this was only suitable for small, fairly simple and robust shapes, which were placed in a cylindrical container or barrel, together with a mass of pebbles or rounded abrasive shapes and a lubricating liquid. The whole assembly was then slowly rotated for some hours. The rubbing and tumbling of the abrasive shapes against the work pieces smoothed them in a rather unselective manner, but the process often inflicted as much damage as it removed. This process has been much improved, sometimes by substituting vibratory or other more gentle motion, so that quite delicate and fragile parts can be treated. Even so, these processes do not produce the best finish, but they have the merit of requiring little labour.

Metal finishing is the last step in the production process, and it is at this stage that a flawless surface must be produced in spite of any surface damage, dents, scratches, etc. which the article may have acquired during previous processes. It is obvious, but

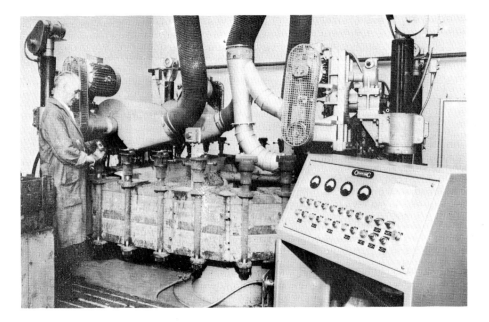

Fig. 9.2 Automatic special purpose mechanical polishing machine.
The circular articles, held on rotating spindles, move under rotating mops set at various angles, so as to polish all parts.

little regarded in practice, that care in the selection of the original material and in the handling during all stages of the manufacturing process can materially reduce the cost of the final smoothing operations.

In conclusion to this section it cannot be too strongly emphasized that although these pre-polishing treatments do not themselves produce a lustrous and polished surface, they, and they only, lay the foundation for a high-quality final finish. The quality of a polished surface lies not in its brilliance or gloss, but in the clarity of the images of surrounding objects reflected in it. Only if the preliminary smoothing has been conscientiously done can clear, sharp reflected images be obtained. Fuzzy, distorted images cannot be corrected by the final 'shining-up' operation, and are a sure sign of insufficient pre-polishing.

9.3.3 Degreasing and etching before electroplating.

Sound, continuous, adherent electroplated coatings can only be obtained on a chemically clean and firm metal foundation. Therefore the degreasing, thorough cleaning, and etching of a metal article before electroplating is a vital preliminary step. Similarly, thorough cleaning is necessary before most wet finishing processes, and also to some extent before painting and other dry finishing processes.

The first part of the preparation is the removal of all traces of grease or oil, so that the metal article is freely wetted by the electroplating solution. Indeed the test for the satisfactory completion of the degreasing process is that the metal surface remains completely wetted all over, without any visible 'water-break', even after dipping in a slightly acid solution.

Metal articles will often be lubricated with oil or grease during their manufacture; on smoothed articles traces of greasy buffing compounds may have been left — even normal handling will transfer traces of greasy or oily films to the articles. Metal surfaces have more affinity for oily and greasy matter than for watery solutions. In particular, certain types of organic compounds, such as fatty acids and fatty alcohols, which are common in natural fats, have certain polar groups attached to long chains of carbon atoms. These polar groups attach themselves tenaciously to a metal surface, whilst the long tails of carbon atoms form a thick greasy barrier to the approach of any other substance to the metal. Deliberate use is made of this property in lubricants by including these polar substances in them. This greasy film can also dissolve other oily, but non-polar, substances and dirt. Sometimes, by the action of heat, light or pressure, the organic material is further changed, or polymerized, on the surface to resinous or tarry materials, which are even more difficult to dislodge. The purpose of the degreasing step is to penetrate, dislodge and remove such films, and the methods used are really not very different in principle from those used domestically for dry-cleaning, clothes washing or dish washing except that they are more thorough.

Removal of heavy films of oil or grease is most conveniently done by dissolving them in an organic solvent. Normal hydrocarbon solvents such as petrol, paraffin and white spirit present too great a fire hazard; their light vapours form an explosive mixture with air, and spread readily. Heavy, non-inflammable chlorinated hydrocarbons are therefore used, principally trichlorethylene (C_2HCl_3) (colloquially known to the plater as 'tri'), and perchlorethylene (C_2Cl_4). An especially effective method of using these solvents is in a solvent-vapour degreasing apparatus. This consists of a deep, open-topped tank or box, at the bottom of which a quantity of trichlorethylene is heated to its boiling point (87°C) by steam coils or electric heaters. A water-cooled coil is arranged around the inside of the tank near the top, to condense the rising vapour, which in any case is much heavier than air. The major part of the tank volume is thus held full of solvent vapour. When a cold, greasy article is hung in this vapour, solvent condenses on it; this dissolves the grease and falls back into the solvent reservoir at the bottom.

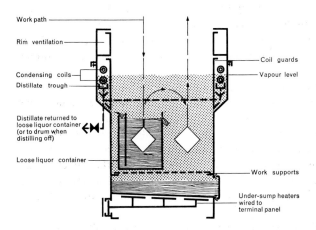

Fig. 9.3 Sectional view of a trichlorethylene degreasing apparatus.
The greasy article is lowered into a container full of warm, freshly distilled solvent, which dissolves most of the grease, and is then transferred into the hot solvent vapour. Clean liquid solvent condenses on it and drips off, removing the last traces of grease.

The article is thus quickly cleansed by being continuously bathed in clean solvent at its boiling point. The oil and grease is not re-vaporized with the solvent. No solvent is consumed but care must be taken that it is not wasted by being drawn out on degreased articles or by disturbance of the vapour, for it is quite expensive. Chlorinated hydrocarbon solvent vapours are dangerously narcotic if inhaled, and they also decompose to poisonous fumes if drawn through a flame. Solvent-vapour degreasing is therefore controlled by statutory regulations, but these are not irksome in a well-conducted plant.

Solvent-vapour degreasing alone does not produce a water-wettable surface and it must be followed by a wet degreasing process. This process works through a combination of saponification (see below), emulsification and change of surface tension, all based on the use of alkaline solutions and all in fact operative in domestic washing. Whereas acids, as a class, are characterized by a high concentration of hydrogen ions, i.e. a low pH number, alkalis are characterized by a high concentration of hydroxyl ions, i.e. by a low concentration of hydrogen ions, and hence a high pH number. It is this high concentration of hydroxyl ions which facilitates metal cleaning. Sodium and potassium hydroxides (soda and potash) are the strongest and harshest alkalis; sodium carbonate (washing soda), trisodium phosphate, disodium hydrogen phosphate, sodium metasilicate (waterglass), sodium cyanide and sodium borate (borax) are milder alkalis.

Alkaline solutions attack and combine with the polar groups of greasy compounds to form soaps which are soluble in water. This action proceeds best in hot solutions, and is called 'saponification'. Soaps, and also a large number of synthetic detergents, have the ability to break up oil and grease into small droplets and to hold them suspended in aqueous solution, without much tendency to recoalesce or to deposit on the solid surface. They do this by coating the exterior of each oil drop with a monomolecular film of the detergent molecules. These molecules are of elongated shape; one end favours contact with water but the other end prefers to be in oil. Thus the molecules arrange themselves in the surface of the oil drops with the water-favouring end outwards and the oil-favouring end inwards. This stabilizes the oil droplets. Soaps and other specially formulated synthetic detergents also have the property of changing both the surface tension and the interfacial tension of liquids. (These are the forces acting at the junction between an aqueous solution and a metal surface or an air surface.)

Degreasing of metallic articles prior to electroplating is therefore accomplished by immersing them for some time in a tank containing the hot, mixed alkaline solution. Typical formulations are given in table 9.2.

The stronger and harsher alkalis are the most effective for saponification, but not necessarily for emulsification; they also tend to stain iron and steel, and attack zinc and aluminium. Potassium hydroxide is even stronger and harsher than sodium hydroxide and is much more expensive. Caustic alkalis rapidly absorb carbon dioxide from the air; they are also not compatible with many useful synthetic detergents. For these reasons most degreasing solutions today are formulated from the milder alkalis, such as sodium carbonate, sodium silicate and sodium phosphate. Sodium cyanide is often also included in cleaners for copper alloys and other non-ferrous metals as it dissolves oxide and sulphide stains chemically.

Alkaline degreasing solutions can conveniently be contained in mild steel tanks, which are not attacked by alkalis. Such tanks are cheap and can easily be heated by external flames or internal steam coils. Handling the articles themselves after the degreasing operation would be likely to re-contaminate them, so it is universal practice first of all to fasten the articles onto wires or jigs of convenient design, by means of

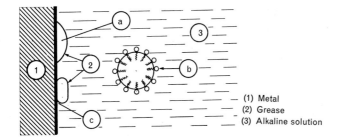

Fig. 9.4 Principle of Alkaline Degreasing.
A metal surface (1) with adherent grease (2) immersed in a hot alkaline solution (3) is cleaned by several actions, viz:
(a) Saponification: The grease, if a fatty acid, reacts with the alkali to form a soluble soap.
(b) Emulsification: Molecules of soap, detergent or wetting agent coat grease droplets and prevent them re-adhering.
(c) Reduction of surface tension: Alkalis, detergents or wetting agents reduce the adherence of the grease to the metal.

which they can be handled and supported in the various tanks, not only for degreasing, but throughout the electroplating processes.

The degreasing action needs some time and 5 min or more immersion is usual; the action is most rapid in an almost boiling solution, but too high a temperature has the disadvantage that the articles dry off when removed before they can be rinsed. Dislodgement of the grease can be encouraged by evolving hydrogen gas electrolytically at the surface by passing a current between the steel tank acting as anode (positive) and the articles as cathode (negative). The magnitude of the current is less important than a copious evolution of gas; 50 to 100 A/ft^2 of cathode area is suitable, for which a voltage of 4 to 6 V suffices.

For small and intricate parts the degreasing action of both organic solvents and aqueous alkaline solutions can be assisted by ultrasonic action. This consists of electronically generated alternate pulses of compression and rarefaction in the liquid, at a frequency above that of the highest pitched audible sound, e.g. 100 000 c/s. The rarefaction pulses are so rapid that they generate bubbles of vacuum in the liquid at the

Table 9.2 Typical alkaline degreasing solutions (preferably used at 180 to 200°F (80 to 95°C)).

		Heavy duty oz/gal (g/l)	Medium duty oz/gal (g/l)	Light duty oz/gal (g/l)
Caustic soda	NaOH	6 (37.5)	2 (12.5)	— —
Sodium carbonate	Na_2CO_3*	4 (25)	4 (25)	— —
Trisodium phosphate	$Na_3PO_4.12H_2O$	1 (6.2)	2 (12.5)	4 (25)
Sodium metasilicate	$Na_2SiO_3.5H_2O$	— —	2 (12.5)	4 (25)
Sodium cyanide	NaCN	— —	— —	2 (12.5)**
Wetting agent		0.25 (1.5)	0.12 (0.75)	0.12 (0.75)

* Anhydrous sodium carbonate, commercially known as soda ash. The decahydrate, $Na_2CO_3.10H_2O$, known as washing soda, may be used in proportionately greater amount.
**Sodium cyanide optional, but particularly valuable with copper alloys.

metal surface, and particularly in holes, crevices and recesses. These collapse equally suddenly during the compression pulses, in fact they 'implode'. The resultant violent disturbance at the metal surface greatly assists in the degreasing action.

After degreasing, the articles must be rinsed very thoroughly to remove all traces of emulsified oils and soaps, which would be precipitated by subsequent acid dips. The articles will then show an unbroken film of water over the whole surface, and should not be allowed to dry off.

Etching. The degreased articles are not yet ready to receive the electroplated coating, because the metal surface may not be sound and it may have superficial oxide films or stains on it. If a clean, sound metal surface can be exposed the subsequent electroplate will bond atomically to it, but any impediment to this intimate contact will lead to poor adhesion, and the plating may peel off later in service. The extreme outer surface of the metal is indeed not representative of the sound metal below, because it may have been disturbed and fragmented by previous manufacturing operations and by the smoothing processes. For this reason, after degreasing, the articles are lightly etched in a dilute acid solution, e.g. by dipping into 10% sulphuric acid. A cyanide solution is often used for lightly etching copper alloys and brass. Where thick and particularly adherent deposits must be formed on steel and the adhesion is very important, an electrolytic anodic etching process may be used. After these etching treatments the articles are again thoroughly rinsed and placed without delay into the electroplating tank. It is advisable for their entry to complete the electrical circuit so that plating starts immediately.

The technique of electroplating is fairly similar for different metals, but naturally the composition of the solution is different, and temperature, current density and other conditions may also vary. The process will therefore be illustrated by describing nickel and chromium plating in some detail, since these are by far the most extensively used electroplating processes, and then outlining processes for other metals.

9.3.4 Principles of electroplating

Electroplating is a method of producing a smooth, compact and fairly uniform film of metal from an aqueous solution of one of its chemical compounds by means of a direct electric current. The current flows from the aqueous solution to the article being electroplated, which must of course itself be electrically conducting, and which forms the 'cathode'. The current enters the aqueous solution through another metallic electrode, called the 'anode'. The rest of the circuit is of course metallic. The anode is usually made of the metal being plated out at the cathode; it dissolves under the action of the current and thus replenishes the solution, so that the net overall effect is to transfer metal from the anode and distribute it more or less uniformly over the cathode, leaving the composition of the solution unchanged. It is, however, incorrect to think of the metal passing at any appreciable speed between the electrodes in combined form. The metal is produced from the solution in one place (the cathode) and returned to it at another (the anode); in continued practical operation if the solution is not stirred it will become unduly concentrated around the anode, and impoverished around the cathode.*

---

* The direction of current mentioned above is the conventional one of flow from the positive to the negative pole; in fact the flow of electrons, which are negatively charged, is in the reverse direction.

METALLIC COATINGS | 681

Electroplating dates from the 1830s, and is closely associated with the name of Michael Faraday. Gold, silver and copper were first electroplated soon after the invention of the Daniell cell, the first reliable source of continuous electric current. Electromagnetic generation of electricity by rotating machinery was not available at that time; dynamos, in fact were first applied commercially in the electroplating industry, but not until 1841.

To understand the electroplating process it is necessary to recall the general principles of electrolysis and the nature of the simpler chemical compounds and of their aqueous solutions. Metals are chemical elements (alloys are mixtures of elemental metals). Metals are distinctive among the elements because the atoms of which they are composed have one or two (occasionally three) outer electrons which are rather loosely attached to the rest of the atom. In the solid metal these outer electrons are shared by all the atoms present, and are very mobile. Their mobility is the cause of the distinctively high electrical and thermal conductivity of metals.

It is convenient in this connection to consider hydrogen as a metal, because its atom also has an easily detached electron; the remainder of the atom is of course so different from that of any metal that hydrogen gas is not actually metallic in character. In chemistry, acids are substances which contain one or more hydrogen atoms attached to another atom or group of atoms. The nature of the latter distinguishes one acid from another, but all acids have hydrogen atoms. These hydrogen atoms are attached to the acid radical in a special way; viz. by transferring the loosely attached electron from the hydrogen to the other atom or group of atoms which comprise the acid radical. The hydrogen atom (now without its electron) is then bound to the acid radical (which has gained the electron) only by the electrostatic force set up by the transfer of the electron. In aqueous solution the hydrogen atom – less its electron, and now called a hydrogen ion – can separate from its original partner acid radical. Thus in the solution there are equal numbers of positively charged hydrogen ions and negatively charged acid-radical ions; the solution as a whole is electrically neutral.

Simple metal salts are derived (directly or indirectly) from acids by replacing the hydrogen ions of the acid by a metal ion; i.e. by a metal atom less its one (or two) loosely attached electrons. In the aqueous solution of a metal salt, the positively charged metal ions can similarly separate from the negatively charged acid radical or anion. The aqueous solution as a whole is still electrically neutral because it contains equal numbers of positive and negative charges. These phenomena are shown diagrammatically in fig. 9.5.

To be more specific, sulphuric acid (H_2SO_4) is an acid which has two hydrogen ions. Copper is a metal the atom of which has two easily removable electrons. Copper sulphate ($CuSO_4$) is the copper salt of sulphuric acid. In aqueous solution the copper ions, each with two positive charges due to the loss of two electrons, separate themselves from the doubly charged negative sulphate ion SO_4^{--}.

If two copper plates, which may be called electrodes, are dipped into the solution of copper sulphate, and a direct current applied to them, one plate will become negatively charged and the other positively charged. The negatively charged plate is called the cathode. When a copper ion touches the cathode, two electrons will transfer themselves from the cathode to the copper ion, and thus turn it back into metallic copper. Simultaneously, at the positive electrode – the anode – a metallic atom on the electrode will slip away into the solution as a copper ion, leaving its two electrons behind. In this way the aqueous solution remains electrically neutral, but two electrons have been

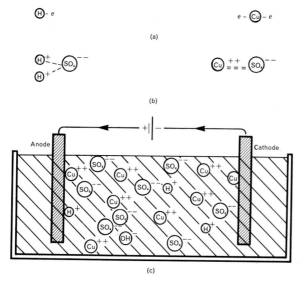

Fig. 9.5 Diagrammatic representation of the ionic nature of acids and metal salts, and of electrolysis.
(a) Atoms of hydrogen gas and metallic copper, showing the detachable electrons, one from hydrogen, two from copper.
(b) Molecule of sulphuric acid and of copper sulphate, showing the electrostatic bond, due to the transfer of electrons from hydrogen or from copper to the sulphate anion.
(c) Representation of the electrolysis of an aqueous copper sulphate solution with copper electrodes.
Seven copper cations (=14 positive charges), three hydrogen ions (=3 positive charges). Total positive charges =17; and eight sulphate anions (=16 negative charges) one hydroxyl ion (=1 negative charge). Total negative charges =17. Thus solution is neutral. There are more hydrogen ions than hydroxyl ions because the solution is slightly acid, but neither is concerned in the electrolysis.
The copper on the cathode (right) has just acquired two electrons and become metallic copper. The copper ion (extreme left) has just left the anode, leaving two electrons behind.
The arrows representing the direction of current flow are to the normal convention. Note that electrons flow in the opposite direction.

able to move down the connecting wire into the cathode, and another two up from the anode along its connecting wire. In other words a current (of electrons) flows in the metallic circuit, and a metallic atom has disappeared from the anode and another has appeared at the cathode.

Ions (whether anions or cations) do not move so freely in a solution as electrons do in a metal, so the rate at which the current can pass from the solution to the cathode and deposit smooth copper is somewhat limited. The amount of current is obviously related to the area of the solution/metal interface, so it is called a current density. Under ordinary conditions it is limited in this case to about 30 A/ft² of cathode area. This may be compared with a safe current of at least 1 000 A/in² of cross-section in a metal. However, this rate of 30 A/ft² of area corresponds to a growth of thickness of the coating of about 0.002 in/h (25 μm). This is fairly typical for most electroplating processes, so that it will be seen that electroplating is essentially a rather slow method of metal coating.

METALLIC COATINGS | 683

Two aspects of this process should be noted; firstly, that the amount of copper transferred from one electrode to the other is exactly proportional to the current passed – one copper atom for each two electrons; secondly, that no chemical work has been done by the current, copper has merely been transferred from one electrode to the other. Everything else is as before. Thus the current density controls the rate of electroplating, and only a small voltage is required to drive the current round. In this case it is about 1 V.

The process described above is a completely practical commercial electroplating process; the copper sulphate solution is usually made slightly acid to decrease its electrical resistance. Apart from its use in electroplating this process is also used for the electrorefining of all copper used for electrical purposes.

However, in general, such simple salts of metals are not always used for electroplating in industry. There are two main reasons for this:
1. The simple salt solutions tend to give rather coarse crystalline deposits, which are more difficult to polish.
2. When a metal is 'baser' (more negative in the electrochemical series, see table 9.1) than the metal of the electroplating solution which it is immersed in, there is a tendency for an exchange reaction to take place, without any applied current. For example, if steel is immersed in the copper sulphate plating solution mentioned above, some of the iron dissolves and is replaced by a film of copper, but this is loose and powdery. It is not a good foundation for electroplating.

This explains why complex salts of a metal are sometimes used for electroplating rather than simple ones.

There is one other principle of electrolysis which must be understood in connection with electroplating. The principal constituent of the aqueous electroplating solution is of course water (H_2O). This itself dissociates to some extent to yield positively charged hydrogen ions, and negatively charged hydroxyl ions (OH^-), thus:

$$H_2O \rightleftharpoons H^+ + OH^-$$

In a copper sulphate solution, although the positively charged hydrogen ions are attracted to the cathode they cannot be discharged there because copper ions are more easily discharged (see table 9.1). In a sodium sulphate solution or in an aluminium sulphate solution, however, it is far easier for the hydrogen ions to be discharged, so sodium or aluminium cannot be electroplated from aqueous solutions. It might appear from table 9.1 that it would be equally impossible to electroplate lead, tin, nickel, cadmium, chromium and zinc from aqueous solutions, but fortunately there is a further overvoltage of about 0.5 to 0.7 V which is necessary to evolve hydrogen gas on these metals, and furthermore, solutions for electroplating these metals are formulated to have plenty of metal ions available, with relatively few hydrogen ions. Thus in fact the metals named can be electroplated from aqueous solutions. However, in all these cases there is a tendency for some hydrogen to be plated out as well as the metal, as is explained in more detail for nickel plating.

If the anode – i.e. the positive electrode – is of a material which cannot dissolve in the solution – for example, if it is of graphite or platinum – current can still pass because of this dissociation of water. Two hydroxyl ions from the water give up one electron each at the anode, and then rearrange themselves to give oxygen gas and water:

$$2OH^- \rightarrow H_2O + O + 2e$$

Thus, if a copper sulphate solution is electrolyzed with a graphite anode, copper is plated out at first on the cathode as usual. However, as the copper in solution becomes used up, some other positively charged ion must be available to balance the negatively charged sulphate ions which are left. The number of hydrogen ions therefore increases to keep the solution as a whole electrically neutral, i.e., the solution becomes acid. Eventually all the copper ions in the solution would be used up. If current were still forced through the solution, it would then be possible for the hydrogen ions to be discharged and form hydrogen gas. The net result of the electrolysis in this case would be the splitting up of water into hydrogen gas at the cathode, and oxygen gas at the anode:

$$H_2O \rightarrow H_2 + \tfrac{1}{2}O_2$$

But now, unlike the case of depositing copper using a copper anode, chemical work has been done in decomposing water, so a higher voltage is required – over 2 V. This type of electrolysis is important in some cleaning processes.

9.3.5 Nickel plating

Nickel plating is by far the most important electroplating process, chiefly because it is the best undercoating immediately below chromium plating. Nickel is a hard, yellowish-white, non-toxic metal which takes a high polish and has considerable resistance to tarnish and corrosion by the weather. After prolonged exposure to damp industrial atmospheres it loses its polish and forms a superficial haze or bloom; but the metal is not deeply corroded and the original lustre can readily be restored by light polishing with a mildly abrasive metal polish. A sufficiently thick coating of nickel protects iron and steel from rusting, but only by excluding the corrosive environment from contact with the ferrous metal; if the nickel is locally defective or absent, the rusting of the steel at the exposed spots is accelerated by its presence on adjacent areas (see choice of metallic coatings, p. 672, and table 9.1).

Soon after the metal became commercially available, about 1870, nickel plating became popular for the protection and embellishment of harness parts and the then fashionable bicycle. Subsequently it was used for all kinds of metal articles. Its use was stimulated by the advent of the motor car, particularly after 1930, when chromium plating became the major decorative, as well as the protective medium on motor cars and many other manufactured products. Out of a total consumption in the UK in 1965 of 36 300 tons of nickel it is estimated that about 5 000 tons were used in electroplating, i.e. approximately one-seventh of the total consumed.

The most usual type of electroplating solution for nickel is the Watts solution. A typical composition is:

Nickel sulphate (as $NiSO_4 \cdot 7H_2O$)	40 oz/g (250 g/l)
Nickel chloride (as $NiCl_2 \cdot 6H_2O$)	6 oz/g (37.5 g/l)
Boric acid (H_3BO_3)	4 oz/g (25 g/l)
Acidity – pH value	3.0 to 5.8
Temperature	35 to 65°C
Current density (depending on temperature and agitation)	15 to 100 A/ft²

These conditions correspond to a rate of deposition of 0.0008 to 0.0053 in/h. The voltage necessary varies with the current density, the temperature and the size of the vat, but is in the range of 3 to 7 V.

In the above formula, the nickel salts are quoted in the form of the hydrated crystalline material commercially available; in solution, of course, this water of crystallization is no longer attached to the salts. The use of two different nickel salts is necessary to ensure the presence of some chloride ions, because with sulphate anions alone the nickel anodes do not dissolve freely, and the solution would become impoverished in nickel during use. The solution is almost saturated with nickel salts, to have the maximum amount of nickel ions available and to achieve a high current density.

In nickel plating the maintenance of a steady, but very slight acidity is most important. Satisfactory nickel plate can only be obtained in the pH range 3.0 to 6.0, but in practice a much closer range is maintained, e.g. pH 5.2 to 5.8. The pH value of the nickel plating solution tends to rise slowly in use, due to a small proportion of current

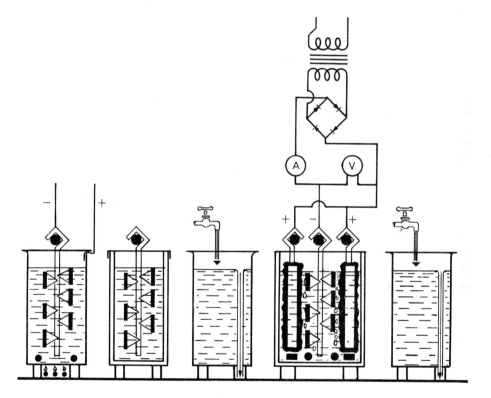

Fig. 9.6 Diagrammatic section of a simple nickel plating plant.
The articles to be plated, hung on suitable jigs, are transferred by hand from tank to tank.
Left to right:
1. Steel tank containing hot alkaline degreasing solution.
   The degreasing action is sometimes assisted by an electric current. After degreasing the articles are rinsed in Tank 3.
2. Lead-linked tank containing cold dilute acid, to etch the articles lightly.
3. Steel rinse tank, with flowing cold water.
4. Nickel plating tank. The lead-lined tank contains a warm solution of nickel and other salts.
   Note the step-down transformer and rectifier, supplying low-voltage direct current from the a.c. mains through an ammeter and voltmeter. The current enters the tank through nickel blocks, called anodes, one on each side, and crosses the solution to the articles. Air bubbling from pipes at the bottom stirs the solution.
5. Final rinse tank; cold water, or hot water to facilitate drying.

discharging hydrogen ions rather than nickel ions, especially at lower pH values of the solution, and must be adjusted from time to time by adding sulphuric acid, or alternatively, if it has become too acid, by adding nickel carbonate. The presence of the boric acid in the solution ensures that these changes of pH value are less sudden than would otherwise be the case.

The nickel plating solution is held in an open-topped, lead-lined or hard rubber-lined tank. The solution is heated by immersed steam coils or electric heaters, or in some other way. A temperature of at least 35°C is usual, but because higher current densities, and hence faster electroplating, can be achieved at higher temperatures, the baths are often operated at temperatures up to 65° or 70°C. The plating solution is stirred or agitated, usually by compressed air, which is blown in through a perforated pipe on the floor of the tank.

The tank is provided with a central metal rod at the top, from which the articles to be plated can be hung by the wires or jigs on to which they were fixed prior to degreasing. This rod is connected to the negative side of the low-voltage direct current supply. Similar rods are arranged along two opposite sides of the tank, and are connected to the positive side of the current supply. On these the nickel anodes are hung by metal hooks. Nickel anodes are usually cast with some care from metal containing oxide and other trace elements to facilitate their dissolution, and these are called depolarized anodes. Nickel (and other electroplating) anodes tend to release tiny metallic fragments into the solution as they dissolve. If these particles should settle on the articles being plated, a rough deposit would result. The anodes are therefore enclosed in heavy cotton twill bags. The electric current and the solution can percolate through the fabric, but the particles are retained. It is nevertheless quite common for nickel plating solutions to be filtered continuously, or from time to time.

The low-voltage direct current is supplied to the tank from the normal mains alternating current supply via a step-down transformer and a static rectifier. It is now usual for each tank to have its own transformer and rectifier; the voltage of the final direct current supply can be adjusted by variation of the transformer ratio, and this serves to adjust the magnitude of the current. A voltage range of 3 to 7 V is sufficient for this. It is unusual to have a switch in the low-voltage circuit, as it is highly desirable that electroplating commence as soon as the article is inserted in the tank, but there is always an ammeter in circuit. The current to be applied must be calculated from the surface area of the articles to be plated. At 35°C with moderate agitation, 15 to 30 A/ft^2 of area can be used, whereas at higher temperatures and with more violent agitation of the solution, current densities of up to 100 A/ft^2 are possible. If these current densities are seriously exceeded the electroplate is dark coloured and powdery; this is called 'burning'. The rate of growth of thickness of the deposit is strictly proportional to the current density, and at 20 A/ft^2 almost exactly 0.001 in (25 μm) per hour. It is thus possible to calculate the time for which the articles must remain in the plating tank to attain any desired thickness of deposit. During this time the current must not be interrupted for any appreciable time, or the plating is liable to be laminated and to peel off. It is, obviously, often difficult to compute the exact surface area of complex articles; in such cases the operator applies the voltage known from experience to produce the desired current density, but this method of current adjustment is not too reliable.

The thickness of nickel plating to be applied must be determined in the light of the type of basis metal and, particularly, with regard to the protection from corrosion which is desired. Although a thin deposit of nickel, e.g. 0.0001 in thick, is visually continuous,

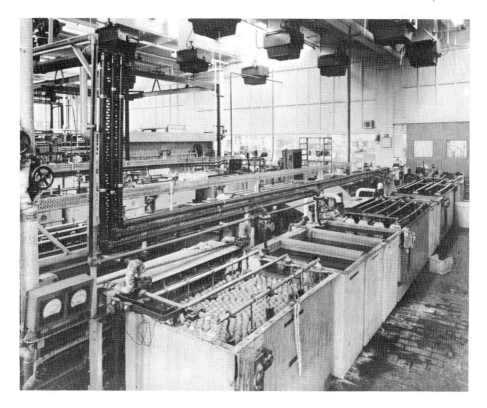

Fig. 9.7 Manually operated plant for nickel and chromium plating. The balls floating on the tank in foreground are to reduce evaporation losses.

it is found that it contains numerous, invisible tiny pores or holes. As the thickness of the plate is increased the number of these pores decreases until at about 0.001 in (25 μm) thickness, the number is negligible. This thickness (0.001 in) is considered to be the minimum reliable thickness if good protection to steel against the weather outdoors is required. Of course if the purpose of the electroplating is mostly decorative, and on a metal not easily corroded, a lesser thickness will suffice. It must, however, be borne in mind that on an article of intricate shape, the current density at any point may differ from the average current density, because of the shape of the electrical field. It will be appreciably less in recesses and appreciably greater on projecting areas. The thickness of plating will accordingly be much less than the reliable minimum in recesses, even though the average thickness is adequate. Some electroplating solutions compensate for this to some extent by a property called throwing power, but unfortunately nickel plating solutions do not have good throwing power. In any case some reduction of thickness must be allowed for in the subsequent polishing operation.

Against these considerations the commercial electroplater has to balance the fact that his plant is occupied longer for thicker deposits and the costs rise proportionately. The inadequacy of thickness of the nickel electroplate does not become apparent until long after it is delivered. For these reasons a British Standard Specification (No. 1224)

Fig. 9.8 Semi-mechanised electroplating plant. Small articles are plated in perforated, slowly rotating barrels, instead of being individually hung on jigs.

and numerous other private and foreign specifications lay down the minimum permissible thickness of nickel plating on any part of an article for various types of service, and describe and specify means of checking this. To achieve safely the minimum thickness specified for steel motor-car parts exposed to the weather (0.0012 in) may involve depositing at least 0.002 in average thickness, i.e. 2 h plating at 20 A/ft^2 or 40 m at 60 A/ft^2. Thinner nickel plating for use under milder conditions of exposure or on other basis metals is also covered in these specifications.

It should be made clear that many articles may be nickel plated at the same time in one nickel plating tank, provided that they do not shield one another from the flow of current from the anodes. All the articles may be put in and withdrawn together, or they may be put in and withdrawn in sequence, provided only that each one gets its proper share of the current and the requisite time.

After the selected period of electroplating the racks or wires carrying the articles are lifted out of the plating tank, thoroughly rinsed in running water to avoid stains and then dried, usually in a current of warm air. The parts may then be detached from the racks or wires. Inevitably, valuable nickel plating solution adheres to the parts or is trapped in recesses, etc. This is called 'drag-out' and is lost during rinsing. Except for this loss the nickel plating solution is little changed in the plating process since the nickel metal plated out from it is automatically replenished from the anodes.

Many other types of nickel plating solution have been advocated, mostly based on nickel sulphate, although nickel chloride and nickel sulphamate baths can be worked more quickly. All these yield matt deposits and are called dull plating baths. The electroplated coating from the Watts bath is smooth but of a matt, pearly milk-white texture and must be mechanically polished if, as is usual, a high lustre is required. This process is called colouring or colour buffing, and is fairly readily accomplished by pressing each part of the surface of the article against a revolving wheel built up of soft cloth discs. The wheel is dressed with very fine lime or similar powder bound in grease, the powder flowing or smearing the surface rather than abrading it.

The major modern change in the practice of nickel plating has been the use of brightening and levelling additions to the nickel plating solution of the Watts bath type. These are complex organic substances which, although present in solution only in small quantities, are adsorbed on the growing electroplate and modify its metallurgical structure. Not only is the size of the crystalline grains in it reduced, but the additions have a specific effect of restraining the growth at any place which tends to grow out above the general level. In this way the electroplate is kept microscopically smooth and hence it is bright and lustrous as plated. It is also much harder and more wear-resisting which is an important advantage in service. The advantage to the electroplater is not only that the cost and labour of mechanical polishing is eliminated, but also that the articles can proceed direct to chromium plating without unracking or rewiring. Rinsing and drying must, however, be carefully carried out to avoid water staining.

The operation and control of nickel plating calls for much care and experience. Temperature and clarity of solution, and the pH value must be constantly checked, e.g. twice daily; analysis for the main constituents, especially nickel, is required at longer intervals. Nickel plating solutions are especially sensitive to trace contamination by traces of copper, zinc and chromium, and by organic materials of a colloidal or glue-like nature. The inorganic contaminants can be eliminated by long continued plating onto a piece of scrap sheet metal at a low current density; organic materials are removed by oxidation with permanganate or by absorption on activated charcoal.

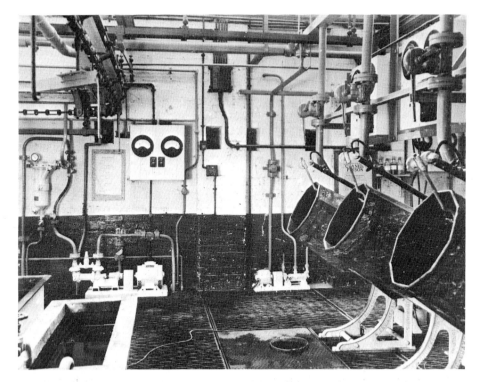

Fig. 9.9 Small articles are sometimes plated in inclined rotating drums partially filled with solution, in which a central anode hangs.

In addition to its use as undercoating to chromium, nickel plating is often used alone for protection of chemical plant and food-processing equipment, or for parts, such as the electrodes of thermionic valves, which must be protected from scaling by the action of heat.

Thicker deposits of electrodeposited nickel are sometimes applied to worn or wrongly machined steel parts to restore them to size or make them oversize. In such cases great attention must be paid to the preliminary etching process, to ensure good adhesion, and to the mechanical properties of the electrodeposited coating. The deposit can be confined to specific areas of the steel part by applying a thick layer of adherent and chemically neutral wax to the remaining area. This is called stopping off and is also used to prevent unwanted plating on the shafts of suspension jigs, etc., in decorative plating.

9.3.6 *Chromium plating*

It is convenient to deal next with chromium plating because it is almost always a direct sequel to nickel plating, and also because the methods of chromium plating are so radically different from all other types of electroplating.

Chromium is a hard, brittle, silvery-white metal, with a faint blue colour, which is not otherwise used unalloyed in industry. It is an important constituent of stainless steel, both of the cutlery type (13%), and of the austenitic ductile type (18:8 nickel: chromium). Chromium confers passivity to corrosion to these alloys because it automatically forms a protective oxide film on their surface. In an alloy with nickel it gives resistance to oxidation at high temperatures, and is thus used for electrical heating elements. It is also an important minor constituent of many special steels and non-ferrous alloys.

Although it would appear from the position of chromium in the electrochemical series (table 9.1) that this metal is about as reactive as zinc, in practice it automatically covers itself with a protective oxide film, and then behaves like a noble metal, such as gold. It is quite unattacked by atmospheric exposure, by most oxidizing chemicals and by acids and waters.

Attempts were made for many years to electroplate chromium from solutions of its simple salts, but these were so difficult to control as to be impractical, chiefly due to the codeposition of oxides and hydroxides.

In 1920 an American research worker named Sargent showed that chromium could be electroplated satisfactorily from chromic acid, if a small, but critical, proportion of of chromic sulphate were also present. It was soon found that it was the sulphate that was critical, although chromium sulphate was indeed formed during the plating process. This surprising solution composition has remained the basis of all chromium plating in spite of considerable research to find alternatives. The surprising aspect arises from the fact in the first place that the chromium in chromic acid solutions is present in the negatively charged anion, and would not be expected to be discharged at the negatively charged cathode, and in the second place, since electroplating is chemically a reduction process, it is unexpected that it should operate with the strongly oxidizing chromic acid.

Experience has shown that chromium plating can be satisfactorily achieved from chromic acid solutions of any concentration provided than an amount of sulphuric acid (or equivalent sulphate) is present equivalent to about one-hundredth part of the

Fig. 9.10 Semi-automatic plating plant.

chromic acid concentration. Beyond the limits of about 50:1 and 200:1 the solution just will not work at all. Hydrofluoric acid or fluorosilicic acid can be substituted for sulphuric acid, but with no particular advantage.

Chromic acid is in fact difficult to isolate as a single chemical substance, although it undoubtedly exists in aqueous solution. The substance commercially sold as chromic acid is the anhydride (CrO_3), i.e. chromic acid less water:

$$CrO_3 + H_2O \rightarrow H_2CrO_4 \rightleftharpoons 2H^+ + CrO_4^-$$

The unexpected type of composition is not the only unusual feature of chromium plating. Unless the current density exceeds a substantial figure (60 to 80 A/ft²), no electroplate whatever is obtained. Even above this minimum the electroplating process is grossly inefficient, i.e. only a minor proportion of the current density is employed usefully in depositing chromium; the remainder uselessly generates hydrogen gas. Thus very high current densities need to be used and even then the rate of growth of the deposit is very slow—at 200 A/ft² it is only 0.0005 in/h. Furthermore the appearance and mechanical properties of the electroplate are very sharply and critically dependent on the temperature of the solution. Below about 45°C the deposit is dark grey, rather hard and brittle, and is considered industrially useless; between 45 and 55°C it is lustrous and bright, glass-hard, intensely brittle and microscopically cracked; above 55°C it is milky-lustrous and unsatisfactory. Finally, the solution cannot be operated satisfactorily with chromium metal anodes, even if sufficiently pure metal were available, which it is not. The bath has therefore to be run with 'insoluble' anodes of lead or of lead-antimony alloy. These soon become coated with an electrically conducting film of lead peroxide, at which the current generates oxygen gas. The oxidizing conditions at the anode maintain the chromic acid in the oxidized state. The mechanism of chromium plating will not be discussed here, except to say that it is exceedingly complex and to some extent not yet fully elucidated.

Two typical formulations are given below. The only important differences are that the more concentrated composition has a lower critical and a lower operating current density, and can be used at a slightly lower temperature to achieve bright plating. At 100 A/ft²—the lowest practical plating current density—the efficiency is only 13% of the theoretical, and the rate of deposition is only about 0.0002 in/h. Chromium plating is mostly used to provide a non-tarnishing top coating to nickel plating; fortunately only a minute thickness is needed for this purpose, so it is the custom only to chromium plate for about 5 m. This gives a thickness of only about ten to twenty-millionths of an inch, but in fact a greater thickness is definitely detrimental because the number and width of the cracks, which are inevitably present, increases so that the protection actually

Table 9.3 Typical chromium plating solution compositions

	For decorative plating	For 'hard' plating
'Chromic Acid' (CrO_3)	64 oz/gal (400 g/l)	40 oz/gal (250 g/l)
Sulphuric acid (H_2SO_4)	0.64 oz/gal (4 g/l)	0.4 oz/gal (2.5 g/l)
Current density (minimum)	100 A/ft²	150 A/ft²
Temperature	A range of about 4° within the limits 45 to 55°C, according to the current density.	

Fig. 9.11 Semi-automatic nickel and chromium plating plant for motor-car silencers or tail mufflers.

decreases with increasing thickness. It cannot be too highly emphasized that the chromium itself is totally unattacked by the weather. Underlying nickel which is exposed at the cracks through the chromium, can of course corrode; indeed its corrosion is greatly stimulated by the close proximity and larger area of the much more 'noble' chromium. The corrosion products of the nickel then spread somewhat over the surface of the chromium adjacent to the cracks, and stain it, but this stain is entirely superficial and can easily be rubbed off, leaving the chromium undamaged.

Much thicker deposits of chromium are widely used in engineering for providing an extremely hard surface to machine parts. For this purpose the more dilute solution is used at a temperature of 50°C and a current density of at least 200 A/ft^2 or more. The deposit is confined to the required areas by coating the remainder of the surface with wax.

Chromium plating is carried out in steel tanks lined with lead or antimonial lead alloy. Since the currents used are very high, very substantial current-carrying bars and jigs are necessary. Because of these high currents, and because of the chemical work done not only in producing chromium at the cathode, but also in decomposing water to hydrogen at the cathode and to oxygen at the anode, a much greater voltage is required for chromium plating than for copper or nickel plating, where metal is only transported across the bath. However, 4 to 7 V suffices. The heavy current has a marked heating effect on the solution so that the means for heating the bath must be sensitive to this, and preferably controlled by thermostat within a narrow range.

The bubbles of oxygen, and especially those of hydrogen which are evolved during the electrolysis create a fine spray-mist of chromic acid solution in the air above the tank. This is irritating to breathe and can cause damage to the respiratory passages. Chromium plating tanks are therefore fume-extracted by ducts on two or more top edges of the tank in such a way as not to obstruct access to it. Another method is to add a proprietary surface-tension lowering chemical which causes the bubbles to be larger so that they do not throw up mist. Chromium plating solution is injurious to the skin and liable to turn minor wounds into festering sores. Operatives are therefore compelled by law to wear rubber gloves, boots and aprons, and to be inspected by a medical officer periodically. With reasonable care, however, chromium plating is not a hazardous operation.

It is not usually necessary to degrease or etch freshly polished or bright nickel plate before chromium plating, in view of the very strongly oxidizing nature of chromic acid and of the very thin coatings applied. Chromium plating tanks need remarkably little attention. Apart from close adjustment of temperature, the main periodical control is the addition of water to replace that lost by evaporation and the addition of chromic acid to replace that decomposed to chromium or lost as spray. Occasional analysis of the sulphate content is advisable since this may be increased by addition of impure water or accidental drag-in of other plating solutions. The amount of trivalent (cationic)

Fig. 9.12 A typical automatic nickel and chromium plating plant for taps and bathroom fittings. The articles are fixed on jigs which are then hung on the conveyor. The machine automatically lifts them from tank and finally returns them, completely plated, to the loading station.

chromium must be held within limits, but this is done by adjustment of the area ratio of anode to cathode.

Advantages are claimed for certain additions or alternatives to sulphuric acid in the solution to achieve so-called 'microcracked' or 'crackless' chromium, or to attain self-regulation of the bath. The simple solution given above, however, gives entirely satisfactory results.

The covering power and throwing power of chromium plating is very poor, and it is difficult to plate highly recessed areas or the edges of holes without special devices.

Experience has shown that good service is only given by chromium plate over an adequate film of nickel, whatever the nature of the basis metal (stainless iron is sometimes directly plated to improve the colour). A process for the direct chromium plating of zinc base die-castings for perambulator parts and similar applications, achieved some success for a time. This was called the Bornhauser solution and used a chromic acid solution partly neutralized with caustic soda and a very low sulphuric acid addition. The solution was operated at about room temperature and at extremely high current densities. The deposit was grey and matt, but could be readily polished to a rather dark lustre.

The disposal of waste chromium plating solutions, or even of the rinsing water presents some problems. Chromic acid, even in minute concentrations, is toxic to most aquatic life, including sewage bacteria. Most local authorities now require it to be dealt with before disposal of the waste liquid. This does not necessarily involve complete elimination of chromium; chemical reduction to the trivalent chromium salt is sufficient.

9.3.7 Electroplating of silver, gold, copper, zinc and cadmium

a. Introduction. It is convenient to deal with the electroplating of all the above metals together, because the same type of solution is used in each case, viz. the double cyanide solution. Cyanides are the salts of the very weak hydrocyanic acid (prussic acid). The cyanides of sodium and potassium are soluble in water, but the cyanides of the heavy metals are insoluble; nevertheless these insoluble cyanides readily dissolve in a solution of sodium or potassium cyanide to form a double salt

$$NaCN + AgCN \rightarrow NaAg(CN)_2 \rightleftharpoons Na^+ + Ag(CN)_2^-.$$

This double salt dissociates in solution to yield sodium ions and a negatively charged double cyanide ion which contains the heavy metal. Although most of the heavy metal is bound up in the anion and not available for plating, this double cyanide ion further dissociates to be in equilibrium with a very small amount of positively charged heavy metal ions, which are available for plating. As they are used up by deposition, more are immediately provided by further dissociation of the double cyanide anion. The advantages of this type of electroplating solution over solutions of simple salts of the metal concerned are twofold; firstly, the deposit is of much finer grain size, and hence is smoother and more easily polished; and secondly, the metal does not plate out chemically onto a more reactive metal, when this is first placed into the plating solution. The basic formulation of all these baths therefore includes the metal cyanide salt and enough sodium cyanide to form the double salt, together with some excess or free sodium cyanide to assist in dissolution of the metal at the anode. Other salts such as sodium carbonate or nitrate are also often present to increase the electrical conductivity of the solution. At one time potassium cyanide was preferred to sodium cyanide because of the higher conductivity it conferred, but it is much more expensive, and the conductivity can be increased equally well by using a slightly higher temperature. In some cases it is more convenient to prepare the solution from salts other than the heavy metal cyanide, e.g. in the case of zinc and cadmium, the metal oxide is used with even more cyanide. These react in solution to form the metal cyanide and sodium hydroxide. Some indicative formulations are given in table 9.4.

Cyanide solutions are of course intensely poisonous if ingested but they are not absorbed through the skin, nor do they fume. The toxic hazard has thus been found by long experience to be remarkably small. Nevertheless, great care must be taken to avoid cyanides being discharged in the effluents from electroplating plants into sewers or streams, as they are so toxic to animal and fish life, and even to sewage bacteria. The reduction of the cyanide content of waste from plants to a very low level is now a statutory obligation and presents some problems.

Cyanide solutions are alkaline and can be safely contained in steel tanks, which are cheap. They do not require much chemical control, except for ensuring that a sufficient

Table 9.4 Typical double cyanide electroplating solutions

Silver		Gold	
Silver cyanide	25 g/l	Gold (as metal, but added as cyanide)	2.1 g/l
Potassium cyanide	37 g/l	Potassium cyanide	15 g/l
Potassium carbonate	25 g/l	Disodium hydrogen phosphate	4 g/l
Free cyanide (approx.)	24 g/l		
Temperature	20–27°C	Temperature	60–80°C
Current density	5–10 A/ft²	Current density	1–5 A/ft²

	Copper (cyanide)	Copper (Rochelle salt)
Copper cyanide	21 g/l	25 g/l
Sodium cyanide	37 g/l	32.5 g/l
Rochelle salt ($KNaC_4H_4O_6 \cdot 4H_2O$)	–	45 g/l
Sodium carbonate	10 g/l	30 g/l
Free cyanide (approx.)	13 g/l	5.5 g/l
pH value	–	12.5
Temperature	45–60°C	50–70°C
Current density	5–10 A/ft²	20–60 A/ft²

	Zinc	Cadmium
Cadmium oxide	–	33 g/l
Zinc oxide	32 g/l	–
Caustic soda	10 g/l	–
Sodium cyanide	56 g/l	88 g/l
pH value	13.5	13
Temperature	30–40°C	18–20°C
Current density	10–20 A/ft²	10–20 A/ft²

excess of cyanide is maintained (free cyanide). This can be done by a simple titration or even by observation of the state of the anodes. On the other hand, cyanide solutions absorb carbon dioxide from the air because of their alkalinity, and form sodium carbonate. This consumes cyanide but does no great harm until the amount of carbonate is so great that it crystallizes out when the bath is cold. Cyanides also tend to oxidize and decompose, especially at high temperatures, so that cyanide plating baths are continually changing their composition.

b. Silver. Silver was one of the first metals to be electroplated from such solutions, and the electroplating industry was thus founded in the 1840s. A great incentive to its development was the ability thereby to provide an attractive appearance and the hygienic handling of foodstuffs without the expense of making a whole article out of the costly metal. For some decades prior to the introduction of silver plating this difficulty had been overcome by the use of Sheffield plate, a composite sheet metal made by rolling silver and copper together. Tableware and decorative articles were then made from the sandwich in such a way that only the silver was visible. Electroplating was able quickly to take over this market and establish itself.

Silver plating has changed little since it was introduced. The solution is traditionally used at room temperature or only slightly warm, although this limits the permissible current density to between 5 and 10 A/ft². As, however, the silver atom is heavy, this corresponds to a rate of deposition of 0.0008 to 0.0016 in/h. The silver electroplate obtained from the simple double cyanide solution is milky-white and matt, but it is very smooth and soft. It can readily be colour buffed or burnished to a very high lustre. Minor additions to the simple silver plating bath have long been known which limit the grain size of the deposit and produce a lustrous or bright coating. These are either organic sulphur compounds, typified by carbon disulphide, or small additions of other heavy metals such as lead or antimony. The increase in lustre lessens the labour of polishing, but the main advantage is that the bright deposit is harder and more wear-resistant.

For household and table articles, silver is usually applied to one of the copper alloys and predominantly to nickel-silver. Nickel-silver is an alloy of copper, zinc and nickel, and is thus essentially a brass, bleached to a yellowish-white colour by the addition of nickel. A minimum of 9% nickel is necessary to eliminate the brassy colour, and 12 to 15% gives a white alloy. With a white alloy eventual wearing through of the silver plate plate is less obvious than if copper or brass were used; nickel-silver is also stronger and more resistant to corrosion.

Silver is, of course, a beautiful white and highly reflective metal, which is intrinsically resistant to most corrosive influences, especially to fruit acids and foodstuffs. It is, unfortunately, readily tarnished to a most disfiguring dark-brown or black sulphide by the action of hydrogen sulphide in the air or by sulphur compounds in foodstuffs. This tarnish is superficial and does not involve much depth of silver. Silver is therefore the ideal electroplate for household and tableware, and now that detarnishing dips are available the tarnish question is not serious. Since the basis metal is always intrinsically corrosion resistant and the conditions of exposure are not severe, thin coatings are sufficient on articles not subject to much abrasive wear. For table articles such as spoons and forks, however, a much thicker coating is necessary to avoid penetration in use. Most reputable silver platers therefore apply relatively thick coatings in excess of 0.001 in to such articles. However, since silver is an expensive metal the tendency is to

698 | THE SURFACE TREATMENT OF METALS

Fig. 9.13 Special scale hand-operated plating plant for precious metals.

express the amount applied by quoting the weight of silver per article or per dozen, which confuses the issue unless the area of electroplate is known. Some ingenuity is often devoted to shielding articles in the plating bath in such a way that a thicker deposit is preferentially applied to places where most wear is to be expected, such as fork prongs and the backs of spoon bowls. In any case silver and all other metals plated from double cyanide electroplating solutions have very good throwing powder, i.e. the thickness of plating is better in recesses than might be expected.

Silver plating is now increasingly used for electronic parts and for some bearings where use is made of the high electrical and thermal conductivity of the metal respectively.

c. Gilding. Gold plating or gilding is also one of the oldest electroplating processes. A double cyanide type bath is used, but to reduce the capital investment this is usually extremely dilute. The consequent limitation on the maximum permissible current density and on the rate of plating is not important since the usual deposit is very thin, and is indeed often limited to that which gives the required colour. For this reason of colour other additions are often made to the bath. Gold does not dissolve very freely as an anode; the cost of providing gold anodes is also considerable, so gold plating baths are often used with insoluble anodes of stainless steel, platinum or graphite until exhausted. Gilding is usually applied to embellish trinkets and cheap jewellery which will be subject only to mild conditions. Recently, however, much thicker and more carefully controlled deposits have been used in the electronics industry, mostly from patented or proprietary solutions.

d. Copper plating from cyanide solutions. Copper plating is most commonly achieved using a double cyanide solution, because the acid copper sulphate solution mentioned earlier in this section cannot be utilized on steel or on zinc base articles owing to the loose metal deposition which takes place by chemical replacement. Copper plating is seldom used alone as a protective or decorative plating since it readily dulls and tarnishes in the atmosphere. Moreover, as a result of its intensely positive nature (see table 9.1) it strongly accelerates the corrosion of steel and most other basis metals, when they are exposed at pores or sites of mechanical damage. Electroplated copper coatings are, however, often used as the initial or undercoat on difficult basis metals such as zinc base alloy die-castings, cast iron, aluminium and also on very irregular and highly recessed articles because of their excellent covering and throwing power. In such cases copper deposits of one to two ten-thousandths of an inch are applied, and then the articles are thoroughly rinsed and transferred immediately to a nickel plating bath. The simple double cyanide bath is slow and is too alkaline for zinc base die-castings. For these purposes a solution containing Rochelle salt–sodium potassium tartrate–is used at a higher temperature and current density. It may be pointed out that the copper in copper cyanide is monovalent, i.e. only one electron is needed to deposit each atom, compared with two electrons per atom from the copper sulphate bath. Thus at the same current density the rate of deposition is twice as great in the cyanide copper bath as in the copper sulphate bath. Copper cyanide baths give a very smooth but matt deposit; additions can be made to provide a lustrous deposit, but this is less necessary since copper is usually only an undercoat.

e. Zinc and cadmium electroplating. Zinc and cadmium are also plated from double cyanide baths. There is less difficulty in dissolving the anodes and it assists if the solution is alkaline, so that caustic soda is often added, or formed automatically by using the metal oxide and more cyanide. The purpose for which these metals are electroplated is almost always to protect iron and steel from rust; they are not usually polished because appearance is secondary. Although zinc is a fairly reactive metal it soon covers itself with a protective film under neutral conditions, so that it is fairly resistant to the weather and to waters. Moreover, by the sacrificial action previously explained it protects the underlying steel even where the deposit is porous or damaged. However, this necessitates a thicker coating to allow for this sacrificial action.

The potential difference between steel and cadmium at exposed points is small but, especially in slightly alkaline solutions, cadmium protects steel by a slight sacrificial action, but it is not so readily corroded as zinc and the corrosion products are less unsightly. Cadmium is more resistant than zinc to weakly acid solutions and is more effective with weakly alkaline solutions such as those used for household washing. But cadmium is very expensive and becoming more so; therefore, although it is preferred for many purposes, particularly for aircraft parts, it is only used where the better service offsets its greater cost. Both zinc and cadmium compounds are somewhat toxic; electroplates of these metals should not be used on articles for use with foodstuffs.

Both zinc and cadmium can be electroplated in a lustrous condition by means of brightening additions. Alternatively the matt deposits from the simple solution can be chemically brightened after completion of the plating by immersion into simple chemical dips. A dip can also be used to preserve a zinc deposit from the superficial but disfiguring 'white rust'–a whitish protective film which zinc soon acquires in the atmosphere. A typical passivating dip is an acid solution of sodium dichromate.

Zinc coatings can also be electroplated on geometrically simple articles, such as sheet and wire, from an acid zinc sulphate solution. This is practised on a large scale by the manufacturers of such semi-fabricated commodities, where the hot-dipping process is impractical.

9.3.8 Tin plating

Tin is a very white metal with high reflectivity in the polished or freshly solidified condition. It has high resistance to corrosion especially against fruit acids, milk products and other foodstuffs, and is not attacked by sulphur compounds. Its salts are not poisonous. The metal is soft, easily melted and readily soft soldered. It is thus ideally suited as a coating metal for steel and copper vessels used with foodstuffs, and has been so used for centuries. The demand has now almost outrun the supply so that it has become a very expensive metal. Although tin is cathodic to steel during corrosion in air, and thus accelerates rusting at pores and breaks, the position is reversed in the absence of air, as in a sealed food can. The tin then sacrificially protects the steel at pores. Tin is an ideal coating for electrical wires and other electrical connections because of its suitability for soldering and its immunity to tarnish by sulphur-bearing rubber insulation. Tin has traditionally been applied to other metals in the molten state, as described later, but it can readily be electroplated either from acid or alkaline solution.

Alkaline plating is carried out in a hot solution of sodium stannate (Na_2SnO_3) formed by dissolving tin oxide in caustic soda. This is another example of an electroplating solution with the main reserve of the metal in the negative anion but in equilibrium with a small concentration of tin cations, from which the actual deposition takes place. The speed of plating, the covering and throwing power are excellent, although the deposit is matt. The rate of deposition is 0.0013 in/h at 20 A/ft^2, but an even faster rate of deposition can be obtained from more concentrated solutions of potassium stannate; this is more expensive. Both solutions can be fed from tin anodes, but great care has to be taken to maintain a high current density on these, or a different compound, sodium stannite is formed, which ruins the bath. Satisfactory anode performance is evident from a characteristic yellow-green appearance.

Acid solutions of tin (stannous) sulphate or fluoborate can be electroplated at very high current densities, but the deposit is rather coarsely crystalline unless complex organic brightening additions are used, e.g. cresol sulphonic acid or β-naphthol and gelatin. The control of this type of solution is rather exacting, and the covering and throwing power are poor. These disadvantages make the acid solution unsuitable for miscellaneous plating but are less detrimental in electroplating moving steel strip to produce electrolytic tin plate. This process has therefore largely replaced hot-dipped tinning. The tin plating is carried out at the steel works which produce large coils of bright cold-rolled steel strip. The plant for plating is laid out in a continuous line, through which the strip passes at high speed. Provision is made to join the end of one coil to the beginning of the next without stopping the line. The strip passes through degreasing, etching, rinsing, plating and drying stages in sequence, the length of each being adjusted to give the requisite time at the strip speed used. Although high current densities are used, e.g. 200 A/ft^2, and only a thin coating is applied (0.00005 in), a considerable length of plating tank is required to allow sufficient time for plating at the high speed of operation, e.g. 100 ft length at 1 000 ft/m. In earlier plants the strip was passed horizontally through a long tank, but as speeds increased, this became impractical and now the strip is passed in vertical loops through very deep tanks. Special precautions

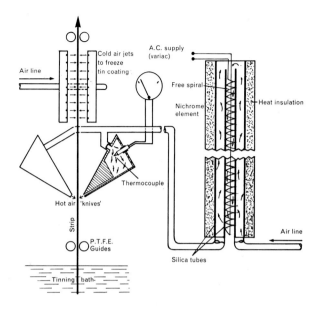

Fig. 9.14 Steel strip is electroplated with a thin layer of tin whilst travelling at high speed to produce "tinplate". The electroplated coating is dull; it is therefore brightened as the strip leaves the plating bath by being melted in a current of hot air.

need to be taken to avoid undue 'drag-out' of solution. The deposit is dull and matt, and somewhat porous. To compete with hot-dipped tin plate the strip therefore passes through a very rapid heating and cooling stage, which melts and 'flow-brightens' the tin. It is finally oiled and cut into the usual standard tin plate sheets. Plant of this type, which must handle steel strip 0.005 to 0.015 in thick and up to 36 in wide at 1 000 ft/m, is enormously expensive to install and operate, and calls for engineering of a high order. Since tin plate is a commodity of world-wide use and very sensitive to price, the whole operation is very finely balanced from an economic point of view.

9.3.9 Electroplating of other metals

Magnesium and aluminium cannot be formed as coatings by electroplating from aqueous solutions as they are too basic. Most other metals in common use can be electroplated, but their industrial demand is limited except for the metals already mentioned. Lead is sometimes plated onto chemical plant or for special purpose bearings. Indium is plated onto the lead bearings. A fluoborate solution is used in both cases. Platinum and palladium coatings are sometimes used for jewellery and for electrical instruments, but of the platinum metals, rhodium is the most utilized as a thin (1 mg/cm^2) tarnish-preventing coating on silver and in thicker layers for electronic switching devices.

9.3.10 Electroplating of alloys

Two or more metals can, with some difficulty, be electroplated simultaneously to give a mixed metal coating. The properties and appearance of such a mixture are, however, often different from that of the metallurgical alloy, which, having been melted, has come into metallurgical equilibrium.

Brass is electroplated from a mixed copper and zinc cyanide solution, and gold alloys from mixed cyanide solutions of gold with copper, silver, zinc or nickel. The objective is usually to obtain a thin plating of a particular colour for purely decorative purposes. Bronze (copper-tin), speculum alloy (copper-tin) and tin-zinc alloys can be electroplated to achieve some of the advantages of each constituent, chiefly good corrosion protection with maintenance of appearance; control is difficult however. A true intermetallic compound of tin and nickel can be electroplated; this has unique properties; although rather dark it is relatively untarnishable and gives good protection.

9.3.11 *Electroplating engineering, operation and control*

It will be clear from the foregoing that in its simplest form an electroplating plant consists of degreasing, etching and rinsing tanks, separate plating tanks for each metal to be plated, each with appropriate and adequate supply of continuous (or unidirectional) electric current. The articles to be plated are fastened to wires or suspension jigs, and these are then lifted into and out of the appropriate tanks in sequence, the times and currents being adjusted at the judgement of the operator. Many electroplating plants are indeed operated in this way. For larger-scale operation it is now, however, much more common to mechanize the whole process to a greater or lesser extent. The various tanks are arranged in line to a common width and depth, but with lengths in proportion to the relative time the articles will spend in them. Overhead mechanisms lift the work from tank to tank, and driving chains progress it laterally through the plant. Not only does this save labour, which may even be reduced only to loading and unloading the machine, but it also ensures that each article receives the predetermined treatment without human judgement. Temperatures, currents, pH values, etc., are automatically controlled and centrally indicated. Such machines are costly to install and maintain, and are uneconomic unless they can be kept fully loaded, but they have high capacity and above all they produce consistent results with a small labour force. This small force is used to supervize, control and inspect the operations and to maintain quality.

Where very large numbers of small parts such as nuts and bolts are to be electroplated a different type of process, called barrel plating, is used. The individual parts are not wired or jigged, but are put in mass into a perforated plastic cylinder which they about half fill. A main negative contact is fitted along the axis of this cylinder, which contacts some of the articles and they in turn contact the others. Thus current is conveyed from one to another throughout the mass. Means are provided for slowly rotating the barrel, which is then handled as a unit and progressed through the various tanks. As the barrel rotates the parts come successively to the surface of the mass and become plated.

In another form of barrel plating, the parts lie at the bottom of an open cup-shaped tub rotatable about an axis at about 45° to the horizontal. There is a negative contact sited centrally in the base, and an anode is hung above the parts. The tub is filled with plating solution and rotated. For removal, the contents are dumped out through a sieve. Barrel plating does not produce such satisfactory deposits as tank plating, for the action on any one part is at best intermittent and some parts may receive an inordinately thin deposit.

The precise procedure for the cleaning and pre-treatment of different metals before electroplating can only be mentioned briefly here. Iron and steel articles can be directly plated with all the metals previously mentioned. However, copper must be deposited from the cyanide rather than from the acid type of solution since otherwise a loose powdery deposit would be produced chemically. This does not apply to nickel-silver or brass, but for silver plating these metals a special 'strike' initial treatment is necessary to circumvent a similar chemical deposition of silver.

Zinc and zinc alloys are electroplated in very large quantities in the form of zinc base alloy pressure die-castings, and are subsequently exposed to severe corrosive conditions, e.g. on the outside of motor cars. The great chemical activity of zinc not only makes the etching and initial plating difficult, but also demands an unusually sound deposit. Any metallic coating is strongly cathodic to zinc, so that severe attack occurs to this base metal where it is exposed at pores, etc. The corrosion products formed there are more bulky than the metal from which they were formed and force up the adjacent plating into blisters. Zinc is invariably given a thin initial plating of copper from a double cyanide solution; the other desired electroplates such as nickel and copper follow.

Aluminium is an even more difficult metal to electroplate. Although aluminium is a very reactive metal, it is normally covered with a protective oxide film, which is rapidly reformed when a fresh metal surface is exposed. The oxide is also insoluble in most acids. It thus cannot be prepared in the normal way by acid dip. The most successful method of preparation is by immersion in a strongly alkaline solution of sodium zincate. The alkali dissolves off the oxide film, and then the naked metal reacts chemically with the zincate to form a thin zinc film on the aluminium. This can then be plated with copper from a double cyanide solution, and any other electroplate subsequently applied. Nevertheless aluminium is even more active than zinc, and thus is anodic to any metallic coating. The electroplating of aluminium and its alloys is therefore not advisable, especially as there are a number of protective and decorative processes, unique to aluminium, which can be applied instead.

Magnesium and its alloys are too reactive to be electroplated.

Control of electroplating processes relies largely on measurements of currents, temperatures and pH values, also on periodic chemical analysis of the solution. However, even today control relies very largely on the skill, knowledge and experience of the operator.

Electroplating plants use large quantities of water, which is expensive to buy and equally expensive to purify before disposal. Considerable attention to the economical use of water and also to recovery of rinsed-off chemicals is worth while. Water supplies suitable for drinking and domestic purposes may be unsuitable for electroplating, because of their content of dissolved salts. These may contaminate solutions and may also cause stains on articles dried off. De-ionizing processes based on ion-exchange resins are available to remove such salts and are often worth while. Large amounts of heat are required to maintain solutions at the required temperature, and for heating air which is afterwards thrown out of the fume extraction ducts. Care in design can often materially reduce such costs.

The inspection of electroplated coatings (and indeed of all metallic coatings) as a method of control by the producer or by the buyer will always include measurements of the thickness of coating. This can be done destructively on sample pieces by metallographic sectioning and direct measurement, or by determining the loss in weight when the coating metal only is selectively dissolved in a suitable solution. The former gives the local thickness, and can therefore be used to find the minimum thickness but the latter gives only the average thickness. The thickness can also be measured nondestructively in certain cases by physical methods. For example, in the case of a non-magnetic coating such as tin on a magnetic base, e.g. steel or vice versa as in the case of nickel on brass, the force of attraction of a permanent magnet can be related to the thickness. In other cases the strength of eddy currents induced in the surface by a high frequency magnetic field can be used, or the lateral thermal conductivity.

The protective capacity of an electroplated (or of any other) coating can only be assessed accurately by its behaviour in actual service. As this is usually too time-consuming and difficult, simulated and accelerated service tests are used, but it is difficult to devise satisfactory tests. Exposure to a continuous mist of salt water, the so-called salt-spray test, has been much used but does not truly simulate atmospheric exposure; nor does any test involving continuous or intermittent immersion in an aqueous solution. A more realistic test is exposure to an atmosphere saturated with water vapour and containing controlled amounts of sulphur dioxide and carbon dioxide, especially if periodical fluctuations of temperature are arranged to induce condensation and re-evaporation of dew. The porosity or cracking of cathodic coatings such as nickel and tin on steel, can be revealed by simple tests, such as boiling in water and observing the rust spots. These are useful for comparing one coating with another but difficult to relate to service behaviour. Equally, it is difficult to measure the adhesion of coatings. A simple test is to cut through the coating in a criss-cross pattern and then try to detach the flakes, e.g. with adhesive tape. The decorative value can be dealt with by a visual inspection, and a practised inspector can often detect other deficiencies, but not, unfortunately, inadequate thickness or poor adhesion. The mechanical properties of the deposit can be assessed by bending and fracturing tests.

Most of the above tests also apply to metallic coatings applied by other methods than electroplating, or even to paint films.

9.3.12 Metal coating by hot dipping

a. Hot tinning. The tin coating of steel and copper sheet and vessels by dipping the cleaned article in molten tin is of considerable antiquity, the suitability of the coating for contact with foodstuffs having long been recognized. The cheap production of tinplate, i.e. thin sheet steel coated with a very thin coating of tin, and its ease of soldering, enabled the enormous food canning industry to develop in the nineteenth century. The use of tin as a coating and for other vital purposes is so extensive that it has threatened to outrun the available supply of the metal, and has already vastly inflated the price. In the UK in 1965, out of a total consumption of 20 717 tons of tin, 10 591 tons, over 50%, was used for hot tinning. This was made up from 9 227 tons for tinplate, 552 tons for tinning copper wire, 88 tons for steel wire, and 724 tons for miscellaneous tinning.

Before World War II tin plate was produced in separate sheets, mostly 20 in by 14 in, broadly as follows. Mild steel plate, about 8 in wide by 4 ft long, was heated and cross-rolled to widen it. It was folded in half and rolling continued first longitudinally, then transversely. As the area increased the sheet was again folded and rolled, until eventually a 'pack' of sixteen folds was obtained. The pack was then rolled to the required gauge, welding of the individual folds being mostly prevented by the oxide scale. The edges of the pack were then sheared off and the sheets separated, stacked and annealed. Pickling and light cold rolling followed to clean, smooth and burnish the surface. A rapid anneal was given in boxes to exclude air, followed by a light pickling operation.

The tinpot was a semi-cylindrical steel vessel filled with molten tin. There was a central division dividing the surface of the tin into an entry and exit side. A molten flux of ammonium chloride floated on the tin on the entry side, and a layer of palm oil on the exit side. The tin was maintained at 300 to 330°C, but the palm oil was cooler (235 to 250°C). The steel sheets, wet from the last pickle, were fed vertically downwards into the molten tin through the flux on the entry side, caught by guide rolls and brought out through the palm oil. Pairs of rolls on the exit side, of which one pair was in the palm oil, distributed the molten tin fairly uniformly, and this tin drew up to a smooth and lustrous surface by surface tension, and then solidified in the palm oil. Thinness of the tin coating could be adjusted by the nip of the rolls. The issuing sheets were cooled and polished with bran or sawdust to remove most, but not all, the palm oil. Some residual oil was essential for lubrication of the sheets during can manufacture. The plates were cut to the standard 14 by 20 in size, critically inspected and packed 112 to a 'basis box'. The amount of coating was, and still is, expressed as the weight of tin distributed over the surface of all the sheets comprising a basis box (31 360 in^2 one side only). A coating of 1 lb per basis box thus corresponds to 0.00006 in thickness. Coatings of 1 to $2\frac{1}{2}$ lbs per basis box (0.00006 to 0.00015 in) are most common, but the process will provide films of 8 ozs to 15 lb per box. The extreme thinness of the usual coating should be noted. The thinner tin coatings are extremely porous, and indeed are not intended so much as a protective coating, as a means of easing soldering, and for decoration. However, the rusting is prevented to some extent by the residual palm oil.

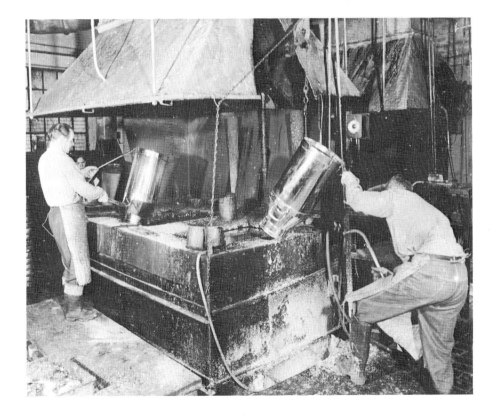

Fig. 9.15 Milk churns being "hot-tinned" by immersion into molten tin.

Today, although a certain amount of hot tinning of sheet is still done, the sheets are cut from cold-rolled steel strip. However, a very considerable proportion of tin plate is produced by the continuous electrolytic tinning of steel strip, as already described.

Copper and steel wire is continuously tinned by drawing through an etching bath, a fluxing bath, and then through molten tin, and through wipers to remove excess tin.

For the hot-dip tinning of fabricated articles of steel or copper, such as cooking utensils, milk churns and kettles, degreasing and pickling are necessary. The articles are then dipped into a fluxing solution of ammonium chloride and are manually dipped into a bath of molten tin, which has recently been skimmed. Considerable care is taken to drain the tin off, especially from the bottom ('list') edge. If necessary the tin is distributed by rubbing with tow or hemp. Sometimes the articles are held in hot palm oil, at a temperature above the melting point of tin to allow excess tin to drain off.

b. Terne plate. Lead is a cheap metal and is relatively inert to corrosion by the weather and by most waters, but it cannot be applied alone to steel by hot dipping because it does not wet or alloy with steel. The addition of quite a small proportion of tin (12 to 25%) provides an alloy which will wet and coat steel by hot dipping. The procedure is similar to hot tinning or galvanizing, but the coating is much thicker (up to 0.001 in). The product is called terne plate and is used for roofing, manufacturing large drums (it can readily be soldered) and for similar purposes. The coating is unattractive in appearance and cannot be used with foodstuffs, but it is an excellent basis for paint.

c. Zinc coating by hot dipping (galvanizing). The zinc coating of steel by hot dipping is confusingly known as galvanizing, although no use is made of electricity in the process. It provides a cheap and effective method of producing a good rust-resisting coating on steel sheet, steel wire and on miscellaneous fabricated steel and cast-iron articles, such as tanks, vessels, hardware, pipes and fittings and structural components, most of which will be exposed to the weather, or to damp air or water. Galvanizing is carried out on a very large scale and, with perhaps the exception of paint, is the most widely used protective coating system. Thus in 1965 in the UK, out of a total consumption of 357 000 tons of zinc, a total of 92 052 tons (25%) was used for galvanizing, comprising sheet and strip 23 836 tons, wire 19 596 tons, tube 13 433 tons, and general 35 187 tons.

Galvanizing consists in essence of dipping the cleaned and fluxed steel into a skimmed bath of molten zinc, and then withdrawing and cooling it as soon as possible. A certain rapidity of processing is necessary because the molten zinc not only wets the steel, but actively reacts with it to form an intermetallic compound which grows rapidly in thickness at temperatures above the melting point of zinc. This alloy layer is intensely brittle, therefore it is essential to keep it as thin as possible; and galvanized articles should not be deformed excessively for fear of fracturing this layer.

Hot-dip galvanizing of steel sheet (corrugated or otherwise) and of steel wire can be mechanized; galvanizing of fabricated parts is mostly done by hand. The parts must first be free from scale, rust and oxide; castings must be free from sand. This is mainly accomplished by pickling in acid, but large castings are sometimes shot-blasted or sand-blasted, followed by a lighter pickling process. It may be necessary to brush off detritus and residues adhering to the surface after pickling. The parts are then thoroughly rinsed and immersed in a fluxing solution, usually an aqueous solution of zinc ammonium chloride. They can be held in this solution without danger of further oxidation or attack until required for galvanizing.

The galvanizing bath is a heavy vessel of welded steel plate, fired from below and from the sides. Although this is often unlined, it may be lined with refractory brick. Such a vessel may hold many tons of molten zinc (spelter) – usually between 30 and 75 tons. During galvanizing the temperature is about 450°C. The surface of the molten metal must be kept clean by constant skimming when galvanization is in progress, or a thin film of ammonium chloride may be present, although this volatilizes to irritating fumes.

During galvanizing the parts are taken from the fluxing solution and well drained, or even dried off, without rinsing, and immersed into the clean molten zinc. They are then withdrawn as quickly as possible, and encouraged to drain by shaking or brushing, special attention being given to the lower edges. The parts are then put aside to cool or quenched in an air blast or in water. The thickness of the zinc coating can be controlled to some extent by the temperature of the zinc and the period of immersion. Coatings are usually expressed by the weight of zinc per unit area; 2 oz zinc per ft^2 = 0.0018 in thickness is a usual coating, but thicknesses of 4 to 6 oz are possible. As a result of surface-tension forces, small re-entrant crevices and angles, such as screw threads, tend to become clogged, whilst the coating may be thinner than average on sharp edges, except at the bottom edge where there may be a thickening.

The surface appearance is irregularly crystalline because the solidification of the zinc proceeds from nuclei, and causes characteristic and decorative 'spangles'. Small additions of aluminium increase the fluidity of the molten bath, and restrain the growth of the alloy layer. Dross and oxide films floating on the bath surface are liable to become entangled in the molten zinc coating and to impair its appearance. The zinc used is often rather impure (about 99%). Moreover, it reacts with the iron immersed in it and with the pot; eventually it becomes saturated with the iron-zinc intermetallic compound, which crystallizes out. These crystals tend to sink to the bottom of the zinc pot, but some may get caught up in the coating, causing unsightly lumps.

Steel sheets are galvanized mechanically by processing them through the molten zinc by power-driven rolls. Steel wire is passed mechanically through the zinc bath, and is usually drawn through a wiper on the exit side.

d. Hot dip aluminizing. Aluminium can also be coated on to iron and steel articles by hot dipping, but the process is considerably more difficult than in the case of tin and zinc, for a number of reasons. Because of the higher melting point of aluminium the process must be carried out at about 700°C, at which temperature it is very difficult to preserve the steel in an oxide-free condition. The aluminium melt itself cannot be kept without an oxide film on its surface, even in an inert atmosphere. The major difficulty in aluminizing, however, is the rapidity of the formation of an intensely brittle aluminium-iron intermetallic compound at the interface. The growth of this layer can be restrained to a limited extent by additions of silicon or beryllium to the melt. A 6% silicon-aluminium alloy is usually used. The steel is freed from oxide by pickling, but must be dried before entry into the molten alloy. Simple and partially successful schemes of providing a thin replacement film of copper on the surface, and then coating it with glycerine, have been used, but for large-scale aluminizing very complicated plant for deoxidizing the sheet and introducing it into the molten bath under a protective gas are employed.

Aluminized coatings are bright and metallic in appearance, although usually marred by streaks and patches of oxide, drawn off the surface of the melt. They will not withstand even slight bending without spalling and cracking, because of the inevitable layer of intermetallic compound at the interface. They have excellent resistance to exposure to the weather, to waters and to damp conditions without much change of appearance. However, their major field of application is in resisting the effect of high temperatures which would oxidize and scale the steel base. Aluminized steel thus finds its chief application for such purposes as motor-car silencers and exhaust systems, for gas-heater heat exchangers and for similar applications where hot products of combustion are concerned. A usual thickness is 0.001 to 0.002 in.

9.3.13 Metal spraying

Protective metal coatings can also be sprayed onto the cleaned and roughened article in the form of molten metal droplets, which congeal and bond immediately on impact. The article does not become unduly heated (even paper can be metal sprayed) and the coating does not wet or alloy with the basis metal, so that the adhesion of the coating is only mechanical. The coating is only an aggregate of flattened quenched droplets, interspersed with oxide films and some porosity, so that it is very brittle.

Nevertheless the process has the special advantage over most of the processes so far described, that it can be applied to structures of any size and, if necessary, *in situ*. It is thus specially suitable for the protection of ships' hulls, structural steel work, bridges and other civil engineering steel parts. The steel surface must be freed from all scale and rust before coating; this is usually done by shot-blasting or sand-blasting, using compressed air. This preparation also roughens the surface and facilitates the mechanical adhesion of the coating.

Metal spraying is chiefly used for applying zinc or aluminium onto steel. Three main methods are used. The most popular and most highly developed utilizes a handheld or machine-held pistol in which a wire of the coating metal is continuously fed into an oxyacetylene flame, where it is melted, and the droplets of molten metal are im-

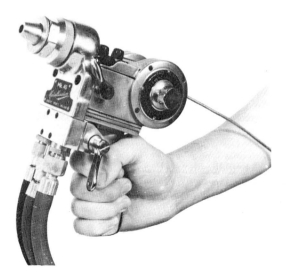

Fig. 9.16 Pistol for spraying molten metal. The pistol is an oxy-gas blowpipe; the metal is continuously fed through the flame, where it melts and is impelled forward as molten droplets.

pelled on to the work piece by compressed air. In another method, the coating metal in the form of powder is blown through a melting flame onto the work by the compressed air. In a third method, pre-melted metal from a reservoir is atomized and blown onto the work piece. The rate of deposition is quite fast, but only a small area is covered at one time. Coatings of at least 0.002 in, and usually of 0.005 in, are applied; such thick coatings are necessary because the deposit is rather porous. The outer surface is rough and unattractive in appearance but is an excellent base for subsequent paint coatings. The porosity is less harmful than in thinner coatings for it soon seals itself with corrosion product.

9.3.14 Fire gilding

This process for producing a gold coating on silver or on copper or its alloys is many thousands of years old, but is still used for the very finest silversmiths' wares, ormolu furniture mounts, buttons for uniforms and for medals and regalia. Metallic gold is dissolved in mercury to form a liquid or paste amalgam: this is uniformly spread or smeared over the areas of the silver or copper article to be gilded. The process has the advantage that it need only be applied to selected areas. This is called parcel gilding. The gold amalgam wets the metal. The whole article is then heated, gently at first, and finally to about 370°C, at which temperature the mercury volatilizes, leaving the gold to alloy with the basis metal. According to the strength of the amalgam and the speed of the heating, the final surface consists of full-coloured pure gold or (on silver) a pale lemon-yellow coloured gold-silver alloy. The coating is well bonded to the basis metal, and resists many years of polishing. The volatilized mercury vapour is extremely poisonous; it is prudent to carry out the heating in an enclosed furnace or oven, from which the fumes are drawn off and cooled by passing over wetted baffles or even by bubbling through water. In this way most of the mercury is recovered. The process is obviously expensive in materials and in labour, but the applications usually justify this.

9.3.15 'Electroless' plating of nickel

Metals can be recovered in metallic form from aqueous solutions of their salts by means of 'reducing' agents. Development of photographic film and the silvering of mirrors are examples of this. The more noble metals such as gold and silver are easily turned out of the combined form; the more electronegative metals require much more powerful reducing agents. It is also essential that the metal deposits on the solid surface as a coherent and continuous film, and highly desirable that it only coats the surface desired, and not the walls of the containing vessel as well. In the silvering of glass mirrors, for example, the silver deposits on the glass but also on the walls of the containing vessel.

A process has been devised using the very powerful reducing agent, sodium hypophosphite, to produce nickel coatings on metals, and has the peculiar and advantageous property of proceeding only on nickel itself or on metals of the platinum group. Once started, it is therefore autocatalytic. This is a purely chemical reduction and nonelectrolytic. It is therefore generally known by the ugly term 'electroless' nickel plating. The solution contains nickel sulphate, chloride or acetate, sodium hypophosphite and an organic acid, such as citric or lactic acid to control the change of pH value, which must be maintained at about pH 4.0. The solution is operated at about 90°C and deposits nickel at the rate of about 0.0008 in/h. In due course the solution becomes spent and must be renewed. The formation of the nickel coating on a different metal such as steel can be ensured by a prior treatment of the surface with a dilute palladium chloride solution; a thin film of palladium is formed by chemical replacement which catalyses the subsequent reduction of the nickel salt by the hypophosphite.

The metallic deposit has an appreciable content of phosphorus. It is smooth and often lustrous, very fine grained and extremely hard (500 Vickers pyramid number – an indentation measure of hardness); it is very brittle. Since the process is chemical and autocatalytic, the thickness of the deposit can be uniform even on the most irregular and re-entrant areas. The process is naturally rather expensive, but it is uniquely suited to coating the interior of complicated chemical plant, tubes, condensers, etc., even after assembly or installation.

9.3.16 *Hot diffusion processes*

There are a number of processes in use on a limited scale in which steel articles are coated with another metal by being heated in contact with the metal or one of its compounds in the absence of air.

a. Calorizing. This process is a means of obtaining an aluminium coating on heavy steel vessels and boxes, furnace parts and other pieces which are used in heat treatment or firing of ceramic goods, or exposed in some other way continually to high temperatures in air. The aluminium coating prevents the steel scaling and wasting away and thus materially prolongs its life. The process of calorizing is carried out by prolonged heating of the steel articles at about 500°C in a box with a mixture of aluminium powder, alumina and ammonium chloride, air being excluded. It probably functions via the formation of aluminium chloride vapour and the reaction of this with the iron in the surface layers.

b. Sherardizing. This is a means of zinc coating small steel articles such as nuts and bolts. The coating is extremely uniform and thus, for example, the functioning of screw threads is not impaired. The articles are tumbled for some hours in zinc dust at a temperature of about 370°C, air being excluded. The coating contains a high proportion of iron-zinc alloy; it is dark-blue in colour and brittle. A normal thickness is 0.5 oz/ft^2. (0.0005 in).

c. Chromizing. The chlorides of a number of metals (and some non-metals) are volatile and will react with steel at elevated temperatures to form iron chloride and deposit the metal in the place of the iron removed. Thus steel articles can be chromized by heating in chromic chloride vapour. An iron-chromium alloy is formed on the surface, which has corrosion-resisting properties similar to those of stainless steel; if carbon is present, the process may be used to effect a surface hardening. The process can also be used to confer resistance to heat oxidation onto steel articles.

9.4 MISCELLANEOUS NON-METALLIC COATING PROCESSES

The more important non-metallic coatings for metal are vitreous enamel and the various types of organic coatings such as paint, enamel and lacquer. Nevertheless there are a number of other processes for producing non-metallic coatings which it is convenient to mention first, since they are either akin to electroplating processes or are preparatory to painting processes.

9.4.1 Metal colouring processes

Although metals dull and tarnish in time, some of the patinas or surface effects so produced are considered to be visually pleasant, especially where patterns of clean and tarnished metal appear side by side due to constant handling. There is extensive literature on methods for the artificial production of such effects on decorative household articles such as light fittings, door furniture and other hardware. The erroneously termed oxidized finish on copper and silver articles is the best known in this class. The mauve, brown or black tarnish which appears in time on copper or silver articles exposed indoors is actually a sulphide film. It can readily be produced artificially on the clean metal surface by immersing it in an aqueous solution of sodium or ammonium polysulphide, known in the trade as liver of sulphur. The coating is formed more uniformly on a freshly electroplated surface. To simulate the effect of constant handling, the coatings are then locally removed or 'relieved' by local rubbing with a cloth or brush loaded with pumice powder. All such tarnish films are neither stable nor wear resistant, so they are invariably further protected by a film of clear lacquer, usually of air-drying nitrocellulose lacquer.

In the hardening of high-carbon steel, it is first quenched from a high temperature into water or oil, and then tempered to improve the toughness by heating to a rather exact and moderate temperature. Before the days of accurate temperature-measuring instruments, this temperature, and thus the degree of temper, was judged by the interference colour formed on the cleaned steel during tempering. When the desired colour was attained, the parts were again quenched. Some of these colours are attractive in appearance, such as the peacock-blue colour suitable for springs; they have moreover become associated with the properties of a particular grade of temper. The colours are caused by interference of light in oxide films of precise thickness determined by the temperature. These films are very thin, but when saturated with oil they have a slight protective value. They are therefore sometimes artificially produced by other chemical oxidation processes, such as heating in a melt of mixed sodium and potassium nitrates at the appropriate temperature. Such films are used, after oiling with, e.g., linseed oil, in tool, watch and instrument manufacture.

The traditional attractive brownish-black coating on firearms and weapons is also an oiled oxide film, but is much thicker and is a mixture of ferric and ferrous oxides. The traditional method of formation is lengthy and tedious; it involves repeated treatments with a complex oxidizing solution and successive heatings. A simpler process of oxide blackening is widely used for springs, clips and similar small steel parts. They are immersed in a very concentrated solution of caustic soda containing sodium nitrate. Owing to the high concentration of caustic soda, this solution can be operated at 150°C. An oxide coating of 0.001 to 0.002 in thickness is obtained in 20 to 30 minutes. This is somewhat porous, but, if kept oiled, it provides a fair protection from rusting.

The Bower-Barff process was formerly used on the inside of radiators and other steel vessels intended to contain water. A thick adherent coating of magnetite scale (Fe_3O_4) was produced. The parts were heated to above 900°C, then steam was introduced for about 20 m to produce oxide, followed by gases containing carbon monoxide for 15 to 25 m to reduce the oxide to magnetite. The magnetite and the steel are said to have similar coefficients of expansion and thus to remain uncracked over considerable changes of temperature.

9.4.2 Phosphating

There are a multitude of separate processes designed to produce thin, crystalline, adherent films of phosphates of iron on the surface of steel articles; these often have an admixture of zinc and manganese phosphates. These films absorb oil readily and the oiled film confers a considerable measure of corrosion protection quite cheaply on iron and steel goods where appearance is not important. Phosphate coatings are also an excellent pre-treatment for paint or stove enamel on steel, ensuring good adhesion and discouraging lateral corrosion under the paint film from points of damage. Basically this phosphating process consists of immersing the steel articles into a hot solution containing ferrous phosphate with some free phosphoric acid, and with additions of zinc and/or manganese phosphates. Phosphoric acid is tribasic, i.e. it has three replaceable hydrogen ions. The acid phosphates, in which only one or two of the hydrogen atoms are replaced, are fairly soluble in water, but the full phosphate, in which all three hydrogens are replaced by iron, is insoluble. As the phosphoric acid in the solution attacks the steel, the local acidity is reduced and a crystalline film of iron phosphate precipitates onto it. The simple process described above is rather slow, taking up to an hour to provide a suitable coating; addition of zinc or manganese phosphates accelerates the process and manganese phosphate ensures a thicker coating. But the action can be speeded up more fully by the addition of accelerators which are slightly oxidizing chemicals such as nitrates or nitrites, or complex organic nitrocompounds. The range of possible formulations and types of coating is very wide, and their chemistry complex, so that the different processes are mostly serviced by specialist companies under proprietary names, such as Coslettizing (the original, now obsolete, process), Bonderising, Granodising, Parkerising, Walterisation, etc. These companies sell the mixed salts or solutions with full instructions for their use, and often provide a supervision and chemical control service. In some cases they supply plant as well. In these circumstances the user is relieved of much of the technical difficulty and control work, and he can integrate the process into his metal manufacturing operations.

Phosphating processes are used very widely in industry for three distinct purposes, for each of which a somewhat different formulation is necessary. These end-uses are:
1. Application of a simple and cheap corrosion preventative for machine and mechanical steel parts. As formed, the phosphate coating is dull matt and grey-black; it is unattractive in appearance and has only slight protective properties. It is, however, moderately rough and absorbent. It is therefore usually treated with an aqueous solution of a black dye and then, after drying, with hot oil.
2. Rather thicker phosphate coatings are used alone or impregnated with oil to provide an exceptionally good lubricant in the processes of heavy pressing, drawing, forming and even extrusion, of steel. Similarly, they are often applied to gears and other sliding parts to reduce friction. The coating not only holds the oil, but itself prevents the metal-to-metal contact which is the cause of scoring and seizure when metal works on metal.
3. The most widespread application of phosphate coatings is as a basis for industrially applied paints, lacquers and enamels. The phosphate coating greatly increases the adhesion of the organic coating to the base, and also largely prevents the lateral extension

of the corrosion from inevitable points of damage, which would otherwise cause blistering of the adjacent paint film.

Phosphate coatings are sometimes applied to zinc-plated or galvanized steel, the zinc coating supplying even more assurance against rusting. Phosphate coatings are also applied to zinc base die-castings as an aid to paint adhesion. Thick coatings are in the range 700 to 2 000 mg/ft^2; light coatings are 150 to 400 mg/ft^2. A process using a solution containing manganese gives thicker coatings, and is prefered for the first two applications. A zinc process is more rapid, and can even be applied by spraying the liquid onto the work.

In the operation of the process, the work must first be descaled and degreased, but strongly acid or strongly alkaline cleaners leave the surface in a condition which encourages a coarsely crystalline phosphate deposit, so these are usually avoided. The articles are then rinsed and hung in the phosphating solution. As the process is non-electrolytic, close spacing is possible, and small parts can be treated in baskets or perforated containers. The solution is usually operated at 80°C or above; treatment times of 3 to 20 m are usual. After phosphating, the parts are rinsed, but residues of the solution inadvertently left in joints, crevices, etc., are not corrosive. A short further dip in 0.5% chromic acid solution is often given for further passivation, with or without a final rinse.

Since very considerable numbers of parts are processed, and these are often very large, e.g. complete car bodies, or very small, phosphating plants are frequently mechanized. In the motor car industry 'Roto-Dip' plants are used in which complete car bodies, each skewered on a horizontal shaft, are passed through an elaborate automatic phosphating plant, and finally dipped in a bath of priming paint, suitably drained, and then stoved. The bodies are rotated continuously to ensure that every area is treated and that solution is not trapped in pockets.

Although the design and formulation of the phosphating process is complex, its control can be achieved by simple titrations with standard alkali solutions, using different indicators. The results are usually expressed as 'points', and additions of an acid phosphate mixture are made, as necessary, to restore these to some pre-determined value.

9.4.3 Treatments specific to aluminium alloys

Aluminium and its alloys owe their excellent resistance to corrosion by the atmosphere, waters, foodstuffs, etc. to an oxide film on the surface of the metal, which is very rapidly reformed when the naked metal is exposed by cutting or deformation. This film is very protective and relatively pore free, but it is very thin (about one millionth of an inch). It is therefore logical to devise means of artificially thickening it to provide even better resistance to corrosion. There are two main types of methods for doing this, one of which is purely chemical, the other electrochemical.

a. MBV treatment. The modified Bauer-Vogel process, known as the MBV process, consists of immersing the aluminium alloy parts for 10 to 30 m in an almost boiling aqueous solution of 5% sodium carbonate and 1% sodium chromate. The oxide film so formed is light-grey to almost black in colour, depending on the aluminium alloy, but is smooth and glossy; it is only a few hundred-thousandths of an inch thick. Nevertheless it has considerable corrosion resistance, and is absorbent to oil or paint so that it is an excellent preparatory process for painting aluminium alloys.

b. Chromating processes. Another chemical process for increasing the thickness of the oxide film employs an acid solution containing phosphoric and chromic acids and sodium fluoride. It is used at more moderate temperatures, (40 to 50°C) and for shorter times than the MBV process. The coating is, green, brown or golden in colour, quite thin and rather soft. It is principally employed as a preparatory process for painting aluminium alloys. Owing to the chromic acid content it should not be used on articles intended for use with foodstuffs. This type of process is mostly used under proprietary names, such as Alodine in the USA, and in GB Alocrom.

c. Anodizing. Anodic oxidation, or anodizing, is the electrochemical method of providing an artificially thickened oxide film on aluminium alloys. It is carried out by making the aluminium alloy anodic (positive) in a solution of chromic, sulphuric, oxalic, or, less commonly, phosphoric acid. In any of these solutions the effect of the current is to convert the surface of the metal to a compact adherent oxide film. This would very soon insulate the metal and stop the current, if the electrolyte did not simultaneously attack the oxide and render it porous. As a result the current can continue to pass and the film can continue to grow in thickness, to the point where the speed of dissolution equals the speed of formation. Chromic acid anodizing was the first of these processes to be developed. A 3% solution is used at 50°C, and the voltage increased gradually up to 50 V. After a few minutes at this voltage the growing resistance of the film practically stops the current. The resulting oxide film is dark and greenish in colour; it has considerable protective capacity, but is an especially good basis for subsequent paint or grease protective systems. Chromic acid anodizing can be applied to the strong copper-containing alloys such as are used in aircraft construction. Residues of the electrolyte trapped in joints, etc. are not harmful and are inhibitory to corrosion. For these reasons the chromic acid anodizing process is very important in the vital protection of aircraft structures and components.

Fig. 9.17 Typical anodising plant for aluminium sheet and sections up to 6ft. long. The work is transferred by overhead crane from tank to tank. The individual tanks contain solutions for cleaning etching, rinsing, anodising, colouring and sealing. A small control laboratory is seen on the left.

Fig. 9.18 Colour anodising of aluminium: a batch of saucepan lids which have been anodised, dyed. The colour is permanently sealed in the anodic film by the action of boiling water. The plastic spheres floating on the water prevent evaporation and heat loss.

In solutions of sulphuric acid, preferably of about 20% concentration, anodizing can be carried out at room temperature; indeed the temperature must not be allowed to rise above 20°C, or the film redissolves. A current density of 15 to 20 A/ft^2 is used which requires 16 to 18 V. Under these conditions the film continues to grow in thickness and 0.001 in is formed in about an hour. On super-purity aluminium, or its alloys with magnesium or zinc, the film is completely transparent. Other alloying elements and impurities, especially iron, render the film milky. Copper-containing alloys cannot be treated. In 8% oxalic acid solutions the process is similar, but a voltage of 50 to 60 V is necessary. The films produced in sulphuric or oxalic acids are porous when first formed; at this stage they are very receptive of dyes from aqueous solutions, or inorganic pigments can be precipitated in the pores. The porosity is subsequently sealed by the relatively simple process of immersing the anodized parts in pure water at 96 to 100°C for 20 to 30 m. In this way coatings which enhance and protect the silvery lustre of the metal, or provide strongly coloured metallic effects are produced. These are used for a multitude of consumer goods, but also for the protection and decoration of aluminium parts on the outside of buildings. The film produced in sulphuric acid is very hard. The throwing power of all anodizing processes is excellent, i.e. the thickness of the anodic film is similar all over the article, regardless of its shape, even inside quite long tubes.

9.4.4 Chemical and electrolytic brightening of metals

Most metals can be superficially brightened by chemical or electrochemical attack in specially formulated solutions; this process is widely used for providing the final lustrous flat surface on metal objects. However, it has only achieved industrial importance in the treatment of brass, stainless steel and, particularly, aluminium. Although these processes are frequently referred to as chemical or electrolytic polishing, this is a misnomer. They will not remove even minor irregularities of surface level or scratches in the way that mechanical polishing will. The actual effect is to convert an already smooth surface which is matt into a brilliantly lustrous one. Even so, unless the metal alloy is very homogeneous, some parts become more severely attacked than others and the reflectivity of the surface after brightening may be lustrous, but will not yield sharp reflected images.

A simple process of this kind has long been used for brass articles under the name of bright dipping. In this the degreased articles are dipped in a mixture of nitric and sulphuric acids containing a small but critical addition of hydrochloric acid. Cuprous chloride is relatively insoluble in this solution, so that when the brass is vigorously attacked by the nitric acid a very thin resistant film of cuprous chloride is formed on the surface, particularly in hollows. This delays the attack there, so that the surface becomes levelled on a micro-scale, i.e. it becomes lustrous. Greatly superior results are obtained from solutions containing sulphuric, nitric and phosphoric acids; the most effective formulations are different for brass and for aluminium. Since these solutions are protected by patent and are critical in composition, they are chiefly marketed and serviced by the licensees, under the trade name of Phosbrite. The solutions are used hot; they are very vigorous in action and evolve unpleasant fumes, so that fume extraction is necessary. The solutions become exhausted fairly speedily in use and are relatively expensive, because of the phosphoric acid content. Nevertheless, in the course of a 2 to 5 m dip they provide a very satisfying lustre on sheet material, pressings and simple assemblies, thereby obviating expensive mechanical polishing.

A radically different chemical immersion brightening process is available for super-purity aluminium (99.99%) and its magnesium alloys, under the name of the Erftwerk process. This utilizes a solution of nitric acid and ammonium bifluoride. A minute trace of lead is also essential to the process. The solution is operated at about 60°C in hard rubber or PVC tanks. The reaction is extremely vigorous, but the fumes are less unpleasant than with Phosbrite; 1 to 2 m immersion only is necessary. The product of the reaction is aluminium fluoride, which precipitates, so that although the solution becomes exhausted it can be continuously regenerated by additions of ammonium bifluoride and occasionally of nitric acid. The process is therefore fairly cheap to operate, and has been widely used in Germany as a component part of the process of producing bright anodized motor-car trim parts.

In these chemical brightening processes the driving force is provided by the chemical energy from the dissolution of the metal in the acid. In electrolytic brightening the driving energy is supplied by an external current. The work pieces are made the anode, i.e. they are connected to the positive side of the current source. This has the advantage, compared with chemical brightening, of being more controllable, and dealing with metals which do not dissolve vigorously in acids, but the disadvantage of requiring more plant (current source, meters, hangers, etc.) and particularly of a varying action on outstanding and re-entrant areas. Comparatively high current densities have to be used, often over 200 A/ft^2.

Electrolytic brightening has achieved considerable industrial importance for stainless steel (usually for the austenitic nickel-chromium 18:8 alloy) which is a hard and difficult metal to polish mechanically. Sulphuric-phosphoric acid mixtures are used, sometimes with the addition of glycerine or chromic acid. Somewhat similar solutions are also used successfully in industry for aluminium alloys. An alkaline electrobrightening process for super-purity and high-purity aluminium alloys (the 'Brytal' process) utilizes a hot alkaline solution of sodium phosphate and sodium carbonate at a moderate current density of about 10 A/ft^2. This process has been widely used for reflectors.

9.4.5 Bright anodizing of aluminium alloys for bright trim

A combined process of chemical (or electrolytic) brightening of high-purity aluminium alloy articles, followed by thin anodizing, has achieved considerable industrial importance. The aluminium alloy in the form of sheet or extrusion is formed into finished parts, taking every care to avoid superficial damage or marking, particularly parts for the exterior decoration of motor cars and domestic appliances, such as fascia panels, name plates, window and windscreen surrounds, wiper parts, radiator grills, etc. These are then chemically brightened by the Phosbrite or Erftwerk process or by an electrolytic brightening process, and are then immediately anodized by the sulphuric acid process. Because the alloy is of high purity, the anodic film is completely transparent and invisible, but it provides lasting protection from tarnish and corrosion by the weather. It is also hard and scratch resistant. The part thus appears to be gleaming, tarnish-resistant bare metal; it matches well with chromium plate. At the present time a large proportion of the bright trim on the more popular makes of motor car, which is generally believed to be chromium plated is in fact bright-anodized aluminium. To the manufacturer this has the advantage of a much more speedy and less laborious finishing process, against which of course must be set the higher cost of the special aluminium alloy over steel, and the greater care necessary in forming; the strength of the finished article is also less. To the car owner it has the advantage that there is no chance of rust appearing because of defective electroplating, or of the plating peeling off.

9.5 VITREOUS ENAMEL

Vitreous-enamel coatings on metal are films of glass, opacified with inorganic pigments. Such coatings have the excellent chemical resistance and inertness of glass, together with both hardness and gloss. This makes them very suitable for coatings on metals to resist the weather, water, chemical solutions, foodstuffs and beverages, as well as heat.

Vitreous-enamel coatings have been used since ancient times on a craft scale for jewellery and decorative table and toilet wares, because of the attractive effects which could be obtained. In these cases they were usually applied to silver or copper. They have also been used on other non-ferrous metals for instrument parts, such as watch dials. These uses, however, are small compared with vitreous enamelling of iron and steel articles. Enamelled steel pots and pans, kettles, basins and jugs were formerly very common, but have now largely been replaced by aluminium, stainless steel and plastics. Vitreous enamel is still much used for baths and other sanitary ware, and for chemical and food handling plant. Another considerable use of enamelled steel is in the construction of signs where the significant lettering or symbolism stands out from the background by the use of a different colour.

Whilst the surface of vitreous enamel withstands exposure to the weather or to water for many years without appreciable change, it suffers from the inherent brittleness of glass. Thus if the vitreous-enamelled sheet is bent, dented or flexed, the enamel cracks, and may even spall off. In such cases the exposed steel rusts; this disfigures the coating over adjacent areas, but, more seriously, it may spread laterally under the adjacent enamel, which becomes prised off. Paint films are much less brittle, but formerly had poorer weather resistance. As the quality of paints has improved in recent years they have tended to supplant vitreous enamel. It may in fact sometimes be quite difficult to differentiate between a well-applied stoved paint finish and a fired vitreous enamel. However, only vitreous enamel will withstand elevated temperatures without change. It is therefore much used for stove parts, cooking appliances and similar articles.

Vitreous enamels are essentially glasses, that is solidified mixtures of fused sodium and other silicates, often with the addition of boro-silicates and phosphates. The chemical resistance of the glass, and hence of the enamel, its co-efficient of thermal expansion, and its softening point can be varied by additions of the silicates of calcium, magnesium, manganese, lead, cadmium and tin, and thus the formulation of a glass for a particular use in vitreous enamelling is a complex exercise in compromise.

In the preparation of vitreous enamel, the ingredients of the glass are fused and reacted together in a crucible, and are then cast into water, where the glass shatters into fragments. The glass needs to be designed with a view to a suitable melting or flowing temperature, and also so that its coefficient of thermal contraction is similar to, if not identical with, that of steel. The glass fragments are then ground in pebble or ball mills with water to a fine powder or slurry and are sieved.

Iron and steel parts must be freed from scale by pickling before enamelling, and must then be thoroughly dried, but a small amount of superficial oxide is not harmful. There are two main methods of applying vitreous enamel. The most commonly used, and that universal for sheet steel products, is a wet process. A carefully formulated charge of glass frit, opacifying oxides – usually titanium oxide (rutile), or tin, antimony or zirconium oxides (which are all white) – together with colouring oxides and suspending agents are milled with water in a pebble mill to give a thin cream-like slurry, which is called the slip. The colouring materials are cadmium, iron or selenium compounds or oxides for reds, chromium compounds for greens, cobalt compounds for blues, and antimony, vanadium and tungsten compounds for yellows. The available colours are rather limited and perhaps crude.

The steel articles are then coated with the slip all over by spraying or sometimes by dipping. The suspending agents added to the slip ensure that it is a colloidal suspension which can be uniformly spread. The article is meanwhile supported on a few spikes called perrits, the marks from which will not be obtrusive. It is usual to dry off the slip, leaving the article covered with a thin film of dry, but loosely adherent mud before it is fired.

The firing is done at 650 to 750°C for 20 to 30 m, during which time the slip melts, reacts to some extent with the steel and draws up by surface tension to a smooth and glossy surface. Small additions of nickel compounds are beneficial. The glass coating becomes hard and rigid long before the assembly is cold. As it cools further from this point to room temperature, both the steel base and the glass coat shrink by thermal contraction, but as the thermal contraction of steel is greater than that of glass, its contraction is resisted by the glass film. The enamel is thus left in a state of compression, which it can withstand better than a tensile stress. For this reason, if the enamel is applied to one side only of a thin steel sheet, this side tends to be convex after cooling. In enamelling sheet it is therefore universal to apply a ground coat all over as a first step, and then a higher grade final coat on the more important surfaces. Each coat is fired separately. Dust or other foreign matter falling on to the enamel, gas evolved from within the metal, and patches where the slip has been accidentally wiped off are the chief defects of the coating after firing. Cracks, dents and spalling due to thermal or mechanical shock are the chief hazards subsequently.

A succession of coats can be applied, each one being fired separately. By applying the second and subsequent coats locally in contrasting colours, through a stencil, or by hand, designs, lettering or signs can readily be produced. Another method is to apply an all over ground coat in a bright colour, and then to spray with the background colour. When this is dried to mud consistency, stencils of the required design are laid on and the mud brushed away. On subsequent firing the design appears in bright colour whilst a new coat is formed elsewhere.

The second method of coating, known as the dry process, is chiefly applied to cast-iron or heavy steel articles, usually for the final coating only. A ground coat is first applied by the wet process and fired. The articles are then heated to a somewhat higher temperature than for the wet process, 800 to 860°C being usual, and are then temporarily removed from the furnace, whilst the dry powdered-frit mixture is dusted on. By skilful application, a uniform coating can be obtained, and initial blemishes can be covered up.

Vitreous-enamel coatings are usually 0.005 in or more thick; they often vary from point to point, tending to be thicker in recesses and sharply re-entrant angles, and sparse on sharp edges. Although the materials for vitreous enamel are not expensive, the plant required, the cost of fuel and above all the amount of skilful and conscientious labour necessary, make vitreous enamelling a relatively expensive process, and one which is not easily mechanized.

Aluminium can also be vitreous enamelled, but because of the much lower melting point of the metal, a special type of glass frit has to be used, and even then the margin between the firing temperature of the enamel and the melting point of the metal is uncomfortably small. Lead or phosphate glasses are used. The process has had some success in the USA for the decoration of aluminium wall panels for buildings, but has not made much headway in GB.

9.6 LACQUER, VARNISH, ENAMEL AND PAINT COATINGS

9.6.1 Introduction

Metal articles are frequently protected and decorated by means of lacquer, varnish, enamel or paint coatings. The final result is that they are covered with a thin, horn-like film of organic material, which may be transparent, translucent or, on the contrary, thickened, opacified and coloured by incorporation of pigment. The material is applied as a more or less viscous liquid and draws up to a smooth, glossy surface because of surface tension forces, and then is caused or allowed to solidify. (It is preferable to speak of solidification in this connection, rather than to drying, which can have several meanings in paint technology.) However the solidification takes place, it occupies some time, during which the article cannot be handled or allowed to touch other articles, and this complicates the industrial application of the process. The final film is required to have the best combination for the intended purpose of a large number of properties, such as: thickness, hardness, adhesion, flexibility, gloss; transparency or opacity, colour, uniformity of colour, resistance to fading; resistance to weather, water, condensation, foodstuffs, washing materials, oils, alcohol, light and heat; and reasonable cost, ease of application, good smoothing and control of film thickness.

Until the 1920s these materials were formulated almost entirely from naturally occurring gums, oils, resins, etc. and the technology was not essentially different from that used in antiquity. Thereafter more and more materials were especially designed and synthesized for the purpose (hence the term synthetics) so that today most paint ingredients are man-made, even if natural products are used as a starting point. This change has made the possible combinations almost limitless, and has confused the former classification. It will only be possible here to give some basic principles of formulation and application, and to cite a few examples.

9.6.2 Nature of lacquers and paints

The term lacquer usually denotes an unpigmented solution of a preformed resin or film-forming material in suitable solvents; it thus solidifies solely by solvent evaporation.

An enamel is a lacquer with opacifying and colouring pigments, which solidifies to a high gloss. In a stove enamel the solidification is assisted by heating, and this may not always be due to acceleration of the evaporation of solvents only.

A varnish is an unpigmented paint. It solidifies by mechanisms other than mere evaporation of solvent, as explained below.

A paint is a pigmented mixture of film-forming oils and resins with volatile solvents and viscosity-reducing fluids, so that is is sufficiently thin to apply and smooth out, but the evaporation of these volatile fluids is *not* the mechanism responsible for the solidification of the paint. This solidification is accomplished by complex chemical reactions. Originally, paint consisted of an unsaturated natural oil, of which linseed oil is typical with lead oxide and other pigments, and a natural hydrocarbon oil diluent. When this was spread out into a thin film, the diluent evaporated off fairly quickly, and then the linseed oil, by oxidation from the air, polymerized slowly to a rigid but rather soft resin. The solidification was accelerated by the presence of the lead pigment, and it was afterwards found that other metals such as manganese and cobalt, in the form of chemical compounds, were even better accelerators or driers. This type of paint is obsolete for industrial purposes.

Thus lacquer, enamel, varnish and paint films finally consisted of an organic resin or polymer, the difference being that in lacquers and enamels the resin has been preformed, whilst in varnishes and paints it is formed *in situ*. Carbon, which is remarkable for its chain-forming capacity, has the capability of linking to other carbon atoms, as well as with other atoms, and can thus form long chains of hundreds of atoms. These are single or branched but in either case they are spiral or zig-zag in shape, so that they lock together. The objective of resin and plastics chemistry is to devise small organic molecules with suitable reactive links which can be made at the right time to polymerize into gigantic molecules. Carbon atoms have four valency bonds each, but in a chain of carbon atoms, two of these may be directed to the next carbon atom. The chain is then unsaturated. The double bond may then be made to open and link up with a different chain, and thus, by repeating the action, to form a very large molecule. Nature solved the problem, and formed natural gums, resins and vegetable oils, which were at first adapted by man for polymerization, but now chemists have been able to start from simple organic compounds and direct the reaction towards their own special purposes. Polymerization into very large, but still separate, molecules provides a resin which is rigid when at room temperature, but at higher temperatures the carbon chains are less inflexible, and can slide over one another. The resin is therefore thermoplastic but becomes rigid again on cooling. In other cases the polymerization can be made to proceed by cross-linking from one chain to another, chiefly by heating. Such resins are thermosetting; once they are cross-linked, they are permanently rigid. Thermosetting resins have the better chemical resistance, but are less soluble in solvents. Therefore thermoplastic resins are mostly used for lacquers, and thermosetting resins are used in, or more often finally formed in, paints. A paint nowadays therefore consists basically of solvents, pigment and the constituents of a film-forming material. The latter solidifies by more sophisticated mechanisms, of which the following are the chief:

1. Polymerization in a short time at room temperature, or more rapidly on heating, of a mixture of two components mixed just before application. Epoxy resins can be made to set in this way by admixture of amines or polyamides. This method, however, has the great disadvantage that the mixture must be applied immediately to prevent it from setting in the pot (short pot-life).
2. Rapid polymerization brought about by heating mixtures of components. Phenolic resins are examples.
3. Rapid polymerization by air oxidation of mixtures of special synthetic resins and drying oils, or of resins modified with drying oils. This is the most widely used system. An example is an alkyd resin modified with dehydrated castor oil.

The presence of a pigment in a paint or enamel stiffens the liquid and allows a much greater thickness to be applied in one coat than would be possible with the organic vehicle alone. A few pigments such as lead oxides, calcium plumbate and chromates, have definite inhibitory properties against corrosion, but most are inert and neutral. They are mostly inorganic oxides or other simple compounds which are very finely ground and then incorporated by grinding with the oily matter.

The solvents of lacquers need to have considerable solvent power and rapid evaporation properties. They are often organic liquids with functional groups other than carbon and hydrogen, since such groups have greater solvent power. In paints the solvent power is less important and they are more diluents. Hydrocarbon fractions are suitable. During the solidification of the paint or lacquer the solvent is almost irretrievably lost; moreover its evaporation presents a serious fire hazard. For this reason there is now a tendency to use water as the carrier.

The final solid resin film of paints or, more particularly, of lacquers may be too brittle, therefore plasticizers, i.e. slowly evaporating oils are incorporated.

The thickness of a paint or lacquer film that can be applied without drops or waviness (sagging) is limited; furthermore the properties required next to the basis metal may be different from those at the top face. It is therefore common practice to apply several separate coats. These usually comprise at least a primer specially designed to adhere to and seal the basis metal, an undercoating, a somewhat heavily pigmented paint, to give thickness and body to the film, and a final gloss coat.

In the manufacture of metal articles it is usually more convenient to stove a paint or lacquer, and thus reduce its drying time, even if it would dry satisfactorily in air. In general the higher the stoving temperature, the more rapid is the solidification reaction. However, too high a temperature may cause yellowing, or even burning, of the paint. It is also difficult to raise the temperature of an irregular article quickly to a uniform temperature.

9.6.3 Methods of application and examples of use in metal manufacture

Tarry and asphaltic coatings are sometimes applied to water and gas pipes and fittings by immersing them in the hot molten liquid. Bitumastic paints, which remain relatively soft and plastic are often applied to metal building components which will be in contact with damp concrete, brickwork, etc.

Polished metal articles are often lacquered with a water-white transparent lacquer to prevent tarnishing and dulling. Examples are decorative silverware, instrument parts, handbag frames and parts of travel goods. Nitrocellulose lacquers–a solution of nitrocotton in alcohols and amyl acetate–together with a plasticizer such as tricresyl phosphate, were formerly much used. These lacquers are very inflammable, and tend to yellow on exposure to light. Cellulose acetate lacquers are better, but even these yellow and disintegrate on outside exposure. For bold exposure to the weather, e.g. to preserve the original appearance of anodized and dyed aluminium curtain walling on modern buildings, cellulose butyrate lacquers must be used. Such lacquers can be applied by spraying, or, for small objects, by dipping and draining. The lacquers can be tinted with a transparent dye, but it is then difficult to get a uniformly coloured film. Thus to simulate brass or gold on handbag frames, cheap locks, etc. made of tin plate it is common to coat with uncoloured lacquer and then dye the lacquer in a water dye solution.

Much aluminium foil is used for packaging, particularly of cheese, milk products and other foodstuffs. The foil is usually thinly lacquered not only to protect it, but also to improve the 'feel'. A lacquer consisting of a dilute solution of a vinyl acetate-chloride copolymer is used in a solvent of ketones with toluene diluent. The straight vinyl chloride-acetate copolymer does not adhere very well and so a variety with a small amount of copolymerized maleic acid is used. This lacquer is applied by continuous roller coating before the foil is cut. In fact, the poor adhesion of the straight chloride-acetate polymer is made use of for strippable coatings applied to metal before pressing or drawing, and subsequently detached.

Tin plate and thin aluminium sheet is elaborately decorated in sheet form before being cut up and formed into cans, bottle tops and containers of all kinds. The sheets are somewhat oily; the first coat is often a size consisting of a very thin and dilute vinyl chloride-acetate copolymer lacquer, applied at 1 to 2 mg/in^2. The second coat is a pigmented paint, often a drying-oil modified alkyd, to give the required opaque ground to receive the printed design, which is applied by lithographic offset printing. A clear, protective gloss coat, usually a vinyl, is then applied. All these coatings are applied by roller coating with intermediate stoving, e.g. 20 m at 150°C. If the can is to be used for wet foodstuffs, particularly vegetables, fish or meat, the tin plate is first roller coated with a rather thick resistant lacquer on the inside. This is applied in a pattern which leaves the seams, which will subsequently be soldered, uncoated. This lacquer is traditionally a phenolic oleoresin, which solidifies by oxidation and polymerization during stoving. It is hard and very resistant to fruit acids, etc. but has the disadvantage of yellowing during stoving. This is acceptable when it is a pleasant golden colour, but

LACQUER, VARNISH, ENAMEL AND PAINT COATINGS | 723

faster stoving at a higher temperature produces a brown colour. If understoved, the lacquer is likely to confer a slight phenolic taste to foodstuffs. Tooth-paste tubes and other preformed impact extrusions are treated similarly but the application machinery is more complicated.

Much aluminium and some steel strip is continuously lacquered at high speed, e.g. 100 ft/m, on very large and costly special purpose machines. It can later be cut up and fabricated into such items as Venetian blind strips, panels for domestic appliances, and cladding panels for modern buildings. A very adherent and flexible paint film is necessary, so that much attention must be given to the pre-treatment. From the speed of throughput and the time necessary for the degreasing, chromating, rinsing and drying, paint-coating and stoving processes, the length of the appropriate section of the plant can be calculated. Even so, very fast-acting processes are necessary; spray degreasing, spray application of Alocrom solution or, alternatively, a very light unsealed anodic oxidation (for aluminium) and roller coating of paint. Reverse roller coating is used if a heavier film is required. (In conventional roller coating the surfaces of the roller and of the strip move at the same speed, and the liquid paint film is merely transferred from roller to strip. In reverse roller coating the surface of the roller moves in the reverse direction to that of the strip, and thus builds up a thicker coating on the strip than that which is carried by the roller.) Stoving for the usual time of about 20 m is out of the question on a continuous plant, so forced stoving at 350°C for 30 to 60 s is done, but the paint must be one that does not yellow under such drastic conditions. In any case once the strip is coated it cannot be physically contacted or supported till it has been stoved or cooled. It is therefore either hung in a long moving catenary through the stoving oven, or supported on jets of hot air in the oven. For simple, flat surfaces such as these, radiant or infrared heating is sometimes used, as it has the advantage of speed of heat transfer and of heating the surface more than the core of the metal.

Fig. 9.19 Roto-dip painting of car bodies. The pink primer and undercoat system gives a film thickness of 1.5 thou. in one single automatic operation. The electrocoat tank is 44 feet (13.5 metres) long, 9 feet (2.1 metres) deep, with a capacity of 11,500 gallons (52,000 litres).

Most metal articles, however, are not painted until after they have been completely fabricated. Their irregular shape then makes it difficult to paint them by mechanized means. Paints can be applied by dipping the whole article in a bath of paint, and draining it carefully. Care has to be taken that paint is not trapped in pockets and folds. Asphaltic paints are often applied in this way to parts such as bicycle frames. These paints are compounded from natural asphalts, e.g. gilsonite, and driers, and are solidified by severe stoving. Motor-car bodies nowadays often receive the first primer coat by dipping. The bodies are skewered on a shaft and carried through degreasing and phosphate pre-treatment, chromic acid dip, and are then dried. They are then dipped mechanically in a fairly heavily pigmented, hard-stoving primer, and stoved.

Intensive research and development is now proceeding in the electrophoretic deposition of paint from aqueous solution for car bodies or, at least, components. In this process the paint is dispersed in the form of minute droplets in the aqueous solution, with suitable dispersing agents. These cause each droplet to acquire a net negative charge, so that in a tank in which the walls act as cathode, metallic articles connected to the positive pole quickly become coated with a film of discharged paint droplets. This process differs from electrodeposition of metals, because it is the individual drops of paint which are charged, and these are several orders of magnitude larger than metallic ions. Thus a film of 0.0005 in can be obtained in 3 m with a current density of about 7 to 10 A/ft^2. The solidification and polymerization of the film is carried out by subsequent stoving.

Most metal parts, however, are painted by spraying the paint as an atomized mist. On contacting the metal surface the droplets coalesce and flow out. In this way a very even coat can be obtained but on small parts a good deal of the paint is lost as overspray which misses the parts altogether. This is not only wasteful, but it needs special arrangements to catch and remove it. The process of electrostatic spraying avoids most of this difficulty of overspray; it provides a useful method of mechanizing paint application to irregular articles. A very high electrical potential (80 000 to 100 000 V) is maintained between the paint spraying device and the work pieces, which are hung on a conveyor and are at earth potential. The paint droplets are inevitably slightly charged and are therefore attracted towards the work pieces. Their line of flight therefore curves towards the work; even the rear of the pieces becomes coated, and little if any paint is lost. Electrostatic spraying will not, however, penetrate into hollow articles or very deep recesses. The paint is sprayed either from conventional air-operated spray guns, or better, by centrifugal atomization from rapidly rotating plates or bells. Elaborate precautions need to be taken to prevent accidental access to the high-voltage part of the apparatus, and to reduce the ignitive energy of any spark which may be inadvertently produced. Although the voltage is extremely high, the current flow is minute and is intentionally limited to only a few hundred micro-amps. In practice the method does not seem to be as hazardous as might be expected. It has even been possible to devise hand-held electrostatic spray guns, in which the gun is at earth potential, and only the nozzle is at high potential. This is connected to the gun by a long insulated tube.

Any kind of paint spraying process requires a fairly thin paint, which means that the solvent's content is high and the dry coat only of limited thickness. The viscosity of a paint can alternatively be kept down by increasing its temperature; processes of hot spraying of paint with consequent reduced solvent content have therefore achieved some popularity. Even then, however, the rapid expansion of the compressed air used to atomize the paint and impel it towards the work, chills the droplets before contact. Thus a process for airless, mechanical hot spraying has been used, although it is relatively complicated.

Although the appearance and serviceability of an article depends largely on the final paint coat, some regard must be had to the system as a whole with respect to the conditions of service. The adhesion of the initial paint coat, and the prevention of underpaint corrosion is of the utmost importance. Steel and zinc surfaces are phosphated, and aluminium surfaces anodized, MBV (see p. 713) treated or chromated, as previously described, but if not, metal surfaces, particularly aluminium, may be treated with an etch primer. This is a very thin wash coating containing phosphoric, and sometimes chromic, acid, a special vinyl resin and zinc chromate. The acids react with the metal to give good adhesion of the resin, and the zinc chromate provides a positive inhibition of corrosion should moisture penetrate to the interface.

The combinations possible in the formulation of paints are infinite, but three of the chief resin systems in wide use today may be mentioned. The most important class of resins are the alkyds, which are made by polymerizing dibasic acids, such as phthalic acid, with polyhydric phenols such as glycerol. They are cheap, water-white, with excellent weather resistance and fair hardness. For use in paints, they are copolymerized with drying oils or with other resins, of which the chief are melamine-formaldehyde and urea-formaldehyde resins. Melamine has the better exterior durability and is used for outdoor finishes, e.g. on car bodies, whereas with urea-formaldehyde better light colours, including white, can be obtained. The latter is therefore preferable for domestic appliances.

Epoxy resins have outstanding chemical resistance, adhesion to metals, flexibility and hardness, but they are rather expensive and are slightly brown in colour. They can be made to solidify by three main types of process:
1. Spontaneously, even at room temperature, by admixture with amines or polyamides. (This method is also utilized in the field of adhesives to stick metals together.)
2. By reaction at elevated temperatures, but without oxidation, in combination with urea or melamine-formaldehyde resins.
3. In esterified form, by oxidation, at elevated temperature.

In general, epoxys are only used where unusually high chemical resistance is required, and hence not usually for weather resistance alone, but often for washing machines and refrigerators, where contact with washing compounds, detergents, foodstuffs, alcohols, etc. may call for exacting resistance. They are also used for industrial equipment, can and drum linings.

Finally, there are the acrylic resins, which are typified by methyl methacrylate (Perspex). These are water-white, very hard and have high chemical resistance, but are mostly thermoplastic and are used dissolved in solvents as lacquers and enamels. In this form their heat resistance is not good, so thermosetting types are coming into use.

For the many engineering components which may be in contact with oil, petrol, etc. a special sort of chemical resistance is required, which is provided by chlorinated rubber.

Solely by way of illustrative example, and by no means in any limiting sense, the following outline systems of complete applications may be given:

a. Motor-car bodies. Primer and undercoat on phosphated coating: An epoxy-ester melamine-formaldehyde paint, or an alkyd melamine-formaldehyde stoving paint

(150°C, 30 m). Top coat: thermosetting acrylic or an alkyd melamine-formaldehyde paint stoved at 125°C for 30 m, or a force-dried acrylic nitrocellulose enamel.

b. Domestic appliances. Primer on phosphate coating on steel, or on etch-primed metals, or on chromated, or MBV treated aluminium: alkyd melamine-formaldehyde (stoved 150°C, 30 m). Finish coat: alkyd urea-formaldehyde plus epoxy resin for special resistance to detergents (stoved 150°C, 30 m), or a force-dried alkyd nitrocellulose enamel.

9.6.4 Some special types of paint

Zinc-rich paints are useful protective coatings for structural steelwork, bridges, boats, etc. They are half-way between a metal coating and a paint. They have a very high pigment ratio of 90 to 95% of fine zinc powder and just enough resistant organic paint vehicle, e.g. chlorinated rubber, to hold this on the steel. The zinc particles touch each other and the underlying steel sufficiently to be in electrical contact, and hence to form a sacrificial metallic coating. This acts to prevent rusting, even at breaks in the film (see p. 673).

Aluminium paints are based on aluminium flake pigment. This is produced by atomizing molten metallic aluminium, and then ballmilling the powder in white spirit and stearic acid. Very thin but relatively extensive metal flakes are produced, each of which is coated with stearic acid. The mill load is filtered off to form a stiff paste, which is diluted with white spirit to a metallic content of about 65%. This paste is used as the pigment in 'leafing' aluminium paints; the vehicle is almost always a solution of indenecourmarone resin in white spirit. When the paint is applied by brushing or spraying, the individual flakes of aluminium tend to rise to the surface of the paint film, to 'de-wet' and to orient themselves on the surface. This is called 'leafing'; it yields a paint with an almost wholly metallic appearance, and is a good barrier to the entry of moisture, and to light. If desired the leafing effect can be prevented to give 'deleafed' aluminium paint. When this dries the multitude of aluminium flakes are orientated parallel to the surface and form a considerable barrier to moisture.

Aluminium flake pigment is used in other paints. For example, a much smaller amount of deleafed aluminium flake in a transparent, but tinted, clear lacquer provides an iridescent effect in polychromatic paints. Aluminium flake in a pigmented, rather thin, paint can be made to form whorls and eddies in the paint as it dries. These resemble hammer marks, and can be used to give an attractive appearance to the rather rough surface of instrument bodies, etc.

Top coat paints may also be formulated to solidify into a wrinkled finish, or even to shrink and crack, revealing a lattice pattern of an undercoat of contrasting colour. (See ch. 11.11. for rust prevention compositions.)

9.6.5 Heat-resisting and non-stick coatings

All normal paints are damaged by continuous high temperatures because of their content of organic resin. Aluminium paint, even of normal formulation, is, however, fairly protective since the aluminium itself is not easily oxidized.

The element silicon has four valencies like carbon, and with some difficulty it can be made to form chains and resin-like substances. These can be formulated into heat-resisting coatings. Their chief application, however, is in non-stick coatings on domestic and industrial cooking utensils. Unfortunately, the non-stick property causes them to have very poor adhesion to the basis metal. For aluminium frying pans and utensils it is usual to rely on mechanical adhesion by severely etching the metal and thus producing undercut cavities, but the appearance is poor.

Another resin with heat-resisting and low frictional properties is polytetrafluoroethylene (PTFE). Although this has a carbon chain backbone, the associated carbon fluorine bonds are very strong and not easily broken by heating. PTFE lacquers also do not adhere at all to metal surfaces, and a special priming system has to be used.

9.7 PLASTIC COATINGS

Metals coated with plastic may well be the constructional materials of the future. Already large quantities of strip steel are laminated with a preformed plastic film, using adhesives. One such product is marketed under the trade name Stelvetite. Plastic is extruded around a metal core for curtain rails. These processes, however, must be considered as methods of manufacture rather than of surface treatment.

Plastics do not melt to a limpid liquid on heating, and cannot be applied in a liquid form. As their name implies, on heating they become plastic and sticky. A very successful method of coating metal, especially wire-work goods, e.g. racks and light structures, chain-link fencing and grills, with polyvinyl chloride or similar thermoplastic material is as follows. The plastic is finely powdered and is placed in a deep box; a perforated bottom plate permits air to be blown up through the plastic powder. This maintains it in a mobile or fluidized condition. The metal articles are preheated to about 150°C and then plunged into the seething mass of powder. The particles which touch the hot metal become softened and sticky; they adhere to it. After withdrawal the metal part is re-heated for a short time to allow the individual plastic particles to coalesce and draw up to a smooth coating by surface-tension forces. The resultant coating is about 0.005 in thick.

Alternatively, colloidal dispersions of partially polymerized polymer in plasticizer, or of completely polymerized thermoplastic material in the minimum of water, consisting of very thick creamy liquids may be used for coating metal objects by dipping. These are called plastisols or organosols. The solidification of the plastic is accomplished by subsequent heating.

9.8 LITERATURE

W. E. BALLARD, *Metal Spraying and the Flame Deposition of Ceramic and Plastics*. Charles Griffen, London, 1952.

W. BLUM and G. B. HOGABOOM, *Principles of Electroplating and Electroforming*. 3rd edn. McGraw-Hill, London, 1949.

A. W. BRACE, *The Technology of Anodising Aluminium*. Robert Draper, Teddington, Middlesex, 1968.

A. K. GRAHAM, *Electroplating Handbook*. Chapman and Hall, London, 1962.

A. A. B. HARVEY, *Paint Finishing in Industry*. 2nd edn. Robert Draper, Teddington, Middlesex, 1967.

N. HEATON, *Outlines of Paint Technology*. Charles Griffen, London, 1956.
I. M. F. SYMPOSIUM, *Nickel-Chromium Plating*. Robert Draper, Teddington, Middlesex, 1961.
H. SILMAN, *Chemical and Electroplated Finishes*. Chapman and Hall, London, 1952.
S. WERNICK and R. PINNER, *The Surface Finishing of Aluminium and its Alloys*. 3rd edn. Robert Draper, Teddington, Middlesex, 1964.
ZINC DEVELOPMENT ASSOCIATION, *Hot-Dip Glavanising*, London, 1953.

CHAPTER 10

Metal Powders and Hard Metals

10.1 INTRODUCTION

The history of metallurgical development is inevitably linked with the development of high temperatures and the means to contain the material at elevated temperature. Of the seven metals of the ancient world – gold, silver, copper, iron, lead, tin and mercury – the first was found native in metallic form, the second frequently so, and the third occasionally. Of the remainder, the last three required comparatively low temperatures to extract and melt, such as could be obtained from a wood or charcoal fire. Copper and iron required higher temperatures, involving a type of furnace body made from clay, and bellows of goatskin or some similar fabric. Even in the absence of high temperatures, small globules of partially reduced metal might have been produced in the hottest portions of a simple wood fire, and washing away of the ash and residue would reveal a type of powdered metal, or a conglomerate which, with minor pressure, would crush to a powder. There is every reason to believe, therefore, that some powdered metals may have been in use long before man learned to attain the higher temperatures necessary to produce a molten bath of metal.

Such crudely produced powder would be of irregular shape and particle size. Provided the material was not excessively brittle, mechanical pressure could be used to flatten the particles so that they either welded together as a larger body or could be secured with adhesive to build up a metallic skin on some substrate. A portion of Greek papyrus found at Thebes, and dated about 300 B.C., describes the manufacture of gold ink by pounding gold leaf with gum. In this case the reverse procedure is involved initially, that of disintegrating a metallic foil produced primarily from a melt.

Powder metallurgy is therefore not a new science, but its applications only became widespread in the early half of the twentieth century. There is every reason to believe that its use and applications will grow, partly for technological reasons and partly from economic pressures.

The bulk production by relatively cheap methods of the more common metals, iron, copper, aluminium, nickel, tin, lead, zinc and their alloys, is not likely to be affected for some time, since the existing capital investment in plant is very high and the quantities involved considerable. For specialist applications, however, especially when the properties are outside the conventional fields of structures and machinery, there is a growing tendency to consider powder methods. The main uses for the technique can therefore be defined as follows:

1. The manufacture of metals which cannot easily be carried out by methods involving conventional melting processes. Into this category may fall the expensive rare metals, such as the platinum group, and the high melting point, or refractory, metals, e.g. tungsten, molybdenum, niobium, vanadium, titanium and zirconium. The rare metals of the platinum group can be melted and fabricated by conventional methods but the value of the material requires adequate recovery from slag formed in melting and from residues produced in separation and refinement. Powder methods involve fewer recovery problems and often less money invested in recirculation as a result. The refractory metals are difficult to melt by virtue of their high melting points and difficult to contain and handle in a crucible when liquid. They are also extremely reactive with atmospheric gases and, although even the powder method requires protection from the atmosphere, the higher temperature of a melting process inevitably increases the rate of attack, and therefore the intrinsic difficulty of adequate protection.

2. The manufacture of components which cannot easily be fabricated by normal methods. Examples include porous bearings and filters, metallic friction materials, cutting tools in very hard materials, magnetic materials and fine wires in refractory metals, e.g. electric lamp filaments. The properties required may dictate the type of structure required, e.g. porosity in bearings to be self-lubricating, and hence the powder method may prove to be the only one feasible.

3. The fabrication of materials which could be made by other means but for which the powder method is more economic, convenient or reliable. Into this category may fall some of the metals mentioned in the first category, such as the precious metals, but it mainly refers to iron, nickel or copper based alloys of small size but intricate shape. The number of operations required to fabricate these small components from bar, sheet or block may be considerable, whereas the powder method allows considerable flexibility in the design of components.

The limitations of powder metallurgy are largely concerned with the preparation of the powder, the pressing of the powder into shape in a die, and the sintering, or high-temperature bonding, of the finished article. Almost all metals and many metallic compounds can be prepared as powder, but the properties may not be ideal, in some cases, for subsequent pressing and sintering. The limitation on pressing involves the size of the article and the ease of forming of the particular powder shape and size. The sintering process almost invariably requires a protective atmosphere, and the atmosphere required varies with the material.

Combinations of materials may also be manufactured by this means, including the latest development of composite materials for strength at high temperature, in which refractory metallic oxides or other compounds are reinforced by metallic fibres or 'mats' to modify the inherent lack of ductility of the compound matrix. Intimate mixtures of refractory oxides and metal powders, known as 'cermets', are also in this category (see ch. Ceramics, vol. II).

10.2 METAL POWDERS

The production of metal powders falls into two main categories: (1) chemical methods whereby the conditions of formation of the metal are adjusted to yield a powder of reasonably uniform size and appropriate shape, and (2) mechanical methods whereby metal produced by more conventional means is distintegrated to powder.

The method chosen for a specific application may depend on economics, but is often decided on grounds of specific properties (as in magnets) or purity. Alloys are difficult to produce by chemical methods since the result is frequently a coprecipitate of the two or more constituents as a mixture, not as a true alloy. In such circumstances, mechanical processes are almost the only feasible ones. Certain refractory metals are produced by chemical reactions which, whilst they may evolve considerable quantities of heat, rarely reach sufficiently high temperatures to melt the metal, and a coke or sponge of very porous material results. Such metallic sponge is friable and easily converted to powder by crushing; production therefore includes both chemical and mechanical methods. Similarly, powders of the metallic compounds used in hard metal tools are obtained by chemical methods, but modified to shape and size, by mechanical means.

10.2.1 Chemical production methods

All chemical compounds can be decomposed into the constituent elements if heated to a sufficiently high temperature. Metallic compounds can often be decomposed by reduction processes in which the non-metallic portion is combined with some other element or compound (the reductant) leaving the metal free in elemental form. The characteristic form of the metal will depend on the physical state of the original compound, the temperature attained during reaction and the rate of reaction. The metallic compound may be solid, liquid (molten) or gas. In certain cases a solution of the compound, usually in water, may be reduced to yield the metal.

Two main principles are involved in assessing a suitable process for the production of a metal by reduction of a compound. The first is the thermodynamic one, that a negative change in free energy must be feasible, i.e. that the total energy of the system must be less after the change than before. The change in free energy involves the heat of reaction. Changes in pressure of gaseous components of a reaction also occur when the reductant or the products are gases. Study of the physical chemistry of a possible reaction allows assessment of these factors and indicates the likelihood of a reaction taking place. Calculations can also be made of the likely yield from the reaction if the reverse reaction, i.e. recombination, is prevented.

Such an approach in the first instance indicates, for example, that aluminium, chromium and titanium powder cannot be prepared by reduction of the oxides with hydrogen gas. On the other hand, the thermodynamic data indicate that copper, iron and tungsten oxides should be reduced either by solid carbon or by hydrogen gas, and that zirconium metal can be made by reduction of zirconium tetrachloride by hydrogen if the temperature can be raised to $2\,000°C$.

The second principle involves not the feasibility of the reaction but the kinetics, or speed, of the reaction. It would be uneconomic, for example, to consider a process (thermodynamically possible) if it proceeded so slowly that high temperatures or pressures were required to be maintained for several days. In many reactions an energy barrier exists between the stable configuration of the initial products of a reaction and the stable state of the products after reaction. To surmount this barrier, energy must be supplied to the system, which, thereafter, proceeds to completion by virtue of the energy generated by the reaction. The initial input of energy is termed the activation energy of the system and can be measured by laboratory experiment. Such an experiment may also indicate which feature of the system controls the rate of reaction once initiated. The lower the activation energy, the more rapid the rate of reaction.

These two principles may assist in the evaluation of more than one possible route for the production of a powder, especially in the economic aspects. Some other factors are relevant, however, the chief ones being the nucleation processes by which the metal powder is formed, or precipitated, and which naturally have a bearing on the size and shape of the particles. The economic aspects of the chemical process may be inverted when the product is examined since a slower reaction may give a finer and far more uniform shape of powder, thereby eliminating expensive grinding, crushing or sorting processes to convert to a workable material.

The selection of a route for the manufacture of metal powder is therefore a critical step in powder metallurgy, and may have a considerable bearing on the properties and behaviour of the final component.

(For more extensive treatment of these aspects, see references 1, 2, 3 and 4.)

a. Reduction of metallic oxides. Many metals are prepared by reduction of their oxides, both for normal molten metal and for production of powder. The oxides can occur naturally or be prepared from more complex compounds by relatively simple chemical methods. The oxides may be crushed or pulverized, and graded by sieving to particular size ranges so as to control the final powder particle size to some extent.

The reductants employed may be gases such as hydrogen, cracked ammonia (ammonia heated to dissociate into nitrogen and hydrogen as a cheap source of hydrogen), carbon monoxide or natural gas. Solid carbon (usually in the form of charcoal) is the oldest form of reductant known to metallurgy and has the merit of relative cheapness and availability. In the presence of air, reduction with solid carbon may be considered as reduction with carbon monoxide since the first steps of the reaction comprise the formation of carbon monoxide which subsequently reduces the metal oxide to metal. The formation of carbon monoxide may be represented by the following reactions.

$$C + O_2 \rightarrow CO_2$$

$$CO_2 + C \rightleftharpoons 2CO$$

The first equation represents the combustion of carbon which supplies the heat needed to form the carbon monoxide in the second reaction. The second reaction is an equilibrium reaction. At a temperature of about 700°C the reaction mixture contains about equal amounts of carbon monoxide and carbon dioxide. When the temperature decreases the latter reaction proceeds to the left with the development of heat and the carbon monoxide content in the gas mixture decreases (at 500°C the carbon monoxide content is about 5%). When the temperature of the system is raised to 800°C the reaction proceeds to the right with the absorption of heat until the gas mixture contains 87% CO and 13% CO_2. At 1000°C the carbon monoxide content is 99.3%.

The required carbon monoxide concentration in the reducing gas mixture for the reduction of metal oxide depends on the kind of metal. In the case of iron oxide the reduction process can be represented by the following simplified equilibrium reaction.

$$FeO + CO \rightleftharpoons Fe + CO_2$$

The reaction is forced to the right when the CO concentration is higher than the equilibrium concentration. Figure 10.1 shows the equilibrium concentrations of CO at various temperatures for both these equilibrium reactions.

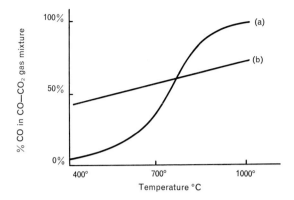

Fig. 10.1 (a) The equilibrium concentration of CO in the reaction of $CO_2 + C \rightleftharpoons 2CO$.
(b) The equilibrium concentration of CO in the reaction of $FeO + CO \rightleftharpoons Fe + CO_2$.

From the above figure can be derived that the temperature in the reduction furnace must be above 700°C.

The reactions mentioned above proceed in the reverse direction on cooling resulting in the reoxidation of iron by carbon dioxide. In addition the decomposition of carbon monoxide to deposit carbon is catalysed by iron surfaces, and soot is readily deposited at temperatures between 300 and 600°C. The carbon diffuses into the iron, causing it to become harder.

It is therefore necessary to cool the reaction mixture rapidly with rapid removal of the gases formed in order to lower the reaction velocity so quickly that no reverse reaction can occur.

Reduction by hydrogen involves the evolution of steam by the reaction.

$$MeO + H_2 \rightleftharpoons H_2O + Me$$

(in which Me represents a metal atom)

The latter reaction is also an equilibrium reaction. In contradistinction to the reduction with CO (in which the CO_2 formed is again converted into CO by the carbon) the steam concentration in the gas mixture increases when the reaction proceeds to the right and the reaction reaches an equilibrium and no further reduction takes place. To convert all the metal oxide quickly into metal it is often necessary to remove the steam continuously from the reaction mixture.

One of the governing factors in any gas/solid reaction is the interface at which the reaction proceeds. A solid oxide particle will react on the surface in the initial stages, but after a skin of metal has been formed the reductant gas must diffuse through the metal before it can reach more oxide. After further reaction, the gaseous products must diffuse in the reverse direction, through the metal, to the surface and there escape into the atmosphere. The over-all speed of the reaction may then depend on the steps by which these diffusion mechanisms occur, quite apart from the aspects so far discussed. Where the same reductant is being used, the diffusion rates will vary from one metal to another, and with temperature, so that every case of metal powder production by gaseous reduction requires to be studied individually. Reduction by solid carbon must depend finally on the same principles of diffusion, since the initial solid carbon/solid oxide interface at the surface is replaced by a solid carbon/solid metal interface. The reactions outlined for carbon monoxide production then become very important.

The reduction of copper oxide by hydrogen is a typical example of the problems involved. The equation for the reaction is

$$CuO + H_2 \rightarrow Cu + H_2O$$

Equilibrium data suggest that even a small hydrogen content, in the presence of steam, will reduce copper oxide. In practice, however, a small amount of moisture in the hydrogen atmosphere suppresses the reaction, and removal of the steam after the initial introduction does not immediately lead to reduction. The reason lies in the production of nuclei which water can suppress, and which require an incubation period of about 60 min at 150°C to form, even in dry hydrogen. Water vapour can 'adsorb' on the surface of the oxide particles, i.e. molecules attach themselves to the surface, and inhibit the reduction. A similar phenomenon occurs with carbon monoxide reduction of cupric oxide, in which carbon dioxide can inhibit reduction.

Hydrogen readily diffuses through solid copper, but water and carbon monoxide do not. When the first skin of copper forms on the oxide, the metal occupies nearly 40% less volume than the oxide from which it is produced and so envelops the oxide tightly. Although hydrogen can diffuse through the skin, water–the product of the reaction–cannot diffuse out, and so the skin must be ruptured to allow the removal of the water. Such a process of fissuring also speeds up the inward movement of the reductant, whether hydrogen or carbon monoxide. At higher temperatures the fissures may sinter together again and slow the reaction once more; this occurs readily above 400°C. Prevention of easy exit for the product of reaction may lead to blowing out of bubbles of gas, and hence a porous metal powder. The ideal conditions for gaseous reduction of cupric oxide are therefore below 400°C for both hydrogen and carbon monoxide reduction.

For the same reasons a temperature above 800°C is not preferred in the reduction of iron oxide by either hydrogen or carbon monoxide. As shown in fig. 10.1, temperatures below 700°C may not occur in reduction of iron oxides with carbon monoxide. The reduction of iron oxides by hydrogen is highly dependent on temperature, and although the presence of some water does not inhibit the reaction at 800°C, the steam may recombine with the metallic iron fairly readily on cooling unless removed beforehand. The reduction of iron oxides by either hydrogen or carbon monoxide is a complex process and experimental work does not always apply in large-scale practice. One additional variable in the case of iron is the existence of three different oxides, with different crystal structures and consequently different conditions of reduction. To summarize the findings of all the investigations made, the following conclusions may be drawn for hydrogen reduction:

1. Iron oxides reduce at a faster rate than the observed rates of diffusion of either iron or oxygen through the phases. The slowest reaction does not, therefore, appear to be the rate-controlling factor.

2. Magnetite (Fe_3O_4) takes longer to reduce than hematite (Fe_2O_3). If magnetite is oxidized to hematite before reduction (in air), then subsequent reduction takes place in the shorter time for hematite.

3. Provided that the particles are not so fine that gas permeability is obstructed, the finer the particles, the greater the rate of reduction.

4. The rate of reduction decreases above 600°C.

5. The use of a fluidized bed (one in which the gas pressure maintains the bed of powdered oxide in a state of fluidity by its passage through the bed from beneath) improves the rate of reduction. The difficulty is in separation of the reduced iron from the oxide still unreduced, but this can be effected by magnetic methods.

One interesting application of carbon reduction occurs in the Mannesmann process in which finely divided high carbon steel powder is heated with iron oxide powder. The carbon in the iron reduces the oxide and escapes as carbon monoxide. Very fine (submicron) iron powder, of particle size less than one-thousandth of a millimetre, is made by decomposition of pure iron compounds, which leave behind a fine reactive oxide, followed by hydrogen reduction. Such submicron powders are suitable for magnetic applications, but the extreme fineness of the powder presents a fire hazard, oxidizing rapidly in air, often with explosion (pyrophoric).

Tungsten metal powder can be produced simply, by reduction of the oxide (WO_3) in hydrogen at temperature above 700°C. The high temperature is used to avoid lower oxides of tungsten, and water is best eliminated from the system. This can be achieved by using dry hydrogen with a high rate of flow to prevent the accumulation of water vapour after reduction. Too much water vapour causes combination with the oxide to form a volatile (easily vaporized) hydrate, which would then be swept out of the system by the hydrogen stream. The tungsten powder tends to agglomerate by sintering at the high temperature, and a two-stage process is often followed, between which the mixed oxide/metal powder is crushed and screened to break up the agglomerates. The presence of some trace impurities affects the particle size of tungsten powder prepared by hydrogen reduction, and analytical control of the original source of tungsten compounds must be continued through each stage of manufacture to prevent contamination (see ref. 5).

Tungsten is a comparatively brittle material and problems arise in the manufacture of fine wires (for electric lamp filaments) from a powder-metallurgy product. To modify the grain recrystallization during sintering, additions may be made to the metal powder and, to ensure uniformity of distribution, the additions ('dope') are best made to the solution of tungsten salts from which the oxide is precipitated prior to the reduction process. The most common addition is about 1% thorium oxide (thoria) added as a hydrate to the tungsten solution. Thoria (ThO_2) is not reduced in the preparation of the tungsten metal powder and remains embedded in the tungsten crystals. The thoria (cubic crystals) causes the tungsten to form cubic crystals which are more easily sintered without grain growth (see also sintering, p. 000).

b. Oxide reduction by a second metal. A number of metallic powders have been produced by reaction between the metallic oxide and, as a reducing agent, another metal whose thermodynamic properties cause an interchange of the oxygen.

$$M_1O + M_2 \rightarrow M_2O + M_1$$

Chromium powder has been produced by the reduction of the oxide (Cr_2O_3) with magnesium, zirconium powder by reduction with calcium or magnesium, thorium by reduction with calcium, and uranium by calcium or magnesium. Such processes often require considerable activation but generate substantial quantities of heat once the reaction starts. This tends to produce very high temperatures in the reaction vessel and to give a melt rather than a powder. By making suitable additions of slag-forming materials, the metallic particles can be kept in suspension until the slag freezes. Some of the heat generated is thus used to form and melt the slag, and the slag physically prevents agglomeration. The difficulty lies in the separation of slag and metal after separation, and the likely contamination of the product by the slag or some side reaction. Only where other methods fail or the product is expensive or difficult to handle (thorium and uranium) would such a course be justified.

c. Reactions in the liquid phase. A number of metals may be precipitated in the powder metallic form by displacement in solution with another metal, often itself in powder form to provide a large surface area and to speed up the reaction. Copper is the classic case, since it can be precipitated from solution by iron in the form of filings or machine turnings. The efficiency of the process is affected by the acidity of the solution and by the rate of dissolution of iron. Too high an acidity dissolves the iron directly, and too low an acidity causes the precipitation of hydrated iron and copper salts in atmospheric air. The solution is normally made to flow over the iron to prevent localized depletion as displacement occurs. Waste copper solutions from copper mining are often treated with iron to recover the last traces of copper, but the product is not of high purity.

It is also possible to precipitate metal powder from aqueous solutions by hydrogen gas, the reaction involving an interchange of electric charge so that the positively charged metallic ions in solution are precipitated and hydrogen gas is converted into hydrogen ions (also positively charged). The reaction may be written as follows:

$$CuSO_4 + H_2 \rightleftarrows H_2SO_4 + Cu$$

which indicates the chemical interchange or, to show the electrochemical charges:

$$\underset{\text{solution}}{Cu^{2+}} + \underset{\text{gas}}{H_2} \rightleftarrows \underset{\text{crystals}}{Cu} + \underset{\text{solution}}{2(H^+)}$$

The reaction is reversible, hence the double arrows, and the chemical kinetics as stated by the law of mass action are such that a high pressure of hydrogen gas (on the left of the equation) will encourage the production of solid copper crystals, i.e. the reaction will proceed from left to right.

The concentration of other ions in solution, especially if they form complexes with the metallic ions, may affect the rate of reaction. The solid/liquid interface is the rate-determining factor in all precipitations from the liquid phase and diffusion of metal ions to the nucleating particles becomes important, particularly when precipitation is caused by gas reduction. Copper is readily precipitated from acid solutions, and the first copper particles precipitated act as nuclei for further growth. Copper, cobalt and nickel form complexes with ammonia, and hence are more difficult to precipitate from ammoniacal solutions. In all cases, the walls of the container may act as nuclei; this does not produce the finest powder since crystals tend to build up on the walls rather than in the solution. Nuclei may be provided, however, in the body of the solution by adding either fine metal powder as 'seeds', or some other nucleating agent to produce a fine colloid of the metal to be precipitated, which then acts as a source of nuclei. Graphite has been used as a seed for cobalt and iron salts have been used to precipitate fine nuclei of nickel in nickel-bearing solutions.

The advantage of precipitation from solutions lies in the use of waste liquors (containing the desired metal) which are by-products from other mining or metallurgical operations. One development of the process, however, does not rely on such economic advantages and concerns the manufacture of composite powders consisting of layers of different metals built up on a nucleus. Such 'core and case' powders find application in powder metallurgy products in which only the outer skin of the particles need be of the more expensive metal, the core acting only as a filler, or in which the core can be of the main material and the case acts as a binder on sintering. An example of the first case is iron coated with nickel for chemical filters, where iron would corrode but nickel does not. The sintered powder consists of pores formed between the nickel coating so that the corrosive liquor sees only the nickel, and the iron core cheapens the cost of material over-all. An example of the second case is the coating of cobalt on tungsten carbide nuclei for hard metal applications, where cobalt is frequently used as the binder for the hard carbide in view of its lower melting point.

A number of reactive metals have been precipitated from molten salts by the addition of a metal reductant. Aqueous solutions would be useless in such cases because of the rapid reaction between the reactive metal and water or other ions. The disadvantage of precipitation from fused salts is the likely contamination by either the salt itself or the metal reductant. Nevertheless, beryllium, zirconium and thorium have been so prepared, and methods have been proposed for chromium. Separation of the metal powder from the molten salt and reaction products can be difficult, and halides are often used so that they can be dissolved out in hot water after cooling the melt. The metal powder is then merely filtered off.

d. Reactions in the vapour phase. Just as precipitation from fused salts avoids the possible reactions between a reactive metal and other ions in the case of solutions, so any reaction in the vapour phase can avoid contamination by excluding extraneous materials as far as possible. Two main methods are available: either reduction of the vapour of a metallic compound by another metal or by hydrogen, or the decomposition of a suitable metallic compound by an input of energy which causes complete precipitation of the metal from the vapour phase.

The use of either method involves one or both of only two contaminating surfaces, the reductant if a metal and the materials of construction of the reaction vessel in all cases. For these reasons, the reactive metals required in a high state of purity have been largely produced from such vapour phase reactions. Where the melting point is very high, the product has most frequently been a powder, since large quantities would be difficult to melt and contain in the molten state. Such a method of powder manufacture is therefore relatively common with the high melting point reactive transition metals such as titanium, zirconium, hafnium and vanadium. Because of the unusual chemistry of their compounds, there are similar methods for the production of iron and nickel.

The most common compounds in use are the halides, since they can be prepared oxygen-free in a high state of purity. The chlorides are frequently used with a metal reductant or hydrogen, and the iodides for spontaneous decomposition on a heated filament. Reduction of chlorides for the production of titanium, zirconium, hafnium and vanadium can be effected by hydrogen, sodium, calcium, magnesium, aluminium and silicon. The last two are eliminated because they tend to alloy with the metal product.

The Kroll process (see reference 6), used for titanium, employs titanium tetrachloride and molten magnesium or sodium to perform the reduction, at temperatures below 900°C:

$$TiCl_4 + 2Mg \rightarrow Ti + 2MgCl_2$$

An atmosphere of inert gas such as argon, at atmospheric pressure, prevents the access of atmospheric gases such as oxygen. The principal problem lies in the separation of the metal powder from the slag remaining, usually carried out by leaching and then vacuum degassing of the powder. The use of sodium reductant has the merit of easy solution of the sodium chloride produced as the slag. Similar reactions have been used for zirconium and vanadium (see references 7 and 8). One other disadvantage of these halide processes applied to certain metals is the tendency for some of the halides to disproportionate, i.e. to decompose into a mixture of two halides or intermediate products, only one of which may react in the reduction process. All likely reactions of this kind require exploration before undertaking a similar process.

The decomposition of halides, and particularly iodides, by an input of energy has been fully investigated by van Arkel (see reference 9). Essentially, this type of process takes place in two stages: the formation of the halide compound, and its subsequent decomposition. The temperature at which the compound is formed is largely a problem of kinetics, i.e. achieving the most rapid rate of reaction to match the rate of decomposition in the second stage, unless, of course, the two processes are carried out separately and not continuously. The dissociation temperature of the compound must be lower than the melting point of the metal, and for this reason chlorides of the lower melting point metals (less than 2 000°C) cannot be used. Iodides are much less restricted, hence their use.

Van Arkel used a sealed chamber in which the iodide vapour decomposed on a hot filament. If the released iodine can be recirculated to react with further material in a separate chamber to form more compound, then the essential feature is the transport of metal from one form to the final decomposition site by means of iodine. The control of decomposition rate and transport can be assisted by an inert carrier gas, which is pumped round the system and plays no part in the reactions. The disadvantages referred to previously, the disproportionation of complex iodides into simpler forms with separate characteristics of decomposition temperature and pressure, can interfere with the efficiency of these processes in practice especially the continuous ones.

Improvements in this type of process have been successfully applied to certain metals, e.g. the addition of hydrogen to assist reduction of the halide, as with zirconium and titanium. To prevent intermediate disproportionation reactions, the dissociation rate may be increased by use of an electric arc, but this is only used in practice when the extra expense is justified by the production of a purer product, e.g. with zirconium and hafnium. Organic compounds have been studied, but less extensively. In practice the disadvantages of using organic compounds centre around the possibility of contamination of the metal by carbon, particularly where the metal readily forms a stable carbide, as in the case of titanium, zirconium and vanadium.

To produce powder metals, the conditions of the decomposition process described must be carefully controlled to produce powder or sponge deposits. The carrier gas technique is very suitable for this, since the carrier molecules help to disperse the metal atoms deposited and hinder their bonding together in a solid mass. Variations in *e. Reduction of metal carbonyl.* The gas carbon monoxide (CO) can combine with a number of metals to produce compounds such as iron pentacarbonyl ($Fe(CO)_5$) and nickel tetracarbonyl ($Ni(CO)_4$). The carbon monoxide is reacted with the metal (after preliminary treatment of the latter to produce some degree of refinement) under pressure but at a comparatively low temperature (200°C). A liquid carbonyl results, which can be stored under pressure until required. The decomposition is achieved by boiling off the liquid and passing the gas into a heated vessel at atmospheric pressure. The decomposition proceeds spontaneously around 100°C:

$$Fe(CO)_5 \rightarrow Fe + 5CO$$

The carbon monoxide is then recovered and recirculated to produce more carbonyl. The resemblance to the van Arkel process will be obvious, the carbon monoxide acting merely as a carrier. Unfortunately, the carbon monoxide itself may produce side-reactions, as described under metallic oxide reduction with carbon monoxide (see p. 732), and this may lead to pick-up of oxygen or carbon, especially in iron. By introducing a reducing gas, such as ammonia, countercurrent to the carbonyl gas in the reaction vessel, the carbon and oxygen content can be substantially reduced, but nitrogen will inevitably result from some reaction with ammonia. Iron powders, particularly, require to be annealed after production to distribute the contaminants uniformly and to reduce high internal stresses produced by their presence. Little information on the nickel carbonyl process is available generally, because of its commercial importance.

f. Electrolytic methods. The use of electrolysis, familiar in plating thin films of metal onto a component, can be adapted to the production of powder by employing an inert cathode, or one from which the metal deposit may be easily stripped, followed by mechanical disintegration. The main advantages are comparatively low cost, high purity, good control of the operating conditions and a product physically suitable for the subsequent pressing and sintering processes. The disadvantages are the difficulty of producing alloys by this method, and the highly active surfaces of the powder which may lead to oxidation when washing it free from the electrolyte and drying it for further use.

Aqueous solutions are obviously cheap, but largely confined to the production of copper, iron, zinc, tin and nickel powders because of the oxidation problem referred to above. The use of organic solvents is possible but more expensive. Fused salts, in which the metal compound is dissolved in a low melting point electrolyte of other metallic salts but containing no water, have been successfully used. A liquid metal cathode, usually mercury, has the advantage of eliminating many of the problems inherent with aqueous electrolytes whilst retaining most of the advantages.

The principle of electrolysis is a very simple one – the use of electrical energy to cause transport of the ions in the electrolyte to individual electrodes, where the desired material can be collected. In the case of an aqueous solution of copper sulphate, the material is ionized and the ions liberated when dissolved:

$$CuSO_4 \rightarrow Cu^{2+} + SO_4^{2-}$$

A separation of the positive (anode) and negative (cathode) electrodes at opposite ends of the bath will cause the copper ions to migrate to the cathode and the sulphate ions to the anode. There they will be neutralized by the charge on the electrodes, and the copper will deposit as metal on the cathode. The conditions of electrolysis – the current density, the temperature and the degree of agitation of the bath – will affect the velocity of transport of the ions, and therefore the speed of deposition. These conditions, together with the nature of the cathode and any additions of organic or other material to the electrolyte, will also affect the nature of the deposit, which may be smooth, rough, spongy or powdery.

The deposition of metal at the cathode causes depletion of metallic ions in that vicinity, hence the transport of further metallic ions from the remainder of the bath is important. Fresh metallic salt must be added to the bath as the metal deposits, otherwise the rate of deposition will fall rapidly towards the end of the process. Where the current density at the cathode changes markedly on depletion of the electrolyte in metallic ions, or when the voltage is varied, then the deposits tend to be spongy or powdery, as with copper, zinc and cadmium. Where this is not so, as with iron, nickel and cobalt, the deposits tend to be compact and strong. The effect of temperature and agitation of the bath is largely one of ion transport, as mentioned above; hence the nature of the deposit can be controlled in part by the conditions of electrolysis. The principal method of transport is the process of diffusion, which can be affected by other salts in the electrolyte, organic additions and the current density. Once the current density at the cathode exceeds the current density in the electrolyte (current carried by the moving ions) then the deposits at the cathode become spongy and, whereas this would be avoided in plating, it is the aim for powder production.

The addition of other materials to the electrolyte – acids, alkalis or neutral salts – provides other current-carrying ions; hence the current density of the electrolyte does not rely on the metal ions alone, thereby causing more rapid depletion of metal ions in the vicinity of the cathode and a more spongy deposit. Barrier layers or films of other material close to the cathode have also been used to the same effect. The basic requirements for powder metal deposits are therefore:

1. High current density.
2. Low metal ion concentration.
3. Addition of other ionized material (neutral to the deposition process).
4. No agitation.
5. Low temperature (to suppress convection and keep diffusion rates low).

The presence of the aqueous electrolyte leads to dissociation of water into its component ions:

$$H_2O \rightarrow 2H^+ + O^-$$

and the hydrogen will automatically be released at the cathode as a gas, along with the metal. The metal deposits may therefore contain occluded hydrogen, and in many cases this leads to embrittlement. Deposits of metal which cannot easily be produced in sponge or powder form may be easily pulverized if they are embrittled by hydrogen. Fortunately, this is the case for iron, which is more easily produced in brittle, compact form and subsequently crushed to powder.

As has been mentioned, after-treatment of the powder obtained may be necessary to prevent oxidation and deterioration during storage of the highly reactive product. Washing to remove traces of electrolyte may employ very dilute acid, followed by neutralization of the acid with dilute ammonia and removal of ammonia by distilled water. Some additions are often made to the final wash water to leave a thin film of water-repellent chemical substance, finally drying in hot air. Powders which may contain substantial quantities of hydrogen, intentionally or accidentally, are frequently annealed to remove hydrogen, and this has the added advantage of removing inherent residual stresses which might interfere with the subsequent pressing of powder into compacts. (For excellent reviews of commercial plant for copper and iron powder production, see references 10 to 13).

The electrolysis of fused salts removes many difficulties of the deposition from aqueous electrolytes, particularly the contamination of the final product. However, the higher temperature necessary to melt the electrolyte and maintain it in the molten condition adds substantially to the cost. At the present time, the process is only justified economically by the specialized requirements of the high melting point transition metals and those of interest to the nuclear energy industry, e.g. titanium, zirconium, tungsten, molybdenum, niobium, beryllium, uranium and plutonium (see references 14 to 16).

The fused salts normally employed consist largely of chlorides of the alkali or alkaline-earth metals, lithium, sodium, potassium, calcium and magnesium. These salts have a high conductance in the molten state and comparatively low melting point (330 to 800°C). Employed as mixtures, they frequently form eutectics in which the melting point is even lower. Many of the chlorides of the metals referred to as suitable for this type of process dissolve in these mixtures. The decomposition by electrolysis must, of course, ensure that the metal desired is the only one deposited, so that the voltage required to decompose, e.g. titanous chloride ($TiCl_3$), must be lower than that for the carrier electrolyte, e.g. a sodium chloride/potassium chloride mixture. In certain cases, alloys have been produced by adding two metallic salts to the carrier, e.g. titanium–aluminium. The electrolytic conditions referred to under aqueous electrolytes apply here also, i.e. current density, diffusion, temperature etc. Disproportionation reactions may also proceed at the cathode, particularly as the metal to be deposited is often one of those which suffer similar breakdown of higher halides in vapour phase deposition processes.

Very little commercial advantage has been taken of fused salt electrolysis for these reasons and the process remains highly specialized, attracting large research and development costs, and limited therefore to applications of expediency or value of the product (see also references 17 to 20).

Liquid metal cathodes, particularly mercury, do not necessarily dissolve the deposited metal, but disperse it in a protective form to prevent oxidation or other contamination. Iron powder is very suitable for this variation of the electrolytic process because the iron can be separated magnetically. Filtration or centrifuging are other feasible methods of separation from mercury (see reference 21).

10.2.2 Mechanical production methods

a. Disintegration of solid metal. As discussed briefly in the introduction, the disintegration of metal by mechanical means is one of the oldest forms of preparing a powder. The process can be applied to solid metal produced by conventional melting and casting processes, but tends to produce a particular geometrical shape of the particles, i.e. a lamellar rather than a more spherical shape. In this form, however, metal powders have important uses, such as the manufacture of paints and finishes. The main use of mechanical methods, however, is in conjunction with the chemical methods which give a spongy or powder product, and where the metal is either inherently brittle or in that condition as a result of the method of production.

The deformation of metals can be briefly described as a combination of three possible processes, namely, the elastic deformation which is recoverable on removal of the applied stress, the plastic deformation which is not recoverable and which leads to work hardening of the metal (i.e. stress must continue to rise at an increasing rate to continue the deformation) and, finally, the process of fracture by which some metal atom bonds are destroyed and portions of metal are detached from the main mass. In normal applications of metals, the proportion of elastic deformation is required to be high and the ratio stress:strain (modulus of elasticity) to be as high as possible. In fabrication of metals by forging, rolling, extrusion or drawing, the plastic stage is required to be extensive, though it must not lead to fracture. To disintegrate a metal, only the fracture stage is really pertinent, and the previous elastic and plastic stages must be passed through as rapidly as possible. To make such a process commercially acceptable, the properties required are, therefore, low elasticity and plasticity; in other words, the material should be brittle. For this reason, soft and ductile metals are not manufactured in powder form by mechanical means unless for specific applications, e.g. aluminium or copper powder for paints. In addition, any prior treatment or process which results in a more brittle form of the metal is to be desired, as in the case of metal embrittled by hydrogen.

A brittle metal may be readily broken by simple impact or crushing processes of the type used to reduce minerals to small size, e.g. by use of stamping mills, roll mills or jaw crushers. Continued crushing of this type would tend to produce lamellar-shaped particles if carried to the particle size normally acceptable for powder metallurgy processes, so that some other method must be introduced for the production of finer particle sizes where lamellar shapes are unsuitable. The rod or ball mill, in which hard metal rods or balls are rotated with the powder inside a cylindrical shell of metal, is very suitable for bulk production of this kind. The rods or balls are lifted by the rotating shell and drop onto the powder particles, crushing them by impact. In addition, the particles are rubbed together like the pebbles on a beach and soon become mainly spherical in shape. For smaller scale, specialized production there are rotating hammer mills, in which pivoted hammers crush by impact, often at relatively high speeds. The impacting of powder particles against each other can also be used in the eddy mill (see Fig. 10.2), in which a stream of particles is caused to flow against a second stream in a small impact chamber by the use of compressed air or inert gas as the fluid medium. The particles leave the impact chamber to be separated by a centrifugal classification chamber, the smallest particles staying close to the centre and the larger particles passing around the periphery of the chamber to be returned to the impact chamber as the second stream meeting the injected new material. The fine powder passes out of the central collector hole in the centrifugal classification chamber only when the particle size has been reduced to dimensions dependent on the fluid pressure and the size of the chamber. Since the process involves no material other than the powder to be reduced, contamination is kept to a minimum, whereas all crushing operations involve contamination to some degree with the material of construction of the mill. Fig. 10.2 represents an eddy mill.

Contamination in all forms of pulverization depends on the mechanical properties of the powder and the lining of the vessel or machine involved. Hard rubber is fre-

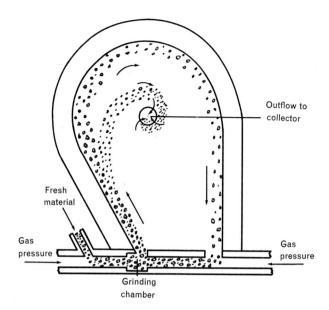

Fig. 10.2 Principle of the eddy Mill.

quently used as a liner in rod and ball mills because the contamination can be burned off in the sintering process later, and the rubber reduces the noise level involved in crushing. Stainless steel can be used to reduce contamination but results in about 0.5% pick-up of iron when crushing hard material. Wear-resisting alloys give lower levels of contamination but are comparatively brittle themselves, and this may occasionally result in fairly large lumps of the lining breaking away and becoming mixed with the powder.

Electrolytically produced powdered metal, e.g. iron, may be given a preliminary crushing treatment in a small jaw or rotary crusher, followed by ball milling to produce fine particle size. Metallic sponges, often softer than electrolytic iron, require only vibration or impact, in swing hammer or eddy mills. Brittle metals such as antimony and bismuth, and many alloys, can be jaw or roll crushed if in lumps above 1 inch in size and then broken to fine powder in ball mills. Occasionally, metals are pulverized at high temperature to take advantage of the decrease in strength which occurs as the material approaches the melting point. Tin and brass powders have been so produced, but the product tends to be more lamellar than from other processes, since the metal is ductile at high temperature in addition to being of lower strength. Stainless steel can be rendered more susceptible to pulverization by attacking the grain boundaries chemically, and then crushing or ball milling. Similar treatments have been applied to nickel-chromium alloys. The possibility of contamination by acids or salts is increased by the embrittling process.

Metals which form hydrides (compounds with hydrogen) or are embrittled by hydrogen can be treated in that gas before crushing. The hydrogen assists in preventing oxidation, but the product is often of fine particle size and very reactive when exposed to air in the compacting processes to follow. Some reactive metals may ignite spontaneously in this condition, i.e. they are pyrophoric.

A similar process to comminution of metal may be involved in the preparation of mixtures of metal powders, where alloys will be formed during the later sintering process. Ball milling helps to smear the softer material over the harder, and ensures better mixing than straightforward blending operations in a tumbler-type mixer.

In all crushing or milling operations, the addition of a liquid dispersing agent may assist the operation, provided that it does not lead to contamination or oxidation and is neither inflammable nor toxic. Organic compounds, often the higher alcohols or hydro carbons, have been claimed to increase the surface area of the final particles by orders of magnitude.

b. Disintegration of liquid metal. This process, often known, incorrectly, as atomization, carries the principle of disintegration to a logical conclusion by utilizing the greatly decreased strength of metal in the liquid state to break it up. Any metal that can be economically melted and pumped through an orifice can be disintegrated by this method. All that is required is to project the liquid stream from an orifice designed to give turbulent rather than laminar flow, at a suitable velocity, and to provide a collecting chamber for the powder which results.

The velocity of the liquid is determined by the formula

$$v = c\{2g(p_1 - p_2)/\rho\}^{1/2}$$

where v is the velocity, c a constant, p_1 the pressure of liquid in the orifice, p_2 the pressure outside the orifice and ρ the density of the liquid.

The constant (c) depends on the orifice and on the inherent properties of the liquid. The relationship between the velocity and viscosity of the liquid affects the type of flow which results, laminar or turbulent. Sir Osborne Reynolds carried out the initial work on this relationship and derived a constant R, the Reynolds number, relating the physical properties involved:

$$R = \rho v d/\eta$$

where ρ is the density of the liquid, v the velocity of the liquid, d the diameter of the orifice and η the absolute viscosity of the liquid.

Laminar, or streamline flow (see fig. 10.3), will persist up to a high velocity of liquid flow provided that the viscosity is high and the density low. At a critical velocity, whatever the viscosity or density, the flow becomes turbulent and hence disintegration of the liquid stream becomes more complete. Molten metals have a much higher density than water and, generally, higher viscosity, but the design of the orifice can be arranged to assist turbulence and hence reduce the velocity at which it occurs. Lower critical velocity means lower pumping pressure, and hence lower cost. High-speed photography has assisted in the study of orifice design and improved the efficiency of atomizing plants.

Once droplets are formed by turbulence at an orifice, it is necessary to prevent coagulation into larger droplets or the production of a molten pool or solid accretion on the floor or walls of the chamber. This can be achieved by several methods but the most important is the use of a high-velocity gas stream impinging on the liquid metal stream, or enveloping it. This assists turbulence and so reduces the liquid velocity at the orifice. It cools the liquid droplets and so prevents them coagulating, either in their fall or on the surrounding chamber. High-velocity liquids (other than the metal stream) have also been used to assist turbulence and prevent coagulation. Since their viscosity is higher

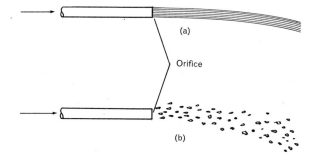

Fig. 10.3 (a) Streamline of laminar flow. (b) Turbulent flow.

than that of a gas, the effect is greater and the cooling efficiency much increased also. Water, aqueous solutions and organic liquids have all been tried, and can be successfully adapted to prevent corrosion and contamination of the powder product (see also references 22 and 23).

Disintegration can also be achieved mechanically, by breaking up a vertical stream of molten metal against a rotating metal plate or cup, or by pouring through vibrating sieves. The latter method has been used for lead shot production for centuries; the first method is of equal longevity in brass flake production for paint used in manuscript illumination and paintings. The Degussa company in Germany used a rotating plate, with radial slots and cutting knives mounted adjacent, to break up a stream of molten metal surrounded by a high-pressure water annulus issuing from the orifice with the metal stream. The product tended to be coarse in particle size and readily oxidized, quite apart from possible freezing onto the knives set into the plate. Atomization at an orifice by turbulent flow is likely to remain the simplest and most economic of liquid metal disintegration processes.

10.2.3 Characteristics of metal powders in common use.

a. *Aluminium.* Considerable quantities of aluminium powder are incorporated in many modern explosives, and atomization gives the cheapest product in quantity for this purpose. The disadvantage of this process is the wide range of particle size which results (see reference 23), the range extending from 350 to 50 μm (1 μm = 10^{-3} mm). Such powder can be classified to separate particles of required size, since the particles are generally spherical in shape. For use as a metal reductant in the preparation of other metals the range of particle size is not important.

The use of aluminium powder in flake or lamellar form is fairly extensive in paints, to give a metallic lustre to highly coloured pigments (automotive industry) and to assist in rust-proofing ferrous metals or in priming coats on timber. The production of flake powder is carried out in ball mills, often with the addition of lubricant such as white spirit to prevent oxidation. The main disadvantage of aluminium is the oxidation of freshly prepared metallic surfaces, and this can occur in both the atomization and ball milling methods. Although the oxide film prevents further rapid oxidation and is highly resistant to abrasion, the rise in temperature which can occur when fine powders oxidize may cause ignition of the remaining metal powder and spread to create an explosion in the whole mass. The use of an oxygen-excluding lubricant is therefore desirable in ball mills so that opening of the vessel does not give rise to an explosive hazard.

Flake aluminium is about 10 μm thick and considerably greater in the approximate diameter of the particle; hence the surface area to volume ratio is high, a condition which assists the rapid rise in temperature and possibility of explosion. The more nearly spherical particles produced by atomization are drawn off from the orifice by vacuum system, and collected in bags, to prevent molten metal droplets reaching the powder and increasing the explosive hazard.

Oxidized flake aluminium particles, containing about 10 to 15% aluminium oxide as a film spread over the particles, have been used in the manufacture of a special material known as sintered aluminium powder (S.A.P.). The mechanical properties of this product are superior to those of aluminium sintered from non-oxidized spherical powder and to those of many fabricated aluminium alloys, particularly at high temperatures (up to 400°C). The explanation of this phenomenon is outside the scope of this chapter, but can be ascribed to the influence of the oxide on the normal deformation processes of the relatively soft aluminium.

b. Copper. Copper is readily produced in powder form by several methods, so that the properties may be all important, depending on the subsequent use of the powder. Reduction of oxide by hydrogen or carbon monoxide produces a porous, soft material (see reduction of metallic oxides, p. 732). The pressing of such a powder requires less force and the sintering characteristics are good as a result of the porosity. The particle size, however, is variable by reason of the chemical process involved and the friability of the metal product. The particles are very irregular in shape too, though this is of less consequence in view of the fragility. The purity of the product in this case is largely dependent on the purity of the original oxide.

Copper powder made by precipitation from aqueous solution contains about 1% impurity, mostly iron, but can be obtained in fine powder form, usually less than 40 to 50 μm and the vast majority of 2 μm size. This process may be capable of considerable improvement in utilizing waste liquors from copper mining or smelting, and has the merit of comparative cheapness. Reports from USA mines indicate improvement to 99.2% purity in copper by gaseous reduction of aqueous solutions, the majority of the particles being less than 40 μm in size and of density 2 to 2.5 g/ml. The particles are obviously very porous, but are comparable with electrolytic powder.

In Europe, the majority of copper powder is produced by electrolysis from aqueous solution, or by atomization. The properties of the powders are very different. Electrolytic copper is produced under conditions which give rise to a spongy or powdery deposit, the higher the concentration of copper in the solution, the larger the particle size. The spongy deposit tends to be dendritic, and therefore angular, which causes the powder produced by comminution to be less free flowing but reasonably dense. Addition of colloidal materials, e.g. glue or corn oil, to the electrolytic bath produces a more brittle deposit, which is readily broken up to a finer particle size by crushing. The washing and drying of electrolytic copper powder affects the resultant properties. Cracked ammonia atmospheres have been used to prevent oxidation, and addition of stearic acid has been used to repel water and act as a lubricant in pressing.

Atomized copper powder suffers from the same disadvantage as aluminium, namely, there is a wide variation in particle size, often from 250 to 10 μm. To obtain a reasonable product for pressing, the larger particles must be removed or crushed. The latter method is not economic with atomized powder as the density is already high and the metal less easily crushed.

c. *Iron.* Iron powder resembles copper in that it can be produced by several methods. It is to be expected that the common process of producing liquid iron, by reduction of oxide with solid carbon, carbon monoxide or hydrogen, could be readily modified for the production of powder. Indeed, the production of very fine iron powder, of less than 1 μm particle size, has been very successful using these gaseous reductants. The metal particles are coarser than the oxide from which they are derived and they increase in size by agglomeration if left for long or reduced at higher temperatures. At low temperatures, the aggregates appear similar in shape to the original oxide crystals, but changes occur at higher temperatures. Since iron oxide particles 0.01 μm in size can be produced, the reduction by gases, even though the resultant metal particles are somewhat larger, is an important process for specialized applications such as the production of magnets.

The industrial production in bulk of iron powders by gaseous reduction depends entirely on a source of pure oxide, cheap fuel and relatively simple plant. In Sweden, the Högariäs process utilizes pure magnetite, its own coal supply, and produces a sponge iron easily crushed. The plant resembles that used in the company's ceramic brick interests, and is consequently cheap. The highest grade powder requires annealing in hydrogen to remove some carbon and oxygen and reduce residual stress in the particles.

The USA Pyron process utilizes mill-scale from steel plants, and reduces it on a continuous belt by hydrogen gas. The only reason for using this normally expensive reductant is the presence of an adjacent chemical plant. Like the Swedish process, however, the product is very uniform in particle size and composition. Similar developments are proceeding with hydro carbon reduction, but costs are bound to play a large part even assuming a good product for pressing can be obtained.

Iron powder produced from carbonyl decomposition shows the characteristic 'onion skin' formation, as layers of metal deposit in succession on the original nucleus. Increase in temperature of decomposition decreases the particle size and apparent density of the powder, presumably by causing more rapid formation of new nuclei. The small particles obtained at high temperature are minute spheres, difficult to remove from the decomposition chamber and less than 1 μm in diameter. The normal range of carbonyl iron powder is from 2 to 15 μm. It is the presence of carbon which detracts from the utility of carbonyl iron powder, but treatment with ammonia, as already mentioned, may only serve to replace carbon by nitrogen, the powder becoming even harder in the process. Nevertheless, large quantities of carbonyl iron powder are used since the content of metallic impurities is extremely low.

Iron powders produced by electrolysis vary according to whether or not they are pulverized after extraction to break up the dendritic deposit which is compact and hard. The use of caustic soda solutions produces iron powder directly. All such iron deposits are dendritic and consequently angular rather than spherical. Comminution may round off the angular shape but in general produces small sections of dendrites rather than spheres.

Iron powder was produced in Germany during world war II by mechanical disintegration of a liquid stream, but the process was not developed further after 1945 except in the USA. Alloy steels have been successfully produced by similar methods but employing air, water or steam jets to break up the liquid, rather than mechanical methods. The shape of particles is changed by employing a denser medium in the disintegrating jet, so that water gives a coke-like spongy particle, easily broken down further in ball mills. The addition of elements such as boron to steel modifies the surface tension, and straight atomization of stainless steel to give spherical particles has been successful. All such methods either add impurity atoms intentionally, to assist the process by modifying the properties of the liquid, or lead to impurities being present as a result of contamination in the process. Provided that no deleterious properties result in the final component, such methods are acceptable, especially if the cost is reduced.

d. Nickel and cobalt. Like iron, nickel can be readily produced by dissociation of the carbonyl, which is more economic when associated with bulk production for other purposes, as is the case with nickel. Nickel does not absorb hydrogen or pick up carbon to the same degree as iron in the carbonyl process, and so this process becomes much more economically viable, since no after-treatment is required and the particle shape and size can be readily controlled by the conditions of reduction.

Where aqueous solutions of nickel or cobalt are available, especially where the two metals occur together in the mineral form, the reduction of such solutions can be justified, and even used as a method of separation. As has already been mentioned, composite powders utilizing an outer coating of nickel or cobalt have been successfully produced by these methods for special applications.

Again, where electrolysis is employed for recovery of small concentrations of nickel or cobalt, the production of powder may be easily achieved by variation of the conditions of electrolysis, a spongy deposit being obtained and readily crushed in ball mills.

The characteristics of these powders naturally vary with the method. The carbonyl powder tends to be spherical, by virtue of the rotation used in the reduction plants, but precipitation from solution and electrolysis tend to produce dendritic structures which break up into smaller angular-shaped and irregular particles. Much depends on the utilization of the powder in later stages or on the economics of the whole metallurgical operation in deciding which process to use.

e. Tungsten. Largely as a result of the specialized uses for tungsten, and its cost, only one type of process has tended to be used for powder production: reduction of the oxide or some tungstate salt with hydrogen or cracked ammonia. Carbon can be used, but the inherent brittle nature of tungsten is intensified by the introduction of carbon, due to the formation of tungsten carbide. The manufacture of tungsten wire for lamp filaments has long been the main outlet for the metal, and the drawing of fine wires is a tricky operation with such a brittle material. The characteristics of the powder depend to a great extent on the characteristics of the original compound undergoing reduction. Finely divided oxide tends to give fine tungsten powder when reduced at low temperature with a high gas velocity. The average size of the spherical particles is 1–5 μm, but very much finer powder can be produced. All that is necessary to coarsen the powder is to increase the temperature and encourage growth of the metal particles during reduction. Other impurities, such as nitrogen or arsenic, may be deleterious and the gas employed must be pure. To prevent build-up of the small traces of impurities present, a proportion of the recirculating gas is bled off from time to time.

Tungsten powder is one of the simplest to produce, but more difficult to process at later stages. The success of the tungsten filament lamp is sufficient evidence of the adaptability of the powder metallurgy method to a high melting point metal.

f. Alloys. Mention has already been made of stainless steel powder production, but a number of other alloys are utilized in powder form. One of the oldest uses of powder is in the manufacture of porous bearings, which can be impregnated with lubricant and so guarantee long life with no maintenance.

Although conditions of electrolysis can often be adjusted to give a deposit of alloys, there are severe limitations to such processes because of the widely differing characteristics of metals under those conditions. As a result, electrolysis is normally considered only for specific alloy compositions, or for the preparation of powders which are not true alloys but consist of layers of the constituent metals deposited in turn by varying the electrolyte and/or the conditions. Such layered powders become true alloys at a later stage, in sintering, by a process of diffusion.

Almost the only truly suitable method for alloy powders is that of atomization of a liquid stream. By this means, although there are problems of contamination and atmosphere control and operational limitations – especially with high melting point materials – at least the alloy composition can be prepared separately in the melting furnace. A wide variety of alloys is now available as atomized powder: steels, brasses (copper–zinc), bronzes (copper–tin), aluminium alloys and tin–lead solders. After production of the powder, classification into particle size is usually necessary for specific applications, all oversize material merely being remelted.

10.2.4 Manufacturing of shaped products from metal powders

The first step in the consolidation of metal powder into component shapes is to cause adherence of millions of small particles into a block, or something approaching the final shape, so that the mass can be handled without disintegrating once more into powder. The requirement is to increase the 'green strength', i.e. the strength before any other process is carried out. The second step is to increase the strength further, by creating more coherent bonds between the atoms of what were originally individual particles, and, at the same time, to remove the inevitable porosity which must result when so many small particles are forced into contact. This step implies movement of metal atoms to rearrange themselves into the maximum number of binding contacts – the process of diffusion, which is assisted by rise in temperature. This step is called 'sintering', because the metal atoms move, albeit over small distances, but do not melt, although in certain cases some material may be added which does melt, to assist the binding. The final step may be a simple fabrication by mechanical means to produce further consolidation or to complete the shaping operation to final size tolerances. Such a step is known as 'coining' after the resemblance to the stamping of embossed designs on coins in special dies.

a. Pressing. The first vital step is theoretically a simple one involving the compression of powder in a die, with a suitable fitting plunger, to cause adherence of the particles and the first elimination of porosity.

Uniformly sized spheres poured into a container with no pressure would not pack together as tightly as theoretically feasible, due to the inevitable entrapment of air. A dense powder like tungsten (specific gravity 19.3) will pack better than a lighter one such as aluminium (specific gravity 2.7) because the weight of the individual powder particles will tend to force air bubbles to escape. The pour density of the powder is a measurement of this loose packing when it is poured into a container from a standard height. The container volume is calibrated by filling with water at a standard temperature; then all that is required is to weigh the powder which just fills the container level at the top. By vibrating the container as the powder is being poured the density increases but may still be below the theoretical density of perfect spheres packed into that volume. The main reason is that powder particles are rarely truly spherical and there will always be a range of particle size. The theoretical close packing of identical spheres should lead to 74% of the volume being occupied (see fig. 10.4(a)), so that the tap density (as the vibrated density is called) can never exceed 74% of the density of the metal concerned. In practice, the tap density may be much less, as mentioned above, and may be only 50% of the true density of the metal. Figures 10.4(b) and (c) illustrate the pour and tap density which are characteristic of metal powders, in contrast to the theoretical close packing. In practice, the range of particle size causes some of the spaces between larger spheres to be filled with smaller particles, but even then the packing cannot be better than that of the theoretical 74%, calculated on the mean particle size. In practice, powders of a wide range of shape still show 37 to 40% porosity, and not more than 66% of theoretical density has been recorded with lead shot shaken to obtain the maximum tap density.

The sizing of powders, to obtain maximum filling of the space available, can be utilized to increase tap density. If the largest spheres, of radius a, are packed as closely as possible, then each has twelve close neighbours, best envisaged as one sphere surrounded by a ring of six in one plane, with three in the plane above and three below. The space between any three such neighbouring spheres can accommodate a sphere of radius $0.414a$, and this secondary sphere creates three more voids in one plane, which can accommodate smaller spheres of radius $0.225a$ (see fig. 10.5). Carrying the analysis further to five sizes of sphere, the radii are $a, 0.414a, 0.225a, 0.175a$ and $0.117a$, and the proportions of each size $1:1:2:8:8$. The space remaining unfilled, i.e. the porosity, would then be 3.9%, but although this is very small it can only be achieved if each sphere goes into its proper place and the sizes are in accordance with the list above. This explains why any attempt to carry out such a synthesis of a commercial powder would not lead to a worth-while return and, in any case, the pressing process packs the material far better than the closest packing of freely moving spheres.

The filling of a pressing die with powder can be accompanied by vibration, in which case the tap density will be higher than the normal pour density. Ultrasonic vibrators attached to the work-table have been used for this purpose. Apart from the assistance of vibration, the minimization of friction between the moving particles can materially improve the packing, and it is common practice to add a lubricant to the powder with this end in view. There is also a considerable frictional force created by movement at the walls of the die, and between die and plunger, when the pressing operation commences. Lubricants, whether added to the powder or coated on the walls of plunger and die, must not contaminate the final product in any way which would affect its properties or spoil the surface finish. A material of low shear strength is required, capable of application to form a thin film, but which is not readily broken, as a film, by movements within the powder or the relative movement between powder and die wall. These properties are possessed to some degree by all substances used as lubricants and accordingly the common organic materials associated with anti-friction qualities are utilized – paraffins (kerosene), alcohols, fatty acids and soaps. The smaller the molecule of the organic compound, the less effective the lubricant, and hence paraffins and alcohols with high molecular weight are the only ones worth consideration.

METAL POWDERS

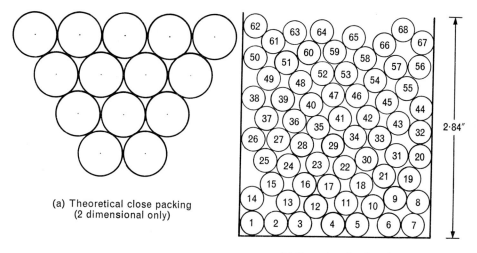

(a) Theoretical close packing (2 dimensional only)

(b) Pour density of spheres.

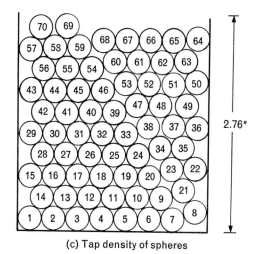

(c) Tap density of spheres

Fig. 10.4 Packing of spherical particles.

As the molecular weight increases the material changes from liquid to solid, and hence some form of solution or suspension in a volatile liquid may be required to allow a thin coat to be applied.

Stearic acid, a fatty acid occurring in animal fats, is commonly used for metal die lubrication because of its adherence to the metal surface. Friction is very low, but the film is fairly strong laterally and not easily removed during the pressing operation. It is frequently suspended or dissolved in alcohol or paraffin to enable the formation of a uniform coating as the evaporation of the solvent commences. For more rigorous requirements, a metallic soap itself can be used since the melting point is higher and the cohesive strength therefore greater. Zinc stearate, melting at 130°C, and lithium stearate (m.p. 220°C) are extensively used.

For lubricating the powder itself, the requirements are slightly different. The tremendous surface area involved in a mass of powder requires a free-running liquid

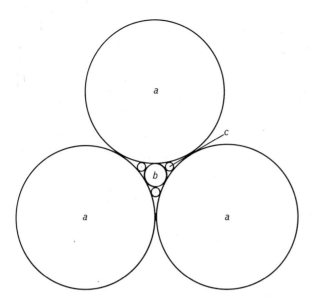

Fig. 10.5 Close packing with three sizes of sphere.

to be dispersed uniformly around each particle, either by dissolution of the lubricant in a large volume of solvent or by heating the lubricant until it softens and readily flows around each particle. In the case of a very porous, spongy powder, the use of a liquid solution of lubricant can give rise to problems in that the solvent may be absorbed into the particles themselves and be difficult to remove later. The lubricant must be blended into the powder to provide good mixing and as uniform a coating as possible. Specially designed mixers and blenders are available for such techniques.

Paraffin wax, camphor, benzoic acid and similar materials have been successfully used for powder lubrication, but stearic acid is not ruled out for this purpose also. The quantity involved is greater in powder lubrication, hence the cheapest successful material tends to be selected. Drying-out of the solvent, if one is used, is almost always necessary to speed the removal of the unwanted liquid once the particles are properly coated. A low-temperature oven is all that is required.

The lubricant, being an addition, decreases the packing density of the powder by occupying part of the volume, but normally this is offset by the better pressing achieved as a result of its presence. The powder does not flow so well to the die through a delivery system, but this is not unduly serious. Quantities used are of the order of 1 to 3% of lubricant, often dispersed in a suitable solvent in the same proportion, e.g. 100 g copper powder might be blended with 2 g paraffin wax dissolved in 100 g carbon tetrachloride.

The lubricant-treated powder is placed in the die, itself being lubricated, and the lubricated plunger is then brought into contact with the top surface of the powder and pressure applied. The movement of the powder during compression is materially assisted by its own lubricant coating in the interior and by the die lubricant at the walls. Flow is required to be smooth, otherwise entrapped air cannot escape easily and larger pockets of air may become entrapped in certain places. Jerky movement may also result in some areas of powder becoming highly compressed but with an area of much lower density between them. This leads to lamination, with the possible separation of these layers at some future date or, at best, a variation in mechanical properties through the final product; neither is to be desired. The density of the material will increase at first by re-stacking of the particles to give closer packing, and will be accompanied by the release of trapped air (or whatever atmosphere is in use in the pressing operation). The porosity may fall at this stage by 10%, depending on the shape, size and type of metal powder, lubricant and die design. Once the particles are re-stacked, further compression must lead to deformation of the particles themselves, and this will occur more readily if the particles are porous or spongy in nature. Spheres will tend to flatten into discs, and angular particles will penetrate each other as they change shape. The metallic material will be extruded into porous areas, and hence more entrapped air (or other gases) will be pressed out of the die. The shape of voids created by first filling the die with powder will therefore change, apart from decreasing in total volume, but porosity will not be entirely eliminated, due to the compressibility of the gas itself and to the deformation of the die as more of the pressure is transmitted to the walls. Generally, it is not possible to reduce the porosity to less than 10% by cold pressing and, even then, only with very high pressures. However, the object of pressing is not solely to remove pores but to improve the handling qualities, or green strength, although the elimination of porosity obviously plays some part in attaining this.

The removal of some pores, the improvement of contact between the metal particles and the extrusion of metal by plastic deformation, all contribute to an increase in the green strength by the creation of more atomic bonds between what were formerly separate, individual particles. As two atoms of a metal are brought closer together, attraction forces are set up between the electrons of each and the positively charged nucleus of the second atom. At the same time, repulsion forces operate between the two electron clouds and the two nuclei, and equilibrium is achieved when the two forces balance – the closest distance of approach, which varies with the atoms involved. There is a state of minimum free energy between the two when this occurs, and any attempt to separate them increases the attraction forces and thus brings them together again. This is the bond set up in cold-welding metals together by deformation, and in powder pressing. The green strength of a compacted powder will depend on the number of such contacts made in pressing, which itself will depend on several factors. The closest distance of approach is very small indeed, about 5×10^{-7} mm, which is far smaller than the degree of smoothness or flatness of two surfaces, however carefully prepared. There are surface 'asperities', or crests on the surface of a metal, which make contact before the main surface, and these asperities will be the first to bond together. Accordingly, the green strength will depend on how these may be flattened by pressure, to allow more bonds to be made. In practice, no metal surface is so clean and pure that contact to give atomic bonding is made readily. All metal surfaces have a film of oxide, however thin, and unless it can be removed this prevents the atoms reaching the closest distance of approach. The removal takes place mechanically in pressing, by rupture of the film and pushing the oxygen atoms into specific sites, thereby exposing pure, clean metal. This process cannot continue indefinitely at room temperature, so for this reason also there is a limit to the number of adhesion bonds. In a vacuum, or a reducing atmosphere which removes oxygen, the oxide film is much less effective in delaying adhesion. Metals such as aluminium, which form coherent, strong oxide films, must be deformed much more to create adhesion between two surfaces than metals, such as gold, with very little oxygen on the surface. The preparation and storage of metal powders can markedly affect the green strength in pressing, for this reason. Soft, spongy powders adhere well because deformation takes place readily on pressing. Hard metals, and those with oxide films, adhere only with extreme pressure.

There is, therefore, a limitation with cold pressing in the maximum green strength attainable. The use of higher temperatures, both die and powder being heated, can assist the deformation by reducing the strength of the powder and increasing the ductility. Furthermore, the metal powder itself work hardens in cold pressing as deformation proceeds, whereas by hot pressing the metal is being continuously annealed and so retains more of its ductility. The disadvantages of hot pressing, however, involve the use of inert atmospheres to prevent excessive oxidation of the powder (which would remove much of the advantage), and the need for much more expensive die materials to resist the temperature without themselves deforming. The furnace is a further expense, and the power consumed. These items are significant in view of the large thermal capacity of the plunger and die, the press platens and the volume of powder involved. Nevertheless, there are many instances where hot pressing is an advantage.

When ejected from the die the pressed compact is released from the pressure, some of which gave rise to elastic deformation of both powder and die. The die contracts therefore, and the compact expands, with the release of this elastic energy. Ejection therefore requires further pressure, in view of the sudden increase in friction between powder compact and die wall. Too sudden a release of pressure, and ejection, can lead to cracking of the compact, whose green strength is still comparatively low, as explained above. Hot pressing does not produce such drastic affects and so may be used with the more fragile powders of high elastic limit at room temperature. (The elastic limit falls with rise in temperature.)

b. Die design. It follows from the preceding section that the behaviour of the die in pressing plays a large part in achieving high pressures and preventing drastic elastic recovery on removal of the load. The uniformity of the pressing operation is also affected by the die design, bearing in mind the friction at the walls and in the body of the powder.

If the plunger presses the powder into a die with one end closed and integral with the side walls, it is obvious that pressure cannot be uniform, since the initial movement is at the plunger face and only gradually extends to the base of the die. Figure 10.6 illustrates this point, and the lower density, and therefore lower green strength, at the base. Ejection is also difficult in such circumstances and, if carried out by pushing on the base, will further compact the base slightly, but leave a lower strength area between top and bottom faces. This problem is overcome by double-ended pressing using two plungers, one of which forms the base but is not attached to the operating part of the machine, the other being the normal top plunger. The powder is pressed by the top plunger with the body of the die supported by stops or springs until the bottom plunger can be used, either against springs, or by removing stops or by inverting, and making the bottom plunger become the top. Figure 10.7 illustrates the principle.

The side pressures on a die are considerable, and it is here that the die itself tends to deform. A 'bolster', or large retaining ring of metal, can be fitted around the die to restrain its movement, or the die walls can be separately supported by springs, cams or levers. Frequently, such devices can be made to assist ejection of the compact by moving in the opposite direction at the appropriate stage. The more complex the component, the more difficult the task of retaining the shape of the die and the dimensional tolerances of the compact.

METAL POWDERS | 755

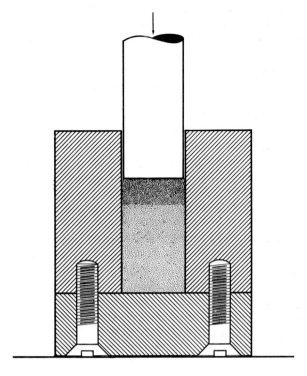

Fig. 10.6 Pressing in single-end die.

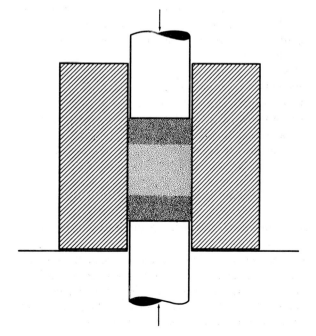

Fig. 10.7 Pressing in a double-end die.

Presses are required to give pressure in the die between 10 000 and 100 000 p.s.i., on average. For small components this is not difficult, but for larger ones the press itself becomes very expensive and the die also. Modern presses for powder metallurgy incorporate feed mechanisms for powder filling and ejection systems for easy removal of the compacts. Vibration of the feed hopper and die during filling and automatic weighing of the powder feed are refinements of considerable value in mass production.

   c. *Isostatic pressing.* The disadvantages of pressing in dies can be overcome by the use of fluid pressure, in which the triaxial stresses in the working fluid act as the die itself, the fluid being virtually incompressible relative to the powder. The pressure achieved is much more uniform and the density of the product equally so. Pressures can be much higher, which is frequently an advantage. Even if initial pressing is carried out in a die, a finishing compaction to give more uniform pressure still remains an advantage. The method is called isostatic to emphasize the uniformity of pressure achieved.

   The powder must be held in some form of plastic container, aided if necessary by supporting metal devices to retain shape in the early stages. A rubber bag surrounding the powder is common, and the whole assembly is placed in the working fluid (see fig. 10.8). Water, oil or gas have been used as the working fluid, and explosive charges have been employed to increase the rate of strain application with great success. Molten metals and glass have been proposed for hot pressing by isostatic methods. The USA Dynapak press is a typical example of the type of machine used for high rates of pressing (see also reference 24).

   d. *Sintering in general.* The pressing of powders, by whatever method, increases the green strength so that the compact can be handled, albeit delicately, for subsequent processing. The adhesion of the individual particles is, at this stage, mechanical and relies on the forces set up between two surfaces known as van der Waals forces after the work of van der Waals on the kinetic theory of gases. The electrons surrounding the atoms in a metal produce an attraction force for the positively charged nucleus of other atoms, and the magnitude of the force varies with the distance apart of the atoms. the electrons also mutually repel each other, and shield each nucleus from the attraction of other electrons. An attraction force remains because the electrons are not stationary and their true position at any point in time depends on a statistical probability. There are always attraction forces of this type, although they are very weak, and it is this attraction which causes some gases to act as fluids rather than undergo complete dispersal in a container. In any element, the van der Waals forces can be expressed as

$$\phi_v = -\frac{3\alpha^2 E}{4r^6}$$

where ϕ_v is the potential energy, α the polarization factor, E the energy factor of the atoms involved, and r the interatomic distance.

   The polarization factor becomes more important in compounds, where the electron distribution is not simple but more directional, and the value of E changes with the element. The important value is r, the distance between atoms, and the sixth power of r indicates the low value of the van der Waals force. The strength of the metal would therefore be very low if no further bonding mechanism were to be involved.

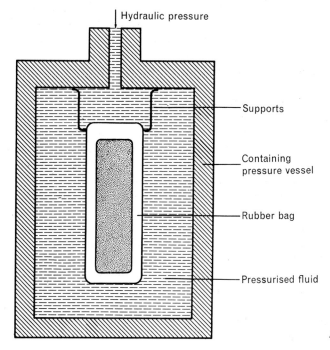

Fig. 10.8 Isostatic pressing.

Within the individual particles of a metal powder the strength of the metal is high because the atoms are arranged on a crystal lattice and some of the electrons are shared amongst all atoms – the electron gas of the theory of metals. To provide stronger adhesion between metallic powder particles, therefore, the bond across the interface between all particles must be converted, as far as possible, into the true metallic bond so that the maximum strength is developed over-all. This is precisely the objective in joining metallic components by welding and other means.

Two barriers exist to prevent a ready bonding across the interface. The first is the presence of oxide films on the surface of the metal, which contribute to the van der Waals force but prevent metallic bonding because their electron structure is not the same as that of the pure metal. The second barrier is the difference in orientation of the crystal lattice from one particle to the next, so that some movement is necessary to allow the electron gas to form uniformly around the atoms in two adjacent particles. This misorientation is present in all metals at grain boundaries, but in metals solidified from the liquid it is comparatively slight and consists of a 'defect' lattice structure which still provides a reasonable cohesive force. To improve the bonding of metal powder particles, therefore, the oxide film must be broken, and the orientation barrier overcome by reorientating some of the particles or parts of the crystal lattice to create a similar situation to the boundary in a normal polycrystalline metal.

The pressing together of metal powders or two metallic surfaces in general achieves, to some degree, both of these objectives. The oxide film is disrupted, at least partially, by the mechanical pressure and the orientation is changed by plastic deformation of the metal. There is therefore some additional binding to the van der Waals forces set up by the pressing operation, but it is localized, and the strength of the compact is still well below the level required for use. The greater the pressure applied, the more the oxide film is disrupted and the greater is the plastic deformation, which affects orientation misfit. The adhesion of powder compacts therefore increases, up to a limit, with rise in compacting pressure.

Apart from improvement in bonding, the packing density of the particles is by no means perfect and voids occur in the compact. To ensure maximum strength the number of contact points should be a maximum, and hence densification plays a large part in the increase in strength after pressing. All these processes are achieved by sintering the compact, a rise in temperature serving to increase the plastic deformation at contact points, thereby aiding the break up of oxide films and reorientation of the crystal lattices. In addition, the movement of atoms by diffusion not only assists in the bonding process but also enables reduction of the void space and densification.

Diffusion occurs in all materials, even in the solid state, by interchange of position of atoms in the crystal lattice. There are always holes in the lattice – vacant lattice lattice sites or 'vacancies'. The number depends upon the treatment of the material in its original formation. When cooled from the liquid, not all the atoms arranging themselves on the lattice can take up the appropriate positions rapidly enough, and the faster the rate of cooling the more difficult this becomes. Quenched in water from a high temperature a metal is always less dense than it should be at room temperature, and the vacancies which exist provide for easy diffusion on reheating, since it is energetically more favourable for an atom to move into a hole in the lattice than to push aside other atoms when changing its position. In powder particles, the number of vacancies varies depending on the method of manufacture, but some are always present. Atomized powder, produced by disintegration of liquid metal into droplets, contains a high proportion of vacancies, for reasons given above. Powder produced by chemical reduction may also contain a high proportion of vacancies due to the chemical reaction itself proceeding by diffusion mechanisms.

Diffusion proceeds more rapidly at high temperatures, as would be expected by reason of the energy required to cause an atom to move. There is a minimum requirement of energy to start the procedure – the activation energy – and the temperature dependence follows an exponential law of the form

$$D = D_0 \exp(-Q/RT)$$

where D is the measured diffusion coefficient, D_0 the diffusion constant for the system, Q the activation energy, R the gas constant and T the temperature.

D_0 varies according to the system which, for powder metallurgy, is self-diffusion, the movement of like atoms. D, the coefficient, varies according to the mechanism by which the atoms move, and Q is therefore a measure of this. Atoms may move along an exposed surface, including the surface of a void (surface diffusion), along a grain boundary with its many lattice defects (grain boundary diffusion) or they may move directly through the lattice (volume diffusion). Of these three possibilities, the activation energy is lowest for the surface (since the minimum resistance is met by the atoms), intermediate for grain boundaries (where a state of disorder exists) and greatest for lattice diffusion. The general view is, therefore, that surface diffusion accounts for the movement of atoms to transport metal from one point to another, especially in voids, the hole then moving towards the surface by a series of steps so that eventually it should disappear at the surface. There is a net movement of metal to the centre of the compact and of voids away from the centre to the surface – in other words, densification. In certain cases it would appear that actual transport of atoms by evaporation from one side of the void and condensation on the other is possible, but the magnitude of this effect is still in dispute.

To summarize, the process of sintering at high temperatures causes a realignment of metal atoms between powder particles so that oxide films are disrupted, thereby increasing points of contact. At the same time, the metal atoms start to rearrange on a common or closely related crystal lattice pattern: the mechanism of recrystallization. These two effects convert the bonding from relatively weak van der Waals forces to strong metallic bonds, and the strength increases rapidly. Simultaneously, the atoms are activated to move with respect to the voids remaining between particles and densification proceeds, albeit more slowly than the formation of the metallic bond. The compact becomes strong, and almost as dense as the theoretical density of the pure metal would indicate.

Some metallic powders contain impurities and, as in the case of alloys, may create phases which melt at the sintering temperature. In such cases, a small proportion of liquid will exist and tend to surround the solid particles or, at least, collect at points between the solid particles. All processes of diffusion will occur more rapidly in the liquid than in the solid, and hence there will be more rapid transport at a given temperature than if no liquid were present. The pores will tend to fill up with liquid as a result, and oxide films will tend to be disrupted on the solid particle unless the phase is a highly oxidizing one, which is unlikely. In the presence of a liquid phase, therefore, sintering can be more rapid at a given temperature, provided that the liquid wets the solid, i.e. there is no great energy barrier between the two which reduces the spreading or movement of the liquid phase. In certain cases, therefore, there may be advantages in allowing certain impurities to remain in the powder or, alternatively, in making additions to the powder to ensure the presence of a liquid phase. However, whilst sintering will be improved and speeded up, the final product must be in part a composite material, and the strength of the interface between the two phases after cooling will be important in meeting specified properties. In many cases, a composite is desirable (see hard metals, p. 774).

e. Loose sintering. Before discussing the practical implications of sintering, it is necessary to remember that metal powder will sinter together without compacting, merely because of the processes involved in diffusion. A powder poured into a container and then heated to a suitable temperature will sinter whether vibrated or not, and will produce a high proportion of perfectly acceptable metallic bonds. In view of the necessary processes contributing to bonding, the very much lower density of loose powder will involve higher temperatures of sintering if the time is to be reduced, or longer times at the temperature acceptable for a pressed compact. Accordingly, the density of loose-sintered material would not be expected to reach that of compacted and sintered metal except with prolonged, and therefore probably uneconomic, sintering conditions. However, for many purposes the density and strength achieved may be perfectly adequate, and a further coining or pressing operation may be possible after sintering if density is important.

The simple approach to loose sintering is therefore to pour the powder into a suitably shaped mould and to raise the temperature in sintering. This permits relatively large compacts – depending only on the size of the sintering furnace – where pressing would require very expensive dies and presses. The powder may be vibrated to increase the density before sintering, if required, especially if the free-flowing characteristics of the powder do not allow it to fill a complex shape adequately. The mould must be compatible with the powder at the temperature of sintering, i.e. there must be no alloying or welding of the two, and the mould must not distort or deform during the sintering process. Graphite moulds have been extensively used where reaction with carbon does not occur in sintering, and have the merit of easy machining. Moulds of cast iron, welded sheet steel and alloy steel can be utilized, and compatibility with the powder can be assured by coating the metal mould with an impermeable skin of oxide, phosphate or other metal by prior treatment.

By controlling the sintering temperature and time, the method of loose sintering can be used to produce a given porosity, and there are many applications of powder products where porosity is desirable, as with porous bearings and filters. The tap density after pouring in powder and vibrating may be as high as 85% of the theoretical density, and the sintering process can be controlled to provide adequate bonding for strength and still retain from 85 to 90% density, with most of the pores remaining interconnected. Although there are limitations to the complexity of shapes produced by loose sintering, the use of this method for bulk production of metal stock, to be worked by rolling, forging or machining at a later stage, has been put forward as economic, considering the immediate advantages of lower temperatures than in melting and casting, the avoidance of macroscopic defects which occur in cast ingots and more efficient control of purity and composition. Ingots of pure carbonyl iron of up to 2 tons have been produced by this method.

In addition to the simple pouring of powder into a mould, methods of improvement of the flow characteristics have been developed to enable more complex shapes to be produced. The metal powder may be mixed with a liquid to form a slurry and poured into an absorbent mould (e.g. plaster of paris) as is done in the pottery industry with finely divided clay slurry (slip-casting). Other techniques have used resins which solidify on evaporation of a suitable solvent, the resin being decomposed and volatilized during sintering. Freezing of the slurry has also been employed, the liquid then being slowly evaporated in a vacuum chamber before sintering. Whilst this latter method produces accurate dimensions in complex shapes, the extra cost of refrigeration and vacuum equipment may prove uneconomic. True slip-casting involves finely divided powders and requires the addition of some reagent to prevent the particles aggregating or flocculating. This type of preparation has been developed for uranium and thorium powder slip-casting, and for many other reactive metals.

Precautions to be taken during the sintering process are the same as for sintering of pressed compacts, and will be described below.

f. Sintering pressed compacts. Since the basic requirement of sintering is a rise in temperature, the precautions necessary are those which apply to any metal heat-treatment process, and involve the same items of plant. However, it is often more economic in metal heat treatment to conduct the operation in air and to remove the outer skin of oxide at a later stage, making due allowance in dimensions for this cleaning-up step. In sintering, one of the prime objects is to disrupt the oxide films on the powder particles to improve the contact of the surfaces, and this would not occur if an oxygen atmosphere were present in the furnace. The use of inert atmospheres or high vacuum is therefore essential.

The furnace used is not required to differ from any other type of furnace designed to operate at the appropriate temperature, but means of handling the compacts in their initial green state and after sintering may be slightly more stringent than when handling bulk solid metal components. Below about 1 100°C, furnaces are commonly heated by gas or electrical resistance methods. Separation of the interior of the furnace from the heat supply is more necessary with gas heating because of the possible effect of the products of combustion on the compacts. Double-walled muffles are widely used for this reason. Between 1 100 and 1 350°C, resistance heating with silicon carbide elements is common, and between 1 350 and 1 700°C, resistors of molybdenum are used, in an atmosphere of hydrogen to prevent oxidation. Above 1 700°C, the use of high-frequency induction heating can be economic and simple in operation, but resistance heating with graphite, tungsten or tantalum elements, in a suitable protective atmosphere, is fairly extensive. The use of high vacuum at very high temperatures causes evaporation of many metallic heating elements, and inert atmospheres are necessary, rather than vacuum, above about 1 800°C.

The sintering process is a function of both temperature and time. If a high temperature can be used to shorten the time involved, mass production methods, using belt conveyors through the furnace, become feasible. Such methods are often utilized in connection with sintering of porous compacts. The heating and cooling times are very important in these cases, as they are with batch processes. Too rapid a heating rate may cause spalling of the compact if the outer surface heats up much more quickly than the interior, and too rapid cooling may have the same effect but in reverse. Vacuum systems cannot be heated or cooled rapidly because of the absence of conduction through the atmosphere, and an inert gas is often used during the cooling cycle after vacuum sintering to speed up the process.

g. Protective atmospheres in sintering. Although a variety of protective atmospheres have been employed, only those involving hydrogen or vacuum are extensively used in practice. Hydrogen is common in laboratory and small-scale processes, to eliminate both oxygen and nitrogen. The advantage of hydrogen is that it can be used again by purification and recycling; the disadvantage is the cost, and the necessary precautions to prevent explosion with air. To start operations with a hydrogen atmosphere involves sweeping out the oxygen in the residual air with a vacuum system or by a flow of nitrogen, before admitting hydrogen. The same technique applies in reverse on cooling down. Hydrogen is available in cylinders but these inevitably contain adsorbed water on the walls and slight traces of oxygen. It is necessary, therefore, to remove the oxygen, usually by combination with some of the hydrogen on a platinum catalyst, and to follow this by drying to remove water from the cylinder together with that formed by the oxygen-hydrogen reaction. Hydrogen can be produced by electrolysis of water, but again the purification and drying process is necessary.

A cheaper source of hydrogen is the atmosphere obtained by cracking ammonia, in which heat dissociates the ammonia into nitrogen and hydrogen, with a doubling of volume. This atmosphere can only be used if there is no reaction between nitrogen and the metal powder, as in the case of brass and steels. Ammonia must be handled with care, but proprietary packaged converters are available for production and control of cracked ammonia furnace atmospheres, which are extensively used in the heat treatment of ferrous metals.

Hydrocarbon gases also can be used in place of hydrogen, but generally require treatment, such as burning with air, to produce a mixture of carbon monoxide, carbon dioxide and hydrogen. Alone, the hydrocarbons tend to decompose in the furnace and deposit soot. By adjusting the air:gas ratio in the range 10:2, the hydrogen content can be varied from a mixture containing as little as 0.05% hydrogen to one of 40%. The range up to 15% hydrogen is exothermic, i.e. the generator supplies sufficient heat by the formation of steam to maintain the reaction. In contrast, with 15 to 40% hydrogen the gas formed is endothermic and requires external heat to the generator. The exothermic gas produces large volumes of steam which are required to be condensed out before use. The endothermic gas has a low moisture content and can form explosive mixtures with air. When formed from coal gas these gas mixtures may contain sulphur, which must be removed by absorption in iron oxide or activated charcoal. All the necessary controls and purification systems are incorporated in the proprietary plant.

The carbon content of these hydrocarbon based atmospheres is important when ferrous metals are involved in sintering. Endothermic gas, with high hydrogen content and carbon mainly as carbon monoxide, will not decarburize steel, but exothermic gas, with carbon as carbon dioxide, will decarburize medium to high carbon steel, especially if the moisture content is high. The presence of residual hydro carbon will cause carburization of iron and steel.

Vacuum equipment for sintering does not differ from that used in other applications. The pressure level required depends entirely on the reaction of the metal concerned with oxygen, nitrogen, moisture or carbon dioxide, all of which are present in air and which can leak into a vacuum system unless carefully leak-tested. Even the best commercial vacuum systems leak to a small degree, in the ultimate by diffusion of the external atmosphere through the furnace chamber walls. The pressure measured on a vacuum gauge should therefore be that of the chamber, and not of the pipeline leading to the vacuum pump. A pressure of 10^{-5} mmHg can mean a substantial draught blowing through the furnace if the chamber is very large, so that the level aimed at should depend on the capacity of the furnace. For most practical purposes, the limit of a mechanical rotary vacuum pump is about 10^{-2} mmHg and this is sufficient to exclude most of the air and therefore prevent undue oxidation. It may be insufficient, however, for ferrous or reactive metals at temperatures above about 800°C, and a diffusion pump must be added to the rotary pump. The pressure attained may then be reduced to 10^{-5} mmHg. As an extra precaution, the inlet and outlet ends of vacuum furnaces may be surrounded by a jacket containing inert gas, and this is also employed when inert atmospheres are used in the main chamber. Often two or more gas supplies are used, the entry point being well inside the inlet and outlet points so that the gas pressure prevents air leaking inwards and any leakage is of the gas itself to the outside. When using hydrogen, gas jets are fitted to doors and seals to burn any escaping gas and provide a blanket or curtain of burning gas to prevent air leaking in.

Air present in powder compacts before sintering, or adsorbed on the surface of any container or other object placed in the furnace, will be pumped away in a vacuum furnace, or swept away in an inert atmosphere. Time should be allowed during the initial heating period for this to occur, as gas adsorbed onto surfaces is difficult to remove. In a vacuum furnace, inert gas is usually provided at the end cooling period to speed up the cooling process. Such a combination materially increases the throughput of a batch-loaded vacuum furnace.

h. Infiltration. The porosity of sintered compacts varies, as described above, with the metal, the type of powder, the sintering conditions of temperature and time, and any subsequent treatment. One of the simplest techniques for reduction of porosity is that of infiltration, whereby a liquid phase is allowed to penetrate the interconnected pores either during sintering or in a further treatment. Economically, infiltration is best carried out during sintering. The material to be added can be placed above or below the compact and, on reaching the appropriate temperature, melts and is absorbed by surface-tension effects into the porous compact. Infiltration is assisted by sintering in vacuum so that the pores are not full of gas and there is no tendency to blow out the infiltrated material once it begins to penetrate.

The material used must be capable of melting at, or just above, the appropriate sintering temperature to give the amount of interconnected porosity desired. It must also wet the skeleton of sintered material, i.e. there must be no high interfacial energy between the infiltration medium and the compact material. Although the materials may alloy together, it is possible to arrange that this alloying is kept to a minimum by reducing the time or temperature of the process. Materials can be chosen which wet each other but are not mutually very soluble, so that little alloying occurs. A perfect example of this is the use of copper to infiltrate iron. The difference in melting points is about 400°C, and the copper can be alloyed to reduce its melting point further and thereby to lower the infiltration temperature. Copper is only slightly soluble in iron, and diffusion into the solid iron is therefore slow. If the temperature is increased, the solubility increases, so that a balance must be preserved between solubility and rapid filling of the pores.

Infiltration is used in the manufacture of tungsten and molybdenum compacts for electrical contacts, the copper being valuable as an infiltration medium in view of its high electrical conductivity. Copper based alloys are infiltrated with lead, or some similar anti-friction metal or alloy, when made into bearings. The use of infiltration is not limited to metals. Oil is used for sintered porous bronze bearings, or anti-friction polymers such as PTFE – both acting as lubricants in this application.

i. Coining. After sintering, the strength of the compact is high and normally little different from that of the same material produced by other methods. Occasionally the sintered powder product is superior in mechanical properties, especially when the metal is high in melting point and difficult to produce by other means without leaving voids or impurities which reduce the strength.

It is obvious that the pores which may remain in a powder compact influence the mechanical strength. By deforming the finished compact in a closed die of the same shape, the pores may be closed up by virtually extruding metal into them. At the same time, the dimensional accuracy of the finished component may be improved, although this is rarely necessary or vitally important. The operation, known as coining because of the similarity with the stamping of coinage from sheet metal, is obviously limited in scope to simple shapes and comparatively small sizes. Nevertheless, the degree of hardening and reduction of porosity obtained is significant. Goetzel (reference 25) quotes a sintered iron compact, prepared from electrolytic iron powder, compacted at 100 000 p.s.i. and sintered at 1 200°C for 1 hr, in which the porosity was 6.7% and the tensile strength 30 000 p.s.i., with 15% elongation. After coining at 100 000 p.s.i., the same pressure as the original compaction, the porosity was reduced to 4.5% and the tensile strength raised to 50 000 p.s.i. at 5% elongation. By a final annealing treatment at 1 200°C for 1 hr, the porosity again fell, to 2.3%, the tensile strength fell to 32 000 p.s.i., and elongation rose to 23%. Such an operation obviously improves the properties, largely by densification, and there is no limit to the working processes which may be applied. Rolling, extrusion and drawing have been carried out on compacts, with similar changes to the coining operation described. Nevertheless, coining or any similar process represents another step and further costs, so that it is not likely to be used except in special circumstances.

Closely related to the coining or working operation is the fabrication of tungsten compacts, particularly for lamp filaments. Tungsten exhibits low ductility and cannot be worked into wire form by heavy deformation, without fracture. The technique used is that of hot swaging, a combination of rolling and forging, with intermediate annealing to restore the small degree of ductility available for the next operation. Automatically, the process not only forms the tungsten into the correct shape but increases the density, as with coining.

In the manufacture of magnetic and electrical alloys, the use of powder methods enables control of the grain size of the material, but the orientation of the grains is often vitally important, as with the silicon-iron alloys used for high permeability in transformer laminations. The vast bulk of such material is required in sheet form, which can best be manufactured by rolling. For special applications, however, powder compacts can be used, and the coining operation may be necessary, with annealing treatments, to control crystal orientation in addition to density.

If any portion of a powder compact component is not required to be in the annealed condition, which, virtually, it is after sintering, a coining or similar operation will be necessary to convey the necessary work hardening.

10.2.5 Applications of powder metallurgy products

The commercial use of powder metallurgy methods may be divided into three main categories:
1. Where other methods are not feasible.
2. To improve specific properties over those achieved by more conventional methods.
3. To lower costs by mass production of simple components or special shapes.

Typical examples in the first category are the production of tungsten wire and the high melting point refractory metals. In the second, the deliberate introduction of porosity, as in bearings and filters, is perhaps the best example, and in the third, the more recent use of iron and steel products in special shapes which would require difficult machining operations from the solid bulk metal. In each category, many of the applications are common to all materials within that class, as with the refractory metals of the first category. There are also examples, however, of specific problems associated with individual metals or combinations and in this case more than one category may be involved, as with tungsten wire. So far as possible, these special cases will be referred to under the main category involved.

a. Feasibility confined to powder metallurgy. The conventional manufacture of metal products by melting and casting in bulk form, with or without subsequent fabrication, accounts for the vast tonnages of ferrous metals, copper, aluminium and alloys based on these and other low melting point non-ferrous metals. Even in these cases, technological development is comparatively recent, e.g. steel could not be melted in quantities above about 120 lb until the mid-nineteenth century and bulk aluminium production had to await the electrical technology associated with large generators. Many commercially useful metals and alloys remained as museum pieces or laboratory samples because of their reactivity with atmospheric gases when melted in air, or the difficulty of attaining a sufficiently high temperature to achieve melting, and the problem of containment of the liquid metal. During the past 25 to 30 years, enormous strides have been made, and many elements in the periodic table have been produced in bulk form for the first time, and fabricated into useful components. The range of useful engineering materials has been greatly extended as a result, and the only real barrier to increased production is that of cost, in view of the very stringent precautions which are often necessary.

Fig. 10.9a Powder metallurgy parts manufactured by S.M.C. (Sterling) Ltd.

Fig. 10.9b Another powder metallurgy part manufactured by S.M.C. (Sterling) Ltd.

To melt a metal successfully the principal requirements are:
1. A high temperature (somewhat above the melting point to ensure a fast reaction).
2. Containment of the liquid metal in a material of higher melting point than the metal and with sufficient strength not to break or distort excessively.
3. Prevention of reactions between the container and the liquid metal, and the metal and atmospheric gases.

The use of fossil fuels, whether solid, liquid or gas, does not allow bulk temperatures much above 1 750°C, and to achieve this regeneration by preheating the air required for combustion by means of the outgoing waste gas is absolutely essential. If pure oxygen is substituted for air, in whole or in part, then this temperature can be exceeded. The use of oxygen, however, increases the velocity of all oxidation reactions, as exemplified by the tonnage oxygen steel-making processes in which the time required has been reduced by an order of magnitude. The metals which could be melted by fossil fuel furnaces, therefore, are limited to melting points below 1 750°C, always assuming that conditions 2 and 3 above can be satisfied. There remains the possibility of electrical heating; even here, resistance heating involves problems with the resistor itself and only induction heating at radio frequencies, and the electric arc, are universally applicable to temperatures above 1 750°C.

The containment of the metal is of even greater importance. Refractory oxides with melting points higher than those of these special metals are very few in number and very expensive to produce. Graphite is particularly useful, and has the merit of being electromagnetically acceptable in an induction circuit if that type of heating is employed. The problem of containment has also been solved by use of the consumable-electrode electric arc furnace. In this, the upper electrode is prepared from compressed and sintered blocks of powder, screwed or welded together. The lower electrode is a water-cooled copper hearth (or mould) into which the molten metal falls when the arc is struck, solidifying almost immediately where it contacts the cooled hearth. The hearth comprises a movable platform which is lowered continuously, producing an ingot of considerable length of the same cross-section as the mould, but in which the metal being melted acts as its own containment – an ideal solution. The size of the ingot produced is limited by the mechanical difficulty of moving the water-cooled hearth within an enclosed vacuum chamber, and by the large height of the consumable electrode protruding (within the chamber) above the hearth. Nevertheless, the product is more homogeneous, and titanium and many other metals and alloys have been successfully produced in bulk by this means.

Reactions between the container and liquid metal are usually less severe than those between atmosphere and liquid metal but, in both cases, contamination of the metal results, often with disastrous effects on the desirable properties. It is unfortunate that most of the high melting point metals and alloys inevitably react with many refractory container materials because they react with oxygen, one of the principal ingredients of refractories. These metals also react with nitrogen and hydrogen and any oxygen-bearing compounds such as water, carbon dioxide and carbon monoxide. If the container problem can be satisfied, the problem of atmosphere remains, and either vacuum or inert gas atmospheres become essential. This, in its turn, creates problems of tracking or glow discharge in arc or induction furnaces and, whilst solutions have been found, they prove expensive in both the development and operational stages.

The powder metallurgy production of these materials avoids many of the above problems by a batch-type production on a smaller scale, with relatively cheap units so that replication of the batch becomes possible and economic. The pressing of powder can be made rapid and one press can serve a number of sintering furnaces. Furnaces can be operated *in vacuo* or with circulating and regenerated inert gas, the small volume involved in each furnace simplifying the operation. Heating can be by radio-frequency induction or by electrical resistance, utilizing the metal powder compact as the resistor by passing current directly through it. Means must be provided to allow for the shrinkage during sintering and yet preserve the electrical contact. This has been done by spring-loaded, water-cooled contacts, or by utilizing a pool of mercury as one contact, with an attached metal strap attached to the specimen at one end and dipping into the mercury. In view of the high vapour pressure of mercury, inert atmospheres must be used, not vacuum, and the mercury vapour inevitably carried over in the gas stream must be condensed out. Fig. 10.10 illustrates the principle of direct-resistance sintering using spring-loaded contacts.

Where components are pressed directly to shape rather than produced as bulk ingots, the direct-resistance heating method is not feasible and indirect resistors or induction heating remains. The majority of high melting point metals and alloys are produced in bulk form for subsequent fabrication to standard shapes, as rod, sheet and tube, rather than directly to shape.

Mention has already been made of the production of tungsten lamp filaments from sintered tungsten powder billets. Tungsten is the metal of highest melting point in the periodic table and is becoming increasingly important as a high-strength metal for heat-resistant applications where the atmosphere can be properly controlled. Since tungsten is somewhat brittle at room temperature, metals of slightly lower melting point, but which nevertheless retain adequate strength at high temperatures, can be substituted. If resistance to chemical attack (other than atmospheric) is also involved, then tantalum and niobium have great potential. Molybdenum is much cheaper than any of the other three but is less heat-resistant and very reactive with oxygen.

All these materials find increasing use in furnace parts, aero-engine applications, chemical plant and electrical contacts. Development of their alloys, for even higher strength or other specific properties, is continuously widening the applications. If a

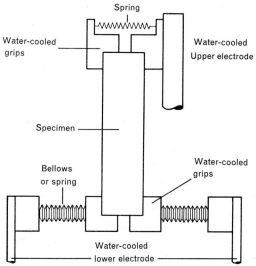

Fig. 10.10 Sintering by direct resistance heating.

skin of oxidation-resistant material can be successfully produced as a coating on these metals, the use in jet propulsion turbines and other heat-resistant applications must advance rapidly.

All can be fabricated into rod, wire or sheet form with little difficulty, apart from the cost of atmosphere control in annealing and working processes. All except tungsten are readily produced in tube form, and all four metals can be welded (with the same stringent precautions) in inert atmospheres. In a similar way, but at slightly lower cost because of the lower melting point, the metals titanium, zirconium and vanadium can be readily produced by powder methods, though here the consumable-electrode electric arc melting process is a very keen competitor. These metals (particularly titanium) find increasing use in aircraft, chemical plant, and the nuclear energy field.

In the nuclear field of fissile materials, uranium, thorium and plutonium have all been successfully manufactured by powder methods. In view of the increasing tendency to move from metallic nuclear fuel to oxides, carbides or similar compounds, the powder metallurgy applications are of less significance, but the powder method is used for the production of the compounds exactly as if a metal were involved. Since the commercial availability of these materials is very strictly controlled, it is unlikely that use of them will reach the normal avenues of industry.

Beryllium is one metal with a similar history to the high melting point metals referred to previously. It is highly reactive, particularly with oxygen, but its melting point is not unduly high (about 1200°C). Nevertheless, the toxic properties of the metal and some of its compounds require strict control of atmosphere and handling. Powder metallurgy methods (being on a smaller scale) can be easily operated in glove boxes or protective cabinets and hence have been extensively used. However, the melting of beryllium is comparatively simple (with appropriate atmosphere control) and the powder method is more likely to be used as a preparatory step to melting, in which small compacts are fed into the electric arc or some similar melting furnace, as practised for titanium alloys. Such methods combine the simplicity of powder metallurgy with the economic advantage of bulk melting procedure.

A number of other metals, not previously regarded as commercially useful, may become more so as a result of research and development based on their specific properties. Chromium has potential in this group, and rhenium, but both suffer from brittle behaviour at room temperature. The powder method may have advantages in these cases if the strict control of purity can best be attained by manufacturing powder, though melting may be incorporated in the final stage, as with beryllium and titanium. Manganese, not dissimilar to chromium in many ways, and the rare-earth metals, may follow in due course as powder products.

b. The attainment of specific properties. Mention has already been made of the manufacture of tungsten wire by swaging the sintered bar. Similar arguments apply to many metals in which the crystal lattice or the difficulty of removing specific impurities renders them brittle at room temperature, and which show little increase in ductility at elevated temperatures. Almost all the metals with a hexagonal crystal lattice are in this category but only beryllium, titanium and zirconium are sufficiently brittle or have sufficiently high melting points to warrant consideration of powder methods. The body-centred cubic lattice metals include many which are comparatively brittle, e.g. tungsten and molybdenum, but others which are relatively ductile, e.g. niobium and iron. Nevertheless, the advantage of the powder method remains, in that the initial product may be simpler and cheaper to produce at the required standard of purity in powder form. It may then be cheaper and more effective to compact and sinter, retaining the high standard of purity, than to melt. From this point onwards, the modification of crystal structure, to homogenize and produce the required orientation in the crystal grains, may be more easily controlled in the sintered compact, with its inherent porosity, than in the cast form with possible segregation and other defect structures. This type of advantage has been reported for molybdenum and platinum, where the powder product is more ductile in wire and sheet form (see ref. 26).

One very important aspect of the powder method is the ability to incorporate additions of material, often in small quantity, to modify properties or to produce a composite rather than a pure material. To distribute a second phase or constituent in a pure metal by casting from the liquid state is not a simple matter. Inevitably, differences in density of the constituents, chemical effects and the convection currents set up in a liquid metal as it solidifies, do not assist homogeneous distribution. The powder method, being entirely in the solid state unless a liquid phase is deliberately introduced, can mix the constituents very uniformly, in the particle size appropriate to the application, and then fix the distribution by sintering. Pressing the compact may have some disturbing influence but much less so than casting. These principles have been incorporated in the manufacture of conventional materials such as silicon-iron for electrical transformers, where additions of manganese sulphide are claimed to affect secondary recrystallization (see ref. 27).

It is in the rapidly developing field of composite materials, however, that the greatest potential must lie. These materials usually incorporate an addition to modify the most disadvantageous property of the bulk material, e.g. the addition of ductile metal artefacts to brittle ceramic bodies, or the addition of ceramics with high-temperature strength to metals which lose strength rapidly at temperatures above half their melting point on the absolute scale. Where the addition is made in a shape comparable to the bulk material, simple mixing is all that is required. When special shapes are involved, e.g. a high length:diameter ratio as in fine wires, this is rarely possible, though extrusion of a body in which powder particles are temporarily held together by an adhesive may allow introduction of fine wires in a manner similar to the production of lettered candy rock in the confectionery trade. Control of the orientation of special shapes is often vital, and the development in this field will certainly include a full review of the potential advantages of powder techniques.

The dispersion of very fine intermetallic compounds in a metallic matrix can produce extremely high strength for engineering constructional purposes. In the main, the most favourable methods so far have relied on a metallic alloy system in which supersaturation of a single phase at room temperature (obtained by quenching after solution treatment at high temperature) allowed an approach to equilibrium by subsequent heat treatment. The use of powder methods may extend this principle by introduction of a variety of compounds which do not dissolve in adequate quantity at high temperature in the matrix, but which nevertheless can be controlled in some form to produce the same result as the coherent precipitate stage of a normal precipitation hardening alloy. This is not a simple matter and may involve several phases to ensure the fine structure required. Such developments are more closely akin to 'cermets', in which the phases are distinct and separate (see ch. Ceramics, vol. II).

The improvement of physical properties other than mechanical strength is more highly developed with the powder method. Magnetic materials affected by crystal orientation have already been referred to with the soft magnetic materials such as silicon-iron. The control of purity, grain size and orientation is much simpler with powder metallurgy than with conventional casting and rolling methods. Pure powdered metals and alloys, free from inclusions, are readily attainable, and control of subsequent fabrication is simpler with compacted and sintered material (see also reference 28). Magnetic cores, for use in alternating current applications, suffer energy losses by reason of eddy currents generated within them. Any increase in the specific resistance of the material can reduce the eddy current loss, and this is the objective of laminating the core material by employing thin sheets pressed together. By using a porous powder compact and incorporating an insulator to fill the pores, the specific resistance can be increased far more than is possible by laminations. The insulator can be mixed with the metal powder or introduced by infiltration methods. The presence of the insulator reduces the magnetic permeability of the bulk material but this is more than offset by the reduction in eddy current loss. This is not a new idea, having been first described by Heaviside in 1887. For a review of the practice in Europe and the USA see reference 29 and publications of the International Nickel Company.

The most recent developments in soft magnetic materials with low eddy current losses are concerned with oxides such as ferrite, the magnetic oxide of iron, and related oxide combinations. These have a higher specific resistance than metals or alloys but are strictly ceramic materials, although manufacture is by powder methods.

'Hard' magnetic materials are those employed as permanent magnets, which require more energy to magnetize them but retain their magnetism with a high field strength over long periods. Most are based on alloys of iron, nickel and cobalt, and incorporate similar principles of dispersion of intermetallic phases as alloys developed for mechanical strength. It follows, therefore, that the application of powder methods for mechanical strength alloys applies also to the hard magnetic materials.

The shape of particles for magnetic applications can affect the properties more than is the case for other physical parameters. Electrolytically deposited iron powder can be made with elongated particle shape in the form of dendrites. If the iron is deposited on a mercury cathode and the temperature of the mercury subsequently raised, the dendrite shape alters, by dissolution of the iron and reprecipitation on the main axis, leaving a rod-shaped particle of excellent magnetic properties. The magnetic properties of these materials are improved by correct orientation of the particles and this can be achieved by pressing in an applied magnetic field. Sintering of the compact would cause some recrystallization and change in orientation, so these magnets are bonded together with non-magnetic material, the volume of the binder serving to keep the particles apart and therefore preventing them from demagnetizing each other. The unsintered powder magnet is now well established and offers scope for further development (see refs. 30 to 32).

The utilization of the porosity in sintered powder compacts to manufacture components involved in gas or liquid flow is one of the earliest applications. Materials involved in separation, i.e. as filters, are generally required to be corrosion-resistant, since chemicals of high reactivity are involved. Alloys based on nickel and copper are most common, with stainless steel as the sole ferrous material. To these may be added pure silver for certain specific requirements. The materials, after mixing, are either pressed to a low density and then sintered, or premixed with a volatile substance, pressed and sintered or loose sintered after vibration to give uniform packing. Among additions which volatilize on sintering and leave behind porosity are ammonium salts and metallo-organic salts such as oxalates, acetates and tartrates. The importance of porosity is here related to interconnected porosity, since there must be permeability of gas or liquid through the filter. The effective porosity may be between 25 and 40%, but the pressing and sintering operations must be carefully controlled, and the cost of production is high. New developments include the mixing of various shapes of particles to control the shape and relative volume of pores, and better use of volatile additions on which some permanent constituent has been plated or coated. Powder rolling, referred to later, has some advantages in bulk production of very fine filters. The use of powder metallurgy to produce porous bodies for effusion cooling of gas-turbine blades has not been as successful as hoped for metallic materials but may improve with the development of ceramic blades with porosity. The manufacture of stainless steel filters is covered by reference33.

Porous bronze bearings, in which 25 to 35% interconnected porosity is customary, are well established for applications where oil feed is undesirable or difficult, although continuous oil feeds are also applied where appropriate. In electric clocks and instruments, the original oil infiltration, about 90% of the porosity available, is sufficient for about 25×10^6 revolutions of the shaft in the bearing, equivalent to several years' continued use. Such bearings are required to preserve close dimensional tolerances and good surface finish on the bearing surface. The materials in common use are bronzes containing about 10% tin, iron-copper mixtures of about 5% copper, iron containing from 0.5 to 1% graphite, and aluminium-tin mixtures. The bronze bearings with 10% tin require as little as 5 min sintering at 800°C to produce the appropriate porosity when simple mixing of the two powders is employed. The process of diffusion at the sintering temperature produces some liquid phase, which rapidly completes the sintering operation and assists the porosity by virtue of the shrinkage resulting from the solidification of any remaining liquid phase. The infiltration with oil is assisted by degassing in vacuum, using hot oil (below 100°C) or gas pressure above the oil. Too high a temperature may cause the oil to oxidize or break down, which could clog the pores in operation.

Bearings have also been produced with other media of infiltration, notably graphite and various polymers. Of the latter, polytetrafluorethylene (PTFE), is the most successful, having a very low coefficient of friction. Polymers can be infiltrated by direct pressure by use of a volatile solvent or by liquefaction before impregnation.

Other uses of powder metallurgy which take advantage of the simplicity of the method are bound to arise, but many will come within the third category of application initially, i.e. the reduction in cost or improvement in the product as compared with conventional methods. After that step, the realization of properties more desirable than those originally acceptable may stimulate the powder method to increased development.

c. Applications involving competitive cost. The advantages of the powder method based on a competitive price must depend largely on complexity of the final product or the number of operations required by conventional methods. If complex machining enters into the list as a final operation, especially on small parts, then the powder method may hold great advantages. Operations involving envelopes of sheet metal can never be competitive by the powder method, and cupping and drawing processes can be readily automated. Small components involving more bulk metal may be highly competitive in powder if dimensional tolerances can be met without too much machining. Against this, the rate of die wear and the tool-making costs may seriously affect the advantages of powder, and little can be said without individual costing exercises on specific components.

Examples of the powder method reducing the cost of production (and occasionally improving the desirable properties) are electrical contacts, brush contacts and similar electrical connectors such as railway pantographs. Copper-nickel electrical contacts can be made by mixing the powders, pressing and sintering under conditions where little diffusion, and therefore alloying, actually takes place. This avoids the increase in electrical resistance which would result from use of the alloy in cast form, but retains sufficient strength. The use of tungsten in contacts can be greatly simplified by powder metallurgy techniques and precious metals such as the platinum group can be used with less waste. Brush materials, which can be regarded as rubbing electrical contacts, require to combine some of the properties of bearings with those of electrical contacts. Silver-graphite mixtures and tin bronzes incorporating graphite are examples. Copper-carbon combinations are used for dynamo brushes and have the advantage that the proportions of each can readily be adjusted to suit particular applications.

Simple mass production, whereby one press can deal with a multiple die, can be utilized to show how competitive the pressing of powder may prove in comparison with stamping from sheet or billet. Components such as washers may prove to be highly competitive, even when the sintering and possible infiltration processes are taken into account.

The production of strip, especially in special metals or alloys, can become competitive with the introduction of powder rolling followed by continuous sintering. Powder rolling consists simply of feeding powder from a hopper into the gap between a pair of rolls set to produce a given degree of compaction by varying the roll pressure. The compacted strip emerges onto a deflector plate and is guided onto a conveyor belt leading to the sintering furnace. Fig. 10.11 illustrates the principle; references 34 and 35 deal with specific aspects. Powder rolling has proved advantageous for porous strip in filter production and in material for gas flow control.

Continuous compaction of powder in block form of geometrical sections has been used, followed by conventional sintering. Distribution of powder on a conveyor belt without pressure, a continuous method of loose sintering, has long been practised in the backing of steel strip with leaded bronze-bearing material. Extrusion of powder containing volatile binder, followed by removal of binder before sintering, is also an old process, which is being used in tungsten wire production. Modern developments of this process utilize a container, with the powder (vibrated to pack well) inside, and

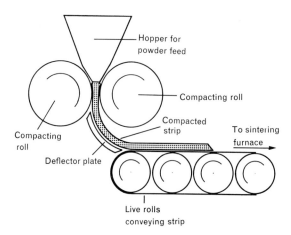

Fig. 10.11 Powder rolling.

direct extrusion without binders. All such processes are bound to be carefully examined in the mass production field as costs of conventional methods increase and automation becomes applied to those processes which can more readily accept control methods. Perhaps the most recent application of powder is to the welding process, whereby the stream of metallic ions in the normal metallic arc process is replaced by a stream of powder injected by gas pressure and sintered *in situ* by generation of heat by any appropriate method. This has the advantage of producing less effect on the parent metal because of lower temperatures and more control over the nature of the material (the powder) forming the weld metal.

10.2.6 Statistics

Production of powder metals in the USA is as follows:

Iron Powder	*1000s of tons**								
	1959	*1960*	*1961*	*1962*	*1963*	*1964*	*1965*	*1966*	*1970 estimated total*
Domestic production	33	30	34	50	57	72	90	100	250
Imported	1	2.5	1.5	1.5	1.2				
Consumption of total				%					
Powder metallurgy components	54.5	52.25	51	58	64	62	65	71	
Welding electrodes	26	26.25	28.2	26.5	23	25	23		
Flame cutting	10	11.5	10.8	7	7	7	7		
Electrical and magnetic	4.5	4.5	5	4	3	2	2		
Miscellaneous	5	5.5	5	4.5	3	4	3		

Copper Powder	*1000s of tons**							
	1960	*1961*	*1962*	*1963*	*1964*	*1965*	*1966*	*1970 estimated total*
Granular copper production	10.6	11	13.3	14.2	16.7	17.6	19.5	40
Flake copper	0.7	0.8	0.95	0.76	0.56	0.94	0.7	
Granular copper alloy powder	5.7	6.5	7.2	8.4	9.05	10.9	10.8	
Flake alloy powder	2.3	2.3	2.4	2.02	1.98	1.9	2.3	

Aluminium Powder	*1000s of tons**			
	1964	*1965*	*1966 estimated total*	*1970 estimated total*
Atomized powder production			11.6 16.2 20	?
Total powder production			23.2 29.0 35	40

*Sources: *Progress in Powder Metallurgy; International Journal of Powder Metallurgy.*

10.3 HARD METALS

10.3.1 Introduction

The cutting of metals to complete a shaping process has been practised since ancient times. The principal requirement for a suitable tool is that it should be harder than the metal being cut, but since the temperature rise at the point of contact can be extremely high, a secondary requirement is the retention of hardness to as high a temperature as possible. Hardened steel, relying on high carbon content, formed the basic cutting tool material for several centuries, improving in heat treatment and reliability during the nineteenth century. F. W. Taylor, the father of scientific management, initiated a development project at the U.S. Steel Corporation in the eighteen-eighties to find a harder and more heat-resistant cutting steel. The result was high-speed steel, so-called because the cutting speed of machine tools could be increased as a result of its use. The tool bit can actually operate at dull red heat with little loss in efficiency or economy. The chief alloying constituent in such steels is tungsten. It was many years after the original use of high-speed steel that the principal agent contributing to the properties was shown to be tungsten carbide.

At the turn of the century, a second alloy type appeared, again incorporating tungsten and other transition metals of high melting point but with only a small proportion of iron, called Stellite. Once again, the presence of intermetallic compounds was responsible for the high hardness at temperature. A natural development then followed: the preparation of pure carbides rather than their incorporation as constituents in a more conventional alloy system. Initially, attempts were made to cast tungsten carbide, e.g. for drawing dies, but the product was inherently brittle. The use of powder metallurgy techniques improved the properties a little but the infiltration or initial admixture of a binder metal finally removed the main problem, that of brittleness. The main constituent of these hard metals remained tungsten carbide, and the cementing or binder metal was one of the iron group, iron, cobalt, nickel, since the melting point of these metals allowed for liquid-phase sintering to bind together the carbide particles, which remained solid during the sintering process. The limitations of the early, simple cemented carbides led to the extensive study of other carbides, and combinations of more than one transition metal carbide have been common for many years.

Tungsten carbide (WC) has a hexagonal crystal lattice arrangement of the tungsten atoms and the inherent brittleness of the carbide is, in part at least, due to this type of lattice. Furthermore, tungsten carbide will oxidize in air at a high enough temperature by the formation of tungsten oxides, which are porous and occupy a greater volume than the metal atoms from which they are derived. This is the principal limitation of tungsten carbide, which alone is perfectly adequate for machining metal which undergoes little plastic deformation when machined, e.g. cast iron, but less so when machining ductile steels, which tend to form long chips at the tool point with a temperature rise as a result.

Titanium carbide (TiC) forms a lattice of cubic type (similar to sodium chloride, NaCl) and when oxidized in air forms a more adherent film of oxide (TiO). For these two reasons, titanium carbide is more suited for those applications involving slightly more oxidizing conditions, or where reduced brittleness is an advantage.

Tantalum carbide is similar to titanium carbide except in oxidation. Molybdenum carbide resembles tungsten carbide more closely. All four of these metals have been extensively used in the production of hard metals, and many combinations of each have been tried.

The cementing metal plays an important part in the properties and production of these hard metals. The discovery, in Germany, that metals of the iron group would slowly dissolve the high melting point metal carbides with a drop in melting point of the resultant alloy formed, provided the stimulus which led to future development. Of the iron group, cobalt has the highest melting point, 1 950°C, and begins to melt when it has dissolved 1.5% tungsten carbide (WC) at 1350° to 1400°C. When cobalt has dissolved about 20% WC, it is completely molten, but on cooling the solubility drops at first to 4% WC and then at room temperature to less than 1% WC, and the excess WC precipitates out as a fine dispersion in the cobalt, or deposits on the adjacent WC particles. An ideal situation then exists: a reasonably high melting point metal–cobalt containing about 1% WC–cements the hard WC particles with a fine dispersion of WC crystals around them merging into a dispersion of WC in the cobalt. Iron dissolves WC readily, but retains much more WC in solid solution on cooling. Nickel also does not reject WC from solution so well as cobalt, but does reject the carbides of titanium, tantalum and molybdenum. When two carbides are mixed, e.g. WC and TiC, cobalt dissolves both and precipitates a solid solution of WC–TiC on cooling after solidification. This solid solution cements the skeleton of WC and TiC particles to give great mechanical strength, and the TiC conveys all the advantages of oxidation resistance to the whole mass. However, TiC is difficult to prepare without the presence of some titanium dioxide (TiO_2), from which it is generally derived, and TiO_2 upsets the dissolution of TiC by cobalt. Reaction between TiO_2 and the carbon atoms present gives rise to some carbon monoxide evolution and the compacts are often porous after sintering as a result.

Numerous proprietary hard metals are available, but all contain one or more of the carbides referred to and one or more of the cementing metals, in varying proportions. Predominantly, however, tungsten carbide and cobalt (Co) are the mainstay of the cemented carbides. To assist in recognition, the trade names originally applied to the various combinations are listed:

WC + Co: Widia; Carboloy; Wimet; Ardoloy
TiC + Ni: Titanit
MoC + Ni: Cutanit
TaC + WC + Ni: Ramet
WC + TiC + TaC + Co: Widia; Carboloy; Wimet; Ardoloy

10.3.2 Production of powder
The manufacture of cobalt and nickel is dealt with in the section on metal powders.

a. Tungsten carbide. Tungsten carbide powder is prepared by first reducing tungsten oxide (WO_3) in pure hydrogen, as described under metal powders. The metal powder is ball milled to break up agglomerates and to reduce the particles to 10 μm or less. The metal powder is then mixed with lamp-black (an amorphous form of carbon) of fine particle size and heated in a furnace either in vacuum or in an atmosphere of hydrogen. Reaction commences slowly at about 850°C, but is rapid at 1 400°C. In practice, even higher temperatures may be used, around 1 600°C.

Hydrogen atmospheres undoubtedly give different rates of reaction from those in a vacuum. Decarburization would occur in hydrogen, if the proportion of carbon present were exactly correct to form tungsten carbide (WC), and in practice an excess of carbon is added as an outer layer around the mixture. The formation of hydro carbons is known to take place, and some propane (C_5H_{12}) is often added to assist the reaction. The presence of halides, notably hydrogen iodide, (HI), has been shown to accelerate carburization, probably by some activation process similar to catalysis. A second carbide of tungsten (W_2C) is undoubtedly formed during the process, and the rate-controlling step is probably surface diffusion of carbon. So long as an excess of carbon is present, therefore, temperature is the only variable which will influence the rate of reaction.

Attempts were made initially to avoid the intermediate production of tungsten metal by reacting tungstic trioxide with carbon directly. The difficulty seems to be control of the final carbon content, since carbon must first reduce WO_3 to W and then combine to form WC. Later methods employed a mixture of hydrogen and methane to reduce WO_3 at 850 to 1000°C. A cheap source of methane (as natural gas) may well revive interest in this method, unless a bulk source of lamp-black as a by-product of coal and petroleum industries maintains the economics of the current methods (see ref. 36).

The final product of WC is ball milled for several hours, or days, before storage. The particle size depends on the size of the original WO_3 and the intermediate tungsten metal. The WC should contain 6% combined carbon, and invariably is accompanied by a small free carbon content of about 0.1 to 0.5%. Tungsten carbide corresponds to 6.12% carbon.

b. Titanium and tantalum carbides. Preparation of these carbides is generally limited to reduction of the oxides with carbon. Commerically pure titanium dioxide ('rutile', TiO_2) is readily available and comparatively cheap, but it is contaminated with impurities such as silica. The oxide is mixed with carbon in the form of lamp-black, often with addition of water, and the wet mixture dried and precarburized by heating to 1700°C in an atmosphere of hydrogen, usually by high-frequency heating. The hydrogen would decarburize the titanium carbide formed, as with tungsten carbide, and a layer of excess carbon is placed on top of the mixture to minimize this. The cool mixture is then ball milled for about 30 min and analyzed. Depending on the carbon content, a second carburization is then carried out by adding either titanium dioxide (TiO_2) (if there is excess carbon) or carbon black (if there is carbon deficiency) and heating to 1800 to 1900°C for 2 h. The final composition aimed at is 17 to 19% total carbon with not more than 1 to 2% free carbon. The final product is washed in hydrochloric acid, and then dried.

Titanium carbide (TiC) can be produced *in vacuo*, and carburization with carbon monoxide gas has been reported. The principal problem with (TiC) is the non-stoichiometric nature of the compound, i.e. there is a range of composition with the general properties of TiC, containing from 14 to 20% by weight of carbon. Lower carbon than the stoichiometric value of 20% by weight implies that some carbon atoms are missing from their crystal lattice positions, leaving vacant lattice sites, a 'defect' structure. Since the compounds titanium oxide (TiO) and titanium nitride (TiN) are isomorphous with TiC, oxygen or nitrogen atoms can occupy these vacant lattice sites with no distortion of the TiC lattice. Since TiC is frequently prepared from titanium dioxide, oxygen is already present and nitrogen may be introduced as impurity in the lamp-black or in the atmosphere. Excess carbon in the original mixture is claimed to assist the removal of nitrogen and oxygen, and a vacuum rather than a hydrogen atmosphere is said to remove reaction products more quickly and so assist oxygen and nitrogen removal. Undoubtedly, no commercial product is likely to have the stoichiometric composition TiC with no oxygen or nitrogen. Attempts to produce a purer product are always more expensive, and perhaps unjustified. Molten aluminium and nickel help to reduce the impurity content, and since neither of these metals forms a stable carbide they do not affect the TiC itself. Their use inevitably adds to the cost of the process.

Tantalum carbide is prepared in the same way as TiC, by reacting the oxide with carbon at 1 500 to 1 600°C.

10.3.3 Compacting and sintering

Tungsten carbide and cobalt powders are ball milled together for up to 60 h in appropriate quantities (about 6 to 15% cobalt) in an endeavour to smear each tungsten carbide particle with a film of cobalt metal. Liquids are often added to the ball mill, even water assisting the action by acting as a lubricant. Benzol, 1% glycerine or a solution of paraffin wax in carbon tetrachloride have all been reported in commercial use as lubricants. The paraffin wax remains to assist in the pressing operation, as does glycerine. Where water or organic liquids are employed in ball milling, they may be dried out by heating the mixture in dry hydrogen at 650°C, largely to reduce any oxide films which may have formed during milling.

Cold or hot pressing is employed for tungsten carbide-cobalt (WC-Co) mixtures. Cold pressing utilizes pressures from 5 to 30 tons per square inch, with hard metal or hardened steel dies in view of the hard constituent of the compact. Die lubrication is essential, and most frequently simple shapes are adopted for subsequent manufacture of tools to simplify the pressing operation and ensure uniform distribution of the powders in the compact. Presintering is carried out on cold-pressed compacts for 30 min at 800 to 1 000°C, depending on the cobalt concentration. This serves only to sinter the cobalt powder and increase the strength and handling qualities. The final sintering operation, at 1 350 to 1 500°C, takes 1 to 2 h, during which the tungsten carbide begins to dissolve in the cobalt forming a solid solution and the compact shrinks from 15 to 25% by volume. Heating is carried out in hydrogen or *in vacuo*, the usual precautions with hydrogen being taken, i.e. covering with a carbon blanket. Carbon tube, molybdenum resistor or radio-frequency heating are most common. Slow cooling follows, to prevent cracking as the tungsten carbide dissolved in the cobalt precipitates out.

Hot pressing uses graphite moulds with either induction heating where the mould acts as a susceptor, or heating by direct current through the mould where the mould acts as the resistor. Pressures from 400 to 2 500 p.s.i. are sufficient, the graphite mould preventing oxidation at the temperature of 1 400°C. No further sintering is necessary but precautions must be taken to prevent squeezing out the binder of cobalt.

Any premachining of the compact to finished size can be carried out after the presintering with cold pressing but not with hot pressing, hence the latter method may be more economic only with simple shapes.

Extrusion of a paste in which WC-Co is held together by organic binders has been successful, and is the only possible method for components of considerable length, which could not be pressed uniformly in the normal way. Up to 24 in lengths have been made for broaches and similar tools. Care is required during sintering to drive off the binder slowly and prevent bubbles from being formed.

Cold-pressed and sintered WC-Co material inevitably contains some porosity, and the sintering process certainly causes some grain growth. Both these factors contribute to brittle fracture in the material under appropriate conditions of stress, and it was this feature which led to the hot-pressing alternative. However, as referred to above, the economic advantage of pressing or extruding bulk material, followed by premachining before final sintering, may outweigh the improved performance of the hot-pressed material.

Titanium carbide (TiC), with nickel binder, is manufactured by similar methods to WO + Co. The TiC powder, usually of 0.5 to 1 μm particle size, is mixed with about 40% nickel powder and either cold or hot pressed. Few values of sintering temperature have been reported but the region of 1 300 to 1 400°C would seem feasible in view of the solubility of TiC in nickel (9.3% eutectic at 1 280°C). Higher temperatures would accelerate the process. Tungsten metal has been added to the nickel to give greater strength in the binder, and inevitably some exchange between tungsten and titanium for the available carbon content would result. However, mixed carbides of WC and TiC, with other possible additions such as tantalum, niobium and vanadium, are fairly common and possess superior properties.

Tantalum carbide alone has been reported, used with cobalt, or cobalt plus nickel binder, pressed at 80 tons per square inch and sintered at 1 350 to 1 400°C for 1 h.

The mixed carbides, particularly WC-TiC, cannot be used to advantage unless the solid solution of the two is achieved either before or after sintering. Pre-prepared solid solution is recommended, either by co-reduction of the oxides with carbon or by mixing the two and presintering (without binder) to cause alloying. The solid solution forms at about 1 600°C and control of carbon content necessitates adjustment by addition of pure WC or carbon. The pre-alloyed material must then be dry milled for 2 h in a mill lined with carbide. Claims to have isolated a definite double carbide compound ($W_2Ti_2C_4$) have never been substantiated, but the solid solution is undoubtedly very hard.

The pre-alloyed carbide is milled with cobalt binder and pressed in the same way as for WC-Co alone. Presintering temperatures are commonly 900 to 1 000°C – the higher the cobalt content, the lower the temperature. After machining to shape, the individual components are sintered finally at 1 475 to 1 525°C for 1 to 2 h. Vacuum is necessary in view of the gas absorption powers of TiC. Hot-pressing methods at from 750 to 1 500 p.s.i. have also been used in Germany. Similar methods can be adapted for all complex carbide compositions containing any combination of WC-Co with titanium, tantalum, niobium or vanadium carbide. Assessment of the advantages of simple carbide or complex in performance is difficult in view of extravagant claims made by adherents to specific types in this lucrative business. Undoubtedly, scientific evidence points to TiC being an advantage when dissolved in WC, but control of carbon content with any proportion of TiC is not easy and materially affects resultant porosity and shrinkage during sintering. The report on the German hard metal industry during world war II (ref. 37) should be read with caution in this respect. The report mentions, for example, 45% TiC plus 45% vanadium carbide, with either 10% or 7% nickel, 3% cobalt binder, being used as a substitute for WC-TiC-Co.

10.3.4. Properties

Measurement of the mechanical properties of hard carbides is not simple. Whilst hardness is relevant in cutting tools, the shock loading and mechanical deformation at high temperatures generally determines the successful performance in practice. The transverse rupture strength of tungsten carbide alone is about 80 000 p.s.i., with hardness over 95 on the Rockwell A scale or over 1750 Vickers diamond hardness. When bound with 20% cobalt, the hardness is still 85 or over on the Rockwell A scale and rupture strength is increased to about 400 000 p.s.i., although pure cobalt shows a hardness of only 120 to 250 Vickers (depending on impurities and method of manufacture) and rupture strength around 25 000 p.s.i. The explanation for the improvement in mechanical strength lies in the dissolution of a portion of the tungsten carbide particles in the cobalt, and subsequent precipitation of a fine dispersion coherent with tungsten carbide and cobalt. This is a classic case of the composite material, in which strength is increased by the fine dispersion, ductility by the ductile binder matrix (cobalt) and over-all properties are enhanced more than a mean calculated on the properties of both constituents. Variation in the cobalt content produces intermediate properties between the extremes of the pure constituents. 3% cobalt gives a rupture strength of about 150 000 p.s.i., and 10% cobalt about 250 000 p.s.i. Hardness correspondingly drops to 91 Rockwell A with 3% cobalt and 88 Rockwell A with 10% cobalt. The properties of cemented carbides can therefore be adjusted with reasonable precision to suit requirements.

The addition of titanium carbide (TiC) to tungsten carbide (WC), in solid solution, does not materially affect the rupture strength, for the same cobalt content, but increases the hardness, e.g. 10% cobalt with equal proportions of the two carbides gives a Rockwell A hardness of 92, with rupture strength 250 000 p.s.i. Replacement of more WC by TiC (in solid solution) depresses the rupture strength slightly but hardly alters the hardness, except in an upward direction. Substitution of tantalum carbide (TaC) for either WC or TiC raises the rupture strength only if the cobalt content is increased, but this slightly lowers the hardness. The behaviour at high temperature improves, and so does the thermal conductivity. Vanadium carbide has a similar influence to tantalum carbide.

The properties of sintered carbides with cobalt binder are therefore capable of adjustment to produce a fairly wide range of specifications, and the claims made for any composition are based more on criteria of performance than on measured properties.

10.3.5 Uses

The obvious use of these materials is in the field of machine tool bits, for increased cutting speed and longer life as compared with alternative steels or alloys. Tool bits are generally formed by brazing carbide components onto a more ductile, hardened steel or alloy supporting-shank, but the range of components has increased remarkably over the past 20 years. Apart from lathe tools (the initial application), drills, rotating files, milling cutters, broaches, mandrels, boring bars, gauge blocks and ball races have been added to components successfully produced and tested in practice.

Dies for extrusion and drawing of metals and alloys have been improved by inserting small components of carbide into a bearing ring of forged steel. Bearings for high-speed motors, guide rings for textile equipment, liners for pumps and valves, radio valve parts and welding rods for hard surface coatings are recent additional uses.

During world war II, the cores of armour piercing bullets, e.g. anti-tank ammunition, were successfully made in tungsten carbide, opening up an avenue for more peaceful application by virtue of the behaviour under pressure. Whilst hard carbides are brittle compared with ductile steels, the high strain rate and temperature developed under conditions applicable to penetration of armour plate served to illustrate the fallacy of associating brittle behaviour with weakness. The high-pressure conditions which occur prevent the development of brittle fractures, which normally propagate very rapidly, and hence a successful performance is achieved.

Hard carbides have been used to support industrial diamonds in larger cutting heads such as oil-well and metal-mining drills, and in dressing tools for grinding wheels on more conventional machines. The cobalt binder is increased to over 25% for this application, the diamonds being set by hand into a steel mould before the carbide–cobalt powder is compacted around them. Both hot- and cold-pressing methods have been used, the hot-pressing producing less shrinkage. The thermal expansion of diamond and carbides is almost the same and bonding between the two is good. Wear on the carbide matrix is much less than on any other backing material and hence there is less likelihood of the diamonds being ripped out in use. The obvious step, to smaller tools incorporating diamond dust with cemented carbide, has enjoyed similar success.

Inserts of conventional ductile alloys into carbide components allow drilling and tapping of the alloy for attachment of the carbide component to a wide variety of parts where brazing might be difficult or impossible. Devices of this type are bound to increase the availability of carbide components and extend their use to more fields of industry. The hard metal aspect of powder metallurgy is bound to expand with the more extensive automation of manufacturing plant, to extend the life of machine components even if initial cost is greater.

10.4 LITERATURE

1. O. KUBASCHEWSKI and E. L. EVANS. *Metallurgical Thermochemistry*, 2nd edn. London, Pergamon, 1958.
2. J. MACKOWIAK. *Physical Chemistry for Metallurgists*. London, Allen & Unwin, 1965.
3. A. H. COTTRELL. *An Introduction to Metallurgy*. London, Edward Arnold, 1967.
4. J. H. HOLLOMON and D. TURNBULL. *Progress in Metal Physics*, vol. 4. London, Pergamon, 1953.
5. K. C. LI and C. Y. WANG. *Tungsten*. New York, Reinhold, 1955.
6. U.S. PATENT 2205854; BRITISH PATENTS 623564 and 658213.
7. S. M. SHELTON, E. D. DILLING and J. H. MCCLAIN. *Progress in Nuclear Energy*, series v, vol. 1, pp. 305–51. London, Pergamon, 1956.
8. E. FOLEY, M. WARD and A. L. HOCK. *Extraction and Refining of the Rarer Metals*, pp. 196–211. London, Inst. Min. and Met., 1957.
9. A. E. VAN ARKEL and J. H. DE BOER. *Z. anorg. chem.* 148 (1925), 345–50.
10. W. M. SHAFER and C. R. HARR. *J. Electrochem. Soc.*, **105**(7) (1958), 413–17.
11. I. LJUNGOERG. *Powder Metallurgy*, **1/2** (1958), 24–32.
12. C. L. MANTELL. *J. Electrochem. Soc.*, **106**(1) (1959), 70–4.
13. E. MEHL. *Powder Metallurgy*, **1/2** (1958), 33–9.
14. H. BLOOM and J. BOCKRIS. *Modern Aspects of Electrochemistry*, No. 2. London, Butterworth, 1959.
15. J. O'M. BOCKRIS et al. *Proc. Roy. Soc.*, A, **255**(1283) (1960), 558–78.
16. W. J. KROLL. *Met. Reviews*, **1**(3) (1956), 291–337.
17. A. D. and M. K. MCQUILLAN. *Titanium*. London, Butterworth, 1956.
18. G. L. MILLER. *Zirconium*, 2nd edn. London, Butterworth, 1957.

19. J. H. BUDDERY and H. J. HEDGER. 'Metallurgy and Fuels', *Progress in Nuclear Energy*, series v, pp. 33–8. London, Pergamon, 1956.
20. B. BLUMENTHAL and R. A. NOLAND *ibid.* pp. 62–80.
21. W. WRIGHT. *Powder Metallurgy*, **4** (1959), 79–89.
22. R. P. FRASER *et al. Brit. Chem. Eng.* **2** (1957), (8), 414–17; (9), 496–501; (10), 536–43; (11), 610–13.
23. J. S. THOMPSON. *J. Inst. Metals*, **74** (1948), 101–32.
24. R. G. MONSEE. *Met. Progress*, **76** (4) (1959), 111–13.
25. C. G. GOETZEL. *Treatise on Powder Metallurgy*. 3 vols. New York, Interscience, 1963.
26. G. L. DAVIS and P. J. BURDON. *Metal Treatment*, **25**(159) (1958), 495–502.
27. J. E. MAY and D. TURNBULL. *Trans. A.I.M.M.E.* **212** (1958), 769–81.
28. C. E. RICHARDS. *Electrical Manufacturing*, **60**(6) (1957), 104–8.
29. C. E. RICHARDS and A. C. LYNCH. *Soft Magnetic Materials*, London, Pergamon, 1953.
30. F. E. LUBORSKY, T. O. PAINE and L. I. MENDELSOHN. *Powder Metallurgy*, **4** (1959), 57–58.
31. R. B. FALK, G. D. HOOPER and R. J. STUDDERS. *J. App. Phys.*, **30**(4) (1959), 132–3.
32. M. W. FREEMAN. *Proc. Met. Powder Assoc.* 1957.
33. D. A. OLIVER *et al. I.S.I. Spec. Rep*, **58** (1956), 180–94.
34. P. E. EVANS and G. C. SMITH. *Powder Metallurgy*, **3** (1959), 26–44.
35. D. K. WORN and R. P. PERKS, *loc. cit.* pp. 45–71.
36. A. E. NEWKIRK and I. ALIFERIS. *J. Am. Chem. Soc.*, **79** (17) (1957) 4629–31.
37. G. COMSTOCK. *Iron Age*, **156**(9) (1945), 36A.

CHAPTER 11

The Corrosion of Metals

11.1 CORROSION IN GENERAL

11.1.1 Introduction

Corrosion is the natural phenomenon in which metals and alloys undergo chemical oxidation: the basic cause of which is their spontaneous tendency to return to their thermodynamic stable state.

The reaction can be represented thus:

$$\text{Stable Metallic Mineral Ore} \xrightarrow[\text{(extraction)}]{\text{(reduction)}} \text{Unstable Massive Metal} \xrightarrow[\text{(oxidation)}]{\text{(corrosion)}} \text{Stable Corrosion Product}$$

In this cycle the free energy of the extracted metal has been raised above that of the mineral ore, where it was at a minimum. Nature then requires the metal to return to its original stable state which it does by losing energy gradually until its free energy is again at a minimum, with the formation of a stable corrosion product. Quite often this latter substance is of very similar chemical composition to the original metallic mineral from which the metal was extracted. Cycles of oxide to oxide, carbonate to carbonate and sulphide to sulphide are found.

Since the thermodynamic equations of the corrosion reaction are not essential to a general understanding of the phenomenon they will not be discussed further (see refs. 1, 3 and 4).

The thermodynamic aspect of corrosion does, however, emphasize that corrosion is a natural phenomenon and that all metals with the exception of gold, iridium, platinum and palladium (which are thermodynamically stable) possess some tendency to change to the oxidized state.

All constructional metals corrode, thus the corrosion engineer's problem is the *rate* at which the oxidation occurs. The rate varies according to the metal and the corrosive environment from being so slow that the reaction may be regarded as not occurring at all to being so fast that the reaction occurs with almost explosive violence. Whether corrosion is detrimental to the use of a particular metal in a specific environment will depend on many factors, e.g. time of exposure, temperature, oxygen concentration of the environment, usage, etc.

N.B. It must also be appreciated that the corrosion of a metal *per se* is not necessarily deleterious (see below).

11.1.2 Definitions

The standard, but not entirely satisfactory definition of corrosion found in almost all glossaries of corrosion terms is: the destruction of a metal by chemical or electrochemical reaction with its environment. This definition seems to imply that corrosion must be, necessarily, a deleterious reaction and since the word 'corrosion' is derived from the Latin verb 'corrodo' – to gnaw to pieces or wear away – this mental picture is understandable. However, in terms of the usage of metals corrosion is not always undesirable; it is well known that a small initial amount of corrosion on a metal surface may be advantageous. The explanation of this apparent contradiction will be discussed in detail later (see film formation, p. 793) but the basic reason is because a limited quantity of corrosion on a metal surface may result in a film of corrosion product at the metal/environment interface which will prevent, stifle or form a barrier against the continuance of the corrosion reaction. Although this is destruction in the sense that these films are formed from the oxidation of a small quantity of the basis metal such corrosion does not lead to harmful failures, and in some cases, but not all, such films are actively encouraged to form to prevent corrosion failure. Thus, it can be appreciated that there is no general definition of corrosion that is generally acceptable unless it is restricted to purely thermodynamic principles.

Erosion, a word which denotes the destruction of a material by mechanical means, e.g. by the abrasive action of water containing air bubbles, solid particles in suspension or cavitation, is often used in corrosion technology. Further, since mechanical action and electrochemical activity can occur simultaneously the terms corrosion–erosion and erosion–corrosion are also found. The definition of such conjoint actions is: the combined effect of corrosion and erosion in which the surface film at the metal/environment interface is continually removed by mechanical means so providing a fresh unprotected metal surface for further electrochemical attack.

11.1.3 Environments

In any definition of corrosion it will be noted that the environment plays a definitive role. Corrosion, beneficial or otherwise, can only occur by a reaction of the metal with some outside agency – the environment. In considering the resistance of a metal to corrosion the conditions under which it is used, where, and also how it is used will have as much or sometimes more importance than the intrinsic corrosion behaviour of the metal or alloy. Corrosion behaviour is, therefore, not a fundamental property of a metal as, for instance, are its mechanical properties.

Many attempts have been made to list the diverse environments in which metals are used, but even should the list be comprehensive the reaction of metals and alloys with any specific environment may take several forms depending on the rate of attack and the character of the reaction product, etc. Such lists, therefore, always have limitations. However, two broad groups can be scheduled but it is, of course, possible for more than one of the factors to be operating at the same time, e.g. moving saline water, or steam erosion. The main environmental influences are as follows:

Locational	*Mechanical*
Wet, dry.	Under conditions of static stress.
Hot, cold.	
Aqueous, atmospheric.	Under fatigue conditions.
Fresh, salt water.	Movement.

Chemicals.   Erosion.
Soil.   Cavitation.
Steam, gases, vapours.   Impingement.
Foodstuffs.   Fretting (see p. 819).
Acid, alkaline or neutral.   Torsion, compression
Oxygen present or absent.   or tensile stress.

Corrosion may occur in either a wet or a dry environment. In dry corrosion the reaction occurs at a gas/metal interface while in wet corrosion the interface is metal/electrolyte solution. In both cases the basic reaction is electrochemical and is effected by the removal from the metal of one or more electrons with the formation of positive ions. In dry corrosion direct oxidation of the metal takes place but in wet corrosion the oxidation is indirect, the metal ions first becoming hydrated. Since these electrochemical reactions occur at a surface they are often modified by the properties of the corrosion product, e.g. whether or not these products adhere to the surface or not and so interfere with the normal electrochemical behaviour of the metal and the environment, also whether or not the corrosion product is or is not soluble in the immediate vicinity of the corrosion cell. Such factors may decrease the amount of corrosion or they may increase it, depending on the conditions.

Since corrosion is the reaction between a metal and its environment it follows that the control of corrosion can be effected by modification of the environment as well as by the use of a corrosion-resistant material.

11.1.4 Cost of corrosion

Corrosion reactions which cause a metal or alloy to fail in service represent financial loss, as does the use of a more expensive material when the corrosive conditions are such that a cheaper, but otherwise suitable metal, would be dangerous.

Financial losses are rarely confined to the price of the failed metal itself, they are usually increased by the cost of the labour of replacement and often by secondary losses such as shut-down or out-of-service expenses which in the shipping industry, for instance, may delay a sailing date. Loss of revenue due to breaks in production, spoilage due to leaks causing contamination of a commercial product, and even explosions or failure of major structures with possible loss of life, all enter into the economics of corrosion and the wastage of metal by overdesign, i.e. heavier gauges of metal scheduled to take care of possible corrosion than would be required by considerations of mechanical properties alone.

What corrosion costs a country per annum is impossible to estimate but it has been stated that the wastage of material resources by corrosion ranks third after war and disease. Figures have been computed in terms of national loss per year, but they are really no more than informed guesses even when the calculations have been made by those experienced in the field of corrosion. There are many imponderables and it is not easy to compare the estimates from country to country. For example, some estimates include the cost of corrosion prevention while others do not, and it can be argued that no figure is realistic unless it includes the cost of research into corrosion science and investigation into the production of more corrosion-resistant alloys.

Corrosion is ubiquitous and when it cannot be eliminated it must be controlled. This raises the problem whether it is better to use a high capital cost corrosion-resistant material with a long trouble-free life or, if it is feasible, to use a cheaper material and accept replacement costs out of income. Formulae are available for estimating which of the two alternatives is the better economic proposition, but even then a decision cannot be reached with certainty because none of the formulae can be loaded with all the factors – some of which may well be unknown for both the metal and the environment. Ultimately, it is not the basic cost of corrosion or corrosion prevention that will affect the attention given to corrosion by industry but the relation of corrosion to profits on which very little information has ever been published.

Nevertheless, to provide some idea of the magnitude of the wastage due to corrosion, however crude it may be, one can quote the figures offered by Uhlig, Vernon and others that in the USA an annual loss of more than $6 billion was not excessive at the prices ruling in the middle and late nineteen-fifties. Similarly, about that time, it was estimated that the UK spent some £100 million merely on paint for protection of steel structures and the Royal Navy are of the opinion that corrosion and corrosion protection costs them approximately £10 million per annum.

11.2 REACTION MECHANISM OF CORROSION

11.2.1 Introduction

In Section 11.1 the ore/corrosion-product cycle is represented and it is shown that the ore is reduced to metal and in the process the metal gains energy with resulting thermodynamic instability. (The energy for this reaction is usually electrical or of chemical origin.) The thermodynamically unstable metal is now poised to take part in some oxidation reaction which will result again in a compound of thermodynamic stability. It must be pointed out, however, that thermodynamics provides no knowledge of the course of the reaction and gives no indication of the *rate* at which the metal is degrading – the kinetics of these reactions may be so slow that for all practical purposes the metal remains stable.

A good example of the ore → metal → metal compound cycle is provided by the behaviour of the metal iron. Many iron ores are found in the oxidized form (oxides, carbonates), and metallic iron is obtained by smelting the ores with carbon as the reducing agent. If the iron so won is allowed to react with moisture containing oxygen, the familiar substance 'rust' – the corrosion product – will be produced. If red rust is chemically analyzed it is found to be an iron oxide of similar composition to some of the natural ores of iron. Of course, the reaction product is not always an oxide, e.g. it may be a sulphide.

11.2.2 The corrosion reaction

A corrosion reaction is always accompanied by a flow of electricity from one metallic area to another with a lower potential, through a non-metallic environment capable of conducting an electric (ionic) current, i.e. an electrolyte solution. The current flow in the metallic portions of the system is electronic and in the electrolyte solution ionic. See fig. 11.1. (An electrolyte is a substance which in solution gives rise to ions. An electrolyte solution is a solution in which the conductance of an electric current occurs by the passage of ions.)

The initial and essential step in the corrosion reaction is the transformation of a metal atom into a metallic compound, with consequent degradation of the metal. The dissolution of a metal is an electrochemical phenomenon (a chemical reaction accom-

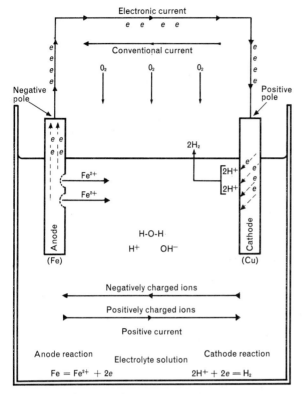

Fig. 11.1 Diagrammatic galvanic cell in which the cathodic reaction is the reduction of hydrogen.

panied by the passage of an electric current) and during the process the metal atom is oxidized by the loss of one or more electrons.

The primary reaction at the region of lower potential – the anode is:

$$M \rightarrow M^{z+} + ze \tag{1}$$

Since this reaction can only occur in the presence of an electron acceptor the anodic reaction must be accompanied by a cathodic reaction, at a region of higher potential, with a consequent flow of electricity (the current is produced because of the chemical energy released during the conversion of the metal to its oxidized form). A number of reactions is possible (see below) at the cathode including:

$$2e + 2H^+ \rightarrow H_2 \tag{2}$$

Combining equations (1) and (2) the basic corrosion equation becomes:

$$M + zH^+ \rightarrow M^{z+} + \tfrac{1}{2}zH_2. \tag{3}$$

The reaction mechanism mentioned above is basically similar to that occurring in a galvanic cell. For this reason the corrosion reaction is usually represented as a corrosion cell.

Before considering the corrosion reaction in greater detail it may assist those who are not electrochemists if some attention is given to a galvanic cell such as is shown in fig. 11.1. This will enable a general picture of the corrosion cell to form in the reader's mind, and will also serve as an introduction to certain terms which are used throughout corrosion technology.

The corrosion cell is often depicted as follows:

Metal A (area x)//Electrolyte Solution//Metal A or Metal B (area y)
Anode, low potential (Corrosion)   *Cathode*, high potential (Protection)

The most simple form of cell to illustrate the component parts of the corrosion cell is the voltaic cell. The bimetallic galvanic cell–iron/copper in oxygenated sodium chloride–has been chosen. It must be stressed, however, that two different metals are not essential for a corrosion cell to form: corrosion can take place on the surface of a single metal, nor need the electrodes be dipping into the same electrolyte solution or into the same concentration of that solution, so long as electrical continuity is maintained.

If a piece of clean, unfilmed pure iron with a homogeneous surface is placed in a solution of sodium chloride (NaCl) in contact with oxygen, a small amount of corrosion will immediately occur at the metal/liquid interface. Similarly, if a piece of clean, unfilmed pure copper is placed in the same solution but in such a way that the metals do not touch, the corrosion reaction at the copper/liquid interface will again be very small. However, if the iron and copper are then connected by an external conducting path the corrosion rate of the iron increases by a very large factor and the corrosion of the copper ceases entirely.

In fig. 11.1, it will be noticed that the cell has four component parts which combine to form the corrosion cell:

1. The iron electrode which leads the *conventional* current into the cell. This is the *anode*, which has a negative polarity and is the electrode where oxidation (corrosion) takes place.
2. The copper electrode which leads the *conventional* current out of the cell. This is the *cathode*, which has a positive polarity and is the electrode at which a process of reduction occurs. Corrosion is prevented at this electrode.
3. The oxygenated saline solution in which the anode and the cathode are immersed. This is the *electrolyte solution* and is the medium that carries the internal ionic electricity flow of the cell.
4. The external, low-resistance, metallic path that carries the *electronic current*, which flows in the opposite direction to the conventional current.

Although reduction and corrosion prevention are the usual reactions at the cathode, with some amphoteric metals 'indirect' corrosion may occur at this electrode. The polarity of the electrodes is immaterial and may be positive or negative depending on the type of cell. In the current *producing* cell (the galvanic cell and corrosion cell described above), the anode is of *negative* polarity, but in the current *consuming* cell (e.g. an electrolysis cell or battery) the polarities are reversed and the anode is *positive*. However, in both cases negative ions travel to the anode and the anode is the electrode which leads the electrons into the external metallic electronic path.

11.2.3 Corrosion of metals in aqueous environments

In general, a corrosion current will flow between any anode/solution interface, and any cathode/solution interface no matter what causes the difference in the potential of the electrodes. The sizes of anodes and cathodes may vary from microscopic to a large area such as the whole wetted area of a ship's hull. The electrodes may be of any size relative to one another but their effectiveness will depend on the various elements of electrical resistance in the ionic half of the cell. Anodes and cathodes may be produced in many ways, e.g. by coupling together two metals of different potential, by differences due to physical inequalities – work hardened/not work hardened, etc. – and by variations in the environment such as concentration gradients in oxygen or electrolyte solution. There may, also, be a spontaneous formation of inseparable anodes and cathodes on a metal surface in which anodic and cathodic areas alternate, as, for example, in the uniform dissolution of metal in acid.

Whatever may be the cause of the production of the anodic and cathodic areas, several specific events take place when such areas of different potential are electrically connected in an electrolyte solution:

1. Metal atoms break away from the anode/solution interface and positively charged metal ions enter the liquid leaving electrons behind in the metal. The number of electrons released will be equivalent to the charge on the ion, e.g.:

$$Ag \rightarrow Ag^+ + e \quad \text{or} \quad Fe \rightarrow Fe^{2+} + 2e$$

2. The released electrons will flow by the external conducting path to the cathode/solution interface – the area of higher potential.

3. When these electrons arrive at this interface they take part in some electrochemical process which may be non-metallic or metallic, e.g.:

$$2H^+ + 2e \rightarrow H_2 \quad \text{or} \quad Cu^{2+} + 2e \rightarrow Cu$$

In the external metallic path there is a flow of current from anode to cathode and, although it is conventional to allude to a flow of current through the electrolyte solution, it must be pointed out that the passage of the charge is not carried by electrons but is maintained by a flow of negative ions (anions) from the cathode to the anode, and by positive ions (cations) from the anode to the cathode. When the charged ion receives its correct quota of electrons at the cathode, it immediately reverts to the atomic state.

The anodic and cathodic reactions must be electrochemically equivalent, and their reaction rates must also be equal. Thus, any condition which affects the cathodic reaction rate will have a corresponding effect at the anode – a most important axiom in corrosion control.

Since the amount of corrosion occurring at the anode is directly related to the flow of electrons in the external circuit, the rate of corrosion can be calculated by the insertion of measuring instruments in this electronic path, and interpreting the readings in terms of Faraday's law.

11.2.4 Reactions at anodes and cathodes

All anodic electrochemical reactions involve processes of oxidation and while there are several ways in which these may be effected, the corrosion engineer is mainly concerned with the oxidation of a metal to its ions as equation (1). (He is sometimes also interested in the more complicated reaction of the oxidation of the metal to a metal oxide which prevents further reaction, i.e. causes passivation of the metal (see passivity, p. 799).)

A number of reactions at cathodes are of interest to the practical corrosion technologist since these frequently provide a means of controlling the corrosion rate:

1. Reduction of hydrogen ions to hydrogen gas:

Acid solution

$$2H^+ + 2e \rightarrow H_2. \tag{3a}$$

Neutral or alkaline solution

$$2H_2O + 2e \rightarrow 2OH^- \tag{3b}$$

2. Reduction of oxygen:

$$O_2 + 2H_2O + 4e \rightarrow 4OH^-. \tag{4}$$

3. Reduction of sulphate (by anaerobic bacteria):

$$4H_2 + SO_4^{2-} \rightarrow S^{2-} + 4H_2O \tag{5}$$

4. Reduction of metal ions:

$$Fe^{3+} + e \rightarrow Fe^{2+}$$

Hydrogen reduction is the usual cathodic process found in acid solutions with a high concentration of H^+ ions, and the corrosion rates of most metals corroding in this way, i.e. where the metal dissolves with the simultaneous evolution of hydrogen (H_2), are high (usually because insoluble corrosion products which might stifle the reaction do not form). In near neutral and alkaline solutions the reaction represented by equation 11.3b is commonly found.

In both cases (equations 11.3a and 11.3b), if the solutions are initially near neutral an excess of OH^- ions will occur. This very often tends to slow the reaction down since the alkali formed in the vicinity of a cathode can act as an inhibitor due to the formation of insoluble hydroxides which will stifle the continuance of the reaction if the anodes come within the influence of this alkali. Further, with amphoteric metals, e.g. aluminium and lead, 'cathodic corrosion' can take place at areas of high pH values since the oxides and hydroxides of these metals are soluble.

The oxygen-reduction reaction (equation 4), often known as the oxygen-absorption reaction, is, in near neutral solutions (natural waters containing oxygen), more commonly found than those described above, e.g. this reaction predominates when iron corrodes in sea-water. The pH of the solution again tends to increase at cathodic areas and this effect is very significant in cathodic protection both for its effect on some paints and in the formation of calcareous scales which may reduce the amount of current required for protection (see cathodic protection, p. 841).

Sometimes, when the products of the anodes and the cathodes meet, they may enter into further mutual reactions. For example, the reaction of hydroxyl ions from the cathodic process with the ferrous ions of the anodic dissolution of corroding ferrous materials:

$$Fe^{2+} + 2OH^- \rightarrow Fe(OH)_2$$

When the limit of the solubility of the ferrous hydroxide is reached a white precipitate will form in the solution. In the presence of oxygen a further reaction will occur:

$$4Fe(OH)_2 + O_2 + 2H_2O \rightarrow 4Fe(OH)_3 \quad \text{or} \quad Fe_2O_3 \cdot H_2O \text{ (Rust)}$$

The oxidation of ferrous hydroxide to ferric oxide can proceed rapidly or slowly. When the reaction is fast the final product is the *alpha* form of hydrous ferric oxide (FeO(OH)) which is similar in structure to goethite mineral, but if the oxidation reaction is slow the *gamma* form of the hydrous oxide is more likely to result, and this is similar in structure to lepidocrocite or limonite ore. Precipitated rust (hydrated ferric oxide) tends to become darker and denser on ageing and eventually dries to the brown ferric oxide (Fe_2O_3); again there are *alpha* and *gamma* forms, of which the best known is haematite – the alpha form. However, if the oxidation of the ferrous hydroxide does not proceed to completion, the precipitated rust will often be black or green in colour due to the presence of ferroso ferric oxide (Fe_3O_4), or the hydrated form ($Fe_3O_4.H_2O$) the familiar mineral magnetite. The presence of green or black corrosion products at the site of corroding ferrous materials usually denotes that the corrosion is of recent origin.

11.2.5 Corrosion rates and polarization

The essential driving force of a corrosion reaction is the potential difference between the anodic and cathodic areas, but the amount of metal dissolved and the rate of dissolution depend on the current flow. Any factor which varies the circuit resistance, the potential of the electrodes, etc., will therefore affect the corrosion rate.

Any change in the potential of an electrode from its equilibrium value is known as polarization. When a metal is immersed in a solution of its own ions there will be an outgoing pressure of metal ions trying to enter the solution, and an inward pressure of metal ions endeavouring to enter the metal. Sooner or later the ingoing and outgoing pressures will come together into equilibrium and the exchange will occur at equal rates. The current, when this condition is reached, is called the exchange current and the potential at which it occurs the equilibrium potential.

If at this situation the metal becomes part of a corrosion cell the equilibrium is upset and the reaction will proceed in a definite direction. If the metal becomes more positive (anodic) it will corrode, and if it becomes more negative (cathodic) corrosion will usually be prevented and a reduction reaction will take place. The cell is said to be polarized and the polarization value is the amount by which the potential has changed from the equilibrium value. (Overvoltage is a synonymous term for polarization – often used in the specific cases of hydrogen or oxygen evolution.) Such changes of potential are also known as anodic or cathodic polarization depending on the electrode affected. Polarization can be due to a number of causes, and it is possible to distinguish at least three:

1. Resistance polarization (or resistance overpotential).
2. Concentration polarization.
3. Activation polarization (or activation overpotential).

Resistance polarization is due to the formation of insoluble protective scales or films on the metal surface (at anodes or cathodes) which promote resistance in the cell circuit. This type of polarization, which results from any voltage drop at the electrode/solution interface, follows Ohm's law.

Concentration polarization results from changes in the concentration of the various reactive ions which are involved at the electrode/solution interfaces. It usually occurs when the electrochemical reactions at the anodes, cathodes or both are so rapid that reactant ions cannot diffuse into or out of the reacting areas sufficiently fast to maintain the reactions. For example, an anode potential can be raised to a more positive value if metal ions accumulate in the immediate vicinity and a cathode will move towards a more negative potential if, due to lack of reactants, the reduction process is terminated or reduced. Concentration polarization will be reduced by an effect, such as stirring, which ensures a continual supply of fresh electrolyte solution to the cathode and/or sweeps any accumulation of metal ions from the anode. The reduction of concentration polarization is one reason why flowing water is more corrosive than calm.

Most electrode reactions require some activation energy before they will proceed. Thus, to produce anodic dissolution a small external potential must be applied to the electrode to assist the passage of ions into solution, e.g. if the electrode becomes part of a galvanic cell. This results in another type of polarization – activation polarization. This type is logarithmically dependent upon the current density. Various relationships have been postulated relating activation polarization with current density. At low current densities the variation in polarization is often roughly linear, but when the irreversibility of the electrode is high the relationship is more complex and the relevant logarithmic equation was first developed by Tafel (see ref. 11).

A graphical method of showing how corrosion rates depend on the polarization of the electrodes was developed by Evans and Hoar of Cambridge University, England (see also corrosion rates – Evans diagrams, p. 797 and 798).

Mention must also be made of the work of M Pourbaix and his associates, though for a detailed study of this work the original literature should be consulted since their potential/pH equilibrium diagrams need careful assessment in relation to practical corrosion problems. (See refs. 12 and 13.)

It is well known that in physical metallurgy equilibrium phase diagrams can be constructed having temperature and composition as ordinates. Pourbaix *et al* have calculated the phases at equilibrium for metal/water from the chemical potentials of the species concerned in the equilibria, and have expressed the data in the form of equilibrium diagrams having pH and potential, E, as ordinates. These diagrams provide a thermodynamic basis for the fundamental study of corrosion reactions. Pourbaix, however, emphasizes that such diagrams have their limitations when applied to practical problems. The separation of such diagrams into zones of corrosion, immunity, and

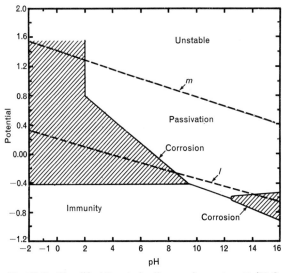

Fig. 11.2 Simplified Pourbaix diagram for system Fe/H$_2$O. Curves *l* and *m* show the reversible hydrogen and oxygen electrodes.

passivity is justified in terms of thermodynamics but since practical conditions often involve other factors besides those expressed on the diagrams they do not provide a panacea for the solution of many corrosion problems (See further ref. 3.)

11.2.6 Film formation and its importance in corrosion reactions

The effect of films of corrosion products formed at anodes and cathodes in determining the progress of the corrosion of the reacting metal has been mentioned frequently in the foregoing sections.

The type of film formed on a metal in any environment is a major factor controlling the corrosion which occurs. The type of film will vary with the environment and any alteration of the conditions at a metal/environment interface may alter the film-forming process with either an increase or decrease in cell resistance and consequent alteration in the progression of the oxidation process. The films are usually oxides or hydroxides though sulphide films, etc. are also found. The type of film found on any specific metal may vary from environment to environment both in physical properties as well as chemical composition.

Film thickness does not appear to be a very important factor in controlling corrosion provided that the film is continuous, impervious, elastic (plastic) and self-healing. If the film has these four properties it will isolate the metal from its environment and so reduce the amount of corrosion occurring, irrespective of its thickness. Examples of thin and thick protective films are the very thin films (a few Å) formed on the stainless steels and the thick complex films which sometimes form on some copper-base alloys in certain atmospheric conditions, e.g. the green patina which forms on copper roofing materials. The rust of mild steel, which is non-adherent and usually porous, is an example of a thick film which is not protective. The corrosion films of some metals and alloys have the very desirable property of being able to reform (repair) themselves after mechanical breakdown. This ability is very important under conditions of corrosion-erosion and an example of an alloy specially developed for resistance to the corrosion-erosion environment in sea water is aluminium brass (76 Cu, 22 Zn, 2 Al). In this alloy the relatively unprotective film of copper oxides usually found on single-phase copper/zinc alloys is replaced by the much more protective and self-healing film of copper oxide/aluminium oxide.

Protective films can also be produced deliberately, by chemical or electrochemical treatment of the metal surface, e.g. anodizing, and all types of passivation; chromate treatment and phosphating are examples of protective film formation.

Some metals, e.g. chromium, titanium and the stainless steels, owe their corrosion resistance to the fact that they form, spontaneously, in air (oxygen) films which comply with the four properties referred to above.

Protective films on metals whether natural or artificial are frequently cathodic to the underlying metal substrate. In consequence a cell:

Cathode	Anode
Filmed metal	Unfilmed metal

will be set up at any break or discontinuity in the film or scale and intense attack on a small area of the basis metal often occurs. One well known type of scale which is liable to accelerate corrosion to a deleterious degree is the rolling- or mill-scale found on rolled ships' plate.

11.2.7 Galvanic series

It has been implicit in the above description of corrosion reactions that when any metal is placed in an electrolyte it assumes an electrical potential in relation to its surroundings. Such metal potentials can, obviously, be placed in some order from, say, noble gold to base sodium. The well-known electromotive series is such an order and gives the *standard electrode potentials* of the various metals. Such tables are constructed from single metal potentials calculated with reference to the standard hydrogen electrode in an electrolyte solution of the metal's own ions at a concentration equal to unit activity – the potential of hydrogen being taken as an arbitrary zero.

It is often thought that, since many instances of corrosion are due to the coupling of two metals with different potentials, the degree of corrosion may be estimated from this table. In terms of the practical corrosion engineer this is not possible; the environment for which the standard electrode potentials are calculated is too peculiar. Such a table cannot even be used to indicate which metal of a couple will corrode if the potentials are grossly different, i.e. are well separated in such a list.

The actual potential of a metal in contact with another metal in an electrolyte solution depends on many variables – polarization effects, existing reactions at the anodes and cathodes, film formation, breakdown and repair and the type of film, protective or non-protective. Thus, the only practical table which can be constructed with any reliability, to indicate the galvanic tendencies of a series of metals, is one made from actual measurements made in the electrolyte solution under investigation in as near as possible the actual service conditions. Such tables for various electrolyte solutions will be found in the literature and the behaviour of metals in sea-water is especially well documented.

The *Nernst equation* which is used to express the exact electromotive force (e.m.f.) of a cell in terms of activities of products and reactants of the cell, has a negative or positive sign, depending on the way in which the reaction is written, although the magnitude is constant. For example, the Daniell cell reaction could be written:

$$Cu^{2+} + Zn \rightarrow Cu + Zn^{2+}$$

or as

$$Cu + Zn^{2+} \rightarrow Cu^{2+} + Zn$$

The convention usually adopted is to write the equation of the cell reaction in the direction in which the reaction actually proceeds, and then to write the cell so that the reduced form of each half-reaction is on the same side as in the cell reaction:

$$Cu^{2+} + Zn \rightarrow Zn^{2+} + Cu$$
$$Zn/Zn^{2+}//Cu^{2+}/Cu$$

Then, by convention, the e.m.f. of the cell is taken as the electrical potential of the electrode on the right-hand side of the cell, and this is arbitrarily assigned a positive value. It is evident that in this convention the anode is always on the left-hand side of the cell, i.e. if the reaction is spontaneous the reduced form on the left-hand side becomes oxidized.

The sign of the potential of a metal is often a matter of variation in the literature and until about 1953 no standard convention was applied. For example:

European convention

$$Cu^{2+} + 2e \quad +0.34 \text{ V } (E^0)$$
$$Zn^{2+} + 2e \quad -0.76 \text{ V } (E^0)$$

US convention

$$Zn^{2+} + 2e \quad +0.76 \text{ V } (E^0)$$
$$Cu^{2+} + 2e \quad -0.34 \text{ V } (E^0)$$

However, in 1953-54 the International Union of Pure and Applied Chemistry adopted the European convention as the standard and the difficulty is now resolved. It is evident that in the European convention the electrons appear on the left-hand side of the reaction, i.e. the reaction is written as a reduction. (A more detailed exposition of this point will be found in refs. 3 and 6). The above is based on L. L. Shrier (ed.) *Corrosion*, London, George Newnes, 1963.

If any corrosion couple is referred to a galvanic series the metal nearer the anodic (base) end of the list will tend to suffer corrosion and the metal nearer the cathodic (noble) end will tend to be protected. In the absence of adhering films the further apart the metals are in the series the greater the chance there is of corrosion of the more anodic member. Many metals may act as either anodes or cathodes of a corrosion cell depending on the couple, e.g. steel is anodic if coupled to copper, but it is cathodic if coupled to aluminium. The anodicity of a metal may vary with the environment, especially in cases of coupled metals which are close together in the table – tin is usually noble to iron in near-neutral solutions but when it (tin) is used in the presence of some fruit juices it may become less noble than iron.

It must be stressed that any galvanic series, though it may indicate anodicities of couples in terms of the probable potentials set up, can never predict the rate at which the corrosion will occur – this depends upon the current flow.

11.2.8 Concentration and differential aeration cells.

Corrosion cells can also be formed in which two identical electrodes are in contact with different environments, e.g. differing oxygen or metal ion concentrations, different temperatures or in areas of varying degree of movement in the electrolyte solution.

Concentration cells due to differences in concentrations of ions sometimes form in liquid environments. An example of such a cell is:

$$Cu/ \text{ Concentrated} \parallel \text{Dilute } /Cu$$
$$CuSO_4 \parallel CuSO_4$$

In such a *reversible* concentration cell, the tendency will be for the solutions to equalize because, in the half-cell of lower concentration, corrosion of the copper (Cu) will occur (anode) and in the half-cell of higher concentration, deposition of copper will take place (cathode). In practice the cell described above is not frequently found, but differences in potential due to the presence of varying concentrations of salt or pH values are common. In such cases the metal in the environment of higher conductivity (higher salt concentration) will be anodic to that in the lower conductivity (lower salt concentration). In this situation it should be noted that the location of the anodes and cathodes is the reverse of that in the reversible concentration cell.

Another variety of concentration cell, which in practice is more important, is known as the *differential aeration cell* and occurs when different metals or different areas of the same metal are in different concentrations of oxygen. Evans (1920-30) showed that if two electrodes of the same metal were immersed in an electrolyte solution and one was aerated and the other starved of oxygen a current was produced, and the aerated electrode became the cathode and the deaerated metal the anode. Such differential aeration cells are capable of causing severe damage at crevices, threaded connections, areas under loose washers, bolt heads, etc. These cells also accelerate pitting under rust or other debris. In all cases the corrosion occurs at the area of lowest oxygen concentration, i.e. *under* the rust, washer or bolt head or *in* the crack, crevice, etc. In other words, the area in contact with the most oxygen becomes an active cathode. In practice, any area shielded from the inward diffusion of oxygen soon becomes lower in oxygen concentration because initial corrosion will soon use up the available oxygen with the production of an active differential aeration cell. Such cells are notorious for initiating pitting on stainless steels, nickel and other 'passive' metals when they are exposed to highly conductive environments such as sea-water.

Corrosion cells can also be initiated by temperature differentials. In an electrolyte solution of copper sulphate ($CuSO_4$), a cell with a hot and cold copper electrode will deposit copper on the hot electrode (cathode) and copper will be dissolved from the cold electrode (anode). Not all metals behave in this way, e.g. with silver the polarities are reversed, and iron in sodium chloride solutions often becomes initially anodic at cold versus hot areas with a reversal with time depending on environmental conditions.

11.2.9 Corrosion rates – Evans diagrams

The current produced in a corrosion cell (corrosion rate) is limited by the resistance of the cell and the polarization of the electrodes. The polarization of an electrode increases with the amount of current flow and the anode tends to assume a more positive potential while the cathode will become increasingly negative. Therefore, polarization moves the two potentials closer together and the driving force of the corrosion reaction is diminished. (The anode is the electrode with the more negative potential, see p. 788. The amount of polarization on an anode may vary from that on its associated cathode and the resistance of the electrolyte solution may also vary from environment to environment. In consequence corrosion reactions may be classified into those which are anodically controlled, those which are cathodically controlled, and, as is more usual, when some polarization occurs on both the anode and cathode areas the situation can be described as mixed control. Resistance control is when the current is not sufficient to induce any significant polarization of the anode or the cathode.

The graphical method of showing how corrosion rates depend on the polarization of the electrodes of the corrosion cell was devised by Evans of Cambridge University and the basic Evans diagram is shown in fig 11.3 and diagrammatic curves for the four types of control are shown in fig. 11.4. The technique as developed by Evans and Hoar was to construct polarization curves by measuring the electrode potential of corroding specimens at different current densities and plotting potential against current. As they used specimens of the same area, this was similar to plotting ΔE against current density. (In all the accompanying diagrams the anodic and cathodic curves are shown as straight lines – in practice this may not be the case.)

In fig. 11.3 the current i' that flows between the anode and the cathode is given by

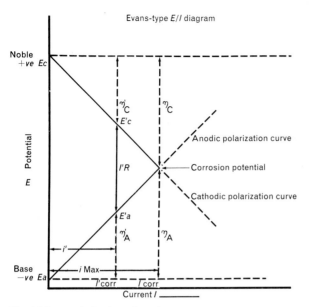

Fig. 11.3 η polarization of anode and cathode i' reaction value.

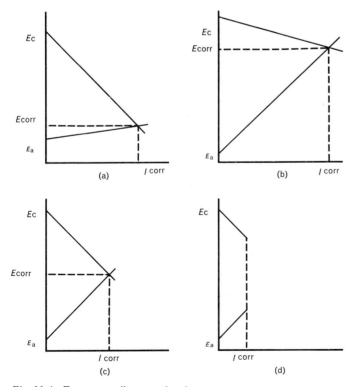

Fig. 11.4 Evans-type diagrams showing:
 I. Cathodic control
 II. Anodic control
 III. Mixed control
 IV. Resistance control

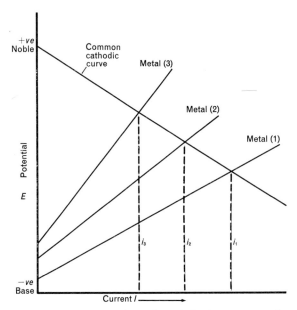

Fig. 11.5 Influence of anodic polarization on corrosion currents. Increase of polarization decreases corrosion current.

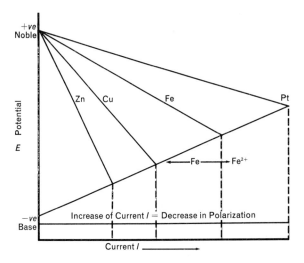

Fig. 11.6 Influence of cathodic polarization ($2H^+ \rightarrow H_2$) on corrosion currents, on platinum (Pt), iron (Fe), copper (Cu) and zinc (Zn) (Diagrammatic).

$E'_c - E'_a/R$, where E'_c and E'_a are the cathodic and anodic polarized potential values for the given current i'. R is the total resistance of the cell circuit (the sum of the resistances due to the electronic and ionic electrolytic paths plus the resistance of any films or reaction products which may interfere in the vicinity of the anode or cathode). In solutions of high conductivity with anodes and cathodes close together (acids) R will be small and the point of intersection of the two polarization curves will represent the maximum corrosion current (i_{max}). The more polarized an electrode reaction becomes, the less corrosion, and this is revealed by an increasing steepness of the polarization curves. This is shown in fig. 11.5 where increasing anodic polarization curves are shown at a common cathodic polarization value – note the decrease in current from i_1 to i_2 to i_3.

Changes in the slope of the cathodic polarization curve are shown in fig. 11.6 which indicate the facility with which the reaction $2H^+ \rightarrow H_2$ takes place on various metal surfaces, i.e. the decrease in current flow in the descending order platinum (Pt), iron (Fe), copper (Cu) and zinc (Zn). In non-oxidizing acids zinc has a higher overpotential than platinum, iron or copper.

11.2.10 Passivity, passivation

A metal exhibits passivity when, although it is exposed to an environment in which it would normally corrode, it remains visibly unchanged for a prolonged period. The passivation of a metal surface is the conversion, by any method, of a reactive surface to a passive surface.

Passivation may be complete or partial. Complete passivation means that a metal surface has been rendered completely resistant to some particular environment. Partial passivation can take two forms: either the passivated surface may corrode, but at a much slower rate than would normally be the case on the untreated surface, or the passivated surface may corrode (usually by pitting) to a greater extent than is normal for the unpassivated surface, due to some areas of the surface being passivated while others are not. Since all passive films are noble to their substrate, the formation of active/passive corrosion cells will be initiated when the passivated surface is discontinuous.

An example of passivation is the behaviour of iron in nitric acid. If a piece of iron is immersed in dilute aqueous nitric acid, vigorous corrosion takes place and the reaction rate increases with any increase in the concentration of the acid. In fuming nitric acid, however, the iron becomes passivated and, after a few moments, corrosion is reduced to a negligible amount. Further, on removing the iron from the fuming acid and again immersing it in a solution of lower concentration of nitric acid (HNO_3), the state of corrosion resistance temporarily remains.

One of the first scientists to observe the passivation phenomenon was Michael Faraday in the middle of the nineteenth century, and it may be said, with some truth, that argument as to the real explanation of the formation and behaviour of passivity films on metals which exhibit this peculiar form of corrosion resistance has not yet been concluded. There are several schools of thought: Evans' oxide film theory, Uhlig's electronic modification, or the theory requiring adsorption of ions or molecules on the surface, put forward by both Uhlig and Kolotyrkin.

It is, however, generally agreed that passivation is due to the formation on the metal surface of a special type of ionic current barrier. This protective film is very thin and, although it may be an oxide, it has very different properties from the normal corrosion films and only forms under specific conditions. It forms when a certain critical current density is exceeded at the anodes of the local corrosion cells of the metal surface. When the current density goes beyond the critical point, a change in the overall potential of the metal occurs and barrier films form with the consequent passivation of the metal surface. Under this condition the only electrochemical reaction now open to such a surface is the very slow growth of oxide film, just sufficient to compensate for the film dissolved by the electrolyte solution.

The passivation of iron in an acid solution is illustrated in fig. 11.7. The anodic polarization curve as shown is obtained by a potentiostatic method in which by means of an electronic instrument called a potentiostat, the potential of the electrode is gradually increased while the corresponding polarizing current is kept at the value necessary to maintain the prevailing potential.

Considering fig. 11.7, the iron is active at small current densities – corroding normally as Fe^{2+}. As the current increases a barrier film begins to form and at $i_{critical}$ the current suddenly drops to $i_{passive}$ of considerably lower value. At this point the thick barrier film dissolves and is replaced by a much thinner film which induces passivity on the metal surface. On further increase in potential the current density remains at the same low value (the iron now corroding as Fe^{3+}) until eventually the equilibrium oxygen-electrode potential is reached and oxygen is evolved. The potential value at $i_{passive}$ is often known as the Flade potential after the German electrochemist who made some of the original studies on passivity. (See ref. 14.)

Some metals, e.g. aluminium, chromium, nickel and the stainless steels, possess a pronounced tendency to passivity, others can be made passive, e.g. iron, by exposure to passivating solutions such as chromates or by anodic passivation at high current densities in sulphuric acid for example. A passivated surface, once produced, often does not need such exacting conditions to maintain passivity as were required to produce the phenomenon. Breakdown of passivity can occur from various causes–any factor which causes complete or partial removal of the film: cathodic depassivation, reactivation of the surface may be initiated by chemicals such as chlorides, heat may also destroy passive films formed in the cold, etc. Passive films can, of course, be broken mechanically and sometimes an environment may be such that the substrate is attacked more violently than the passive film and 'undermining' of the passive film occurs.

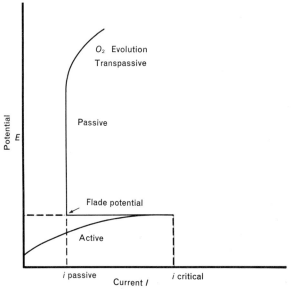

Fig. 11.7 Potentiostatic anodic curve for polarization of iron in oxidizing acid solution. (Diagrammatic.)

11.2.11 Dry corrosion

Corrosion reactions at temperatures above which the environment cannot contain moisture–high-temperature oxidation–are often very complicated. Such reactions usually occur at metal/gas or metal/vapour interfaces where the chances of environmental variations and/or contamination are likely to be more frequent and severe than would be the case in liquids. Another characteristic factor which affects these reactions is that the initial oxidation and the accompanying reduction must occur at one and the same place with consequent probability of critical interference with the progress and rate of the reaction.

If the compound formed at the seat of the corrosion is, for example, volatile the corrosion rate will be very different from that which occurs when the corrosion products form a film. In the latter event a barrier to further corrosion will be formed since all reactants both anodic and cathodic must pass through it. Thus, not only is the type of film a controlling factor–whether or not it is a conductor–but also its behaviour, for example, the rate at which it builds up, whether or not it cracks and if it does whether it heals again. The kinetics of dry corrosion are very involved and although some further consideration of the variation in corrosion rates during high-temperature oxidation is given in section 11.3.8, more detailed accounts (see refs. 1, 3 and 4) should be consulted for more definitive information. (See also, on the fundamental principles of corrosion. (refs. 1 to 6).)

11.3 FORMS OF CORROSION IN THE ABSENCE OF STRESS

11.3.1 General corrosion

General or uniform corrosion is distributed more or less evenly over the whole surface and it usually denotes a condition in which the anodes and cathodes of the corrosion cells cannot be distinguished and in which any area may be an anode at one moment of time and a cathode at the next.

It is, probably, the most acceptable form of corrosion, since very often the service life of a metal undergoing general attack can be calculated within design limits because loss of strength is likely to be directly proportional to the thickness of the metal lost.

11.3.2 Pitting corrosion

When corrosive attack is highly localized it is usually referred to as pitting. This type of corrosion takes many forms and is found in all aqueous environments; no constructional metal is entirely immune to this form of corrosion under some condition or another.

Pitting is a most dangerous type of corrosion reaction and it has two major characteristics. It is very local and the rate of attack is fast. Pits may occur singly at well dispersed locations or they can be so closely grouped together as to resemble local areas of general attack. Pits vary considerably in shape and they can penetrate into the metal at various angles and to various depths depending on the condition of the metal and the environment. If they follow grain boundaries very penetrative attack can be found with the metal honeycombed with cracks and fissures. Pitting is especially dangerous in stressed metal (see also corrosion in the presence of stress, p. 813).

Some conditions which encourage pitting, e.g. salt, metal ion or oxygen concentration cells are well understood, but some of the cases of rapid pitting found under specialized conditions on various passive materials – usually highly resistant to corrosion – are not so well documented.

Pitting will be encouraged by deposits on the surface of the metal, an insufficient amount of inhibitor, the plating out of a noble metal on one less noble, at discontinuities in protective films (either natural or artificially produced) and in crevices, laps, joints, etc.

The basic condition of all pitting is the presence of a relatively small active anode surrounded by a much larger active cathode. This configuration ensures that polarization in such a corrosion cell will remain small and the amount of metal lost will come from a small area but will be large in relation to that area.

Pitting, once it has begun, is usually very difficult to arrest, although some alloys (aluminium among them) are reported to be self-stifling, in which case penetration to failure can be controlled by assessing the metal thickness in terms of rate and degree of penetration, and mechanical strength in relation to the metal remaining after corrosion has stopped. Some improvement in susceptibility to pitting can be effected by alloying; resistance to such attack has been improved in some of the copper-base alloys and the stainless steels by the addition of small amounts of inhibiting elements.

11.3.3 Galvanic corrosion

Although corrosion is fundamentally galvanic (see principles of corrosion, p. 787), corrosion engineers usually allude to corrosion which occurs at the junction of two different metals as galvanic corrosion. In galvanic attack it must be remembered that while the anodic metal is corroded the cathodic member of the couple is protected – corrosion is decreased at this electrode. In any case of galvanic corrosion there will be the corresponding half of the cell where 'non-corrosion' is taking place and this may be a desirable effect as, for example, in cathodic protection.

When dissimilar metals are in electrical contact with one another in an electrolyte solution, a current will flow and the magnitude of that current will depend on the potential difference of the metals and the environment, and also on the basic electrochemical habit of the metals concerned. In Section 11.2 tables of metal potentials and galvanic series were discussed but it is, perhaps advisable to emphasize again that such tables, except possibly those for sea-water, are often not very satisfactory as design data unless the conditions match those of the practical application very closely.

11.3.4 Deposit attack

Deposit attack is a form of pitting. It develops in conducting solutions which contain foreign matter – loose debris, sand, coke, dead organisms, skeletons of dead matter, fouling, etc. – which settle on the surface of the metal. These deposits encourage corrosion in various ways: they prevent the formation of protective films and in consequence corrosion cells – active area/non-active area – are likely to be formed. The deposits can also act as screens between two areas of metal and thus cause concentration difference cells which may be due to varying oxygen concentrations or to differences in salt or metal ion concentrations. In all these cases the anode (corroding area) will be underneath the deposit. Should the deposit be associated with the decomposition of organic matter, it is possible for the biological activity of the bacteria moulds fermenting the environment to use up the available oxygen causing an area of oxygen deficiency with the formation of differential aeration cells. Further, the biological decomposition of organic matter both plant and animal may, on occasion, release certain organic substances which can act as active depolarizers of any corrosion cells in the vicinity with considerable acceleration of corrosion (see section 11.3.11).

The pits associated with deposit attack are often shallow and widespread. If, however, the deposit occurs in a pipe through which an electrolyte is flowing, turbulence may also be promoted and a corrosion-erosion condition set up when pitting may be rapid and grooves may be cut in the metal. In many cases the rate at which such pitting proceeds increases with time and rarely 'stops off'.

11.3.5 Crevice corrosion

Crevice corrosion as the name suggests refers to cases of accelerated corrosion which occur at the junction of two materials – not necessarily galvanically incompatible and of which only one need be a metal – wherever there are crevices. These will usually be found at lap joints, loose rivets, under badly fitting washers, and at loose bolts, etc. The most common corrosion reaction at such locations is the differential aeration mechanism, since any liquid in the crevice is likely to contain less oxygen than the liquid outside the crevice. Also, as soon as any incipient corrosion occurs in the crack, the oxygen present will soon be used up and replacement by inward diffusion will be slow. Thus, a differential aeration cell is formed and corrosion at the anode (inside the crevice) will be severe since its associated cathode (the area outside the crevice) will be large and will not easily polarize due to the presence of ample oxygen.

Crevice corrosion is also dangerous because it is possible for corrosion products to build up in the space to such an extent that they may force the crack open with possible disruption of the joint. Crevice corrosion is often found in threaded joints which split open the union with consequent leakage or even complete failure of the joint.

Some of the more noble constructional alloys, such as the stainless steels and the high nickel alloys are prone to crevice corrosion and deposit attack due to the fact that passive surfaces are usually very active cathodes. This should be borne in mind when scheduling such alloys for bolted or riveted assemblies in active electrolytes such as sea-water.

11.3.6 Selective corrosion (attack)

The term selective attack covers an area of corrosion which is difficult to cover adequately in a few short paragraphs, especially if, as is often the case, the corrosion is associated with stress (see also corrosion in the presence of stress, p. 813).

Selective attack often follows intergranular or intercrystalline paths. In such cases selective attack is really another form of pitting which is very specifically localized. The extent and direction of this attack is controlled by small, often microscopic, galvanic cells which occur due to differences in potential between grain boundaries and their associated grains. The amount of metal lost is often extremely small, yet the loss of strength may be considerable, since the metal can become honeycombed with a network of fine cracks.

The delineation of such corrosion is often difficult without laboratory equipment. In the field, rough tests may be made by bending the metal and observing if any cracks open up or grains drop away from the surface. Also, the metal will often lose its characteristic metallic ring.

It is rare for constructional metals to be completely uniform, and the processes of manufacture sometimes lead to certain constituents being concentrated at grain boundaries, especially if the material has been cast. Segregation of elements or of metal compounds may also be found. The presence of such inequalities will frequently cause the formation of corrosion couples. The presence in the metal surface of dirt, scale, intermetallic compounds, etc., can give rise to more than one type of corrosive attack. For instance, if the inclusion is more noble than the surrounding material, e.g. a graphite flake, localized corrosion will be rapid, since the cathodicity of the carbon will cause rapid and severe pitting. If, on the other hand, the inclusion is anodic to the basis material, the corrosion will stop as soon as the anodic particle is consumed and thus, the amount of corrosion will be limited to the size of the particle. If the inclusions are interconnecting the loss of mechanical strength may be serious.

Cases of selective corrosion are frequently found in some cast irons, some copper-aluminium alloys; the preferential corrosion of nickel has been detected in the copper-nickel alloys and some of the high-alloy solders are not immune to selective corrosive attack. Intergranular corrosion is often found where there is precipitation of a phase at grain boundaries which may happen when a metal is heated after fabrication in uncontrolled conditions, such as local welding.

11.3.7 Dezincification

Dezincification is a form of selective attack which occurs in the brasses. It takes various forms depending on the environment and the composition of the brass but in general terms the strong copper-zinc alloy (brass) loses zinc and becomes a weak mass of porous copper.

The exact mechanism of the reaction has been the subject of much enquiry and has, perhaps, never been completely resolved. Some investigators have suggested that the attack begins with an overall solution of the brass (both copper and zinc ions going into solution), followed by redeposition of the copper. Other workers have put forward the thesis that there is actual selective solution of the zinc with the copper left behind *in situ*. Laboratory experiments have indicated that both theories are possible and that temperature, pH and oxygen concentration play a part in controlling the reaction. Thus, it is reasonable to suppose that under practical conditions both mechanisms may occur simultaneously and that rise in temperature, low oxygen tensions and the presence of chlorides encourage this type of attack. Warm sea-water is a very active promoter of dezincification in all uninhibited brasses (see below).

Dezincification occurs in two different forms, firstly a type of pitting in which an area of the brass undergoes deep, rapid penetrative attack with the formation of a pit plugged with a porous mass of copper–this is usually known as 'plug' dezincification–and, secondly a general attack of the surface layer of the brass which gradually transforms from brass to copper, thickening as time goes on–this is called 'layer' dezincification. In brass tubes carrying a liquid, plug dezincification can be very detrimental, since the plugs tend to drop out leaving holes through which the medium carried can escape. Layer dezincification is often less serious since although the liquid side of the tubes may be attacked, with subsequent loss of strength and possible sagging as the transformation from brass to copper proceeds, it may be a long time before actual penetration takes place.

It was discovered by Bengough and May many years ago that the addition of a few hundredths of 1% of arsenic when added to alpha, single phase brass will ensure almost 100% inhibition from this type of corrosion. Other inhibiting additions, antimony (Sb) and phosphorus (P), have also been proposed but reference to the literature suggests that arsenic is the preferred inhibitor–British Standard Specifications require alpha brasses to be inhibited with 0.02 to 0.06% arsenic (As).

Dezincification also occurs in the two-phase brasses but it must be emphasized that, to date, there is no addition which will inhibit dezincification in these materials to an order of magnitude corresponding to the effect of arsenic in single-phase copper-zinc alloys (see also corrosion of copper and its alloys, p. 832).

11.3.8 Corrosion-erosion

A metal surface can be attacked and destroyed by the combined action of electrochemical corrosion and mechanical abrasion, and with the exception of such obvious instances as sand erosion, or erosion due to slurries and other liquids carrying abrasive particles where the deterioration is almost entirely erosion, there are two types of metal destruction, due to the conjoint action of erosion and corrosion, which are commonly encountered by the corrosion engineer. They are known as impingement attack and cavitation damage.

Although both processes contain elements of corrosion and erosion, the proportions of the two reactions are not the same. Impingement attack is usually mainly electrochemical corrosion accelerated by erosive action; while cavitation damage is usually mainly mechanical damage with some electrochemical component which varies from very small if the liquid is severely cavitating to a relatively large amount if the mechanical action is incipient.

It is, perhaps, necessary to point out that 'cavitation' refers to the hydromechanical formation and collapse of 'cavities' in the water or liquid and not to the pits formed during the process of cavitation. The pitting is the damage and not the cavitation itself.

a. Impingement attack. Impingement attack is caused by a liquid–usually water–striking a metal surface. For significant abrasive action to occur on most metals, air bubbles (entangled air) must be present and there is usually a critical speed of impingement at which deterioration becomes unacceptable. A few metals, e.g. copper, can undergo impingement attack by water containing no air bubbles and at relatively low water speeds (3 to 4 ft/s).

The basic reaction is as follows: the air bubbles erode away any protective surface films and a local injury of the surface is produced with the formation of a corrosion cell, filmed metal (cathode)//unfilmed metal (anode). This area is, then, prevented from healing and/or the corrosion 'stopping off' due to the continual action of the moving air-containing water. The point of impingement thus becomes a very active anodic area and since the water is usually well aerated any cathode associated with the broken surface (anode) will be actively and continuously depolarised which will assist in maintaining the corrosion cell and accelerating the electrochemical component of the reaction. It is not easy to determine the relative amounts of corrosion and erosion but since many instances of impingement attack may be prevented by the application of cathodic protection (see the control of corrosion by design, p. 847) the electrochemical component must, in such cases, be large.

Pits formed by the impingement of moving aerated water have a very characteristic shape. They are smooth, with rounded contours and are usually free of corrosion product. They will be found to be undercut on the downstream side, semicircular in shape with a slightly raised centre. They look like the imprints of horseshoes and the direction of the liquid flow is usually shown by the fact that the 'horse' appears to be walking upstream.

Copper, most of the brasses, the copper-nickels, the copper-nickel-iron alloys, some steels and bronzes will suffer this type of attack, especially in strong electrolyte solution of cavities. In other words, the liquid 'boils', and vapour cavities (areas of partial on the alloy and the design of the system. Tables will be found in ref. 10 for sea-water but designers should consult the manufacturers before accepting such figures as other than general guides to metal resistance.

Impingement attack can occur at any area where there is a turbulent aerated electrolyte solution–in pipes, tubes, on flat surfaces, at valves, branch connections, changes in pipe diameter, and at partial obstructions, etc.

Impingement attack can be reduced by cathodic protection, removal of high velocity and turbulent flow and by using alloys which form protective films which can resist the specific conditions, e.g. aluminium brass in condensers where the average water speed is of the order of 8 ft/s.

b. Cavitation damage. Cavitation damage is usually associated with materials in rapidly moving fluid environments where the hydromechanical process of cavitation occurs. The cavities which promote the attack are formed in the liquid at regions of low pressure caused by irregularities in the liquid flow, often initiated by rotation, vibration or any alteration in the rate of flow which increases the relative velocity of the fluid towards the surface over which it is flowing. At such areas the liquid is subjected to tensile forces greater than its cohesive strength and thus it splits open with the formation of cavities. In other words, the liquid 'boils', and vapour cavities (areas of partial vacuum) are formed. When the liquid returns to normal pressure the cavities collapse extremely rapidly, producing a high acceleration of their liquid envelopes; as the walls of a cavity collide, a shock wave is produced which 'hammers' any adjacent surface. The number of cavities which may collapse is very large on even relatively small areas and it has been calculated that some 2 000 000 cavities may collapse in one second with such devastating effect that pits of the order of several-thousandths of one inch deep can be eroded in aluminium in a few minutes. The cavitation damage cycle is as follows:
1. Formation in the liquid of extremely low pressure areas.
2. Formation of vapour 'cavities'.
3. Abrupt change of pressure in the liquid.
4. Collapse of cavities with the production of shock waves involving pressures in the region of 100 ton/in^2.

Although the occurrence of cavitation damage without any electrochemical component has been demonstrated by the way in which glass, plastics, etc., can be eroded by cavitation produced in non-ionic fluids, the attack is slow; much slower than that found in conducting liquids. This suggests that in electrolyte solutions corrosion plays some part in accounting for the amount of metal lost. That the conductivity of the water is important has been shown in experiments which reveal that in chloride solutions, the attack on steel is some 50% greater than in distilled water, yet 25% sulphuric acid (H_2SO_4) does not increase the damage over that found in the distilled water.

Cavitation damage is a well-known cause of deterioration in ships' propellers, on rudders, etc. It is common in hydraulic turbines, in which components it is sometimes unavoidable. Cavitation damage is also found in diesel engine cooling systems, where, although the water velocity is low, the severe vibrations which occur in cylinder liners can set up cavitation in the surrounding liquid with consequent pitting on their watersides.

The onset and severity of cavitation can be influenced by the properties of the liquid under consideration, e.g. air content and temperature. The amount of air (dispersed air, not dissolved air) in a liquid has been shown to affect the susceptibility of a liquid to cavitation. Small bubbles of about 50μ m diameter constitute active accelerators of the onset of cavitation; on the other hand, the injection of much larger bubbles has been used successfully to prevent cavitation damage. Such air bubbles since they are compressible tend to absorb the energy of the pressure waves caused by the cavities collapse. The effect of temperature on cavitation is somewhat involved but it can be accepted that, in the range of temperature rise found in natural waters, a rise in temperature will increase cavitation. Thus, cavitation damage can be expected to be more in summer than in winter and more in the warmer oceans than in the colder.

The pits caused by cavitation are markedly different from those produced by impingement attack. The pits are sharp, irregular and deep, often attacking one constituent of a casting more than another. The exact sequence of the pit and subsequent crack development is often confused. Surface deformation of the metal surface usually occurs due to the 'hammer' effect before the actual disruption of the material. This hammering may lower the fatigue or corrosion-fatigue resistance and eventual cracking may be associated with fatigue phenomena.

As mentioned above, cavitation can be relatively mild with little release of energy. For example, on some modern well-designed propellers it is probable that the cavitation does no more than continually break the metal's surface oxide film with attacks of the order of that found with impingement conditions. Godfrey has recommended that such attack be called *cavitation corrosion* to distinguish it from the more violent, and more mechanical, *cavitation erosion*.

11.3.9 Stray-current corrosion

Stray-current corrosion occurs mainly on buried or submerged metallic structures, though it can, of course, occur at any point where a direct current leaves a metal surface to follow some electrolytic path, e.g. down an electrolyte flowing in a pipe.

It is different from other forms of corrosion in that the activating current is external to the corrosion cell. An important source of stray currents is leakage of electricity from electric railways and similar systems.

Although stray-current corrosion is not possible, theoretically, with alternating current, many a.c. systems contain some d.c. component and thus the chance of stray-current corrosion occurring in an a.c. system should not be dismissed. Work by the US Bureau of Standards indicates that lead cable sheathing can be damaged by alternating current at commercial frequencies, but that the deterioration is only about 1% of that which would be found with a direct current of the same magnitude. (In view of the remarks below, even this percentage may be very deleterious.)

Stray-current electrolysis can cause very severe corrosion since, if the metal dissolves according to Faraday's law–which postulates that when I ampere passes the amount of corrosion which takes place will be I/F gram-equivalents per second–the following amounts of metal will be dissolved in one year: 20 lb of iron (Fe), 23 lb of divalent copper (Cu), 75 lb of lead (Pb) and 24 lb of zinc (Zn), etc.

The reasons why stray currents are so deleterious is because the rate of attack is controlled by the value of the current flowing and little or no polarization effects are found at the high driving voltages usually encountered. Current of this type will tend to disrupt any protective coatings which may be present and thus attacks will often become concentrated at small areas such as pores and discontinuities. Further, stray-current corrosion tends to be preferential, e.g. at grain boundaries.

One essential towards the prevention of stray-current corrosion is the bonding of components so that any wild currents remain in the metal and are not allowed to escape. In ships it is often unwise to use the hull as an indiscriminate earth; properly fitted busbars are to be preferred.

Welding machines are also likely to set up stray-current leaks unless they are properly earthed; this is especially true if the welding is taking place on a ship afloat. Most countries issue specifications giving the correct connections for earthing welding machines when ships are used alongside or are moored to each other, possibly by electrical conductors such as wire hawsers. Basically, the regulations call for one generator source for each ship, isolated from the shore or another ship.

Stray currents are also found in electroplating plants and, in consequence, the design of the connections of current-carrying cables needs care.

11.3.10 *Corrosion of metals at high temperatures*

Corrosion may occur in other conditions besides those which may be generically termed wet.

At temperatures above about 100°C the presence of moisture is unlikely, and for reactions occurring in the absence of liquid water the expressions 'hot corrosion' or 'dry corrosion' are used. Some investigators, however, prefer to refer to corrosion taking place in such environments as high-temperature oxidation.

In wet corrosion the progress of oxidation is affected by the formation, extent, character and tendency to breakdown or repair of the oxide films or scales formed by the corrosion reaction. Similarly, under conditions of high-temperature oxidation (corrosion) the most important influence on the corrosion rate is the formation, type and persistence of the compound scales which form and must form on the metal surface, unless they are volatile.

As mentioned earlier, all constructional metals have a thermodynamic spontaneity to form compounds with their environment, and their free energy characteristics indicate that their affinity for oxygen will be high. It is well known, however, that little or no oxidation will occur for quite a long period if a highly polished oxide-free metal surface is exposed to dry air. The reason for this contradiction lies in the fact that although there is a rapid initial reaction between the metal and oxygen, this film soon isolates the metal from the dry air environment and continuation of the corrosion reaction is virtually stopped unless it is stimulated by a rise in temperature.

The permeability of such films is very important, and the scale which forms between the metal and the high temperature oxygen-containing environment has some similarity with the electrolyte in the wet reaction: since, for corrosion to occur, the ions of the metal and/or those of the environment must migrate through it.

a. Equations of oxidation. When a metal is attacked by oxygen gas the corrosion product–the metal oxide–will, in the majority of cases, build up as a surface layer on the metal. This layer may be such that access of oxygen to the metal surface is, at some point, prevented, or it may be of such a structure that inward diffusion of oxygen never stops. In this latter case where the continual access of oxygen is not prevented, the oxide film, x, thickens at an undiminished rate with time, t, and the rate is unaffected by the thickness of the surface layer; the growth law is:

$$\frac{dx}{dt} = k_1$$

which on integration gives a linear law (see fig. 11.8). Examples are, magnesium (Mg), calcium (Ca), molybdenum (Mo), etc.

If an oxide forms a film between the metal and the environment of such a variety that the rate of diffusion of the reactants (metal outwards to the oxygen, and oxygen inwards to the metal, or a combination of both) is the controlling step in the oxidation, the oxidation must decrease as the thickness of the film increases. In other words, the effectiveness of the oxide barrier must be directly proportional to the thickness and the growth law will be represented by:

$$\frac{dx}{dt} = \frac{k_2}{x}$$

which on integration gives a parabolic law (see fig. 11.8). The carbon and low-alloy steels are among those metals which follow this law. Since, in this case the oxide film grows until it completely separates the metal and the environment, the growth will proceed in two stages: firstly, a surface reaction at the oxygen/oxide and the oxide/metal interfaces, and secondly, a diffusion of material through the oxide. The rate of growth will be controlled by whichever of the two reactions is the slower.

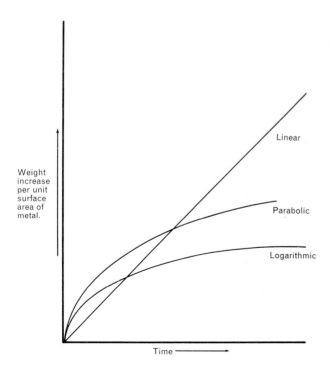

Fig. 11.8
The main oxidation laws.

Of course, in the initial stages of the oxidative film formation when the film is very thin the second stage will be absent.

A third process of oxide-film formation is where the metal oxidizes very rapidly at first, followed by a slowing down to a constant rate at which the thickness does not increase. Here the film formation follows a logarithmic law:

$$x = k_1 \log(1 + k_2 t)$$

relating thickness with time (see fig. 11.8). This equation applies to relatively thin protective films which form when some metals oxidize at low temperatures, e.g. copper, tin, aluminium.

It must be pointed out that these rate laws do not always apply in practice with the regularity which may be found in laboratory experiments and that metals vary in the law they follow depending on the environmental circumstances. Further, some metals change their growth rate factors as oxidation proceeds and the composition of the films alter. Oxidation rates are also altered by the physical breakdown of the oxide films and the sequence of attack may proceed in steps rather than continuously. For example, copper at 800°C follows a smooth and continuous rate of oxide growth, while at 500°C the oxide formation rate slows down and then suddenly begins again at an increased rate, followed by a further slow down, etc.

The amount of oxide formed from the oxidation of an area of metal—whether the volume of the oxide is greater or less than that of the metal consumed—has been the subject of considerable enquiry and in 1923 Pilling and Bedworth argued that the oxide layer would be continuous if its volume was greater than that of the metal destroyed in forming it. It was suggested that such an oxide would be impervious and so restrain the access of gaseous oxygen. On the other hand, if the oxide was less voluminous than the metal destroyed, the layer would allow the gaseous oxygen to reach the metal surface continuously. This simple law, though of some validity, is now known not to fit all the facts, e.g. many metals whose oxide : metal ratio is greater than unity are found to be porous at elevated temperatures, and some metals for which the ratio is less than unity only form a porous oxide layer at temperatures greater than some critical value which is characteristic of the metal. However, the Pilling and Bedworth rule can be used to predict whether the oxide layer will be in lateral compression (oxide : metal ratio greater than one) or in lateral tension (ratio less than one).

In the main, the practical significance of hot corrosion is controlled by the kinetics of the reactions rather than by the thermodynamic stability of the reactants or the products involved. It must be stressed that reaction rates found in laboratory experiments are frequently not applicable to the more complicated practical environments. Two major factors affect such results: stress, and thermal cycling. In the first case, attack somewhat similar to stress-corrosion cracking may be initiated and stresses may also break the films with varying degrees of acceleration of general attack. Thermal cycling may also lead to film breakdown or flaking with acceleration in the rate of corrosion.

The behaviour of the various metals and alloys at high temperatures are too variable to be discussed in detail here and any specific problem must be referred to the literature where many excellent tables are given of the behaviour of metals at high temperatures and the compositions most likely to be satisfactory in the high-temperature environments found in operating plant.

11.3.11 Biological activity and corrosion.
Life processes lead to many forms of fermentation and the production of a large range of organic and inorganic compounds some of which have been shown to play a part in initiating or sustaining the corrosion reaction.

As early as 1891 it was suggested that compounds of nitrogen, especially ammonia, produced by bacterial action could accelerate the corrosion of lead in some natural waters. In the same decade a Dutch bacteriologist (Beijerinck) demonstrated that sulphates could be reduced to sulphides (H_2S) by the action of bacteria, and in the early 1920s two of his compatriots (von Wolzogen Kühr and van der Vlugt) drew attention to the possibility of a connection between the acceleration of corrosion under anaerobic conditions and the bacterial reduction of sulphides. About the same time three Australians (Grant, Bate and Myers), investigating corrosion problems in heat exchangers cooled by the highly polluted waters of Sydney Harbour, demonstrated experimentally that the pitting rate in condenser tubes was considerably increased when ammonia produced by bacterial action was present in the water, especially during shut-down periods when the bacteria fermented the polluted water which lay stagnant along the bottom of the tubes.

Since few corrosion investigators are trained bacteriologists and not many bacteriologists had any interest in corrosion it was some years before the incidence of bacteria in metallic corrosion was studied in any systematic detail.

However, stemming from the pioneer work of the two Dutchmen the 1930s saw considerable activity in the interpretation of the role of bacterial action on the corrosion of metals. In the UK Bunker and Rogers studied the problem, as did Hadley and others in the USA, and it is now accepted by both corrosion engineers and corrosion scientists that biological activity can play a role, sometimes a major role, in the corrosion of metals. Also, to bacteria must be added various other organisms such as moulds, lichens, etc., which are capable of producing corrosive substances, e.g. acid exudates, carbon dioxide, etc., produced in the course of their normal metabolic cycles.

The corrosion reaction can be accelerated by bacteria etc. in various ways, e.g. by the production of such compounds as hydrogen sulphide (H_2S), carbon dioxide (CO_2), ammonium ions (NH_4^-), sulphuric acid (H_2SO_4), etc. which are directly corrosive; by the production of organic compounds which can act as catalysts or depolarizers in corrosion reactions and, as is the case in the anaerobic corrosion of iron, where the activities of the sulphate-reducing bacteria become involved as an integral part of the corrosion reaction.

In some cases bacterial colonies or moulds can live on a metal surface, and the necessary concentration of the corroding agent may be found only in the immediate vicinity of the colony and may be undetectable in the surrounding environment; in others, the whole environment may become contaminated. Further, a bacterial colony growing on a metal surface can set up an oxygen screen, with the consequent formation of a differential aeration cell which may be more than usually active if the colony also provides some added corrosion accelerator. In such cases local attack may be very severe.

Probably the best known example of the association of bacterial action and corrosion is found where those bacteria which reduce sulphates are active. The most important of these organisms is known as *Desulfovibrio desulfuricans*, an organism which can only live in the absence, or at very low concentrations, of oxygen. Certain harbours and rivers around the world are highly contaminated with sulphides produced from the activities of these particular bacteria–Hamburg, Los Angeles and the lower reaches of the Thames are well known examples. These organisms can live in either fresh or salt water and one can expect to find them active in any stagnant (low oxygen) locality which contains soluble organic matter and sulphates. Those littoral areas which receive waste products from fish-processing plant, meat-packing factories or sugar refineries are usually heavily contaminated with hydrogen sulphide, the same is also true of fitting-out basins and many non-tidal docking areas. Instances are on record of the serious corrosion of ships' hulls before they have ever been to sea.

Besides the production of hydrogen sulphide as a direct-action corrodant, it has been suggested that the sulphate-reducing bacteria can become involved in the corrosion reaction itself. (A mechanism involving the cathodic depolarization of an active corrosion cell by the removal of hydrogen from a cathode by the bacteria, as an integral part of their metabolic energy cycle.)

The theory stems from the analysis of the corrosion products which are formed when iron or steel corrode anaerobically in the presence of hydrogen sulphide. Direct corrosion by hydrogen sulphide (H_2S) should be according to the equation:

$$Fe + H_2S \rightarrow FeS + H_2 \tag{1}$$

on the other hand, cathodic polarization predicts the following equation:

$$4Fe + 4H_2O + SO_4^{2-} \rightarrow 3Fe(OH)_2 + FeS + FeS + 2OH^- \tag{2}$$

Analysis of corrosion products usually found at the site of the anaerobic corrosion of iron are found to be in closer agreement with equation (2) than with (1). Thus it has been deduced that bacterial removal of cathodic hydrogen produced during an anaerobic corrosion reaction can take place with consequent increase in the corrosion rate. The removal of hydrogen atoms from a cathode has been shown to be associated with the presence of the enzyme dehydrogenase, and organisms rich in this compound (as are the sulphate-reducing organisms) have been shown experimentally to be able to use the hydrogen evolved from cathodic reactions in anaerobic corrosion. This reaction appears to take place in direct relation to the amount of enzyme present. Thus, the corrosion of a ferrous metal in an environment of very low oxygen concentration such as waterlogged soil or mud will, in the presence of *Desulfovibrio desulfuricans*, be accelerated even in the absence of oxygen. The cathodic hydrogen will be used up in the formation of S^{2-} ions (cathodic depolarization); these ions will then react with the Fe^{2+} ions going into solution at the anode with the formation of black ferrous sulphide and the corrosion will, therefore, be further stimulated by anodic depolarization. Under these conditions it is possible for the corrosion of iron to be more severe in anaerobic than in aerobic environments. It may also be mentioned that when metals acquire sulphide films there is always the possibility, since these films are cathodic to the underlying metal, that local corrosion may occur at breaks in the films.

Other bacteria which produce direct-acting corrodants are those which excrete nitrogen compounds, ammonia, etc., and those which produce acids: *Thiobacillus thiooxidans* can synthesize sulphuric acid and there are many bacteria which decompose cellulose, with the production of carbon dioxide and organic acids which will cause corrosion. For example, metal pipes lagged with cellulose-containing compounds may be attacked.

The acceleration of corrosion by the products of bacterial metabolism is sometimes due to the presence of redox systems. One particular system that has been studied is that of the sulphur compounds cystine and cysteine. Sulphydryl compounds of this type are capable of reversible oxidation/reduction reactions, e.g. cysteine in alkaline solution and in the presence of certain metal catalysts (such as copper) and oxygen can be oxidized to cystine and the cystine can be reduced again to cysteine by hydrogen; and it has been shown that the cathodic hydrogen of the corrosion reaction can effect this reduction.

$$\text{Cystine (oxidized disulphide)} + H_2 \rightarrow \text{Cysteine}$$
$$\text{Cysteine (reduced mercaptan)} + O_2 \rightarrow \text{Cystine}$$

The reaction is complicated, and the presence of certain enzymes may be necessary, but it has been shown experimentally that sea-water which contains sulphydryl group compounds is excessively corrosive to copper-base alloys and that the addition of small quantities (a few ppm) of cystine or its related compound glutathione will increase the corrosiveness of saline (near neutral) waters towards alloys containing copper.

The control of biological action in relation to corrosion is difficult since it is not easy to inhibit bacterial action in large quantities of water, particularly since many of the indigenous soil and water bacteria are very resistant to most of the common disinfectants, unless they are used in concentrations which become uneconomic. Chlorine has been used in condensers in land power stations and some dyestuffs, though capable of selectively destroying the sulphate-reducing bacteria, can only be used in closed systems. The adjustment of the pH of an environment to a range lethal to the offending organisms has been used as well as aeration or rise in temperature, but again, these palliatives are only likely to be feasible if the volume of water is small. Cathodic protection can be used to reduce such corrosion but in some cases the current density required for full protection may be considerably more than that required in uncontaminated water.

11.4 CORROSION IN THE PRESENCE OF STRESS

11.4.1 Introduction

In general, it can be said that all tensile stresses, static or alternating, tend to increase the probability of a corrosion reaction resulting in a metal failure; similarly, corrosion tends to increase susceptibility to failure where stresses are operating. Stresses other than tensile vary in their effects, e.g. compressive stresses tend to decrease the probability of failure by stress-corrosion cracking. In the majority of cases the two phenomena – corrosion and stress – acting together, irrespective of their ratio, produce more serious consequences than would be expected from the combined effects of both.

Almost all mixed metal structures will contain constructional materials that are susceptible to stress corrosion or corrosion fatigue, though for some metals specific environments are necessary for these types of corrosion.

The essential factors from the practical aspect of corrosion plus stress are the uncertainty of time to failure and the that in corrosion fatigue (see p. 817) *there is no fatigue limit*. Alterations in environment from benign to injurious may alter the time to failure from years to minutes, also if the stress is, for some reason, reduced or relieved or the corrosion stopped, the behaviour of a metal prone to stress corrosion may alter in character with consequent variation in the interval before the onset of unacceptable behaviour. Further, the amount of corrosion necessary to initiate failure under stress varies considerably. It may be very small and it may have occurred before the application of the stress; alternatively, it may be very large, with the result that the corrosion reduces the metal to a size that cannot support the desired load.

In general, the stress level necessary to cause failure by stress corrosion is roughly that of the yield point of the metal; but it is by no means unknown for failure to occur in the presence of corrosion well below this figure.

Corrosion fatigue, the condition where a metal fails due the presence of corrosion and cyclic stresses, can be considered as similar to normal fatigue. The damaging effects are accelerated, however, more by chemical than mechanical agencies and with the significant exception, already mentioned, that corrosion fatigue has no fatigue limit as is normally found in a metal fatiguing in air.

Fretting corrosion is another type of metal failure which may be basically failure by fatigue or corrosion fatigue but which is found in certain specific and characteristic locations.

The technology of metal failure entailing corrosion and stress is very complex and any attempt to put forward the various theories relating to the phenomenon, if ambiguity and indecisiveness are to be avoided, would need more space than can be given here. Considerable investigation into this type of failure has been made, and the literature on the subject is voluminous but excellent surveys will be found in refs. 1, 3 (vol. 1, ch. 8), 4 and 5. A further difficulty in the field of stress and corrosion is that the nomenclature used to describe the various reactions is often confusing to a reader unfamiliar with corrosion. Some of the terms used are specific to environments, e.g. season cracking, while the expressions stress corrosion and stress-corrosion cracking may mean different things to the corrosion engineer and the corrosion scientist. Again, the reason why certain otherwise ductile metals suffer local brittle fracture, in the presence of mild corrosion and low stress while maintaining their general ductility, is a problem that needs to be studied from the specific literature.

Normally, stress-corrosion cracking and corrosion fatigue failures are initiated by the formation of pits on the metal surface. Some investigators have endeavoured to relate the configuration of the pits to the various types of conjoint action but there seems little certainty about such possibilities – most probably due to the lack of sufficient data to make statistical estimates viable.

Examples of the 'time to failure' parameter are provided by the case of stressed brass in mercury salts and the aluminium-magnesium alloys. In the copper-base alloy, time to failure in mercurous nitrate is a matter of seconds while some of the aluminium alloys may not become susceptible to stress-corrosion failure before considerable metallurgical changes (precipitation) have occurred.

In stress-corrosion cracking it is usually considered that internal stresses are more likely to be dangerous than those which are applied – many manufacturing processes may produce such effects. For example, non-uniform cold work, unequal cooling rates, changes in metallurgical structure during or after casting, deformation by riveting, thermal treatments, ageing and composition, are all capable of affecting the possibilities of stress and corrosion producing a deleterious condition. The fact that some failures are transgranular and some are intergranular offers obvious differences which have been suggested as diagnostic aids. However, it does not appear that there is sufficient evidence to render such differences specific. While it is not easy to dogmatize, the following conditions are usually found in stress-corrosion cracking:

1. The damage is caused by the conjoint activity of tensile stress and some electrochemical process.
2. The cracks show no signs of ductile tearing.
3. The development of cracks can be arrested by removing the electrochemical component of the reaction.
4. If the stress level is increased after corrosion has stopped further cracking will be ductile.
5. The phenomenon can be divided into two parts – a relatively long induction (corrosion) period followed by a short period of crack propagation.
6. Cracks may be transgranular or intergranular but they are, usually, predominantly one or the other. No generalizations can be offered on what conditions favour one type of cracking more than another; a change in environment may suffice to change the mode of cracking.
7. The average rate of crack propagation in many metals appears to have a characteristic value of the order of 0.5 cm/h.
8. The environments for stress corrosion cracking are often highly specific. For example, certain ions (Cl^-, NO_3^-, OH^-) appear necessary for transgranular cracking. In the brasses, cracking is intergranular at pH 6.3 to 7.7, while at lower and higher pH values the cracking is transgranular. The cracking of mild steel due to nitrates also has a narrow pH range.

11.4.2 The behaviour of certain metals towards stress-corrosion cracking

   a. Aluminium alloys. Pure aluminium is not susceptible to stress-corrosion cracking which occurs nevertheless in the aluminium-zinc-magnesium, aluminium-magnesium and the aluminium-copper series of aluminium alloys. In cast aluminium alloys the incidence of stress-corrosion cracking is rare, though it has been reported in the high (10%) magnesium-containing materials when used in sea-water – though there is some evidence that failures are usually associated with compositional or manufacturing inequalities.

   b. Magnesium alloys. Some wrought magnesium alloys are susceptible to stress-corrosion cracking but the cast materials usually have good resistance to this type of attack. It is impossible to relate stress-corrosion cracking to any definite schedule of behaviour for these alloys. The failure of both aluminium and magnesium by stress-corrosion cracking is often closely associated with the presence, or absence, of separate phases at the grain boundaries, and while these phases are usually anodic to the grains, and will suffer preferential attack, their presence and extent can depend on many metallurgical factors, while the degree of attack can vary considerably with the environmental conditions.

   c. Copper-base alloys. Although the required environments may be different and sometimes very specific, most copper-base materials will exhibit a tendency to stress-corrosion cracking if the conditions are suitable. When the deleterious stresses are produced during manufacture the failures are often known as season cracking.

Among the most important specific environments are the presence of ammoniacal compounds, both organic and inorganic, and of mercury and solutions of its salts.

It has been shown that the mechanisms of the stress-corrosion cracking of brass in the two above environments are different. Mercury, and its salts, appears always to produce failure by intercrystalline cracking, while in an environment containing ammonia or one of its compounds the cracking may be intercrystalline, transcrystalline or a mixture of both. The cracking of internally stressed brass by mercurous nitrate is so specific that it is often used as a diagnostic for the presence of built-in stresses in manufactured brass articles.

Zinc is reported to increase susceptibility to stress-corrosion cracking of the brasses, high zinc alloys being more likely to crack than the low zinc materials. Failure by stress-corrosion cracking may be reduced in the 70/30 type alpha brasses by the addition of 1% silicon.

The aluminium-bronzes, tin-bronzes, nickel-silvers and the copper-nickel alloys can all undergo stress-corrosion cracking if the conditions, both manufacturing and environmental, are just right – what these 'right' conditions are may be difficult to evaluate and certainly cannot be tabulated. Electrolytic, tough pitch and oxygen-free-high-conductivity copper have been reported as immune to season cracking, but phosphor-copper and copper containing about 0.4% arsenic and 0.2% phosphorous are suspect.

d. Mild steels. Certain corrosive solutions will encourage stress-corrosion cracking in mild steels under the following conditions if all are present at the same time: a susceptible structural condition of the steel; a tensile stress of sufficient magnitude; and certain specific corrosive environments. The most important factor in sensitizing the plain carbon steels seems to be the carbon content – the susceptibility decreases with increasing carbon content. The most usual environments likely to produce stress-corrosion cracking in these steels are solutions in which sodium hydroxide or nitrates are present. In the latter case the cation and the pH have some effect on the tendency for a solution to encourage stress-corrosion cracking. Ammoniacal vapours, hydrogen sulphide and carbon dioxide, have also been shown to encourage this type of attack at specific concentrations and temperatures; chlorides do not appear to increase the risk of stress-corrosion cracking occurring.

In an environment of sodium hydroxide and static stress, low-carbon steels undergo a type of attack known as caustic cracking, caustic embrittlement or boiler embrittlement (embrittlement referring to the type of crack and not to a general loss of ductility). For caustic cracking to occur the stress level must approach the yield point of the metal and the concentration of the caustic and the degree of temperature must be high. This type of cracking is found in alkaline boiler waters or waters which have been rendered alkaline by treatment. It has been stated that such cracking may be prevented if a certain ratio of sodium sulphate to sodium hydroxide is maintained. The addition of sodium nitrate, trisodium phosphate and tannin have also been proposed as inhibitors of the effect of the caustic alkali.

e. High tensile steels. In these steels the possibility of hydrogen playing a role in cracking is well known, and since the hydrogen may arise from a corrosion reaction some failures may be broadly classed as stress-corrosion cracking.

Steels increase in susceptibility to hydrogen cracking in direct relation to increases in tensile strength, and failure from this type of phenomenon has become more probable with the introduction of the modern steels which can resist 90 to 130 ton/in^2; although means of combating the trouble by careful control of pickling baths, etc., is receiving considerable attention, users of these steels are recommended to consult the modern specifications.

The ductility of hydrogen-embrittled steel may be restored by ageing in the atmosphere or by soaking in hot water; both treatments tend to cause the hydrogen to diffuse out of the metal. If, however, the occluded atomic hydrogen accumulates in any cavity, molecular hydrogen may be formed and then its outward diffusion will be prevented.

In hydrogen embrittlement there is a critical load stress below which failure does not occur and this type of attack is found less frequently at very low (−220°C) or at high (+200°C) temperatures.

It must be mentioned that surface metallic coatings which are anodic to the basis steel can encourage hydrogen embrittlement by galvanic action at pores, and there is always the possibility of hydrogen evolution with subsequent cracking if such steels are subjected to uncontrolled cathodic protection.

f. Stainless steels. Stress-corrosion cracking in the austenitic stainless steels has been found in hot water, steam, hot chloride solutions, hot caustic liquids, pickling baths, organic acids, some fruit juices and perspiration, etc.; it has been suggested, however, that the presence of chlorides or caustic alkali may be essential.

The stress-corrosion cracking of austenitic stainless steels is a serious problem and each environment must be considered peculiar. Some substances, e.g. phosphates, can act as inhibitors, but rises in temperature tend to increase susceptibility.

The cracks in the austenitic types are usually transgranular and tend to spread over the surface with little penetration; 18 : 8 stainless steels are also susceptible to stress-corrosion cracking if sea-water evaporates on warm surfaces. The possible serious deterioration of all stainless steels in sea-water if often insufficiently appreciated.

The martensitic stainless steels are also susceptible to stress-corrosion cracking, and the phenomenon has been reported in environments containing sulphides, chlorides and in marine and industrial atmospheres.

11.4.3 Corrosion fatigue

Corrosion fatigue has been defined as the type of failure which takes place when a metal is subjected to cyclic stressing in an environment which continuously corrodes the bare metal, i.e. if the metal is able to react with the environment at areas not protected by oxide or other natural or artificial films. It is important to note that the natural protective films which form on some metals in some environments and play such an important part in the prevention of many corrosion reactions, often lose their protective value under conditions of alternating stress – they tend to crack and admit the corrodant to the basis metal. In consequence corrosion fatigue is possible in almost any metal under the conditions defined above. For example, aluminium is quite corrosion resistant to normal atmospheres under static conditions since it easily forms an excellent protective oxide film. However, when alternating stresses are present it is possible for the oxide film to be disrupted more quickly than it can be repaired and, in such an event, corrosion fatigue is not only possible but likely.

All engineers are familiar with the standard fatigue curves which are used for the design of structures or components which must resist alternating stresses. Such curves exhibit a fatigue limit, the limit of stress below which failure does not normally occur no matter how many reversals may take place. In corrosion fatigue there is no fatigue limit and as long as corrosion is active failure will eventually occur provided that the alternating stress, whatever its magnitude, is maintained.

Failures due to corrosion fatigue may, or may not, show visible corrosion but they will usually show a large number of cracks at the point of failure in addition to the one where failure actually occurred. When the stress system is uniaxial the cracks are generally parallel to each other and in a plane perpendicular to the direction of the stress. In a torsion failure a group of cracks emanating from one point is more usual and they are often cruciform in configuration with arms at 45° to the torsion axis. Cracking in pipes, due to cyclic thermal stresses, usually makes parallel circumferential cracks with further cracks at right angles, parallel to the pipe axis.

Another characteristic of corrosion fatigue is the exudation of corrosion products from the cracks which sometimes stain the surface of the metal in the area of the failure. Corrosion fatigue cracks are often transgranular, but not invariably, thus the diagnosis of corrosion fatigue from the presence of transgranular cracking may not necessarily be correct.

Corrosion-fatigue failures may be roughly divided into two classes: those which are due to regular reciprocating stresses such as are found in static machinery and whose mean oscillation is often small and where failure is likely to be repetitive; and those found in such large objects or components as ships, moving vehicles, etc. where the operational conditions which caused the failure, e.g. the passage of a ship through a particular sequence of seas and waves, may never occur again.

The results of corrosion fatigue are extremely dangerous and can be catastrophic, and the methods of eliminating this type of damage are different for the failure in the continuously vibrating object or component from the occasional and isolated failure due to some peculiar circumstance. In the first case probably the most important remedial factor will be by attention to design to remove the stresses followed by removal of the corrosive environment, while in the second case protection from the environment may be a sufficient remedy. With regard to design, not only should alternating stresses be maintained at a minimum but also cracks and crevices suitable for initiating corrosion should be avoided or sealed. In addition, if inhibitors are used care must be taken to ensure that the amount of inhibitor is sufficient to maintain the passivity and that local non-passivated areas do not occur with the formation of active-passive cells, with consequent pitting which may act as stress raisers for the occurrence of stress corrosion phenomenon.

As mentioned above, alternating stresses can disrupt protective films; thus, remedial measures which include the use of material whose corrosion resistance depends on the presence of oxide films, need to be carefully considered before use. Further, any artificial protective films will need to be more plastic when used in the presence of alternating stress than under static conditions.

Compressive stresses are useful as inhibitors of corrosion fatigue, and such stresses can be induced on metal surfaces by chemical treatment (nitriding), metallurgical processes (heat treatment) or by mechanical fettling (shot-peening). In the exclusion of a corrosive environment by electroplating it must also be remembered that tensile stresses may be induced in the basis material by the plating itself, and that corrosion at discontinuities in a coating may be deleterious. For example, zinc plating produces a coating in compression which will tend to reduce the likelihood of corrosion fatigue, but since very active corrosion may occur between zinc and some basis metals at pores or cracks with the production of hydrogen – which may enter the metal surface – the resistance to corrosion fatigue may be lowered. Spraying with pure aluminium has been used effectively to protect susceptible high-strength aluminium alloys. Many organic coatings and modern plastics can be used very satisfactorily as barrier coats, as long as they remain plastic and bonded to the vibrating base. In general, if the problem is one solely of corrosion fatigue, with no component of ordinary fatigue, the most simple way of preventing failure is the removal, if operationally possible, of contact between the metal and the corrosive environment.

From the above it will be noted that the prevention of corrosion fatigue, similarly of stress-corrosion cracking, is extremely complicated and where either type of attack is a recurring problem the authoritative literature must be consulted.

11.4.4 Fretting corrosion

Fretting corrosion is a form of damage which occurs when two closely fitting surfaces are subjected to slight relative oscillatory slip. The surfaces become pitted and a finely divided corrosion product (oxide detritus) is usually formed. Nearly all constructional metals are subject to fretting if the correct conditions are present. It is most commonly found in vibrating machinery and the vibrations may be indigenous to the machine or they may be induced from some other source of movement, e.g. in moving vehicles which vibrate as a normal consequence of their operation.

Fretting is known to engineers by a variety of names and many of the names contain the words blood or cocoa. These stem from the fact that fine rust formed at the point of slip in ferrous materials appears to 'bleed' from the area, with characteristic discoloration and red streaks.

Fretting is found where shrink and press fits, bolted parts, splined couplings, keyways, tracks of all types of ball race, are incorporated in a machine or component. Sometimes the pits formed in fretting will initiate fatigue failure.

Fretting fatigue is affected by many different conditions among which are: the amount of oxygen and moisture in the environment (the amount of damage increases as these increase), the temperature (above 50°C its effect is reported to be slight), load (increase in load), amplitude of slip, and number of vibratory cycles are all reported to increase attack. Hardness is beneficial and, in general, attack of this type decreases with increasing degrees of hardness.

Although lubricants can reduce the tendency to fretting they are not so successful on oscillating surfaces as on normal sliding surfaces. The remedies for fretting seem to fall into two groups according as to whether the surfaces were, or were not, intended to slip. If the surfaces should not move in relation to each other then, either the vibrations must be removed, or the friction between the surfaces increased. This latter requirement may be accomplished by plating the opposing surfaces or by inserting some suitable elastic substance between them to absorb the motion. Where relative motion is a necessary requirement, the oscillatory movement in the system should again be brought to as low a value as possible and great attention paid to the lubrication, the type of oil and its access to the relevant surfaces. Phosphated and oil-impregnated surfaces have been used to reduce this type of attack.

It is perhaps worth mentioning that, although the fretting of ferrous materials is usually indicated by a reddish exudation, with other metals the colour is different, e.g. black for aluminium.

11.5 CORROSION OF METALS
11.5.1 Ferrous metals. (wrought)

All ferrous materials will corrode – some very freely – and the rate of attack will depend on the composition of the alloy, the overall environment, the local conditions and the amount of protection given to the metal, e.g. by paint, metallic coatings or cathodic protection, etc., or some combination of the various methods.

The constructional steels can be roughly divided into three broad groups:
1. Those ferrous materials which contain no deliberately added alloying elements and which corrode very freely in most environments unless protected in some way. Examples: plain carbon steels, ordinary cast irons, wrought iron.
2. Those steels which contain small percentages of alloying elements, the total elements (other than iron) usually not exceeding 2%. These steels corrode freely in the presence of chlorides, but under some conditions, expecially in the atmosphere, corrode at a much slower rate than those in group 1. This group contains those alloys which contain small amounts of copper (0.5%) or chromium (1%).
3. The stainless steels which contain large percentages of chromium, nickel and molybdenum. Steels in this class remain virtually unattacked in some environments; though they are by no means completely resistant to corrosion under some conditions, e.g. they are very sensitive to concentration and differential aeration cells, especially in chloride solutions.

The high-alloy steels resistant to high temperature oxidation will be considered separately but many of these steels while resistant to high temperatures may not always be resistant to corrosive waters and atmospheres.

a. Group 1 (plain carbon steels). This group contains the common constructional steels: those plain iron-carbon alloys to which no elements are deliberately added above those present in the pig and scrap from which they are normally manufactured.

In general, neither the manufacturing process nor the normal variations in the carbon, manganese and silicon contents of these alloys has any major effect on their overall corrosion resistance. It is true, however, that those steels which contain relatively large percentages of these three elements are slightly more resistant to atmospheric attack than those with low percentages. Copper, on the other hand, has a definite effect on the resistance of this class of steel to atmospheric rusting and an increase in the copper content from 0.01 to 0.05% may increase the resistance by two to three times.

The corrosion resistance of these steels can also be considerably affected by surface condition, e.g. if, during fabrication, their surfaces become contaminated by oxide rolling- or mill-scales. Since mill-scale is cathodic to the underlying metal and is usually discontinuous and full of cracks, severe local corrosion is likely if such scaled steel is exposed to any strong electrolyte such as sea-water. Thus it is usual to use only de-scaled material in marine environments.

Rusting in the atmosphere. In air the rusting of these steels depends to a large extent on the climatic conditions and the impurities which are present in the air and also, to some extent, on the mass of the metal under consideration, since the rate at which a mass of metal will react to differences in ambient temperatures will affect condensation rates.

The presence of moisture, rain, spray, etc., will be of considerable importance in considering corrosion rates. It has also been shown that water vapour in the air, even at relative humidities below saturation, can be effective in causing the rusting of constructional steels especially in the presence of such impurities as sulphur dioxide. In fact, in any given humidity the presence of impurities, usually sulphur dioxide and dust, will increase the corrosion rate to a marked degree.

The effect of temperature on the rusting of mild steel in air is complex. Rise in temperature should increase reaction rates but if the effect is to dry the environment the corrosion rates may be reduced, and as a general rule the moisture and contamination contents of the air are more important than rises in atmospheric temperature.

In table 11.1 some figures from the researches of The British Iron and Steel Research Association (BISRA) are quoted showing variations in the rate of rusting of mild steel in the open air. (These figures, since they are obtained from small coupons of ingot iron, may not be definite for large areas and masses but they are useful in showing the variations in corrosion rates between different environments.) The highest rate 0.0244 ipy (inches/year) at Lagos (W. Africa) beach represents the destruction of $^1/_8$ in material in under three years. It will be noticed that the corrosion rates for GB have a general relationship to the amount of industrial pollution, which tends to increase corrosion even more than marine localities. It will also be noticed that areas of low humidity also show low corrosion rates, and in an area such as Khartoum where the air is clean as well as of low humidity, the corrosion rate is exceptionally small.

Experiments carried out by the British Iron and Steel Research Association and Larrabee in the USA (see refs. 15 and 16) have shown that the orientation of the steel has a considerable effect on the corrosion rates, e.g. the underside of a horizontal surface even though it may be protected from rain can at times remain wet for long periods because it may be protected from the sun, drying winds, etc. In consequence the lower surface may corrode faster than the topside. Further, if the steel is vertical the corrosion rate will be affected by the direction of the prevailing wind. Larrabee has also shown that the climatic conditions at the time of initial exposure can influence future rusting rates. For example, steel initially exposed in the autumn corroded about 20% more than similar steel first exposed in the spring. The rate of rusting with time has also been studied by BISRA and their graphs show that rusting rates fall in the first year or so and then become fairly constant – similar results have been obtained by the American investigators.

Table 11.1 Rates of rusting of mild steel outdoors in different localities (based on BISRA data).

Type of atmosphere	Corrosion rates (inches/year)
GB, rural or suburban	0.0021 (average 3 sites)
GB, marine.	0.0026 (average 2 sites)
GB, industrial	0.0053 (average 5 sites)
Khartoum, rural	0.0001
Delhi, rural	0.0003
USA, rural	0.0017
Germany, rural	0.0021
Singapore, marine	0.0006
Nigeria, marine	0.0011
Lagos, surf beach	0.0244
USA, industrial	0.0043

The above remarks refer to the corrosion of bare steel in outdoor conditions; indoors where there is no precipitation and no direct sunlight, the rates of rusting depend almost entirely on the condensation of moisture and contamination of the local atmosphere.

Rusting in waters. Here again, so many factors affect the corrosion rates that generalized figures are not of much value, particularly, in industrial service, where the operating conditions vary considerably with effects on the corrosion of these steels which cover wide variations.

However, when totally immersed in natural waters the structural steels of group 1 corrode at roughly the same rate and in sea-water the average rate, in any but the most extreme environments, is of the order of 0.005 to 0.006 ipy.

Surface condition is also important and for use in waters the removal of manufacturing oxide scales is mandatory.

The composition of the water can affect corrosion rates, i.e. salinity and pH, etc. Saline and acid soft waters will be more corrosive than near neutral hard waters. Alkaline waters are usually the least aggressive and in some cases the corrosion of mild steels can be controlled by artificially increasing the pH of a water, e.g. boiler waters. Further, since oxygen can act as a cathodic depolarizer in aqueous corrosion reactions its concentration in any locality will affect the corrosion rates. In some cases, e.g. boiler tubes, the corrosion rate can be considerably reduced by removal of oxygen as well as by increasing the pH.

The presence of organic matter, fouling and decomposing organisms can also affect the corrosion rates in waters, as can the presence or absence of calcareous scales on the metal surface whose presence will be controlled by the amount of available carbon dioxide and the alkalinity of the waters.

The effect of ambient temperatures is more marked in liquids than in air, mainly because it affects the solubility of the corrosion products, precipitation of scales, etc. The rate of flow of a liquid will also affect corrosion rates and particularly turbulent flow which may occur at changes in direction of flow, speed, etc., or around deposits, weirs or projections into the water stream. In addition the type of corrosion found, e.g. pitting, general attack, presence or absence of stress, etc., will affect corrosion rates. Thus, it is impossible to offer generalized figures for the corrosion of mild steel in waters without considerable knowledge of the service conditions.

The corrosion of the constructional steels when buried in soil also varies considerably. In general, dry sandy soils will be less corrosive than waterlogged clays where the presence of the sulphate-reducing bacteria may prevent such steels being used without considerable protection.

If detailed information is required, reference should be made to the many publications of US Steel, BISRA, the Office Technique pour l'Utilization de l'Acier, the US Bureau of Standards, etc., and the works of Hudson, Larabee, Chandler, Stanners, etc.

b. Group 2 (low-alloy steels). The structural steels of low or medium carbon content can be improved by the addition of small amounts of chromium, which, although primarily added to increase mechanical properties have the added advantage that the addition also has some effect on corrosion resistance, particularly in relation to rusting in the atmosphere.

These so called 'low-alloy steels' often contain other added elements such as molybdenum, manganese and nickel and, in particular, copper which in combination with chromium produces a steel which has the best corrosion resistance to atmospheric conditions of all the cheap constructional steels.

On the other hand, the steels in group 2 are not usually significantly more resistant to corrosion by natural waters than the steels of group 1 unless they contain approximately 3% chromium. This is also true for resistance to soils and, again it is not until considerably more chromium than 1% is added that resistance increases to any marked extent.

In the atmosphere, low-alloy steels containing about 1% chromium plus 0.5% copper show a substantial advantage in industrial atmospheres and they are frequently used for locations where protection after erection will be difficult and infrequent. Another advantage of these low-alloy steels is that the type of rust formed (depending on the elements added and their amounts) is more compact and adherent, and appears to accept paint coatings better than does the rust formed on ordinary mild steel. Further, the paint coats tend to last longer and do not spall so easily. Little advantage with regard to corrosion will be gained, however, when these steels are used in enclosed areas.

c. Group 3 (stainless steels). Chromium steels. Although the general, as distinct from atmospheric, corrosion resistance of steels begins to show significant improvement when additions of chromium reach about 3% for 'stainless' attributes, at least 12% chromium is required and amounts up to 20% are found, especially if resistance to high-temperature oxidation is required.

The chromium stainless steels fall into two metallurgical groups—ferritic and martensitic—and common analyses of chromium and carbon are 13 Cr: 0.20 to 0.30C and 16 to 17 Cr: 0.10 to 0.15C respectively. These steels are on the whole not so corrosion resistant as the chromium-nickel steels with the possible exception of those which contain chromium and molybdenum of the order of 20Cr: 2Mo.

The martensitic types are not fully stainless though in the hardened and polished condition they are resistant to the specific environment of domestic cutlery, for which purpose the 13Cr is the commonly used alloy. (The suggestion that modern detergents have some deleterious effect on the corrosion resistance of these steels appears to have some validity: probably due to the fact that residual grease films are very efficiently removed.) In air, martensitic steels are considerably more corrosion resistant than the alloys in groups 1 and 2 but they are not as resistant as the chromium-nickel austenitic types. Similarly, in natural waters and in sea-water, their resistance is not as good as is found in those alloys which contain nickel. The plain chromium alloys are suitable for many industrial processes, e.g. in processes involving ammonia, petroleum, many chemicals and sodium compounds. All the stainless types of ferrous materials, however, which depend for their corrosion resistance on the formation of highly resistant surface oxide films, are susceptible to oxygen concentration cells and areas where such cells might be produced—at laps, weld imperfections, under loose washers and bolts, etc.— should not be incorporated or creep into designs. Also if these steels are used in reducing environments corrosion may be severe. A useful corrosion-resistant steel is the composition which lies between the martensitic and ferritic alloys. This steel has a nominal composition of 16Cr: 2.5Ni and has a martensitic-ferritic structure; it can be given a hardness of 380 Brinell with a maximum stress of 85 ton/in^2 and an elongation of 15%. For use in strong electrolytes, however, care must be taken that carbide precipitation is absent at grain boundaries.

When the chromium content of these iron-chromium materials reaches 16 to 17% the steels become essentially ferritic in structure. These alloys are relatively easy to fabricate and their resistance to atmospheric corrosion is better than that of the martensitic types. They are frequently used for automotive trim and railway rolling stock. The addition of molybdenum is sometimes made for specific service requirements where high corrosion resistance is required of steels of this type.

Austenitic (18:8) chromium (Cr):nickel (Ni) stainless steels. These steels which were originally manufactured to a 18Cr:8Ni specification now usually contain rather more nickel (9 to 10%); and molybdenum up to 3% is often added to obtain the maximum corrosion resistance especially for use in the presence of chlorides.

An early trouble with these steels was intergranular attack associated with carbide precipitation but this problem should no longer be of much moment since its control is well understood. Carbide precipitation may be controlled by maintaining the carbon content of these alloys at a very low figure (0.02 to 0.03%) or by the addition of titanium or niobium as stabilizing elements. Which of these methods should be used, and the percentages to be added are matters for the metallurgists rather than the corrosion engineer whose responsibility only lies in ensuring that the stabilized types of austenitic stainless steel are used where appropriate. There are many publications on stainless steel from which the information may be obtained – not the least important of which are the publications offered by the stainless steel makers.

The chromium:nickel stainless steels are expensive and are usually only economic for use as structural materials where the corrosion resistance of the bare metal is a prime consideration. They are used extensively for decorative trim, and are usually satisfactory unless the atmosphere is badly contaminated with chlorides when it may be necessary to maintain some form of protection of the 'wax' type. These steels behave well in natural waters as long as oxidizing conditions are maintained. Care should be taken when these alloys are used in sea water that the formation of differential aeration cells does not occur, e.g. the presence of unsealed nooks and crevices.

These alloys find considerable use in chemical engineering and they are highly resistant to all oxidizing acids, but with reducing acids their behaviour may not be satisfactory and before they are designed into plant using such acids, information should be sought from the manufacturers as to the best alloy to use and which service conditions are likely to encourage attack. They are not attacked by ammonia and they are resistant to caustic alkalis at temperatures below 100°C. Their resistance to salt solutions is also a matter for expert advice though it may be taken as a general rule that if there is insufficient oxygen to maintain passivity their behaviour will be suspect, especially in acid chlorides. Hypochlorites can cause pitting and in the presence of moisture all halogens are corrosive to stainless steel. They are excellent for most food processing plant provided care is taken that contact with hypochlorite sterilizing solutions is of short duration. They are also now used in some areas of atomic power plants.

11.5.2 Ferrous metals (cast)

The term cast iron is given to a very wide range of ferrous-base casting materials whose total carbon content exceeds 1.7% except in what are known as the 'high-silicon irons' in which the carbon content rarely exceeds a nominal figure of 0.85%.

Since the corrosion resistance of the true cast irons is associated with the amount and type of carbon present and with the structure of the matrix it is convenient to classify them in terms of their microstructures:

1. *White iron.* All the carbon is in solid solution and the matrix is pearlitic.
2. *Grey iron.* Most of the carbon is in the form of flakes. This is the form usually meant by the term cast iron and the matrix is commonly pearlitic.
3. *Nodular, ductile, spheroidal iron.* Most of the carbon is present as nodules or spherules and the alloys solidify with a pearlitic matrix which may be transformed to ferritic to develop full ductility by suitable annealing.
4. *Malleable iron.* The carbon is present as nodules but these materials can be produced in either pearlitic or ferritic structure.

The corrosion behaviour of the cast irons is markedly different from that of the wrought ferrous metals. When steel corrodes in water and the corrosion product (rust) is precipitated away from the corroding surface, the steel wastes away and its dimensions become smaller with time. With cast iron, however, serious corrosion may occur without any apparent alteration in size or shape of the component. The reason for this behaviour is that cast irons contain in their structure various microconstituents, which are usually not present in steel, and which are less severely attacked than the general matrix. The most important of these substances are graphite, phosphide eutectic and carbides. When cast iron corrodes, these compounds are left behind as a skeleton with the interstices filled with corrosion products. This structure although it retains its original shape, and in many cases its original dimensions, has no strength; and it is not surprising that corrosion failures in cast iron are frequently unsuspected and catastrophic. In cast iron the configuration of the graphite is most important. The finer the graphite particles and the more that they are interlocked, the denser will be the graphite layer on the surface; and the corrosion resistance of the material will be greater than if the flake structure is course and/or individual. Thus, the spheroids of nodular cast iron allow more corrosion to occur than do some of the other structural types. It has also been shown that the ferritic materials do not form such resistant films as do the materials of pearlitic structure.

In most cases the plain cast irons are not selected for use on the basis of their corrosion resistance, but the high-alloy cast irons, e.g. the austenitic grey irons, the high-silicon irons and the high-chromium irons, are often used on the basis of their enhanced corrosion properties.

Small variations in the compositions of the common cast irons have little effect on the corrosion resistance except that additions of nickel (up to 2%) are reported to enhance the corrosion resistance of the nodular types.

Cast iron is usually used in the atmosphere and most cast iron components are massive and in consequence, with reasonable maintenance, failure due to atmospheric corrosion is likely to be infrequent. In liquids and waters, as is usual with ferrous metals, their corrosion rate is associated with oxygen concentrations and while it is difficult to generalize bare cast iron will usually resist corrosion to no better degree than will mild steel. However, some cast irons do form protective films which tend to make them more resistant to soils but it can be accepted that if an environment is aggressive to constructional steel, it will also be aggressive to cast iron of the plain and low-alloy compositions. With cast iron, in contrast to the plain carbon steels, it has been shown that manufacturing processes can affect corrosion resistance (e.g. spun pipe, compared with vertically cast pipe) and it must also be mentioned that the presence of the sulphate-reducing bacteria in soil can seriously affect the corrosion rate of cast iron.

Grey cast iron suffers a type of attack known as graphitization. This occurs only in the grey iron and the ductile types. Iron undergoing this type of corrosion has the disadvantage that the graphitization forms a very cathodic film and this tends to accelerate attack; and, while superficially the metal appears to be unattacked and the final dimensions are roughly the same as that of the original component, the graphitized metal is so weak and soft that it can easily be cut with a knife.

In moving water, the common cast irons are not particularly resistant though if they form a hard protective film in the environment under consideration they may, at low velocities, have some advantage; they certainly have an advantage in handling slurries. The cast irons do not have appreciable resistance to cavitation damage and if they are used for ships' propellers they must be protected by cathodic protection for them to behave satisfactorily. The unalloyed cast irons are rapidly attacked by hydrochloric, nitric and most organic acids, and also by phosphoric, and hydrofluoric acid. They are slightly more resistant to sulphuric acid at concentrations of the order of 65% though below this concentration they are attacked fairly rapidly. The cast irons, with certain exceptions, (see below) are reasonably resistant to alkalis and neutral solutions.

The high alloy cast irons come into an entirely different category with regard to corrosion resistance. The austenitic cast irons are basically nickel cast irons with nickel contents from 13.5 to 36%. They also may contain copper and chromium for special resistance, and are available in the ductile form. The high-silicon cast irons contain between 14 and 18% silicon and the high-chromium cast irons have 12 to 35% chromium additions. Table 11.2 gives a rough guide to the behaviour of the high-alloy cast irons but much more specific information will be found in refs. 3 and 7, from which most of the information in the table was abstracted.

The galvanic compatibility of the cast ferrous materials is very variable and much depends on the environment and the type of protective film formed on the alloy. It must be stressed that their enhanced resistance to corrosion is often vitiated by their

Table 11.2 Approximate behaviour of high-alloy cast irons

Environment	Austenitic	High silicon	High chromium
Hydrochloric acid,			
Sulphuric acid	Good[1]	Excellent[3]	Excellent[4]
Nitric acid	Poor	Excellent	Excellent
Organic acids	Moderate	Good	Good
Phosphoric	Poor	Good	Good
Alkalis	Good	Poor	Poor
Salts	Good	Good	Good
Abrasion	Good	Good	Good[5]
Cavitation	Good[2]	Good	Good
Soils	Good	Good	Excellent

All assessments are relative and the best resistance in any group may refer to special alloys.
1. Hydrochloric acid only up to certain concentrations. Sulphuric acid good except at high temperatures (38°C) in aeration.
2. Improved by the addition of chromium.
3. For hydrochloric acid 3% molybdenum additions are recommended.
4. Exceptionally resistant to sulphuric acid but attack is a function of temperature. Not good in free sulphur trioxide.
5. Not suitable for chloride slurries, but good for mine waters.

brittleness and this precludes their use in some industries, e.g. shipbuilding where their sensitivity to shock outweighs their resistance to attack in many applications.

11.5.3 Ferrous metals (high-temperature)

The progress of high-temperature oxidation or corrosion is more often controlled by the kinetics of the reaction than by the thermodynamics of the stability of the reactants, the presence of stress and thermal cycling assuming considerable importance since cracking or flaking of the surface scales invariably leads to increased attack. While the corrosion will be considerably affected by the presence or absence of oxygen, the contamination of the reacting atmosphere plays a significant part in many practical corrosion problems at high temperatures. Further, at high temperatures where the presence of moisture is possible, deterioration of the metal can be brought about by the breakdown of water into oxygen and hydrogen when both elements may be reactive.

Steam, carbon dioxide, sulphur compounds, general combustion products, ash, halogens, and oil decomposition products, may all affect the corrosion rates and the severity of attack when metals are subjected to high-temperature environments. Some high-temperature processes take place in the presence of refractories and molten metals and salts. In the latter instances, the presence of galvanic corrosion can occur and its possibility must not be disregarded in the diagnosis of high-temperature failures.

Resistance to high-temperature corrosion is almost entirely controlled by the type, thickness and characteristics of the oxide film which is present on the surface before the alloy is submitted to heat, or which forms in the early stages of the heating and then, by its behaviour as the heating period increases, is continuous or intermittent. The technology of these films and their growth laws (see corrosion of metals at high temperatures p. 808) is involved and refs. 1 and 3 should be consulted for more definitive techcriptions of their behaviour.

For ease of presentation the ferrous alloys (following the scheme used by Edwards (see reference (3) will be divided into two groups: those with less than 12% total alloy content; and those with more than this amount; roughly a division between stainless and non-stainless materials.

The low-alloy materials find many industrial applications at temperatures where the heat does not exceed 600 to 650°C. These alloys contain many alloying elements, not all added to increase oxidation resistance; and Edwards (see ref.3) lists aluminium, carbon, chromium, molybdenum and silicon as the most successful modifying elements. Nickel, at the percentages usually found in these steels (of the order of 3%), has no particular effect on the heat resistance though it may be added for other reasons. Table 11.3 is based on the data provided in *Steels in Modern Industry* (see ref. 17). It must, however, be stressed that the figures offered are only relative and may be raised or lowered depending on the degree of oxidation promoted by the environment.

Many low-alloy steels have originally had their alloying elements scheduled in terms of mechanical properties for use at elevated temperatures rather than for resistance to corrosion. This has rendered much data on their behaviour somewhat difficult to disentangle and although there is a vast amount of literature on the behaviour of steels at high temperatures little of it has been correlated outside the steel manufacturers and their research organizations. Specific problems still need to be investigated. The relationship between low-alloy steels and heat is well summed up by Edwards (see ref. 3).

'There is sufficient evidence to indicate that low-alloy steels have a considerable part

Table 11.3 Mild and low-alloy steels for high-temperature services

Type of steel	General applications	Maximum service temperature
Mild steel (fully 'killed')	Steam pipes, fittings and super-heater tubes for boilers, etc, of low service requirements	454°C (850°F)
Low carbon: molybdenum	For service conditions similar to above as regards oxidation but has elevated temperature resistance 50–75% better	510°C (950°F)
0.75 to 1.0 chromium: 0.4 to 1.0 molybdenum	For high-temperature steam services, high-pressure tubes, valves bolts etc.	538°C (1000°F)
High silicon	Oil pipe still-furnace tubes under mild conditions. High silicon content advantageous	593°C (1100°F)
Nickel: chromium: molybdenum 0.75 to 1.0 chromium	The chromium content of this steel is rather low but it is useful for components in high-temperature steam engineering under moderate conditions	538°C (1000°F)
2 to 3 chromium: 0.5 to 1.0 molybdenum	Just sufficient chromium to give useful scale resistance for marked improvement to high-temperature conditions. Used in hot oil piping and general steam engineering. The 3% chromium alloy used in ammonium plant and hydrogenation processes. Also used for turbines when modified by nickel, tungsten or vanadium	600°C (1112°F)
4 to 6 chromium: 0.5 to 1.0 molybdenum	Widely used in oil plant similar to above steel, but where corrosive sulphur-bearing oils are processed. High-temperature steam and hydrogenation plants, pressure vessels.	650°C (1202°F)
7 to 9 chromium 0.5 to 1.0 molybdenum	High oxidation resistance used where oxidation and corrosion are severe. Similar applications to above.	700°C (1292°F)
8 chromium: 3 silicon	Resistance to scaling good. Tendency to brittleness at normal temps. Retains properties at high temperature well. Internal combustion engines, soot blowers, gas-producers, ore roasting, etc.	800°C (1472°F)

to play in construction equipment in which heat resistance is important. They are, however, limited to an absolute maximum temperature of 600 to 650°C, this limit being controlled by the alloy content of the steel and the conditions of service, the alloying elements of major importance are Cr, Al, and Si and a knowledge of the working atmosphere is imperative if the full value of these steels is to be exploited.'

Many of the high-alloy steels were originally developed with resistance to high-temperature oxidation very much in mind and the major alloying elements used are chromium and nickel. There is a considerable range of these materials from electrical heating elements and resistance wires through the chemical engineering industry to steam engineering and their use as containers for liquid salts and metals. In their selection for any particular high-temperature service their resistance to creep, which is a very important factor in the use of metals at the temperatures these steels are required to resist, must be borne in mind. Typical heat-resisting steels are the following (see reference 3.):
1. 13 chromium, limited use as a scale-resistant material, mainly for steam valves and turbine blades.
2. 21 chromium and 29:2 chromium:nickel, limited but important applications. These low-nickel ferritic steels have excellent corrosion/oxidation resistance but their strength is not as great as the austenitic chromium:nickel steels and they are not so easy to fabricate.
3. 18:9 chromium:nickel:titanium and 18:11 chromium:nickel:niobium steels are austenitic and have good oxidation resistance to at least 800°C (1472°F), and the 23:12 chromium:nickel:tungsten steel is commonly used to 1000°C (1832°F). A steel containing 24:21 chromium: nickel can be used to 1100°C (2012°F) and has better hot strength and improved fabrication response over most of the austenitic types.
Of the above types the 18:11 chromium:nickel:niobium has the best creep response.

Probably the main environments in which these steels are employed involve atmospheres of air, flue gas, carbon dioxide, and the many oxidizing gases of industrial plant operation, all of which may affect the choice of alloy.

Further, the severity of many high-temperature locations is modified by the presence or absence of combustion ash, particularly that formed from some fuel oils, e.g. ash containing vanadium pentoxide. This has become a pressing problem with the advent of the gas turbine. Other atmospheres which require special heat-resistant steels are those containing sulphates, especially if contaminated with chlorides (possibly due to seawater in a fuel oil) sulphides: may also be deleterious. High-alloy steels are also used in industrial plant handling molten salts and metals. No specific compositions can be offered since the resistance to attack varies with the salt and the composition of the steel. The resistance of these steels also varies in liquid metals and they are not recommended for use with zinc, cadmium, aluminium, antimony or copper.

When cast iron is subjected to high-temperature oxidation a surface scale is formed which adheres to the metal surface and acts as a barrier against further oxidation: thus, the deterioration can be expected to decrease with time, and the degree of adherence of the film becomes an important factor in the resistance of these irons to high temperatures. The adherence of the film can be influenced by the addition of such elements as aluminium, chromium and silicon. The rate of oxidation will also be affected by the structure of the matrix and the type of the graphite – fine flakes tend to restrict the inward diffusion of oxygen.

Due to the formation of these scales and the changes which may occur in the structure of the iron during heating, the cast irons are very prone to show dimensional distortion, the degree of which will depend on the temperature and whether the heating is continuous or cyclic. Therefore cast irons for high-temperature use must be chosen after considering the maximum temperature to be resisted, the heating cycle, and whether the component can undergo 'growth' without deleterious effects – including the possibility of incipient cracking of the scale. The International Nickel Company have shown that some advantage over the more conventional cast irons is provided by the 'ductile' types. Cast irons used up to 400°C (750°F) are often stabilized with chromium, molybdenum and vanadium in combination, sometimes, with nickel and copper. For higher temperatures, the high-alloy austenitic high-silicon or high-chromium are usually used. Temperatures of 999°C (1830°F) may be resisted by those irons which contain of the order of 35% chromium. For high-temperature steam service nickel and chromium are common additions and the high-alloy austenitic types are frequently used. The cast irons are often successfully used for handling molten metals but the varieties of iron which are most suitable depend on the molten metal and the temperature of the operation.

11.5.4 Aluminium and its alloys

Aluminium can resist a wide variety of environments and aluminium and its alloys lend themselves to many engineering, domestic, industrial and food processing applications. Service environments which call for lightness, good thermal and electrical conductivity, hygienic non-toxic properties, etc. combined with good corrosion resistance frequently exploit the various aluminium alloys with great success. Aluminium alloys have been used as architectural materials, transmission lines and culinary metals for over half a century and since World War II the use of this material has been considerably extended even to such a corrosive environment as sea-water in which some of the modern alloys can be used in the totally unprotected condition.

Aluminium, a relatively base metal in the e.m.f. series, has the property of forming thin inert protective films on its surface which have strong self-healing properties when damaged, and in many environments the initial corrosion diminishes rapidly with time as the film builds up. The corrosion behaviour of aluminium is, thus, very dependent upon whether this film is damaged or removed and the rate at which it will be repaired. Environments which affect the corrosion behaviour of aluminium can be roughly divided into: those which attack the oxide film to such an extent as to render aluminium unsuitable; those which cause localized breakdown (pitting) of the oxide film; and those where the oxide film is not attacked. Environments of the first type include the strong alkalis, the strong acids, mercury compounds and a wide range of inorganic chemicals. Environments causing local breakdown of the oxide film include many natural waters, aqueous solutions of the salts of heavy metals, especially copper and some inorganic salt solutions. The final group includes many organic acids, organic compounds generally, petroleum products, food stuffs, alcohol, etc.

There are many aluminium alloys and one of the commonest causes of failure by corrosion is the use of the wrong alloy for the environment. This trouble may be aggravated by the fact that all aluminium materials look alike, although their corrosion resistant properties may vary enormously, and for a designer to schedule 'aluminium' can lead to considerable trouble (see below) in some environments, especially marine.

Since aluminium is a comparatively base metal, when its film is broken, destroyed or has not formed, it will be anodic to most of the other constructional metals and all bimetallic joints should be insulated, in some way, to obviate the chance of galvanic couple action. Further, because the corrosion resistance of aluminium is dependent on an oxide film, crevices, cracks and other configurations which exclude the entry of oxygen must be avoided. Indoors, where relative humidities are low, these precautions against galvanic attack may, of course, be relaxed.

Aluminium can suffer general dissolution, pitting, intercrystalline corrosion, stress-corrosion cracking (usually limited to the high-strength alloys), filiform corrosion, poultice corrosion, deposition attack, layer corrosion, etc., or combinations of these types of attack. Most of the corrosion, unless the alloy is entirely unsuitable, can be prevented by the use of the correct material, alteration in the design of the component, or by the use of a suitable barrier coat and, in some cases by the use of inhibitors.

Most aluminium alloys can be used satisfactorily in air both indoors and outside. Although aggressive industrial atmospheres may necessitate consideration of the use of special surfaces such as anodizing, rural atmospheres usually offer no problem. Corrosive attack decreases with time and a very protective film eventually builds up which allows materials, where the initial corrosion does not effect the aesthetic look of the alloy, to be used without paint or any other barrier coat. The problem of the initial corrosion can be overcome by waxing or continual polishing though the colour of the initially corroded surface may not be as attractive as the original metal.

Some aluminium alloys are available which resist many fresh waters although variations in corrosion rate may occur due to the presence of pollution, solid matter and relatively very high pH natural waters. Waters, of any type, which contain traces of metal compounds of copper, lead, tin, nickel, cobalt, magnesium, etc. are extremely unsuitable environments for aluminium. All the heavy metals tend to plate out on the surface of aluminium, even in the presence of the natural film, with the formation of very active aluminium-heavy metal couples which rapidly develop into pits. Sea-water can be used in contact with several aluminium alloys *but the high-strength copper-containing alloys must in no circumstances be used in contact with sea-water.* In sea-water, also, pollution may encourage pitting and if anti-fouling paints are used on aluminium alloys they must be well insulated from the metal by a good barrier coat – the most successful of such coatings are those based on zinc chromate. Paints containing lead or mercury should not be used on bare aluminium and it is probably good practice to avoid using paints containing mercury even if they are applied over a barrier.

One way of overcoming corrosive attack on aluminium in contact with aggressive environments is the use of clad alloys. The cladding is usually an aluminium alloy containing 1 to 2% zinc which is anodic to the strong underlying structural aluminium and it thus provides a built-in cathodic protection system, and such material has been known to pipe sea-water for over ten years without failure. Aluminium is frequently used successfully for piping and storing high purity water and since aluminium is not attacked by dissolved oxygen or carbon dioxide it is well suited for the piping and handling of steam condensates provided that the steam is not entraining alkaline boiler compounds.

Aluminium can be used underground but rates of attack vary considerably and made-up ground containing cinders is very aggressive to aluminium alloys. It is usually advisable to protect the surface of buried aluminium unless the soil is known from previous experience to be benign. Both high temperature, in air and liquids, and the velocity of a liquid environment can affect the corrosion behaviour of the various aluminium metals. Steam up to about 250°C (480°F) can be satisfactorily handled by aluminium alloys; above this figure reactions between steam and aluminium may occur. Sintered aluminium powder (SAP) has been used at steam temperatures of 500°C (930°F) when about 10% nickel has been added to the SAP.

Aluminium alloys, again with the exception of the high-strength copper-containing alloys, are suitable for handling building materials and are not attacked by concrete, mortar, plaster so long as the materials are dry and remain so. One exception to this generalization of good behaviour in the building trade is the sensitivity of these alloys to magnesium oxychlorides (used for floor coverings); these materials must be isolated from aluminium, usually by bituminous compounds.

Limitations of space preclude lists of the various aluminium alloys and the variety of conditions in which they can be purchased. The technology of the use of aluminium is relatively new and many fabricators do not realize that its use is somewhat different from steel with regard to the use of the correct alloy in the correctly heat-treated condition, coupled with certain differences in constructional techniques. Storage and marking of alloys are also more critical. Welding, too, needs a different approach from steel particularly with reference to fluxes. However, all the aluminium companies of the Western World maintain extensive information services which will advise, at no cost, which are the correct alloys for any specific environment and how to use the material to the best advantage.

11.5.5 Copper and its alloys

From the point of view of corrosion resistance the copper-base alloys probably comprise the most important family of metals. The group covers a wide range of corrosion resistant materials, some of outstanding quality, many of which have excellent workability (either hot or cold), high thermal and electrical conductivities and attractive mechanical properties which are maintained to moderately high temperatures.

Copper is not a very reactive metal and in the e.m.f. series it is near the noble end. It can, therefore, be expected to have relatively good general corrosion resistance and by alloying many compositions have been developed which are considerably more corrosion resistant than the parent metal.

The types of corrosion found in the copper-base materials are generally similar to those found with other metals, though the corrosion rates for any specific type of attack may, of course, be different. For example, copper is not so sensitive to differential aeration as some metals and the brasses can be affected by certain specific corrosion reactions. Some of these latter alloys undergo a selective type of attack in which zinc is lost with a general weakening of the metal, (see dezincification, p. 804), and in the presence of internal stresses and specific corrodants, e.g. ammonia, they can suffer a destructive type of fracture called season cracking. Of the whole group, the brasses are generally the least corrosion resistant and in many instances the modern complex bronzes are taking their place when corrosion resistance is a prime factor.

a. Copper. Copper is found in several grades. The purest commercially available is usually designated OFHC (oxygen-free high-conductivity). Tough pitch, also of high conductivity, containing about 0.04% oxygen is, however, the material usually specified for electrical conductors. Phosphorus deoxidized copper is used for pipe and tube, and although its electrical conductivity is slightly less than that of the above types, it has the advantage, in hot reducing atmospheres, of being less susceptible to embrittlement than tough pitch. Arsenical coppers containing about 0.4% arsenic are used where high strength at elevated temperatures is required.

Coppers vary little in their corrosion resistance. They behave well in the atmosphere, both rural and marine, although the presence of hydrogen sulphide in industrial air can increase the normal corrosion rate quite considerably. Their resistance to acids, both organic and mineral, varies with the acid: they do not have satisfactory resistance to oxidizing acids and the rate of attack will be affected by temperature and the amount of aeration. The coppers are corroded by caustic alkalis, all compounds of ammonia, oxidizing salts, halogens and sulphides. Chlorides, sulphates and nitrates are resisted to a moderate degree depending on the film formation. The coppers are resistant to many types of fresh water and corrosion rates are, generally, of the order of 0.0001 to 0.001 ipy or less, depending on the hardness–soft and acid waters are sometimes aggressive towards the standard coppers. When copper is used with potable waters, distilled water or carbonated water, it is often good practice to tin the copper surface in contact with the water. Copper may be used in sea-water if the liquid is calm, fouling is absent and the water speed is less than 3 ft/s. On the whole arsenical copper is not more corrosion resistant than the others but beryllium copper may have some advantage in localities of flowing water.

b. Brasses. The brasses are essentially binary alloys of copper and zinc, in which the zinc content can vary from about 10 to 45%. However, many other elements are added to obtain specific properties including corrosion resistance. The brasses are usually divided into three groups: the alpha (single phase brasses with zinc content up to about 37%); the alpha plus beta (two phase alloys containing more than about 37% and less than about 50% zinc); and the beta brasses (alloys which contain roughly equal quantities of copper and zinc). These latter alloys, some of which are alloyed with aluminium, etc. are of very high strength but they are of little interest to the corrosion engineer. They tend to be subject to season cracking and will fail from stress-corrosion in saline solutions. They are, however, found quite commonly as brazing materials and although their melting point is low, their high zinc content tends to make them not very corrosion resistant and, they should not be used for brazing in sea-water.

The alpha brasses cover a wide range of useful materials–they are ductile and can be easily cold worked. This group contains the well-known materials Cartridge Brass (70 copper:30 zinc), Admiralty Brass (70 copper:29 zinc:1 tin) and Aluminium Brass (76 copper:22 zinc:2 aluminium). These alloys have similar general resistance to corrosion as has copper especially in the atmosphere, but some are considerably more resistant to chloride solutions than copper. It should be noted, however, that if dezincification is to be avoided these alloys must contain 0.02 to 0.06% arsenic.

The alpha-beta brass family are essentially hot-working alloys, and although they are essentially alloys of copper and zinc, other elements such as lead, to improve machining, and tin, to increase corrosion resistance, are often added. This group contains such famous materials as Muntz Metal (60 copper:40 zinc) and Naval Brass (60 copper:39 zinc:1 tin) as well as the so-called manganese bronzes. The former two alloys, useful though they are, should not be used in strong electrolytes such as seawater since they will dezincify freely and in contrast to the behaviour of the alpha alloys when inhibited with arsenic they cannot be inhibited against dezincification–as of 1969. No attention should be paid to the name Naval Brass which has long outlived its original association with sea-water.

The alloys of alpha plus beta, or sometimes beta, structure also cover the group of materials which are often known as manganese bronzes though high-tensile brass is the better designation. These materials usually contain many other elements besides copper and zinc, e.g. aluminium, iron, nickel, tin, etc. These alloys are found in both cast and wrought forms and their corrosion resistance varies over wide limits but in general they do not behave well in strong electrolytes such as sea-water.

resistance. This point should be appreciated by designers and only alloys produced by manufacturers used to handling these materials should be used. Sometimes aluminium bronzes undergo dealuminification, which is a somewhat similar phenomenon to dezincification, and it is probable that conditions which favour the one will favour the other. Most of these alloys are less susceptible to cracking than the brasses and although they are not especially prone to stress-corrosion failure they are not immune.

Other brasses which have good corrosion resistance are those which contain nickel – the nickel-aluminium bronzes – those which contain substantial amounts of manganese (several per cent) and those which contain relatively large amounts of silicon either alone or in combination with aluminium.

The bronzes, as a whole, form a very important group of corrosion-resistant materials. However, since their behaviour to different environments varies with their composition, designers and corrosion engineers should consult more detailed sources of information for specific service uses. Since published evidence of the behaviour of these alloys is not extensive in the general literature, the publications of the various trade research associations and development organizations will offer the best areas of enquiry.

d. Gunmetals. Some copper-base alloys contain copper, tin and zinc. These alloys are known as gunmetals. Some gunmetals also contain lead added to improve casting and pressure tightness qualities. These materials all have high corrosion and erosion characteristics, especially towards sea-water. Two well known alloys are Admiralty Gunmetal (copper 88 : tin 10 : zinc 2) and Eighty-five-three-fives (copper 85 : tin 5 : zinc 5 : lead 5). On the whole, the higher copper alloy has the best corrosion resistance though the second material is probably more frequently used because of its better casting qualities and pressure tightness – it is also cheaper. There are several newer alloys sponsored by the British Non-ferrous Metals Research Association and the International Nickel Company which contain other percentages of copper, tin and lead or, in some cases, nickel. These alloys are claimed to have a better combination of corrosion resistance and fabrication properties than the older alloys.

e. Cupro-nickels and cupro-nickel-iron alloys and nickel-copper alloys. There are three important groups of copper-base alloys whose major alloying element is nickel. Firstly, the cupro-nickels which contain 20 to 30% nickel with a few tenths of 1% of both iron and manganese as obligatory modifiers for the best corrosion resistance. Secondly, the cupro-nickel-iron alloys, which are found commercially in two ranges: the 5% and the 10% nickel with the remainder copper, also with an obligatory requirement that each nickel percentage contains a specific iron content which varies between 1 and 2%. Lastly, those nickel-copper alloys where the nickel content usually exceeds that of the copper.

The 70 : 30 cupro-nickel alloy properly modified with iron and manganese is probably the best condenser tube alloy now available (1969). It will withstand, in properly designed and relatively turbulent free sea-water cooling systems, velocities of well over 15 ft/s. It is also found in sheet form where its high strength and ductility makes it a very desirable corrosion resistant material; and the cupro-nickels are the most resistant of all the copper-base alloys towards corrosion by alkali solutions. If, in terms of cost, designers schedule the 80:20 alloy, it must be remembered that the overall corrosion resistance of this alloy is less than that of the 70 : 30; and as regards resistance

Most of the brasses are not very resistant to impingement attack and cavitation corrosion with the exception of Aluminium Brass which is the most resistant of all the high brasses to corrosion-erosion.

The general resistance of the high-copper brasses towards acids is better than copper. Those alloys which contain above about 15% zinc, however, are not suitable for use with acids and, in general, the brasses should not be used in contact with nitric and hydrochloric acids though they are less attacked by hydrofluoric acid in calm low-oxygen environments. The high-copper brasses have some resistance to sulphuric acid but not to phosphoric acid. Organic acids can be handled if the aeration is low but most oxidizing ions will accelerate the corrosion of the brasses. The high-copper alloys can resist moderate concentrations of sodium and potassium hydroxide but are virulently attacked by all solutions containing ammonium compounds.

Natural waters, except sea-water, can be handled by red brass as long as the water is relatively free from chlorides, and copper and its alloys are frequently successfully used underground. The brasses are not very suitable for high temperature environments unless they contain aluminium (2 to 5%) when resistance to atmospheric oxidation can be expected up to 1470°F (800°C).

The brasses may undergo stress-corrosion cracking and corrosion fatigue in certain environments, and these types of attack usually increase with increase in zinc content.

c. Bronzes. The bronzes were originally binary alloys of copper and tin, but many modern bronzes contain numerous other elements and some, the aluminium bronzes, contain substantially no tin.

The phosphor bronzes, whose tin contents range from about 2 to 10%, are an important group of this class of materials (the amount of phosphorus rarely exceeds 0.4%). The tin content of the wrought alloys is usually restricted to about 8%, when selective attack due to equilibrium problems will be absent. With the cast materials, however, the tin content may be higher and maximum corrosion resistance will only be obtained if they are fully homogenized, e.g. by soaking at about 700°C for long periods. The homogenized alloys, both wrought and cast, have good general corrosion resistance, especially in sea-water and, on the whole, if the tin is not out of equilibrium, the higher percentage tin alloys have the best corrosion behaviour. They have been used extensively in pulp mills for resistance to acid sulphides and sulphurous acid solutions (see ref. 7). Their behaviour in the atmosphere is roughly comparable with the high brasses.

Alloys, containing of the order of 10% aluminium and no tin, are now well established as a series of alloys which have very good corrosion resistance both towards strong electrolytes and for service at moderately high temperatures – for which purpose they are among the best of the copper-base alloys. Their corrosion resistance is due to their ability to form tough and very protective films of oxide, which though thin, are very adherent and heal rapidly when damaged. Unfortunately, this oxide film tends to make jointing and repair of these materials difficult. However, successful jointing techniques, especially by welding, have been developed and such difficulty should no longer deter designers from scheduling their use. One of the largest applications of these materials is for marine propellers and many special compositions which contain nickel and iron as well as aluminium are found. This group of alloys is used in both the wrought and cast forms and they can vary from an alpha to a duplex structure, and in the latter case the presence of out-of-phase constituents can seriously affect their corrosion to corrosion-erosion by sea-water, considerably less. These alloys sometimes suffer attack under oxygen screens such as may be formed by debris if the water speeds through condensers are allowed to fall too low. They also undergo what is known as 'hot-spot' corrosion which is due to thermogalvanic effects which may require design alteration for their elimination.

The copper-nickel-iron alloys were originally developed for the sea-water services of warships where lightness combined with good corrosion resistance is a vital necessity. The 5% nickel alloy has been used extensively for piping sea-water in cooling systems etc., and while it is considerably more resistant to corrosion and corrosion-erosion than copper and will, in non-turbulent conditions, withstand an average flow speed of 5 to 7 ft/s, it is recommended that unless the system is substantially free from the possibility of impingement attack, the 10% alloy is to be preferred. These alloys which contain substantial amounts of iron in copper must be manufactured under controlled heat-treatment conditions for maximum corrosion and mechanical properties to be maintained.

The range of applications of the nickel-copper alloys – Monels, Inconels, Nimonics, Hastelloys and Nichrome, etc. – is very wide, and the corrosion resistance of the high-nickel-copper alloys to atmospheric, liquid and high temperature environments is outstanding. However, the nickel-copper alloys are not all acceptable for all environments but with careful selection, preferably guided by the maker's experience, the designer can be assured that there is, probably, no other group of alloys which can be scheduled for high-duty corrosion service with such confidence. It must be mentioned, of course, that the alloys are expensive and their use for resistance to corrosion will be considerably affected by economics. It is also true that they, like most high nickel-containing materials can suffer attack due to the formation of active-passive cells if their surfaces are, locally, screened from access to oxygen (see also nickel, p. 839).

11.5.6 Lead and its alloys

In considering the corrosion resistance of lead and its alloys it is important that the correct grade or alloy be used since some of the alloying elements added to lead to improve mechanical properties do not always also improve corrosion resistance.

High purity lead has good corrosion resistance to a wide range of chemical compounds and some natural waters. Lead is characterized by its ability to form anodic corrosion products which mechanically insulate the basis metal from the corrosive environment if the conditions are right by the formation of lead sulphate ($PbSO_4$) triplumbic tetroxide or red lead (Pb_3O_4), lead carbonate ($PbCO_3$), etc. Further, since lead is amphoteric, cathodic reactions are sometimes responsible for corrosion and plumbates are often soluble in alkalis. On the whole the many additions made to lead to improve mechanical properties, e.g. antimony and tellurium, do not enhance the corrosion resistance of the basis metal to any great extent and such alloys are rarely chosen from the point of view of corrosion resistance alone.

Lead and its alloys have excellent resistance to corrosion by the atmosphere over a wide range of humidities and the presence of sulphur dioxide (SO_2), sulphurtrioxide (SO_3), hydrogen sulphide (H_2S), and carbon dioxide (CO_2) do not cause acceleration of attack. Further, the insulating films formed are able to prevent much galvanic action at couples; this might not be expected by the position of lead in the e.m.f. series.

These materials are not considered as suitable for use with distilled water as tin. Their use for piping domestic waters will depend on the plumbosolvency, since lead compounds are usually poisonous, even in trace quantities if ingested over long periods. Waters having a few degrees of hardness can be handled safely since protective scales of carbonates or sulphates, etc., are formed. They must not be used, however, for soft acid waters stemming, for instance, from peaty moorlands.

Many lead alloys are used for coatings of underground cables, pipes, etc., and, due to their service purpose stray current corrosion is a common type of attack. It is frequently the practice to cover lead in contact with soils with tar or bitumen impregnated wrappings and bare and wrapped lead structures are often protected by cathodic protection systems which are very successful providing they are properly designed and monitored.

Lead-coated metals are commonly found in the chemical industry. They are used for handling sulphuric acid, phosphoric acid and chromic acid and the corrosive sulphite wastes from paper mills. Lead is not resistant to acetic acid and nitric acid, nitrates and nitrites. Its behaviour towards hydrochloric acid is reported to be variable (see ref. 3) but on the whole it is advisable not to use lead in contact with this acid. In general, lead is not very resistant to organic acids, and lead covered cable should not be used in contact with materials which are likely to exude organic acids. This is true for some woods, especially oak whose acetic acid content can cause lead coverings to disintegrate. The corrosion of lead by acids is often accelerated by aeration and movement of oxygenated liquids may alter corrosion rates.

Alkalis also attack lead although, perhaps, not as severely as those acids which are corrosive to lead; again aeration of the chemical can affect the corrosion rate.

Corrosion by neutral solutions very much depends on the type of protective scale which forms, and rates of attack range from high, where the film is not protective, to negligible where the corrosion products are continuous, non-porous and adherent. Sea-water can be handled successfully by lead and its alloys since calcareous scales are usually formed which soon become protective.

Because of its ability to form very protective scales, lead-coated metal is frequently used in chromium plating baths and the associated pipelines, etc. It is a frequent routine to age the protective film artificially by making the surfaces which will be in contact with the plating solutions anodic for a short period before they are put into service so that an adequate dioxide film will be present. As with all protective coatings there is always the danger that if the coating is damaged severe local corrosion will take place at any area where the basis metal is exposed.

Lead and lead-alloy anodes are also often found in other electroplating processes as well as for the positive grids in accumulators. Various special lead alloys are used for these purposes and more authoritative sources should be consulted for specific information.

The corrosion rates of lead in some electrolytes can be reduced by the addition of inhibitors, e.g. cobalt sulphate in sulphuric acid.

Lead and lead alloy anodes are common in cathodic protection systems, e.g. lead-silver alloys, and recently lead alloys containing microelectrodes of platinum inserted into the surface have been shown to be suitable for high current densities – of the order of 25 A/ft^2.

11.5.7 Magnesium

Magnesium is the lightest of the constructional metals; it has a high chemical activity and it has little intrinsic corrosion resistance. It does not corrode to any extent in dry unpolluted atmospheres but contamination or saline moisture in the air renders attack almost certain, although it may not proceed further than surface discoloration. Magnesium has poor resistance to mineral acids except hydrofluoric and chromic. Most organic acids attack magnesium but some fatty acids are capable of producing magnesium salts which form protective corrosion films. Fresh waters will attack magnesium especially if they are agitated and the protective film is not allowed to form. The corrosion of magnesium in salt solutions depends on the ions in solution; this is exemplified in sea-water which is less corrosive to magnesium than pure sodium chloride of the same strength. Its resistance to alkalis, again, depends on the formation of the protective film and in some caustic solutions its corrosion resistance is greater than that of the aluminium alloys. Outside the specialized usages for aircraft, missiles and guided weapons, the main interest of the corrosion engineer in magnesium and its alloys is for its use as a sacrificial anode in cathodic protection. Magnesium has a high driving voltage *vs* protected steel, a low electrochemical equivalence and good anodic polarization; characteristic properties which indicate its usefulness as such material. The potential of pure magnesium in dilute chloride solution is 1.70 V *vs* the copper sulphate half cell. It is, thus, anodic to the common constructional metals, steel, aluminium, copper, zinc and lead.

The most common magnesium anode materials for cathodic protection services are either very high-purity metals containing limiting percentages of copper, iron and nickel (approx. 0.02%, 0.003%, and 0.002% respectively) or alloys containing of the order of 6% aluminium and 3% zinc.

If it is used as a constructional metal some form of protective coating is almost obligatory. Chemical or electrochemical treatments which oxidize and passivate the surface are frequently applied. Extraneous coatings of paint, plastics or stoving enamels, even a coat of grease, will also be found used to form an impervious coating so that the metal is isolated from its environment. With the exception of stoved enamels, applied protective coatings are usually more satisfactory if instead of being applied directly to the bare metal, some chemical film has first been formed, e.g. by a chromate treatment. It may be mentioned that die lubricants used in pressure die-casting can initiate corrosion, especially lubricants containing graphite, and they should always be removed from articles formed in this way.

11.5.8 Tin and its alloys

The corrosion of tin in the atmosphere is slow, varying from about 0.000 11 ipy in marine atmospheres through industrial atmospheres to the low rural rate of about 0.000 02. Pure tin is very resistant to both hot and cold distilled water and when totally immersed in sea-water the corrosion resistance approaches that found in the atmosphere; if, however, conditions are such that the protective film is broken, corrosion rates may be greater.

In organic acids and dilute non-oxidizing mineral acids, the corrosion of tin is mainly controlled by the amount of oxygen present. Phosphoric and chromic acids form protective layers but nitric acid will freely attack tin even in the absence of oxygen.

Tin is attacked by alkalis the rate increasing sharply at about pH 9 to 10 and at alkalinities over pH 12, the use of tin is not recommended. The corrosion rate is, again, much affected by the presence of oxygen and the temperature.

Tin alloys such as solders suffer corrosion in some supply waters, the rate increasing with temperature and any increase in softness.

By the nature of their use many solders are often present in their environment as a small metallic area surrounded by a much larger area of another metal. Thus, if the solder is anodic to the adjacent metal considerable couple action may be expected if the environment is a good electrolyte and scale forming ingredients are absent.

Babbitt bearing metals (alloys of lead, tin, antimony and copper) usually suffer little corrosion and are usually resistant to the weak acids and sulphur compounds found in oils. Instances have occurred, however, with some modern additive oils in the presence of water where very hard crusts of tin oxide have formed on the bearing surface with consequent damage if this hard compound breaks down. The real cause and progress of this type of attack has not, at the time of writing (1969), been successfully resolved.

Babbitt metals are not considered suitable for bearings to which sea-water has access; and for bearings operating under sea-water it is usual to use an alloy of 70 tin: 1.5 copper: balance, zinc.

11.5.9 Nickel and its alloys

Nickel is one of these metals which has the ability to form, in the presence of oxygen, passive protective films on its surface which can be very resistant to many environments and high temperatures.

Nickel is resistant to most atmospheric corrosion though it does not, usually, retain its brightness and many nickel alloys are more resistant than nickel itself.

Nickel is normally passive in fresh and distilled water and in the latter corrosion rates are very low, of the order of 0.000001 ipy. Nickel is less resistant to sea-water with corrosion rates of about 0.003 ipy, and in such environment its copper-containing alloys are more usually used since nickel, like the stainless steels, is liable to become pitted at active-passive areas formed by oxygen concentration cells.

Nickel has a relatively high resistance to solutions of alkalis and is used satisfactorily for chemical plant handling caustic liquids.

Nickel resists non-oxidizing acids but corrodes freely in oxidizing acids and solutions of oxidizing salts. It is used for handling fruit juices, wines and milk products and it is usually considered non-toxic for handling most foods.

Important nickel-base alloys are the Monel compositions of the International Nickel Company. Monel (70 nickel: 30 copper) is found in both the cast and wrought forms. 'K' Monel (containing 2.75% aluminium) is a precipitation hardening alloy and 'S' Monel (containing 4% silicon) has great resistance to erosion. From the point of view of corrosion resistance, these alloys may be considered together though, in general and especially in sea-water, Monel has the lowest corrosion rate. The resistance to corrosion and corrosion-erosion in all natural waters is excellent. They behave well in non-oxidizing organic and inorganic salts at alkaline, neutral and mildly acid pH but they are not suitable for handling oxidizing acids or solutions of oxidizing salts. They have good resistance to relatively dilute alkalis but they are not so resistant as pure nickel to solutions containing over 40% caustic alkali. They are frequently used for food-processing plant and chemical plant, such as textile treatment vats, where corrosion may be severe.

Certain nickel-base alloys contain substantial amounts of chromium. Typical compositions are 80 nickel: 20 chromium, primarily used for heating elements and high-temperature service and 76 nickel: 15 chromium: 7 iron (Inconel Alloy). Both these alloys, and others containing large percentages of chromium, are used for handling nitric acid, mineral acid mixtures, many alkalis and hydrogen sulphide. Many of these materials are available in both the cast and wrought state and some containing boron are used as hard facing alloys for use at high temperatures in steam plant, etc. There are also austenitic-type stainless steel compositions which are also obtainable in wrought and cast form and they are very resistant to corrosion by steam condensates, natural waters and acid mine waters; these highly alloyed nickel-base materials are among the few metals which will resist corrosion by water used for scrubbing furnace gases.

Most of these alloys depend on their surface film (passive) of oxide for their corrosion resistance, and hence they must have continual access to oxidizing conditions if the film is to remain protective. Thus, they are liable to suffer corrosion at areas of low aeration or at oxygen shields, such as fouling, or at crevices where differential oxygen cells can form. Nickel-base alloys containing more than 40 to 50% nickel are immune from stress-corrosion cracking in waters containing traces of chlorides, and at high temperatures and pressures. For this reason Inconel (International Nickel Alloy) is frequently used in atomic-energy cooling systems and modern high efficiency steam-raising plants (see also refs. 3 and 7).

11.5.10 Zinc

Metallic coatings of zinc are probably the most widely used method of preventing corrosion by imposing a barrier between the metal and the environment in which it is operating. About one-half of the world production of zinc is used in the form of coatings for the prevention of the corrosion of ferrous structures.

Zinc and zinc alloys have good resistance to corrosion in the atmosphere and most natural waters, though any coating will need to be thicker for waters containing chlorides. The corrosion resistance of zinc, as with many other metals, is due to its ability to form protective coatings of oxides, hydroxides and some basic salts; it is, however not as sensitive to differential aeration conditions as is the case with those alloys which form passive films. Also, since zinc forms an amphoteric oxide, the formation of such layers is affected by pH at both ends of the scale. The corrosion rate increases rapidly at below pH 6 and above pH 12.5; but in the pH range 6 to 12 the film is stable and its corrosion rate is relatively low.

The American Society for Testing Materials reports the following figures for the corrosion of zinc in various atmospheres: industrial, 0.000 25 ipy; marine, 0.000 06 ipy; rural, 0.000 04 ipy; and 0.000 007 ipy for desert conditions.

Zinc forms protective films in hard waters and corrosion rates are low but in distilled water, especially if it contains carbon dioxide, serious corrosion of the pitting type can occur, which may lead to penetration of the zinc and, further, the water may pick up sufficient zinc for the water to become toxic. Waters containing chlorides also tend to promote pitting though BISRA tests have shown that steel coated with aluminium, cadmium, lead or tin behaves less well in sea-water than zinc coatings. In considering the relative behaviour and economics of coatings it is important to remember that thickness and ease of application may play their part in the scheduling of the correct material for a particular service use.

Zinc coatings are very frequently used in contact with soils and their behaviour ranges from good to bad depending on the type of soil. The National Bureau of Standards (USA) exposed zinc coated specimens at a variety of sites covering a pH range from about 3 to 10 and resistivities of 62 to 17 800 ohm cm, and the results indicated that poor aeration, high acid and soluble salt content caused the most severe attack. The behaviour of zinc-coated steel pipe compared with bare pipe showed that the uncoated pipe corroded about five times as rapidly as the coated material and rates of pitting, where it occurred, were at an even higher differential.

Zinc coatings are an excellent base for paint and paint films used in conjunction with a zinc coating may behave better than the sum of the lives of the coatings if estimated separately. Paints adhere well if applied directly to sprayed coats but galvanized and other types of coating often require some surface preparation, such as a pre-paint primer to roughen the surface.

Temperature has a marked effect on the corrosion of zinc in water. The corrosion rate is least at about 20°C but increases by a factor of about 100 at temperatures between 55 and 65°C and then falls again to about six times the low temperature rate at 100°C.

The resistance of zinc to many chemical environments depends on the pH (see above) and the purity of the metal. All common acids both organic and mineral attack zinc but only the stronger solutions of alkali; and, on the whole, commercial zinc is likely to be more susceptible to attack than high-purity material. Hydrocarbons and other non-acidic organic liquids can be handled in galvanized tanks, pipes, etc. but zinc is not suitable for handling food stuffs.

The use of zinc in domestic housing has a technology of its own. Galvanized storage tanks, small-bore piping for underground services in farms are, in the majority of cases, very successful and satisfactory. However, galvanized equipment occasionally fails – and fails very quickly. The reasons may be diverse, water, temperature, type of application, purity of the zinc, to mention only a few; and since these failures are likely to be characteristic it is advisable to refer them to the relevant research organizations dealing with zinc.

Zinc is widely used for sacrificial anodes in cathodic protection (see cathodic protection, below).

11.6 CATHODIC PROTECTION

11.6.1 Introduction

The use of cathodic protection to control the corrosion of pipe lines, ships' hulls, etc., is becoming so common that any account of corrosion must contain some reference to the method, though it must be pointed out with some insistence that cathodic protection is a *specialized* technology, full of imponderables, and that successful, economic and viable systems are only likely to be designed by technological experts.

The essential factor in cathodic protection is the application of a countercurrent sufficiently large to neutralize all the local corrosion currents on the surface of a corroding area. As long ago as 1824 Sir Humphrey Davy and his assistant Michael Faraday demonstrated that all local corrosion – which is represented by the sum of all the corrosion cells on any area – could be swamped out by connecting the corroding surface to a secondary area which was anodic to the area to be protected thus, making the local corroding area an overall current receiver, i.e. cathodic. In other words, all the local anodes are polarized to the same potential as the local cathodes so that no corrosion cells can be formed. (fig. 11.9)

To produce this condition in practice the counter e.m.f. may be provided by a metal which is more electronegative than the metal to be protected, e.g. magnesium and steel (sacrificial protection), or by applying an opposing external current from an electrical

842 | THE CORROSION OF METALS

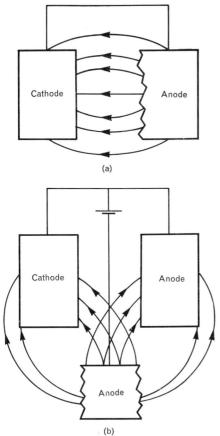

Fig. 11.9
(a) Anode corroding due to current flow to Cathode.
(b) Anode of I prevented from corroding due to current flow from external anode which blanks corrosion on both components of I.

supply via an auxiliary anode which may be metallic or non-metallic as long as it conducts electrons (impressed-current protection).

When a cathodic protection system is inaugurated there is a tendency for the pH of natural electrolytes to increase in the vicinity of the protected structure. Under certain conditions this is useful because ferrous structures tend to corrode less at high pH values; and also because, in waters containing calcium carbonate or magnesium hydroxide (e.g. sea-water), there is a deposition of scales which have a protective effect and thus the amount of current necessary for protection is reduced. On the other hand, if the alkali produced is excessive oil-based, paints will be attacked and if a metal, such as aluminium, is overprotected corrosion may be increased and the cathodic protection vitiated.

11.6.2 Current required for protection

The amount of current required to effect protection will vary from metal to metal and it will also be affected by differences in environment, the geometry of the structure to be protected and the resistance of the circuit. For example, bare clean steel in calm sea-water (cathodic protection can only be applied in an environment capable of conducting an electric current) will need a current density of about 10 mA/ft² for full protection, but this figure will fall as the calcareous scale builds up to about 3 mA/ft², and on a painted structure the average value will be even less. A newly painted ship with an undamaged paint coat may require only one-tenth of the current required for bare steel at its lowest requirement.

Since the amount of current required to protect any particular structure cannot be computed from theoretical considerations – the variables are imponderable – the most satisfactory and effective method of estimating the necessary criterion for the correct current density, to protect any structure, is to measure the potential of the metal, to be protected, in its environment relative to some *standard reference electrode*.

The absolute reference electrode is the well-known hydrogen half-cell in which hydrogen at 1 atm pressure is bubbled into a solution of the metal's own ions at unit activity of hydrogen ions. Such an electrode is obviously not practicable under field conditions. It is thus replaced either by the copper sulphate or the silver chloride electrodes. Particulars of the reference electrodes and their manufacture can be found in books on cathodic protection (see ref. 18). The potentials of these electrodes *vs* the standard hydrogen electrode at 25°C are, respectively, -0.318 V and -0.222 V, if the silver/silver chloride electrode is made up with sea water the potential *vs* the hydrogen electrode is -0.250 V. In cathodic protection technology it is general practice to use the copper/copper sulphate electrode for monitoring buried structures and silver/silver chloride electrode for marine applications. High-purity zinc–bare metal–can also be used as a reference electrode and it is frequently found in automatic systems designed for ships' hulls.

Satisfactory protection of ferrous metals in sea-water is usually accomplished if the structure is polarized to *more* than -0.75 V against silver or *less* than $+0.30$ V against zinc and the practical values used in the field are of the order of -0.80 to -0.85 V *vs* silver and $+0.20$ to $+0.25$ V *vs* zinc, and to avoid alkali damage the potential should not be allowed to become more negative than -0.90 V or less positive than 0.15 V against silver and zinc respectively. In soils the recommended value *vs* the copper half-cell is -0.85 V but when the protection is required for soils which contain sulphate reducing bacteria a value of -0.90 V is often used. As with ships' hulls the amount of current required to protect a buried surface depends on many parameters, e.g. the type and quality of any protective coating, the nature of the soil, particularly with regard to its conductivity, the nature of the electrolyte present, and the amount of oxygen.

11.6.3 Anode materials

The choice of system, 'sacrificial anodes' or 'impressed-current', for any particular

Table 11.4 Potentials of reference electrodes in sea-water

Metal	vs Silver/Silver Chloride	vs Zinc
Magnesium	-1.50 V	-0.45 V
Zinc	-1.05 V	0.00 V
Iron	-0.75 V	$+0.30$ V
Copper	-0.25 V	$+0.80$ V
Silver	0.00 V	$+1.05$ V
Aluminium	Potentials vary with the alloy from about -1.0 V to -1.4 V *vs* silver/silver chloride	

system is beyond the scope of this account, but their major differences are as follows.

Galvanic anode systems require no electrical power or equipment. If properly designed they are relatively fool-proof and require little supervision. The current cannot be connected in the wrong direction – which produces an increase in corrosion instead of a reduction. Installation is simple (though correct design of the structure of the system is important) and the anode leads are themselves cathodically protected. On the other hand, the current available depends on the anode area which may entail unacceptable weight, e.g. on large ships. Also the cost factor may be larger than with impressed current systems – electrical energy cannot be produced as cheaply from galvanic anodes as from generators – and high currents require large leads to reduce resistance losses.

Sacrificial anodes are usually constructed from magnesium, zinc or aluminium, and in each case the composition of the alloys used for this purpose is of critical analysis. These anodes are required to give long life without losing their efficiency or corroding unevenly. Certain impurities in the metal, especially with magnesium and zinc anodes, require to be kept to certain minimum quantities. The correct types of materials are available commercially and alloys of magnesium, zinc or aluminium not specifically developed for cathodic protection systems should not be used under any circumstances.

Impressed current systems can be designed to give any (large or small) protecting current provided that the anode material is suitable for the requirements and remains intact. It is usually found, however, that impressed-current systems are only used above certain limiting current requirements, i.e. above that which cannot be economically supplied by small galvanic anodes. The leads can be of any size so long as the resistance losses can be overcome by an increase in the applied voltage. A continuous d.c. electrical power supply must be available and the connections to anodes and structure to be protected must ensure that the current flows in the correct direction. The *positive* terminal of the d.c. supply (red) must be connected to the anodes and the *negative* terminal (black) to the ships' hull or other structure to be protected. In general, more technical supervision will be required. The anode leads must be well insulated since any exposure of the lead to the environment will encourage preferential corrosion at the point of leakage.

These are the main advantages and disadvantages of the two systems but this account does not claim to be all embracing and it is recommended that one of the organizations specializing in cathodic protection should be consulted before any system is fitted.

The anode materials for impressed-current systems will require different properties from those for use with galvanic systems. In the first place, the degree of permanence must be considered; it may be uneconomic to use a material the permanence of which exceeds the required life of the structure. In theory, any electronic conductor may be used for a 'permanent' anode in an impressed-current array. The following materials are commonly employed: silicon-iron, lead-silver alloys, graphite, platinized titanium or tantalum, scrap ferrous metal, and in certain cases aluminium, e.g. for potable waters since the corrosion products of aluminium are non-toxic. For further information see ref. 18.

11.6.4 Anode shields

An important component in cathodic protection systems, especially in ships' hulls, is the anode shield. Cathodic protection systems using large anodes for either galvanic or impressed-current protection need some dielectric shield to spread the current from the immediate vicinity of the anodes. With galvanic anodes, high-duty non-alkali paint correctly applied is often sufficient but for impressed-current systems more permanent and better dielectric qualities are necessary. Neoprene sheet is a commonly used material though many modern installations use some of the newer plastics which have high insulating properties and are easy to apply. Care must be taken that all joints are shingled so that they are not undercut by movement of the ship and for anodes fitted on the top of bilge keels, anode shields are very important if good protection is to be obtained. It is often good practice to protect the edges of the insulation by tack-welded metal strips.

11.7 CORROSION CONTROL

11.7.1 Introduction

Although all the usual constructional metals are subject to corrosion under some conditions, much of the attack can be kept within acceptable economic limits by selected methods of control. For instance, since corrosion is a reaction between a metal and its environment the obvious method is to prevent the environment from having any contact with the corrodible material. This is, however, only one method and it is usual to schedule corrosion control into three broad groups: control of the metal, which will include alloying for specific forms of attack, e.g. the addition of arsenic to brass as an inhibitor of dezincification; control of the environment, including its exclusion from the metal surface; and control by design, e.g. drain holes to allow run-off of moisture. The order of importance of the method will vary from case to case.

While the choice of the metal is usually the first consideration in corrosion control it is more often than not true that the best material from the corrosion viewpoint cannot be chosen because of other factors such as economics, mechanical properties, toxicity, user attraction, etc.

Excluding the problem of design, if the choice of metal leaves some unacceptable possibility that it will corrode in the relevant service and environment, three methods of control are usually available. These are: the environment must be treated to render it less corrosive, e.g. boiler feed-water treatment; the environment must be prevented from reaching the metal surface, e.g. paint on mild steel; or the application of cathodic protection (see cathodic protection, p. 841).

Treatment of the environment is usually somewhat specific to the particular environment and will include such methods as the use of inhibitors, the control of pH, or the addition to the environment of passivators or special chemicals which will act as protective film formers.

11.7.2 Corrosion inhibitors

The definition of a corrosion inhibitor is: 'a chemical substance which when added in small amounts to a corrosive environment effectively decreases the corrosion rate at any metal surface with which it is in contact'. Inhibitors work at the sites of the anodic or the cathodic reactions or sometimes both.

a. Anodic inhibitors. These are often compounds which form essentially insoluble salts with the metals they are required to protect. Sodium and potassium hydroxides, chromates, nitrites, phosphates, carbonates and benzoates are found as inhibitors of the corrosion of iron in waters. It must be stressed that the concentration of the inhibitor must be such that there is sufficient present, at all times, to stifle attack over all the contacting metal surface. If too little inhibitor is added, corrosion may occur at local areas where concentration cells may be set up with consequent severe local corrosion which may produce more attack than would have occurred in the absence of the inhibitor (see safe and dangerous inhibitors, below).

b. Cathodic inhibitors. Two general types of cathodic inhibitor are used: those which affect the oxygen adsorption reaction, and those which affect the hydrogen evolution reaction. The two reactions are as follows:

$$O + H_2O + 2e \rightarrow 2OH^-$$
$$2H^- + 2e \rightarrow H_2$$

Salts of magnesium, nickel and zinc, which form almost insoluble hydroxides, are inhibitors of the oxygen adsorption reaction. They react with the OH^- ions to form insoluble deposits on the cathodes and a barrier is so formed which prevents the access of oxygen which is required to depolarize the corrosion reaction. An example of the stifling of the hydrogen evolution reaction is the addition to the environment of salts of arsenic which when they adsorb onto the metal surface increase the hydrogen overvoltage and thus prevent the evolution of hydrogen.

c. Safe and dangerous inhibitors. An inhibitor is said to be safe if it reduces all the corrosion on the required surface without increasing the amount of corrosion on adjacent areas, which may remain unprotected. Dangerous inhibitors are those which tend to increase the rate of attack on unprotected areas. They are frequently found among the anodic inhibitors though they are the most efficient inhibitors if they are used at the correct concentration. Chromates, phosphates, soluble oils and other anodic inhibitors are capable of increasing corrosion if used in insufficient quantity or the original concentration is allowed to fall – the attack usually takes the form of pitting. Sodium benzoate, however, usually causes general attack if used in too low a concentration while zinc sulphate, a cathodic inhibitor, may accelerate corrosion if used in too high a concentration.

d. Adsorption inhibitors. The use of organic chemicals to modify liquids so that the corrosion rate is reduced is a very wide subject. Some compounds are able to film both the anodic and the cathodic areas of the corrosion reaction. Others may operate on the anodic or cathodic areas respectively.

In using inhibitors, the operator must be familiar with the technology of their use; concentrations, sometimes of narrow limits, must be maintained and in some cases it is necessary to use some measuring reaction to maintain concentration, e.g. titration in boiler feed-water treatments.

e. Passivators. Passivators are chemical compounds which, when added to an environment, are capable of changing the electrochemical potential of the metal surface with which they come in contact to a more cathodic-noble value. Cathodic inhibitors do not usually react in this way but some anodic inhibitors may. Again, the concentrations of these substances and the maintenance of the correct value as the passivator is used up in service require specialized operational techniques.

f. Temporary protectives. Sometimes especially during erection periods, it is necessary to prevent corrosion occurring while the metal awaits (usually in the open) its final location in a building or other large structure. There are many of these products most of which are petroleum based and they may be deposited as either hard or soft films. Their use would seem to be too infrequent and many corrosion troubles have been known to stem from initial corrosion begun when the metal was lying awaiting use. They are, admittedly, a nuisance to use since the complete removal of the compound may be a necessity, especially in such components as tubes which will be subjected to heat on installation. In such cases failure to remove the protective film will often result in a burst tube.

11.7.3 Volatile corrosion inhibitors

These materials are in a class by themselves compared with the orthodox inhibitors added to corrosive liquids. They are solids, many of them with patented compositions, which vaporize relatively slowly, producing inhibiting ions which, in the presence of oxygen and moisture, can act as stifling agents in the corrosion reaction.

Successful volatile corrosion inhibitors (VCIs) have the following characteristics: they must be stable under the required service conditions; they must be soluble in water; they must be sufficiently volatile to maintain the necessary inhibiting concentration; and their diffusion characteristics must be requisite for the job in hand.

They should not be used without adequate care; for although nearly all ferrous metals can be protected by these substances other metals, either in the massive state or as alloying elements may be attacked by some VCI compounds.

Volatile compounds used to inhibit the atmospheric corrosion of metals are found in two groups: basic water-repellent compounds (usually amines), which are used in steam condensate lines, and the volatile nitrites, carbonates, benzoates, etc., used for the prevention of corrosion in packaged or stored materials. These latter are usually solid amine salts which are added to packages or enclosed areas on trays, in cloth bags or coated on paper and they are used almost exclusively for ferrous metals.

VCIs are only effective if they are chosen with regard to their specific behaviour and properties, and although if properly used they are excellent, since they are cheap, convenient and easy to use, they should not be put into service without adequate attention to their specific technology.

11.7.4 Control of corrosion by design

Many corrosion engineers find it regrettable that some designers using metals are very often not familiar with the corrosion characteristics of their materials and pay scant heed to what can be done to prevent future corrosion problems by the common sense application of relatively minor design conventions. If possible the prevention of corrosion should begin at the drawing board stage in the design of buildings, chemical, industrial plant and ships, etc. The following are some of the precautions which should be undertaken. They are not to be accepted as all embracing and they are in no special order of priority since that will vary from structure to structure.

Since moisture, or water, is the basic initiator of corrosion, all structures should be designed so that water or moisture is not entrapped in pockets, etc. where it will lie and collect. Particularly if the pocket will also attract dirt and debris. Thus, crevices and ledges should be avoided and channels and gutters should be fitted with drain holes.

Although painting may be easy during erection, designers should remember to allow access after structures are built so that repainting can be also easy and old rust and debris can be satisfactorily removed. If metal plates are shingled they should be fitted so that moisture drains away from the laps.

Any component which is scheduled to contain liquid should be designed so that it can, if necessary, be completely drained. If the metal is in contact with flowing liquids, care must be taken to avoid water speeds and turbulence above the capacity of the metal to withstand. Changes in diameter, junctions and valve sitings, etc. should be carefully designed so that local turbulence is not set up. The fitting of flanges and washers should not allow 'proud' barriers to project into the water stream.

Stray currents should be eliminated as far as possible and electricians should be instructed on the probability that earthing to water-borne structures may promote corrosion. Laggings can cause corrosion and those which are successful with ferrous metals may not be equally so with other metals such as aluminium. The possibility of metallic couples setting up corrosion must not be neglected.

Those alloys which rely on passive films for their corrosion resistance should not be designed into a structure so that access of oxygen to maintain the passive film is restricted. The problems of possible stress-corrosion cracking and corrosion fatigue should be considered, especially in the high-strength alloys.

The use of cathodic protection should always be borne in mind: it can be designed into many internal locations and if the correct technology is applied – usually through consultants – anodes of the correct size and shape can assist, materially, in the prevention of corrosion in relatively inaccessible locations. The designer should always bear in mind that galvanic attack is not the only form which corrosion takes, and although galvanic attack can be serious it is often true that the more subtle types of corrosion such as crevice attack or the presence of differential cells can be more dangerous since they encourage pitting and corrosion usually occurs in hidden places.

11.7.5 Control of the metal

In choosing the correct metal for any specific environment the first consideration will usually be adequate strength which is a relatively straightforward problem. To correlate strength with corrosion resistance is a much more difficult task.

It is first of all necessary to select a material which has the correct level of corrosion resistance compatible with the economics of the project. It has to be considered whether the initial material will last, but not appreciably outlast, the life of the structure or component, or whether a less corrosion-resistant material should be fitted and the consequent replacement cost accepted out of future income.

Corrosion control, so far as the selection of the metal is concerned, should be based on accurate costing and the designer's job should be to achieve absence of corrosion for the service life of the metal for a minimum of outlay in relation to initial material, maintenance and/or replacement schedules.

Another consideration in designing for corrosion resistance will be the availability of materials. It is sometimes necessary to accept substitutes, in which case if the second choice material is of a different corrosion characteristic from the original choice relevant changes in design – such as reworking metallic couples – may be necessary. Criteria of corrosion resistance often offer considerable problems since corrosion rates vary from environment to environment and very often reliable data for the service requirement under consideration is not available. In such cases laboratory trials are often initiated. These 'accelerated' tests can be most useful but if the laboratory conditions do not closely correspond with those of the service required they can be misleading. For example, manufacturers' corrosion rates computed on total immersion may be quite unrealistic if the metal is to be used in an intermittent environment, at a different rate of flow or at a different oxygen concentration. Further, corrosion rates expressed in 'inches per year' or 'milligrams per square decimetre per day' are not very useful if the metal is prone to pitting or the effects of stress in the presence of corrosion. The extrapolation of short-term tests to long-time service also offers difficulties. and such estimated corrosion rates will only be of practical value if the experimental rates were very high or very low. If, as frequently occurs, the accelerated tests give rather indeterminate figures the designers will, understandably, tend to work to the highest rates, thus increasing the amount of metal used, wasting material and adding to the capital cost. It is, therefore, often economic, though initially expensive, to make adequate *ad hoc* tests under service conditions in the actual locality where the metal is to be used before the final choice is made.

11.7.6 Paint

Although the prevention of corrosion by coats of paint is a complete technology of its own there are several very important points with which the corrosion engineer should be familiar.

Perhaps the most important axiom is that a good paint badly applied to an unprepared surface will not behave as well, and will be far less economic, than a poor or even bad paint properly applied over a well-prepared surface. In other words, the behaviour of an anti-corrosive paint is almost entirely controlled by the quality of the surface preparation prior to painting and the efficiency of the application.

The second most important point is that the primer should have anti-corrosive properties, the correct porosity and an acceptable drying time.

Thirdly, since it is probable that the anti-corrosive paint will not have much decorative value, it must be chosen so that it is compatible with any subsequent coats.

It must be accepted that all environments – atmospheric, marine, fresh water, soil, etc. – have their special problems and only consultation with those experienced in each field will ensure satisfactory and economic paint barriers against the ravages of corrosion.

Paint is essentially a pigment dispersed in a solution of an organic binding medium. The type of binder will decide the basic physical and chemical properties of the paint. In a finishing paint the pigment is there to provide colour, but in a primer (most anti-corrosion paints are primers) it should contribute to the durability of the whole system. The most commonly used pigments in anti-corrosion paints used on ferrous metals are red lead, zinc chromate, calcium plumbate and zinc dust. For non-ferrous metals the most commonly used barrier against corrosion is, probably, zinc chromate and it should be noted that lead-based paints must never be used on aluminium. Calcium plumbate is frequently recommended for use on galvanized articles. Those paints containing zinc dust, since they have a built-in cathodic protection system, are becoming deservedly popular for all metals. However, the paints must be porous and they will only function satisfactorily in an electrolyte solution. e.g. in saline waters.

The function of the third major component of a paint – the volatile solvent – is to aid manufacture and application and to give some control of drying time.

Paints vary considerably in their composition and 'painting' properties and it is rarely that a single paint possesses all the requirements of an anti-corrosion barrier. It is, therefore, usually necessary to schedule a system: primer, intermediate coat or coats, with, possibly, a final coat of varnish. The primer should adhere well–it is the bond between the metal and subsequent coats–as well as provide most of the protection. The finishing coat should protect the primer and add any aesthetic qualities which may be necessary.

The question of adherence is most important and no matter what may be the method of surface preparation for the reception of the primer the final criterion will be what is called the 'anchor pattern'–the surface must have a satisfactory 'key'. The metal-paint interface must be rough, but not so rough that economic paint coat thicknesses do not cover the hills as well as the valleys of the surface configuration. This requirement may be rather less stringent with the 'etch primer' type of preparation since adherence of the primer, in such cases, is usually chemical as well as mechanical.

Finally, it must be mentioned that some pigments promote corrosion (see above, lead-aluminium). Also primers containing lamp black and graphite can cause intense local corrosion since the graphite particles can act as active local cathodes. Further, some anti-fouling paints can be actively corrosive. Heavy metals such as copper and, especially, mercury should not be applied directly to metal surfaces, particularly to aluminium and its alloys.

11.8 LITERATURE

1. U. R. EVANS. *The Corrosion and Oxidation of Metals–Scientific Applications.* London, Edward Arnold, 1960.
2. U. R. EVANS. *An Introduction to Metallic Corrosion.* London, Edward Arnold, 1963
3. L. L. SHREIR (Ed.). *Corrosion.* London, George Newnes, 1963.
4. N. D. TOMASHOV. *Theory of Corrosion and Protection of Metals.* (Translated from the Russian and edited by Boris H. Tytell, Isidore Geld and Herman S. Preiser.) New York and London, Macmillan and Collier-Macmillan, 1966.
5. H. H. UHLIG (Ed.). *The Corrosion Handbook.* New York and London, John Wiley and Chapman and Hall, 1948.
6. H. H. UHLIG. *Corrosion and Corrosion Control.* New York and London, John Wiley, 1963.
7. F. L. LAQUE and H. R. COPSON (Ed.). *Corrosion Resistance of Metals and Alloys.* New York and London, Reinhold Publishing Corp. and Chapman and Hall, 1963.
8. F. N. SPELLER. *Corrosion–Causes and prevention.* New York and London, McGraw-Hill, 1951.
9. G. BUTLER and H. C. K. ISON. *Corrosion and its Prevention in Waters.* London, Leonard Hill, 1966.
10. T. HOWARD ROGERS. *The Marine Corrosion Handbook.* Toronto, McGraw-Hill, 1960 and *Marine Corrosion,* London, George Newnes, 1968
11. J. TAFEL. *Zeitschrift der Physikalischen Chemie.* **50** (1904) 641.
12. M. POURBAIX. *Thermodynamics of dilute aqueous solutions.* London, Arnold, 1949.
13. M. POURBAIX. *Atlas of electrochemical equilibria in aqueous solutions.* Oxford, Pergamon, 1966.
14. F. FLADE. *Zeitschr. Physik. Chem.* **76** (1911) 513.

15, 16. VISRA and C. C. LARRABEE. *Corrosion* **9** 1953 and *Trans. Electrochem. Soc.* **85** (1944).
17. W. J. BURLING SMITH. *Steels in Modern Industry* (ed. W. E. BENBOW). London, Iliffe, 1953.
18. J. H. MORGAN. *Cathodic Protection.* London, Leonard Hill, 1959.

CHAPTER 12

Composite Materials

12.1 PRINCIPLES OF COMPOSITE MATERIALS

12.1.1 Definitions

A composite material may, in a broad sense, be considered to be any material which is composed of two or more physically distinct components, each of which, by virtue of its own particular properties, fulfils a major role in the functioning of the material.

This definition covers an immense field. Indeed, many useful alloys, including the steels, consist of a metallic matrix (or ground mass) reinforced by particles of intermetallic compounds. Galvanized or painted steelwork are also composite materials, the steel providing strength and stiffness and the zinc or paint the protection from the environment. Within these limits there is a much more closely defined range of materials which are being developed to meet the present and future engineering demands for improved physical and mechanical properties. For the purpose of this chapter composites will be considered to be those materials which have been formed from two or more components and in which the components retain to a large extent their original geometry and properties.

This definition now excludes materials such as the conventional alloys, painted metals and the like. For the sake of completeness, however, the definition should include the term 'man-made'. This now excludes those natural composite materials such as wood and bone which consist of strong fibres bonded together by a relatively soft matrix.

The present chapter is limited to composite materials involving metals and some composites which directly compete with metal composites (especially the reinforced plastics). Other important composites outside the metal field are various laminates of different plastics, laminates of plastics with textiles or wood, laminates of different glasses, laminates of strong materials such as wood and plastics with heat or sound insulating materials etc. Wire reinforced concrete, although an important material involving metals, is discussed in vol. II under concrete since the concrete retains to a large extent its original geometry and properties.

The aim in producing a composite material must always be to make use of the advantageous properties of all the components to yield a final material which possesses as few of their disadvantages as possible. One of the basic advantages of composites is that

12.1.2 The properties of materials

Before examining the concepts underlying composite materials and the desirable properties which can be attained, it is necessary to consider in general terms the components of these systems, what each class of material has to offer, and their limitations. Indeed, if these limitations did not exist, there would be no case for composites.

a. Metallic materials. Metals and alloys are the most widely used engineering materials, for good reasons. Firstly, metals can be alloyed to produce materials with tensile strengths up to 224 000 t.s.i. and above for some of the high-strength steels. Furthermore, many metals also possess a high resistance to alternating stress (fatigue) and to compressive, torsional and shear stresses. Unfortunately the strong alloys are usually based on metals with high densities, so that specific strength or strength-to-weight ratio is limited, limiting their application.

A further advantage is that most metals will plastically deform prior to complete failure. This capacity for plastic deformation can be made use of to work materials to a required shape and, perhaps more important, allows the stress at the root of a small crack or other defect to be redistributed, preventing catastrophic failure.

Most materials have useful strengths at temperatures up to about half the melting point; beyond this the strength falls drastically. In this respect metals are inferior to the ceramics because their melting points are relatively low. Moreover, below the melting temperature, metals under stress have a tendency to deform slowly (creep), while in many cases oxidation resistance is poor at elevated temperatures.

It is obviously impossible in a few sentences to generalize the limitations and advantages of metals. For instance, tungsten has a melting point of 3 850°C (higher than many ceramic materials) and chromium and aluminium are extremely resistant to oxidation due to their adherent protective oxides; on the other hand, metals such as beryllium and chromium lack the ductility mentioned previously. Nevertheless, from the foregoing, certain limitations of metals and the need for stronger and more temperature-resistant materials can be recognized.

b. Ceramics. Unlike the situation in metals where, when atomic bonds are broken, it is relatively easy for the atoms to recombine with different neighbours, new bonds cannot easily be made in ceramics. This means that ceramics cannot readily undergo plastic deformation and that even very small cracks in the surface of the ceramic will propagate, causing failure. This difference between metals and ceramics can be demonstrated by comparing the size of crack which will propagate in various materials at a stress of, say, 100 000 p.s.i. Thus a strong steel can tolerate cracks 1 inch long and aluminium can tolerate cracks about 0.016 inch long, but in glass cracks only 0.001 inch in length will propagate rapidly. This inability to accommodate even minute cracks constitutes one of the major disadvantages of ceramics and is associated with lack of ductility and, unless the surface is perfect, a low bulk strength.

On the other hand, the nature of the atomic bonding does ensure that if defects are absent very high strength can be obtained, and the hardness and elastic modulus*

---

* There are in fact, several elastic moduli. In the present context the tensile modulus (Young's modulus) is implied.

(ratio of stress to strain, upon which the rigidity depends) are both high. In addition, the kind of atomic bonding present in ceramics is responsible for their high melting points. Therefore, at temperatures above which most metals have negligible strength many ceramics still retain a measure of strength and stiffness. This is particularly true of materials such as carbon and boron.

Again there are exceptions to these generalizations; the behaviour of glass with respect to crack propagation is similar to that of a ceramic but it has a low melting point and its elastic modulus is only equal to that of aluminium. In spite of this, glass is an extremely useful component of the most widely used composite – the glass-fibre reinforced plastics.

c. *Plastics.* The plastics are organic compounds in which individual molecules are linked to form chains. Such substances are known as polymers (e.g. polyvinyl chloride (PVC), polyesters etc.). Although at present plastics are based on carbon it is possible that plastics based on other elements will appear in the future.

A major advantage of the polymers is that the type of atomic bonding is such that the molecules can slide over each other as do the atoms in a metal so that cracks propagate less easily than in the ceramics. A distinct limitation of the polymers, however, is their poor resistance to heat – they soften and char at quite moderate temperatures. However, the softening phenomenon is a useful property in that plastics can be readily moulded under low pressure by the application of moderate heat.

Their strength and stiffness are rather low, although this is somewhat offset by the low density of the material, so that there is widespread and increasing use of plastics for purposes which do not call for high strength. As a class, plastics have good corrosion resistance, although owing to the fact that the molecules are based on carbon they do soften or swell in carbon based solvents such as carbon tetrachloride, trichloroethylene etc.

12.1.3 Concepts employed in composite materials

From the foregoing it is clear that combinations of the three classes of materials would, theoretically, yield a variety of properties depending upon the choice of materials and their relative proportions. Various methods of classifying the composites have been used. One depends upon the property required, e.g. strength, stiffness or resistance to crack propagation, or upon the application to be met. Another classification depends on the matrix material. Probably the simplest is that based on the shape of the reinforcing component or phase, that is, on whether it consists of particles, fibres or layers. This classification is used in this chapter. The properties of the matrix and the second phase are obviously of first importance in determining those of the composite. The properties of the common surface (or interface) between the matrix and the second phase are also vital, and can markedly affect the performance of the whole composite.

It is not the purpose of this contribution to discuss in detail the many theoretical treatments given to composites; however, the basic concepts underlying the various types of composite materials will be discussed in general terms in order to demonstrate how the choice of components is made and how these components can be used to best advantage.

a. *Particulate composites.* There are two methods of employing particles of a second phase within a continuous matrix, the difference between the two lying essentially in the size of the second phase particle and the role which it fulfils.

856 | COMPOSITE MATERIALS

The simplest method is that known as aggregate hardening, in which particles of the two systems are distributed as shown in fig. 12.1(a). This type of composite may consist of two continuous intermingled matrices or of particles of one phase distributed throughout the other. Particles several microns* in diameter are used and the properties are, in general, proportional to the ratio of the constituents.

For instance, the cermets, which are metal ceramic mixtures, may have compositions between 30 and 90% of ceramic, depending on the properties required. Thus a material to withstand high temperatures may have a high proportion of ceramic while one destined for an application involving some measure of impact loading would have much less.

In systems of this type, the final solution is generally a compromise based on how much of one property can be sacrificed for an improvement in another. Similar consi-

* 1 micron, or micrometre (abbreviated as μ or μm) is a thousandth of a millimetre.

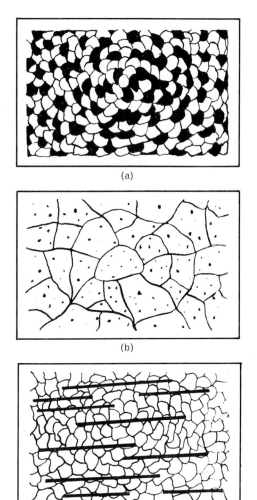

Fig. 12.1 Different methods of strengthening a matrix. (a) Aggregate; (b) dispersion; and (c) fibre.

derations apply in the case of the other important type of aggregate strengthening, in which powders are used as fillers in plastics to produce either a more rigid, or in some cases merely a cheaper, material than the unfilled material.

In order to describe the second type of particulate composite – the dispersion-hardened metals – it is necessary to examine in greater detail one of the properties of metals discussed previously, namely, the capacity to deform plastically. All metals consist of regular arrays of atoms forming a geometrical pattern called a lattice. This lattice, repeated in three dimensions, constitutes the crystal. Metals contain irregularities or dislocations in their crystal structure and it is by the generation and movement of these defects under stress that metals deform plastically, eventually to fracture. Any hindrance to the generation or movement of dislocations will mean that, to produce a given degree of plastic strain, the stress on the system with hindrances will need to be greater than on the 'pure' metal. Thus the introduction of hindrance to dislocation movement results in a strengthening. Other dislocations, produced by cold-working or by second phase particles such as the intermetallic compounds found in conventional alloys, may constitute the hindrance.

One method of improving the strength of a metal is to dissolve a second metal in it and then cause this second metal to fall out of solution as a precipitate of very fine particles. This can be done if the solubility of the second metal is greater at higher temperatures than at low and if enough of the second metal is added. The process is known as precipitation hardening. Dispersion hardening (fig. 12.1(b)) somewhat resembles precipitation hardening; the difference between the two is in the nature and mode of introduction of the second phase. Precipitates are relatively easy to produce by suitable heat treatment but, in general, because they are intermetallic compounds they lack stability at high temperatures. On the other hand the dispersed phases are usually stable oxides such as alumina or thoria, but they present problems of incorporation into the material in a suitable form.

The distance between particles is of vital importance in determining the strength of particulate composites; the closer the particles are together the more difficult it is to force a dislocation between them. Thus, the strengthening achieved will increase with the amount of second phase present and with a given quantity of second phase maximum strength will be achieved when it is dispersed as finely as possible. In conventional precipitation-hardened alloys, the free path between precipitate particles is of the order of a few hundred ångström units (1 Å = 10^{-10} m) but this is difficult to achieve in dispersion-hardened systems where the second phase is added separately. Nevertheless, the particle size employed in dispersion hardening is one or two orders of magnitude lower than that used in aggregate hardening.

The second difference between the aggregate-hardened and dispersion-hardened particulate composites is the amount of second phase. Although it is possible to tailor the composition of dispersion-hardened systems to a far greater extent than that of precipitate-hardened systems, the amount of second phase is usually limited to 20%. Beyond this level problems of fabrication arise.

Essentially then, there are the two distinct systems of particulate composites: one has properties roughly approximating to the sum of the component properties, while in the other the second phase is a tool used to modify the properties of the matrix.

b. Fibre reinforced systems. Aggregate hardening employs ceramics finely dispersed within a ductile matrix so that there is no continuous easy path for a crack to pass through the material. The second method of reinforcing a matrix consists of introducing fibres (fig. 12.1(c)). It can be applied to metals and plastics and indeed to other materials such as rubber and cement. The theory of the strengthening action of the fibres is based on the assumption that when the system is loaded both components (i.e. both fibres and matrix) are equally strained. The fibres are usually chosen to have a high strength and to be less extensible than the matrix (i.e. to stretch less than the matrix for a given tensile stress). The simplest system considered is one consisting of continuous fibres unidirectionally aligned in the direction of stress. On stressing, the fibres and matrix being equally strained, the less extensible component, i.e. the fibres, carries the majority of the stress on the system. For this condition to apply there must obviously be sufficient cohesion or bonding between the components to ensure the simultaneous straining of the fibres and matrix.

For maximum strength the fibres should be as strong as possible and much less extensible than the matrix and the fibre content should be as high as possible.

In the case of metal matrices the theory has to be modified to accommodate plastic behaviour of the matrix.

The foregoing remarks apply to composites containing continuous fibres all lying in the direction of stress. If these conditions are not present two effects are immediately apparent. For all matrices the strength is reduced if the fibres are not continuous. Fig. 12.2(a) shows this for fibres with a length/diameter ratio varying from ∞ to 8. Cohesion between the fibres and the matrix becomes all the more important because, for a given cohesive or bond strength, the shorter the fibre the greater is the tendency to pull out from the matrix. The strength is also much reduced if the fibres are aligned at an angle to the direction of stress (fig. 12.2(b)). This is due to a change in mode of

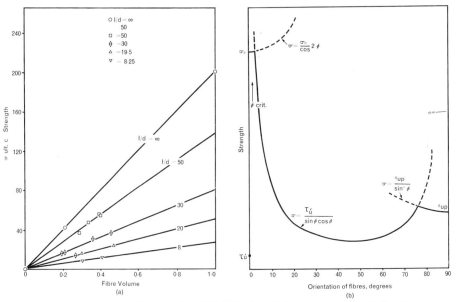

Fig. 12.2 Effect on composite strengths of (a) Fibre length due to diameter ratio (for the copper tungsten system at 600°C) and (b) fibre orientation.

failure. Fibre failure only occurs when fibres are aligned in the direction of stress or very nearly so; with increasing misalignment shear failure in the matrix and eventually tensile failure of the matrix become the dominant modes.

Because the fibres in a fibre reinforced composite are usually stronger and less extensible than the matrix, the composite is stronger and stiffer than the matrix from which it is derived. This, and the fact that ceramic fibres retain a high proportion of their strength and stiffness to high temperatures, explains why ceramic-fibre reinforced metals are of particular interest for aerospace applications.

The role of the common surface or interface between the fibres and the matrix is of particular importance in fibre reinforced systems. Cohesion of fibre and matrix at the interface is essential if both are to take part in resisting stress. It becomes even more important in materials reinforced with short fibres or with fibres having a range of strengths. In the latter case, when a weak fibre breaks, the interfacial bond permits the broken parts of the fibre to continue to function and resist stress.

Too strong a bond is, however, undesirable as it may impair the resistance to crack propagation. If a crack develops perpendicular to the applied stress and the fibre/

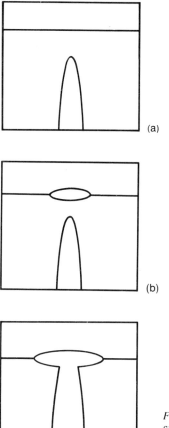

Fig. 12.3 Deflection of a propagating crack by a weak fibre/matrix interface. (a) Crack advancing towards fibre; (b) Weak interface, giving lack of cohesion in crack path; (c) and (c) crack stopped at fibre interface.

matrix interface is weak in tension, the applied stress causes failure at the interface and so causes the crack propagation to cease. This mechanism is demonstrated in fig. 12.3. On the other hand, if the material is subjected to compressive stress perpendicular to the fibres, a strong interface is essential to resist buckling. It is therefore necessary to compromise.

 c. Planar composites. These consist of 'sandwiches', the layers of which are comprised of the components of the composite. One component usually acts as reinforcement to the other. This type of reinforcement – called planar – can be used for laminated composites with controlled crack propagation properties and for surface protection. In both instances the planar reinforcement acts as a barrier – in the former a barrier to the crack, in the latter a barrier to prevent the atmosphere, perhaps an oxidizing atmosphere in the case of the refractory metals, reaching the underlying metal.

 Tough composites. Many ferrous materials suffer a ductile to brittle transition as the temperature decreases. That is, the energy necessary to cause complete failure in an impact is reduced markedly over a fairly narrow temperature range. Several factors such as specimen size, notch effect and speed of loading affect the transition temperature. Recently, attempts have been made to reduce the transition temperature by using composite materials.

 Arresting a crack by opening up a weak interfacial region, as discussed for fibre reinforced systems, is also possible in planar reinforcements where the crack is travelling perpendicular to the weak interface.

 A second method of improving the resistance to brittle failure makes use of the known size effect, that is, the phenomenon that in general the smaller the specimen the lower the transition temperature. Thus, by producing a laminated material, it should be possible to produce in each lamination conditions equivalent to those in very thin homogeneous specimens.

 Oxidation-resistant coatings. The refractory metals – tungsten, molybdenum, niobium and tantalum – all have high melting points. In addition, all have a high modulus of elasticity, a large proportion of which, together with strength, is retained at high temperatures. Although they appear useful engineering materials for high temperatures, the refractory metals all suffer in some measure from high density and brittleness, and they lack oxidation resistance. Attempts to produce oxidation-resistant alloys based on these metals have not been entirely successful so that more recent efforts have been concentrated upon oxidation-resistant coatings. Major improvements have been achieved by diffusing such elements as silicon, aluminium, titanium and chromium to produce a surface alloy which not only resists oxidation but is completely impervious to the passage of oxygen into the underlying metal.

12.1.4 Problems associated with composites

The problems associated with the production and utilization of specific composite materials are discussed in appropriate sections of this chapter. At this stage, however, it is possible to isolate certain inherent difficulties which most composite systems have in common.

 a. Uniformity of structure. With composite materials it is generally a case of the strength lying in the weakest link in the chain. Any areas deficient in one or other of the components may constitute a major weakness and lead to premature failure. It is a relatively simple matter to ensure that homogeneous materials or heterogeneous alloys formed by heat treatment are macroscopically uniform. Furthermore, any defects such as pores or surface cracks which may occur are readily recognizable by well-understood, easily operated inspection techniques.

This is not so with composites, where one of the first steps in the manufacture is usually to obtain a uniform distribution of the fine particles or fibres within the matrix. In the case of fibrous composites, in order to obtain maximum effect the fibres also have to be uniformly aligned. In addition, any non-uniform regions are most difficult to discover and very careful inspection must be applied to ensure that any defects are detected. In planar composite systems uniformity of both the distribution and interfacial conditions are also of vital importance.

b. Compatibility. The problems of compatibility between components must be solved before any useful composite material can be developed. Two general points should be stressed in this respect:

(i) The whole problem of wetting and bonding between materials of different species is complex, and in some ways obscure. Ideal compatibility lies somewhere between two extremes: at the one end, complete interaction between the components results in loss of integrity of matrix or reinforcement; at the other, the matrix component will not even wet the reinforcement and the two components are quite separate.

(ii) The degree of bonding necessary is dependent upon the nature of the composite, the properties required and the environment to which the material will be subjected.

c. Cost. One of the major problems with composite materials is cost, which stems from the cost of the components and the cost of incorporating the fine fibre or particles into the composite material or of introducing the planar reinforcements where required. The benefits of fine particles and fibres have already been stressed, but the finer the reinforcement the greater the processing necessary to achieve it. In many cases this is a magnification of an already high cost of a material in bulk form.

Some idea of cost of fibre reinforced systems can be gained by consideration of the costs of various types of fibre. Glass fibres are readily available in diameters of the order of a few tenths of thousandths of an inch. However, as we strive for better properties in a glass, so the cost increases. Boron, graphite and boron nitride all have advantages as reinforcing members and are being produced commercially in the USA, but costs per pound for these materials can be up to $700, $550, $175, respectively. Silicon carbide, perhaps the most recent of the continuous fibres, costs $2100 per pound.

This high cost is one reason for the limited commercial application of the more exotic composites; obviously the applications which justify such high costs are limited and will in the early stages be concerned with military and aerospace transport systems, where specific strength and stiffness are at a premium.

d. Engineering design. Composite materials are in many ways unique and careful design must be employed to ensure that the properties are used to their best advantage. The dangers of blanket acceptance of design methods suitable for conventional materials cannot be too strongly emphasized.

Anisotropy*, often a common phenomenon in composites, particularly those reinforced with fibres, must be fully understood and compensated for.

Often in obtaining the superior properties of composites others have to be sacrificed. This must be borne in mind and compensated for in design. Failure to take note of these simple concepts often results in perfectly good materials being written off as inherently unsuitable, when in fact the only fault is in unenlightened design.

12.2 PARTICULATE COMPOSITES

12.2.1 Dispersion-hardened systems

a. Method of manufacture. Dispersion-hardened systems consist of ultrafine particles dispersed in a matrix. In the case of metallic systems the matrix is a metal and the disperse phase a ceramic, usually an oxide. The aims of all processes of manufacture must be to obtain the ceramic in the required particle size ($\sim 2\,\mu$m) and incorporate it uniformly into the matrix with as little change in shape and size as possible. Several methods have been employed to produce these systems and are described here.

Oxide coating. The particles which ultimately comprise the matrix are superficially oxidized (naturally or artificially) and then sintered together. The thin film of oxide on each particle then constitutes the dispersion-hardening phase. For example, the natural film of oxide (alumina) which forms on aluminium powder becomes the ceramic reinforcement in the sintered composite. This sintered aluminium powder (S.A.P.), consisting as it does of alumina dispersed in an aluminium matrix, is the material which aroused interest in dispersion-hardened systems. Here the metal is reinforced with its own oxide which forms naturally. The aluminium powder is cold-pressed and sintered in vacuum to remove hydrogen porosity, or hot-pressed, to form a compact with a density of 90% theoretical. The compact is finally extruded (i.e. squeezed through an aperture smaller than the cross-sectional area of the original compact) to achieve greater densification.

Powder mixing. The simplest method involves the direct mixing of the finest powders of the metal and the oxide. The major problem is the production of the fine powders and techniques have to be employed to prevent the particles agglomerating during milling as a result of heating or electrostatic attraction. Extremely fine (0.1 μm) powders of nickel have been obtained by causing the absorption of certain polyvalent ions on the powders; this prevents agglomeration during milling.

The fine powder is then pressed, sintered, and subjected to a working operation, usually hot extrusion, to give final densification to almost theoretical density. As the amount of oxide increases, so it becomes more difficult to attain theoretical densities.

Internal oxidation. This method can be employed only with those alloys consisting of a solid solution of one metal in another, one of which forms an oxide more readily than the solvent metal does so that it oxidizes preferentially. The alloy, preferably in the powder form, is exposed to oxygen, which diffuses inwards and oxidizes the more reactive metal. The choice of alloy is limited to those in which the diffusion of oxygen inwards is more rapid than the diffusion of solute outwards, otherwise an oxide scale is formed.

---

* An isotropic substance has properties which are the same in all directions. An anisotropic substance has different properties in different directions

The resulting dispersion is uniformly distributed, provided that fine powders are used. After internal oxidation the powder is pressed and sintered to a final compact. (Fig. 12.4 is an electronmicrograph of a typical structure.)

Reduction. This is essentially the reverse of internal oxidation in that particles consisting of a solid solution of oxides are formed, after which the oxide of the more noble metal is reduced to leave particles of metal plus oxide. These are then pressed and sintered. The major problems are removal of the by-product, usually steam, and of ensuring complete reduction of the matrix metal.

Solution methods. A uniform dispersion can be obtained by coating fine oxide powders with a solution of a salt of the metal, followed by heating to decompose the salt. Again, the powders are pressed and sintered.

b. Properties of dispersion-hardened systems

Stability of structure at elevated temperatures. When metals are cold-worked, dislocations are generated so that further dislocation movement is hindered and the material thereby strengthened. When these metals are subsequently heated the hardening induced by cold-work is removed by processes of recovery and recrystallization due to the movement of dislocations and grain boundaries at elevated temperatures. These phenomena do not occur in dispersion-hardened systems until considerably higher temperatures than those needed for recrystallization of the metal matrix alone. It is claimed that the dispersed phase prevents the normal processes by pinning the original boundaries and dislocations. This retention of hardness is an important quality of dispersion-hardened systems for use at elevated temperatures.

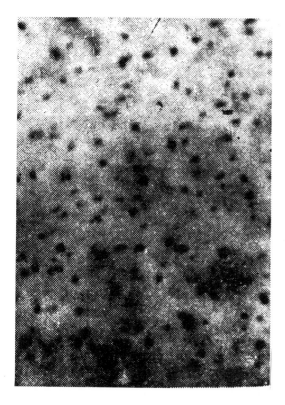

Fig. 12.4 Copper–0.2% magnesium internally oxidised (× 120 000)

864 | COMPOSITE MATERIALS

Strength at high temperatures. Stress-rupture curves (fig. 12.5) are often used to demonstrate the elevated temperature properties of metals. The curves show the time that elapses before rupture when a material is stressed at a given temperature and different stresses. There are three major differences between these curves for dispersion-hardened systems and those of conventional alloys: a higher stress can be withstood, the stress does not fall off so much with time (compare the curves at 450°C in fig. 12. 5) and there is no sudden change of direction in the curves.

This all-round improvement in long-term properties at elevated temperature has been attributed to:

(i) Hindrance of dislocations by particles.
(ii) Hindrance of dislocation by the dense dislocation fields formed by cold-working, and retained despite heating.
(iii) Inactivation of dislocation sources by particles.

Effect of melting. Some dispersion-hardened systems demonstrate a unique property. When they are heated to temperatures well above the melting point of the matrix, they are still able to resist stress and have a finite strength. Furthermore, despite the fact that the content of dispersion is insufficient to form a continuous network, on subsequent testing at room temperature these composites retain a large portion of their original strength. The fact that this treatment does not destroy all the strength has led to the suggestion that the property could be used advantageously during welding in that any regions of local melting would retain their integrity and could be subsequently worked to their original strength.

c. Limitations of dispersion-hardened systems

Fabrication difficulties. Fabrication of dispersion-hardened systems still presents major problems. Ideally one would like to mix the oxide powder into molten metal and cast the components to shape. However, as yet no successful liquid metal techniques

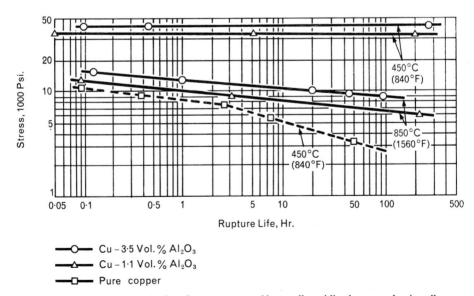

Fig. 12.5 Stress rupture properties of pure copper and internally oxidised copper alumina alloys.

have been developed. Density differences between the components can cause problems of segregation while some methods involving liquid metal have failed due to growth of the particle by solution and diffusion in the liquid or by direct agglomeration. A major difficulty in all these techniques is, however, the reluctance of the metal to wet the particles.

Even when both components are powders and extrusion is the final operation, it is extremely difficult to obtain densities of the order of 100%. It is very difficult to obtain a fully uniform product with all methods involving mixing particles of different species.

Thus, dispersion-hardened systems can show batch-to-batch and inter-batch variations. In this respect systems such as S.A.P. and the internally oxidized systems have a significant advantage by virtue of the mode of formation of the oxide particles.

Low room-temperature strength. The strength of dispersion-hardening and precipitation-hardening systems is inversely proportional to the distance between the particles. It is much easier to produce a system with a fine dispersion of particles by heat treatment than by mixing powders. Thus dispersion-hardened systems are generally weaker at room temperatures than precipitation-hardened systems of the same particle content. Furthermore, any porosity will also serve directly to weaken the system.

Low ductility. Although strength increases with increased volume fraction of oxide, the ductility shows a marked decline. Several mechanisms for this fall-off in ductility have been demonstrated. Pores have been shown to play a major role in initiating cracking. This has been shown in S.A.P. where, unless the sintering stage is carried out in vacuum to remove hydrogen pores, a fourfold decrease in ductility can result.

In other cases, particularly hard particles which resist deformation have been shown to build up sufficient stress at the interface to cause premature cracking. Agglomerates of oxide, particularly in the grain boundary regions of the matrix, are also a source of cracks.

Growth of pores at elevated temperatures. The importance of a uniform structure is also demonstrated by the resistance to creep of dispersion-hardened systems (see 12.2.1c). In one such system, Ni–MgO, it was shown that the best creep resistance occurred at an optimum level of oxide. The fall-off beyond this maximum was attributed to the increased difficulty of removing the pores at higher oxide contents. Although the voids in this case apparently had limited effect on room-temperature strength, at an elevated temperature under load they grew, eventually causing complete failure.

Growth of particles. As the strength of the system is dependent upon the size of the particles it is essential that these do not grow during stressing at elevated temperatures. Ceramic oxides do not readily grow in a metal matrix. Nevertheless, growth of alumina particles in nickel and cobalt matrices has been observed, accompanied by marked reduction in strength and hardness. This difficulty must be overcome before such alloys find application at elevated temperatures.

d. *Application.* So far, commercial exploitation has been largely confined to the S.A.P. type of alloys, and the use of this material is limited on account of its low room-temperature strength and its high cost, due in some measure to the small batch size for manufacture.

However, one of the major reasons for the lack of commercial success of dispersion-hardened systems is that, although the useful range of temperature may have been increased, the increases are largely into fields already served by alloys based on other metals. A major advance must surely come with the development of dispersion-hardened super alloys capable of extended use at temperatures well beyond those withstood by conventional alloys.

12.2.2 Cermets

a. The principles of cermet formation. The aim of the cermet systems is to combine the advantages of a ceramic, usually an oxide or a carbide, with those of a metal, by producing what is virtually a mechanical mixture of the components. Depending upon the relative proportions of ceramic and metal, either may form a continuous matrix, with the other component as a dispersed phase, or there may in fact be two interlocking continuous phases.

In marrying the ceramic and metal, the aim is to capitalize on the properties of the ceramic, such as strength, stiffness, oxidation resistance at high temperatures or hardness and abrasion resistance, and use the metal to overcome the problems introduced by the brittleness of the ceramic. However, this is by no means the full story; other properties of some special ceramics can be exploited by the use of cermets.

In manufacturing the cermets, particles of appropriate size must be properly intermingled and bonded together. This is a matter of compromise. Absence of chemical interaction between the components would result in failure along the interface. Conversely, excessive chemical interaction between the components may result in one or other being completely destroyed.

The ideal situation, one where there is partial solubility between the components, is found in the tungsten carbide/cobalt system, where there is a good bond between the components and the material has a high resultant strength. In the case of cermets based on oxides, this condition does not exist and artificial aids to bonding have to be provided. These rely on the provision of a material which will bond with both components and so form a transition region between the two. One approach used in systems employing alumina particles is to allow the surface of the metal phase to oxidize and to rely on bonding between the two oxide phases, either by the formation of a special reaction product or by mutual solubility.

The desirable structure depends ultimately on the application to be met, but generally a uniformly fine dispersion of ceramic particles in a continuous metal phase is considered suitable. The uniform dispersion gives good strength while the separation of the ceramic particles restricts the passage of cracks through the system.

Bearing in mind these two guiding principles of adequate bonding and uniform microstructure, the various methods of manufacture will be considered.

b. Methods of manufacture

Pressing and sintering. A mixture of the metal and ceramic powders, both of the order of 25 μm in size, is ball-milled until the particles of both phases range between 10 μm and 1 μm in size and then compacted (i.e. pressed to size and shape) at room temperature. To provide the compact with strength a lubricant can be added to the original mixture or the compact may be hydrostatically pressed in rubber moulds after initial pressing. The compact is machined roughly to shape, sintered and finished by diamond grinding. In the case of oxide systems bonding is aided, as previously discussed, by allowing the metal surface to oxidize during the firing operation.

Hot-pressing. After the same initial steps of powder preparation as for cold-pressing, the compact can be hot-pressed, i.e. pressed and sintered simultaneously. Resulting properties are roughly the same as achieved by cold-pressing and sintering, but this is not a popular method as the lives of the expensive graphite dies which have to be used are short.

Slip casting. This method, widely used in the ceramics industry, has not achieved widespread commercial acceptance for cermets, but is nevertheless useful for the production of hollow components. The powder mixture is made into an aqueous slurry with an electrolyte to hold the particles in suspension. This is cast into a plaster of Paris mould and allowed to remain there so that the mould absorbs liquid to leave a layer of powder on the mould wall. After a predetermined time to allow a sufficient thickness to be built up, the excess slurry is poured out and the specimen allowed to dry and harden. After removal, the component is sintered to achieve densification.

Infiltration techniques. The ceramic powder is cold-pressed with an organic binder and given a preliminary sinter to achieve sufficient strength for handling. The organic binder volatilizes or is burnt off, leaving a network of voids. The skeleton is machined to size and infiltrated with molten metal, either by dipping the compact or allowing the metal to penetrate by capillary action, often with a superimposed force to assist complete infiltration. The two major problems are those of ensuring complete penetration without loss of the integrity of the components and geometrical stability of the skeleton during infiltration. The latter is overcome by confinement in a mould and the former by careful choice of components and control of time and temperature of infiltration.

Resistance sintering. This is a modification of the hot-pressing technique which overcomes the need for expensive dies. The mixed powder or a cold-pressed compact is resistance heated by a low-voltage high amperage current and simultaneously pressed at high pressure for a very short time (a small fraction of a second).

Plastic forming. Plasticizing additives are added to the powder mixture which is then formed against plaster dies or extruded into rods or tubes. The required shape is then sintered to produce densification.

c. *Properties of cermets*

Strength and stiffness at temperature. The quest for high-temperature strength led to the initial thinking behind the introduction of ceramics into metal matrices. As with all properties of mixtures, the strength depends upon the relative proportions of the two components which is dictated primarily by the balance of properties required for the application.

Alumina base cermets such as 23% Al_2O_3/77% Cr and 15% Al_2O_3/25% Cr/60% W can be used at service temperatures up to 3000°F (1650°C) while the 1000 h stress-rupture life of a 70% Al_2O_3/30% Cr cermet is some 16000 p.s.i. Provided the limitations of cermets, which are discussed later, can be overcome, this family of materials represents a useful extension of the temperature range over that for which metal and conventional alloys are currently used.

Coupled with strength, materials based on alumina have a high modulus of elasticity at room temperature which is retained over a substantial temperature range.

Chemical stability. The oxide or carbide phases are extremely stable even at high temperatures so that, provided a suitable choice of metal matrix is made, cermets can be produced with good chemical stability and oxidation resistance. Titanium carbide/nickel cermets demonstrate excellent resistance to mineral acids and sodium hydroxide, while for more exotic applications platinum and cobalt have been used with titanium carbide. Alumina cermets can be produced to resist oxidation and are particularly useful in that they are not wetted by liquid metals, while cermets based on 83% Cr_3C_2/15% Ni/2% W have exceptional resistance to oxidation at 1 800°F in air.

Hardness. Cermets in which the hardness is derived principally from the ceramic phase are employed in cutting tools and materials subjected to abrasive wear.

d. Limitations of cermets

Cost. Cermets are expensive materials; not only are the components expensive but the methods of fabrication are time-consuming and often require expensive capital equipment.

Brittleness. Cermets have low impact resistance due to the high proportion of the ceramic phase. Attempts have been made to overcome this deficiency by cladding the surface with a metal such as nickel or by incorporating metal reinforcements in the form of mesh or fibres.

Thermal shock resistance. Although better than that of ceramics the thermal shock resistance (i.e. resistance to rapid heating or cooling) of cermets is low. Again, reinforcements can assist.

Uniformity of product. Cermets tend to be made in small batches and the properties of a particular material may show batch-to-batch variations. Standards of inspection of cermets must be high, for even very small undetected defects which would cause no trouble in a metal component may cause failure of a cermet.

e. Applications.

Combination of the above-mentioned advantages with properties derived from the metal, e.g. thermal conductivity, or from special ceramics such as uranium dioxide enable cermets to be used for specialized applications, although because of the expense the actual tonnage of cermets is low. Major fields of application are summarized in table 12.1.

Table 12.1 Applications of cermets

Application	Components	Cermet
Molten metal devices	Flow control devices	Cr/Al_2O_3
	Thermocouple sheath	Cr/Al_2O_3
	Pouring spouts	Cr/Al_2O_3
	Reactive metal pots	Ti/ZrO_2
Furnaces	Tubes	Cr/Al_2O_3
Electronics	Thermionic cathodes	Nickel/barium carbonate; tungsten/thoria
Jet engines	Flame holder	Cr/Al_2O_3
	Turbine blades and valves	
Wear resistance	Bearings and seals	Ni/Cr_3C_2, $Cr/Mo/Al_2O_3/TiO_2$
	Gauge blocks	Ni/Cr_3C_2
	Brake linings and clutches	
Nuclear applications	Fuel elements	Stainless steel/ UO_2; $Cr/Al_2O_3/UO_2$
	Control rods	Stainless steel/boron carbide

In conclusion it must be re-emphasized that a considerable amount of the work on cermets is still in the experimental or pilot plant stage; as yet application is limited.

12.2.3 Filled plastics

Particulate reinforcement was applied to the plastics initially as a means of cheapening the material, by 'diluting' the plastic with a relatively cheap powder such as calcined clay. However, it was soon realized that the operation had significant technical merit.

Although the strength was relatively unaffected by the presence of the filler, the stiffness was increased by a factor of three, the dimensional stability was increased by virtue of a lowering of the coefficient of thermal expansion and the heat destruction temperature was raised. Thus, several of the limitations of the plastics as a class of material were overcome.

Typical applications include bodies for hand tools such as electric drills.

12.3 FIBRE COMPOSITES

12.3.1 Glass-fibre reinforced plastics

a. Principles of glass reinforced plastics. The aim of fibre reinforced plastics is to combine the stiffness and strength of fibrous materials, discussed previously, with the advantageous properties of plastics, i.e. corrosion resistance, low density and mouldability. The majority of reinforced plastics produced today are glass reinforced epoxy or polyester resins, both of which are thermosetting (i.e. they set as a solid shape under heat and pressure).

Glass fibres have also been used with phenolics, silicones, polystyrene and polyvinyl chloride. The application of these composites is limited, however, by the high moulding pressures required for all such systems, the high cost of silicones, the inherent colour of the phenolics and the heat-softening characteristics of all thermoplastics (thermoplastics soften when heated and become rigid again on cooling).

Glass fibres are the obvious choice as reinforcing agents, principally because of the relative ease with which high strengths can be obtained in fibres a few microns in diameter.

Glass can be produced either as continuous threads, sixty of which are bundled into a 'roving', or as staple fibre. Depending upon the properties required, the rovings can be used as they are, woven into fabrics or bonded into a mesh to form a mat. It is possible to produce composites with a range of strengths according to the glass content and nature of the reinforcement (see fig. 12.6). Furthermore, by choice of the lay-up of the roving it is possible to control the directional nature of the properties in a component. With respect to fig. 12.6, it should be noted that generally the stronger the reinforcement the more difficult and expensive is the manufacturing process.

In order to utilize the strength of the fibre it is essential that there is a good bond between fibre and resin. Unfortunately, there is a low affinity between the two so that a dressing must be applied to the glass to promote the bonding. This is usually an inorganic compound which copolymerizes with the resin.

Further interfacial problems may arise due to the shrinkage of the resin on setting, which can cause high stress in the resin–fibre interfacial region and may lead to defects in service. The epoxy resins have lower shrinkage than the other resins and so have

870 | COMPOSITE MATERIALS

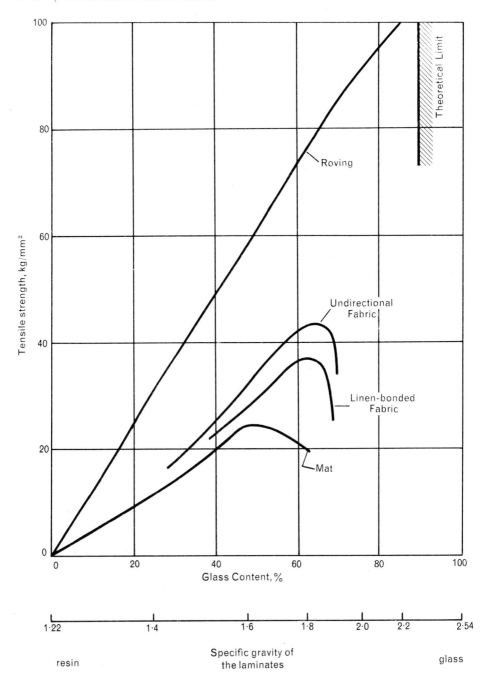

Fig. 12.6 Tensile strength of various types of glass reinforced plastic as a function of glass content.

marked advantages as matrix materials. Moreover, unlike some plastics, neither epoxy nor polyester resins have curing by-products, which can lead to porosity in the material.

b. Methods of manufacture. Several methods of manufacture are available; in all cases certain basic steps must be taken to ensure a good product.

(i) The fibre content, distribution and direction must be chosen commensurate with cost, properties required and ease of manufacture.

(ii) Sufficient pressure must be used to ensure removal of all air bubbles and penetration of the resin between the fibres.

(iii) The final stage must be hardening of the resin.

The most widely used methods of manufacture are as follows.

Hand lay-up. A smooth wooden or plaster mould precoated with lacquer and release agents (which prevent sticking of the resin to the mould) is given a fine layer of resin and the component subsequently built up by applying layers of glass-fibre material or fabric precut to shape. Between each layer of glass a measured quantity of resin is added and uniformly distributed with rollers or brushes, pressure being applied to remove air bubbles. To ensure a smooth finished surface the last layer of glass is added dry, pressed into the excess resin from the previous addition and coated with a thin film of polythene.

This is a flexible method in that rovings can be incorporated into highly stressed regions of the component, but wherever sharp curves occur, it is advisable to use mats. After lay-up the resin is hardened by allowing it to stand at room temperature.

Although the method is cheap and can be used for large components (e.g. dinghies, motor-car bodies), it has disadvantages in that it is not easy to ensure uniform distribution of the glass and the resin content is high. Thus its application to highly stressed components is limited.

Several manufacturing methods are used which represent relatively minor modifications to the hand lay-up method. One system employs two-part moulds to ensure greater uniformity of wall thickness. For lowly stressed components, instead of mats or fabrics, the fibres are sprayed into the mould to produce layers of loose felt and the resin, catalyst and hardener are sprayed into the felt with a gun. This is a rapid method but only suitable for high resin contents.

Pressure methods. Pressure methods are often used to produce large components with higher glass content and better surface finish that can be achieved by hand lay-up. The application of fibre and resin are the same as the hand lay-up but pressure is applied after build-up of the component and prior to hardening. Methods of applying pressure include vacuum, autoclaves, pressure bags and flexible plungers.

Compression moulding. This method, which employs two-part moulds within a press, is normally used for good quality articles. For short production runs the moulds are usually filled resins and the pressure is applied cold, while for long runs metal moulds are more economical and it is possible to apply pressure and heat simultaneously. In this way up to 70% glass can be incorporated.

Pre-impregnation methods. Mats or fabrics are pre-impregnated with resin by passing through baths containing a solution of the resin. The solvents are removed by drying and the material is worked into mouldings. This is an expensive method but does ensure extremely uniform distribution of fibres and resins and is therefore used for high quality products. In building up the structure the impregnated fibres are cut to shape and great care is taken to ensure uniform distribution of overlaps. After preforming, the article is hardened under pressure.

Filament winding. Filament winding or wrapping methods are used for the production of hollow bodies such as tubes and pressure vessels. Filament winding of pre-impregnated rovings is an expensive method requiring sophisticated equipment but the fibres can be accurately laid in position to resist anticipated service stresses most efficiently. After building up the structure, the component is finished by methods employed for impregnated fibres.

Spinning. In the spinning method, rolled up mats or fabrics are placed inside a tubular mould, resin is introduced and the mould rotated so that the resin is forced between the fibres. The resulting component has a very smooth surface finish.

In addition to these methods, rods of glass reinforced plastics can be produced by extrusion of pre-impregnated rovings through dies.

c. Properties of glass reinforced plastics

Low density. Both components of glass reinforced plastics are of low density: the resin has a specific gravity of 1.2, the fibre about 2.5 depending on the type of glass used. This is a beneficial property when components have to be fixed, lifted, transported or used in transportation vehicles.

Strength. The strength of the glass fibres normally used is about 157 000 p.s.i. and, by incorporating high fractions of fibres carefully aligned, it is possible to produce materials with strengths of up to 112 000 p.s.i. in the direction of the fibres. This strength, coupled with low density, enables materials to be fabricated within a wide range of specific strengths.

Corrosion resistance. The plastics have extremely good corrosion resistance when exposed to carbon base solvents. Thus, by ensuring that the glass is entirely surrounded by the plastic, it is possible to produce highly corrosion-resistant materials.

Impact resistance. Although neither material in itself possesses exceptional resistance to impact, the combination produces a shock-resistant material by virtue of the crack stopping mechanisms discussed previously. Arrest of the crack results in any impact damage being localized.

Surface finish. By careful moulding methods a good smooth finish can be obtained; because of its smoothness and weathering resistance this surface can be easily maintained in excellent condition. Pigments can be added to produce decorative finishes where required.

Ease of fabrication. The thermosetting epoxy and polyester resins normally used are readily formable at low temperatures. Also glass fibres, unless woven into stiff fabrics, can be laid to a complexity of shapes and designs. Thus the composite material can be readily fabricated to desired shapes with desired properties.

Electrical properties. Glass reinforced plastics are good electrical insulators unless water has been absorbed in weak interfacial regions.

d. Limitations of glass reinforced plastics. Glass reinforced plastics are extremely worth-while materials provided that the components are designed to suit the material. In this respect, having outlined the advantages it is well worth considering how the materials are limited.

Strength. Strength is dictated essentially by three factors relating to the fibre: strength, content and orientation with respect to stress. Glass fibres have good strength provided the surface is undamaged; unfortunately the surfaces can be easily damaged, for instance by other fibres. This need to prevent fibre/fibre contact limits the fibre content to 70% in a carefully aligned composite, while in a randomly oriented composite only 30% fibres can be incorporated. Thus we are faced with a highly anisotropic composite at one end of the scale and an isotropic but weaker one at the other. Fortunately the materials can be made so that the fibres are oriented to give the best resistance to the applied stresses. Nevertheless, these limitations have to be recognized when designing with reinforced plastics.

Stiffness. The elastic modulus of glass fibres is 10×10^6 p.s.i. and that of the plastic much lower, so that the moduli of the composites are fairly low - less than 6 or 7×10^6 p.s.i. For many applications it is of little use having materials which can withstand high stresses if this stress results in high degrees of strain. This again has to be recognized and taken into account in design.

Heat resistance. Epoxy and polyester resins char or flow at temperatures around 200°C, but this disadvantage can be somewhat offset by the use of newer resins, e.g. high-temperature polyimide resins. However, glass itself suffers marked deterioration in strength at temperatures of about 400°C so that this will impose a maximum temperature in the use of glass reinforced systems.

Corrosive environments. Both the glass fibre and the fibre/resin interface are weakened by water absorption, although this can be overcome if the matrix forms a continuous envelope around the fibres. Any cracks formed in service may allow water penetration and subsequently deterioration of the composite.

Fatigue. The fatigue characteristics of fibre reinforced plastics are good, but at very high rates of cycling (i.e. many alternations per second) the heat generated by internal friction in the resin cannot be conducted away, and high temperatures can be developed in these circumstances.

e. Applications of glass reinforced plastics. Half a million tons of reinforced plastics are produced in the world per annum. This figure is growing continually; figures for the USA indicate an 11 to 13% annual increase from the 160 000 tons used there in 1965. The major applications for reinforced plastics can be classified as follows.

Boat building. The US Navy have been using glass reinforced plastics for more than 20 years, during which time some 3 000 boats have been built, taking advantage of the durability, low maintenance costs and ease of repair. It is claimed that failures which have occurred can be traced to bad design and misuse of the material, rather than any inherent deficiencies.

At the 1966 London Boat Show, 55% of the craft were made from reinforced plastics. This demonstrates how the material has taken over from wood for the manufacture of small boats. At the other end of the scale, a recent US Navy survey of large boat construction in glass reinforced plastics yielded the following conclusions:

(i) Construction of boats up to 83 ft in length has been achieved and is a practical and economical proposition.

(ii) Many fabrication systems are available for selection.

(iii) Construction of boats up to 200 ft in length is entirely feasible.

Among countries developing large boats are the United Kingdom, USA, Japan, France, Sweden and South Africa.

Storage tanks. Glass reinforced plastics have found widespread application for storage tanks where the corrosion resistance, weathering capability and light weight are advantages. Often the tanks are supported within a steel or concrete framework.

Specific applications of interest include frozen food containers and warm-water storage tanks, where glass reinforced plastics are used in a sandwich construction with polyurethane foam as an insulating filler, and silos for agricultural usage, e.g. food and fertilizer stores.

A developing application in the USA is for gasolene tanks; recently a 10 000 gal tank was constructed. As metal tanks rust away they are now being replaced with glass-polyester tanks which, in addition to being much lighter, have increased resistance to soil corrosion.

The building industry. Although perhaps not making the major inroads that were predicted, glass reinforced plastics are nevertheless very useful materials for the building industry.

As roofing materials, flat and corrugated sheets have found application by virtue of their low density, ease of fixing, weathering capability, translucence and maintenance-free service. Similar attributes have led to the development of modular panels for industrialized building systems and as monolithic facing sheets. Reinforced plastics are also used for decorative faces on buildings and recently complete bathroom units, incorporating bath, toilet, wash-basin and shower compartment, have been produced.

Transport. The high specific strength of these materials is of great interest in all forms of transportation where maximum pay-load is required. Although the aircraft industry probably initiated much of the interest in the materials, far more reinforced plastics are currently used for surface transport. Compressor blades for some engines, the radome for the Concorde and many interior fittings are examples of aircraft applications.

In road transport, reinforced plastics compete successfully with aluminium, wood and zinc-coated steel. Applications include storage tankers, lorry cab bodies, translucent roofs on vans and insulated trucks, plus, to a limited extent, some passenger transport bodies.

These four groups represent the major usage of reinforced plastics but additional applications worthy of note include vaulting poles, hovercraft hulls, as a reinforcement for pvc piping and as moulds for casting concrete components.

12.3.2 Plastics reinforced with other fibres than glass

Although glass is by far the most widely used reinforcing fibre for plastics its limitations have led to the quest for other fibrous materials. Essentially, the aim is to introduce a fibre which has improved mechanical properties but which still retains the manufacturing capabilities of glass. For this reason, the main fibres of interest are those which can be produced in continuous form so that they can either be filament wound or formed into mats. Within this sphere, newer reinforced plastics include the following.

a. Wire reinforced plastics. Steel wires can be obtained with a consistent strength of the order of 448 000 p.s.i. and a modulus three times that of glass (30×10^6 p.s.i.). Furthermore, there is considerably less variation in strength of the metal wire than is found in glass fibre and of course the surface is not easily damaged. For high duty applications such as pipes and pressure vessels a wire sheet consisting of many parallel hand-drawn wires resin-bonded to a polyester tissue has been produced.

Filament winding techniques have been used to build up pressure vessels and pipes which combine high strength with the corrosion resistance of the resin. These are still in the developmental stage but suggested applications include suspension bridge cables, pipe reinforcements and panel laminates.

b. Modified glass reinforced plastics. To overcome the relatively low strength and stiffness of glass fibres several attempts have been made to develop glasses with improved properties. Unfortunately, any compositional modification usually impairs the viscosity–temperature relationship so necessary for easy drawing.

Glass with a modulus of 12×10^6 p.s.i. and strength of 5×10^5 p.s.i., despite its increased cost over the normal type of glass, has found some application for reinforcing plastics for rocket motor casings for US missile systems.

A high beryllia content glass has been developed with a modulus of 17×10^6 p.s.i.

c. Boron reinforced systems. Boron filaments are now commercially available in the USA and have been service tested for reinforcing epoxy resins in certain aircraft components, such as horizontal stabilizers, and for several other aerospace and hydrospace applications.

These filaments are prepared by depositing elemental boron from a boron tribromide/hydrogen mixture on to a thin tungsten filament heated to 1 200°C. The boron, which melts at 2 050°C, has a density of only 0.095 lb/in³, a strength of 400 000 p.s.i. and an elastic modulus of 60×10^6 p.s.i. Some properties of an epoxy system reinforced with 70% boron are given in table 12.2.

d. Carbon-fibre reinforced systems. Carbon fibres are prepared by an initial pre-oxidation of rayon or polyacrilonitrile, followed by graphitization (i.e. conversion to carbon) under controlled temperature and atmosphere for a predetermined time. The fibres are stressed during graphitization to encourage the crystallites to align themselves parallel to the fibres. This material has half the density of boron but a tensile strength of the same order and an elastic modulus of about 50×10^6 p.s.i.

Although some forms of carbon cause inhibition of the curing of polyester resins it is reported that no inhibition has been caused by the graphite fibres. Two methods of fabrication have been employed in an effort to obtain maximum packing. In the first, fibres are added to the resin held in a mould, pressure is applied and some resin squeezed out and allowed to leak away. The second method involves impregnation in dilute resin solution, evaporation of the solvent, pressing and subsequent hot-moulding by conventional methods.

No design details are available but experimental studies on the material have yielded the following test results.

(i) Tensile strength is up to 105 000 p.s.i. with 40% fibre, with an elastic modulus of 22×10^6 p.s.i., giving a specific stiffness (stiffness/weight ratio) three times that of steel.

(ii) Impact strength is equivalent to that of glass reinforced plastics and is retained even after immersion in boiling water.

Table 12.2 Properties of Boron Reinforced Epoxy

Compressive strength	300 000 p.s.i.
Flexure strength	2.5 to 3.5×10^5 p.s.i.
Shear strength	13×10^3 p.s.i.
Young's modulus	40×10^6 p.s.i.

(iii) Damping (absorption of vibration) is high, because of the internal friction of the resin matrix, while the fatigue strength is good.

(iv) Immersion in water has little effect on stiffness and there is only 5% loss in strength.

(v) The coefficient of friction is low and usually consistent at 0.25 to 0.35. This is true of all graphite/resin systems, indicating that the graphite fibres are in fact bearing the major portion of the load. (The coefficient of friction of low frictional PTFE is actually increased by the addition of graphite.) Wear is at least two orders of magnitude less than with unreinforced material.

Clearly graphite reinforced plastics eliminate many of the problems of glass reinforced plastics, but at a price. Therefore, at least at first, application will be limited. Likely areas of interest include aircraft components, especially rotating members such as compressor blades and helicopter rotors, where specific stiffness is important. Space programmes may well make extensive use of the materials because of the potential weight savings.

Graphite reinforced nylons are also likely materials of the future, perhaps as bearing materials where low coefficient of friction and low wear rate will be sought.

e. Boron nitride reinforced phenolics. This fibre is also becoming commercially available in the USA. It has a relatively low modulus (10×10^6 p.s.i.) and strength (20 000 p.s.i.) but excellent abrasion resistance and is being considered for reinforcing phenolic materials for rocket nozzles and nose cones.

12.3.3 Fibre reinforced metals in general

One of the newer approaches to composite materials has been the application of the concepts of fibre reinforcement to metals. Many papers have been devoted to the theoretical assessment of the properties obtainable in a fibre reinforced metal if the fibre properties could be utilized. During the last five to ten years much research effort has been devoted to the problems of incorporating fibres into metals.

The gains to be made are so great that, in addition to work aimed at developing commercially useful systems, much effort has been devoted to understanding fully the factors which control strength and mode of failure. At this stage much is known about fibre reinforced metals but as yet no commercially viable systems have been developed. Metal wires, continuous non-metallic filaments and whiskers* have all been used in model studies and, in a limited number of cases, for direct attempts at reinforcement of commercial alloys. Table 12.3 lists the strength and moduli of some of the fibres which have been used in these efforts.

The review of fibre reinforced metals is dealt with under the headings devoted to each class of reinforcing fibre. However, the limitation of fibre reinforced metals as a class are first considered.

12.3.4 Problems of fibre reinforced metals

a. The growth and harvesting of strong filaments in sufficiently large quantities for reinforcing components is time-consuming and expensive, even when relatively cheap raw materials can be used.

b. Incorporation of the filaments into the matrix with suitable orientation and interfacial conditions is essential.

*Long thin crystals with a high degree of surface perfection and high strength.

c. Conventional fabrication and joining techniques will in most cases be of no use for these materials. New techniques will have to be developed.

d. When the composite is loaded, the stress concentration at fibre ends or at fibre breaks may initiate fatigue cracks.

e. A weak interface may be desirable as a crack arresting mechanism, but may lead to creep of the matrix.

f. The difference in the thermal expansion of the components may result in severe stresses being set up during manufacture and/or during service.

12.3.5 Metals reinforced with metal wire

As a reinforcing agent metal wires have the advantages of ready availability in diameters of the order of 0.001 inch, uniform mechanical properties, easy handling in that minor surface damage is not so likely to produce drastic strength reductions, and availability in continuous lengths. The disadvantages as far as practical reinforcement is concerned lie in the fact that metals are generally denser than ceramic fibres.

With regard to commercial exploitation of metal wires, two possibilities exist: (i) the use of brittle but low density beryllium wires; and (ii) the use of refractory metals where the matrix serves as a binder and as an oxidation-resistant sheath for the fibres.

a. Methods of manufacture

Infiltration. This method is applicable in those systems where the molten matrix will wet the fibres; the system copper/tungsten is ideal in this respect and has been used extensively in a model system because liquid copper will wet the tungsten without damaging the fibres. The fibres are held in a suitable refractory container and the molten metal is usually allowed to infiltrate the system under vacuum.

In this way fully aligned composites with both continuous and semi-continuous fibres have been produced; however, due to the need to hand lay the fibres only small test-pieces have been obtained.

Other systems have been produced in which the fibres are pressed into a random felt before infiltration.

Powder and fibre mixing. This method has the advantage of lower fabrication

Table 12.3

Fibre	Elastic modulus $\times 10^6$ p.s.i.	Ultimate tensile strength, $\times 10^3$ p.s.i.
Metal		
Steel	30	600
Tungsten	40	580
Filament		
Graphite	50	400
Boron	60	400
Silica	10	600
Whiskers		
Alumina	65	1700
Silicon carbide	70	1600

temperatures than the liquid metal techniques so that the interaction between components is reduced and any cold-work in the wires (which improves the strength) is not entirely removed by heat.

The fibres are chopped to size and are either intimately mixed with the matrix powder to produce a random array or, in some cases, are carefully laid between layers of powder. After the desired distribution has been obtained the mixture is then hot-pressed or cold-pressed and sintered. In one case a random composite of molybdenum in a titanium alloy was subsequently extruded to produce a rod with aligned fibres.

Hot-rolling. Composites of aluminium alloy and steel have been produced by hot-rolling a sandwich consisting of continuous strands of steel wire between aluminium alloy sheets.

Electroforming. The disadvantages of the high temperature in liquid metal techniques and pressure in powder techniques, both of which may cause fibre damage, can be overcome by using the electroforming technique. As an example, tungsten wires were wound on to a stainless-steel mandrel submerged in an electroplating bath containing nickel sulphamate solution, so that the nickel was deposited on to the wire. By controlling the speed of rotation of the mandrel it was possible to produce nickel/tungsten composites with various fibre contents.

b. Properties

Strength. Much of the work with metal-wire systems has been devoted to investigating the tensile strength of the composites produced. This work demonstrates that in fully aligned systems, providing the manufacturing technique does not cause serious damage to the fibres and does produce a dense composite, the fibre strength can be realized. Essentially, tensile strength has been used as a measure of the suitability of a manufacturing process for a particular system.

Fatigue resistance. The effect of fatigue crack propagation was investigated in the rolled sandwich construction. Two points of major interest emerged: not only was the rate at which a fatigue crack propagated considerably reduced, but the length of crack which the matrix could accomodate before the crack became unstable and propagated rapidly was increased.

Stress-rupture properties. The time required for failure at a given stress and temperature of titanium reinforced with 10% molybdenum was increased a thousand-fold at 540°C for a stress of 20 000 p.s.i.

c. Possible practical systems

(i) Beryllium fibres. Attempts to reinforce aluminium alloys were largely unsuccessful owing to interaction, destruction and coalescence of beryllium, and the formation of a brittle interface.

(ii) Refractory metal wires. Refractory metal wires have been shown to increase the tensile strength and stress-rupture properties of titanium, nichrome (a nickel/chromium alloy) and steel matrices. However, owing to the relatively high density of the refractory metal, particularly tungsten, the gain in specific strength (strength/weight ratio) is only marginal over conventional alloys and at this stage appears unlikely to give sufficient incentive to proceed with these materials.

(iii) The ability of fibres to arrest cracks could be a valuable asset for certain components but the composites so far developed for this purpose probably have insufficient strength perpendicular to the fibres to be of commercial interest.

12.3.6 Whisker reinforcements in metals

Whiskers of ceramics such as alumina, silicon nitride and boron carbide are considered by many to be potentially the best reinforcing members. Indeed, on the basis of their specific strength and stiffness this is certainly true.

However, major problems still remain in the growth, harvesting and alignment of these very small particles while the perfect surfaces may present serious problems to the bonding with the metal matrix.

a. Methods of manufacture

Infiltration. Using silver as the matrix, composites some few hundredths of an inch in diameter have been produced containing up to 30% aligned whiskers. In these structures, the whiskers were carefully selected and aligned by hand (see fig. 12.7 (a), showing a cross-section of such a composite).

Powder methods. Whiskers are normally produced as a mat, in many ways resembling cotton wool. This can be carefully broken and the whiskers dispersed in a high viscosity liquid, filtered and redispersed in water plus a wetting agent. After screening the whiskers are redispersed with a metal powder in a liquid and the mixture is filtered to produce a composite felt which is then hot-pressed to the solid.

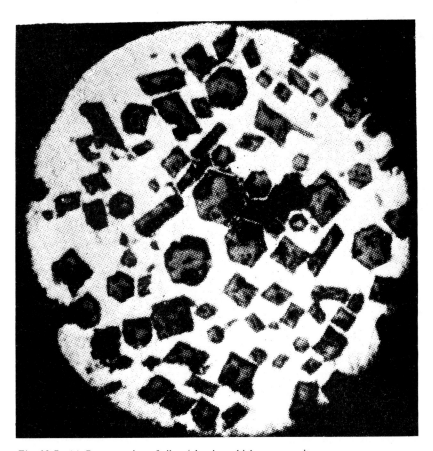

Fig. 12.7 (a) Cross-section of silver/alumina whisker composite.

Similar techniques have been used in which, prior to sintering, the felt is extruded to align the whiskers.

b. Properties. Tensile strength has been the principal property used to assess the manufacturing process. Fig. 12.7(b) demonstrates how the strength at elevated temperatures of these very small samples can be increased above that of the matrix both when unreinforced and when dispersion-hardened.

c. Possible applications. Although many speculations for fibre reinforced metal have been based upon whisker properties, the production of large components of whisker reinforced super alloys still seems a long way from realization. Aside from the problems already outlined, discontinuous fibre systems have inherent disadvantages over the continuous type. Problems of stress concentration of fibre ends and the need to develop a strong bond to prevent the fibres being pulled from the matrix prior to fracture have yet to be investigated.

12.3.7 Continuous fibre reinforcements in metals

This is probably the type of system most likely to succeed in producing components suitable for use at high temperatures. To date much of the work has been devoted to model studies with silica fibres; these fibres, as outlined earlier, suffer disadvantages of limited temperature resistance and low stiffness, and are therefore most unlikely to be of major practical use for reinforcing metals. Preliminary work has been devoted to compatibility studies between metals and the boron and graphite fibres now becoming available. Details of these studies have not yet been published, thus this section is of necessity devoted to the work on the system of aluminium reinforced with silica fibres.

a. Methods of manufacture. The first stage involves drawing the silica rod into

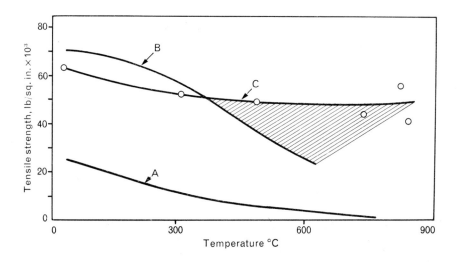

Fig. 12.7 (b) Strength of silver and silver composites at elevated temperature. A = Unreinforced, B = dispersion hardened, C = fibre strengthened.

fibre and passing it through molten aluminium to provide a continuous coating. The aluminium serves the dual purpose of matrix and protection of the silica during subsequent handling. By careful control of the coating variables, the interfacial interaction can be controlled to produce the required degree of bonding.

After coating, the fibres are either bundled and hot-pressed, or filament wound into sheets which can be cut to any desired shape and size, and then hot-pressed.

b. Fabricational methods. Three methods of fabrication have been studied.

Filament winding. Filament winding and direct hot-pressing has enabled rings of the material to be produced, indicating a possible method for production of pressure vessels and similar tanks.

Flat components. The effect of fibre orientation on tensile strength has been demonstrated to be in close agreement with predicted values (fig. 12.2(b)). Following on from this it has been possible to build up thin plates with desired degrees of anisotropy from unidirectionally aligned monofilament layers.

Joining. One of the problems with fibre reinforced systems is to produce a joint with strength commensurate with the remainder of the system. Successful joining has been accomplished by pressure welding and filament winding the fibres around bushed inserts.

c. Properties. The material has been used extensively as a model on which to study mechanical properties, mode of failure and possible component manufacturing routes.

Strengths. Room-temperature tensile strengths of the order of 224 000 p.s.i. can be obtained with this material. The stress-rupture properties of the material demonstrate that up to 500°C there is no fall-off in stress to rupture with time. This behaviour is very similar to that of the dispersion-hardened systems and unlike the metal matrices, the strength of which falls off drastically at these temperatures.

Fracture characteristics. The effect of the fibres in arresting a fatigue crack has been demonstrated to be very similar to the results with aluminium–steel wire sandwich specimens, and the crack arrest by delamination at a weak interface has been observed.

d. Possible applications. As the model work has shown, filament winding techniques can be usefully employed to produce engineering structures. Methods of fabrication have been developed to make use of the high fibre strength and much is known which enables fracture characteristics to be controlled.

Therefore, once the problems of compatibility between the matrices and graphite or boron fibres have been solved, it can be expected that for some applications the resulting composite material will have sufficient technical advantages to outweigh the extra expense.

12.3.8 Glass reinforced cement

An interesting new addition to the range of fibre reinforced composites is glass reinforced cement. In many ways this material, which has 20% fibre, lies intermediate between the ferro-cements, i.e. steel mesh embedded in cement and the glass reinforced plastics.

The major problem with the fabrication of this type of material is that of preventing interaction between the glass and cement. Two approaches have been adopted. In Russia, where the majority of work appears to have been carried out, the technique is to employ conventional glass fibres in a super sulphate cement. Attempts are being made in other countries, including the United Kingdom and China, to develop specially alkali-resistant glasses which can be used in conventional cements.

a. Manufacture. Not surprisingly, many of the manufacturing techniques successfully developed for reinforced plastics have been adopted for glass reinforced cement (e.g. hand lay-up, filament winding). After the component has been fabricated, however, an additional stage-curing-must be carried out. In the case of the high alumina cements this is done at 16°C in high humidity by covering with a wet polythene cloth for 24 h, followed by three days immersion in water.

b. Properties and applications. Glass reinforced cement is claimed to have good strength, which is of course dependent upon the fibre content, is easily repaired and maintained and is watertight and, most, important, readily fabricated.

Only sparse information is available on the application of this material, but there are claims from the USSR that it has been used for a factory roof and wall panels, and as a watertight covering for reinforced concrete water tanks.

12.4 PLANAR COMPOSITES

12.4.1 Oxidation-resistant coatings

Of the refractory metals, molybdenum and niobium are generally considered to be the most likely to find widespread application in aero-engine components because they have a lower density than tantalum and tungsten. Of these two, niobium has the better oxidation resistance and is generally more ductile and easy to fabricate. Even so, the oxidation resistance of both the metal and the alloys developed from it leave much to be desired at temperatures of 1 300°C in oxygen-bearing atmospheres, while molybdenum oxidizes rapidly at temperatures above 700°C. Thus there is a need in both cases for a coating material.

a. Requirements for a coating. The coating must be oxidation-resistant and mechanically and thermally stable over a wide range of temperatures; it may also have to withstand erosion and impact damage. Furthermore, one of the problems with niobium is that the interstitial elements may penetrate the metal, causing deep subsurface embrittlement–therefore the coating must be impervious. The optimum thickness of coating will be different for different applications and environments, but will generally be limited to a few thousandths of an inch.

b. Coatings

Silicide coatings. Silicide coatings have been developed for both molybdenum and niobium base alloys. Although the impact resistance was fairly low, the coatings on molybdenum had good oxidation resistance and mechanical strength.

Niobium alloys have been protected with niobium disilicide coatings, produced by reduction of silicon tetrachloride vapour at 1 200°C or by pack cementation from a silicon powder, and with molybdenum disilicide coatings, produced by vapour plating molybdenum and subsequently converting to silicide.

Modified silicide coatings. In an effort to produce coatings with a better oxidation resistance for service at 1300°C, silicide coatings with traces of iron and chromium have been developed. Although oxidation resistance was improved, the coatings contained cracks which developed by oxidation at the root during thermal cycling.

Composite coatings. Diffusion-bonded silicide coated with a glass overlay has proved to be a good coating at temperatures up to 1300°C. The glaze acts as a barrier and a filler for any minor cracks and has excellent resistance to impact, thermal shock and creep.

Another composite coating for niobium alloys consists of sprayed alumina subsequently infiltrated by a glaze.

c. Potential applications. In the USA satisfactory tests on coated refractory metals have been reported on turbine components such as blades and nozzle guide vanes, while in the UK there is general interest for turbojets, ram jets and air breathing engines.

12.4.2 Plastic-coated metals

Within the strictest sense of the original definition, plastic-coated metals may be regarded as composite materials in that they retain the stiffness and strength of the metal and the corrosion resistance and low frictional resistance of the plastic surface.

The various types of coating and method of application are listed in table 12.4.

12.4.3 Crack-restricting systems

The experimental work carried out on the composite specimens designed to stop the cracks have so far been limited to demonstrating that the concepts outlined in section 12.1.3 c. are justified.

Laminated specimens were produced from mild-steel plates by brazing or soldering and by explosively bonding* with copper layers. The brittle to ductile transition temperature was lowered from $-15°C$ for the homogeneous mild-steel specimens to $-120°C$ for the laminates. The progress of the crack was monitored during the tests and it was

---

* Explosive bonding is a process in which the plates are welded together by placing them in contact and detonating an explosive on one side against an anvil.

Table 12.4

Plastic	Method of Application	Advantages
Polyvinyl chloride (PVC)	Gluing sheets Liquid dipping	Easily shaped by conventional methods Thick coatings
Polytetrafluoroethylene (PTFE)	Sintering	Low friction for bushes and slideways
Nylon Epoxide Phenolics	Dipping into fluidized beds and stoving	Thick coatings

observed that, although at low temperatures the first steel sheet failed in a brittle manner, the second did not fail until after appreciable deformation and delamination at the interface. Thus, the composite was considerably tougher than the homogeneous specimen.

It is well known that, as the thickness of a homogeneous specimen decreases, so the transition temperature is decreased. This behavior was borne out in the crack divider type of laminate to the extent that the transition temperatures for the laminated specimens are close to those of the homogeneous specimens of the same thickness as the laminates.

These tests indicate that toughness can be built into a laminated material, either by arresting or dividing the crack. Further work will no doubt be directed towards developing useful applications for this type of material.

12.4.4 Honeycomb structures

a. General concept. One further family of laminated materials which are often included within the broadest definition of composite materials are the honeycomb/sandwich structures. These are not really composite structures but are nevertheless worth brief comment in a chapter of this nature.

The aim of the honeycomb structure is to build a material with a light-weight honeycomb core sandwiched between skins of high-strength material; the low-density core is used to restrict the bending or buckling of the thin skin material. In this way the advantages of both components are incorporated into the final structure, in exactly the same fashion as with a composite material. The prime advantages of the honeycomb sandwich are therefore: (i) high strength-to-weight ratio, (ii) high stiffness-to-weight ratio and (iii) smooth external surfaces, all of which make these materials eminently suitable for aircraft construction.

In addition, an important bonus property which the designer gladly accepts when using these materials is its fail-safe mechanism. That is, once a fatigue crack starts, the rate of propagation in this type of structure is low enough to ensure that the loss in strength suffered by the component before the crack is discovered by routine inspection is small.

In honeycomb and indeed in all sandwich structures, the adhesives used for bonding the components are the key to a successful final structure. Adhesives are outside the scope of this chapter, but this point cannot be too strongly stressed.

b. Types of honeycomb. The most widely used honeycomb structures incorporate an aluminium core between either steel or high-strength aluminium-alloy skins. The core can be made in one of two ways:

  i. Aluminium foil is formed into corrugated ribbon and adhesive applied to the crests of the corrugations. The foil is then built into blocks of honeycomb, oven-cured and sliced to the required dimensions.

  ii. Adhesive is applied to flat foil in evenly pitched stripes and the foil cut to length and laid into a block with alternate sheets offset at half the pitch of the adhesive strips. The block of sheets is then hot-pressed to cure, cut at right-angles to the adhesive stripes and finally expanded by pulling the outer layers of the foil apart.

Stainless-steel honeycombs are used for high-temperature applications but here, of course, adhesive is no use and brazing or welding is necessary for adequate bonding. Other core materials used are glass fabric, resin-impregnated paper and cotton fabrics.

An extension of the concept is to fill the holes of the honeycomb with an expanded resin material, for instance, polyurethane.

This leads on to a further development in sandwich structures - the use of rigid foam core with glass reinforced plastic skins. This structure, again barely a composite material, uses the rigidity of the expanded material to resist buckling in its own right.

c. Application. Materials of this type are mostly used in transport and in civil engineering applications, or in very limited quantities in very specialized systems.

Some of the major applications are listed in table 12.5.

In all the applications in the table, it should be borne in mind that the structures, although ideal for flat components, can be easily machined so that tapered and curved structures can be made.

12.4.5 Metal laminates

An important and growing field of composite materials is that of metal laminates, in which two metals or a metal and non-metal in sheet form are joined together to give a better combination of properties than either of the two individual metals. A typical example is stainless steel laminated to aluminium, where the chemical inertness of the stainless steel is associated with the lower cost, better themal conductivity and easier forming of the aluminium. Laminates of this type are used for such applications as the manufacture of domestic utensils and for decorative trim on automobiles.

Another form of metal laminate is one in which the metal is laminated to one of its alloys. For example, a sheet of aluminium alloy of high strength but low corrosion resistance may be covered on both sides with a sheet of the pure metal, which has lower strength but higher corrosion resistance. Bimetallic strips in which the two

Table 12.5

Type of Structure	Core	Skin	Application
True honeycomb structure	Glass fabrics	Resin-impregnated glass fibres	Container for airborne radar equipment
	Aluminium	High-strength aluminium alloys	Fuselage shells, aerofoil panels, helicopter rotor-blades, aircraft doors and floors
	Aluminium	Reinforced plastics	Bases for production tools, checking plates
Expanded plastic core structure	PVC foam	Glass/polyester	Beams for boat construction
	Polyurethane	Glass/laminates	Building panels; making use of insulating properties of the foam
	Phenolic foam	Glass reinforced plastic	Building panels

metals have different coefficients of thermal expansion are widely used as switching devices in electrical circuits. When the strip is heated, the uneven expansion causes it to bend in one direction and operate the switch. Another use of bimetallic laminates is to provide a means of joining two metals which are difficult to join by normal processes. Aluminium can be joined to steel by the insertion of an aluminium/steel laminate. The steel portion is welded to the steel and aluminium portion to the aluminium by conventional techniques.

Metals are also laminated to non-metals such as wood or plastics. Facings of wood veneers are applied to sheet metal for decorative purposes and metal is laminated to plywood to give extra strength or hardness. Metal/wood laminates were widely used at one time in aircraft construction but this use is now less important than formerly. Another laminating technique used in aircraft construction is to join together a number of sheets of metal of different outline to give a thick member of varying cross-section. This is a cheaper and easier way of making complex structural sections than cutting them from a solid block of metal. To be classed as laminates it is essential that the two materials should be in sheet form. Coatings applied in other ways, for example, by spraying, hot dipping, painting etc., serve a similar function, but the resultant product is not classed as a laminate.

Laminates are usually made by metallurgical bonding or by the use of adhesives. Many metals can be made to stick together provided the contact surfaces are clean and sufficient pressure is applied. Consequently, laminates can be made by passing suitably prepared sheets of metal through rolling mills. This cannot be done, however, when laminating metal to non-metals and in such cases adhesives are employed. Once again, however, adequate preparation of the surfaces is essential to give satisfactory adhesion.

Another property of the aluminium honeycomb is its ability to absorb energy. Thus it is extremely useful for decelerating moving bodies and is therefore used in arresting devices. By judicious choice of honeycomb density it is possible to design devices over a wide range of deceleration rates.

Specific applications depending upon this property include impact/deceleration testing devices for instruments and aircraft crash-landing devices.

12.5 THE FUTURE OF COMPOSITES

12.5.1 Production levels

Much of this chapter has been devoted to those materials which, to the design engineer, are still in the embryo stage. Many of the materials quoted are as yet only probable or possible solutions to the quest for stronger, stiffer, lighter, more temperature-resistant materials. It has therefore been impossible to quote production figures for many composite materials or to attempt to predict future production; indeed most of the chapter is of necessity restricted to the results of research studies.

What then of future production of composites? Certainly the only major tonnage material at present is fibre reinforced plastic. This is expected to grow from its present level of some half a million tons per year. Plastics reinforced with the new fibres, graphite and boron, are likely to become the next major composite material, if only for the great advantage of stiffness that they offer and because of the fund of knowledge on building components from glass reinforced plastics which should form the basis for engineering the new materials.

However, most of these materials that involve the use of small components, whether they be particles or fibres, will be expensive even if only by virtue of the price of the reinforcements. All composites require fabrication steps involving mixing or cladding, so again much expense is likely to be incurred. It is clear, therefore, that composite production will, at least in the early stages, be directed towards specific applications for aerospace and military uses where the desired properties are at a premium.

Widespread use will only be achieved if major technical gains can be made, coupled with economic advantages.

12.5.2 Research

The last five to ten years have been exciting ones for those engaged in materials research. It is virtually impossible to estimate the amount of money spent during this period. Perhaps the biggest contributors, the US Air Force, spent several million dollars in 1966, while both the US Navy and British government organizations have spent considerable sums on sponsored and 'in house' research during this period.

During this time, reinforcements have been developed and model and potential practical systems evaluated so that much is now known of the best methods to tailor materials to meet the demands of different environments of stress and temperature.

The next few years' research could well be the real test for composite materials. During this period there will probably be a reorientation of effort towards developing:

   a Economical and reliable production of fibres, particles and coatings.

   b. Economical and consistent fabrication techniques for large-scale production of components.

   c. Designs utilizing the advantages of composites but bearing in mind and accommodating any disadvantages.

   d. Evaluation of components.

The interdisciplinary approach has been successful in combining ceramics, metals and plastics to achieve the best combination of properties. Full utilization will be dependent upon further integration of the work of production engineers and design engineers with the materials scientists.

12.6 LITERATURE

Many review articles on composite materials have appeared in recent years. For further information the reader is referred to the following articles, all of which give extensive reference to the topics covered in this chapter.

1. J. R. TINKLEPAUGH and W. B. CRANDALL, Eds. *Cermets*. New York, Reinhold, London, Chapman and Hall. 1960.
2. C. G. SMITH. 'Dispersion hardened systems' in *Powder Metall.*, **11** 1963.
3. H. HAFERKAMP. 'Glass reinforced plastics' in *Engrs Dig.*, **24** 1963, 81.
4. L. N. PHILLIPS. 'Carbon reinforced plastics' in *Trans. Plast. Inst., Lond.*, 1967.
5. D. CRATCHLEY. 'Fibre reinforced metals in *Metall. Rev.*, **1** 1965.
6. J. E. RESTALL. 'Oxidation resistant coatings' in *Metals and Materials*, 1967.
7. J. D. EMBORY *et al*. 'Crack resisting systems' in *Trans. Am. Inst. min. (metall.) Engrs*, **239** 1967, 144.

Index

Abrasives, for metal working, 571
Abrasive material, 570-1
Abrasive wheel sawing, 570
Absolute reference electrode, 843
Absorption coefficient, 76
Accelerated hammer, 483
Accumulator, 264
　hydraulic, 484, 487
Accumulator press system, 475
Acid, ionic nature of, 681, 682
Acid components of steel-making, 152-3
Acid steel, 161
Acrylic resins, 654
Activation energy, 731
　and diffusion, 758
Additives,
　for investment castings, 387
　for moulding sand, 367-8
　for steel-making, 151-2
Adherence of paint, 850
Adherence surfaces, nature of, 646
Adhesion, theories of, 644-46
Adhesive joining of metals, 644-64
Adhesive locking, 587
Adhesive materials, 653-64
Adhesive methods of bonding, 585-6
Adhesives for metals, 586
Adit, 92
ADL process, 150
Admiralty Brass, 833
Admiralty Gunmetal, 835
Adsorption inhibitors, 846
Adsorption theory of adhesion, 645
Aerofall, 139-40
Aerospace industry,
　fibre reinforced metals in, 859
　titanium in, 315
Aeroweb honeycomb structures, 644, 661
Age of Metals, 81, 82
Age-hardening of steel, 196-7, 219

Aging, 32
　of aluminium alloys, 72
Agglomeration, 117
　of iron ores, 143
Aggregate hardened metals, 856-7
Agricola, 83
Air-curing moulding, 377
Air-setting moulding, 377
Aircraft construction, anodizing for, 714
Aircraft industry,
　mechanical joints in, 587
　molybdenum in, 333
AJAX method, 162
Alkyd resins, 721, 725
Allemontite, 265
Allotropy, 192
Allowances, in pattern, 351-4
　distortion, 353-4
　machining, 352
　shrinkage, 351-2
Alloying addition, beryllium as, 239
Alloying of steel,
　effect on hardenability, 223-5
　effects of, 203
　elements for, 202-3
Alloy powders, characteristics of, 749
Alloy theory, 21-6
Alloys, 12-4
　for atomic energy applications, 430
　casting,
　　classification of, 430-1
　　microstructure of, 433-5
　　properties of, 436-9
　　selection of, 431-2
　clad, 831
　compounds in, 13
　freezing of, 406-8
　fusible, 268
　for high-temperature applications, 430
　long freezing range, 406-7, 411, 417-8

890 | INDEX

Alloys *(cont.)*
  manufacture of, by powder methods, 764
  microstructure of, 21-3, 28, 71-2
  properties of, 71-2
  short freezing range, 406-7, 411, 417
  structure of, 13
  temperatures for casting, 390
  used in dentistry, 430
  wear resistant, 272
  zone forming, 441-2
Alternating current for arc welding, 604
Alternating current systems, stray-current corrosion and, 807
Alumina cermets, 867, 868
Aluminium, 243-52
  adhesive bonding of, 646, 647
  alloying of, 247
  available forms of, 248-9
  brazing of, 634
  brightening of, 716
  casting of, 249
  as coating for steel, 710
  corrosion of, 830-2
  electroplating of, 683, 703
  extraction of, 244-7
  extrusion of, 475
  flake, 746
  fluxes for welding, 605
  history of, 243-4
  in honeycomb structures, 884
  laminated to stainless steel, 885
  occurrence of, 244
  painting and lacquering of, 722-3
  properties of, 247-8
  protection of, 671
  radiographs of, 76
  reinforced with silica fibres, 880
  sheathing with, 474
  soldering of, 640
  solubility of copper in, 24, 25
  statistics for, 251-2
  in titanium, 313
  uses of, 250-1
  and vitreous enamel, 719
  welding of, 613, 615, 622
  working of, 249-50, 463
Aluminium alloys, 248, 426
  bright anodizing of, 717
  brightening of, 716, 717
  cast, 441-2
  corrosion of, 830-2
  hardening of, 32-3
  mechanical properties of, 457
  microstructure of, 435
  and stress-corrosion cracking, 815
  surface treatment of, 713-5
  welding of, 613, 622
Aluminium Brass, 833, 834
  in corrosion-erosion, 793

Aluminium bronzes,
  cast, properties of, 442-3
  corrosion of, 834-5
Aluminium foil, lacquering of, 722
Aluminium oxide, as abrasive, 571
Aluminium powder, 249
  characteristics of, 745-6
  production of, for paint, 742
Aluminized coatings, properties of, 708
Alluvial mines, 91
Alnico steel, 208
Alocrom, 714
Alodine, 714
Amalgamation, 118
Ammonia,
  and corrosion, 811
  cracked, as atmosphere for sintering, 761
  for nitriding steel, 31-2
Ammonium chloride, fluxing solution of, 706
Ammonium polysulphide, for metal colouring, 711
Anchor pattern, 850
Angle joints, 652-3
Anisotropic phases, examination of, 67
Anisotropy, 494-5, 862
Annealing, 32-6, 461-2
  full, 34-5
  isothermal, 35
  of steel, 216
  structural changes in, 34
Anode, 681
  definition of, 788
  depolarized, 686
  materials for cathodic protection, 843-4
  reactions at, 789
  shields for cathodic protection, 844-5
Anodic inhibitors, 845-6
Anodicity, 795
Anodizing, 714-5
  bright, 717
Anti-fouling compositions, 292
Antimonial lead, 266
Antimony, 265-7
  alloys of, 266-7
  extraction of, 266
  history of, 265
  occurrence of, 265
  ore treatment, 265-6
  properties of, 266
  in silver plating, 697
  statistics for, 267
  uses of, 267
Arc blow, 603
Arc-casting process for molybdenum, 331-2
Arc length,
  control systems for, 620
  in welding, 602
Arc plasma welding, 619-22
  applications of, 621

Arc plasma welding *(cont.)*
  equipment for, 619-20
  process of, 619
  properties of joints by, 621-2
Arc welding, 601-23
Architecture, copper in, 291
Ardoloy, 775
Argon in metal inert gas welding, 610
Argyrodite, 255
van Arkel, 311, 316, 738
Arsenical antimony, 265
Assel mill, 515-7, 519
Atmospheres, for heat treatment, 37
Atom, structure of, 1
Atomic energy, alloys for use in, 430
Atomic fuel, cladding for, 319
Atomic hydrogen welding, 622-3
Atomization, 744-5
Austempering, 30-1, 219
Austenite, 26, 192-5
  isothermal transformations of, 213
  rate of transformation of, 213
Austenitizing temperature, 195
Autogenous grinding, 105, 139
Automatic bar machines, 544-6
Automation, development of, 527
Azurite, 285

Babbitt, Isaac, 259
Babbitt metals, corrosion and, 839
Back-stand emerying, 675
Bacteria, and corrosion, 811-3
Bain, E. C., 210
Bainite, 30-1, 213
Balanced steel, 185
Ball mill, 103-4, 742, 743
Band saws, 569
Bar,
  drawing of, 492-3
  rolling of, 470, 510
Barium titanate, for ultrasonic testing, 74
Barrel plating, 702
Barrel polishing, 675
Base circle of gear, 561
Basic components for steel-making, 153
Basis box, 705
Battery, 673
Bauer-Vogel process, modified, see MBV process
Bauxites, 244
Bayer process, 244-5, 246-7
  residues of, for gallium, 253
Bayerite, 646
Bearings,
  lead in, 264
  porous, 730, 749
    infiltration and, 763
    and powder metallurgy, 771

  and sintering, 760
  tin in, 258-9
Bells, manufacture of, 582
Bench worl, in core-making, 397
Bending, 496-501
  machines for, 499-501
  press, 497-8
  press-brake, 497
  section, 499-500
Beneficiation, 139, 244
  of zirconium ore, 316-7
Benzoic acid, for lubrication, 752
Benzol, for lubrication, 777
Bernoulli's theorem, 411-2
Bertrandite, 236
Beryllium, 236-9
  alloys of, 237
  extraction of, 236
  fabrication of, 236-7
  fibres for reinforcement of aluminium, 878
  history of, 236
  joining of, 238
  occurrence of, 236
  production of, by powder methods, 737, 741, 768, 769
  properties of, 237-8
  statistics for, 239
  uses of, 238-9
  wires for reinforcing metals, 877
Berzelius, 310, 316, 320
Bessemer converter, 163-5
Bessemer process, 155-6
Bessemer, Sir Henry, 130, 154-6
Betterton process, 263
Betts electrolytic process, 263, 295
Bevel gear generating, 565
Bevel gears, 563
Bicycle frames, manufacture of, 635
Billet, 182
  rolling of, 508-10, 512
Binders,
  for cores, 402-3
  ethyl silicate, 385-6
  phosphate, 386-7
Binding, acetate process of, 387
Bingham Canyon copper mine, 91
Biological activity and corrosion, 811-3
Biological aspects of molybdenum, 334-5
Bisphenol A, 656
Bismuth, 267-9
  alloys of, 268
  extraction of, 267-8
  history of, 267
  occurrence of, 267
  properties of, 268
  statistics for, 268-9
  structure of, 3-4
Bismuth glance, 267
Bismuthinite, 267

892 | INDEX

Blacksmith welding, 597-8
Blake type jaw crusher, 101-2
Blast furnace, 144-9
  charge in, 147
  development of, 127
  lining of, 146
  physico-chemical processes in, 146-7
  slag in, 147-8
  temperature of air blast in, 148
Blasting in underground mining, 92
Blister steel, 129
Block-caving, 286-7
Bloom, 182
  rolling of, 508, 511
Blooming, history of, 125-6
Blow squeeze moulding, 373
  shell, machines for, 381-2
Blowholes and steels, 184
Boat building, glass-fibre reinforced plastics in, 873
Bobbing, 675
Body-centred cubic structure, 14-5
  notch-brittleness of, 55
Boehmite, 646
Boiler embrittlement, 816
Boisbaudran, Lecoq de, 253
Bolster, 754
Bolting, 586-7
Bolts, tightening of, 587
Bond, 571
Bone, chemical, 3-5, 12, 13
Bonding, metallurgical surface, 585, 632-41
Bonding surfaces, pretreatment of, 646
Bordeaux Mixture, 292
Borehole exploration, 86-88
Boring, 538-46
  machines for, 542
  tool for, 538-40
Bornhauser solution, 695
Bornite, 285
Boron nitride reinforced phenolics, 876
Boron reinforced systems, 875
Borts, 87
Bower-Barff process, 711-2
Bragg, Sir Laurence, 195
Brass flake, production of, 745
Brass powder, production of, 743
Brass wire, 290
Brasses, 306
  brightening of, 716
  corrosion of, 804-5, 833-4
  electroplating of, 702, 703
  etching of, 680
  extrusion of, 475
  impingement attack and, 806
  machining of, 292
  mechanical properties of, 72, 457
  melting of, 426
  micrographs of, 66
  microstructure of, 21-2
  stress-corrosion cracking and, 816
  working temperatures of, 463
Braze welding, 632-3
Brazing, 582, 633-8
  of beryllium, 238
  materials for, 633-4
  and welding, distinction between, 582
Brearley, H., 202, 206
Brewing vats, 290-1
Bright dipping, 716
Brightening,
  chemical, 716
  electrolytic, 716-7
Brinell hardness number, 51
Brinell hardness test, 51-2
Briquettes, molybdenum, 331
Britannia metal, 267
British Iron and Steel Research Association, 131
British Steel Corporation, 131
Brittle behaviour of materials, 54
Brittle failure, resistance to, 860
Brittleness of cermets, 868
Broaches, 558-9, 564
Broaching, 558-61
  machines for, 559-60
Bronze Age, 81, 82, 123
  metal cutting and, 523
  metal deformation in, 455
  tin and, 256
Bronze welding, 585, 633
Bronzes, 259
  corrosion of, 834-5
  electroplating, 702
  impingement attack and, 806
  melting of, 426
Brucite, 240
Brytal process, 717
Buffing, 675
Building industry,
  aluminium in, 250
  glass-fibre reinforced plastics in, 874
Bullets, 779-80
Bullhead, 508
Bunsen, 240
Burgers vector, 19
Burning, in electroplating, 686
Butt joints, 642-3, 652-3

Cable-sheathing press, vertical hydraulic, 474
Cadmium, 307-8
  electroplating of, 683, 695-7, 699
  powder, 740
  pretreatment of, for adhesive bonding, 647
Cadmium copper, 307
Caesium-137, as source in radiography, 77

Calamine, 301
Calcium aluminate cements, 387
Calorizing, 710
Camphor, for lubrication, 752
Camshafts, economic aspects of, 37
Capacitor discharge welding, see Percussion welding,
Capped rimmed steel, 184
Capstan lathes, 544
Carat, 299
Carbides in steels, 227-8
Carboloy, 775
Carbon, see also Diamond and Graphite,
  for carburizing steel, 31
  and hardness of steel, 201
  in iron, 13-4
  as reductant, 732, 735
  removal from steel, 189
  and resins, 720-1
Carbon arc welding, 622
Carbon dioxide moulding process, 378
Carbon dioxide flux welding, 617
Carbon dioxide welding, 615-7
Carbon disulphide, in silver plating, 697
Carbon-fibre reinforced systems, 875-6
Carbon monoxide, as reductant, 732-3, 734
Carbon steel, see Steel, carbon,
Carbon tetrachloride, in lubricants, 777
Carbo-nitriding process for steel, 31
Carbonyl gas process for refining nickel, 277-8
Carburizing, 31, 232
Carburizing flame, 625
Cartridge Brass, 833
Case-hardening, 31, 231
Cassiterite, 257
Cast iron, 208-9
  corrosion of, 825-7, 829-30
  flake graphite, properties of, 433, 440
  grey, 35, 208
    corrosion of, 825-6
    micrograph of, 197-8
    properties of, 433, 440
    volume change on solidification of, 416
  high-alloy, corrosion of, 825-6
  malleable, 35-6, 191, 440
    blackheart, 35-6
    corrosion of, 825-6
    early production of, 124
    whiteheart, 35
    melting of, 427-8
    microstructure of, 434
    nodular, see Cast iron, spheroidal graphite
    pretreatment of, for adhesive bonding, 647
  for rolls, 471-2
  spheroidal graphite, 209, 440
    contraction of, 352
    corrosion of, 825-6
    vitreous enamel for, 719
  white, 35, 209, 440

  corrosion of, 825-6
  shrinkage allowance for, 351-2
Cast ingot, structure of, 457-8
Cast metals, 432-3, 457
Cast components, 341-3
Casting, Ch. 5
  advantages of, 341-3
  centrifugal, 357, 393-5
  choice of method, 357-9, 360-1
  classification of processes, 356-7
  continuous, of steel, 185-8
  definition of, 341
  economics of, 358
Casting alloys,
  classification of, 430-1
  microstructure of, 433-5
  properties of, 436-9
  selection of, 431-2
Castings,
  basic steps in making, 344-6
  centrifugal, 357
  centrifuging of, 395
  chemical composition of, 453-4
  defective, reclamation of, 626
  defects within, 449
  evaluation of, 432, 446-54
  feeding of, 415-20
  inquiries from buyers of, 358-9
  investment, 381-7
  pouring of, 404-15
  properties of, 431, 450-3
  shape factor of, 418
  soundness and unsoundness of, 449, 451
Catalysts,
  for phenolic resins, 660
  platinum group metals as, 283
  for polyurethane adhesives, 662
Cathode, 681
  definition of, 788
  reactions at, 789-91
'Cathode ray', 629
Cathodic inhibitors, 846
Cathodic protection, 841-5
  anode materials for, 843-4
  anode shields for, 843-4
  current required for, 842-3
  principles of, 841-2
  use of magnesium for, 242-3, 838
Caustic cracking, 816
Caustic embrittlement, 816
Caustic soda, molten, for descaling, 674
Cavitation, 806
  damage, 805, 806-7
  environment and, 807
Cell,
  concentration, 795-6
  corrosion, 787, 796
  Daniell, reaction in, 794
  differential aeration, 796, 803

894 | INDEX

Cell (cont.)
  galvanic, 673, 787-8
  reversible concentration, 795
  unit, 14
  voltaic, 788
Cellulose,
  in core sands, 403
  in moulding sand, 368
Cement-bonded moulds, 374
Cement moulding, drawback of, 377
Cementation, 129
Cemented carbides, 336, 537
Cementite, 195, 197-8
Centre-type cylindrical grinder, 572-3
Centreless cylindrical grinders, 573-4
Centrifugal casting, 393-5
Centrifuging of castings, 395
Ceramic moulding, 388
Ceramic-shell moulds, 387
Ceramics, 537, 854-5
Cereal in core sand, 403
Cereals in moulding sand, 368
Cermets, 856, 866-9
  applications of, 868-9
  limitations of, 868
  methods of manufacture, 866-7
  and powder metallurgy, 730
  principles of formation, 866
  properties of, 867-8
Chalcocite, 285
Chalcolithic Age, 285
Chaplets, 401
Chalcopyrite, 276, 286
Charge,
  for blast furnace, 147
  for LD converter, 167
  for open hearth furnace, 158-60, 162
Charpy V-notch method, 54
Chatter, 532
Checkers, 157
Chemical brightening of metals, 716
Chemical composition of castings, 453-4
Chemical compounds, nature of, 681
Chemical industry,
  lead-coated metals for, 837
  molybdenum in, 333
  use of tantalum in, 326
  titanium in, 315
Chemical stability of cermets, 867-8
Chemical treatment of ores, 117-8
Chemisorption, 645
Chills, 419
Chip formation, 528-9
Chip friction coefficient, 529
Chip friction force, 529
Chip thickness ratio, 530
Chisel edge angle, 548
Chlorimet, 445
Chlorinated rubber, 725

Chromating processes for aluminium, 714
Chromic acid, 691, 714
Chromic chloride, 710
Chromic sulphate, 691-2
Chromite, 327
Chromium, 327-9
  alloying of, 329
  commercially pure, 328
  electroplating of, 683
  history of, 327
  mining of, 327
  occurrence of, 327
  powder, 735, 768
  pretreatment of, for adhesive bonding, 647
  production of, 327-8
  properties of, 328, 690-1
  statistics for, 329
  uses of, 329
Chromium plating, 329, 690-5
  plant, 687, 693, 694
  precautions in, 694
  solution,
    composition of, 691-2
    disposal of waste, 695
    temperature of, 692
  tanks for, 693
  undercoating for, 684
  uses of, 693
Chromizing, 329, 710
Chucks, lathe, 542-3
Cinnabar, 308
Circular pitch of gear, 561
Circular saws, 569-70
Clad plate, 598, 600
Clapper box, 556
Classification of iron ore, 140
Classifier,
  cyclone, 107-8
  hydraulic, 107
  for particle size, 106-7
Clays, for moulding, 363-4, 367
Clay-water-silica bond in mould material, 364
Cleaning of castings, 345
Cleavage, stresses in, 649
Climb milling, 553
Close-packed hexagonal lattice, 14-5
Close-plating, 293
Coatings, 669
  asphaltic, 722, 724
  gold, 300
  heat-resisting, 726-7
  lead for, 264-5
  manganese, 339
  metallic, 670-710
  nickel as, 684
  for niobium, 325
  non-metallic, 670, 725-6
  non-stick, 726-7
  oxidation-resistant, 860, 882-3

INDEX | 895

Coatings (cont.)
  paint, 671
  plastic, 727
  of platinum group metals, 284
  preparation for, 673-6
  produced by reducing agents, 709-10
  tarry, 722
  surface, for cores, 403-4
  tin, 259-60
  vitreous enamel, 671, 717-9
  zinc, 303-5
Coating processes, non-metallic, 710-27
Coal, in moulding sand, 368
Coal mining, comparison with ore mining, 89-90
Cobalt, 269-73
  alloys of, 272-3
  extraction of, 270-1
  history of, 269
  mining of, 269-70
  occurrence of, 269
  production of powder, 736-7, 740, 748
  properties of, 271-2
  statistics for, 273
  uses of, 272-3
Cobalt-60, as source in radiography, 77
Cobalt, tungsten carbide mixtures, 775, 777-8
Cobble, 510
Coefficient, absorption, 76
Cogging, 467
Coggling mills, 183
Coinage,
  copper in, 292
  gold as, 297, 300
  nickel in, 278-9
  silver as, 294
Coining, 749, 763-4
Cold chamber die-casting process, 392-3
Cold forging, 490-1
Cold-heading machine, multistation, 490-1
Cold pilgering, 518, 520
Cold pressure welding, 599-600
Cold-setting moulding, 377
Cold working, 196, 458, 459-61
Collapsibility, 366-7
Collectors, in flotation processes, 116
Colour buffing, 689
Colouring processes, 711-2
Columbite, 321-2
Columbium, see Niobium,
Columnar grains, 405-6, 457
Compaction of green sand moulds, 369-73
Compo, composition of, 375
Compo and chamotte moulds, 374-6
Compo moulding, drawback of, 377
Components, powder method of manufacture of, 730
Composite materials, Ch. 12
  classification of, 855
  compatibility in, 861
  concepts employed in, 855-60
  cost of, 861
  definition of, 853
  future of, 886-7
  manufacture by powder method, 730
  and powder metallurgy, 769
  principles of, 853-62
  problems associated with, 860-2
  literature on, 887
Composite coatings for metals, 883
Composite powders, 737
Composition, relation between microstructure and, 70-2
Compound, intermediate, in alloys, 13
  intermetallic, 24
  interstitial, 13
  valence, 13
Compression moulding, 871
Compressive force, 529
Compressive stress, 450, 452
Computer, Quantimet Image Analysing, 68
Concentration cell, 795-6
Concentration of iron ore, 140-2
Concentrator, Humphreys spiral, 140-1
Condensation, polymerization, 365
Conduit tubing, electrical, 498
Cone piercer, 513, 516
Conical forms, grinding of, 573
Constitutional undercooling, 406
Consumable guide technique of electro-slag welding, 631
Consumable patterns, 350-1
Contamination,
  in arc welding, 602
  in metal pulverization, 742-3
Continuity, law of, 411-2
Continuous broaching machine, 559-60
Continuous fibre reinforcements in metals, 880-1
Continuous mills for tube, 515
Continuous tandem rolling mill, 466
Contour squeezing, 371
Contouring control systems, 579
Contraction of metal in casting, 351
Control,
  of corrosion, 813, 845-50
  of corrosion reaction, 796-7
  structural, of casting, 433
  unit, in numerical control, 577-8
Controlled arc welding, 613
Convention,
  for cell reactions, 794
  sign, for potental of metal, 795
Converters, 155, 162-72
  comparison of Bessemer and LD, 169-70
  copper, 288
  using gas with low nitrogen, 165
  Rotor, 171-2
  side-blown, 423, 428-9

896 | INDEX

Conveyors for mining, 95
Coolant for grinding, 572
Cooling of steel, 213-4, 215
Cooperite, 281
Co-ordination number, 3
Cope, 348
Cope and drag pattern plates, 349-50
Copper, 284-93
  for brazing, 633
  castings of, 292
  chiek mines, 286
  corrosion of, 832-3
  as corrosion inhibitor, 822
  electrolysis of, 681-3, 684
  electrolytic, 746
  electrorefining of, 683
  extraction of, 287-9
  extrusion of, 475
  fabrication of, 289-92
  gilding of, 709
  history of, 284-5
  Melting of, 426
  microstructure of, 71-2
  mining of, 286-7
  OFHC, 832
  ores, leaching of, 118
    principal, 285-6
  plating from cyanide solutions, 695-7, 699
  pretreatment of, for adhesive bonding, 647
  properties of, 289, 457
  solubility in aluminium, 24, 25
  statistics for, 293
  stress-rupture curves for, 864
  uses of, 289-92, 763
  welding of, 614, 626
  wire, tinning of, 706
  working temperatures of, 463
Copper alloys,
  for brazing, 633-4
  cast, 442-3
  corrosion of, 832, 833-6
  cost comparison of, 38-9
  degreasing of, 678
  etching of, 680
  gilding of, 709
  impingement attack and, 806
  melting of, 426
  microstructure of, 71-2, 435
  pretreatment of, for adhesive bonding, 647
  for small components, 730
  and stress-corrosion cracking, 815-6
Copper glance, 285
Copper mine, Bingham Canyon, 91
Copper oxide, reduction of, 734
Copper powder,
  atomized, 746
  characteristics of, 746
  for paint, 742
  production of, 734, 736, 739, 740

Copper sulphate, electrolysis of, 739-40
Copper-tungsten fibre systems, 877
Copy turning, 541
Core blowers, 398
Core boxes, 355-6, 397
  for core blowers, 398-9
Core driers, 355-6
Core finishing, 400
Core-making, 345, 356-8
  machines for, 397-9
  methods of, 396-9
Core materials, 362-6, 401-4
Core oils, 402
Core ovens, 400
Core prints, 345, 355, 400
Core sands, 402
  properties of, 401-2
Core setting, 400-1
Core shooters, 399
Core shooting, 373
Cores, 396-404
  applications of, 401-2
  buoyancy forc in, 400-1
  function of, 396
Coring diamond drilling, 87-8
Corrosion, Ch. 11
  in absence of stress, 801-13
  in aqueous environments, 788-9
  and biological activity, 811-3
  control of, 845-50
  cost of, 785-6
  crevice, 594, 803
  definition of, 784
  deposit attack, 802-3
  dry, 801
  and environment, 784-5
  fretting, 814, 819-21
  galvanic, 802
  high temperature, 808-11, 827, 828
  and induction brazing, 637
  inhibitors, 845-7
  literature on, 851
  natural inhibition of, 672-3
  nature of, 671-3, 674
  oxidation equations for, 808-11
  in percussion welding, 596-7
  phenomenon of, 783
  phosphating as preventative of, 712
  pitting, 801-2
  in presence of stress, 813-21
  rates of, 796-9
    measurement of, 789
    and polarization, 791-3
  reaction mechanism of, 786-801
  resistance to, 313, 615, 872-3
  selective, 803-4
  stray-current, 807-8
  in welded austenitic stainless steel, 606
Corrosion cell, 787, 796

Corrosion currents, influence of polarization on 798-9
Corrosion-erosion, 784, 805-7
Corrosion fatigue, 60, 814, 817-9
Corrosion reaction, 786-8
　control of, 796-7
　importance of film formation in, 793
　ways of accelerating, 811
Cort, Henry, 128-9
COR-TEN, 203-4
Cost,
　of cermets, 868
　of corrosion, 785-6
　of melting, 422
Counter-drilling, 546
Counterblow hammer, 483
Covalent bond, 3-4, 5, 12, 13
Cracking,
　caustic, 816
　detection of, 73-4
　in freezing metal, 409
　season, 816
　stress-corrosion, 814-7
Crack-resisting systems, 883-4
Cracks,
　in corrosion fatigue, 818
　propagation of, 854, 855
Creep, 56
　curves, 57-9
　in dispersion-hardened systems, 865
　mechanism of, 58-60
　resistance to, in alloy steels, 230
　tension tests for, 56-7
　testing, 55-60
Crevice corrosion, 803
　in mechanical joints, 588
　in spot welds, 594
Cronstedt, Axel, 274
Crookesite, 253
Cross-linking, 366
　in phenolic based resins, 660
　in polyurethane, 662
Crown wheel, 563
Crucibles, development of, 130
Crushing of ores, 101-3, 139
Crusher,
　cone, 103-4
　gyratory, 103
　hammer, 103
　jaw, 101-2
Cryolite, 101-2
Crystal structures, metallic, 14-6
Crystallography, 14
Cupellation, 294, 298
Cupola, 423, 425, 427
Cuprite, 285
Curing,
　agents for, 657
　of silicate adhesives, 664

Current density,
　in cathodic protection, 842-3
　in chromium plating, 692
　in nickel plating, 682
　in tinplating, 700
Cut and cutting force, 531
Cut and fill stoping, 93-4
Cutanit, 775
Cutters, types of, 552
Cutting, Ch. 7
　action in milling, 553
　economics of, 534-5
　fluids for, 532, 537-8, 539
　force, 529, 530-2
　general, 527-38
　literature on, 579
　principles of, 527-32
　tool materials, 536-7
　tools for, 539-40
Cutting ratio, 530
Cutting speeds,
　for broaching, 560-1
　and cutting force, 531
　for drilling, 551
　for milling, 554-5
　for reaming, 552
　for shaping and planing, 558
　for tapping, 552
　and tool life, 553
　for turning, 546
Cyanidation, 118-9
Cyanide method for gold extraction, 294-5, 298
Cyanide solutions for electroplating,
　formulation of, 695-6
Cyaniding, 232
Cyanoacrylates, 654-6
Cyclone, 140
Cylindrical joints, 651-2
Cysteine, 813
Cystine, 813

Damage, cavitation, 805, 806-7
Daniell cell reaction, 794
Darby Abraham, 127
Data processing, 577
Davis, A. V., 244
Davy, Sir Humphrey, 240, 243, 841
Dealuminification, 835
Decarburization, 35
Decay, rate of, 76
Deep drawing, 493-5
Deep hole drilling, 549-50
Deformation, Ch. 6, 16-9, 742
　elastic, 455-6
　history of, 455
　literature on, 67
　mechanism of, 458-9
　mechanical, products of, 505-21

Deformation *(cont.)*
  plastic, 16, 455-6, 857
    mechanisms of, 58-60, 857
    reasons for, 457-8
    recent developments of, 501-5
Degreasing, 676-80
  agent for, 647
  alkaline, 678-9
  apparatus for, 677-8
  solvent-vapour, 677-8
  solvents used for, 677
  wet, 678-9
Dehydrogenase, 812
De Levaud process of pipe manufacture, 395
Density, see also Pour density and Tap density
  of glass-fibre reinforced plastics, 872
Dental amalgams, 259
Dentistry, alloys used in, 430
Deoxidants for steel, 183-4
Deoxidation of steel, 183-5
Deposit attack, 802-3
Depth of cut, in shaping and planing, 558
Design,
  of castings, 452-3
  data, 577
  factor in controlling corrosion, 847-8
  of joints, 641-4, 648-53
  methods for composites, 861-2
Desulfovibrio desulfuricans, 812
Detergents, synthetic, action of, 678
Detritus, removal of, in drilling, 86-7
Deville, Henry St. Claire, 244
Dextrine, in moulding sand, 368
Dezincification, 804-5
Diamagnetism, 268
Diametral pitch of gear, 561
Diamond,
  as abrasive, 571
  as cutting tool material, 537
  industrial, and hard carbide, 780
  structure of, 3-4
Diamond drilling, 87-8
Diamond-square sequence for rolling billet, 509-10, 512
Diaphragm squeezing, 371
Die,
  definiton of, 473
  design of, 754-6
  forging, 485-6
  manufacture of, 779
  metal, 390
  pressing, filling of, 750-1
Die casting, 345, 356-7, 392-3
  molybdenum in, 334
  of zinc alloys, 306
Die castings, 356
Differential aeration cell, 796, 803
Diffraction, 69
Diffusion in metals, 22-3
  processes, hot, 710
  in solid state, 758-9
  in steels, 210
  temperature dependence of, 758
Diffusion bonding, 585
Dip brazing, 635
Dip-transfer welding, 616
Direct current for arc welding, 603-4
Direct-oxygen process of steel-making, 155
Direct-reduction of iron ore, 149
Direct steel-making, 189-90
Disc-type surface grinder, 574-5
Disintegration,
  of liquid metal, 744-5
  of solid metal, 741-4
Dislocations, 19
  edge, 460
  pile up of, 58, 460-1
  theory of, 459-60
  and work hardening, 49
Dispersion-hardened systems, 857, 862-3
  applications of, 865-6
  limitations of, 864-5
  manufacture of, 862-3
  properties of, 863-4
Disproportionation, 738
Dissimilar joints, welding of, 600, 601, 602
Distilling columns, copper, 291
Distortion allowance, 353-4
Dividing head attachment for milling machines, 554
Dolomite, 240
Domestic appliances,
  surface treatment of, 726
  aluminium in, 251
Dope, 735
Doré metal, 295
Double lap joint, 651
Double scarf joint, 651
Down milling, 553
D process, 379-80
Drag, 348
Drag-out, in electroplating, 689
Draw bench, 492-3
Draw-die forming, 499
Drawing, 491-3
  mandrel bar, 518-9
  plug, 519
  principles of, 491-2
  products of, 492
Drawing ratio, limiting, 492
Drawn sections, 511-3
Driers for cores, 402
Drilling, 546-52
  definition of, 546
  diamond, 87-8
  location of hole for, 551
  machines for, 548-51
    difference between, 546

Drilling *(cont.)*
  percussive, 86-7
  performance and accuracy of, 550-1
  rotary, 87-8
  speeds and feeds for, 551
  tools for, 547-8
Drills, 547-8
Dry corrosion, 808
Dry sand moulds, 373-7
Dry strength, 366
Ductility, 48, 452
  assessment of, 606
  of dispersion-hardened systems, 865
  testing of, 41-50
Dump box, 355-6, 397
Dump box method of shell moulding, 380-1
Duplex process of steel-making, 162, 428
Duralumin, 72-3, 244, 475
Dwight, A. S., 143
Dwight-Lloyd-McWane process, 149
Dynamo brushes, 772
Dynapak machine, 504

Earing, 494-6
Earth's crust, 135
Economic aspects of metal selection, 37-40
Economics of metal cutting, 534-5
Eddy current testing of metals, 78-9
Eddy mill, 742-3
Edge dislocation, 19-20
Eighty-five-three-fives, 835
Elastic deformation, 455-6
Elastic limit, 455
Elastic strain, 450
Electric arc furnace, 173-6
  power source for, 174-5
  basic, 175-6
  acid, 176
  charge for, 175
  slag in, 176
Electric furnace method of steel-making, 155
Electric storage battery, 264
Electrical industry, aluminium in, 251
Electrical methods of prospecting, 85
Electrical properties of glass-fibre reinforced plastics, 872
Electrical separation of ores, 114-5
Electrochemical series for metals, 672
Electrode potentials, standard, 794
Electrodes,
  for arc welding, 602-3
  reference, 843
Electroforming of metal-fibre reinforced metals, 878
'Electroless' plating of nickel, 709-710
Electrolysis,
  of fused salts for metal powder production, 741

  principles of, 681, 682, 739-40
Electrolyte solution, 786, 788
Electrolytic brightening of metals, 716-7
Electrolytic methods for extraction of silver, 295
Electrolytic methods for metal powder production, 739-41
Electrolytic process,
  Betts, 263
  for production of magnesium, 240-1
  for refining zinc, 303
  for tinplate production, 259-60
Electrolytic production of manganese, 338
Electrolytic refining,
  of copper, 289
  of nickel, 277-8
Electrolytic tinplate, 700-1
Electromagnetic forming, 503-4
Electromagnetic testing of metals, 78-9
Electro-slag refining of steel, 180-1
Electro-slag welding, 630-2
Electron beam, production of, 628-9
Electron beam melting of niobium, 323
Electron beam welding, 585, 628-30
Electron diffraction, 69
Electron microscopy, 68-9
Electron probe microanalysis, 70
Electroplated coatings, inspection of, 703-4
Electroplated nickel silver, see EPNS,
Electroplating, 670, 671, 702-4
  of alloys, 701-2
  control of, 703
  degreasing before, 676-80
  etching before, 680
  history of, 681
  mechanization of, 702
  precautions in, 696
  principles of, 680-4
  use of lead in, 837
  water requirements of, 703
Electroplating plant,
  for precious metals, 698
  semi-mechanized, 688-691
Electroplating solutions, double cyanide, typical, 696
Electrostatic separation of ores, 114-5, 141
Electrostatic spraying of paint, 724
Electrothermal refining process for zinc, 303
Electrum, 299
Elongation, percentage, 47
Embrittlement, hydrogen, 674
Emerying, back-stand, 675
EMMA instrument, 70
Emulsification, 678-9
Enamel, *see also* Vitreous enamel,
  coatings, 719-26
  definition of, 720
  phosphate base for, 712-3
End thrust in rolling, 470-1

Endogeoeous freezing, 408
Endothermic reaction, 147
Endurance limit, 60
Engineering, use of aluminium in, 251
Environment, and corrosion, 784-5, 788-9
Epichlorhydrin, 656
Epigenetic deposits, 83
EPNS, 279, 297
Epoxide 828 Thiokol LP-3 co-polymer, 659
Epoxide polyamide systems, 658
Epoxy based resins, pretreatment for, 646-8
Epoxy phenolics, 658
Epoxy polysulphides, 658-9
Epoxy resins, 656-9, 869-71, 721
   solidification of, 725
Epoxy silicone, 659
Equi-axed crystals, 457
Equi-axed grains, 406
Equilibrium diagrams, 25, 26
Equilibrium potential, 791
Erftwerk process, 716, 717
Erosion, definition of, 784
Erosion-corrosion, 784
Etch primer, 725
Etchant solution, 65
Etching, 65, 680
   of steel, 197
   thermal, 67
Ethyl silicate, in investment moulding, 385-6
Ethylene diamine, 661
Eutectic alloy, 408-9
Eurectoid, 192, 201
Evans diagrams, 796-9
Exchange current, 791
Exogeneous freezing, 408
Exothermic compounds, addition to riser, 419-20
Exothermic reaction, 147
Exploration,
   borehole, 86-8
   in mining, 83-6
Explosion welding, 600-1
Explosive forming, 502
Extensometer, 43, 56-7
External grinder, 572
Extreme pressure oils, 538
Extruded section, 249, 510-1
Extrusion, 473-9
   composition and uses of tools for, 476
   direct, 473
   hydrostatic, 504-5
   impact, 478-9
   indirect, 473-4
   multihole die, 476
   principles of, 473-4
   tool design, 476-7
   tool materials, 475-6
   for tube-shell production, 513-4
   typical temperatures and pressures for, 475

Fabrication, see also individual elements,
   cost of, 38-9
   of dispersion-hardened systems, 864-5
   of glass-fibre reinforced plastics, 972
Face-centred cubic lattice, 14-5
Face width of gear, 561
Faraday, M., 799, 841
Faraday's law, 807
Fatigue, 60-2
   in bolting, 586
   corrosion, 60, 814, 817-9
   cumulative damage in, 60-1
   definition of, 60
   of glass-fibre reinforced plastics, 873
   mechanism of, 63
   resistance to, 878
   in spot welds, 594
   and stress reversal, 61-3
   tests, 60-3
   typical fracture, 63
Fatigue limit, 818
Fatty oils as cutting fluids, 538
Feed rate, 546, 551-2, 555, 558
Feedback in numerical control, 577
Feeder, see Riser,
Feeding of castings, 415-20
Feldmann, H. D., 490
Ferrite, 34, 192-5, 197-8
Ferrochromium, 327-8
Ferromanganese, 148, 151, 337-8
Ferromolybdenum, 331
Ferroniobium, 322, 325
Ferrotungsten, 335
Ferrous hydroxide, oxidation of, 790-1
Ferrous metals,
   cast, corrosion of, 824-7
      galvanic compatibility of, 826-7
      properties of, 433-40
   cathodic protection of, 843
   costs of, 39-41
   high temperature corrosion of, 827-30
   melting of, 427-9
   sawing of, 570
   and sintering, 762
   wrought, corrosion of, 821-4
Fettling time, 161
Fibre composites, 869-82
Fibre failure, 859
Fibre-matrix interface, importance of, 859-60
Fibre reinforced metals, 876
Fibre reinforced systems, 857-60
   cost of, 861
Fibres,
   costs of, 861
   silica, 880
   strength and moduli of, 876-7
   unidirectional, in metal matrix, 858
Field ion microscopy, 70-1
Filament winding techniques, 872, 874, 881

INDEX 901

File hardness test, 51
Filled plastics, 869
Filler metal,
   for braze welding, 632
   for brazing, 633
Fillet joints, 641-2
Film formation, importance in corrosion reactions, 793
Films, protective, 793
Filters, 730, 737, 760, 770-1
Finishing, 467
Finniston, H. M., 204
Fire cracker welding, 604
Fire gilding, 709
Fire-refining of copper, 288
Fire-setting, 83
Flade potential, 800
Flame, oxyacetylene, 623-4, 625, 633
Flame brazing, 636
Flash welding, 592-3
Flaw, size detected by ultrasonic tests, 75
Flocculation, 119
Floor moulding, 374
Flotation, see Froth flotation,
Flotation reagents, 117
Flow, grain, 481-2
Flow of molten metal, 410-2
Fluid-sand process, 378
Fluidity, 408
Fluidized bed, 735
Fluidizing, 150
Fluoroscopy, 77
Fluorosilicic acid, 692
Flux,
   for soldering, 639, 640-1
   for steel-making, 152
   for welding, 584, 598, 602, 605, 607
Flux arc welding processes, 584-5
Flux dip brazing, 635-6
Foil,
   aluminium, 249
   rolling of, 471
   technique for electron microscope, 68
Follow board, 350
Food industry, soldering in, 640
Foot wall, 93
Force, upsetting, 592
Forces at tool edge, 529
Forge welding, 583-4, 597-8
Forging, 480-91, 581
   closed die, 480-1
   cold, 490-1
   free-form, 480
   open-die, 480
   operations of, 485-90
   rotary, 517-8, 520
Forging machines, 484-5, 488
Forging tools, 485, 488
Forgings, drop, 486, 489

Forming,
   draw-die, 499
   electromagnetic, 503-4
   of gears, 564
   high rate, 501-4
   high-voltage discharge, 502-3
   roll, 498-500
   using chemical explosives, 502
Foundries, classification of, 344
Foundry industry, 343-4
Fractional distillation, 153
Fracture characteristics of metal reinforced with continuous fibres, 881
Fracture stress, 47
Free energies of liquid and solid phases, 405
Freezing of alloys, 406-8
Fretting corrosion, 814, 819-21
Friction in rolling, 463
Friction sawing, 570
Friction welding, 599-600
Forth flotation,
   of copper ores, 287
   for iron ores, 141-2
   separation of ores by, 115-7
   for tin ores, 262
   for zinc ores, 302
Fuel oil, in moulding sand, 368
Full-mould process, 377-8
Furnace, 36
   air, 423, 428
   blast, see Blast furnace,
   choice of, 36, 421-2
   consumable electrode vacuum arc, 317
   crucible, 422, 425, 426
   cupola, 423, 425, 427
   direct arc, 428
   electric, for steel-making, 172-7
   electric arc, 173-6, 327, 423
   holding, 421-2
   induction, 177, 426, 428-9
   induction melting, 422-3, 424
   melting, 422-5
   metallurgical features of, 426-9
   open-hearth, 154, 157-62, 423
   oxygen, comparison of, 172
   puddling, 128-9
   reverberatory, 423
   shaker hearth, 36
   side-blown converter, 423, 428-9
Furnace brazing, 636-7
Fusion welding, 584-5, 601-32

Gahn, J. G., 337
Galena, 262
Galling, 587
Gallium, 252-5
Galvanic anode systems for cathodic protection, 844

Galvanic cell, 787-8
Galvanic corrosion, 802
Galvanic series, 794-5
Galvanized articles, corrosion of, 841
Galvanized iron sheet, joining of, 633
Galvanizing, 305, 706-7
Gamma radiography, 77
Gang box, 355
Gangue, 98
Gas cylinders, connections to, 624
Gas welding, 584, 623-6
Gated patterns, 348
Gates, 414-5
Gating design, 413-6
Gating ratio, 415
Gating system, 409-10
Gasket, compressible, 588-9
Gauges, strain, 78-9
Gears, 561-2
 cutting of, 560-8
 finishing processes for, 567-8
 generating of, 564-6
 types of, 562-3
Geochemical prospecting for minerals, 84-5
Geology, in mining, 84
Geophysical prospecting for minerals, 85-6
Germanite, 253
Germanium, 255-6
Gettering, 319
Gilding, 30 , 695-7, 698-9
 fire, 709
 parcel, 709
Glass, 333-4, 718
Glass fibres, 869
Glass-fibre reinforced cement, 881-2
Glass-fibre reinforced plastics, 869-74
Glutathione, 813
Goethite, 135
Gold, 297-301
 alloys of, 299, 702
 available forms, 299
 cyanidation of, 118-9
 extraction of, 82, 298
 history of, 81-2, 297
 occurrence of, 297-8
 pretreatment of, for adhesive bonding, 647
 properties of, 298-9
 refining of, 298
 statistics for, 300-1
 uses of, 299-300
Gold ink, 729
Gold mine, Western Deep Levels, 92
Gold plating, see Gilding,
Goodeve, Sir Charles, 130
Grade of grinding wheel, 572
Grading of moulding sand, 367
Graef, R., 171
Grain, definition of, 459
Grain boundaries, 49-50, 460

Grain flow, 481-2
Grain refinement in single phase alloys, 71
Grain size,
 and deformation, 461-2
 factors affecting, 405
 and hardenability, 222
 and quenching in alloy steels, 228
 in steel, 195-6
 in welds, 622
Graphite, 368, 766, 771
Graphite moulds, 390
Graphitization, 825-6, 875
Gravitational method of prospecting, 85
Gravity die castings, 390-1
Gravity hammer, 482-3
Gravity methods of ore separation, 107-12
Grease, removal of, from metal, 677-80
Green sand moulds, 366-9
 compaction of, 369-73
Green sand moulding, procedure for, 368-9
Green strength, 366, 749, 753-4
Greenockite, 307
Gregor, Rev. William, 310
Grinding, 570-6
 ball-less, 105, 139
 history of, 523
 of iron ore, 139-40
 of ores, 103-5
Grinding machines, 572-3
Grinding wheels, 571-2
 for sawing, 570
Grooves in butt joints, 642-3
Growth of grains, 405
Growth of particles in dispersion-hardened
 systems, 865
Guinier-Preston zones, 442
Gun drill, 549-50
Gunmetals, corrosion of, 835
Glycerine, as lubricant, 777

Hack saws, 568-9
Hadfield, Sir Robert, 202
Haematite, 135
Halides, decomposition of, 738
Hall, C. M., 244
Hall-Hercult reduction process, 245
Hammers, 481-4
Hammer welding, 598
Hand lay-up, 871
Hanging wall, 93
Hard metals, 774-80
 compacting of, 777-8
 composite powders in, 737
 literature on, 780-1
 properties of, 778-9
 proprietary, 775
 sintering of, 777-8
 uses of, 779-80

Hard soldering, 638
Hardenability, 27, 221-31
  improvement of, 226
  measurement of, 221
  mechanism of, 224
Hardening, 32
  profile of, 225-6
  secondary, of steel, 230
  surface, of steel, 31, 231-2
Hardness, 50
  of cermets, 868
  in dispersion- hardened systems, 863
  of steel, 201, 212-3, 219-32
  testing of, 50-3
Hardness values, comparative, 53
Harris process, 263
Hastelloy, 445, 836
Hatchett, C., 321
Heat resistance of glass-fibre reinforced plastics, 873
Heat-resistant applications, metals for, 767-8
Heat treatment, 19-37
  of cast alloys, 441-3
  post-weld, 605, 610, 632
  principles of, 21-6
  of steels, 209-19
  of zirconium, 318
Helical gears, 562-3
Helium in inert gas welding, 610-1
Helix angle of drills, 547
Heroult, P. L. T., 173, 244
High-frequency resistance welding, 597
High-pressure moulding, 373
High-rate forming, 501-4
High temperature, alloys for use in, 272, 430
High-temperature oxidation, 808-11
High-tension separator, 114-5
High-voltage discharge forming, 502-3
H-irn process, 150
Hjelm, P. J., 329
Hobbing, 565-6
Höganäs process, 150, 747
Hoists, in mining, 96-7
Homogenization, 33
Honeycomb structures, 884-5
Honing, 570-1, 575-6
Hooker process, 478-9
Horizontal broaching machine, 559-60
Horizontal spindle-type surface grinder, 574
Hot-box process, 881, 400
Hot-chamber die-casting process, 392-4
Hot corrosion, 808
Hot dip aluminizing, 707-8
Hot dip galvanizing, 305
Hot dipping, 260, 670, 671, 704-8
Hot pressing for cermet manufacture, 866-7
Hot-rolling of metal fibre reinforced metals, 878
Hot-spot corrosion, 836
Hot strength, 366

Hot swaging, 764
Hot tearing, 409
Hot tinning, 704-6
Hot working, 196, 458, 462
Humphreys spiral concentrator, 112, 140-1
Hydraulic bond in mould material, 364-5
Hydraulic classifier, 107
Hydraulic press, 484-5, 486
Hydro carbon fluids in core sands, 403
Hydro carbon gases, 761-2
Hydrocyclone classifier, 107-8, 112
Hydrofluoric acid, in plating solution, 692
Hydrogen,
  as atmosphere for sintering, 761
  embrittlement by, 674, 740, 743
  as a metal, 672, 681
  as reductant, 732-3, 734-5
  reduction of, at cathode, 790
  use in liquid phase reactions, 736
Hydrogen sulphide as corrodant, 812
Hydrogoethite, 135
Hydrometallurgical methods of treating ores, 117-8
Hydrophilic, 116
Hydrophobic, 116
Hydrostatic extrusion, 504-5
HyL process, 150
Hypereutectoid, 201
Hypo-eutectoid, 201

Ilmenite, 135, 311
Image Analysing Computer, Quantimet, 68
Impact extrusion, 478-9
Impact properties, temperature dependence of, 55-6
Impact resistance, 54, 872
Impact testing, 53-5
Impingement attack, 805-6
Impressed current systems for cathodic protection, 842, 844
Inclusions in steel, 221
Inconel alloy, 445, 836, 839-40
Indentation hardness, 51-3
Indexing, 554
Indium, 252-5, 701
Induction brazing, 637-8
Induction electric furnace, 177, 422-4, 426, 428-9
Induction welding, 585
Infiltration, 762-3
  for manufacturing composite materials, 867, 877, 879
In-gates, 414-5
Ingot moulds, 182
Inhibiting additions against dezincification, 804-5
Inoculation effects in casting, 409
Inorganic adhesives, 664

904 | INDEX

Inorganic binders, 402, 403
Instrumentation in steel-making, 190-1
Insulators, addition to riser, 419-20
Internal gears, 564, 565
Internal grinder, 573
International Iron and Steel Institute, 131
International Nickel Company, 274, 276
Interferometry for metallographic examination, 67
Intersecting shaft gears, 563
Investment casting process, 381-7
   materials for, 383-4, 385-7
   principles in, 381-4
   sequence of operations in, 382-3
Investment moulds, comparison with sand moulds, 385
Involute, 561
Ionic bond, 5, 12, 13
Iridium, see Platinum group metals
Iridium-192, as source in radiography, 77
Iron, see also Cast iron and Pig iron,
   behaviour in nitric acid, 799
   coating of, 684, 707-8
   corrosion of, 790-1, 812
   corrosion inhibitors for, 846
   corrugated, 498
   effect of coining on, 763
   and electrochemical series, 673
   electroplating of, 702-3
   enamelling of, 718
   forms of, 192
   galvanized sheet, joining of, 633
   galvanizing of, 706-7
   history of, 82
   infiltration of, 763
   passivation of, 799-800
   production of, by sintering, 760
   protection from rust, 699
   pure, 191
   slip systems in, 17
   solid, production of, 149-51
   with tungsten carbide, 775
   use in copper powder production, 736
   wrought, 191
Iron and steel, Ch. 3
   history of, 123-31
   industry, future development of, 188-9
   materials needed by, 136
   literature on, 232
   phase diagram of, 191-3
   statistics for, 123-4
Iron alloys, see also Steel,
   manufacture of small components, 730
Iron Age, 81, 82, 123-4, 523
Iron-carbon system, 13-4, 25-6
Iron ore, 131-51
   classification of, 137-8
   composition and mining of, 134-8
   concentration of, 140-2

Kiruna mine, 96, 97
   preparation of, 139-44
   separation of minerals in, 113-4
   sintering and pelletizing of, 142-4
   statistics of production, 132-3
   supplies, 131-4
   terminology for, 136-7
   transport of, 132-4
Iron oxide, reduction of, 732-3, 734-5
Iron-oxide boil, 160
Iron powder,
   bulk production of, 747
   from carbonyl decomposition, 739, 747
   characteristics of, 747-8
   production of, 734-5, 737
   shape of particles, 770
ISF process, 302-3
Isoprene molecule, 663
Isotopes, radioactive, use of in non-destructive testing, 77
Isothermal treatments of steel, 28-31
Izod impact testing method, 54-5

Jackson, A., 162
Jewellery, use of gold for, 299
Jig borers, 542-4
Jig for ore separation, 111
Jigging, 140
Joggled lap joint, 650
Joining of metals, Ch. 8
   adhesive, 644-64
   classification of methods of, 581-6
   fibre reinforced systems, 881
   literature on, 665-7
   mechanical, 581-2, 586-9
   metallurgical, 582-5
Joints,
   design of, 648-53
   metallurgical, design of, 641-4
   stresses in, 648-9
Jolt machines, for cores, 397
Jolting machines, 371-2
Jolt-squeeze machines, 372
Junghaus, S., 186

Kaldo converter process, 169-71
Katanga-Shituru process, 270
Killed spirits, 639
Killed steel, 185
Kiruna iron ore mine, 96
Klaproth, M. H., 310, 316, 327
Knee-type milling machines, 553-4
Kroll process, 312, 738
Kroll, W. J., 311, 316
Krupp-Renn process, 149

Laboratory trials for corrosion control, 848-9
Lacquers, 719-26
　cellulose butyrate, 722
　definition of, 720
　methods of application of, 722-5
　nature of, 720-2
　nitrocellulose, 722
　phosphate base for, 712-3
　solvents of, 721
Laminar flow, 410, 744-5
Laminates, 883-4
　metal, 885-6
Lamination, 752-3
Lamp filaments, tungsten for, 336, 748, 764
Langer, Dr. C., 274
Lap, 575
Lap joints, 641, 649-51
Lapping, 570-1, 575-6
Lapping compounds, 575
Larrabee, C. C., 820
Lateritic ores, 275-6
Lathes, 541-6
　accuracy of, 542
　attachments for, 542
　comparison with drilling machines, 546
Lattice,
　cubic, 14-5
　hexagonal, 14-5
　space, 14
Lattice directions, 15-7
Law of continuity, 411-2
Layer dezincification, 804
LD-AC process, 167-9
LD process, 165-9
　comparison with Kaldo process, 170-1
Leaching, 118, 294
Lead, 261-5
　corrosion of, 836-7
　creep in, 56
　electroplating of, 683, 701
　history and occurrence of, 261-2
　ores of, 262
　pretreatment of, for adhesive bonding, 647
　and production of terne plate, 706
　properties of, 263
　smelting and refining of, 262-3
　statistics for, 265
　uses of, 264, 474, 697
Lead alloys, 264-5, 445-6, 836-7
Lead shor production, 745
Leafing, of aluminium paint, 726
Lewis, W., 280
Liberation of ores, 99-102
Light microscopy for metallographic
　examination, 64-5, 66
Lignine in core sands, 403
Lime boil, 160
Limit,
　endurance, 60
　fatigue, 818
　of resolution, 68
Limiting drawing ratio, 492
Limonite, 135
Lip relief angle, 548
Liquation for treatment of antimony ore, 266
Liquid penetrant test for crack detection, 73-4
Liquid phase reactions for metal powder
　production, 736-7
Lithium stearate, for lubrication, 751
Little, A. D., 150
Liver of sulphur, 711
Livingstonite, 308
Lloyd, R. L., 143
Load-extension diagrams, 43-6, 48
Loading methods, 452
Loam moulding, 376-7
Locating points in patterns, 355
Lockalloy, 237
Locomotives in mining, 95-6
Loctite adhesives, 654
Logarithmic strain, see True strain,
Looping mill, 467
Loose patterns, 347-8
Lorandite, 253
'Lost-wax' process, 381
Lubricants,
　for ball milling, 777
　for filling metal die, 750-1
　for metal powders, 751-2
　molybdenum in, 333
　in pre-polishing metals, 675
Lubrication, 492, 712
Lüders lines, 505

Machinability, 535-6
　of cast iron, 208
　of steel, 220-1
Machine movements, control of, 577
Machine tool bits, 779
Machine tools,
　development of, 524
　numerically controlled, 576-9
Machining allowance, 352, 354
Magmas, 83
Magnesite, 240
Magnesium, 240-3, 426
　coatings for, 838
　corrosion of, 837-8
　extrusion of, 475
　history and occurrence of, 240
　pretreatment of, for adhesive bonding, 647
　production of, 240-1
　properties of, 241, 242
　statistics for, 243
　uses of, 242-3
　welding of, 617-8
Magnesium alloys, 241-2

906 | INDEX

Magnesium alloys *(cont.)*
   cast, properties of, 443-4
   corrosion of, 815, 838
   microstructure of, 435
Magnetic cobbing, 141
Magnetic cores, and powder metallurgy, 770
Magnetic dust test for crack detection, 73-4
Magnetic materials, 272, 770
Magnetic prospecting for minerals, 85
Magnetic response of austenitic stainless steels, 606
Magnetic separation of ores, 112-4, 141
Magnetic separators, 113-4
Magnetite, 135
Magnets, permanent, steels for, 208
Malachite, 285
Malleability, 452
Malleablizing, 35-6
Mandrel bar drawing, 64-5
Manganese, 200, 337-9
Manganese bronzes, 443, 833
Manganese phosphate in phosphating, 712
Manganin, 339
Mannesmann piercing, 513-5
Mannesmann process, 735
Marquenching, 219
Martempering, 29-30, 219
Martensite, 27, 196
Master pattern, 350
   allowances for, 352
Match, 350
Match-plate patterns, 348-9
Materials, see also Composite materials,
   for brazing, 633-4
   properties of, 854-5
   for soldering, 639
Matthey, G., 283
MBV treatment for aluminium, 713
Mechanical handling of molten steel, 181-2
Mechanical joining methods, 581-2, 586-9
Mechanical joints, for beryllium, 238
Mechanical press, 484
Mechanical properties, reasons for determination of, 452-3
Mechanical testing, 40-63, 447
Media, for quenching steel, 215
Melting, 420-9
   arc, of titanium, 312
   batch processes of, 422-3
   continuous processes, 423
   effect of, in dispersion hardened systems, 864
   for making castings, 345
   vacuum arc, of niobium, 323
Mercury, 308-10
Metal arc welding, 601-7
   applications of, 605
   equipment for, 604-5
   process of, 601-4
   properties of joints, 605-7

Metal finishing, 670, 675-6
Metal inert gas welding, 610-5
   applications of, 613-4
   equipment for, 613
   process of, 610-2
   properties of joints, 614-5
Metal ion reduction at cathode, 790
Metal laminates, 885-6
Metal powders, 730-73
   characteristics of, 745-9
   flow of, 760
   literature on, 780-1
   manufacture of shaped products from, 749-64
   preparation of mixtures of, 743-4
   production of,
      chemical methods for, 731-41
      choice of method for, 731
      mechanical methods for, 741-5
   products, applications of, 764-73
Metal spraying, 708-9
Metal/wood laminates, 886
Metallic bonding, 5, 13
Metallic materials, properties of, 854
Metalliferous mining, see Ore mining,
Metalliferous ore, 97, 98
Metallographic examination, 63-72, 197
Metallurgical joining methods, 582-5, 590-1
Metallurgical joints, design of, 641-4
Metallurgical surface bonding, 585, 632-41
Metallurgy, powder, see Powder metallurgy,
Metals,
   basic costs of, 37-8
   cast in foundry, 429-46
   cast and wrought, comparison of, 457
   casting of, Ch. 5
   cold working of, 459-61
   continuous-fibre reinforced, 880-1
   corrosion of, Ch. 11
   cutting of, Ch. 7, 774
   deformation of, Ch. 6, 16-9, 742
   disintegration of, to powder, 741-45
   economic aspects of, 37-40
   examination of surface relief features, 65
   general, Ch. 1
   hard, 737, 774-80
   heat treatment of, 19-37
   joining of, Ch. 8
   melting of, 345, 420-9, 766
   molten, flow of, 410-2
   non-destructive testing of, 72-9
   non-ferrous, Ch. 4, 437-9, 441-6
   physical and mechanical properties of, 5-11
   reinforced with metal wire, 877-8
   solidification of, 404-6
   structure of, 1-19, 21-2
   surface treatment of, Ch. 9
   work-hardening capacity of, 53
Metastability, 25
Methane, liquid, storage of, 589

Methyl methacrylate, see Perspex,
Meyer analysis, 53
Microanalysis, electron probe, 70
Microhardness measurement, 52, 68
Microporosity, 408
Microscope, metallurgical, 64
Microscopic techniques, for metallographic examination, 65-8
Microscopy, electron, 68-9
  field ion, 70-1
  high-temperature, 67-8
  light, 64-5, 66
Microstructure, 21-2, 70-2
  of casting alloys, 433-5
  of steel, 195, 201
M.i.g. welding, see Metal inert gas welding,
Mills,
  coggling, 183
  grinding and crushing, 103-5, 742-3
Miller index, 15-6
Miller process, 298
Millers, 558
Milling, 552-5
  speeds and feeds for, 554-5
Milling cutters for forming, 564
Milling machines, 525, 553-4
Mineral, definition of, 97
Mineral deposits, sampling of, 88-9
Mineral engineering, 97
Mineral oils as cutting fluids, 538
Mineral processing, 97
Mineral technology, 97
Mineral treatment, major stages of, 101
Mineralogy, 98-9
Minerals,
  common iron, 135
  geochemical prospecting for, 84-5
  geophysical prospecting for, 85-6
  magnetic characteristics of, 11, 2-3
  properties and separation techniques for, 99-100
Mines,
  alluvial and open-cast, 91-2
  chief copper, 286
  early, 83
  molybdenum, 330
Mining, see also Stoping and Ore mining, 89-90
  early, 83
  haulage in, 94-7
  statistics for, 90
  supports in, 94
Mirrors, use of silver in, 296
Missiles, molybdenum in, 333
Modified glass-fibre reinforced plastics, 875
Modified silicide coatings, 883
Modulus of elasticity, Young's, 456
Moebius process, 295
Mohr scale of hardness, 50-1
Moissan, H., 330

Molasses, in moulding sand, 368
Molybdenite, 329
Molybdenum, 329-35
  biological aspects of, 334-5
  casting of, 331-2
  compounds of, 334
  history of, 329-30
  infiltration and, 763
  manufacture by powder method, 730
  production of, 330-2
  properties of, 332
  uses of, 333-4, 761
  welding of, 630
Molybdenum carbide, and hard metals, 774
Molybdenum powder, 331, 741
  advantages of, 769
  uses of, 767-8
Molybdic oxide, 330, 331
Mond, L., 274
Monel metals, 444-5, 836, 839
Motor-car industry, 204, 587, 724-6
Mould blowers, 372-3
Mouldability, 366
Moulding, 345, 356-8
  ceramic, 388
  choice of method of, 357-9
  classification of, 357
  compo and chamotte, 375-6
  floor and pit, 374
  high-pressure, 373
  investment, 381-7
  limitations on method, 360-1
  loam, 376-7
  precision, 381-91
  sand, 359-81
  shell, 380-1
Moulding machines, 370-3
Moulding mixtures for investment moulds, 383-4
Moulding sands,
  bonds in, 364-6
  flow diagram for use of, 368-9
  properties and ingredients of, 366-8
Moulds,
  carbon dioxide, 378
  cement-bonded, 374
  ceramic-shell, 387
  comparison with cores, 396
  comparison of sand and investment, 385
  compo and chamotte, 374-6
  dry sand, 373-7
  freezing in, 411
  graphite, 390
  green sand, 366-9
  materials for, 362-6
  permanent, 390-1
  plaster, 388-9
  self-curing, 377
  for sintering, 760

Multiple spot welding, 594
Multispindle drills, 549-50
Mushet, R. F., 156, 202
Muntz Metal, 833

Nails, production of, 491
Nasmyth, J., 129
Natural moulding sand, 367
Naval Brass, 292, 833
Necking, 456
Neoprene rubber, 660, 663, 845
Neoprene rubber cements, 663
Nernst equation, 794
Neutral, flame, oxyacetylene, 623, 625
New Jersey process for zinc extraction, 302
Newton's law of motion, 412
Nichrome, 836, 878
Nickel, 274-80
  corrosion of, 839
  'electroless' plating of, 709-10
  electroplating of, 683
  history of, 274
  as a metal coating, 684
  mining and smelting of, 275-7
  ore bodies, 274-5
  pretreatment of, for adhesive bonding, 647
  properties of, 279
  refining of, 277-8
  statistics for, 280
  with tungsten carbide, 775
  uses of, 278-9
  welding of, 605, 607
Nickel alloys, 279
  for brazing, 634
  cast, properties of, 444-5
  corrosion of, 836, 839-40
  creep in, 56
  manufacture of small components, 730
  welding of, 605, 607
Nickel chloride for nickel plating solutions, 684, 689
Nickel plating, 684-90
  plant, 685, 687, 693, 694
  solution, 689
  tank, 686
  thickness of, 686-7
  uses of, 690
Nickel powder,
  characteristics of, 748
  production of, 737, 739, 740, 748
Nickel-silver, 475, 697, 703
Nickel sulphamate, for nickel plating solutions, 689
Nickel sulphate for nickel plating solutions, 684, 689
Nickel tetracarbonyl, for nickel powder production, 739
Nickel-titanium carbide mixtures, 778

Nimonic alloys, 207, 836
Niobium, 321-6
  alloys of, 324-5
  consolidation of, 322-4
  fabrication and coatings for, 325
  history and occurrence of, 321-2
  joining of, 325
  manufacture of, by powder method, 730
  mining and extraction of, 322
  powder, 741, 767-8
  properties of, 324
  statistics for, 326
  uses of, 325
Nitriding, 31-2, 232
Nitrile rubber, 663
Noble metals, 673
Nominal strain, 450
Nominal stress, 452
Non-coring diamond drilling, 87
Non-destructive testing, 72-80
  of castings, 447-8
  electromagnetic, 78-9
  radioactive isotopes in, 77
  strain gauges for, 78-9
Non-ferrous alloys,
  cast, properties of, 437-9, 441-6
  fatigue in, 60-2
  hardening of, 32
  machining of, 536
Non-ferrous metals, Ch. 4
  melting of, 426-7
  properties of, 235
Non-flux arc welding processes, 585
Non-intersecting shaft gears, 563
Non-metallurgical adhesive methods of joining, 585-6
Non-metals, comparison with metals, 1-12
Normal thrust force, 529
Normalizing, 35, 216, 627-8
Nose radius, 540
Notch bend test, 54
Botch-brittleness, 54
Nuclear industry, molybdenum in, 334
Nucleating agent, 404-5
Nucleation, 404-5
Nucleation-growth process, 24-5
Numerical control,
  advantages of, 576-7
  essential features of, 577
  types of system, 577-9
Numerically controlled machine tools, 576-9
Nuts, self-locking, 587

Oblique cutting, 527-8
Oersted, H. C., 243-4
Oil,
  for cutting fluid, 538
  for infiltration of bearings, 771

Oil *(cont.)*
  in production of tinplate, 705
  removal of, from metal, 677-80
  self-curing drying, 377
Oleoresin, phenolic, 722-3
Olivine, as mould material, 365
Open-cast mines, 91-2
Open-cast mining, 90, 95
Open-hearth furnace,
  acid, 161-2
  basic, 158-61
  chemical action of, 161
  lining of, 157
  raw materials for, 158-9
Open-hearth process of steel-making, 155-6
Open stoping, 93
Ore, 97, 98
  diluted, 89
Ore bodies, nickel, 274-5
Ore boil, 160
Ore/corrosion product cycle, 783
Ore deposits, underground mining of, 92
Ore dressing, 97
Ore grades, estimation of, 88-9
Ore mining, 81-97
  comparison with coal mining, 89-90
  history of, 81-2
  literature on, 97
  underground, 92
Ores, see also Iron ore, Ch. 2
  crushing of, 101-3
  grinding of, 103-5
  liberation of, 99-102
  literature for, 121
  processing of, 97-121
    control of, 120
    disposal of product and tailing, 119-20
    hydrometallurgical methods of, 117-8
    sampling in, 120
  separation of, 105-117
    electrical, 114-5
    by flotation, 115-7
    by gravity methods, 107-12
    magnetic, 112-4
  typical, treatment of, 99
Organic binders for cores, 402-3
Organic bonds in mould material, 365-6
Organosols, 727
Orowan, E., 19
Orthogonal cutting, 528
Osmiridium, 284
Osmium, see Platinum group metals,
Over-aging, 32
Overburden, 90
Overvoltage, see Polarization,
Oxalic acid, for anodizing, 714-5
Oxidation, equations of, in corrosion, 808-11
Oxidation-resistant coatings, 860, 882-3
Oxidizing flame, oxyacetylene, 623

Oxyacetylene blowpipe for spraying paint, 708-9
Oxygen-absorption reaction at cathode, 790
Oxygen reduction at cathode, 790
Oxygen, use in steel-making, 153, 162

Packaging, use of aluminium in, 251
Packing of spherical particles, 749-51
Paint, 720
  aluminium, 726
  bitumastic, 722
  coatings of, 719-26
  methods of application of, 722-5
  nature of, 720-2
  phosphate base for, 712-3
  pretreatment of surface for, 725
  thickness of, 721-2
  use of, for corrosion control, 849-50
  zinc as base for, 841
  zinc-rich, 305, 726
Palladium, see also Platinum group metals, 701
Palladium chloride, for pretreatment of steel, 710
Paraffin wax, for lubrication, 752, 777
Parallel shaft gears, 562-3
Parkes process, 263, 295
Particulate composites, 855-7, 862-9
Parting line, 348, 355
Parting off, 541
Passivation, 799-800
Passivators, 846
Passivity, 799-800
Paste, aluminium, 249
Patenting of steel, 216-7
Pattern draft, 352
Pattern making, 345, 354-5
Patterns, 346-56
  allowances in, 351-4
  types of, 346-51
  wax, 382
Pattinson's process, 295
Peacock ore, 285
Pearlite, 34, 195, 197-8
Pearlitic steel, 35
Peeling stress, 649
Pegmatite, 253
Pelletizing, 117, 142-4
Pellets, iron ore, performance of, 144
Pendulum rolling mill, 466-7
Penetratmeter, 77
Pentlandite, 276
Perchlorethylene, as grease solvent, 677
Percussing welding, 596-7
Percussive drilling, 86-7
Periodic table of elements, 1-3
Perlite, in moulding sand, 368
Permanent moulds, 390-1
Permeability, 366

Perspex, 725
Petroforge, 504
Pewter, 259
pH value, 678, 685-6, 790
Phase change, 25, 351
Phase contrast technique for metallographic examination, 67
Phase diagram of iron and steel, 191-3
Phenol formaldehyde, in core-making, 403
Phenolic resin based adhesives, 659-61
Phenolic resins, 658, 721
Phosbrite process, 716, 717
Phosphating, 712-3
Phosphor bronzes, 39, 834
Phosphoric acid, 712, 714
Photographic industry, silver in, 296
Pickling, 673-4
Pickling inhibitors, 674
Piercers, 513-7
Pig iron,
  composition of, 208
  elements present in, 148
  history of, 125
  manufacture of, 131-51
Pigments,
  in anti-corrosion paints, 849
  causing corrosion, 850
Pilger process for tube, 517-8, 520
Piling, 51-2
Pillar drills, 548
Pinion cutter, 566
Pipe, in casting, 449
Pipe butt welds, 643-4
Pipeline welding, 617, 621
Pit moulding, 374, 375-6
Pitch, in moulding sand, 368
Pitch circle, 561
Pitch surfaces, 561
Pitting corrosion, 801-2
Pitting, in cavitation damage, 807
  in impingement attack, 805-6
Plain turning, 540
Planar composites, 860, 882-6
Planar-type milling machines, 554-5
Planers, 558
Planetary grinder, 573
Planetary rolling mill, 465-6
Planing, 556-8
  gears, 564
  machines for, 557-8
  speeds and feeds for, 558
Plaster moulds, 388-9
Plaster of paris,
  hydration of, 365
  in investment moulding, 387
  for moulds, 388-9
Plastic-coated metals, 883
Plastic coatings, 727
Plastic deformation, 455-6

and ceramics, 854
mechanism of, 58-60, 857
Plastic forming of cermets, 867
Plastic strain, 450
Plasticizers, 721
Plastics, properties of, 855
Plastisols, 727
Plate, aluminium, 249
  joints between edges of, 642-3
  production of, 505-6
Platinum, see also Platinum group metals,
  electroplating of, 701
  powder, 769
Platinum group metals, 280-4
  alloys and available forms, 282-3
  extraction and refining of, 281-2
  history and occurrence of, 280-1
  properties of, 282
  statistics for, 284
  uses of, 283-4
Plug dezincification, 804
Plug drawing, 519
Plug-rolling process for tube, 514, 518
Plunge grinding, 572
Plutonium powder, 741, 768
Pneumatic separators for ores, 112
Point angle of drills, 547
Poisson's ratio, 450
Polanyi, M., 19
Polarization,
  and corrosion rates, 791-3
  influence on corrosion currents, 798-9
Polarized light for metallographic examination, 67
Pole lathe, primitive, 523-4
Polishing machine, mechanical, 675, 676
Polyacrilonitrile, 875
Polyamide thermoplastic resins, 661
Polyamides, 658, 661
Polyester resins, 869, 871
Polyimide resins, 664, 873
Polymerization, 720-1
  condensation, 365
  of epoxy resins, 657
  of phenolic based resins, 660
  of polyurethane, 661
Polymers, 855
Polystyrene, expanded, 377
Polytetrafluorethylene, see PTFE,
Polyurethane, 661-2, 885
Polyvinyl acetate, 662-3
Polyvinyl chloride, 260
Pores in dispersion-hardened systems, 865
Porosity, 362, 457, 770-1
Port Colborne process, 270-1
Positioning control systems, 578
Potassium cyanide, 695-6
Potassium stannate, 700
Potential/pH equilibrium diagrams, 792-3

INDEX | 911

Potentiostat, 800
Pour density, 749-51
Pourbaix, M., 792
Pouring, of castings, 345, 404-15
Pouring basin, 413
Powder metallurgy, see also Metal powders, Ch. 10
  advantages of, 766-7
  applications of, 764-73
  applied to steel, 190
  and composite materials, 769
  history of, 729
  limitations of, 730
  literature on, 780-1
  main uses for, 729-30
  for reducing cost, 771-3
  statistics of, 773
Powder rolling, 772
Power source for welding, 603, 608, 613
  carbon dioxide welding, 616
  tungsten inert gas welding, 618-20
Precipitation, 24
Precipitation hardening, 32, 219, 857
Precision moulding processes, 381-91
Pre-impregnation methods for manufacture of glass-fibre reinforced plastic, 871-2
Pre-polishing of metal, 675-6
Presses, 475, 484-5, 486
Press bending, 497-8
Press-brake bending, 497
Pressing, 493-5, 749-54
  comparison with spinning, 495-6
  and green strength, 753-4
  isostatic, 756, 757
  lubricants for, 750-2
  principles of, 749-50
  technique of, 752-3
Pressure, welding methods involving, 583-4, 589-601
Pressure angle of gear, 561
Pressure die casting, see Die casting
Pressure methods for manufacture of glass-fibre reinforced plastic, 871
Pressure vessel manufacture, 632, 643
Pressure welding, 599-600
Pretreatment, surface, for adhesive bonding, 646-8
Printed circuits, copper in, 291
Printing, 266, 446
Process annealing, 33, 216
Process control for ore treatment, 120
Processing of metalliferous ores, 97-121
Producibility, 358-9
Production, mass
  development of techniques, 526
  and powder metallurgy, 772
Production milling machines, 553
Projection welding, 595-6
Proof stress, 47

Properties, mechanical,
  relation between microstructure and, 70-2
  utilization of, 40
Prospecting, 83-6
Prospecting drill, 84
Protection, sacrificial, 673
PTFE,
  for infiltration, 771
  lacquers, 727
Puckering, 612
Pulsed arc welding, 612
Purofer process, 150
Push-bench process for tube formation, 65
PVC, 260
Pyrites, 135
Pyrochlore, 321-2
Pyron process, 747
Pyrrhotite, 276

Quality control,
  of castings, 453
  hardness testing for, 50
  mechanical testing for, 40-1
Quantimet Image Analysing Computer, 68
Quartz, for ultrasonic testing, 74
Quartz sand, volume change in, 363-4
Quenchant, choice of, 36-7
Quenching, 25, 26-8, 215-6

Rack-type cutter, 566, 567
Radial drills, 548-9
Radioactive isotopes, use of in non-destructive testing, 77
Radioactivity method of prospecting, 85
Radiographic sensitivity, 77
Radiography, 75-7
Rail section, rolling of, 508, 510
Railway track, welding of, 593, 626-7
Rake angles, 528, 539-40
Rake face, 528
Ram-type bending machine, 499-500
Ramet, 775
Rates of corrosion, 791-3, 796-9
Raw materials for casting, 421
Rayon, 875
Reaction mechanism of corrosion, 786-801
Reamers, 548
Reaming, 546-52
  accuracy of, 551
  speeds and feeds for, 551-2
Recrystallization, 33-4, 461
Recrystallization temperature, 462, 463
Rectification, 153
Rectifier systems for spot welding, 594
Redox systems, 813
Redrawing operations, 494-5
Reducing flame, oxyacetylene, 625

Reductants for metallic oxides, 732
Reduction plant for aluminium, 246-7
Reduction methods for metal production, 731-9
Redux bonded aluminium alloy, 649-50
Redux process, 661
Reference electrode, 843
Refining,
  history of, 126
  secondary, of steel, 177-81
Refractories, 152
Refractoriness, 366
Refractory metals,
  coatings for, 860
  as wires for reinforcing metals, 878
Refractory permanent moulds, 391
Relief angles, 540
Remelt ingot, aluminium, 248
Replica technique for electron microscope, 68-9
Resin bonding of beryllium, 238
Resin systems, 725
Resins, 720-1
  acrylic, 725
  for shell moulding, 380
Resistance brazing, 634
Resistance sintering of cermets, 867
Resistance welding, 583, 589-97
Resolution, limit of, 68
Retrograde solid solubility, 441
Reusability of moulding sand, 367
Reverse flush drilling, 88
Reversible concentration cell, 795
Reynolds number 744
Rhenium powder, 768
Rhodium, see also Platinum group metals
  electroplating of, 701
Rhokana process, 270
Rickard, T. A., 81
Rigging, 346
Rimmed steel, 184
Rimming, 184
Riser, 416, 418
Riser neck, 418-9
Riser shape, 417-8
Riser size, 418
Riveting, 581, 588
RN process, 149
Roberts-Austen, Sir William, 131
Rockwell B scale hardness, 53
Rockwell hardness number, 53
Rockwell hardness test, 51, 52-3
Rod,
  aluminium, 249, 470
  drawing of, 492-3
  rolling of, 510
Rod mill, 103-4, 742, 743
Roll forming, 498-500
Roll welding, 598
Rolled gold, 300

Rolled zinc, 305
Rolling, 462-72
  of billet, 508-10, 512
  of bloom, 508, 511
  of foil, 471
  of rod and bar, 470, 510
  of section, 470, 508-10
  of sheet and strip, 469, 9-70
  temper, 505
Rolling mill,
  continuous tandem, 466
  layout of, 466-8
  looping, 467
  pendulum, 466-7
  planetary, 465-6
  principle types of, 464
  Sendzimir, 465
Rolls,
  design of, 467-71
  manufacture of, 471-2
  principal parts of, 467-9
Root mean square deviation, 532
Rose, Heinrich, 321
Rossi, Irving, 186
Rotary bending machine, 500-1
Rotary cutter for gear shaving, 567
Rotary drilling, 87-8
Rotary forging, 517-8, 520
Rotating hammer mill, 742
Rotor oxygen converter, 171-2
Roughing, 467
Roving, 869
Rubber based adhesives, 663-4
Runner, 414
Running and feeding systems, 354, 409-10
Rustenburg mines, 281
Rusting,
  in the atmosphere, 822
  rate of in mild steel, 820, 822
  in waters, 820-1
Ruthenium, see Platinum group metals,
Rutile, 311

Sacrificial anodes for cathodic protection, 844
Sacrificial protection, 242-3, 841
SAE 1045, 226
SAE 3140, 226
Salt bath brazing, 635-6
Salt-spray test, 704
Sampling, 89, 120
Sand mixtures for D process, 379
Sand moulding, 359-81
Sand moulds, comparision with investment moulds, 385
Sand slingers for cores, 398
Sand-buffing, 675
S.A.P., see Sintered aluminium powder,
Saponification, 678

Sawing, 568-70
Scale, on unalloyed cobalt, 271-2
Scarf joint, 650-1
Scheele, K. W., 329, 335, 337
Scheelite, 335
Scleroscope, Shore, 53
Scratch hardness, 50-1
Screening, 106, 142
Screw cutting, 541
Screw dislocation, 19-20
Scurfing, 675
Seam welding, 594-5
Seamless tube, 513
Season cracking, 513
Sea-water, effect on aluminium alloys, 831
Secondary hardening of steel, 230
Section,
    bending of, 499-500
    drawn, 511-3
    extruded, 510-1
    heavy, welding of, 631
    rolling of, 470, 508-10
Section sensitivity, 433
Segregation in steel, 185
Seismic prospecting for minerals, 85
Selective corrosion, 803-4
Selenium, use in xeroradiography, 77
Self-adjusting arc welding, 613
Self-curing moulds, 377
Self-locking nuts, 587
Sendzimir-type rolling mill, 465
Sensitive drills, 549
Sensitivity,
    radiographic, 77
    section, 433
Separation,
    of ores, 105-17, 140-1
    techniques, 99-100
Separators, 112-5
Sequential control system, 577-8
Series spot welding, 594
Shape factor of casting, 418
Shape of tool, and cutting force, 531
Shapers, 558
Shaping, 556-8
    of gears, 564-5
    machines for, 556-7
    speeds and feeds for, 558
Shaw process, 389-90
Shear angle, 528, 530
Shear force, 529
Shear plane, 528
Shear stress, 450, 452
    critical, 49
    for slip, 18-9
Sheet,
    deep drawing and pressing of, 493-5
    joints between edges of, 642-3
    production of, 505, 506

    rolling of, 469-70
Sheffield Compo, 374-5
Shell core machines, 397
Shell core-making, 400
Shell formation for tube, 513-6
Shell moulding, 380-1
Sherardizing, 305, 710
Shielding in arc welding, 604
Shipbuilding, welding in, 609
Shore scleroscope, 53
Shot-blasting, 674
Shot-peening, 231
Shrink rule, 352
Shrinkage allowance, 351-2, 353
Shrinkage stoping, 93
Siemens-Martin process, 155-6
Siemens, Sir William, 130-1
Silica, 135, 363-4
Silica fibres, 880
Silica flour in moulding sand, 368
Silicate adhesives, 664
Silicates, as mould materials, 363, 365
Silicide coatings for metals, 882-3
Silicon, in heat-resisting and non-stick coatings, 726-7
Silicon bronzes, 443
Silicon carbide, 571, 761
Silicon iron, and powder method of manufacture, 769
Silicone adhesives, 664
Silver, 293-7
    alloys and available forms, 296, 634
    in composites, 879
    extraction and refining, 294-5
    gilding of, 709
    history and occurrence of, 293-4
    properties of, 295-6
    statistics for, 297
    uses of, 296-7
Silver plating, 697-8
Simple lap joint, 650
Sinking, 51-2
Single point forming, 564
Sintered aluminium powder, 746, 831, 862
    for cutting tools, 537
    ductility, 865
Sintering, 117, 756-62
    definition of, 749
    by direct resistance heating, 767
    of galena, 262
    of iron ores, 142-3
    loose, 759-60
    of pressed compacts, 760-1
    principles of, 756-9
    vacuum, of niobium, 322-3
    of zinc ores, 302
Size of particles, classification of, 106-7
Size tolerances for castings, 352-3
Skeleton patterns, 349-50

INDEX

Skin-dried moulds, 374
Skull cutting, 476-7
Slab, 182
Slag, 147-8, 152, 604
Slingers, 372-3
Slip, 16-9, 459, 460
Slip casting, 160, 867
Slip systems, 17-8
Slotting, 556-8
Sluice, 109-110
Slurry, 383-4, 389
Smelting,
 flash, 287-8
 history of, 82, 123-4
 of iron ore, 145-6
Soaking pits, 182
Soaps, action of in degreasing, 678
Société Le Nickel, Le Havre refinery of, 275
Sodium, electroplating of, 683
Sodium carbonate, 678, 679
Sodium chloride, 5, 12
Sodium cyanide, 678, 679, 695
Sodium hypophosphite, 709-10
Sodium phosphate, 678, 679
Sodium polysulphide, 711
Sodium silicate, 678, 679
Sodium silicate process, see Carbon dioxide moulding process,
Sodium stannate, 700
Sodium zincate, 703
Soft soldering, 582, 638
Soil, use of zinc coatings in, 840-1
Soldering, 582, 638-41
Soldering iron, 638
Solders,
 corrosion of, 838-9
 pretreatment of, for adhesive bonding, 647
 soft, 264
 tin, 258
Solid solubility, 13, 441
Solid solution, 13, 23-4
Solidification of metals, 184, 404-6, 416-7
Soluble oils as cutting fluids, 538
Solubility, solid, 13, 441
Solution, solid, 13, 23-4
Solution treatment, 32
Sorbite, 217
Sorby, H. C., 64, 131
Soundness of casting, 449
Space lattice, 14, 459
Specifications, standard, for castings, 448-9
Specimens,
 of different geometry, tensile testing of, 42-5
 thickness of, for electron microscopy, 68
Speculum metal, 256, 702
Sperrylite, 281
Sphalerite, 253, 301
Spheroidization, 35, 216
Spiegeleisen, 148, 151, 337

Spinning, 495-6, 872
Split box, 355-6
Spot welding, 593-4
Spray steel-making, 189
Spray-transfer mode of metal inert gas welding, 611-2
Spraying of paint, 724-5
Spring washer, 589
Sprue, 413-4
Spur gears, 562-3
Squeeze machines, 370-1
Standard electrode potentials, 794
Stainless steels, 204-6
 austenitic, 205-6, 229, 606
 brightening of, 716-7
 chromium in, 690-1
 corrosion of, 606, 823-4
 corrosion-resisting, welding of, 609, 610
 ferritic, 205, 206
 galling of, 587
 in honeycomb structures, 884
 martensitic, 205, 206
 pretreatment of, for adhesive bonding, 647-8
 and protection, 671
 stress-corrosion cracking and, 817
 welding of, 595, 614, 621
Stainless steel powder, production of. 743, 748
Stainless W, 206
Stand, 466
Standard metre, 283
Standard reference electrode, 843
Standard of resistance, international, 309
Standard specifications for castings, 448-9
Stannous sulphate, 700
Stearic acid, for lubrication, 751, 752
Steel-making, 151-91
 chemistry of, 153
 direct, 189-90
 instrumentation in, 190-1
 in oxygen furnaces, 165-72
 spray, 189
Steel powders, 747-8
Steels, see also Iron and steel and Stainless steels, 191-209
 age-hardening of, 219
 alloy, 200, 203-4
 anaerobic corrosion of, 812
 annealing of, 216
 austenitic, fatigue in, 60-2
 carbon, 200-2
  annealing of, 34-5
  classification of, 199
  microstructure of, 21-3, 28
 carbon removal from, 189
 carburizing of, 31
 castings of, 434, 440
 classification of, 197-200
 coating of, 684, 706-10, 723
 cold extrusion of, 479

Steels *(cont.)*
corrosion of, 809, 812, 821-2
corrosion rates of, 820
corrosion resistant, machining of, 536
and creep resistance, 230
deoxidation of, 183-5
descaling of, 674
effect of elements in, 178
electroplating of, 702-3
enamelling of, 718
etching of, 680
examples of use of, 183
ferritic, fatigue in, 60-2
grain size of, 195-6
hardenability of, 221-31
hardening of, 26-32
hardness of, 219-32
heat resisting, 536, 620-1, 829
heat treatment of, 209-19
high carbon, 197-8, 201-2
high-speed, 537, 774
high-strength low-alloy, 203-4
impact properties of, 55-6
impingement attack and, 806
influence of working on, 196-7
isothermal treatment of, 28-31
low-alloy,
 corrosion of, 822-3, 828
 for high temperature service, 827-8
low carbon, 50, 200-1
for making rolls, 471-2
maraging, 207-8
medium carbon, 201
medium-carbon alloy, 204
metallurgical examination of, 197
mild, 200
 for high temperature service, 827-8
 pretreatment of, for adhesive bonding, 647
 rate of rusting, 820, 822
 welding of, 626
 working temperatures of, 463
nitriding of, 31-2
normalizing of, 216
patenting of, 216-7
and phase change recrystallization, 33-4
phosphating of, 712-3
powder metallurgy and, 190
preparation for use, 181-3
protection from rust, 699
quenching of, 26-8, 215-6
reinforced with metal fibres, 878
secondary hardening of, 230
special, 177-81, 202-8
stress-corrosion cracking and, 816-7
structure of, 192-7
tempering of, 217-9, 229-30
tool, 207, 230, 272, 536-7
types of, 151
for use at high temperature, 206-7

vitreous enamel for, 719
welding of, 594
Steel wire, tinning of, 706
Steelwork, integrated, 190
Stellites, 272, 336, 774
Stelvetite, 727
Sterling silver, 296
Stibnite, 265
Stiefel disc piercer, 513, 516
Stiffening joints, 651
Stiffness, 867, 873
Stitch welding, 593-4
Stock core machine, 399
Stone Age, 81, 455
Stoping, 92-4
Storage, 589, 873-4
Stoving, 721-2
Strain, 450, 456
 true, 46
Strain gauges, 78-9
Strategic-Udy process, 149
Stray-current corrosion, 807-8
Streamline flow, 410, 744-5
Strength,
 of cermets, 867
 of dispersion-hardened systems, 864, 865
 of materials, testing of, 41-50
 of reinforced metals, 878, 881
 of reinforced plastics, 869-70, 872-3
Stress, 46-7, 450, 452
 corrosion in presence of, 813-21
 critical shear, 18-9, 49
 definition of, 456
Stress-corrosion cracking, 814-5
 behaviour of metals to, 815-7
Stress-relieving in steel, 217
Stress reversal and fatigue, 61-3
Stress-rupture curves for dispersion-hardened
 systems, 864
Stress-rupture properties of metal-fibre
 reinforced metals, 878
Stress-rupture test, 57-9
Stress-strain diagrams, 43-6, 50, 452-3, 456
Stresses in joints, 648-9
Stressing, cyclic, 817
Stretch-forming, see Pressing,
Strip,
 deep drawing and pressing of, 493-5
 production of, 505, 506-9
 production of welded tube from, 597
 rolling of, 469-70
Strip mill, 506-7, 508-9
Structure,
 of composites, 860-1
 of dispersion-hardened system, 863
 of steel, 192-7
Stückofen, 125
Studs, attachment of, 596
Sublevel stoping, 93

Submerged arc welding, 607-10
Sudbury ore body, 274, 275, 276, 281
Sulphate reduction at cathode, 790
Sulphide ores of nickel, 275, 276-8
Sulphite binders for cores, 403
Sulphuric acid, 681-2, 691-2, 714-5
Superconducting alloys, 325
Superheating, 410
Supersaturation, 25
Surface bonding, metallurgical, 585, 632-41
Surface finish, 532-3, 872
  by grinding, 575-6
Surface hardening of steels, 231-2
Surface pretreatment for adhesive bonding, 646-8
Surface treatment, 669, 670-1
Surfacing, 541
Swage, 485
Sweep patterns, 350
Swing of lathe, 542
Syngenetic deposits, 83
Synthetic moulding sand, 367

Tafel, J., 792
Tailing disposal from mineral treatment plant, 119-120
Tailing pound, 120
Tantalite, 321-2
Tantalum, 326-7
  history and occurrence of, 321-2
  powder, 767-8
  uses of, 326, 761
  welding of, 630
Tantalum carbide, 774, 777, 778
Tap density, 750-1
Taper turning, 541
Tapered single lap joint, 651
Tapping, 546-52
  speeds for, 552
  tools for, 548
Tartar emetic, 267
Taylor, F. W., 774
Taylor, G. I., 19
Taylor's law, 533
Technologies, interdependence of, in production foundry, 347
Teeming, 182
Television tube, 629
Temper rolling, 505
Temperature,
  effect of, on adhesives, 645
  effect of on dispersion-hardened system, 863-4, 865
  limits of, for hot working, 462, 463
  of metal in casting, 421
  tempering, 218
Tempering of steel, 28-9, 217-9
Tempering water, 367

Tensile load, ultimate, 46
Tensile stress, 450, 452
  ultimate, 47
Tensile testing, 41-50
  atomic description of, 48-9
  literature on, 80
Tensile testing machine, 42
Terminology in iron ore technology, 136-7
Terne plate, 706
Test,
  fatigue, 60-3
  for crack detection, 73-4
  notch bend, 54
  stress-rupture, 57-9
  tension creep, 56-7
  ultrasonic, 73-5
Testing,
  of castings, 446-8
  creep, 55-60
  hardness, 50-3
  impact, 53-5
  mechanical, 40-63
  non-destructive, 72-80
  tensile, 41-50
Thallium, 252-5
Thermal shock resistance of cermets, 868
Thermal stability, 366
Thermit powder, 627
Thermit reaction, 338
Thermit welding, 584, 626-8
Thermoplastic resins, 721
Thermosetting phenolic resins, 660-1
Thermosetting plastics, in core-making, 403
Thermosetting resins, 721
Thickening process for products from ores, 119
Thiobacillus thio-oxidans, 812
Thiokol LP 33, 659
Thomas process, 165
Thomas, Sidney Gilchrist, 130-1, 154, 156
Thompson, J. L., 274
Thorium ores, prospecting for, 85
Thorium oxide, in tungsten production, 735
Thorium powder, 735, 736, 737, 768
Thread grinders, 575
Three-roll piercer, 513, 517
Threshold of spray transfer, 612
Throwing power,
  of anodizing processes, 715
  of electroplating solutions, 687-9, 695, 698, 699
Ticonal, 208
T.i.g. welding, see Tungsten inert gas welding
Time-temperature-transformation diagram, see TTT diagram
Tin, 256-61
  corrosion of, 838
  electroplating of, 683
  powder, 739, 743
Tin alloys, 258-9

INDEX | 917

Tin alloys *(cont.)*
  cast, properties of, 446
  corrosion of, 838-9
  electroplating of, 702
  use of, in soldering, 638
Tin fluoborate, 700
Tinplate,
  application of paint and lacquer to, 722-3
  production of, 259-60, 704-5
  uses of, 260
Tinplating, 700-1
Titanit, 775
Titanium, 310-5
  reinforced with metal fibres, 878
  welding of, 630
Titanium alloys, 313-4, 315, 446
Titanium carbide, 774, 776-7
Titanium carbide-nickel cermets, 867-8
Titanium carbide-nickel mixtures, manufacture of, 778
Titanium carbide-tungsten carbide mixtures, 778
Titanium clad mild-steel plate, 600
Titanium powder, 737, 738, 741, 768, 769
Titanomagnetite, 311
Tolerances in castings, 352-3
Tonnage steel, 130
Tool, cutting, 538-40, 548, 556
  forces at edge, 529
  life of, 532-4
  material of, and cutting force, 532
  materials for, 536-7
  shank, 540
  shape of, and cutting force, 531
Tool bits, 779
Tool steel materials, development of, 526-7
Tooth height and profile of gear, 561
Torch brazing, see Flame brazing,
Touch welding, 604
Tough composites, 860
Toughness, 54
Transformer cores, steels for, 208
Transition elements, 3
Transport, of iron ore, 132-4
Transport industries, 250-1, 874
Tributyl tin oxide, 260
Trichlorethylene, as grease solvent, 677
Tröpenas converter, 165
True strain, 450, 456
True stress, 452, 456
TTT diagram, 30, 213-4
  for alloy steels, 223
  for carbide alloys of steel, 229
Tube, 513
  drawing of, 492-3, 518-9, 521
  production of, 513-21
Tube-in-strip, roll bonding of, 598
Tubular rivet, 588
Tumbling mill, 105

Tundish, 182
Tungsten, 335-7
  fabrication of, for lamp filaments, 764
  infiltration and, 763
  manufacture by powder method, 730
  and powder metallurgy, 772
  pretreatment of, for adhesive bonding, 648
  uses of, 336, 761
  wire, production of, 772-3
Tungsten carbide,
  and hard metals, 774
  pretreatment of, for adhesive bonding, 648
  powder, production of, 775-6
Tungsten carbide-cobalt cermet system, 866
Tungsten carbide-cobalt mixtures, 777-8
Tungsten carbide-titanium carbide mixtures, 778
Tungsten inert gas welding, 617-22
Tungsten powder,
  characteristics of, 748
  production of, 735, 741
  uses of, 767-8
Tup, 482
Turbulence, 410-1, 744-5
Turbulent flow, 410
Turning, 538-46
  speeds and feeds for, 546
Turret lathes, 526, 544-5
Twinning, 17
Twist drill, 547
Tysland-Hole process, 149

Ultimate tensile load, 46
Ultimate tensile stress, 47
Ultrasonic degreasing, 679-80
Ultrasonic testing of metals, 73-5
Ultrasonic welding, 600
Underground mining, 92, 95
Underground use,
  aluminium for, 831
  lead alloys for, 837
Unit cell, 14
Unsoundness of casting, 449, 451
Up milling, 553
Upsetting force, 592
Uranium ores, prospecting for, 85
Uranium powder, 768
  production of, 735, 736, 741
Urea formaldehyde, in core-making, 403

Vacancies, 58, 459, 758
Vacuum arc remelting of steel, 179
Vacuum conditions for electron beam welding, 629
Vacuum degassing of steel, 178-9
Valence, 1
Van der Waals forces, 756
Vanadium, 319-21

Vanadium *(cont.)*
    production by powder method, 730, 737, 738
    in steel, 228
Vapour phase reactions for metal powder production, 737-9
Varnish, definition of, 720
    coatings, 739-26
Vertical broaching machine, 560
Vertical spindle-type surface grinder, 574
Vertical welding, 602, 622, 630-1
Vickers hardness test, 51, 52
Vinyl resin, thermoplastic, 661
Viscosity of adhesives, 645-6
Vitreous bond, 571
Vitreous enamel, for coatings, 717-9
Vitreous silica, 362-4
Voids, 71
Voltaic cell, 788
Vulcanization, 663

Washburn core, 418
Washers, 589
Water,
    composition of, and rusting in, 820
    and electrolysis, 683-4
    joining of supply pipes for, 639-40
    in moulding sand, 367
    in steel-making, 153
    use of zinc in, 840
Watson, Sir William, 280
Watts solution, 684-5
Wax patterns, 382
Wear resistant alloys, 272
Weld, definition of, 582
Welded tube, 513
Welding,
    and brazing, distinction between, 582
    and dispersion-hardened systems, 864
    methods of, involving pressure, 583-4, 589-601
    and powder metallurgy, 773
    of steel, 221
Western Deep Levels gold mine, 92
Weston, D., 139
Whisker reinforcements in metals, 879-80
Whitney, Eli, 525
Wiberg-Soderfors process, 150
Widia, 775
Wilinson, I., 127
Wilkinson, J., 127
Wilkinson, W., 127
Wilm, A., 244
Wimet, 775
Wire, drawing of, 492
Wire reinforced plastics, 874-5
Wöhler, F., 244, 320, 328
Wöhler fatigue test, 60-1
Wohlwill process, 298
Wolfram, see Tungsten,

Wolframite, 335
Wood, C., 280
Wood flour in core sands, 403
Woods metal, 268
Woodz steel, 125
Work function of oxide interface, 618
Work hardening, 46, 459
    capacity of metals for, 53
    and drawing, 492
    mechanism of, 49, 460-1
Worm and wheel pair, 563
Wrought and cast metals, comparison of, 39, 457

Xanthates, 117
Xeroradiography, 77
X-ray diffraction crystallography, 69
X-ray examination of metals, 76-7
X-ray tube, 629

Yield point, 46, 455
Yield stress, 47
Young's modulus of elasticity, 46, 456

Zinc, 301-6
    corrosion of, 840-1
    deformation in, 17
    electroplating of, 683, 695-7, 699-700, 703
    history and occurrence of, 301
    micrograph of, 66
    ores, preparation of, 301-3
    pretreatment of, for adhesive bonding, 648
    properties and uses of, 303-6
    refining of, 303
    statistics for, 306
Zinc alloys, 304-5, 306
    cast, properties of, 445
    copper plating of die-castings of, 699
    corrosion of, 840-1
    electroplating of, 703
    pretreatment of, for adhesive bonding, 648
Zinc blende, 301
Zinc phosphate, 712
Zinc plating, 305, 699
Zinc stearate, for lubrication, 751
Zircaloy, 318
Zircon, 316, 365
Zirconium, 316-9
    alloys of, 318
    heat treatment and fabrication of, 318-9
    history and occurrence of, 316
    mining of, 316-7
    powder, 730, 735, 737, 738, 741
    production of, 317, 731
    properties of, 316, 317-8
    uses and statistics for, 319
Zone forming alloys, 442-3
Zone refining, of germanium, 255